# Comprehensive
# Toxicology

# Comprehensive Toxicology

*Editors-in-Chief*

**I. Glenn Sipes**
*University of Arizona, Tucson, AZ, USA*

**Charlene A. McQueen**
*University of Arizona, Tucson, AZ, USA*

**A. Jay Gandolfi**
*University of Arizona, Tucson, AZ, USA*

**Volume 14:**

CELLULAR AND MOLECULAR TOXICOLOGY

*Volume Editors*

**John P. Vanden Heuvel, Ph.D.**
*Penn State University, University Park, PA, USA*

**Gary H. Perdew, Ph.D.**
*Penn State University, University Park, PA, USA*

**William B. Mattes, Ph.D.**
*Pharmacia and Upjohn, Kalamazoo, MI, USA*

**William F. Greenlee, Ph.D.**
*CIIT Centers for Health Research, Research Triangle Park, NC, USA*

2002

ELSEVIER

Amsterdam—London—New York—Oxford—Paris—Shannon  Tokyo

# Contents

# Preface

Toxicology is the study of the nature and actions of chemicals on biological systems. In more primitive times, it really was the study of poisons. Evidence of the use of venoms and poisonous herbs exists in both the Old World and the New World. Early recorded history documents the use of natural products as remedies for certain diseases, and as agents for suicides, execution, and murders. In those days, the "science of toxicology" certainly operated at the upper limit of the dose response curve. However, in the early 1500s, it was apparent to Paracelsus, that "the dose differentiates a poison and a remedy". Stated more directly, any chemical can be toxic if the dose is high enough. In addition, with respect to metals, he recognized that the effects they produce may differ depending upon the duration of exposure (i.e., acute vs. chronic). Clearly, the two most important tenets of toxicology were established during that time. The level of exposure (dose) and the duration of exposure (time) will determine the degree and nature of a toxicological response.

Since that time the discipline of toxicology has made major advances in identifying and characterizing toxicants. It is the scientific discipline that combines the elements of chemistry and biology. It not only addresses how chemicals affect biological systems, but how biological systems affect chemicals. Due to this broad definition, toxicology has always been an interdisciplinary science. It uses advances made in other disciplines to better describe a toxic event or to elucidate the mechanism of a toxic event. In addition, toxicology contributes to the advancement of other disciplines by providing chemicals with selective or specific mechanisms of action. These can be utilized to modify biochemical and physiological processes that advance our understanding of biology and the treatment of disease.

During the last quarter century, toxicology has made its greatest advances. Reasons for this include a demand by the public for protection of human and environmental health, an emphasis on understanding cellular and molecular mechanisms of toxicity, and use of such data in safety evaluation/risk assessment. Clearly, the explosion of information as a result of the new biology (molecular and cellular) provided the conceptual framework as well as tools to better elucidate mechanisms of toxicity and to understand interindividual differences in susceptibility to toxicants. Due to this new information, more sensitive end points of toxicity will be identified. This is critical because the real need in toxicology is to understand if/how exposure to trace levels of chemicals (i.e., low end of the dose–response curve) for long periods of time cause toxicity.

Thus, the time is appropriate for the development and publication of an extensive and authoritative work on toxicology. The concept of *Comprehensive Toxicology* was developed at Pergamon Press in 1989–1990. After market research documented a need for such a work, discussions took place to identify an Editor-in-Chief. The project was too large for one individual and the current team of co-Editors-in-Chief was identified. These individuals along with scientific and support staff from Pergamon Press (now an imprint of Elsevier Science, Ltd.), developed a tentative series of volumes to be edited by experts. These Volume Editors then solicited the Authors, who provided the chapters that appear in these several thousand pages. It is to the Authors that we are indebted, because without them, this project would still be a concept, not a product.

Although the title of this work is *Comprehensive Toxicology*, we certainly realize that it does not meet the literal definition of comprehensive. Due to space limitations, certain critical tissues are not included (skin, eye). Also, the time it takes to receive, edit, and print the various chapters limits inclusion of the latest information. Our goal was to provide a strong foundation. Similarly, our understanding of toxic mechanisms is much greater in certain tissues (liver, kidney, lung) than in others (cardiovascular, nervous system, hematopoietic). Thus, the scope of these chapters will differ. They reflect the state of the science and the specializations of the Editors/Authors. Not only will the reader find information on toxicology, but superb chapters on key physiological and biochemical processes in a variety of tissues. In all cases, a wealth of scientific material is presented that will be of interest to a wide audience.

As co-Editors-in-Chief we have appreciated the opportunity to be involved in this project. We are grateful to the very supportive staff at Elsevier and at the University of Arizona who have assisted us with the project. We trust that you will consider *Comprehensive Toxicology* a milestone in the literature of this very old, but very dynamic discipline.

I. Glenn Sipes          Charlene A. McQueen          A. Jay Gandolfi
*Tucson*                    *Tucson*                    *Tucson*

# Contributors to Volume 14

Professor M.E. Andersen
Colorado State University, International Center for Risk Assessment, Department of Environmental Health, Environmental Health Building, Fort Collins, CO 80532, USA

Dr. S.P.Anderson
GlaxoSmithKline Research & Development, 5 Moore Drive, Research Triangle Park, NC 27709, USA

Professor G.K. Andrews
Department of Biochemistry, University of Kansas Medical Center, 39th and Rainbow Blvd, Kansas City, KS 66160-7421, USA

Dr. D.A. Bell
Environmental Genomic Section, National Institute of Environmental Health Sciences, C3-03, PO Box 12233, Research Triangle Park, NC 27709, USA

Dr. D. Bloom
Department of Pharmacology, Baylor College of Medicine, One Baylor Plaza, Houston, TX 77584, USA

Dr. S.M. Boyle
University of Washington, Department of Environmental Health, 4225 Roosevelt Way N.E. #100, Seattle WA 98105, USA

Professor C.A. Bradfield
University of Wisconsin, McArdle Laboratory for Cancer Research, 1400 University Avenue, Madison, WI 53706, USA

Dr. C.-C Chang
246 Natl. Food Safety and Toxicology Center, Dept. Pediatrics and Human Development, Michigan State University, East Lansing, MI 48824, USA

Dr. J.C. Corton
CIIT Centers for Health Research, P.O. Box 12137, 6 Davis Drive, Research Triangle Park, NC 27709-2137, USA

Professor M. Costa
New York School of Medicine, Dept. of Environmental Medicine, 550 First Avenue, New York, NY 10016, USA

Dr. P. Daniels
Department of Biochemistry, University of Kansas Medical Center, 39th and Rainbow Blvd, Kansas City, KS 66160-7421, USA

Dr. S. Dhakshinamoorthy
Department of Pharmacology, Baylor College of Medicine, One Baylor Plaza, Houston, TX 77584, USA

Professor C.J. Elferink
Department of Pharmacology and Toxicology, Room 1.101, Pharmacology Building, 301 University Boulevard, University of Texas Medical Branch, Galveston, TX 77555-1031, USA

K.C. Fertuck
Michigan State University, Department of Biochemistry and Molecular Biology, 223 Biochemistry Building, Wilson Road, East Lansing, MI 48824-1319, USA

M.R. Fielden
Michigan State University, Department of Biochemistry and Molecular Biology, 223
Biochemistry Building, Wilson Road, East Lansing, MI 48824-1319, USA

Dr. A. Gaikwad
Department of Pharmacology, Baylor College of Medicine, One Baylor Plaza, Houston, TX
77584, USA

Dr. W.F. Greenlee
CIIT Centers for Health Research, P.O. Box 12137, 6 Davis Drive, Research Triangle Park,
NC 27709-2137, USA

Professor J.D. Groopman
Johns Hopkins University, School of Hygiene & Public Health Environ. Health Sciences, 615
N. Wolfe Street, Baltimore,  MD 21205-2179, USA

Dr. W. Hanneman
Colorado State University, International Center for Risk Assessment, Department of
Environmental Health, Environmental Health Building, Fort Collins, CO 80532, USA

Christoph Helma
Machine Learning Lab, University of Freiburg, Georges Koehler Allee 079, d-79110
Freiburg/BR, GERMANY

Dr. E. Holmes
Biological Chemistry, Biomedical Sciences Division, Faculty of Medicine, Imperial College
Of Science, Technology and Medicine, South Kensington, London, SW72AZ UK

Dr. W.R. Howard
Department of Biochemistry, University of Kentucky, 800 Rose St., Lexington, KY 40536-
0298, USA

Dr. P.E. Jackson
Johns Hopkins University, School of Hygiene & Public Health Environ. Health Sciences, 615
N. Wolfe Street, Baltimore,  MD 21205-2179, USA

Professor A.K. Jaiswal
Department of Pharmacology, Baylor College of Medicine, One Baylor Plaza, Houston, TX
77584, USA

Professor. T.W. Kensler
John Hopkins University, School of Hygiene & Public Health Environ. Health Sciences, 615
N. Wolfe Street, Baltimore,  MD 21205-2179, USA

Dr. S.A. Kliewer
GlaxoSmithKline, Nuclear Receptor Discovery Research, Venture 116, 5 Moore Drive,
Research Triangle Park, NC  27709, USA

Dr. M. Kumar
Center for Molecular Toxicology and Carcinogenesis, Department of Veterinary Science, 226
Fenske Laboratory, Penn State University, University Park, PA 16802, USA

Dr. J.M. Kyriakis
Diabetes Research Laboratories, Massachusetts General Hospital, Bldg 149, 8th Floor,
Charlestown, MA 02129, USA

Professor J.C Lindon
Biological Chemistry, Biomedical Sciences Division, Faculty of Medicine, Imperial College
Of Science, Technology and Medicine, South Kensington, London, SW72AZ UK

J.B. Matthews
Michigan State University, Department of Biochemistry and Molecular Biology, 223
Biochemistry Building, Wilson Road, East Lansing, MI 48824-1319, USA

Dr. W.B. Mattes
Pharmacia and Upjohn, Predictive & Mechanistic Toxicology, Mail Stop 7228-300-523, 301
Henrietta Street, Kalamazoo,  MI 49007-4940, USA

Dr. J.T. Moore
GlaxoSmithKline, Nuclear Receptor Discovery Research, Venture 116, 5 Moore Drive, Research Triangle Park, NC 27709, USA

Professor J.K. Nicholson
Biological Chemistry, Biomedical Sciences Division, Faculty of Medicine, Imperial College Of Science, Technology and Medicine, South Kensington, London, SW72AZ UK

Professor D.J. Noonan
Department of Biochemistry, University of Kentucky, 800 Rose St., Lexington, KY 40536-0298, USA

Dr. M.L. O'Brien
Department of Biochemistry, University of Kentucky, 800 Rose St., Lexington, KY 40536-0298, USA. [Current Address: Dr. M.L. Twaroski, US Food and Drug Administration, CFSAN/Office of Food Additive Safety, 200 C Street, SW, HFS-225, Washington, DC, 20204, USA].

Professor C.J. Omiecinski
University of Washington, Department of Environmental Health, 4225 Roosevelt Way N.E. #100, Seattle WA 98105, USA

Professor S. Orrenius
Karolinska Institute, Division of Toxicology, Institute of Environmental Medicine, Box 210, SE-171 77, Stockholm, SWEDEN

Dr. W.B. Pennie
Syngenta, Alderley Park, Macclesfield, Cheshire SK10 4TJ, UK

Professor G.H. Perdew
Center for Molecular Toxicology and Carcinogenesis, Department of Veterinary Science, 226 Fenske Laboratory, Penn State University, University Park, PA 16802, USA

Professor J.M. Peters
Center for Molecular Toxicology and Carcinogenesis, Department of Veterinary Science, 226 Fenske Laboratory, Penn State University, University Park, PA 16802, USA

Professor T.R. Rebbeck
University of Pennsylvania School of Medicine, 904 Blockley Hall, 423 Guardian Drive, Philadelphia, PA 19104-6021, USA

Dr. L. Recio
CIIT Centers for Health Research, P.O. Box 12137, 6 Davis Drive, Research Triangle Park, NC 27709-2137, USA

Dr. M.D Reily
Pfizer Global Research and Development, 2800 Plymouth Road, Ann Arbor, MI 48105-1047, USA

Dr. D.G. Robertson
Pfizer Global Research and Development, 2800 Plymouth Road, Ann Arbor, MI 48105-1047, USA

Dr. J.D. Robertson
Karolinska Institute, Division of Toxicology, Institute of Environmental Medicine, Box 210, SE-171 77, Stockholm, SWEDEN

Dr. J. E. Sutherland
New York School of Medicine, Dept. of Environmental Medicine, 550 First Avenue, New York, NY 10016, USA [Current Address: Charles River Laboratories, Discovery and Development, 57 Union Street, Worcester, MA 01608 USA]

Dr. D.J. Tomso
Environmental Genomic Section, National Institute of Environmental Health Sciences, C3-03, PO Box 12233, Research Triangle Park, NC 27709, USA

Professor J.E. Trosko
246 Natl. Food Safety and Toxicology Center, Dept. Pediatrics and Human Development, Michigan State University, East Lansing, MI 48824, USA

Professor J.P. Vanden Heuvel
Center for Molecular Toxicology and Carcinogenesis, Department of Veterinary Science, 226 Fenske Laboratory, Penn State University, University Park, PA 16802, USA

Dr. J.A.Walisser
University of Wisconsin, McArdle Laboratory for Cancer Research, 1400 University Avenue, Madison, WI 53706, USA

Professor C.L. Walker
UTMD Anderson Cancer Center, Department of Carcinogenesis, P.O. Box 389, Park Road 1C, Smithville, TX 78957, USA

Professor M.K. Walker
College of Pharmacy, 2502 Marble NE, University of New Mexico, Albuquerque, NM, 87131-1066, USA

Dr. J.-S. Wang
John Hopkins University, School of Hygiene & Public Health Environ. Health Sciences, 615 N. Wolfe Street, Baltimore, MD 21205-2179, USA

Dr. T.M. Willson
GlaxoSmithKline, Nuclear Receptor Discovery Research, Venture 116, 5 Moore Drive, Research Triangle Park, NC 27709, USA

Professor F.A. Witzmann
IUPU-Columbus, Molecular Anatomy Laboratory, 4601 Central Avenue, Columbus, IN 47203, USA [Current Address: Indiana University School of Medicine, Department of Cellular and Integrative Physiology, 635 Barnhill Drive, MS405, Indianapolis, IN 46202, USA]

Dr. X. Yuan
Department of Biochemistry, University of Kentucky, 800 Rose St., Lexington, KY 40536-0298, USA [Current Address: Millenium Pharmaceuticals, 38 Sidney St., Cambridge, MA 02139, USA]

Professor T.R. Zacharewski
Michigan State University, Dept. of Biochemistry, 402 Biochemistry Building, Wilson Road, East Lansing, MI 48824-1319, USA

# Introduction

**Volume 14: Cellular and Molecular Toxicology (edited by J. P. Vanden Heuvel, G. H. Perdew, W. B. Mattes and W. F. Greenlee)**

The growth of Toxicology as a scientific discipline has been driven to a large extent by the use of extremely powerful molecular and cell biology techniques. The overall aim of this volume is to demonstrate how these advances are being used to elucidate causal pathways (or linkages) for potential adverse health consequences of human exposure to environmental chemicals or radiation. Detailed knowledge of the molecular mechanisms of action of these agents on biological systems provides the basis for understanding (and predicting) tissue- and species-specificity, and can be used in the development of science-based approaches for human risk assessment. Specific areas to be covered include receptor theory, cell cycle regulation, cell signaling and the molecular stress response, genetic responses to environmental signals, endocrine disruptors, genetic determinants of susceptibility, development and use of high throughput technologies for screening toxicant action on gene arrays, and bioinformatics. Prototype agents representing broad classes of environmental agents will be used throughout.

A unique feature of this volume will be its illustration of how carefully-designed studies of the molecular mechanisms of chemical action provide not only understanding of the potential toxicity of the chemical under investigation, but also new insights into the functioning of the biological system used as an experimental model. In this context, chemicals continue to serve as invaluable research tools to probe biological function through carefully controlled study of the nature of their perturbation. Thus molecular toxicology bridges both basic and applied research. Each chapter in this volume will contain a listing of major peer-reviewed articles and reviews, and useful web-sites. In addition, each section will contain a broad introductory chapter that outlines the subsequent chapters. The *Introductory and Overview* chapters are designed to be stand-alone units, and may be packaged as a textbook in graduate level courses. This volume is intended for research scientists and regulators with an appropriate academic background in contemporary biology and also can be used as a text in upper division undergraduate courses and graduate training programs.

J.P. Vanden Heuvel, G.H. Perdew, W.B. Mattes and W.F. Greenlee (Eds.)
Comprehensive Toxicology, Vol. xiv

## 14.1 BASIC CONCEPTS

# 14.1.1
# Introduction to Molecular Toxicology

## JOHN P. VANDEN HEUVEL
### Penn State University, University Park, PA, USA

Toxicology as a discipline has been growing rapidly, due in large part to the proliferation of chemicals being produced and deployed, and the increased public awareness of the risks these products pose to humans and their environment. The Society of Toxicology, the oldest professional group in this field, has had steady increases in its membership since its inception in 1961. The number of journals devoted to toxicology and the number of articles published in these and related journals have burgeoned as a result, in particular within the last ten years. In addition to the public concern, another reason for the growth of this field may be the embrace of molecular biology by the toxicologist.[1] Within the Society of

Toxicology, a molecular biology specialty section has been developed, and over the past ten years it has influenced the direction and approaches of the discipline as a whole. "How do genetic differences lead to variation in human responses to drugs?", "What are the cellular mediators of toxicity?", "How can we detect and treat human cancers?", are all questions being posed by today's molecular toxicologist.[2,3] A key to answering these and other questions are the molecular probes and tools that have been developed which increase speed, sensitivity and accuracy in defining health risks of chemicals. The molecular toxicologist has forced a paradigm shift in how toxicology is performed, moving from endpoints such as high-dose overt toxicity to more meaningful low-dose biological effects.[1] Recent advances in high-throughput and whole-genome assays have also had an impact on toxicology, causing it to relinquish some of its reductionism approaches and adopt a more global understanding of cell biology.

Molecular toxicology is a rapidly advancing field and, although it is a new area of expertise, has an often under-appreciated past. Certainly, we cannot do justice to this history nor can we foresee the promise to come. However, adding prominence to molecular toxicology and delineating its role in the toxicological and basic biological sciences is the purpose of Volume 14 of the *Comprehensive Toxicology* series. This chapter has several less ambitious goals. First, the tenets of classical toxicology will be compared and contrasted to that of molecular toxicology. Although some basic concepts of toxicology will be re-introduced, the reader is directed to Volume 1 of this series[4] as well as several other excellent texts[5–10] for a more detailed description of classical toxicology. An equally important goal of this unit is to encourage the molecular biologist to understand and acknowledge the contributions that toxicology have made to the development of their discipline. Next, the integration of molecular toxicology into a less reductionist approach with an emerging concept in biology, *complexity theory*, will be explored. Throughout this chapter, the difficulties, challenges and promise of assimilating molecular toxicology into the risk and health assessment hierarchy will be discussed.

### 14.1.1.1   BASIC PRINCIPLES AND TERMS OF MOLECULAR TOXICOLOGY

#### 14.1.1.1.1   Scope

Of course, the term *toxicology* literally means "study of poisons", and can be defined as the branch of science that is concerned with the harmful effects of chemicals on organisms. The essence of the science of toxicology is to distinguish the fine-line between tolerable and unacceptable risks to humans and other organisms in the manufacturing, handling, use and disposal of chemical agents. A toxicologist must be well versed in many areas of research in order to achieve this determination of risk. In fact, investigative toxicology encompasses the disciplines of anatomy, physiology, biochemistry, chemistry, immunology, pharmacology, cell and molecular biology as well as pathology and laboratory medicine.

Depending on the source and fate of the chemical(s) in question, the toxicologist may also require knowledge of human and veterinary medicine, agricultural and nutritional sciences and perhaps inorganic chemistry and engineering. It is generally believed that only through rigorous basic scientific research can responsible regulation of potentially hazardous compounds be achieved. The main goals of toxicology include:[5] (1) To elucidate the toxic properties of chemicals; (2) To evaluate the hazards of chemicals to organisms in relation to the concentration of these substances in the environment (risk estimation); (3) To advise society on measures to control or prevent the harmful effects of chemicals (hazard control).

To achieve these goals an interdisciplinary approach and a broad appreciation of many aspects of biological sciences must be observed. As is the case with most modern sciences, a large degree of specialization has evolved among toxicologists, each contributing to the basic mission. A *Descriptive Toxicologist* is concerned directly with toxicity testing for safety evaluation and regulatory requirements. The appropriate toxicity tests are designed to yield information that can be used to evaluate the risk posed to humans and the environment by exposure to specific chemicals. Similarly, a *Regulatory Toxicologist* is responsible for deciding, on the basis of data provided by descriptive and mechanistic toxicologists, whether a chemical poses a sufficiently low risk to be marketed for the stated purpose. A *Mechanistic Toxicologist* is a person who wishes to identify and understand the mechanisms by which chemicals exert toxic effects on living organisms. The mechanistic information is very useful in all areas of applied toxicology. Also an understanding of mechanisms of toxic action contributes to the knowledge of basic physiology, pharmacology, cell biology and biochemistry. There are also three specialized areas of toxicology, that of forensic, clinical and environmental toxicology. *Forensic Toxicologists* are hybrids of clinical chemists and basic toxicologists and study medical-legal aspects of harmful effects of chemicals on humans and animals. *Clinical Toxicologists* are concerned with diseases caused by, or uniquely associated with, toxic substances. Efforts are directed at treating patients poisoned by drugs or chemicals. Last, an *Environmental Toxicologist* focuses on the impacts of chemical pollutants in the environment on biological organisms.

A *Molecular Toxicologist* can be affiliated with each of the specializations described above, but is mainly a mechanistic toxicologist who uses the latest biochemical and molecular tools to understand how a chemical results in toxicity (see Figure 1). The key to this sub-discipline is the powerful tools, but the goals are the same as other branches of toxicology, to detect, treat and assess the risk posed by chemical exposure to living organisms. The use of molecular biology in toxicology research has been fully embraced by the mechanistic, forensic and clinical toxicologist and is enjoying rapid growth among environmental toxicologists. The key challenge that faces the regulatory agencies is to understand and fully assimilate the information being presented to them by the molecular toxicologists, a task that has been somewhat slow and daunting.

*Molecular toxicology* focuses its attention on how harmful effects of chemicals are manifested at the cellular and molecular level. Having detailed molecular information on how chemicals result in toxicity may help achieve all three of the goals stated above. Elucidation of toxicity, risk estimation and hazard control are all improved by the sensitivity, accuracy and speed of modern technologies that dominate molecular toxicology. There are additional benefits, or objectives, of molecular toxicology that may be added to the previous list: (4) To achieve an appreciation for the complexity of cell and molecular biology and how chemicals may perturb the delicate and intricate order of living systems; (5) To develop more accurate screens and therapies for the toxic effects of chemicals, not the least of which is cancer, based on the knowledge of molecular events associated with particular diseases. It has been a relatively recent occurrence that scientists have been able to achieve these goals for a handful of chemicals and it is hoped that detailed mechanistic information of many xenobiotics will follow. Also, the astounding pace that technology is advancing may result in other goals being developed for molecular toxicology research that are unfathomable at this time.

### 14.1.1.1.2 History

Toxicology is one of the most ancient forms of science (see References 5–8 for a more extensive description). Egyptian papyrus scrolls from 1550 B.C. showed an extensive knowledge of toxic and curative properties of natural products. The toxic properties of chemicals have been exploited, such as the case with Socrates who was executed with an extract of hemlock. The formal, causal relationship between amount of chemical and extent of toxic response is often attributed to the Swiss physician Paracelsus (1493–1541), who drew attention to the dose-dependency of the toxic effects of substances with the saying

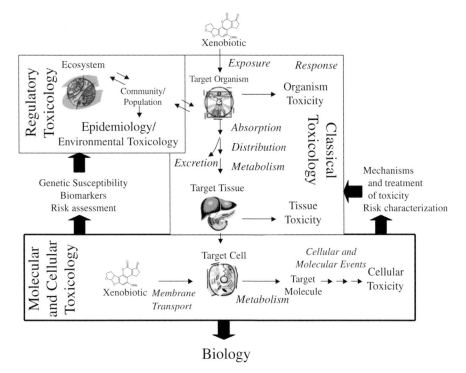

**Figure 1** How molecular toxicology complements other areas of toxicology research.

"Everything is a poison ... it is only the dose that makes it not a poison." Subsequently, Bonaventura Orfila was the first scientist to systematically test the toxicity of chemicals and developed methods for chemical analysis of poison in the tissue and body fluids (published *Trait des Poison*, 1814). This classical toxicology approach is still evident today and has taken on increased vigor since World War II when an increasing range of chemicals appeared on the market.

Perhaps the beginning of molecular toxicology can be traced to the advent of the *receptor theory*. As discussed in detail in Chapter 14.1.3, the concept of a specific cognate protein that is responsible for the effects of chemicals has been a founding principle for many disciplines including endocrinology and pharmacology. The receptor theory has been developed by a succession of eminent biologists and chemists: Claude Bernard (1813–1878) elucidated the actions of carbon monoxide and curare and coined the phrase "Locus of drug effect" to describe the specificity of the toxic response; Paul Erhlich (1854–1915) adapted this concept to "Selectivity of drug action" and stated "Corpora non agunt nisi fixati" or an agent cannot act unless bound; J. N. Langley (1852–1926) felt there was a "Receptive Substance" in the cell and developed the first model for pharmacological antagonism; A. J. Clark (1885–1941) believed that complexes are formed according to some law based on the concentration of the drugs and the affinity for the receptive substances. These ideas developed into a quantitative description of selectivity and saturability and the *law of mass action* that is still the prevailing model for most bimolecular interactions.

Although much of the history of molecular toxicology is inextricably connected to that of basic biochemistry, there have been defining moments. The first occurred in the area of chemical carcinogenesis. Researchers such as Rous and Kidd (1941), Mottram (1944) and later Boutwell (1964) used chemicals to define the etiology of skin cancer in what has become the initiation-promotion, or two-stage, model of carcinogenesis.[11] The second occurred more recently in regard to the molecular effects surrounding 2,3,7,8-tetrachlorodibezo-*p*-dioxin (TCDD) toxicity. Several researchers, included Poland (1973),[12,13] speculated that a specific intracellular receptor existed for dioxin and related compounds. This hypothesis was proven correct and the protein was later cloned and deemed the *Arylhydrocarbon Receptor* (AhR). This marked the first xenobiotic receptor and has opened the door to a burgeoning and productive field.

Molecular toxicology has played an extremely important role in history of other sciences as well. Various toxins such as curare, strychnine and botulism toxin have been used extensively to study and dissect specific cellular and physiological events. The initiation-promotion paradigm, as mentioned above, initially developed to explain how mutagenic compounds and hormone-like chemicals result in tumors, has formed the backbone of carcinogenesis research. Many modern chemotherapeutic agents are overtly toxic to cancerous cells at doses that are well tolerated by normal cells. These agents could not have been developed without an understanding, at the molecular level, of how a chemical causes its toxic effects. That many biotechnology companies have extensive lists of specific enzyme inhibitors and activators attests to the fact that detailed knowledge of how chemicals affect cellular processes is important in modern biochemical research. Thus, many areas in the biological science owe a measure of gratitude to the mechanistic and molecular toxicologist for developing the tools required to probe specific cellular processes.

### 14.1.1.1.3 Classification of Toxic Agents

There are many ways in which a chemical may be grouped, or classified, into a particular family of related compounds. Although this is a reductionist approach, it may convey useful mechanistic information if the classification scheme is based on detailed scientific knowledge. First, a chemical is often classified based on origin. A *toxin* refers to toxic substances that are produced naturally, such as curare. *Toxicant* is a term reserved for a substance that is of anthropogenic origin and includes compounds such as dioxin as well as most drugs. Further classification may exist of *use* (food additive, drug, pesticide), *source* (animal, plant), *chemistry* (amine, hydrocarbon) or *physical state* (gas, liquid). The chemical may be grouped based upon toxic profile such as *target organ* (liver, kidney), *effect* (necrosis, mutagenesis) or *potency* (very toxic, nontoxic). To a molecular toxicologist, the best means of classification is *biochemical mechanism of action* (AchE inhibitor) or *target molecule* (estrogen receptor). This type of information can immediately result in a useful assessment of the risk of a chemical. For example, knowing an agent is an estrogen receptor antagonist may suggest target organs, potential toxic effects, as well as potency. Most modern drug discovery is geared toward classification of chemicals by their targets. Libraries are often screened through high-throughput, target-based assays to uncover novel therapeutic agents.

### 14.1.1.1.4 Toxicokinetics in the Context of Cellular Toxicology: Membrane Transport

Insight into the molecular target of a chemical is meaningless unless the agent can reach that site in the organism. That is, adverse or toxic effects are not produced unless that chemical or its metabolic products reach the appropriate site in the body at a sufficient concentration and length of time. *Toxicokinetics* focuses on these time-dependent, whole organism, transport and toxicity of chemicals. In molecular toxicology, toxicokinetics is relevant at several levels, including understanding the concentration at the site of entry, within distribution (plasma), and at the target cell. Particularly germane is the concentration of chemical (or active metabolites) at the target molecule itself, be it a receptor, enzyme or DNA. The classic toxicology tetrad of absorption-distribution-metabolism-excretion (ADME) has a somewhat limited scope in most molecular biology studies, at least for those performed in cell culture or *in vitro*. However, one must always place the results obtained in molecular toxicology back to the physiological context. If a chemical results in a particular response in a cell culture model, is that concentration of xenobiotic attainable *in vivo* and is the metabolic fate similar?

In the molecular toxicology realm, perhaps the most important toxicokinetic consideration is passage of the substance across biological membranes. Crossing membranes is an important determinant of the cellular effect for many xenobiotics. Transport across the membrane may be accomplished by several different means (Figure 2). *Passive diffusion* is the most common route by which xenobiotics pass through membranes. Passive diffusion is non-specific and unsaturable and is fueled by the difference in concentration on both sides of the membrane, and on the lipophilicity of the compound concerned. In contrast, *Filtration* requires transport through pores and aqueous channels in the cell membrane and is restricted to hydrophilic substances with a low molecular mass. Unlike diffusion, filtration utilizes a pressure gradient in addition to an osmotic gradient. Most aqueous pores allow molecules with molecular mass less than 100 daltons to pass. *Facilitated diffusion* and *active transport* both involve a carrier protein for transport across the membrane. Facilitated diffusion takes place due to a difference in concentration on both sides of the membrane, in the direction of the lowest concentration and does not require energy. Active transport may occur against a concentration gradient and requires energy for transport. Both processes are substrate-specific and saturable since they require a specific interaction between the carrier molecule and the xenobiotic. This is often the case for amino acids and sugars, or for xenobiotics that are chemically related to these compounds. *Pinocytosis* takes place by invagination of the cell membrane around the foreign entity. This process is of particular importance in the removal of larger particles, for example of particles from the alveoli in the lungs.

As stated above, xenobiotics generally cross membranes by passive diffusion. Passive diffusion largely takes place across the lipid component of the cell membrane, and will only transport the lipid-soluble, non-ionized form of the molecule. The rate of transport across the membrane is directly proportional to the difference in concentration between the non-ionized molecules on either side of the membrane. The ratio of charged to uncharged molecule is, of course, described by the Henderson–Hasselbach equation that relates ionization to $pK_a$ and pH. In most cases, however, there will be a dynamic system, with the concentration on both sides of the membrane continuously changing as a result of removal of molecules by the bloodstream, metabolism, ionization, as well as binding to proteins. In addition to charge, the transport rate is also dependent upon molecule size, lipid solubility of the molecule (hydrophobicity) and, to some extent, temperature. Non-ionized molecules with a high (oil–water) *partition coefficient* ($P_{oct}$) will generally diffuse rapidly. The partition coefficient of a substance refers to the ratio of the amount of that substance that dissolves in an organic solvent not mixable with water and the amount that dissolves in an aqueous solution. Molecules with a very high partition coefficient will easily penetrate into a membrane, and having done so, leave it less easily.

### 14.1.1.1.5 Spectrum of Toxic Effects

In therapeutics, each drug produces a number of effects, but only one is associated with the primary objective of the therapy; all other effects are considered undesirable or side effects. The effects that are always undesirable are referred to as *adverse*, *deleterious* or *toxic* effects of the drug. In classical toxicology, there exists a spectrum of toxic endpoints that may be examined to characterize this untoward

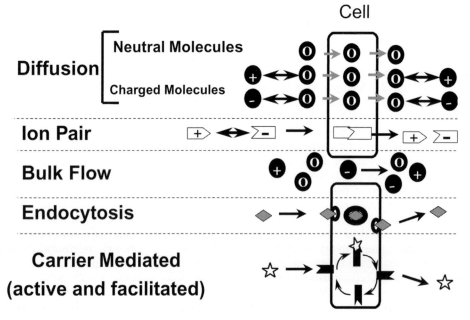

**Figure 2** Basic membrane transport mechanisms utilized by xenobiotics. The charge of the molecules is listed as +, − or o (neutral). See text for details.

effect. *Allergic Reactions* are immunologically mediated adverse reaction to a chemical resulting from previous sensitization to the chemical or to a structurally similar one. *Idiosyncratic reactions* refer to a genetically determined abnormal reactivity to a chemical. The response is usually qualitatively similar to that observed in other individuals but may take the form of extreme sensitivity/insensitivity to the chemical. (Note this is not the same as a reaction that occurs with low frequency.) Of course, with *in vivo* studies, the responses may be *Immediate* or *Delayed*. Immediate responses are those that occur rapidly after a single administration of a substance whereas delayed toxic effects are those that occur after a lapse of some time. Cancer is classic example of a delayed (20–30 years in humans) toxic response. The toxic effects may be *reversible versus irreversible*. If a chemical produces pathological injury to a tissue, the ability of that tissue to repair itself will determine whether the effect is reversible or irreversible. *Local effects* refer to those that occur at the site of first contact between the biological system and the toxicant. *Systemic effects* require absorption and distribution of a toxicant from its entry point to a distant site at which deleterious effects are produced. Most chemicals that produce systemic toxicity do not cause a similar degree of toxicity in all organs; instead they elicit their major toxicity at one or two sites. These are referred to as *target organs* of toxicity. Often the target organ is not the site of highest concentration of that chemical.

The spectrum of toxic responses takes on a completely different flavor when the studies are performed at the cellular and molecular level. In this case, the object is to examine the *Direct Effect* of a chemical on the *Target Cell*. Thus, the molecular toxicologist is dependent on results obtained via classical toxicology to determine the predominant target and whether the chemical itself, a metabolite or a secondary product from another tissue, is responsible for the effect.

#### 14.1.1.1.6 Interaction of Chemicals

There are a number of ways that different chemicals may interact with each other. Since many of the terms are misused, they warrant a brief review. (The numbers in parenthesis are arbitrary effects seen with the drugs given separately. For example $2 + 2 = 4$ indicates that xenobiotic A and xenobiotic B have a quantified effect of "2" and when given together will have an effect of "4".) Two chemicals are *Additive* when the combined effect of the two chemicals is equal to the sum of the effects of each agent given alone $(2 + 2 = 4)$. A *Synergistic* effect is noted when the combined effect of the two chemicals is far greater than the sum of the effects of each agent given alone $(2 + 2 = 10)$. *Potentiation* occurs when one substance has no toxic effect but when added to another chemical makes it much more toxic $(0 + 2 = 10)$. *Antagonism* comes in four different forms and occurs when two chemicals administered together interfere with each other's action or one interferes with the action of the other $(2 + 2 = 1; 4 + 0 = 2)$: *Functional antagonism* occurs when two chemicals counterbalance each other by producing opposite effects on the same physiological function. Since the sympathetic and parasympathetic nervous systems have opposing physiological function, chemicals differentially activating these pathways will be functional antagonists of each other; *Chemical antagonism* occurs when a chemical reaction between two chemicals produces a less toxic product. For example dimercaprol (BAL) chelates metals such as arsenic and is used to treat acute poisoning cases; *Dispositional antagonism* results when the disposition ("ADME") is altered or the concentration or duration of action at the target site is diminished. Charcoal and ipecac syrup are useful as dispositional antagonists. Increasing metabolism (if the resultant product is less toxic; cytochrome P450 inducer phenobarbital) or decreasing metabolism (if products are more toxic, cytochrome P450 inhibitor piperonyl butoxide) may affect toxicity and are other instances of dispositional antagonism; Finally, *Receptor antagonism* (or *Pharmacologic antagonism*) occurs when two chemicals bind to the same receptor and produce less of an effect than the two when given separately $(2 + 2 = 2$ or $2 + 0 = 0)$. This form of antagonism is perhaps the most important type of interaction of chemicals in molecular toxicology research. Being able to specifically inhibit an effect at the cognate macromolecule allows for detailed studies into the biological role of that receptor. Also knowing that interaction of the chemical with a specific receptor occurs results in a better understanding of how that antagonist results in alteration of toxicity.

### 14.1.1.1.7 Dose-Response

The characteristics of exposure to a chemical and the spectrum of effects are a correlative relationship referred to as the *dose-response* (DR) *relationship*. This relationship is the most fundamental and important concept in classic toxicology and is also a key component of molecular toxicology studies. The DR relationship will be described in detail in the following chapter, and will only be briefly discussed herein. There are basically two types of DR relationships: That which describes the response of an individual to varying doses of a chemical, often referred to as a *Graded* DR because the measured effect is continuous over a wide range of doses; and, one which

characterizes the distribution of responses to different doses in a population of individuals or a *Quantal* DR. Both types of DR relationships are often expressed as log-normal distributions, and the shape of the two curves are virtually identical. The quantal DR relationship has gained prominence in the epidemiology arena and is used by molecular toxicologist examining polymorphisms in drug metabolizing enzymes, for example. However, the graded DR has a more universal role in cell and molecular biology and will be our primary focus.

#### 14.1.1.1.7.1 Assumptions in Deriving the Dose-Response Relationship

In order to gain the full use of a graded DR relationship, one must adhere to certain assumptions. The most obvious assumption is that the response being measured is a direct result of the chemical being administered. This assumption is not always easy to ascertain and causality must be backed up by mechanistic data. Second, the magnitude of the response is related, in some manner, to the dose or concentration of the chemical. This premise is comprised of three sub-assumptions: There is a macromolecule or target site with which the chemical interacts to produce the response; The production of the response and the degree of the response are related to the concentration of the agent at the target site; The concentration at the target site is related to the dose administered. The third assumption is that there exists a quantifiable method of measuring and expressing toxicity. The ideal response to measure would be one that is closely associated to molecular events, since this will be the type of effect that adheres to the assumptions listed above in most cases.

#### 14.1.1.1.7.2 Basics of the Dose-Response Relationship

Dose-response data are typically depicted graphically in two dimensions, by plotting a measured effect on the ordinate (dependent variable) and the dosage or a function of dose (e.g., log dose) on the abscissa (independent variable). The chemical that results in a measurable effect is often referred to as the *agonist*. Since the agonist effect is a function of dose and time, a graded DR curve is a time-independent relationship. The agonist's actions may be quantified at several levels of analysis: molecular, cellular, tissue, organ, organ system, or organism. Thus, the conditions of study, as well as the mathematical approach to data analysis, determine the precise means of defining the xenobiotic-induced effect and preclude the existence of any single characteristic relationship between the intensity of drug effect and drug dosage. Effects are frequently recorded at time of peak-effect, or under steady-state conditions. Figure 3 illustrates the four variable features of a typical DR curve: *Potency* (location of curve along the dose axis), *efficacy* (greatest attainable response of the system under measurement), *slope* (change in response per unit dose), and biologic variation (variation in magnitude of response among test subjects in the same population given the same dose of drug).

The shape of the DR curve varies depending on how the data is expressed (Figure 3). A sigmoidal DR curve results when the magnitude of effect observed is plotted versus the logarithm of agonist concentration. However, when the amount of agonist is plotted on a linear scale, a hyperbolic curve results. The shape of the DR curve is particularly important in risk assessment where "safe" levels of exposure to xenobiotics is to be ascertained. By examining a sigmoidal curve, one may predict a *threshold* dose, or a concentration of chemical that must be exceeded before toxicity is observed. The concept of threshold has very little practical application in molecular toxicology for the following reasons. First, a threshold may be an artifact of the manner in which the data is visualized, as is clearly evident in Figure 3. Second, determining a threshold is often dictated by the sensitivity of the assays. In modern molecular toxicology the indices measured are exquisitely sensitive, and are only

apt to become more so. Theoretically, the measurement of an effect elicited by one molecule of a xenobiotic is possible. Last, the assumptions stated above for the interpretation of a DR curve preclude the existence of a threshold. The most meaningful DR relationships are those that use molecular events as the response; however, thresholds are only of use in assessing complicated, multifaceted events such as lethality or cancer. Thus, finding a threshold concentration may have some practical purpose in risk assessment, but this practice has little value in molecular toxicology and in fact has little mechanistic basis.

Perhaps the best use of DR relationships is to compare the potency of a series of agonists. In fact, determining a rank-order potency for a series of chemicals in eliciting a particular response is a key criteria in evaluating toxicity. That is, the potential for toxicity of an under-studied chemical may be estimated by comparing its potency to that of xenobiotics with known toxicity. The relative potency is usually reported as a comparison of $EC_{50}$ values, or concentrations that result in one-half maximal response. Comparison of agonists and partial agonists based on DR curves is described in detail in Chapter 14.1.3.

### 14.1.1.1.8   Variation in Toxic Response

Differences in toxic responses among species and within populations represent a colossal challenge to the toxicologist. For example, how can the risk posed to humans by a given chemical be ascertained given that the only data available is from laboratory animals? Equally difficult is trying to understand if

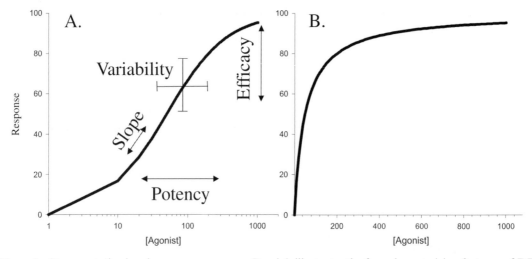

**Figure 3**   Representative log dose-response curves. Panel A illustrates the four characterizing features of DR curves, potency, efficacy, slope and variability. Panel B examines the shape of the DR curve when plotted on an arithmetic scale. The data in both panels are identical (adapted from Reference 24).

subpopulations of individuals exist who are at increased danger to chemical exposure. These challenges are perhaps where molecular toxicology may be best utilized to improve health decisions (see Figure 4). By understanding the details of how a chemical results in toxicity within a sensitive species or cell, one can then determine if another cell type has the same potential for response. The three types of variation of response[5] are described below.

### 14.1.1.1.8.1 Selective Toxicity

Selective toxicity means that a chemical produces injury to one kind of living matter without harming another. The living matter that is injured may be referred to as *uneconomic* or undesired; the protected matter is referred to as *economic* or desirable form. The terms economic and uneconomic may refer to different species of animals, as in pesticides for example, or may refer to differences in cells within the same organism. For example, cancer chemotherapeutic agents cause preferential toxicity in cancer (uneconomic) cells while leaving normal (economic) cells relatively unscathed. Drugs and other chemicals used for selective toxicity produce the effects in two ways: The chemical is equally toxic to both economic and uneconomic, but is accumulated by the uneconomic cells; The chemical reacts fairly specifically with a cytologic or biochemical feature that is absent or does not play an important role in the economic form.

### 14.1.1.1.8.2 Species Differences

Both quantitative and qualitative differences may exist in response to chemicals. A good example of a quantitative difference in response exists for dioxin where the lethal dose ($LD_{50}$) in guinea pigs differs by 1000-fold from hamsters. The difference in the potency of dioxin can be attributed to the incongruity in the affinity of the cognate receptor (AhR) for the chemical between the two species. This shows the validity of understanding the detailed mechanism of action of chemicals when extrapolating between species. One could predict the possibility of a particular toxic endpoint, in this case lethality, in a test subject by examining the existence and then the affinity of the AhR. Note, that this is not a fool-proof approach and confidence would increase if the proper molecular and biochemical events were also observed in the test subject. Obviously,

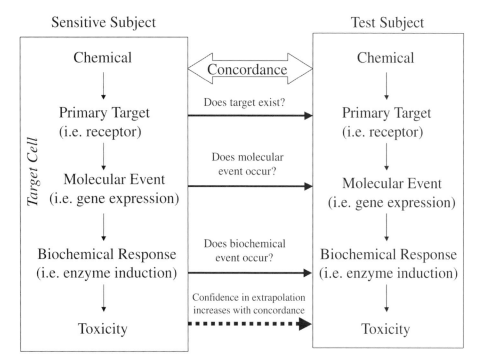

**Figure 4** How detailed mechanistic information may assist in evaluating risk and examining variation in toxic response. The information gathered from a sensitive subject (cell, individual or species) may assist in determining if toxicity is likely to occur in another subject (test subject). Molecular toxicology may assist in determining the sequence of events that result in a toxic event and may also be used for examining concordance between the sensitive and test subjects. With increasing concordance comes increasing confidence in predicting toxicity.

identifying the mechanistic basis for species differences in response to chemicals is an important part of toxicology, especially in extrapolating animal data to humans.[5]

### 14.1.1.1.8.3 *Individual Differences in Response*

The examination of hereditary differences in a single gene (polymorphisms), and their role in idiosyncratic reactions to chemicals or in differences in responsiveness among populations has become a major area of research within molecular toxicology. The genetics of the toxic response is described in detail in the third section of the present volume. Once again, by understanding how a chemical manifests its toxicity, the effect of harboring a polymorphic gene that participates in the cascade can be better appreciated.

### 14.1.1.2 REDUCTIONISM VERSUS COMPLEXITY

In the above discussion, we have focused on reducing a chemical response to a series of linear events, ultimately culminating in toxicity, and on using this reductionism to perform risk-assessment. This approach utilizes simple *Boolean logic* and has also been applied to explain monogenic diseases, or those responses that obey Mendel's laws. In many of the subsequent chapters a similar approach will be used to place specific proteins, enzymes and genes into a somewhat rigid, easy to conceptualize, circuit. Although this is certainly a necessary and productive exercise, it may result in an under-appreciation of complex biological systems and how chemicals perturb cellular processes. Thus, the major challenge now is to decipher the polygenic and multifactorial etiology of toxic responses, such as cancer and auto-immune diseases.

One reason for taking the minimalist approach in the past has been limitations in the technology available. Until recently, only a small subset of molecular parameters could be examined concurrently, thereby limiting the scope of investigative studies. However, technological developments, as outlined in Section 5 of the present volume (*Novel Technologies and Predictive Assays*), has resulted in a revolutionary transition. Robust data acquisition, the human and other genome projects, genomic and proteomic technologies, now allow for a huge pool of information to be gathered from a single experimental situation.[14] It is now possible to uncover crucial clues to the underlying molecular changes in genetic networks that result in a toxic response.

It has become obvious that the development of cancer, or any complex toxic response, is a non-linear process and requires a less rigid premise for mechanistic studies. The new mathematics of *non-linear dynamics*, which includes chaos and complexity theory, have led to paradigm shifts in many unrelated disciplines such as economics, meteorology and seismology and is taking hold in biology and toxicology. (The reader is directed to several excellent reviews[14–17] and books[18–22] of *chaos, complexity* and non-linear dynamics in the biological sciences.) Our current understanding of toxicology has resulted from a conventional view of the disease process. In this paradigm, the alteration in the function of a gene product, or several gene products, leads to toxicity. Applying chaos theory and complexity to toxicology, however, leads to a different perception and the toxic response can be viewed as a complex adaptive system[23] (see Figure 5). In fact, small changes in genetic response may result in a robust toxic response, as a result of the linkage of many genes, and is characteristic of non-linear systems. The abrupt changes seen in non-linear systems are termed *emergent properties* and is also a hallmark of chaotic systems.[23] In other words, a normal cell is more than the sum of its parts, and disrupting the network of gene products by the addition of a xenobiotic results in increased chaos, disorder and a complex adaptive response.

Redefining our perception of toxicology as a complex adaptive response may lead to a deeper understanding of disease states, and possibly result in novel insight into cell biology. In this scenario, every gene when perturbed by a xenobiotic is a potential disease or toxic response target. This results in the concept of 'reverse toxicology' whereby pathogenic pathways are derived from the knowledge of the structure and function of every gene. Certainly, the completion of the human genome project will facilitate the advancement of this concept. By going from sequence (genomics) to function (proteomics, metabonomics) we will gain insight into basic mechanisms of major functions such as cell proliferation, differentiation and development, which are perturbed by many xenobiotics.

### 14.1.1.3 CONCLUSIONS

Toxicology research over the past several decades has contributed enormously to our understanding of basic cell biology as well as

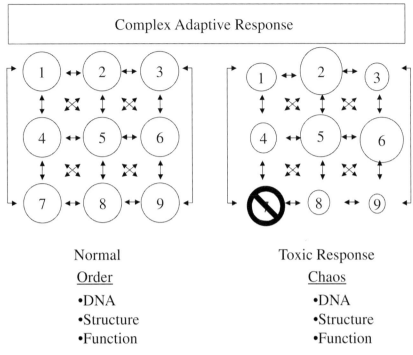

**Figure 5** Self-organization and interactive network of genes. Each unit (1–9), representing an element (cell, gene or gene product), interacting with neighboring elements. The linkage and interactions obey a simple set of rules known as nonlinear dynamics. A small change in one gene product (in this case gene 7) results in a new "emergent state", and perhaps a large and unexpected toxicological response (adapted from Reference 23).

mechanisms of disease. The subdiscipline of molecular toxicology has a relatively short history, but it has had a major impact in how toxicology research is being conducted. Rapid advances in the scientific tools available have increased the penetration of molecular biology in toxicology, and in fact these technologies have driven toxicology to new and exciting heights. These advances provide an unprecedented opportunity for toxicologists to examine detailed cell biology without losing site of the ultimate goals, to improve the health of humans and our environment. Research at the interface of molecular biology and toxicology is the major emphasis of the present volume.

## 14.1.1.4 REFERENCES

1. E. Marshall, 'Toxicology goes molecular.' *Science*, 1993, **259**, 1394–1398.
2. H. Vainio, 'Molecular approaches in toxicology: change in perspective.' *J. Occup. Environ. Med.*, 1995, **37**, 14–18.
3. J. L. Stevens and L. J. Marnett, 'Defining molecular toxicology: a perspective.' *Chem. Res. Toxicol.*, 1999, **12**, 747–748.
4. J. Bond, 'General Principles', Elsevier, New York, NY, 1997.
5. L. J. Casarett *et al.*, 'Casarett and Doull's Toxicology: The Basic Science of Poisons', 5th edn., McGraw-Hill, Health Professions Division, New York, 1996.
6. B. Ballantyne, T. C. Marrs and T. L. M. Syversen, 'General and Applied Toxicology', 2nd edn., Grove's Dictionaries, New York, NY, 1999.
7. I. G. Sipes, C. A. McQueen and A. J. Gandolfi, 'Comprehensive Toxicology', Pergamon, New York, 1997.
8. M. J. Derelanko and M. A. Hollinger, 'CRC Handbook of Toxicology', CRC Press, Boca Raton, 1995.
9. R. A. Lewis, 'Lewis' Dictionary of Toxicology', Lewis Publishers, Boca Raton, FL, 1998.
10. K. Stine and T. M. Brown, 'Principles of Toxicology', CRC Lewis Publishers, Boca Raton, FL, 1996.
11. H. C. Pitot, 'Fundamentals of Oncology', 2nd edn., M. Dekker, New York, 1981.
12. A. Poland and E. Glover, 'Studies on the mechanism of toxicity of the chlorinated dibenzo-*p*-dioxins.' *Environ. Health Perspect*, 1973, **5**, 245–251.
13. A. Poland and E. Glover, '2,3,7,8-Tetrachlorodibenzo-*p*-dioxin: a potent inducer of aminolevulinic acid synthetase.' *Science*, 1973, **179**, 476–477.
14. Z. Szallasi, 'Bioinformatics. Gene expression patterns and cancer.' *Nat. Biotechnol.*, 1998, **16**, 1292–1293.
15. P. B. Persson, 'Modulation of cardiovascular control mechanisms and their interaction.' *Physiol. Rev.*, 1996, **76**, 193–244.
16. E. D. Schwab and K. J. Pienta, 'Cancer as a complex adaptive system.' *Med. Hypotheses*, 1996, **47**, 235–241.

17. D. A. Rew, 'Modelling in surgical oncology—Part III: massive data sets and complex systems.' *Eur. J. Surg. Oncol.*, 2000, **26**, 805–809.
18. E. Mosekilde and L. Mosekilde, North Atlantic Treaty Organization. Scientific Affairs Division. 'Complexity, Chaos, and Biological Evolution', Plenum Press, New York, 1991.
19. R. Lewin, 'Complexity: Life at the Edge of Chaos', 1st edn., Collier Books, New York, 1993.
20. R. Robertson and A. Combs, 'Chaos Theory in Psychology and the Life Sciences', Lawrence Erlbaum Associates, Mahwah, NJ, 1995.
21. B. H. Kaye, 'Chaos and Complexity: Discovering the Surprising Patterns of Science and Technology', Vch, Weinheim, New York, 1993.
22. J. Gleick, 'Chaos: Making a New Science', Penguin, New York, NY, U.S.A., 1988.
23. D. S. Coffey, 'Self-organization, complexity and chaos: the new biology for medicine.' *Nat. Med.*, 1998, **4**, 882–885.
24. L. S. Goodman, A. Gilman and J. G. Hardman, *et al.* 'Goodman & Gilman's the Pharmacological Basis of Therapeutics', 9th edn., McGraw-Hill, Health Professions Division, New York, 1996.

J.P. Vanden Heuvel, G.H. Perdew, W.B. Mattes and W.F. Greenlee (Eds.)
Comprehensive Toxicology, Vol. xiv

# 14.1.2
# Exposure-Dose-Response: A Molecular Perspective

## MELVIN E. ANDERSEN and WILLIAM H. HANNEMAN
### Colorado State University, Ft. Collins, CO

**14.1.2.1  PUBLIC HEALTH CONSIDERATION IN TOXICOLOGY AND RISK ASSESSMENT**

From a public health perspective the main contributions of toxicity testing and toxicology research are to assess, with minimum uncertainty, the risks posed to human populations by exposure to specified concentrations of potentially toxic compounds. Toxicology studies also evaluate risks posed by exposure of environmental species or risks of environmental degradation. In what way do these estimates of expected risks serve the public? Firstly, they provide measures to determine if the compounds are themselves likely to lead to harm in a specific population. Secondly, they provide a means to compare one chemical with another, i.e., to assess relative risks, thereby aiding in selecting particular chemicals for a particular societal use. Thirdly, these health risk estimates can be useful in a larger context to compare risks from different activities, i.e., from chemical exposure to disinfection byproducts versus risks of discontinuation of water disinfection, a point of controversy with disinfection byproducts in drinking water. These various risk assessment activities are only as rigorous as the data that are available for the decision-making process and require contributions from a variety of biological disciplines, including toxicology and molecular biology.

Over the past decades, many practicing toxicologists have considered chemical risk assessment to be a pseudo-scientific process, primarily determined by public policy that masquerades as science, rather than representing a process well grounded in firm biological and toxicological principles. Many toxicologists are uncomfortable when their work is linked to risk assessment or even when asked how their work will influence the risk assessment process. This chapter discusses the integration of molecular toxicology and toxicogenomic information via an exposure-dose-paradigm, provides a brief synopsis of the emerging activities to increase the scientific basis of chemical risk assessment, and notes the areas where molecular toxicology will likely contribute to refinements of the current risk assessment process. The societal need for quality risk assessment will only be achieved as practicing toxicologists take an increasing responsibility in discussing the manner in which their data will be applied to improve risk assessment.

### 14.1.2.1.1   The Exposure-Dose-Response Paradigm

Providing a risk assessment context for studying toxicological issues related to public health requires a more interdisciplinary approach than is available from the in-depth studies of a molecular mechanism by which a chemical interacts with cellular constituents. In assessing real world risks, the entire continuum, including exposure, absorption, distribution, cellular interactions, and ultimately impaired health, needs to be considered in context. This contextual basis for studying the health consequences of exposures to toxicants is referred to here as an exposure-dose-response paradigm (Figure 1). Frequently, the main issue that confronts both the molecular toxicologist and the risk assessor is the problem of how individual mechanistic studies fit into this larger exposure-dose-response continuum. Risk assessments have to consider the integration of all the steps in this exposure-dose-response continuum to make definitive statements of the risks at various levels of exposure. Risk assessment then requires the melding of various disciplines to arrive at a product that is generally larger than the contributions of the individual portions. What are the disciplines

that contribute to a formal risk assessment? They include exposure assessment, toxicity testing, mechanistic and molecular toxicology, pathology, statistics, and mathematical modeling. In some recent examples of chemical risk assessments, such as with methylene chloride[1] and vinyl chloride,[2] all these disciplines are utilized to assess appropriate exposure guidelines for various human populations. The ongoing multi-volume reassessment of the risks of exposure to 2,3,7,8-TCDD[3] by the US Environmental Protection Agency (US EPA) outlines the broad set of considerations that are necessary in arriving at estimates of human risk from toxicant exposures.

To a large extent, the emphasis placed on an exposure-dose-response continuum for organizing toxicology data is directly related to the need to have the product of all the individual studies be useful for chemical risk assessment. The marriage of risk assessment methods and molecular toxicology requires close co-operation between risk assessment professionals and molecular toxicologists. The former will be challenged to keep abreast of the increase in understanding of biological processes afforded by the new genomic and proteomic tools available for studying biological function and toxicological problems. This task is formidable even for experts in the area. For the molecular toxicologists, the challenge is to understand the exposure-dose-response paradigm and to articulate how individual pieces of their work link together to inform the risk assessment process for determining expected effects on public health.

### 14.1.2.1.2   Risk Assessment and Toxicological Research

Risk assessment for chemical hazards in the workplace and in the general environment have become increasingly formalized over the past several decades. In early stages in the 1950s animal studies were used to determine No-Observed Effect Levels (NOELs). The US Food and Drug Administration (FDA) derived acceptable daily intakes (ADIs) by dividing animal NOELs by 100.[4] The factor of 100 consisted of two safety factors of 10 each, intended in a general way to account for (1) differences in sensitivity of humans compared to animals; and (2) variation in sensitivity of individuals in a heterogeneous human population compared to more homogeneous sensitivity in inbred animal stains. These ADIs were usually established based on organ or organism level responses that were clearly adverse to health. An underlying premise in this approach

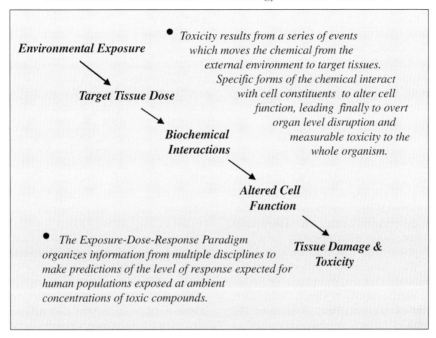

**Environmental Exposure**

• *Toxicity results from a series of events which moves the chemical from the external environment to target tissues. Specific forms of the chemical interact with cell constituents to alter cell function, leading finally to overt organ level disruption and measurable toxicity to the whole organism.*

**Target Tissue Dose**

**Biochemical Interactions**

**Altered Cell Function**

• *The Exposure-Dose-Response Paradigm organizes information from multiple disciplines to make predictions of the level of response expected for human populations exposed at ambient concentrations of toxic compounds.*

**Tissue Damage & Toxicity**

**Figure 1**   The major components of the exposure-dose-response paradigm for organizing toxicology research and testing for risk assessment applications.

was the existence of a threshold dose, i.e., the belief that there were concentrations or exposure levels below which the risk of adverse health effects was zero.

Modern toxicology has continuously incorporated research and testing strategies from the clinical and biological sciences. In the 1960s toxicologists borrowed methodologies from the medical community to assess organ level damage in animals treated with toxicants. Plasma enzymes and urinary enzymes were monitored to assess damage to kidneys and liver; BSP or inulin clearances assessed the functional consequences of chemical exposures on these organs. These responses, impaired organ function in the intact animal, were clearly detrimental to integrated function in the intact animal.

The 1970s brought a focus on the biology of cancer and a shift of testing and research resources in toxicology toward chemical carcinogenesis. The recognition of the role of mutagenesis as a requisite precursor to carcinogenesis brought attention to a variety of alternative test methods that could be used to assess mutagenesis in cells and simpler organisms. These alternative systems provided rapid screening of the mutagenic potential of many compounds. However, it was not clear how these data could be used to establish exposure standards. One obvious use for *in vitro* cell system data was in priority setting for long term carcinogenicity studies in animals. If a compound had high exposures in human populations and showed evidence for mutagenicity, it would be a higher priority compound for lifetime bioassay testing than a compound with similar exposure that lacked mutagenic activity. Another application is in product development. Compounds that proved strongly mutagenic in short-term assays could be dropped from further consideration, a sieving process.

Another contribution of the 1970s was development of risk assessment methods for carcinogens.[5] Animal studies provided information of the incidence of tumors at specific doses in test animals, usually rats and mice. How could these results be used to predict human risks at much lower doses by other routes of administration? Two extrapolations were introduced; one predicted the shape of the dose response curve at low levels of incidence, the second adjusted the expected responses for different species. The low dose extrapolation utilized a mathematical model of carcinogenesis, the linearized multi-stage (LMS) model. This model predicts some measure of carcinogenicity at every dose, no matter how small the dose. Interspecies extrapolation, justified based on studies with various chemotherapeutic compounds, was calculated on a surface area adjustment for dose. This body surface extrapolation regards humans as more sensitive to toxic responses than are the smaller rodent species.

Philosophically, this cancer risk assessment approach was quite distinct from that with the non-cancer endpoints. Threshold approaches formed the basis of ADIs; however, carcinogens were treated as if they had no threshold. Although the cancer methods were primarily intended for carcinogens that interacted with DNA via mutational processes, these extrapolation tools were quickly applied to all chemical carcinogens. The LMS model and body surface area-dose adjustments were defined as defaults to be used for DNA-damaging carcinogens, but came to be routine for all chemicals found to cause cancer. It bears emphasis that the understanding of the multiple molecular mechanisms of chemical carcinogenesis was much less developed in the mid-1970s when these initial cancer risk assessment paradigms were under development.

The concept of dose was also being refined by scientists who borrowed methods from the field of clinical pharmacokinetics (PK) to assess the relationship between exposure, sometimes called administered dose, and the concentrations of active chemical/metabolites at target tissues. The emphasis on compartmental pharmacokinetic models arose mainly due to the high doses used in many animal tests, doses at which capacity limited processes, i.e., metabolism, tubular excretion in kidney, *etc.*, became saturated. Scientists in the chemical industry were the first to apply these PK methods to many commodity chemicals and to discuss the need to relate toxicity to target tissue dose rather than simple measures of dietary composition or ppm of inhaled chemical in the air. Work with vinyl chloride carcinogenesis showed a clear relationship between metabolized dose of this compound and liver cancer.[6]

The 1980s provided several important developments for toxicology. Among them were increasing use of *in vitro* cell systems for assessing chemical interactions in living cells and the first applications of molecular techniques emerging from the new field of molecular biology. Another advance was the increasing sophistication applied to assessing how chemicals caused their effects, i.e., the mode of action of chemicals in biological systems. In addition, quantitative, mechanistic tools were increasingly being developed to assist in analysis of dose-response relationships. The earlier successes of compartmental PK models for unraveling dose-response curves continued in the development of physiologically based pharmacokinetic (PBPK) modeling to permit extrapolations across route, dose level, and species.[7] At the same time, mechanistic models for carcinogenesis, primarily the Moolgavkar–Venzon–Knudson (MVK) model,[8] provided a biological framework for considering the roles of mutation, cell birth and cell death, and cell differentiation within quantitative structures. The MVK model provided information on the role of promotion, defined as a growth advantage of preneoplastic versus normal cells, in giving rise to various shapes of dose response curves. The MVK model for cancer was an early example of what is now referred to as biologically based dose response (BBDR) models. These scientific contributions, coupled with the new quantitative tools in PBPK and BBDR modeling, provided pressure to apply this growing information to improve the scientific basis of chemical risk assessment.

### 14.1.2.1.3 Evolution of Risk Assessment Methods

The 1980s and 1990s also provided a continuous evolution of risk assessment methods with an increasing emphasis on introduction of "best available" science into the risk assessment process. Several advances in this area deserve mention in explaining the development of risk assessment procedures and preparing the stage for introduction of molecular toxicology studies into the process. The major contributions were:

- Publication of Risk Assessment in the Federal Government in 1983[9]
- Development of Guidelines for Various Toxic Responses by US EPA
- Refinement of Reference Dose/Reference Concentration Methods
- Publication of Science and Judgment in Risk Assessment in 1994[10]
- US EPA's Proposed Carcinogen Risk Assessment Guidelines in 1996.[11]

#### 14.1.2.1.3.1 Risk Assessment in the Federal Government: Managing the Process

This publication[9], also called the Red Book, provided a consistent structure for risk assessment. It attempted to create a set of consistent "inference guidelines" for use by federal regulatory agencies in the risk assessment process. This publication emphasized the need to separate the scientific and the policy aspects of risk assessment. It defined risk assessment as: " ... the use of the factual data base to define the health effects of exposures of individuals or populations to hazardous materials or situations." It also organized the risk assessment processes into four areas. Hazard identification is the determination of the effects of the chemical in exposed animals or people.

Dose-response assessment evaluates the exposure conditions under which these effects are observed. Exposure assessment estimates the amounts of the chemical present in the workplace, home, and general environment. The fourth component, risk characterization combines information on the exposures and dose-response assessment to estimate the risk level for specific individuals, groups, or populations. The recommendations of this report had widespread effects on both attempts to develop new carcinogen assessment guidelines[11] and on formulation of non-cancer risk assessment guidelines.[12–14]

#### 14.1.2.1.3.2 Guidelines for Studying Various Toxic Responses by US EPA

Subsequent to the Red Book, US EPA developed several guidance documents. These documents covered Superfund Sites,[15] exposure assessment,[16] specific types of toxic processes—cancer,[11,17] reproductive[18] developmental,[19] neurotoxicity,[20] and mixtures.[21] As the fields of toxicology and risk assessment continued to develop, early versions of guidance documents were revised. In general, the guidance documents for different toxic endpoints focused on hazard identification more than on dose response assessment. Future revisions of these documents will surely provide many opportunities for discussing areas where molecular endpoints, toxicogenomics and proteomics might be evaluated for specific target organ responses.

#### 14.1.2.1.3.3 Reference Dose (RfD) and References Concentration (RfC) Methods

Noncancer response data for orally administered compounds are examined empirically to derive a dose with some particular response rate, a no-adverse-effect level (NOAEL), a low-adverse-effect level (LOAEL), or a benchmark dose (BMD).[18–20,22] The BMD, NOAEL or LOAEL are further modified to account for several factors, including interspecies extrapolation, interindividual variability, extrapolation of subchronic-to-chronic exposure duration, the starting point for the risk calculation (i.e., is it a NOAEL, LOAEL or BMD) and the quality/breadth of the data base. Individual uncertainty factors (UCFs) with values of 10 or 3 are multiplied together leading to an aggregate uncertainty factor which that may be as large as 3000. Based on professional judgment, these calculations can

also be changed by application of a modifying factor (MF).

$$RfD = NOAEL\ (BMD)/$$
$$(UCF1 \times UCF2UCF3 \times UCF4 \times UCF5 \times MF)$$
$$(1)$$

In general, the RfD determinations have not had an explicit procedure built in for evaluating the impact of mode of action or toxicokinetics on risk evaluations. Furthermore, there has been little guidance provided describing how available data might be used to set appropriate values for uncertainty factors, though such proposals have been discussed in the toxicology literature.[23,24]

Inhalation exposure guidelines, referred to as reference concentrations (RfCs), account in a general fashion for deposition of gases and particulate matter within the respiratory tract.[25] The delivered dose for each region of the respiratory tract has been defined with default equations based on general knowledge of particle deposition and regional gas deposition characteristics within specific regions of the respiratory tract. For the RfC estimation, a BMD or NOAEL exposure concentration is first adjusted based on duration of exposure to account for continuous exposure and then corrected based on the respiratory tract defaults in order to calculate a human equivalent concentration (HEC). This adjustment accounts for dosimetry differences between test animals and humans. After these so-called "above the line" corrections, other uncertainty factors are applied as noted with the RfD calculation in equation (1).

Noncancer endpoint risk assessments have been based on the belief that there is a level of exposure, a threshold, below which the risks should be negligible. The difficulty in practice is defining objectively how these thresholds should be established.[26] The concept of a threshold may be biologically plausible. However, it is very difficult, if not impossible, to differentiate between the observance of no response because a dose is below a threshold, and the failure to observe a response due to being below the limit of detection in a typical laboratory study with a limited number of animals. The evaluation of the mechanistic basis of thresholds and the more complete elucidation of the dose response for these effects will be a major emphasis in future toxicological research. Such studies must be based on improved knowledge of the interrelationship of molecular, cellular, organ and organism level responses.

In looking at the historical development of noncancer dose-response approaches, there is the appearance that a new uncertainty factor is proposed in response to each newly identified deficiency in the current risk paradigm, without regard to how these adjustments fit into the existing strategy. An example of this proliferation in uncertainty factors is in the language of H.R.1627, The Food Quality Protection Act of 1996,[27] where a new safety factor of 10 was proposed to protect children and infants. The text reads, "In the case of threshold effects, for purposes of clause (ii)(I) an additional tenfold margin of safety for the pesticide chemical residue shall be applied for infants and children to take into account potential pre- and post-natal toxicity and completeness of the data with respect to exposure and toxicity to infants and children." Protecting children and infants is a commendable goal. If a policy decision is made to apply this uncertainty factor routinely, it needs to be explained as policy and not as a formal part of the dose-response assessment process. When various policy related uncertainty factors are included and treated independently (i.e., multiplied together without clear justification), the ability to conduct comparative risk ranking across endpoints or to develop a common metric for comparing relative risks becomes virtually impossible.

#### 14.1.2.1.3.4 Science and Judgment in Risk Assessment

In 1994, there was publication of a second important risk assessment oriented-document, *Science and Judgment in Risk Assessment.*[10] This work was commissioned by the authorizing legislation for the Clean Air Act (NAS/ NRC, 1994). Section 112(0) of the Clean Air Act Amendments of 1990 directed the EPA (1) to arrange for the National Academy of Sciences to review the methods used by EPA to determine the carcinogenic risk associated with exposure to hazardous air pollutants from sources subject to Section 112; (2) to evaluate methods used for estimating the carcinogenic potency of hazardous air pollutants; and for estimating human exposures to these air pollutants; and (3) to evaluate risk assessment methods for noncancer health effects for which safe thresholds might not exist. This report expanded concepts from the 1983 Red Book,[9] primarily in emphasizing that the interaction between research and risk assessment is an iterative process, whereby the conduct of specific risk assessments uncovers research needs for improving the process. Though primarily commissioned to address cancer

risk assessment, its recommendations are also pertinent for noncancer endpoints. One of its strongest recommendations was that the US EPA should provide a clear definition of the assumptions supporting the application of defaults and the explicit codification of the criteria for departure from these defaults. Default approaches, such as use of the linearized multistage model for low dose extrapolation and surface area adjustment for interspecies extrapolation, and default values, such as for uncertainty factors, are those options used in the absence of convincing scientific knowledge on which of several competing models and theories is correct.

In the "Validation: Methods and Models" section of the 1994 report, several recommendations were made with respect to developing specific models that influence the conduct of risk assessments across multiple endpoints. They were:

EPA should continue to explore and, when scientifically appropriate, incorporate pharmacokinetic models of the link between exposure and biologically effective dose (i.e., dose reaching the target tissue).

EPA should continue to use the linearized multistage model as a default option but should develop criteria for determining when information is sufficient to use an alternative extrapolation model.

EPA should develop biologically based quantitative methods for assessing the incidence and likelihood of noncancer effects in human populations resulting from chemical exposure. These methods should incorporate information on mechanisms of action and differences in susceptibility among populations and individuals that could affect risk.

#### 14.1.2.1.3.5 US EPA's Revised Carcinogen Assessment Guidelines

The revised carcinogen guidelines[11] advocate an approach in which evaluation of tumor incidence data in the range of observation and extrapolation to low doses are conducted as two distinct steps. After evaluation in the observable range, low-dose extrapolations can be conducted using several different options including linear, non-linear, or margin-of-exposure (MOE) approaches. These guidelines emphasize incorporation of mode of action, encourage use of pharmacokinetic information, and specify the default procedures. The preferred option for the low dose extrapolation is a biologically based dose response model.

In the proposed revisions to the cancer risk assessment guidelines, the provision for non-linear modes of actions and MOE calculations together with use of precursor effects, are in essence moving toward a dose response assessment strategy for non-genotoxic carcinogens similar to that used with these noncancer endpoints. However, the draft cancer guidelines emphasize mode of action in structuring the dose response assessment; in risk assessment for noncancer endpoints, this central theme, informing the various extrapolations based on knowledge of mode of action and toxicokinetics, has not been well developed.

## 14.1.2.2 MOLECULAR TOXICOLOGY AND RISK ASSESSMENT

### 14.1.2.2.1 Molecular Toxicology and the Challenge of Relevance

An important emphasis in the broad area of toxicology in the past 10 years is the emergence of concern for compounds that interfere with endocrine function.[28] The concern was heightened by publication of *Our Stolen Future*.[29] Endocrine active compounds (EACs) can alter responses to endogenous hormones affecting adult function, reproduction, and development. The interest in the potential public health consequences of these compounds coincided with the burgeoning knowledge and methodologies for assessing biological consequences of hormones on cellular function.

Hormones are messengers within the body passing information from one portion of the organism to another. They include compounds that interact with cell surface receptors and compounds that interact with intracellular receptors that regulate gene transcription. Mechanistic knowledge about the function of cytosolic and cell surface receptors increases almost daily. Additionally, the creative suite of new methods for assessing molecular and cellular responses has also increased remarkably. As the biological sciences provide new tools for research, these tools become available for refining toxicity research. This theme with molecular methods, toxicogenomics and proteomics, follows the other advances throughout the past 40 years in the field of toxicology. Do new biomedical tools automatically improve our ability to solve public health questions arising from chemicals in the environment? Without doubt, we are able to assess molecular

interactions at lower levels of exposure and to assess cellular level responses *in vitro* in a variety of natural and bio-engineered organisms. Research efforts utilizing these methods, however, need to be placed in a risk perspective to provide data useful for assessing public health consequences of exposures. Otherwise, the field of molecular toxicology stands in jeopardy of creating a huge hazard identification database that will inflame concerns about chemicals without providing exposure-dose-response perspective on adversity of the observed alterations or perspective on the linkage between these alterations and health consequences in target populations. From the point of view of a risk assessor or risk manager, there are a variety of disciplines that must contribute if the risk assessment is intended to have a strong basis in mechanistic biology (Figure 2). These individual disciplines can remain independent and Balkanized, as noted by the stark gridlines between them. In these cases, the integration process responsibility falls to the risk manager who is unlikely to be trained in the disciplines of molecular and cellular biology. The subsequent jumble of information represents a formidable obstacle in creating the assessment.

A major problem in "handing the data off" for interpretation by the risk assessor is that mechanistic data may fail to have a spokesperson arguing for their quantitative importance and noting how they might influence risk assessment. Molecular toxicology data collected and published, even in the most prestigious journal, do not automatically convey information about how they fit into the exposure-dose-response paradigm. Data do not talk for themselves. Scientists that collect the data need to serve as the spokesperson for their application and take responsibility for their inclusion in assessing risks of the chemicals in human populations. As noted earlier the challenges with application of molecular level data generally are two-fold. The first question is in how these observations relate to portions of the exposure-dose-response continuum that are immediately upstream and immediately downstream. The second question with molecular level observations is the relationship of precursor mechanistic endpoints to more obviously adverse responses at the organism level. The exposure-dose-response paradigm should be depicted by the more tightly integrated picture of a linked chain, connecting independent areas of research/testing to create a more seamless understanding of the responses of target organisms to toxic chemical exposures (Figure 3). Having provided the background on risk assessment methodologies, the remainder of the chapter discusses research

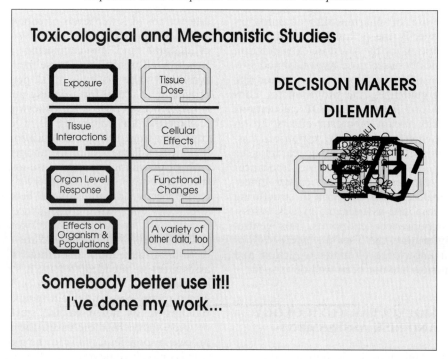

**Figure 2** In a not uncommon method of using information in risk assessment contributions are passed on in the absence of interpretive context from the practicing scientist to the risk assessor who may not be knowledgeable about the applications of all the cutting edge science.

strategies that introduce molecular methods within this exposure dose response structure.

### 14.1.2.2.2 Designing Molecular Toxicology Studies to Make a Difference

Default risk assessment methods represent a convenience for estimating dose response in the low dose regions in the absence of biological knowledge. The revolution in molecular toxicology is a subsidiary revolution based on the enormous growth of knowledge of genomics and the nascent growth of proteomics. More than ever before the science of toxicology, especially at the level of altered morbidity or mortality in exposed populations, is now a study of perturbation of normal biological systems. In the past, the normal system has been the black box of biology. Toxicologist established dose-response relationships as if toxicology were some fundamental property of nature rather than a secondary property expressed as an alteration in a dynamic biological environment. New light is shining on virtually every critical process related to cellular function, intercellular interactions, and organism level integration of distributed functions. Instead of focusing on individual proteins or messenger RNA species, genomic and proteomic methods permit assessment of the suite of genes of batteries of protein products

that are coordinately regulated and the manner in which toxicants alter their expression. In the weekly magazine *Science* throughout the year 2000, there appeared a series of articles, one per month, called Pathways of Discovery. The article for March, 2000 was "Genomics: Journey to the Center of Biology".[30] Toward the end the authors note the overall goals of studies on genomics, goals that apply equally to toxicology and risk assessment.

> "The long-term goal is to use this information to reconstruct the complex molecular circuitry that operates within the cell—to map out the network of interacting proteins that determines the underlying logic of various cellular biological functions including cell proliferation, responses to physiologic stresses, and acquisition and maintenance of tissue-specific differentiation functions. A longer term goal, whose feasibility remains unclear, is to create mathematical models of these biological circuits and thereby predict these various types of cell biological behavior."

In toxicology the questions we ask are how to reconstruct the normal circuitry and examine the manner in which excesses of various exogenous chemicals lead to physiological stress, alterations in gene batteries and resulting degradation of function. This manifesto for

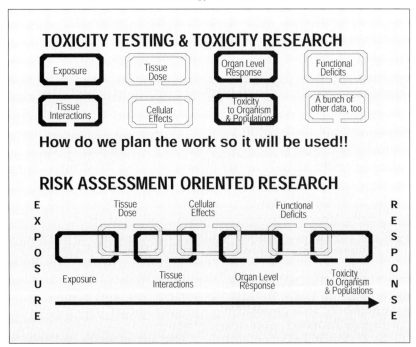

**Figure 3** To achieve a risk assessment focus the contributions from individual disciplines need to be designed to fit into the exposure-dose-response perspective, considering the adjacent disciplines and the relationship of all the steps to dose.

biology points to an integration of the multiple interacting biological cassettes that create normal cellular function and provides the main targets for assessing the actions of toxic compounds. The approaches to studying these changes in batteries of coordinately controlled genes include observational assessment of the changes in gene/protein expression after dosing with toxic compounds. This approach resembles studies in cancer in which the final aggressive transformed cell is compared with the initial normal cell to see the differences in characteristics between the two states. This strategy may place groups of compounds into categories in relation to their modes of action. It may tell little about the molecular alterations that move the cell from the initial to the final state. Uncovering the circuitry involved in these transitions will tell more about the molecular targets for the toxic responses and allow improved dose response assessment of the toxic actions of the compounds.

EACs are examples of compounds that interact with the normal cell circuitry to mimic or antagonize the actions and functions of normal signaling systems. Excess of the EAC or deficiencies of the natural hormone alter hormone system function leading to impaired health. Impaired health covers a wide range of responses—loss of viability, impaired performance, altered reproductive success, and delayed maturation. The actions of many signaling

cassettes—cell surface receptors, cytosolic transcriptional factors, kinase/phosphatase cascades—are now more completely understood than just a few years ago. Studies of the EACs almost all have to ask what is known about normal function and how is the normal function perturbed by exogenous compounds (Figure 4). In a very real way, this strategy is a perturbation theory approach to molecular and cellular toxicology. A re-orientation to a health based approach to toxic action rather than an emphasis on final pathology may provide new avenues to evaluate dose-response relationships in toxicology.

### 14.1.2.2.3 Maturation of the Exposure-Dose-Response Paradigm

The exposure-dose-response depiction in Figure 1 captures an ideal situation. Paramount to incorporation of molecular toxicological study results in risk assessments would be an understanding of normal function of target macromolecules within the cell. The extent to which the normal function was understood both qualitatively and quantitatively would determine the extent to which the impact of the perturbation could be assessed at various doses and in various species. Although this ideal may still be some distance off, it is no longer a fantasy. Some examples of candidates

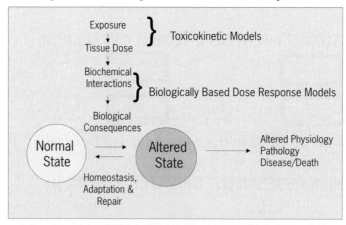

**Figure 4** The maturing exposure-dose-response paradigm for toxic responses is related to perturbations of the normal control processes in the cell by toxicant exposures.

for this approach would be EACs that affect kinetics of native hormones, such as estradiol, testosterone, or thyroid hormones. Such toxicants need to be evaluated based on their impact on time courses of endogenous hormones. One obvious research need with EACs is to adequately characterize the kinetics and the perturbation of the kinetics of endogenous hormones along with determination of the kinetics of the EAC. Continued efforts to more fully characterize the biological control mechanisms that regulate the hypothalamic-pituitary-gonadal feedback axes should be helpful in elucidating the dose response for estrogenic/anti-androgenic EACs, a topic of intense debate in the regulatory community.

### 14.1.2.2.4 Exploring the Molecular Basis of Dose-Response Curves

Current risk assessments are increasingly being organized around two primary concepts—mode of action and tissue dosimetry associated with particular modes of action.[31] Most toxicological data are utilized to come to one simple conclusion about the nature of the toxic response. Is the response expected to be low-dose linear or is it expected to have a threshold? If it is non-linear, the RfD/RfC approaches are employed to provide some justification of the uncertainty factor required for adequate protection to the general population. The nature of the response may lead to reduction or enhancement of the net uncertainty factors. However, these application of uncertainty factors, while embedded in a set of individual uncertainty factors that are legitimate concerns, are employed in a highly subjective manner. They do not have a clear relationship to the shape of the dose response

curve in low dose regions, i.e., those regions of most interest from a public health perspective.

One advance with molecular markers, such as induction of cytochrome P450 1A1 (CYP 1A1) by 2,3,7,8-TCDD[32] or adduct formation by DNA-reactive carcinogens,[33] is extension of the dose response region to lower ranges using these molecular markers. However, these measurements also lead to questions about the linkage between precursor effects and clearly adverse responses of the organism. Should the observation of a 1% increases in CYP 1A1 mRNA after 2,3,7,8-TCDD treatment be considered as adverse as higher dose observations of liver pathology with single cell necrosis? If these two endpoints, minimal induction or cell necrosis, were to be used for risk assessment, what types of low dose extrapolation would be applied and should risk assessments for cancer and non-cancer endpoints utilize different uncertainty factors?

A second important contribution of research that clarifies the mode of action is the ability to define the proper measure of target tissue dose, i.e., the proper dose metric. A mode of action statement conveys both the manner in which the compound alters the biology to cause the effect as well as the nature of the compound (parent compound, metabolite, DNA-or-Protein adduct, *etc.*) that is associated with the initial interactions leading eventually to an adverse response of the organism.[31] Once the appropriate tissue dose metric is identified, compartmental or PBPK models can be developed to predict these dose metrics in target tissues for various dose levels, dosage routes, and in humans. Pharmacokinetic modeling has become a mainstay of low dose and inter-species extrapolations for both carcinogenic and non-carcinogenic responses. The utility of these models depends on the strength of the association of a particular dose metric with a well-supported mode of action.

### 14.1.2.2.5 Mechanistic/Computational Concepts in a Molecular/Cellular Context

Despite the marvelous increase in sophistication of the methods available for studying cellular and molecular level responses, we are still faced with the same question, what is the shape of the dose response curve at low incidence levels for adverse effects? Will the new technology answer these questions or will we simply continue to collect more and more information to inform hazard identification without providing insight for dose response in the intact animal or human? This question is central to the expectations of the potential impact of molecular toxicology on improving human health risk assessments.

Probably the most significant contribution that may arise from the combination of molecular toxicology coupled with a perturbation theory of toxicity is the ability to understand the molecular basis of dose response relationships for toxic compounds and provide biologically plausible methods for low dose risk assessment. Our dose response assessment tools have largely been empirical or driven by defaults. The commentary in *Science*[30] talked about the long-term goal of creating mathematical models of biological circuits that describe responses to physiologic (also read *chemical*) stress and the acquisition and maintenance of tissue-specific differentiation function. A similar goal pervades the desire to create biological models of dose response curves for toxicity based on alterations in cell and tissue function. While most of our dose response assessment models have smooth continuous changes in response with dose, the real world of cells demonstrate a more complex variety of interacting circuitry. We describe chemical processes by statistical methods, average behaviors of molecules, since the numbers of particles involved in most reactions and interactions are large. At the cellular level, behaviors are stochastic. A cell either divides or it does not divide. A challenge in formulating the mathematical models of cellular functions is the requirement to grasp the manner in which continuous changes of chemical variables within the cellular machinery leads to dichotomous, discontinuous responses, such as apoptosis, proliferation, differentiation, or activation of global cellular circuitry by exposure to chemicals.

Stochastic, non-linear models of cellular level responses may provide the basis for developing tools that will predict threshold behaviors toward toxic exposures or predict dose regions where the proportionate response to increasing dose varies considerably from the dose response structure at high doses of toxicants. Of course, some of the models applied for assessing cancer risks are stochastic models of cell division, cell death, and cell mutation. The MVK model (Figure 5) represents a stochastic model of a biological process.[8] As noted in perturbation approaches to biological-toxicological responses, the models have to be initially set to adequately describe tumor incidence in the control animals. In deriving the BBDR models, it is necessary to evaluate the effect of dose on intrinsic biological parameters of the model. The effects can be described empirically, as has been done, or mechanistically. Mechanistically, the relationship between dose and proliferation or dose and apoptosis are unlikely to be simple continuous functions. The control of biological circuitry and the transition between different states of the cellular circuitry in response to exogenous signaling molecules will underlie the dose response manifestations of the toxic responses.

### 14.1.2.2.6 Applying these Mechanistic Ideas to a Toxic Endpoint with Tumor Promotion

Many liver tumor promoters appear to cause effects by increasing proliferation of hepatocytes transiently with longer term-adaptation to the exposures. The adaptation, with phenobarbital, involves elaboration of specific anti-proliferative cytokines that restrain proliferation.[34] Cells resistant to cyto-inhibition are presumed to derive a growth advantage and grow out to preneoplastic foci under the selection pressure from the promoter. The dose response relationship for carcinogenesis requires characterization of the dose of promoter required to increase the proliferation of hepatocytes. Phenobarbital, in common with a large number of liver tumor promoters, has a receptor-mediated mode of action. These promoters interact with protein receptors that serve as transcriptional modulators to alter expression of batteries of genes in the hepatocytes. Phenobarbital interacts via the CAR (constitutive androstane receptor).[35] Other promoters act via the Ah receptor (2,3,7,8-TCDD), the PPAR, peroxisome proliferation activating receptor (trichloroacetic acid), or the pregnane-X receptor. All of these receptors, in

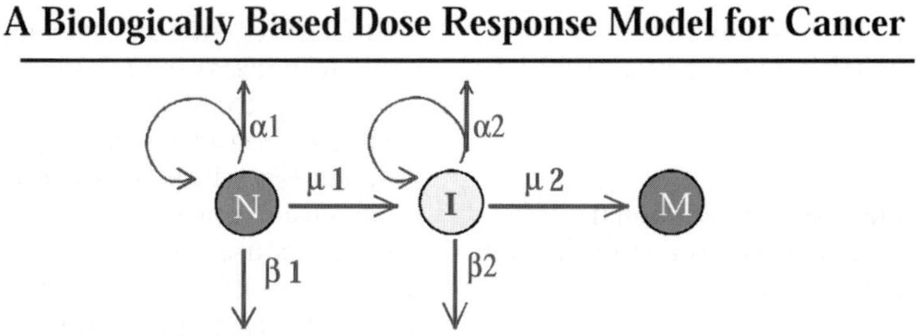

## A Biologically Based Dose Response Model for Cancer

**Figure 5** The development of cancer proceeds in two steps from the conversion of a normal cell (N) to an intermediate cell type (I) and the conversion of that intermediate cell to a malignant cell genotype (M). The expected incidence of cancer is a function of the birth ($\alpha$), death ($\beta$), and mutation rates ($\mu$) in the model. The linkage of these biological parameters, especially the birth and death rates, with dose of an exogenous compound may be highly non-linear and be associated with alterations in suites of gene products in the cells.

concert with the toxic compounds, act to increase expression of batteries of genes, leading to several alterations in expression of many individual gene products, including specific CYP P450 genes.

Among these promoters, 2,3,7,8-TCDD has received a great deal of attention in the past decade as US EPA has re-evaluated the risks of exposure to this environmental contaminant.[3] PBPK and protein induction models describe dioxin kinetics in the body, including binding to the Ah receptor with activation of specific gene products. These models have also described the induction of specific genes through interactions of the Ah receptor-TCDD complex with DNA-response elements for the Ah receptor and various partnering molecules (see this volume, Chapter 14.2.3). Like most mathematical models of biological systems, these BBDR models with dioxin represent a significant simplification of the individual molecular processes. Simplifications are necessary to attain a computationally tractable model and may be useful to gain insights about the major processes involved in creating a dose response behavior. However, it is important that the simplifications retain the important biological aspects of the responses.

With these inducers, the response of cells in the liver and of cells *in vitro* do not appear to follow a continuous response pattern where a 50% induction in the total liver is reflected by a 50% induction in all hepatocytes. The induction of individual cells occurs almost in an all-or-none fashion. Cells are either induced or remain in a basal state.[36,37] As dose of the inducer increases, the induction slowly fills in the entire liver acinus with centrilobular hepatocytes showing induction at lower doses than do the periportal hepatocytes. This response, a stochastic response of a cell to the receptor-ligand complex, is not yet well understood. The molecular substrate that causes this

switch-like behavior leads to a qualitative alteration in response over a narrow range of dose moving the cell from one state to a new one. These responses appear to represent a reversible differentiation to a new stable state (a new phenotype) of the hepatocyte.

The evaluation of these induction responses of hepatocytes leads to a set of interesting concepts for future exposure-dose-response assessments. They include switching, cellular circuits, and multiple stable states of the cells, in addition to our old concerns regarding the relationship of molecular level responses and the ultimate expression of toxicity. Chaos and complexity advocates have discussed concepts of stable attractors in complex systems. In particular, mammalian cells may exist in a suite of differentiated forms that represent stable attractors for the overall behavior of the genetic content of the cell.[38] Limited modeling of stable states for hormone responsiveness of cells has been pursued for estrogen responsive actions.[39,40] The two states of hepatocytes in a basal or induced state following tumor promoters may represent two stable attractors. The increasing concentration of the receptor ligand complex may alter the concentration of a limited set of initial gene products that move the circuitry from that for one stable attractor to a second stable attractor. After a period of time the overall content and behavior changes, consistent with the new stable state. The new state determines the pathological or physiological consequences of induction for the cell while the dose response of the process is more likely determined by some of the early interactions of the ligand and the receptor molecules. Non-linear switching modules exist for receptor auto-regulation,[40] kinase/phosphatase cascades,[41] $Ca^{++}$-mediated nerve cell signaling for long term potentiation,[42] among multiple possibilities. Although the non-linear characteristics of these switches are evident, none has been examined in sufficient detail to provide an understanding of the molecular basis of the switch and its influence on the dose-response curve at low doses.

Another area for interaction of dose-response modeling and molecular toxicology is the characterization of the relationship between molecular responses and ultimate toxic action. With 2,3,7,8-TCDD the dose response curves for more simple molecular responses tended to behave in a linear manner. For responses at the organ and organism level the responses were less likely to be linear with apparent Hill-equation slope factors of greater than 1.0.[3,43] Unraveling the control circuitry and the consequences of the molecular changes for the overall responses of the organism are important for providing the context to interpret many of the molecular changes that are measurable at much lower doses than the doses that are associated with overtly toxic responses.

In addition, the control of gene products by these transcriptionally active xenobiotics and their cognate receptors has been treated as if they were independent processes. However, the responses are in some way co-ordinated by the cell machinery to insure smooth induction of a battery of gene products. From a modeling vantage, the induction of a suite of distant genes, with individual promoter regions, in mammals simply by increasing the concentration of the inducer represents a competitive process, not a co-ordinated response. The response elements on the multiple distributed genes would compete for available inducer complex. In the induction models, either for regional induction in the liver with 2,3,7,8-TCDD[37] or with progesterone induced maturation of Xenopus oocytes,[41] the nonlinear response is described empirically, not mechanistically. Further studies of the early events following treatment and a focus on control of multiple gene batteries will provide more biological insights into these dose response behaviors.

New methods for modeling the control of gene batteries in normal system may utilize Boolean networks[38] or apply neural network models[44] for expression of multiple gene families. The contributions from groups developing more quantitative tools to assess physiological coordination of multiple cellular activities, e.g., the physiome project,[45] may significantly expand the mathematical dose-response modeling approaches used by toxicologists and risk assessors for integrating the exposure-response-dose paradigm into a health-based perturbation paradigm for assessing toxic potential of compounds.

### 14.1.2.2.7 Toxicokinetic Models for Extrapolation

The second linchpin of dose response assessment has become PBPK and other models for predicting tissue dose of active compounds for various exposure conditions. In truth, it is difficult to tell where the TK models end and the response models begin. The improvements in analytical chemistry in general and now the revolution in proteomics provide a broader array of potential endpoints for evaluation in these TK models including circulating compound/metabolites, and protein adducts in red blood cells, plasma proteins, saliva, hair, *etc.* Other applications of protein mass spectrometry will likely provide measures of function of adducted proteins and estimates of altered protein processing associated with exposures, but not directly related to adduct formation. PBPK models will need to provide more biological detail on protein processing, red blood cell control, and hair growth to link adduct concentrations and both external exposure (to better support exposure assessment) and tissue dosimetry within the test animal. As shown in Figure 4, the goal of both the toxicokinetic and the biologically based dose response models is to link exposures with a perturbation of the normal system that, when intense enough, leads to an altered physiology and overt pathological sequelae. Several chapters at the end of the book discuss the use of protein and DNA adducts for exposure assessment.

### 14.1.2.3 OPPORTUNITIES AND CONCLUSIONS

In retrospect, breakthroughs in biomedical technologies have continuously spurred new discoveries of modes of action and new approaches to measuring and defining toxicology over the past 50 or 60 years. These new technologies are in turn followed by changes in our basic understanding of biological and toxicological processes. Today's molecular toxicologists take advantage of a marvelous array of new technologies that have the potential to provide new discoveries in toxicology in a more integrative context. The recent practice of studying one gene at a time is fading fast; screening strategies are emerging that permit evaluation of the interplay between hundreds or even thousands of genes at both the *in vitro* and *in vivo* levels.

The turn of the millennium has coincided with major developments in widespread genome sequencing (human and mouse) and technical platforms to support "large-scale" functional gene analysis. The availability of sequence information for thousands of genes (and, in most cases the coding regions of these genes) has allowed investigators to construct large-scale gene microarrays for semi-quantitative measurement of the transcriptional activity of thousands of genes during chemical exposures.[46] These arrays can be either "broad spectrum" or custom designed to profile particular tissues (brain, liver, *etc.*) or specific toxicological pathways (AhR, PPAR, *etc.*). The utilization of this technology has become a distinct discipline, "toxicogenomics". Advances in this disciplinary activity will lead to a number of scientific windfalls (i.e., rapid fingerprints of chemical toxicity, a greater appreciation of molecular basis of toxicity, and lastly an enhanced ability to extrapolate between *in vitro* and *in vivo* approaches in the context of risk assessment). Toxicogenomics opens a window to examine multiple molecular responses (at low levels of exposure). However, the accretion of this data on sets of expressed gene/protein also raises questions regarding their interpretation with respect to both hazard assessment and risk assessment. Respectively, are the changes in expression adverse responses and do they serve as markers for adverse responses in human populations? Interpretation of these changes in gene and protein in a public health perspective will require merging toxicogenomic results with more traditional studies designed to understand toxicity at the physiological, pathological and biochemical levels.

A primary challenge for linking molecular toxicology and toxicogenomics, (especially the large-scale methods for assessing altered gene expression) with risk assessment is to avoid the collection of large bodies of data that serve only as hazard identification. Such uses of new methods tend to inflame public concern about potential risks of chemical exposures without providing any contextual analysis of the "actual risks" posed by exposures at low doses. Part of the solution to this is the design of studies that take advantage of the exposure-dose-response paradigm. For molecular markers, it is a focus on the steps immediately preceding and downstream of the markers in the exposure-dose-response continuum. Other questions arise in relation to the relationship of the doses used in cellular studies to doses that would be present in exposed persons.

Such concerns are applicable with any new methodology brought to bear on issues related to protection of public health from toxic compound exposures. The converse is the opportunities provided by new technologies to make advances on primary issues that have resisted resolution with other tools. As the methods of molecular toxicology, toxicogenomics and proteomics are brought to bear in integrated research programs, new perspectives will emerge regarding the steps involved in chemical–organism interactions and how these interactions progress to overtly adverse health consequences. From the personal perspective of the authors of this chapter, the possibility of explaining the molecular basis of non-linear dose response curves would be a significant contribution of molecular methods to public health in relation to toxic chemical exposures. With this information, it would be possible to predict the shape of the dose response curve at low doses in test animals and in human populations. Such insights would provide much more accurate risk assessments than estimates derived from arbitrary application of uncertainty factors and avoid the proliferation of uncertainty factors for each new concern and significantly increase the value of mechanistic toxicology data in protecting public health. These are exciting times for advancing our knowledge of the impact of xenobiotics on biological function.

Considerations of the integration of these molecular toxicology tools with public health needs for quantitative risk assessment will greatly enhance the value of these molecular toxicology studies and their influence on public health decision-making with toxic compounds.

## 14.1.2.4  REFERENCES

1. Occupational Safety and Health Administration (OSHA), 'Occupational exposure to methylene chloride; final rule.' *Fed. Reg.*, 1997, **62**(7), 1493–1619.
2. US Environmental Protection Agency. 'Toxicological review of vinyl chloride.' EPA/635R-00/004, 2000.
3. US EPA 'Exposure and Human Health Reassessment of 2,3,7,8-Tetrachlorodibenzo-*p*-dioxin (TCDD) and Related Compounds. Part II. Health Assessment for 2,3,7,8-Tetrachlorodibenzo-*p*-dioxin (TCDD) and Related Compounds.' NCEA-1-0835. May 2000. Science Advisory Board Review Draft, 2000.
4. A. J. Lehman and O. G. Fitzhugh, '100-Fold Margin of Safety.' *Assoc. Food Drug Off. U.S.Q. Bull.*, 1954, **18**, 33–35.
5. R. E. Albert, R. E. Train and E. Anderson, 'Rationale developed by the Environmental Protection Agency for the assessment of carcinogenic risks.' *J. Natl. Cancer Inst.*, 1977, **58**, 1537–1541.
6. P. G. Watanabe and P. J. Gehring, 'Dose-dependent fate of vinyl chloride and its possible relationship to oncogenicity in rats.' *Environ. Health Perspect*, 1976, **17**, 145–52.
7. H. W. Leung, 'Development and utilization of physiologically based pharmacokinetic models for toxicological applications.' *J. Toxicol. Environ. Health*, 1991, **32**, 247–67.

8. S. H. Moolgavkar and A. G. Knudson Jr., 'Mutation and cancer: A model for human carcinogenesis.' *J. Natl. Cancer Inst.*, 1981, **66**, 1037–1052.

9. NAS/NRC (National Academy of Sciences/National Research Council). 'Risk Assessment in the Federal Government: Managing the Process,' Washington, DC., NAS Press, 1983.

10. NAS/NRC. (National Academy of Sciences/National Research Council). 'Science and Judgment in Risk Assessment,' NAS Press, Washington, DC, 1994.

11. USEPA (US Environmental Protection Agency). Proposed Guidelines for Carcinogen Risk Assessment. Office of Research and Development, Washington, DC, USEPA 600-P-92-/P-92/003C, 1996.

12. USEPA (US Environmental Protection Agency). Integrated Risk Information System (IRIS), US Environmental Protection Agency (www.epa.gov/iris/index.html), 1996.

13. M. L. Dourson and J. F. Stara, 'Regulatory history and experimental support of uncertainty (safety) factors.' *Reg. Toxicol. Pharmacol.*, 1983, **3**, 224–238.

14. D. G. Barnes and M. Dourson, 'Reference dose (RfD): description and use in health risk assessments.' *Regul. Toxicol. Pharmacol.*, 1988, **8**, 471–486.

15. USEPA (US Environmental Protection Agency). Risk Assessment Guidance for Superfund: Human Health Evaluation Manual (Part A). Washington, DC, USEPA/5~0/1-89/002, 1989.

16. USEPA (US Environmental Protection Agency). 'Exposure Factors Handbook.' Office of Health and Environmental Assessment, Washington, DC, USEPA/600/8-89/043, 1989.

17. USEPA (US Environmental Protection Agency). 'Guidelines for carcinogen risk assessment.' *Fed. Regist.*, 1986, **51**, 33992.

18. USEPA (US Environmental Protection Agency). 'Guidelines for Reproductive Toxicity Risk Assessment.' Office of Research and Development, Washington, DC, USEPA/630/R-96/009, 1986.

19. USEPA (US Environmental Protection Agency). 'Guidelines for Developmental Toxicity Risk Assessment.' *Fed. Regist.*, 1991, **56**(234), 63798–63826.

20. Office of Research and Development, Washington, DC, USEPA/600/90/066F. USEPA (US Environmental Protection Agency). 'Guidelines for Neurotoxicity Risk Assessment.' *Fed. Regist.*, 1995, **60**, 52032–52056.

21. USEPA (US Environmental Protection Agency). 'Guidelines for Health Risk Assessment of Chemical Mixtures.' *Fed. Regist.*, 1986, **51**, 34014.

22. M. L. Dourson, 'Methods for establishing oral reference doses,' 'Risk assessment of essential elements', eds. W. Mertz, C. O. Abernathy and S. S. Olin, Washington, DC, ILSI Press, 1994.

23. A. G. Renwick, 'Safety factors and the establishment of ADI.' *Food Add. Contarn.*, 1991, **8**, 135–150.

24. H. A. Barton and S. Das, 'Alternatives for a risk assessment on chronic noncancer effects from oral exposure to trichloroethylene.' *Regul. Toxicol. Pharmacol.*, 1996, **24**, 269–285.

25. USEPA (US Environmental Protection Agency). Methods for Derivation of Inhalation Reference Concentrations and Application of Inhalation Dosimetry, 1994.

26. G. P. Daston, 'Do thresholds exist for developmental toxicants? A review of the theoretical and experimental evidence,' in: 'Issues and Reviews in Teratology', Volume 6, ed. H. Kalter, Plenum Press, New York, 1993.

27. The Food Quality and Protection Act. *Public Law*, 1996, 104–170.

28. R. J. Kavlock and G. T. Ankley, 'A perspective on the risk assessment process for endocrine-disruptive effects on wildlife and human health.' *Risk Anal.*, 1996, **16**, 731–739.

29. T. Colburn, D. Dumanoski and J. P. Myers, 'Our Stolen Future: Are we Threatening Our Fertility, Intelligence, and Survival? A Scientific Detective Story,' Penguin Books, New York, USA, Inc., 1996.

30. E. S. Lander and R. A. Weinberg, 'Genomics: Journey to the center of biology.' *Science*, 2000, **287**, 1777–1782.

31. M. E. Andersen and J. E. Dennison, 'Mode of Action and tissue dosimetry in current and future risk assessments.' *Sci. Total Environment*, 2001, in press.

32. J. P. Vanden Heuvel et al., 'Dioxin-responsive genes: Examination of dose-response relationships using quantitative reverse transcriptase-polymerase chain reaction.' *Cancer Res.*, 1994, **54**, 62–68.

33. J. A. Swenberg et al., 'DNA adducts: effects of low exposure to ethylene oxide, vinyl chloride and butadiene.' *Mutat. Res.*, 2000, **464**, 33–86.

34. R. L. Jirtle, S. A. Meyer and J. S. Brockenbrough, 'Liver tumor promoter phenobarbital: A biphasic modulator of hepatocyte proliferation.' *Prog. Clin. Biol. Res.*, 1991, **369**, 209–216.

35. D. J. Waxman, 'P450 gene induction by structurally diverse xenochemicals: central role of nuclear receptors CAR, PXR, and PPAR.' *Arch. Biochem. Biophys.*, 1999, **369**, 11–23.

36. A. M. Tritscher et al., 'Dose-response relationships for chronic exposure to 2,3,7,8-tetrchlorodibenzo-*p*-dioxin in a rat tumor promotion model: quantification and immunolocalization of CYP1A1 and CYP1A2 in the liver.' *Cancer Res.*, 1992, **52**, 3436–3442.

37. M. E. Andersen et al., 'Regional hepatic CYP1A1 and CYP1A2 induction with 2,3,7,8-tetrachlorodibenzo-*p*-dioxin evaluated with a multi-compartment geometric model of hepatic zonation.' *Toxicol. Appl. Pharmacol.*, 1997, **144**, 145–155.

38. S. Kaufman, 'At Home in the Universe', Oxford University Press, New York, 1995.

39. Z. Simon, A. Vlad and G. I. Michalas, 'Positive gene control and steroid hormone receptor protein interactions. Mathematical model.' *Studia Biophysica*, 1988, **125**, 137–142.

40. D. J. Shapiro et al., 'Estrogen regulation of gene transcription and mRNA stability.' *Recent Prog. Horm. Res.*, 1989, **45**, 29–58.

41. J. E. Ferrell Jr. and E. M. Machleder, 'The biochemical basis of an all-or-none cell fate switch in Xenopus oocytes.' *Science*, 1998, **280**, 895–898.

42. U. S. Bhalla and R. Iyengar, 'Emergent properties of networks of biological signaling pathways.' *Science*, 1999, **283**, 381–387.

43. L. F. McGrath, P. Georgopoulos and M. A. Gallo, 'Application of a biologically-based RfD estimation method to tetrachlorodibenzo-*p*-dioxin (TCDD) mediated immune suppression and enzyme induction.' *Risk Anal.*, 1996, **16**, 539–548.

44. J. Vohradsky, 'Neural network model of gene expression.' *FASEB J.*, 2001, **15**, 846–854.

45. The Physiome Project, University of Washington (www.physiome.org).

46. A. Puga, A. Maire and P. Medvedovic, 'Transcriptional signature of dioxin in human hepatoma HepG2 cells.' *Biochem. Pharmacol.*, 2000, **60**, 1129–1142.

J.P. Vanden Heuvel, G.H. Perdew, W.B. Mattes and W.F. Greenlee (Eds.)
Comprehensive Toxicology, Vol. xiv

# 14.1.3
# Receptor Theory and the Ligand-Macromolecule Complex

## JOHN P. VANDEN HEUVEL
*Penn State University, University Park, PA, USA*

---

## 14.1.3.1   INTRODUCTION

A key concept in both pharmacology and toxicology is that the chemical must achieve an adequate concentration at its site of action for a response to be seen. For many chemicals, the ultimate site of action is a cognate protein or "receptor." The main criteria for the operational term "receptor" are the functions of recognition and transduction. By this definition, a receptor must recognize a distinct chemical entity and translate information from that entity into a form that the cell can interpret, and alter its state accordingly. This altered state may be a change in permeability, activation of a guanine nucleotide regulatory protein or an alteration in the transcription of DNA. To differentiate a receptor from an enzyme, the recognition unit should not chemically alter the ligand and, to differentiate from a binding protein, a receptor must produce a biochemical change and transmit the signal. Often the receptor is a protein and a single component of a large complex of macromolecules that may include other proteins, RNA and DNA.

The quantification of the interaction between a macromolecule and a xenobiotic is often called *receptor theory*. As will be discussed subsequently, the development of receptor theory predates many of the modern techniques of molecular biology and coincides with development of analytical biochemistry and basic enzymology. The study of structure-activity relationships and modifications of chemicals to fit the active site of the macromolecule have become standard in the pharmaceutical industry and are an important part of modern pharmacology and toxicology. Also, the basic tools of quantification and characterization of a bimolecular interaction is applicable to many disciplines including enzymology (Michaelis complex between enzyme and substrate), immunology (formation of antibody–antigen complex), pharmacology (drug-receptor complex) and toxicology (xenobiotic-receptor complex).

Much of the conceptual framework regarding receptor theory evolved from pharmacology and the investigation of drug action. Consequently, the historical account of the development of receptor theory contains many references to drugs as opposed to hormones, neurotransmitters and xenobiotics. The term *ligand* (*L*) is used interchangeably with drug in this case, and simply denotes a chemical that binds with affinity and specificity to a receptor. Drugs and certain xenobiotics presumably bind to receptors designed for interaction with endogenous hormones and neurotransmitters. By way of definition, agonists are analogous to endogenous hormones and neurotransmitters, in the sense that they elicit a biological effect, although the effect elicited may be stimulatory or inhibitory. In contrast, antagonists are defined as agents that block receptor-mediated effects elicited by hormones, neurotransmitters, or agonist drugs by competing for receptor occupancy. Antagonists may not have an endogenous, physiological counterpart in the strict sense of a competitive inhibitor of receptor occupancy.

## 14.1.3.2   HISTORICAL PERSPECTIVE

The general treatises of "receptor theory" have a long history that dates back at least 100 years and for all intents and purposes mark the advent of contemporary pharmacology and toxicology. Modern scientists often take it for granted that drugs elicit their effects via interaction with specific receptors. However, this concept was not always evident and evolved as a result of the remarkable insights of early scientists exploring a number of fundamental living processes. In particular, pioneers such as Claude Bernard (1813–1878), Paul Erhlich (1854–1915), John Langley (1852–1926) and A. J. Clark (1885–1941) are responsible for much of our understanding of quantifiable receptor responses. For an extensive review of this historical perspective, the reader is directed to a book of great value *Cell surface receptors: A short course on theory and methods* by L. E. Limbird.[1]

### 14.1.3.2.1   Receptors and Chemical Selectivity

Although Claude Bernard did not talk about receptors *per se*, he demonstrated that the effects of a drug or toxin (in his case curare) depended on its ability to access a particular site of action, what he called a *locus* of drug effect. Bernard concluded from a detailed set of experiments with curare and frog muscle contractions that this chemical did not affect the sensory nerves, but instead altered motor nerve function. Through isolating particular nerves, applying curare followed by electrical stimulation, Bernard was able to show a locus of effect whereby some sites were resistant to curare's blockade of nerve transmission. His experiments revealed the existence of a neuromuscular junction prior to the demonstration of the muscular endplate as a discrete anatomic structure. Also, these studies showed that an unknown component of particular cells made them more/less sensitive to chemical insult.

The receptor concept is generally attributed to Paul Erhlich and expanded the idea of chemical "selectivity." His most acclaimed studies (which formed the basis of a Nobel Prize in Medicine in 1908) dealt with immunochemistry and antibody–antigen interactions. Erhlich demonstrated that antigen–antibody interactions could be described as direct chemical encounters. He postulated that mammalian cells possess entities or "side chains" that can associate with certain chemical groups on the antigen, and thus serve as the basis for attaching the antigen to the cell. In addition to antigens, he also felt that small molecules possessed distinct domains for binding to the target cell. His own studies established that the arsenoxide group of organic arsenicals was essential for the lethal effects of these agents. Target cells need to first "bind" the arsenical, and cells that were unable to recognize certain constituents were insensitive to toxicity. Inherent in his "side chain theory" was the concept of specific cell surface receptors as the basis for targeting bioactive agents to the appropriate responsive cell. Erhlich conjectured that the normal function of cellular side chains was the binding of cell nutrients, and that the affinity of toxic substances for these groups was the fortuitous analogy between the structure of the exogenous toxic substance and the endogenous nutrient. All of his studies form the basis of quote "*Corpora non agunt nisi fixata*" (Agents cannot act unless they are bound).

J. N. Langley studied the chemical basis for autonomic transmission and neuromuscular communication and extended the work of Bernard. Langley demonstrated that nicotine could chemically stimulate the frog gastrocnemius muscle even following severing and degeneration of its motor nerves, and that this action could be blocked by curare. The mutually antagonistic effects of curare and nicotine led Langley to conclude that nicotine

and curare act on the same site within the nerve or muscle which he called a *receptive substance*. He postulated that complexes are formed between the receptive substance and the drugs according to some law by which the relative concentration of the drugs and their affinity for each other are critical factors. Thus, Langley first stated the concept of drug-receptor interactions which predated the algebraic description of these interactions, the mass-action law, often attributed to A. J. Clark (see below).

### 14.1.3.2.2 Law of Mass Action and Occupancy Theory

As described above, the concept of receptor-mediated events dates back to the late 1800s and helped explain, qualitatively, the selectivity and saturability of drug action. However, it was not until A. J. Clark (1885–1941), performed his research that the quantitative nature of receptor theory emerged. Based on studies of the antagonism between acetylcholine and atropine, Clark postulated that drugs combine with their receptors at a rate dependent on the concentration of drug and receptor. Similarly, the resulting drug-receptor complex breaks down at a rate proportional to the number of complexes. These statements implied that drug-receptor interactions obey the *law of mass action* that was used to describe the adsorption of gases onto metal surfaces and other physical–chemical binding isotherms. Based on these principles, a mathematical expression can be provided to describe drug-receptor interactions:

Rate of combination $= k_1[L][R]$

Rate of dissociation $= k_{-1}[LR]$

Where $k_1$ = rate constant for combination (also called $k_{on}$); $k_{-1}$ = rate constant for dissociation also called $k_2$ or $k_{off}$);

L = concentration of drug;

$R$ = concentration of unoccupied receptor;

$LR$ = concentration of drug/receptor complex.

At equilibrium (or apparent equilibrium, steady state), the rate of combination equals rate of dissociation, or

$$k_1[L][R] = k_{-1}[LR]$$

Which can be arranged to

$$\frac{k_{-1}}{k_1} = \frac{[L][R]}{[LR]} = K_d \qquad (1)$$

The term $K_d$ is defined as the *equilibrium dissociation constant* and is usually expressed as a concentration. (The *equilibrium association constant*, $K_a$, is equivalent to $K_d^{-1}$, but is of limited use.) Equation (1) relates the concentration of drug applied, $L$, to the proportion of receptors occupied. Fractional occupancy ($Y$) is the proportion of all receptors that are bound to ligand at equilibrium.

$$\text{Fractional Occupancy} = Y = \frac{[LR]}{[R]_{\text{Total}}}$$

$$= \frac{[LR]}{[R] + [LR]}$$

Which after some algebra (multiply the numerator and denominator by $[L]$ and divide by $[LR]$), can be described as follows.

$$Y = \frac{[L]}{[L] + K_d} \qquad (2)$$

It is important to note that the algebraic relationship describing fractional receptor occupancy as a function of drug concentration (Equation (2)) is analogous to the quantitative relationships between enzyme and substrate introduced by Michaelis and Menton.

Subsequently, Clark extended his hypothesis that drug-receptor interactions obeyed mass action law by postulating that the fraction of receptors occupied, $Y$, was directly proportional to the response of the tissue. Some early data of Clark and others describing concentration-response relationships in various model systems were consistent with this postulate. However, certain data conflicted with this simple relationship between occupancy and effect. First, the slope of the concentration-response relationships reported was often steeper than predicted from Equation (2). Second, a number of examples existed where the application of extremely high concentrations of drug did not elicit a maximal response. These latter findings suggested that even maximal, saturating occupancy of a receptor does not necessarily translate into a maximal physiological effect.

The lack of maximal response in the presence of maximal occupancy proved to be troubling to early researchers and required the addition of a new concept. Ariëns introduced the term *intrinsic activity* to describe the ability of drug to elicit its pharmacological effect. He expressed the relationship between the effect ($E_L$) elicited by drug $L$ and the concentration of drug receptor complexes as follows.

$$E_L = \alpha[LR]$$

He defined $\alpha$ as the "intrinsic activity" of the particular drug. The intrinsic activity of a drug was meant to be a constant that determines the effect elicited per unit of LR complex formed. This early definition of intrinsic activity addressed the observation that maximal receptor occupancy by some agonists did not elicit a maximal response. However, this conceptualization still could not explain dose-response relationships that were steeper than predicted by mass action law. Obviously, the law of mass action and fractional occupancy were an oversimplification of the biological response and would require refinement. An important concept that developed from the work of Ariëns and Clark was that of *efficacy*, as described subsequently.

### 14.1.3.2.3    The Concept of Efficacy

It was Stephenson (1954) who introduced a major conceptual advance in the understanding of the quantitative relationship between receptor occupancy and receptor-elicited effects.[1] Stephenson argued that A. J. Clark's experimental findings were not in accord with a linear relationship between occupancy and effect. Although Stephenson concurred that Equation (2) is the probable relationship between the concentration of drug introduced and the concentration of drug-receptor complexes formed, he felt that there was no experimental justification for extending this relationship to the response of the tissue. He postulated three principles governing receptor-mediated functions that could explain the previously anomalous observation that agonist response curves were often steeper than the dose-response relationships that would be predicted by simple mass action law. In addition, his postulates offered an explanation for the observed progressive variation in the agonistic properties of a homologous series of drugs:

(1) An agonist when occupying only a small proportion of the receptors can produce a maximum effect.

(2) The response is a function of receptor occupancy, although not linearly proportional.

(3) Different drugs may have varying capacities to initiate a response and consequently occupy different proportions of the receptors

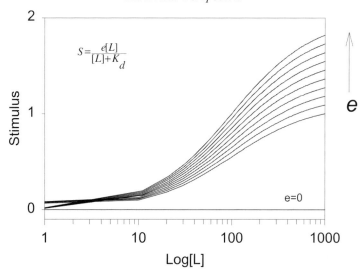

**Figure 1** Dose-stimulus relationship with drugs of increasing efficacy (*e*).

when producing equal responses. This property was referred to as the *efficacy* of the drug and is a property of the *L* and the *LR* complex. A pure antagonist would have an efficacy of zero.

Stephenson followed these postulates with the following derivations:

*S* = stimulus to the tissue from the ligand *L*

*S* = *eY* (the stimulus is proportional to fractional occupancy)

Where *e* is *efficacy* as described above and *Y* is fractional receptor occupancy

$$S = \frac{e[L]}{[L] + K_d} \quad \text{(From Equation (2) for } Y\text{)} \quad (3)$$

*R* = response of a tissue and *R* = *f*(*S*)

To test the validity of his postulates regarding various efficacies for different agonists, Stephenson studied a series of alkyl-trimethyl-ammonium salts on contraction of the guinea pig ileum. He noted that the short alkyl chain lengths (e.g., butyl-trimethylammonium) behaved as an agonist like acetylcholine, whereas the longer chain length homologues acted like atropine, an antagonist. He interpreted this antagonism as a property expected for a drug with low efficacy. Thus, the drug produces a response much less than maximum even when occupying all or nearly all of the receptors. However, because a drug with low efficacy can nonetheless occupy the receptors, it decreases the response elicited by a drug with high efficacy when added simultaneously. Stephenson termed these low efficacy drugs that possessed properties intermediate between agonists and antagonists *partial agonists*.

The concept of efficacy may be best described by examining a dose-relationship for a hypothetical series of drugs (see Figure 1) with efficacies ranging from 1 to 1000. Of course, an efficacy of zero would result in no stimulus, and hence no response. The response from efficacies greater than one would be a function of the stimulus; an increase in *e* results in an increase in the maximal stimulus without a major change in the shape of the curve.

### 14.1.3.2.4 The Concept of Spare Receptors

The law of mass action implies that the formation of a ligand-receptor complex is reversible. However, it was noted that certain antagonists did not follow this rule and were in fact irreversible, i.e., they form covalent bonds with a receptor. (The types of receptor antagonism will be discussed in detail in Section 14.1.3.4.3.) This was shown when extensive washing of a tissue preparation was unable to reduce the antagonist's effect. A receptor that is occupied by an irreversible antagonist is permanently disabled; this property of irreversible antagonists was used to examine a phenomenon know as *spare receptors*. Research performed by Furchgott (1967)[3] and others, showed that there was a *receptor reserve* and that 100% occupancy was not required for maximal response. Consider a simplified model as shown in Figure 2 in which drugs A, B and C stimulate the same receptor. Although all three agonists can elicit a maximal response, each requires a different percentage of receptors to be occupied (consider each drug-receptor complex to have a different intrinsic activity or efficacy). Upon irreversible

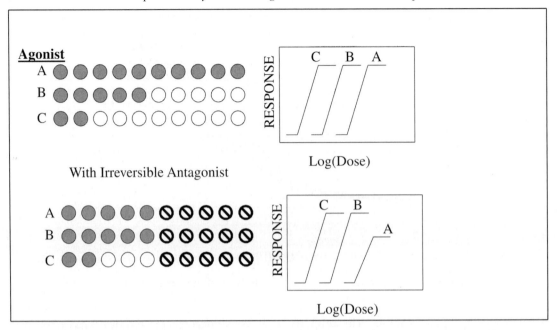

**Figure 2**  Hypothetical model for examining the existence of spare receptors. Three agonists (A, B, C) were examined in the absence (top) or presence (bottom) of an irreversible antagonist. In the case of drug A, there is no spare receptor and full occupancy is required for maximal induction. With drugs B and C only 50% and 20% occupancy is required. When one-half of the receptors are inhibited, only drug A's effects are reduced. Greater than 80% inhibition would be required to affect the maximal induction elicited by agonist C.

inactivation of a population of the receptors, it becomes possible to determine the fractional occupancy and rank the efficacies for the three agonists. If 50% of the receptors are inactivated, the maximal response of the agonist A is decreased by 50% whereas drugs B and C are unaffected. If 75% of the receptors are irreversibly inhibited, the response of A is decreased by 75%, B's effects are decreased by 50% and C still elicits a maximal response. Using this model system, Nickerson (1956) demonstrated that occupancy of only 1% of the histamine receptor population of guinea pig ileum was required to elicit maximal contractile response, suggesting the existence of a large receptor reserve for histamine receptors in this tissue. However, it is probably worth noting that receptor reserves were not always so dramatic. For example, Furchgott (1967) noted that for epinephrine and the muscarinic receptor there was only a shift of a half log unit, if anything, before a decrease in the maximum response was observed following exposure to the alkylating agent dibenamine. Goldstein (1974)[1,3] has offered a tenable teleological explanation for such a phenomenon, i.e., that sensitivity to low drug concentrations and low affinity is achieved by the spare receptor capacity. In circumstances where the response desired is to be rapid in onset and in termination, as in

neurotransmission, a spare receptor capacity provides a mechanism for obtaining a response at a very low concentration of an agonist that nonetheless has a relatively low affinity for the receptor. The low affinity (i.e., high $K_d$) of the drug assures its more rapid rate of dissociation, since $K_d = k_{-1}/k_1$. If, alternatively, sensitivity to low concentrations of agonist were achieved by a high affinity of the drug for the receptor, then the rate of reversal of the effect would necessarily be slow.

### 14.1.3.3   QUANTITATION OF RECEPTOR OCCUPATION

#### 14.1.3.3.1   The Law of Mass Action

Much of the discussion above showcases the difficulty in extrapolating receptor occupancy to responses elicited, especially in light of the existence of partial agonists, steep dose-response curves and spare receptors. However, the assessment of receptor occupation *per se*, follows the law of mass action in many instances. We will re-examine the law of mass-action as outlined by Clark and others and subsequently explain how these derivations may be used experimentally. Most analyses of radioligand binding experiments are based this simple model:

$$L + R \underset{k_{-1}}{\overset{k_1}{\rightleftharpoons}} LR$$

where $L$ is the ligand, $R$ is the receptor and $LR$ is the ligand-receptor complex.

The model is based on these simple ideas:

(1) Binding occurs when ligand and receptor collide due to diffusion, and when the collision has the correct orientation and enough energy. The rate of association (number of binding events per unit of time) equals $[L][R]k_1$, where $k_1$ is the association rate constant in units of $M^{-1}min^{-1}$.

(2) Once binding has occurred, the ligand and receptor remain bound together for a certain amount of time influenced by the affinity of the receptor and ligand for one another. The rate of dissociation (number of dissociation events per unit time) equals $[LR]k_{-1}$, where $k_{-1}$ is the dissociation rate constant expressed in units of $min^{-1}$.

(3) After dissociation, the ligand is the same as before binding. This is to differentiate a receptor from an enzyme.

(4) Equilibrium is reached when the rate at which new $LR$ complexes are formed equals the rate at which these complexes dissociate. Equilibrium may be more appropriately termed *apparent equilibrium* or *steady state*, since the achievement of true equilibrium is often not possible in pharmacologic systems.

At equilibrium, $LR$ complexes form at the same rate that they dissociate:

$$[L][R]k_1 = [LR]k_{-1}$$

which may be rearranged to define the equilibrium dissociation constant $K_d$.

$$\frac{[L][R]}{[LR]} = \frac{k_{-1}}{k_1} = K_d$$

The $K_d$, expressed in units of moles/liter or molar, is the concentration of ligand that occupies half of the receptors at equilibrium (see below). A small $K_d$ means that the receptor has a high affinity for the ligand whereas a large $K_d$ means that the receptor has a low affinity for the ligand. Often $K_d$, the equilibrium dissociation constant, is confused with $k_{-1}$, the dissociation rate constant. However, they are obviously not the same, as can be seen by the fact that their units are different.

The law of mass action predicts the fractional receptor occupancy at equilibrium as a function of ligand concentration. Fractional occupancy $(Y)$ is defined as the fraction of all receptors that are bound to ligand:

$$Y = \frac{[LR]}{[R]_{\text{TOT}}} = \frac{[LR]}{[R] + [LR]}$$

$$= \frac{[L]}{[L] + K_d} \qquad \text{(See Equation (2))}$$

This equation predicts the following. When no ligand is available, the occupancy equals zero. When the concentration of ligand is very high (many times $K_d$), the fractional occupancy approaches (but never reaches) 100%. When $[L] = K_d$, the fractional occupancy is 50%. Equation (2) predicts that the approach to saturation as ligand concentration increases is quite slow. When the ligand concentration equals four times its $K_d$, it will only occupy 80% of the receptors at equilibrium. The occupancy rises to 90% when the ligand concentration equals 9 times the $K_d$. It takes a concentration equal to 99 times the $K_d$ to occupy 99% of the receptors at equilibrium.

### 14.1.3.3.2 Assumptions Inherent in the Law of Mass Action

The law of mass action is a convenient and simple model to describe a bimolecular interaction. However, in order to coerce a complex biological interaction into a simple model, several assumptions or criteria must be met. Although termed a "law", the law of mass action is simply a model based on these assumptions:

(1) *All receptor sites are equally accessible to ligand.* In the simplest model, all receptor sites are considered to have equal affinity for ligand and to be independent. That is, occupancy of some receptor sites does not alter the binding to other, unoccupied sites (*cooperative binding*).

(2) *The two reactants (receptor and ligand) are either free or bound.* The model ignores any states of partial binding. Also, the measured products do not include degraded, metabolized or other unavailable forms of drug or receptor. Importantly, *non-specific binding* must be accounted for such that the concentration of the free or bound ligand is measured correctly. Non-specific binding includes any non-receptor site for ligand binding that would diminish free concentration of the chemical being examined.

(3) *Binding is reversible.* The association between a ligand and receptor depends only on the interaction of ligand and receptor and dissociation only on the breakdown of the ligand receptor complex.

**Table 1**   Considerations when choosing an isotope for labeling of ligand in receptor binding assays.

| Isotope | Specific activity (Ci/mmol) | Half-life | Other considerations |
|---|---|---|---|
| $^3H$ | 29.4 | 12.3 years | Bioactivity usually unchanged by tritiation; can introduce $> 1$mole $^3H$/mole ligand to increase specific activity |
| $^{125}I$ | 2125 | 60.2 days | Require tyrosine or unsaturated cyclic system in ligand structure to achieve incorporation of $^{125}I$; high specific radioactivity especially useful when receptor availability limited |
| $^{32}P$ | 9760 | 14.2 days | Short half-life makes $^{32}P$ difficult choice |
| $^{35}S$ | 4200 | 86.7 days | Good sensitivity; used in metabolic labeling |
| $^{14}C$ | 0.064 | 5568 years | Poor sensitivity |

From Reference 3.

The discussions that follow regarding the incubation and plotting methods are dependent on the assumptions. If it is not possible to meet all the criteria above, two choices exist. If the first assumption is not satisfied, there are more complicated model systems to describe ligand-receptor complexes that are cooperative. The reader is directed to Limbird (1996),[1] *Cell Surface Receptors* for description of these models as they are beyond the scope of the present chapter. If the assumptions 2 and 3 are not met, the only option is to analyze your data in the usual way, but interpret the results as an empirical description of the data without attributing rigorous thermodynamic meaning to the $K_d$ values and rate constants. In other words, the data must be considered as highly suspect.

### 14.1.3.3.3   Assessment of Receptor Occupation

#### 14.1.3.3.3.1   Choice of Label

In order to examine the binding of ligand and receptor, the former must be easily quantified. The most common means to quantify a ligand is to add an isotope to form a *radioligand*. One of the most critical considerations in the successful application of radioligand binding techniques is the choice of an appropriate isotope for radiolabeling the ligand. Each common isotope has certain characteristics to take into consideration (see Table 1). For example, tritium is a popular choice because of its long half-life. However, tritium exchanges with water and may require re-purification. Carbon-14 ($^{14}C$) has benefits of ease of incorporation into organic ligands, but suffers from low specific activity. Iodine-125 ($I^{125}$) is the most common choice for radioligand products and affords examination of

low affinity binding and low concentration of receptor. However, the incorporation of iodine into a drug may affect its affinity for a biological receptor. Taken together, the choice of isotope dependent on sensitivity issues (receptor density, counting efficiency, availability of tissue) as well as ease of synthesis.

Another option that is gaining popularity is to examine a drug or chemical's fluorescence, either natural or added exogenously. Whether this method can be performed depends on the ligand under question, but when available represents a sensitive and rapid means to examine receptor-ligand interactions.

#### 14.1.3.3.3.2   The Incubation

Critical components of the incubation include the labeled ligand, a source of receptor (crude membrane preparation, recombinant protein *etc.*) and a suitable buffer. Also, the incubation must proceed for a sufficient amount of time, at an appropriate temperature, until enough ligand-receptor complex is formed to be detectable. In general, increasing the concentration of either or both of the retactants $L$ and $R$ can increase the overall rate of association. These manipulations also will increase the amount of ligand-receptor complex formed and thus lengthen the time required to reach equilibrium. Optimal incubation conditions are determined empirically for each system and are chosen to maximize complex formation will minimizing the degradation of $L$ and $R$.

The source of receptor is dependent on the system under study and may include crude extracts, membrane preparations, cytosol, or nuclear extracts. Typical concentrations of receptors in crude preparations are $1 \times 10^{-12}$ to $1 \times 10^{-9}$ M (femtomoles to picomoles per mg tissue). However, in many instances the receptor is expressed in bacteria, yeast or insect cells, if possible. This allows for a much higher concentration of receptor than seen *in vivo*, often at higher purity (>1 mg/ml). However, receptors that are multiprotein complexes or require post-translational modifications are difficult to examine using recombinant proteins.

### 14.1.3.3.3.3 Separation of Bound from Free Ligand

To determine the quantity of *LR* accumulated, one must have a method for determining the incubation that permits the resolution of *L* from *LR*. When receptors are embedded within membranes, the free and bound forms of ligand may be easily separated by filtration or centrifugation. The speed of separation is an important consideration in determining which method to use to separate *L* from *LR*. If the rate of dissociation is slow in comparison to the rate of separation, the amount of *LR* formed can be determined accurately. This is the case for high affinity ligands (low $K_d$). However, if the rate of ligand dissociation is high, the rate of separation may become an important consideration for accurately measuring the amount of ligand-receptor complex.

Consider the following equation (for a derivation of Equation (4), see L. Limbird, 1996).[1]

$$K_d = \frac{[L][R]}{[LR]} = \frac{k_{-1}}{k_1} = \frac{6.93 \times 10^{-1} \text{ M sec}}{t_{1/2}} \quad (4)$$

where $t_{1/2}$ is the half-life of the *LR* complex (amount of time required for *LR* to decrease by 50%). If a 10% loss of *LR* is considered acceptable during the separation process, then a 0.15 times the half-life is the allowable time required. (Note that these calculations are based on a $k_1$ of $10^6 \text{ M}^{-1} \text{ sec}^{-1}$ and that $k_{-1}$ is equivalent to $\ln 2/t_{1/2}$.) As shown in Table 2, with low affinity ligand-receptor interactions,

**Table 2** Relationship between equilibrium dissociation constant $K_d$ and allowable separation time of *LR*.

| $K_d$ | Allowable separation time (0.15 $t_{1/2}$) |
|---|---|
| $10^{-12}$ | 1.2 days |
| $10^{-11}$ | 2.9 hr |
| $10^{-10}$ | 17 min |
| $10^{-9}$ | 1.7 min |
| $10^{-8}$ | 10 sec |
| $10^{-7}$ | 0.1 sec |
| $10^{-6}$ | 0.01 sec |

From Reference 3.

the separation of *L* from *LR* must exceeding rapid. In the proceeding section, the basic separation techniques will be briefly discussed. For a more detailed description of these methods see Levitski (1984)[2] or Limbird (1996).[1]

#### (i) Equilibrium Dialysis

The dialysis chamber consists of two compartments separated by a membrane to which the receptor preparation is added to one side and the radiolabeled drug to the other. After the solution reaches equilibrium, samples from both chambers are taken and the radioactivity measured. The amount of ligand-receptor complex formed is determined based on the assumption that the free drug is equal on the protein and non-protein containing compartments.

| Protein compartment | $LR + L$ |
|---|---|
| -Non-protein compartment | $- \quad L$ |

| Difference | $LR$ |
|---|---|

(the concentration of bound radioligand)

The advantages of equilibrium dialysis include the fact that equilibrium is not disturbed when samples are taken and therefore is an accurate estimate of *LR*. Also, this method is useful for low affinity receptor-ligand interactions. Equilibrium dialysis is no longer a frequently utilized method in receptor binding assays. This is predominantly due to the long dialysis time required to reach equilibrium, thus increasing the risk of protein or ligand breakdown and decreasing throughput.

#### (ii) Centrifugation

The following techniques require that the ligand-receptor complex be formed and is at equilibrium with the free ligand. There are several variations of centrifugal separation of ligand-receptor complexes. If the receptor is membrane bound, the ligand-receptor complex may be centrifuged and a tight pellet formed. The amount of radioactivity retained in the pellet represents *LR* and the free ligand is found in the supernatant. One minor variation is to layer the receptor-containing solution on a high density solution (sucrose or oil). This alteration results in less free drug being trapped in the pellet, but also makes the assay more cumbersome and increases the risk of *LR* dissociating during the centrifugation.

In the case of soluble receptors, either the receptor or the drug may be precipitated by centrifugation. In the former situation, an antibody specific to the receptor may be coupled to a large molecule such as sepharose. Upon centrifugation, the pellet will contain *LR* and the supernatant *L*, as described above for membrane bound receptors. Alternatively, agents such as polyethyleneglycol (PEG), or Sephadex G-50 may be used to non-specifically precipitate proteins. Alternatively, the radioligand may be precipitated with the aid of dextran-coated charcoal, leaving the *LR* in the supernatant.

Typically, centrifugation can pellet the bulk of the particulate matter within 5 sec, and as shown Table 2 is appropriate for receptors with a $K_d$ of $10^{-8}$ M and lower. This method is relatively fast and many replicates may be examined in one assay. However non-specific binding and trapping of free ligand is often a major limitation.

### (iii)   Vacuum Filtration

The incubation is terminated by pouring the mixture through a filter under vacuum. The proteins (*R* and *LR*) are retained by the filter and the free radioactivity flows through. This procedure can handle a large number of samples and the filtration devices are commercially available. The filters are generally made from glass, nylon or nitrocellulose which have high affinity and capacity for protein binding. The amount of *LR* formed is determined by counting the amount of radioactivity present on the filter after washing with a suitable buffer. Filtration usually takes 2–5 seconds per wash, which means that the minimum affinity of a receptor for a ligand that can be identified by this technique is $10^{-8}$ M. As with other methods, to slow down the dissociation rate of the drug, the washing of the drug-receptor complex is performed with cold buffer. Nonspecific binding and trapping of radioligand to the filter are potential problems associated with vacuum filtration.

### (iv)   Scintillation Proximity Assays

A newer type of assay is the scintillation proximity assay, which does not require the free and the bound drug to be separated. In this assay, the receptor is attached to a synthetic bead which contains a scintillin (a material that emits photons when it absorbs radioactive emissions). In most instances the receptor is bacterially expressed, purified and chemically attached to the bead, although it is possible to adhere membrane bound or crude receptor preparations. When a radioactive ligand is added to the medium, only the portion that is bound to the receptor will be close enough (in proximity to) the scintillin and

thereby produce a photon emission. Therefore, the *LR* is assessed as the number of light units produced and is a true equilibrium binding assay. Scintillation proximity assays have become quite popular because of their high-throughput capabilities. However, the production of the synthetic beads containing the receptor of interest may be an arduous task for many laboratories.

### 14.1.3.3.3.4   Nonspecific Binding

In addition to binding to the receptors of physiological interest, radioligands bind to nonreceptor sites. Any interaction with a non-receptor site would be considered *nonspecific* binding and is found at excess over true, specific interactions with the receptor. When performing radioligand binding experiments, a measure of both total and nonspecific binding is performed, and specific (receptor) binding calculated as the difference (see Figure 3). The binding of radioligand is examined in the presence of excess, saturating amount of unlabeled ligand. This technique assumes that the 100-fold excess of nonradioactive ligand saturates the specific, receptor binding but has no effect on the nonspecific sites. A useful rule-of-thumb is to use the unlabeled compound at a concentration equal to 100 times its $K_d$ for the receptor, if that is known from previous experiments. The drug used in excess may be the compound as the radioligand, but unlabeled, or it may be a chemical that is known to bind to the same receptor. Ideally, you should get the same results defining nonspecific binding with a range of concentrations of several ligands. Nonspecific binding is generally proportional to the concentration of radioligand (within the range it is used) and will not plateau at high concentration. However, as shown in Figure 4 and discussed subsequently, the specific binding is saturable and has a hyperbolic shape when examined on an arithmatic plot.

### 14.1.3.3.3.5   Additional Criteria

The binding of a ligand to a receptor is an *in vitro* experiment performed under very controlled situations. The desire is to be able to extend this data to a physiologically relevant situation. The following are criteria that may be used to assess whether the data generated may indeed to extrapolated to a real-world (i.e., a real cell) situation.

• The specific binding of *L* should be saturable, since a finite number of receptors are expected in a biological preparation.

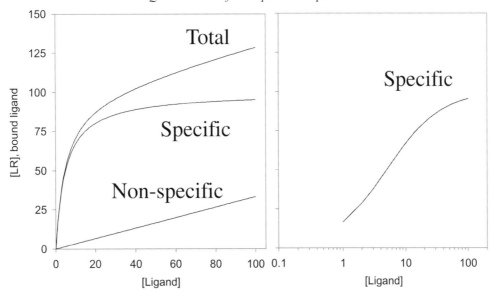

**Figure 3** Hypothetical binding curves. The left figure shows total and nonspecific binding. The difference between total and nonspecific binding is the specific binding and if this is plotted on a log-dose scale, has a sigmoidal shape (right panel). If an insufficient number of concentrations are used, often the plateau is not evident in such a representation.

• The specificity of agents in competing for *L* for binding to *R* should parallel the specificity of physiological effect via the putative receptor of interest. The use of competition binding will be discussed subsequently.

• The kinetics and *LR* of binding should be consistent with the time course of the biological effect elicited by *L* and there should be consistency between $K_d$ values obtained using steady state incubations and those determined through kinetic experiments.

$$[LR] = \frac{[L][R]_{tot}}{K_d + [L]}$$

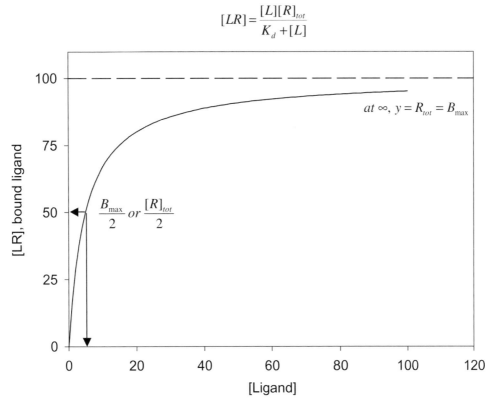

**Figure 4** The hyperbolic relationship of ligand binding isotherms.

#### 14.1.3.3.4 Plotting Methods

As shown in Figure 4, the relationship between receptor occupancy ($[LR]$) and drug concentration is hyperbolic when a saturable receptor population binds to ligand in a freely reversible, bi-molecular interaction. The reason for the shape of the curve is discussed below. Often the data is examined by linear transformation of the data. The advantage of such transformations is that $K_d$, and receptor density ($B_{max}$ or $R_{tot}$) can be easily determined. However, if the drug concentrations examined are insufficient, linear transformations may give a distorted view of the binding phenomena. In this section, both non-linear and linear transformations will be discussed. All plotting methods must meet the following criteria: (1) Assumptions of the law of mass action, as described above, are met; (2) The incubation and separation techniques are appropriate; and (3) Nonspecific binding has been adequately determined.

##### 14.1.3.3.4.1 Plotting Saturation Binding Data

Saturation binding experiments measure specific binding at equilibrium at various concentrations (often 6–12) of the radioligand to determine receptor number and affinity. Use of drug concentrations that allow binding to **approach saturation** is crucial for accurate examination of the interaction between drug and receptor. The next consideration is that of non-specific binding. As mentioned above, a good rule-of-thumb is to use a concentration of unlabeled ligand at 100 times the $K_d$. When examining binding in the presence of $100 \times K_d$, the amount of $[LR]$ detected is considered *nonspecific*, and as shown in Figure 3 is linear with respect to free ligand concentration. *Total* binding is the amount of $[LR]$ detected in parallel samples without this competitor included in the incubation. The difference, (*total – nonspecific*) is the *specific binding*, and is the data that should conform to the law of mass action. It is this binding that we will examine in more detail.

One of the criteria stated above for a physiologically relevant receptor is that the binding sites are saturable. To assess saturability, the characteristics of binding as a function of increasing concentrations of radioligand are determined. The saturation binding curve can described by Equation (2)

$$Y = \frac{[LR]}{[R]_{TOT}} = \frac{[L]}{[L] + K_d} \quad \text{(See Equation (2))}$$

which can be rearranged to the following,

$$[LR] = \frac{[L][R]_{tot}}{K_d + [L]} \quad (5)$$

Equation (5) is the equation of a rectangular hyperbola $y = \frac{ax}{b+x}$. An important consequence of this equation representing a rectangular hyperbola is that the horizontal asymptote is $R_{tot}$ or $B_{max}$. Thus $B_{max}$ can only be obtained at infinite concentrations of $L$. This will be of importance when linear transformations are discussed. Other important pieces of information can be obtained from Equation (5). The term $K_d$ is a measure of the affinity of the receptor for ligand and is the concentration of ligand that occupies one-half of the maximal binding sites. This can be easily derived from Equation (5) if $[LR]$ is replaced by $\frac{[R]_{tot}}{2}$ and solving for $K_d$ (the result being $K_d = [L]$). The estimation of $B_{max}$ and $K_d$ can be solved by either linear transformation or by non-linear regression, which is described below.

##### 14.1.3.3.4.2 Scatchard Plots

The most common linear transformation of binding data is the Scatchard plot. In this plot, the $X$ axis is specific binding (usually labeled "bound") and the $Y$ axis is the ratio of specific binding to concentration of free radioligand (usually labeled "bound/free") (Figure 5). Some of the terms used in Scatchard plots are:

$B =$ Bound ($LR$)    Concentration of ligand in the incubation that is specifically bound to the receptor at equilibrium.

$F =$ Free ($L$)    Concentration of free ligand present in the incubation at equilibrium. As we will discuss in Section 14.1.3.3.6. Ligand depletion, this is often estimated by the concentration of the drug added to the incubation.

$K_d$    Equilibrium dissociation constant. In the Scatchard plot, the slope of the line is equal to $K_d^{-1}$.

$B_{max}$ ($R_{TOT}$)    Maximum number of binding sites in the incubation at equilibrium, or total receptor concentration. $B_{max}$ is expressed in the same units as the drug concentration (i.e., molarity).

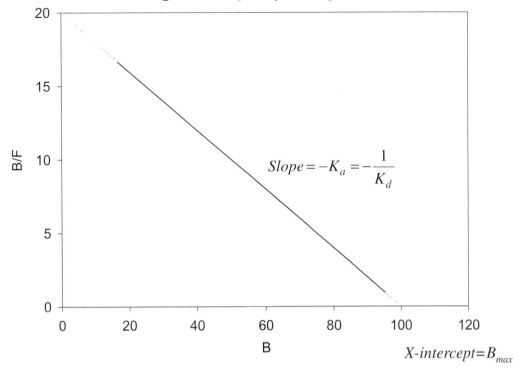

**Figure 5** Scatchard Analysis. The slope of B/F versus B is equal to $-1/K_d$ whereas the $X$-intercept estimates the total binding sites ($B_{max}$ or $LR_{max}$).

The reason for plotting $B/F$ as the $y$-axis and $B$ as the $x$-axis comes from rearranging equation Equation (5) (and substituting $B$ for $LR$, $F$ for $L$ and $B_{max}$ for $R_{tot}$):

$$B = \frac{(F)(B_{max})}{K_d + F} \text{ becomes } (B)(K_d) + (B)(F)$$
$$= (F)(B_{max})$$

Dividing by $F$, and rearranging results in Equation (6), in the form of $y = mx + b$.

$$(B/F)(K_d) + B = B_{max}$$
$$(B/F) = \frac{B_{max}}{K_d} - \frac{B}{K_d}$$
$$(B/F) = \frac{B_{max}}{K_d} - \frac{B}{K_d}$$
$$(B/F) = -\frac{1}{K_d}(B) + \frac{B_{max}}{K_d}$$

When making a Scatchard plot, you have to choose units for the $Y$ axis. One choice is to express both free ligand and specific binding in counts per minute (cpm) so the ratio bound/free is a unitless fraction. The advantage of this choice is that you can interpret $Y$ values as the fraction of radioligand bound to receptors. If the highest $Y$ value is large (greater than 0.10), then the free concentration will be substantially less than the added concentration of radioligand, and the standard analyses are not appropriate. You should either revise your experimental protocol or use special analysis methods that deal with ligand depletion (see 14.1.3.3.6). The disadvantage is that you cannot interpret the slope of the line without performing unit conversions. An alternative is to express the $Y$ axis as fmol ligand bound per mg protein per concentration (nM). While these values are hard to interpret, they simplify calculation of the $K_d$ which equals the reciprocal of the slope. The specific binding units cancel when you calculate the slope. The negative reciprocal of the slope is expressed in units of concentration (nM) which equals the $K_d$.

### 14.1.3.3.4.3 Double Reciprocal Plot

As the name implies, a double reciprocal plot is simply the reciprocal of the amount of specifically bound ligand versus the reciprocal of the free ligand concentration. The $y$-intercept yields the reciprocal of the value of maximal binding ($1/B_{max}$) while the $x$-intercept represents the negative reciprocal of the equilibrium dissociation constant ($-1/K_d$). Note that this is identical to the Lineweaver–Burke plot (Figure 6).

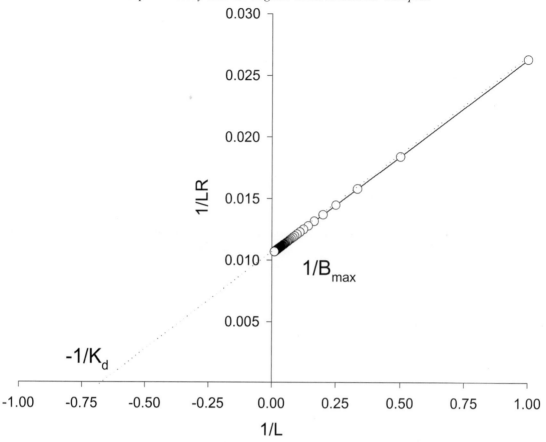

**Figure 6**   Double reciprocal plot. When the inverse of specific binding $(1/LR)$ is plotted versus inverse of ligand concentration $(1/L)$, a straight line is obtained. The $y$-intercept is $1/B_{max}$ and the $X$-intercept in $-1/K_d$.

#### 14.1.3.3.4.4   Hill Plot

The Hill plot was originally used to describe the cooperativity of binding of oxygen to hemoglobin and has been adapted to examine ligand-receptor interactions. The general equation from which the derivation is based:

$$[LR] = \frac{[L]^n \times [LR]_{max}}{K_d + [L]^n} \qquad (7)$$

The difference between Equations (5) and (7) is the inclusion of $n$ or the Hill Coefficient. The Hill Coefficent is the maximum number of ligand molecules bound to each molecule of receptor and is also a measure of cooperativity between binding sites. When a ligand binds to a single class of noncooperative receptors, this binding is described by the law of mass action and $n = 1$. If $n$ is greater than 1, evidence supports positive cooperatively while $n$ less than 1 indicates negative cooperativity or multiple classes of binding sites.

Equation (7) can be rearranged to the following linear equation,

$$\mathrm{Log}\frac{[LR]}{[LR]_{max} - [LR]} = n\,\mathrm{Log}[L]$$
$$- n\,\mathrm{Log}(K_d) \qquad (8)$$

Therefore a plot of $\log[LR]/([LR]_{max} - [LR])$ as a function of $\log[L]$ yields a line whose slope is $n$ and whose intercept on the ordinate is $- n \log K_d$ (see Figure 7).

In addition to the Hill plot, the cooperativity between binding sites may be examined by Scatchard analysis. In this case the plot is non-linear and concave when positively cooperative and convex when negatively cooperative (see Figure 8). Both the Hill plot and the Scatchard may be used to determine if your experiments are fulfilling the criteria listed in Section 14.1.3.3.2, in particular if the binding sites are of a single population and independent of each other.

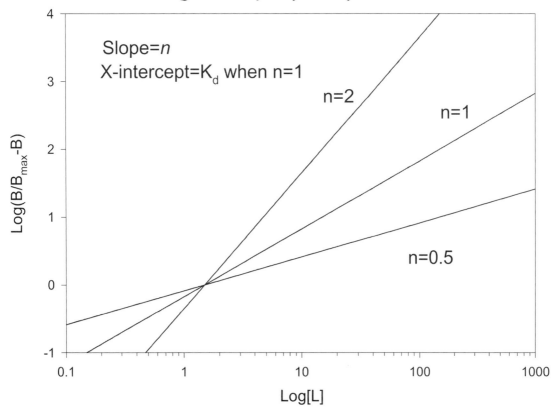

**Figure 7** Hill plot. When the log (Specific binding/(Maximum specific binding-specific binding)) is plotted versus the log [Ligand] a straight line with a slope equal to 1 is obtained, if the law of mass action is being followed. If positive cooperativity is observed, the slope is >1; in the case of negative cooperativity the slope is <1.

### 14.1.3.3.4.5 *Linear Regression and Analysis Error*

While Scatchard, double reciprocal and Hill plots are very useful for visualizing data, they may not be the most accurate way to analyze data. Linear regression assumes that the data is normally (gaussian) distributed and that the standard deviation is the same at every concentration of ligand. This is not the case with data transformed in the Scatchard plot. A second problem is value of $X$ (bound) is used to calculate $Y$ (bound/free). Since the assumptions of linear regression are violated, the $B_{max}$ and $K_d$ you determine by linear regression of Scatchard transformed data are likely to be further from their true values than the $B_{max}$ and $K_d$ determined by nonlinear regression.

With many receptor binding experiments, there is insufficient data points throughout the binding isotherm. This results in a clustering of data points when the Scatchard analysis is performed. For example, if there are few data points at the beginning of the curve and many in the plateau region the Scatchard transformation will give undo weight to the data point collected at the lowest concentration of radioligand (the upper left points on the Scatchard plot). Although it is inappropriate to analyze data by performing linear regression on a Scatchard plot, it is often helpful to display data as a Scatchard plot. As mentioned above, visually interpreting data with Scatchard plots can be of diagnostic value, but it may not result in the most accurate values of $K_d$ and $B_{max}$.

### 14.1.3.3.4.6 *Fitting a Curve to Determine $B_{max}$ and $K_d$*

Equilibrium specific binding at a particular radioligand concentration equals fractional occupancy times the total receptor number ($B_{max}$ or $R_{tot}$):

Specific Binding $= [LR]$

$$= \frac{[L]B_{max}}{K_d + [L]} \text{ (See Equation (5))}$$

This equation describes a rectangular hyperbola or a binding isotherm. As before, $[L]$ is the concentration of free radioligand, and is

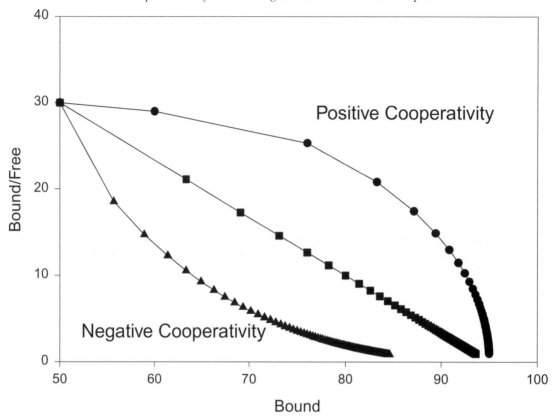

**Figure 8** Non-linear Scatchard plots. If the law of mass action is being observed, a plot of B/F versus B will be a straight line. However, in the case of cooperative binding, the Scatchard plot is non-linear.

plotted on the $X$ axis. $B_{max}$ is the total number of receptors expressed in the same units as the $Y$ values (i.e., cpm, sites/cell or fmol/mg protein) and $K_d$ is the equilibrium dissociation constant (expressed in the same units as $[L]$ usually nM). Typical values might be a $B_{max}$ of 10–1000 fmol binding sites per milligram of protein and a $K_d$ between 10 pM and 100 nM.

To determine the $B_{max}$ and $K_d$, fit specific binding data to Equation (5) using nonlinear regression. There are currently many resources available for performing non-linear regression and many statistics and plotting programs will perform the analysis. An excellent source for a detailed description of non-linear regression may be found at the GraphPad website.*

### 14.1.3.3.4.7 *Evaluation of Binding Data: Have the Assumptions been Met?*

In order for the binding data to result in physiologically relevant values of $K_d$ and $B_{max}$, the law of mass action must be followed. Also,

*Introduction to nonlinear regression, Copyright 1999 by Graph-Pad Software, Inc.
http://www.graphpad.com/curvefit/introduction.htm

the choice of ligand, separation procedures and incubations must be appropriate. When examining your own data, or evaluating binding data within a manuscript, it is advantageous to evaluate the quality of studies. Here are some guidelines that may be used for evaluation of studies on receptor–ligand interaction.

*Has the law of mass action been followed?* In general, there are some benchmarks that can be used to assess whether the law of mass action is being followed. First, the Scatchard plot must be linear and the Hill plot should have a slope equal to one. If a non-linear Scatchard is observed, more than one population of receptor exists, cooperativity is an issue and $K_d$ is not a constant. Second, *ligand depletion* should not occur such that the amount of drug added to the incubation is an appropriate approximation of free drug, or $L$. The amount of binding (total binding) in an incubation should be less than 10% of the amount added. See Section 14.1.3.3.6 for details on what to do if you exceed this value. Third, the incubation must be at steady state, or pseudo-equilibrium. To decrease the time to steady-state, one of the constituents should be at much lower concentration than the other. Since the ligand is the constituent that is measured, decreasing its concentration may affect sensitivity. Therefore, the receptor concentration should be limiting. ($[R] \ll K_D$; $[R] \ll [L]$). Note, this will also decrease the risk of ligand depletion.

*Is the binding physiologically relevant?* The physiological relevance often cannot be addressed without conducting other studies in addition to receptor binding. However, there are a few issues that can be examined. First, the formation of the ligand–receptor complex must appear to be saturable. Note, as described above, you will never achieve 100% binding, but complete saturation should be approached. If a plateau is not observed, either non-specific binding has not been properly accounted for, or a non-specific interaction is being examined. Second, the specificity of *L* for *R* should parallel the physiological effects, if known. That is, the $K_d$ should be within an achievable concentration at the receptor site. Examining a series of chemicals and measuring their affinity for a particular receptor may also address physiological relevance. Affinity and potency of effect should have a similar rank order within this group of chemicals. This may be easier to examine by competition binding experiments, as described subsequently.

*Have the experiments been properly designed and executed?* Several important issues regarding the design and implementation of a ligand binding study was discussed in 14.1.3.3.3. The following questions should be asked when interpreting receptor characterization studies.

(1) *Was the choice of radioligand and source of receptor appropriate?* In particular, the source of receptor is often an issue. Bacterial and yeast expression systems are used frequently to produce soluble receptors. Often, these purification methods result in receptor preparations that recapitulate the *in vivo* situation. However, if post-transcriptional processing is required for receptor function, these systems may not be appropriate.

(2) *Was the separation of bound and free drug performed appropriately?* In particular was the time of separation adequate for the $K_d$ of the receptor-ligand interaction (see Table 1).

(3) *Was nonspecific binding measured correctly?* Remember the rule of thumb that the concentration of competitor should be 100 times the $K_d$ of the drug receptor complex.

### 14.1.3.3.5 Competitive Binding Experiments

#### 14.1.3.3.5.1 *Why Use a Competitive Binding Curve?*

Competitive binding experiments measure the binding of a single concentration of labeled ligand in the presence of various concentrations of unlabeled ligand. To quantitate the potency of ligands in competing for the receptor, the $IC_{50}$ (concentration that inhibits 50% of the specific radioligand binding) value is determined for each competitor. Competitive binding experiments have several advantages to direct binding assays and have been used for the following types of experiments:

• *Validate a direct binding assay.* Competition binding experiments are an excellent way to examine the physiological significance of a direct binding assay. In this type of analysis the radioligand is competed with ligands whose potencies are known from functional experiments. Demonstrating that these ligands bind with the expected potencies, or at least the expected order of potency, helps prove that your radioligand has identified the correct receptor.

• *Determine whether a ligand binds to the receptor.* You can screen thousands of compounds to find those that bind to the receptor simply by examining if they are able to compete with a known ligand. This can be faster and easier than other screening methods. In fact, this is the most common way the pharmaceutical companies are identifying novel ligands for important receptor systems.

• *Investigate the interaction of low affinity ligands with receptors.* Binding assays are only useful when the radioligand has a high affinity ($K_d < 100$ nM). A radioligand with low affinity generally has a fast dissociation rate constant, and will not stay bound to the receptor while you wash the filters or pellet the complex.

#### 14.1.3.3.5.2 *Performing the Experiment*

The experiment is done with a single concentration of radioligand, usually equal to the $K_d$ of the ligand for the receptor. A higher concentration will increase the sensitivity of the assay and decrease counting error but will also increase non-specific binding and time to equilibrium. As with direct binding experiments, the incubation should reach equilibrium. However, the reaction is more complicated with the presence of a competitor, often at high concentration. To assure that a steady-state has been reached, the incubation proceeds for 4–5 times the half-life of the radioligand for receptor dissociation as determined in an off-rate experiment. In order to have a compete profile of the competition, typically 12–24 concentrations of unlabeled compound spanning about six orders of magnitude are examined.

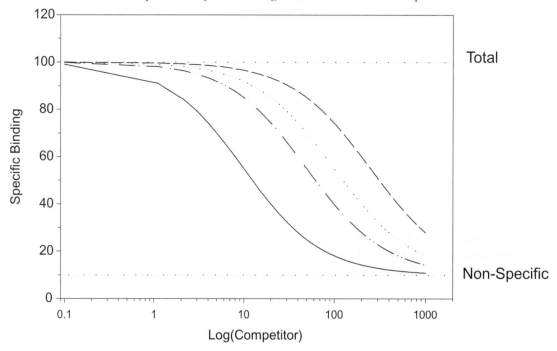

**Figure 9** Competition binding experiments. Total binding of a single concentration of radioligand is examined following incubation with graded concentrations of competitor. Shown are four competitors with different affinity for the receptor. The plateau at the top of the curve represents total binding whereas the bottom plateau is non-specific binding. The data may be analyzed using non-linear regression to calculate IC$_{50}$ values. $[LR]_I = NS + \frac{\text{Total}-NS}{1+10^{\log[I]-\log IC_{50}}}$.

### 14.1.3.3.5.3 *Analyzing Competitive Binding Data*

Visual inspection of competition of radioligand binding reveals a sigmoidal curve, with the most potent ligands inhibiting specific binding at lower concentration than less potent chemicals (Figure 9). The plateau at the top of the curve, the radioligand binding in the absence of the competing unlabeled drug, represents total binding. Note that total binding is not the same as $B_{max}$ since the receptor is not fully saturated (if using a $K_d$ concentration of radioligand). The bottom of the curve is a plateau equal to nonspecific binding (NS) with the difference between the top and bottom plateaus being specific binding. The IC$_{50}$ is the concentration of unlabeled drug that blocks half the specific binding.

If the labeled and unlabeled ligand compete for a single binding site, the steepness of the competitive binding curve is determined by the law of mass action. As shown in Figure 10, the curve descends from 90% specific binding to 10% specific binding with an 81-fold increase in the concentration of the unlabeled drug. More simply, nearly the entire curve will cover two log units (100-fold change in concentration). This generalization can be determined in terms of $[L]$, $K_d$ and the fractional saturation, $Y$.

$$Y = \frac{[L]}{K_d + [L]}$$

The fractional occupancy at 90% and 10% saturation will be

$$0.9 = \frac{S_{0.9}}{K_d + S_{0.9}} \quad \text{and} \quad 0.1 = \frac{S_{0.1}}{K_d + S_{0.1}}$$

Solving these equations simultaneously, and introducing the term cooperativity index $(R_s)$, or the change in drug concentration from 90% to 10%:

$$R_s = \frac{S_{0.9}}{S_{0.1}} = \frac{\frac{0.9}{0.1}}{\frac{0.1K_d}{0.9K_d}} = 81$$

When the interaction between competitor and receptor does not proceed via the law of mass action, the value of $R_s$ will not equal 81 (positive cooperativity $R_s < 81$; negative cooperativity $R_s > 81$.

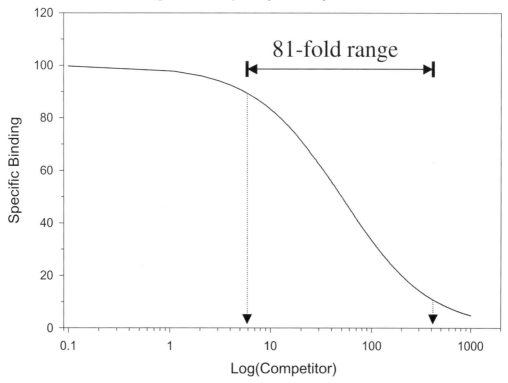

**Figure 10** Analyzing assumptions of the law of mass action. The law of mass action predicts that the concentration of competitor that results in 90% and 10% inhibition cover a 81-fold (roughly two orders of magnitude) range.

Competitive binding curves are described by this equation:

$$[LR]_I = NS + \frac{\text{Total} - NS}{1 + 10^{\log[I] - \log IC_{50}}} \quad (8)$$

where $[LR]_I$ (*Y*-axis) is the amount of binding measured at each $[I]$ or competitor (*X*-axis). Nonlinear regression is used to fit your competitive binding curve to determine the $\log(IC_{50})$. In order to determine the best-fit value of $IC_{50}$, the nonlinear regression problem must be able to determine the 100% (total) and 0% (nonspecific) plateaus.

Alternatively, values of $IC_{50}$ can also be ascertained by logit-log plot (or pseudo-Hill).

$$\log \frac{[LR]_I}{[LR] - [LR]_I} = n \log[I] + n \log IC_{50} \quad (9)$$

Where $[LR]$ is the amount of binding in the absence of inhibitor and $n$ is Hill coefficient. Plotting the $\log([LR]_I/[LR] - [LR]_I)$ versus $[I]$ will yield a straight line with slope equal to $n$ and an *x*-intercept equal to $IC_{50}$.

#### 14.1.3.3.5.4  Calculating the $K_I$ from the $IC_{50}$

The $IC_{50}$ value is not equivalent to $K_d$ for the competitor (or $K_I$) and is dependent on the amount of radioligand in the incubation. Therefore, the $IC_{50}$ value can vary between studies whereas $K_I$ is a constant. The $K_I$ can be calculated from $IC_{50}$ using the Cheng and Prusoff method

$$K_I = \frac{IC_{50}}{1 + \frac{[L*]}{K_d}} \quad (10)$$

where $K_I$ is the equilibrium dissociation constant for the inhibitor and $K_d$ is the equilibrium dissociation constant for the radioligand $L*$. Several assumptions were made in the derivation of the Cheng and Prussof equation that must be met.

(1) The radioligand and the inhibitor must interact with receptor according to the law of mass action. That is, the binding of both chemicals should be reversible and to a single population of $R$. Whether the law of mass action has been met can be determined from an pseudo-Hill plot where the slope equals $-1$.

(2) The concentration of $L$ added should equal the amount free. There should not be ligand depletion ($[LR] < 10\%$) and the concentration of receptor is much less than $K_d$.

(3) The incubation has reached equilibrium for radioligand and all concentrations of the competitor.

### 14.1.3.3.6   Ligand Depletion

The equations that describe the law of mass action include the variable $[L]$ which is the free concentration of ligand. All the analyses presented so far depend on the assumption that a very small fraction of the ligand binds to receptors (or to nonspecific sites) so you can assume that the free concentration of ligand is approximately equal to the concentration added. In some experimental situations, the receptors are present in high concentration and have a high affinity for the ligand, so that assumption is not valid. A large fraction of the radioligand binds to receptors so the free concentration of radioligand is quite a bit lower than the concentration added. If radioligand depletion occurs, alternative analysis is required. For example, the free concentration of ligand can be measured in every tube. Alternatively, the free concentration in each tube by subtracting the number of cpm of total binding from the cpm of added ligand. One problem with this approach is that experimental error in determining specific binding also affects the calculated value of free ligand concentration.

### 14.1.3.4   QUANTITATION OF RECEPTOR RESPONSES

#### 14.1.3.4.1   Why Examine Receptors Based on Responses?

The interaction between the receptor and ligand is the first step in the elicitation of a biological response. In fact, there are three components of a receptor system: the ligand, the receptor and the *effector* ($E$) (see pathway below). The effector may be an enzyme, an ion or a transcription factor and it transmits the biophysical interaction of ligand and receptor into a biochemical or molecular signal. Physiological receptors are linked to the *signal transduction* apparatus of the cell. Ultimately, the dose-response of a cell to a ligand is more complicated than the direct binding assay may predict. The ultimate response will be proportional to the amount of ligand-receptor complex formed, but is not necessarily a direct proportionality. Therefore, it is often difficult to determine $K_d$

values for a ligand-receptor interaction based solely on the response of that cell.

$$L + R \underset{k_{-1}}{\overset{k_1}{\rightleftarrows}} LR + E \underset{k_{-2}}{\overset{k_2}{\rightleftarrows}} LRE \overset{k_3}{\longrightarrow} Effect$$

In addition to adding complexity, the effector system amplifies the response. Therefore, detection of a receptor-mediated event is often easier to examine than is direct binding *per se*. This is the major reason why receptors are characterized by their responses prior to the implementation of direct binding assays. Sensitive assays for detecting influx or efflux of $Na^+$, $K^+$, $Ca^{2+}$ or $Cl^-$, activation of adenylate cyclase, phosphorylation of a receptor or effectors, alterations in mRNA expression, have been developed. In addition, binding of a chemical to a receptor says nothing about the response this ligand may elicit. That is, antagonists and agonists may bind with the same affinity to a receptor (in fact antagonists are often more avidly associated) but are coupled to the effector molecular in a different manner. Therefore receptor responses are often more physiologically relevant than are direct binding assays.

Prior to the advent of modern molecular pharmacology, which has allowed for cloning of receptors without knowledge of their function or ligands, the existence of a cognate receptor for a particular chemical was based on cellular responses. Some of the criteria that lead investigators to believe that a given effect was receptor-mediated included:

(1) *Specificity of response.* It is often the specificity of a drug, hormone or xenobiotic that persuades an investigator that the effects are receptor-mediated. Rank order of potency for effect of a series of agonists should be characteristic of a particular receptor.

(2) *Existence of antagonists.* Antagonists can selectively block receptor-mediated events. The order of potency of antagonists should be characteristic of a particular receptor.

(3) *Protection against irreversible receptor blockade.* In the presence of a ligand, the irreversible antagonist cannot reach its site of action.

(4) *Cross-desensitization.* If you can down-regulate the receptor, ligands will have less of an effect. Genetically-altered "Knockout" mice are the ultimate desensitized model.

Examination of receptor responses offers a mechanism by which the physiological relevance of a ligand–receptor interaction can be confirmed. If the four criteria listed above can be fulfilled, the receptor system has been adequately characterized and the chemical(s) under question elicit effects via a cognate protein.

##### 14.1.3.4.2 Assumptions for Determining $K_d$ Values for Receptor-Ligand Interactions in Intact Tissue Preparations

As mentioned above in the context of $IC_{50}$ values versus $K_I$, $K_d$ values are more useful than are $EC_{50}$. Both $EC_{50}$ and $IC_{50}$ are dependent on the nuances of the incubation conditions whereas $K_d$ is only dependent on the receptor and the ligand. However, there are many assumptions that must be made in order to determine a $K_d$ from a cellular response.

(1) *The response of the tissue or cell should be due solely to the interaction of an agonist with one type of receptor.* Thus the response on which $K_d$ values are based should not be a composite on effects of two different receptor populations nor should it be the result of nonspecific or non-receptor mediated effects of the agonist.

(2) *The altered sensitivity to an agonist observed in the presence of an antagonist should be due to competition for a shared recognition site.* This holds true for pharmacological antagonism of the competitive and irreversible variety (see Section 14.1.3.4.4).

(3) *The response obtained following addition of agonist should be measured at a time when the maximal response is elicited.* Biological preparations especially suited for determination of $K_d$ values for receptor-ligand interactions are those that maintain a maximal response over a reasonable length of time.

(4) *When agonists or competitive antagonists are added to the tissue incubation, the concentration of ligand free in solution should be maintained at a known level.* Tissue uptake, degradation and other losses must be prevented or overcome by continual re-addition of ligand.

(5) *The experimental design should contain proper controls to permit measurement of, or correction for, any changes in tissue sensitivity over time.* Internalization of a membrane-bound receptor, or proteolysis of soluble receptors within the timeframe of the experiments may alter the sensitivity of the cell to the ligand.

##### 14.1.3.4.3 Full Agonists

The determination of a $K_d$ for a full antagonist based on responses is not as straightforward as Clark's theory of occupancy. Complicating factors such as the existence of spare receptors and efficacy issues will impact the calculations. Clark's theory of occupancy may be used for examining responses if examining a full agonist (efficacy $e$ is 100%) and the receptor requires full occupancy for maximal response. The key feature of this formulation is that response is proportional to the number of receptors occupied.

$$L + R \underset{k_{-1}}{\overset{k_1}{\rightleftharpoons}} LR \overset{k_e}{\longrightarrow} \text{Effect}$$

Thus, if $k_e$ (the rate constant for coupling, or of the effector system) is known, for any given $[LR]$ the extent of the response can be predicted.

$$\text{Response}(\Delta) = f([LR]) = f\left(\frac{[L][LR]_{max}}{K_d + [L]}\right)$$

The maximal response $\Delta_{max}$ will occur at $[LR]_{max}$

$$\frac{\Delta}{\Delta_{max}} = \frac{LR}{LR_{max}} = \frac{[L]}{K_d + [L]} \qquad (11)$$

As with receptor occupancy, half-maximal response will occur at $[L]$ equal to $K_d$. If it is assumed that occupancy and response are proportional, the same graphical means to examine $K_d$ as used in receptor binding may be employed. That is, a plot of fractional response $(\Delta/\Delta_{max})$ versus log[Agonist] will result in a hyperbolic curve whereas $\log(\Delta/(\Delta_{max} - \Delta))$ versus log[Agonist] represents a straight line.

However, as mentioned in 14.1.3.2, this simple relationship between occupancy and effect does not adequately predict many biological responses. In particular, dose-response relationships tend to have a steeper slope than predicted and the existence of spare-receptors shows that maximal responses do not require full occupancy. Stephenson and Furchgott formalized a non-linear relationship between receptor occupancy and biological response,[1]

$$\text{Response} = f(S) = fe\frac{LR}{LR_{max}}$$

$$\text{Efficacy } (e) = \varepsilon \bullet LR_{max}$$

$$\text{Response} = f\left(\varepsilon \bullet LR_{max}x\frac{LR}{LR_{max}}\right) = f(\varepsilon[LR])$$

$$(12)$$

where $e$ is efficacy, $\varepsilon$ is intrinsic efficacy and $S$ is the stimulus applied to the tissue or cell. The introduction of the term $\varepsilon$ effectively divides efficacy $(e)$ into two components, a ligand-

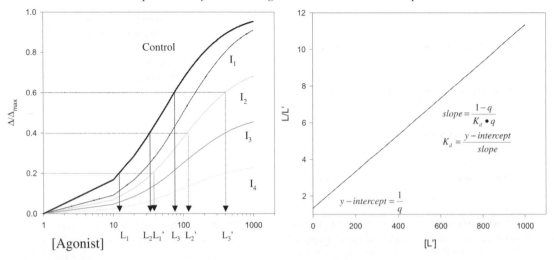

**Figure 11**   Determination of $K_d$ for receptor-agonist interactions using irreversible receptor blockade.

dependent ($e$) and a tissue-dependent ($LR_{max}$) term.

In order to use this relationship to examine $K_d$ for a ligand–receptor interaction, the experiments are similar to those used to determine the spare receptor reserve (discussed in 14.1.3.2.4). Dose-response data for a particular agonist before and after irreversible receptor blockade is obtained. In effect, what these experiments are doing is changing $LR_{max}$ in a preparation, and hence altering the efficacy ($e$) of the drug-receptor complex. There are several ways to irreversibly inactivate a receptor, and for many receptors these agents have been described. If an irreversible antagonist has not been characterized, chemical cross-linking may be utilized. In this instance, a highly reactive side-chain such as an -azido group may be added to the ligand; Upon binding to the receptor and activation with UV light, the ligand will covalently attach to the protein. With peptide ligands, bifunctional reagents such as disuccinimidyl suberimidate, may result in irreversible blockade. Non-specific means, such as alkylation or proteolysis have been employed. All means of irreversible inhibition must affect the receptor complex and not affect the effector signaling pathways.

Substituting Equation (12) into that of (11) results in the following relationship:

$$\frac{\Delta}{\Delta_{max}} = fe\frac{[L]}{K_d + [L]}$$

$$= f\varepsilon\left(LR_{max} \bullet \frac{[L]}{K_d + [L]}\right) \quad (13)$$

If the receptor preparation is treated with an irreversible antagonist, the concentration of total active receptors is reduced to a fraction

($q$) of the original $LR_{max}$. Since efficacy is equal to $\varepsilon LR_{max}$, in the presence of irreversible blockade, $e = q\varepsilon LR_{max}$, and the following equation is derived:

$$\frac{\Delta'}{\Delta'_{max}} = fe\frac{[L]'}{K_d + [L]'}$$

$$= f\varepsilon\left(qLR_{max} \bullet \frac{[L]'}{K_d + [L]'}\right) \quad (14)$$

where $\Delta'$, $\Delta'_{max}$, $[L']$ represent values after irreversible blockade of the receptor. When comparing equal responses before and after inhibition, for example $EC_{50}$ values, it is assumed that the stimulus applied to the tissue is the same ($S = S'$).

$$\varepsilon LR_{max}\frac{[L]}{K_d + [L]} = \varepsilon qLR_{max}\frac{[L']}{K_d + [L']}$$

eliminating $\varepsilon LR_{max}$                                      (15)

$$\frac{[L]}{K_d + [L]} = q\frac{[L']}{K_d + [L']}$$

The advantage of comparing equal responses is that the unknown $LR_{max}$ is eliminated and the function $f$ relating occupancy to response is no longer required. Equation (15) can be rearranged to the linear equation:

$$\frac{[L']}{[L]} = \frac{1}{q} + \frac{[L'](1 - q)}{K_d \bullet q} \quad (16)$$

A plot of $[L']/[L]$ versus $[L']$ results in a straight line with a $y$-intercept equal to $1/q$ and a slope equal to $(1 - q)/K_d \bullet q$. The $K_d$ value can be calculated as ($y$-intercept $-1$)/slope (see Figure 11).

#### 14.1.3.4.4 Quantitation of Pharmacologic Antagonism

The use of antagonists to block receptor responses is a classic approach in pharmacology and antagonism is a major mechanism by which chemicals cause toxicity. There are five basic types of antagonism: (1) *Functional antagonism*. This type of inhibition often results from stimulating two pathways with opposite function (i.e., sympathetic *versus* parasympathetic). Functional antagonism results from action via two distinct receptors and cannot be examined by the methods described herein; (2) *Competitive antagonism*. A competitive antagonist is binding to the receptor in roughly the same physical space as the agonist, precluding the latter from eliciting a biological response. A competitive antagonist is reversible, with a rate of dissociation that is relevant to the time-frame of the experiments; (3) *Irreversible antagonism*. As with the competitive antagonist, the irreversible antagonist binds to the same region of the receptor as the agonist. However, in this instance, the rate of dissociation is very low or non-existent; (4) *Noncompetitive antagonism*. Receptors are complicated macromolecules that may interact with small molecules in a variety of ways. A non-competitive antagonist is defined as a chemical that binds to the receptor at a site other than the agonist-binding pocket, yet it's binding precludes the binding of the agonist. This is often caused by the antagonist producing a conformational change in the binding pocket; and (5) *Mixed antagonism*. A combination of any of the type of antagonism listed above.

As stated above, antagonism is an important mechanism by which drugs exert beneficial responses and how certain toxicants produce their hazardous effects. Therefore, it is appropriate to use the methods described below to *characterize the chemical* and determine which type of inhibition is occurring. Knowing the mechanism of action may aid in designing more potent drugs or in ameliorating toxicity. However, antagonists are very useful tools for *characterizing the receptor* as well. This will be the focus of the present section. To characterize a receptor using antagonists, such blockade must fulfill certain requirements:

(1) Blockade must be selective for the family of agonist in this receptor type.

(2) Blockade of responses in different tissues by a series of antagonists should result in a similar rank order of potency.

(3) Blockade should be sufficiently rapid to allow for precise quantitative measurement.

(4) An identical value for $K_I$ in antagonizing different agonists that act at the same receptor should be obtained.

(5) Values for the dissociation constants $K_I$ derived from studies assessing tissue response should be identical to antagonist dissociation constants obtained in studies of binding to a particular type of receptor.

##### 14.1.3.4.4.1 Functional Antagonism

Functional antagonism is defined as antagonism of tissue response that is unrelated to blockade at drug receptors but instead represents blockade of tissue response at a site distal to receptors. This precludes its precise quantitation using the methods described. May affect second messengers or depress cellular excitability and/or energy status. Alternatively, may produce an opposing action through a different receptor.

##### 14.1.3.4.4.2 Competitive Antagonism

Competitive antagonists are ligands that compete with agonists, usually for a common binding site in a receptor. The interaction of agonist $(L)$, competitive antagonist $(I)$ with receptor $(R)$, is described using the following scheme.

$$
L + \quad
\begin{array}{c}
I \\
+ \\
R \\
{\scriptstyle Ki}\uparrow\downarrow \\
IR
\end{array}
\quad \overset{Kd}{\underset{}{\rightleftarrows}} LR \overset{Ke}{\longrightarrow} \text{Effect}
$$

Note that the inhibitor–receptor complex is ineffectual, and cannot couple occupancy with a biological response. The formation of the ligand–receptor and the inhibitor-receptor complex may be described by the Clark equations and have equilibrium dissociation constants of $K_d$ and $K_I$ respectively. As a result of $IR$ formation, there are fewer receptors available for $LR$ formation.

$$
\frac{\Delta}{\Delta_{\max}} = \frac{LR}{LR_{\max}} = \frac{[LR]}{[R] + [LR] + [IR]}
$$
$$
= \frac{\frac{[L]}{K_d}}{1 + \frac{[L]}{K_d} + \frac{[I]}{K_I}} \tag{17}
$$

This can be rearranged (divide by $[L]/K_d$):

$$
\frac{\Delta}{\Delta_{\max}} = \frac{1}{1 + \frac{K_d}{L}\left(1 + \frac{[I]}{K_I}\right)} \tag{18}
$$

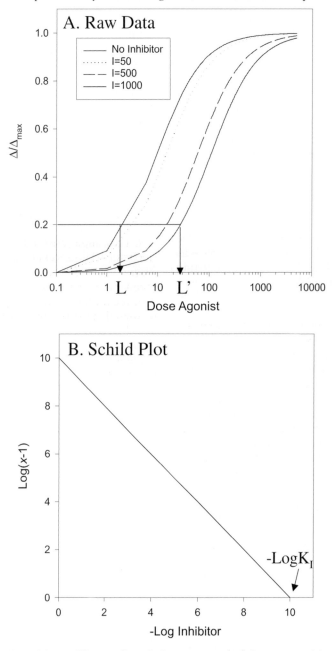

**Figure 12**  Determination of the equilibrium dissociation constant $(K_I)$ for a competitive antagonist. **Panel A**. The fractional response was calculated using Equation (17) with a $K_d$ of 10, $K_I$ of 10 and varying the inhibitor concentration from 0 to 1000. **Panel B**. Schild plot with log(dose ratio $-1$) versus concentration of inhibitor. The $K_I$ may be calculated from either the $x$- or $y$-intercept.

As you can see, the extent of antagonism depends on the agonist and antagonist concentration, as well as their dissociation constants, $K_d$ and $K_I$. In the presence of competitive inhibitor $I$, fractional occupation $[LR]/[LR]_{\max}$ decreases as a consequence of an increase in the apparent equilibrium dissociation constant of the agonist $L$. This decrease is from a value of $K_d$ in the absence of inhibitor to a value of $K_d(1 + [I]/K_I)$ in the presence of $I$.

The increase in apparent dissociation constant yields an equation with the same shape dose-response curve, only shifted to the right with no effect on maximal response (see Figure 12).

Since the competitive antagonist acts by excluding the agonist from binding, it does not change the relationship between occupation and response. At equal fractional occupancy, equal responses are observed. When $\Delta/\Delta_{\max}$ in the absence of competitor equals

$\Delta'/\Delta_{max}$ of the presence of competitor $I$, then

$$\frac{[L]}{[L]+K_d} = \frac{[L']}{[L']+K_d\left(1+\frac{[I]}{K_I}\right)} \tag{19}$$

where $L$ and $L'$ are the concentration required to generate equivalent responses $\Delta/\Delta_{max}$ and $\Delta'/\Delta_{max}$ in the absence and presence of $I$. This equation can be simplified,

$$\frac{[L']}{[L]} - 1 = \frac{[I]}{K_I}$$

and by taking the logarithms of this equation and substituting $x$ for $[L']/[L]$, the Schild equation is derived:

$$\log(x-1) = \log[I] - \log K_I \tag{20}$$

Note that the $K_d$ is removed from the equation and hence is not required to calculate $K_I$. The term *dose ratio* ($x$) is applied to the expression $[L]/[L']$. Dose ratios depend on the $[I]$ and $K_I$. (Must assume that the dose ratio is independent of agonist used and that a single receptor type is being examined.) The determination of dose-ratios is identical to that described above for full agonists (Section 14.1.3.4.3) where concentrations in the presence and absence of inhibitor result in equivalent responses. The Schild equation does not assume there is a linear relationship between occupancy and effect. However, the Lineweaver–Burke plot does. Therefore it is not appropriate to calculate a $K_I$, but is of diagnostic value. As shown in Figure 13, double reciprocal plots in the presence of increasing amounts of competitive inhibitor intersect at $1/B_{max}$. Non-competitive inhibitors will result in a plot with the intersection occurring at $-1/K_d$, as discussed below.

#### *14.1.3.4.4.3 Noncompetitive Antagonism*

A noncompetitive antagonist interacts with the receptor, however this interaction takes place at a site different than that of the agonist. Upon binding to the noncompetitive antagonist, the affinity of the receptor for the agonist is altered, possibly the result of a conformational change in the protein structure. The interaction of agonist ($L$), noncompetitive antagonist ($I$) with receptor ($R$), is described using the following scheme.

$$\begin{array}{ccccccc} & I & & I & & \\ & + & & + & & \\ L\,+ & R & \overset{Kd}{\longleftrightarrow} & LR & \overset{Ke}{\longrightarrow} & \text{Effect} \\ & {\scriptstyle Ki}\uparrow\downarrow & & {\scriptstyle Ki}\uparrow\downarrow & & \\ & IR & \longleftrightarrow & LIR & & \end{array}$$

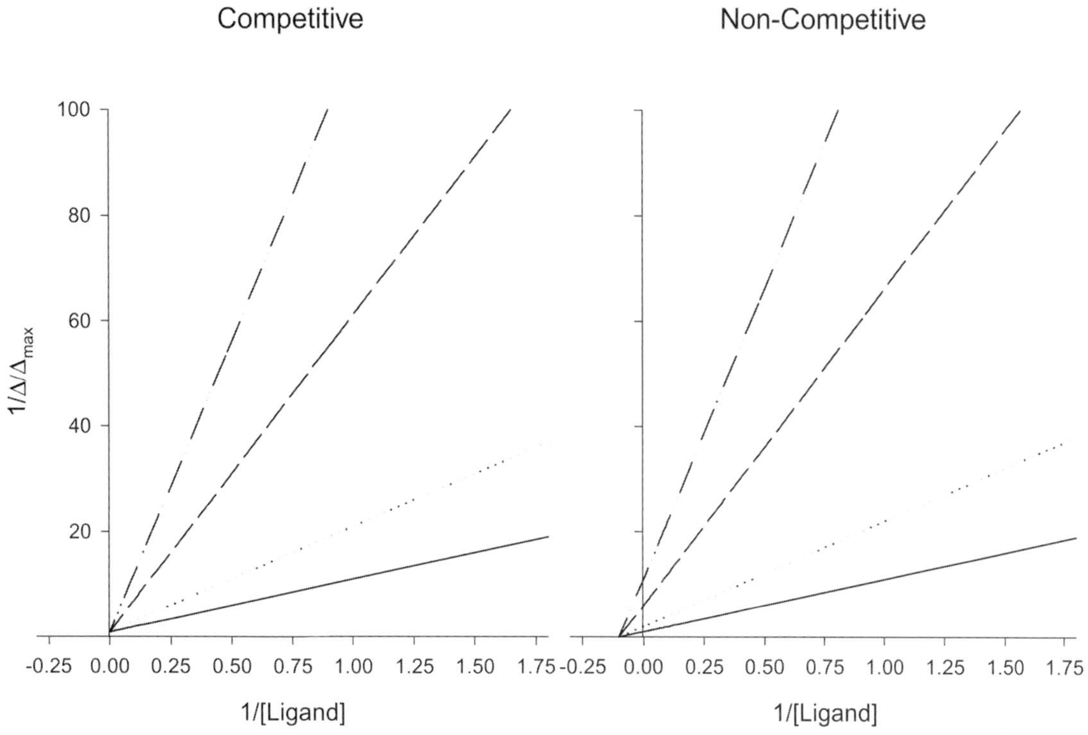


**Figure 13** Comparison of double-reciprocal plots for competitive and noncompetitive antagonists.

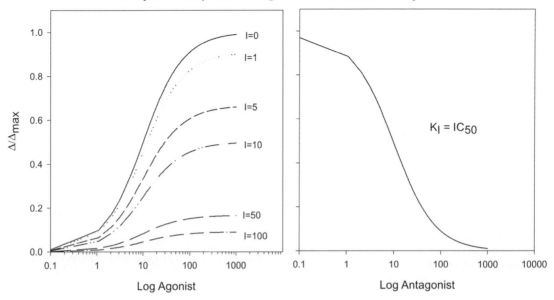

**Figure 14**   Determination of the equilibrium dissociation constant $(K_I)$ for a noncompetitive antagonist.

An antagonist of this sort produces quantitative changes in the agonist dose-response curves: The maximal response is decreased while the $EC_{50}$ of the agonist does not change (Figure 14). The antagonist does not alter the concentration of bound agonist $(LR + LRI)$ but the $LRI$ complex is nonfunctional. The response of the agonist may be derived as follows:

$$\frac{\Delta}{\Delta_{max}} = \frac{k_e[LR]}{[R]_{Total}}$$

$$= \frac{k_e[LR]}{[R] + [LR] + [IR] + [ILR]} \quad (21)$$

Which can be rearranged to:

$$\frac{\Delta}{\Delta_{max}} = \frac{k_e[L]}{([L] + K_d)\left(1 + \frac{[I]}{K_I}\right)}$$

The response is decreased by a factor whose magnitude is increased by $[I]$ and low values of $K_I$ (i.e., high affinity). For non-competitive antagonists $K_I = IC_{50}$, the concentration of antagonist that inhibits the response by 50%.

#### 14.1.3.4.4.4   *Irreversible and Pseudoirreversible Antagonists*

Some types of antagonists produce irreversible antagonism of agonist-mediated responses. This is most commonly observed when an antagonist produces a covalent modification in a receptor. Irreversible antago-

nists will yield curves that are identical to noncompetitive antagonist if the response is proportional to the number of receptors occupied. (However, remember spare receptors.) Note that if the $k_{-1}$ is slow enough, competitive antagonists may appear to be irreversible. Noncompetitive, irreversible, and pseudoirreversible antagonism share a common property in that they are non-surmountable antagonists, i.e., increased agonist does not cause maximal response.

#### 14.1.3.4.4.5   *Mixed Antagonism*

Mixed antagonism is observed when antagonists have multiple types of interaction with receptor systems. For example, the antagonist may initially block receptors competitively but over time may become irreversible. Two other possibilities exist if the $LRI$ complex is able to produce a response or if the $K_I$ for $R$ does not equal that of $LR$.

#### 14.1.3.4.5   **Partial Agonists**

A partial agonist has affinity, but less than maximal efficacy. The only way to determine the $K_d$ for a partial agonist (designated $K_{dp}$) is to compare it to a full agonist (Figure 15). However, many assumptions will be made to simplify the comparisons. First, it is assumed that the concentration of full agonist $(A)$ resulting in a response is small relative to $K_{da}$. We can therefore simplify Equation (2),

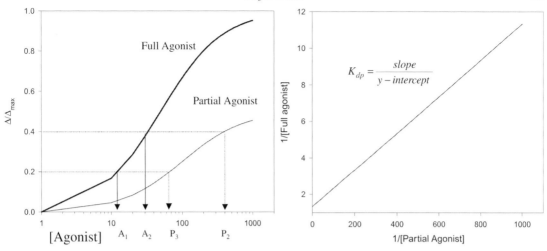

**Figure 15**  Determination of the equilibrium dissociation constant ($K_{dp}$) for a partial agonist.

for a full agonist to:

$$Y_A = \frac{[A]}{K_{da}}$$

For a partial agonist, Equation (2) cannot be simplified and remains:

$$Y_p = \frac{[P]}{[P] + K_{dp}}$$

To determine the $K_{dp}$ of a partial agonist, a dose-response curve for the full agonist $A$ and the partial agonist $P$ is obtained. As is the case in the previous examples, comparisons of $[A]$ and $[P]$ that elicit equal responses are used.

$$S = e_A Y_a$$
$$S = e_P Y_p$$

At equal response

$$e_A \frac{[A]}{K_{da}} = e_P \frac{[P]}{[P] + K_{dp}}$$

which can be arranged to the following linear equation

$$\frac{1}{[A]} = \left[\frac{e_A K_{dp}}{e_P K_{da}}\right] \frac{1}{[P]} + \frac{e_A}{e_P K_{da}} \quad (22)$$

Thus, a plot of $1/[A]$ versus $1/[P]$ should be linear with the slope equal to $\frac{e_A K_{dp}}{e_P K_{da}}$ and a $y$-intercept of $\frac{e_A}{e_P K_{da}}$. The $K_{dp}$ can be determined by dividing the slope by the $y$-intercept. It is important to stress that an assumption was made regarding the $K_{da}$ versus the concentration of a full agonist that results in a response.

If this is not the case, the mathematics become much more difficult.[1]

### 14.1.3.5  CONCLUSIONS

The quantification of the interaction between a macromolecule (receptor) and a xenobiotic (ligand) is called *receptor theory* and is often overlooked in modern molecular toxicology. The proper use of receptor theory can be used to examine structure-activity relationships and modifications of chemicals to fit the active site of the macromolecule. Also, the basic tools of quantification and characterization of a bimolecular interaction is applicable to many disciplines including enzymology (Michaelis complex between enzyme and substrate), immunology (formation of antibody–antigen complex), pharmacology (drug–receptor complex) and toxicology (xenobiotic–receptor complex). The shape of the dose-response curve, the most important relationship examined in toxicology, is dictated by receptor theory. Therefore, by applying the equations described herein, taking into account all assumptions and limitations, the estimation of risk posed by chemical exposure may be more accurately determined.

### 14.1.3.6  REFERENCES

1. L. E. Limbird, 'Cell Surface Receptors: A Short Course on Theory and Methods,' 2nd edn, Kluwer Academic Publishers, Boston, MA, 1996.
2. A. Levitzki, 'Receptors: A Quantitative Approach,' Benjamin/Cummings Pub. Co., Menlo Park, CA, 1984.

3. W. B. Pratt, P. Taylor and A. Goldstein, 'Principles of Drug Action: The Basis of Pharmacology,' 3rd edn, Churchill Livingstone, New York, 1990.

4. H. Motulsky, 'The GraphPad Guide to Analyzing Radioligand Binding Data,' GraphPad Software, Inc., 1996.

J.P. Vanden Heuvel, G.H. Perdew, W.B. Mattes and W.F. Greenlee (Eds.)
Comprehensive Toxicology, Vol. xiv
2002 Elsevier Science BV. All rights reserved.

# 14.1.4
# Control of Gene Expression

## JOHN P. VANDEN HEUVEL
*Penn State University, University Park, PA, USA*

## 14.1.4.1  INTRODUCTION

The survival of a living system depends on its ability to communicate with and respond to its surroundings. In general, how a gene responds to incoming signals depends on the physical state of that gene as well as cell-specific factors such as transcription and translation machinery. Signaling may be initiated by xenobiotics directly, by interacting with cell surface or soluble receptors (discussed in 14.2.1), or indirectly, through a paracrine or autocrine event. Many forms of toxicity, not the least of which is cancer, result from altered signaling and hence abhorrent gene expression control.[1]

In order to understand how a chemical alters gene expression, we must have an appreciation for how genes are selectively expressed in cells and the mechanisms by which genes are affected by external stimuli. Signaling events are the major focus of Section 14.2 (Xenobiotic Receptor Systems) and Section 14.4 (Alterations in Cell Signaling by Xenobiotics), and will not be discussed in the present chapter. Rather, we will concentrate on the end result of signaling, be it via kinases, transcription factors or receptors, namely the molecular mechanisms by which a gene is turned "on" or "off". Although much of what we know about eukaryotic gene expression was derived from prokaryotic studies (operons) we will concentrate on multicellular organisms. For a good basic overview of transcriptional control of gene expression, the reader is directed to *Molecular Biology of the Cell.*[2] In addition, there have been several excellent, recent review articles on eukaryotic transcriptional control.[3-9] Post-transcriptional gene regulation is, in general, more difficult to examine experimentally but is a very important determinant of cellular events. We will briefly discuss mRNA processing and stability, with emphasis on events altered by xenobiotics. Translational and post-translational regulation of gene control will not be examined in detail in this section, but is discussed in many chapters throughout the current volume. In addition, the reader is directed to review articles in these broad categories of gene expression, that of post-transcriptional[10-15] and post-translational[16-23] control mechanisms.

## 14.1.4.2  AN OVERVIEW OF GENE CONTROL

A xenobiotic can affect the expression of a particular gene in many ways, as shown in Figure 1.[2] First, and perhaps most important from a xenobiotic standpoint, is controlling how and when a given gene is transcribed (*transcriptional control*). This is the predominant, but not the sole, manner in which ligands for nuclear hormone receptors (NHRs) and other soluble receptors affect gene expression. Second, a chemical may control how the primary RNA transcript is spliced or otherwise processed (*RNA processing control*). Alternatively spliced transcripts may result in different protein products or may result in mRNA with different rates of turnover or translation. Although not studied in great detail, there are several examples of chemically-induced splice variants. Third, a chemical may select which completed mRNAs in the nucleus are exported to the cytoplasm (*RNA transport control*). To date, there are not any examples of this means of regulation by xenobiotics, although the potential exists. Fourth, chemicals may affect gene expression by selecting which mRNAs in the cytoplasm are translated by ribosomes (*translational control*). Similar to RNA processing control, there are few concrete examples of chemicals specifically regulating the expression of a gene via this pathway. However, many chemicals affect the expression of ribosomal proteins that may in turn affect translation efficiency. Fifth, certain mRNA molecules may be stabilized or destabilized in the cytoplasm (*mRNA degradation control*). This may be the most under-appreciated area in chemically induced alteration in mRNA levels. There are many examples of soluble receptor ligands that affect the stability of mRNA thereby resulting in alteration in corresponding protein levels. Last, a specific protein molecule may be selectively activated, inactivated, stabilized or compartmentalized after it has been made (*protein activity control*). Alteration in protein stability is a potentially important way in which certain xenobiotics affect cytochrome P450 apoprotein concentration. Also, receptor internalization (membrane bound receptors) and ligand-induced degradation (soluble receptors) are important means of effecting the responsiveness of a tissue to a particular chemical.

Certainly, affecting protein activity is an important mechanism by which chemicals result in their biological and toxicological effects. However, often it is the mRNA levels that are being examined following treatment of a particular cell with the chemical of interest. This is due to the fact that mRNA levels are the predominant manner in which altered gene expression is being manifested by a ligand for a soluble receptor. Equally important is the fact that the examination of mRNA levels is at a much higher level of sophistication than are equivalent protein measurements. For

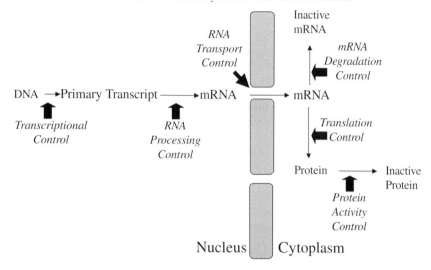

**Figure 1** Control of gene expression. Adapted from Reference 2. In the present section, the focus is on mRNA accumulation and hence transcription, mRNA processing and stability will be described.

example, microarray techniques make the concurrent examination of thousands of mRNA transcript levels feasible. Other high throughput quantitation means, such as real time and competitive RT-PCR, allow for detailed examination of when and how a transcript is affected by a xenobiotic. Equivalent methods for protein and enzyme activity are being developed (see Section 14.5, this volume), but are not readily available at this time. For these reasons, mRNA control will be the focus of the subsequent discussion.

### 14.1.4.3 GENE CONTROL IN EUKARYOTES AT THE mRNA LEVEL

As mentioned above, the ability of chemicals to affect the expression of fully processed mRNA, which may be translated into an active protein, comes under regulation at several points:

(1) *Chromatin structure.* The physical structure of the DNA, as it exists compacted into *chromatin*, can affect the ability of transcriptional regulatory proteins (*transcription factors* and *soluble receptors*) and RNA polymerases to find access to specific genes and to activate transcription from them. The presence of histones and CpG methylation may affect accessibility of the chromatin to RNA polymerases and transcription factors.

(2) *Transcription initiation and elongation.* This is the most important mode of control for eukaryotic gene expression and alteration by xenobiotics. Specific factors that exert control include the strength of *promoter elements*

within the DNA sequences of a given gene, the presence or absence of *enhancer sequences* (which enhance the activity of RNA polymerase at a given promoter by binding specific transcription factors), and the interaction between multiple activator proteins and inhibitor proteins.

(3) *Transcript processing and modification.* Eukaryotic mRNAs must be capped and polyadenylated, and the introns must be accurately removed. Several genes have been identified that undergo tissue-specific patterns of alternative splicing, which generate biologically different proteins from the same gene. As mentioned above, this a potential means by which chemicals can affect the activity of the protein encoded by a particular transcript.

(4) *RNA transport.* A fully processed mRNA must leave the nucleus in order to be translated into protein.

(5) *Transcript stability.* Unlike prokaryotic mRNAs, whose half-lives are all in the range of 1–5 min, eukaryotic mRNAs can vary greatly in their stability. Certain unstable transcripts have sequences (predominately, but not exclusively, in the 3'-non-translated regions) that are signals for rapid degradation.

### 14.1.4.3.1 Chromatin Structure

#### 14.1.4.3.1.1 *Structure of the Nucleosome*

Chromatin is a term designating the structure in which DNA exists within the nucleus of cells. The structure of chromatin is determined and stabilized through the interaction of the DNA with DNA-binding proteins. There are two classes of DNA-binding proteins. The *histones* are the major class of DNA-binding proteins involved in maintaining the compacted structure of chromatin. There are five different histone proteins identified as H1, H2A, H2B, H3 and H4. The other class of DNA-binding proteins is a diverse group of proteins called simply, *non-histone proteins.* This class of proteins includes the various transcription factors, polymerases, soluble receptors and other nuclear enzymes. In any given cell there are greater than 1000 different types of non-histone proteins bound to the DNA. The class of non-histone proteins that will be discussed in more detail in Chapter 14.2.1 are the activator proteins.

The binding of DNA by the histones generates a structure called the *nucleosome.* The nucleosome core contains an octamer protein structure consisting of two subunits each of H2A, H2B, H3 and H4. The assembly of the nucleosome depends on many histone dimerization events, beginning with the H3/H4 heterodimer and subsequent tetramer (H3/H4)$_2$. A left-handed superhelical ramp of proteins is then assembled, with heterodimers linked end-to-end: H2/H2B-H4/H3-H3/H4-H2B/H2A.[24] With the stability of the individual heterodimers, the H2B-H4 interface is the likely to be the initial site of disruption of the nucleosome core.

The nucleosome core contains approximately 160 bp of DNA, in two helical loops of 80 bp each. The initial histone DNA association begins when the (H3/H4)$_2$ tetramer forms a stable complex with approximately 120 bp of DNA. Domains within the histones contain α-helix structures (histone fold domains) that are key sites of DNA contact. The histone fold of each heterodimer is capable of interacting with approximately three-turns of the DNA helix (30 bp). The interaction with DNA utilizes arginine residues in the histone fold, which penetrate into the minor groove, as well as substantial hydrophobic interactions. The core nucleosome conformation is similar for most genes, regardless of the DNA sequence itself.

In addition to histone-fold domains, a critical region of these DNA binding proteins is the tail domain. This tail, if fully extended, can project into the helices of DNA in the nucleosome. In addition, the histone tails are involved in internucleosomal contact and formation of higher order chromatin structure (coiled-coil). Post-translational modification of the histone tails may evoke a change in the chromatin fiber, affecting the accessibility to other modifying proteins (see below).

Histone H1 occupies the internucleosomal DNA and is identified as the linker histone and participates in higher order chromatin structures. The tails of the linker histone bind to DNA within and in between the nucleosome cores. Histone H1 linkers contain numerous basic residues, which may neutralize the acidic backbone of the DNA and aid in the folding of nucleosomal arrays. The H1-chromatin interaction is relatively weak and removal of H1 may result in destabilization of high order chromatin structure.

The linker DNA between each nucleosome can vary from 20 to more than 200 bp. These nucleosomal core structures would appear as beads on a string if the DNA were pulled into a linear structure. The nucleosome cores themselves coil into a solenoid shape which itself coils to further compact the DNA. These final coils are compacted further into the characteristic chromatin seen in a karyotyping spread. The protein-DNA structure of chromatin is stabilized by attachment to a non-histone protein scaffold called the nuclear matrix.

### 14.1.4.3.1.2    *Transcription States*

Nucleosomes are a very stable physical structure and under physiological conditions fold into higher order structures of high concentration.[24] In spite of this stability and coiled-coil structure, many metabolic processes are capable of modifying this structure *in vivo*. In fact, coiling of DNA around histones to form the nucleosomal structure is an important regulator of transcriptional control of gene expression,[6] as outlined in Figure 2. The physical state of a gene can be categorized as repressed, basal or induced,[3] each of which is associated with a different structure of the nucleosome. A repressed gene is likely to be encased in chromatin to such an extent that the transcription machinery cannot access the promoter DNA. A basally expressed gene might have a more permissive structure of the chromatin that allows for a low level of gene expression. Last, an induced gene is likely to have an open chromatin structure and be bound by transcriptional activators and efficient recruitment of the transcriptional machinery.

Unlike prokaryotes, which utilize repressor proteins to maintain an "off-state" of genes, eukaryotic genes are repressed by the nucleosome complex.[6,25] This repression occurs in one of three manners.[6] First, nucleosomes may block the DNA binding sites for activator proteins (enhancer sequences, Enh), such as receptors, transcription factors or DNA modifying enzymes. Second, higher-ordered folding into the solenoid configuration may repress

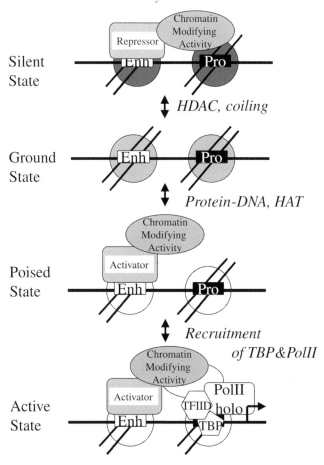

**Figure 2** Transcriptional states in eukaryotic cells are regulated by chromatin structure. Activators and repressors of transcription interact with enhancer sequences of DNA (Enh). Recruitment of chromatin modifying activities by activators and repressors results in altered chromatin structure, which affects whether a gene will be expressed. In the active state, the PolII machinery (TFIID and polII holoenzyme) are associated with the transcription start site via TATA-binding protein (TBP). See text for details. Adapted from Reference 25.

entire chromosomal domains. Third, interactions of nucleosomes with additional proteins to form heterochromatin may result in a hereditably-repressed gene. The significance of nucleosomal structure is that the binding of the TATA-binding protein (TBP) to the promoter (Pro) is prevented, and hence the polymerase machinery (i.e., PolII) is absent from the gene context.[25] TBP and PolII will be described in more detailed below.

An important regulatory mechanism that involves chromatin structure is the recruitment of chromatin modifying enzymes by activator and repressor proteins. For example, a DNA-binding repressor may inhibit the transcription machinery by recruiting histone deacetylase (HDAC) activity to the promoter[26] resulting in a *silent state*. Similarly, activator proteins may recruit proteins that acetylate histone tails (HAT), thereby resulting in template activation (*poised state*[25]). These aspects of gene regulation will be described in more detail in

Chapters 14.2.8 and 14.2.9 in regard to co-activator and co-repressor proteins.

### 14.1.4.3.1.3 Regulation of Chromatin Structure

That chromatin structure is important in gene regulation is underscored by its importance in development.[26] The activation of developmentally regulated genes may occur as a result of sequential changes in chromatin structure.[25] Chromatin modifying enzymes such as Swi/Snf may be recruited to a gene by Swi5, at a particular time in development. Only after this association, and presumably a change in DNA structure, other factors become associated with the gene, usually closer to the transcription initiation site. Recent studies have shown that coactivators and corepressors mediate communication between upstream regulatory proteins and RNA pol II. As discussed in a subsequent

chapter (14.2.8), these transcription coregulators carry DNA modifying activities, in
particular histone acetylation and deacetylation
capabilities.[26] Ultimately, the end result of this
complex, tiered approach to transcription
regulation is that a particular gene is expressed
in a proper temporal and spatially-localized
manner and is sensitive to small differences in
levels of an extracellular signaling molecule.
Three means by which chromatin structure may
be modified are briefly described below and
include acetylation, other posttranslational
modification (phosphorylation/methylation/
ubiquitinization) of the core histones[24] and
DNA methylation.

### 14.1.4.3.1.3.1  Consequences of acetylation of core histones

Of the means by which transcription may be
regulated by post-translation modification,
none has received as much attention as histone
acetylation. This is predominantly due to the
fact that the co-activator complex contains
proteins with histone acetyltransferase (HAT)
activity while co-repressors contain histone
deacetylase (HDAC) functions. Histones are
modified on specific lysine residues in the tail
region and are the targets of HAT and HDAC
containing coregulators. The acetylation state
of histones affects chromatin structure on
several levels.[24] First, acetylation of histone
tails may reduce the stability of interaction
with nucleosomal DNA. Acetylated histones
wrap DNA less tightly and are more mobile
than are hypoacetylated histone tails. Second,
acetylation may disrupt the protein–protein
interactions between histones, in particular the
H3–H4 junction. Third, acetylated histones are
less able to interact with adjacent nucleosomal
arrays, thereby decreasing the stability of the
compacted fiber. Last, acetylation of the core
histones decreases the association of the linker
histone H1, further destabilizing the higher
order structure. Ultimately the end result of
acetylating histone is an increase in transcription factor access to nucleosomal DNA. This
fact may be observed by DNase sensitivity
assays where regions of sensitivity correlate
with histone acetylation.

There are a total of 28 lysines in the tails of a
core histone octamer, of these only about one-
half need to be acetylated to produce increased
transcription. A portion of the decreased
histone-DNA interaction may be charge neut-
ralization that results from this post-translational modification. However, it is evident that
other modifications of histones are also playing
a role, some of which are described below.

Also, the initial acetylation of histones may
start a complex series of protein–protein
interactions with other chromatin modifying
enzymes.

A number of co-activators that are recruited
to activator proteins have intrinsic HAT
activity. These include the p160 family, important in nuclear receptor (NR) function, and the
more general coactivators CBP/p300 and
PCAF.[27] The p160 family includes steroid
receptor coactivator-1 (SRC-1), transcription
intermediary factor-2 (TIF2), glucocorticoid
receptor interacting protein (GRIP), pCIP and
many others. These proteins associate with the
NR's activator function-2 (AF2) domain, a
latent domain that is revealed upon ligand
binding. Thus, in the presence of hormone, the
activated receptor binding to DNA and the
HAT activity is recruited to the histone core.

HDAC activity is equally important in the
regulation of gene expression.[28] Deacteylation
results in an increase in histone/DNA interaction, internucleosomal interactions and higher
order DNA structure (reversal of HAT activity). Similar to HAT-containing proteins,
HDAC proteins are often parts of large
complexes of proteins. Several co-repressors
are associated with histone deacetylase 1
(HDAC1) including Sin3, N-CoR and
SMRT. Transcriptional regulation by co-
repressors complexes can be blocked by
chemical HDAC inhibitors (Tricostatin A,
trapoxin), showing the role of this enzymatic
activity in gene control. The recruitment of
HDAC to DNA by NRs is often the reverse to
that of HAT proteins. That is, co-repressors
bind to NRs in the absense of ligand and are
released upon addition of hormone.

### 14.1.4.3.1.3.2  Consequences of phosphorylation, ubiquitination and methylation of core histones

In contrast to the many studies on the
structural and functional consequences of
histone acetylation, the impact of other post-
translational modifications of the core histones is much less characterized.[24] In response
to phorbol esters, histone H3 is rapidly
phosphorylated on serine residues within its
basic N-terminal domain. In addition, H3
phosphorylation is correlated with mitotic and
meiotic chromatin condensation. The phosphorylated residue (Serine 10) is located
within the basic N-terminal domain of histone
H3 and may interact with the ends of DNA in
the nucleosomal core. Phosphorylation of
histones might be expected to have structural

consequences comparable with acetylation, presumably the result of change in protein charge and decreased histone-DNA binding. This may be one mechanism by which phorbol esters affect the chromatin structure and increase the rate of transcription of c-fos and c-jun. However, protein–protein interactions may also be affected by histone phosphorylation. Proteins containing exposed sulfhydryl groups accumulate on the chromatin surrounding fos and jun following phorbol ester treatment. The proteins containing exposed sulfydryl groups include both non-histone proteins, such as RNA polymerase, and molecules of histone H3 with exposed cysteine residues. Phosphorylation and acetylation of histone H3 might act in concert to cause these changes. There is likely to be an important link between cellular signal transduction pathways and chromatin targets for post-translational modification.

Ubiquitination of histones, in particular H2A, is associated with transcriptional activity. Ubiquitin is a 76 amino acid peptide that is attached to the C-terminal tail of approximately one histone H2A in every 25 nucleosomes in an inactive gene. This increases to one nucleosome in two for the transcriptionally active gene. Enrichment in ubiquitinated H2A is especially prevalent at the 5′-end of transcriptionally active genes. Since the C-terminus of histone H2A contacts nucleosomal DNA at the dyad axis of the nucleosome, ubiquitination of this tail domain might be anticipated to disrupt higher order chromatin structures.

Lysine residues are also targeted for modification by methylation. Most methylation in vertebrates occurs on histone H3 at Lys9 and Lys27 and histone H4 at Lys20, not known to be sites of acetylation. The consequence of this modification has not been elucidated.

### 14.1.4.3.1.3.3 Consequences of DNA methylation

Mammalian cells possess the capacity to modify their genomes via the covalent addition of a methyl group to the 5-position of the cytosine ring within the context of the CpG dinucleotide.[29] Approximately 70% of the CpG residues in the mammalian genome are methylated, although this does not mean that the event is nonspecific. The distribution of CpG is mainly in the 5′ untranslated region (UTR) of genes whereas the majority of the genome is CpG-poor. Certain regions of the genome that possess the high CpG frequency are termed *CpG islands*. These islands are often

protected from methylation in normal cells by mechanisms that remain unclear and may be hypermethylated in certain tumors. The critical genes that are inappropriately hypermethylated, and hence silenced, are tumor suppressor genes including retinoblastoma protein (Rb) and adenomatous polyposis coli (APC). DNA methylation has been shown to be essential for normal development, X-chromosome inactivation and imprinting.

A strong correlation between DNA hypermethylation, transcriptional silence and tightly compacted chromatin has been established in many different systems. At least a major portion of chromatin remodeling appears to be accomplished through acetylation and deacetylation of the histone tails, as described above. Inactive regions of DNA that were demonstrated to be heavily methylated were enriched in hypoacetylated histones. A major advancement in understanding the link between DNA methylation, histone hypoacetylation, and gene silencing came from the discovery of the methyl-binding protein MeCP2. This protein is known to be involved in transcriptional repression of methylated DNA, and through the bridging protein Sin3 recruits the histone deacetylase, HDAC1.

Only one DNA methyl transferase (DNMT) protein has been examined in detail. DNMT1 is a large enzyme ($\sim 200 \, \text{kDa}$) composed of a C-terminal catalytic domain with homology to bacterial cytosine-5 methylases and a large N-terminal regulatory domain with several functions, including targeting to replication foci. Disruption of Dnmt1 in mice results in abnormal imprinting, embryonic lethality, and greatly reduced levels of DNA methylation. Targeting of DNMT1 to replication foci via the N-terminal domain is believed to allow for copying of methylation patterns from the parental to the newly synthesized daughter DNA strand. Inhibition of DNMT1 activity through the use of antisense, knockout, pharmacologic means, or a combination of the latter two, inhibits tumor cell growth and induces differentiation. The enzymatic removal of 5-methylcytosine from DNA has also been described but has been far less extensively studied. One mechanism identified involves a 5-methylcytosine DNA glycosylase activity.

Hypomethylation of growth-related genes is associated with their overexpression and this could favor overgrowth of preneoplastic liver tissue. Decreasing the amount of S-adenosyl methionine (SAM), the methylating agent in the enzymatic process described above, is a model system in for the development of liver tumors. Treatment with exogenous SAM prevents the development of neoplastic lesions. This is associated with recovery of DNA methylation and inhibition of growth-related gene expression. The effect of feeding SAM on prenoplastic cell growth is abolished by 5-azacytidine, a hypomethylating agent, indicating the involvement of DNA methylation in this process.

### 14.1.4.3.2  Control of Eukaryotic Transcription Initiation and Elongation

The general configuration the upstream regulatory region of a gene is depicted in Figure 3 and is comprised of enhancer (Enh) or operator sequences as well as promoter (Pro) elements. There are many nuances to this configuration including, but not limited to: multiple, overlapping and competing enhancer sequences; enhancers within an intron, 3' untranslated region (UTR) or far upstream elements; alternative promoter sites, and TATA-less promoters. In the following discussion, we will focus on relatively simply systems of eukaryotic transcription initiation and elongation.

#### *14.1.4.3.2.1  Binding of Activators to Enhancer Sequences*

There are many other regulatory sequences in mRNA genes that bind various transcription factors (see Table 1). These regulatory sequences are predominantly located upstream (5') of the transcription initiation site, although some elements occur downstream (3') or even within the genes. The activator proteins themselves contain DNA binding motifs (see Chapter 14.2.1, this volume) that specifically recognize certain enhancer sequences. The number and type of regulatory elements to be found varies with each mRNA gene. Different combinations of transcription factors also can exert differential regulatory effects upon transcriptional initiation. The various cell types each express characteristic combinations of transcription factors; this is the major mechanism for cell-type specificity in the regulation of mRNA gene expression.

The general approach to examine transcriptional enhancer elements within a 5' regulatory region has been to identify protein-binding sites using electrophoretic mobility shift assays (EMSA, or *gel shift*) or DNase fooprinting assays. Subsequently, these elements are tested in reporter assays where the DNA element is placed upstream of a basal promoter and an

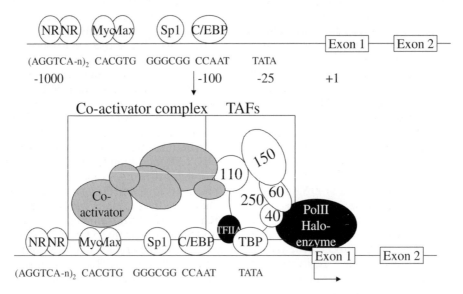

**Figure 3**  Structure of the upstream region of a typical eukaryotic mRNA gene that hypothetically contains two exons and a single intron. The diagram indicates the TATA-box and CCAAT-box basal elements at positions -25 and -100, respectively. The transcription factor TFIID has been shown to be the TATA-box binding protein, TBP. Several additional transcription factor binding sites have been included and shown to reside upstream of the two basal elements and of the transcriptional start site. The location and order of the variously indicated transcription factor-binding sites is only diagrammatic and not indicative as being typical of all eukaryotic mRNA genes. There exists a vast array of different transcription factors that regulate the transcription of all 3 classes of eukaryotic gene encoding the mRNAs, tRNAs and rRNAs. C/EBP = CCAAT-box/enhancer binding protein. NR = nuclear receptor.

easily quantified gene-coding region (i.e., luciferase, alkaline phosphatase). Finally, the element is mutated within the context of this reporter construct and is tested in transient transfection experiments. This approach has been extremely well characterized and utilized and will be described in more detail in several subsequent sections.

A recent review has suggested that synthetic enhancer constructs provide information whether a transcription factor has the *potential* to affect gene expression, but provides little information regarding the contribution of that factor to the regulation of the natural gene.[30] Mechanisms by which context-dependent transcription activation can be achieved include: (1) *Activator-induced DNA bending may influence promoter activity*. In general, the position of the Enh element has little bearing on transcriptional activity. However, there are instances where the distance between the Enh and Pro affects the rate of transcription. In addition, the general transcription factor Sp1 can achieve context-dependent transcriptional activation by stabilizing an intrinsic DNA bend; (2) *DNA-sequence specific protein–protein interactions influence response to activator*. Cooperative interactions between DNA bound transcription factors can result in context specific gene regulation. For example, protein-protein interactions between NF-κB and SP1 can result in cooperative DNA binding and synergistic activation of promoters; (3) *The structure of the core promoter can influence the response to activator*. Certain activator proteins prefer promoters containing a particular promoter element. For example, Sp1 shows more of an enhancing effect from the TdT promoter than from a TATA box; (4) *The cellular environment can influence the response to activator*. This is particularly germane to reporter assays performed in transformed cell lines. The loss of tumor suppressor genes or increase in oncogenes can affect the activity of certain activator proteins. For example, c-myc, which is commonly amplified in transformed cell lines, can associate with YY1 and inhibit its transcriptional regulation ability; (5) *Overlapping binding sites can influence response to activator*. Within many genes, the presence of overlapping elements makes it difficult to determine which activator is mediating the response. For example, transcription factors that binding GC rich regions, such as Sp1 and early growth response-1 (Egr-1), bind to overlapping sites. In this case the binding is mutually exclusive.

### 14.1.4.3.2.2 Promoter Elements, TBP and TAFs

Almost all eukaryotic mRNA genes contain basal promoters of two types and any number of different transcriptional regulatory domains (see diagram below). The basal promoter elements are termed *CCAAT-boxes* (pronounced "cat") and *TATA-boxes* because of their sequence motifs. The TATA-box resides 20–30 bases upstream of the transcriptional start site and is similar in sequence to the prokaryotic Pribnow-box (consensus $TATAT/_AA T/_A$, where $T/_A$ indicates that either base may be found at that position). Numerous proteins identified as TFIIA, B, C, etc. (for transcription factors regulating RNA pol $II$), have been observed to interact with the TATA-box. The CCAAT-box (consensus $GG T/_C CAATCT$) resides 50–130 bases upstream of the transcriptional start site. The protein identified as C/EBP (for CCAAT-box/Enhancer Binding Protein) binds to the CCAAT-box element.

The general transcription machinery assembles over the core promoter and drives transcription at the initiator site (+1). The presence of the TATA-binding protein (TBP) at the TATA box appears to be a pivotal step in transcriptional regulation. (For a review of TBPs and its function, see Reference 3.) TBP becomes associated with the TATA box, partially the result of activator proteins associating with enhancer elements. In fact, the preinitiation complex is not assembled in the cell unless facilitated by activator proteins. As described above, activator proteins recruit a variety of chromatin remodeling factors that may be required for TBP binding. Therefore, it appears that the core promoter sequence is hidden within nucleosomes, thus preventing assembly of the general transcription machinery.

TBP and TBP-related factors are somewhat unique among DNA binding proteins in that they bind to the minor groove of DNA. The binding of TBP to TATA box is high affinity ($Kd \approx 1\,nM$), although it appears that binding to DNA *per se* is not sufficient to explain the activity of this protein. For example, TBP is required for many TATA-less promoters and is responsible for RNA polI and RNA pol III transcription. This may be achieved by binding to non-TATA elements or by association with TBP-associated factors (TAFs). Once TBP is bound to DNA, it forms the scaffolding by which other factors assemble. Another important aspect of TBP's action is that it is able to bend DNA, which may bring upstream factors in contact with the promoter or with downstream elements.

TBP is associated with a variety of complexes. Of particular importance is TFIID, which is a TBP/TAF$_{II}$ complex specific for promoters of mRNA genes. The SL1 and TFIIB complexes are specific for pol I and III, respectively. TFIID is composed of approximately 14 distinct TAF$_{II}$ subunits plus TBP. The major functions of TAFs in the TFIID complex include serving as activator binding sites, mediate core-promoter recognition and provide essential catalytic activity. A number of TAF$_{II}$ subunits such as TAF$_{II}$40 and TAF$_{II}$110 are targets for activator proteins such as p53 and Sp1. Nuclear receptors interact with a number of TAF$_{II}$ including TAF$_{II}$28, TAF$_{II}$30 and TAF$_{II}$135. The "bridging" function of coactivators may be attributed to similar TAF interactions. TFIID occupies approximately 80 bp of DNA surrounding the transcription start site. However the specificity of this interaction may be determined by the composition of TAFs within the TFIID complex. This was determined by mutating certain TAFs in yeast and examining which genes were affected.[31] Unlike deletion of PolII, TAF$_{II}$s are not universally required for transcription and the expression of a characteristic subset of genes is affected by mutation. Last, many TAF$_{II}$ such as TAF$_{II}$250 and TAF$_{II}$145 have intronic histone acetylase activity, and some subunits such as TAF$_{II}$250, have serine/threonine kinase functions. In addition to bridging the activator to the RNA polymerase, TFIID may also contain chromatin-remodeling activities.

**Table 1**  Representative transcription factors response elements.

| Factor | Sequence motif | Comments |
|---|---|---|
| c-Myc and Max | CACGTG | c-Myc first identified as retroviral oncogene; Max specifically associates with c-Myc in cells. |
| c-Fos and c-Jun | TGA$^C$/$_G$T$^C$/$_A$A | Both first identified as retroviral oncogenes; associate in cells, also known as the factor AP-1. |
| CREB | TGACG$^C$/$_T$$^C$/$_A$$^G$/$_A$ | Binds to the cAMP response element; family of at least 10 factors resulting from different genes or alternative splicing; can form dimers with c-Jun. |
| SP1 | GGG GCG GGG | Binds to GC rich sequences found within promoters. Can act as a basal promoter as well as upstream activator. |
| c-ErbA; also TR (thyroid hormone receptor) | GTGTCAAAGGTCA | First identified as retroviral oncogene; member of the steroid/thyroid hormone receptor superfamily; binds thyroid hormone. |
| c-Ets | $^G$/$_C$$^A$/$_C$GGA$^A$/$_T$G$^T$/$_C$ | First identified as retroviral oncogene; predominates in B- and T-cells. |
| GATA | $^T$/$_A$GATA | Family of erythroid cell-specific factors, GATA-1 to -6. |
| c-Myb | $^T$/$_C$AAC$^G$/$_T$G | First identified as retroviral oncogene; hematopoietic cell-specific factor. |
| MyoD | CAACTGAC | Controls muscle differentiation. |
| NFκB and c-Rel | GGGA$^A$/$_C$TN$^T$/$_C$CC[1] | Both factors identified independently; c-Rel first identified as retroviral oncogene; predominate in B- and T-cells. |
| ARE | RTGACNNNGC | Anti-oxidant response element. |
| SRF (serum response factor) | GGATGTCCATATTAGGACATCT | Exists in many genes that are inducible by the growth factors present in serum. |
| HSE | CNNGAANNTCCNNG | Heat shock. HSTF binds. |
| GRE | TGGTACAAATGTTCT | Glucocorticoid receptor. |
| RARE | ACGTCATGACCT | Binds to elements termed RAREs (retinoic acid response elements) also binds to c-Jun/c-Fos site. |
| MRE | CGNCCCGGNCNC | Cadmium response element. |
| PPRE | AGGCCAAGGTCA | Peroxisome proliferator response element. |
| TRE | TGACTCA | Phorbol ester. |

The list is only representative of the hundreds of identified factors, some emphasis is placed on several factors that exhibit oncogenic potential.
[1]N signifies that any base can occupy that position.

### 14.1.4.3.2.3  RNA Polymerases and Transcription Initiation

Transcription is the mechanism by which a template strand of DNA is utilized by specific RNA polymerases to generate one of the three different classifications of RNA. All RNA polymerases are dependent upon a DNA template in order to synthesize RNA. The resultant RNA is, therefore, complimentary to the template strand of the DNA duplex and identical to the non-template strand. The non-template strand is called the coding strand because its sequences are identical to those of the mRNA.

Transcription of the different classes of RNAs in eukaryotes is carried out by three different polymerases (RNA polymerase I, II, III). The capacity of the various polymerases to synthesize different RNAs was shown with the toxin α-amanitin (see Table 2). At low concentrations of α-amanitin, synthesis of mRNAs are affected but not rRNAs nor tRNAs. At high concentrations, both mRNAs and tRNAs are affected. These observations have allowed the identification of which polymerase synthesizes which class of RNAs. *RNA pol I* synthesizes the rRNAs, except for the 5S species. This class of RNAs are assembled, together with numerous ribosomal proteins, to form the ribosomes. Ribosomes engage the mRNAs and form a catalytic domain into which the tRNAs enter with their attached amino acids. The proteins of the ribosomes catalyze all of the functions of polypeptide synthesis. *RNA pol II* synthesizes the mRNAs and some small nuclear RNAs (snRNAs) involved in RNA splicing. These classes of RNAs are the genetic *coding* templates used by the translational machinery to determine the order of amino acids incorporated into an elongating polypeptide in the process of translation. RNA pol III synthesizes the 5S rRNA and the tRNAs. This class of small RNAs form covalent attachments to individual amino acids and recognize the encoded sequences of the mRNAs to allow correct insertion of amino acids into the elongating polypeptide chain. The vast majority of eukaryotic RNAs are subjected to post-transcriptional processing. The most complex controls observed in eukaryotic genes are those that regulate the expression of RNA pol II-transcribed genes, the mRNA genes.

Transcription proceeds in the following general scheme. After activators and TBP/TAFs bind to the 5′ regulatory region of he gene, the RNA polII haloenzyme and general transcription machinery is recruited. Bending of the TATA-box DNA around TBP confers a context for interaction with TFIID. This large protein complex becomes the scaffolding upon which the rest of the transcription machinery can assemble. Similarly, TFIIH includes an ATP-dependent helicase activity that can unwind the promoter around the transcription start site. Working together, they open the DNA double helix and polII proceeds down one strand working in the 3′ to 5′ direction. As it moves down this strand, polII assembles ribonucleotides into the strand of RNA following the rules of base pairing. Synthesis of the RNA proceeds in the 5′ to 3′ direction. The primary transcript produced is called heterogeneous nuclear RNA (hnRNA) and it will undergo processes, as described below, including adding a 5′ cap, poly-A tail and splicing. The mammalian polII large subunit contains an essential, multifunctional carboxy terminal domain (CTD) that is a target for many kinases and phosphatases and is comprised of 52 seven amino acid repeats.[32,33] This particular region of polII is an important point of convergence of mRNA transcription initiation, elongation as well as mRNA processing.

### 14.1.4.3.2.4  Transcription Elongation

The transition from transcription initiation to elongation is loosely defined in eukaryotes. One key step must be breaking the initial ties to the promoter and the accessory initiation factors, but it may also include conversion of RNA polII to an elongation-competent form.[34] This conversion involves phosphorylation of the CTD, as described above, followed by association with other proteins, such as TFIIS, TFIIF,[32] elongin and positive transcription elongation factor b (P-TEFb).[35] The CTD is phosphorylated by TFIIH, which itself is composed of cyclin dependent kinase 7 and cyclin H. Other

**Table 2**  Eukaryotic nuclei have three RNA polymerases.

| Enzyme | Location | Product | Relative activity | Inhibitor |
|---|---|---|---|---|
| RNA pol 1 | Nucleolus | Ribosomal RNA | 50–70% | Actinomycin D |
| RNA pol 2 | Nucleoplasm | Nuclear RNA | 20–40% | α-amanitin |
| RNA pol 3 | Nuceloplasm | Transfer RNA | ~10% | Species Specific |

cyclin- and cell cycle-dependent kinases may also phosphorylate polII. During the pre-initiation and initiation stage of transcription, several key proteins bind to the hypophosphorylated CTD. Subsequently, this region of polII becomes hyperphosphorylated, thereby allowing transcription elongation to occur.

Transcription elongation is discontinuous and may encounter a *pause*, an *arrest* or may reach *termination*. During a pause state, RNA polII temporarily stops RNA synthesis for a finite period of time. An arrested elongation complex is unable to resume transcript elongation without the aid of accessory factors. Both transcription pause and arrest may be intrinsic (due to RNA sequence or structure) or extrinsic (DNA binding proteins or nucleotide concentration).[34] Transcription termination is the result of RNA polII becoming catalytically inactive. It is important to note that once the elongation apparatus is terminated, or falls off the template, it cannot simply resume its function; Transcription initiation must be restarted at the promoter. Pausing or arresting the elongation process allows for more control over the amount of transcript being produced. Termination may be important in negating the expression of mutant mRNAs or may be closely tied to DNA damage.

There are several mechanisms by which elongation may be temporary stalled or arrested.[34] Intrinsic factors such as stretches of Ts in the DNA template may result in a pause in elongation. Chromatin structure may affect the rate of elongation as well. Extrinsic factors such as phosphorylation of CTD causes polII to stall. For example, an oncogene induced by ionizing radiation, c-abl, negatively regulates transcription by increasing phosphorylation of CTD.[32] This cessation of transcriptional activity results in p53 activation and hence stimulates cell cycle arrest. RNA polII phosphorylation status changes during the course of the cell cycle and in response to external stimuli such as growth factors, stress, mitogens and DNA damaging agents. In fact, transcription elongation is carefully coordinated with DNA replication and mitogenesis and there is evidence that RNA polymerases encounter DNA replication machinery during transcription.[34]

### 14.1.4.3.3 Posttranscriptional Processing of RNAs

#### 14.1.4.3.3.1 *Introduction to RNA Processing*

Post-transcriptional processing is a very important process in eukaryotic mRNA accu-

mulation. When transcription of bacterial rRNAs and tRNAs is completed they are immediately ready for use in translation with no additional processing being required. Translation of bacterial mRNAs can begin even before transcription is completed due to the lack of the nuclear-cytoplasmic separation that exists in eukaryotes. An additional feature of bacterial mRNAs is that most are *polycistronic*, which means that multiple polypeptides can be synthesized from a single primary transcript. This does not occur in eukaryotic mRNAs. In contrast to bacterial transcripts, eukaryotic RNAs (all 3 classes) undergo significant post-transcriptional processing and are transcribed from genes that contain *introns*. The sequences encoded by the intronic DNA must be removed from the primary transcript prior to the RNAs being biologically active. The process of intron removal is called *RNA splicing*. Additional processing occurs to eukaryotic mRNAs. The 5′ ends of all eukaryotic mRNAs are *capped* with a unique 5′ to 5′ linkage to a 7-methylguanosine residue ($m^7G$). Messenger RNAs also are polyadenylated (poly(A)) at the 3′ end.

Although not discussed herein, eukaryotic tRNAs and rRNAs are extensively processed. In addition to intron removal in tRNAs, extra nucleotides at both the 5′ and 3′ ends are cleaved, the sequence *5′-CCA-3′* is added to the 3′ end of all tRNAs and several nucleotides undergo modification. There have been more than 60 different modified bases identified in tRNAs. Both prokaryotic and eukaryotic rRNAs are synthesized as long precursors termed *preribosomal RNAs*. In eukaryotes a 45S preribosomal RNA serves as the precursor for the 18S, 28S and 5.8S rRNAs. In the following section, we will limit our discussion to eukaryotic mRNA and will emphasize the important connections between transcription initiation, pre-mRNA elongation and mRNA processing.

#### 14.1.4.3.3.2 *Capping of mRNA*

The addition of 7-methylguanine to RNA is the first post-transcriptional event to occur, and it proceeds as a result of three enzymes, a phosphatase, a guanyl transferase and a methylase. The end result is the addition of a "cap" with a unique 5′ to 5′ linkage (Figure 4). Although the capping enzymes are distinct from the transcriptional machinery, they are coupled to the transcription process.[36] The guanyltransferase and methylase bind to the phosphorylated form of the polII CTD and carry out their enzymatic functions before the

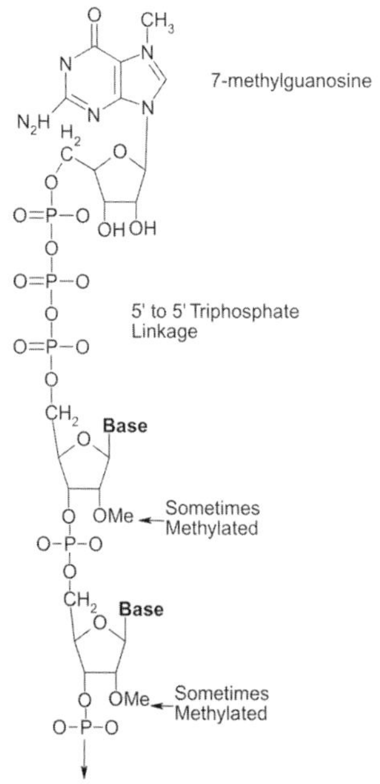

**Figure 4** Structure of the 5′-Cap of Eukaryotic mRNAs.

transcript has reached 30 nucleotides long. The addition of the 5′ cap may serve two distinct functions. First, it may signal the switch from transcription initiation to that of transcript elongation. Stalling of transcription elongation may result from defective capping of mRNA. Second, due to the 5′ to 5′ linkage of the 7-methylguanine at the capped end of the mRNA, the transcript is protected from exonucleases and, more importantly, is recognized by specific proteins of the translational machinery.

### 14.1.4.3.3.3  *Splicing of RNAs*

The next RNA processing step to occur is that of intronic splicing. Most eukaryotic genes contain introns, with approximately 90% of the gene being comprised of noncoding sequences. Splicing of nuclear mRNAs requires the formation of an internal "lariat structure" and is catalyzed by specialized RNA-protein complexes called small nuclear ribonucleoprotein particles (snRNPs, pronounced "snurps").[37] The RNAs found in snRNPs are identified as U1, U2, U4, U5 and U6. The genes encoding these snRNAs are highly conserved in vertebrate and insects and are also found in yeasts and slime molds indicating their importance. Analysis of a large number of mRNA genes has led to the identification of highly conserved consensus sequences at the 5′ and 3′ ends of essentially all mRNA introns

(Figure 5). The U1 RNA has sequences that are complimentary to sequences near the 5′ end of the intron. The binding of U1 RNA distinguishes the GU at the 5′ end of the intron from other randomly placed GU sequences in mRNAs. The U2 RNA also recognizes sequences in the intron, in this case near the 3′ end. The addition of U4, U5 and U6 RNAs forms a complex identified as the spliceosome that then removes the intron and joins the two exons together.

As with mRNA capping, splicing is also closely connected with transcription and may in fact occur simultaneous to mRNA elongation.[36] The splicesome is a large complex that includes the snRNPs described above, as well as splicing regulatory proteins (SR proteins). In general, SR proteins possess RNA-binding domains (RRMs) that target the protein to exon enhancer sequences and arginine/serine rich region (RS) that may provide for protein–protein interaction. *In vitro* splicing reactions can be affected by the phosphorylation status of the polII CTD and SR proteins interact directly with the heptad repeats of this protein. Overexpressing various SR proteins affects the splicing patterns initiated at specific promoters. This is consistent with a model in which SR proteins interactions with CTD are set up early in the transcription initiation process and affect splicing patterns as the mRNA chain is elongated.[36]

Alternative splicing is regulated in a cell- and developmental stage-specific manner and is a versatile mechanism of gene expression regulation.[13] As can be seen in Figure 5, the splice recognition sequence contains very little information and accurately predicting a splice recognition site is very difficult. Therefore, recognition of the appropriate splice site by the splicesome is a daunting problem. This fact is underscored by the numerous genetic diseases that arise from the loss of splice recognition, as discussed below. In addition, coordinated changes in the splicing patterns of pre-mRNAs is an integral component of gene expression programs involved in cell differentiation and apoptosis.[13] A summary of four genes that are affected by differential splicing is shown in Figure 6 (adapted from Reference 13).

### 14.1.4.3.3.4  *Polyadenylation*

A specific sequence, AAUAAA, is recognized by the endonuclease activity of *cleavage polyadenylation stimulatory factor* (CPSF) that cleaves the primary transcript approximately 11–30 bases 3′ of the sequence element. A stretch of 20–250 A residues is then added to

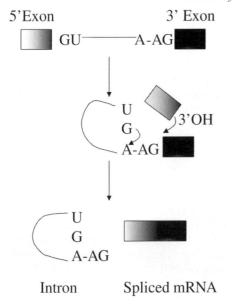

5'Exon                        3' Exon

U
G
3'OH
A-AG

U
G
A-AG

Intron          Spliced mRNA

**Figure 5**  Splicing of mRNA. The splicing of pre-mRNA involves two successive reactions. In the first reaction, the 2'OH of a specific adenosine at the branch point near the 3' end of the intron attacks the 5' splice site. The reaction releases the 5' exon and leaves the 5' end of the intron joined by the 2' to 5' phosphodiester bond to the branch point (lariat structure). In the second step, the 3'OH of the 5' exon intermediate attacks the 3' splice, producing the intron containing lariat and the spliced mRNA.

the 3' end by the *polyadenylate polymerase* activity. The polyA tail has been conclusively shown to play a role in mRNA stabilization and decay, as will be discussed in 14.1.4.3.5. However the addition of the 5' cap and polyA tail may also be involved in nuclear export of the transcript (discussed in 14.1.4.3.4).

The CTD of polII affects the polyadenylation of mRNA and the processing of mRNA regulates transcriptional elongation. Shortly after transcription initiation, CPSF binds to polII and in carried along during the elongation process. Deletion of the CTD inhibits the polyadenylation process, once again linking transcription to mRNA processing. The recognition of the polyA signal site triggers the termination of the transcription process. Without a polyA signal site, long RNA transcripts are produced as the result of polII remaining attached to the elongating transcript.[36]

### 14.1.4.3.3.5  Clinical Significances of Alternative and Aberrant mRNA Processing

The presence of introns in eukaryotic genes would appear to be an extreme waste of cellular energy when considering the number of nucleotides incorporated into the primary transcript only to be removed later as well as the energy utilized in the synthesis of the splicing machinery. However, the presence of introns can protect the genetic makeup of an organism from genetic damage by outside influences such as chemical or radiation. An additionally important function of introns is to allow alternative splicing to occur, thereby, increasing the genetic diversity of the genome without increasing the overall number of genes. By altering the pattern of exons, different proteins can arise from the processed mRNA from a single gene. Alternative splicing can occur either at specific developmental stages or in different cell types. This process of alternative splicing has been identified to occur in the primary transcripts from at least 40 different genes. Depending upon the site of transcription, the calcitonin gene yields an RNA that synthesizes calcitonin (thyroid) or calcitonin-gene related peptide (CGRP, brain). Even more complex is the alternative splicing that occurs in the α-tropomyosin transcript. At least 8 different alternatively spliced α-tropomyosin mRNAs have been identified. Abnormalities in the splicing process can lead to various disease states. Many defects in the β-globin genes are known to exist leading to β-thalassemias. Some of these defects are caused by mutations in the sequences of the gene required for intron recognition and, therefore, result in abnormal processing of the β-globin primary transcript. Patients suffering from a number of different connective tissue diseases exhibit humoral auto-antibodies that recognize cellular RNA-protein complexes. Patients suffering from systemic lupus erythematosis have auto-antibodies that recognize the U1 RNA of the spliceosome.

### 14.1.4.3.4  RNA Transport

As mentioned briefly above, a major difference between eukaryotic and prokaryotic mRNA production is the fact the former segregates RNA and protein synthesis. This segregation allows for a level of regulation of gene expression not available to prokaryotes.[38] Although all RNAs must be transported, our focus will remain with mRNA. Prior to mRNA export from the nucleus, the introns must be removed, and the addition of the m[7]G cap at the 5' end and polyA at the 3' end may enhance the rate of transport. The general scheme for mRNA nuclear export is shown in Figure 7.

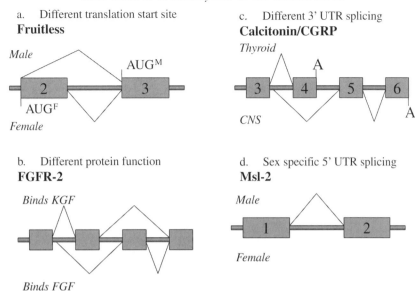

**Figure 6** Biological consequences of differential splicing. Four examples are given of how differential splicing affects the expression of a particular gene. **Panel a.** *Alternative start site.* In the drosophila fruitless gene, male and female splicing patterns give rise to different open reading frames. **Panel b.** *Different protein function.* Mutually exclusive use of exons in the fibroblast growth factor receptor 2 (FGFR-2) changes the binding capacity of the resulting protein. **Panel c.** *Different 3' polyA signal.* The tissue specific splicing of the calcitonin gene results in different poly adenylation signals in the 3' untranslated region (UTR). The resultant protein has different function in the two tissues, acting as a calcium homeostatic protein in the thyroid and as a vasodilator in the central nervous system. **Panel d.** *Sex specific retention of a 5' intron.* Inclusion of the intron in the male specific lethal-2 (msl-2) gene is seen in females, resulting in retention in the nucleus and translational repression.

In the nucleus, the pre-mRNA is associated with heterogeneous nuclear ribonucleoproteins (hnRNPs), a family of approximately 20 proteins.[38] Some hnRNPs such as hnRNPA1 contain both nuclear localization (NLS) and export (NES) sequences, while others including hnRNPC are retained in the nucleus. Several other proteins are needed for efficient nuclear export of mature mRNA. Tap and its coregulator p15 bind to mRNA are a heterodimer and shuttle between the nucleus and the cytoplasm. The carboxy terminus of Tap is recognized by several nucleoporins and transportin, which are part of the nuclear pore complex (NPC). The function of p15 in this complex is less clear, but its presence may increase Tap binding to mature mRNA. The helicase Dbp5 is also required for polyA mRNA transport and may function by unwinding mRNA as it travels through the NPC. This may also assist in the removal of non-shuttling hnRNPs for retention in the nucleus.

The end result of this complicated mechanism of nuclear export is that only properly processed RNA will be allowed to exit the nucleus for later translation. Retroviruses have evolved ways to express both sliced and un-spliced transcripts in the cytoplasm by circumventing some of these controls.[38] The regulation of mRNA transport by chemicals, or by disease states such as cancer, has not been studied to a large extent. However, it is potentially a plausible manner to regulate gene expression.

#### 14.1.4.3.5  mRNA Stability

##### 14.1.4.3.5.1  Introduction to mRNA Stability

Modulation of mRNA stability is an important posttranscriptional mechanism of modulating gene control. For example, inherently unstable mRNAs can reach steady-state levels more rapidly than their stable counterparts. This property would allow for cytoplasmic concentrations and translation to be altered quickly in response to a change in transcription rate.[39] The half-life of mRNAs is often controlled in a cell cycle-dependent manner, and important genes such as c-myc are controlled to a large extent by transcript stability. The importance of mRNA decay in gene expression is underscored by the fact that there are large differences in half-lives of transcripts ranging from 15 min to several

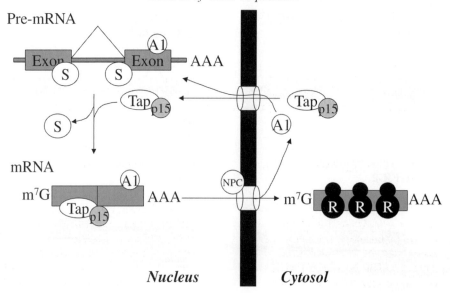

**Figure 7**   Model of mRNA export. Pre-mRNA containing introns are associated with RNPs, such as RNPA1 (A1) and the splicesome complex (S). These pre-mRNA are retained in the nucleus since they cannot access the RNA export factors. The mRNA export factors such as Tap and p15, become associated with the mRNA after the splicesome complex is released. The complex containing the transport factors docks with the nuclear pore complex (NPC) and exits the nucleus. In the cytosol, the mRNA is bound by other protein factors and is a substrate for the ribosomal (R) translation machinery.

hundred hours.[40] This variation may result in many-fold differences in the levels of specific proteins being expressed. Despite the importance of mRNA stability in gene expression, these processes have been largely ignored. The predominant reason for this oversight is the fact that the tools available for examining transcription, such as reporter assays, have not become as universally accepted for the study of mRNA stability and thus examining decay rates can be quite difficult.

### 14.1.4.3.5.2   Cis-acting Elements in mRNA Stability

Similar to transcription, mRNA decay is a precise process dependent on a variety of specific cis-acting sequences and trans-acting factors.[41] Entry into the pathways of mRNA decay is triggered by at least three types of initiating event: poly(A) shortening, arrest of translation at a premature nonsense codon, and endonucleolytic cleavage. However, other potential mRNA sequences may play a role, as outlined in Figure 8. After polyA shortening or premature arrest of translation and endonuclease cleavage, the cap is removed and exonuclease activity proceeds. In general the mRNA 5' cap remains unchanged after mRNA transport and the 5' to 5' phosphodiester bond

makes the mRNA resistant to most ribonucleases.

Most eukaryotic mRNAs contain a 3' polyA tract, which are gradually shortened in the cytoplasm in a transcript- and cell-specific manner. For some mRNA, the shortening of the polyA tail is the rate-determining event in transcript decay. Shortening by 3' to 5' exonuclease is followed by decapping at the 5' end and entry of exonucleases that proceed in the 5' to 3' direction. Exonuclease digestion in the 3' to 5' direction is, in general, inhibited by a polyA tail of at least 30 nucleotides. Deadenylation-independent decapping occurs in some mRNAs that contain nonsense codons. This mechanism, which occurs by the protein decapping protein 1 (Dcp1), exists to ensure that aberrant mRNAs are not translated.

The presence of specific sequence elements can affect the rate of decay of certain mRNAs. These rate-modifying elements include adenosine + uridine and hairpin structures in the 3' and 5'UTR or in the coding region. The best-characterized destabilization sequence, the AU rich element (ARE), is characterized as a pentamer of AUUUA, repeated once or several times in the 3' UTR. AREs are found in many labile cytokines, transcription factors and oncogenes. The 3' terminal stem loop, represents a stable structure that affects the stability of cell cycle regulated histone genes. These transcripts are not polyadenylated.

Evidence that poly(A) stimulates translation initiation, that some destabilization sequences must be translated in order to function, and that premature translation termination promotes rapid mRNA decay supports a close linkage between the elements regulating mRNA decay and components of the protein synthesis apparatus.

### 14.1.4.3.5.3 Trans-*Acting* Factors *in mRNA Stability*

Subsequent to transport via the nuclear export machinery, mRNA are bound by the cytoplasmic cap-binding complex, eIF4F, which is composed of three subunits eIF4E, eIF4A and eIF4G. The subunit eIF4E interacts directly with the $m^7G$ cap while eIF4A contains helicase activity and eIF4G interacts with the polysomal complex. Other proteins bind with high affinity to the 5′ region of mRNA including iron regulatory proteins (IRPs) that block entry of the small ribosomal subunit and repress translation. The decay of mRNA commences while it is still attached to the translation apparatus, indicating how eIF4F and IRPs affect mRNA turnover.

An important functional mutation in a gene is one that adds a premature stop codon. These mutant transcripts must be recognized and removed before they lead to disease. Nonsense mutations in BRCA1 and NF1 lead to human diseases including breast cancer and neurofibromatosis. Several proteins that may affect nonsense mediated mRNA decay are Upf1p, a helicase, and Upf2p and Upf3p, RNA binding proteins. Upf3p or Upf2p recognizes the nonsense mutation and bind to the transcript that must be removed. Subsequently Upf1p interacts with the mRNA and unwinds the RNA, making it more accessible to nuclease digestion.

The AU binding proteins (AUBPs or AUFs), are 30–45 kDa proteins that appear to be involved in both stabilization and destabilization of mRNA.[42] The human AUBP HuR has been postulated to chaperone mRNA from the nucleus and decreases susceptibility to nucleases. AU-binding factor 1 (AUF1) has been implicated in many cytokine-regulated genes[39] and binds to AU containing transcripts in the nucleus or cytoplasm. This protein interacts with two other RNA binding proteins eIF4G and polyA binding protein (PABP).

PABP is a highly conserved 71 kDa protein that binds with high affinity to polyA sequences. Stabilization of mRNA by PABP requires current translation and affects the turnover of transcripts that rely on polyadenylation dependent pathways. One model suggests that PABP holds the 5′ and 3′ end of the

**A.**

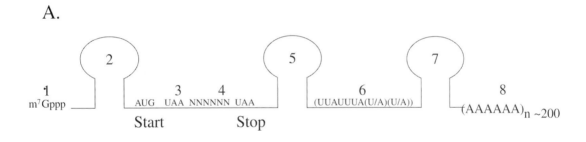

**B.**

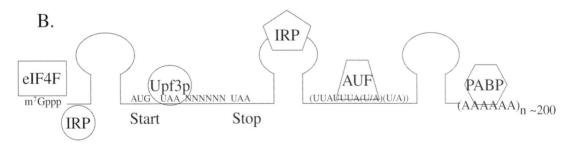

**Figure 8** Elements of mRNA stability. **Panel A**. *Cis-acting elements*. (1) 5′ cap; (2) 5′UTR secondary structure; (3) Premature (nonsense) stop codon; (4) Open reading frame sequence; (5) 3′ UTR sequences; (6) AU-rich elements (AREs); (7) Histone 3′ terminal stem loop; (8) PolyA + tail. **Panel B**. *Trans-acting factors*. See text for details.

transcript in close proximity, presumably by protein-protein interactions with eIF4G.

#### 14.1.4.4  COMBINATORIAL GENE CONTROL

An important concept about gene expression, and the effects of xenobiotics or any external stimuli, is that often the events observed are cell-, species-, sex-, developmentally-specific. This is due to the fact that combinatory gene expression is the norm in eukaryotes (see Figure 9). In general, a combination of multiple gene regulatory factors, rather than a single protein, determines where and when a gene is transcribed. A single gene may be necessary but not sufficient to alter the phenotype of a cell. A good example of combinatorial gene control comes from muscle cell differentiation.[2] Myogenic proteins (MyoD, myogenin; both are helix-loop-helix proteins) are key to causing certain fibroblasts to differentiate into muscle cells. It appears that myoD regulates myogenin. If you remove myogenin, the cells will not differentiate and if you overexpress this gene in fibroblasts they will convert to muscle. However, other cell types are not converted to muscle by myogenic proteins. This suggests that some cells have not accumulated the other regulatory proteins required.

As we will be discussed below, multiple gene regulatory proteins can act in combination to regulate the expression of an individual gene. But, combinatorial gene expression means much more: Not only does each gene have multiple "inputs", each regulatory proteins contributes to the control of many genes ("outputs"). Although some regulatory genes, such as MyoD and myogenin, are specific to a specific cell type, more typically production of a given regulatory protein is switched "on" in a variety of cell types. With combinatorial control, a given gene regulatory protein does not have a single, simply definable function. It is the combination of gene products that conveys the information. The consequence of adding a new gene is dependent on the past history of the cell. (The metaphor often used is proteins are like words and phenotype of a cell is the language. Words require context to convey their meaning.)

#### 14.1.4.5  EXAMINING GENE CONTROL

##### 14.1.4.5.1  Introduction

Differential gene expression is a key component of many complex phenomena including

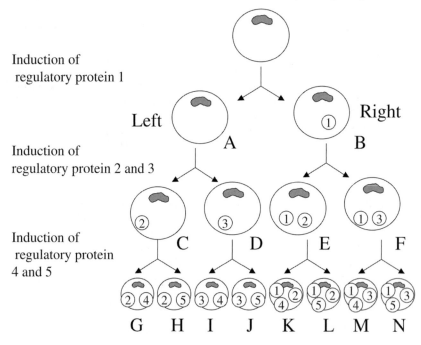

**Figure 9**  The importance of combinatorial gene control for development. A highly schematic illustration showing how combinations of a few gene regulatory proteins can generate many different cell types during development. In this simple scheme a "decision" to make one of a pair of different gene regulatory proteins is made after each cell devision. The "decision" may be influenced by xenobiotics. Also, this scheme is similar to what is seen in chemical-induced cancer. The letters under each cell show a putative "phenotype" of that cell.

cellular-development, -differentiation, -maintenance and -injury or -death. In fact, the subset of genes that are being expressed determine to a large extent the phenotype of that cell. Also, a loss of control of differential gene expression underlies many disease states, not the least of which being cancer. The identification of genes that are being expressed in one cell type versus another (i.e., control versus treated; tumor versus normal) can help in explaining the function of those genes as well as lend insight into the system being examined. For this reason, the identification of differentially expressed genes has been pursued for diverse stimuli such as responses to biological programs (developmental and circadian cues), physical agents (i.e., UV irradiation, X-rays) and chemical agents (hormones and xenobiotics). In fact, over 300 examples of these pursuits can be found in the literature[43] and have resulted in a much greater understanding of cellular biology and our responses to physical and chemical insult.

In the following, we will discuss approaches that are used to examine gene control as well as identify the genes being regulated by key activator proteins. The general concepts of model development and molecular approaches for identifying genes responsive to receptor-mediated xenobiotics and hormones will be discussed.

### 14.1.4.5.2 Basic Methods

For specific procedures on the analysis of gene expression the reader is directed toward laboratory manuals such as *Current Protocols in Molecular Biology*[44] and *Molecular Cloning*.[45] Popular methods for examining gene expression are outlined in Table 3. The basic approach that many take for examining altered gene expression is as follows.[46] First, an mRNA for a specific gene is shown to accumulate to a different extent in two populations of cells. In the past, this was most often observed by Northern blotting, but high-throughput microarray assays has made genome-wide analysis possible. As stated above, the accumulation of an mRNA species may be the result of several possible mechanisms including transcription, mRNA stabilization *etc*. Therefore, it is often of interest to determine how a gene is being regulating and transcription initiation assays, such as nuclear run-on assays, are performed. If it is determined that transcription initiation is not the primary cause of mRNA accumulation, other possibilities are examined. If transcription initiation is the predominant

mechanism at play, specific *cis-* and *trans-*acting factors are explored.

With any laboratory procedure, one must be cognizant of limitations of the procedures and technical expertise. A critical factor that is often overlooked, however, is the model system. That is, when gene expression is examined in different populations of cells, often proper controls are not exercised. In the subsequent discussion, the development of a cellular model system to examine differential gene expression will be explored.

### 14.1.4.5.3 Differential Gene Expression

#### 14.1.4.5.3.1 Introduction

Three factors that contribute to this complexity and difficulty in examining differentially expressed genes are: (1) Genes are not present at the same abundance; (2) The intensity of response varies greatly from gene-to-gene; (3) There are many ways to alter gene expression. We will briefly discuss these parameters as they pertain to examining altered gene expression, and how these factors impinge on developing an optimal model system. Much of the following has been described in more detail in reference[47] as it pertained to cloning xenobiotic-induced genes.

#### 14.1.4.5.3.2 mRNA Abundance

The mammalian genome of $3 \times 10^9$ base pairs has enough DNA to code for approximately 300,000 genes, assuming a length of 10,000 bp per gene.[48] Obviously, not every gene is expressed per cell (also, not every segment of DNA may be associated with a gene product). In fact, in a mammalian cell, hybridization experiments have shown that approximately 1–2% of the total sequences of nonrepetitive DNA are represented in mRNA.[48] Thus, if 70% of the total genome is non-repetitive, 10,000–15,000 genes are expressed at a given time.

The average number of molecules of each mRNA per cell is called its *representation* or *abundance*. Of the 10,000–15,000 genes being expressed, the mass of RNA being produced per gene is highly variable. In fact, usually a few sequences are providing a large proportion of the total mass of mRNA. Hybridization and kinetic experiments between excess mRNA and cDNA in solution identifies several components of mRNA complexity. The majority of the mass of RNA (50%) is being accounted for by a component with few mRNA species. In

**Table 3** Methods for examining gene expression at the mRNA level.

| Process | Method | Comments |
|---|---|---|
| RNA accumulation | Northern Blot Analysis | RNA is resolved on a gel, transferred to nylon and probed with a radioactive DNA or RNA specific for gene of interest. Relatively insensitive and nonquantitative. |
| | RNase protection assay (RPA) | Radioactive probe specific for gene of interest is added to mRNA in solution. RNase digests non-hybridized RNA and products are resolved on gel. More sensitive and quantitative than Northerns. |
| | Dot/Slot blots, Microarrays | Probes are placed on matrix and labeled mRNA is hybridized and analyzed for expression. High throughput and quantitative. |
| | Reverse-transcriptase PCR | RNA is reverse transcribed and the cDNA is amplified using PCR and gene-specific primers. Very sensitive but quantitation requires standardization. |
| Transcription rate | Nuclear run-on assays | Nuclei are isolated and incubated with radioactive nucleotides. The newly transcribed, radioactive mRNA is hybridized to dot/slot blots or microarray. The standard for examining rate of transcription. |
| | Reporter assays | DNA response elements are cloned upstream of a promoter element and reporter gene. This plasmid is transfected into an appropriate cell and the amount of reporter activity is used to indicate rate of transcription. This is an indirect measure as the reporter mRNA must be processed, transported, stabilized and translated and therefore may not measure transcription directly. |
| Enhancer/Promoter analysis | DNase footprinting | DNA containing the regulatory region of the gene of interest is radiolabeled and incubated with appropriate proteins (*in vivo* or *in vitro*). Unprotected DNA is digested with Dnase and the "footprinted" DNA is resolved on a gel. |
| | Reporter assays | See above for description. The response elements may be mutated to show their contribution to gene expression. |
| mRNA turnover | Pulse-chase experiments | Cells are grown in culture in the presence of radioactive nucleotides (pulse) followed by incubation with an excess of cold nucleotides. RNA is extracted at various times after the chase and the loss of transcript is assessed by dot or slot blot analysis. Standard method for examination of mRNA turnover |
| | Inhibition of transcription | Transcription is inhibited in cell culture using the appropriate chemical treatment (see Table 2). The amount of mRNA the accumulates is indicative of turnover. Indirect measure of turnover. Polymerase inhibitors must be properly titrated. |
| | Reporter assays | See above for description, with the exception that the 3′UTR is cloned downstream of the reporter. The 3′UTR response elements may be mutated to show their contribution to gene expression. |
| Protein-DNA interaction | Electrophoretic mobility shift assays | A labeled DNA response element is incubated with protein source and the complex resolve on a gel. The protein-DNA complex is "retarded" in mobility relative to the probe alone. |
| Protein-RNA interaction | Electrophoretic mobility shift assays | A labeled RNA response element is incubated with protein source and the complex resolve on a gel. The protein-RNA complex is "retarded" in mobility relative to the probe alone. |

fact, approximately 65% of the total mRNA may be accounted for in as few as 10 mRNA species. The remaining 35% of the total RNA represents the remaining 10,000–15,000 genes being expressed in that tissue. Of course, the genes present in each category may be present in very different amounts and represent a continuum of expression levels. For means of this discussion, we will divide the three major components[49] into abundant, moderate and scarce representing approximately 100,000 copies, 5000 copies and <10 copies per cell respectively.

There are several reasons for discussing the components of mRNA. First, when doing a differential screen (i.e., subtractive hybridization, differential display, microarray) what is actually being compared is two *populations* of mRNA and you are examining the genes that overlap or form the intersection between groups. When comparing two extremely divergent populations, such as liver and oviduct, as much as 75% of the sequences are the same[48] equating to 10,000 genes that are identical and approximately 3000 genes that are specific to the oviduct. This suggests that there may be a common set of genes, representing required functions, that are expressed in all cell types. These are often referred to as housekeeping or constitutive genes. Second, there are overlaps between all components of mRNA, regardless of the number of copies per cell. That is, *differentially expressed genes may be abundant, moderate or scarce*. In fact, the scarce mRNA may overlap extensively from cell to cell, on the order of 90% for the liver to oviduct comparison. However, it is worthy to note that a small number of differentially expressed genes are required to denote a specialized function to that cell, and the level of expression does not always correlate with importance of the gene product.

As to be discussed subsequently, the key to developing an effective model for the study of differential gene expression may be to keep the differences in the abundant genes to a minimum. This is due to the fact that a small difference in expression of a housekeeping gene, say 2 fold, will result in a huge difference in the number of copies of that message from cell type-to-cell type (i.e., an increase of 10,000 copies per cell). Also, it is important to have a screening method that can detect differences in the scarce component. If the two populations to be compared have little difference in the abundant genes and you have optimized your screening technique to detect differences in the scarce population, the odds of identifying genes that are truly required for a specialized cellular function have increased dramatically.

### 14.1.4.5.3.3 *Intensity of Response*

A basic pharmacologic principle is that drugs and chemicals have different *affinities* for a receptor and the drug–receptor complex will have different *efficacies* for producing a biological response, i.e., altering gene expression. A corollary of this principle states that not every gene being affected by the same drug–receptor complex will have identical dose–response curves. That is, when comparing two responsive genes, the affinity of the drug–receptor complex for the DNA response elements found in the two genes and the efficacy of the drug–receptor–DNA complex at effecting transcription could be quite different. In fact, similar DNA response elements may cause a repression or an induction of gene expression, depending on the context of the surrounding gene. Therefore, when comparing two populations of mRNAs (i.e., control versus treated), there may be orders of magnitude differences in the levels of induction and repression regardless of the fact that all the genes are affected by the same drug–receptor complex.

Needless-to-say, the extent of change is important in the *detection* of these differences, but not the *importance* of that deviation. For technical reasons, it is often difficult to detect small changes in gene expression (< 2 fold). However, a two-fold change in a gene product may have dramatic effects on the affected cell, especially if it encodes a protein with a very specialized or non-redundant function. Also, the detection of a difference between two cell populations is easier if the majority of the differences are in scarce mRNAs. Once again, this is due to technical aspects of analyzing gene expression whereby the change from 500 to 1000 copies per cell is a dramatic effect compared to a change from $1 \times 10^5$ to $2 \times 10^5$, an effect that may be virtually unnoticed.

### 14.1.4.5.3.4 *Specificity of Response*

The last factor we will discuss regarding the complexity of mRNA species, is that regulation of gene expression is multi-faceted. The analysis of differential gene expression is most often performed by comparing steady-state levels of mRNA. That is, the amount of mRNA which accumulates in the cell is a function of the rate of formation (transcription) and removal (processing, stability, degradation). If differences in protein products are being compared, add translation efficiency, processing and degradation to the scenario. With all the possible causes for altered gene expression, the specificity of response must be questioned. Is the difference in mRNA or protein observed an important effect on expression or is it secondary to a parameter in your model system you have not controlled or accounted?

In the best-case situation, the key mechanism of gene regulation that results in the end-point of interest should be known. At least, one should have criteria in mind for the type of response that is truly important. With most receptor systems, early transcriptional regulation may predominate as this key event. However, the other modes must be acknowledged, at least when novel responses are being characterized. By assuming that the key event is mRNA accumulation, the true initiating response, such as protein phosphorylation or processing, may be overlooked. Also, the extent and diversity of secondary events, i.e. those that require the initial changes in gene expression, may far exceed the primary events. The amplification of an initial signal (i.e., initial response "gene A" causes regulation of secondary response "gene B") can confuse the interpretation of altered mRNA accumulation. Once again, one must have a clear understanding of whether a primary or secondary event is the key response and design your model accordingly.

## 14.1.4.6 SUMMARY

There are three key nuances regarding the regulation of gene expression by exogenous chemicals. (1) The phenotype of a cell is dictated by which genes are being expressed at that particular time. For example, different cell types in a multicellular organism arise because they synthesize and accumulate different sets of RNA and protein molecules. Similarly, a tumor cell expresses a different subset of genes than the normal cell from which it arose.[1] Understanding how a chemical regulates gene expression is particular germane to carcinogenesis, but it may also apply to other forms of chemical-induced toxicity as well. (2) A cell can change the expression of its genes in response to external signals, such as xenobiotics, in a *cell-*, *species-* and *sex-dependent* manner. For example, many environmental pollutants regulate the expression of cytochrome P450s in liver cells. This may be considered an adaptive response as cytochrome P450s are involved in many detoxification pathways. However, the subset of genes regulated in the liver may be quite different than that seen in the kidney or lung. This illustrates the general feature of cellular specialization. (3) Gene expression can be regulated at many different steps in the pathway from DNA to RNA to Protein. This is an extension of the first two points, whereby the phenotype of a cell, and the cell-type dependent regulation of gene expression, can be modulated by a xenobiotic at many points in this complicated cascade.

## 14.1.4.7 REFERENCES

1. S. M. Fisher, 'Chemical Carcinogen and Anticarcinogen 1st edn,' eds. G. T. Bowden and S. M. Fisher, Elsevier Science Inc, New York, NY, 1997, pp. 349–382.
2. B. Alberts, D. Bray, J. Lewis *et al.*, 'Molecular Biology of the Cell 3rd edn,' eds. Garland Publishing, Inc., New York, NY, 1994, pp. 401–476.
3. B. F. Pugh, 'Control of gene expression through regulation of the TATA-binding protein.' *Gene*, 2000, **255**, 1–14.
4. N. Dillon and P. Sabbattini, 'Functional gene expression domains: defining the functional unit of eukaryotic gene regulation.' *Bioessays*, 2000, **22**, 657–665.
5. L. Huang, R. J. Guan and A. B. Pardee, 'Evolution of transcriptional control from prokaryotic beginnings to eukaryotic complexities.' *Crit. Rev. Eukaryot. Gene. Expr.*, 1999, **9**, 175–182.
6. R. D. Kornberg, 'Eukaryotic transcriptional control.' *Trends Cell. Biol.*, 1999, **9**, M46–49.
7. J. W. Chin, J. J. Kohler, T. L. Schneider *et al.*, 'Gene regulation: protein escorts to the transcription ball.' *Curr. Biol.*, 1999, **9**, R929–932.
8. R. G. Roeder, 'Role of general and gene-specific cofactors in the regulation of eukaryotic transcription.' *Cold Spring Harb. Symp. Quant. Biol.*, 1998, **63**, 201–218.
9. G. C. Franklin, 'Mechanisms of transcriptional regulation.' *Results Probl. Cell. Differ.*, 1999, **25**, 171–187.
10. K. F. Wilson and R. A. Cerione, 'Signal transduction and post-transcriptional gene expression.' *Biol. Chem.*, 2000, **381**, 357–365.
11. A. V. Philips and T. A. Cooper, 'RNA processing and human disease.' *Cell Mol. Life Sci.*, 2000, **57**, 235–249.
12. T. Pederson, 'Movement and localization of RNA in the cell nucleus.' *Faseb. J.*, 1999, **13**, Suppl 2, S238–242.
13. C. W. Smith and J. Valcarcel, 'Alternative pre-mRNA splicing: the logic of combinatorial control.' *Trends Biochem. Sci.*, 2000, **25**, 381–388.
14. S. Maas and A. Rich, 'Changing genetic information through RNA editing.' *Bioessays*, 2000, **22**, 790–802.
15. A. Herbert and A. Rich, 'RNA processing and the evolution of eukaryotes.' *Nat. Genet.*, 1999, **21**, 265–269.
16. F. I. Comer and G. W. Hart, 'O-GlcNAc and the control of gene expression.' *Biochim. Biophys. Acta*, 1999, **1473**, 161–171.
17. J. K. Hood and P. A. Silver, 'In or out? Regulating nuclear transport.' *Curr. Opin. Cell Biol.*, 1999, **11**, 241–247.
18. U. Shinde and M. Inouye, 'Intramolecular chaperones: polypeptide extensions that modulate protein folding.' *Semin. Cell Dev. Biol.*, 2000, **11**, 35–44.
19. M. S. Sheikh and A. J. Fornace Jr, 'Regulation of translation initiation following stress.' *Oncogene*, 1999, **18**, 6121–6128.

20. G. Reuter and H. J. Gabius, 'Eukaryotic glycosylation: Whim of nature or multipurpose tool?' *Cell Mol. Life Sci.*, 1999, **55**, 368–422.

21. N. L. Weigel, 'Steroid hormone receptors and their regulation by phosphorylation.' *Biochem J.*, 1996, **319**, 657–667.

22. G. G. Kuiper and A. O. Brinkmann, 'Steroid hormone receptor phosphorylation: is there a physiological role?' *Mol. Cell Endocrinol.*, 1994, **100**, 103–107.

23. R. J. Davis, 'Transcriptional regulation by MAP kinases.' *Mol. Reprod. Dev.*, 1995, **42**, 459–467.

24. A. P. Wolffe and J. J. Hayes, 'Chromatin disruption and modification.' *Nucleic Acids Res.*, 1999, **27**, 711–720.

25. K. Struhl, 'Fundamentally different logic of gene regulation in eukaryotes and prokaryotes.' *Cell*, 1999, **98**, 1–4.

26. M. Mannervik, Y. Nibu, H. Zhang *et al.*, 'Transcriptional coregulators in development.' *Science*, 1999, **284**, 606–609.

27. C. Bevan and M. Parker, 'The role of coactivators in steroid hormone action.' *Exp. Cell Res.*, 1999, **253**, 349–356.

28. J. L. Workman and R. E. Kingston, 'Alteration of nucleosome structure as a mechanism of transcriptional regulation.' *Annu. Rev. Biochem.*, 1998, **67**, 545–579.

29. K. D. Robertson and P. A. Jones, 'DNA methylation: past, present and future directions.' *Carcinogenesis*, 2000, **21**, 461–467.

30. C. J. Fry and P. J. Farnham, 'Context-dependent transcriptional regulation.' *J. Biol. Chem.*, 1999, **274**, 29583–29586.

31. M. R. Green, 'TBP-associated factors (TAFIIs): Multiple, selective transcriptional mediators in common complexes.' *Trends Biochem. Sci.*, 2000, **25**, 59–63.

32. J. W. Conaway, A. Shilatifard, A. Dvir *et al.*, 'Control of elongation by RNA polymerase II.' *Trends Biochem. Sci.*, 2000, **25**, 375–380.

33. D. B. Bregman, R. G. Pestell and V. J. Kidd, 'Cell cycle regulation and RNA polymerase II.' *Front. Biosci.*, 2000, **5**, D244–257.

34. S. M. Uptain, C. M. Kane and M. J. Chamberlin, 'Basic mechanisms of transcript elongation and its regulation.' *Annu. Rev. Biochem.*, 1997, **66**, 117–172.

35. J. W. Conaway and R. C. Conaway, 'Transcription elongation and human disease.' *Annu. Rev. Biochem.*, 1999, **68**, 301–319.

36. N. Proudfoot, 'Connecting transcription to messenger RNA processing.' *Trends Biochem. Sci.*, 2000, **25**, 290–293.

37. A. Newman, 'RNA splicing.' *Curr. Biol.*, 1998, **8**, R903–905.

38. B. R. Cullen, 'Nuclear RNA export pathways.' *Mol. Cell Biol.*, 2000, **20**, 4181–4187.

39. J. M. Staton, A. M. Thomson and P. J. Leedman, 'Hormonal regulation of mRNA stability and RNA-protein interactions in the pituitary.' *J. Mol. Endocrinol.*, 2000, **25**, 17–34.

40. A. Jacobson and S. W. Peltz, 'Tools for turnover: methods for analysis of mRNA stability in eukaryotic cells (editorial).' *Methods*, 1999, **17**, 1–2.

41. A. Jacobson and S. W. Peltz, 'Interrelationships of the pathways of mRNA decay and translation in eukaryotic cells.' *Annu. Rev. Biochem.*, 1996, **65**, 693–739.

42. B. T. Kren and C. J. Steer, 'Posttranscriptional regulation of gene expression in liver regeneration: role of mRNA stability.' *Faseb. J.*, 1996, **10**, 559–573.

43. J. S. Wan, S. J. Sharp, G. M. Poirier *et al.*, 'Cloning differentially expressed mRNAs.' *Nat. Biotechnol.*, 1996, **14**, 1685–1691.

44. F. M. Ausubel, R. Brent, R. E. Kingston *et al.* 'Current Protocols in Molecular Biology,' John Wiley and Sons Inc, New York, NY, 1994.

45. J. Sambrook, E. F. Fritsch, T. Maniatis, eds. 'Molecular Cloning: A laboratory manual,' Cold Spring Harbor Press, Cold Spring Harbor, NY, 1989.

46. J. P. Vanden Heuvel, 'PCR Protocols in Molecular Toxicology,' eds. J. P. Vanden Heuvel, CRC Press, Boca Raton, FL, 1997, pp. 41–98.

47. J. P. Vanden Heuvel, 'Toxicant Receptor Interactions: Modulation of Signal Transduction and Gene Expression,' eds. M. S. Denison, W. G. Helferich, Taylor and Francis, Philadelphia, 1998, pp. 217–235.

48. B. Lewin, 'Genes IV,' Oxford University Press, Oxford UK, 1990.

49. J. O. Bishop, J. G. Morton, M. Rosbash *et al.*, 'Three abundance classes in HeLa cell messenger RNA.' *Nature*, 1974, **270**, 199–204.

J.P. Vanden Heuvel, G.H. Perdew, W.B. Mattes and W.F. Greenlee (Eds.)
Comprehensive Toxicology, Vol. xiv

14.2  XENOBIOTIC RECEPTOR SYSTEMS

# 14.2.1
# Introduction and Overview

JOHN P. VANDEN HEUVEL
*Penn State University, University Park, PA, USA*

---

### 14.2.1.1   INTRODUCTION

Activator proteins, as defined in Chapter 14.1.4, have the ability to bind to DNA in the enhancer region of target genes and ultimately regulate gene expression. These proteins respond to external stimuli such as xenobiotics either directly (receptors) or indirectly (downstream transcription factors, kinases) to regulate gene expression. The outline of the present section is shown in Figure 1. The predominant focus of the present chapter is to describe the basic structure/function of transcription factors and introduce *soluble xenobiotic receptors*. These latter activator proteins often play key roles in mediating toxic responses from a variety of xenobiotics, in particular lipophilic compounds (reviewed in Reference 1). The majority of xenobiotic ligands for these receptors are tumor promoters and/or epigenetic carcinogens. (See Volume 12 of this series for classification of chemical carcinogens.) Therefore, xenobiotic

receptor systems are regulators of cellular proliferation, apoptosis and differentiation. These diverse and complicated processes are affected by alteration in the expression of networks of genes, ultimately resulting in a change in cellular phenotype. Soluble xenobiotic receptors are also an active area of research in drug discovery where the beneficial effects of xenobiotics may be exploited and drug–drug interactions may be examined in detail.

The toxicologic and pharmacologic effects of ligands for xenobiotic receptor systems are dependent on the control of gene expression. The reader is directed to Chapter 14.1.4 of this volume to the basic tenets of gene control and the role of activator proteins. Receptors act as ligand-activated transcription factors, and are potent transcriptional modulators, although they may also affect post-transcriptional events. Binding to specific DNA sequences in genes that are regulated by xenobiotics is only part of the story. The pathways leading to DNA recognition by a xenobiotic receptor are complicated in

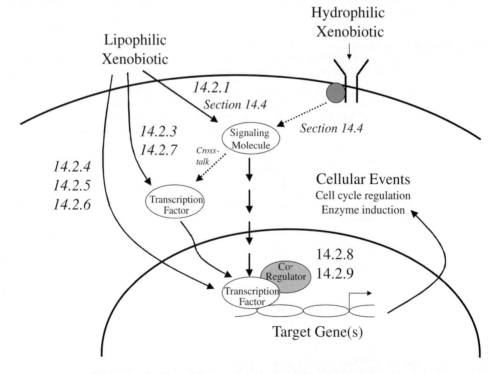

**Figure 1**   Outline of Section 14.02, *Xenobiotic receptor systems.*

that they often require additional events including receptor-ligand binding (see Chapter 14.1.3), protein conformational changes, translocation into the nucleus and protein–protein interactions. In the present section, the mechanism of action of several xenobiotic receptor systems will be examined in detail, documenting the sequence of events initiated by the ligand-receptor complex and ultimately resulting in a toxic or pharmacologic event.

In order to affect the transcription of target genes, activator proteins such as nuclear hormone receptors (NHRs) or the arylhydrocarbon receptor (AhR) must be able to interact with DNA modifying enzymes or other transcription coregulators. As described in Section 14.1.4, chromatin structure affects the rate of gene transcription and activator proteins recruit histone modifying enzymes. Alternatively, activator proteins may associate with large, multiprotein complexes that bring RNA polymerase in proximity to the transcription initiation site. These aspects of environmental influence on gene expression will be described in the last two chapters of the present section. In particular, the consequences of recruiting common coregulator proteins in the modification of the toxic response initiated by the xenobiotic receptors will be described.

## 14.2.1.2 STRUCTURAL MOTIFS IN EUKARYOTIC TRANSCRIPTION FACTORS

Activator proteins, or more specifically transcription factors, are extremely varied in their structure and function. This holds true for xenobiotic receptors as well, although certain types of domains have been conserved and are found in many such proteins. Thus, certain motifs or domains are used in a variety of proteins to impart a particular function to that protein. For an overview of these domains, the reader is directed to *Molecular Biology of the Cell*.[2] The classification of domains found in transcription factors[3] is summarized in Table 1. The most common motifs found in xenobiotic receptors are introduced, below.

### 14.2.1.2.1 Basic Domain

The DNA binding, basic domain is typically associated with leucine zippers or helix-loop-helix dimerization domains (see below) and has a high percentage of basic amino acids. Examples of basic leucine zipper (bZIP) proteins include fos and jun while basic helix-loop-helix (bHLH) proteins are typified by

USF, myc and AhR. Proteins containing basic domains generally function as dimers (homo or hetero) and bind to DNA elements that are symmetric, such as inverted repeats. This binding occurs in the major groove of the DNA helix. Basic region dimers are sufficient for DNA binding and do not require the ZIP or HLH regions for this function. In solution, the basic region is a random coil, but it adopts an α-helical structure upon DNA binding.

### 14.2.1.2.2 Zinc Fingers

The zinc finger domain is a DNA-binding motif consisting of specific spacing of cysteine and histidine residues that allow the protein to bind zinc atoms. The metal atom coordinates the sequences around the cysteine and histidine residues into a finger-like domain. The finger domains can "interdigitate" into the major groove of the DNA helix. The spacing of the zinc finger domain in this class of transcription factor coincides with a half-turn of the double helix. The classic example is the RNA pol III transcription factor, TFIIIA. Proteins of the steroid/thyroid hormone family of transcription factors also contain zinc fingers. There are several basic types of Zn finger proteins, each which differ in the coordination of Zn, either with cysteine (C) or histidine (H). Examples are shown in Table 1 and include $C_4$, $C_2H_2$ and $C_6$ coordination of Zn within the finger.

### 14.2.1.2.3 *Helix-Loop-Helix (HLH)*

The HLH domain is used by a wide range of transcription factors and is predominantly involved in protein dimerization. The HLH motif is composed of two regions of α-helix separated by a region of variable length that forms a loop between the α-helices. This motif is quite similar to the helix-turn-helix motif found in several prokaryotic transcription factors such as the cyclic AMP receptor protein (CRP) involved in the regulation of the *lac* operon.[2] The α-helical domains are structurally similar and are necessary for protein interaction with sequence elements that exhibit a twofold axis of symmetry. This class of transcription factor most often contains a region of basic amino acids located on the N-terminal side of the HLH domain (termed bHLH proteins) that is necessary for the protein to bind DNA at specific sequences. The HLH domain is necessary for homo- and heterodimerization. Examples of bHLH proteins include MyoD (a myogenesis inducing transcription factor) and c-Myc (originally identi-

**Table 1**  Transcription factor classification.

| Superclass | Class | Motif | Examples |
|---|---|---|---|
| Basic domains | Leucine Zipper (bZIP) | [KR]-x(1,3)-[RKSAQ]-N-x(2)-[SAQ](2)-x-[RKTAENQ]-x-R-x-[RK] | *Fos, Jun, Maf, Nrf-1,* CREB, C/EBP |
| | Helix-loop-helix (bHLH) | [DENSTAP]-[KR]-[LIVMAGSNT]-{FYWCPHKR}-[LIVMT]-[LIVM]-x(2)-[STAV]- [LIVMSTACKR]-x-[VMFYH]-[LIVMTA]-{P}-{P}-[LIVMRKHQ]. | E2A, MyoD, Myogenin, Twist, **AhR, Arnt** |
| | bHLH-ZIP | See above | c-myc, Mad, Max, USF, SREBP, E2F |
| | NF-1 | No consensus motif | NF-1A, NF-1B |
| | RF-X | No consensus motif | RF-X |
| | nHSH | No consensus motif | AP-2 |
| | Cys$_4$ of nuclear receptor type | C-x(2)-C-x-[DE]-x(5)-[HN]-[FY]-x(4)-C-x(2)-C-F-x-R | GR, MR, AR, **PPAR, LXR, FXR/SXR, PXR, CAR, ER** NGFI-B, HNF-4 |
| | Diverse Cys$_4$ zinc fingers | C-x-[DN]-C-x(4,5)-[ST]-x(2)-W-[HR]-[RK]-x(3)-[GN]-x(3,4)-C-N-[AS]-C | GATA-1, GLN3 |
| Zinc-finger domains | Cys$_2$His$_2$ zinc fingers | C-x(2,4)-C-x(3)-[LIVMFYWC]-x(8)-H-x(3,5)-H | TFIIIA, Sp1, Egr-1, Krüppel, Ikaros, WT1, HIV-EP1, **MTF-1** |
| | Cys$_6$ zinc cluster | [GASTPV]-C-x(2)-C-[RKHSTACW]-x(2)-[RKHQ]-x(2)-C-x(5,12)-C-x(2)-C-x(6,8)-C | Gal4, Leu3 |
| | Zinc fingers of alternating composition | No consensus motif | Baf1, Byr3 |
| Helix-turn-helix | Homeo domain | [LIVMFYG]-[ASLVR]-x(2)-[LIVMSTACN]-x-[LIVM]-x(4)-[LIV]-[RKNQESTAIY]- [LIVFSTNKH]-W-[FYVC]-x-[NDQTAH]-x(5)-[RKNAIMW] | HoxA1, Ftz, Cad, Ubx, HNF1, Msh, Pit-1, Oct-1Lim-1, ATBF1 |
| | Paired Box | R-P-C-x(11)-C-V-S | Pax-1, Pax-2, Poxn |
| | Fork head/ winged helix | [KR]-P-[PTQ]-[FYLVQH]-S-[FY]-x(2)-[LIVM]-x(3,4)-[AC]-[L 1M] | Fkh, Slp1, ILF, HFH-1, HNF-3α |
| | Heat shock factors | No consensus motif | HSF |
| | Tryptophan clusters | W-[ST]-x(2)-E-[DE]-x(2)-[LIV] | Myb, GL1, Ets, Elk-1, Ets-2Erg-1, IRF |
| | TEA domain | G-R-N-E-L-I-x(2)-Y-I-x(3)-[TC]-x(3)-R-T-[RK](2)-Q-[LIVM]-S-S-H-[LIVM]-Q-V | TEF-1, Sd |
| β-scaffold factors with minor groove contacts | Rel-homology domain (RHR) | No consensus motif | *NF-κB,* RelA, IκB, NF-AT |
| | STAT | No consensus motif | Stat1, Stat2, Stat5 |
| | p53 | No consensus motif | p53 |
| | MADS box | R-x-[RK]-x(5)-I-x-[DNGSK]-x(3)-[KR]-x(2)-T-[FY]-x-[RK](3)-x(2)-[LIVM]-x-K(2)-A-x-E-[LIVM]-[STA]-x-L-x(4)-[LIVM]-x-[LIVM] (3)-x(6)-[LIVMF]-x(2)-[FY] | MEF-2, ZEM, Glo, MCM1 |
| | β-barrel α-helix | No consensus motif | E2, EMNA-1 |

**Table 1** Continued.

| Superclass | Class | Motif | Examples |
|---|---|---|---|
| | TATA-binding proteins | No consensus motif | TBP |
| | HMG | No consensus motif | Sox-2, SRY, TCF, UBF |
| | Heteromeric | No consensus motif | CP1A, CBF-4 |
| | CCAAT factors | No consensus motif | |
| | Grainhead | No consensus motif | CP2, Grainyhead |
| | Cold-shock domain factors | [FY]-G-F-I-x(6,7)-[DER]-[LIVM]-F-x-H-x-[STKR]-x-[LIVMFY]. | DbpA, Ybx-3 |
| Other | Runt | No consensus motif | Runt, PEBP2, AML1 |
| | Copper fist | No consensus motif | AMT1, ACE1 |
| | HMGI(Y) | No consensus motif | HMGI(Y) |
| | Pocket domain | No consensus motif | Rb, CBP, p107 |
| | E1A-like factors | No consensus motif | E1A |
| | AP2/EREBP-related | No consensus motif | AP2, EREBP, R1AV1 |

Derived from Reference 3. **Bold** type is a xenobiotic receptor, *italics* a receptor affected by xenobiotics, indirectly.

fied as a retroviral oncogene). Several HLH proteins do not contain the basic region and act as transcriptional repressors. These HLH proteins repress the activity of other bHLH proteins by forming heterodimers with them and preventing DNA binding.

#### 14.2.1.2.4 Leucine Zipper

Similar to the HLH motif, the leucine zipper domain is necessary for protein dimerization. Leucine zipper motifs are generated by a repeated distribution of leucine residues spaced seven amino acids apart within α-helical regions of the protein. These leucine residues have their R-groups protruding from the α-helical domain in which the leucine residues reside. The protruding R-groups are thought to "interdigitate" with leucine R groups of another leucine zipper domain, thus stabilizing homo- or heterodimerization. The leucine zipper domain is present in many DNA-binding proteins, such as c-Myc, and C/EBP.

#### 14.2.1.2.5 Activation Domains

An activation domain (AD) (or transactivation domain, TAD) is that part of a sequence in a protein that when tethered to a promoter leads to greatly elevated levels of transcription. An AD does not bind directly to DNA, but must be associated with this activity. The criteria for defining an AD are to attach this region to a heterologous DNA binding domain (i.e., lexA, Gal4) and assess its ability to regulate a reporter gene. The general mechanism of action of ADs is that they serve as a contact site with coactivator proteins and aid in the recruitment of the transcription machinery. Not all transcription factors contain an AD; however, these particular proteins interact with other factors that do contain an activation domain.

There are several varieties of ADs known and a transcription factor can have one or more such domain of the same or different classes. For example, nuclear hormone receptors often contain two ADs (called activation function 1 and 2, AF1 and AF2) one of which being ligand-dependent, the other ligand-independent. Activation domains share little homology even in a class except for the preponderance of a certain amino acid. This does not mean that no homology exists or that they do not adopt a common structure. *"Acidic" activation domains* contain a preponderance of aspartic acid or glutamic acid residues. The structure of acidic ADs is that of an amphipathic α-helix. Examples include steroid hormone receptors, Gal4 and VP16. *Glutamine-rich activation domains*, such as those seen in Sp1 and Oct1, are specific for higher eukaryotes and do not function in yeast. Similarly, *Proline-rich activation domains* only function in higher eukaryotes. This particular AD is found in CTF/NF1, Oct2 and AP-2. *Serine/Threonine-rich activation domains* may also serve as transcriptional activators and are sites of protein phosphorylation.

### 14.2.1.3 DIRECT XENOBIOTIC SIGNALING PATHWAYS

As stated above, nature has reused common structural motifs in a variety of proteins for the purpose of regulating gene expression in a specific manner. By using a combination of DNA binding, protein–protein interface and activation domains, a protein can be designed that regulates a specific set of genes at a particular time or in a particular cellular context. Adding a sensing or a responding domain to the protein results in an additional level of specificity. This domain can act like a switch; when a particular event is sensed, the activator protein is turned "on" or "off" (transcriptionally active or inactive). This sensing domain has been used successfully in the hormone or growth factor receptors where gene batteries are regulated in the presence of a physiologically controlled ligand. However, sensing domains may also allow xenobiotics to inappropriately regulate gene expression if these exogenous chemicals are able to mimic a natural effector molecule. Of interest to the following discussion are *ligand-binding domains* that are a direct contact site between the xenobiotics and the responding protein. If a transcription factor contains a ligand-binding domain, it contains the two functions required to be defined a receptor, *recognition* or binding of ligand and signal *transduction* (see Chapter 14.1.3). There are several important examples of proteins that meet the definition of a xenobiotic receptor (Table 2). The most common type of protein of this class are nuclear hormone receptors and include peroxisome proliferators-activated receptors (PPARs), estrogen receptor (ER) and pregnane-X-receptor (PXR). Dioxin and related chemicals bind directly to the aryl hydrocarbon receptor (AhR), a member of the bHLH, Per-Arnt-Sim (PAS) family of transcription factors. Heavy metals interact with metal transcription factor, a $C_2H_2$ zinc finger protein. Although not transcription factors, protein kinase C and protein phosphatases may be considered the cognate receptors for phorbol esters and okadaic acid, respectively. Last, jun, fos and Nrf are leucine zipper proteins that are activated by antioxidants. Since it has not yet been conclusively shown that anti-oxidants bind directly to these transcription factors, fos/jun/Nrf may not be true xenobiotic receptors and hence will be discussed in Section 14.4 as signaling molecules.

A common feature of the xenobiotics shown in Figure 1 is that they are lipophilic compounds that can diffuse across the cell membrane with moderate ease. The interaction with the xeno-

**Table 2** Xenobiotics and endogenous compounds whose effects are mediated directly by nuclear receptors, transcription factors or signaling molecules.

| Ligand | Receptor | Class |
|---|---|---|
| Polycyclic aromatic hydrocarbons | Ah receptor | bHLH-PAS[a] |
| Phenobarbital | CAR | NHR[b] |
| Dexamethasone | PXR/SXR | NHR |
| Fibrates, fatty acids | PPARα | NHR |
| Cholesterol | LXRα | NHR |
| Bile acids | FXR | NHR |
| Thryoid hormone | TR | NHR |
| Estrogen/Xenoestrogens | ER | NHR |
| Benzoates | BXR | NHR |
| Metals | MTF | $C_2H_2$ ZFP[c] |
| Antioxidants[1] | Fos/jun/Nrf | bZIP[d] |
| Phorbol esters | PKC | Kinase |
| Okadaic acid | PP-1, PP2A | Phosphatase |

[a]Basic helix-loop-helix Per-Arnt-Sim.
[b]Nuclear hormone receptor ($C_4$ zinc finger protein).
[c]Zinc finger proteins.
[d]Leucine zipper.
[1]Not definitive xenobiotic receptors.

biotic receptor may occur predominantly in the cytosol (AhR) or in the nucleus (ER, PXR). Another common feature of these xenobiotics is that they act as tumor promoters. The ability of xenobiotic receptors to regulate gene expression makes them potent regulators of the cell cycle and apoptosis. In addition, many cytochrome P450 enzymes are regulated by xenobiotic receptors. In addition to affecting the metabolism of endogenous compounds, enzyme induction is an important aspect of drug–drug interactions and toxicity (Table 3).[4]

### 14.2.1.4 NUCLEAR HORMONE RECEPTORS

#### 14.2.1.4.1 Introduction

Unlike receptors found on the cell surface, members of the nuclear hormone receptor (NHR) superfamily are restricted to metazoan organisms such as nematodes, insects, and vertebrates. (There are several excellent reviews on steroid hormone receptors, including References 4–19). These proteins are intracellular transcription factors that directly regulate gene expression in response to lipophilic molecules. They affect a wide variety of functions, including fatty acid metabolism, reproductive development, and detoxification of foreign substances. As described above, many of the NHRs act as ligand-inducible transcription factors, responding to endogenous and exogenous chemicals. However, the vast majority of known receptors do not have an identifiable physiologically relevant ligand, and are deemed *orphan receptors.*

To date, over 300 NHRs have been cloned. Early classification of these receptors was based on ligands, DNA binding properties or other functional characterization. Recently a more systematic classification has been proposed, based on sequence similarity, much like

that employed for cytochrome P450s. Phylogenic analysis has shown six subfamilies (NR1-6) with various groups and individual genes.[7] As discussed below, most NHRs have the same basic structure. The most highly conserved region is the $C_4$ zinc finger domain, which as described above, is a DNA binding motif.

#### 14.2.1.4.2 Structure to Function

##### 14.2.1.4.2.1 Functional Domains

NHRs generally follow a standard blueprint, as shown in Figure 2. While most NHRs have all of these elements, there are a number of exceptions, some of which will be discussed below. The N terminus of the NHR, sometimes called the modulator, hypervariable or A/B domain, has transactivation activity, termed activation function 1 (AF-1). This acidic AD is ligand-independent, or constitutively functional. The A/B domain's sequence and length are highly variable between receptors (i.e., GR versus RXR) and among receptor subtypes (RXRα versus β). In addition, this region is the most frequent site of alternative splicing and secondary start sites and contains a variety of kinase recognition sequences. For these reasons, it is thought that the variable N-terminal sequences may be responsible for the receptor-, species-, and cell type-specific effects as well as promoter context-dependent properties of NHR transactivation.[20]

NHRs bind to hormone response elements (HREs) in their target promoters through the DNA binding domain (DBD) or C domain (see Figure 3). Composed of two zinc fingers, the DBD is the most conserved region within the NHR superfamily. The first zinc finger contains the proximal- or P-box region, an alpha helix that is responsible for high-affinity recognition of the "core half-site" of the response element. Located within the second

**Table 3** Xenobiotic receptors and cytochrome P450 (CYP) induction.

| Xenobiotic receptor | Prototypic CYP induced | Core DNA response element |
|---|---|---|
| Aryl hydrocarbon receptor (AhR) | 1A1, 1A2, 1B1 | TNGCGTG |
| Constitutive androstane receptor (CAR) | 2B1, 2B2 | AGGTCA, DR4 |
| Pregnane-X-receptor (PXR) | 3A1, 3A2, 3A23 | AGGTCA, DR3,ER6 |
| Peroxisome proliferators-activated receptors (PPARs) | 4A1, 4A2, 4A3 | AGGTCA, DR1 |
| Liver-X-receptor (LXR) | 7A1 | AGGTCA, DR4 |
| Farnesol-X-receptor (FXR) | 7A1 | AGGTCA, IR1 |
| Protein kinase C/AP1 | Affect CYP induction by other receptors | |

Adapted from Reference 4.

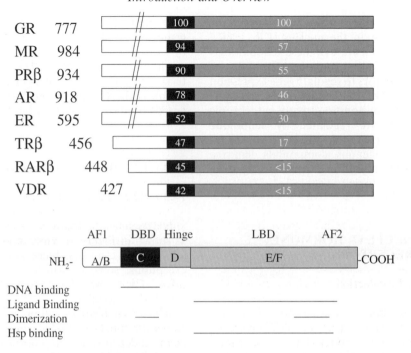

**Figure 2**   Basic structure of nuclear hormone receptors. Top panel: Comparison of several nuclear hormone receptors. The percent identity of the DNAbinding-(C-)domain and the ligand binding-domain (DEF) domains are shown, as well as the length of the proteins. Bottom panel, putative function of the various NHR domains. Abbreviations GR, Glucocorticoid receptor; MR, mineralocorticoid receptor; PRβ, progesterone receptorβ; AR, androgen receptor; ER, estrogen receptor; TRβ, thyroid hormone receptorβ; RARβ, retinoic acid receptorβ; VDR, Vitamin D receptor; hsp, heat shock protein.

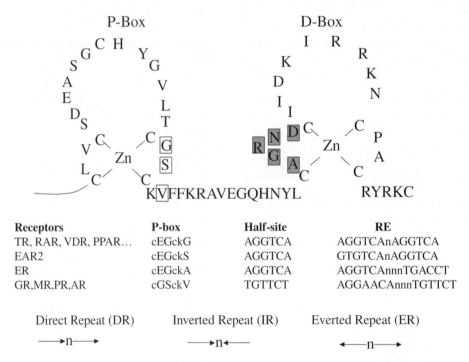

**Figure 3**   Sequence specific recognition of DNA by nuclear receptors. The core half-site recognized by a NHR is based on amino acid residues within the P-box (boxed residues in figure, capital letters in text). The helical structure of the P-box provide contacts with the major groove of the DNA helix. The shaded residues in the D-box are important for interactions with the phosphate groups of the DNA helix as well as dimerization. Adapted from References 21, 56, 57.

zinc finger is the distal or D-box, an α-helix which lies perpendicular to P-box helix, and is a site that mediates receptor dimerization. NHRs bind to DNA as heterodimers, homo-dimers, or monomers, depending on the class of NHR. The steroid hormone receptors GR, PR, ER, AR and MR (receptors for gluco-corticoid, progesterone, estrogen, androgen and mineralocorticoids, respectively) bind to DNA as homodimers and recognize palindro-mic response elements.[21] Thyroid, retinoid, vitamin D and peroxisome proliferator recep-tors (TR, RAR, VDR and PPAR), as well as most orphan receptors, bind to DNA as a heterodimer with retinoid-x-receptor (RXR). However, the three dimensional structure of the RXR heterodimer complex produces dif-ferent DNA binding affinities. Response ele-ments may be direct repeats ($DR_x$, AGGTCA-$N_x$-AGGTCA, where N is any nucleotide and $x$ is any number of residues from 0 to 10), everted repeats ($ER_x$, ACTGGA-$N_x$-AGGTCA) or inverted repeat ($IR_x$, AGGTCA-$N_x$-ACTGGA).

Immediately adjacent to the DNA binding domain is the D or hinge domain. This particular region has an ill-defined function. The hinge domain contains the carboxy-term-inal extension (CTE) of the DBD, which may be involved in recognizing the extended 5′ end of the HRE. The D-domain appears to allow for conformational changes in the protein structure following ligand binding. Also, this region may contain nuclear localization signals and protein–protein interaction sites.

The sequence of the ligand binding domain (LBD) or E/F domain varies substantially between NHRs, but they all share a common structure of 11–13 α-helices organized around a hydrophobic binding pocket. Residues within the binding pocket confer specificity, determin-ing whether the LBD will accept steroid hormones, retinoid compounds or the host of xenobiotic ligands that affect receptor func-tion. Ligand-dependent activation requires the presence of activation function 2 (AF-2), located at the extreme C terminus of the NHR. LBDs also contain nuclear localization signals, protein interaction with dimerization motifs for heat shock proteins, coregulators and other transcription factors.

### 14.2.1.4.2.2 Exceptions to the Rule

There are several NHRs that do not follow the basic structure to function paradigm. For example, the constitutive androstane receptor (CAR) allows constitutive transcription of the target gene in the absence of ligand. Transcrip-tional activity is be suppressed by binding to ligands androstenol and androstanol, unlike other NHRs examined. Another important exception to the classic model of NHR activa-tion comes from receptors that do not require DNA binding.[22] DAX-1 and SHP lack DBDs completely, and transgenic mice bearing a GR that does not bind DNA are viable and fertile even though at least some GR functions are crucial for survival. NHRs can also have important biological effects without ligand binding. Most NHRs are known to be phos-phoproteins, and it is now known that many of these proteins are activated by crosstalk with other signal transduction pathways such as those responding to EGF and TGF-α.[8]

### 14.2.1.4.2.3 Basic Mechanism of Action

The mechanism of action of nuclear hor-mone receptors can take one of two basic forms, that of steroid hormone receptors (SHRs) or that of retinoid/thyroid/Vitamin D receptors (Figure 4).[23] In the absence of ligand, the transcriptionally inactive SHRs MR, PR, GR, AR and ER are sequestered in a large complex comprising the receptor, heat shock protein-90 (Hsp90), Hsp70, FKBP52/51 and possibly other proteins.[23] The cellular localiza-tion of this inactive complex is somewhat controversial and cytoplasmic or nuclear loca-lization may be observed depending on the cell type and the conditions examined; however, the central dogma is that SHRs are cytosolic in the un-liganded form. One consequence of hor-mone binding to the receptor is a distinct conformational change in receptor structure (discussed below). This conformational change marks the beginning of the signal transduction process. In the case of the GR subfamily (GR, AR, MR, PR), hormone binding elicits a dissociation of hsps and the release of a monomeric receptor from the complex. Genetic analysis and *in vitro* protease digestion experi-ments indicate that the conformational changes in receptor structure induced by agonists are similar but distinct from those produced by antagonists. However, both conformations appear to be incompatible with hsp binding.

The TR, RAR and VDR receptors do not avidly interact with hsps and are localized predominantly in the nucleus in the absence of ligand. Some unliganded NHRs of this class may interact with DNA and act as transcrip-tion repressors. This may be the result of interaction with co-repressor proteins (described in Chapter 14.2.8). An interesting exception to this observation is CAR, which is transcriptionally active in the absence of its

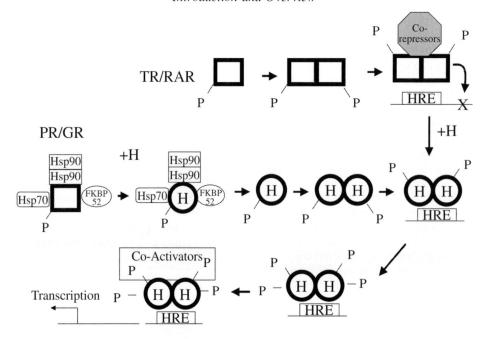

**Figure 4**  Basic mechanism of action of nuclear hormone receptors. See text for details. Abbreviations used: H, hormone; P, phosphorylated residue; HRE, hormone response element.

ligand. Hormone induced conformational changes also occur upon activation of this class of NHR, suggesting that alteration of receptor shape by ligands is a key step in the activation pathway.

Evidence suggests that receptors of the GR subfamily (SHRs) cooperatively bind to DNA as homodimers. The TR, RAR, VDR, PPAR and most of the orphan receptors form heterodimers with other members of the intracellular receptor superfamily. TR, RAR, PPAR and VDR can utilize RXRs as partners for heterodimer formation. The DNA site of contact depends on certain sequences within the C-domain, namely the proximal (P-box) and distal (D-box) zinc finger motifs (see description of the C-domain above). The P-box determines the half-site recognized, while the D-box determines the spacing between half-sites. Following activation, the SHRs receptors are capable of interacting with DNA, and both classes of NHRs (SHRs and TR/RAR) can now recruit co-activators. The DNA bound NHR complex is now a substrate for general transcription apparatus and the initiation of transcription commences.

### 14.2.1.4.3   Ligands and Activators

#### 14.2.1.4.3.1   *Natural and Xenobiotic Ligands*

Ligands for NHRs are as varied as the proteins themselves, as can be observed in

Figure 5. The reader is directed to the chapters contained in the present volume for more details on specific NHR ligands. However, a few generalized comments can be made. All ligands are lipophilic and can easily transverse the plasma membrane as well as the nuclear membrane, if required. The affinity ($K_d$) of the ligand–receptor complex is generally in the nM range, but can vary from pM to μM. It should be noted however, that the concentrations of each natural ligand should approach their $K_d$ to be considered a physiologically relevant ligand (see Chapter 14.1.3 for a discussion of physiologically relevant ligand–receptor complexes). This is of particular importance when considering reclassifying (or *adopting*) an orphan receptor in the process of *reverse endocrinology* (14.2.1.4.3.3). Xenobiotic agonists and antagonists share structure features with the natural effector molecules, although the similarities are often cryptic (note the PPARγ ligands PGJ$_2$ and Troglitazone). Some receptors, such as PPARγ, have a large ligand-binding cavity that allows for the association of a variety of endogenous ligands.[24]

#### 14.2.1.4.3.2   *Ligand-Induced Activation*

Structural studies of empty and ligand-bound LBDs have led to the "mousetrap" model of NHR activation[16,25] (Figure 6). The ligand is attracted to the trap, the receptor's electrostatic potential, and a conformational

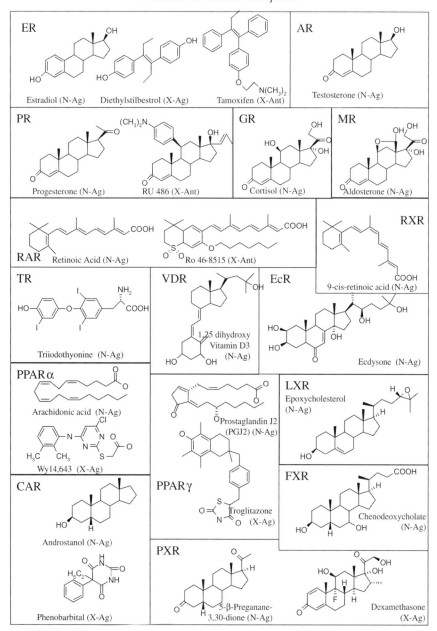

**Figure 5** Natural and xenobiotic ligands of nuclear hormone receptors. Abbreviations used; N, natural ligand; X, xenobiotic; Ag, agonist; Ant, antagonist.

change takes place, preventing the ligand's exit. In the same way that the sprung mousetrap is more stable than the primed trap, ligand binding to the NHRs ligand-binding domains stabilizes their structures relative to the unliganded receptor. The ligand forms an integral part of the hydrophobic core of the liganded LBD. This structural change is different for ligands that are full agonists *versus* those that are partial agonists or antagonists. Much attention is focused on the accessibility of the AF-2 domain to accessory proteins. The AF-2 domain can serve as an activator of transcription when excised from the rest of the protein and linked to a heterologous DNA-binding domain. In this model, binding of the ligand molecule induces a conformational change in the LBD, whereby the AF-2 sequences fold back against the binding pocket, obstructing the opening and causing rearrangements in adjacent helices. In the process, a new surface is revealed that recruits specific transcriptional coactivators. This model may explain why receptor antagonists block transactivation; these compounds do not induce the proper conformational rearrangements in the LBD, interfering with the formation of the transcriptional activation complex.

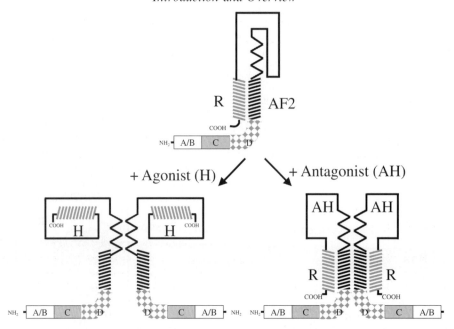

**Figure 6** Ligand-dependent NHR activation. Abbreviations used: R, regulatory domain; AF2, activation function 2.

### 14.2.1.4.3.3   *Reverse Endocrinology*

The increasing use of bioinformatics and the spread of genome projects has lead to the discovery of hundreds of proteins which share structural characteristics with NHRs. By some estimation, the Human Genome Project alone will reveal that there are as many as 2000 NHRs. When a NHR is discovered without any knowledge of its natural ligand, it was dubbed an orphan nuclear receptor (ONR). Efforts to understand ONR function and identify their physiological ligands (a process known as *reverse endocrinology*)[13] have led to the discovery of novel metabolic pathways involving the PPAR, liver X receptor (LXR), and farnesoid X receptor (FXR), new developmental systems involving the benzoate X receptor (BXR), novel classes of ligands (benzoates, terpenoids), and alternative mechanisms for NHR receptor regulation and function.[26]

The cloning of the intracellular receptor cDNAs and their reconstruction into ligand responsive transcription units has allowed for the identification of new agonist/antagonists. The "cis-trans" or reporter assay consists of transfecting a receptor expression vector and a vector carrying a receptor-responsive reporter into mammalian cells. Induction of the reporter gene activity (i.e., luciferase, chloramphenicol acetyl transferase, β-galactosidase) reflects the transcriptional activity of the transfected receptor induced by hormone. The identification of receptor antagonists is more complex as it is sometimes difficult to distinguish true receptor antagonism from toxicity. The domain structure of NHRs allows for construction of reporter assays without knowledge of the DNA binding domain of the protein. A common means to assess ligand activation of a receptor is to perform a "finger-swap" whereby the LBD for the novel NHR is placed downstream of a heterologous DBD of another, well-characterized protein. Examples of DBD used include the C-domain of GR as well as the transcription factor Gal4. Activation of this chimeric protein by agonists for the novel LBD would regulate a reporter that contains the appropriate DNA binding site (i.e., a GRE for proteins containing the C-domain of GR). This type of approach allows for rapid screening of chemicals and may be used to find both xenobiotic and endogenous ligands.

### 14.2.1.4.3.4   *Ligand-Independent Activation*

An emerging concept for intracellular receptors is that they function not only as transducers of nuclear effects of steroids, hormones and nutrients, but also as key points of convergence of multiple signal transduction pathways. The first realization of this "crosstalk" between signaling cascades came from the observation that transcriptional activity of PR, ER, TR and COUP-TF is modulated by the neurotransmitter dopamine.[21,23] The specific activation of the dopamine D1 receptors results in activation of at least two distinct signal transduction processes in the cell.

Mutation of a specific serine site close to the carboxyl terminus of the NHR can prevent dopamine activation, with no effect on the ability of the ligand to activate. The two activation pathways may differentially phosphorylate this serine. A similar phenomena is seen with other signaling pathways, in particular polypeptide growth factors such as insulin-like growth factor (IGF-1), transforming growth factor α (TGFα) and epidermal growth factor (EGF).

## 14.2.1.5 BASIC HELIX-LOOP-HELIX PROTEINS

### 14.2.1.5.1 Introduction

The helix-loop-helix (HLH) family contains over 240 transcriptional regulatory proteins and is found in a variety of organisms ranging from yeast to humans (reviewed in Reference 27). In eukaryotes, these proteins play key roles in developmental processes, including neurogenesis, myogenesis, hematopoiesis, and pancreatic development. Although a systematic nomenclature has not been developed for HLH proteins, a classification scheme based upon tissue distribution, dimerization capabilities, and DNA-binding specificities has been devised.[28] A more comprehensive alignment has also been utilized based solely on phylogenetics.[29] From a toxicological standpoint, of interest are the basic HLH (bHLH) and the bHLH Per-Arnt-Sim (bHLH-PAS) proteins. Many important transcription factors including myc, Max, sterol-responsive element-binding protein (SREBP) are bHLH family members and lack the PAS domain. The bHLH-PAS family includes the aryl hdrocarbon receptor (AhR), AhR nuclear translocator (Arnt) and hypoxia inducible factor-1α (HIF1α). This latter group of proteins will be discussed in detail in Chapters 14.2.2, 14.2.3 and 14.4.4. The goal of the following review is to place AhR, Arnt and HIF1α into a broader perspective and describe the structure to function relationships of bHLH proteins in general.

### 14.2.1.5.2 Structure to Function

#### 14.2.1.5.2.1 Functional Domains

##### 14.2.1.5.2.1.1 Basic Helix-Loop-Helix and DNA Binding

The basic HLH proteins are characterized by common possession of highly conserved bipartite domains for DNA binding and protein–protein interaction.[29] A motif of mainly basic residues permits helix-loop-helix proteins to bind to a consensus hexanucleotide E-box (CANNTG). A second motif of primarily hydrophobic residues, the HLH domain, allows these proteins to interact and to form homo- and/or heterodimers. The dimerization motif contains about 50 amino acids and produces two amphipathic α-helices separated by a loop of variable length. Additionally, some basic helix-loop-helix (bHLH) proteins contain a leucine zipper (LZ) dimerization motif characterized by heptad repeats of leucines that occur immediately C-terminal to the bHLH motif.

The solution structure of of several HLH proteins, including Myc, Max and E47 have been solved.[27] A number of interesting features were revealed from the crystal structure of E47. The E47 dimer forms a parallel, four-helix bundle that allows the basic region to contact the major groove. In addition to the basic region, residues in the loop and helix 2 also make contact with DNA. The HLH dimer is centered over the E box, with each monomer interacting with either a CAC or CAG half-site. A glutamate present in the basic region of each subunit makes contact with the cytosine and adenine bases in the E-box half-site. An adjacent arginine residue stabilizes the position of the glutamate by direct interaction with these nucleotides and additionally the phosphodiester backbone. Both the glutamate and the arginine residues are conserved in most bHLH proteins, consistent with a role in specific DNA binding.

##### 14.2.1.5.2.1.2 Per-Arnt-Sim (PAS) Domain

The unique feature of bHLH-PAS proteins is the PAS domain, named for the first three proteins identified with this motif, *Drosophila* Period (Per), human Arnt, and *Drosophila* Single-minded (Sim).[30] The PAS domain is 260–310 amino acids long and can be subdivided into two well-conserved regions, PAS-A and PAS-B, separated by a poorly conserved spacer.[31] Within both the A and B regions lies a copy of a 44-amino acid repeat referred to as the PAS repeat. The repeat begins with a nearly invariant Phe residue and terminates with a His X X Asp motif. Overall, the PAS domain is not well conserved with other family members sharing less than 25% identical in amino acid sequence.

It is not surprising, given the diversity in sequence, that the PAS domain can mediate a number of biochemical functions. The most definitive function of PAS is protein–protein interaction.[31] The PAS domain is used to interact with other members of the bHLH-PAS family of proteins. In mammalian cells, Arnt functions as a common dimerization partner with other bHLH-PAS proteins such as HIF and AhR. In this respect, Arnt functions much like RXR in the NHR arena. These heterodimers bind a sequence element related to the XRE (core GCGTG) or CME (core ACGTG). Other protein–protein interactions utilize the PAS domain, such as hsp90 binding.

The PAS domain may represent a small molecule interaction domain or a "sensing domain". To date, only one bHLH-PAS family member may be considered a true xenobiotic receptor. AhR interacts with halogenated aromatic hydrocarbons such as TCDD (dioxin) through its PAS domain, making this region similar to the E/F or LBD of NHRs described above. All of the high affinity AhR ligands identified to date are planar, hydrophobic molecules and include several classes of HAHs (PCDDs, PCDFs, PCBs *etc.*) as well as PAHs (benzopyrene, flavones). Although the affinity of binding of TCDD and other HAHs is in the picomolar to nanomolar range, PAHs bind with significantly lower affinity (nanomolar to micromolar). The differences in affinity correlate well with observed differences in toxic potency. Three dimensional molecular volume mapping studies suggest that the ligand binding properties of AhR can accommodate planar ligands with maximal dimensions of $14\,\text{Å} \times 12\,\text{Å} \times 5\,\text{Å}$ (Figure 8). In addition to

limitations on the physical characteristics of AhR ligands, high affinity binding of chemicals to the AhR are also critically dependent on key electronic and thermodynamic properties of the ligand. Although several naturally occurring AhR ligands have been identified, no high affinity physiologic endogenous ligand has been found. Recent observations demonstrating AhR-mediated responses that occur in the absence of exogenous ligand suggest that endogenous ligands must exist. Several dietary compounds such as indole-3-carbazole and the oxidized carotinoids, canthaxanthin and astaxanthin, as well as indoles (including trytophan) have been shown to be AhR ligands. The phenotype of the Ahr null mouse (described in Chapter 14.2.3), in the absence of xenobiotic treatment, also supports a biologically relevant means of receptor activation, such as a ligand.

bHLH-PAS proteins control both developmental and physiological aspects of oxygen tension.[31] Hypoxia inducible factor (HIF) and related proteins control the response to oxygen levels, including vascular branching. In the initial characterization, HIF was shown to consist of two subunits, HIF-1α, a Sim-related bHLH-PAS protein, and HIF-1β, which was later identified as Arnt. The binding site for HIF, the hypoxia response element (HRE), contains a core ACGTG sequence, which is similar to the E-box described above. Other bHLH-PAS proteins related to HIF-1α, such as endothelial PAS domain protein 1 (EPAS1), also form heterodimers with Arnt and are also believed to play roles in controlling the physiological response to oxygen levels. How

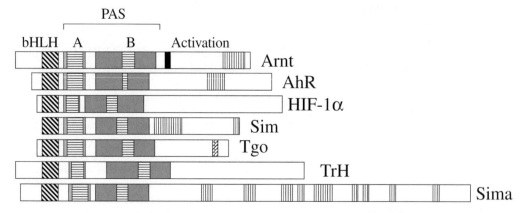

**Figure 7**  The structure of bHLH-PAS proteins is highly conserved. Shown are representations of eukaryotic bHLH-PAS proteins, aryl hydrocarbon receptor nuclear translocator (Arnt), aryl hydrocarbon receptor (Ahr) and hypoxia inducible factor-1α, (HIF-1α) relative to *Drosophila* proteins Single-minded (Sim), Tango (Tgo), Trachaeless (TrH) and Similar (Sima) proteins. The bHLH domain (cross-hatch) is near the amino terminus followed closely by the PAS domain. PAS consists of two conserved regions: A (light gray) and B (dark gray) separated by a variable spacer. Within each PAS region is a 44-amino acid PAS repeat (horizontal lines). The carboxyl termini of these bHLH-PAS proteins function as transcriptional activation domains. Shown in blocks with vertical lines are glutamine repeats associated with activation function.

the PAS domain is sensing changes in oxygen tension are not clearly understood.

#### 14.2.1.5.2.2 *Basic Mechanism of Action*

The classification scheme based on Massari and Murre[27,28] divides HLH into seven distinct classes, based partially on the mechanism of action. In Table 4, the means by which these HLH-containing proteins regulate gene expression is summarized. In the interest of space, the detailed molecular mechanism of action of HLH proteins is not discussed in this section. The reader is directed to Chapters 14.2.2, 14.2.3 and 14.4.4 for more details on the mechanism of action of the Class VII HLH proteins (Ahr, Arnt, HIF). Also, there are several reviews on the mechanism of action of oncogenic HLH proteins such as c-myc.[32–35]

### 14.2.1.6 C₂H₂ ZINC FINGER PROTEINS

#### 14.2.1.6.1 Introduction

The C₂H₂ zinc finger proteins (ZFPs) are an extremely large superfamily of nucleic acid binding proteins, with thousands of different members in vertebrates.[36] Members of the ZFP family include the transcription factors Sp1, TFIIIA, Egr-1 and WT. Recent work, as outlined in Section 14.2.7, has shown that the xenobiotic receptor metal transcription factor-1 (MTF-1) also contains C₂H₂ motifs. Additional structural features within these proteins define classes, or ZFP subgroups.[36] Class 1 ZFPs contain a Krüppel associated box (KRAB) domain and often contain 10 or more C₂H₂ motifs. The KRAB domain is similar to the HLH domain described above, in that it is a protein–protein interface motif. Human chromosome 19 contains over 40 KRAB-ZFPs in a large cluster with similar exon/intron organization. In addition, the zinc finger motifs are found on a single exon within the KRAB-ZFPs. Therefore, it appears that many Class 1 ZFP proteins arose from gene amplification events somewhat late in evolution. In contrast, Class 2 ZFPs are highly conserved in evolution. These proteins often contain fewer C₂H₂ motifs and do not contain KRAB domains. Examples of Class 2 ZFPs include SP1, TFIIIA, Gli and Krox20. In these proteins the zinc finger modules are contained on separate exons. These proteins often play key roles as housekeeping genes and in cellular differentiation. MTF-1 appears to most resemble Class 2 ZFPs since it does not have a KRAB domain, it contains 6 zinc finger motifs, and it is conserved evolutionarily. It differs from other ZFPs in that is a latent nucleotide binding protein that requires ligand (zinc or other metals) to become active.[37,38]

The basic function of C₂H₂ zinc finger motifs in either Class 1 or 2 ZFPs is to bind to RNA or DNA. ZFPs may regulate gene expression as transcription factors, chromatin remodeling proteins and RNA processing factors. In this section we will focus simply on the nucleic acid association aspect of ZFP function and redirect attention of the molecular biology of ligand binding and gene regulation of MTF-1 to a subsequent chapter.

#### 14.2.1.6.2 Structure to Function

The main structural feature of ZFPs is the coordination of zinc in the invariant cysteine and histidine residues to form a zinc-finger fold (Figure 9). Similar to C₄ motifs, the C₂H₂ finger contains a β-hairpin and an α helix folded around a zinc ion.[39] These domains are DNA-binding structures that make contact with the major groove of DNA. Formation

**Table 4** Mechanism of action of HLH containing proteins.

| *bHLH family* | *Examples* | *Hetero/homo 1dimers* | *DNA binding* |
|---|---|---|---|
| I | E12, E47, "E-box" | Both | E-box A (CA*G*CTG) |
| II | Myogenin MyoD | Hetero (with I) | E-box A (CA*G*CTG) |
| III | c-myc, SREBP (Contain LZ¹ motif) | Both | E-box B (CA*C*GTG) |
| IV | Mad, max (Contain LZ motif) | Both (with III) | E-box B (CA*C*GTG) |
| V | Id (No basic region) | Both | Transcription repressor of class I and II |
| VI | Hairy | Both | E-box B (CA*C*GTG) |
| VII | AhR, Arnt, HIF (PAS) | Hetero with Arnt | DRE GCGTG Ahr/Arnt HREACGTGHif/Arnt Others E-box A or B for others |

¹Leucine zipper.
Adapted from Reference 27–29.

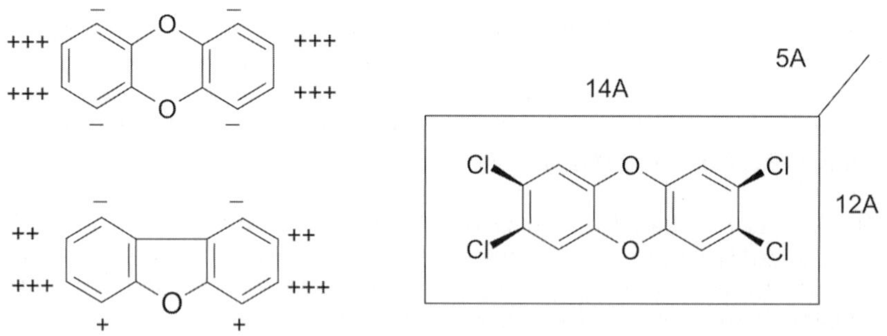

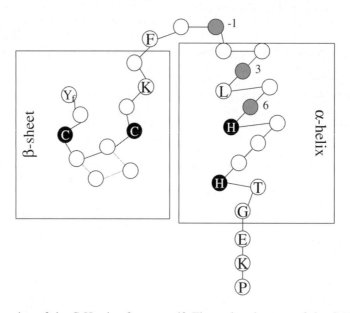

**Figure 8** Basic structure of AhR ligands. Ligands of the arylhydrocarbon receptor include the polychlorinated dibenzo-*p*-dioxins (PCDDs, top left), polychlorinated dibenzofurans (PCDFs, bottom left), polychlorinated biphenyls (PCBs) and various chlorinated and nonchlorinated compounds (not depicted). The affinity of PCDDs and PCDFs for AhR is affected by the position of the halogen atoms (+ indicated increased affinity, − the opposite). A subjective definition of the binding pocket is shown in the right panel where a planar structure of 14 Å × 12 Å × 5 Å has the maximal activity. Adapted from Reference 58.

of base-specific contacts involves at least six different amino acids within or immediately adjacent to the zinc finger helix. Similar to the studies outlined above for the P-box of NHRs, the DNA binding specificity of the zinc finger modules can be altered by site directed mutagenesis. As shown in Figure 9, sites −1, 3 and 6 (relative to the α-helix) confer selectivity of the protein for DNA. For example, altering the residue 6 of ZIF268 from a T to an R changes the binding preference from TGG to GGG.[40] Although each zinc finger domain typically recognizes 3 bp of DNA, variation in helical presentation can allow for recognition of a more extended

site. Also, in contrast to most transcription factors that rely on dimerization of protein domains for more complex DNA contact site recognition, repeating the zinc finger domain allows for the recognition of longer asymmetric sequences of DNA by a ZFP.

There are no sequence characteristics that distinguish DNA binding zinc fingers from RNA binding elements.[36] In addition, no direct contact sites on either the protein or the ribonucleic acid have been shown. RNA recognition is primarily driven by the tertiary structure of the RNA and phosphate contacts appear to predominate. Proteins such as TFIIIA contain both DNA and RNA binding

**Figure 9** Representation of the $C_2H_2$ zinc finger motif. The major elements of the $C_2H_2$ zinc finger motif ("zinc finger fold") is shown. The β-sheet and α helix are shown. Invariant cysteines and histidines, used in zinc coordination, are shaded black and other moderately conserved residues are indicated. The residues shaded in gray are direct DNA contact sites. Changing these residues alters the selectivity to the NNN nucleotide sequence it recognizes. Adapted from References 36 and 40.

ability, but they are manifested by different sets of $C_2H_2$ motifs. In this particular instance, the RNA binding activity is essential for 5S RNA storage and nucleocytoplasmic transport. The tumor suppressor gene WT1 co-immunoprecipitates with a number of spliceosomal proteins and may function in RNA splicing, a function that requires its $C_2H_2$ motifs.[41]

In addition to being nucleic acid association sites, $C_2H_2$ domains are involved in protein–protein interactions. For example, Ikaros uses separate $C_2H_2$ motifs for DNA binding and for protein binding. The last two zinc fingers mediate Ikaros homodimerization, while the first four are involved in sequence-specific DNA binding. Site-directed mutagenesis has shown that both protein and nucleotide binding require the zinc coordinating cysteines and histidines. Heterodimerization with other $C_2H_2$ motifs as well as with $C_4$ and bZIP domains has also been observed.[39]

## 14.2.1.7 PROTEIN KINASES AND PHOSPHATASES

### 14.2.1.7.1 Introduction

Certain protein kinases and phosphatases are true xenobiotic receptors, although they differ significantly from that of NHRs, bHLH-PAS and $C_2H_2$ zinc finger proteins. The function of ligand recognition and signal transduction still exists in these proteins, although they are not DNA binding transcription factors in their own right. Certain xenobiotics alter the activity of protein kinase C (PKC) and protein phosphatases 1 and 2 (PP1 and PP2) through mimicking endogenous co-factors. Ligand binding in this case results in altered phosphorylation of serine/threonine residues on substrate proteins. The substrate proteins (either directly or indirectly acted upon by the kinase/phosphatases) are the transcription factors that act as activator proteins to affect gene expression. The kinase/phosphatase target's activity is altered by phosphorylation status and can be any of the types of proteins described above.

An important question can be raised at this point, why are PKCs and PP1 and 2 considered receptors whereas other enzymes, such as cytochrome P450s (Cyps) are not? Xenobiotic substrates for Cyps may affect gene expression through the displacement of endogenous substrates from the binding pocket or by affecting intracellular oxygen tension. The criteria of specific binding and signal transduction have been met. However, in these instances the substrate is altered by the macromolecule,

thereby violating the definition of receptor set forth in Chapter 14.1.3.

The classification of PKCs and phosphatases as xenobiotic receptors, and their structure and function has been described in a section from a prior volume of this series (12.15)[42] and hence will not be recapitulated in detail herein. However, basic aspects of structure/function and ligand recognition will be described.

### 14.2.1.7.2 Protein Kinase C

The discovery of the interaction of phorbol esters with PKC represented a major breakthrough in understanding the action of these tumor promoting chemicals.[42] PKCs play a fundamental role in signaling mechanisms that result in cellular proliferation, apoptosis and differentiation as well as insulin signaling, actin remodeling and immune function.[43] As a family, PKCs are ubiquitous protein kinases that phosphorylate serines and threonines in a variety of proteins. PKC was first termed "phospholipid-activated calcium-dependent protein kinase" because of its dependency for activation on membrane phospholipids and calcium. Thus far, at least 12 forms of the PKC family have been identified in mammalian cells with 11 different genes and one splice variant.

#### 14.2.1.7.2.1 Structure to Function

The mammalian isoforms are divided into three basic families, based on primary sequence as well as activation profiles (see Figure 10). The conventional PKCs (cPKCs) include alpha, beta (splice variants 1 and 2), and gamma. Novel PKCs (nPKC) are represented by delta, epsilon, eta, and theta. The PKCs zeta and lambda comprise the atypical PKCs (aPKCs). Each class has distinct structural differences and functions. Conventional PKCs are activated by phospholipids such as phosphatidylserine (PS) in a calcium-dependent manner. In addition, cPKCs bind to diacylglycerol (DAG), which increases the affinity of the enzyme for calcium. Novel PKCs are activated by DAG and PS, but no longer require calcium. Only PS is required to activate the atypical PKCs.

All members of the PKC family contain a regulatory and a catalytic domain. Comparison of the sequence of all the PKC isoforms has shown that there are four highly conserved regions (C1–C4) and five variable regions (V1–V5). The regulatory domain (V1–C2) contains the phorbol ester-binding site, the calcium binding site (cPKCs only), the DAG binding

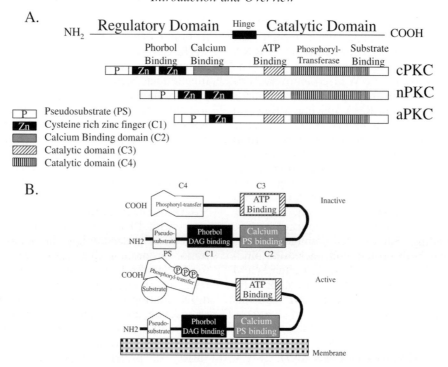

**Figure 10** Schematic of the primary structure and function of the PKC family. Adapted from References 42 and 43.

site, a pseudosubstrate region, and one or two cysteine-rich, zinc finger regions. The catalytic domain (C3-V5) contains the substrate-recognition site and the ATP binding site. The hinge or V3 region is flexible and thus allows the regulatory section to move away from the active site of the catalytic section during enzyme activation. The hinge region is accessible to protease, particularly calpain, when the enzyme is in the unfolded or activated state.

Much has been learned about the regulation of PKCs from their crystal structure (reviewed in Reference 43). The ligand-binding cavity (C1) is a globular structure comprised of two β-sheets. Unlike ligand binding to NHRs, phorbol ester binding to C1 does not result in a significant conformation change, but instead alters the hydrophobicity of the surface and results in membrane targeting. The hydrophobic surface is lacking in the PKCs that do not respond to phorbol esters (aPKCs). The C2 domain is rich in β-sheets and forms a novel calcium-binding pocket. Divalent ion binding to this region results in a conformational change and exposure of lysine residues on the surface. In PKCs that lack calcium regulation (nPKCs, aPKCs) there is a lack of residues required for coordinating metal ions and divalent ion binding. The autoinhibitory, pseudosubstrate domain interacts with the kinase domain via electrostatic interactions with acidic residues.

### 14.2.1.7.2.2   Substrates

Unlike activator proteins such as NHRs and AhR, PKCs result in altered gene expression via phosphorylation of substrate proteins, which in turn affect downstream signaling events. Many proteins are phosphorylated by PKC. The consensus phosphorylation motif is described as RXX$\underline{S}$/$\underline{T}$XRX where X indicates any amino acid. Basic residues surrounding the T and S phosphorylation site, as well as at positions $-6$, $-4$ and $-2$, are preferred by all PKCs, and there are isoform-specific preferences in substrate sequence.[43] Identification of substrate targets has greatly contributed to understanding the function of PKCs. Myristoylated alanine-rich C-kinase substrate (MARCKS) is a substrate for most PKCs. MARCKS is an acidic protein of 80 kilodaltons and possess a very basic 25 amino acid domain that is both the site of PKC phosphorylation and calmodulin binding. This same domain, when dephosphorylated, interacts with actin and causes actin cross-linking. Conversely when phosphorylated by PKC, MARCKS is unable to perform this function. Thus, it appears that MARCKS is a mediator for PKC-controlled cell motility. There are also several related proteins, Mac-MARCKS (F52 or MRP), neurogranin, and GAP-43 (neuromodulin) that are substrates.

Important signaling molecules have also been shown to be substrates for PKC isozymes.[42] PKC phosphorylates the epidermal growth factor (EGF) receptor on serine residues, which induces its internalization and down-regulation. Similarly, the insulin receptor is phosphorylated by PKC *in vivo*. PKC phosphorylates calmodulin-binding domains of certain proteins, which inhibit calmodulin binding. Notably, PKCα phosphorylates Ser499 and Ser259 and activates Raf-1, the upstream activator of mitogen-activated protein (MAP) kinase. Activation of PKC has also been associated with induction of immediate early genes and activation of transcription factors.[43] PKC (in particular epsilon and zeta) increases the expression of c-fos and c-jun, as well as junB and egr-1. The activity of the AP-1 transcription factor is also affected by PKC-dependent phosphorylation, while nuclear factor κB (NFκB) is activated by PKCzeta. Finally, PKC may also play a role in activation of nuclear hormone receptors including the estrogen receptor.[44] Obviously PKCs play in important role in signaling through a variety of pathways and is a key regulator of cellular proliferation, differentiation and apoptosis.

### 14.2.1.7.2.3 *Ligands*

The mechanism by which DAG and phorbol esters can activate the cPKC and nPKC is through binding to the regulatory domain, causing an unfolding of the enzyme and exposure of the catalytic site,[42] as described above. Several lipids, including arachidonic acid and phosphatidylinositol-3,4,5-triphosphate, can enhance activation of the aPKC isoforms through an unknown mechanism. The physiological ligand for PKCs is DAG, formed via hydrolysis of inositol phospholipids and phosphatidylcholine by several forms of phospholipase C. Phorbol esters such as TPA mimic DAG (in the cPKC and nPKC isoforms), thus dysregulating a normally highly controlled pathway. (The structure of TPA is shown in Figure 11.) Phorbol esters are more potent than DAG due to a longer half-life and the fact they decrease or abolish the requirement for calcium. The complete activation of PKC by phorbol esters is also associated with translocation of the enzyme from the cytosol to cellular membranes. Activation of PKC by physiological stimulators like DAG, however, causes only limited translocation. This has suggested that specific physiological signals may activate only a subset of isoforms. Bryostatin binds to and activates PKC but this results in only a small number of the responses that are elicited by TPA. This appears to be due to differential activation of PKC-α, -δ, and -ε by bryostatin compared to TPA.

### 14.2.1.7.3 Protein Phosphatases

As mentioned above, PKCs regulate gene expression by affecting the phosphorylation status a substrate proteins, i.e., by adding a phosphate group onto serines or threonines. Phosphatases are equally adept at affecting gene expression, although the response is in the opposite direction of that of kinases. The protein serine/threonine phosphatases (PPs) comprise a family of at least four major

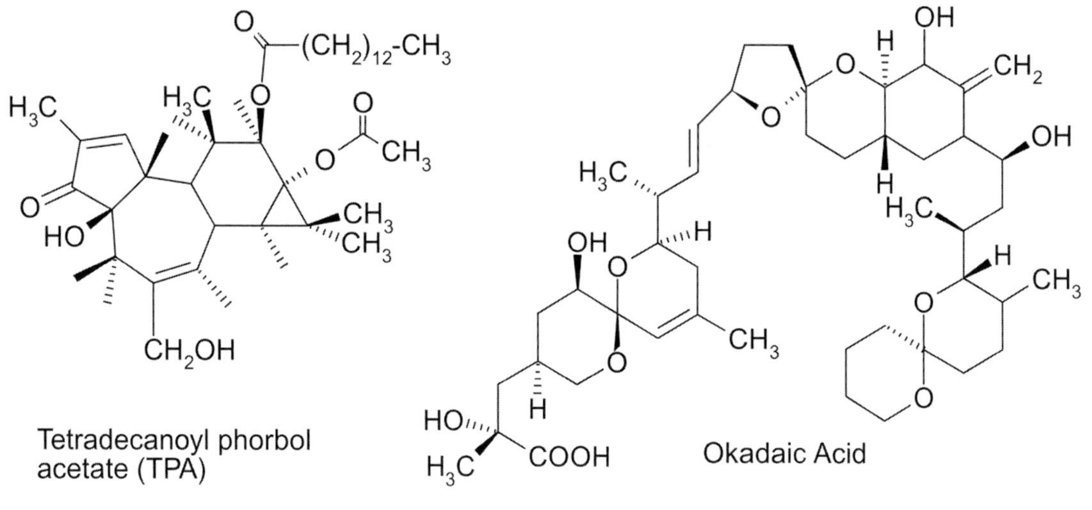

Tetradecanoyl phorbol acetate (TPA)

Okadaic Acid

**Figure 11** Ligands for PKCs and phosphatase.

enzyme types with a variety of novel forms.[45] The PPs are functionally classified, as a systematic nomenclature has not been devised. Type 1 PPs preferentially dephosphorylate the β-subunit, and type 2 PPs the α-subunit, of phosphorylase kinase. The type 2 PPs are further subdivided into types 2A, 2B (also known as calcineurin) and 2C based on dependence on divalent cations. PP2A activity is independent of divalent cation activity whereas PP2B and PP2C require $Ca^{2+}$ and $Mg^{2+}$ respectively.

Protein phosphatases-1 and -2A and 5 (PP-1, PP-2A, PP-5) are the major protein phosphatases in the cytosol of mammalian cells that perform this dephosphorylation of phospho-serine and -threonine residues. The phosphatase that has been characterized to the highest extent is PP-2A, and will be the focus of the present section. Importantly, not only are PP2A and other phosphatases targets of viral products and toxins, but their substrates include many important kinases including PKC. It stands to reason that phosphatases have been implicated in the regulation of cellular metabolism, DNA replication, transcription, RNA splicing, cell-cycle progression, differentiation and oncogenesis.[46]

### 14.2.1.7.3.1 *Structure to Function*

PP2A is structurally complicated in that a single catalytic subunit (C) can associate with a variety of regulatory (A) and targeting (B) subunits. The regulatory subunits modulate activity, substrate specificity and subcellular localization.[46] The C subunit in bound constitutively to a scaffold, or regulatory subunit (A) or PR65. The PR65-C heterodimer can then complex with a wide variety of B regulatory subunits. The binding of one B subunit precludes the binding of others. Three gene families encode for the B subunit including PR55, PR61 and PR72. Two isoforms of the C unit (α and β) and two isoforms of PR65 are known, whereas the B subunit has several isoforms and splice variants. Theoretically, greater than 50 trimeric PP2A haloenzymes may be formed by different combinations of subunits A, B and C.[46]

The B subunits of PP2A appear to have a number of functions.[47] First, they contain the targeting information that directs the heterotrimer to distinct intracellular locations. For example, B55α directs PP2A to microtubules. Second, PP2A B subunits determine the substrate specificity of the enzyme. For example, a PP2A heterotrimer with a 72-kDa (PR72) regulatory subunit dephosphorylates casein kinase I sites on SV40 large T antigen, whereas PP2A with a 55-kDa regulatory B subunit dephosphorylates cyclin-dependent kinase sites. Third, tissue-specific and developmentally regulated patterns of the B subunit gene expression likely determine what substrates are dephosphorylated in specific tissues. This has been in *Drosophila*, where flies with mutant PR55 alleles have both imaginal disc and cellular (mitotic) abnormalities. Last, the regulatory subunits of PP2A may act as receptors of second messengers, nutrients and toxins. For example the lipid ceramide as well as the Vitamin E activate PP2A, perhaps acting through the B subunit. It is of interest that at least three distinct DNA tumor viruses encode proteins that regulate PP2A activity by binding to or displacing the endogenous B subunit, the most important being SV40 small T antigen.

Classically, the protein phosphatases have been classified into four types based on their sensitivity to inhibitors, requirement for cations, and *in vitro* substrate specificity (PP-1, PP-2A, PP-2B (calcineurin), and PP-2C) (see Figure 12). The catalytic subunits of three of these phosphatases (PP-1c, PP-2A, and PP-2B) constitute a single gene family, termed the PPP gene family.[48] However, additional protein

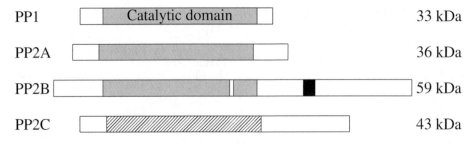

**Figure 12** Structure of the catalytic subunits of common protein phosphatases. PP1, PP2A and PP2B share a conserved catalytic domain (gray shading, PPP family) while PP2C is unrelated (cross-hatched). PP2B is calcium/calmodulin regulated and the calmodulin binding site is indicated (black shading). PP2B is relatively insensitive to okadaic acid, compared to PP1 and PP2A, partially due to differences in the catalytic domain, L7 loop (indicated with a gap in the catalytic domain of PP2B). Adapted from References 45.

serine/threonine phosphatase catalytic subunits possessing PPP family structures have been recently identified, including PP-4, PP-5, PP-6, and PP-7. This suggests that the PPP family of phosphatases share similar protein folding and catalytic mechanisms. Some of the amino acids important for catalysis and involved in interactions with the marine toxins such as okadaic acid, including the L7 loop.[49]

### 14.2.1.7.3.2 Substrates

Much that is known about the signaling pathways of PPs is the result of the examination of the tumor promoter okadaic acid (see Figure 11). This marine toxin specifically inhibits PP-1 and PP-2A, with little effect on the other major phosphatases. The effect of inhibiting phosphatases is to cause a net increase in the level of phosphorylated proteins, an effect very similar to the increased phosphorylation that occurs from the activation of protein kinases such as PKC.[42] Although some differences exist, increased phosphorylation is believed to be the basis for the similar biological effects of phosphatase inhibitors and phorbol esters including inflammation and tumor promotion. Not surprising, several substrates for PP-1 and PP-2A are the same proteins that are substrates for PKC. PP-1 and PP-2A can dephosphorylate the PKC-mediated phosphorylation of MARCKS. Other intermediate filament proteins, including the keratins and vimentin, are also hyperphosphorylated by okadaic acid. Like TPA, okadaic acid also increases the phosphorylation of serine and threonine residues on the EGF receptor.

Some of the most important targets of PPs may be kinases themselves. In fact, over 30 protein kinase activities are modulated by PP2A.[46] Theses include kinases of the AGC subgroup such as protein kinase B, p70 S6 kinase, protein kinase C, $Ca^{2+}$-calmodulin dependent kinases, mitogen activated protein kinase (MAPK), IκB kinase and cyclin-dependent kinases. As described above, cPKCs have a basal level of phosphorylation that allows for subsequent activation by DAGs. A PP55 containing PP2A trimer can dephosphorylate PKCα at these basal sites and decreases DAG sensitivity. Thus PPs can affect kinase function on two levels, de-phosphorylating the same targets acted upon by the kinase, or alternatively by affecting the activity of these enzymes.

### 14.2.1.7.3.3 Ligands

Okadaic acid has been an important tool in the analysis of PP function. A product of dinoflagellate metabolism, okadaic acid is the causative agent of diarrhetic shellfish poisoning. This toxin potently and specifically inhibits PP-1 and PP-2A, although with different $IC_{50}$ concentrations (Table 5). Despite sharing similar structure and catalytic mechanisms, PP-2B is inhibited to a much lesser extent. Structural modification, such as adding hydroxyl groups throughout the structure of okadaic acid change its ability to affect enzyme activity, showing that the interaction has distinct structural requirements.[49] Although both rapidly bind and potently inhibit PP-1 and PP-2A, only microcystins form a covalent linkage with PP-1 or PP-2A. Microcystins form this covalent link with and Cys-273 of PP-1, placing microcystins very close to the catalytic center of PP-1 (Tyr-272 coordinates one of the catalytic center water molecules[49]). Evidence suggests that okadaic acid binds in the same site as microcystin-LR, since the latter can compete with okadaic acid for phosphatase binding sites. Comparison of the structures of the C subunit of PP-1 and PP-2B have

**Table 5** Inhibition of protein phosphatases (PPs) by toxins.

| Phosphatase | Okadaic acid $IC_{50}$ (nM) | Inhibitor | $IC_{50}$ for PP2A (nM) |
|---|---|---|---|
| PP1 | 100 | Okadaic Acid | 2.0 |
| PP2A | 1.0 | Calyculin A | 7.0 |
| PP2B | > 5000 | Nodularin | 1.0 |
| PP2C | None | Microcystin-LR | 2.0 |
| PP3 | 5.0 | Tautomycin | 20 |
| PP4 | 0.2 | Fumonisin B1 | 30,000 |
| PP5 | < 1.0 | Cantharidin | 160 |
| PP6 | Unknown | Thyrsiferyl | 1,600 |
| PP7 | None | Motuporin | 0.1 |
| | | Fostriecin | 40 |

Adapted from Reference 5.

suggested that PP-2B possesses a cryptic marine toxin binding site comprised of amino acids that are important for interaction with the toxins.

Recent studies have shown that Vitamin E ($\alpha$-tocopherol) can bind directly to, and activate PP2A *in vitro*.[46] This may help explain the effects of VitE on inactivation of PKC$\alpha$ in smooth muscle cells. Lipid-like second messengers, such as ceramide are also activators of PP2A.[50] Depending on the cell type, ceramide can induce differentiation, cell proliferation, growth arrest, inflammation or apoptosis. A number of direct cellular targets for ceramide have been identified, including a ceramide-activated protein kinase, PKC$\zeta$, as well as, PP2A. Ceramide activation of PP2A requires the catalytic subunit, as it activates ABC and AB'C, as well as AC complexes. Recently, ceramide was found to specifically activate a mitochondrial PP2A, which rapidly and completely induced the dephosphorylation and inactivation of Bcl2, a potent anti-apoptotic protein.[50]

### 14.2.1.8  MODIFIERS OF XENOBIOTIC SIGNALING PATHWAYS, COREGULATORS

#### 14.2.1.8.1  Introduction

As discussed in Chapter 14.1.4, transcriptional regulation in eukaryotes is achieved through multiprotein complexes assembled at the enhancer and promoter regions of target genes. In the case of protein-coding genes, RNA polymerase II (Pol II) and its associated general transcription factors (TFIIA, B, D, E, F and H) are sufficient for recognition and low levels of accurate transcription from common core promoter elements. These components of the basal transcription machinery are ubiquitous and, although not static, are also not under dynamic control by xenobiotics. By contrast, a second class of transcription factors, the transcriptional activators described above, typically assemble at distal enhancer sites and are often responsible for cell-type and gene specific regulation of gene expression. Despite the complexity of this basal transcription machinery ($>40$ proteins), its response to activators on specific target genes is still dependent on additional factors called coregulators (coactivators and corepressors). This represents a third class of transcription factors, whose purpose is to integrate the transcriptional activators with the Pol II basal machinery.[51] Coactivators are explicitly defined as factors that are required for the function of DNA-binding activators, but not for basal transcription *per se*, and do not show site-specific DNA binding by themselves. Co-repressors serve the opposite role in that they prevent the integration of the activator protein with that of the basal transcription factors.

With the exception of the kinases and phosphatases, all the xenobiotic receptors described have been shown to require coregulators for transcriptional activity. It is important to note that the targets of PKC and PP2A, such as AP1 and c-myc, also require coregulators, so even these signaling cascades are not devoid of the involvement of coactivators and corepressors. In fact, this sharing of essential transcription apparatus may be a point of cross-talk between nuclear receptors and other signaling pathways.

Coregulator function has been best characterized with the nuclear hormone receptors. The first functional coactivator discovered was steroid receptor coactivator-1 (SRC-1, reviewed in Reference 52) and it appears to be a general coactivator for NHRs, as well as several receptors not within this family. SRC-1 enhances transactivation of steroid hormone-dependent target genes via interaction with AF-2 domains of NHRs. Since the discovery of SRC-1, an explosion of putative co-activators has occurred. These coactivators include the SRC-1 related proteins, TIF2 and GRIP1, and other putative and unrelated co-activators such as ARA70, Trip1, RIP140, and TIF1, to name a few. Another co-activator, CREB-binding protein (CBP), is similar to SRC-1 in the broad specificity of activator proteins it is capable of affecting. CBP and SRC-1 interact and synergistically enhance transcriptional activation of several receptor systems, perhaps via a ternary complex consisting of CBP, SRC-1, and liganded receptors. Similarly, co-repressors, such as SMRT and N-CoR, have been identified for xenobiotic receptors. Several type II NHRs, which include TR and RAR, can inhibit basal promoter activity of certain genes. Corepressor proteins mediate the silencing of target gene transcription by unliganded receptors.

The general scheme of how coregulators affect activator protein function is as follows. Upon binding of agonist the receptor changes its conformation in the ligand-binding domain that enables recruitment of coactivator complexes, which may serve two distinct functions. First, coactivators serve as a bridge between the receptor and the basal transcriptional machinery. This is the result of multiple protein–protein interactions, ultimately leading to recruitment of TBP and RNA polII. Second, coactivators may alter chromatin structure via histone acetylase (HAT) activity. The addition of acetyl groups to histones causes charge repulsion with DNA, resulting in a less compacted, more accessible region of DNA. In contrast, binding of antagonists induces a different conformational change in the receptor. The conformational change may fail to dislodge the associated corepressors. The activator protein-corepressor complex results in a nonproductive interaction with the basal transcriptional machinery. Corepressors are also able to remodel chromatin via association with histone deacetylase (HDAC) activity.

Detailed information on xenobiotic coregulators is given in Chapter 14.2.8. In the present section, the emphasis will be on structure/function relationships of the most common coregulators. It is interesting to note that certain of the motifs described above for activator proteins have also been co-opted for use in coregulators.

### 14.2.1.8.2 Structure to Function

The most characterized group of HAT containing coactivators is the p160 family, which is typified by SRC-1 (see Figure 13).[6,53,54] At least three sub-families of p160 family members exist and include SCR-1/NcoA-1, TIF2/GRIP1/NoCoA-2 and p/CIP, ACTR/RAC3/AIB1/ TRAM-1. Members of the p160 family have basic HLH motifs, although they are not able to bind to DNA directly. In addition, the protein–protein interaction domain used in AhR and Arnt, the PAS domain, is found in this family. An important structural feature of SRC-1 and many other coactivators is the multiple uses of amphipathic LXXLL helical motifs, important in the interaction with NHRs. CBP and p300 are highly related HAT coactivators that interact with SRC-1 and also bind directly to NHRs in a ligand-dependent manner. In addition to LXXLL motifs, p300/CBP share in common with the p160 family the HAT activity and glutamine rich regions. Unlike p160, p300 and CBP contain cysteine/histidine rich regions (C/H) as a point of contact with proteins such as TBP and p53. Corepressors are typified by silencing mediator of retinoid and thyroid receptors (SMRT) and the related protein nuclear hormone receptor corepressor (NCoR). SMRT and NCoR interact with unliganded NHRs such as TR and RAR. A region in NHRs called the CoR box is found in

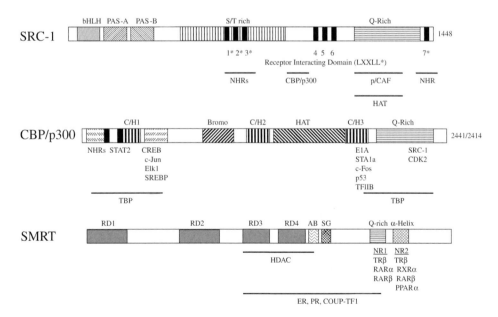

**Figure 13** Structural features of receptor coregulators. The structure of SRC-1 is representative of a class of coactivators including NCoA-1, NcoA-2, TIF-2, GRIP1, ACTR, p/CIP and TRAM-1. CBP and p300 are similar to each other, but quite different than SRC-1 in their structure function. The corepressor SMRT is similar to NcoR in structure (not shown) but differs from the coregulators signficantly, in particular with the repressor domains (RD) as well as recruitment of histone deacetylase (HDAC) activity. Lines below each figure show interaction with other factors. Domains are indicated above each figure. Abbreviations: bHLH, basic helix loop helix; PAS, Per-Arnt Sim domain; S/T, serine/threonine rich; HAT, histone acetyltransferase; Q-rich, glutamine rich; RID 1–7, receptor interacting domain; *, RID which contains LXXLL motif; C/H, cysteine/histidine rich; Bromo, bromodomain; RD, repressor domain, AB, acidic basic; SG, serine/glycine rich. Adapted from References 6, 53, 54.

the hinge region between the DNA binding and ligand binding motifs. The corepressors are components of a complex that contains Sin3, a protein with four paired amphipathic helical domains, and HDAC activity.

### 14.2.1.9  CONCLUSIONS

Xenobiotic receptors are an important class of activator protein that mediate the toxicity of a number of important chemicals including dioxin, peroxisome proliferators, xenoestrogens, heavy metals and a variety of clinically-relevant drugs. To perform the functions of *recognition* and *transduction*, the requisite for being a true receptor, these proteins have co-opted functional motifs used in other activator proteins, but have also including a xenobiotic binding or *sensing* domain. The nuclear hormone receptors (NHRs) represent the largest and most important family of xenobiotic receptors. To perform the function of DNA binding, the NHR signature $C_4$ zinc finger motif is used. NHRs include two activation domains, AF1 and AF2, which serve as points of coactivator recruitment in the unliganded and liganded receptor, respectively. Multiple protein–protein interactions as well as kinase/phosphatases recognition sites add complexity to the gene regulation by NHRs. The ligand-binding domain of NHRs is the point of interaction with xenobiotics, often allowing a wide structural variety of chemicals to associate. Upon ligand binding, the conformation of the entire protein is altered causing release of corepressors, addition of coactivators and ultimately transcriptional activity. The aryl hydrocarbon receptor (AhR) has a mechanism of action that is remarkably similar to that of NHRs, although they share no structural similarity. The AhR is a basic helix-loop-helix (bHLH) protein that contains two per-arnt-sim (PAS) domains. The basic region of the protein is used in DNA binding whereas the HLH and PAS domains are used to associate with its heterodimerization partner Arnt. The ligand-binding region is also contained within the PAS domain and it can recognize a variety of polycyclic compounds. Upon ligand binding, a conformation change occurs and coactivators, many of which are the same as those used by NHRs, are recruited. A third class of xenobiotic receptor is represented by metal transcription factor-1 (MTF-1). This protein is a member of the $C_2H_2$ zinc finger family of transcription factors and is activated by a wide range of heavy metals including lead and zinc. MTF-1 is much less characterized that the NHRs or AHR so little is known about coregulator interaction and other functional motifs. Finally, the atypical xenobiotic receptors protein kinase C (PKC) and phosphatases 2A (PP2A) are regulated by important tumor promoters such as phorbol esters and okadaic acid. Unlike the receptors described above, PKC and PP2A exert their effects by altering the phosphorylation of substrate proteins. By themselves they are not DNA binding proteins, although they can alter the function of downstream transcription factors.

Detailed molecular information on the structure/function of xenobiotic receptors is required for an understanding of how many chemicals cause both toxicity and benefit. The myriad of receptor-mediated effects of chemicals is testimony to the importance of this area, but is perhaps most striking in regard to tumor promotion. Most of the chemicals described in this chapter (dioxins, peroxisome proliferators, hormones, metals, phorbol esters) result in tumors without directly damaging DNA. A generally held belief is the aforementioned tumor promoters are affecting the cell cycle in target tissues through the inappropriate activation of xenobiotic receptors. That is, a protein such as the estrogen receptor has evolved to respond to its natural effector, estrogen, and the response is short-lived and the target genes regulated in a temporally appropriate manner. Xenoestrogens may regulate gene expression at an inappropriate time or may have a sustained effect due to slow chemical turnover. Other important health issues such as susceptibility to disease that results from mutation or polymorphisms of these receptor systems, require a precise understanding of key regulatory regions of the xenobiotic receptors. The list of xenobiotic receptors has been growing steadily and the study of these systems will remain an important area of research in cellular and molecular mechanisms of toxicity for many years to come.

### 14.2.1.10  REFERENCES

1. M. S. Denison and W. G. Helferich, 'Toxicant-Receptor Interactions: Modulation of signal transduction and gene expression', Taylor and Francis, Philadelphia, PA, 1998.
2. B. Alberts, D. Bray, J. Lewis *et al.*, 'Molecular Biology of the Cell, 3rd edn,' eds. Garland Publishing, Inc., New York, NY, 1994, pp. 401–476
3. E. Wingender, X. Chen, R. Hehl *et al.*, 'TRANSFAC: an integrated system for gene expression regulation.' *Nucleic Acids Res.*, 2000, **28**, 316–319.
4. D. J. Waxman, 'P450 gene induction by structurally diverse xenochemicals: central role of nuclear receptors CAR, PXR, and PPAR.' *Arch. Biochem. Biophys.*, 1999, **369**, 11–23.

5. M. Beato, 'Transcriptional control by nuclear receptors.' *Faseb. J.*, 1991, **5**, 2044–2051.

6. T. N. Collingwood, F. D. Urnov and A. P. Wolffe, 'Nuclear receptors: Coactivators, corepressors and chromatin remodeling in the control of transcription.' *J. Mol. Endocrinol.*, 1999, **23**, 255–275.

7. N. R. N. Committee, 'A unified nomenclature system for the nuclear receptor superfamily.' *Cell*, 1999, **97**, 161–163.

8. L. Di Croce, S. Okret, S. Kersten *et al.*, 'Steroid and nuclear receptors. Villefranche-sur-Mer, France, May 25–27, 1999.' *Embo. J.*, 1999, **18**, 6201–6210.

9. P. F. Egea, B. P. Klaholz and D. Moras, 'Ligand-protein interactions in nuclear receptors of hormones.' *FEBS Lett.*, 2000, **476**, 62–67.

10. H. Escriva, M. C. Langlois, R. L. Mendonca *et al.*, 'Evolution and diversification of the nuclear receptor superfamily.' *Ann. N. Y. Acad. Sci.*, 1998, **839**, 143–146.

11. L. P. Freedman, 'Strategies for transcriptional activation by steroid/nuclear receptors.' *J. Cell Biochem.*, 1999, **Suppl**, 103–109.

12. K. B. Horwitz, T. A. Jackson, D. L. Bain *et al.*, 'Nuclear receptor coactivators and corepressors.' *Mol. Endocrinol.*, 1996, **10**, 1167–1177.

13. S. A. Kliewer, J. M. Lehmann and T. M. Willson, 'Orphan nuclear receptors: Shifting endocrinology into reverse.' *Science*, 1999, **284**, 757–760.

14. V. Laudet, 'Evolution of the nuclear receptor superfamily: early diversification from an ancestral orphan receptor.' *J. Mol. Endocrinol.*, 1997, **19**, 207–226.

15. M. Manteuffel-Cymborowska, 'Nuclear receptors, their coactivators and modulation of transcription.' *Acta. Biochim. Pol.*, 1999, **46**, 77–89.

16. J. W. Schwabe, 'Transcriptional control: How nuclear receptors get turned on.' *Curr. Biol.*, 1996, **6**, 372–374.

17. R. Sladek and V. Giguere, 'Orphan nuclear receptors: an emerging family of metabolic regulators.' *Adv. Pharmacol.*, 2000, **47**, 23–87.

18. S. Tenbaum and A. Baniahmad, 'Nuclear receptors: structure, function and involvement in disease.' *Int. J. Biochem. Cell Biol.*, 1997, **29**, 1325–1341.

19. W. Wahli and E. Martinez, 'Superfamily of steroid nuclear receptors: Positive and negative regulators of gene expression.' *Faseb. J.*, 1991, **5**, 2243–2249.

20. P. Escher and W. Wahli, 'Peroxisome proliferator-activated receptors: Insight into multiple cellular functions.' *Mutat. Res.*, 2000, **448**, 121–138.

21. M. J. Tsai and B. W. O'Malley, 'Molecular mechanisms of action of steroid/thyroid receptor superfamily members.' *Annu. Rev. Biochem*, 1994, **63**, 451–486.

22. N. L. Weigel and Y. Zhang, 'Ligand-independent activation of steroid hormone receptors.' *J. Mol. Med.*, 1998, **76**, 469–479.

23. D. P. McDonnell, E. Vegeto and M. A. Gleeson, 'Nuclear hormone receptors as targets for new drug discovery.' *Biotechnology (NY)*, 1993, **11**, 1256–1261.

24. R. V. Weatherman, R. J. Fletterick and T. S. Scanlan, 'Nuclear-receptor ligands and ligand-binding domains.' *Annu. Rev. Biochem.*, 1999, **68**, 559–581.

25. G. J. Maalouf, W. Xu, T. F. Smith *et al.*, 'Homology model for the ligand-binding domain of the human estrogen receptor.' *J. Biomol. Struct. Dyn.*, 1998, **15**, 841–851.

26. B. Blumberg and R. M. Evans, 'Orphan nuclear receptors – new ligands and new possibilities.' *Genes Dev.*, 1998, **12**, 3149–3155.

27. M. E. Massari and C. Murre, 'Helix-loop-helix proteins: Regulators of transcription in eucaryotic organisms.' *Mol. Cell Biol.*, 2000, **20**, 429–440.

28. C. Murre, G. Bain, M. A. van Dijk *et al.*, 'Structure and function of helix-loop-helix proteins.' *Biochim. Biophys. Acta*, 1994, **1218**, 129–135.

29. W. R. Atchley and W. M. Fitch, 'A natural classification of the basic helix-loop-helix class of transcription factors.' *Proc. Natl. Acad. Sci. USA*, 1997, **94**, 5172–5176.

30. J. R. Nambu, J. O. Lewis, K. A. Wharton, Jr. *et al.*, 'The *Drosophila* single-minded gene encodes a helix-loop-helix protein that acts as a master regulator of CNS midline development.' *Cell*, 1991, **67**, 1157–1167.

31. S. T. Crews, 'Control of cell lineage-specific development and transcription by bHLH- PAS proteins.' *Genes Dev.*, 1998, **12**, 607–620.

32. M. A. Peters and E. J. Taparowsky, 'Target genes and cellular regulators of the Myc transcription complex.' *Crit. Rev. Eukaryot. Gene. Expr.*, 1998, **8**, 277–296.

33. P. Steiner, B. Rudolph, D. Muller *et al.*, 'The functions of Myc in cell cycle progression and apoptosis.' *Prog. Cell Cycle Res.*, 1996, **2**, 73–82.

34. M. D. Cole and S. B. McMahon, 'The Myc oncoprotein: a critical evaluation of transactivation and target gene regulation.' *Oncogene*, 1999, **18**, 2916–2924.

35. C. V. Dang, L. M. Resar, E. Emison *et al.*, 'Function of the c-Myc oncogenic transcription factor.' *Exp. Cell Res.*, 1999, **253**, 63–77.

36. T. Pieler and E. Bellefroid, 'Perspectives on zinc finger protein function and evolution-an update.' *Molecular Biology Reports*, 1994, **20**, 1–8.

37. D. C. Bittel, I. V. Smirnova and G. K. Andrews, 'Functional heterogeneity in the zinc fingers of metalloregulatory protein metal response element-binding transcription factor-1.' *J. Biol. Chem.*, 2000, **275**, 37194–37201.

38. S. J. Langmade, R. Ravindra, P. J. Daniels *et al.*, 'The transcription factor MTF-1 mediates metal regulation of the mouse ZnT1 gene.' *J. Biol. Chem.*, 2000, **275**, 34803–34809.

39. J. P. Mackay and M. Crossley, 'Zinc fingers are sticking together.' *Trends Biochem. Sci.*, 1998, **23**, 1–4.

40. D. J. Segal, B. Dreier, R. R. Beerli *et al.*, 'Toward controlling gene expression at will: selection and design of zinc finger domains recognizing each of the 5'-GNN-3' DNA target sequences.' *Proc. Natl. Acad. Sci. USA*, 1999, **96**, 2758–2763.

41. D. Kennedy, T. Ramsdale, J. Mattick *et al.*, 'An RNA recognition motif in Wilms' tumour protein (WT1) revealed by structural modelling.' *Nat. Genet.*, 1996, **12**, 329–331.

42. S. M. Fisher, 'Chemical Carcinogens and Anticarcinogens, 1st edn,' eds. G. T. Bowden and S. M. Fisher, Elsevier Science Inc., New York, NY, 1997, pp. 349–382.

43. A. Toker, 'Signaling through protein kinase C.' *Front. Biosci.*, 1998, **3**, D1134–1147.

44. N. L. Weigel, 'Steroid hormone receptors and their regulation by phosphorylation.' *Biochem. J.*, 1996, **319**, 657–667.

45. M. J. Peirce, M. R. Munday and P. T. Peachell, 'Role of protein phosphatases in the regulation of human mast cell and basophil function.' *Am. J. Physiol.*, 1999, **277**, C1021–1028.

46. T. A. Millward, S. Zolnierowicz and B. A. Hemmings, 'Regulation of protein kinase cascades by protein phosphatase 2A.' *Trends Biochem. Sci.*, 1999, **24**, 186–191.

47. B. McCright, A. M. Rivers, S. Audlin *et al.*, 'The B56 family of protein phosphatase 2A (PP2A) regulatory subunits encodes differentiation-induced phosphoproteins that target PP2A to both nucleus and cytoplasm.' *J. Biol. Chem.*, 1996, **271**, 22081–22089.

48. S. Orgad, N. D. Brewis, L. Alphey *et al.*, 'The structure of protein phosphatase 2A is as highly conserved as that of protein phosphatase 1.' *FEBS Lett.*, 1990, **275**, 44–48.

49. J. F. Dawson and C. F. Holmes, 'Molecular mechanisms underlying inhibition of protein phosphatases by marine toxins.' *Front. Biosci.*, 1999, **4**, D646–658.

50. V. Janssens and J. Goris, 'Protein phosphatase 2A: a highly regulated family of serine/threonine phosphatases implicated in cell growth and signalling.' *Biochem. J.*, 2001, **353**, 417–439.

51. S. Malik and R. G. Roeder, 'Transcriptional regulation through mediator-like coactivators in yeast and metazoan cells.' *Trends Biochem. Sci.*, 2000, **25**, 277–283.

52. H. Shibata, T. E. Spencer, S. A. Onate *et al.*, 'Role of co-activators and co-repressors in the mechanism of steroid/thyroid receptor action.' *Recent Prog. Horm. Res.*, 1997, **52**, 141–164.

53. J. D. Chen, 'Steroid/nuclear receptor coactivators.' *Vitam. Horm.*, 2000, **58**, 391–448.

54. C. Leo and J. D. Chen, 'The SRC family of nuclear receptor coactivators.' *Gene*, 2000, **245**, 1–11.

55. A. H. Schonthal, 'Role of PP2A in intracellular signal transduction pathways.' *Front. Biosci.*, 1998, **3**, D1262–1273.

56. B. M. Forman and H. H. Samuels, 'Interactions among a subfamily of nuclear hormone receptors: the regulatory zipper model.' *Mol. Endocrinol.*, 1990, **4**, 1293–1301.

57. D. J. Mangelsdorf and R. M. Evans, 'The RXR heterodimers and orphan receptors.' *Cell*, 1995, **83**, 841–850.

58. J. P. Vanden Heuvel and G. Lucier, 'Environmental toxicology of polychlorinated dibenzo-p-dioxins and polychlorinated dibenzofurans.' *Environ. Health Perspect.*, 1993, **100**, 189–200.

J.P. Vanden Heuvel, G.H. Perdew, W.B. Mattes and W.F. Greenlee (Eds.)
Comprehensive Toxicology, Vol. xiv

# 14.2.2
# The PAS Protein Superfamily

JACQUELINE A. WALISSER and CHRISTOPHER A. BRADFIELD
*McArdle Laboratory for Cancer Research, University of Wisconsin School of Medicine, Madison, WI, USA*

## 14.2.2.1 INTRODUCTION

An organism responds to changes in its environment by attempting to preserve physiological homeostasis. This adaptive response typically includes a sensory mechanism to detect the change and a molecular response to adapt to the altered environment. Many responses to environmental challenge occur at the level of gene expression, meaning that regulation of transcription, mRNA translation, and protein expression are fundamental components of cellular adaptation. Common mechanisms of environmental adaptation include alterations in the levels of metabolic enzymes, cytokines, and transcription factors. For example, in the classical inductive response, a foreign chemical upregulates its own metabolism and decreases its biological half-life.[1–4] While maintenance of physiological homeostasis is one potential outcome of adaptation, it is often not the only consequence. Xenobiotic-induced alterations in gene expression can also lead to a variety of adverse effects. Therefore, from a toxicological viewpoint, understanding gene-environment interactions can often explain both physiological defenses against chemicals, as well as the molecular mechanisms underlying their toxicity.

A major objective of this chapter is to present the observation that members of the Per-Arnt-Sim (PAS) superfamily of proteins play important roles as environmental sensors and mediate transcriptional responses to a number of environmental stimuli. Additionally, over stimulation of certain PAS pathways has been shown to produce deleterious effects. In this short chapter, we will attempt to summarize the various roles of PAS proteins in environmental response and highlight similarities in their molecular mechanisms. As a brief introduction to this superfamily we will begin with a discussion of the members and the general features of the PAS domain. We will follow with a discussion of the aryl hydrocarbon receptor (AHR), a protein that is central to the mechanism by which animals adapt to environments contaminated with planar aromatic hydrocarbons (PAHs). Next we will describe the signal transduction pathway that allows organisms to adapt to changes in atmospheric and cellular oxygen (the hypoxia inducible factor or "HIF" system). We will also outline the pathway that entrains an animal's activity to its illuminated environment (the circadian response pathway). Finally, we will describe recent observations that suggest that rudimentary PAS domains are found in a number of light and oxygen sensors of prokaryotes and plants.

It is hoped that these model pathways can provide a framework for use in understanding the biology of the less well-understood members of this emerging superfamily, as well as those to be characterized in the future. For additional viewpoints on these various pathways, the reader is also referred to a number of excellent reviews.[5–10] Moreover, detailed descriptions of both the AHR and HIF pathways are described in detail in following chapters.

## 14.2.2.2  THE PAS SUPERFAMILY

The PAS domain is found in a rapidly growing number of proteins (Figure 1 and http://mcardle.oncology.wisc.edu/bradfield/). The PAS domain is commonly described as a region of homology among family members. It typically encompasses 250–300 amino acids and contains a pair of highly degenerate 50 amino acid subdomains termed the "A" and "B" repeats (Figure 2).[11–13] The term "PAS" originates from the first letter of the three founding members of the family, *PER*, *ARNT* and *SIM*, each of which was identified through genetic means (Figure 2). The PER protein, the product of the *Drosophila Period* (*per*) gene, was discovered as a result of its involvement in the regulation of circadian rhythms in flies.[14,15] The ARNT gene product was originally identified from mammalian cells as a protein that was essential for induction of CYP1A1/CYP1A2 activity in response to PAH expo-

sure.[11] SIM, the product of the *Drosophila Single-minded locus*, was identified through its role as a regulator of midline cell lineage.[12,13]

### 14.2.2.3  CLASSIFICATION OF PAS PROTEINS

In higher eukaryotes, the PAS domain functions as a surface for interactions with other PAS proteins, as well as interactions with cellular chaperones such as the 90 kDa heat shock protein (Hsp90).[16,17] In the case of the AHR, the PAS domain can also function as a binding surface for small molecule ligands (e.g., certain PAHs and dioxins).[18–20] Most of the vertebrate PAS proteins that have been cloned also contain basic helix-loop-helix (bHLH) motifs immediately N-terminal to their PAS domain (Figure 1 and Figure 2). The HLH domains participate in dimerization between two bHLH-PAS proteins and they position the basic regions to allow specific contacts within the major groove of target DNA enhancers.[21,22] Consistent with their activities as signal transduction molecules, most PAS proteins have transactivation domains (TADs) within their C-terminal variable regions. Despite the relative conservation of the bHLH and PAS domains, and their apparent functional similarities, most PAS proteins show little sequence homology in their C-terminal sequences.

Members of the PAS superfamily can be divided into subclasses that specify their functional similarity and evolutionary relatedness. We refer to these classes as α, β and γ class (Figure 3(A)). The α-class is by far the largest and may be the most poorly defined. In addition to sequence similarities in their PAS domains, another characteristic of α-class proteins is that they often act as sensor molecules. That is, many α-class proteins are directly influenced by the binding of small molecules (i.e., the AHR) or stabilized by decreases in oxygen tension (i.e., HIF1α, HIF2α and HIF3α). The β-class is a smaller subdivision and can be best characterized by the fact that they all form heterodimers with members of the α-class (i.e., ARNT, ARNT2, MOP3 and MOP9).[23–28] The pairing of α- and β-class PAS protein lies at the heart of how these proteins signal, since it is the heterodimer that positions the basic regions of the respective partners within the major groove of DNA, thus providing DNA binding specificity. The γ-class PAS proteins were discovered by virtue of their activity as coactivators for a number of steroid receptors.[29–31] Given that the bHLH-PAS domains of the γ-class PAS proteins are

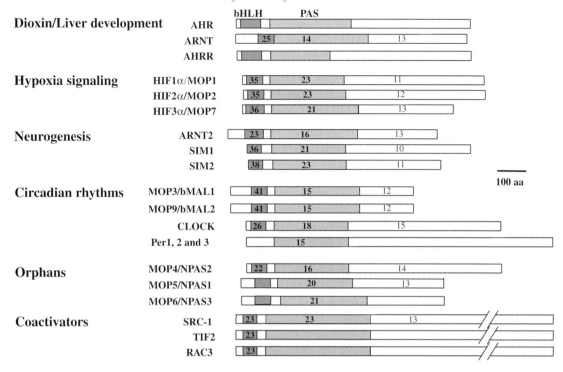

**Figure 1** Members of the PAS superfamily. At the present time, the vertebrate PAS superfamily is know to have at least 21 members. The members of this family are grouped according to their primary biological function. The PAS and the basic helix-loop-helix (bHLH) functional domains are shaded. When given, the numbers in the domains reflects the amino acid identity with the AHR.

not required for their activity as coactivators, they form a special subgroup where the function of their PAS domain remains unknown. For a discussion of this subclass of PAS proteins see recent reviews.[32,33]

The pairing rules that govern how bHLH-PAS proteins heterodimerize are a bit more complicated than simple α–β partnerships (Figure 3(B)). For example, ARNT and ARNT2 serve as high affinity partners for the AHR and HIFαs, but do not appear to dimerize with CLOCK or MOP4. One of the primary distinctions between ARNT and ARNT2 is where they are expressed. That is, ARNT2 may be the principle partner in the central nervous system, while ARNT is expressed primarily in non-neuronal tissues.[24,25,34] Conversely, MOP3 and MOP9 act

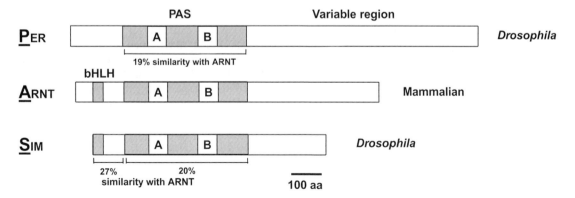

**Figure 2** The founding members of the PAS superfamily. Domain structures of the founding PAS proteins, PER, ARNT and SIM are shown. The name PAS stems from the first letters of PER, ARNT and SIM. The basic region (b), helix-loop-helix (HLH), PAS and C-terminal variable region are labeled on the *top*. The A and B repeat regions are shown within the PAS domain as *white boxes*. The percentage amino acid similarities of SIM and PER, compared with ARNT, are labeled *beneath* their respective domains. See text for details.

**A.  Phylogenetic relationships**

**B.  Pairing rules of α- and β-class members**

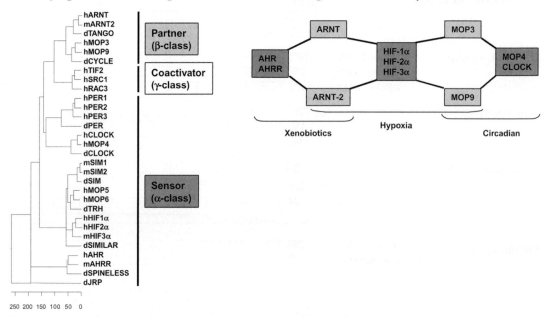

**Figure 3** Classification and pairing rules of the bHLH-PAS family members. (A) A phylogenetic tree based upon the nucleotide sequence similarity in the PAS domain was constructed using Clustal alignment.[79] The axis below denotes sequence distance. The PAS proteins are classified into sensors (α-class), partners (β-class) and coactivators (γ-class) as indicated on the right. Species prefix "h" represents human, "m" represents mouse and "d" represents *Drosophila*. (B) Scheme describing the pairing rules of bHLH-PAS proteins. The α-class proteins are shown in dark gray boxes. These are the sensors. The β-class proteins are shown in light gray boxes. These are general partners for the α-class sensors. All β-class partners bind to the HIFs. The ARNTs bind to the AHR and AHRR, but not MOP4 or CLOCK. MOP3 and MOP9 also bind to MOP4 and CLOCK, but not to the AHR or AHRR. See text and Gu *et al.*[10] for more details.

as β-class partners for the CLOCK, MOP4, and the HIFαs, yet do not dimerize appreciably with the AHR. Although the *in vivo* partners for many less well defined PAS proteins still remain to be elucidated (e.g., MOP5/NPAS1, MOP6/NPAS3), the above observations indicate that pairing of α–β proteins is controlled by tissue specific expression levels of the relevant partners, as well as subtle differences in selectivity that may result from slight differences in the structure of bHLH or PAS domains.[35–37]

### 14.2.2.4  PAS PROTEIN SIGNAL TRANSDUCTION PATHWAYS

#### 14.2.2.4.1  The AH Receptor Pathway

The AHR plays important roles in how vertebrates respond to a number of important environmental contaminants. Environmental PAHs bind to the AHR, leading to the induction of a battery of xenobiotic metabolizing enzymes (XMEs) collectively referred to as the AH gene battery.[38] These XMEs include the cytochromes P4501A1, 1A2 and 1B1, as well as the glutathione S transferase Ya subunit and quinone:oxidoreductase.[39–44] As a result of the upregulation of this gene battery, PAHs are metabolized to more polar and readily excretable products. Since agonist exposure leads to upregulation of XMEs and increases in the metabolism and excretion of PAHs, this pathway easily falls under the rubric of an "adaptive response." Not all of the outcomes of the AHR pathway are positive. Potent agonists of the AHR, such as the halogenated dioxins, stimulate the AH gene battery, but also lead to a number of toxic endpoints such as hepatomegaly, liver damage, porphyria, endocrine disruption, thymic involution, chloracne, cancer, birth defects, a severe wasting syndrome, and death.[45–47] Although the AHR is known to transduce the dioxin signal resulting in these toxic outcomes, the exact molecular mechanism by which dioxin exposure causes toxicity is less clear (see Chapter 14.2.3 for more details).

The AHR signal transduction pathway provided a "first look" into how PAS proteins sense environment, transduce a signal to the nucleus and produce an adaptive response

(Figure 4). In the absence of agonists, the AHR resides in the cytoplasm as a complex with cellular chaperones. Upon exposure to agonist (e.g., dioxin), the AHR binds the chemical, translocates to the nucleus, sheds its chaperones and dimerizes to the β-class protein, ARNT. As a consequence of heterodimerization, the AHR-ARNT complex is able to recognize and bind to specific DNA sequences called dioxin response elements (DREs). These DREs lie upstream of genes in the AH gene battery. As the result of this interaction, transcription of this gene battery is initiated, resulting in expression of gene products with metabolic activity toward the chemical insult.

### 14.2.2.4.2 The Hypoxia Response Pathway

The mechanism by which vertebrates respond to their oxygenated environment is also mediated by PAS heterodimers. The hypoxia response pathway is mediated by a PAS protein heterodimer that was originally referred to as hypoxia inducible factor, or HIF.

The HIF complex was later found to be comprised of a subunit of the α-class protein HIF1α and the β-class ARNT (also referred to as HIF1β) (see Chapter 14.4.4). Under normoxic conditions, HIF1α is rapidly degraded in the cell and is thus unable to dimerize with ARNT and transduce a signal. However, when cellular oxygen tension drops below 3%, the HIF1α protein stability increases and the protein translocates to the nucleus where it binds to ARNT (Figure 5). The HIF1α-ARNT heterodimer subsequently binds to specific DNA sequences called hypoxia response elements (HREs). The binding of the HIF1α-ARNT heterodimer to these response elements initiates transcription of a battery of hypoxia inducible genes involved in mediating a response to low oxygen tension. This gene battery includes the peptide hormone, erythropoietin (EPO),[48] the angiogenic factors, vascular endothelial growth factor (VEGF), platelet-derived growth factor (PDGF) and fibroblast growth factor (FGF),[49–51] and genes encoding glycolytic enzymes such as aldolase A, phosphoglycerate kinase 1, lactate dehydrogenase A, phosphofructokinase L and glucose trans-

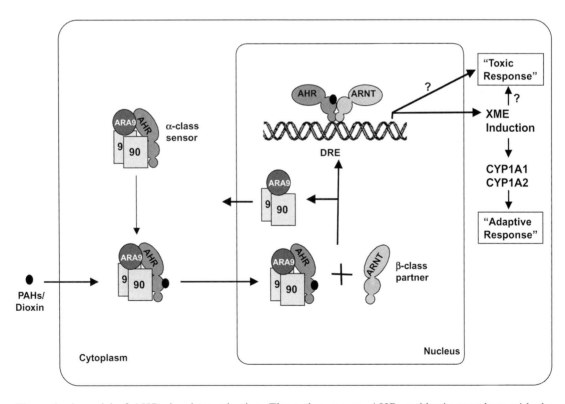

**Figure 4** A model of AHR signal transduction. The α-class sensor, AHR, resides in cytoplasm with the cellular chaperones, a dimer of Hsp90 and ARA9.[80,81] Upon activation by ligand, AHR translocates from cytoplasm into the nucleus and exchanges the chaperones for its β-class transcriptional partner, ARNT. The AHR-ARNT heterodimer then binds to the dioxin response element (DRE) and activates transcription of downstream genes belonging to the AH gene battery. Among the activated target genes, the cytochrome P450 isozymes are involved in the adaptive response to PAH exposure.

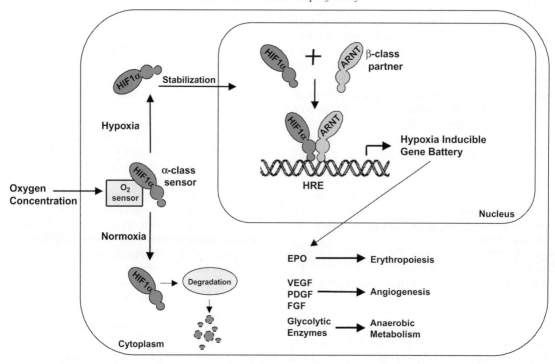

**Figure 5**  A model of the hypoxia response pathway. HIF1α is the α-class protein that responds to changes in oxygen tension, although it is not the direct oxygen sensor. Under normoxic conditions, HIF1α is rapidly degraded. When a cell is exposed to a hypoxic environment, the HIF1α protein is stabilized and it translocates to the nucleus where it dimerizes with its β-class partner, ARNT. The HIF1α-ARNT heterodimer then binds to the hypoxia response elements (HREs) and activates the transcription of downstream target genes. The target genes of the hypoxia-inducible gene battery include erythropoietin (EPO), vascular endothelial growth factor (VEGF), platelet derived growth factor (PDGF), fibroblast growth factor (FGF) and a series of glycolytic enzymes. Upregulation of these genes is an adaptive response to the change in environmental oxygen concentrations.

porters such as GLUT-1.[52–55] Less is known about the signaling pathways for the HIF1α homologues, HIF2α and HIF3α, including the exact identities of their β-class partners. Although their role in oxygen sensing and gene regulation is probably similar, *in situ* hybridization studies indicate each of the HIFαs act in different cell types and possibly at different stages in development.[34]

### 14.2.2.4.3   The Circadian Response Pathway

PAS proteins also play important roles in the maintenance of circadian and other biological rhythms. Daily changes in light and dark impose a significant environmental challenge that requires continual physiological and behavioral adaptation. Thus, it is not surprising that a number of biological activities are regulated by circadian rhythms with oscillations that have a period of approximately 24 hours. These rhythms are regulated by a master clock that operates with remarkable precision and persistence. However, this internal clock still requires continual fine-tuning by external

light. Moreover, in situations where circadian time is altered (e.g., transworld travel) more significant adaptation is required and the clock is reset by external cues. Circadian rhythms are maintained through input pathways that transmit environmental cues to the circadian clock, the clock itself that generates the biological rhythm, and output pathways that transmit the clock's information to the rest of the organism.[6,7,56]

The regulation of circadian rhythms has both a sensory component and an adaptive component. In the circadian pathway, the PAS heterodimer is composed of the (α-class protein CLOCK and the β-class partner MOP3 (Figure 6).[57–59] The CLOCK-MOP3 heterodimer binds to response element sequences termed M34RE (or a circadian responsive E-box). As a result, the CLOCK-MOP3 heterodimer positively regulates the levels of circadian responsive gene products including PER and TIM.[60] In return, PER and TIM negatively feedback to regulate the CLOCK/MOP3 complex either by binding to one member of the complex and disrupting its function or by indirectly influencing the signaling of the MOP3/CLOCK

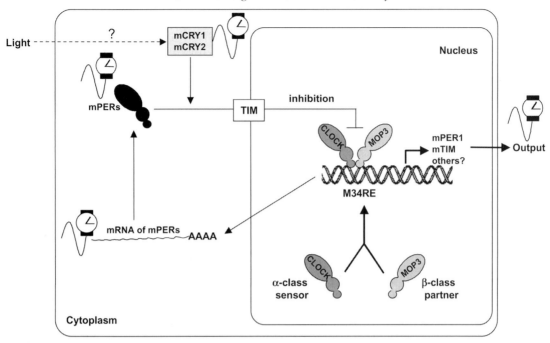

**Figure 6** A model of the mammalian circadian response pathway. The mammalian and fruit fly circadian response pathways may be slightly different thus the model of the mammalian pathway is shown here. The expression of PERs and cryptochromes (CRYs) are circadian regulated (indicated by a clock icon). CLOCK and MOP3 form a heterodimer. The heterodimer binds to the responsive element M34RE (stands for the MOP3 and MOP4 responsive element) containing the core sequence of 5'-CACGTG-3' and activates the transcription of downstream target genes such as mPER1 and mTIM. The mammalian CRYs (CRY1 and CRY2) are proposed to function in entrainment of behavioral rhythms.[82–85] It is unknown whether or how CRYs respond to light, but mCRYs interact with mPERs and help in translocating mPERs from cytoplasm into nucleus. PER, TIM and CRY can block the CLOCK-MOP3 dependent transcriptional activation, and therefore complete the feedback inhibition loop of PER. See text for details.

complex through interactions with the basal transcriptional machinery.[61–64] A key feature of this system is that there is a lag between the transcriptional induction of *per* and *tim* on one hand and the nuclear translocation of the repressor proteins they encode on the other. This lag creates a temporal separation between phases of induction and repression of gene expression, which is required to generate an oscillation. The molecular and genetic details of the circadian system will not be detailed further in this context. However, an excellent description of the circadian system can be found in recent reviews.[65,66]

### 14.2.2.4.4 Primordial PAS Proteins in Prokaryotes

It has recently been observed that PAS proteins are also found in a variety of prokaryotes. Although prokaryotes are not known to harbor bHLH domains, they do express proteins that contain local homology to PAS domains.[67,68] These PAS proteins are best thought of as having domains akin to the region surrounding one of the repeats found in their vertebrate homologues. Members of the prokaryotic PAS family are also involved in environmental adaptation. These roles include oxygen sensing (Dos, Aer, and FixL) and negative phototropism (PYP).[69–72]

Unlike their eukaryotic relatives, these prokaryotic proteins often harbor histidine kinase activity and transduce their signals via phosphorylation cascades that lead to activation of transcription factors.[73]

### 14.2.2.4.5 Primordial PAS Proteins in Plants

Structural homology with the PAS domain has also been identified in several photoreceptors in plants. In *Arabidopsis*, these proteins include NPH1, phytochromes from PhyA to PhyE, and the *phytochrome interacting factor* PIF3.[74,75] The blue-light photoreceptor NPH1 is required for directional growth towards light.[76] This protein has two repeats of approximately 110 amino acids, referred to as *light, oxygen, voltage sensor domains* ("LOV" domains).[77] The LOV domains share sequence homology to the PAS domains and have been shown to function as the binding sites for the chromophore flavin mononucleotide (FMN).[77] Five types of phytochromes (from PhyA to PhyE) have been identified. They respond mainly to light at the red/far-red region of the spectrum and are involved in many aspects of plant development.[75] These phytochromes contain two repeats homologous to the typical PAS domain and a histidine kinase like domain at their carboxy-terminus.[75] In keeping with the importance of homotypic interactions in PAS protein function, a yeast two-hybrid screen using the carboxy-terminal region of PhyB as the bait, lead to the identification of a photochrome-interacting factor, PIF3. This protein was shown to harbor a bHLH-PAS domain and to participate in the signaling pathways of both PhyA and PhyB.[78] The presence of both bHLH and PAS domains as well as its involvement in PhyA and PhyB signaling suggest that PIF3 is a transcription factor.[78]

## 14.2.2.5  SUMMARY

PAS domains have been identified in dozens of signal transduction molecules and various forms have been found in animals, plants and prokaryotes. In this review, we have attempted to summarize this rapidly expanding research area by providing a brief description of three signal transduction pathways that utilize PAS protein heterodimers to drive their transcriptional output. We have also used this review to develop the idea that most eukaryotic PAS proteins can be classified based upon functional similarities, as well as predicted phylogenetic relationships. We described the α-class proteins, which often act as sensors of environmental signals, and the β-class proteins, which typically act as broad-spectrum partners directing these heterodimers to their genomic targets. The α-class sensors, such as the AHR, the HIFαs and possibly even CLOCK detect changes in the environment and regulate an adaptive response. The β-class partners, such as the ARNTs, MOP3 and MOP9 dimerize with a broad spectrum of sensors and are essential for the transcriptional output of a number of these biologically important pathways. The sensors often act by directly detecting environmental change (AHR and ligand binding) or by transducing a signal from an upstream sensor (HIF1α from low $O_2$ tension). The mechanisms by which the signal transduction occurs in response to environmental cues are quite diverse and can involve ligand binding, protein stability or subcellular locali-

zation. It is now easy to predict that the PAS proteins arise from one of the largest families of signal transduction molecules encoded by the mammalian genome, rivaling the size of the steroid receptor superfamily. The elucidation of dioxin, hypoxia and circadian signal transduction pathways has provided valuable information about this superfamily and has allowed us to make a number of generalizations about PAS protein function. Analysis of the functional roles of vertebrate, plant and prokaryotic PAS proteins suggest that these proteins represent a primary mechanism by which organisms sense and adapt to environmental change. Furthermore, we are just beginning to appreciate the concept that a developing embryo is a microcosm of environmental adaptation and, as such, many of these pathways are utilized for normal development. This should not be a surprise since ontogeny of complex organisms is also a cellular struggle to adapt to environmental change.

## 14.2.2.6  ACKNOWLEDGMENTS

This work was supported by The National Institutes of Health (Grants R01-ES05703 and P30-CA07175) and a Post-doctoral Fellowship from the Natural Sciences and Engineering Research Council of Canada.

## 14.2.2.7  REFERENCES

1. A. H. Conney, E. C. Miller and J. A. Miller, 'The metabolism of methylated aminoazo dyes. V. Evidence for induction of enzyme synthesis in the rat by 3-methylcholanthrene.' *Cancer. Res.*, 1956, **16**, 450–459.
2. A. H. Conney, J. R. Gillette, J. K. Inscoe *et al.*, 'Induced synthesis of liver microsomal enzymes which metabolize foreign compounds.' *Science*, 1959, **130**, 1478–1479.
3. H. Remmer and H. J. Merker, 'Drug-induced changes in the liver endoplasmic reticulum: Association with drug-metabolizing enzymes.' *Science*, 1963, **142**, 1657–1658.
4. D. W. Nebert and J. E. Jones, 'Regulation of the mammalian cytochrome p1450 (CYP1A1) gene.' *Int. J. Biochem.*, 1989, **21**,
5. S. T. Crews, 'Control of cell lineage-specific development and transcription by bHLH-PAS proteins.' *Genes and Development*, 1998, **12**, 607–620.
6. J. C. Dunlap, 'Molecular bases for circadian clocks.' *Cell*, 1999, **96**, 271–290.
7. P. L. Lowrey and J. S. Takahashi, 'Genetics of the mammalian circadian system: Photic entrainment, circadian pacemaker mechanisms, and posttranslational regulation.' *Annual Review of Genetics*, 2000, **34**, 533–562.
8. J. P. Whitlock, Jr., 'Induction of cytochrome P4501A1.' *Annual Review of Pharmacology and Toxicology*, 1999, **39**, 103–125.

9. G. L. Semenza, 'Regulation of mammalian $O_2$ homeostasis by hypoxia-inducible factor 1.' *Annual Review of Cell and Developmental Biology*, 1999, **15**, 551–578.

10. Y.-Z. Gu, J. Hogenesch and C. Bradfield, in 'The PAS superfamily: Sensors of environmental and developmental signals', eds. Academic Press, 2000, pp. 519–561.

11. E. C. Hoffman, H. Reyes, F. F. Chu *et al.*, 'Cloning of a factor required for activity of the Ah (dioxin) receptor.' *Science*, 1991, **252**, 954–958.

12. F. R. Jackson, T. A. Bargiello, S. H. Yun et al., 'Product of per locus of *Drosophila* shares homology with proteoglycans.' *Nature*, 1986, **320**, 185–188.

13. J. R. Nambu, J. O. Lewis, K. A. Wharton, Jr. *et al.*, 'The *Drosophila* single-minded gene encodes a helix-loop-helix protein that acts as a master regulator of CNS midline development.' *Cell*, 1991, **67**, 1157–1167.

14. P. Reddy, A. C. Jacquier, N. Abovich *et al.*, 'The period clock locus of *D. melanogaster* codes for a proteoglycan.' *Cell*, 1986, **46**, 53–61.

15. Y. Citri, H. V. Colot, A. C. Jacquier *et al.*, 'A family of unusually spliced biologically active transcripts encoded by a *Drosophila* clock gene.' *Nature*, 1987, **326**, 42–47.

16. M. Denis, S. Cuthill, A. C. Wikstrom *et al.*, 'Association of the dioxin receptor with the Mr 90,000 heat shock protein: A structural kinship with the glucocorticoid receptor.' *Biochem. Biophys. Res. Comm.*, 1988, **155**, 801–807.

17. G. H. Perdew, 'Association of the Ah receptor with the 90-kDa heat shock protein.' *J. Biol. Chem.*, 1988, **263**, 13802–13805.

18. K. M. Burbach, A. Poland and C. A. Bradfield, 'Cloning of the Ah-receptor cDNA reveals a distinctive ligand-activated transcription factor.' *Proc. Natl. Acad. Sci. USA*, 1992, **89**, 8185–8189.

19. K. M. Dolwick, H. I. Swanson and C. A. Bradfield, '*In vitro* analysis of Ah receptor domains involved in ligand-activated DNA recognition.' *Proc. Natl. Acad. Sci. USA*, 1993, **90**, 8566–8570.

20. A. Poland, D. Palen and E. Glover, 'Analysis of the four alleles of the murine aryl hydrocarbon receptor.' *Mol. Pharmacol.*, 1994, **46**, 915–921.

21. T. Kadesch, 'Consequences of heteromeric interactions among helix-loop-helix proteins.' *Cell Growth and Differentiation*, 1993, **4**, 49–55.

22. C. Murre, G. Bain, M. A. van Dijk *et al.*, 'Structure and function of helix-loop-helix proteins.' *Biochimica et Biophysica Acta*, 1994, **1218**, 129–135.

23. H. Reyes, S. Reisz-Porszasz and O. Hankinson, 'Identification of the Ah receptor nuclear translocator protein (Arnt) as a component of the DNA binding form of the Ah receptor.' *Science*, 1992, **256**, 1193–1195.

24. E. Maltepe, B. Keith, A. M. Arsham *et al.*, 'The role of ARNT2 in tumor angiogenesis and the neural response to hypoxia.' *Biochem. Biophy. Res. Comm.*, 2000, **273**, 231–238.

25. G. Drutel, A. Heron, M. Kathmann *et al.*, 'ARNT2, a transcription factor for brain neuron survival?' *Europ. J. Neurosci.*, 1999, **11**, 1545–1553.

26. J. B. Hogenesch, Y.-Z. Gu, S. Jain *et al.*, 'The basic helix-loop-helix-PAS orphan MOP3 forms transcriptionally active complexes with circadian and hypoxia factors.' *Proc. Natl. Acad. Sci. USA*, 1998, **95**, 5474–5479.

27. J. Hogenesch, Y.-Z. Gu, S. Moran, K. Shimomura *et al.*, 'The basic helix-loop-helix-PAS protein MOP9 is a brain specific heterodimeric partner of circadian and hypoxia factors.' *J. Neuroscience*, 2000, **20**, 1–5.

28. H. I. Swanson, W. K. Chan and C. A. Bradfield, 'DNA binding specificities and pairing rules of the Ah receptor, ARNT, and SIM proteins.' *J. Biol. Chem.*, 1995, **270**, 26292–26302.

29. H. Li, P.J. Gomes and J.D. Chen, 'RAC3, a steroid/nuclear receptor-associated coactivator that is related to SRC-1 and TIF2.' *Proc. Natl. Acad. Sci. USA*, 1997, **94**, 8479–8484.

30. J. J. Voegel, M. J. Heine, C. Zechel *et al.*, 'TIF2, a 160 kDa transcriptional mediator for the ligand-dependent activation function AF-2 of nuclear receptors.' *EMBO J.*, 1996, **15**, 3667–3675.

31. S. A. Onate, V. Boonyaratanakornkit, T. E. Spencer *et al.*, 'The steroid receptor coactivator-1 contains multiple receptor interacting and activation domains that cooperatively enhance the activation function 1 (AF1) and AF2 domains of steroid receptors.' *J. Biol. Chem.*, 1998, **273**, 12101–12108.

32. C. K. Glass, D. W. Rose and M. G. Rosenfeld, 'Nuclear receptor coactivators.' *Curr. Opin. Cell Biol.*, 1997, **9**, 222–232.

33. L. Xu, C. K. Glass and M. G. Rosenfeld, 'Coactivator and corepressor complexes in nuclear receptor function.' *Curr. Opin. Gene. Dev.*, 1999, **9**, 140–147.

34. S. Jain, E. Maltepe, M. M. Lu *et al.*, 'Expression of ARNT, ARNT2, HIF1 alpha, HIF2 alpha, and Ah receptor mRNAs in the developing mouse.' *Mech. Dev.*, 1998, **73**, 117–123.

35. Y. D. Zhou, M. Barnard, H. Tian *et al.*, 'Molecular characterization of two mammalian bHLH-PAS domain proteins selectively expressed in the central nervous system.' *Proc. Natl. Acad. Sci. USA*, 1997, **94**, 713–718.

36. J. B. Hogenesch, W. K. Chan, V. H. Jackiw *et al.*, 'Characterization of a subset of the basic helix-loop-helix-PAS superfamily that interacts with components of the dioxin signaling pathway.' *J. Biol. Chem.*, 1997, **272**, 8581–8593.

37. E. W. Brunskill, D. P. Witte, A. B. Shreiner *et al.*, 'Characterization of npas3, a novel basic helix-loop-helix PAS gene expressed in the developing mouse nervous system.' *Mech. Dev.*, 1999, **88**, 237–241.

38. D. W. Nebert and H. V. Gelboin, 'Substrate-inducible microsomal aryl hydroxylase in mammalian cell culture I. Assay and properties of induced enzyme.' *J. Biol. Chem.*, 1968, **243**, 6242–6249.

39. M. S. Denison, J. M. Fisher and J. P. Whitlock, Jr., 'Inducible, receptor-dependent protein-DNA interactions at a dioxin-responsive transcriptional enhancer.' *Proc. Natl. Acad. Sci. USA*, 1988, **85**, 2528–2532.

40. K. W. Jones and J. P. Whitlock, Jr., 'Functional analysis of the transcriptional promoter for the CYP1A1 gene.' *Molec. Cell. Biol.*, 1990, **10**, 5098–5105.

41. T. H. Rushmore, R. G. King, K. E. Paulson *et al.*, 'Regulation of glutathione S-transferase Ya subunit gene expression: identification of a unique xenobiotic-responsive element controlling inducible expression by planer aromatic compounds.' *Proc. Natl. Acad. Sci.*, 1990, **87**, 3826–3830.

42. L. V. Favreau and C. B. Pickett, 'Transcriptional regulation of the rat NAD(P)H:quinone reductase gene. Identification of regulatory elements controlling basal level expression and inducible expression by planar aromatic compounds and phenolic anitoxidants.' *J. Biol. Chem.*, 1991, **266**, 4556–4561.

43. T. Iyanagi, M. Haniu, K. Sogawa *et al.*, 'Cloning and characterization of cDNA encoding 3-methylcholanthrene inducible rat mRNA for UDP-glucuronosyltransferase.' *J. Biol. Chem.*, 1986, **261**, 15607–15614.

44. F. J. Gonzalez, R. H. Tukey and D. W. Nebert, 'Structural gene products of the Ah locus. Transcriptional regulation of cytochrome P1-450 and P3-450

mRNA levels by 3-methylcholanthrene.' *Mol. Pharmacol.*, 1984, **26**, 117–121.

45. C. L. Wilson and S. Safe, 'Mechanisms of ligand-induced aryl hydrocarbon receptor-mediated biochemical and toxic responses.' *Toxicol. Path.*, 1998, **26**, 657–671.

46. A. Poland and J. C. Knutson, '2,3,7,8-tetrachlorodibenzo-*p*-dioxin and related halogenated aromatic hydrocarbons: examination of the mechanism of toxicity.' *Ann. Rev. Pharm. Toxi.*, 1982, **22**, 517–554.

47. J. V. Schmidt and C. A. Bradfield, 'Ah receptor signaling pathways.' *Annu. Rev. Cell Dev. Biol.*, 1996, **12**, 55–89.

48. M. A. Goldberg, S. P. Dunning and H. F. Bunn, 'Regulation of the erythropoietin gene: evidence that the oxygen sensor is a heme protein.' *Science*, 1988, **242**, 1412–1415.

49. J. A. Forsythe, B. H. Jiang, N. V. Iyer *et al.*, 'Activation of vascular endothelial growth factor gene transcription by hypoxia-inducible factor 1.' *Mol. Cell. Bio.*, 1996, **16**, 4604–4613.

50. A. P. Levy, N. S. Levy, S. Wegner *et al.*, 'Transcriptional regulation of the rat vascular endothelial growth factor gene by hypoxia.' *J. Biol. Chem.*, 1995, **270**, 13333–13340.

51. K. Kuwabara, S. Ogawa, M. Matsumoto *et al.*, 'Hypoxia-mediated induction of acidic/basic fibroblast growth factor and platelet-derived growth factor in mononuclear phagocytes stimulates growth of hypoxic endothelial cells.' *Proc. Natl. Acad. Sci. USA*, 1995, **92**, 4606–4610.

52. G. L. Semenza, P. H. Roth, H. M. Fang *et al.*, 'Transcriptional regulation of genes encoding glycolytic enzymes by hypoxia-inducible factor 1.' *J. Biol. Chem.*, 1994, **269**, 23757–23763.

53. G. L. Semenza, B. H. Jiang, S. W. Leung *et al.*, 'Hypoxia response elements in the aldolase A, enolase 1, and lactate dehydrogenase A gene promoters contain essential binding sites for hypoxia-inducible factor 1.' *J. Biol. Chem.*, 1996, **271**, 32529–32537.

54. H. Li, H. P. Ko and J. P. Whitlock, 'Induction of phosphoglycerate kinase 1 gene expression by hypoxia. Roles of Arnt and HIF1α.' *J. Biol. Chem.*, 1996, **271**, 21262–21267.

55. J. M. Gleadle and P. J. Ratcliffe, 'Induction of hypoxia-inducible factor-1, erythropoietin, vascular endothelial growth factor, and glucose transporter-1 by hypoxia: evidence against a regulatory role for Src kinase.' *Blood*, 1997, **89**, 503–509.

56. D. P. King and J. S. Takahashi, 'Molecular genetics of circadian rhythms in mammals.' *Annu. Rev. Neurosci.*, 2000, **23**, 713–742.

57. N. Gekakis, D. Staknis, H. B. Nguyen *et al.*, 'Role of the CLOCK protein in the mammalian circadian mechanism.' *Science*, 1998, **280**, 1564–1569.

58. J. B. Hogenesch, W. C. Chan, V. H. Jackiw *et al.*, 'Characterization of a subset of the basic helix-loop-helix-PAS superfamily that interact with components of the dioxin signaling pathway.' *J. Biol. Chem.*, 1997, **272**, 8581–8593.

59. M. Ikeda and M. Nomura, 'cDNA cloning and tissue-specific expression of a novel basic helix-loop-helix/PAS protein (BMAL1) and identification of alternatively spliced variants with alternative translation initiation site usage.' *Biochem. Biophy. Res. Comm.*, 1997, **233**, 258–264.

60. N. Gekakis, L. Saez, A. M. Delahaye-Brown *et al.*, 'Isolation of timeless by PER protein interaction: defective interaction between timeless protein and long-period mutant PERL.' *Science*, 1995, **270**, 811–815.

61. T. K. Darlington, K. Wager-Smith, M. F. Ceriani *et al.*, 'Closing the circadian loop: CLOCK-induced transcription of its own inhibitors per and tim.' *Science*, 1998, **280**, 1599–1603.

62. K. Kume, M. J. Zylka, S. Sriram *et al.*, 'mCRY1 and mCRY2 are essential components of the negative limb of the circadian clock feedback loop.' *Cell*, 1999, **98**, 193–205.

63. C. Lee, K. Bae and I. Edery, 'PER and TIM inhibit the DNA binding activity of a Drosophila CLOCK-CYC/dBMAL1 heterodimer without disrupt formation of the heterodimer: a basis for circadian transcription.' *Mol. Cell. Biol.*, 1999, **19**, 5316–5325.

64. A. M. Sangoram, L. Saez, M. P. Antoch *et al.*, 'Mammalian circadian autoregulatory loop: a timeless ortholog and mPer1 interact and negatively regulate CLOCK-BMAL1-induced transcription.' *Neuron*, 1998, **21**, 1101–1113.

65. D. C. Chang and S. M. Reppert, 'The circadian clocks of mice and men.' *Neuron*, 2001, **29**, 555–558.

66. K. Wager-Smith and S. A. Kay, 'Circadian rhythm genetics: from flies to mice to humans.' *Nature Genetics*, 2000, **26**, 23–27.

67. J. L. Pellequer, R. Brudler and E. D. Getzoff, 'Biological sensors: More than one way to sense oxygen.' *Curr. Biol.*, 1999, **9**, R416–418.

68. B. L. Taylor and I. B. Zhulin, 'PAS domains: internal sensors of oxygen, redox potential, and light.' *Microbiol. Mole. Biol. Rev. (Washington, DC)*, 1999, **63**, 479–506.

69. V. M. Delgado-Nixon, G. Gonzalez and M. A. Gilles-Gonzalez, 'Dos, a heme-binding PAS protein from *Escherichia coli*, is a direct oxygen sensor.' *Biochemistry*, 2000, **39**, 2685–2691.

70. S. I. Bibikov, L. A. Barnes, Y. Gitin *et al.*, 'Domain organization and flavin adenine dinucleotide-binding determinants in the aerotaxis signal transducer Aer of *Escherichia coli*.' *Proc. Natl. Acad. Sci. USA*, 2000, **97**, 5830–5835.

71. M. David, M. L. Daveran, J. Batut *et al.*, 'Cascade regulation of nif gene expression in Rhizobium meliloti.' *Cell*, 1988, **54**, 671–683.

72. J. L. Pellequer, K. A. Wager-Smith, S. A. Kay *et al.*, 'Photoactive yellow protein: a structural prototype for the three-dimensional fold of the PAS domain superfamily.' *Proc. Natl. Acad. Sci. USA*, 1998, **95**, 5884–5890.

73. P. G. Agron, E. K. Monson, G. S. Ditta *et al.*, 'Oxygen regulation of expression of nitrogen fixation genes in Rhizobium meliloti.' *Res. Microbiol.*, 1994, **145**, 454–459.

74. G. C. Whitelam and K. J. Halliday, 'Photomorphogenesis: Phytochrome takes a partner.' *Curr. Biol.*, 1999, **9**, R225–227.

75. C. Fankhauser and J. Chory, 'Photomorphogenesis: Light receptor kinase in plants.' *Curr. Biol.*, 1999, **9**, R123–126.

76. J. M. Christie, P. Reymond, G. K. Powell *et al.*, 'Arabidopsis NPH1: a flavoprotein with the properties of a photoreceptor for phototropism.' *Science*, 1998, **282**, 1698–1701.

77. J. M. Christie, M. Salomon, K. Nozue *et al.*, 'LOV (light, oxygen, or voltage) domains of the blue-light photoreceptor phototropin (nph1): Binding sites for the chromophore flavin mononucleotide.' *Proc. Natl. Acad. Sci. USA*, 1999, **96**, 8779–8783.

78. M. Ni, J. M. Tepperman and P. H. Quail, 'PIF3, a phytochrome-interacting factor necessary for normal photoinduced signal transduction, is a novel basic helix-loop-helix protein.' *Cell*, 1998, **95**, 657–667.

79. D. G. Higgins and P. M. Sharp, 'CLUSTAL: A package for performing multiple sequence alignment on a microcomputer.' *Gene*, 1988, **73**, 237–244.

80. L. A. Carver, J. J. LaPres, S. Jain *et al.*, 'Characterization of the Ah Receptor-associated Protein, ARA9.' *J. Biol. Chem.*, 1998, **273**, 33580–36159.

81. J. J. LaPres, E. Glover, E. E. Dunham *et al.*, 'ARA9 Modifies Agonist Signaling through an Increase in Cytosolic Aryl Hydrocarbon Receptor.' *J. Biol. Chem.*, 2000, **275**, 6153–6159.

82. A. R. Cashmore, J. A. Jarillo, Y. J. Wu *et al.*, 'Cryptochromes: blue light receptors for plants and animals.' *Science*, 1999, **284**, 760–765.

83. E. A. Griffin, Jr., D. Staknis and C. J. Weitz, 'Light-independent role of CRY1 and CRY2 in the mammalian circadian clock.' *Science*, 1999, 286, 768–771.

84. K. Kume, M. J. Zylka, S. Sriram *et al.*, 'mCRY1 and mCRY2 are essential components of the negative limb of the circadian clock feedback loop.' *Cell*, 1999, **98**, 193–205.

85. R. J. Lucas and R. G. Foster, 'Circadian clocks: A cry in the dark?' *Curr. Biol.*, 1999, **9**, R825–828.

J.P. Vanden Heuvel, G.H. Perdew, W.B. Mattes and W.F. Greenlee (Eds.)
Comprehensive Toxicology, Vol. xiv

# 14.2.3
# The Aryl Hydrocarbon Receptor: A Model of Gene-Environment Interactions

JACQUELINE A. WALISSER and CHRISTOPHER A. BRADFIELD

*McArdle Laboratory for Cancer Research, University of Wisconsin School of Medicine, Madison, WI, USA*

## 14.2.3.1 ENVIRONMENTAL TOXICANTS AND THE ADAPTIVE RESPONSE PATHWAY

Compounds such as benzo(a)pyrene and benzanthracene, are examples of the polycyclic aromatic hydrocarbons (PAHs) that are widely distributed in the environment. These compounds can be formed from incomplete combustion of organic material (e.g., forest fires, diesel exhaust, cigarette smoke and charbroiling of food).[1] In general, the toxicity of PAHs is related to their metabolic activation to electrophiles and their propensity to react with cellular macromolecules. Consequences of this metabolic activation include the formation of DNA and protein adducts, DNA mutations, and cell death. One way in which vertebrates adapt to the exposure to these environmental chemicals is through the induction of battery of xenobiotic metabolizing enzymes (XMEs) that metabolize PAHs to more polar and excretable products. The most well understood aspect of this adaptive response is the induction of the microsomal cytochrome P450 dependent monooxygenases such as CYP1A1, CYP1A2, and CYP1B1.[2–5] Prior to our ability to discriminate between these three gene products, the collective activity of these xenobiotic metabolizing enzymes was referred to as "aryl hydrocarbon hydroxylase" (AHH) based upon their capacity to hydroxylate aromatic hydrocarbons.[6]

## 14.2.3.2 THE AH RECEPTOR MEDIATES AN ADAPTIVE RESPONSE

The idea that a receptor was involved in mediating the adaptive response to PAHs is now a well-established one. Early on it was observed that the inductive response to PAHs was polymorphic among strains of mice. For example, it was determined that some murine strains (e.g., C57BL/6) were highly responsive to PAH induction of XMEs whereas other strains (e.g., DBA) were relatively nonresponsive.[6] Classical genetics involving crosses, backcrosses and intercrosses of these mouse lines indicated that a single autosomal locus played a primary role in controlling the inducibility of XMEs. This locus was termed *Ah*, for *a*ryl *h*ydrocarbon responsiveness.[7–9] Thus, the responsive strains were designated as encoding the $Ah^b$ allele, whereas non-responsive strains encoded the $Ah^d$ allele. It was later learned that halogenated aromatic compounds, such as 2,3,7,8-tetrachlorodibenzo-*p*-dioxin (dioxin), were more potent inducers of the XME response than were the PAHs.[10] This greater potency suggested an increased binding affinity for an *Ah* locus encoded receptor, the aryl hydrocarbon receptor (AHR),[10] and provided direction for the synthesis of high affinity radioligands and the pharmacological investigation of this receptor.[11–16]

The development of radioligands and purified receptor preparations provided initial insights into the mechanism of PAH/dioxin signal transduction and XME induction. Reversible radiolabeled ligands allowed the demonstration of a saturable, high affinity receptor present in target cells.[17,18] Competitive binding studies allowed the correlation of congener binding affinity to biological response.[11,12,19] Such structure-activity relationships are still a central aspect of any proof that the AHR mediates a given biological response.[20,21] Saturation binding isotherms with [125]I-labeled congeners have demonstrated that the AHR has a remarkably high binding affinity for halogenated agonists. That is, the $K_D$ for dioxin binding to the AHR from responsive alleles is approximately $1 \times 10^{-12} \, M$[15]. Furthermore, studies with these ligands demonstrated that the nonresponsive $Ah^d$ alleles encoded a receptor with a 2–10 fold lower binding affinity for agonists as compared to AHRs encoded by responsive $Ah^b$ alleles.[22–29] Radiolabeled ligands also provided the tool that led to the purification of the AHR and the biochemical characterization of the receptor.[15,30,31] Early on, these ligands were used to demonstrate that agonist exposure induced a change in the oligomeric state of the AHR.[32–35] Furthermore, this change was coincident with a shift in cellular localization of the receptor from the cytoplasm to the nuclear compartment and upregulation of CYP1A1.[35–37] Much of our understanding of the AHR signal transduction pathway has come from studying XME induction and expression of *Cyp1a1* (reviewed in Reference 3).

Thus, the three important proofs that the *Ah* locus is mediating a given biological response are: (1) the genetic proof where binding affinity and biological effect co-segregate with the responsive and nonresponsive alleles; (2) the pharmacological proof, whereby ligand binding affinity for the AHR correlates with biological response; and (3) the biochemical proof, demonstrating the translocation of the ligand-bound receptor to the nucleus and transactivation of classic target genes such as *Cyp1a1*.

## 14.2.3.3 THE AH RECEPTOR SIGNAL TRANSDUCTION PATHWAY

An important step in understanding AHR signal transduction was the elucidation of the molecular features of the protein. Protein sequence information from the purified protein lead to the molecular cloning of the receptor's cDNA and revealed that it was a member of the bHLH-PAS superfamily.[38,39] AHR was also shown to have amino acid sequence homology with aryl hydrocarbon nuclear translocator (ARNT). The ARNT cDNA was cloned as the result of a genetic screen designed to identify gene products that played roles in AHR signal transduction in mouse hepatoma cells.[40] In one class of signaling mutants identified in this screen, the AHR was present and bound ligand normally, but did not attain an increased affinity for the nuclear compartment. A human gene fragment encoding the ARNT protein rescued this loss-of-function mutation. Further experiments demonstrated that the ARNT protein was required to direct the ligand-activated AHR to specific regulatory elements upstream of genomic targets like the *Cyp1a1* gene.[40,41] The realization that the AHR and ARNT were structurally related bHLH-PAS proteins shed light on the model of AHR signal transduction and provided the first example of a bHLH-PAS heterodimer (Figure 1).

A model of AHR signal transduction that accounts for its role as a sensor and mediator of environmental adaptation can be summarized as follows (Figure 2). In the absence of ligand, AHR resides in the cytoplasmic compartment bound to a variety of molecular chaperones, including a dimer of the 90 kDa heat shock protein (Hsp90), the AH receptor associated protein, ARA9 (also known as AIP1 and XAP2), and p23.[42–45] Hsp90 is a well recognized cellular chaperone. ARA9 displays structural similarity to the immunophilin, FKBP52, including the binding region for FK506 and a region that harbors multiple tetratricopeptide repeat (TPR) domains. p23 is an acidic protein often associated with Hsp90.[45] The interaction of AHR with these

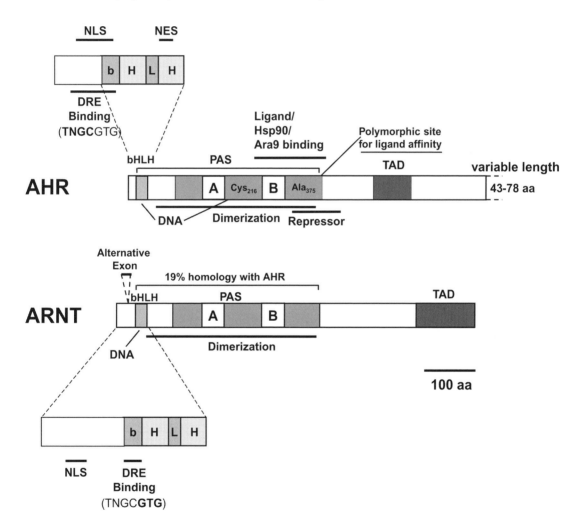

**Figure 1**   The molecular structures of AHR and ARNT. The basic region (b), helix-loop-helix (HLH), PAS domain with A and B repeats, as well as the transactivation (TAD) domains of AHR and ARNT are labeled. For AHR, the regions that have been shown to play a role in nuclear localization (NLS), DNA binding (DNA), PAS protein dimerization, ligand/Hsp90/Ara9 binding and repression of AHR activity are labeled.[69,142] $Cys_{216}$ is marked for its role in DNA binding.[143] $Ala_{375}$ is marked for its importance for high affinity ligand binding.[22] AHR binds to the 5′-TNGC-′3 half-site of the DRE. The C-terminal end "variable length" represents the length of different *Ah* alleles ($Ah^b$ and $Ah^d$) in various mouse strains.[22] AHR also contains a leucine-rich nuclear export signal (NES) in the $NH_2$-terminal region indicating that the AHR may undergo nucleocytoplasmic recycling.[50] For ARNT, the location of an alternative exon, the regions that are involved in nuclear localization (NLS), DNA binding and PAS protein dimerization are marked. ARNT binds to the 5′-GTG-′3 half-site of the DRE. The percentage similarity of the ARNT PAS domain with that of AHR is indicated. See text for details.

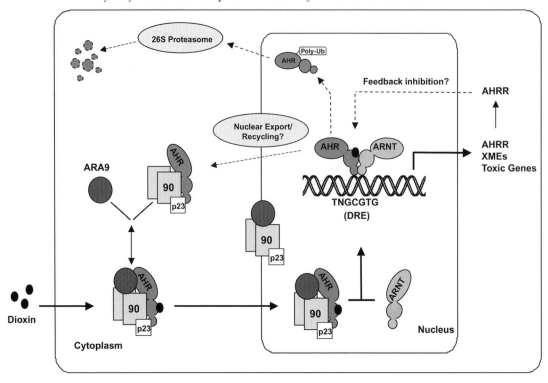

**Figure 2**  A model of AHR signal transduction. AHR normally resides in cytoplasm with the chaperones Hsp90, ARA9 and p23, which stabilize the receptor and increase ligand responsiveness. Upon activation by its ligand, AHR translocates from cytoplasm into the nucleus and exchanges its chaperones for ARNT. The AHR-ARNT heterodimer then binds to the DRE with the core sequence of TNGCGTG and activates transcription of downstream target genes (XMEs, AHRR and toxic genes). Ligand-activated AHR is degraded through a proteasome pathway or may undergo nuclear export and recycling within the cytoplasm. See text for details. This pathway is also described and periodically updated at: http://www.biocarta.com/pathfiles/acrPathway.asp

co-chaperones is thought to stabilize the receptor in the cytosolic fraction of cells and increase the number of receptors available to bind ligand.[46–49] Nuclear localization signals (NLS) have been identified in the $NH_2$-terminal regions of both AHR and ARNT. In the absence of ligand, the AHR-chaperone complex is present primarily in the cytoplasm. Upon ligand binding, the AHR undergoes a conformational change that exposes the NLS and results in translocation of the receptor to the nucleus.[36,50] In contrast, ARNT appears to be constitutively present in the nucleus.[36,51,52]

Concomitant with the movement of the AHR to the nucleus are changes in the size of the receptor complex that are believed to be the result of the shedding of its cellular chaperones in exchange for binding a second PAS protein, ARNT.[33,35,53,54] AHR-ARNT heterodimerization is mediated by both the PAS domain and the HLH region (Figure 1). Within the nucleus, the AHR-ARNT heterodimer becomes competent to bind specific "dioxin response elements" (DREs) and drive transcription from adjacent target promoters.[37,53,55–61] The basic

regions of the bHLH domains within the AHR-ARNT heterodimer carry out recognition of the specific DNA elements. Molecular analysis, coupled with the identification of consensus DREs from known target genes, has defined the core DRE as 5'-TNGCGTG-3'.[56,58,62–64] Evidence generated *in vitro* indicates that the AHR basic region binds to the TNGC half site, while the ARNT basic region binds to the GTG half site.[62,63]

The protein sequences required to drive transcription appear to reside in the C-terminal halves of both AHR and ARNT, in regions called the transactivation domain (TAD).[65–68] Some laboratories have proposed that these two proteins contribute differently to transcriptional activation.[69,70] One model of transcriptional activation by the AHR-ARNT heterodimer suggests that interactions with multiple upstream DREs facilitates the disruption of local chromatin structure, allowing downstream promoter elements to bind their respective transcription factors and initiate transcription.[70–72] Recent evidence also suggests a variety of coactivators, such as RIP140,

SRC-1, and the retinoblastoma protein (Rb), play roles in the transcriptional activity of the AHR-ARNT complex.[73-75]

Upregulation of DRE-regulated genes forms the basis for the adaptive response mediated by the AHR signal transduction pathway. The target genes of this adaptive response include a battery of xenobiotic metabolizing enzymes (XMEs) such as CYP1A1, CYP1B1, CYP1A2, the glutathione S transferase Ya subunit and quinone oxidoreductase.[76-81] Since the XMEs ultimately metabolize the compounds that caused the induction, AHR-ARNT mediated transcriptional activation can be seen as an adaptive response to an environmental exposure.

### 14.2.3.4 REGULATION OF AH RECEPTOR PATHWAY

The AHR pathway has been shown to be attenuated by two mechanisms. The first mechanism involves the ligand dependent transcription of a dominant negative bHLH-PAS protein known as the AH receptor repressor (AHRR) (Figure 2).[82] The AHRR is a PAS protein that contains only a single PAS repeat and has been found to inhibit AHR signal transduction.[82] The primary mechanism whereby AHRR inhibits AHR signal transduction is though its activity as a transcriptional repressor. AHRR has been show to directly inhibit gene expression from linked promoters. Thus, it has been suggested that AHRR may be a part of a negative feedback loop to down-regulate or attenuate an activated AHR pathway. The feedback inhibition idea is based upon the observations that the AHRR promoter is driven by a functional DRE and that the level of the AHRR mRNA is up-regulated by agonists of the AHR.[82] In addition, when the AHRR is expressed it can act as a constitutive partner that may compete for ARNT dimerization and DRE binding. As a consequence of its repressor activity and competition, AHRR significantly inhibits dioxin induced transcription.

The second mechanism whereby AHR's signal can be down-regulated is through the proteasome degradation pathway.[83] It has been demonstrated that the ligand activated AHR is rapidly proteolyzed, leading to a decrease in receptor number immediately following agonist exposure in many cell types.[84-86] Recent evidence suggests that activated nuclear receptor undergoes poly-ubiquitination, which acts as a recognition signal for the 26 S proteasome complex.[87,88] In contrast, ARNT does not appear to be down-regulated by the ubiquitin degradation pathway.[87] This agonist-dependent down-regulation of AHR may be a mechanism of attenuating extended periods of transcription that could potentially be deleterious to the cell.

### 14.2.3.5 TOXICITY OF AH RECEPTOR AGONISTS

The AHR pathway provides an example of how overstimulation of an adaptive response can have deleterious effects. In addition to PAHs, most vertebrates are also exposed to halogenated aromatic hydrocarbons, like dioxin.[89,90] Exposure to dioxins can lead to a number of toxic endpoints such as hepatomegaly, liver damage, porphyria, endocrine disruption, thymic involution, chloracne, cancer, birth defects, a severe wasting syndrome and death.[91-93] Application of the pharmacological and genetic proofs outlined above indicates that the AHR is directly involved in mediating many, if not all of these toxic endpoints.[7,94] The adaptive response model depicted in Figure 2 appears to accurately describe the up-regulation of XMEs by dioxins, however, it has yet to be proven that this model explains any of dioxin's toxic actions. Although this may ultimately be proven true, it is important to emphasize that the pharmacological and genetic proofs for AHR involvement do not necessarily implicate the ARNT protein or DRE mediated gene expression in any toxic mechanism. Other than a role for the AHR, little is understood about the molecular mechanisms that underlie most of dioxin's toxic effects. For more information on dioxin toxicity we refer you to the website, http://dioxins-r-us.ucdavis.edu/Dioxin.HTML

#### 14.2.3.5.1 Models of DRE-Dependent Toxicity

Various models have been proposed to explain the toxic effects of AHR mediated dioxin toxicity. The simplest model of dioxin toxicity is a reflection of the adaptive response model and suggests that the AHR-ARNT heterodimer directly regulates the expression of genes involved in toxicity through ARNT dependent, DRE driven gene expression. In fact, the genes involved in toxicity could be those already known to be members of the AHR regulated gene battery. In this regard, it has been proposed that induction of the CYP-monooxygenases could be cytotoxic due to increased oxidative stress related to redox cycling.[95-98] It is also possible that some of these XMEs have important endogenous substrates. Therefore, alterations in the production/degradation of XMEs could lead to inappropriate cellular levels of certain physiologically active substances leading to toxicity. Alternatively, these "toxic genes" could be among a battery of less-characterized dioxin-responsive genes. For example, the expression of ecto-ATPase, major histocompatibility complex Q1b, epidermal growth factor receptor, tissue plasminogen activator inhibitor-2, cyclooxygenase-2, interleukin 1β, interleukin 2 and δ-aminolevulinic acid synthetase have all been shown to be influenced by dioxin exposure and many of these genes have been shown to harbor DREs.[99–105] Finally, the genes involved in toxicity may not yet have been discovered, or have been discovered but have not yet been linked to regulation by the AHR/ARNT complex.

There is literature support for alternate homotypic interactions involving AHR mediated transcription that could also account for toxicity. Dioxins may induce the formation of the AHR-ARNT complexes to high enough levels that interaction with nonconsensus DREs is possible. Recently, the AHR/ARNT complex has been shown to interact with estrogen response elements (EREs) or sequences referred to as "imperfect DREs" (iDREs) in a number of estrogen responsive genes.[106] Such interactions could explain the endocrine disruptive effects of dioxin. Alternatively, it is conceivable that AHR may be able to partner with other PAS proteins besides ARNT and that these alternate heterodimers may transcribe a unique set of genes. In fact, a close structural homologue of the ARNT molecule (ARNT2) has been described and proposed to act as an alternate partner for AHR in the central nervous system.[107,108] This assertion is based upon the observations that ARNT2 dimerizes with the AHR *in vitro* and that the resultant complex is capable of driving transcription from a DRE linked promoter in a heterologous expression system. Other ARNT homologues are the PAS proteins, MOP3 (also called BMAL1)[109–111] and MOP9 (also called BMAL2).[112] These PAS proteins have relatively weak affinity for the AHR and, as such, it is unlikely that these heterodimers have biological relevance *in vivo*. Based upon dimerization affinity and tissue specific expression patterns, it appears that ARNT may be the most prominent bHLH-PAS partner of the AHR, with the role of ARNT2 in this pathway still to be understood. A summary of the potential mechanisms of DRE-dependent toxicity in response to dioxin is shown in Figure 3.

### 14.2.3.5.2 Models of DRE-Independent Toxicity

AHR-mediated toxicity may also involve mechanisms that are quite distinct from its classical role as a DRE-binding transcription factor. Several DRE independent mechanisms of dioxin toxicity have been proposed that suggest AHR's direct physical interaction/interference with other signaling pathways (Table 1). In a homotypic interference model, dioxin toxicity could occur if the activated AHR sequestered partners like ARNT from other PAS proteins. For example, ARNT is also a partner with another bHLH-PAS protein called the hypoxia-inducible factor 1-α (HIF1α), which is involved in sensing and responding to low oxygen environments (discussed in detail in Chapter 14.4.4). HIF1α partners with ARNT and together they bind to specific DNA sequences called hypoxia response elements (HREs) that drive hypoxia responsive target genes. In the cross-talk interference model, ligand-bound AHR could compete with HIF1α for ARNT thereby compromising hypoxia signaling.[113–116] However, recent evidence suggests that ARNT is not limiting in many cell types.[113] In heterotypic interference models, the AHR could influence the signaling activities of non-PAS proteins through direct interactions. Suggested heterotypic partners for the AHR include molecules such as Rb, NFκB, c-Src or the estrogen receptor (ER) (Table 1). It has been proposed for each of these factors that interactions could lead to upregulation or downregulation of cellular pathways involved. Finally, given the recent discovery that many bHLH-PAS proteins can act as coactivators, it is possible that AHR is a coactivator for some unknown signal transduction pathway (Table 1). That is, AHR may still be found to participate in gene expression without directly contacting DREs or without interacting with ARNT.

In summary, the molecular details of AHR mediated dioxin toxicity are yet to be elucidated. Numerous mechanisms for AHR mediated toxicity have been proposed. It may be that exposure to dioxin and subsequent toxicity is the result of some combination of these mechanisms. However, given the diverse effects of dioxin on an organism, the proven involvement of the AHR in this toxicity and its role as a DRE-dependent transcription factor, it seems most probable that dioxin toxicity can be explained by transcription of known, unrecognized, or yet to be discovered toxic genes. Active investigation coupled with the expanding knowledge of human and mouse

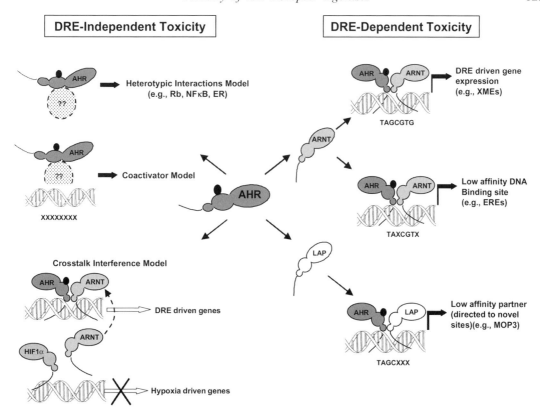

**Figure 3** DRE dependent and independent models of AHR-mediated toxicity: A number of mechanisms have been proposed to explain AHR mediated toxicity of dioxins. Ligand activated AHR is capable of interacting with regulatory elements driving a variety of genes. The gene batteries that play a role in cytotoxicity are unknown but may include DRE driven genes, such as the XME gene battery or as yet unidentified genes. AHR may also bind to low affinity DREs or to a low affinity partner (LAP) and induce expression of an alternate battery of genes that may be involved in toxicity. Alternately, several DRE independent mechanisms have been proposed and include direct heterotypic interactions of AHR with various non-PAS proteins, the cross-talk interference model and the possibility that AHR is acting as a coactivator in other signaling pathways. See text for details.

**Table 1** Interaction/interference models of dioxin toxicity. DRE-independent models of dioxin toxicity mediated by the AHR have been proposed by a number of laboratories. This table summarizes some of the current models of AHR interaction with ARNT, RB, NFκB, CSrc/kinase, ceramide and the estrogen receptor. See text for details.

| Interaction/interference model | Suggested effect | Reference |
|---|---|---|
| Retinoblastoma (Rb) | Direct interaction between AHR and the Rb protein may mediate effects of dioxin on cell cycle regulation. | 75,144,145 |
| NFκB | Direct interaction and mutual functional repression between AHR and NF-κB may provide a mechanism for the PAH-induced toxic responses. | 146,147 |
| cSrc/kinase | Dioxin can exert rapid, pleiotropic effects through the AHR-associated kinase to alter functions of many proteins through a cascade of protein phosphorylations. | 148 |
| Estrogen receptor (ER) | The antiestrogenic/estrogenic effects of dioxin may be the result of AHR directly interacting with ERα, COUP-TF, and ERRα1. | 149 |
| Ceramide model | AHR modulates aspects of ceramide signaling associated with the induction of apoptosis but not cell cycle arrest. | 150 |
| Cross-talk model | AHR and HIF1α compete for ARNT occupancy and therefore, toxicity could result from a dioxin induced decreased availability of ARNT. | 113–116 |

genomics will undoubtedly provide an answer to this question in the future.

### 14.2.3.6 PHYSIOLOGICAL ROLE OF AH RECEPTOR

Due to its role in mediating responses to environmental contaminants, AHR biology has been studied extensively from a toxicological viewpoint. However, it is hard to imagine that the AHR evolved solely as a defense against PAHs or related environmental toxicants. A phylogenetic survey indicates that the AHR arose over 450 million years ago, with functional orthologs in marine, aquatic, avian and mammalian species.[40,117–121] This observation suggests that the AHR has conferred a selective advantage throughout vertebrate evolution, in a variety of chemical environments and before environmental pollution by anthropogenic compounds. In addition, the AHR is expressed in a variety of tissues and at developmental time points that are inconsistent with a singular role as part of a metabolic defense against environmental chemicals.[122–124] Based on the above observations, we can conclude that the AHR is an important player in normal development. Although the underlying developmental mechanism is unclear, it is tempting to consider its established role as an environmental sensor for guidance. Given that the embryonic system can be viewed as a continually changing environment, the AHR may be sensing a variety of developmental cues and mediating related physiological transitions.

#### 14.2.3.6.1 Targeted Disruption of the *Ah* Locus in Mice

Evidence to support an important physiological role for the AHR has come from gene targeting experiments in the mouse, where the *Ah* null alleles have been generated.[125–128] As predicted, AHR null mice fail to show up-regulation of XMEs in response to receptor agonists and are resistant to the classical toxic endpoints of dioxin exposure.[126,129,130] Interestingly, these mouse lines commonly show defects in liver development, decreased animal weights and poor fecundity.[125,126] The decreased liver size in *Ah* null animals appears to be the result of smaller hepatocyte size, which is apparently the consequence of inadequate vascular perfusion.[131] Studies carried out to visualize the vascular architecture of the *Ah* null mouse clearly demonstrated the presence of a patent (open) ductus venosus (DV) and

decreased blood flow within liver lobules (Figure 4). The DV is a fetal vascular structure that shunts approximately 50% of the blood from the umbilical vein past the liver during fetal life. In wild-type animals, this structure closes around the time of birth, but fails to do so in the *Ah* null animals. Additional vascular abnormalities were noted in other organs (Figure 4), including the kidney and the eye, suggesting a broader role for the AHR in resolution of fetal vascular development. Thus, in addition to its adaptive role in mediating responses to environmental chemicals and its role in mediating the toxicity of dioxin, the AHR also appears to play an important physiological role in development.

#### 14.2.3.6.2 Endogenous Ligands of the AH Receptor

The concept that an endogenous ligand exists for the AHR has been a prevailing theme in this area of biological research and this idea is consistent with the conservation of the receptor throughout evolution. In recent years a number of studies have been suggestive that an endogenous ligand for the AHR exists.[132–134] However, little information exists as to the possible identity of the cellular factor(s) that activate the receptor. A number of studies have proposed candidate endogenous ligands for the AHR, yet no study has been conclusive in this regard. Several studies suggest that oxidized products of the amino acid, tryptophan, which are formed after exposure to ultraviolet light, induce AHR transformation, binding to its specific DNA recognition site and CYP1A1 induction in mammalian cells.[135–138] Other studies point to bilirubin and biliverdin, the degradation products of heme, as endogenous ligands of the AHR.[139,140] A structurally different type of endogenous ligand was suggested in the studies of lipoxin $A_4$, an acyclic, negatively charged compound formed from arachadonic acid.[141] Lipoxin $A_4$ functions as an anti-inflammatory agent to dampen neutrophil recruitment to a site of inflammation. Although it is tempting to speculate that an endogenous ligand for the AHR exists, we must also accept the formal possibility that the receptor may be constitutively active at a particular time point or in particular tissues during the developmental process.

### 14.2.3.7 SUMMARY

This chapter has endeavored to illustrate the AHR as a model of a PAS superfamily member

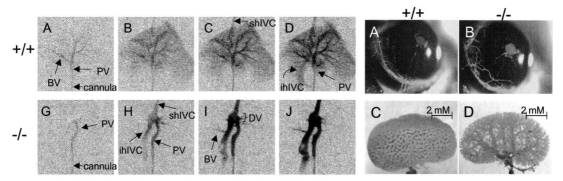

**Figure 4** Vascular defects in *Ah* null organs. Left: Patent ductus venosus (DV) and decreased liver perfusion in *Ah* null mice. Continuous X-ray images were obtained over approximately 10 s as contrast was injected into the portal vein of *Ah* +/+ (A–D) and −/− (G–J) mice. Serial radiographs are from left to right, portal vein (PV), infrahepatic inferior vena cava (ihIVC), suprahepatic inferior vena cava (shIVC), and branching vessels (BV). Right. *Ah* −/− mice have expansion of limbal vessels and altered kidney vascular structure. A and B: Limbal vessels were photographed after mice were injected with fluorescein isothiocyanate-dextran and eyes were epi-illuminated at 488nm. C and D: Latex corrosion casts of the kidney vasculature showing the difference in the extent of vascularization of the *Ah* −/− compared to the *Ah* +/+ mice.

that is involved in sensing environmental change and mediating a genetic response to those environmental alterations. The interaction between genes and the environment is an important consideration in toxicology since many responses to toxicants are likely to involve changes in gene expression. We began with a discussion of the adaptive response to PAHs that included the upregulation of XMEs. We then went on to discuss the involvement of the AHR in the toxic response associated with dioxin exposure. While the mechanism of the adaptive response pathway is well accepted, our understanding of AHR's role in the toxic response pathway is less certain. Finally, we framed the developing fetus as a model of environmental change and used that model to illustrate an important endogenous role for the AHR. The combined biological and toxicological investigations of the AHR will continue to shed light on the role of the receptor as a mediator of environmental change.

## 14.2.3.8 ACKNOWLEDGMENTS

This work was supported by The National Institutes of Health (Grants R01-ES05703 and P30-CA07175) and a Post-doctoral Fellowship from the Natural Sciences and Engineering Research Council of Canada.

## 14.2.3.9 REFERENCES

1. S. A. Skene, I. C. Dewhurst and M. Greenberg, 'Polychlorinated dibenzo-*p*-dioxins and polychlorinated dibenzofurans: The risks to human health. A review.' *Human Toxicol.*, 1989, **8**, 173–203.

2. A. H. Conney, E. C. Miller and J. A. Miller, 'The metabolism of methylated aminoazo dyes. V. Evidence for induction of enzyme synthesis in the rat by 3-methylcholanthrene.' *Cancer Res.*, 1956, **16**, 450–459.

3. J. P. Whitlock, Jr., 'Induction of cytochrome P4501A1.' *Ann. Rev. Pharm. Toxicol.*, 1999, **39**, 103–125.

4. L. C. Quattrochi, T. Vu and R. H. Tukey, 'The human CYP1A2 gene and induction by 3-methylcholanthrene.' *J. Biol. Chem.*, 1994, **269**, 6949–6954.

5. W. Li, P. A. Harper, B. K. Tang *et al.*, 'Regulation of cytochrome P450 enzymes by aryl hydrocarbon receptor in human cells: CYP1A2 expression in the LS180 colon carcinoma cell line after treatment with 2,3,7,8-tetrachlorodibenzo-*p*-dioxin or 3-methylcholanthrene.' *Biochem. Pharm.*, 1998, **56**, 599–612.

6. D. W. Nebert and H. V. Gelboin, 'The *in vivo* and *in vitro* induction of aryl hydrocarbon hydroxylase in mammalian cells of different species, tissues, strains, and development and hormonal states.' *Arch. Biochem. Biophys.*, 1969, **134**, 76–89.

7. J. E. Gielen, F. M. Goujon and D. W. Nebert, 'Genetic regulation of aryl hydrocarbon hydroxylase induction.' *J. Biol. Chem.*, 1972, **247**, 1125–1137.

8. D. W. Nebert, F.M. Goujon and J.E. Gielen, 'Aryl hydrocarbon hydroxylase induction by polycyclic hydrocarbons: Simple autosomal dominant trait in the mouse.' *Nat. New Biol.*, 1972, **236**, 107.

9. A. Poland and E. Glover, 'Genetic expression of aryl hydrocarbon hydroxylase by 2,3,7,8-tetrachlorodibenzo-*p*-dioxin: evidence for a receptor mutation in genetically non-responsive mice.' *Mol. Pharmacol.*, 1975, **11**, 389–398.

10. A. Poland and E. Glover, 'Comparison of 2,3,7,8-tetrachlorodibenzo-*p*-dioxin, a potent inducer of aryl hydrocarbon hydroxylase, with 3-methylcholanthrene.' *Mol. Pharm.*, 1974, **10**, 349–359.

11. A. Poland and E. Glover, 'An estimate of the maximum *in vivo* covalent binding of 2,3,7,8-tetrachlorodibenzo-*p*-dioxin to rat liver protein, ribosomal RNA, and DNA.' *Cancer Res.*, 1979, **39**, 3341–3344.

12. A. Poland, W. F. Greenlee and A. S. Kende, 'Studies on the mechanism of action of the chlorinated dibenzo-*p*-dioxins and related compounds.' *Ann. N. Y. Acad. Sci.*, 1979, **320**, 214–230.

13. A. Poland, E. Glover, F. H. Ebetino *et al.*, 'Photoaffinity labeling of the Ah receptor.' *J. Biol. Chem.*, 1986, **261**, 6352–6365.

14. A. Poland, P. Teitelbaum and E. Glover, '[125I]2-iodo-3,7,8-trichlorodibenzo-*p*-dioxin-binding species in mouse liver induced by agonists for the Ah receptor: characterization and identification.' *Mol. Pharmacol.*, 1989, **36**, 113–120.

15. C. A. Bradfield, A. S. Kende and A. Poland, 'Kinetic and equilibrium studies of Ah receptor-ligand binding: use of [125I]2-iodo-7,8-dibromodibenzo-*p*-dioxin.' *Mol. Pharm.*, 1988, **34**, 229–237.

16. L. Poellinger, R. N. Kurl, J. Lund *et al.*, 'High-affinity binding of 2,3,7,8-tetrachlorodibenzo-*p*-dioxin in cell nuclei from rat liver.' *Biochem. Biophys. Acta*, 1982, **714**, 516–523.

17. A. Poland, E. Glover and A. S. Kende, 'Stereospecific, high affinity binding of 2,3,7,8-tetrachlorodibenzo-*p*-dioxin by hepatic cytosol.' *J. Biol. Chem.*, 1976, **251**, 4936–4946.

18. T. A. Gasiewicz, W. Ness and C. G. Rucci, 'Ontogeny of the cytosolic receptor for 2,3,7,8-tetrachlorodibenzo-*p*-dioxin in rat liver, lung, and thymus.' *Biochem. Biophys. Res. Comm.*, 1984, **118**, 183–190.

19. J. A. Goldstein, 'The structure-activity relationship of halogenated biphenyls as enzyme inducers.' *Ann. NY Acad. Sci.*, 1979, **320**, 164–171.

20. M. A. Denomme, K. Homonoko, T. Fujita *et al.*, 'Effects of substituents on the cytosolic receptor-binding avidities and aryl hydrocarbon hydroxylase induction potencies of 7-substituted 2,3-dichlorodibenzo-*p*-dioxins. A quantitative structure-activity relationship analysis.' *Mol. Pharm.*, 1985, **27**, 656–661.

21. A. S. Kende, F. H. Ebetino, W. B. Drendel *et al.*, 'Structure-activity relationship of bispyridyloxybenzene for induction of mouse hepatic aminopyrine *N*-demethylase activity. Chemical, biological, and X-ray crystallographic studies.' *Mol. Pharm.*, 1985, **28**, 445–453.

22. A. Poland, D. Palen and E. Glover, 'Analysis of the four alleles of the murine aryl hydrocarbon receptor.' *Mol. Pharmacol.*, 1994, **46**, 915–921.

23. P. E. Thomas, R. E. Kouri and J. J. Hutton, 'The genetics of aryl hydrocarbon hydroxylase induction in mice: A single gene difference between C57BL/6J and DBA/2J.' *Biochem. Genetics*, 1972, **6**, 157–168.

24. P. E. Thomas and J. J. Hutton, 'Genetics of aryl hydrocarbon hydroxylase induction in mice: Additive inheritance in crosses between C3H/HeJ and DBA/2J.' *Biochem. Genet.*, 1973, **8**, 249–257.

25. W. F. Greenlee and A. Poland, 'An improved assay of 7-ethoxycoumarin O-deethylase activity: Induction of hepatic enzyme activity in C57BL/6J and DBA/2J mice by phenobarbital, 3-mthylcholanthrene and 2,3,7,8-tetrachlorodibenzo-*p*-dioxin.' *J. Pharmacol. Exp. Therap.*, 1978, **205**, 596–605.

26. S. W. Bigelow and D. W. Nebert, 'The murine aromatic hydrocarbon responsiveness locus: a comparison of receptor levels and several inducible enzyme activities among recombinant inbred lines.' *J. Biochem. Toxicol.*, 1986, **1**, 1–14.

27. M. Ema, N. Ohe, M. Suzuki *et al.*, 'Dioxin binding activities of polymorphic forms of mouse and human aryl hydrocarbon receptors.' *J. Biol. Chem.*, 1994, **269**, 27337–27343.

28. P. A. Harper, C. L. Golas and A. B. Okey, 'Ah receptor in mice genetically "nonresponsive" for cytochrome P4501A1 induction: cytosolic Ah receptor, transformation to the nuclear binding state, and induction of aryl hydrocarbon hydroxylase by halogenated and nonhalogenated aromatic hydrocarbons in embryonic tissues and cells.' *Mol. Pharm.*, 1991, **40**, 818–826.

29. A. B. Okey, L. M. Vella and P. A. Harper, 'Detection and characterization of a low affinity form of cytosolic Ah receptor in livers of mice nonresponsive to induction of cytochrome P1–450 by 3-methylcholanthrene.' *Mol. Pharmacol.*, 1989, **35**, 823–830.

30. C. A. Bradfield, E. Glover and A. Poland, 'Purification and *N*-terminal amino acid sequence of the Ah receptor from the C57BL/6J mouse.' *Mol. Pharmacol.*, 1991, **39**, 13–19.

31. G. H. Perdew and A. Poland, 'Purification of the Ah receptor from C57BL/6J mouse liver.' *J. Biol. Chem.*, 1988, **263**, 9848–9852.

32. M. Denis, S. Cuthill, A. C. Wikstrom *et al.*, 'Association of the dioxin receptor with the Mr 90,000 heat shock protein: A structural kinship with the glucocorticoid receptor.' *Biochem. Biophys. Res. Comm.*, 1988, **155**, 801–807.

33. T. A. Gasiewicz, C. J. Elferink and E. C. Henry, 'Characterization of multiple forms of the Ah receptor: Recognition of a dioxin-responsive enhancer involves heteromer formation.' *Biochemistry*, 1991, **30**, 2909–2916.

34. R. D. Prokipcak, M. S. Denison and A. B. Okey, 'Nuclear Ah receptor from mouse hepatoma cells: effect of partial proteolysis on relative molecular mass and DNA-binding properties.' *Arch. Biochem. Biophys.*, 1990, **283**, 476–483.

35. A. Wilhelmsson, S. Cuthill, M. Denis *et al.*, 'The specific DNA binding activity of the dioxin receptor is modulated by the 90 kd heat shock protein.' *EMBO J.*, 1990, **9**, 69–76.

36. R. S. Pollenz, C. A. Sattler and A. Poland, 'The aryl hydrocarbon receptor and aryl hydrocarbon receptor nuclear translocator protein show distinct subcellular localizations in Hepa 1c1c7 cells by immunofluorescence microscopy.' *Mol. Pharmacol.*, 1994, **45**, 428–438.

37. R. H. Tukey, R. R. Hannah, M. Negishi *et al.*, 'The Ah locus: Correlation of intranuclear appearance of inducer-receptor complex with induction of cytochrome $P_1$-450 mRNA.' *Cell*, 1982, **31**, 275–284.

38. K. M. Burbach, A. Poland and C. A. Bradfield, 'Cloning of the Ah-receptor cDNA reveals a distinctive ligand-activated transcription factor.' *Proc. Natl. Acad. Sci. USA*, 1992, **89**, 8185–8189.

39. M. Ema, K. Sogawa, N. Watanabe *et al.*, 'cDNA cloning and structure of mouse putative Ah receptor.' *Biochem. Biophys. Res. Comm.*, 1992, **184**, 246–253.

40. E. C. Hoffman, H. Reyes, F. F. Chu *et al.*, 'Cloning of a factor required for activity of the Ah (dioxin) receptor.' *Science*, 1991, **252**, 954–958.

41. H. Reyes, S. Reisz-Porszasz and O. Hankinson, 'Identification of the Ah receptor nuclear translocator protein (Arnt) as a component of the DNA binding form of the Ah receptor.' *Science*, 1992, **256**, 1193–1195.

42. L. A. Carver, J. J. LaPres, S. Jain *et al.*, 'Characterization of the Ah receptor-associated Protein, ARA9.' *J. Biol. Chem.*, 1998, **273**, 33580–36159.

43. Q. Ma and J. P. Whitlock, Jr., 'A novel cytoplasmic protein that interacts with the Ah receptor, contains tetratricopeptide repeat motifs, and augments the transcriptional response to 2,3,7,8-tetrachlorodibenzo-*p*-dioxin.' *J. Biol. Chem.*, 1997, **272**, 8878–8884.

44. B. K. Meyer, M. G. Pray-Grant, J. P. Vanden Heuvel *et al.*, 'Hepatitis B virus X-associated protein 2 is a subunit of the unliganded aryl hydrocarbon receptor core complex and exhibits transcriptional enhancer activity.' *Mol. Cell. Biol.*, 1998, **18**, 978–988.

45. A. Kazlauskas, L. Poellinger and I. Pongratz, 'Evidence that the co-chaperone p23 regulates ligand responsiveness of the dioxin (Aryl hydrocarbon) receptor.' *J. Biol. Chem.*, 1999, **274**, 13519–13524.

46. C. Antonsson, M. L. Whitelaw, J. McGuire *et al.*, 'Distinct roles of the molecular chaperone hsp90 in

modulating dioxin receptor function via the basic helix-loop-helix and PAS domains.' *Mol. Cell. Biol.*, 1995, **15**, 756–765.

47. I. Pongratz, G. G. Mason and L. Poellinger, 'Dual roles of the 90-kDa heat shock protein hsp90 in modulating functional activities of the dioxin receptor.' *J. Biol. Chem.*, 1992, **267**, 13728–13734.

48. L. A. Carver, V. Jackiw and C. A. Bradfield, 'The 90-kDa heat shock protein is essential for Ah receptor signaling in a yeast expression system.' *J. Biol. Chem.*, 1994, **269**, 30109–30112.

49. M. L. Whitelaw, J. McGuire, D. Picard *et al.*, 'Heat shock protein hsp90 regulates dioxin receptor function *in vivo*.' *Proc. Natl. Acad. Sci. USA*, 1995, **92**, 4437–4441.

50. T. Ikuta, H. Eguchi, T. Tachibana *et al.*, 'Nuclear localization and export signals of the human aryl hydrocarbon receptor.' *J. Biol. Chem.*, 1998, **273**, 2895–2904.

51. N. G. Hord and G. H. Perdew, 'Physicochemical and immunocytochemical analysis of the aryl hydrocarbon receptor nuclear translocator: characterization of two monoclonal antibodies to the aryl hydrocarbon receptor nuclear translocator.' *Mol. Pharm.*, 1994, **46**, 618–626.

52. H. Eguchi, T. Ikuta, T. Tachibana *et al.*, 'A nuclear localization signal of human aryl hydrocarbon receptor nuclear translocator/hypoxia-inducible factor 1β is a novel bipartite type recognized by the two components of nuclear pore-targeting complex.' *J. Biol. Chem.*, 1997, **272**, 17640–17647.

53. M. R. Probst, S. Reisz-Porszasz, R. V. Agbunag *et al.*, 'Role of the aryl hydrocarbon receptor nuclear translocator protein in aryl hydrocarbon (dioxin) receptor action.' *Mol. Pharmacol.*, 1993, **44**, 511–518.

54. C. J. Elferink and J. P. Whitlock, Jr., 'Dioxin-dependent, DNA sequence-specific binding of a multiprotein complex containing the Ah receptor.' *Receptor*, 1994, **4**, 157–173.

55. K. M. Dolwick, H. I. Swanson and C. A. Bradfield, '*In vitro* analysis of Ah receptor domains involved in ligand-activated DNA recognition.' *Proc. Natl. Acad. Sci. USA*, 1993, **90**, 8566–8570.

56. K. E. McLane and J. P. Whitlock, Jr., 'DNA sequence requirements for Ah receptor/Arnt recognition determined by *in vitro* transcription.' *Receptor*, 1994, **4**, 209–222.

57. E. S. Shen and J. P. Whitlock, Jr., 'Protein-DNA interactions at a dioxin-responsive enhancer. Mutational analysis of the DNA-binding site for the liganded Ah receptor.' *J. Biol. Chem.*, 1992, **267**, 6815–6819.

58. A. Lusska, E. Shen and J. P. Whitlock, Jr., 'Protein-DNA interactions at a dioxin-responsive enhancer. Analysis of six bona fide DNA-binding sites for the liganded Ah receptor.' *J. Biol. Chem.*, 1993, **268**, 6575–6580.

59. N. Matsushita, K. Sogawa, M. Ema *et al.*, 'A factor binding to the xenobiotic responsive element (XRE) of P-4501A1 gene consists of at least two helix-loop-helix proteins, Ah receptor and Arnt.' *J. Biol. Chem.*, 1993, **28**, 21002–21006.

60. G. G. Mason, A. M. Witte, M. L. Whitelaw *et al.*, 'Purification of the DNA binding form of dioxin receptor. Role of the Arnt cofactor in regulation of dioxin receptor function.' *J. Biol. Chem.*, 1994, **269**, 4438–4449.

61. H. I. Swanson, K. Tullis and M. S. Denison, 'Binding of transformed Ah receptor complex to a dioxin responsive transcriptional enhancer: Evidence for two distinct heteromeric DNA-binding forms.' *Biochem.*, 1993, **32**, 12841–12849.

62. H. I. Swanson, W. K. Chan and C. A. Bradfield, 'DNA binding specificities and pairing rules of the Ah receptor, ARNT, and SIM proteins.' *J. Biol. Chem.*, 1995, **270**, 26292–26302.

63. S. G. Bacsi, S. Reisz-Porszasz and O. Hankinson, 'Orientation of the heterodimeric aryl hydrocarbon (dioxin) receptor complex on its asymmetric DNA recognition sequence.' *Mol. Pharmacol.*, 1995, **47**, 432–438.

64. K. Sogawa, R. Nakano, A. Kobayashi *et al.*, 'Possible function of Ah receptor nuclear translocator (Arnt) homodimer in transcriptional regulation.' *Proc. Natl. Acad. Sci. USA*, 1995, **92**, 1936–1940.

65. S. Jain, K. M. Dolwick, J. V. Schmidt *et al.*, 'Potent transactivation domains of the Ah receptor and the Ah receptor nuclear translocator map to their carboxyl termini.' *J. Biol. Chem.*, 1994, **269**, 31518–31524.

66. H. Li, L. Dong and J. P. Whitlock, Jr., 'Transcriptional activation function of the mouse Ah receptor nuclear translocator.' *J. Biol. Chem.*, 1994, **269**, 28098–28105.

67. Q. Ma, L. Dong and J. P. Whitlock, Jr., 'Transcriptional activation by the mouse Ah receptor. Interplay between multiple stimulatory and inhibitory functions.' *J. Biol. Chem.*, 1995, **270**, 12697–12703.

68. M. L. Whitelaw, J. A. Gustafsson and L. Poellinger, 'Identification of transactivation and repression functions of the dioxin receptor and its basic helix-loop-helix/PAS partner factor Arnt: Inducible versus constitutive modes of regulation.' *Mol. Cell. Biol.*, 1994, **14**, 8343–8355.

69. B. N. Fukunaga, M. R. Probst, S. Reisz-Porszasz *et al.*, 'Identification of functional domains of the aryl hydrocarbon receptor.' *J. Biol. Chem.*, 1995, **270**, 29270–29278.

70. H. P. Ko, S. T. Okino, Q. Ma *et al.*, 'Dioxin-induced CYP1A1 transcription *in vivo*: the aromatic hydrocarbon receptor mediates transactivation, enhancer-promoter communication, and changes in chromatin structure.' *Mol. Cell. Biol.*, 1996, **16**, 430–436.

71. L. Wu and J. P. Whitlock, Jr., 'Mechanism of dioxin action: Ah receptor-mediated increase in promoter accessibility *in vivo*.' *Proc. Natl. Acad. Sci. USA*, 1992, **89**, 4811–4815.

72. S. T. Okino and J. P. Whitlock, Jr., 'Dioxin induces localized, graded changes in chromatin structure: implications for Cyp1A1 gene transcription.' *Mol. Cell. Biol.*, 1995, **15**, 3714–3721.

73. M. B. Kumar, R. W. Tarpey and G. H. Perdew, 'Differential recruitment of coactivator RIP140 by Ah and estrogen receptors. Absence of a role for LXXLL motifs.' *J. Biol. Chem.*, 1999, **274**, 22155–22164.

74. M. Kumar and G. Perdew, 'Nuclear receptor coactivator, SRC-1 interacts with Q-rich subdomain of the AHR and modulates its transactivation potential.' *The Toxicologist*, 2000, **54**, 985.

75. C. J. Elferink, N.-L. Ge and A. Levine, 'Maximal aryl hydrocarbon receptor activity depends on an interaction with the retinoblastoma protein.' *Mol. Pharmacol.*, 2001, **59**, 664–673.

76. K. W. Jones and J. P. Whitlock, Jr., 'Functional analysis of the transcriptional promoter for the CYP1A1 gene.' *Mol. Cell. Biol.*, 1990, **10**, 5098–5105.

77. M. S. Denison, J. M. Fisher and J. P. Whitlock, Jr., 'Inducible, receptor-dependent protein-DNA interactions at a dioxin-responsive transcriptional enhancer.' *Proc. Natl. Acad. Sci. USA*, 1988, **85**, 2528–2532.

78. T. H. Rushmore, R. G. King, K. E. Paulson *et al.*, 'Regulation of glutathione S-transferase Ya subunit gene expression: identification of a unique xenobiotic-responsive element controlling inducible expression

by planer aromatic compounds.' *Proc. Natl. Acad. Sci.*, 1990, **87**, 3926–3830.

79. L. V. Favreau and C. B. Pickett, 'Transcriptional regulation of the rat NAD(P)H:quinone reductase gene. Identification of regulatory elements controlling basal level expression and inducible expression by planar aromatic compounds and phenolic anitoxidants.' *J. Biol. Chem.*, 1991, **266**, 4556–4561.

80. T. Iyanagi, M. Haniu, K. Sogawa *et al.*, 'Cloning and characterization of cDNA encoding 3-methylcholanthrene inducible rat mRNA for UDP-glucuronosyltransferase.' *J. Biol. Chem.*, 1986, **261**, 15607–15614.

81. F. J. Gonzalez, R. H. Tukey and D. W. Nebert, 'Structural gene products of the Ah locus. Transcriptional regulation of cytochrome P1-450 and P3-450 mRNA levels by 3-methylcholanthrene.' *Mol. Pharmacol.*, 1984, **26**, 117–121.

82. J. Mimura, M. Ema, K. Sogawa *et al.*, 'Identification of a novel mechanism of regulation of Ah (dioxin) receptor function.' *Genes Dev.*, 1999, **13**, 20–25.

83. N. A. Davarinos and R. S. Pollenz, 'Aryl hydrocarbon receptor imported into the nucleus following ligand binding is rapidly degraded via the cytosplasmic proteasome following nuclear export.' *J. Biol. Chem.*, 1999, **274**, 28708–28715.

84. M. J. Lees and W. M. L., 'Multiple roles of ligand in transforming the dioxin receptor to an active basic helix-loop-helix/PAS transcription factor complex with the nuclear protein Arnt.' *Mol. Cell. Biol.*, 1999, **19**, 5811–5822.

85. H. I. Swanson and G. H. Perdew, 'Half-life of aryl hydrocarbon receptor in Hepa 1 cells: evidence for ligand-dependent alterations in cytosolic receptor levels.' *Arch. Biochem. Biophys.*, 1993, **302**, 167–174.

86. B. L. Roman, R. S. Pollenz and R. E. Peterson, 'Responsiveness of the adult male rat reproductive tract to 2,3,7,8-tetrachlorodibenzo-*p*-dioxin exposure: Ah receptor and ARNT expression, CYP1A1 induction, and Ah receptor down-regulation.' *Toxicol. Appl. Pharm.*, 1998, **150**, 228–239.

87. B. J. Roberts and M. L. Whitelaw, 'Degradation of the basic helix-loop-helix/Per-ARNT-Sim homology domain dioxin receptor via the ubiquitin/proteasome pathway.' *J. Biol. Chem.*, 1999, **274**, 36351–36356.

88. Q. Ma, A. J. Renzelli, K. T. Baldwin *et al.*, 'Superinduction of CYP1A1 gene expression. Regulation of 2,3,7, 8-tetrachlorodibenzo-*p*-dioxin-induced degradation of Ah receptor by cycloheximide.' *J. Biol. Chem.*, 2000, **275**, 12676–12683.

89. A. K. Liem, P. Furst and C. Rappe, 'Exposure of populations to dioxins and related compounds.' *Food Additives Contam.*, 2000, **17**, 241–259.

90. D. E. Lilienfeld and M. A. Gallo, '2,4-D, 2,4,5-T, and 2,3,7,8-TCDD: an overview.' *Epidem. Rev.*, 1989, **11**, 28–58.

91. C. L. Wilson and S. Safe, 'Mechanisms of ligand-induced aryl hydrocarbon receptor-mediated biochemical and toxic responses [see comments].' *Toxicol. Path.*, 1998, **26**, 657–671.

92. A. Poland and J. C. Knutson, '2,3,7,8-tetrachlorodibenzo-*p*-dioxin and related halogenated aromatic hydrocarbons: examination of the mechanism of toxicity.' *Ann. Rev. Pharmacol. Toxicol.*, 1982, **22**, 517–554.

93. J. V. Schmidt and C. A. Bradfield, 'Ah receptor signaling pathways.' *Annu. Rev. Cell Dev. Biol.*, 1996, **12**, 55–89.

94. A. Poland and E. Glover, '2,3,7,8-Tetrachlorodibenzo-*p*-dioxin: Segregation of toxicity with the Ah locus.' *Mol. Pharm.*, 1980, **17**, 86–94.

95. D. W. Nebert, D. D. Petersen and A. J. Fornace, Jr., 'Cellular responses to oxidative stress: the [Ah] gene battery as a paradigm.' *Environ. Health Perspect.*, 1990, **88**, 13–25.

96. E. A. Hassoun, F. Li, A. Abushaban *et al.*, 'The relative abilities of TCDD and its congeners to induce oxidative stress in the hepatic and brain tissues of rats after subchronic exposure.' *Toxicology*, 2000, **145**, 103–113.

97. N. Z. Alsharif, T. Lawson and S. J. Stohs, 'Oxidative stress induced by 2,3,7,8-tetrachlorodibenzo-*p*-dioxin is mediated by the aryl hydrocarbon (Ah) receptor complex.' *Toxicology*, 1994, **92**, 39–51.

98. H. G. Shertzer, D. W. Nebert, A. Puga *et al.*, 'Dioxin causes a sustained oxidative stress response in the mouse.' *Biochem. Biophys. Res. Comm.*, 1998, **253**, 44–48.

99. B. V. Madhukar, K. Ebner, F. Matsumura *et al.*, '2,3,7,8-Tetrachlorodibenzo-*p*-dioxin causes an increase in protein kinases associated with epidermal growth factor receptor in the hepatic plasma membrane.' *J. Biochem. Toxicol.*, 1988, **3**, 261–277.

100. L. Gao, L. Dong and J. P. Whitlock, Jr., 'A novel response to dioxin. Induction of ecto-ATPase gene expression.' *J. Biol. Chem.*, 1998, **273**, 15358–15365.

101. L. Dong, Q. Ma and J. P. Whitlock Jr., 'Down-regulation of major histocompatibility complex Q1b gene expression by 2,3,7,8-tetrachlorodibenzo-*p*-dioxin.' *J. Biol. Chem.*, 1997, **272**, 29614–29619.

102. T. R. Sutter, K. Guzman, K. M. Dold *et al.*, 'Targets for dioxin: Genes for plasminogen activator inhibitor-2 and interleukin-1β.' *Science*, 1991, **254**, 415–418.

103. A. Poland and E. Glover, 'Chlorinated dibenzo-*p*-dioxins: Potent inducers of δ-aminolevulinic acid synthetase and aryl hydrocarbon hydroxylase.' *Mol. Pharmacol.*, 1973, **9**, 736–747.

104. M. S. Jeon and C. Esser, 'The murine IL-2 promoter contains distal regulatory elements responsive to the Ah receptor, a member of the evolutionarily conserved bHLH-PAS transcription factor family.' *J. Immunol.*, 2000, **165**, 6975–6983.

105. A. Puga, A. Hoffer, S. Zhou *et al.*, 'Sustained increase in intracellular free calcium and activation of cyclooxygenase-2 expression in mouse hepatoma cells treated with dioxin.' *Biochem. Pharmacol.*, 1997, **54**, 1287–1296.

106. C. M. Klinge, J. L. Bowers, P. C. Kulakosky *et al.*, 'The aryl hydrocarbon receptor (AHR)/AHR nuclear translocator (ARNT) heterodimer interacts with naturally occurring estrogen response elements.' *Mol. Cell. Endocrin.*, 1999, **157**, 105–119.

107. K. Hirose, M. Morita, M. Ema *et al.*, 'cDna cloning and tissue-specific expression of a novel basic helix-loop-helix/Pas factor (arnt2) with close sequence similarity to the aryl hydrocarbon receptor nuclear translocator (arnt).' *Mol. Cell. Biol.*, 1996, **16**, 1706–1713.

108. G. Drutel, M. Kathmann, A. Heron *et al.*, 'Cloning and selective expression in brain and kidney of ARNT2 homologous to the Ah receptor nuclear translocator (ARNT).' *Biochem. Biophys. Res. Comm.*, 1996, **225**, 333–339.

109. M. Ikeda and M. Nomura, 'cDNA cloning and tissue-specific expression of a novel basic helix-loop-helix/PAS protein (BMAL1) and identification of alternatively spliced variants with alternative translation initiation site usage.' *Biochem. Biophys. Res. Comm.*, 1997, **233**, 258–264.

110. J. B. Hogenesch, W. K. Chan, V. H. Jackiw *et al.*, 'Characterization of a subset of the basic-helix-loop-helix-PAS superfamily that interacts with components of the dioxin signaling pathway.' *J. Biol. Chem.*, 1997, **272**, 8581–8593.

111. S. Takahata, K. Sogawa, A. Kobayashi *et al.*, 'Transcriptionally active heterodimer formation of

an Arnt-like PAS protein, Arnt3, with HIF-1a, HLF, and clock.' *Biochem. Biophys. Res. Comm.*, 1998, **248**, 789–794.

112. M. Ikeda, W. Yu, M. Hirai *et al.*, 'cDNA Cloning of a Novel bHLH-PAS trascription factor superfamily gene, BMAL2: its mRNA expression, subcellular distribution, and chromosomal localization.' *Biochem. Biophys. Res. Comm.*, 2000, **275**, 493–502.

113. R. S. Pollenz, N. A. Davarinos and T. P. Shearer, 'Analysis of aryl hydrocarbon receptor-mediated signaling during physiological hypoxia reveals lack of competition for the aryl hydrocarbon nuclear translocator transcription factor.' *Mol. Pharm.*, 1999, **56**, 1127–1137.

114. W. K. Chan, G. Yao, Y. Z. Gu *et al.*, 'Cross-talk between the aryl hydrocarbon receptor and hypoxia inducible factor signaling pathways. Demonstration of competition and compensation.' *J. Biol. Chem.*, 1999, **274**, 12115–12123.

115. K. Gradin, J. McGuire, R. H. Wenger *et al.*, 'Functional interference between hypoxia and dioxin signal transduction pathways: competition for recruitment of the Arnt transcription factor.' *Mol. Cell. Biol.*, 1996, **16**, 5221–5231.

116. J. E. Kim and Y. Y. Sheen, 'Inhibition of 2,3,7,8-tetrachlorodibenzo-*p*-dioxin (TCDD)-stimulated Cyp1a1 promoter activity by hypoxic agents.' *Biochem. Pharm.*, 2000, **59**, 1549–1556.

117. M. E. Hahn, A. Poland, E. Glover *et al.*, 'Photoaffinity labeling of the Ah receptor: phylogenetic survey of diverse vertebrate and invertebrate species.' *Arch. Biochem. Biophys.*, 1994, **310**, 218–228.

118. K. M. Burbach, A. Poland and C. A. Bradfield, 'Cloning of the Ah receptor cDNA reveals a distinctive ligand-activated transcription factor.' *Proc. Natl. Acad. Sci. USA*, 1992, **89**, 8185–8189.

119. M. E. Hahn, S. I. Karchner, M. A. Shapiro *et al.*, 'Molecular evolution of two vertebrate aryl hydrocarbon (dioxin) receptors (AHR1 and AHR2) and the PAS family.' *Proc. Natl. Acad. Sci. USA*, 1997, **94**, 13743–13748.

120. R. L. Tanguay, C. C. Abnet, W. Heideman *et al.*, 'Cloning and characterization of the zebrafish (Danio rerio) aryl hydrocarbon receptor.' *Biochimica et Biophysica Acta*, 1999, **1444**, 35–48.

121. R. S. Pollenz, H. R. Sullivan, J. Holmes *et al.*, 'Isolation and expression of cDNAs from rainbow trout (Oncorhynchus mykiss) that encode two novel basic helix-loop-Helix/PER-ARNT-SIM (bHLH/PAS) proteins with distinct functions in the presence of the aryl hydrocarbon receptor. Evidence for alternative mRNA splicing and dominant negative activity in the bHLH/PAS family.' *J. Biol. Chem.*, 1996, **271**, 30886–30896.

122. B. D. Abbott, M. R. Probst, G. H. Perdew *et al.*, 'AH receptor, ARNT, glucocorticoid receptor, EGF receptor, EGF, TGF alpha, TGF beta 1, TGF beta 2, and TGF beta 3 expression in human embryonic palate, and effects of 2,3,7,8-tetrachlorodibenzo-*p*-dioxin (TCDD).' *Teratology*, 1998, **58**, 30–43.

123. S. Jain and C. A. Bradfield, 'Developmental profiles of basic helix-loop-helix PAS proteins in mice.' *Mech. Dev.*, 1998, **73**, 117–123.

124. A. Kuchenhoff, G. Seliger, T. Klonisch *et al.*, 'Arylhydrocarbon receptor expression in the human endometrium.' *Fertility Sterility*, 1999, **71**, 354–360.

125. P. Fernandez-Salguero, T. Pineau, D. M. Hilbert *et al.*, 'Immune system impairment and hepatic fibrosis in mice lacking the dioxin-binding Ah receptor.' *Science*, 1995, **268**, 722–726.

126. J. V. Schmidt, G. H. Su, J. K. Reddy *et al.*, 'Characterization of a murine Ahr null allele:

127. involvement of the Ah receptor in hepatic growth and development.' *Proc. Natl. Acad. Sci. USA*, 1996, **93**, 6731–6736.

127. J. Mimura, K. Yamashita, K. Nakamura *et al.*, 'Loss of teratogenic response to 2,3,7,8-tetrachlorodibenzo-*p*-dioxin (TCDD) in mice lacking the Ah (dioxin) receptor.' *Genes to Cells*, 1997, **2**, 645–654.

128. G. P. Lahvis and C. A. Bradfield, 'Ahr null alleles: distinctive or different?' *Biochem. Pharmacol.*, 1998, **56**, 781–787.

129. F. J. Gonzalez, P. Fernandez-Salguero, S. S. Lee *et al.*, 'Xenobiotic receptor knockout mice.' *Toxicol. Lett.*, 1995, **82–83**, 117–121.

130. P. M. Fernandez-Salguero, D. M. Hilbert, S. Rudikoff *et al.*, 'Aryl-hydrocarbon receptor-deficient mice are resistant to 2,3,7,8-tetrachlorodibenzo-*p*-dioxin-induced toxicity.' *Toxicol. Appl. Pharmacol.*, 1996, **140**, 173–179.

131. G. Lahvis, S. Lindell, R. Thomas *et al.*, 'Portosystemic shunts and persistent fetal vascular structures in Ah-receptor deficient mice.' *Proc. Natl. Acad. Sci. USA*, 2000, **97**, 10442–10447.

132. C. M. Sadek and B. L. Allen-Hoffmann, 'Cytochrome P450IA1 is rapidly induced in normal human keratinocytes in the absence of xenobiotics.' *J. Biol. Chem.*, 1994, **269**, 16067–16074.

133. C. Y. Chang and A. Puga, 'Constitutive activation of the aromatic hydrocarbon receptor.' *Mol. Cell. Biol.*, 1998, **18**, 525–535.

134. J. J. Willey, B. R. Stripp, R. B. Baggs *et al.*, 'Aryl hydrocarbon receptor activation in genital tubercle, palate, and other embryonic tissues in 2,3,7,8-tetrachlorodibenzo-*p*-dioxin-responsive lacZ mice.' *Toxicol. Appl. Pharmacol.*, 1998, **151**, 33–44.

135. R. K. Sindhu, S. Reisz-Porszasz, O. Hankinson *et al.*, 'Induction of cytochrome P450IA1 by photooxidized tryptophan in Hepa 1c1c7 cells.' *Biochem. Pharmacol.*, 1996, **52**, 1883–1893.

136. A. Rannug, U. Rannug, H. S. Rosenkranz *et al.*, 'Certain photooxidized derivatives of tryptophan bind with very high affinity to the Ah receptor and are likely to be endogenous signal substances.' *J. Biol. Chem.*, 1987, **262**, 15422–15427.

137. W. G. Helferich and M. S. Denison, 'Ultraviolet photoproducts of tryptophan can act as dioxin agonists.' *Mol. Pharmacol.*, 1991, **40**, 674–678.

138. Y. D. Wei, U. Rannug and A. Rannug, 'UV-induced CYP1A1 gene expression in human cells is mediated by tryptophan.' *Chemico-Biological Interactions*, 1999, **118**, 127–140.

139. C. J. Sinal and J. R. Bend, 'Aryl hydrocarbon receptor-dependent induction of cyp1a1 by bilirubin in mouse hepatoma hepa 1c1c7 cells.' *Mol. Pharmacol.*, 1997, **52**, 590–599.

140. D. Phelan, G. M. Winter, W. J. Rogers *et al.*, 'Activation of the Ah receptor signal transduction pathway by bilirubin and biliverdin.' *Arch. Biochem. Biophys.*, 1998, **357**, 155–163.

141. C. M. Schaldach, J. Riby and L. F. Bjeldanes, 'Lipoxin A4: a new class of ligand for the Ah receptor.' *Biochemistry*, 1999, **38**, 7594–7600.

142. B. N. Fukunaga and O. Hankinson, 'Identification of a novel domain in the aryl hydrocarbon receptor required for DNA binding.' *J. Biol. Chem.*, 1996, **271**, 3743–3749.

143. W. Sun, J. Zhang and O. Hankinson, 'A mutation in the aryl hydrocarbon receptor (AHR) in a cultured mammalian cell line identifies a novel region of AHR that affects DNA binding.' *J. Biol. Chem.*, 1997, **272**, 31845–31854.

144. A. Puga, S. J. Barnes, T. P. Dalton *et al.*, 'Aromatic hydrocarbon receptor interaction with the retinoblastoma protein potentiates repression of E2F-depen-

dent transcription and cell cycle arrest.' *J. Biol. Chem.*, 2000, **275**, 2943–2950.

145. N.-L. Ge and C. J. Elferink, 'A direct interaction between the aryl hydrocarbon receptor and retinoblastoma protein. Linking dioxin signaling to the cell cycle.' *J. Biol. Chem.*, 1998, **273**, 22708–22713.

146. Y. Tian, S. Ke, M. S. Denison *et al.*, 'Ah receptor and NF-κB interactions, a potential mechanism for dioxin toxicity.' *J. Biol. Chem.*, 1999, **274**, 510–515.

147. D. Kim, L. Gazourian, S. Quadri *et al.*, 'The RelA NF-κ B subunit and the aryl hydrocarbon receptor (Ah) cooperate to transactivate the c-myc promoter in mammary cells.' *Oncogene*, 2000, **19**, 5498–5506.

148. A. Blankenship and F. Matsumura, '2,3,7,8-Tetra-chlorodibenzo-*p*-dioxin-induced activation of a protein tyrosine kinase, pp60src, in murine hepatic cytosol using a cell-free system.' *Mol. Pharmacol.*, 1997, **52**, 667–675.

149. C. M. Klinge, K. Kaur and H. I. Swanson, 'The aryl hydrocarbon receptor interacts with estrogen receptor alpha and orphan receptors COUP-TFI and ERRα1.' *Arch. Biochem. Biophys.*, 2000, **373**, 163–174.

150. J. J. Reiners, Jr. and R. E. Clift, 'Aryl hydrocarbon receptor regulation of ceramide-induced apoptosis in murine hepatoma 1c1c7 cells. A function independent of aryl hydrocarbon receptor nuclear translocator.' *J. Biol. Chem.*, 1999, **274**, 2502–2510.

J.P. Vanden Heuvel, G.H. Perdew, W.B. Mattes and W.F. Greenlee (Eds.)
Comprehensive Toxicology, Vol. xiv

# 14.2.4
# Peroxisomes, Peroxisome Proliferators and Peroxisome Proliferator-Activated Receptors (PPARs)

JEFFREY M. PETERS and JOHN P. VANDEN HEUVEL
*Pennsylvania State University, University Park, PA, USA*

## 14.2.4.1   PEROXISOMES: HISTORY AND FUNCTION

Peroxisomes were first described in mouse kidney cells as "microbodies" characterized as subcellular organelles with single lipid membranes surrounding a granular matrix of soluble proteins.[1] Subsequently, peroxisomes were shown to be present in many cell types in numerous organisms including plants and mammalian cells.[2-4] The distribution and relative size of peroxisomes in mammals is variable. Cells containing considerable amounts of peroxisomes include liver, kidney, adrenal cortex, and sebaceous cells while peroxisome are rarely found in smooth muscle or fibroblasts. In rat liver, peroxisomes account for approximately 2.5% of total protein[5] and are approximately 1.5% of the cell volume.[6]

Urate oxidase, D-amino acid oxidase and catalase were some of the first enzymes identified that were localized in peroxisomes.[7,8] The term "peroxisome" was first used to describe these organelles, based on the biochemical property of peroxide formation by the enzymes associated with peroxisomes.[7] Since this time, numerous enzymes have been localized within peroxisomes and include many that catalyze reactions central to lipid metabolism. Among the general groups of enzymes are those involved in fatty acid activation (Acyl CoA synthases), β-oxidation, carnitine acyltransferase, bile acid synthesis, glycerolipid synthesis, cholesterol synthesis, fatty acid elongation, metabolism of oxygen or reactive oxygen species and nucleotide binding proteins (see Reference 9 for review).

Biological roles of peroxisomes in respiration, gluconeogenesis, thermogenesis, purine catabolism and lipid metabolism have all been described. There are similar enzymes to those found in peroxisomes in other cellular compartments, including the cytoplasm and mitochondria. This is particularly true for the fatty acid metabolizing enzymes, where similar steps in β-oxidation occur in both peroxisomes and mitochondria (see Figure 1). It is thought that peroxisomes preferentially oxidize long chain fatty acids prior to further mitochondrial oxidation to modulate energy production, since peroxisomal oxidation yields $H_2O_2$ and heat, rather than mitochondrial oxidation which produces energy in the form of NADH or $FADH_2$. A key component of the peroxisome is the enzyme catalase. This enzyme detoxifies hydrogen peroxide generated from the first and rate-limiting enzyme in the peroxisomal β-oxidation pathway, acyl-CoA oxidase. The relationship between catalase, acyl-CoA oxidase and cancer will be discussed below under the auspices of oxidative stress.

### 14.2.4.2  PEROXISOME PROLIFERATION

The phenomenon of peroxisome proliferation was first reported by several groups in the mid-1960s.[10–12] These investigators demonstrated significant increases of hepatic peroxisomes in both size and number in response to administration of the hypolipidemic drug clofibrate in rats. This peroxisome proliferation accompanied hepatomegaly in these animals. Peroxisomes typically occupy less than 2% of the cytoplasmic volume, while treatment with fibrate drugs increase this volume to as much as 25%. In addition to peroxisome proliferation and hepatomegaly, peroxisomal fatty acid oxidation is induced, and long term administration of hypolipidemic drugs causes hepatocarcinogenesis,[13] as will be discussed in more detail below. A number of structurally diverse compounds were later identified that share the morphological and biochemical response of clofibrate and deemed "peroxisome proliferators." Peroxisome proliferators (PPs) are a diverse class of chemicals and include the fibrate class of hypolipidemic drugs (clofibrate, ciprofibrate), WY-14,643 (often used as the prototypical peroxisome proliferator), commercially used plasticizers (phthalates), steroids (dehydroepiandrosterone), and dietary fatty acids or derivatives (Figure 2). Further, starvation or feeding high levels of fatty acid can induce peroxisome proliferation, although the magnitude of the effect is substantially lower than that found with treatment with xenobiotics.

During the mid 1980s, it was suggested that the biological effects induced by PPs were mediated by an unidentified cytosolic receptor that modulated gene expression to cause the pleiotropic effect of these chemicals.[14] There were several reasons for this hypothesis. First, the peroxisome proliferation response was tissue-, species-, and sex-dependent. Second, although PPs are structurally diverse, structure activity relationships were identified, and most chemicals able to cause peroxisome proliferation, were found to share a fatty acid-like structure. Last, PPs regulate a battery of genes involved in fatty acid metabolism with a similar dose-response relationship. Based on the hydrophobic nature of PPs, it was hypothesized that the heretofore-unidentified receptor would be a nuclear hormone receptor. In 1990, a receptor termed peroxisome proliferator-activated receptor (PPAR) was cloned from mouse liver using the estrogen receptor DNA binding domain as a probe. Discovery of PPAR has proven to be pivotal to the understanding of how peroxisome proliferators regulate gene expression and ultimately result in toxicity and cancer.[15]

### 14.2.4.3  PEROXISOME PROLIFERATOR-ACTIVATED RECEPTORS (PPARS)

Since the initial discovery in mice, PPARs have been cloned in several species, including

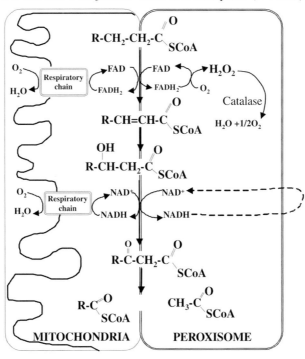

**Figure 1** Comparison of the fatty acid β-oxidation pathway in the mitochondria versus the peroxisome.

humans, rodents, amphibians, teleosts and cyclostoma.[16] The cloning of three distinct PPARs from *Xenopus*,[17] lead to the realization that a subfamily of these receptors existed. The expression of PPARα, β(δ) and γ varies widely from tissue-to-tissue. In numerous cell types from either ectodermal, mesodermal, or endodermal origin, PPARs are co-expressed, although their concentration relative to each other varies widely.[18] PPARα is highly expressed in cells that have active fatty acid oxidation capacity including hepatocytes,

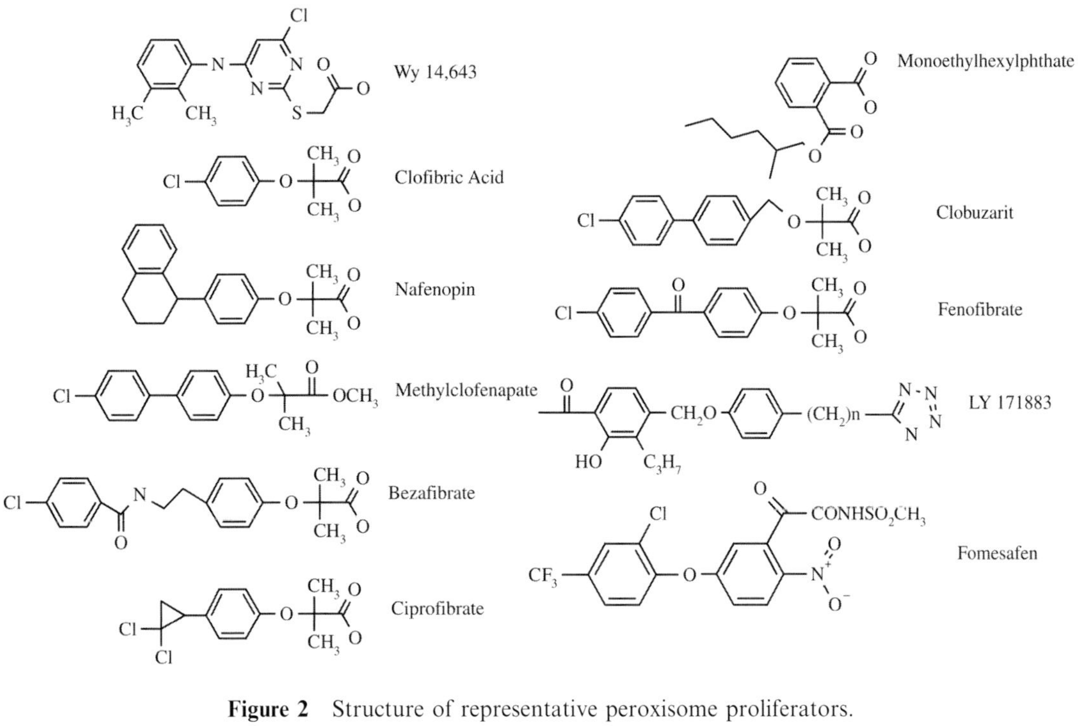

**Figure 2** Structure of representative peroxisome proliferators.

cardiomyocytes, enterocytes, and the proximal tubule cells of kidney. PPARβ is expressed ubiquitously and often at higher levels than PPARα and γ. PPARγ is expressed predominantly in adipose tissue and the immune system and actually exists as two distinct protein forms γ1 and γ2, which arise by differential transcription start sites and alternative splicing.[19]

Each PPAR subtype in every species examined thus far has the same basic mechanism of action, although there may be slight differences. One must keep in mind that subtle differences in structure and function may lend clues as to the biological niche filled by PPARα, β and γ and potential species differences. The steps involved fall into the basic categories of ligand binding, heterodimerization/DNA binding and cofactor recruitment will be discussed, followed by a review of domains within PPAR that are involved in each process.

### 14.2.4.3.1 Regulation of Gene Expression by PPAR

#### 14.2.4.3.1.1 Ligand-dependent and -independent Activation

The most-accepted and studied mechanism of action of PPAR is shown in Figure 3 and is very similar to that of other type 2 steroid hormone receptors (SHRs; for example Vitamin D receptor, thyroid hormone receptor).[20] These receptors are, by-and-large, ligand-activated transcription factors with PPARs being activated by xenobiotic-PPs as well as endogenous fatty acids and their metabolites.[15,17,21,22] The term "activation" denotes an altering in the three-dimensional structure of the receptor complex such that it is able to regulate gene expression. The physical alteration that is initiated by ligand binding may include events such as loss of heat shock proteins and chaperones, nuclear translocation, and protein turnover. To a large extent, the components of the pre- and post-liganded receptor complex have not been examined for any of the PPARs, so the precise sequences of events that result in "activation" are not known. However, in various reporter gene expression systems, the formation of a transcriptional active complex upon addition of ligand is evident and conformational changes of PPARα and β have been observed using limited proteolysis.[23,24]

A ligand-independent mechanism of regulating the activity of SHRs is a relatively new area of study. This type of activation is most often associated with kinase-dependent processes

and has been extensively studied for the estrogen receptor.[25,26] Several labs have shown that PPARα and PPARγ are phosphoproteins and that mitogen-activated protein kinase (MAPK, in particular ERK2), can modulate PPAR activity.[27] However, whether this is true ligand-independent modulation remains to be seen and may be particularly difficult to examine in light of the relatively high concentration of endogenous ligands present within the cell (see below).

#### 14.2.4.3.1.2 Endogenous PPAR Ligands

Some clues to the biological function of PPARs can be ascertained by examining endogenous and dietary activators and ligands. As mentioned above, many PPs and fatty acids are PPAR activators. The xenobiotic PPs, such as the fibrate hypolipidemic drugs, are structurally similar to endogenous and dietary fatty acids and their metabolites, containing a carboxylic acid functional group and a hydrophobic tail. Until recently, there has been some debate as to whether PPs and fatty acids activate PPAR through a direct, physical interaction. However, most receptor activators have been demonstrated to bind to PPAR subtypes with reasonable affinity and several naturally occurring ligands of these receptors have been identified (reviewed in Reference 28).

Many mono- and polyunsaturated fatty acids bind directly to PPARα at physiological concentrations and cause transcriptional activation. Long-chain unsaturated fatty acids such as linoleic acid, polyunsaturated fatty acids (PUFAs) including arachidonic, eicosapentaenoic, and linolenic acids, as well as the branched-chain fatty acid phytanic acid, bind to PPARα with reasonable affinity (μM range[28]). Recently, conjugated linoleic acid (9 *cis*, 11*trans* CLA), a dietary fatty acid, has been identified as a potent PPARα ligand with a $K_d$ in the low nM range.[29] Eicosanoid metabolites of the linoleic acid cascade have a higher affinity for PPARα than does the parent compound. The eicosanoids 8(S)-hydroxyeicosatetraenoic acid (8(S)-HETE) and leukotriene B4 (LTB4) are relatively potent PPARα ligands. Despite the higher affinity of these compounds in *in vitro* systems, it is unclear whether the concentration of 8S-HETE or LTB4 is sufficient to cause activation of PPARα *in vivo*. This has lead some to speculate that PPARα has evolved to respond to the cumulative amount of intracellular fatty acids.[28]

In contrast to PPARα, PPARγ has a preference for PUFAs over mono- or un-

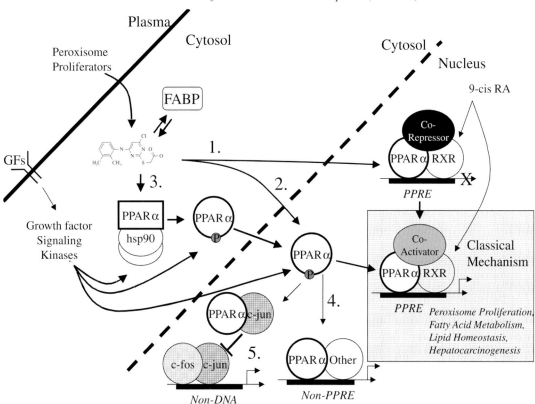

**Figure 3** Mechanism of Gene Regulation by Peroxisome Proliferators via PPARα. The most recognized mechanism by which peroxisome proliferators regulated gene expression is through a PPARα/RXR heterodimeric complex binding to a peroxisome proliferator-response element (PPRE) (classical mechanism, boxed and shaded region). However, there are the possibility of several variations on this theme: (1) The peroxisome proliferator interacts with PPARα that preexists as a DNA complex with associated corepressors proteins. The interaction with ligand causes release of the corepressor and association with a coactivator, resulting in the classical mechanism; (2) The peroxisome proliferator interacts with PPARα as a soluble member of the nucleus. The binding of ligand results in RXR heterodimerization, DNA binding and coactivator recruitment; (3) In this scenario, PPARα exists in the cytosol, perhaps complexed to heat shock protein 90 and/or other chaperones. Binding of peroxisome proliferator causes a conformational change and translocation into the nucleus. Scenarios 4 and 5 require regulation of gene expression via non-classical mechanisms; (4) PPARα is capable of interacting with, and forming DNA binding heterodimers with, several nuclear receptors including the thyroid hormone receptor. The binding site for this non-RXR heterodimer need not be the classic DR-1 motif found in the PPRE; (5) PPARα may participate in the regulation of gene expression without binding to DNA. By association with transcription factors such as c-jun or p65, PPARα diminishes the ability of AP1 or NFκB to bind to their cognate DNA sequences, respectively. Also shown in this scheme are two means to modify the peroxisome proliferator response. Most importantly, growth factor signaling has a pronounced affect on PPARα via post-translational modification. PPARα is a phosphoprotein and its activity is affected by insulin. Phosphorylation may have an impact on all schemes described above. Another modulator of PPARα activity is sequestering of peroxisome proliferators by fatty acid binding protein (FABP).

saturated fatty acids, although even the PUFAs are not considered potent activators. Lipoxygenase products of arachidonic acid, 9-HODE and 13-HODE are slightly more efficacious than PUFAs. The most widely studied endogenous PPARγ activator is 15-deoxy-$^{12,14}$-prostaglandin $J_2$ (PGJ$_2$). This prostaglandin acts similarly to the potent synthetic PPARγ ligands, the thiazolidindiones, in inducing adipocyte differentiation.

The PPARβ subtype has a fatty acid pre-ference that is similar to that of PPARα, although the amount of activation is much less. PUFAs, methyl palmitate and dihomo-γ-linoleic acid are known fatty acid ligands of PPARβ; Eicosapentaenoic acid may be the most potent activator of this chemical class. The eicosanoids PGA$_1$ and PGD$_2$ activate PPARβ in the low μM concentration range. A synthetic molecule similar in structure to the prostaglandin PGI$_2$, carbaprostacyclin, is one of the most efficacious PPARβ ligands yet described.

Taken together, it is apparent that PPARs have evolved to respond to fatty acids and their metabolites. Each subtype has preferences for certain branches of the linoleic and arachidonic acid metabolic cascades, although there is a fair amount of overlap. In addition, the apparent affinity constants ($K_d$) for most of these chemicals are within the range known to exist in cells or in serum. Therefore, it is appropriate to remove PPARs from the list of orphan receptors and consider fatty acids to have hormone-like activity. In fact, it has been suggested that fatty acids and their metabolites are primitive, ancestral hormones that predate glucocorticoids, androgens and estrogens as signaling molecules.[30]

### 14.2.4.3.1.3 Heterodimerization and DNA Binding

PPARs function in a manner very similar to that of the Vitamin D, retinoic acid and thyroid hormone receptors.[20] In this scenario, the activated PPAR binds to DNA as a heterodimeric complex with retinoid-X-receptor (RXR; NR2B[31]). The PPAR/RXR complex controls gene expression by interacting with specific DNA response elements (peroxisome proliferator response elements, PPREs) located upstream of target genes.[32] Genes containing PPRE motifs include acyl-CoA oxidase (*ACO*[32]), peroxisomal bifunctional enzyme (PBE or BIF[33]), liver fatty acid-binding protein (*L-FABP*[34]), and microsomal *CYP4A*,[35] although many more have been described. The role of altered gene expression in the overall biological function of the PPARs will be discussed in a subsequent section.

The natural PPREs for several genes have been examined[36] and a partial list is shown in Figure 4. Members of the RXR-interacting subgroup of SHRs typically bind to DNA elements containing two copies of direct repeat arrays spaced by 1–6 nucleotides (DR1-DR6). The idealized consensus binding site (AGGTCA) is similar for most members of this class with the specificity dictated by the number of nucleotides between half-sites as well as the 5′ flanking elements.[36] In the case of PPAR, a DR1 motif is preferred with PPAR interacting with the 5′ repeat and RXR ($\alpha$, $\beta$ or $\gamma$) binding to the 3′ motif.[37] This response element, called a peroxisome proliferator-response element or PPAR-response element (PPRE) is similar for all three PPAR subtypes.[36–38] Interestingly, the consensus PPRE is also recognized by HNF,[39] COUP-TF,[40,41] ARP-1,[42] RAR,[42] RZR[43] and TAK-1;[44] the interaction of these transcription factors with

the PPRE may inhibit PPAR's ability to activate gene expression.

As discussed above, the association of PPAR and RXR$\alpha$ directs the complex to bind to DR1 motifs. However, if PPAR interacts with another transcription factor, this complex may no longer associate with DNA and the interacting protein may be removed from its normal site of action. For example, the crosstalk between PPAR and thyroid hormone receptor is due, at least in part, to direct association of the two proteins. Peroxisome proliferators may inhibit thyroid hormone's ability to regulate gene expression by increasing the formation of an inactive PPAR$\alpha$/TR complex.[45,46] Similarly, PPAR$\alpha$ inhibits the effects of C/EBP$\alpha$,[47] LXR$\alpha$,[48] and GHF-1[49] through a non-DNA binding mechanism. PPAR$\gamma$ exerts some of its anti-oxidant effects by blocking the association of NF-$\kappa$B, AP1, STAT1$\alpha$ and Ets to their cognate DNA elements.[50] A classic example of how peroxisome proliferators may regulate gene expression via this route was given by Sakai *et al.* 1995,[51] with reference to glutathione S-transferase $\pi$ (GST-P) expression. Peroxisome proliferators regulate GST-P expression by decreasing activity at the AP-1 site of this gene, presumably the result of a PPAR$\alpha$/jun dimerization.[51] Steroid hormone receptors and interaction with the AP-1 complex are well recognized[52,53] and may have important consequences for the ability of PPARs to modulate expression of growth regulatory genes.

### 14.2.4.3.1.4 Target Genes

Table 1 provides a summarized list of mRNAs that are altered as a result of treatment with PPs. As noted, administration of PPs causes changes in numerous mRNAs and many of these genes are central to lipid metabolism. For example, mRNAs encoding peroxisomal, mitochondrial and microsomal fatty acid metabolizing enzymes are all increased by PPs. Further, liver mRNAs encoding proteins involved in lipid transport including fatty acid binding protein, fatty acid transporters, lipoprotein lipase, and apolipoproteins are also regulated by these chemicals. Expression of mRNAs encoding other lipid-related proteins that are altered as a result of PPs include HMG CoA synthase, lecithin:cholesterol acyl transferase, stearoyl-CoA desaturase 1, fatty acid synthase, S14 and malic enzyme which are involved in cholesterol metabolism and lipogenesis. In addition to modifying gene expression essential to lipid metabolism, PPs can also influence mRNAs

**Table 1**  Peroxisome proliferator-induced mRNAs in rodents (Target genes).

| | Tissue | Influence[1] | Reference | Null mouse?[2] | Reference |
|---|---|---|---|---|---|
| *Peroxisomal* | | | | | |
| ACS | Liver | Increased | 124,242 | No | |
| AOX | Liver | Increased | 243 | Yes | 68 |
| BIEN | Liver | Increased | 243 | Yes | 68 |
| THIOL | Liver | Increased | 244 | Yes | 68 |
| PTE-Ia/Ib | Liver | Increased | 245 | No | |
| | | | | | |
| *Mitochondrial* | | | | | |
| MCAD | Heart | Increased | 85 | Yes | 85 |
| LCAD | Liver | Increased | 80 | Yes | 80 |
| MTE | Liver | Increased | 245 | No | |
| HMG CoA Syn | Liver | Increased | 246 | No | |
| | | | | | |
| *Microsomal* | | | | | |
| CYP4A | Liver | Increased | 247 | Yes | 68 |
| | | | | | |
| *Lipid transport* | | | | | |
| FABP | Liver | Increased | 128,129 | Yes | 68 |
| FAT | Liver | Increased | 78 | Yes | 78 |
| FATP | Liver | Increased | 78,124 | Yes | 78 |
| CPT-I | Liver, heart | Increased | 80,85 | Yes | 80 |
| LPL | Liver | Increased | 248 | No | |
| Apo AI | Liver | Decreased* | 132 | Yes | 70 |
| Apo AII | Liver | Decreased | 132 | Yes | 70 |
| Apo AIV | Liver | Decreased | 132 | No | |
| Apo CII | Liver | Decreased | 130 | No | |
| Apo CIII | Liver | Decreased | 133 | Yes | 70 |
| | | | | | |
| *Lipid related* | | | | | |
| LCAT | Liver | Decreased | 249 | No | |
| SCD-1 | Liver | Increased | 250 | No | |
| FAS | Liver | Increased | 251 | No | |
| S14 | Liver | Decreased | 251 | No | |
| ME | Liver | Increased | 252 | Yes | 80 |
| | | | | | |
| *Cell cycle* | | | | | |
| CDK-1 | Liver | Increased | 74 | Yes | 74 |
| CDK-4 | Liver | Increased | 74 | Yes | 74 |
| Cyclin B1 | Liver | Increased | 74 | Yes | 74 |
| Cyclin D | Liver | Increased | 74 | Yes | 74 |
| c-myc | Liver | Increased | 165,167 | Yes | 74 |
| c-jun | Liver cells | Increased | 164,168 | No | |
| c-fos | Liver cells | Increased | 164,168 | No | |
| egr-1 | Liver cells | Increased | 164,168 | No | |
| junB | Liver cells | Increased | 164,168 | No | |
| ERK1 | Liver cells | Increased | 170 | No | |
| ERK2 | Liver cells | Increased | 170 | No | |
| HGF | Liver | Decreased | 173 | No | |
| TNFα | Liver | Increased | 253 | No | |
| | | | | | |
| *Acute phase* | | | | | |
| α2U | Liver | Decreased | 72,86 | Yes | 72,86 |
| Ceruloplasmin | Liver | Decreased | 82 | Yes | 82 |
| Haptoglobin | Liver | Decreased | 82 | Yes | 82 |
| Fibrinogen | Liver | Decreased | 86,93 | Yes | 86,93 |
| α1-acid GP | Liver | Decreased | 86 | Yes | 86 |
| | | | | | |
| *Other* | | | | | |
| 17β-HSD IV | Liver | Increased | 254 | Yes | 91 |
| COX-2 | Liver cells | Increased | 169 | No | |
| Transferrin | Liver | Decreased | 255 | No | |
| GOT | Liver | Decreased | 100 | Yes | 100 |

*continued*

**Table 1** Continued.

| | Tissue | Influence[1] | Reference | Null mouse?[2] | Reference |
|---|---|---|---|---|---|
| GPT | Liver | Decreased | 100 | Yes | 100 |
| CTE | Liver | Increased | 245 | No | |
| ZFP-37 | Liver cells | Both† | 98 | Yes | 98 |
| HBV | Liver | Increased | 92 | Yes | 92 |
| PK | Liver | Decreased | 76 | Yes | 76 |

Abbreviations: ACS: Acyl-CoA synthase; AOX: Acyl-CoA oxidase; BIEN: enoyl-CoA hydratase/3-hydroxyacyl-CoA dehydrogenase (bifunctional enzyme); THIOL: 3-ketoacyl CoA thiolase; PTE: peroxisomal thioesterase; MCAD: medium chain acyl-CoA dehydrogenase, LCAD: long chain Acyl-CoA dehydrogenase; MTE: mitochondrial thiolesterase; HMG CoA Syn: 3-hydroxy-3-methylgluatary-CoA synthase; CYP4A: cytochrome P450 4A subfamily, including CYP4A1, CYP4A2, CYP4A3, CYP4A6; FABP: fatty acid binding protein; FAT: putative fatty acid transporter/CD36; FATP: fatty acid transporter; CPT-I: carnitine palmitoyl transferase-I, LPL: lipoprotein lipase; Apo AI: apolipoprotein AI; Apo AII: apolipoprotein AII; Apo AIV: apolipoprotein AIV; Apo CII: apolipoprotein CII; Apo CIII: apolipoprotein CIII; LCAT: lecithin:cholesterol acyltransferase; SCD-1: stearoyl-CoA desaturase 1; FAS: fatty acid synthase; S14: spot 14; ME: malic enzyme; CDK-1: cyclin dependent kinase-1; CDK-4: cyclin dependent kinase-4; ERK1: extracellular signal-regulated kinase-1 (mitogen-activated protein kinase); ERK2: extracellular signal-regulated kinase-2 (mitogen-activated protein kinase); HGF: hepatocyte growth factor; TNFα: tumor necrosis factor-α; α2U: α-2-urinary globulin; α1-acid GP: α1-acid glycoprotein; 17β-HSD IV: 17β-hydroxysteroid dehydrogenase type IV; COX2: cyclooxygenase 2 (prostaglandin H synthase); GOT: aspartate aminotransferase; GPT: alanine aminotransferase; CTE: cytosolic thioesterase; ZPP37: zinc-finger protein 37; HBV: hepatitis B virus; PK: pyruvate kinase.
[1]Effect of PPARα activator.
[2]whether or not this change in mRNA level has been evaluated in the PPARα-null mouse.
*Species difference. In rodents Apo AI is decreased, in humans it is increased.
†Induced in rats, repressed in mice.

encoding proteins that regulate cell proliferation or the acute phase response. Lastly, PPARα can mediate alterations in mRNAs associated with steroid metabolism (17β-HSD), prostaglandin synthesis (COX2), iron metabolism (transferrin), nitrogen metabolism (GOT, GPT) and hepatitis B viral replication. While it is clear that PPARα is critical in the regulation of lipid metabolism, and that this receptor likely regulates cell proliferation/apoptosis underlying PP-induced hepatocarcinogenesis, the precise role for some of these alterations in gene expression are not understood. For many

of the target genes that regulate lipid metabolism, functional PPREs have been identified in their respective promoter region, including peroxisomal fatty acid metabolizing enzymes, fatty acid transporter, apolipoproteins, and lipoprotein lipase. Identification of PPREs in target genes is an ongoing research area that will likely confirm the presence of functional responsive elements for many of the target genes listed in Table 1 if this has not been established. However, it is also possible that alterations in mRNA expression are an indirect effect induced by a PPARα-specific target gene

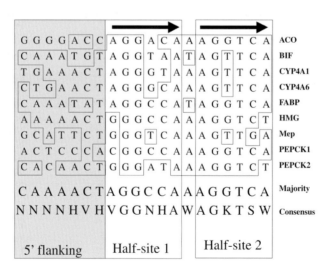

**Figure 4** Natural PPREs for peroxisome proliferator responsive genes. (Adapted from Reference 31). The shaded region corresponds to the 5′ flanking region, whereas the open box with the arrows above represents the two half-sites of the DR-1. PPAR interacts with half-site 1 while RXR associates with half-site 2. Alignments were performed using Multi-alignment Editor in Genetool (Biotools Inc., Edmonton, AB). The majority sequence is defined as the nucleotide most prevalent, while the consensus allows for ambiguity (IUPAC nomenclature). In addition, nucleotides which contain the sequence found in the majority motif are boxed.

that in turn regulates expression of a subsequent mRNA. For example, while it is clear that CDK and cyclin expression are increased by PPs, data demonstrating the presence of a functional PPRE is lacking. This may be true for other mRNA species listed in Table 1.

Additionally, there is indirect evidence suggesting alterations in other proteins from PP administration, yet confirmation of corresponding increased mRNA is lacking. (Table 1 does not include such PP-regulated genes.) For example, *in vitro* data suggests that Cu/Zn superoxide dismutase (SOD) is modulated by PPARα.[54] Further, analysis of enzyme activities suggest that acyl CoA carboxylase[55] and enzymes that regulate phospholipid levels[56] are also modulated by this class of chemical.

### 14.2.4.3.1.5 Cofactor Recruitment

Steroid hormone receptor co-activators and co-repressors (collectively referred to as cofactors) are an important class of SHR-interacting proteins. Cofactors directly interact with nuclear receptors and repress (co-repressors) or enhance (co-activators) their transcriptional activities; the ability of cofactors to affect gene expression is receptor, tissue and gene specific.[57] Classically it is suggested that in the absence of ligand, nuclear receptors may bind to co-repressors, which decreases receptor activity. Ligand binding induces a conformational change in the nuclear receptor that favors binding to co-activators. The receptor/coactivator complex can then activate gene transcription by means of recruiting chromatin-modifying enzymes (i.e., acetyl transferases) or by forming a "bridge" with the preinitiation complex at the hormone regulated promoter.[57] DNA-bound receptors function as large multiprotein complexes, the constituents and function are dependent on the cell and promoter context. PPAR cofactors have been identified

by a variety of techniques including yeast two-hybrid, transient transfections and biochemical means. A partial list of PPAR cofactors is given in Table 2.

### 14.2.4.3.2 PPAR Structure/Function

Similar to other transcription factors, PPARs are comprised of different domains, each of which confers a particular activity or function to the overall protein. In PPAR there are four functional domains, starting from the amino terminus of the protein, called A/B, C, D and E/F (see Figure 5)[16] as described below: The A/B domain is a poorly conserved region ("hypervariable") whose function has not been clearly defined. Based on what is known from other SHRs, this may be the region that contains the activation function-1 (AF-1), a sequence that is involved in ligand-independent activation. This region contains a large number of consensus phosphorylation sites, including those of MAPK, casein kinase-2 and protein kinase C.[27] The C domain or DNA binding domain (DBD) is the highest conserved region and contains two $C_4$ zinc finger motifs. The first zinc finger is also called the proximal or P-box and forms an α-helix that interacts with the major groove of DNA. The PPAR P-box (CEGCKG) is identical to that of other SHRs that recognize the AGGTCA DNA half-site. The distal, or D-box, is also helical in structure but is aligned perpendicular to the P-box. This region is involved in dimerization with other SHR DBDs as well as recognition of spacing elements within the PPRE. The primary function of the D or hinge domain appears to be allowing for a flexible connection between the DBD and the E/F domain. The site contains the carboxy-terminal extension (CTE) of the DBD, which may be involved in recognizing the extended 5′ end of the PPRE.[58] The fourth region is the E/F or ligand binding domain

**Table 2** List of known PPAR cofactors.

| Co-activator | Reference | Co-repressor | Reference |
|---|---|---|---|
| PBP | 256 | NcoR | 257 |
| SRC-1 | 258–261 | SMRT | 44 |
| CBP/P300 | 262 | | |
| RIP140 | 44,263 | | |
| PGC-1 | 264 | | |
| TRAP220 | 265 | | |
| PRIP | 266 | | |
| RAP250 | 267 | | |

Abbreviations used: PBP, PPAR binding protein; SRC, steroid receptor co-activator; CBP, CREB binding protein; RIP140, receptor interacting protein; PGC-1, PPARγ coactivator; TRAP220, thyroid receptor associated protein; PRIP, PPAR interacting protein; RAP250, receptor associated protein; NcoR, nuclear receptor coactivator; SMRT, silencing mediator of retinoid and thyroid receptors.

**A.**

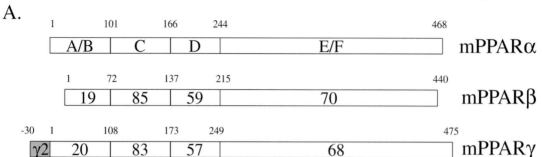

**B.**

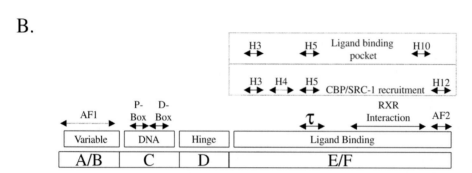

**Figure 5** Basic structure of the mouse PPARs. **Panel A.** Structure and functional domains of PPARα, β and γ. A/B is the hypervariable region containing the putative activation function-1 (AF-1) domain. PPARγ2 contains a 30 amino acid region that arises from differential promoter use and splicing. The C-domain is the most conserved and contains the DNA binding motif. The D-domain (hinge) is believed to allow for conformational change following ligand binding. The E/F domain contains the ligand-binding region of PPAR. Alignments and percent similarities were performed with MegAlign (DNAStar, Madison WI). **Panel B.** Detailed functional domains of PPARs. Above the outline for the hypothetical PPAR are the structural features of PPARs that have been deduced by sequence comparisons and crystallography. The AF-1 domain has not been fully characterized although it is known to reside in the A/B domain. The DNA binding motif contains two $C_4$ zinc fingers, the proximal (P-box) and distal (D-box) boxes, which confer DNA binding and heterodimerization, respectively. Much has been learned about PPAR structure/function from recent crystallographic studies. PPAR E/F contains 13 alpha helices (H1-H12, H2′) and 4 short β strands[268] and helices 3, 5 and 10 forms the ligand-binding pocket. RXR interacts along several helices including H7-H10. The coactivator CBP interacts with H3–5 and H12 while SRC-1 associated with H3, H5 and H12. The τ1 domain contains a leucine-zipper-like heptad repeat.[269] The AF-2 domain is highly conserved among all PPARs and is intimately associated with ligand-induced transcriptional events.

(LBD). The crystal structure of PPARβ and PPARγ has been solved in the presence or absence of ligand and RXR.[59,60] Because of the relatively high conservation in this region, the basic structure of PPARα may be deduced from what has been learned form the other subtypes. The overall structure of the PPAR LBD is 13 α helices (H1-H12 plus H2′) and a four stranded β-sheet.[16] In addition to binding ligand, the LBD mediates hormone-dependent activation (AF-2) and is the major interface for RXR and cofactor interactions.

One of the major interests in developing a detailed picture of the functional regions of PPARs is to understand mutations, polymorphisms and splice variants in human populations that may contribute to diseases of energy metabolism such as obesity, atherosclerosis, diabetes and cancer. A truncated form of hPPARα is formed as a result of differential splicing and introduction of a premature stop codon (PPARα$_{tr}$,[61]). The PPARα$_{tr}$ lacks a portion of the D domain and the entire E/F domain and acts as a dominant negative form of the receptor. A naturally occurring variant of PPARα (PPARα 6/29) also acts as a dominant negative and contains different residues at positions 71, 123 and 444.[62] A leucine to valine mutation in human PPARα (L162V) has been identified by three laboratories and is associated with hyperapobetalipoproteinemia, hyperlipidemia and altered transactivation by PPARα ligands.[63–65] Four single amino acid substitutions have been described in human PPARγ, including P12A, P115Q mutations, which have little effect on insulin sensitivity,[66] and two substitutions (at residues 290 and 467) which result in a dominant negative PPARγ and severe insulin resistance.[67]

### 14.2.4.3.3 PPARα-Null Mouse

In 1995, the PPARα-null mouse was first described.[68] Targeted disruption of the murine PPARα gene was not developmentally lethal thus providing an excellent model to examine the role of PPARs in homeostasis, toxicity and cancer. In the initial report, PPARα-null mice were described as refractory to peroxisome proliferation, hepatomegaly, and increased expression of mRNAs encoding peroxisomal and microsomal fatty acid metabolizing enzymes.[68] Subsequently, this mouse model has been used extensively to examine the role of the PPARα in PP hepatocarcinogenesis, lipid homeostasis and alterations in gene expression elicited from a variety of stimulus such as feeding experimental diets, fasting and cold acclimation.[69–119] Combined, these data have provided definitive *in vivo* evidence that the PPARα has a critical role in modulating gene expression to maintain lipid homeostasis. Further, these reports conclusively demonstrate that the PPARα is the only member of this receptor subfamily capable of mediating

peroxisome proliferation and, most importantly, hepatocarcinogenesis induced by PPs. Hence, the term peroxisome proliferator-activated receptor best describes the PPARα, yet will continue to be used for the other two PPAR isoforms due to significant sequence homology. Since PPARβ and PPARγ are relatively uninfluenced by PPs, the remainder of this chapter will focus on the role of the PPARα in mediating the carcinogenic effects of these chemicals. A summary of the known roles of PPARα, as determined from the null mouse model, is shown in Figure 6.

### 14.2.4.3.4 Physiological Roles of the PPARα

PPARα regulates genes encoding proteins that regulate lipid homeostasis, including fatty acid uptake, activation and oxidation. For excellent reviews on the physiological roles for this receptor, refer to the following references.[120–123] Activation of PPARα increases expression of lipoprotein lipase, fatty acid binding protein and fatty acid translocase, all

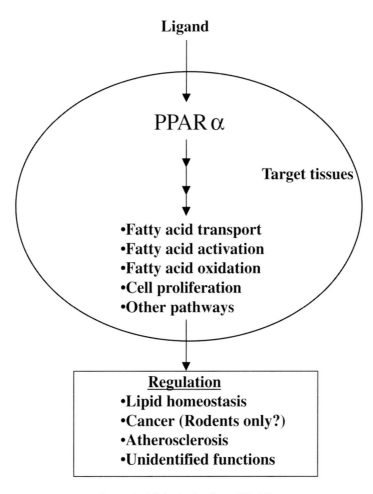

**Figure 6** Biological roles of PPARα.

of which may contribute to increased intracellular uptake of fatty acids.[68,78,124–129] Expression of hepatic apolipoproteins that influence lipoprotein uptake can also be altered by PPARα activation, such as apolipoproteins CII, CIII, AI, AII and AIV.[130–133] Long-chain fatty acids are typically activated into CoA thioesters by acyl CoA synthase, another gene product that is regulated by PPARα.[124,126] Interestingly, thioesterases which catalyze the hydrolysis of acyl CoAs to the corresponding free fatty acid and CoA are also responsive to PPARα activation.[77] Transport of fatty acids across the mitochondrial membrane by CPT is also under control of PPARα.[80,85,134] Oxidation of fatty acids is significantly influenced by PPARα as peroxisomal, microsomal and mitochondrial fatty acid oxidizing enzymes are all induced by treatment with PPs.[68,80] Combined, PPARα has a central role in regulating lipid homeostasis by modulating gene expression in response to both exogenous and endogenous chemicals. For example, many of the above changes in gene expression that modulate fatty acid uptake, activation and oxidation are facilitated by the PPARα in response to both treatment with PPs, dietary fatty acids and fasting. However, alterations in mRNAs encoding lipid-related enzymes such as UCPs, CPTs and lipogenic proteins observed with dietary fatty acids and fasting are independent of PPARα since these changes are found in both wild-type and PPARα-null mice.[75,79,94]

In addition to being a key regulator of fatty acid metabolism, PPARα has also been shown to modulate other physiological functions. It is thought that this receptor regulates the inflammatory response induced by leukotriene B4 by altering catabolism of this endogenous ligand.[90] PPARα also inhibits interleukin-1 (IL-1) signaling by interacting with NF-κB mediated cyclooxygenase-2 induction.[135] Given the influence of PPARα in modulating inflammation and fatty acid catabolism, it is not surprising that there is evidence demonstrating a direct role of this PPAR subtype in the molecular mechanisms contributing to atherosclerosis.[136] Because of the central role of this receptor in lipid-related pathways, it is likely that other lipid dependent functions will be modified as a result of PPARα activation.

### 14.2.4.4 PEROXISOME PROLIFERATOR HEPATOCARCINOGENESIS

Long term administration of PPs causes hepatocarcinogenesis in rodent models.[13,137] The carcinogenic effect of these chemicals is only seen when there is significant peroxisome proliferation (i.e., greater than fourfold increase in peroxisomal acyl CoA oxidase activity).[138] Due to limited peroxisome proliferation, hepatocarcinogenesis is not likely to occur in response to physiological activation of PPARα-dependent transcription such as that observed with fasting or feeding a diet rich in fatty acids. PPs that cause hepatocarcinogenesis include the fibrate class of hypolipidemic drugs,[13] WY-14,643,[139] dehydroepiandrosterone,[140] di(2-ethylhexyl) phthalate (DEHP)[141] and trichloroethylene.[142] The incidence of hepatocarcinogenesis in wild-type mice fed the potent PP WY-14,643 for 11 months is 100% while PPARα-null mice are refractory to this effect.[73] Microscopic examination of livers from PPARα-null mice fed WY-14,643 also failed to identify any preneoplastic lesions. Thus, it is clear that the carcinogenic effect of these chemicals is mediated by the PPARα, however the specific mechanisms underlying this effect have not been fully elucidated. Further, it is likely that a species difference in sensitivity to PPs exists, and this will be discussed later. A number of possible mechanisms for PP-induced hepatocarcinogenesis have been hypothesized, including: (1) increased oxidative damage to intracellular macromolecules (DNA, protein, lipids) that contribute to gene mutations indirectly; (2) increased cell replication in the presence of damaged cells; and (3) inhibition of apoptosis, resulting in the persistence of damaged cells. Each of these possibilities will be discussed and are outlined in Figure 7.

#### 14.2.4.4.1 Oxidative Stress

One of the first hypotheses to explain the carcinogenic effect of these chemicals is that peroxisome proliferation causes oxidative damage, resulting from increased intracellular concentration of $H_2O_2$ produced by peroxisomal acyl CoA oxidase (ACO).[143] Hydrogen peroxide in the presence of metals facilitates free radical formation, which may in turn damage macromolecules. Oxidative stress is said to occur as a result of the large induction of ACO caused by PPs,[144,145] in the absence of significant increases of catalase.[144,146] Evidence demonstrating intracellular oxidative damage resulting from PPs is inconsistent. Peroxide-modified lipids have been detected in hepatocytes from rats treated with either WY-14,643, clofibrate or DEHP.[146,147] However, more sensitive measures of oxidative damage, including ethane exhalation and hepatic levels of esterified F2-isoprostanes, are not markedly influenced by administration of WY-14,643.[148,149] Evidence showing oxidative damage to DNA resulting from PPs is also inconsistent. DNA damage in the form of 8-hydroxydeoxyguanosine (8-OH-dG) has been reported in liver from rats after long-term treatment with PPs, including DEHP.[150,151] However, the increase in 8-OH-dG residues are small (<2-fold), not sustained with chronic DEHP treatment,[151,152] and the correlation between 8-OH-dG levels and tumor multiplicity is weak[145,152]. One study suggested that the increase in 8-OH-dG levels resulting from PPs is due in part to increased mitochondrial DNA,[153] although this study may be flawed due to the absence of antioxidants in the preparations.[154] In contrast to the studies demonstrating oxidative damage to DNA, no measurable increase in 8-OH-dG or thymidine glycol was reported in rat liver after administration of nafenopin.[155]

Despite evidence that PPs cause intracellular oxidative damage to lipids or DNA, most investigations support the idea that PPs are non-genotoxic carcinogens.[138,154,156,157] Numerous assays designed to determine the genotoxicity of chemicals have been utilized to assess the effect of PPs, including mutation, genotoxicity, DNA damage, chromosomal damage and cell transformation assays. Nonetheless, there is one report suggesting that H$_2$O$_2$ produced from overexpression of ACO in NIH3T3 cells causes cell transformation.[158] While this suggests that oxidative damage can lead to cell transformation *in vitro*, this has not been clearly demonstrated *in vivo*.

Interestingly, spontaneous peroxisome proliferation, induction of many PPARα target genes, and hepatocarcinogenesis are all found in ACO-null mice after one year of age.[159] This phenotype is remarkably similar to that of rodents fed PPs, suggesting that in the absence

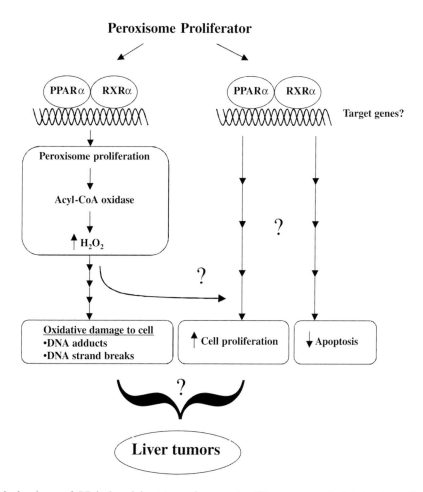

**Figure 7** Mechanisms of PP-induced hepatocarcinogenesis. PPs cause marked induction of peroxsiomal Acyl-CoA oxidase in the absence of substantial catalase induction resulting in increased intracellular H$_2$O$_2$ which could cause oxidative damage to cells. PPs cause increased cell proliferation of hepatocytes, which in the absence of apoptosis, could cause proliferation of cells that may contain damaged DNA thus contributing to tumor formation.

of fatty acid oxidation by peroxisomal acyl CoA oxidase, a fatty acid metabolite accumulates and increases PPARα dependent transcription. Thus, ACO may function to metabolize this unidentified fatty acid metabolite to prevent accumulation that ultimately would result in peroxisome proliferation and hepatocarcinogenesis. These data suggest that ACO functions as a tumor suppressor and PPARα as an oncogene.[159]

### 14.2.4.4.2  Cell Cycle Control

Alterations in cell cycle control is another postulated mechanism of PP-induced carcinogenesis. A significant increase in liver weight is always found in rodents fed PPs[139,160] and this is due in large part to increased cell proliferation as measured by [3]H-thymidine or bromodeoxyuridine incorporation in hepatocytes.[145,161,162] Since hepatomegaly and increased expression of cell cycle control proteins are not found in PPARα-null mice treated with PPs,[73,74] the mitogenic effect of these chemicals clearly requires this receptor. Perhaps PPs, acting through PPARα, alter the expression of a particular subset of genes that in turn affects the rate of proliferation of cells. Several early studies indicated that the degree of peroxisome proliferation correlated poorly with the relative hepatocarcinogenicity of PPs. However, a strong correlation was observed between the relative tumor promoting activity of the PPs DEHP and WY-14,643, with the ability to induce a persistent increase in replicative DNA synthesis.[145]

Although most genes known to be regulated by these chemicals are involved in fatty acid metabolism (reviewed in References 120–123), there are several PP-responsive genes with a link to cell cycle control. Induction of the oncogenes c-Ha-ras,[163] jun[164] and c-myc[74,165–167] by PPs has been reported and the ability to induce these genes correlates well with tumor promoting potential.[164,168–170] For example, the tumor promoters WY-14,643, clofibrate, ciprofibrate and DEHP were inducers of c-fos, c-jun, junB egr-1, and NUP475 whereas the weak-carcinogenic PP dehydroepiandosterone (DHEA) was ineffective.[168] In addition, an immediate early gene (IEG) critically involved in lipid metabolism, tumor promotion and inflammation, COX-2, is also regulated by PPs.[169] A novel IEG involved in neuronal differentiation, rZFP-37, is a PP-regulated gene in rodent liver.[98] Both rZFP37 and c-myc have been demonstrated to be controlled by PPARα,[84,98] using the null mouse model. Many of the genes described above are critical in the progression

of the cell cycle, in particular the $G_1$ to S transition. PP induced expression of growth regulatory genes precedes entry of the cell into S phase.[168] Also, expression of cell cycle control proteins associated with cell replication (cyclins, cyclin dependent kinases) are increased as a result of treatment with PPs.[171,172]

Hepatocyte growth factor (HGF) is another gene product that is altered as a result of PP treatment and thus could influence hepatocyte cell proliferation. Administration of PPs to rats causes down-regulation of hepatic HGF mRNA after short-term feeding WY-14,643[173,174] and continues to decline to less than 50% of control levels after forty weeks of feeding.[173] The reduction in liver HGF mRNA levels is found in non-tumor portions of cancerous livers in PP-treated rats, although a more striking decrease is found in the tumor regions of the same livers.[173,175] Since HGF can also inhibit colony formation in soft agar of neoplastic or preneoplastic liver cells from WY-14,643-fed rats,[173,175] this suggests that decreased HGF expression resulting from PP treatment could contribute to the mechanisms underlying PP-induced hepatocarcinogenesis. While HGF can function to stimulate hepatocyte cell proliferation in response to partial hepatectomy,[176,177] the previous observations and others indicate that HGF may also function as a growth suppressor in hepatocarcinoma cells.[178,179] Recent studies using PPARα-null mice demonstrated that PP-induced reduction of HGF does indeed require a functional PPARα.[180] However, these studies also showed that hepatomegaly and expression of mRNAs encoding cell cycle control proteins resulting from PP feeding are not prevented in two different lines of transgenic mice over-expressing liver HGF at substantially high levels.[180] Thus, the precise role of reduced HGF expression in the pleiotropic response resulting from PP treatment is unclear.

Consistent with PPs role as hepatic tumor promoters, these chemicals decrease the rate of programmed cell death[181,182] thereby altering the balance between mitosis/apoptosis, a key mechanism in carcinogenesis.[183] Recently, PPARα has been shown to be an essential component of PPs repression of cell death.[62,184] These researchers showed that overexpression of a dominant negative PPARα, (thereby abolishing PPARα activity) abrogates the suppression of apoptosis induced by PPs.[62] Additionally, hepatocytes from PPARα-null mice are also refractory to suppression of apoptosis caused by PPs.[184] Combined, there is evidence that PPARα can regulate the response to PPs by preventing programmed cell death that potentially results in cell proliferation of "damaged" cells that would normally undergo apoptosis.

### 14.2.4.4.3 Indirect Mechanism

A related hypothesis is that hepatic cell proliferation is influenced by changes secondary to oxidative- or growth factor-signal transduction. For example, intracellular $H_2O_2$ produced from significant induction of ACO caused by PPs may modulate NF-κB activity[185] that in turn can alter cell proliferation. This hypothesis has not been examined in great detail, although there is evidence that anti-oxidants such as selenium or Vitamin E can decrease PP-induced hepatocarcinogenesis.[186,187]

Growth factors released from extrahepatic tissues may represent important secondary mitogeneic stimuli to heptocytes. Specifically, Kupffer cell-derived growth factors such as TNFα may contribute to mitogenesis in liver resulting from treatment with PPs.[188] A critical role for the resident macrophages in the mitogenic response of hepatocytes to PPs has been suggested by, (1) inactivation of Kupffer cells by methyl palmitate or dietary glycine prevents Wy-14,643-induced release of TNFα[189,190] and hepatocyte proliferation;[191] and (2) Wy-14,643 directly activates superoxide production in isolated Kupffer cells.[191,192] It is unclear how Kupffer cells and parenchymal cells interact. Parenchymal cells seperated from non-parenchymal cells and treated with PPs show increased activity of acyl-CoA, but DNA synthesis is unchanged.[193] Further, purified parenchymal cells cultured in medium conditioned from isolated Kupffer cells and treated with WY-14,643 exhibit increased DNA synthesis. This supports the hypothesis that both cytokines released from Kupffer cells, and PPARα are required for the proliferative response and cancer induced by PPs. Combined, these observations support the hypothesis that PPs activate Kupffer cells which may affect release of paracrine factors such as TNFα.

The role of Kupffer cells in PP-induced tumors is uncertain. Clearly, the hepatocarcinogenic effect of PPs requires a functional PPARα;[73] however, Kupffer cells do not express this receptor subtype.[194] Additionally, activation of these macrophage-like cells is independent of the PPARα since Wy-14,643-induced superoxide production is similar in wild-type and PPARα-null mice.[194] Therefore, the production of TNFα by Kupffer cells is independent of PPARα, whereas the cell proliferation requires this protein. In order for this particular combined theory to be correct, the mitogenic response of hepatocytes to TNFα must be dependent on PPARα. This hypothesis has not been adequately tested, although this cytokine does indeed affect the transcriptional activity of PPARα.[27] This is similar to the ligand-independent activation of PPAR described above. Importantly, PPs have mitogenic effects when given directly to primary hepatocytes in culture,[195] adding further uncertainty as to the role of extrahepatic growth factors in PP-induced hepatocarcinogenesis. Lastly, TNFα-null mice are not refractory to PP-induced cell proliferation,[196] suggesting that the role of Kupffer cells in the mechanisms underlying peroxisome proliferator-induced hyperplasia may involve cytokines/chemical mediators other than TNFα.

### 14.2.4.5 SPECIES DIFFERENCES

There is considerable debate concerning the mechanisms by which PPs cause liver tumors in rodent models and whether these chemicals represent a human cancer risk. It is well established that long-term administration of PPs results in hepatocarcinogenesis in rats and mice.[14,138,197,198] In contrast, liver tumors are not found in Syrian hamsters after long-term administration of PPs.[199] Similar investigations in non-human primates indicate that this species are also refractory to the PP-induced liver cancer,[200] although these experiments were not carried out over the lifespan of the animals. Consistent with this idea, hepatic peroxisome proliferation is not found in non-human primates treated with these chemicals.[138,198,201] Liver biopsies from humans treated with gemfibrozil, clofibrate or fenofibrate for periods ranging from 27 to 94 months did not reveal convincing evidence of peroxisome proliferation, supporting the hypothesis that humans are less responsive to PPs.[202–205] Further support of this idea is provided by two limited epidemiological studies that showed no evidence of increased cancer risk as a result of fibrate therapy.[206,207] While these observations suggest that a significant species difference in the response to PPs exists, it is critical to delineate the mechanisms of this putative difference in order to increase the confidence of risk assessments made for this class of chemicals.

Short-term *in vivo* studies evaluating the mechanisms of PP-induced alterations have also revealed marked species differences. Rats and mice typically respond to these chemicals by peroxisome proliferation, increased hepatic cell replication, and large increases in hepatic expression of peroxisomal and microsomal lipid metabolizing enzymes. While rats and mice are highly responsive to PPs, Syrian hamsters exhibit an intermediate phenotype, and guinea pigs and dogs appear to be unresponsive.[200] Examination of typical PP-induced alterations in non-human primates also suggests that this species is refractory to these changes.[208] Replicative DNA synthesis is significantly increased in the liver of rats or mice treated with PPs but the increase in cell replication is not observed in similarly treated Syrian hamsters.[199] *in vitro* analysis using cells from different species support many of the *in vivo* observations. Cultured rodent hepatocytes respond to PPs by increased replicative DNA synthesis, suppression of apoptosis, increased expression of marker mRNAs and proteins, including peroxisomal ACO, and peroxisome proliferation.[209–217] PPs do not stimulate cell proliferation in cultured guinea pig hepatocytes.[218] Human and non-human primate hepatocytes cultured in the presence of PPs do not exhibit increased replicative DNA synthesis, suppression of apoptosis, increased expression of marker mRNAs and proteins including peroxisomal ACO, or peroxisome proliferation.[209–217] Thus, *in vivo* and *in vitro* data strongly support the idea that humans are refractory to many of the effects induced by PPs in rodents. However, a reduction in serum lipids which is the primary use for these chemicals in humans, occurs in all species tested to date. Since this effect is clearly mediated by the PPARα in mice,[70] this suggests that humans have a functional PPARα capable of modulating gene expression in the same mechanism as rodents.

Transient transfections with a human PPARα cDNA show that this receptor is capable of transactivating reporter constructs, providing evidence that the human isoform is functional.[22] Thus it is not surprising that PPREs have been described in human genes that are transcriptionally regulated by the PPARα, including apo C-III,[219] lipoprotein lipase,[38] apo A-I,[220] apo A-II,[221] carnitine palmitoyltransferase-I[134] and acyl CoA oxidase.[222] Because the hypolipidemic effects of PPs are mediated by the PPARα in rodents and possibly humans, and the human genome encodes for PPARα and contains genes with apparent PPREs, it is of great interest to determine why human cells appear to be refractory to these chemicals. Several possible explanations have been postulated to account for this disparity. One possible theory why humans appear to be insensitive to the adverse effects of PPs is that there are differences in the expression levels or function of the expressed protein. Indeed, PPARα mRNA in human liver RNA samples are reported to be less than 10-fold, as compared to a limited set of mouse RNA controls.[223] Interestingly, guinea pig liver also contains significantly less PPARα as compared to mouse liver.[224] This suggests that the level of PPARα in the human liver may not be sufficient to activate the plethora of target genes that are altered by PPs in rodent models, yet may be expressed at levels capable of modulating lipid homeostasis. However, given the relatively small number of human liver samples examined, further quantification of PPARα mRNA and protein, in addition to other transcription factors, would be helpful to serve as positive controls.

While it appears that lower levels of PPARα expression may in part contribute to the species differences between rodents and humans, expression of truncated or mutant PPARα have also been described.[61,64,65,223,225] This suggests that in addition to lower expression levels of human PPARα, variants may also convey a relative insensitivity to PPs in humans. However, since most humans are responsive to fibrate therapy, the hypothesis that altered PPARα protein accounts for the species difference is likely not true in all circumstances.

Another postulate to explain the species difference in sensitivity is that some of the human target genes contain mutations or polymorphisms in the regulatory region. For example, it is reported that a PPRE upstream of the human acyl CoA oxidase gene is inactive in cells transiently transfected with PPARα expression vector, and this PPRE in a reporter construct.[226] However, another group had previously found that this PPRE is capable of being transactivated in similar reporter assays[222] and corrected the original reported sequence of the 5' regulatory region of this gene.[227] Subsequently, it was shown using site directed mutagenesis of the rat ACO PPRE that the human PPRE sequence is not capable of being transactivated in reporter assays.[228] Due to the disparity in results, further analysis of this and other PPARα target genes in humans should be evaluated before drawing the conclusion that mutations or polymorphisms in PPREs contribute to the species difference in sensitivity to PPs. It is worth noting that differences in the formation of PPARα-RXRα/ACO PPRE complexes in rat and human cell extracts has also been suggested as a related hypothesis and deserves further investigation.[229]

The hepatotoxic effects of PPs are well documented and are primarily the result of this class of chemicals interacting with the PPARα. Hepatomegaly, the hallmark of liver toxicity occurs rapidly in rodent liver due to increased cell replication and/or inhibition of apoptosis. The use of PPARα-null mice provided definitive evidence that this effect is mediated by this subtype.[68,71,73,74,97] Considerably less is known about the role of this receptor in mediating other toxic effects in liver and other tissues. A number of lesions induced by PPs have been described, including nephropathy, testicular lesions, mononuclear cell leukemia, and spongiosis hepatis.[97,230,231] While the biochemical mechanisms underlying the nephropathy and testicular lesions are unclear, they are mediated by PPARα.[97] Mononuclear cell leukemia induced by PPs is found only in one strain of rats and may not be relevant in other species, including humans.[231] Whether spongiosis hepatis found in liver of phthalate-treated rodents is mediated by PPARα has not been determined. Lastly, at least two different PPs are developmentally toxic resulting in teratogenesis. Trichloroethylene administration during pregnancy results in cardiac abnormalities in the developing conceptus[232] and it is not known if this is mediated by PPARα. In contrast, DEHP exposure during development in the mouse is also teratogenic, resulting in severe neural tube defects in term fetuses, and this effect is independent of the PPARα.[71] DEHP teratogenicity is due to the induction of a maternal zinc deficiency, a phenomenon that occurs with a host of structurally unrelated compounds.[233,234]

Interestingly, pretreatment with PPs can greatly diminish the toxic effects of a number of chemicals including cerium (a rare earth metal),[235] 2,3,7,8-tetrachlorodibenzo-*p*-dioxin[236] and acetaminophen.[237] It has been suggested that the protection against acetaminophen toxicity provided by pretreatment with PPs may be due to increased levels of hepatic glutathione which potentially traps the acetaminophen electrophile, thus preventing covalent binding to macromolecules.[238] In support of this hypothesis, biliary excretion of 3-glutathionyl-acetaminophen is increased twofold and adduct formation of hepatic proteins is reduced in mice pretreated with clofibrate.[239] In contrast, *in vitro* evidence suggests that clofibrate-induced hepatoprotection against acetaminophen is not the result of altered glutathione regulation.[240] Thus, the mechanisms underlying the effect of PPs in preventing hepatotoxicity of acetaminophen are unclear. While it has recently been shown that this protective effect requires the PPARα,[241] the specific changes in gene expression that mediate this phenomenon have not been identified.

## 14.2.4.6 SUMMARY AND CONCLUSION

In summary, PPs are a diverse class of chemicals that are capable of producing a consistent pleiotropic response. The effects observed in rodent liver include peroxisome proliferation, hepatomegaly, regulation of gene expression, cell cycle control and ultimately, carcinogenesis. The majority of biological effects induced by PPs are mediated by the PPARα, and include physiological, toxicological and carcinogenic pathways. Acting as transcription factors, PPARα functions in classic nuclear receptor fashion by ligand binding, heterodimerization, recruitment/dissociation of co-activators/co-repressors, and modulation of transcription after binding to specific responsive elements of target genes. Definitive evidence of the biological and toxicological role of this receptor has been provided by extensive analysis of PPARα-null mice. The mechanisms underlying PP-induced hepatocarcinogenesis are not clear although oxidative stress, increased cell proliferation and/or repressed apoptosis are likely to have roles in this process. Whether or not the hepatocarcinogenic effect of PPs routinely observed in rodent models is relevant to humans is a controversial issue and requires ongoing molecular and cellular research. Given the apparent lower levels of PPARα expression, potential mutant forms of this receptor subtype, and putative differences in PPREs in target genes, humans appear to be somewhat refractory to the carcinogenic effect of these chemicals.

## 14.2.4.7 REFERENCES

1. J. Rhodin, in 'Correlation of ultrastructural organisation and function in normal and experimentally changed proximal tubule cells of the mouse kidney', Karolinska Institute, Stockholm, Aktiebolaget Godvil, 1954, p. 76.
2. H. Beevers, 'Microbodies in higher plants.' *Ann. Rev. Plant Physiol.*, 1979, **30**, 159–179.
3. A. B. Novikoff, P. M. Novikoff, C. Davis *et al.*, 'Studies on microperoxisomes. V. Are microperoxisomes ubiquitous in mammalian cells?' *J. Histochem. Cytochem.*, 1973, **21**, 737–755.
4. Z. Hruban, E. L. Vigil, A. Slesers *et al.*, 'Microbodies: constituent organelles of animal cells.' *Lab. Invest.*, 1972, **27**, 184–191.
5. F. Leighton, B. Poole, H. Beaufay *et al.*, 'The large-scale separation of peroxisomes, mitochondria, and lysosomes from the livers of rats injected with triton WR-1339. Improved isolation procedures, automated

analysis, biochemical and morphological properties of fractions.' *J. Cell Biol.*, 1968, **37**, 482–513.

6. E. R. Weibel, W. Staubli, H. R. Gnagi *et al.*, 'Correlated morphometric and biochemical studies on the liver cell. I. Morphometric model, stereologic methods, and normal morphometric data for rat liver.' *J. Cell Biol.*, 1969, **42**, 68–91.

7. C. de Duve, 'Functions of microbodies (peroxisomes).' *J. Cell Biol.*, 1965, **27**, 25A.

8. C. De Duve and P. Baudhuin, 'Peroxisomes (microbodies and related particles).' *Physiol. Rev.*, 1966, **46**, 323–357.

9. G. P. Mannaerts and P. P. Veldhoven, in 'Metabolic role of mammalian peroxisomes', eds. G. Gibson and B. Lake, Taylor and Francis, Washington D.C., 1993, pp. 19–62.

10. R. Hess, W. Staubli and W. Reiss, 'Nature of the hepatomegalic effect produced by ethyl-chlorphenoxyisobutyrate in the rat.' *Nature*, 1965, **208**, 856–858.

11. G. E. Paget, 'Experimental studies of toxicity of atromid with particular reference to the fine structural changes in the livers of rodents.' *J. Artheroscler. Res.*, 1963, **3**, 729–736.

12. D. J. Svoboda and D. Azarnoff, 'Response of hepatic microbodies to a hypolipidemic agent, ethyl-chlorophenoxybutyrate (CBIP).' *J. Cell Biol.*, 1966, **30**, 442–450.

13. J. K. Reddy, S. Rao and D. E. Moody, 'Hepatocellular carcinomas in acatalasemic mice treated with nafenopin, a hypolipidemic peroxisome proliferator.' *Cancer Res.*, 1976, **36**, 1211–1217.

14. J. K. Reddy and N. D. Lalwai, 'Carcinogenesis by hepatic peroxisome proliferators: evaluation of the risk of hypolipidemic drugs and industrial plasticizers to humans.' *Crit. Rev. Toxicol.*, 1983, **12**, 1–58.

15. I. Issemann and S. Green, 'Activation of a member of the steroid hormone receptor superfamily by peroxisome proliferators.' *Nature*, 1990, **347**, 645–650.

16. P. Escher and W. Wahli, 'Peroxisome proliferator-activated receptors: insight into multiple cellular functions.' *Mutat. Res.*, 2000, **448**, 121–138.

17. C. Dreyer, G. Krey, H. Keller *et al.*, 'Control of the peroxisomal beta-oxidation pathway by a novel family of nuclear hormone receptors.' *Cell*, 1992, **68**, 879–887.

18. O. Braissant, F. Foufelle, C. Scotto *et al.*, 'Differential expression of peroxisome proliferator-activated receptors (PPARs): tissue distribution of PPAR-α, -β, and -γ in the adult rat.' *Endocrinology*, 1996, **137**, 354–366.

19. L. Fajas, D. Auboeuf, E. Raspe *et al.*, 'The organization, promoter analysis, and expression of the human PPARγ gene.' *J. Biol. Chem.*, 1997, **272**, 18779–18789.

20. R. M. Evans, 'The steroid and thyroid hormone receptor superfamily.' *Science*, 1988, **240**, 889–895.

21. M. Gottlicher, E. Widmark, Q. Li *et al.*, 'Fatty acids activate a chimera of the clofibric acid-activated receptor and the glucocorticoid receptor.' *Proc. Natl. Acad. Sci. USA*, 1992, **89**, 4653–4657.

22. T. Sher, H. F. Yi, O. W. McBride *et al.*, 'cDNA cloning, chromosomal mapping, and functional characterization of the human peroxisome proliferator activated receptor.' *Biochemistry*, 1993, **32**, 5598–5604.

23. J. Berger, P. Bailey, C. Biswas *et al.*, 'Thiazolidinediones produce a conformational change in peroxisomal proliferator-activated receptor-γ: binding and activation correlate with antidiabetic actions in db/db mice.' *Endocrinology*, 1996, **137**, 4189–4195.

24. P. Dowell, V. J. Peterson, T. M. Zabriskie *et al.*, 'Ligand-induced peroxisome proliferator-activated receptor alpha conformational change.' *J. Biol. Chem.*, 1997, **272**, 2013–2020.

25. G. Bunone, P. A. Briand, R. J. Miksicek *et al.*, 'Activation of the unliganded estrogen receptor by EGF involves the MAP kinase pathway and direct phosphorylation.' *Embo. J.*, 1996, **15**, 2174–2183.

26. M. K. El-Tanani and C. D. Green, 'Two separate mechanisms for ligand-independent activation of the estrogen receptor.' *Mol. Endocrinol.*, 1997, **11**, 928–937.

27. J. P. Vanden Heuvel, 'Peroxisome proliferator-activated receptors (PPARS) and carcinogenesis.' *Toxicol. Sci.*, Review, 1999, **47**, 1–8.

28. T. M. Willson, P. J. Brown, D. D. Sternbach *et al.*, 'The PPARs: from orphan receptors to drug discovery.' *J. Med. Chem.*, 2000, **43**, 527–550.

29. S. Y. Moya-Camarena, J. P. Vanden Heuvel, S. G. Blanchard *et al.*, 'Conjugated linoleic acid is a potent naturally occurring ligand and activator of PPARα.' *J. Lipid. Res.*, 1999, **40**, 1426–1433.

30. C. Sumida, 'Fatty acids: ancestral ligands and modern co-regulators of the steroid hormone receptor cell signalling pathway.' *Prostaglandins Leukot. Essent. Fatty Acids*, 1995, **52**, 137–144.

31. S. A. Kliewer, K. Umesono, D. J. Noonan *et al.*, 'Convergence of 9-*cis* retinoic acid and peroxisome proliferator signalling pathways through heterodimer formation of their receptors.' *Nature*, 1992, **358**, 771–774.

32. J. D. Tugwood, I. Issemann, R. G. Anderson *et al.*, 'The mouse peroxisome proliferator activated receptor recognizes a response element in the 5′ flanking sequence of the rat acyl CoA oxidase gene.' *Embo. J.*, 1992, **11**, 433–439.

33. B. Zhang, S. L. Marcus, F. G. Sajjadi *et al.*, 'Identification of a peroxisome proliferator-responsive element upstream of the gene encoding rat peroxisomal enoyl-CoA hydratase/3-hydroxyacyl-CoA dehydrogenase.' *Proc. Natl. Acad. Sci. USA*, 1992, **89**, 7541–7545.

34. I. Issemann, R. Prince, J. Tugwood *et al.*, 'A role for fatty acids and liver fatty acid binding protein in peroxisome proliferation?' *Biochem. Soc. Trans.*, 1992, **20**, 824–827.

35. A. S. Muerhoff, K. J. Griffin and E. F. Johnson, 'The peroxisome proliferator-activated receptor mediates the induction of CYP4A6, a cytochrome P450 fatty acid omega-hydroxylase, by clofibric acid.' *J. Biol. Chem.*, 1992, **267**, 19051–19053.

36. C. Juge-Aubry, A. Pernin, T. Favez *et al.*, 'DNA binding properties of peroxisome proliferator-activated receptor subtypes on various natural peroxisome proliferator response elements. Importance of the 5′-flanking region.' *J. Biol. Chem.*, 1997, **272**, 25252–25259.

37. A. Jpenberg, E. Jeannin, W. Wahli *et al.*, 'Polarity and specific sequence requirements of peroxisome proliferator-activated receptor (PPAR)/ retinoid X receptor heterodimer binding to DNA. A functional analysis of the malic enzyme gene PPAR response element.' *J. Biol. Chem.*, 1997, **272**, 20108–20117.

38. K. Schoonjans, J. Peinado-Onsurbe, A. M. Lefebvre *et al.*, 'PPARα and PPARγ activators direct a distinct tissue-specific transcriptional response via a PPRE in the lipoprotein lipase gene.' *EMBO. J.*, 1996, **15**, 5336–5348.

39. C. Nishiyama, R. Hi, S. Osada *et al.*, 'Functional interactions between nuclear receptors recognizing a common sequence element, the direct repeat motif spaced by one nucleotide (DR-1).' *J. Biochem. (Tokyo)*, 1998, **123**, 1174–1179.

40. S. L. Marcus, J. P. Capone and R. A. Rachubinski, 'Identification of COUP-TFII as a peroxisome pro-

liferator response element binding factor using genetic selection in yeast: COUP-TFII activates transcription in yeast but antagonizes PPAR signaling in mammalian cells.' *Mol. Cell Endocrinol.*, 1996, **120**, 31–39.

41. K. S. Miyata, B. Zhang, S. L. Marcus *et al.*, 'Chicken ovalbumin upstream promoter transcription factor (COUP-TF) binds to a peroxisome proliferator-responsive element and antagonizes peroxisome proliferator-mediated signaling.' *J. Biol. Chem.*, 1993, **268**, 19169–19172.

42. H. Nakshatri and P. Bhat-Nakshatri, 'Multiple parameters determine the specificity of transcriptional response by nuclear receptors HNF-4, ARP-1, PPAR, RAR and RXR through common response elements.' *Nucleic Acids Res.*, 1998, **26**, 2491–2499.

43. C. J. Winrow, J. P. Capone and R. A. Rachubinski, 'Cross-talk between orphan nuclear hormone receptor RZRα and peroxisome proliferator-activated receptor alpha in regulation of the peroxisomal hydratase-dehydrogenase gene.' *J. Biol. Chem.*, 1998, **273**, 31442–31448.

44. Z. H. Yan, W. G. Karam, J. L. Staudinger *et al.*, 'Regulation of peroxisome proliferator-activated receptor alpha-induced transactivation by the nuclear orphan receptor TAK1/TR4.' *J. Biol. Chem.*, 1998, **273**, 10948–10957.

45. T. Miyamoto, A. Kaneko, T. Kakizawa *et al.*, 'Inhibition of peroxisome proliferator signaling pathways by thyroid hormone receptor. Competitive binding to the response element.' *J. Biol. Chem.*, 1997, **272**, 7752–7758.

46. J. Hunter, A. Kassam, C. J. Winrow *et al.*, 'Crosstalk between the thyroid hormone and peroxisome proliferator-activated receptors in regulating peroxisome proliferator-responsive genes.' *Mol. Cell Endocrinol.*, 1996, **116**, 213–221.

47. A. N. Hollenberg, V. S. Susulic, J. P. Madura *et al.*, 'Functional antagonism between CCAAT/Enhancer binding protein-α and peroxisome proliferator-activated receptor-γ on the leptin promoter.' *J. Biol. Chem.*, 1997, **272**, 5283–5290.

48. K. S. Miyata, S. E. McCaw, H. V. Patel *et al.*, 'The orphan nuclear hormone receptor LXR alpha interacts with the peroxisome proliferator-activated receptor and inhibits peroxisome proliferator signaling.' *J. Biol. Chem.*, 1996, **271**, 9189–9192.

49. R. M. Tolon, A. I. Castillo and A. Aranda, 'Activation of the prolactin gene by peroxisome proliferator-activated receptor-α appears to be DNA binding-independent.' *J. Biol. Chem.*, 1998, **273**, 26652–26661.

50. M. Ricote, J. T. Huang, J. S. Welch *et al.*, 'The peroxisome proliferator-activated receptor-gamma is a negative regulator of macrophage activation.' *Nature*, 1998, **391**, 79–82.

51. M. Sakai, Y. Matsushima-Hibiya, M. Nishizawa *et al.*, 'Suppression of rat glutathione transferase P expression by peroxisome proliferators: interaction between Jun and peroxisome proliferator-activated receptor alpha.' *Cancer Res.*, 1995, **55**, 5370–5376.

52. A. S. Mohamood, P. Gyles, K. V. Balan *et al.*, 'Estrogen receptor, growth factor receptor and protooncogene protein activities and possible signal transduction crosstalk in estrogen dependent and independent breast cancer cell lines.' *J. Submicrosc. Cytol. Pathol.*, 1997, **29**, 1–17.

53. X. K. Zhang, K. N. Wills, M. Husmann *et al.*, 'Novel pathway for thyroid hormone receptor action through interaction with jun and fos oncogene activities.' *Mol. Cell Biol.*, 1991, **11**, 6016–6025.

54. H. Y. Yoo, M. S. Chang and H. M. Rho, 'Induction of the rat Cu/Zn superoxide dismutase gene through

the peroxisome proliferator-responsive element by arachidonic acid.' *Gene*, 1999, **234**, 87–91.

55. M. E. Maragoudakis and H. Hankin, 'On the mode of action of lipid-lowering agents. V. Kinetics of the inhibition *in vitro* of rat acetyl coenzyme A carboxylase.' *J. Biol. Chem.*, 1971, **246**, 348–358.

56. H. Mizuguchi, N. Kudo, T. Ohya *et al.*, 'Effects of tiadenol and di-(2-ethylhexyl)phthalate on the metabolism of phosphatidylcholine and phosphatidylethanolamine in the liver of rats: comparison with clofibric acid.' *Biochem. Pharmacol.*, 1999, **57**, 869–876.

57. K. B. Horwitz, T. A. Jackson, D. L. Bain *et al.*, 'Nuclear receptor coactivators and corepressors.' *Mol. Endocrinol.*, 1996, **10**, 1167–1177.

58. T. Lemberger, O. Braissant, C. Juge-Aubry *et al.*, 'PPAR tissue distribution and interactions with other hormone-signaling pathways.' *Ann. NY Acad. Sci.*, 1996, **804**, 231–251.

59. H. E. Xu, M. H. Lambert, V. G. Montana *et al.*, 'Molecular recognition of fatty acids by peroxisome proliferator-activated receptors.' *Mol. Cell.*, 1999, **3**, 397–403.

60. R. T. Nolte, G. B. Wisely, S. Westin *et al.*, 'Ligand binding and co-activator assembly of the peroxisome proliferator-activated receptor-gamma.' *Nature*, 1998, **395**, 137–143.

61. P. Gervois, I. P. Torra, G. Chinetti *et al.*, 'A truncated human peroxisome proliferator-activated receptor alpha splice variant with dominant negative activity.' *Mol. Endocrinol.*, 1999, **13**, 1535–1549.

62. R. A. Roberts, N. H. James, N. J. Woodyatt *et al.*, 'Evidence for the suppression of apoptosis by the peroxisome proliferator activated receptor alpha (PPAR alpha).' *Carcinogenesis*, 1998, **19**, 43–48.

63. D. M. Flavell, I. Pineda Torra, Y. Jamshidi *et al.*, 'Variation in the PPARalpha gene is associated with altered function *in vitro* and plasma lipid concentrations in Type II diabetic subjects.' *Diabetologia*, 2000, **43**, 673–680.

64. A. Sapone, J. M. Peters, S. Sakai *et al.*, 'The human peroxisome proliferator-activated receptor alpha gene: identification and functional characterization of two natural allelic variants (In Process Citation).' *Pharmacogenetics*, 2000, **10**, 321–333.

65. M. C. Vohl, P. Lepage, D. Gaudet *et al.*, 'Molecular scanning of the human PPARa gene. Association of the l162v mutation with hyperapobetalipoproteinemia (in process citation).' *J. Lipid Res.*, 2000, **41**, 945–952.

66. S. Kersten, B. Desvergne and W. Wahli, 'Roles of PPARs in health and disease.' *Nature*, 2000, **405**, 421–424.

67. I. Barroso, M. Gurnell, V. E. Crowley *et al.*, 'Dominant negative mutations in human PPARγ associated with severe insulin resistance, diabetes mellitus and hypertension.' *Nature*, 1999, **402**, 880–883.

68. S. S. Lee, T. Pineau, J. Drago *et al.*, 'Targeted disruption of the alpha isoform of the peroxisome proliferator-activated receptor gene in mice results in abolishment of the pleiotropic effects of peroxisome proliferators.' *Mol. Cell Biol.*, 1995, **15**, 3012–3022.

69. J. M. Peters, Y. C. Zhou, P. A. Ram *et al.*, 'Peroxisome proliferator-activated receptor alpha required for gene induction by dehydroepiandrosterone-3 beta-sulfate.' *Mol. Pharmacol.*, 1996, 50, 67–74.

70. J. M. Peters, N. Hennuyer, B. Staels *et al.*, 'Alterations in lipoprotein metabolism in peroxisome proliferator-activated receptor alpha-deficient mice.' *J. Biol. Chem.*, 1997, **272**, 27307–27312.

71. J. M. Peters, M. W. Taubeneck, C. L. Keen *et al.*, 'Di(2-ethylhexyl) phthalate induces a functional zinc

deficiency during pregnancy and teratogenesis that is independent of peroxisome proliferator-activated receptor-alpha.' *Teratology*, 1997, **56**, 311–316.

72. K. Motojima, J. M. Peters and F. J. Gonzalez, 'PPAR alpha mediates peroxisome proliferator-induced transcriptional repression of nonperoxisomal gene expression in mouse.' *Biochem. Biophys. Res. Commun.*, 1997, **230**, 155–158.

73. J. M. Peters, R. C. Cattley and F. J. Gonzalez, 'Role of PPAR alpha in the mechanism of action of the nongenotoxic carcinogen and peroxisome proliferator Wy-14,643.' *Carcinogenesis*, 1997, **18**, 2029–2033.

74. J. M. Peters, T. Aoyama, R. C. Cattley *et al.*, 'Role of peroxisome proliferator-activated receptor alpha in altered cell cycle regulation in mouse liver.' *Carcinogenesis*, 1998, **19**, 1989–1994.

75. B. Ren, A. P. Thelen, J. M. Peters *et al.*, 'Polyunsaturated fatty acid suppression of hepatic fatty acid synthase and S14 gene expression does not require peroxisome proliferator-activated receptor alpha.' *J. Biol. Chem.*, 1997, **272**, 26827–26832.

76. D. A. Pan, M. K. Mater, A. P. Thelen *et al.*, 'Evidence against the peroxisome proliferator-activated receptor alpha (PPARalpha) as the mediator for polyunsaturated fatty acid suppression of hepatic L-pyruvate kinase gene transcription.' *J. Lipid Res.*, 2000, **41**, 742–751.

77. M. C. Hunt, P. J. Lindquist, J. M. Peters *et al.*, 'Involvement of the peroxisome proliferator-activated receptor alpha in regulating long-chain acyl-CoA thioesterases.' *J. Lipid Res.*, 2000, **41**, 814–823.

78. K. Motojima, P. Passilly, J. M. Peters *et al.*, 'Expression of putative fatty acid transporter genes are regulated by peroxisome proliferator-activated receptor alpha and gamma activators in a tissue- and inducer-specific manner.' *J. Biol. Chem.*, 1998, **273**, 16710–16714.

79. S. Kersten, J. Seydoux, J. M. Peters *et al.*, 'Peroxisome proliferator-activated receptor alpha mediates the adaptive response to fasting.' *J. Clin. Invest.*, 1999, **103**, 1489–1498.

80. A. Aoyama, J. M. Peters, N. Iritani *et al.*, 'Altered constitutive expression of fatty acid-metabolizing enzymes in mice lacking the peroxisome proliferator-activated receptor alpha (PPARα).' *J. Biol. Chem.*, 1998, **273**, 5678–5684.

81. D. L. Kroetz, P. Yook, P. Costet *et al.*, 'Peroxisome proliferator-activated receptor alpha controls the hepatic CYP4A induction adaptive response to starvation and diabetes.' *J. Biol. Chem.*, 1998, **273**, 31581–31589.

82. S. P. Anderson, R. C. Cattley and J. C. Corton, 'Hepatic expression of acute-phase protein genes during carcinogenesis induced by peroxisome proliferators.' *Mol. Carcinog.*, 1999, **26**, 226–238.

83. T. B. Barclay, J. M. Peters, M. B. Sewer *et al.*, 'Modulation of cytochrome P450 gene expression in endotoxemic mice is tissue specific and peroxisome proliferator-activated receptor-α dependent.' *J. Pharmacol. Exp. Ther.*, 1999, **290**, 1250–1257.

84. M. A. Belury, S. Y. Moya-Camarena, H. Sun *et al.*, 'Comparison of dose-response relationships for induction of lipid metabolizing and growth regulatory genes by peroxisome proliferators in rat liver.' *Toxicol. Appl. Pharmacol.*, 1998, **151**, 254–261.

85. J. M. Brandt, F. Djouadi and D. P. Kelly, 'Fatty acids activate transcription of the muscle carnitine palmitoyltransferase I gene in cardiac myocytes via the peroxisome proliferator-activated receptor alpha.' *J. Biol. Chem.*, 1998, **273**, 23786–23792.

86. J. C. Corton, L. Q. Fan, S. Brown *et al.*, 'Down-regulation of cytochrome P450 2C family members and positive acute-phase response gene expression by

87. P. Delerive, K. De Bosscher, S. Besnard *et al.*, 'Peroxisome proliferator-activated receptor alpha negatively regulates the vascular inflammatory gene response by negative cross-talk with transcription factors NF-kappaB and AP-1.' *J. Biol. Chem.*, 1999, **274**, 32048–32054.

88. F. Djouadi, C. J. Weinheimer, J. E. Saffitz *et al.*, 'A gender-related defect in lipid metabolism and glucose homeostasis in peroxisome proliferator-activated receptor alpha-deficient mice.' *J. Clin. Invest.*, 1998, **102**, 1083–1091.

89. F. Djouadi, J. M. Brandt, C. J. Weinheimer *et al.*, 'The role of the peroxisome proliferator-activated receptor alpha (PPAR alpha) in the control of cardiac lipid metabolism.' *Prostaglandins Leukot. Essent. Fatty Acids*, 1999, **60**, 339–343.

90. P. R. Devchand, H. Keller, J. M. Peters *et al.*, 'The PPARalpha-leukotriene B4 pathway to inflammation control.' *Nature*, 1996, **384**, 39–43.

91. L. Q. Fan, R. C. Cattley and J. C. Corton, 'Tissue-specific induction of 17 beta-hydroxysteroid dehydrogenase type IV by peroxisome proliferator chemicals is dependent on the peroxisome proliferator-activated receptor alpha.' *J. Endocrinol.*, 1998, **158**, 237–246.

92. L. G. Guidotti, C. M. Eggers, A. K. Raney *et al.*, 'In vivo regulation of hepatitis B virus replication by peroxisome proliferators.' *J. Virol.*, 1999, **73**, 10377–10386.

93. M. Kockx, P. P. Gervois, P. Poulain *et al.*, 'Fibrates suppress fibrinogen gene expression in rodents via activation of the peroxisome proliferator-activated receptor-alpha.' *Blood*, 1999, **93**, 2991–2998.

94. T. C. Leone, C. J. Weinheimer and D. P. Kelly, 'A critical role for the peroxisome proliferator-activated receptor alpha (PPARalpha) in the cellular fasting response: the PPARalpha-null mouse as a model of fatty acid oxidation disorders.' *Proc. Natl. Acad. Sci. USA*, 1999, **96**, 7473–7478.

95. T. Nakajima, Y. Kamijo, N. Usuda *et al.*, 'Sex-dependent regulation of hepatic peroxisome proliferation in mice by trichloroethylene via peroxisome proliferator-activated receptor alpha (PPARα).' *Carcinogenesis*, 2000, **21**, 677–682.

96. M. E. Poynter and R. A. Daynes, 'Peroxisome proliferator-activated receptor alpha activation modulates cellular redox status, represses nuclear factor-kappaB signaling, and reduces inflammatory cytokine production in aging.' *J. Biol. Chem.*, 1998, **273**, 32833–32841.

97. J. M. Ward, J. M. Peters, C. M. Perella *et al.*, 'Receptor and nonreceptor-mediated organ-specific toxicity of di(2-ethylhexyl)phthalate (DEHP) in peroxisome proliferator-activated receptor alpha-null mice.' *Toxicol. Pathol.*, 1998, **26**, 240–246.

98. J. P. Vanden Heuvel, P. Holden, J. Tugwood *et al.*, 'Identification of a novel peroxisome proliferator responsive cDNA isolated from rat hepatocytes as the zinc-finger protein ZFP-37.' *Toxicol. Appl. Pharmacol.*, 1998, **152**, 107–118.

99. F. Gueraud, J. Alary, P. Costet *et al.*, 'In vivo involvement of cytochrome P450 4A family in the oxidative metabolism of the lipid peroxidation product trans-4-hydroxy-2-nonenal, using PPARalpha-deficient mice.' *J. Lipid. Res.*, 1999, **40**, 152–159.

100. A. D. Edgar, C. Tomkiewicz, P. Costet *et al.*, 'Fenofibrate modifies transaminase gene expression via a peroxisome proliferator activated receptor alpha-dependent pathway.' *Toxicol. Lett.*, 1998, **98**, 13–23.

101. P. Costet, C. Legendre, J. More *et al.*, 'Peroxisome proliferator-activated receptor alpha-isoform defi-

ciency leads to progressive dyslipidemia with sexually dimorphic obesity and steatosis.' *J. Biol. Chem.*, 1998, **273**, 29577–29585.

102. C. L. May, T. Pineau, K. Bigot *et al.*, 'Reduced hepatic fatty acid oxidation in fasting PPARalpha null mice is due to impaired mitochondrial hydroxymethylglutaryl-CoA synthase gene expression.' *FEBS Letters*, 2000, **475**, 163–166.

103. M. Turunen, J. M. Peters, F. J. Gonzalez *et al.*, 'Influence of peroxisome proliferator-activated receptor alpha on ubiquinone biosynthesis.' *J. Mol. Biol.*, 2000, **297**, 607–614.

104. K. Watanabe, H. Fujii, T. Takahashi *et al.*, 'Constitutive regulation of cardiac fatty acid metabolism through peroxisome proliferator-activated receptor alpha associated with age-dependent cardiac toxicity.' *J. Biol. Chem.*, 2000, **275**, 22293–22299.

105. L. G. Komuves, K. Hanley, A. M. Lefebvre *et al.*, 'Stimulation of PPARalpha promotes epidermal keratinocyte differentiation *in vivo*.' *J. Invest. Dermatol.*, 2000, **115**, 353–360.

106. K. Hanley, L. G. Komuves, D. C. Ng *et al.*, 'Farnesol stimulates differentiation in epidermal keratinocytes via PPARalpha.' *J. Biol. Chem.*, 2000, **275**, 11484–11491.

107. Y. J. Wan, Y. Cai, W. Lungo *et al.*, 'Peroxisome proliferator-activated receptor alpha-mediated pathways are altered in hepatocyte-specific retinoid X receptor α-deficient mice.' *J. Biol. Chem.*, 2000, **275**, 28285–28290.

108. S. Kersten, S. Mandard, N. S. Tan *et al.*, 'Characterization of the fasting-induced adipose factor FIAF, a novel peroxisome proliferator-activated receptor target gene.' *J. Biol. Chem.*, 2000, **275**, 28488–28493.

109. T. Hashimoto, W. S. Cook, C. Qi *et al.*, 'Defect in peroxisome proliferator-activated receptor alpha-inducible fatty acid oxidation determines the severity of hepatic steatosis in response to fasting.' *J. Biol. Chem.*, 2000, **275**, 28918–28928.

110. C. Chen, G. E. Hennig, H. E. Whiteley *et al.*, 'Peroxisome proliferator-activated receptor alpha-null mice lack resistance to acetaminophen hepatotoxicity following clofibrate exposure.' *Toxicol. Sci.*, 2000, **57**, 338–344.

111. F. Casas, T. Pineau, P. Rochard *et al.*, 'New molecular aspects of regulation of mitochondrial activity by fenofibrate and fasting.' *FEBS Lett.*, 2000, **482**, 71–74.

112. D. D. Patel, B. L. Knight, A. K. Soutar *et al.*, 'The effect of peroxisome-proliferator-activated receptor-alpha on the activity of the cholesterol 7alpha-hydroxylase gene.' *Biochem. J.*, 2000, **351**(3), 747–753.

113. P. Delerive, P. Gervois, J. C. Fruchart *et al.*, 'Induction of ikappa balpha expression as a mechanism contributing to the anti-inflammatory activities of peroxisome proliferator-activated receptor-alpha activators.' *J. Biol. Chem.*, 2000, **275**, 36703–36707.

114. N. Macdonald, K. Barrow, R. Tonge *et al.*, 'PPAR-alpha-dependent alteration of GRP94 expression in mouse hepatocytes.' *Biochem. Biophys. Res. Commun.*, 2000, **277**, 699–704.

115. H. Lee, W. Shi, P. Tontonoz *et al.*, 'Role for peroxisome prolifcrator-activated receptor alpha in oxidized phospholipid-induced synthesis of monocyte chemotactic protein-1 and interleukin-8 by endothelial cells.' *Circ. Res.*, 2000, **87**, 516–521.

116. A. Hermanowski-Vosatka, D. Gerhold, S. S. Mundt *et al.*, 'PPARalpha agonists reduce 11beta-hydroxysteroid dehydrogenase type 1 in the liver.' *Biochem. Biophys. Res. Commun.*, 2000, **279**, 330–336.

117. G. Lazennec, L. Canaple, D. Saugy *et al.*, 'Activation of peroxisome proliferator-activated receptors (PPARs) by their ligands and protein kinase A activators.' *Mol. Endocrinol.*, 2000, **14**, 1962–1975.

118. J. F. Louet, F. Chatelain, J. F. Decaux *et al.*, 'Long-chain fatty acids regulate liver carnitine palmitoyltransferase I gene (L-CPT I) expression through a peroxisome-proliferator-activated receptor alpha (PPARalpha)-independent pathway.' *Biochem. J.*, 2001, **354**, 189–197.

119. J. Dallongeville, E. Bauge, A. Tailleux *et al.*, 'Peroxisome proliferator activated receptor alpha is not rate limiting for the lipoprotein lowering action of fish oil.' *J. Biol. Chem.*, 2001, **276**, 4634–4639.

120. B. Desvergne, I. J. A, P. R. Devchand *et al.*, 'The peroxisome proliferator-activated receptors at the cross-road of diet and hormonal signalling.' *J. Steroid Biochem. Mol. Biol.*, 1998, **65**, 65–74.

121. B. Desvergne and W. Wahli, 'Peroxisome proliferator-activated receptors: nuclear control of metabolism.' *Endocr. Rev.*, 1999, **20**, 649–688.

122. K. Schoonjans, B. Staels and J. Auwerx, 'The peroxisome proliferator activated receptors (PPARs) and their effects on lipid metabolism and adipocyte differentiation.' *Biochim. Biophys. Acta*, 1996, **1302**, 93–109.

123. W. Wahli, O. Braissant and B. Desvergne, 'Peroxisome proliferator activated receptors: transcriptional regulators of adipogenesis, lipid metabolism and more.' *Chem. Biol.*, 1995, **2**, 261–266.

124. G. Martin, K. Schoonjans and A. M. Lefebvre *et al.*, 'Coordinate regulation of the expression of the fatty acid transport protein and acyl-CoA synthetase genes by PPARalpha and PPARgamma activators.' *J. Biol. Chem.*, 1997, **272**, 28210–28217.

125. B. Staels and J. Auwerx, 'Perturbation of developmental gene expression in rat liver by fibric acid derivatives: lipoprotein lipase and alpha-fetoprotein as models.' *Development*, 1992, **115**, 1035–1043.

126. K. Schoonjans, M. Watanabe, H. Suzuki *et al.*, 'Induction of the acyl-coenzyme A synthetase gene by fibrates and fatty acids is mediated by a peroxisome proliferator response element in the C promoter.' *J. Biol. Chem.*, 1995, **270**, 19269–19276.

127. J. P. Vanden Heuvel, P. F. Sterchele, D. J. Nesbit *et al.*, 'Coordinate induction of acyl-CoA binding protein, fatty acid binding protein and peroxisomal β-oxidation by peroxisome proliferators.' *Biochem. Biophys. Acta*, 1993, **1177**, 183–190.

128. R. M. Kaikaus, W. K. Chan, N. Lysenko *et al.*, 'Induction of peroxisomal fatty acid beta-oxidation and liver fatty acid-binding protein by peroxisome proliferators. Mediation via the cytochrome P-450IVA1 omega-hydroxylase pathway.' *J. Biol. Chem.*, 1993, **268**, 9593–9603.

129. P. Besnard, A. Mallordy and H. Carlier, 'Transcriptional induction of the fatty acid binding protein gene in mouse liver by bezafibrate.' *FEBS Lett.*, 1993, **327**, 219–223.

130. Y. Andersson, Z. Majd, A. M. Lefebvre *et al.*, 'Developmental and pharmacological regulation of apolipoprotein C-II gene expression. Comparison with apo C-I and apo C-III gene regulation.' *Arterioscler. Thromb. Vasc. Biol.*, 1999, **19**, 115–121.

131. L. Berthou, R. Saladin, P. Yaqoob *et al.*, 'Regulation of rat liver apolipoprotein A-I, apolipoprotein A-II and acyl-coenzyme A oxidase gene expression by fibrates and dietary fatty acids.' *Eur. J. Biochem.*, 1995, **232**, 179–187.

132. B. Staels, A. van Tol, T. Andreu *et al.*, 'Fibrates influence the expression of genes involved in lipoprotein metabolism in a tissue-selective manner in the rat.' *Arterioscler. Thromb.*, 1992, **12**, 286–294.

133. B. Staels, N. Vu-Dac, V. A. Kosykh *et al.*, 'Fibrates downregulate apolipoprotein C-III expression inde-

pendent of induction of peroxisomal acyl coenzyme A oxidase. A potential mechanism for the hypolipidemic action of fibrates.' *J. Clin. Invest.*, 1995, **95**, 705–712.

134. C. Mascaro, E. Acosta, J. A. Ortiz *et al.*, 'Control of human muscle-type carnitine palmitoyltransferase I gene transcription by peroxisome proliferator-activated receptor.' *J. Biol. Chem.*, 1998, **273**, 8560–8563.

135. B. Staels, W. Koenig, A. Habib *et al.*, 'Activation of human aortic smooth-muscle cells is inhibited by PPARalpha but not by PPARgamma activators.' *Nature*, 1998, **393**, 790–793.

136. B. P. Neve, J. Fruchart and B. Staels, 'Role of the peroxisome proliferator-activated receptors (PPAR) in atherosclerosis.' *Biochem. Pharmacol.*, 2000, **60**, 1245–1250.

137. J. K. Reddy, D. L. Azarnoff and C. E. Hignite, 'Hypolipidaemic hepatic peroxisome proliferators form a novel class of chemical carcinogens.' *Nature*, 1980, **283**, 397–398.

138. J. Ashby, A. Brady, C. R. Elcombe *et al.*, 'Mechanistically-based human hazard assessment of peroxisome proliferator-induced hepatocarcinogenesis.' *Hum. Exp. Toxicol.*, 1994, **13**, S1–117.

139. J. K. Reddy, M. S. Rao, D. L. Azarnoff *et al.*, 'Mitogenic and carcinogenic effects of a hypolipidemic peroxisome proliferator, [4-chloro-6-(2,3-xylidino)-2-pyrimidinylthio]acetic acid (Wy-14, 643), in rat and mouse liver.' *Cancer Res.*, 1979, **39**, 152–161.

140. M. S. Rao, V. Subbarao, A. V. Yeldandi *et al.*, 'Hepatocarcinogenicity of dehydroepiandrosterone in the rat.' *Cancer Res.*, 1992, **52**, 2977–2979.

141. J. A. Popp, L. K. Garvey and R. C. Cattley, '*In vivo* studies on the mechanism of di(2-ethylhexyl)phthalate carcinogenesis.' *Toxicol. Ind. Health*, 1987, **3**, 151–163.

142. N. Motohashi, H. Nagashima and J. Molnar, 'Trichloroethylene. II. Mechanism of carcinogenicity of trichloroethylene.' *in vivo*, 1999, **13**, 215–219.

143. J. K. Reddy and M. S. Rao, 'Oxidative DNA damage caused by persistent peroxisome proliferation: its role in hepatocarcinogenesis.' *Mutat. Res.*, 1989, **214**, 63–68.

144. M. R. Nemali, N. Usuda, M. K. Reddy *et al.*, 'Comparison of constitutive and inducible levels of expression of peroxisomal beta-oxidation and catalase genes in liver and extrahepatic tissues of rat.' *Cancer Res.*, 1988, **48**, 5316–5324.

145. D. S. Marsman, R. C. Cattley, J. G. Conway *et al.*, 'Relationship of hepatic peroxisome proliferation and replicative DNA synthesis to the hepatocarcinogenicity of the peroxisome proliferators di(2-ethylhexyl)phthalate and [4-chloro-6-(2,3-xylidino)-2-pyrimidinylthio]acetic acid (Wy-14,643) in rats.' *Cancer Res.*, 1988, **48**, 6739–6744.

146. J. G. Conway, K. E. Tomaszewski, M. J. Olson *et al.*, 'Relationship of oxidative damage to the hepatocarcinogenicity of the peroxisome proliferators di(2-ethylhexyl)phthalate and Wy-14,643.' *Carcinogenesis*, 1989, **10**, 513–519.

147. D. S. Marsman, T. L. Goldsworthy and J. A. Popp, 'Contrasting hepatocytic peroxisome proliferation, lipofuscin accumulation and cell turnover for the hepatocarcinogens Wy-14,643 and clofibric acid.' *Carcinogenesis*, 1992, **13**, 1011–1017.

148. J. G. Conway and J. A. Popp, 'Effect of the hepatocarcinogenic peroxisome proliferator Wy-14,643 *in vivo*: no increase in ethane exhalation or hepatic conjugated dienes.' *Toxicol. Appl. Pharmacol.*, 1995, **135**, 229–236.

149. M. S. Soliman, M. L. Cunningham, J. D. Morrow *et al.*, 'Evidence against peroxisome proliferation-induced hepatic oxidative damage.' *Biochem. Pharmacol.*, 1997, **53**, 1369–1374.

150. H. Kasai, Y. Okada, S. Nishimura *et al.*, 'Formation of 8-hydroxydeoxyguanosine in liver DNA of rats following long-term exposure to a peroxisome proliferator.' *Cancer Res.*, 1989, **49**, 2603–2605.

151. A. Takagi, K. Sai, T. Umemura *et al.*, 'Relationship between hepatic peroxisome proliferation and 8-hydroxydeoxyguanosine formation in liver DNA of rats following long-term exposure to three peroxisome proliferators; di(2-ethylhexyl) phthalate, aluminium clofibrate and simfibrate.' *Cancer Lett.*, 1990, **53**, 33–38.

152. R. C. Cattley and S. E. Glover, 'Elevated 8-hydroxydeoxyguanosine in hepatic DNA of rats following exposure to peroxisome proliferators: relationship to carcinogenesis and nuclear localization.' *Carcinogenesis*, 1993, **14**, 2495–2499.

153. P. J. Sausen, D. C. Lee, M. L. Rose *et al.*, 'Elevated 8-hydroxydeoxyguanosine in hepatic DNA of rats following exposure to peroxisome proliferators: relationship to mitochondrial alterations.' *Carcinogenesis*, 1995, **16**, 1795–1801.

154. J. Doull, R. Cattley, C. Elcombe *et al.*, 'A cancer risk assessment of di(2-ethylhexyl)phthalate: application of the new U.S. EPA Risk Assessment Guidelines.' *Regul. Toxicol. Pharmacol.*, 1999, **29**, 327–357.

155. M. E. Hegi, D. Ulrich, P. Sagelsdorff *et al.*, 'No measurable increase in thymidine glycol or 8-hydroxydeoxyguanosine in liver DNA of rats treated with nafenopin or choline-devoid low-methionine diet.' *Mutat. Res.*, 1990, **238**, 325–329.

156. S. M. Galloway, T. E. Johnson, M. J. Armstrong *et al.*, 'The genetic toxicity of the peroxisome proliferator class of rodent hepatocarcinogen.' *Mutat. Res.*, 2000, **448**, 153–158.

157. P. A. Lefevre, H. Tinwell, S. M. Galloway *et al.*, 'Evaluation of the genetic toxicity of the peroxisome proliferator and carcinogen methyl clofenapate, including assays using Muta Mouse and Big Blue transgenic mice.' *Hum. Exp. Toxicol.*, 1994, **13**, 764–775.

158. S. Chu, Q. Huang, K. Alvares *et al.*, 'Transformation of mammalian cells by overexpressing $H_2O_2$-generating peroxisomal fatty acyl-CoA oxidase.' *Proc. Natl. Acad. Sci. USA*, 1995, **92**, 7080–7084.

159. C. Y. Fan, J. Pan, R. Chu *et al.*, 'Hepatocellular and hepatic peroxisomal alterations in mice with a disrupted peroxisomal fatty acyl-coenzyme A oxidase gene.' *J. Biol. Chem.*, 1996, **271**, 24698–24710.

160. J. K. Reddy, D. L. Azarnoff and C. R. Sirtori, 'Hepatic peroxisome proliferation: induction by BR-931, a hypolipidemic analog of WY-14,643.' *Arch. Int. Pharmacodyn. Ther.*, 1978, **234**, 4–14.

161. S. R. Eldridge, L. F. Tilbury, T. L. Goldsworthy *et al.*, 'Measurement of chemically induced cell proliferation in rodent liver and kidney: a comparison of 5-bromo-2′-deoxyuridine and [3H]thymidine administered by injection or osmotic pump.' *Carcinogenesis*, 1990, **11**, 2245–2251.

162. P. I. Eacho, T. L. Lanier, C. A. Brodhecker, 'Hepatocellular DNA synthesis in rats given peroxisome proliferating agents: comparison of WY-14,643 to clofibric acid, nafenopin and LY171883.' *Carcinogenesis*, 1991, **12**, 1557–1561.

163. M. Cherkaoui Malki, P. Passilly, B. Jannin *et al.*, 'Carcinogenic aspect of xenobiotic molecules belonging to the peroxisome proliferator family.' *Int. J. Mol. Med.*, 1999, **3**, 163–168. (Record as supplied by publisher.)

164. B. J. Ledwith, S. Manam, P. Troilo *et al.*, 'Activation of immediate-early gene expression by peroxisome proliferators *in vitro*.' *Mol. Carcinog.*, 1993, **8**, 20–27.

165. T. L. Goldsworthy, S. M. Goldsworthy, C. S. Sprankle *et al.*, 'Expression of myc, fos and Ha-ras associated with chemically induced cell proliferation in the rat liver.' *Cell Prolif.*, 1994, **27**, 269–278.

166. O. Bardot, M. C. Clemencet, M. C. Malki *et al.*, 'Delayed effects of ciprofibrate on rat liver peroxisomal properties and proto-oncogene expression.' *Biochem. Pharmacol.*, 1995, **50**, 1001–1006.

167. R. T. Miller, S. E. Glover, W. S. Stewart *et al.*, 'Effect on the expression of c-met, c-myc and PPAR-alpha in liver and liver tumors from rats chronically exposed to the hepatocarcinogenic peroxisome proliferator WY-14,643.' *Carcinogenesis*, 1996, **17**, 1337–1341.

168. B. J. Ledwith, T. E. Johnson, L. K. Wagner *et al.*, 'Growth regulation by peroxisome proliferators: opposing activities in early and late G1.' *Cancer Res.*, 1996, **56**, 3257–3264.

169. B. J. Ledwith, C. J. Pauley, L. K. Wagner *et al.*, 'Induction of cyclooxygenase-2 expression by peroxisome proliferators and non-tetradecanoylphorbol 12,13-myristate-type tumor promoters in immortalized mouse liver cells.' *J. Biol. Chem.*, 1997, **272**, 3707–3714.

170. C. L. Rokos and B. J. Ledwith, 'Peroxisome proliferators activate extracellular signal-regulated kinases in immortalized mouse liver cells.' *J. Biol. Chem.*, 1997, **272**, 13452–13457.

171. X. Ma, D. A. Stoffregen, G. D. Wheelock *et al.*, 'Discordant hepatic expression of the cell division control enzyme p34cdc2 kinase, proliferating cell nuclear antigen, p53 tumor suppressor protein, and p21Waf1 cyclin-dependent kinase inhibitory protein after WY14,643 ([4-chloro-6-(2,3-xylidino)-2-pyrimidinylthio]acetic acid) dosing to rats.' *Mol. Pharmacol.*, 1997, **51**, 69–78.

172. J. A. Rininger, G. D. Wheelock, X. Ma *et al.*, 'Discordant expression of the cyclin-dependent kinases and cyclins in rat liver following acute administration of the hepatocarcinogen [4-chloro-6-(2,3-xylidino)-2-pyrimidinylthio] acetic acid (WY14,643).' *Biochem. Pharmacol.*, 1996, **52**, 1749–1755.

173. Y. Motoki, H. Tamura, R. Morita *et al.*, 'Decreased hepatocyte growth factor level by Wy-14,643, non-genotoxic hepatocarcinogen in F-344 rats.' *Carcinogenesis*, 1997, **18**, 1303–1309.

174. Y. Motoki, H. Tamura, T. Watanabe *et al.*, 'Wy-14,643, a peroxisome proliferator, inhibits compensative cell proliferation and hepatocyte growth factor mRNA expression in the rat liver.' *Cancer Lett.*, 1999, **135**, 145–150.

175. T. Suga, Y. Motoki, H. Tamura *et al.*, 'Involvement of hepatocyte growth factor on hepatocarcinogenesis induced by peroxisome proliferators.' *Cell Biochem. Biophys.*, 2000, **32**, 221–228.

176. T. Nakamura, H. Teramoto and A. Ichihara, 'Purification and characterization of a growth factor from rat platelets for mature parenchymal hepatocytes in primary cultures.' *Proc. Natl. Acad. Sci. USA*, 1986, **83**, 6489–6493.

177. T. Nakamura, K. Nawa and A. Ichihara, 'Partial purification and characterization of hepatocyte growth factor from serum of hepatectomized rats.' *Biochem. Biophys. Res. Commun.*, 1984, **122**, 1450–1459.

178. G. Shiota, D. B. Rhoads, T. C. Wang *et al.*, 'Hepatocyte growth factor inhibits growth of hepatocellular carcinoma cells.' *Proc. Natl. Acad. Sci. USA*, 1992, **89**, 373–377.

179. G. Shiota, H. Kawasaki, T. Nakamura *et al.*, 'Inhibitory effect of hepatocyte growth factor against FaO hepatocellular carcinoma cells may be associated with changes of intracellular signalling pathways mediated by protein kinase C.' *Res. Commun. Mol. Pathol. Pharmacol.*, 1994, **85**, 271–278.

180. A. Kiss, R. Ortiz-Aguayo, R. Sharp *et al.*, 'Evidence that reduction of hepatocyte growth factor (HGF) is not required for peroxisome proliferator-induced hepatocyte proliferation.' *Carcinogenesis*, 2001, In press.

181. A. C. Bayly, R. A. Roberts and C. Dive, 'Suppression of liver cell apoptosis *in vitro* by the non-genotoxic hepatocarcinogen and peroxisome proliferator nafenopin.' *J. Cell Biol.*, 1994, **125**, 197–203.

182. R. A. Roberts, 'Non-genotoxic hepatocarcinogenesis: suppression of apoptosis by peroxisome proliferators.' *Ann. NY Acad. Sci.*, 1996, **804**, 588–611.

183. R. A. Roberts, D. W. Nebert, J. A. Hickman *et al.*, 'Perturbation of the mitosis/apoptosis balance: a fundamental mechanism in toxicology.' *Fundam. Appl. Toxicol.*, 1997, **38**, 107–115.

184. S. C. Hasmall, N. H. James, N. Macdonald *et al.*, 'Suppression of mouse hepatocyte apoptosis by peroxisome proliferators: role of PPARalpha and TNFalpha.' *Mutat. Res.*, 2000, **448**, 193–200.

185. Y. Li, L. K. Leung, H. P. Glauert *et al.*, 'Treatment of rats with the peroxisome proliferator ciprofibrate results in increased liver NF-κB activity.' *Carcinogenesis*, 1996, **17**, 2305–2309.

186. H. P. Glauert, M. M. Beaty, T. D. Clark *et al.*, 'Effect of dietary selenium on the induction of altered hepatic foci and hepatic tumors by the peroxisome proliferator ciprofibrate.' *Nutr. Cancer*, 1990, **14**, 261–271.

187. H. P. Glauert, M. M. Beaty, T. D. Clark *et al.*, 'Effect of dietary Vitamin E on the development of altered hepatic foci and hepatic tumors induced by the peroxisome proliferator ciprofibrate.' *J. Cancer Res. Clin. Oncol.*, 1990, **116**, 351–356.

188. M. L. Rose, I. Rusyn, H. K. Bojes *et al.*, 'Role of Kupffer cells in peroxisome proliferator-induced hepatocyte proliferation.' *Drug Metab. Rev.*, 1999, **31**, 87–116.

189. M. L. Rose, D. Germolec, G. E. Arteel *et al.*, 'Dietary glycine prevents increases in hepatocyte proliferation caused by the peroxisome proliferator WY-14,643.' *Chem. Res. Toxicol.*, 1997, **10**, 1198–1204.

190. M. L. Rose, D. R. Germolec, R. Schoonhoven *et al.*, 'Kupffer cells are causally responsible for the mitogenic effect of peroxisome proliferators.' *Carcinogenesis*, 1997, **18**, 1453–1456.

191. M. L. Rose, I. Rusyn, H. K. Bojes *et al.*, 'Role of kupffer cells and oxidants in signaling peroxisome proliferator-induced hepatocyte proliferation.' *Mutat. Res.*, 2000, **448**, 179–192.

192. M. L. Rose, C. A. Rivera, B. U. Bradford *et al.*, 'Kupffer cell oxidant production is central to the mechanism of peroxisome proliferators.' *Carcinogenesis*, 1999, **20**, 27–33.

193. W. Parzefall, W. Berger, E. Kainzbauer *et al.*, 'Peroxisome proliferators do not increase DNA synthesis in purified rat hepatocytes.' *Carcinogenesis*, 2001, **22**, 519–523.

194. J. M. Peters, I. Rusyn, M. L. Rose *et al.*, 'Peroxisome proliferator-activated receptor alpha is restricted to hepatic parenchymal cells, not Kupffer cells: implications for the mechanism of action of peroxisome proliferators in hepatocarcinogenesis.' *Carcinogenesis*, 2000, **21**, 823–826.

195. W. G. Karam and B. I. Ghanayem, 'Induction of replicative DNA synthesis and PPAR alpha-dependent gene transcription by Wy-14, 643 in primary rat hepatocyte and non-parenchymal cell co-cultures.' *Carcinogenesis*, 1997, **18**, 2077–2083.

196. J. W. Lawrence, G. K. Wollenberg and J. G. DeLuca, 'Tumor necrosis factor alpha is not required for WY14,643-induced cell proliferation.' *Carcinogenesis*, 2001, **22**, 381–386.

197. P. Bentley, I. Calder, C. Elcombe *et al.*, 'Hepatic peroxisome proliferation in rodents and its significance for humans.' *Food Chem. Toxicol.*, 1993, **31**, 857–907.

198. B. G. Lake, 'Mechanisms of hepatocarcinogenicity of peroxisome-proliferating drugs and chemicals.' *Annu. Rev. Pharmacol. Toxicol.*, 1995, **35**, 483–507.

199. B. G. Lake, J. G. Evans, M. E. Cunninghame *et al.*, 'Comparison of the hepatic effects of nafenopin and WY-14,643 on peroxisome proliferation and cell replication in the rat and Syrian hamster.' *Environ. Health Perspect.*, 1993, **5**, 241–247.

200. R. C. Cattley, J. DeLuca, C. Elcombe *et al.*, 'Do peroxisome proliferating compounds pose a hepatocarcinogenic hazard to humans?' *Regul. Toxicol. Pharmacol.*, 1998, **27**, 47–60.

201. B. G. Lake, 'Peroxisome proliferation: current mechanisms relating to nongenotoxic carcinogenesis.' *Toxicol. Lett.*, 1995, **83**, 673–681.

202. F. A. De La Iglesia, J. E. Lewis, R. A. Buchanan *et al.*, 'Light and electron microscopy of liver in hyperlipoproteinemic patients under long-term gemfibrozil treatment.' *Atherosclerosis*, 1982, **43**, 19–37.

203. S. Blumcke, W. Schwartzkopff, H. Lobeck *et al.*, 'Influence of fenofibrate on cellular and subcellular liver structure in hyperlipidemic patients.' *Atherosclerosis*, 1983, **46**, 105–116.

204. P. Gariot, E. Barrat, P. Drouin *et al.*, 'Morphometric study of human hepatic cell modifications induced by fenofibrate.' *Metabolism*, 1987, **36**, 203–210.

205. M. Hanefeld, C. Kemmer and E. Kadner, 'Relationship between morphological changes and lipid-lowering action of *p*-chlorphenoxyisobutyric acid (CPIB) on hepatic mitochondria and peroxisomes in man.' *Atherosclerosis*, 1983, **46**, 239–246.

206. M. R. Law, S. G. Thompson and N. J. Wald, 'Assessing possible hazards of reducing serum cholesterol.' *British Medical Journal*, 1994, **308**, 373–379.

207. J. K. Huttunen, O. P. Heinonen, V. Manninen *et al.*, 'The Helsinki Heart Study: an 8.5-year safety and mortality follow-up (see comments).' *J. Intern. Med.*, 1994, **235**, 31–39.

208. IARC. Peroxisome Proliferation and its Role in Carcinogenesis. IARC, Lyon, France, 1995.

209. C. R. Elcombe, 'Species differences in carcinogenicity and peroxisome proliferation due to trichloroethylene: a biochemical human hazard assessment.' *Arch. Toxicol. Suppl.*, 1985, **8**, 6–17.

210. S. Duclos, J. Bride, L. C. Ramirez *et al.*, 'Peroxisome proliferation and beta-oxidation in Fao and MH1C1 rat hepatoma cells, HepG2 human hepatoblastoma cells and cultured human hepatocytes: effect of ciprofibrate.' *Eur. J. Cell Biol.*, 1997, **72**, 314–323.

211. M. C. Cornu-Chagnon, H. Dupont and A. Edgar, 'Fenofibrate: metabolism and species differences for peroxisome proliferation in cultured hepatocytes.' *Fundam. Appl. Toxicol.*, 1995, **26**, 63–74.

212. C. E. Perrone, L. Shao and G. M. Williams, 'Effect of rodent hepatocarcinogenic peroxisome proliferators on fatty acyl-CoA oxidase, DNA synthesis, and apoptosis in cultured human and rat hepatocytes.' *Toxicol. Appl. Pharmacol.*, 1998, **150**, 277–286.

213. V. Goll, E. Alexandre, C. Viollon-Abadie *et al.*, 'Comparison of the effects of various peroxisome proliferators on peroxisomal enzyme activities, DNA synthesis, and apoptosis in rat and human hepatocyte cultures.' *Toxicol. Appl. Pharmacol.*, 1999, **160**, 21–32.

214. C. R. Elcombe, D. R. Bell, E. Elias *et al.*, 'Peroxisome proliferators: species differences in response of primary hepatocyte cultures.' *Ann. NY Acad. Sci.*, 1996, **804**, 628–635.

215. N. Bichet, D. Cahard, G. Fabre *et al.*, 'Toxicological studies on a benzofuran derivative. III. Comparison of peroxisome proliferation in rat and human hepatocytes in primary culture.' *Toxicol. Appl. Pharmacol.*, 1990, **106**, 509–517.

216. S. C. Hasmall, N. H. James, N. Macdonald *et al.*, 'Species differences in response to diethylhexylphthalate: suppression of apoptosis, induction of DNA synthesis and peroxisome proliferator activated receptor alpha-mediated gene expression.' *Arch. Toxicol.*, 2000, **74**, 85–91.

217. S. C. Hasmall, N. H. James, N. Macdonald *et al.*, 'Suppression of apoptosis and induction of DNA synthesis *in vitro* by the phthalate plasticizers monoethylhexylphthalate (MEHP) and diisononylphthalate (DINP): a comparison of rat and human hepatocytes *in vitro*.' *Arch. Toxicol.*, 1999, **73**, 451–456.

218. J. A. Styles, M. Kelly, N. R. Pritchard *et al.*, 'A species comparison of acute hyperplasia induced by the peroxisome proliferator methylclofenapate: involvement of the binucleated hepatocyte.' *Carcinogenesis*, 1988, **9**, 1647–1655.

219. R. Hertz, J. Bishara-Shieban and J. Bar-Tana, 'Mode of action of peroxisome proliferators as hypolipidemic drugs. Suppression of apolipoprotein C-III.' *J. Biol. Chem.*, 1995, **270**, 13470–13475.

220. N. Vu-Dac, K. Schoonjans, B. Laine *et al.*, 'Negative regulation of the human apolipoprotein A-I promoter by fibrates can be attenuated by the interaction of the peroxisome proliferator-activated receptor with its response element.' *J. Biol. Chem.*, 1994, **269**, 31012–31018.

221. N. Vu-Dac, K. Schoonjans, V. Kosykh *et al.*, 'Fibrates increase human apolipoprotein A-II expression through activation of the peroxisome proliferator-activated receptor.' *J. Clin. Invest.*, 1995, **96**, 741–750.

222. U. Varanasi, R. Chu, Q. Huang *et al.*, 'Identification of a peroxisome proliferator-responsive element upstream of the human peroxisomal fatty acyl coenzyme A oxidase gene.' *J. Biol. Chem.*, 1996, **271**, 2147–2155.

223. C. N. Palmer, M. H. Hsu, K. J. Griffin *et al.*, 'Peroxisome proliferator activated receptor-alpha expression in human liver.' *Mol. Pharmacol.*, 1998, **53**, 14–22.

224. A. R. Bell, R. Savory, N. J. Horley *et al.*, 'Molecular basis of non-responsiveness to peroxisome proliferators: the guinea-pig PPARalpha is functional and mediates peroxisome proliferator-induced hypolipidaemia.' *Biochem. J.*, 1998, **332**, 689–693.

225. J. D. Tugwood, T. C. Aldridge, K. G. Lambe *et al.*, 'Peroxisome proliferator-activated receptors: structures and function.' *Ann. NY Acad. Sci.*, 1996, **804**, 252–265.

226. N. J. Woodyatt, K. G. Lambe, K. A. Myers *et al.*, 'The peroxisome proliferator (PP) response element upstream of the human acyl CoA oxidase gene is inactive among a sample human population: significance for species differences in response to PPs.' *Carcinogenesis*, 1999, **20**, 369–372.

227. U. Varanasi, R. Chu, Q. Huang *et al.*, 'Identification of a peroxisome proliferator-responsive element upstream of the human peroxisomal fatty acyl coenzyme A oxidase gene.' *J. Biol. Chem.*, 1998, **273**, 30842.

228. K. G. Lambe, N. J. Woodyatt, N. Macdonald *et al.*, 'Species differences in sequence and activity of the

peroxisome proliferator response element (PPRE) within the acyl CoA oxidase gene promoter.' *Toxicol. Lett.*, 1999, **110**, 119–127.

229. C. Rodriguez, V. Noe, A. Cabrero *et al.*, 'Differences in the formation of PPARalpha-RXR/acoPPRE complexes between responsive and nonresponsive species upon fibrate administration.' *Mol. Pharmacol.*, 2000, **58**, 185–193.

230. A. W. Lington, M. G. Bird, R. T. Plutnick *et al.*, 'Chronic toxicity and carcinogenic evaluation of diisononyl phthalate in rats.' *Fundam. Appl. Toxicol.*, 1997, **36**, 79–89.

231. D. J. Caldwell, 'Review of mononuclear cell leukemia in F-344 rat bioassays and its significance to human cancer risk: A case study using alkyl phthalates.' *Regul. Toxicol. Pharmacol.*, 1999, **30**, 45–53.

232. P. D. Johnson, B. V. Dawson and S. J. Goldberg, 'Cardiac teratogenicity of trichloroethylene metabolites.' *J. Am. Coll. Cardiol.*, 1998, **32**, 540–545.

233. M. W. Taubeneck, G. P. Daston, J. M. Rogers *et al.*, 'Altered maternal zinc metabolism following exposure to diverse developmental toxicants.' *Reprod. Toxicol.*, 1994, **8**, 25–40.

234. M. W. Taubeneck, G. P. Daston, J. M. Rogers *et al.*, 'Tumor necrosis factor-alpha alters maternal and embryonic zinc metabolism and is developmentally toxic in mice.' *J. Nutr.*, 1995, **125**, 908–919.

235. M. Salas, B. Tuchweber, K. Kovacs *et al.*, 'Effect of cerium on the rat liver: an ultrastructural and biochemical study.' *Beitr. Pathol.*, 1976, **157**, 23–44.

236. B. Tuchweber and M. Salas, 'Prevention of CeCl3-induced hepatotoxicity by hypolipidemic compounds.' *Arch. Toxicol.*, 1978, **41**, 223–232.

237. F. A. Nicholls-Grzemski, I. C. Calder and B. G. Priestly, 'Peroxisome proliferators protect against paracetamol hepatotoxicity in mice.' *Biochem. Pharmacol.*, 1992, **43**, 1395–1396.

238. J. E. Manautou, D. J. Hoivik, A. Tveit *et al.*, 'Clofibrate pretreatment diminishes acetaminophen's selective covalent binding and hepatotoxicity.' *Toxicol. Appl. Pharmacol.*, 1994, **129**, 252–263.

239. J. E. Manautou, A. Tveit, D. J. Hoivik *et al.*, 'Protection by clofibrate against acetaminophen hepatotoxicity in male CD-1 mice is associated with an early increase in biliary concentration of acetaminophen-glutathione adducts.' *Toxicol. Appl. Pharmacol.*, 1996, **140**, 30–38.

240. F. A. Nicholls-Grzemski, I. C. Calder, B. G. Priestly *et al.*, 'Clofibrate-induced *in vitro* hepatoprotection against acetaminophen is not due to altered glutathione homeostasis.' *Toxicol. Sci.*, 2000, **56**, 220–228.

241. C. Chen, V. M. Silva, J. C. Corton *et al.*, 'PPARalpha knockout mice are not protected against acetaminophen (APAP) hepatotoxicity by clofibrate (CFB) pretreatment.' *Toxicol. Sci. (Supp)*, 2000, **54**, 42.

242. K. Schoonjans, B. Staels, P. Grimaldi *et al.*, 'Acyl-CoA synthetase mRNA expression is controlled by fibric-acid derivatives, feeding and liver proliferation.' *Eur. J. Biochem.*, 1993, **216**, 615–622.

243. J. K. Reddy, S. K. Goel, M. R. Nemali *et al.*, 'Transcription regulation of peroxisomal fatty acyl-CoA oxidase and enoyl-CoA hydratase/3-hydroxyacyl-CoA dehydrogenase in rat liver by peroxisome proliferators.' *Proc. Natl. Acad. Sci. USA*, 1986, **83**, 1747–1751.

244. M. Hijikata, J. K. Wen, T. Osumi *et al.*, 'Rat peroxisomal 3-ketoacyl-CoA thiolase gene. Occurrence of two closely related but differentially regulated genes.' *J. Biol. Chem.*, 1990, **265**, 4600–4606.

245. M. C. Hunt, S. E. Nousiainen, M. K. Huttunen *et al.*, 'Peroxisome proliferator-induced long chain acyl-CoA thioesterases comprise a highly conserved novel multi-gene family involved in lipid metabolism.' *J. Biol. Chem.*, 1999, **274**, 34317–34326.

246. N. Casals, N. Roca, M. Guerrero *et al.*, 'Regulation of the expression of the mitochondrial 3-hydroxy-3-methylglutaryl-CoA synthase gene. Its role in the control of ketogenesis.' *Biochem. J.*, 1992, **283**, 261–264.

247. M. S. Rao and J. K. Reddy, 'Peroxisome proliferation and hepatocarcinogenesis.' *Carcinogenesis*, 1987, **8**, 631–636.

248. F. Heller and C. Harvengt, 'Effects of clofibrate, bezafibrate, fenofibrate and probucol on plasma lipolytic enzymes in normolipaemic subjects.' *Eur. J. Clin. Pharmacol.*, 1983, **25**, 57–63.

249. B. Staels, A. van Tol, G. Skretting *et al.*, 'Lecithin:cholesterol acyltransferase gene expression is regulated in a tissue-selective manner by fibrates.' *J. Lipid. Res.*, 1992, **33**, 727–735.

250. C. W. Miller and J. M. Ntambi, 'Peroxisome proliferators induce mouse liver stearoyl-CoA desaturase 1 gene expression.' *Proc. Natl. Acad. Sci. USA*, 1996, **93**, 9443–9448.

251. D. B. Jump, S. Ren, S. Clarke *et al.*, 'Effects of fatty acids on hepatic gene expression.' *Prostaglandins Leukot. Essent. Fatty Acids*, 1995, **52**, 107–111.

252. R. Hertz, R. Aurbach, T. Hashimoto *et al.*, 'Thyromimetic effect of peroxisomal proliferators in rat liver.' *Biochem. J.*, 1991, **274**, 745–751.

253. H. K. Bojes, D. R. Germolec, P. Simeonova *et al.*, 'Antibodies to tumor necrosis factor alpha prevent increases in cell replication in liver due to the potent peroxisome proliferator, WY-14,643.' *Carcinogenesis*, 1997, **18**, 669–674.

254. J. C. Corton, C. Bocos, E. S. Moreno *et al.*, 'Rat 17 beta-hydroxysteroid dehydrogenase type IV is a novel peroxisome proliferator-inducible gene.' *Mol. Pharmacol.*, 1996, **50**, 1157–1166.

255. R. Hertz, M. Seckbach, M. M. Zakin *et al.*, 'Transcriptional suppression of the transferrin gene by hypolipidemic peroxisome proliferators.' *J. Biol. Chem.*, 1996, **271**, 218–224.

256. Y. Zhu, C. Qi, S. Jain *et al.*, 'Isolation and characterization of PBP, a protein that interacts with peroxisome proliferator-activated receptor.' *J. Biol. Chem.*, 1997, **272**, 25500–25506.

257. P. Dowell, J. E. Ishmael, D. Avram *et al.*, 'Identification of nuclear receptor corepressor as a peroxisome proliferator-activated receptor alpha interacting protein.' *J. Biol. Chem.*, 1999, **274**, 15901–15907.

258. S. Jain, S. Pulikuri, Y. Zhu *et al.*, 'Differential expression of the peroxisome proliferator-activated receptor gamma (PPARgamma) and its coactivators steroid receptor coactivator-1 and PPAR-binding protein PBP in the brown fat, urinary bladder, colon, and breast of the mouse.' *Am. J. Pathol.*, 1998, **153**, 349–354.

259. J. DiRenzo, M. Soderstrom, R. Kurokawa *et al.*, 'Peroxisome proliferator-activated receptors and retinoic acid receptors differentially control the interactions of retinoid X receptor heterodimers with ligands, coactivators, and corepressors.' *Mol. Cell Biol.*, 1997, **17**, 2166–2176.

260. G. Krey, O. Braissant, F. L'Horset *et al.*, 'Fatty acids, eicosanoids, and hypolipidemic agents identified as ligands of peroxisome proliferator-activated receptors by coactivator-dependent receptor ligand assay.' *Mol. Endocrinol.*, 1997, **11**, 779–791.

261. Y. Zhu, C. Qi, C. Calandra *et al.*, 'Cloning and identification of mouse steroid receptor coactivator-1

(mSRC-1), as a coactivator of peroxisome proliferator-activated receptor gamma.' *Gene. Expr.*, 1996, **6**, 185–195.

262. P. Dowell, J. E. Ishmael, D. Avram *et al.*, 'p300 functions as a coactivator for the peroxisome proliferator-activated receptor alpha.' *J. Biol. Chem.*, 1997, **272**, 33435–33443.

263. E. Treuter, T. Albrektsen, L. Johansson *et al.*, 'A regulatory role for RIP140 in nuclear receptor activation.' *Mol. Endocrinol.*, 1998, **12**, 864–881.

264. P. Puigserver, Z. Wu, C.W. Park *et al.*, 'A cold-inducible coactivator of nuclear receptors linked to adaptive thermogenesis.' *Cell.*, 1998, **92**, 829–839.

265. C. X. Yuan, M. Ito, J. D. Fondell *et al.*, 'The TRAP220 component of a thyroid hormone receptor-associated protein (TRAP) coactivator complex interacts directly with nuclear receptors in a ligand-dependent fashion.' *Proc. Natl. Acad. Sci. USA*, 1998, **95**, 7939–7944.

266. Y. Zhu, L. Kan, C. Qi *et al.*, 'Isolation and characterization of peroxisome proliferator-activated receptor (PPAR) interacting protein (PRIP) as a coactivator for PPAR.' *J. Biol. Chem.*, 2000, **275**, 13510–13516.

267. F. Caira, P. Antonson, M. Pelto-Huikko *et al.*, 'Cloning and characterization of RAP250, a novel nuclear receptor coactivator.' *J. Biol. Chem.*, 2000, **275**, 5308–5317.

268. R. T. Gampe, V. G. Montana, M. H. Lambert *et al.*, 'Structural basis for autorepression of retinoid X receptor by tetramer formation and the AF-2 helix.' *Genes Dev.*, 2000, **14**, 2229–2241.

269. A. Gorla-Bajszczak, C. Juge-Aubry, A. Pernin *et al.*, 'Conserved amino acids in the ligand-binding and tau(i) domains of the peroxisome proliferator-activated receptor alpha are necessary for heterodimerization with RXR.' *Mol. Cell Endocrinol.*, 1999, **147**, 37–47.

J.P. Vanden Heuvel, G.H. Perdew, W.B. Mattes and W.F. Greenlee (Eds.)
Comprehensive Toxicology, Vol. xiv

# 14.2.5
# The Physiological Role of Pregnane-X-Receptor (PXR) in Xenobiotic and Bile Acid Homeostasis

JOHN T. MOORE, TIMOTHY M. WILLSON and
STEVEN A. KLIEWER
*GlaxoSmithKline Research and Development, Research Triangle Park, NC, 27709, USA*

## 14.2.5.1 DISCOVERY OF THE NUCLEAR RECEPTOR PXR

Nuclear receptors (NRs) comprise a large family of ligand modulated transcription factors that include the steroid, retinoid, and thyroid hormone receptors.[1–4] NRs are generally modular in nature, containing distinct functional domains that are evolutionarily conserved and serve as hallmark features that distinguish the superfamily. Two of the most extensively studied domains are the DNA

Binding Domain (DBD) and the Ligand Binding Domain (LBD). The centrally positioned DBD contains two C4-type zinc fingers and targets the receptor to specific DNA response elements within regulatory regions of genes. The LBD is a COOH-terminal domain that interacts directly with the hormone, contains a hormone-dependent transcription activation motif which can recruit specific coactivator proteins upon hormone binding,[5] and ultimately serves as a molecular switch to activate transcription of target genes. In short, NR activation by ligand culminates in the modulation of target gene expression. By transducing the binding of small, lipophilic hormones into transcriptional responses, NRs play critical roles in the regulation of cell division, cell differentiation, organ physiology, metabolism, and homeostasis.[6,7] Because nuclear receptors afford the opportunity to modulate these processes with synthetic ligands, they represent attractive pharmacological targets.

The approach used to study NRs has essentially "reversed" in the last decade[8] (Figure 1). Historically, new hormones were discovered through analysis of their effects on physiological or developmental processes. The purified hormone was subsequently used to identify its partner receptor. Recently, recognition that NRs harbor conserved domains has led to a rapid increase in the number of identified receptors. Cloning efforts over the last decade and a half have expanded the NR superfamily to approximately 50 members in humans, the majority of which are still "orphan" in that they lack associated ligands. The cloning of receptors based upon sequence similarity alone has ushered in a new era in which the process begins with the receptor that is then used to search for previously unknown ligands. Rapid, high throughput ligand screening methodologies combined with increases in available chemical diversity make this "reverse endocrinology" approach a very efficient means to link receptors to selective ligands and in turn to dissect the biology of new receptors. We have utilized such a reverse endocrinology approach to study the nuclear receptor PXR (*Pregnane X Receptor*, because natural and synthetic C21 steroids were among the most efficacious natural ligands initially identified for this receptor).

PXR was first identified in 1997 when a sequence appeared in the mouse EST databases (Accession # AA277370) representing the COOH-terminus of a novel orphan nuclear receptor. This sequence tag led to the isolation of the corresponding full-length mouse PXR cDNA, a clone with an open reading frame

predicting an approximately 45 kilodalton protein with primary structural features typical of a nuclear receptor family member.[9] Since that time, full-length cDNAs representing PXR orthologs from the human,[10-12] rat,[13] rabbit,[14] as well as rhesus monkey, dog and pig have been obtained. Figure 2 shows the overall size and sequence similarity between these orthologs and their similarity to the Vitamin D receptor, the most closely related member of the family that has a known ligand. The LBD sequences between members of the PXR subfamily are surprisingly divergent, the significance of which will become apparent later in this chapter.

The identification of PXR prompted the challenge to link this orphan to a biological function. To begin this process, we developed assays to identify selective chemical tools for PXR. This reverse endocrinology approach has led to important insights into PXR function.

## 14.2.5.2  PXR LIGANDS PROVIDE LINK TO CYP3A BIOLOGY

Historically, many NR ligands have been identified in cell-based assays where NR effects on transcription are monitored indirectly by effects on transcription of a reporter gene. Receptor/reporter plasmid combinations are typically co-introduced into a relevant cell line via transient transfection. For PXR, we utilized a receptor/reporter pair consisting of either (A) full-length PXR in combination with a reporter gene driven by PXR response elements (Figure 3); or (B) a chimeric LBD sequence fused with the yeast Gal4 DBD in combination with a reporter gene driven by Gal4 response elements. The latter was used initially since its use did not require defining PXR response elements.

Using these cell-based assays, we and others[9,11-13] found that PXR is a promiscuous receptor, activated by a large and structurally diverse set of compounds. Compounds which activated human PXR in cell-based assays included rifampicin, lovastatin, trans-nonaclor, clotrimazole, RU486, phenobarbital, and the bisphosphonate ester, SR12813 (Figure 4). SR12813 is among the most potent of these compounds, activating human PXR with a half-maximal effective concentration ($EC_{50}$) of $\sim 800$ nM. Lovastatin and clotrimazole activated human PXR with an $EC_{50}$ value of roughly 1–5 $\mu$M, and RU486 activated human PXR with an EC50 value of $\sim 10\,\mu$M.

From the growing list of PXR agonists, a pattern began to emerge. Many of the compounds that were shown to activate PXR

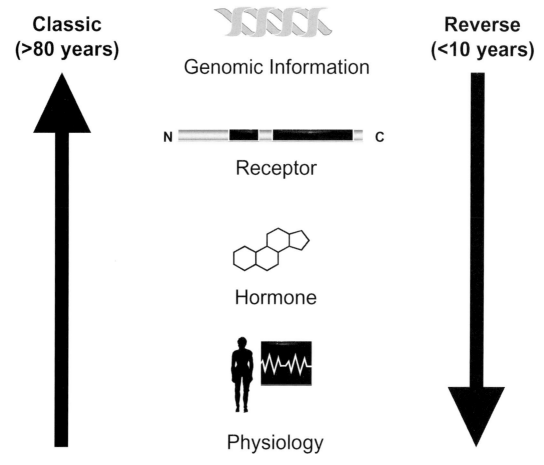

**Figure 1** Reverse endocrinology: The classic approach to endocrinology involved linking a physiological response to a particular hormone, which subsequently initiated the search for a cognate receptor. Because of the rapid increase in the availability of DNA sequence information, this approach has essentially reversed in recent years. It is now feasible to identify a large number of receptors from sequence data alone, which then provide the basis of a search for cognate ligands and ultimately for physiological relevance. This approach has been termed "reverse endocrinology".

(including all of the compounds listed above) were already established inducers of cytochrome P450 3A (CYP3A) activity. This effect had been demonstrated both *in vivo* and in primary cultures of human, rat, and/or rabbit hepatocytes.[15–17] The $EC_{50}$ values noted above were consistent with the concentrations known to induce CYP3A activity in clinical and laboratory studies. The potential link between PXR and CYP3A regulation was exciting because of the important role that CYP3A plays in drug interactions and because it potentially explained a conundrum that has existed in the field since the 1980s.

### 14.2.5.3 CYP3A AND DRUG INTERACTIONS

Drug interactions result when administration of one drug alters the properties of a second co-administered drug. CYP3A monooxygenase

activity catalyzes hydroxylation of susceptible substrates and subsequent increased clearance from the kidney,[18] thus altering drug pharmacodynamic profiles. Because of its abundance in liver and intestine and because of its broad substrate specificity, CYP3A-related drug interactions are of major significance in the clinic. It is estimated that CYP3A4 (the predominant CYP3A subtype in adult humans) is involved in the metabolism of $> 50\%$ of all drugs in use today.[19,20] CYP3A4 substrates include contraceptive steroids, cyclosporin analogs, and HIV protease inhibitors.

CYP3A activity limits the utility of many therapeutically valuable drugs.[19,21] For example, rifampicin (an inducer of CYP3A) would be a valuable tool to treat infectious diseases associated with HIV but is usually contraindicated because patients are simultaneously being treated with HIV protease inhibitors. Identifying the factors responsible for CYP3A

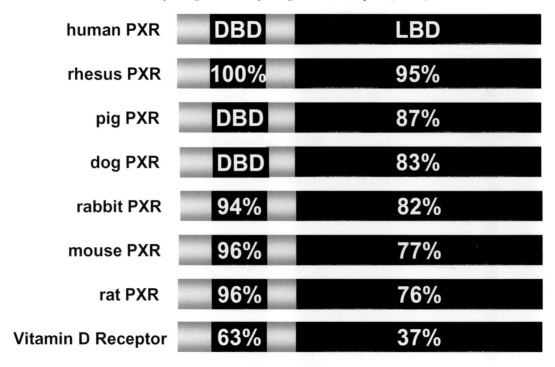

**Figure 2**   PXR homologs: The overall size and sequence similarity between various mammalian PXR DNA binding domain (DBD) sequences and ligand binding domain (LBD) sequences are shown. The DBD sequences of the dog and the pig PXRs are not yet available. Their similarity to the Vitamin D receptor, their closest homolog with a known natural ligand, is also shown.

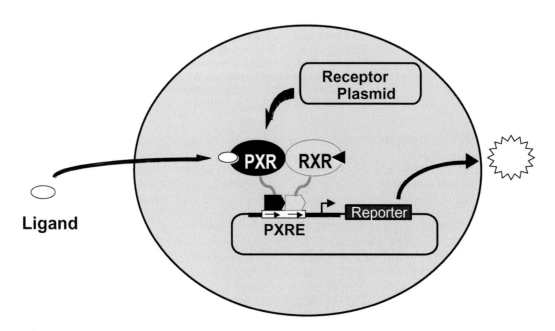

**Figure 3**   Transient transfection assay: Nuclear receptor ligand activation can be monitored in a cellular context by co-transfecting relevant receptor/reporter combinations. Full-length receptor assays require knowledge of functional response elements for that receptor. The response elements are typically inserted proximal to a minimal promoter (e.g., thymidine kinase) and a reporter gene (e.g., chloramphenical acetyl transferase). Alternatively, a more generic assay can be generated by fusing a nuclear receptor ligand binding domain (LBD) to the yeast Gal4 DNA binding domain (DBD) which is used in combination with a reporter construct containing Gal4 response elements. One advantage of this type of assay is that it eliminates possible interference from endogenous receptors.

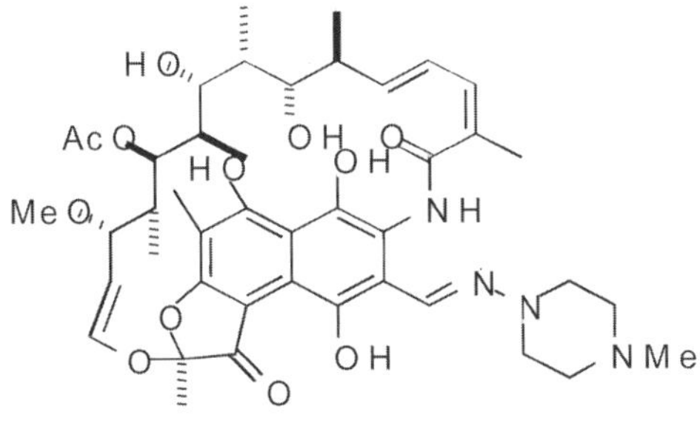

rifampicin 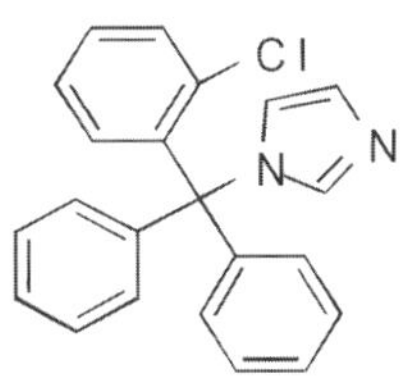                                clotrimazole

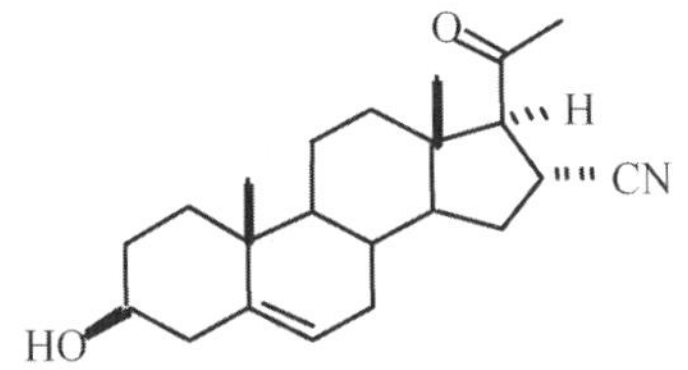

pregnenolone 16α-                              RU486
carbonitrile

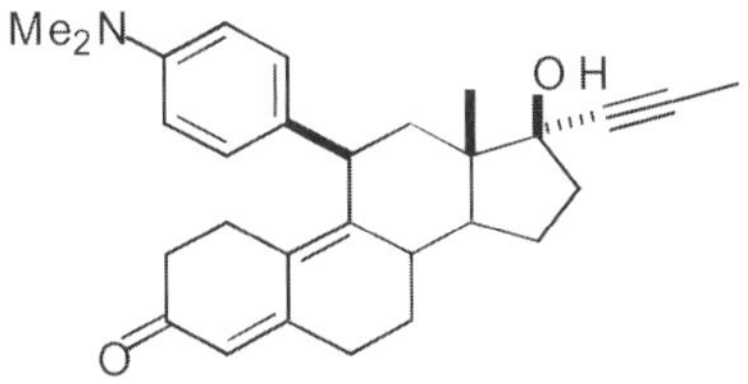

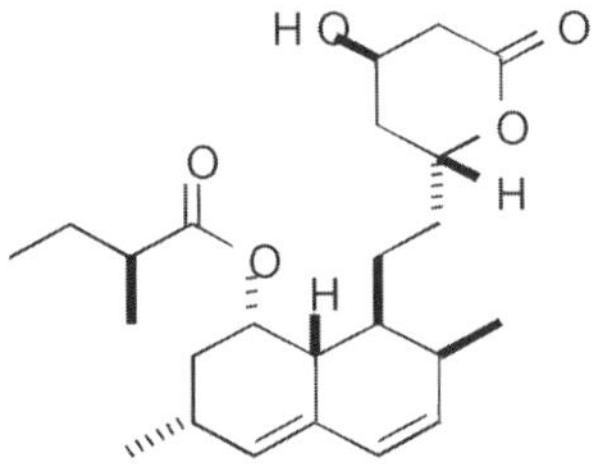

lovastatin 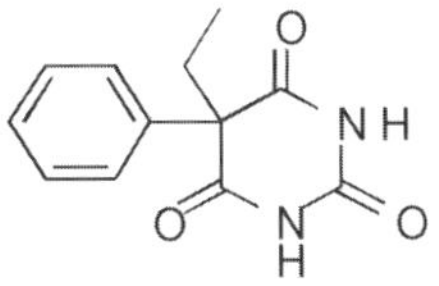                               phenobarbital

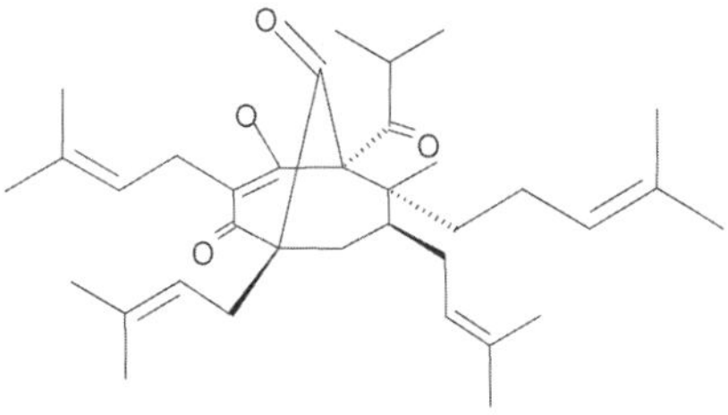

hyperforin 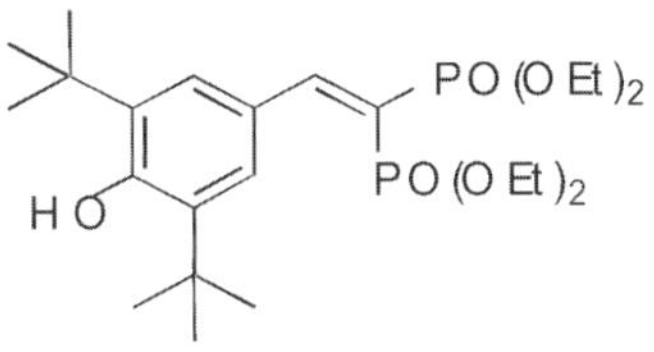                                SR12813

**Figure 4**   Structures of xenobiotics that activate PXR in cotransfection assays.

induction would represent a major milestone towards overcoming these types of limitations.

Early attempts to uncover the molecular details of CYP3A induction produced a puzzling observation. CYP3A expression was found to be induced not only by glucocorticoid receptor (GR) agonists (e.g., dexamethasone), but also by glucocorticoid receptor antagonists such as RU486.[22] Furthermore, the DNA elements within the *Cyp3a1* (mouse) and *Cyp3a4* (human) gene promoters that were responsive to dexamethasone were not GR binding sites.[17,23,24] These data indicated that another factor other than GR was mediating the effect of GR agonists and antagonists. Indeed, it appeared that this missing regulatory factor was a nuclear receptor because the DNA elements shown to be necessary for *Cyp3a* gene induction resembled orphan nuclear receptor hormone-response elements.[4]

#### 14.2.5.4 SOLIDIFYING THE CONNECTION BETWEEN PXR AND CYP3A

Additional analysis of PXR provided evidence that this new orphan receptor could be the missing factor necessary for CYP3A induction. As noted above, hormone response elements within the *Cyp3a* gene provide an important clue.

Nuclear receptor hormone response elements are typically composed of one or two copies of a TGA(A/C)CT core motif. For dimeric receptors, two copies of this "half-site" comprise a single functional response element. The relative orientation of the two half-sites and their spacing determine which nuclear receptors can bind to the response element. For example, half-sites organized as direct repeats with spacers of 3, 4, or 5 nucleotides serve as response elements for the Vitamin D receptor, thyroid hormone receptor, and retinoic acid receptor, respectively.[4] Two different types of response elements were seen in different *Cyp3a* gene promoters. Within the human *Cyp3a4* promoter, the region shown to be essential for dexamethasone/rifampicin responsiveness contained two nuclear receptor half-sites organized as an everted repeat and separated by six nucleotides (ER6).[17] In contrast, in the mouse *Cyp3a1* promoter, dexamethasone/ PCN responsiveness mapped to a region containing two nuclear receptor half-sites organized as a direct repeat and separated by three nucleotides (DR3).[23,24] These response elements were used in electrophoretic mobility shift assays (EMSA) to determine whether PXR bound to these sites. Since several of the nuclear receptors are known to bind response elements as heterodimers with the 9-*cis* retinoic acid nuclear receptor RXR,[4,25] EMSA was performed in the presence of RXR. These experiments revealed that PXR did not bind any of the tested response elements as a homodimer, but instead bound to both the ER6 and DR3 response elements as a heterodimer with RXR.[9] Thus, the PXR/RXR heterodimer is capable of binding to response elements previously shown to be essential for xenobiotic induction of the *Cyp3a* gene (Figure 5). It is additionally interesting that PXR is capable of binding to such an architecturally diverse set of response elements.

Expression data also support the link between PXR and CYP3A.[9,10] CYP3A is expressed predominantly in liver and intestine. PXR regulation of CYP3A expression would predict that PXR expression overlaps with CYP3A expression. In order to determine the expression of PXR, total RNA was prepared

from different tissues from mouse and human, rabbit, rat, and mouse and examined by Northern blotting experiments.[9,10,14] Although subtle interspecies differences were noted (e.g., weak expression of PXR was seen in kidney and stomach in rodents but not in humans and rabbits), PXR was found to be most abundantly expressed in liver and intestine in all species tested. Thus, the expression pattern of PXR bolsters the link between CYP3A and PXR, because the tissues where PXR is expressed mimics those where CYP3A is induced.

Though the human, rabbit, rat and mouse PXR are activated by several of the same compounds, each receptor is pharmacologically distinct. Indeed, it is the pharmacological differences between PXRs from different species that provide one of the strongest links between PXR and CYP3A induction. One of the classic interspecies differences with regard to CYP3A induction involves pregnenolone 16-α carbonitrile (PCN) and rifampicin in humans and rabbits versus rats. Rifampicin is an efficacious inducer of CYP3A mRNA in human and rabbit hepatocytes, but not in rat hepatocytes. Conversely, PCN has marked effects on CYP3A mRNA levels in rat hepatocytes but only modest effects in human or rabbit hepatocytes (Figure 6(a)). The availability of the PXR functional assay allowed the examination of interspecies differences in PXR activation.[14] When these compounds were tested against rat, rabbit, and human PXRs in the cell-based reporter assays, PCN was found to be a much more potent inducer of rat PXR than its human or rabbit orthologs (Figure 6(b)). Conversely, rifampicin was a much more potent activator of human and rabbit PXR than rat PXR (Figure 6(b)). Thus, the interspecies differences in CYP3A induction by PCN and rifampicin are accounted for by differential activation of PXR.

The preceding studies demonstrate the ability of various CYP3A inducers to activate PXR. One surprise from this work is the remarkable structural and chemical diversity of the compounds that activate this receptor. Given this observation, it was important to rule out the possibility that PXR activators were regulating the receptor's activity in the cell-based assays indirectly through another yet unidentified receptor(s). The availability of a potent PXR activator (SR12813) provided the chemical tool necessary to develop a binding assay with which to test whether the diverse compounds that activate PXR do so via direct interactions. A competition type scintillation proximity assay (SPA) was developed with the radioligand [$^3$H]-SR12813 (Figure 7). Data

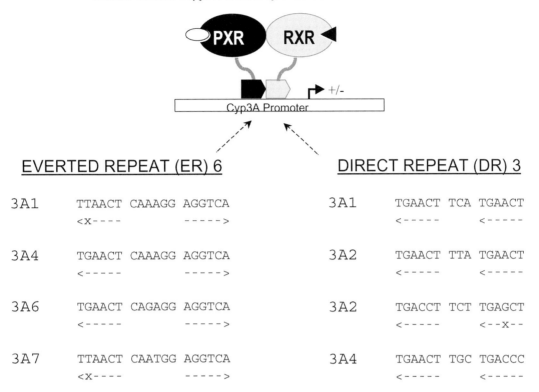

**Figure 5** PXR response elements. PXR binds to response-elements in CYP3A promoters. Electrophoretic mobility shift assays showed that PXR binds to the DR3 element of the *Cyp3a1* gene promoter (mouse) and the *Cyp3a4* gene promoter (human) as a PXR:RXR heterodimer. Other DR3 and ER6 motifs present in known *Cyp3a* promoters are indicated.

from the binding assay revealed that all of the compounds that activated PXR in the cell-based assay also competed with [³H]-SR12813.[14] The IC$_{50}$ data from the SPA were in good agreement with EC$_{50}$ values obtained in cell-based reporter assays. Thus, these data indicated that PXR is activated by direct binding of the structurally diverse set of ligands identified in the cell-based functional assays. The combined set of data suggests that PXR binds a diverse set of compounds and transduces this signal to the *Cyp3a* gene.

## 14.2.5.5 GENETIC STUDIES SUPPORT A ROLE OF PXR IN XENOBIOTIC METABOLISM

Further support for the role of PXR in *Cyp3a* gene regulation and in xenobiotic metabolism have been derived from analysis of mice lacking a functional Pxr allele. Such "*PXR* knock-out" mice have been generated by our group[26] and by the Evans' group[27] and

both support the proposed role of PXR in xenobiotic metabolism.

In each study, PXR knock-out strains were created by disrupting the DNA binding domain of the wild-type allele and subsequently generating mice homozygous for the defective copy.[26,27] Both groups observed that the PXR knock-out mice were viable, fertile, bred with normal Mendelian distribution, and did not exhibit any overt phenotypic changes. Extensive serum analysis did not reveal any significant changes in a number of parameters, including free- and HDL-cholesterol, steroid hormones (progesterone, estradiol, testosterone, and corticosterone), triglycerides, transaminases, albumin, total bilirubin, and total bile acids.

Initial experiments focused on the effects of the PXR knock-out on the regulation of the mouse *Cyp3a* gene. As described above, it is established that PCN, phenobarbital, and dexamethasone strongly induce expression of mouse CYP3A (specifically the CYP3A11 isoform). To determine the PXR dependence of this effect, these compounds were tested in

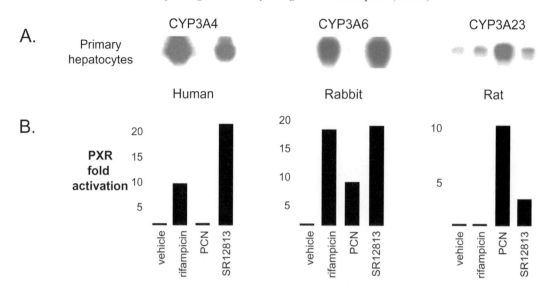

**Figure 6** Species-specific activation of CYP3A mRNA expression in hepatocytes. Panel A represents Northern blots of total RNA from hepatocytes treated with vehicle, rifampicin, pregnenolone 16a-carbonirtrile, or SR12813, and hybridized to a labeled DNA probe representing either human CYP3A4, rabbit CYP3A6, or rat CYP3A23. Compounds were tested at a10mM concentration. Panel B shows that the induction of CYP3A mRNA expression correlates with species-specific activation of PXR in cell-based transient co-transfection assays. These assays were performed in CV-1 cells using either full-length human, rabbit, or rat PXR in combination with a (CYP3A1 DR3)2-tk-CAT reporter vector.

PXR knock-out mice.[26,27] In marked contrast to the results seen in wild-type mice, PCN and dexamethasone did not induce CYP3A11 mRNA in the knock-out strain. These data demonstrate that PXR is required for induction of CYP3A11 expression by PCN and dexamethasone *in vivo*.

To further extend their analysis, Xie *et al.*[27] generated "humanized" transgenic mice by introducing human PXR (hPXR) into a mouse PXR (mPXR) knock-out mouse genetic background, deriving a hPXR-transgenic strain. As predicted, the hPXR transgenic mouse strain mice did not respond to PCN. In contrast, the *Cyp3a* gene was readily induced by the human-specific inducer rifampicin in these same mice. Thus, hPXR alone can reconstitute *Cyp3a* xenoregulation and confer a human *Cyp3a* induction profile.

Additionally, Xie *et al.* generated mice containing a constitutively active hPXR allele by introducing a gene into the PXR knock-out strain which encodes hPXR fused at the amino terminus to the transcriptional activation domain of viral VP16. In this VP16-hPXR strain, the activated form of hPXR causes constitutive upregulation of *Cyp3a* gene

expression. To directly test the hypothesis that PXR is important for protection against xenobiotics, Xie *et al.*[27] explored their potential resistance to xenobiotic toxicants, such as tribromoethanol and zoxazolamine, in the VP16-hPXR strain. Resistance to such anesthetics is a hallmark of an activated xenobiotic system and a traditional test for drug–drug interaction. Indeed, tribromoethanol and zoxazolamine efficiently induced anesthesia and paralysis in wild-type mice as expected from earlier studies.[28] The wild-type mice slept for an average of 32.8 min in response to tribromoethanol, but the VP16-hPXR mice were virtually resistant to the drug. Similarly with zoxazolamine treatment, wild-type mice averaged more than 60 min of paralysis, and the transgenics averaged less than 5 min. Therefore, sustained activation of hPXR and induction of CYP3A result in enhanced protection against xenotoxic compounds. Thus, the combined data from both groups[26,27] using PXR knock-out mice establish PXR as the central mediator for inducible expression of CYP3A and add new insight into the essential role of PXR in xenobiotic metabolism.

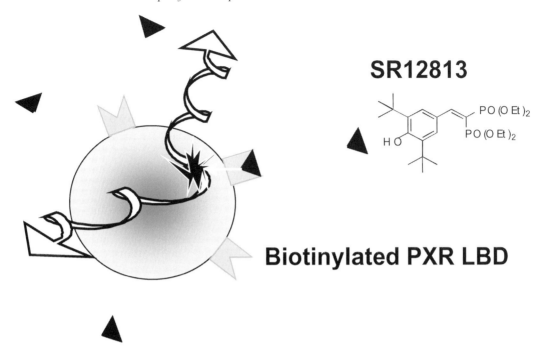

**Figure 7** Schematic representation of the scintillation-proximity assay (SPA). In the PXR SPA, bacterially expressed PXR ligand-binding domain (shown above as structures on bead surface) is affixed to a scintillant-containing bead and incubated with [³H]-SR12813 (black triangles). When bound to PXR on the bead, the radioligand is in close enough proximity to the scinitillant in the bead so that radioactive decay results in energy transfer to the scintillant and a subsequent flash of light. Binding of non-radioactive ligands can be measured by virtue of their competition with the radioligand PXR binding sites on the bead.

### 14.2.5.6 IMPLICATIONS OF PXR IN DRUG DEVELOPMENT

The discovery of PXR has important implications in drug development. The availability of human PXR cell-based functional assays and the high throughput SPA provide valuable tools for the rapid identification of compounds that will induce CYP3A expression. Thus, these assays provide the means to carry out high throughput prospective assays to identify compounds that would interact with other drugs. The removal of compounds that activate PXR from the drug discovery process has the potential to accelerate the development process and result in safer medicines.

Linking PXR activation to CYP3A induction also has practical applications with regard to the use of non-human hepatocytes for preclinical evaluation of compounds. Due to interspecies variability, induction studies performed in non-human hepatocytes have not reliably predicted CYP3A induction in human hepatocytes. These differences have complicated the development of predictive non-human assays. Consequently, analysis of compound effects on CYP3A expression has been largely restricted to time-consuming and costly assays involving human liver tissue. Thus, access to an *in vitro* human PXR assay to predict CYP3A effects is an important advance for predictive biometabolism studies. Additionally, comparative functional studies using rabbit, rat, mouse and human PXR clones as well as clones from other species will increase the ability to evaluate metabolism data from relevant non-human pharmacological model systems.

### 14.2.5.7 SPECIFIC EXAMPLE—PXR AND ST. JOHN'S WORT

Some of the implications of this new found knowledge concerning the importance of PXR in *Cyp3a* gene regulation can be illustrated by considering the example of St. John's wort. St. John's wort (*Hypericum perforatum*) is an ancient herbal remedy that has gained widespread popularity as "nature's Prozac." Several recent reports have indicated that St. John's wort promotes the metabolism of coadministered drugs, including the HIV protease inhibitor indinavir, the immunosuppressant cyclosporin, and oral contraceptives.[29–33] Because each of these drugs is metabolized by

CYP3A4, these findings suggested that St. John's wort might induce CYP3A4 expression. This led to the investigation of whether St. John's wort is an activator of PXR.[34]

Initially, extracts of hypercium were tested for PXR activation using the standard cell-based reporter assay described above. Three different commercial preparations of St. John's wort were extracted with ethanol and the extracts used to treat transfected cells. All three extracts at optimal dilution resulted in PXR activation comparable to that achieved with rifampicin. Subsequently, isolated components of hypercium extracts were tested for activity in the PXR transfection assay. Most of the compounds had little or no PXR activity. In contrast, hyperforin (Figure 4) was found to be an efficacious activator of PXR. Full dose response analysis showed that hyperforin activated PXR with an EC50 of 23 nM. Thus, hyperforin is the most potent PXR activator reported to date. Furthermore, the concentration of hyperforin required to activate PXR is well below the levels achieved in individuals taking the standard regimen of St. John's wort.

These findings indicate that St. John's wort should be used with caution by people taking other medications, especially those metabolized by CYP3A4. The data illustrate the value of testing herbal remedies as well as drug candidates in PXR binding and activation assays as a means for assessing their potential to induce hepatic monooxygenase activity and interact with other drugs. Moreover, insight into PXR inducibility creates the opportunity to screen for compounds that retain antidepressant activity but fail to induce CYP3A4. St. John's Wort is one of an ever-increasing number of specific examples that illustrate the emerging importance of PXR in pharmacology and drug development.

## 14.2.5.8   BROADER UNDERSTANDING OF PXR FUNCTION GAINED FROM PXR KNOCK-OUT STUDIES

One of the next challenges for PXR research is to expand the scope of understanding of this gene to uncover broader aspects of PXR function. The availability of PXR knock-out mice provides the opportunity to examine the effects of PXR loss on regulation of genes other than *Cyp3a*. For example, it had been previously established that PCN treatment of rats decreases expression of CYP7A1, the enzyme responsible for the first and rate-limiting step in the metabolism of cholesterol

to bile acids.[35] Staudinger *et al.*[26] examined whether PXR plays a role in this negative regulation of CYP7A1 expression by utilizing the PXR knock-out strain. Interestingly, CYP7A1 mRNA expression in the PXR-null mice was dysregulated in two respects: First, the basal level of CYP7A1 mRNA was decreased $\sim 2$-fold in PXR-null animals; second, CYP7A1 mRNA steady-state levels were refractory to PCN treatment. These data demonstrate a role for PXR in the regulation of both basal CYP7A1 expression and its repression by PCN.

The effects of PCN on the expression of genes involved in the transport of bile acids were also examined. One such transporter is the organic anion transporter-2 (OATP2). OATP2 is a basolateral (sinusoidal) transporter that can mediate hepatocellular uptake of a wide range of amphipathic substrates, including bile acids and xenobiotics.[36] PCN was found to be a potent inducer of OATP2 mRNA in wild-type mice, but not in PXR-null mice.

Since CYP7A1 and OATP2 are involved in bile acid biosynthesis and transport, Staudinger *et al.* examined whether bile acids might directly regulate PXR activity. Transfected cells were treated with various bile acids and their taurine or glycine conjugates. Notably, the secondary bile acid lithocholic acid (LCA) was an efficacious activator of human and mouse PXR. Furthermore, 3-keto-LCA (a major metabolite of LCA in rats), was also an efficacious activator of both the human and mouse PXR ($IC_{50} = 15 \mu M$ in the binding assay). LCA is a toxic secondary bile acid, which in rodents causes cholestasis, a pathogenic state characterized by decreased bile flow and the accumulation of bile constituents in the liver and blood.[37] LCA levels of $5–10 \mu M$ have been reported in the livers of cholestatic patients and in rat models of biliary cholestasis.[38] In mice, it was also shown that LCA treatment induces CYP3A11 and OATP2 expression *in vivo* in wild type animals but not in PXR-null mice. The data demonstrate that LCA can stimulate CYP3A11 and OATP2 expression through a PXR-dependent mechanism *in vivo*. These findings suggest that PXR may function as a receptor for LCA and/or its metabolites under certain pathophysiological conditions. Interestingly, LCA is eliminated in the urine of rodents and humans following hydroxylation at the 6-position, a reaction that is catalyzed by CYP3A family members.[39,40]

Based upon these data, Staudinger *et al.* propose that PXR functions as a physiological sensor of LCA and perhaps other toxic bile acids in the liver and coordinately regulates gene expression so as to reduce their concentrations. According to this model, PXR activation represses levels of CYP7A1, which blocks further bile acid biosynthesis, and induces OATP2 and CYP3A levels, which results in increased hepatocellular uptake and metabolism of LCA and other bile acids. A prediction of this model is that PCN and other potent PXR ligands might protect the liver from the toxic effects of LCA. Consistent with this model, it was previously shown that PCN treatment dramatically reduces mortality and blocks the formation of lesions in the liver and common bile duct of LCA-treated rats.[28] A schematic representation of this model is shown in Figure 8.

In humans, the urine of patients suffering from cholestasis contains elevated levels of 6α-hydroxylated bile acids, including the LCA metabolite hydrodeoxycholic acid, the production of which is CYP3A4-mediated.[39–40] These findings suggest that increased 6α-hydroxylation is a relevant mechanism for reducing the levels of toxic bile acids in humans. Elevated levels of 6α-hydroxylated bile acids are also observed in the urine of healthy subjects treated with the PXR ligand rifampicin.[41] Notably, rifampicin has been used successfully in the treatment of pruritus associated with intrahepatic cholestasis and, in some instances, has been reported to induce remission of cholestasis,[42] though the molecular basis for these clinical effects has remained obscure. Based upon this data, it is possible that the anti-cholestatic effects of rifampicin may be mediated through PXR, and that potent PXR ligands may be more efficacious in the treatment of cholestasis, a severe hepatic disease for which there is no known cure. These studies form the beginning of a broader understanding

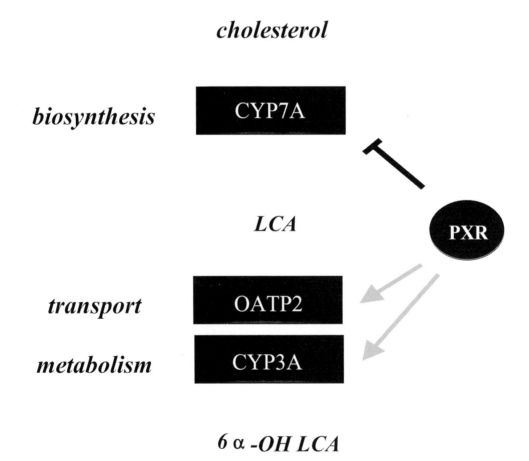

**Figure 8**  Schematic model of PXR regulation of bile acid metabolism: PXR is responsible for the regulation of genes involved in metabolism and transport of bile acids. PXR negatively regulates the *Cyp7a* gene, thus limiting the first and rate-limiting step in cholesterol conversion to bile acids. PXR also up-regulates the organic anion transporter-2 gene (*Oatp2*) and the *Cyp3a* gene, which are involved in the transport and metabolism of bile acids, respectively. The coordinate regulation of this set of genes has the net effect of reducing the intrahepatic concentration of the toxic bile acid, lithcholic acid (LCA) by increasing production and ultimately clearance of the 6α-hydroxylated metabolite.

of PXR biology, and demonstrate its role in the metabolism of both exogenous and endogenous compounds (Figure 9). Further information is expected to be forthcoming from differential gene expression, yeast two-hybrid, and other genetic approaches.

### 14.2.5.9 CONCLUSIONS/ PERSPECTIVES

PXR now joins several other nuclear receptors in being intimately linked to *Cyp* gene regulation. The glucocorticoid, progesterone, androgen, estrogen, PXR, retinoid, Vitamin D, PPAR, CAR, LXR, FXR, and HNF4 receptors all induce specific P450 isoforms. It is clear that these nuclear receptor-regulated P450s play indispensable roles in steroid hormone, bile acid, and fatty acid metabolism. Additionally, PXR highlights the role that nuclear receptors play in regulating *Cyp* genes involved in the response to hormones and toxins. The close connection between the *Cyp* and nuclear receptor gene families reflects their shared functional convergence on compounds from the mevalonate pathway. A large number of compounds from this pathway (including cholesterol, bile acids, oxysterols, and steroids) are both products of *Cyp* gene metabolism as well as ligands for nuclear receptors. The fact that many nuclear receptors regulate *Cyp* gene expression provides an important feedback loop capable of responding to dietary and environmental cues that impinge on these

biologically central metabolites. This interwoven relationship between CYPs and nuclear receptors is clearly an ancient partnership expanded through evolution to regulate homeostasis.

Detailed analysis of NR/CYP interaction has revealed unexpected complexity. For example, the *Cyp7a* gene is known to be regulated by at least eight nuclear receptors, the set comprising a complex regulatory circuit. Much of the regulation involves NR:NR crosstalk, such as when ligand activation of FXR causes induction of the short heterodimer partner (SHP) which subsequently causes repression of *Cyp7a* transcription through its interactions with LRH-1.[43] As we learn more details of PXR biology, the intricacies of how this NR interacts within the overall framework of CYP regulation will ultimately unfold. For example, it has recently been shown that dexamethasone activates the *Cyp3a* gene partly by direct interaction with PXR as described above and partly through activating GR that causes increased expression of PXR. Also, Moore *et al.*[44] have demonstrated ligand cross-talk between the related nuclear receptor CAR and PXR. Their results demonstrate a surprising degree of overlap in the compounds that are capable of binding and regulating the activities of these two receptors. Since CAR and PXR can bind to some of the same response elements and are likely to share many target genes, the net effect of a xenobiotic on gene transcription will often be complex and depend on its effects on both CAR and

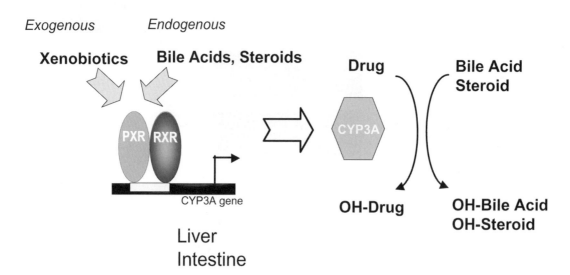

**Figure 9** Model for PXR involvement in regulation of CYP3A expression. PXR binds as a heterodimer with RXR to a response element in the *Cyp3a* gene promoter. Activation of PXR by various exogenous or endogenous ligands results in increased levels of CYP3A enzyme activity that, in turn, results in the hydroxylation of susceptible drugs, steroids, and bile acids.

PXR. An exciting next step in understanding PXR/CAR activity will be to use crystallographic data to derive the structures of their respective ligand binding pockets. This work should explain how these receptors are activated by such structurally and chemically diverse compounds as well as provide further insight into how the PXR/CAR functional overlap is reflected in their structures.

Increases in technology should aid in the unraveling of these complex regulatory pathways as well as the broader implications of the biology of receptors such as PXR. Ultimately, finding selective chemical tools (i.e., the reverse endocrinology approach) will continue to increase our understanding of orphan NRs. These advances in understanding should make these receptors more tractable in pharmacology, either as drug targets or as activities to monitor during drug development.

## 14.2.5.10 ACKNOWLEDGMENTS

The authors would like to thank the large group of individuals at GlaxoSmithKline who contributed to this work. We would also like to acknowledge Jodi Maglich for her contribution of unpublished data for this review.

## 14.2.5.11 REFERENCES

1. M. Beato, P. Herrlich and G. Schutz, 'Steroid hormone receptors: many actors in search of a plot.' *Cell*, 1995, **83**, 851–857.
2. P. Kastner, M. Mark and P. Chambon, 'Nonsteroid nuclear receptors: what are genetic studies telling us about their role in real life?' *Cell*, 1995, **83**, 859–869.
3. D. J. Mangelsdorf and R. Evans, 'The RXR heterodimers and orphan receptors.' *Cell*, 1995, **83**, 841–850.
4. V. Giguere, 'Orphan nuclear receptors: from gene to function.' *Endocrine Rev.*, 1999, **20**, 689–725.
5. R. B. Lanz and S. Rusconi, 'A conserved carboxy-terminal subdomain is important for ligand interpretation and transactivation by nuclear receptors.' *Endocrinology*, 1994, **135**, 2183–2195.
6. R. Evans, 'The steroid and thyroid hormone receptor superfamily.' *Science*, 1988, **240**, 889–895.
7. P. J. Willey and D. J. Mangelsdorf, 'Nuclear orphan receptors: The search for novel ligands and signaling pathways.' *Hormones and Signaling*, 1998, **1**, 307–358.
8. S. A. Kliewer, J. M. Lehmann and T. M. Willson, 'Orphan nuclear receptors: Shifting endocrinology into reverse.' *Science*, 1999, **284**, 757–760.
9. S. A. Kliewer, J. T. Moore, L. Wade *et al.*, 'An orphan nuclear receptor activated by pregnanes defines a novel steroid signaling pathway.' *Cell*, 1998, **92**, 73–82.
10. J. M. Lehmann, D. D. McKee, M. A. Watson *et al.*, 'The human orphan nuclear receptor PXR is activated by compounds that regulate CYP3A4 gene expression and cause drug interactions.' *J. Clin. Invest.*, 1998, **102**, 1016–1023.
11. G. Bertilsson, J. Heidrich, K. Svennson *et al.*, 'Identification of a human nuclear receptor defines a new signaling pathway for CYP3A induction.' *Proc. Natl. Acad. Sci. USA*, 1998, **95**, 12208–12213.
12. B. Blumberg, W. Sabbagh, H. Juguilon *et al.*, 'SXR, a novel steroid and xenobiotic-sensing nuclear receptor.' *Genes Dev.*, 1998, **12**, 3195–3205.
13. H. Zhang, E. LeCulyse, L. Liu *et al.*, 'Rat pregnane X receptor: Molecular cloning, tissue distribution, and xenobiotic regulation.' *Arch. Biochem. Biophys.*, 1999, **368**, 14–22.
14. S. A. Jones, L. B. Moore, J. L. Shenk *et al.*, 'The pregnane X receptor: A promiscuous xenobiotic receptor that has diverged during evolution.' *Mol. Endocrinol.*, 2000, **14**, 27–39.
15. T. A. Kocarek, E. G. Schuetz and P. S. Guzelian, 'Regulation of phenobarbital-inducible cytochrome P450 2B1/2 mRNA by lovastatin and oxysterols in primary cultures of adult rat hepatocytes.' *Appl. Pharmacol.*, 1993, **120**, 298–307.
16. T. A. Kocarek, E. G. Schuetz, S. C. Strom *et al.*, 'Comparative analysis of cytochrome P4503A induction in primary cultures of rat, rabbit, and human hepatocytes.' *Drug Met. Dispos.*, 1995, **23**, 415–421.
17. J. L. Barwick, L. C. Quattrochi, A. S. Mills *et al.*, '*Trans*-species gene transfer for analysis of glucocorticoid-inducible transcriptional activation of transiently expressed human CYP3A4 and rabbit CYP3A6 in primary cultures of adult rat and rabbit hepatocytes.' *Mol. Pharmacol.*, 1996, **50**, 10–16.
18. D. W. Nebert and F. J. Gonzalez, 'P450 genes: structure, evolution, and regulation.' *Ann. Rev. Biochem.*, 1987, **56**, 945–993.
19. P. Maurel, in 'Cytochromes P450: Metabolic and Toxicological Aspects', ed., C. Ioannides, CRC Press, Inc., Boca Raton, FL, 1996, pp. 241–270.
20. F. P. Guengerich, 'Cytochrome P-450 3A4: Regulation and role in drug metabolism.' *Ann. Rev. Pharmacol. Toxicol.*, 1999, **39**, 1–17.
21. P. S. Guzelian, in '*Microsomes and Drug Oxidations*', eds. J. O. Miners, D. J. Burkitt, R. Drew and M. McManus, Taylor and Francis, London, 1988, pp. 148–155.
22. E. G. Schuetz and P. S. Guzelian, 'Induction of cytochrome P-450 by glucocorticoids in rat liver. II. Evidence that glucocorticoids regulate induction of cytochrome P-450 by a nonclassical receptor mechanism.' *J. Biol. Chem.*, 1984, **259**, 2007–2012.
23. J. M. Huss, S. I. Wang, A. Astrom *et al.*, 'Dexamethasone responsiveness of a major glucocorticoid-inducible CYP3A gene is mediated by elements unrelated to a glucocorticoid receptor binding motif.' *Proc. Natl. Acad. Sci. USA*, 1996, **93**, 4666–4670.
24. L. C. Quattrochi, A. S. Mills, J. L. Barwick *et al.*, 'A novel *cis*-acting element in a liver cytochrome P450 3A gene confers synergistic induction by glucocorticoids plus antiglucocorticoids.' *J. Biol. Chem.*, 1995, **270**, 28917–28923.
25. X. K. Zhang, B. Hoffmann, P. B. Tran *et al.*, 'Retinoid X receptor is an auxiliary protein for thyroid hormone and retinoic acid receptors.' *Nature*, 1992, **355**, 441–446.
26. J. L. Staudinger, B. Goodwin, S. A. Jones *et al.*, ' The nuclear receptor PXR is a lithocholic acid sensor that protects against liver toxicity'. *Proc. Natl. Acad. Sci. USA*, 2001, **98**, 3369–3374.
27. W. Xie, J. L. Barwick, M. Downes *et al.*, 'Humanized xenobiotic response in mice expressing nuclear receptor SXR', *Nature*, 2000, **406**, 435–439.
28. H. Selve, 'Prevention of catatoxic steroids of lithocholic acid-induced biliary concrements in the rat.' *Proc. Soc. Exp. Biol. Med.*, 1972, **141**, 555–558.
29. S. S. Chatterjee, S. K. Bhattacharya, M. Wonnemann *et al.*, 'Hyperforin as a possible antidepressant

component of hypericum extracts.' *Life Sciences*, 1998, **63**, 499–510.

30. M. E. Muller, A. Singer, M. Wonnemann *et al.*, 'Hyperforin represents the neurotransmitter reuptake inhibiting constituent of hypericum extract.' *Pharmacopsychiatry*, 1998, **1**, 16–21.

31. S. T. Kaehler, C. Sinner, S. S. Chatterjee *et al.*, 'Hyperforin enhances the extracellular.' *Neuro. Lett.*, 1999, **262**, 54–59.

32. G. Laakmann, C. Schule, T. Baghai *et al.*, 'St. John's wort in mild to moderate depression; the relevance of hyperforin for the clinical efficacy.' *Pharmacopsychiatry*, 1998, **1**, 54–59.

33. A. Biber, H. Fischer, A. Romer *et al.*, 'Oral bioavailability of hyperforin from hypercium extracts in rats and human volunteers.' *Pharmacopsychiatry*, 1998, **1**, 36–43.

34. L. B. Moore, B. Goodwin, S. A. Jones *et al.*, 'St. John's wort induces hepatic drug metabolism through activation of the pregnane X receptor.' *Proc. Natl. Acad. Sci. USA*, 2000, **97**, 7500–7502.

35. Y. C. Li, D. P. Wang and J. Y. Chiang, 'Regulation of cholesterol 7a-hydroxylase in the liver: Cloning, sequencing, and regulation of cholesterol 7a-hydroxylase mRNA.' *J. Biol. Chem.*, 1990, **265**, 12012–12019.

36. B. Noe, B. Hagenbuch, B. Stieger *et al.*, 'Isolation of a multispecific organic anion and cardiac glycoside transporter from the rat brain.' *Proc. Natl. Acad. Sci. USA*,1997, **94**, 10346–10350.

37. K. Miyai, W. W. Mayr and A. L. Richardson, 'Acute cholestasis induced by lithocholic acid in the rat: A freeze-fracture replica and thin section study.' *Lab. Invest.*, 1975, **32**, 527–535.

38. K. D. Setchell, C. M. Rodrigues, C. Clerici c *et al.*, 'Bile acid concentrations in human and rat liver tissue and in hepatocyte nuclei.' *Gastroenterology*, 1997, **112**, 226–235.

39. T. K. Chang, J. Teixeira, G. Gil *et al.*, 'The lithochilic acid 6b-hydroxylase cytochrome P-450, CYP3A10, is an active catalyst of steroid hormone 6b-hydroxylation.' *Biochem. J.* 1993, **291**, 429–433.

40. Z. Araya and K. Wikvall, '6a-Hydroxylation of taurochenodeoxycholic acid and lithocholic acid by CYP3A4 in human liver microsomes.' *Biochem. Biophys. Acta*, 1999, **1**, 47–54.

41. H. Wietholtz, H. U. Marschall, J. Sjovall *et al.*, 'Stimulation of bile acid 6a-hydroxylation by rifampicin.' *J. Hepatology*, 1996, **24**, 713–718.

42. E. L. Cancado, R. M. Leitao, F. J. Carrilho *et al.*, 'Unexpected clinical remission of cholestasis after rifampicin therapy in patients with normal or slightly increased levels of gamma-glutamyl transpeptidase.' *Am. J. Gastroenetrol.*, 1998, **93**, 1510–1517.

43. B. Goodwin, S. A. Jones, R. R. Price *et al.*, 'A regulatory cascade of the nuclear receptors FXR, SHP-1, and LRH-1 represses bile acid biosynthesis.' *Mol. Cell*, 2000, **6**, 517–256.

44. L. B. Moore, D. J. Parks, S. A. Jones *et al.*, 'Orphan nuclear receptors constitutive androstane receptor and pregnane X receptor share xenobiotic and steroid ligands.' *J. Biol. Chem.*, 2000, **275**, 15122–15127.

J.P. Vanden Heuvel, G.H. Perdew, W.B. Mattes and W.F. Greenlee (Eds.)
Comprehensive Toxicology, Vol. xiv

# 14.2.6
# Regulation of Phenobarbital Responsiveness Via the Constitutive Androstane Receptor (CAR)

CURTIS J. OMIECINSKI and SEAN M. BOYLE
*University of Washington, Seattle, WA, USA*

## 14.2.6.1 INTRODUCTION

In mammalian organisms, during development and under basal conditions, a repertoire of >30,000 genes undergo differential transcription. Activation of specific genes in response to chemicals, pathogenic infection, or environmental stressors, requires a highly integrated signal transduction process that signals the transcriptional machinery to direct the expression patterns of appropriate genes.[1] The activation of a given gene will depend on the simultaneous interplay of particular combinations of nuclear proteins that localize to their promoter regions and other regulatory DNA elements, such as transcriptional enhancers. Most DNA enhancer elements contain distinct sets of transcription factor binding sites. Variation in the arrangement of these sites provides the potential to create unique and context specific DNA-protein complexes.[2,3] Cooperative interplay between the proteins in these complexes and with other nuclear factors, such as coactivators and corepressors, can lead to a high level of discrimination in gene activation and to a marked level of transcription synergy.[1–4]

It is noteworthy that several classes of environmental and therapeutic substances are recognized for their capacity to markedly modulate the transcriptional status of mammalian biotransformation enzymes. These enzymes include certain glutathione-*S*-epoxide transferases, UDP-glucuronosyl transferases, epoxide hydrolases, aldehyde dehydrogenases, and the cytochrome P450s or CYPs.[5–9] The CYPs constitute a very important phase I enzyme network and principally catalyze the oxidation of a wide variety of chemicals, including pharmaceuticals. Typically, the biotransformation process tends toward detoxification, with the resulting metabolites being more water-soluble and exhibiting increased likelihood to undergo further reactions *via* phase II conjugation pathways. However, a large number of procarcinogens and other environmental toxins are bioactivated by the xenobiotic metabolizing CYPs.[10]

Since many relevant substances are either bioactivated or detoxified by CYP mediated metabolism, it is likely that variation in expression patterns and levels of CYPs broadly impact the outcome of chemical exposures.[11] It is further anticipated that certain interindividual differences in CYP expression may lead to altered risk for the development of toxicities, such as certain cancers, birth defects and adverse drug reactions.[10–12] This chapter will focus on the phenomenon whereby phenobarbital (PB) and "PB-like" agents modulate expression of the mammalian biotransformation system, in particular with respect to the effects of these agents on CYP gene expression pathways. This topic has been the subject of several recent reviews, to which the reader is directed for further information.[5,6,13]

## 14.2.6.2　THE CYTOCHROME P450S

The CYP monooxygenases are believed to have evolved from an ancestral gene 3.5–4 billion years ago.[14] The CYPs are intimately involved in functionalization reactions representing the first phase of xenobiotic detoxification, as well as being integral parts of many biosynthetic reactions in the cell. Over 1000 CYP genes have been characterized across many species of animals, plants, and microbes.[14] It has been proposed that evolution of more recent CYP genes coincided with terrestrial colonization of plant-eating animals 400 million years ago. One theory proposes that an animal–plant war began as plants synthesized toxic compounds to discourage predators, and animals responded to this selective pressure by evolving multiple CYP genes in order to detoxify the novel toxins.[14,15] Different CYP family members tend to exhibit substrate preferences, however extensive overlap in substrate specificity does exist.[16] CYPs are expressed in most mammalian tissues, however, the liver is responsible for the bulk of chemical biotransformation.[8,17]

Although many CYPs have been characterized, at least 17 CYP gene families exist in mammals and can be functionally divided into two main classes: CYPs that are involved in the biosynthesis of steroids, bile acids and fatty acids, and those that metabolize foreign chemicals, or xenobiotics.[14,16,18] Of the latter CYPs, the CYP1, CYP2, and CYP3 families are principally responsible for drug metabolism. In humans, CYP3A (i.e., CYP3A4 and CYP3A5), and CYP2C (i.e., CYP2C8, CYP2C9, CYP2C18, CYP2C19) are the most abundant, accounting on average for 30% and 20% of the total liver CYPs, respectively; with CYP1A2 (13%), CYP2E1 (7%), CYP2A6 (4%), CYP2D6 (2%), and CYP2B6 (0.4%) also participating as the major CYP forms involved in drug and chemical biotransformation.[19,20] Several web sites exist that catalog and update the information available related to the CYP superfamily. The reader is referred to internet sites maintained by David Nelson, (http://drnelson.utmem.edu/CytochromeP450. html) and the gene locus link at NIH/NCBI (http://www.ncbi.nlm.nih.gov/LocusLink/list. cgi?Q = P450&ORG = Hs&V = 0).

Individuals can differ in the relative levels of CYPs that are expressed constitutively; similarly, a number of members of these principle CYP subfamilies are markedly inducible upon exposure to chemicals. For example, in rats, phenobarbital (PB) treatments can induce both CYP2B1 and CYP2B2 levels in liver up to 50–100 fold.[18,21] PB inducible responses have been documented to occur for human CYP2B6 in primary hepatocyte culture.[22] The human and rodent CYP3A4 genes are similarly markedly responsive to prototypical PB-like inducers.[23,24] CYP gene induction can manifest both beneficial and detrimental effects on xenobiotic metabolism. Understanding the mechanisms that account for these induction responses may greatly facilitate both therapeutic and prophylactic intervention in disease states. The remaining portion of this chapter will review the mechanistic information available that accounts for the induction response to the PB-like inducing agents.

### 14.2.6.3  MECHANISMS OF PB INDUCTION: GENERAL CONSIDERATIONS

Several mechanisms may be operative in the induction process, including post-transcriptional stabilization of mRNA, post-translational stabilization of the enzyme, post-translational protein modifications such as changes in phosphorylation status, and/or direct transcriptional activation of the respective genes. The principle mechanism responsible for the PB inducible response involves direct gene activation at the transcriptional level.[25] With respect to induction of biotransformation enzyme systems, often the same chemical that elicits the gene induction signal is also a substrate for the gene product. Therefore, in theory, induction of enzyme activities allows an organism to quickly respond to toxicant exposures by facilitating enhanced elimination of the foreign substances.

Prototypical inducing agents include the polyaromatic and polychlorinated hydrocarbons, ethanol and organic solvents, peroxisome proliferator compounds such as the phthalate esters, dexamethasone, and several sedative-hypnotic medications.[5,7–9] These different 'classes' of inducers tend to impact the expression levels of the CYP1A, CYP2E, CYP4A, CYP3A, and CYP2B subfamilies of P450, respectively. In the latter case, PB serves as a model agent for other barbiturates and a variety of xenobiotic compounds such as chlordanc, DDT, certain PCBs, *etc.*, that exhibit profound inductive effects on the biotransformation system.[5,26,27] The PB induction response occurs in most mammalian species, including humans, and is principally manifested in the liver.[8] Details of the other processes that relate to the various inducer pathways are the subject of other chapters in this text.

### 14.2.6.4  CYP2B GENES

The CYP2B genes are typically expressed in the liver at very low basal levels. Similarly low levels of CYP2B mRNA can be detected in selected other tissues.[21,28] When an inducer is present, mRNA and protein levels increase markedly, up to 100-fold over basal levels. Although the induction response is relatively liver-specific, reports indicate that PB also enhances CYP expression within intestinal enterocytes and in the brain.[29,30] However, in these latter tissues the induced levels of expression are far lower than those occurring in liver. Within the liver, inducible CYP2B expression is most pronounced in the centrilobular region, i.e., those liver hepatocytes in closest proximity to the central vein.[31] The underlying mechanisms controlling tissue- and regio-specific expression and induction responses remain relatively poorly understood.

Through study of the rodent PB-inducible P450 genes, in particular the mouse Cyp2b10 and the rat CYP2B1 and CYP2B2, a great deal of progress has been made regarding the molecular mechanisms involved in the PB induction response. The principle features of the 5'-flanking promoter regions of these genes, as well as the orthologous human CYP2B6 gene, appear quite similar.[6,32] The hallmark structural features are described below and indicated in Figure 1.

The CYP2B genes contain the typical DNA elements (*cis* elements) that serve as binding sites for transcription factors (*trans* elements) found in many eukaryotic genes (see Figure 1). The respective CYP2B promoter contains *cis* elements such as the TATAA box and other elements required for RNA polymerase complex formation. This region is referred to as the proximal promoter region and is involved in directing the basal transcriptional levels of the gene.[33] The proximal region is just upstream of the transcription start site, where the structural DNA begins its transcription into heterogeneous RNA for subsequent splicing into mature mRNA.

Initially, experiments conducted in the PB-inducible bacteria *Bacillus megaterium* revealed the existence of a seventeen-base pair operon termed the Barbie box,[34] that responded to PB by increasing the transcription of CYP102 and CYP106 genes.[35] This response is mediated by both positive and negative acting transcription factors that bind to a *cis* element[36] on the operon. Although putative barbie box consensus elements were subsequently identified in several mammalian PB-responsive genes, including certain CYP genes,[37] further investigations demonstrated that the Barbie box does

## Hypothetical Accessory Factors

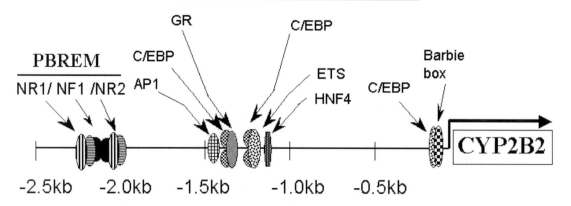

**Figure 1** Structural rendition of the rat CYP2B2 gene. The relative position of the PBREM, other potential nuclear accessory factors, and the core promoter region of the gene are indicated. The basic features of CYP2B gene architecture are similar in the rat, mouse and human.

not appear to direct PB-mediated induction of CYP genes in eukaryotes.[5,38–39]

A basal transcription element, or BTE, has also been identified in the proximal region of the promoter in several PB inducible CYP genes.[40,41] Both the BTE and the Barbie box elements may be involved in basal transcriptional regulation, but are not likely critical in directing PB induction.[42] Additional DNA elements exist further upstream of the proximal promoter region of the CYP2B1/2 and Cyp2b10 genes that appear to interact *in vitro* with a variety of transcription factors. For example, a functional CCAAT/enhancer binding protein (C/EBPα) element has been characterized as important for the constitutive transcription in the CYP2B1 gene.[41,43] Further upstream, nuclear factor kappa B (NFκB) sites have been identified in the CYP2B gene promoters from rat, mouse and human that may be involved in gene repression.[44] This potential mode of regulation may be important since a number of inflammatory processes, cytokines, and viral exposures have been reported to repress the PB induction process.[45,46] A glucocorticoid response element (GRE) was characterized in the CYP2B2 promoter approximately 1.3 kb 5′ of the transcription start site[47] that may be involved in the apparent glucocorticoid dependence of the PB induction response.[24,48–49] Other transcription factors, such as AP1, may also be involved in regulating the CYP2B genes.[50]

An enhancer is a region of the promoter that functions to markedly activate the transcription of the gene. An enhancer element, termed the PB responsive enhancer module, or PBREM, has been localized up to 2.3 kb upstream of the transcriptional start sites of

the CYP2B1, CYP2B2, Cyp2b10, and CYP2B6 genes. The PBREM appears to be intimately involved in the induction, or up-regulation response, subsequent to exposures to the "PB-class" of agents.[32,51]

### 14.2.6.5 THE PHENOBARBITAL RESPONSIVE ENHANCER MODULE: PBREM

Transgenic studies in mice were the first to implicate the existence of the upstream PB enhancer region, using rat CYP2B2 promoter constructs as mouse "transgenes".[52] Using both primary hepatocyte culture models,[51,53–54] and direct *in situ* injection of DNA into the liver,[55] investigators successfully delineated a 51 bp PBREM approximately 2.3 kb upstream of the core promoters of the rat CYP2B2, mouse Cyp2b10, and human CYP2B6[56] PB-inducible genes. By definition, an enhancer fragment of DNA can function in either orientation. In this respect, it is noteworthy that the PBREM region is in the opposite orientation in the human CYP2B6 gene compared to that of the corresponding rat or mouse genes.[56] With the discovery of the enhancer region, the anatomy of the enhancer and its regulatory factor interactions are being delineated.

As indicated in Figure 2, the PBREM is composed of two nuclear receptor sites, NR1 and NR2. These sites are termed "NR" because the response element exhibits an imperfect direct repeat nucleotide sequence known to bind nuclear receptors. Nuclear receptors have been identified that interact

# Enhanceosome
# Complex

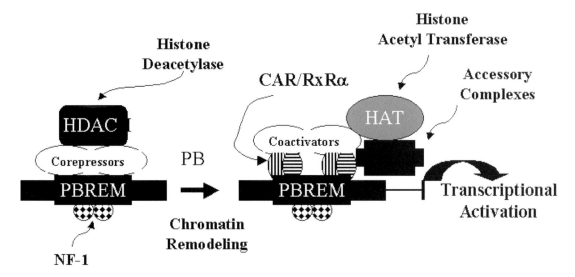

**Figure 2** A putative scheme for transcriptional activation mediated by the PBREM. In this scheme, the PBREM region is bound to nuclear factor 1 and corepressor factors. As a result of the protein context, histone deacetylases have been recruited that function to close the nucleosomal structure and restrict access to other transcription factors. The presence of a PB-like inducer causes the translocation of the CAR-RXRα complex to the nucleus that interacts with the NR-1 and/or NR-2 regions of the PBREM. The interaction favors further interactions with nuclear coactivators proteins, the recruitment of histone acetyl transferases, and remodeling of the local chromatin environment. The remodeling process further results in the recruitment of transcriptional accessory proteins, culminating in the activation of gene transcription.

with the NR regions within the PBREM and are discussed in Section 14.2.6.7. The NR sites within the PBREM flank a core nuclear factor-1 (NF1) motif.[57] NF1 is a liver-enriched nuclear factor involved in the regulation of many genes expressed in the liver. The potential role of the NF1 element in the PB induction process has been studied in various models. The NF1 region is marked by a strong DNase I footprint, and coincides with a DNase I hypersensitivity region.[42] Although NF1 itself does not appear to possess PB-mediated transactivation activity in transfected cells, mutation of the NF1 site within CYP2B genes reduced the PB induction response in *in vitro* transfection assays.[42,58–59] Studies of NF1 interaction on chromatinized templates indicate that NF1 binds independent of other nuclear factors to the PBREM but may enhance the nuclear factor activity during the induction response.[60] However, results from studies using CYP2B2 transgenic mouse models that contained loss-of-function mutations in the NF1 motif indicated no substantial loss of PB inducibility with the mutant NF1 transgenes, arguing against a critical role for

the factor in directing the PB induction response *in vivo*.[61]

## 14.2.6.6 NUCLEAR RECEPTORS

It has become well established that an array of nuclear receptor proteins serve as key transducers of chemical inducer signals.[13] This important area of biology has been the subject of a number of recent reviews.[13,62–64] Like the CYPs, the nuclear hormone receptors (NHRs) are encoded by a superfamily of genes. In any particular mammalian species, $> 70$ different proteins are included in this array that function to coordinate and integrate a large variety of ligand-regulated and ligand-independent signaling events and serve as critical modulators of the cellular transcriptional machinery.[65] To further illustrate this point, *C. elegans* expresses $\sim 250$ nuclear receptor proteins, representing the largest expressed protein family in this organism.[66,67]

As illustrated in Figure 3, the proteins in the NHR family share common structural features, including the general conservation of six

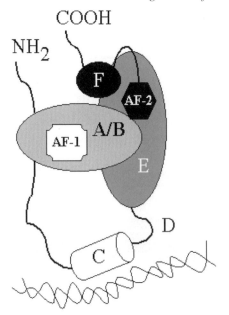

**Figure 3** Modular structure of a typical nuclear hormone receptor. The DNA binding domain C is interposed between activation function and ligand binding domains. Although not indicated, nuclear receptors often heterodimerize with the retinoid X receptor protein by virtue of heterodimerization domains. The figure is adapted from Reference 63.

functional domains.[68,69] The N-terminal A/B domain of these receptors is only weakly conserved and contains a non-ligand regulated transcriptional activation function (AF-1 domain). The C domain is the most conserved domain. It contains a highly conserved P-box and two zinc fingers necessary for forming the DNA binding domain. The D-domain is a hinge region that exhibits little conservation. The E domain contains the ligand binding domain and the ligand-dependent activation function AF-2 domain. The F domain is hypervariable and its function is not well defined.[65]

There are two general classes of nuclear receptors, the steroid receptors, represented by estrogen, androgen, mineralocorticoid, glucocorticoid, and progesterone receptors; and the nonsteroid receptors. Examples of nonsteroid receptors include the thyroid hormone receptor, pregnane X receptor (PXR), constitutive androstane receptor (or, constitutively active receptor) (CAR), the human steroid and xenobiotic receptor (SXR) and the retinoid X receptors (RXRs) that include RXRα, RXRβ and RXRγ.[65,70] Key differences exist between these classes of receptors. Steroid receptors usually bind DNA as homodimers, while nonsteroid receptors typically dimerize with RXR and bind DNA as heterodimers.[71] With

the estrogen receptor being an exception, most steroid receptors are localized to the cytosol in their non-ligand bound forms and translocate to the nucleus upon ligand binding.[72] The nonsteroid receptors most often are constitutively nuclear, although CAR appears to be an exception.[73] It is also apparent that steroid receptors will not actively repress gene expression unless an antagonist is present, such as tamoxifen in the case of the estrogen receptor (ER).[74] In contrast, a number of nonsteroid receptors actively repress transcription in their non-ligand bound form through recruitment of corepressor complexes.[75] A unique characteristic of NHRs in comparison to most other transcription factors is that they bind chromatinized DNA templates with similar affinity as naked DNA templates, a property that enables the receptors to nucleate transcriptional activation events within the chromosome.[76]

### 14.2.6.7 THE CAR NUCLEAR RECEPTOR

The Negishi group was the first to identify the interaction of the previously identified nuclear receptor CAR within the PBREM region of the mouse *Cyp2b10* gene.[53–54,59] CAR is most abundantly expressed in the liver.[77] In earlier studies, CAR was shown to heterodimerize with RXRα (see Figure 4) and transactivate through imperfect direct repeat DR-4 and DR-5 elements, although the list has since expanded to include DR-3 and everted repeat ER-6 elements.[77,78] Initial results obtained from experiments using combined transient transfection of RXRα and CAR, in combination with a heterologous promoter/ reporter construct containing the PBREM, supported the hypothesis that CAR confers PB responsiveness to the PBREM and that this element likely facilitates the induction process for other PB-inducible genes. This conclusion has been further solidified recently by generation of a CAR knockout mouse model that exhibits none of the induction responses typically seen with PB exposure.[79] Interestingly, CAR is capable of conferring PB responsiveness to the human *CYP3A4* gene, indicating a combinatorial regulation of CYP3A4 by both CAR and PXR (SXR is the human ortholog of PXR).[56] The latter observation, coupled with the information that CAR/RXRα complexes can interact with a variety of DNA target elements,[78] strongly implied that CAR may have a broad role in the regulation of CYP genes in the liver. The possibility of an endogenous ligand binding to and inhibiting the activation function of CAR

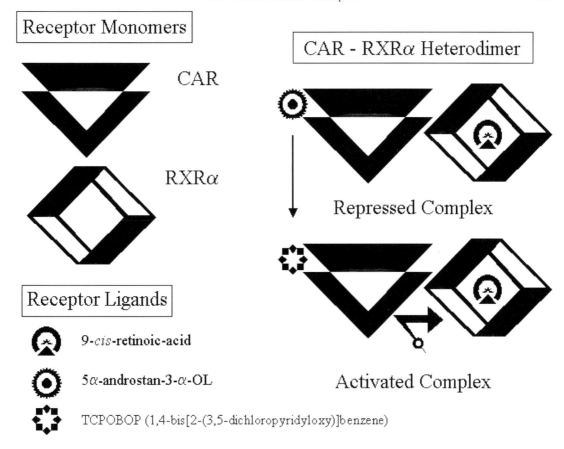

**Figure 4** Ligand-receptor interactions involving the constitutive androstane receptor, CAR, also known as the constitutively active receptor, and the retinoid X receptor, RXR. The figure illustrates a proposed scheme whereby CAR and RXR dimerize, and upon interaction with the inducer ligand, TCPOBOP, the negative androstane agonist is displaced, giving rise to an activated receptor complex.

is supported by findings that sterol biosynthesis inhibitors are potent inducers of CYP2B genes.[80,81]

Despite the evidence that CAR is the PB responsive NHR, it is noteworthy that the experiments performed to date indicate that PB is not a typical ligand activator of CAR, in that it does not appear to bind human CAR at concentrations up to 1mM.[82] These observations imply that PB activates CAR through a separate mechanism.[73,82] However, an extended range of substances have been identified that interact as ligands with human CAR,[82] supporting an important role for this receptor as a component of an integrated cellular regulatory network. The receptor likely functions as a regulatory sensor for both xenobiotic chemicals as well as endogenous lipid-derived mediators.[82] Some marked differences exist with respect to the ability of CAR receptors from different species to interact with their respective ligands. For example, 1,4-bis[2-

(3,5-dichloropyridyloxy)]benzene (TCPOBOP) is a potent activator of mouse CAR but appears to lack activity for human CAR.[78,82] Of further note, PB does not facilitate the recruitment of steroid receptor coactivator-1 (SRC-1) in *in vitro* coactivator recruitment assays, an observation that does not hold true for the more efficacious mouse CYP2B inducer, TCPOBOP. The latter inducer does facilitate coactivator recruitment and will function as a typical ligand in a radioligand-binding assay.[78,82] Specifically, CAR has been shown to interact with the SRC-1, an interaction that is mediated through the LXXLL motif in SRC-1.[78,82] CAR has been used as "bait" in a yeast-2-hybrid screen. Outside of the expected partners, RXR and SRC-1, CAR associates with TBP7, a protein component of the 26S proteosome,[83] and SHP (small heterodimer partner).[84] The association of CAR with the 26S proteosome implies that the association may be important in the exchange of coactivator complexes.[85] SHP appears to act as a transcriptional repressor of a diverse array of NHRs, including CAR.[84] The latter observation is supported by the recent finding that SHP, through interaction with LRH-1, mediates repression of the LXR activated gene, CYP7A,[86,87] as well as RXR transactivation.[88] The adapter proteins that mediate the association of the complexes with the ligand binding domain of the NHRs contain LXXLL motifs.[89–91] These motifs are necessary and sufficient for interaction of coactivators with ligand bound NHRs.[92] These observations raise the interesting possibility that CAR is capable of interaction with one or more mammalian coactivator complexes in the process of transcriptional activation.

Initially, when CAR was cloned, it was determined to be a constitutively active transcription factor and a permanent resident of the nuclear compartment; an observation made in a cell line.[77] Studies to identify ligands for CAR revealed that the NHR is negatively regulated by the pheromone 5α-androstan-3-α-OL,[93] indicating that CAR transcriptional activity may be inhibited by the pheromone in the absence of PB. However, there are a number of problems with this hypothesis, foremost is that the levels of androstanes needed to inhibit CAR are considerably higher than those found in the blood.[94] Additional work from the Negishi group, performed with primary hepatocytes, indicated that CAR is localized to the cytosol in the absence of inducer and translocates to the nucleus (an okadaic acid-inhibited process) upon inducer treatment.[73] Thus, according to this scheme, PB functions in as yet an unknown manner to facilitate the nuclear translocation of CAR and thus activate transcription in an indirect manner.

### 14.2.6.8   RETINOID X RECEPTOR: RXR

As indicated previously and indicated in Figure 3, CAR appears to function biologically in liver through hetero-dimerization with RXRα. In the adult liver, RXRα is the most abundant of the three RXR receptors, suggesting an important regulatory role as a regulator of gene expression in this organ.[70] The importance of RXRα is quite clear in a broader biological context in that it engages as a dimer partner with a large number of other nuclear receptors, including the retinoic acid receptors, Vitamin D receptor, thyroid hormone receptor, PPARα, PXR, LXR, FXR, and SXR.[13,71] In this context, RXR has been referred to as a "master regulator".[71] The levels of RXR

within the cell appear to be regulated themselves by various effectors, including exposures to dexamethasone.[95]

### 14.2.6.9   DO OTHER NUCLEAR FACTORS IMPINGE ON THE PB INDUCTION PROCESS?

As mentioned earlier, a number of other nuclear factors have been implicated as participants in the PB transcription response mechanism. For example, C/EBPα,[41,43,96] NFκB,[44,97] the glucocorticoid receptor[47,58] and activator protein-1 (AP-1)[50] have all been implicated to interact with elements found in CYP2B promoters. HNF-4, another liver enriched transcription factor, appears to bind to the rat CYP2B2 PBREM *in vitro*.[61] Hepatic nuclear factor 4 (HNF-4) is an orphan receptor, a term that describes nuclear receptors that lack known ligands,[98] and has been implicated as an important regulator of liver-specific gene responses.[99–100] However, preliminary studies using recombinant adenovirus-mediated over expression of HNF-4, or a dominant negative form of HNF-4, have failed to demonstrate substantial modulatory effect on CYP2B1 transcription (S. M. Boyle and C. J. Omiecinski, unpublished results). In addition, other nuclear factors may interact with the PBREM, or surrounding sequences, to modulate the induction response.[58,61] For example, putative GREs have been identified within the PBREM, together with an accessory factor element to the 3′ side of NR2.[58] Recently, a third NR site, NR-3, was also detected.[60] It appears likely that factors in addition to CAR and RXR functionally interact within the PBREM region and surrounding sequences to direct the transcriptional activation responses triggered by PB-like inducers. The nature of these interactions, together with their cooperative interplay, remains to be fully elucidated.

Estrogen is another potentially important modulator of the PB induction response. It was reported that relatively high concentrations of 17α-ethinylestradiol or diethylstilbestrol,[101] and a number of proestrogenic organochlorine pesticide compounds,[29,101,102] exhibit PB inductive effects in rat liver and hepatocyte culture. Recently, estrogen also was reported to activate the Cyp2b10 gene in the mouse through interaction with CAR and stimulating the subsequent translocation of CAR into the nucleus.[103] Interestingly, progesterone inhibited these effects.[103] In contrast, the PB induction response can be effectively inhibited by a number of agents, including high concentrations of PB and dexamethasone (DEX).[104] Furthermore, 9-*cis* retinoic acid, a known ligand for RXRα, inhibits induction of the CYP2B1/2 genes.[105] Various cytokines and inflammatory stimuli have been demonstrated to inhibit PB induction,[106] as well as the expression of other CYP pathways,[107] including other inducible systems.[45,107] Exposure to baculovirus, an insect virus, is also capable of inhibiting the PB induction response in primary cultures of rat hepatocytes.[46] Recently, reactive oxygen species have also been shown to repress PB induction in cultured hepatocytes.[108]

### 14.2.6.10 CHROMATIN REMODELING

When the milieu of transcription factors bind their respective DNA response elements of a promoter, activation domains of these transcription factors interact with other proteins involved in assembling the transcriptional complex, known as the enhanceosome.[109,110] The appropriate interaction of transcriptional activators is thought to result in the localized remodeling of chromatin, driven largely by post-translational modifications of histone proteins within the core nucleosomal structures, thereby enhancing the further recruitment of accessory factors that in turn couple with the RNA polymerase II to drive transcription.[111] The reader is referred to recent reviews for additional details on this process.[112–114] A simple model schematizing an enhanceosome assembly process with respect to PB induction of the CYP2B genes is illustrated in Figure 4.

The most common *in vitro* analysis used to determine chromatin remodeling is the DNase I footprinting assay. If a segment of the regulatory region of a naked DNA template is incubated *in vitro* with a nuclear protein extract, the interaction of transcription factors with their cognate recognition elements will typically confer protection to those occupied regions of DNA from the endonuclease activity of DNase I. However, in the case of the CYP2B genes, DNase I footprint analysis typically fails to detect differences in footprint patterns with nuclear protein isolated from control versus PB-treated animals.[33,43,61] For example, with these analyses, avid NF1 interactions are detected within the PBREM in both untreated and treated extracts.[61] Using *in vivo* footprint analysis, where nuclei are typically subjected to a chemical cleavage procedures prior to the isolation of the DNA,[115] Kemper and colleagues have reported a detectable alteration in the footprint within the boundary of the PBREM region.[116]

Recently, these investigators have employed a chromatin assembly method (involving the use of a *Drosophila* embryo S-190 protein extract), to analyze a segment of DNA containing the *cis* elements of the PBREM from the rat CYP2B1 enhancer.[60] When this DNA region was assembled with histones to model its *in vivo* chromatin context, it was shown that after chromatin complexation, NF1 was bound more strongly than to naked templates and enhanced the transactivation of a reporter gene mediated by CAR/RXR.[60] Another method used to detect changes in chromatin structure is the chromatin immunoprecipitation (ChIP) assay. This method uses antibodies directed against modified histones or to transcription factors implicated in the transcriptional regulation of the gene under study.[117,118] The protein-DNA complex is immunoprecipitated and the relative levels of specific DNA promoters, or enhancers, present in the immunoprecipate are identified and quantified using PCR.[117,118] Using this assay, PB-activated chromatin remodeling events within the PBREM have been detected (S. A. Auerbach and C. J. Omiecinski, unpublished results).

### 14.2.6.11 CELLULAR DIFFERENTIATION AND SIGNALING

Despite major advances in identifying key receptor pathways and transcriptional events involved in the PB induction process, many additional details of this regulatory response remain to be elucidated. These include details of the signaling processes necessary for determining the differentiation status of the cell.

The PB induction process is tightly coupled to a highly differentiated hepatic phenotype. Although some evidence for a PB induction response has been obtained in intestinal enterocytes and in certain brain regions, the magnitude of the expression response in the latter cell types is substantially less than that occurring in the liver.[30] Historically, it has been difficult to model the PB responses that occur in the intact liver. Hepatoma cell lines are largely refractory to PB induction; moreover, as these cells become "immortalized", they rapidly and permanently lose many differentiation features, often including the capacity to express other genes that are responsible for xenobiotic metabolism.[24,26,119]

Research performed in our laboratory has established the importance of a variety of defined conditions that enable cultured hepatocytes to exhibit most features inherent in the differentiated adult liver phenotype, including *in vivo*-like responsiveness to PB inducers.[120,121] Appropriately maintained cultures of primary hepatocytes retain differentiated features and are therefore valuable models for assessing liver-specific responses to pharmacological and toxicological agents. Important parameters required for the maintenance of these responses include the provision of extracellular matrix contacts for cultured hepatocytes, use of a serum-free media formulation, and inclusion of physiological concentrations of insulin and dexamethasone.[122] Hepatocytes are quite sensitive to their culture environment such that sub-optimal conditions activate prototypical stress pathways, including stimulation of SAPK/JNK (stress activated protein kinase/Jun-N-terminal kinase) and MAPK (mitogen activated protein kinase) phosphorylation, with the resultant nuclear recruitment of the stress-associated transcription factors, AP-1 and NF-κB.[122] For example, elevated concentrations of several commonly used PI3 (phosphatidyl-inositol-3) kinase inhibitors enhance hepatocyte cytotoxicity and result in the elevated expression CYP2E1 mRNA.[122] Stress-related responses arising from chemical or oxidant exposures, or sub-optimal levels of extracellular matrix, insulin, and/or dexamethasone are associated with a marked down-regulation of several hepatocyte-enriched nuclear transcription factors, factors that are critical to the maintenance of the hepatic phenotype, including maintenance of PB inducibility (J. S. Sidhu and C. J. Omiecinski, unpublished data).

### 14.2.6.12 SUMMARY

Liver metabolism clearly contributes the bulk of xenobiotic metabolism capacity in a mammalian organism, and the liver is likewise the principle organ responsive to the inductive effects of chemical exposures.[8,10,17,123] For substances undergoing biotransformation, it is likely that an individual's capacity for tolerance to chemical injury will be determined in large part by prior exposures to inducing agents, and the magnitude of inductive response mounted within target tissues. Inter-individual variation in induction responses can lead to drastic differences in response to pharmaceutical exposures.[8,11,12] In the clinical setting, induction of drug metabolism can accelerate the clearance of a pharmaceutical agent, thereby reducing its drug's half-life and altering its therapeutic efficacy. Variation in expression of certain P450s may exacerbate

alternate pathways of metabolism for therapeutic agents leading to severe toxicological side effects, including toxicities associated with the use of cancer chemotherapeutics.[124–126] Similarly, exposures to inducing agents may predispose adverse drug–drug interactions. Recent studies have indicated that more than 2.2 million hospitalized Americans suffer adverse drug reactions each year and that approximately 100,000 die unintentionally from administration of medications.[127–129] The situation is further complicated, since as more people take more drugs, the potential for toxic drug interactions increases. The CYP superfamily of enzymes is arguably the most important system directing the metabolic fate of many drugs and protoxins.[10,17,124,127] It is generally accepted that medications that alter CYP gene expression are particularly likely to provoke drug–drug interactions and to promote other adverse consequences associated with drug usage.[8,119,129] Through improved understanding of the molecular mechanisms that regulate the expression of these enzymes, including the PB induction response, it may be possible to optimize therapeutic design or pharmacologically manipulate the metabolic system in order to avoid any number of toxicological consequences of biotransformation.

### 14.2.6.13 REFERENCES

1. B. Lemon and R. Tjian, 'Orchestrated response: a symphony of transcription factors for gene control.' *Genes Devel.*, 2000, **14**, 2551–2569.
2. T. K. Kim and T. Maniatis, 'The mechanism of transcriptional synergy of an *in vitro* assembled interferon-beta enhanceosome.' *Mol. Cell.*, 1997, **1**, 119–129.
3. M. Carey, 'The enhanceosome and transcriptional synergy.' *Cell*, 1998, **92**, 5–8.
4. Y. S. Lin, M. Carey, M. Ptashne *et al.*, 'How different eukaryotic transcriptional activators can cooperate promiscuously.' *Nature*, 1990, **345**, 359–361.
5. B. Kemper, 'Regulation of cytochrome P450 gene transcription by phenobarbital.' *Prog. Nucl. Acid Res. Mol. Biol.*, 1998, **61**, 23–64.
6. T. Sueyoshi and M. Negishi, 'Phenobarbital response elements of cytochrome P450 genes and nuclear receptors.' *Ann. Rev. Pharm. Tox.*, 2001, **41**, 123–143.
7. F. J. Gonzalez, 'The molecular biology of cytochrome P450s.' *Pharm. Rev.*, 1988, **40**, 243–288.
8. U. Fuhr, 'Induction of drug metabolising enzymes: pharmacokinetic and toxicological consequences in humans.' *Clin. Pharmacokinet.*, 2000, **38**, 493–504.
9. M. S. Denison and J. P. Whitlock, 'Xenobiotic-inducible transcription of cytochrome P450 genes.' *J. Biol. Chem.*, 1995, **270**, 18175–18178.
10. F. P. Guengerich, 'Metabolism of chemical carcinogens.' *Carcinogenesis*, 2000, **21**, 345–351.
11. D. L. Eaton, 'Biotransformation enzyme polymorphism and pesticide susceptibility.' *Neurotoxicology*, 2000, **21**, 101–111.

12. D. W. Nebert, 'Polymorphisms in drug-metabolizing enzymes: what is their clinical relevance and why do they exist? [editorial; comment].' *Am. J. Hum. Genet.*, 1997, **60**, 265–271.

13. D. J. Waxman, 'P450 gene induction by structurally diverse xenochemicals: central role of nuclear receptors CAR, PXR, and PPAR.' *Arch. Biochem. Biophys.*, 1999, **369**, 11–23.

14. D. R. Nelson, 'Cytochrome P450 and the individuality of species.' *Arch. Biochem. Biophys.*, 1999, **369**, 1–10.

15. D. W. Nebert and R. A. McKinnon, 'Cytochrome P450: evolution and functional diversity.' *Prog. Liv. Dis.*, 1994, **12**, 63–97.

16. D. R. Nelson, L. Koymans, T. Kamataki *et al.*, 'P450 superfamily: update on new sequences, gene mapping, accession numbers and nomenclature.' *Pharmacogenet*, 1996, **6**, 1–42.

17. C. J. Omiecinski, R. P. Remmel and V. P. Hosagrahara, 'Concise review of the cytochrome P450s and their roles in toxicology.' *Toxicol. Sci.*, 1999, **48**, 151–156.

18. C. J. Omiecinski, R. Ramsden, A. Rampersaud *et al.*, 'Null phenotype for cytochrome P450 2B2 in the rat results from a deletion of its structural gene.' *Mol. Pharmacol.*, 1992, **42**, 249–256.

19. T. Shimada, H. Yamazaki, M. Mimura *et al.*, 'Interindividual variations in human liver cytochrome P450 enzymes involved in the oxidation of drugs, carcinogens and toxic chemicals: studies with liver microsomes of 30 Japanese and 30 Caucasians.' *J. Pharm. Exp. Ther.*, 1994, **270**, 414–423.

20. J. H. Lin and A. Y. Lu, 'Interindividual variability in inhibition and induction of cytochrome P450 enzymes.' *Ann. Rev. Pharm. Tox.*, 2001, **41**, 535–567.

21. C. J. Omiecinski, 'Tissue-specific expression of rat mRNAs homologous to cytochromes P-450b and P-450e.' *Nucl. Acids. Res.*, 1986, **14**, 1525–1539.

22. L. Gervot, B. Rochat, J. C. Gautier *et al.*, 'Human CYP2B6: expression, inducibility and catalytic activities.' *Pharmacogenet.*, 1999, **9**, 295–306.

23. C. Hassett, E. M. Laurenzana, J. S. Sidhu *et al.*, 'Effects of chemical inducers on human microsomal epoxide hydrolase in primary hepatocyte cultures.' *Biochem. Pharm.*, 1998, **55**, 1059–1069.

24. J. S. Sidhu and C. J. Omiecinski, 'Modulation of xenobiotic-inducible cytochrome P450 gene expression by dexamethasone in primary rat hepatocytes.' *Pharmacogenet.*, 1995, **5**, 24–36.

25. J. P. Hardwick, F. J. Gonzalez and C. B. Kasper, 'Transcriptional regulation of rat liver epoxide hydratase, NADPH-cytochrome P-450 oxidoreductase, and cytochrome P-450b genes by phenobarbital.' *J. Biol. Chem.*, 1983, **258**, 8081–8085.

26. D. J. Waxman and L. Azaroff, 'Phenobarbital induction of cytochrome P-450 gene expression.' *Biochem. J.*, 1992, **281**, 577–592.

27. J. M. Rice, B. A. Diwan, H. Hu *et al.*, 'Enhancement of hepatocarcinogenesis and induction of specific cytochrome P450-dependent monooxygenase activities by the barbiturates allobarbital, aprobarbital, pentobarbital, secobarbital and 5-phenyl- and 5-ethylbarbituric acids.' *Carcinogenesis*, 1994, **15**, 395–402.

28. B. G. Lake, J. A. Beamand, A. C. Japenga *et al.*, 'Induction of cytochrome P-450-dependent enzyme activities in cultured rat liver slices.' *Food Chem. Tox.*, 1993, **31**, 377–386.

29. P. G. Traber, J. Chianale, R. Florence *et al.*, 'Expression of cytochrome P450b and P450e genes in small intestinal mucosa of rats following treatment with phenobarbital, polyhalogenated biphenyls, and organochlorine pesticides.' *J. Biol. Chem.*, 1988, **263**, 9449–9455.

30. B. Schilter, M. R. Andersen, C. Acharya *et al.*, 'Activation of cytochrome P450 gene expression in the rat brain by phenobarbital-like inducers.' *J. Pharm. Exp. Ther.*, 2000, **294**, 916–922.

31. T. Oinonen and K. O. Lindros, 'Zonation of hepatic cytochrome P-450 expression and regulation.' *Biochem. J.*, 1998, **329**, 17–35.

32. P. Honkakoski and M. Negishi, 'Regulatory DNA elements of phenobarbital-responsive cytochrome P450 CYP2B genes.' *J. Biochem. Mol. Toxicol.*, 1998, **12**, 3–9.

33. K. M. Sommer, R. Ramsden, J. Sidhu *et al.*, 'Promoter region analysis of the rat CYP2B1 and CYP2B2 genes.' *Pharmacogenet.*, 1996, **6**, 369–374.

34. J. S. He and A. J. Fulco, 'A barbiturate-regulated protein binding to a common sequence in the cytochrome P450 genes of rodents and bacteria.' *J. Biol. Chem.*, 1991, **266**, 7864–7869.

35. L. O. Narhi and A. J. Fulco, 'Phenobarbital induction of a soluble cytochrome P-450-dependent fatty acid monooxygenase in Bacillus megaterium.' *J. Biol. Chem.*, 1982, **257**, 2147–2150.

36. Q. Liang, J. S. He and A. J. Fulco, 'The role of Barbie box sequences as *cis*-acting elements involved in the barbiturate-mediated induction of cytochromes P450BM-1 and P450BM-3 in Bacillus megaterium.' *J. Biol. Chem.*, 1995, **270**, 4438–4450.

37. J. S. He and A. J. Fulco, 'A barbiturate-regulated protein binding to a common sequence in the cytochrome P450 genes of rodents and bacteria.' *J. Biol. Chem.*, 1991, **266**, 7864–7869.

38. G. C. Shaw, C. C. Sung, C. H. Liu *et al.*, 'Evidence against the Bm1P1 protein as a positive transcription factor for barbiturate-mediated induction of cytochrome P450$_{BM1}$ in *Bacillus megaterium*.' *J. Biol. Chem.*, 1998, **273**, 7996–8002.

39. C. N. Palmer, M. C. Gustafsson, H. Dobson *et al.*, 'Adaptive responses to fatty acids are mediated by the regulated expression of cytochromes P450.' *Biochem. Soc. Trans.*, 1999, **27**, 374–378.

40. D. Foti, D. Stroup and J. Y. Chiang, 'Basic transcription element binding protein (BTEB) transactivates the cholesterol 7 alpha-hydroxylase gene (CYP7A).' *Biochem. Biophys. Res. Commun.*, 1998, **253**, 109–113.

41. Y. Park and B. Kemper, 'The CYP2B1 proximal promoter contains a functional C/EBP regulatory element.' *DNA Cell Biol.*, 1996, **15**, 693–701.

42. S. Liu, Y. Park, I. Rivera-Rivera *et al.*, 'Nuclear factor-1 motif and redundant regulatory elements comprise phenobarbital-responsive enhancer in CYP2B1/2.' *DNA Cell Biol.*, 1998, **17**, 461–470.

43. P. V. Luc, M. Adesnik, S. Ganguly *et al.*, 'Transcriptional regulation of the CYP2B1 and CYP2B2 genes by C/EBP- related proteins.' *Biochem. Pharmacol.*, 1996, **51**, 345–356.

44. S. H. Lee, X. Wang and J. DeJong, 'Functional interactions between an atypical NF-κB site from the rat CYP2B1 promoter and the transcriptional repressor RBP-Jkappa/CBF1.' *Nucl. Acids Res.*, 2000, **28**, 2091–2098.

45. E. T. Morgan, 'Regulation of cytochromes P450 during inflammation and infection.' *Drug Metab. Rev.*, 1997, **29**, 1129–1188.

46. N. B. Beck, J. S. Sidhu and C. J. Omiecinski, 'Baculovirus vectors repress phenobarbital-mediated gene induction and stimulate cytokine expression in primary cultures of rat hepatocytes.' *Gene Ther.*, 2000, **7**, 1274–1283.

47. A. K. Jaiswal, T. Haaparanta, P. V. Luc *et al.*, 'Glucocorticoid regulation of a phenobarbital-inducible cytochrome P-450 gene: the presence of a functional glucocorticoid response element in the 5′-

flanking region of the CYP2B2 gene.' *Nucl. Acids Res.*, 1990, **18**, 4237–4242.

48. E. G. Schuetz, W. Schmid, G. Schutz, *et al.*, 'The glucocorticoid receptor is essential for induction of cytochrome P-4502B by steroids but not for drug or steroid induction of CYP3A or P-450 reductase in mouse liver.' *Drug Metab. Dispos.*, 2000, **28**, 268–278.

49. P. M. Shaw, M. Adesnik, M. C. Weiss *et al.*, 'The phenobarbital-induced transcriptional activation of cytochrome P-450 genes is blocked by the glucocorticoid-progesterone antagonist RU486.' *Mol. Pharmacol.*, 1993, **44**, 775–783.

50. A. L. Roe, R. A. Blouin and G. Howard, '*In vivo* phenobarbital treatment increases protein binding to a putative AP-1 site in the CYP2B2 promoter.' *Biochem. Biophys. Res. Commun.*, 1996, **228**, 110–114.

51. E. Trottier, A. Belzil, C. Stoltz *et al.*, 'Localization of a phenobarbital-responsive element (PBRE) in the 5′-flanking region of the rat CYP2B2 gene.' *Gene*, 1995, **158**, 263–268.

52. R. Ramsden, K. M. Sommer and C. J. Omiecinski, 'Phenobarbital induction and tissue-specific expression of the rat CYP2B2 gene in transgenic mice.' *J. Biol. Chem.*, 1993, **268**, 21722–21726.

53. P. Honkakowski, R. Moore, J. Gynther *et al.*, 'Characterization of phenobarbital-inducible mouse Cyp2b10 gene transcription in primary hepatocytes.' *J. Biol. Chem.*, **271**, 9746–9753.

54. P. Honkakoski, R. Moore, K. A. Washburn *et al.*, 'Activation by diverse xenochemicals of the 51-base pair phenobarbital-responsive enhancer module in the CYP2B10 gene.' *Molec. Pharmacol.*, 1998, **53**, 597–601.

55. Y. Park, H. Li and B. Kemper, 'Phenobarbital induction mediated by a distal CYP2B2 sequence in rat liver transiently transfected *in situ*.' *J. Biol. Chem.*, 1996, **271**, 23725–23728.

56. T. Sueyoshi, T. Kawamoto, I. Zelko *et al.*, 'The repressed nuclear receptor CAR responds to phenobarbital in activating the human CYP2B6 gene.' *J. Biol. Chem.*, 1999, **274**, 6043–6046.

57. S. Liu, Y. Park, I. Rivera-Rivera *et al.*, 'Nuclear factor-1 motif and redundant regulatory elements comprise phenobarbital-responsive enhancer in CYP2B1/2.' *DNA Cell Biol.*, 1998, **17**, 461–470.

58. C. Stoltz, M. H. Vachon, E. Trottier *et al.*, 'The CYP2B2 phenobarbital response unit contains an accessory factor element and a putative glucocorticoid response element essential for conferring maximal phenobarbital responsiveness.' *J. Biol. Chem.*, 1998, **273**, 8528–8536.

59. P. Honkakoski, I. Zelko, T. Sueyoshi *et al.*, 'The nuclear orphan receptor CAR-retinoid X receptor heterodimer activates the phenobarbital-responsive enhancer module of the CYP2B gene.' *Mol. Cell Biol.*, 1998, **18**, 5652–5658.

60. J. Kim, G. Min and B. Kemper, 'Chromatin assembly enhances binding to the CYP2B1 PBRU of NF-1, which binds simultaneously with CAR/RXR and enhances CAR/RXR-mediated activation of the PBRU.' *J. Biol. Chem.*, 2000, **276**, 7559–7567.

61. R. Ramsden, N. B. Beck, K. M. Sommer *et al.*, 'Phenobarbital responsiveness conferred by the 5′-flanking region of the rat CYP2B2 gene in transgenic mice.' *Gene*, 1999, **228**, 169–179.

62. R. Kumar and E. B. Thompson, 'The structure of the nuclear hormone receptors.' *Steroids*, 1999, **64**, 310–319.

63. A. C. Steinmetz, J. P. Renaud and D. Moras, 'Binding of ligands and activation of transcription by nuclear receptors.' *Annu. Rev. Biophys. Biomol. Struct.*, 2001, **30**, 329–359.

64. U. Savas, K. J. Griffin and E. F. Johnson, 'Molecular mechanisms of cytochrome P-450 induction by xenobiotics: An expanded role for nuclear hormone receptors.' *Molec. Pharmacol.*, 1999, **56**, 851–857.

65. D. J. Mangelsdorf, C. Thummel, M. Beato *et al.*, 'The nuclear receptor superfamily: the second decade.' *Cell*, 1995, **83**, 835–839.

66. L. Stein, P. Sternberg, R. Durbin *et al.*, 'WormBase: network access to the genome and biology of *caenorhabditis elegans*.' *Nucl. Acids Res.*, 2001, **29**, 82–86.

67. C. J. Francoijs, J. P. Klomp and R. M. Knegtel, 'Sequence annotation of nuclear receptor ligand-binding domains by automated homology modeling.' *Protein Eng.*, 2000, **13**, 391–394.

68. D. Robyr, A. P. Wolffe and W. Wahli, 'Nuclear hormone receptor coregulators in action: diversity for shared tasks.' *Molec. Endocrinol.*, 2000, **14**, 329–347.

69. J. M. Wurtz, W. Bourguet, J. P. Renaud *et al.*, 'A canonical structure for the ligand-binding domain of nuclear receptors.' *Nat. Struct. Biol.*, 1996, **3**, 87–94.

70. D. J. Mangelsdorf, U. Borgmeyer, R. A. Heyman, Jr. *et al.*, 'Characterization of three RXR genes that mediate the action of 9-*cis* retinoic acid.' *Genes Devel.*, 1992, **6**, 329–344.

71. D. J. Mangelsdorf and R. M. Evans, 'The RXR heterodimers and orphan receptors.' *Cell*, 1995, **83**, 841–850.

72. T. Ylikomi, J. M. Wurtz, H. Syvala *et al.*, 'Reappraisal of the role of heat shock proteins as regulators of steroid receptor activity.' *Crit. Rev. Biochem. Mol. Biol.*, 1998, **33**, 437–466.

73. T. Kawamoto, T. Sueyoshi, I. Zelko *et al.*, 'Phenobarbital-responsive nuclear translocation of the receptor CAR in induction of the CYP2B gene.' *Molec. Cell Biol.*, 1999, **19**, 6318–6322.

74. R. M. Lavinsky, K. Jepsen, T. Heinzel *et al.*, 'Diverse signaling pathways modulate nuclear receptor recruitment of N-CoR and SMRT complexes.' *Proc. Natl. Acad Sci. USA*, 1998, **95**, 2920–2925.

75. I. Hu and M. A. Lazar, 'Transcriptional repression by nuclear hormone receptors.' *Trends Endocrinol. Metab.*, 2000, **11**, 6–10.

76. M. Beato and K. Eisfeld, 'Transcription factor access to chromatin.' *Nucl. Acids Res.*, 1997, **25**, 3559–3563.

77. M. Baes, T. Gulick, H. S. Choi *et al.*, 'A new orphan member of the nuclear hormone receptor superfamily that interacts with a subset of retinoic acid response elements.' *Molec. Cell Biol.*, 1994, **14**, 1544–1551.

78. I. Tzameli, P. Pissios, E. G. Schuetz *et al.*, 'The xenobiotic compound 1,4-bis[2-(3,5-dichloropyridyloxy)]benzene is an agonist ligand for the nuclear receptor CAR.' *Molec. Cell Biol.*, 2000, **20**, 2951–2958.

79. P. Wei, J. Zhang, M. Egan-Hafley *et al.*, 'The nuclear receptor CAR mediates specific xenobiotic induction of drug metabolism.' *Nature*, 2000, **407**, 920–923.

80. T. A. Kocarek, J. M. Kraniak and A. B. Reddy, 'Regulation of rat hepatic cytochrome P450 expression by sterol biosynthesis inhibition: inhibitors of squalene synthase are potent inducers of CYP2B expression in primary cultured rat hepatocytes and rat liver.' *Mol. Pharmacol.*, 1998, **54**, 474–484.

81. T. A. Kocarek and A. B. Reddy, 'Negative regulation by dexamethasone of fluvastatin-inducible CYP2B expression in primary cultures of rat hepatocytes: role of CYP3A.' *Biochem. Pharmacol.*, 1998, **55**, 1435–1443.

82. L. B. Moore, D. J. Parks, S. A. Jones *et al.*, 'Orphan nuclear receptors constitutive androstane receptor and pregnane X receptor share xenobiotic and steroid ligands.' *J. Biol. Chem.*, 2000, **275**, 15122–15127.

83. H. S. Choi, W. Seol and D. D. Moore, 'A component of the 26S proteasome binds on orphan member of

the nuclear hormone receptor superfamily.' *J. Steroid Biochem. Mol. Biol.*, 1996, **56**, 23–30.

84. W. Seol, H. S. Choi and D. D. Moore, 'An orphan nuclear hormone receptor that lacks a DNA binding domain and heterodimerizes with other receptors.' *Science*, 1996, **272**, 1336–1339.

85. C. K. Glass and M. G. Rosenfeld, 'The coregulator exchange in transcriptional functions of nuclear receptors.' *Genes Devel.*, 2000, **14**, 121–141.

86. B. Goodwin, S. A. Jones, R. R. Price *et al.*, 'A regulatory cascade of the nuclear receptors FXR, SHP-1, and LRH-1 represses bile acid biosynthesis.' *Mol. Cell*, 2000, **6**, 517–526.

87. T. T. Lu, M. Makishima, J. J. Repa *et al.*, 'Molecular basis for feedback regulation of bile acid synthesis by nuclear receptors.' *Mol. Cell*, 2000, **6**, 507–515.

88. Y. K. Lee, H. Dell, D. H. Dowhan *et al.*, 'The orphan nuclear receptor SHP inhibits hepatocyte nuclear factor 4 and retinoid X receptor transactivation: two mechanisms for repression.' *Mol. Cell Biol.*, 2000, **20**, 187–195.

89. D. M. Heery, S. Hoare, S. Hussain *et al.*, 'Core LXXLL motif sequences in CBP, SRC1 and RIP140 define affinity and selectivity for steroid and retinoid receptors.' *J. Biol. Chem.*, 2001, **276**, 6695–6702.

90. E. Treuter, L. Johansson, J. S. Thomsen *et al.*, 'Competition between thyroid hormone receptor-associated protein (TRAP) 220 and transcriptional intermediary factor (TIF) 2 for binding to nuclear receptors. Implications for the recruitment of TRAP and p160 coactivator complexes.' *J. Biol. Chem.*, 1999, **274**, 6667–6677.

91. C. X. Yuan, M. Ito, J. D. Fondell *et al.*, 'The TRAP220 component of a thyroid hormone rece.' *Proc. Natl. Acad Sci. USA.*, 1998, **95**, 7939–7944.

92. E. M. McInerney, D. W. Rose, S. E. Flynn, Sr. *et al.*, 'Determinants of coactivator LXXLL motif specificity in nuclear receptor transcriptional activation.' *Genes Devel.*, 1998, **12**, 3357–3368.

93. B. M. Forman, I. Tzameli, H. S. Choi *et al.*, 'Androstane metabolites bind to and deactivate the nuclear receptor CAR-beta.' *Nature*, 1998, **395**, 612–615.

94. D. B. Gower and B. A. Ruparelia, 'Olfaction in humans with special reference to odorous 16-androstenes: their occurrence, perception and possible social, psychological and sexual impact.' *J. Endocrinol.*, 1993, **137**, 167–187.

95. J. M. Pascussi, L. Drocourt, J. M. Fabre *et al.*, 'Dexamethasone induces pregnane X receptor and retinoid X receptor-alpha expression in human hepatocytes: synergistic increase of CYP3A4 induction by pregnane X receptor activators.' *Molec. Pharmacol.*, 2000, **58**, 361–372.

96. S. C. Dogra and B. K. May, 'Liver-enriched transcription factors, HNF-1, HNF-3, and C/EBP, are major contributors to the strong activity of the chicken CYP2H1 promoter in chick embryo hepatocytes.' *DNA Cell Biol.*, 1997, **16**, 1407–1418.

97. D. M. Roth and M. Karin, 'The NF-κB activation pathway: a paradigm in information transfer from membrane to nucleus.' *Science STKE*, 1999, **5**, 1–16.

98. D. B. Mendel and G. R. Crabtree, 'HNF-1, a member of a novel class of dimerizing homeodomain proteins.' *J. Biol. Chem.*, 1991, **266**, 677–680.

99. J. Li, G. Ning and S. A. Duncan, 'Mammalian hepatocyte differentiation requires the transcription factor HNF-alpha.' *Genes Dev.*, 2000, **14**, 464–474.

100. F. M. Sladek, 'Orphan receptor HNF-4 and liver-specific gene expression.' *Receptor*, 1994, **4**, 64.

101. T. A. Kocarek, E. G. Schuetz and P. S. Guzelian, 'Regulation of cytochrome P450 2B1/2 mRNAs by Kepone (chlordecone) and potent estrogens in primary cultures of adult rat hepatocytes on Matrigel.' *Toxicol. Lett.*, 1994, **71**, 183–196.

102. D. Blizard, T. Sueyoshi, M. Negishi *et al.*, 'Mechanism of induction of cytochrome P450 enzymes by the proestrogenic endocrine disruptor pesticide-methoxychlor: Interactions of methoxychlor metabolites with the constitutive androstane receptor system.' *Drug Metab. Dispos.*, 2001, **29**, 781–785.

103. T. Kawamoto, S. Kakizaki, K. Yoshinari *et al.*, 'Estrogen activation of the nuclear orphan receptor CAR (constitutive active receptor) in induction of the mouse Cyp2b10 gene.' *Molec. Endocrinol.*, 2000, **14**, 1897–1905.

104. J. S. Sidhu and C. J. Omiecinski, 'Modulation of xenobiotic-inducible cytochrome P450 gene expression by dexamethasone in primary rat hepatocytes.' *Pharmacogenet.*, 1995, **5**, 24–36.

105. H. Yamada, T. Yamaguchi and K. Oguri, 'Suppression of the expression of the CYP2B1/2 gene by retinoic acids.' *Biochem. Biophys. Res. Commun.*, 2000, **277**, 66–71.

106. M. A. Clark, B. A. Bing, P. E. Gottschall, *et al.*, 'Differential effect of cytokines on the phenobarbital or 3-methylcholanthrene induction of P450 mediated monooxygenase activity in cultured rat hepatocytes.' *Biochem. Pharm.*, 1995, **49**, 97–104.

107. J. H. Parmentier, P. Kremers, L. Ferrari *et al.*, 'Repression of cytochrome P450 by cytokines: IL-1 beta counteracts clofibric acid induction of CYP4A in cultured fetal rat hepatocytes.' *Cell Biol. Toxicol.*, 1993, **9**, 307–313.

108. K. I. Hirsch-Ernst, K. Schlaefer, D. Bauer *et al.*, 'Repression of phenobarbital-dependent CYP2B1 mRNA induction by reactive oxygen species in primary rat hepatocyte cultures.' *Molec. Pharmacol.*, 2001, **59**, 1402–1409.

109. A. P. Wolffe and D. Guschin, 'Review: chromatin structural features and targets that regulate transcription.' *J. Struct. Biol.*, 2000, **129**, 102–122.

110. T. N. Collingwood, F. D. Urnov and A. P. Wolffe, 'Nuclear receptors: coactivators, corepressors and chromatin remodeling in the control of transcription.' *J. Mol. Endocrinol.*, 1999, **23**, 255–275.

111. F. J. Dilworth, C. Fromental-Ramain, K. Yamamoto, *et al.*, 'ATP-Driven chromatin remodeling activity and histone acetyltransferases act sequentially during transactivation by RAR/RXR *in vitro*.' *Mol. Cell.*, 2000, **6**, 1049–1058.

112. M. Grunstein, 'Histone acetylation in chromatin structure and transcription.' *Nature*, 1997, **389**, 349–352.

113. M. Vignali, A. H. Hassan, K. E. Neely *et al.*, 'ATP-dependent chromatin-remodeling complexes.' *Molec. Cell. Biol.*, 2000, **20**, 1899–1910.

114. A. P. Wolffe and J. J. Hayes, 'Chromatin disruption and modification.' *Nucl. Acids Res.*, 1999, **27**, 711–720.

115. P. R. Mueller and B. Wold, '*In vivo* footprinting of a muscle specific enhancer by ligation mediated PCR.' *Science*, 1989, **246**, 780–786.

116. J. Kim and B. Kemper, 'Phenobarbital alters protein binding to the CYP2B1/2 phenobarbital-responsive unit in native chromatin.' *J. Biol. Chem.*, 1997, **272**, 29423–29425.

117. M. H. Kuo and C. D. Allis, '*In vivo* cross-linking and immunoprecipitation for studying dynamic protein: DNA associations in a chromatin environment.' *Methods*, 1999, **19**, 425–433.

118. C. Crane-Robinson, F. A. Myers, T. R. Hebbes *et al.*, 'Chromatin immunoprecipitation assays in acetylation mapping of higher eukaryotes.' *Meth. Enzymol.*, 1999, **304**, 533–547.

119. E. G. Schuetz, D. Li, C. J. Omiecinski *et al.*, 'Regulation of gene expression in adult rat hepatocytes cultured on a basement membrane matrix.' *J. Cell Physiol.*, 1988, **134**, 309–323.

120. J. S. Sidhu, F. M. Farin and C. J. Omiecinski, 'Influence of extracellular matrix overlay on phenobarbital-mediated induction of CYP2B1, 2B2, and 3A1 genes in primary adult rat hepatocyte culture.' *Arch. Biochem. Biophys.*, 1993, **301**, 103–113.

121. J. S. Sidhu, F. M. Farin, T. J. Kavanagh *et al.*, 'Effect of tissue-culture substratum and extracellular matrix overlay on liver-selective and xenobiotic inducible gene expression in primary rat hepatocytes.' *In Vitro Toxicol.*, 1994, **7**, 225–242.

122. J. S. Sidhu, F. Liu, S. M. Boyle *et al.*, 'PI3K inhibitors reverse the suppressive actions of insulin on CYP2E1 expression by activating stress-response pathways in primary rat hepatocytes.' *Molec. Pharmacol.*, 2001, **59**, 1138–1146.

123. H. Glatt, I. Gemperlein, G. Turchi *et al.*, 'Search for cell culture systems with diverse xenobiotic-metabolizing activities and their use in toxicological studies.' *Mol. Toxicol.*, 1987, **1**, 313–334.

124. J. H. Lin and A. Y. Lu, 'Inhibition and induction of cytochrome P450 and the clinical implications.' *Clin. Pharmacokinet.*, 1998, **35**, 361–390.

125. W. E. Evans and M. V. Relling, 'Pharmacogenomics: translating functional genomics into rational therapeutics.' *Science*, 1999, **286**, 487–491.

126. S. L. MacLeod, S. Nowell, J. Massengill *et al.*, 'Cancer therapy and polymorphisms of cytochromes P450.' *Clin. Chem. Lab. Med.*, 2000, **38**, 883–887.

127. D. W. Nebert, 'Pharmacogenetics and pharmacogenomics: why is this relevant to the clinical geneticist?' *Clin. Genet.*, 1999, **56**, 247–258.

128. E. S. Vesell, 'Advances in pharmacogenetics and pharmacogenomics.' *J. Clin. Pharmacol.*, 2000, **40**, 930–938.

129. U. A. Meyer, 'Pharmacogenetics and adverse drug reactions.' *Lancet*, 2000, **356**, 1667–1671.

J.P. Vanden Heuvel, G.H. Perdew, W.B. Mattes and W.F. Greenlee (Eds.)
Comprehensive Toxicology, Vol. xiv

# 14.2.7
# Mechanisms of Metal-Induction of Metallothionein Gene Expression

## PATRICK J. DANIELS and GLEN K. ANDREWS
*University of Kansas Medical Center, Kansas City, KS, USA*

## 14.2.7.1 INTRODUCTION

### 14.2.7.1.1 Transition Metals and Biological Processes

Inorganic substances are essential components of many intracellular biochemical pathways. Indeed, metals compose 17 of the 30 required elements for sustaining life. The 31 transition metals lie between scandium and copper in the periodic table. Transition metals, such as zinc and copper, while nutritionally essential, are also toxic in high concentrations. Others, such as cadmium, have no known physiological roles in higher eukaryotes, are extremely toxic in low concentrations, and can accumulate in the food chain. A common characteristic of some transition metals (e.g. copper and iron) is their function as redox centers, due to their large number of oxidation states. Others, such as zinc and cadmium, are not redox active. Transition metals serve multiple functions. They serve as cofactors for hundreds of enzymes, and are critical elements in several signal transduction pathways, including those involving protein kinases, cellular adhesion molecules and hormones. Furthermore, they function in protein–DNA, protein–RNA and protein–protein interactions, which often result in altered gene expression.[1]

Regulation of gene expression by transition metals has been demonstrated in organisms ranging from bacteria to mammals (Table 1). These metals regulate genes involved in protection from metal toxicity, such as the bacterial mercury resistance proteins encoded by the MER genes.[2] Transition metals may also regulate genes involved in the homeostasis of essential metals, such as the yeast ZIP (zinc and iron regulated proteins) genes, which encode zinc importers.[3]

Proteins that sense metals and signal changes in gene expression are known as metalloregulatory proteins (Table 1).[1,2] These proteins can regulate gene transcription, and/or mRNA stability and translation. Transition metals are ubiquitous in our diet and environment and have a great impact on gene expression and, therefore, cellular function.

### 14.2.7.1.2   Overview of Metallothioneins (MT)

In prokaryotes, lower eukaryotes, plants and throughout the animal kingdom, one of the most intensely-studied examples of metal-regulation of gene transcription is that of the metallothionein (MT) genes.[1,4–7] The small molecular weight MTs ($\sim 7$ kDa in mammals) are defined by their metal-thiolate bonds contributed by cysteine-rich regions.[8,9] In mammalian MTs (Figure 1) these cysteine-rich metal-binding clusters are distributed in $\alpha$ and $\beta$ domains of the protein[10] and consist of cysteine-cysteine, cysteine-X-cysteine or cysteine-X-Y-cysteine motifs, where X and Y are amino acids other than cysteine. The cysteine residues are often flanked by basic amino acids, and these motifs are essential for the high-affinity binding of metals. In fact, MTs adopt their specific, biologically unique tertiary structure only upon metal binding. The presence of these cysteine-motifs is conserved across evolution of the MTs from bacteria to mammals.[4,11]

MTs are the most abundant intracellular metal-binding proteins in higher eukaryotes and they function to chelate transition metals.[8] Multiple isoforms of the protein are often present. The complexity of the MT gene family varies among organisms (1 in bacteria, 4 in mice, 16 in humans) and levels of expression of individual genes varies among tissues in higher eukaryotes.[7] MT genes are apparently not essential genes,[12–14] but critical functions of

**Table 1**   Metal-sensing transcription factors.

| Metal | Factor | System | Genes |
|-------|--------|--------|-------|
| Hg | MerR | Bacterial | Hg-ion reductase |
| Fe | Fur(p) | Bacterial | Iron transporters |
|    | Dtx(p) | | Toxins |
|    | FecRp-Fec1p | | Fe(III) citrate transport |
|    | Aft1p | Fungal | Fe(III)/Cu(II) reductase |
| Heme | Hap1p | Fungal | Respiratory proteins |
| Cu | Ace1p | Fungal | Metallothionein |
|    | Amt1p | | |
|    | Mac1p | | Copper transporters |
|    | | | Fe(III)/Cu(II) reductase |
| Zn | SmtBp | Bacterial | Metallothionein |
|    | Zap1p | Fungal | Zinc transporters |
|    | MTF1 | Higher eukaryotes | Metallothionein Zinc exporter $\gamma$GCS |

Metalloregulatory proteins control gene expression, and mediate ancient genetic responses to transition metals. Transition metals often directly interact with specific metal-sensing transcription factors, which, in turn, activate or repress the expression of genes which encode proteins that detoxify (reduce, transport, store) the metal. For details see References 2, 26, 30, 65, 125, 126. This chapter focuses on regulation of the metallothionein genes which encode proteins that chelate transition metal ions, and are usually associated in the cell with the essential metals zinc and copper or the nonessential metal cadmium.

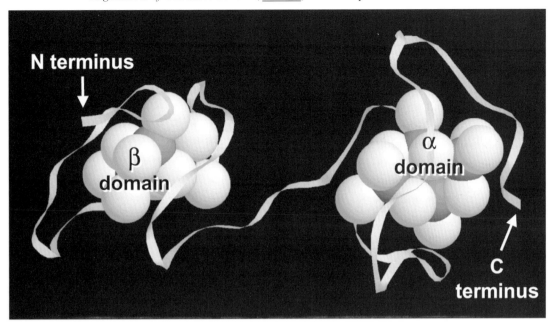

**Figure 1** Structure of the mammalian metallothionein-metal complex. In this model, derived from the crystal structure of rat MT-I, the amino acid backbone is displayed as a ribbon, the 20 cysteine thiols are space filled, and the seven atoms of zinc and cadmium are shown (only six are visible in this orientation). The α-domain of the protein (the carboxyterminal half of the protein) is wrapped around four cadmium atoms, whereas the β-domain is wrapped around two zinc and one cadmium atom. These metal-thiolate clusters are separated by a hinge region. Remarkably, MT adopts this specific, biologically unique, tertiary structure only upon metal binding.

MT in protection against metal toxicity or deficiency, as well as oxidative stress have been demonstrated. For example, mouse MT-I and -II can provide a biologically important reservoir of zinc under zinc-limiting conditions,[15,16] and protect the animal against cadmium toxicity and oxidative stress.[5,17–19] In contrast, MT predominantly functions to protect against copper toxicity in yeast and the *Drosophila*.[14,20–22]

MT can sequester reactive oxygen and hydroxyl radicals, and provide for zinc, copper, or cadmium exchange with other proteins.[23] Although primarily a metal-binding cytoplasmic protein, MT can translocate to the nucleus and may protect DNA from oxidative damage and participate in zinc exchange with zinc-dependent transcription factors.[23,24] These specific actions relate to the broader fields of tumorigenesis and apoptosis.

There are several examples, among the prokaryotes, and lower and higher eukaryotes, where the metalloregulatory proteins which mediate metal-induction of MT gene expression have been identified and studied in some detail.[2,7,25,26] The remainder of this chapter will focus on what is currently understood about the molecular mechanisms of metal-regulation of MT gene expression, in various organisms.

### 14.2.7.2 REGULATION OF METALLOTHIONEIN (SMTA) GENE EXPRESSION IN BACTERIA

#### 14.2.7.2.1 Description of the Bacterial (SMTA) Metallothionein Locus

The best-defined bacterial metallothionein is expressed in the Cyanobacteria *Synechoccus PCC 7942* and *PCC 6301*. In *Synechoccus PCC* the SMT locus contains both the SMTA and a divergently transcribed gene SMTB. SMTA encodes a class II metallothionein (56 amino acids) which uses eight cysteine and two histidine residues to coordinate three $Zn^{2+}$ ions.[27,28] SMTA gene expression is induced by copper, cadmium, and more potently by zinc, and this induction is dependent on SmtBp. Mutation deletion of the SMT locus reduces tolerance to zinc, as well as cadmium.[29]

In the SMT locus, a 100 bp operator-promoter is located between the SMTA and SMTB genes (Figure 2). It is characterized by divergent promoters for each gene and a degenerate 12-2-12 hyphenated inverted repeat that overlaps the transcriptional start site of SMTA.[30] This operator-promoter is bound by three distinct metal-activated protein com-

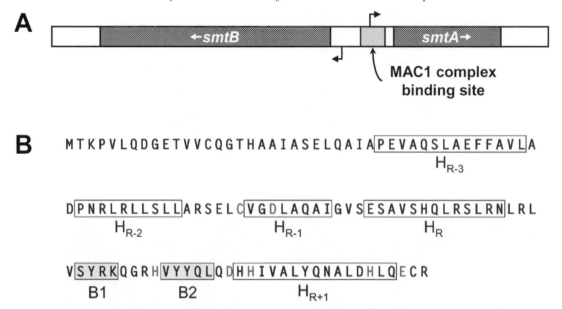

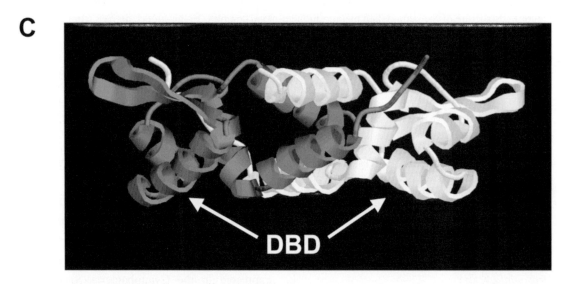

**Figure 2** Structure of the bacterial metallothionein (SMT) locus, and of the zinc-inhibited repressor SmtBp which regulates the transcription of this locus. (A) Organization of the SMT locus.[30] The SMTA and SMTB coding regions, the MAC1-complex binding site, and transcription start sites (*arrows*) are indicated. (B) Primary structure of SmtBp, a component of the MAC1 complex, including the α-helix $H_R$ postulated to be involved in DNA-binding, other α- helices, and β-sheets (*B1*, *B2*).[32] (C) Model of SmtBp dimer, including putative DNA-binding domain (*DBD*), based on information deposited with the Brookhaven Protein Data Bank (entry number 1smt) and data from.[32] Zinc ions inhibit its DNA-binding, and relieve transcriptional repression of SMTA.

plexes (MAC1, MAC2, MAC3). In the absence of zinc, MAC1 (which contains SmtBp) binds to a region (−15 bp to +24 bp, relative to the transcription start site of SMTA) spanning the 12-2-12 inverted repeat, but dissociates from this region in zinc-replete conditions.[30,31] Thus, MAC1 functions to repress SMTA gene expression in zinc-limiting conditions.[30]

### 14.2.7.2.2 SmtBp: a Zinc-Inhibited Repressor of SMTA

SmtBp, a component of MAC1, shows strong similarity to prokaryotic and eukaryotic winged-helix proteins, with a putative helix-turn-helix DNA binding motif (Figure 2). It functions as an elongated dimer, with each monomer containing 122 amino acids. Based on a conservation of residues at the amino-terminus, it is associated with a family of bacterial metalloregulatory proteins including ArsR, a repressor of arsenic-resistant operons, and Cadc, a cadmium-dependent repressor of the expression of cadmium efflux proteins. It is atypical in that metal ions inhibit rather than promote its DNA-binding.[30–32] SmtBp also binds cobalt and nickel, but with lower affinity than zinc.[33]

Although, zinc has not been co-purified with SmtBp, crystallographic and optical spectrographic evidence suggest that one or two cysteine thiolate ligands in combination with a mixture of carboxylate- and imidazole-containing ligands mediate zinc or cobalt binding.[33] The precise amino acid residues required for zinc-binding *in vivo* have not yet been determined. The putative DNA-binding motif is the helix-loop-helix domain (Figure 3) comprising residues 73–85 of each monomer.[32]

Based on metal-binding studies of other bacterial metalloregulatory proteins, the information concerning the tertiary structure of SmtBp suggests that a bend in the DNA helix or a conformational change in the protein is required for DNA-binding. The working hypothesis is that the conformation required for SmtBp binding to DNA and that required for binding to zinc are mutually exclusive, which would explain zinc's inhibitory effect on SmtBp interaction with the SMTA operator-promoter.[30,32] However, definitive proof of this hypothesis awaits further studies.

## 14.2.7.3 METAL-REGULATION OF COPPER METALLOTHIONEIN (CUP1) GENE EXPRESSION IN YEAST

### 14.2.7.3.1 Description of the Yeast (CUP1) Metallothionein Locus

MT gene function and structure have been studied in detail in *Saccharomyces cerevisiae* and *Candida glabrata*,[21,34–39] as has the mechanism of metal-regulation of expression of these genes.[40–45] Yeast tolerate a wide range of copper concentrations in the environment due to the transcriptional regulation, as well as the selective amplification of the CUP1 gene locus.[20,21] The CUP1 locus is located on chromosome VIII in *S. cerevisiae* and encodes a protein of 61 amino acids which is cysteine-rich (12 residues).[21,40] Yeast MTs are analogous in structure to the MTs of higher eukaryotes.[21,38] However, yeast MT genes are induced by copper and silver, but not by zinc

or cadmium.[35] These proteins chelate eight $Cu^{+1}$ ions exclusively through cysteinyl thiolates and serve to protect the cell against the toxicity of copper, not against cadmium or zinc.[20]

### 14.2.7.3.2 Ace1p/Amt1p: A Copper-Dependent, Positive Regulator of CUP1

Expression of the yeast MT genes is activated by a single, metalloregulatory transcription factor encoded by the ACE1 gene in *S. cerevisiae*[40,46] and the AMT1 gene in *C. glabrata*.[41,42,47] ACE1 gene is located on chromosome VII and the ACE1 protein activates CUP1 gene expression. Copper directly and rapidly activates the specific DNA-binding activity of Ace1p/Amt1p. The activated protein binds as a monomer to several sites in the upstream activation sequence in the proximal region of the CUP1 promoter, resulting in increased transcription of this locus (Figure 3).[41,42]

ACE1 encodes a 225 amino acid polypeptide characterized by an amino-terminal half, which is rich in cysteine residues and positively charged amino acids,[48] and a carboxy-terminal acidic transactivation domain. The 122 amino acid amino terminus contains 11 cysteine residues which are critical for copper- or silver-activated DNA-binding. Amt1p also shares this structural motif.[47] Both the DNA-binding and metalloregulatory functions of these factors are located in the aminoterminal domain, and the binding of either silver or copper causes a major conformational change in the protein.[43] Copper binding to Ace1p/Amt1p is cooperative[49,50] and on metal-binding, polynuclear copper- (or silver)-thiolate clusters are formed which serve to organize and stabilize the DNA-binding conformation.[49–51] The Ace1p/Amt1p aminoterminal peptide actually contains two independent, contiguous submodules, one of which binds a single Zn atom (residues 1–42 in Amt1p) and the other (residues 41–110 in Amt1p) which binds to DNA as a tetracopper cluster.[49,52,53]

DNA-binding involves interactions with both the minor and major grooves of the DNA. Ace1p binding spans one and a half turns of the DNA helix[54] and both Ace1p and Amt1p binding involves base-specific contacts within the major groove at a GCTG core sequence and within the minor grove in an A/T-rich region.[54–56] The tetracopper cluster or "copper fist" is essential for the major groove interactions, whereas residues 37–39 stabilize DNA-binding using minor groove contacts.[56]

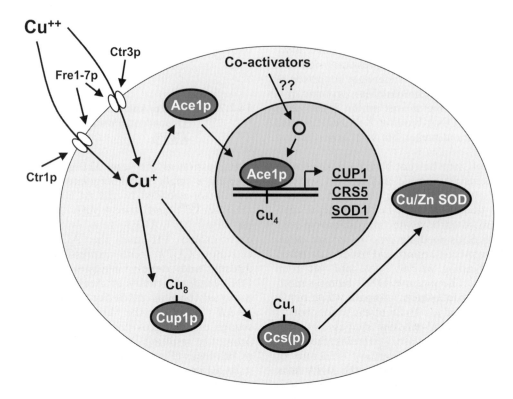

**Figure 3** Overview of yeast copper-metallothionein (CUP1, CRS5) regulation and copper utilization. The yeast copper-metallothionein (Cup1p) binds copper (8 atoms/molecule) and protects the cell from the toxicity of free copper. CUP1 gene expression is regulated by the positive transacting protein Ace1p in *Saccharomyces cerevisiae*. Ace1p binds to copper, causing the formation of a tetracopper cluster and the resulting protein conformation that binds to DNA elements in the promoters of these genes. Ace1p does not appear to require several of the general co-activators of transcription that have been described in yeast. Although not detailed in this diagram, copper also regulates the expression of the copper transporter genes (CTR1 and CTR3) and the Fe (III)/Cu(II) reductase gene (FRE3) (proteins are shown). The latter protein(s) reduce $Cu^{++}$ and $Fe^{+++}$ at the cell surface and are essential for copper and iron transport. The metalloregulatory protein Mac1p regulates these genes in response to copper. Copper levels are carefully controlled in the cell, and no free copper ions are present under normal physiological conditions. Instead, copper is carried by copper chaperones, such as Ccs(p) which specifically delivers copper to SOD1. No copper chaperone for Ace1p has been described.

The transactivation domain of these proteins is less well understood, but transcriptional activation of CUP1 by Ace1p/Amt1p can occur in the absence of several general transcription factors (TF), which are important for the expression of many other genes in yeast. For example, Ace1p function is independent of TFIIE, which regulates TFIIH kinase and the phosphorylation of the carboxyl-terminal domain of RNA polymerase II.[57] Ace1p function is independent of TAF17, a component of the TFIID and the SAGA/histone acetyltransferase complexes.[58] In addition, Ace1p function is independent of TFIIA, a protein which complexes with TFIID at the TATA box and which can interact with the transactivation domains of several transcriptional activators.[59] Whether Ace1p/Amt1p requires specific co-activators of transcription is unknown, but the general co-activator SRB/mediator is not essential.[60] The symmetrical placement of nucleosomes may be important for the activation of transcription by Amt1p,[61] which suggests that a specific chromatin structure of Ace1p/Amt1p-responsive promoters may play a role in regulation of gene expression.

In summary, the reversible and cooperative interaction of copper with the metalloregulatory protein Ace1p/Amt1p modulates its DNA-binding structure. Occupancy of four copper-binding sites stabilizes the DNA-binding form of the protein leading to increased transcription of the yeast MT genes (Figure 3). These genes encode efficient copper-chelating proteins (CUP1, CRS5) which protect the cell from excess copper. Remarkably, Ace1p/Amt1p co-regulates expression of the protective copper/zinc superoxide dismutase-1 (SOD1) gene[62] and its own gene[56,61,63] which augments the response to elevated copper. In addition, yeast respond to copper by repressing the transcription of genes (CTR1, CTR3 and FRE1) which, respectively, encode the high affinity copper transporters and the Fe (III)/Cu(II) reductase required for the system to function.[25,64] Copper-mediated repression of these genes occurs through Mac1p which binds copper, forming a polythiolate cluster, which may inhibit the transactivation domain.[25,65] Yeast cells tightly control copper levels, and essentially no free copper is present under normal physiological conditions.[66] Under normal conditions copper ions may be chaperoned throughout the cell by specific copper-chaperones. For example, the CCS gene product in yeast chaperones copper, once it enters the cell, to SOD1.[67]

### 14.2.7.4   METAL-REGULATION OF METALLOTHIONEIN (MTL) GENE EXPRESSION IN NEMATODES

#### 14.2.7.4.1   Description of Nematode (MTL) Metallothionein Locus

Relatively few mechanistic studies have been directed towards understanding metal-regulation of MT gene expression in invertebrates. However, MT-induction by metal ions (zinc, copper, cadmium) has been described in many invertebrate species. The most advanced studies of the regulation of MT gene expression in invertebrates have been directed toward the nematode *Caenorhabditis elegans* and the fruitfly *Drosophila melanogaster*.[11,22] This latter organism is discussed below in the section on higher eukaryotes.

In *C. elegans*, cadmium induces the expression of the two MT genes (MTL-1 and MTL-2), as well as HSP-70, collagens, rRNAs, pyruvate carboxylase, DNA gyrase, and β-adrenergic receptor kinase.[68] Although Mtl-1p (75 residues) and Mtl-2p (63 residues) contain the cysteine-motifs that are characteristic of

mammalian MTs, they exhibit poor alignment of the putative α and β domains and of cysteine residues with other eukaryotic MTs.[69]

Mtl-1p and Mtl-2p can bind cadmium or zinc, and MTL-1 and MTL-2 gene expression is induced by micromolar concentrations of cadmium.[69–71] In contrast, zinc-induction of these *C. elegans* genes has not been described. Remarkably, the cadmium-induction of MTL-1 and MTL-2 is restricted to intestinal cells at post-embryonic stages of development.[70] Expression of several non-metal-inducible genes, including the six vitellogenic genes, the cysteine protease CPR-1, and the gut protease GES-1 are also restricted to the gut.[72–74]

The upstream regulatory elements (URE) in the promoters of these genes all contain heptameric elements **CTGATAA** which are identical to the binding sites for the GATA family of transcription factors. There are two GATA elements in the MTL-1 promoter and five in the MTL-2 promoter. One GATA element in MTL-1, and two in MTL-2, are required for MTL gene expression.[75] However, these elements have not been linked to metal-induction of MTL gene expression.[76]

#### 14.2.7.4.2   GATA-binding Factors and the MTL Locus

The recent sequencing of the entire *C. elegans* genome and the versatile genetics of this organism provide a unique system for analysis of MT gene regulation.[77] GATA-binding factors in *C. elegans* include ELT-1p, ELT-2p, ELT-3p and END-1p. Similar to MTL-1 and MTL-2, the expression of ELT-2, is restricted to intestinal epithelial cells. Both the GATA promoter elements and ELT-2p are required for transcription of MTL-1 and MTL-2. However, ETL-2p induces MT transcription in the absence of metals, suggesting that other factors, possibly metal-sensors, participate in cadmium-induction of MTL gene expression.[75] Furthermore, no putative homologues of metal-response element binding protein-1 (MTF-1), a key metalloregulatory protein for MT gene expression in higher eukaryotes[78] (discussed below), are present in *C. elegans*. Thus, metal-regulation of MTL genes in this species appears to be unique in its cell specificity and molecular mechanisms. Elucidation of these mechanisms awaits identification of *cis*- and *trans*-acting factors involved in metal-induction of MTL genes.

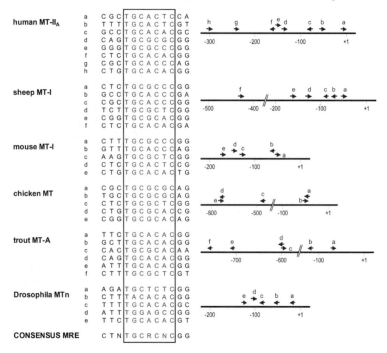

**Finger 1**  YQCTFEGCPRTYSTAGNLRTHQKTHRGEYT

**Finger 2**  FVCNQEGCGKAFLTSYSLKIHVRVHTKEKP

**Finger 3**  FECDVQGCEKAFNTLYRLKAHQRLHTGKT

**Finger 4**  FNCESEGCSKYFTTLSDLRKHIRTHTGEKP

**Finger 5**  FRCDHDGCGKAFAASHHLKTHVRTHTGEKP

**Finger 6**  FFCPSNGCEKTFSTQYSLKSHMKGHD

### Sequence identity between species

|            | Fugu | Chicken | Mouse | Human |
|------------|------|---------|-------|-------|
| Drosophila | 68%  | 68%     | 67%   | 67%   |
| Fugu       |      | 92%     | 92%   | 92%   |
| Chicken    |      |         | 97%   | 97%   |
| Mouse      |      |         |       | 99%   |

**Figure 4**  The zinc-finger domain of the positive acting, zinc-dependent transcription factor MTF-1 is highly conserved among higher eukaryotes. (A) The proximal promoter region ( + 1 designates the transcription start site) of metallothionein genes from higher eukaryotes (insects to mammals) contains multiple copies of metal response elements (MRE$_{a-h}$) which represent binding sites for the metalloregulatory protein MTF-1. The arrows above each promoter indicate the sense or antisense orientation of each MRE. The 12 bp MRE sequence is well conserved and the core bases, which are essential for MTF-1-binding, are highly conserved in functional MREs. (B) The highly conserved six zinc-finger domain of MTF-1 is shown here as a consensus sequence which was derived by comparisons of this domain of the protein from the five indicated species of mammals, birds, fish, and insects. Shown are the cysteine and histidine residues which coordinate a zinc atom in similar C$_2$H$_2$-type zinc fingers. The underlined amino acids are present at that position in 4 out of the 5 species examined, and the amino acid sequence identity in this domain is well over 90% between fish and man. The *Drosophila* sequence is more divergent (67% identity). The reversible binding of zinc to specific zinc fingers of MTF-1 apparently modulates DNA-binding activity of this transcription factor.

## 14.2.7.5 METAL-REGULATION OF *MT* GENE EXPRESSION IN HIGHER EUKARYOTES

### 14.2.7.5.1 Description of MT Gene Loci in Higher Eukaryotes

The metal-inducible MT genes from higher eukaryotes encode proteins of 60–63 amino acids in length which contain 20 cysteine residues and no aromatic amino acids.[8] The placement of cysteine residues is highly conserved, and the majority of other changes in amino acid sequence are conservative. These proteins generally can bind seven zinc or cadmium ions or up to 12 copper ions. Although MT in most mammalian tissues is isolated as a zinc$_7$-complex, it is found as a cadmium$_7$-complex under toxicological conditions.[8,9] MT is also isolated as a copper/zinc-complex from animals with inherited disorders of copper metabolism, such as the Menke's and Wilson's diseases in humans[79] or the LEC rat.[80] Furthermore, in *Drosophila melanogaster*, MT is naturally found complexed with copper.[81]

The MT gene loci in higher eukaryotes are diverse in structure. Two MT genes are found in *Drosophila*,[81–83] and the entire genome of *D. melanogaster* has been sequenced. In contrast, a single MT gene is apparently found in several species of birds,[84] although no avian genomes have yet been sequenced. In the mouse four linked MT genes have been cloned (MT-I to -IV)[85] and currently much of the mouse genome has been sequenced, which suggests that this is the entire gene family. The MT-I and MT-II genes are expressed in many cell-types[4] and in most cultured cells, whereas the MT-III and MT-IV genes show a very restricted cell-type specific pattern of expression.[86] In humans, the MT gene locus contains over 16 members, which can be differentially expressed and regulated.[87–92] Despite the diversity of structure in MT loci, fruit flies, fish, birds and mammals, each share in common the fact that they possess an MT gene(s) which is dramatically regulated by the metal ions zinc, cadmium and/or copper.

### 14.2.7.5.2 Metal-Response Element Binding Protein-1 (MTF-1): A Zinc-Dependent, Positive Regulator of MT Genes in Higher Eukaryotes

Essential for the induction of these MT genes (fly, fish, bird, mammal) by metal ions are DNA motifs termed metal response elements (MRE), which are present in multiple copies in the proximal promoters of these genes. The MRE confers response predominantly to zinc and cadmium, in transfected mammalian cells,[93,94] whereas in *Drosophila* they confer response predominantly to copper.[22] Differences in metal uptake or efflux may account, in part, for this effect. Metal responsiveness is dependent on the MRE core consensus sequence **TGCRCNC**[93–95] which is found within an extended consensus sequences of 12 bp (Figure 4). The MRE is a binding site for the transcription factor MTF-1. MTF-1 was first cloned from the mouse,[96] and MTF-1 homologues have since been identified in humans,[97,98] *Drosophila*,[22] pufferfish,[99] and the chicken (Jiang and Andrews, unpublished data). Homozygous disruption of the mouse MTF-1 gene in cultured cells eliminates heavy metal-induced MT gene expression, as well as basal expression of this gene,[78] and RNA-inhibition experiments suggest that *Drosophila* MTF-1 mediates copper regulation of those MT genes.[22]

It is important to note that, in contrast to the MT genes and their metalloregulatory factors discussed earlier in this chapter, which are nonessential genes that function to protect the organism from metal toxicity and/or from metal deficiency, the mouse MTF-1 gene is essential for embryonic development.[100] Embryos homozygous for MTF-1 null mutations die on day 14 of development. The liver fails to develop properly (or degenerates) in these mice, while development of the nervous system and visceral yolk sac is not impaired.[100–102] MTF-1 clearly regulates the expression of certain essential genes, in addition to the nonessential MT genes. These essential genes include the zinc transporter-1 (ZnT1) gene[103] and the γ-glutamylcysteine synthetase heavy chain gene.[100] In addition, other recent studies suggest that the alpha-fetoprotein, C/EBPα, lipocalin genes and placental-derived growth factor genes are also potentially regulated by MTF-1.[127,129]

MTF-1 is a zinc-finger transcription factor in the Cys$_2$His$_2$ family. The six zinc-finger domain has been highly conserved during evolution (Figure 4), while significant divergence has occurred in the remainder of the protein.[99] Although, the precise mechanisms by which MTF-1 activates MT gene expression in response to metals remain unknown, this observation is consistent with the concept that the zinc-finger domain of MTF-1 is critical for both its metalloregulatory and DNA-binding functions in response to zinc.[26]

Several lines of evidence suggest that MTF-1 directly senses zinc. Treatment of mammalian cells with zinc *in vivo* causes a rapid, dramatic increase in DNA-binding activity of MTF-1 measured *in vitro*,[104,105] and this is accompanied by the nuclear translocation of MTF-1[106] and the concomitant occupancy of MREs in the mouse MT-I promoter *in vivo*.[107] MTF-1 is more sensitive to metal chelators than are other zinc-finger transcription factors[96,104] and the DNA-binding activity of native and recombinant MTF-1 can be reversibly modulated by zinc.[78,98,104] Finally, mouse MTF-1 can function as a zinc-sensor to activate MRE-driven gene expression in yeast and *Drosophila* cells,[108] and *Drosophila* MTF-1 can activate gene expression in response to zinc in transfected mammalian cells.[22] In contrast to these studies of zinc, other transition metals (particularly cadmium) which are potent inducers of MT gene expression do not activate the DNA-binding activity of mammalian MTF-1 *in vitro*.[105,106,109]

Zinc-activation of the DNA-binding activity of MTF-1 involves reversible interactions with $Zn^{2+}$ ions[96] and these interactions occur with the zinc-finger domain.[104] The molecular mechanisms of this interaction remain to be defined, and the exact contributions of each of the zinc fingers to DNA binding and metal activation remains controversial. The temperature requirement for this activation step suggests that zinc-binding induces a conformational change in MTF-1.[104,110] There is structural heterogeneity in the zinc-binding sites in the isolated MTF-1 zinc finger domain,[111,112] and about half of the fingers exhibit high-affinity zinc-binding, while the other half exhibit lower affinity zinc-binding. Deletion mapping of the zinc-finger modules of mouse MTF-1, in the context of the entire protein, suggest that fingers 2, 3, and 4 constitute the core DNA-binding domain,[108] whereas mapping of the isolated zinc finger domain suggests that the first 4 fingers are essential for DNA binding and that fingers 5 and 6 stabilize this binding.[111,112] This is consistent with the fact that a single zinc-finger of this type can interact tightly with three or four bases in DNA.[113] In the context of the intact protein, MTF-1 fingers 5 and 6 appear to be unnecessary for DNA-binding *in vitro* or metal-dependent function *in vivo* in transfected cells,[108,128] yet these fingers are highly conserved during evolution (Figure 4). Finally, finger 1 of MTF-1 may constitute a unique zinc-sensing domain which, in cooperation with the transactivation domains (discussed below), produces a zinc-sensing metalloregulatory transcription factor.[108] Further validation of this model awaits the crystal structures of active DNA-bound MTF-1 and inactive MTF-1.

In contrast to the above studies, it has also been suggested that the MRE-binding activity of MTF-1 may be constitutively active,[96,111,114] while a genetic study using transfected cells suggested that the function of MTF-1 may be inhibited in the cell by a zinc-sensitive inhibitor.[114] Inactivation of the putative inhibitor could lead to heightened expression of the MT-I gene. Furthermore, it remains to be determined how cadmium and copper affect MTF-1 activity leading to the activation of MT gene expression. Cadmium treatment of mammalian cells causes only a modest increase in MTF-1 binding activity *in vivo* and this metal has no effect *in vitro*.[105,106,109,115] It is conceivable that these metals may cause the redistribution of zinc in the cell,[114] utilize specific co-activators of MTF-1 and/or activate signal transduction cascades that impinge on MTF-1 to affect MT gene transcription.

The transactivation domains of MTF-1 are less well characterized than the zinc-finger domain, but intramolecular interactions are important for optimal MTF-1 function.[108,116,117,128] The carboxyl-termini of human and mouse MTF-1 contain three transactivation domains which are acidic, proline-rich and serine-threonine-rich, respectively.[116] Although the VP16 transactivation domain can function in concert with the zinc-finger domain of MTF-1 to produce a weak metal-responsive factor in transfected cells,[114,116] the transactivation domain of the zinc-finger protein Sp1 cannot perform this function.[108]

The transactivation domains of MTF-1 contain several potential sites for phosphorylation by known kinases, but functional phosphorylation of MTF-1 has not been demonstrated. In addition, it remains to be determined whether MTF-1 interacts with specific co-activators in the cell. Inhibition of histone deacetylase activity renders cultured cells hypersensitive to metal-induction of MT gene expression,[118] which suggests that the threshold for sensitivity to metals may have different set points in different cells based on nucleosome structures. That MTF-1 molecules may interact is suggested by the findings that; two or more MREs cooperate to confer metal-responsiveness;[93] and a single pallindromic MRE directs metal-regulation in avian MT promoters.[119] However, dimerization of MTF-1 has not been demonstrated.

MTF-1 also cooperates with other transcription factors to regulate MT gene expression. The basic helix-loop-helix protein upstream stimulatory factor-1 may play a role in the cadmium activation of MT-I gene expression in cultured cells,[120] and this protein cooperates with MTF-1 to regulate the high level expression of <u>MT</u> genes in the visceral endoderm of the early mouse embryo.[102] Whether or not these proteins directly interact with each other is under investigation. Finally, methylation of the mouse <u>MT-I</u> gene represses its activity and inducibility by metal ions,[121] although methylation of the MRE does not appear to inhibit the *in vitro* binding of MTF-1.[123]

In summary, current models of the mechanisms by which metals regulate MT gene expression suggest that the reversible interactions of zinc with the metalloregulatory protein MTF-1 modulates its DNA-binding structure. Reversible occupancy of perhaps a single zinc-binding site in a zinc-finger may stabilize the DNA-binding form of the protein, leading to increased transcription of <u>MT</u> genes. These genes encode efficient zinc, cadmium and copper-chelating proteins which protect the cell against excess metal or from periods of zinc deficiency.

Studies of MTF-1 are currently underway to further test this model and to determine how cadmium and copper utilize MTF-1 to activate <u>MT</u> gene expression. In addition, MTF-1 has recently been shown to mediate metal-induction of the mouse ZnT-1 gene,[103] which encodes a transmembrane protein involved in the efflux of zinc from the cell, thus providing protection against metal toxicity.[124] Furthermore, MTF-1 regulates the expression of the mouse γ-glutamylcysteine synthetase heavy chain gene.[100] γ-Glutamylcysteine synthetase controls the rate limiting step in glutathione (GSH) biosynthesis. GSH, which is present in mM concentrations in cells, effectively chelates large amounts of zinc. Furthermore, both MT and GSH function as antioxidants. As mentioned above, several other genes may also be regulated by MTF-1.[127] Clearly, MTF-1 coordinates the expression of several genes which protect the cell from environmental insults.

## 14.2.7.6 REFERENCES

1. J. M. DeMoor and D. J. Koropatnick, 'Metals and cellular signaling in mammalian cells.' *Cell. Mol. Biol.*, 2000, **46**, 367–381.
2. T. V. O'Halloran, 'Transition metals in control of gene expression (see comments).' *Science*, 1993, **261**, 715–725.
3. M. L. Guerinot and D. Eide, 'Zeroing in on zinc uptake in yeast and plants.' *Curr. Opin. Plant Biol.*, 1999, **2**, 244–249.
4. G. K. Andrews, 'Regulation of metallothionein gene expression.' *Prog. Food Nutr. Sci.*, 1990, **14**, 193–258.
5. C. D. Klaassen, J. Liu and S. Choudhuri, 'Metallothionein: An intracellular protein to protect against cadmium toxicity.' *Annu. Rev. Pharmacol. Toxicol.*, 1999, **39**, 267–294.
6. D. J. Thiele, 'Metal-regulated transcription in eukaryotes.' *Nucleic Acids Res.*, 1992, **20**, 1183–1191.
7. A. T. Miles, G. M. Hawksworth, J. H. Beattie *et al.*, 'Induction, regulation, degradation, and biological significance of mammalian metallothioneins.' *Crit. Rev. Biochem. Mol. Biol.*, 2000, **35**, 35–70.
8. J. H. R. Kagi and A. Schaffer, 'Biochemistry of metallothionein.' *Biochemistry*, 1988, **27**, 8509–8515.
9. J. H. R. Kagi, 'Overview of metallothionein.' *Methods Enzymol.*, 1991, **205**, 613–626.
10. M. Nordberg and G. F. Nordberg, 'Toxicological aspects of metallothionein.' *Cell. Mol. Biol.*, 2000, **46**, 451–463.
11. R. Dallinger, 'Metallothionein research in terrestrial invertebrates: Synopsis and perspectives.' *Comp. Biochem. Physiol. (C)*, 1996, **113C**, 125–133.
12. A. E. Michalska and K. H. A. Choo, 'Targeting and germ-line transmission of a null mutation at the metallothionein I and II loci in mouse.' *Proc. Natl. Acad. Sci. USA*, 1993, **90**, 8088–8092.
13. B. A. Masters, E. J. Kelly, C. J. Quaife *et al.*, 'Targeted disruption of metallothionein I and II genes increases sensitivity to cadmium.' *Proc. Natl. Acad. Sci. USA*, 1994, **91**, 584–588.
14. L. T. Jensen, W. R. Howard, J. J. Strain *et al.*, 'Enhanced effectiveness of copper ion buffering by *CUP1* metallothionein compared with CRS5 metallothionein in *Saccharomyces cerevisiae*.' *J. Biol. Chem.*, 1996, **271**, 18514–18519.
15. T. P. Dalton, K. Fu, R. D. Palmiter *et al.*, 'Transgenic mice that over-express metallothionein-I resist dietary zinc deficiency.' *J. Nutr.*, 1996, **126**, 825–833.
16. G. K. Andrews and J. Geiser, 'Expression of metallothionein-I and -II genes provides a reproductive advantage during maternal dietary zinc deficiency.' *J. Nutr.*, 1999, **129**, 1643–1648.
17. R. D. Palmiter, 'The elusive function of metallothioneins.' *Proc. Natl. Acad. Sci. USA*, 1998, **95**, 8428–8430.
18. J. S. Lazo, S. M. Kuo, E. S. Woo *et al.*, 'The protein thiol metallothionein as an antioxidant and protectant against antineoplastic drugs.' *Chem. Biol. Interact.*, 1998, **112**, 255–262.
19. J. S. Lazo, Y. Kondo, D. Dellapiazza *et al.*, 'Enhanced sensitivity to oxidative stress in cultured embryonic cells from transgenic mice deficient in metallothionein I and II genes.' *J. Biol. Chem.*, 1995, **270**, 5506–5510.
20. R. K. Mehra and D. R. Winge, 'Metal ion resistance in fungi: Molecular mechanisms and their regulated expression.' *J. Cell. Biochem.*, 1991, **45**, 30–40.
21. M. Karin, R. Najarian, A. Haslinger *et al.*, 'Primary structure and transcription of an amplified genetic locus: the CUP1 locus of yeast.' *Proc. Natl. Acad. Sci. USA*, 1984, **81**, 337–341.
22. B. Zhang, D. Egli, O. Georgeiv *et al.*, 'The *Drosophila* homolog of mammalian zinc-finger factor MTF-1 activates transcription in response to copper and cadmium.' *Mol. Cell. Biol.*, 2001, **21**, 4505–4514.
23. G. Roesijadi, 'Metal transfer as a mechanism for metallothionein-mediated metal detoxification.' *Cell. Mol. Biol.*, 2000, **46**, 393–405.
24. M. G. Cherian and M. D. Apostolova, 'Nuclear localization of metallothionein during cell proliferation and differentiation.' *Cell. Mol. Biol.*, 2000, **46**, 347–356.
25. D. R. Winge, 'Copper-regulatory domain involved in gene expression.' *Prog. Nucleic Acid Res. Mol. Biol.*, 1998, **58**, 165–195.

26. G. K. Andrews, 'Regulation of metallothionein gene expression by oxidative stress and metal ions.' *Biochem. Pharm.*, 2000, **59**, 95–104.

27. J. W. Huckle, A. P. Morby, J. S. Turner *et al.*, 'Isolation of a prokaryotic metallothionein locus and analysis of transcriptional control by trace metal ions.' *Mol. Microbiol.*, 1993, **7**, 177–187.

28. M. J. Daniels, J. S. Turner-Cavet, R. Selkirk *et al.*, 'Coordination of $Zn^{2+}$ (and $Cd^{2+}$) by prokaryotic metallothionein-Involvement of His-imidazole.' *J. Biol. Chem.*, 1998, **273**, 22957–22961.

29. J. S. Turner, A. P. Morby, B. A. Whitton *et al.*, 'Construction of $Zn^{2+}/Cd^{2+}$ hypersensitive cyanobacterial mutants lacking a functional metallothionein locus.' *J. Biol. Chem.*, 1993, **268**, 4494–4498.

30. J. S. Turner, P. D. Glands, A. C. R. Samson *et al.*, '$Zn^{2+}$-sensing by the cyanobacterial metallothionein repressor SmtB: Different motifs mediate metal-induced protein-DNA dissociation.' *Nucleic Acids Res.*, 1996, **24**, 3714–3721.

31. A. P. Morby, J. S. Turner, J. W. Huckle *et al.*, 'SmtB is a metal-dependent repressor of the cyanobacterial metallothionein gene *smtA*: Identification of a Zn inhibited DNA-protein complex.' *Nucleic Acids Res.*, 1993, **21**, 921–925.

32. W. J. Cook, S. R. Kar, K. B. Taylor *et al.*, 'Crystal structure of the cyanobacterial metallothionein repressor SmtB: A model for metalloregulatory proteins.' *J. Mol. Biol.*, 1998, **275**, 337–346.

33. M. L. VanZile, N. J. Cosper, R. A. Scott *et al.*, 'The zinc metalloregulatory protein *Synechococcus* PCC7942 SmtB binds a single zinc ion per monomer with high affinity in a tetrahedral coordination geometry.' *Biochemistry*, 2000, **39**, 11818–11829.

34. G. I. Naumov, E. S. Naumova, H. Turakainen *et al.*, 'A new family of polymorphic metallothionein-encoding genes *MTH1* (*CUP1*) and *MTH2* in *Saccharomyces cerevisiae*.' *Gene*, 1992, **119**, 65–74.

35. T. R. Butt, E. J. Sternberg, J. A. Gorman *et al.*, 'Copper metallothionein of yeast, structure of the gene, and regulation of expression.' *Proc. Natl. Acad. Sci. USA*, 1984, **81**, 3332–3336.

36. R. K. Mehra, J. R. Garey, T. R. Butt *et al.*, '*Candida glabrata* metallothioneins. Cloning and sequence of the genes and characterization of proteins.' *J. Biol. Chem.*, 1989, **264**, 19747–19753.

37. R. K. Mehra, J. R. Garey and D. R. Winge, 'Selective and tandem amplification of a member of the metallothionein gene family in *Candida glabrata*.' *J. Biol. Chem.*, 1990, **265**, 6369–6375.

38. R. K. Mehra, E. B. Tarbet, W. R. Gray *et al.*, 'Metal-specific synthesis of two metallothioneins and gamma-glutamyl peptides in Candida glabrata.' *Proc. Natl. Acad. Sci. USA*, 1988, **85**, 8815–8819.

39. R. K. Mehra, J. L. Thorvaldsen, I. G. Macreadie *et al.*, 'Disruption analysis of metallothionein-encoding genes in *Candida glabrata*.' *Gene*, 1992, **114**, 75–80.

40. J. Welch, S. Fogel, C. Buchman *et al.*, 'The *CUP2* gene product regulates the expression of the *CUP1* gene, coding for yeast metallothionein.' *EMBO J.*, 1989, **8**, 255–260.

41. P. Zhou, M. S. Szczypka, T. Sosinowski *et al.*, 'Expression of a yeast metallothionein gene family is activated by a single metalloregulatory transcription factor.' *Mol. Cell. Biol.*, 1992, **12**, 3766–3775.

42. J. M. Huibregtse, D. R. Engelke and D. J. Thiele, 'Copper-induced binding of cellular factors to yeast metallothionein upstream activation sequences.' *Proc. Natl. Acad. Sci. USA*, 1989, **86**, 65–69.

43. P. Furst, S. Hu, R. Hackett *et al.*, 'Copper activates metallothionein gene transcription by altering the conformation of a specific DNA binding protein (published erratum appears in *Cell* 1989 Jan 27; **56**(2):following 321).' *Cell*, 1988, **55**, 705–717.

44. V. C. Culotta, T. Hsu, S. Hu *et al.*, 'Copper and the ACE1 regulatory protein reversibly induce yeast metallothionein gene transcription in a mouse extract.' *Proc. Natl. Acad. Sci. USA*, 1989, **86**, 8377–8381.

45. P. Zhou and D. J. Thiele, 'Copper and gene regulation in yeast.' *Biofactors*, 1993, **4**, 105–115.

46. D. J. Thiele, 'ACE1 regulates expression of the Saccharomyces cerevisiae metallothionein gene.' *Mol. Cell Biol.*, 1988, **8**, 2745–2752.

47. P. Zhou and D. J. Thiele, 'Isolation of a metal-activated transcription factor gene from *Candida glabrata* by complementation in *Saccharomyces cerevisiae*.' *Proc. Natl. Acad. Sci. USA*, 1991, **88**, 6112–6116.

48. M. S. Szczypka and D. J. Thiele, 'A cysteine-rich nuclear protein activates yeast metallothionein gene transcription.' *Mol. Cell. Biol.*, 1989, **9**, 421–429.

49. J. A. Graden, M. C. Posewitz, J. R. Simon *et al.*, 'Presence of a copper(I)-thiolate regulatory domain in the copper-activated transcription factor Amt1.' *Biochemistry*, 1996, **35**, 14583–14589.

50. F. Casas, Jr., S. Hu, D. Hamer *et al.*, 'Characterization of the copper- and silver-thiolate clusters in N-terminal fragments of the yeast ACE1 transcription factor capable of binding to its specific DNA recognition sequence.' *Biochemistry*, 1992, **31**, 6617–6626.

51. C. T. Dameron, D. R. Winge, G. N. George *et al.*, 'A copper-thiolate polynuclear cluster in the ACE1 transcription factor.' *Proc. Natl. Acad. Sci. USA*, 1991, **88**, 6127–6131.

52. J. L. Thorvaldsen, A. K. Sewell, A. M. Tanner *et al.*, 'Mixed $Cu^+$ and $Zn^{2+}$ coordination in the DNA-binding domain of the AMT1 transcription factor from Candida glabrata.' *Biochemistry*, 1994, **33**, 9566–9577.

53. R. A. Farrell, J. L. Thorvaldsen and D. R. Winge, 'Identification of the Zn(II) site in the copper-responsive yeast transcription factor, AMT1: A conserved Zn module.' *Biochemistry*, 1996, **35**, 1571–1580.

54. C. Buchman, P. Skroch, W. Dixon *et al.*, 'A single amino acid change in CUP2 alters its mode of DNA binding.' *Mol. Cell Biol.*, 1990, **10**, 4778–4787.

55. A. Dobi, C. T. Dameron, S. Hu *et al.*, 'Distinct regions of Cu(I).ACE1 contact two spatially resolved DNA major groove sites.' *J. Biol. Chem.*, 1995, **270**, 10171–10178.

56. K. A. Koch and D. J. Thiele, 'Autoactivation by a *Candida glabrata* copper metalloregulatory transcription factor requires critical minor groove interactions.' *Mol. Cell. Biol.*, 1996, **16**, 724–734.

57. H. Sakurai and T. Fukasawa, 'Activator-specific requirement for the general transcription factor IIE in yeast.' *Biochem. Biophys. Res. Commun.*, 1999, **261**, 734–739.

58. Z. Moqtaderi, M. Keaveney and K. Struhl, 'The histone H3-like TAF is broadly required for transcription in yeast.' *Mol. Cell*, 1998, **2**, 675–682.

59. S. Chou, S. Chatterjee, M. Lee *et al.*, 'Transcriptional activation in yeast cells lacking transcription factor IIA.' *Genetics*, 1999, **153**, 1573–1581.

60. D. K. Lee, S. Kim and J. T. Lis, 'Different upstream transcriptional activators have distinct coactivator requirements.' *Genes Dev.*, 1999, **13**, 2934–2939.

61. K. A. Koch and D. J. Thiele, 'Functional analysis of a homopolymeric (dA-dT) element that provides nucleosomal access to yeast and mammalian transcription factors.' *J. Biol. Chem.*, 1999, **274**, 23752–23760.

62. M. T. Carri, F. Galiazzo, M. R. Ciriolo *et al.*, 'Evidence for co-regulation of Cu,Zn superoxide dismutase and metallothionein gene expression in yeast through transcriptional control by copper via the ACE 1 factor.' *FEBS Lett.*, 1991, **278**, 263–266.

63. P. Zhou and D. J. Thiele, 'Rapid transcriptional autoregulation of a yeast metalloregulatory transcription factor is essential for high-level copper detoxification.' *Genes Dev.*, 1993, **7**, 1824–1835.

64. M. M. Pena, K. A. Koch and D. J. Thiele, 'Dynamic regulation of copper uptake and detoxification genes in Saccharomyces cerevisiae.' *Mol. Cell Biol.*, 1998, **18**, 2514–2523.

65. D. R. Winge, 'Copper-regulatory domain involved in gene expression.' *Adv. Exp. Med. Biol.*, 1999, **448**, 237–246.

66. T. D. Rae, P. J. Schmidt, R. A. Pufahl *et al.*, 'Undetectable intracellular free copper: The requirement of a copper chaperone for superoxide dismutase.' *Science*, 1999, **284**, 805–808.

67. V. C. Culotta, L. W. J. Klomp, J. Strain *et al.*, 'The copper chaperone for superoxide dismutase.' *J. Biol. Chem.*, 1997, **272**, 23469–23472.

68. V. H. C. Liao and J. H. Freedman, 'Cadmium-regulated genes from the nematode Caenorhabditis elegans—Identification and cloning of new cadmium-responsive genes by differential display.' *J. Biol. Chem.*, 1998, **273**, 31962–31970.

69. L. W. Slice, J. H. Freedman and C. S. Rubin, 'Purification, characterization, and cDNA cloning of a novel metallothionein-like, cadmium-binding protein from Caenorhabditis elegans.' *J. Biol. Chem.*, 1990, **265**, 256–263.

70. J. H. Freedman, L. W. Slice, D. Dixon *et al.*, 'The novel metallothionein genes of Caenorhabditis elegans. Structural organization and inducible, cell-specific expression.' *J. Biol. Chem.*, 1993, **268**, 2554–2564.

71. C. You, E. A. Mackay, P. M. Gehrig *et al.*, 'Purification and characterization of recombinant Caenorhabditis elegans metallothionein.' *Arch. Biochem. Biophys.*, 1999, **372**, 44–52.

72. T. Blumenthal, M. Squire, S. Kirtland *et al.*, 'Cloning of a yolk protein gene family from Caenorhabditis elegans.' *J. Mol. Biol.*, 1984, **174**, 1–18.

73. C. Britton, J. H. McKerrow and I. L. Johnstone, 'Regulation of the Caenorhabditis elegans gut cysteine protease gene cpr-1: requirement for GATA motifs.' *J. Mol. Biol.*, 1998, **283**, 15–27.

74. B. P. Kennedy, E. J. Aamodt, F. L. Allen *et al.*, 'The gut esterase gene (ges-1) from the nematodes Caenorhabditis elegans and Caenorhabditis briggsae.' *J. Mol. Biol.*, 1993, **229**, 890–908.

75. L. H. Moilanen, T. Fukushige and J. H. Freedman, 'Regulation of metallothionein gene transcription—Identification of upstream regulatory elements and transcription factors responsible for cell-specific expression of the metallothionein genes from Caenorhabditis elegans.' *J. Biol. Chem.*, 1999, **274**, 29655–29665.

76. F. Kugawa, H. Yamamoto, S. Osada *et al.*, 'Metallothionein genes in the nematode Caenorhabditis elegans and metal inducibility in mammalian culture cells.' *Biomed. Environ. Sci.*, 1994, **7**, 222–231.

77. R. K. Wilson, 'How the worm was won. The C. elegans genome sequencing project.' *Trends Genet.*, 1999, **15**, 51–58.

78. R. Heuchel, F. Radtke, O. Georgiev *et al.*, 'The transcription factor MTF-I is essential for basal and heavy metal-induced metallothionein gene expression.' *EMBO J.*, 1994, **13**, 2870–2875.

79. C. D. Vulpe and S. Packman, 'Cellular copper transport.' *Annu. Rev. Nutr.*, 1995, **15**, 293–322.

80. N. Sugawara, C. Sugawara, M. Katakura *et al.*, 'Copper metabolism in the LEC rat: Involvement of induction of metallothionein and disposition of zinc and iron.' *Experientia*, 1991, **47**, 1060–1063.

81. G. Maroni, A. S. Ho and T. Laurent, 'Genetic control of cadmium tolerance in Drosophila melanogaster.' *Environ. Health Perspect.*, 1995, **103**, 1116–1118.

82. W. Stephan, V. S. Rodriguez, B. Zhou *et al.*, 'Molecular evolution of the metallothionein gene Mtn in the melanogaster species group: Results from Drosophila ananassae.' *Genetics*, 1994, **138**, 135–143.

83. F. Bonneton, L. Théodore, P. Silar *et al.*, 'Response of Drosophila metallothionein promoters to metallic, heat shock and oxidative stresses.' *FEBS Lett.*, 1996, **380**, 33–38.

84. G. K. Andrews and L. P. Fernando, in 'Metallothionein in Biology and Medicine', eds. C. D. Klaassen and K. T. Suzuki, CRC Press, 1991, pp. 103–120.

85. R. D. Palmiter, E. P. Sandgren, D. M. Koeller *et al.*, in 'Metallothionein III: Biological Roles and Medical Implications', eds. K. T. Suzuki, N. Imura and M. Kimura, Birkhauser Verlag, 1993, pp. 399–406.

86. L. Liang, K. Fu, D. K. Lee *et al.*, 'Activation of the complete metallothionein gene locus in the maternal deciduum.' *Mol. Reprod. Devel.*, 1996, **43**, 25–37.

87. M. Karin and R. I. Richards, 'Human metallothionein genes—primary structure of the metallothionein-II gene and a related processed gene.' *Nature*, 1982, **299**, 797–802.

88. E. Lambert, P. Kille and R. Swaminathan, 'Cloning and sequencing a novel metallothionein I isoform expressed in human reticulocytes.' *FEBS Lett.*, 1996, **389**, 210–212.

89. F. A. Stennard, A. F. Holloway, J. Hamilton *et al.*, 'Characterisation of six additional human metallothionein genes.' *Biochim. Biophys. Acta Gene Struct. Expression*, 1994, **1218**, 357–365.

90. N. W. Shworak, T. O'Connor, N. C. W. Wong *et al.*, 'Distinct TATA motifs regulate differential expression of human metallothionein I genes MT-I$_F$ and MT-I$_G$.' *J. Biol. Chem.*, 1993, **268**, 24460–24466.

91. A. Heguy, A. West, R. I. Richards *et al.*, 'Structure and tissue-specific expression of the human metallothionein IB gene.' *Mol. Cell Biol.*, 1986, **6**, 2149–2157.

92. A. F. Holloway, F. A. Stennard and A. K. West, 'Human metallothionein gene MT1L mRNA is present in several human tissues but is unlikely to produce a metallothionein protein.' *FEBS Lett.*, 1997, **404**, 41–44.

93. G. W. Stuart, P. F. Searle, H. Y. Chen *et al.*, 'A 12-base-pair DNA motif that is repeated several times in metallothionein gene promoters confers metal regulation to a heterologous gene.' *Proc. Natl. Acad. Sci. USA*, 1984, **81**, 7318–7322.

94. G. W. Stuart, P. F. Searle and R. D. Palmiter, 'Identification of multiple metal regulatory elements in mouse metallothionein-I promoter by assaying synthetic sequences.' *Nature*, 1985, **317**, 828–831.

95. V. Cizewski Culotta and D. H. Hamer, 'Fine mapping of a mouse metallothionein gene metal response element.' *Mol. Cell. Biol.*, 1989, **9**, 1376–1380.

96. F. Radtke, R. Heuchel, O. Georgiev *et al.*, 'Cloned transcription factor MTF-1 activates the mouse metallothionein I promoter.' *EMBO J.*, 1993, **12**, 1355–1362.

97. E. Brugnera, O. Georgiev, F. Radtke *et al.*, 'Cloning, chromosomal mapping and characterization of the human metal-regulatory transcription factor MTF-1.' *Nucleic Acids Res.*, 1994, **22**, 3167–3173.

98. F. Otsuka, A. Iwamatsu, K. Suzuki *et al.*, 'Purification and characterization of a protein that binds to metal responsive elements of the human metallothionein IIA gene.' *J. Biol. Chem.*, 1994, **269**, 23700–23707.

99. A. Auf der Maur, T. Belser, G. Elgar *et al.*, 'Characterization of the transcription factor MTF-1 from the Japanese pufferfish (Fugu rubripes) reveals evolutionary conservation of heavy metal stress response.' *Biol. Chem.*, 1999, **380**, 175–185.

100. Ç. Günes, R. Heuchel, O. Georgiev *et al.*, 'Embryonic lethality and liver degeneration in mice lacking the metal-responsive transcriptional activator MTF-1.' *EMBO J.*, 1998, **17**, 2846–2854.

101. P. Lichtlen, O. Georgiev, W. Schaffner *et al.*, 'The heavy metal-responsive transcription factor-1 (MTF-1) is not required for neural differentiation.' *Biol. Chem.*, 1999, **380**, 711–715.

102. G. K. Andrews, D. K. Lee, R. Ravindra *et al.*, 'The transcription factors MTF-1 and USF1 cooperate to regulate mouse metallothionein-I expression in response to the essential metal zinc in visceral endoderm cells during early development.' *EMBO J.*, 2001, **20**, 1114–1122.

103. S. J. Langmade, R. Ravindra, P. J. Daniels *et al.*, 'The transcription factor MTF-1 mediates metal regulation of the mouse ZnT1 gene.' *J. Biol. Chem.*, 2000, **275**, 34803–34809.

104. T. D. Dalton, D. Bittel and G. K. Andrews, 'Reversible activation of the mouse metal response element-binding transcription factor-1 DNA binding involves zinc interactions with the zinc-finger domain.' *Mol. Cell. Biol.*, 1997, **17**, 2781–2789.

105. S. Koizumi, K. Suzuki, Y. Ogra *et al.*, 'Transcriptional activity and regulatory protein binding of metal-responsive elements of the human metallothionein-IIA gene.' *Eur. J. Biochem.*, 1999, **259**, 635–642.

106. I. V. Smirnova, D. C. Bittel, R. Ravindra *et al.*, 'Zinc and cadmium can promoter the rapid nuclear translocation of MTF-1.' *J. Biol. Chem.*, 2000, **275**, 9377–9384.

107. T. P. Dalton, Q. W. Li, D. Bittel *et al.*, 'Oxidative stress activates metal-responsive transcription factor-1 binding activity—Occupancy *in vivo* of metal response elements in the metallothionein-I gene promoter.' *J. Biol. Chem.*, 1996, **271**, 26233–26241.

108. D. C. Bittel, I. Smirnova and G. K. Andrews, 'Functional heterogeneity in the zinc fingers of the metalloregulatory transcription factor, MTF-1.' *J. Biol. Chem.*, 2000, **275**, 37194–37201.

109. D. Bittel, T. Dalton, S. Samson *et al.*, 'The DNA-binding activity of metal response element-binding transcription factor-1 is activated *in vivo* and *in vitro* by zinc, but not by other transition metals.' *J. Biol. Chem.*, 1998, **273**, 7127–7133.

110. T. P. Dalton, W. A. Solis, D. W. Nebert *et al.*, 'Characterization of the MTF-1 transcription factor from zebrafish and trout cells.' *Comp. Biochem. Physiol. (B)*, 2000, **126**, 325–335.

111. X. H. Chen, A. Agarwal and D. P. Giedroc, 'Structural and functional heterogeneity among the zinc fingers of human MRE-binding transcription factor-1.' *Biochemistry*, 1998, **37**, 11152–11161.

112. X. H. Chen, M. Chu and D. P. Giedroc, 'MRE-binding transcription factor-1: Weak zinc-binding finger domains 5 and 6 modulate the structure, affinity, and specificity of the metal-response element complex.' *Biochemistry*, 1999, **38**, 12915–12925.

113. E. J. Rebar, H. A. Greisman and C. O. Pabo, 'Phage display methods for selecting zinc finger proteins with novel DNA-binding specificities.' *Methods Enzymol.*, 1996, **267**, 129–149.

114. R. D. Palmiter, 'Regulation of metallothionein genes by heavy metals appears to be mediated by a zinc-sensitive inhibitor that interacts with a constitutively active transcription factor, MTF-1.' *Proc. Natl. Acad. Sci. USA*, 1994, **91**, 1219–1223.

115. S. Koizumi, H. Yamada, K. Suzuki *et al.*, 'Zinc-specific activation of a HeLa cell nuclear protein which interacts with a metal responsive element of the human metallothionein-II$_A$ gene.' *Eur. J. Biochem.*, 1992, **210**, 555–560.

116. F. Radtke, O. Georgiev, H.-P. Müller *et al.*, 'Functional domains of the heavy metal-responsive transcription regulator MTF-1.' *Nucleic Acids Res.*, 1995, **23**, 2277–2286.

117. H. P. Müller, E. Brugnera, O. Georgiev *et al.*, 'Analysis of the heavy metal-responsive transcription factor MTF-1 from human and mouse.' *Somat. Cell Mol. Genet.*, 1995, **21**, 289–297.

118. G. K. Andrews and E. D. Adamson, 'Butyrate selectively activates the metallothionein gene in teratocarcinoma cells and induces hypersensitivity to metal induction.' *Nucleic. Acids. Res.*, 1987, **15**, 5461–5475.

119. K. L. Shartzer, K. Kage, R. J. Sobieski *et al.*, 'Evolution of avian metallothionein: DNA sequence analyses of the turkey metallothionein gene and metallothionein cDNAs from pheasant and quail.' *J. Mol. Evol.*, 1993, **36**, 255–262.

120. Q. W. Li, N. M. Hu, M. A. F. Daggett *et al.*, 'Participation of upstream stimulatory factor (USF) in cadmium-induction of the mouse metallothionein-I gene.' *Nucleic Acids Res.*, 1998, **26**, 5182–5189.

121. M. W. Lieberman, L. R. Beach and R. D. Palmiter, 'Ultraviolet radiation-induced metallothionein-I gene activation is associated with extensive DNA demethylation.' *Cell*, 1983, **35**, 207–214.

122. Deleted in revision.

123. F. Radtke, M. Hug, O. Georgiev *et al.*, 'Differential sensitivity of zinc finger transcription factors MTF-1, Sp1 and Krox-20 to CpG methylation of their binding sites.' *Biol. Chem. Hoppe Seyler*, 1996, **377**, 47–56.

124. R. D. Palmiter and S. D. Findley, 'Cloning and functional characterization of a mammalian zinc transporter that confers resistance to zinc.' *EMBO J.*, 1995, **14**, 639–649.

125. M. Serpe, A. Joshi and D. J. Kosman, 'Structure-function analysis of the protein-binding domains of Mac1p, a copper-dependent transcriptional activator of copper uptake in *Saccharomyces cerevisiae*.' *J. Biol. Chem.*, 1999, **274**, 29211–29219.

126. A. J. Bird, H. Zhao, H. Luo *et al.*, 'A dual role for zinc fingers in both DNA binding and zinc sensing by the Zap1 transcriptional activator.' *EMBO J.*, 2000, **19**, 3704–3713.

127. P. Lichtlen, Y. Yang, T. Belser *et al.*, 'Target gene search for the metal-responsive transcription factor-1.' *Nucleic Acids Res.*, 2001, **29**, 1–10.

128. S. Koizumi, K. Suzuki, Y. Ogra *et al.*, 'Roles of the zinc fingers and other regions of the transcription factor human MTF-1 in zinc-regulated DNA binding.' *J. Cell. Physiol.*, 2000, **185**, 464–472.

129. C. J. Green, P. Lichtlen, N. T. Huynh *et al.*, 'Placental growth factor gene expression is induced by hypoxia: A central role for metal transcription factor-1.' *Cancer Res.*, 2001, **61**, 2696–2703.

J.P. Vanden Heuvel, G.H. Perdew, W.B. Mattes and W.F. Greenlee (Eds.)
Comprehensive Toxicology, Vol. xiv

# 14.2.8
# Modulation of Soluble Receptor Signaling by Coregulators

MOHAN KUMAR and GARY H. PERDEW
*Penn State University, University Park, PA, USA*

## 14.2.8.1 INTRODUCTION

An important aspect of molecular toxicology is the ability of a wide array of xenobiotics to alter gene transcription through soluble receptors. Over the past twenty years mechanistic studies have examined the ability of these receptors to be activated by ligands and subsequently bind to enhancer elements upstream of a promoter for a specific gene. In recent years the focus on mechanisms of transcriptional regulation has shifted to the recruitment of coregulators to the transactivation domain (TAD) of enhancer proteins. In this chapter coregulators are defined as proteins or protein complexes that are recruited either directly or indirectly to a TAD, and either enhance or repress transactivation potential. Thus, regulation of transcriptional activation involves several layers of regulation including activation of enhancer binding factors, competition for overlapping enhancer sites, release of corepressors, and subsequent recruitment of coactivators. There are a number of transactivation domains found in transcription factors, including; acidic, glutamine-rich, proline/serine/threonine rich, activation function-1 (AF-1) and AF-2 domains (see Chapter 1.4). The level of specificity imparted by a given TAD is only beginning to be understood. The binding of oligomeric coactivator complexes to a transcription factor leads to the recruitment of histone acetyltransferases (HAT) which results in acetylation of histone and subsequent remodeling of histone-DNA interactions. This HAT activity may also modulate transcription through the acetylation of non-histone proteins including targets such as p53 and Transcription factor$_{II}$F.[1] In addition, other enzymatic activities may also be associated with modulation of transcription such as protein methylation.[2]

Highly ordered chromatin structures inhibit gene transcription, presumably by limiting access to transcription factors and RNA polymerase II machinery binding to target genes.[3,4] Over the past few years a growing list of coactivators have been cloned, some of which contain intrinsic histone acetyltransferase (HAT) activity.[5–9] In contrast, several corepressors have been characterized to have histone deacetylase (HDAC) activity and are also capable of interacting with proteins with potent histone deacetylase activity.[10,11] Gene transcription mediated by certain steroid receptors such as the thyroid hormone receptor (TR) and peroxisome proliferator activated-receptor γ (PPARγ), appear to depend on a balance between the HAT and HDAC activities. A number of recent reviews have been published on coregulators.[12–15] In this chapter a broad approach will be taken to look at the role of coregulators in modulating transcriptional activity and how these factors may play a significant role in toxicology and carcinogenesis through modulation of soluble receptor activity.

## 14.2.8.2 COACTIVATORS

Several classes and sub-classes of coactivators have been identified based on their ability to directly interact with nuclear receptors (NRs) as well as their potential to enhance transcriptional activity of NRs. These include the switch/sucrose non-fermentable (SWI/SNF) family of proteins and coregulators with HAT activity (Table 1). Both classes of proteins are capable of facilitating chromatin remodeling and may potentially interact with basal transcription factors, resulting in either formation and/or stabilization of a pre-initiation complex at the promoters of cognate genes. Originally identified as regulators of gene expression in yeast,[16,17] SWI/SNF proteins have been observed to modulate ligand-dependent transcriptional activation of the glucocorticoid receptor (GR).[18] Some of the SWI class of proteins, including SwI2p, are part of a large chromatin-remodeling complex in yeast.[19] The *Drosophila* and human homologues of SwI2p are Brahma[20] and Brahma-related gene 1 (BRG-1),[21] respectively. BRG-1 is part of large NR-interacting complexes and is required for ER activity.[22] Several coactivators, which are capable of directly interacting with ligand bound receptors, have been cloned and subsequently identified to contain HAT activity.[9] Some of these coactivators are members of the p160 sub-family, which includes steroid receptor coactivator-1 (SRC-1), transcriptional intermediary factor 2 (TIF2), and coactivator p300/CBP/cointegrator-associated protein (p/CIP).[5,23–28] CREB (cAMP response element binding protein)-binding protein (CBP)/ p300 coactivators represent a separate class of HAT coactivators that are often referred to as cointegrators. A common thread through most studies with coactivators is the fact that an increase in expression of a given coactivator leads to an enhanced level of transcriptional activity by a given soluble receptor or transcription factor. It would appear that the level of coactivators is rate-limiting and thus contributes to the level of response obtained by activation of a soluble receptor, at least at a high level of receptor activation. These studies would also suggest that transcription factors probably compete for the available coactivator

**Table 1**

| Coregulator | Target receptors | Function |
| --- | --- | --- |
| Coactivators | | |
| SRC-1/NCoA-1 | PR, ER, GR, RAR, RXR, TR, PPAR, AhR, NF-κB | Weak HAT activity, acetylates histones H3 and H4 Interacts with CBP, pCAF and some basal transcription factors, TBP, TFIIB |
| TIF2/GRIP1/NcoA-2 | ER, GR, PR, AR, RAR, RXR ER, PR, RAR, TR, p53 | Interacts with CBP, has HAT activity |
| p/CIP/ACTR/RAC3/AIB1 | | |
| pCAF | RAR, RXR | HAT activity, acetylates some basal transcription factors Interacts with CBP, RNAP II |
| CBP/p300 | ER, RAR, RXR, TR, p53, myoD | Co-integrator, HAT activity, acetylates all four histones Interacts with SRC-1, pCAF, RNAP II, with SRC-1 synergistically enhances receptor activity, required for pre-initiation |
| RIP140 | ER, PPARα, TRβ, AhR, LXRα | Weakly stimulates some receptor activities, but represses at high levels of expression Represses TR activity |
| ARA70 | AR | Specifically interacts with AR and increases activity |
| Sug1/Trip1 | TR, RAR, RXR | Component of 26S proteosome in yeast |
| PBP/DRIP205 | PPAR | Interacts with PPAR and enhances its activity, part of DRIP complex, anchors DRIP complex to VDR and TR |
| PGC-1 | PPARγ, TR | Modulates PPARγ and TR transcription activity on the uncoupling protein (UCP-1) promoter Up-regulation of UCP-1 and key mito-chondrial enzymes of the respiratory chain |
| Secondary Coactivators | | |
| CARM1 | | Methylates H3 histones *in vitro*, binds to the carboxyl terminus of p160 coactivators, enhances transcriptional activity of NRs only when p160 coactivators were expressed |
| Coactivator complexes | | |
| TRAP complex | TR | Increases transcription from a T3-response element driven template |
| DRIP complex | Vitamin D₃ Receptor | HAT activity, chromatin remodeling |
| Corepressors | | |
| NcoR/SMRT | TR, ER, PR, RAR, RXR | Recruits histone deacetylase complex |
| TIF1α, TIF1β | RAR | Recruits histone deacetylase complex |

pool, especially in cells with high levels of transcriptional activity (e.g., tumor cells). Little is known about the transcriptional and post-translational regulation of coactivators and only a few studies have been published examining this aspect of coactivator activity. For example, expression of receptor interacting protein 140 (RIP140) mRNA is ubiquitous, although the expression is induced by estrogens in estrogen receptor (ER)-positive MCF7 breast cancer cell lines, but not in Ishikawa endometrial cells, indicating cell-specific regulation of cofactor expression.[29]

Clearly the temporal and cell-specific expression of coregulators is an area that will require extensive further investigation in order to understand the role of coregulators in regulating gene transcription. In the sections to follow what is known about the most studied coregulators will be explored.

### 14.2.8.2.1 SRC-1

SRC-1 was the first coactivator to be identified and was cloned as a PR-LBD

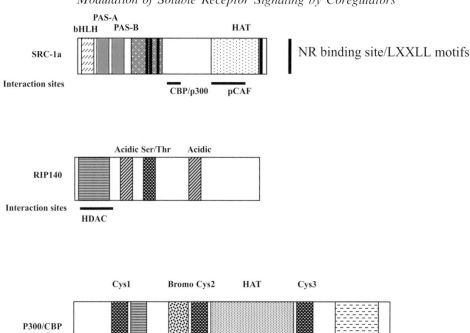

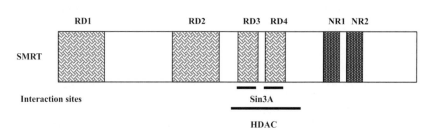

**Figure 1** Schematic representation of the domain structure of nuclear receptor coactivators belonging to distinct families.

interacting protein using a yeast two-hybrid screen of a human B-cell cDNA library.[30] SRC-1 interacts with the progesterone receptor (PR)-LBD in a ligand-dependent manner and modulates the transcriptional activity of PR. Further studies have shown that SRC-1 can interact with several other NRs, including ER,[31–34] TR,[30,31,35] androgen receptor (AR),[36,37] and PPARs[38–40] *via* their AF-2 TAD. Other non-NR family transcription factors that associate with SRC-1 include the *Ah* receptor (AhR)[41] and NF-κB.[42] The ability of SRC-1 to enhance transcriptional activity of a number of transcription factors suggests a pleiotropic role for SRC-1. In addition, in the case of ER[34] and AR,[36,37] SRC-1 modulates the activity of the ligand-independent AF-1 domain.[33]

SRC-1 and several other members of the SRC-1 family of coactivators have both a basic-helix-loop-helix (bHLH) domain and

periodicity/aryl hydrocarbon receptor nuclear translocator (ARNT)/ single-minded (PAS) domain[43] (Figure 1), which are found in the AhR, ARNT, and hypoxia inducible factor α (HIF1α). Although the function of these domains in coregulators remains to be determined, it has been suggested that they may be involved in either intermolecular or intramolecular interaction.[43] In the case of the AhR and ARNT, the bHLH and PAS domains are involved in heterodimerization. The transcription enhancer factor 4 (TEF4) directly interacts with the bHLH-PAS domain of SRC-1, this domain is required for SRC-1 to enhance TEF4-mediated transcription.[44] However, hetero- or homodimerization of bHLH-PAS coactivators has not been demonstrated. SRC-1 also contains a receptor interaction domain (RID) that interacts with several nuclear receptors[30,45] *via* three LXXLL amino acid motifs, which are required for binding with NRs and may also impart specificity to protein–protein interactions.[40,46,47] SRC-1 also contains an activation domain (AD), with three additional LXXLL motifs. Significantly, SRC-1 contains a HAT domain, which is involved in acetylation of histones H3 and H4, suggesting a role for SRC-1 in chromatin remodeling.[9] SRC-1 is also capable of interacting with p300/CBP and pCAF[25,45,48,49] along with certain basal transcription factors such as TATA binding factor (TBP) and transcription factor IIB, which suggests a possible adapter role for SRC-1.[50,51] It has also been shown that SRC-1 may be acetylated by CBP/p300, which may regulate SRC-1 function, but the significance of this finding will require additional investigation.[52]

Disruption of the SRC-1 gene leads to no obvious phenotype and animals grow at a rate similar to wild type mice.[53] However, steroid receptor target organs including the uterus, testis, prostate, and mammary gland exhibited decreased growth and development upon challenge with steroid hormones. One possible reason for the lack of a dramatic phenotype may be due to the observed increase in the expression of TIF2 in SRC-1 null mice. TIF2 is a coactivator that is highly homologous to SRC-1. The possible role of SRC-1 in thyroid hormone action was tested in SRC-1 null mice.[54] These animals exhibited resistance to thyroid hormone (RTH), which leads to a 2.5-fold elevation of serum thyrotropin levels. This study demonstrates for the first time that a factor other than thyroid receptor can modulate RTH. This opens up the possibility that mutations in coactivator genes could contribute to a disease state in humans.

### 14.2.8.2.2 TIF2

TIF2 was cloned as an ER- and retinoic acid receptor (RAR)-interacting protein by screening using the Far-Western blotting technique[55] while the mouse homologue, GRIP1, was identified in a yeast two hybrid screen using GR-ligand binding domain as bait.[56] TIF2 is related to SRC-1, with 30% identical amino acid sequence homology, and exhibits the highest homology in the PAS A-bHLH domain. TIF2 interacts with several NRs, including ER, PR and RAR, in a ligand-dependent manner and potentiates their transcriptional activity.[37] TIF2 contains several standard NR box motifs (LXXLL), along with an auxiliary NR box (NIDaux) that is required for interaction with a subset of NRs, including GR, AR, TRα, and RARα.[57] TIF2 is expressed in a wide range of tissues. A unique

function has yet to be established for this coregulator.

### 14.2.8.2.3 pCIP

pCIP was biochemically identified using a part of CBP as bait in pull-down assays.[58] Examination of the amino acid sequence for pCIP revealed 30% and 36% homology with SRC-1 and TIF2, respectively. The pCIP cDNA has also been cloned from a number of other laboratories and has been labeled RAC3, AIB1, ACTR, SRC3, and TRAM-1.[5,23,26,27,59] A number of soluble receptors have been shown to interact with pCIP, including RAR, ER, TR, and PR.[28] Disruption of the pCIP gene in mice revealed an unexpected phenotype, which included stunted growth, delayed puberty, decreased mammary gland development, and reduced female reproductive function.[60,61] The decreased mammary gland development appears to be due to decreased estrogen serum levels, while retardation of growth may be due to decreased levels of insulin-like growth factor-1. Transcriptional activity in mouse embryonic fibroblasts (MEFs) was examined using gene chips and revealed a small subset of genes that were down regulated, these target genes appear to be sufficient to lead to the growth phenotype observed.[60] Examination of members of the p160 family (e.g., SRC-1, TIF2, and pCIP) in transfection assays would suggest that they have overlapping activities. However, studies in null mice have so far indicated that each member has, at least in part, unique biological activities. Another interpretation of these studies may be that, due to developmental and tissue specific expression of each family member, there is a unique requirement for a given coactivator at a specific stage of growth and development. Support for this later hypothesis can be seen upon examination of the differential expression of pCIP and SRC-1 in tissue sections.[61,62]

### 14.2.8.2.4 PBP

In a screen for PPARγ cofactors a 165 kDa protein, identified as a unique PPARγ-binding protein, was designated as PPAR-binding protein (PBP).[63] Independently, two other laboratories have also cloned this cDNA and designated it as Vitamin D receptor (VDR)-interacting protein 205 (DRIP205) and thyroid receptor-associated protein 220 (TRAP220).[64,65] PBP interacts in a ligand dependent manner with a variety of steroid receptors, including PPARγ, PPARα, RXRα, RARα, TRβ1, and ER. PBP is widely expressed early during development and is found at differing levels in a variety of tissues.[63] Disruption of the PBP gene leads to embryonic lethality, the defects that appear to lead to lethality are impaired placental vasculature, neural development, and heart failure.[66,67] Studies with PBP null mouse embryonic fibroblasts indicate that the absence of PBP expression leads to decreased TRα, RXRα, PPARγ-mediated transactivation potential. In addition, the PBP null fibroblasts are defective in cell cycle regulation. These results clearly suggest that PBP may be a coactivator of central importance.

### 14.2.8.2.5 SRA

Using the AF-1 domain of the progesterone receptor as bait in the yeast two-hybrid cloning assay, a putative coactivator, termed steroid receptor RNA activator (SRA), was cloned.[68] The inability to find an open reading frame lead the investigators to hypothesize that SRA's activity as a coactivator was due to the ability of the RNA to directly interact with steroid receptors and/or coactivator proteins. Several experimental approaches were taken to clearly demonstrate that SRA is functioning as an RNA coactivator. The first was the use of stop codons within a putative open reading frame, which did not alter SRA's coactivation potential. Secondly, generation of anti-peptide antibodies against a sequence in the putative open reading frame failed to recognize a protein in cellular extracts. Finally, the use of cycloheximide during a transient transfection of an SRA construct into HeLa cells failed to influence the level of enhanced reporter gene RNA, while cyclohexamide blocked any SRC-1 mediated enhancement of transactivation potential. Taken together, these results would appear to establish that SRA's coactivator activity is mediated by the expression of RNA. The ability of SRA to mediate enhanced transactivation potential was tested with a number of type I and type II steroid receptors. The PR, GR, ER, and AR transactivation potential were all enhanced by SRA expression. In contrast, SRA had no effect on TRβ, RARγ, RXRγ, or PPARγ activity. It will be interesting to test whether SRA can alter transactivation potential of other transcriptional enhancers. Fractionation of soluble extracts from T-47D breast carcinoma cells revealed that SRA is found in large oligomeric complexes, which contain SRC-1. In addition, immunoprecipitation of lysates from cells co-expressing SRA and SRC-1, along with TR and RXR with TR antibody, suggested that SRA enters into a TR complex only in the presence of SRC-1. Examination of SRA RNA expression in human tissues revealed that SRA is expressed at high levels in liver and skeletal muscle, with a relatively low level of expression detected in other tissues. Interestingly, a survey of SRA RNA expression in established cell lines indicated that SRA is widely expressed in the cell lines tested and appears to be highly expressed in some breast carcinoma cell lines.[68] Using RT-PCR, the level of SRA RNA in a number of human breast tumor tissue samples was examined, results indicated that the expression levels of SRA varied considerably from sample to sample.[69] The level of SRA correlated with tumor grade, while the relationship of SRA expression with ER or PR expression was more complex. Taken together, these studies show that SRA is an RNA coactivator that is widely expressed in tumor tissue and that it enhances type II steroid receptor transactivation potential, in conjunction with SRC-1, through binding to AF-1 domains.

### 14.2.8.2.6 RIP140

Receptor interacting protein 140 (RIP140) was identified as an ER-interacting nuclear protein using a Far-Western assay.[70] It interacts with the ER AF2 domain only in the presence of agonist and not in the presence of anti-estrogens. It was found to potentiate ER transactivation potential by a modest two-fold, when transfected into cells at the appropriate vector levels.[70] Unlike other coactivators, addition of increasing amounts of RIP140 vector leads to repression of ER mediated transactivation potential. This appears to indicate that either RIP140 also exhibits corepressor activity at higher expression levels, or at high levels RIP140 sequesters a factor that is required for transactivation. Subsequent studies have demonstrated that RIP140 can modulate transcriptional activity of a number of soluble receptors including; GR, RXR, PPARα, RAR, TR2, VDR, and AhR.[71–74] RIP140 also interacts with TRβ in a ligand-dependent manner. RIP140 and SRC-1 both interact weakly with a L454V TRβ mutant relative to WT TRβ. Interestingly, this mutant form of TRβ is associated with a syndrome of resistance to thyroid hormone.[75] RIP140 also increases ER and RAR transactivation several-fold in yeast, and this effect was enhanced at sub-optimal doses of ligand.[76]

Although originally identified as a coactivator, subsequent experiments have suggested a more complex regulatory role for RIP140. RIP140 was found to compete with SRC-1 for interaction with NRs and down-regulated a SRC-1 mediated increase in reporter gene transcriptional activity. This suggests that RIP140 may play a role in moderating NR AF-2 activity by competing with SRC-1.[73] In a separate report, RIP140 blocked enhanced transactivation of ER by TIF-2, also indicating a possible negative regulatory role for RIP140.[72] In addition, RIP140 was found to repress TR2 transactivation potential.[77] Significantly, RIP140, when fused to the binding domain of GAL4, was able to repress a GAL4 reporter gene by itself, further suggesting a role of RIP140 as a corepressor protein.[77] Additional evidence supporting the concept that RIP140 can act as a co-repressor is the ability of RIP140 to bind to histone deacetylase (HDAC), which will lead to repression of gene transactivation potential.[78] HDAC binds to RIP140 at an N-terminal domain, located near where the ER binds, while RIP140 interacts with RAR and RXR at a C-terminal domain (Figure 1). Thus, considering that RIP140 has multiple interacting domains, the ability of RIP140 to recruit HDAC may be context and receptor specific. A null RIP140 mouse has been generated that should be useful in examining the role of RIP140 in receptor-mediated signaling. The phenotype of the RIP140 null mice is a complete failure of mature follicles to release the oocyte at ovulation which results in female infertility.[79] This is the first demonstration that a coregulator is required for a specific biological function in adult mice.

RIP140 interacts with PPARα and liver X receptor α (LXRα) in the absence of ligand, in contrast to the requirement for ligand with most NRs that have been studied.[80] In the case of PPARα/RXR and LXRα/RXR, RIP140 appeared to decrease their transactivation potential, further indicating that RIP140 may play a role as co-regulator.[80] In addition RIP140 was found to interact with the GR in a ligand-dependent fashion, yet down-regulated GR activity.[72] This suppression is relieved by over-expression of the p160 class of coactivators, suggesting that RIP140 may compete for transcription factor binding with p160 coactivators, thus moderating/regulating transcriptional activity. However, the repressive effects are selective in that RIP140 actually slightly increased signaling through NFκB subunit p65 (RelA) and the pituitary-specific factor Pit-1/homeodomain transcription factor Pbx complex and only slightly repressed

signaling through the Pbx1/HOXB1 and AP-1 proteins.[72] This argues against general squelching as the mechanism of RIP140 repression. RIP140 also inhibits ER/Pit-1 and Pit-1/TR mediated transcriptional synergy at the pituitary-specific prolactin promoter, but not at the growth hormone promoter.[81] This promoter context-specific synergistic interplay between transcription factors and coactivators illustrates the complexity of the type of responses that are mediated by RIP140.

### 14.2.8.2.7  pCAF

pCAF exists in a complex with p300/CBP, possesses intrinsic HAT activity, and is the human homolog of yeast coactivator/adapter protein GCN5.[82,83] pCAF, along with p300/CBP and TAF$_{II}$250, can direct the acetylation of several members of the basal transcriptional machinery, including transcription factors IIEβ and IIF.[84] Interestingly, unlike other coactivators, pCAF interacts with the DNA binding domain of NRs independent of p300/CBP binding, suggesting a more direct role for pCAF in NR transcription.[82] Indeed, pCAF is found in a complex of more than 20 distinct polypeptides, some of which are identical to the TAFs.[85] Similar to transcription factor IID, pCAF complexes contain several histone-fold components. Significantly, the human RNA Polymerase complex contains several HAT proteins, including pCAF, SWI/SNF complex, Srb proteins and CBP. In addition, p300 and pCAF interact directly with RNA Polymerase II and pCAF also interacts with the elongation-competent, phosphorylated form of RNA Pol II, further suggesting that pCIP has a central role in transcription.[86]

### 14.2.8.3  COINTEGRATORS—CBP/P300

Although originally identified as a CREB interacting protein,[87] CBP has been subsequently shown to be involved in ligand-dependent transcriptional activation *via* direct, or indirect, interaction with several transcription factors, including; p53,[88–90] STAT2,[91,92] myoD,[93,94] and several NRs. CBP has been shown to bind other coactivators like SRC-1[95] and also exists in stable preformed complexes with RNA Pol II.[86] Another highly related protein, p300, was originally identified as one of the key cellular targets of the adenovirus E1A oncoprotein.[96] CBP or p300 have been suggested to be limiting, common cointegrators for distinct but convergent signaling pathways, with NR competing with AP-1 for the available

CBP/p300 pool.[25] Although found to be in distinct preformed complexes,[97] CBP and SRC-1 synergistically increase transactivation potential of ER and PR.[98] CBP contains several distinct domains, which are involved in interacting with different proteins, again indicating a possible central role for CBP[25] (Figure 1). CBP/p300 not only function as transcriptional adapters but also possess a novel HAT activity capable of acetylating all four histones in nucleosomes, suggesting that CBP/p300 is involved in chromatin remodeling.[6] The role of p300 in transcriptional complexes was examined in an *in vitro* system using ER. The cointegrator p300 appeared to be required for preinitiation, however its role in subsequent rounds of transcription-reinitiation was found to be less important, suggesting a multistep process for transcriptional activity.[99]

Targeted disruption of the p300 and CBP gene in mice results in embryonic lethality between days 9 and 11.[100–102] Embryonic fibroblasts isolated from null p300 mice exhibited severely retarded proliferative rates. This confirmed that p300 and CBP play a central role in the cell. Surprisingly, mice heterozygous for CBP or p300 exhibited significant phenotypes, while mice with double heterozygosity for p300 and CBP resulted in embryonic lethality.[100] A number of abnormalities have been detected in the skeletal system of CBP heterozygous mice, the defects observed were also dependent on the genetic background.[101] A comparison of monoallelic inactivated p300 and CBP mice revealed that only CBP heterozygous mice had defects in hematopoietic differentiation, along with hematologic malignancies in older animals.[102] This latter observation clearly indicates that CBP has tumor suppressor activity and the level of CBP expression is important. Results from this report also indicate that CBP and p300, despite being highly related, have distinct functions. In addition, studies in embryonic carcinoma F9 cells have also shown that CBP and p300 are involved in distinctly different biochemical processes.[103] A strikingly similar disease state compared with the CBP heterozygous mice is seen in humans with Rubenstein–Taybi syndrome, patients exhibit growth and mental retardation, facial abnormalities, broad thumbs, and increased risk for certain cancers.[104] Genetic analysis of these individuals revealed breakpoints in chromosome 16p13.3, these breakpoints occurred in the CBP gene, resulting in humans heterozygous for the CBP gene. Examination of CBP heterozygous mice revealed some of the same abnormalities detected in Rubinstein–Taybi syndrome patients.[101,105] These studies may also indicate

that other genetic abnormalities (e.g., point mutations) in CBP/p300 could contribute to a wide range of diseases.

### 14.2.8.4   TISSUE SPECIFIC COACTIVATORS

#### 14.2.8.4.1   PGC-1

PPARγ coactivator 1 (PGC-1) is a novel coactivator cloned from a brown fat cDNA library whose expression is induced by cold exposure in brown fat and skeletal muscles, which are key thermogenic tissues.[106] PGC-1 modulates PPARγ and TR transcription activity on the uncoupling protein (UCP-1) promoter. Over-expression of PGC-1 leads to up-regulation of UCP-1 and key mitochondrial enzymes of the respiratory chain. In addition, it increases the cellular levels of mitochondrial DNA, linking it to adaptive thermogenesis.[106] Significantly, PGC-1 has little inherent transcriptional activity, but upon binding to PPARγ, PGC-1 undergoes a conformational change that enables recruitment of SRC-1 and CBP, resulting in increased transcriptional activity.[107] *In vitro*, PGC-1 interacts with PPARα even in the absence of exogenous ligand. However, PPARα interaction with PGC-1 requires the presence of a ligand, suggesting diverse mechanisms of recruitment of the same coactivator to different receptors.[108] The hPGC-1 is expressed in a tissue-dependent manner with significant expression detected in heart, kidney, brown fat, and brain. Interestingly, PGC-1 contains two potentially novel motifs, a SR domain that may interact with RNA polymerase and a RNA binding motif.[106] Further work on the SR domain of PGC-1 has shown that it is involved in direct coupling of transcription and RNA processing only after PGC-1 is bound to the promoter region.[109] In addition, the SR domain is capable of binding to RNA splicing factors, and thus mediates mRNA processing. PGC-1 was also found to interact with other receptors like ERα in a ligand-independent manner,[110] the significance of which will require further investigation.

#### 14.2.8.4.2   ARA Coregulators

AR associated protein 70 (ARA70) was identified as AR interacting protein in a yeast two-hybrid screen and was able to enhance AR-mediated transcriptional activity several-fold in the presence of testosterone in human prostate cancer cells.[111] Interestingly, ARA70 was highly selective for AR, suggesting that it is an AR-specific coactivator.[111] However, a more recent report found that ARA70 only weakly enhances AR-mediated transactivation potential.[112] ARA 70 is able to directly interact and enhance transcription of the PPARγ, thus casting doubt as to how specific a coactivator ARA70 is for the AR.[113] Examination of tissue expression of ARA70 reveals mRNA expression in a wide range of tissues, this also suggests that ARA70 may play a role in the regulation of transcriptional activity of other factors.[111] There is some evidence that ARA70 coactivators may recruit proteins with HAT activity such as CBP and associated co-factors, consistent with its role as a coactivator.[114] Other ARA coactivators, ARA54, ARA55, and ARA160 also increase ligand-dependent AR transactivation, whereas SRC-1 failed to significantly increase AR transactivation.[114,115] Just how specific these coactivators are for the AR will require further investigation. A testis-specific AR coregulator has been identified, termed AR interacting protein 3 (ARIP3), and belongs to a novel family of nuclear proteins.[116] ARIP3 is able to both enhance and repress AR-mediated reporter activity, depending on the level of transient expression, similar to what is seen with RIP140 expression. The level of ARIP3 specificity for AR-mediated transcriptional enhancement relative to other NR needs to be further assessed.

### 14.2.8.5 COREPRESSORS

The co-repressors SMRT (silencing mediator of retinoid and thyroid hormone receptor) and NCoR (nuclear receptor co-repressor) interact with several unliganded NRs, including RAR and TR bound to RXR, which can bind their cognate response elements in the absence of a ligand and repress their transcriptional activation.[117–122] Some hormone receptors, including the ER and PR, recruit co-repressors when bound to antagonists.[121,123] NCoR and SMRT, in turn, recruit mSin3 and HDACs (Figure 1).[10,11] Sin3, a protein with four amphipathic α-helices, may serve as an adapter between receptors and the core HDAC complex.[124] Purification of the core mSin3 complex revealed a total of seven polypeptides, including SAP30. It has been suggested that SAP30 and mSin3 serve as a bridge between the NRs and the core mSin3 complex.[123,125] The activities of co-repressors, like those of co-activators, are regulated by the cAMP pathway[121] and by proteolysis.[126] A core motif in NRs, called the CoR box, in the hinge region between the DNA binding domain and the LBD, has been identified as a requirement for interaction with co-repressors. An additional motif at the C-terminus has been suggested to be critical for the release of co-repressors upon ligand binding.[118]

Another class of co-repressors is the Transcriptional Intermediary Factor or TIF1, family of co-repressors. Although TIF1α interacts with ligand-bound RAR, TIF1α and TIFβ have been shown to bind to the Krüppel associated box (KRAB) repression domain in $C_2H_2$ zinc-finger proteins and with heterochromatin proteins, suggesting that NRs may target the assembly of a microdomain of heterochromatin.[127,128] TIF1 proteins have been proposed to be involved in ligand-dependent transcriptional repression and/or in converting an inactive heterochromatin-like structure to an active chromatin structure.[129] It also appears that KRAB and TIF1 repression requires the recruitment of deacetylase complexes. These studies underscore the complexity of NR regulation as a balance between the release of a corepressor and subsequent recruitment of a coactivator complex.

### 14.2.8.6 SECONDARY COACTIVATORS

A number of NR/transcription factor-interacting coactivators have been cloned and shown to also interact with specific basal transcription factors. However, this is little evidence for secondary coactivators, which are defined as proteins that do not bind to response element bound transcription factors, but instead interact with coactivators complexed with a transcription factor. One exception to this statement is the coactivator-associated arginine methyltransferase (CARM1), a secondary coactivator which is capable of binding to the carboxyl terminal AD2 domain of p160 coactivators. CARM1 was identified and found to enhance transcriptional activity of NRs only when p160 coactivators were expressed. CARM1 methylates H3 histones *in vitro*, thus introducing a functional and structural complexity to coactivator complexes.[130] A protein highly related to CARM1 is the enzyme arginine methyltransferase 1 (PRMT1), this protein also binds to p160 coactivators and thus is also considered a secondary coactivator.[131] In addition, coexpression of CARM1 and PRMT1 leads to synergistic activation of NRs. This synergistic effect may be due to differential substrate specificity of each methyltransferase.[2,130] Examination of tissue distribution reveals that PRMT1 is widely distributed and is the predominant Type I protein methyltransferase, while CARM1 is predominantly expressed in heart, kidney, and testis.[130,132] This suggests that methyltransferases may exhibit tissue specific effects on transcription. However, as this emerging field continues to develop it is quite probable that additional secondary coactivators will be identified.

### 14.2.8.7 LXXLL MOTIFS

Most coactivators contain several copies of a signature motif, LXXLL, that are necessary for interaction with NRs.[46] Mutation of the critical leucine residues disrupts recruitment of these coactivators to most NRs. Co-crystal structures of ligand bound LBD of TR,[133] PPAR[38] and ER[134] with p160 coactivator peptides have been resolved. These LXXLL motifs also play a critical role in imparting specificity of interaction, with flanking amino acid residues modulating both specificity and affinity.[33] While ER only requires a single LXXLL motif, TR, RAR, PR, and PPARα require different combinations of three motifs in p160 coactivators. For example, in the case of PPARα, different LXXLL motifs are required in response to binding to distinct ligands.[38] Using peptide inhibition and yeast two-hybrid assays, it was determined that an eight amino acid LXXLL-containing sequence is required for efficient binding to NRs, with the non-conserved residues altering their affinity and selectivity (Figure 2).[135] The presence of multiple LXXLL motifs would appear to allow a given coactivator to interact with a greater number of transcription factors, and also perhaps interact with more than one protein at a time.

### 14.2.8.8 OLIGOMERIC COREGULATOR COMPLEXES

A number of large preformed oligomeric coactivators have been isolated from cellular extracts. These complexes appear to vary in their subunit composition, with some complexes being quite similar and others having little subunit overlap. The role of each subunit in these large complexes is only beginning to be understood. Here we will examine what is known about two of the most studied coregulator complexes in a growing list of approximately six complexes that have been biochemically isolated.[15] Whether these complexes are recruited to distinct transcription factor families, or serve a more generalized role, will require additional studies. In addition, whether the subunit composition of these complexes varies from tissue to tissue also needs to be explored.

#### 14.2.8.8.1 TRAP Complex

The hTRα was found to bind to a large complex, labeled the thyroid hormone receptor associated complex (TRAP), containing at least 13 distinct proteins, ranging in size from 20 to 240 kDa. The TRAP complex enhanced transcription from a T3-response element driven template.[136] TRAP220 (PBP), a component of the complex, contains LXXLL motifs and interacts with several other nuclear receptors in a ligand-dependent manner and targets the entire complex to NRs.[65] In fact, it has been suggested that a single molecule of TRAP220 interacts with both molecules of the NR heterodimer.[137] Interestingly, the TRAP-mediated enhancement of TR activity did not require TFIID proteins, indicating functional redundancy. In addition, SRC-1 and CBP are not a part of the TRAP complex, indicating the presence of distinct multi-subunit complexes.[138]

#### 14.2.8.8.2 DRIP Complex

Using Vitamin $D_3$ receptor LBD as an affinity matrix, a 15-protein complex containing proteins ranging from 65 to 250 kDa was purified and named the VDR-interacting protein complex, or DRIP complex.[139] This complex appears to be identical to ARC coactivator complex and quite similar to the TRAP complex.[15] The DRIP complex contains HAT activity, suggesting a chromatin remodeling function, but the complex also selectively enhanced VDR activity on a naked DNA template.[139] The DRIP complex shares components with the RNA Pol II holoenzyme complex. However, studies suggest that the DRIP and RNA Pol II holoenzyme complex represent distinct complexes.[140] Interaction between the two complexes is induced by ligand binding, suggesting a two-step transcriptional activation of cognate genes, comprised of recruitment of DRIP or DRIP-like complexes to NRs, followed by recruitment of RNA Pol II transcriptional machinery.[140] DRIP205 (PBP), a component of the complex, anchors the complex to VDR and TR in a ligand-dependent manner. PBP also interacts with other NRs in a ligand-dependent manner.[141] Interestingly, PBP, is a component of several coactivator complexes, including DRIP, ARC, CRSP, and TRAP complexes.[142,143] The role of the other subunits in

```
    -2 -1 1 2 3 4 5 6 7 8 9 10
  747-Q  L  L  R  Y  L  L  D  K  D  E  K  -758    SRC1 Motif 3
  688-K  I  L  H  R  L  L  Q  E  G  S  P  -699    SRC1 Motif 2
 1433-S  L  L  Q  Q  L  L  T  E           -1441   SRC1 Motif 4
→ 498-T  L  L  Q  L  L  L  G  H  K  N  E  -509    RIP140 L6     HIGH AFFINITY
  817-G  L  L  S  R  L  L  R  Q  N  Q  D  -828    RIP140 L8
  934-N  V  L  K  Q  L  L  L  S  E  N  C  -945    RIP140 L9
  131-T  L  L  A  S  L  L  Q  S  E  S  S  -142    RIP140 L2

→ 711-T  V  L  Q  L  L  L  G  N  P  K  G  -722    RIP140 L7
   19-T  Y  L  E  G  L  L  M  H  Q  A  A  -30     RIP140 L1
  265-S  Q  L  A  L  L  L  S  S  E  A  H  -276    RIP140 L4
  631-H  K  L  V  Q  L  L  T  T  T  A  E  -642    SRC1 Motif 1   LOW AFFINITY
   67-K  Q  L  S  E  L  L  R  G  G  S  G  -78     CBP LXM-1
  355-Q  Q  L  Q  V  L  L  L  H  A  H  K  -366    CBP LXM-2
 2066-S  A  L  Q  D  L  L  R  T  L  K  S  -2077   CBP LXM-3
  183-S  H  L  K  T  L  L  K  K  S  K  V  -194    RIP140 L3
  378-S  L  L  L  H  L  L  K  S  Q  T  I  -389    RIP140 L5
```

**Figure 2** Sequences of LXXLL motifs showing differential affinities for the estrogen and retinoid receptors. The sequences of LXXLL motifs derived from SRC-1, RIP140, or CBP, and their respective amino acid numbers are shown. The sequences are aligned with reference to the first conserved leucine residue. The high and low affinity grouping was derived from yeast two-hybrid data using each receptor tested. However, differential affinities for these LXXLL motifs between the ER and retinoid receptors does exist. The arrows denote two LXXLL motifs that only vary in their flanking amino acid residues, yet show significant differences in binding to receptors. This figure was adapted from the work of Heery *et al.*[135]

the DRIP complex is less clear. However, both PBP and DRIP150 interact with the GR receptor,[144] while hSur-2 (DRIP130) interacts with E1A.[145] These studies may suggest that at least part of the rationale for the multiple subunits of DRIP is to allow interaction with a wide range of transcription factors. Nevertheless, the distinct role of each subunit will require extensive additional studies.

There is also evidence for other large, but distinct, complexes of different types of coregulators. Small pools of p300, BRG-1, and pCAF co-purify with SRC-1 in steady state complexes, although CBP and SRC-1 do not exist in the same complexes.[97] It is interesting to note that promoters coupled to different enhancer elements (RARE, CREB and STAT) require different combinations of coactivators for maximal coactivation at each promoter.[83] This may suggest a possible mechanism for cell and promoter-specific activity of genes.

### 14.2.8.9 MECHANISMS OF COREGULATOR EXCHANGE

Transcription can be viewed as a two step process involving first a derepression of ordered chromatin structure, which subsequently allows greater access to DNA, followed by recruitment of the large RNA polII enzymatic machinery complexes. The presence of multiple oligomeric coregulator complexes would suggest that these complexes have some level of specificity that may be promoter or context specific. In addition, it is probable that these complexes are in a dynamic equilibrium which allows different coregulator complexes to bind and perform specific remodeling activities, followed by binding of another complex that is able to exhibit different remodeling activities. The presence of different remodeling complexes on a given enhancer/ promoter DNA sequence can be viewed as either sequential, combinatorial, or parallel, as depicted in Figure 3.[12] These models are not mutually exclusive and, for many genes, all of these models may be utilized to obtain efficient chromatin remodeling.

### 14.2.8.10 AH RECEPTOR AND COACTIVATORS

The vast majority of the studies performed have used NR as targets for the examination of coactivator activity. This raises the question as to whether essentially all enhancer binding transcription factors are capable of recruiting a common set of coregulators. Several coactivators, corepressors, and transcription factors/ enhancers interact with the AhR/ARNT heterodimer and modulate dioxin-responsive gene transcriptional activity. The coactivator, ER associated protein 140 (ERAP140) and a corepressor, SMRT interacts with the AhR and ARNT and altered the AhR-mediated transactivation in the breast tumor cell line MCF-7.[146] However, these studies failed to show a detailed analysis characterizing these protein–protein interactions. Studies examining the role of RIP140 and SRC-1 have indicated that both of these coactivators are capable of enhancing AhR-mediated transactivation potential, thus demonstrating that coactivators that interact with NRs can also interact with the AhR.[41,71] Both of these coactivators interact with the Q-rich domain of the AhR and not with ARNT. Yet another coactivator, CBP, was shown to interact with ARNT preferentially,[147] although whether CBP is involved in AhR/ARNT transcriptional complexes in cells remains to be determined. These studies suggest that the AhR interacts with many of the coregulators that have been initially shown to bind to the NR, which would suggest that the AhR may compete with the NR for the available coactivator pool. Considering that the AhR and ARNT have TADs that appear to be quite different from NR, it remains to be determined whether they are capable of interacting with a distinct group of coactivators.

The presence of characterized regulatory elements of the CYP1A1 promoters would suggest that several transcription factors might regulate CYP1A1 transcription. However, their exact role(s) in the induction of dioxin responsive genes needs to be further explored. The enhancer region of the mouse CYP1A1 contains six DREs that are inaccessible to the AhR/ARNT heterodimer, due to nucleosomal configuration of the DNA in the absence of TCDD.[148–152] Interestingly, AhR has been found in a complex with histone H4, similar to that seen with the GR and PR, which are suggested to bind nucleosomes.[153] This interaction is thought to be the initial step in chromatin remodeling of the enhancer region, which should pave the way for further localized chromatin remodeling, binding of additional transcription factors, and assembly of a pre-initiation complex. The nuclear factor Sp1 interacts with the AhR through the bHLH-PAS domain in *in vitro* binding experiments.[154] Binding of either Sp1 or AhR/ARNT to their respective elements appears to facilitate binding of the other factor in the process of formation of an active transcriptional complex. Indeed, the presence of Sp1 enhances the DRE-CYP1A1 promoter driven reporter gene 15-fold over the activation provided by AhR/ ARNT heterodimer alone.[154]

Two non-DNA binding transcription factors have also been shown to interact with the AhR and modulate its ability to alter transcription. The retinoblastoma tumor suppres-

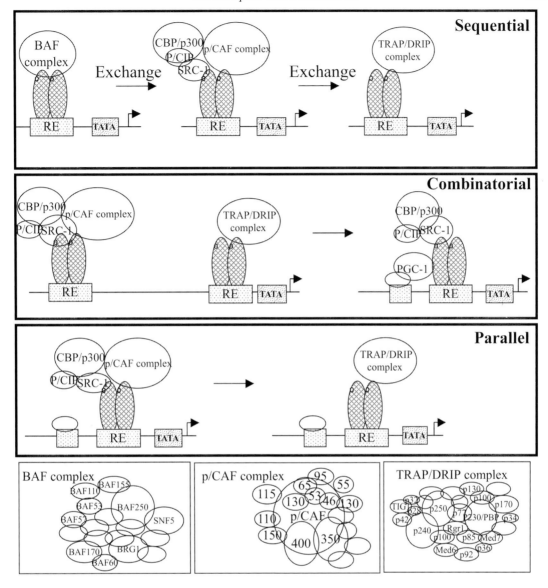

**Figure 3** Models of coactivator complex role in transcriptional activation. In the first model, one coactivator complex is recruited and acetyltransferase activity occurs. After this complex leaves another complex binds that performs a distinct set of remodeling activities. In at least some promoters the presence of multiple enhancer sites results in recruitment of different coactivator complexes at the same time, as is shown in the combinatorial model. The third model considers the possibility that different coactivator complexes can lead to enhanced transcription on the same promoter. This figure has been adapted from a review by Glass and Rosenfeld.[12]

sor protein, pRb, which is involved in controlling cell cycle progression through G1 and promoting differentiation, interacts with the AhR *via* the glutamine-rich sub-domain of the AhR through residues 589–774 (see Chapter 4.6).[155] The Rb protein also interacts presumably through an LXCXE motif present in the N-terminal 364 amino acid residues of the AhR. The interaction of AhR with Rb may provide a mechanism for the cell cycle arrest of 5L rat hepatoma cells at G1 stage upon TCDD treatment.[155] However, the transcription of p27 (Kip1) cyclin/cdk inhibitor is induced by AhR, which has also been suggested to lead to suppression of 5L hepatoma cell proliferation.[156] RelA, a component of the transcription factor NF-kB, which is activated by cytokines, has also been reported to interact with the AhR. This interaction is presumed to down-regulate the expression of TCDD-responsive genes.[157] Clearly, additional studies are needed to establish the functional significance of RelA and pRb in regulating AhR transcriptional activity. In addition, how these protein–protein interactions modulate coactivator or corepressor interaction with the AhR/

ARNT heterodimer will also require further investigation.

## 14.2.8.11  CANCER AND COREGULATORS

For many years the role of steroid hormone receptors and transcription factors in modulating the carcinogenesis process has been documented. So it is not surprising that coactivators, which present an additional layer of regulation on transcriptional activity, would also appear to play an important role in the carcinogenesis process. Perhaps the most studied aspect of coactivator activity, in terms of carcinogenesis, is its role in modulating estrogen receptor activity in breast tumor cells. The ER coactivator AIB1 was the first shown to be amplified in breast and ovarian tumor tissue, this would suggest that tumors would have altered transcriptional activity through dysregulation of coactivators.[23] More recent studies have shown a frequent amplification or overexpression of AIB1 in ovarian and gastric tumors.[158] Importantly, amplification of AIB1 was associated with ER positively in sporadic ovarian carcinomas.[159] The coactivator PBP is either amplified and/or overexpressed in ~50% of primary and established breast cell lines tested, with 24% of breast tumors actually containing an amplified PBP gene.[160] Yet another nuclear receptor coactivator, activating signal cointegrator-2 (ASC-2), has also been found to be amplified in human cancers.[161] Thus, as a more complete picture emerges from examining the amplification or overexpression of coactivators in tumors, it would appear likely that they play an important role in tumor progression.

Tamoxifen, a mixed ER antagonist/agonist is at least initially effective at inhibiting the growth of some breast tumors. However, at some point in the treatment regimen tumors often resume growth, the reason for this apparently acquired resistance to tamoxifen could be due to multiple mechanisms. In addition, tamoxifen may actually behave as an agonist in certain tumors. One possible explanation for these results may be due to the balance between expression of coregulatory proteins that can be recruited to tamoxifen/ER complexes. The yeast two-hybrid approach was utilized to clone proteins that bind to ER or ER/tamoxifen complexes, or to a mutated AF-2 domain.[119,162] This approach cloned the protein REA (repressor of estrogen activity) which has been identified to selectively inhibit estradiol/ER-mediated transactivation potential.[162] In contrast, REA had little effect on the transactivation potential of PR, RAR, or Gal4-VP16. REA is able to effectively inhibit SRC-1 mediated enhancement of ER transactivation potential, suggesting that REA compete for binding to the AF-2 domain of the ER. REA has no significant sequence homology with other known corepressors such as N-CoR or SMRT and REA also differs from these corepressors in that it does not interact in the presence of ligand. In addition, the presence of REA appears to enhance the inhibitory properties of antiestrogens. N-CoR and SMRT suppress the agonistic activity of the tamoxifen-occupied ER.[119] Yet another coregulator named L7/SPA increases tamoxifen-occupied ER-mediated transcriptional activity, but is unable to enhance agonist mediated ER activity.[119] These studies would suggest that the response to an antagonist is more than simply preventing agonist binding to the ER, but also involves altered recruitment of coregulators that can in turn lead to agonist or antagonist activity, independent of the available coregulatory complexes in tumor cells.[163]

Yet another proposed mechanism of tumor progression is the modification of protein activities important in histone remodeling processes, such as an alteration in the balance between HATs and HDACs.[164] In tumor cells this can be accomplished either by gene amplification, chromosomal deletions, and or point mutations. This, in turn, can lead to the silencing of normally activated genes or activation of normally silenced genes. A more targeted mechanism of altered histone remodeling is the generation of fusion proteins between a coactivator with HAT activity and a transcription factor. A prime example of these processes is the alteration of CBP or p300 function during tumorigenesis as seen in a number of tumor types.[165] As this area continues to develop coactivators may be a useful target for the development of new drugs to treat a wide variety of diseases.

## 14.2.8.12  CONCLUSIONS AND FUTURE DIRECTIONS

A wide range of studies has examined the ability of coregulators to modulate soluble receptor-mediated transactivation potential. In the case of coactivators, it is clear that the level of expression of the total pool of coactivators is rate-limiting. This, in turn, would suggest that in a given cell type there may be some level of competition for the available coactivator pool. Considering that a growing number of coregulators can be recruited to a soluble receptor (e.g., ER, PPARs, AhR, *etc.*), a combination of the relative concentration of individual coactivators and their relative affinities for a given TAD would appear to be what determines which coactivator is recruited. The vast majority of studies that examine coregulator function have utilized heterologous reporter vectors to measure transactivation potential in transient transfection assays. Future studies should begin to examine the ability of coregulators to modulate the level of transcription of endogenous genes. In addition, studies that examine the influence of coregulator expression on a battery of genes regulated by a specific receptor would be particularly insightful. The cell-specific expression of coactivators and corepressors is being slowly delineated, but their role in cell line-, tissue- and developmentally-specific expression of genes needs to be explored in detail. Yet another area that has only begun to be explored is the transcriptional and post-translational regulation of coregulators. For example, several studies have indicated that both the ER and certain coactivators are targets for growth factor activated serine kinases.[166,167] The overall picture emerging is that coregulators, along with various transcription factors and receptors, are key to a given tissue's response to exposure to a xenobiotic. In addition, dysregulation of coactivator function through disease processes probably play an important role in the aberrant transcriptional activity seen in a disease state such as cancer.

## 14.2.8.13  REFERENCES

1. D. E. Sterner and S. L. Berger, 'Acetylation of histones and transcription-related factors.' *Mol. Biol. Rev.*, 2000, **64**, 435–459.
2. M. R. Stallcup, D. Chen, S. S. Koh *et al.*, 'Co-operation between protein-acetylating and protein-methylating co-activators in transcriptional activation.' *Biochem. Soc. Trans.*, 2000, **28**, 415–418.
3. J. T. Kadonaga, 'Eukaryotic transcription: an interlaced network of transcription factors and chromatin-modifying machines.' *Cell*, 1998, **92**, 307–313.
4. K. Struhl, 'Histone acetylation and transcriptional regulatory mechanisms.' *Genes Dev.*, 1998, **12**, 599–606.
5. H. Chen, R. J. Lin, R. L. Schiltz *et al.*, 'Nuclear receptor coactivator ACTR is a novel histone acetyltransferase and forms a multimeric activation complex with P/CAF and CBP/p300.' *Cell*, 1997, **90**, 569–580.
6. A. J. Bannister and T. Kouzarides, 'The CBP co-activator is a histone acetyltransferase.' *Nature*, 1996, **384**, 641–643.
7. V. V. Ogryzko, R. L. Schiltz, V. Russanova *et al.*, 'The transcriptional coactivators p300 and CBP are histone acetyltransferases.' *Cell*, 1996, **87**, 953–959.
8. X. J. Yang, V. V. Ogryzko, J. Nishikawa *et al.*, 'A p300/CBP-associated factor that competes with the adenoviral oncoprotein E1A.' *Nature*, 1996, **382**, 319–324.
9. T. E. Spencer, G. Jenster, M. M. Burcin *et al.*, 'Steroid receptor coactivator-1 is a histone acetyltransferase.' *Nature*, 1997, **389**, 194–198.
10. L. Nagy, H. Y. Kao, D. Chakravarti *et al.*, 'Nuclear receptor repression mediated by a complex containing SMRT, mSin3A, and histone deacetylase.' *Cell*, 1997, **89**, 373–380.
11. T. Heinzel, R. M. Lavinsky, T. M. Mullen *et al.*, 'A complex containing N-CoR, mSin3 and histone deacetylase mediates transcriptional repression.' *Nature*, 1997, **387**, 43–48.
12. C. K. Glass and M. G. Rosenfeld, 'The coregulator exchange in transcriptional functions of nuclear receptors.' *Genes Dev.*, 2000, **14**, 121–141.
13. C. K. Glass, D. W. Rose and M. G. Rosenfeld, 'Nuclear receptor coactivators.' *Curr. Opin. Cell Biol.*, 1997, **9**, 222–232.
14. N. J. McKenna, J. Xu, Z. Nawaz *et al.*, 'Nuclear receptor coactivators: multiple enzymes, multiple complexes, multiple functions.' *J. Steroid Biochem. Mol. Biol.*, 1999, **69**, 3–12.
15. C. Rachez and L. P. Freedman, 'Mechanisms of gene regulation by Vitamin D(3) receptor: a network of coactivator interactions.' *Gene*, 2000, **246**, 9–21.
16. M. Stern, R. Jensen and I. Herskowitz, 'Five SWI genes are required for expression of the HO gene in yeast.' *J. Mol. Biol.*, 1984, **178**, 853–868.
17. L. Neigeborn and M. Carlson, 'Genes affecting the regulation of SUC2 gene expression by glucose repression in *Saccharomyces cerevisiae*.' *Genetics*, 1984, **108**, 845–858.
18. S. K. Yoshinaga, C. L. Peterson, I. Herskowitz *et al.*, 'Roles of SWI1, SWI2, and SWI3 proteins for transcriptional enhancement by steroid receptors.' *Science*, 1992, **258**, 1598–1604.
19. C. L. Peterson, A. Dingwall and M. P. Scott, 'Five SWI/SNF gene products are components of a large multisubunit complex required for transcriptional enhancement.' *Proc. Natl. Acad. Sci. USA*, 1994, **91**, 2905–2908.
20. O. Papoulas, S. J. Beek, S. L. Moseley *et al.*, 'The *Drosophila* trithorax group proteins BRM, ASH1 and ASH2 are subunits of distinct protein complexes.' *Development*, 1998, **125**, 3955–3966.
21. K. Zhao, W. Wang, O. J. Rando *et al.*, 'Rapid and phosphoinositol-dependent binding of the SWI/SNF-like BAF complex to chromatin after T lymphocyte receptor signaling.' *Cell*, 1998, **95**, 625–636.
22. J. DiRenzo, Y. Shang, M. Phelan *et al.*, 'BRG-1 is recruited to estrogen-responsive promoters and cooperates with factors involved in histone acetylation.' *Mol. Cell. Biol.*, 2000, **20**, 7541–7549.
23. S. L. Anzick, J. Kononen, R. L. Walker *et al.*, 'AIB1, a steroid receptor coactivator amplified in breast and ovarian cancer.' *Science*, 1997, **277**, 965–968.
24. H. Hong, K. Kohli, M. J. Garabedian *et al.*, 'GRIP1, a transcriptional coactivator for the AF-2 transactivation domain of steroid, thyroid, retinoid, and Vitamin D receptors.' *Mol. Cell. Biol.*, 1997, **17**, 2735–2744.
25. Y. Kamei, L. Xu, T. Heinzel *et al.*, 'A CBP integrator complex mediates transcriptional activation and AP-1 inhibition by nuclear receptors.' *Cell*, 1996, **85**, 403–414.
26. H. Li, P. J. Gomes and J. D. Chen, 'RAC3, a steroid/nuclear receptor-associated coactivator that is related to SRC-1 and TIF2.' *Proc. Natl. Acad. Sci. USA*, 1997, **94**, 8479–8484.
27. A. Takeshita, G. R. Cardona, N. Koibuchi *et al.*, 'TRAM-1, A novel 160-kDa thyroid hormone receptor activator molecule, exhibits distinct properties from steroid receptor coactivator-1.' *J. Biol. Chem.*, 1997, **272**, 27629–27634.

28. J. Torchia, D. W. Rose, J. Inostroza *et al.*, 'The transcriptional co-activator p/CIP binds CBP and mediates nuclear-receptor function.' *Nature*, 1997, **387**, 677–684.

29. C. M. Chan, A. E. Lykkesfeldt, M. G. Parker *et al.*, 'Expression of nuclear receptor interacting proteins TIF-1, SUG-1, receptor interacting protein 140, and corepressor SMRT in tamoxifen-resistant breast cancer.' *Clin. Cancer Res.*, 1999, **5**, 3460–3467.

30. S. A. Onate, S. Y. Tsai, M. J. Tsai *et al.*, 'Sequence and characterization of a coactivator for the steroid hormone receptor superfamily.' *Science*, 1995, **270**, 1354–1357.

31. H. Shibata, T. E. Spencer, S. A. Onate *et al.*, 'Role of co-activators and co-repressors in the mechanism of steroid/thyroid receptor action.' *Recent Prog. Horm. Res.*, 1997, **52**, 141–164; discussion 164–165.

32. E. M. McInerney, M. J. Tsai, B. W. O' Malley *et al.*, 'Analysis of estrogen receptor transcriptional enhancement by a nuclear hormone receptor coactivator.' *Proc. Natl. Acad. Sci. USA*, 1996, **93**, 10069–10073.

33. A. Tremblay, G. B. Tremblay, F. Labrie *et al.*, 'Ligand-independent recruitment of SRC-1 to estrogen receptor beta through phosphorylation of activation function AF-1.' *Mol. Cell.*, 1999, **3**, 513–519.

34. P. Webb, P. Nguyen, J. Shinsako *et al.*, 'Estrogen receptor activation function 1 works by binding p160 coactivator proteins.' *Mol. Endocrinol.*, 1998, **12**, 1605–1618.

35. S. A. Onate, V. Boonyaratanakornkit, T. E. Spencer *et al.*, 'The steroid receptor coactivator-1 contains multiple receptor interacting and activation domains that cooperatively enhance the activation function 1 (AF1) and AF2 domains of steroid receptors.' *J. Biol. Chem.*, 1998, **273**, 12101–12108.

36. P. Alen, F. Claessens, G. Verhoeven *et al.*, 'The androgen receptor amino-terminal domain plays a key role in p160 coactivator-stimulated gene transcription.' *Mol. Cell. Biol.*, 1999, **19**, 6085–6097.

37. H. Ma, H. Hong, S. M. Huang *et al.*, 'Multiple signal input and output domains of the 160-kilodalton nuclear receptor coactivator proteins.' *Mol. Cell. Biol.*, 1999, **19**, 6164–6173.

38. R. T. Nolte, G. B. Wisely, S. Westin *et al.*, 'Ligand binding and co-activator assembly of the peroxisome proliferator-activated receptor-γ.' *Nature*, 1998, **395**, 137–143.

39. S. Westin, R. Kurokawa, R. T. Nolte *et al.*, 'Interactions controlling the assembly of nuclear-receptor heterodimers and co-activators.' *Nature*, 1998, **395**, 199–202.

40. E. M. McInerney, D. W. Rose, S. E. Flynn *et al.*, 'Determinants of coactivator LXXLL motif specificity in nuclear receptor transcriptional activation.' *Genes Dev.*, 1998, **12**, 3357–3368.

41. M. B. Kumar and G. H. Perdew, 'Nuclear receptor coactivator SRC-1 interacts with the Q-rich subdomain of the AhR and modulates its transactivation potential.' *Gene Expr.*, 1999, **8**, 273–286.

42. K. A. Sheppard, K. M. Phelps, A. J. Williams *et al.*, 'Nuclear integration of glucocorticoid receptor and nuclear factor-κB signaling by CREB-binding protein and steroid receptor coactivator-1.' *J. Biol. Chem.*, 1998, **273**, 29291–29294.

43. C. Leo and J. D. Chen, 'The SRC family of nuclear receptor coactivators.' *Gene*, 2000, **245**, 1–11.

44. B. Belandia and M. G. Parker, 'Functional interaction between the p160 coactivator proteins and the transcriptional enhancer factor family of transcription factors.' *J. Biol. Chem.*, 2000, **275**, 30801–30805.

45. J. J. Voegel, M. J. Heine, M. Tini *et al.*, 'The coactivator TIF2 contains three nuclear receptor binding motifs and mediates transactivation through CBP binding-dependent and -independent pathways.' *EMBO J.*, 1998, **17**, 507–519.

46. D. M. Heery, E. Kalkhoven, S. Hoare *et al.*, 'A signature motif in transcriptional co-activators mediates binding to nuclear receptors.' *Nature*, 1997, **387**, 733–736.

47. J. Leers, E. Treuter and J. A. Gustafsson, 'Mechanistic principles in NR box-dependent interaction between nuclear hormone receptors and the coactivator TIF2.' *Mol. Cell. Biol.*, 1998, **18**, 6001–6013.

48. B. Hanstein, R. Eckner, J. DiRenzo *et al.*, 'p300 is a component of an estrogen receptor coactivator complex.' *Proc. Natl. Acad. Sci. USA*, 1996, **93**, 11540–11545.

49. T. P. Yao, G. Ku, N. Zhou *et al.*, 'The nuclear hormone receptor coactivator SRC-1 is a specific target of p300.' *Proc. Natl. Acad. Sci. USA*, 1996, **93**, 10626–10631.

50. A. Takeshita, P. M. Yen, S. Misiti *et al.*, 'Molecular cloning and properties of a full-length putative thyroid hormone receptor coactivator.' *Endocrinology*, 1996, **137**, 3594–3597.

51. M. Ikeda, A. Kawaguchi, A. Takeshita *et al.*, 'CBP-dependent and independent enhancing activity of steroid receptor coactivator-1 in thyroid hormone receptor-mediated transactivation.' *Mol. Cell. Endocrinol.*, 1999, **147**, 103–112.

52. H. Chen, R. J. Lin, W. Xie *et al.*, 'Regulation of hormone-induced histone hyperacetylation and gene activation via acetylation of an acetylase.' *Cell*, 1999, **98**, 675–686.

53. J. Xu, Y. Qiu, F. J. DeMayo *et al.*, 'Partial hormone resistance in mice with disruption of the steroid receptor coactivator-1 (SRC-1) gene.' *Science*, 1998, **279**, 1922–1925.

54. R. E. Weiss, J. Xu, G. Ning *et al.*, 'Mice deficient in the steroid receptor co-activator 1 (SRC-1) are resistant to thyroid hormone.' *EMBO J.*, 1999, **18**, 1900–1904.

55. J. J. Voegel, M. J. Heine, C. Zechel *et al.*, 'TIF2, a 160 kDa transcriptional mediator for the ligand-dependent activation function AF-2 of nuclear receptors.' *EMBO J.*, 1996, **15**, 3667–3675.

56. H. Hong, K. Kohli, A. Trivedi *et al.*, 'GRIP1, a novel mouse protein that serves as a transcriptional coactivator in yeast for the hormone binding domains of steroid receptors.' *Proc. Natl. Acad. Sci. USA*, 1996, **93**, 4948–4952.

57. H. Hong, B. D. Darimont, H. Ma *et al.*, 'An additional region of coactivator GRIP1 required for interaction with the hormone-binding domains of a subset of nuclear receptors.' *J. Biol. Chem.*, 1999, **274**, 3496–3502.

58. J. Torchia, C. Glass and M. G. Rosenfeld, 'Co-activators and co-repressors in the integration of transcriptional responses.' *Curr. Opin. Cell Biol.*, 1998, **10**, 373–383.

59. C. S. Suen, T. J. Berrodin, R. Mastroeni *et al.*, 'A transcriptional coactivator, steroid receptor coactivator-3, selectively augments steroid receptor transcriptional activity.' *J. Biol. Chem.*, 1998, **273**, 27645–27653.

60. Z. Wang, D. W. Rose, O. Hermanson *et al.*, 'Regulation of somatic growth by the p160 coactivator p/CIP.' *Proc. Natl. Acad. Sci. USA*, 2000, **97**, 13549–13554.

61. J. Xu, L. Liao, G. Ning *et al.*, 'The steroid receptor coactivator SRC-3 (p/CIP/RAC3/AIB1/ACTR/TRAM-1) is required for normal growth, puberty, female reproductive function, and mammary gland development.' *Proc. Natl. Acad. Sci. USA*, 2000, **97**, 6379–6384.

62. S. Misiti, N. Koibuchi, M. Bei *et al.*, 'Expression of steroid receptor coactivator-1 mRNA in the developing mouse embryo: a possible role in olfactory epithelium development.' *Endocrinology*, 1999, **140**, 1957–1960.

63. Y. Zhu, C. Qi, S. Jain *et al.*, 'Isolation and characterization of PBP, a protein that interacts with peroxisome proliferator-activated receptor.' *J. Biol. Chem.*, 1997, **272**, 25500–25506.

64. C. Rachez, B. D. Lemon, Z. Suldan *et al.*, 'Ligand-dependent transcription activation by nuclear receptors requires the DRIP complex.' *Nature*, 1999, **398**, 824–828.

65. C. X. Yuan, M. Ito, J. D. Fondell *et al.*, 'The TRAP220 component of a thyroid hormone receptor-associated protein (TRAP) coactivator complex interacts directly with nuclear receptors in a ligand-dependent fashion.' *Proc. Natl. Acad. Sci. USA*, 1998, **95**, 7939–7944.

66. Y. Zhu, C. Qi, Y. Jia *et al.*, 'Deletion of PBP/PPARBP, the gene for nuclear receptor coactivator peroxisome proliferator-activated receptor-binding protein, results in embryonic lethality.' *J. Biol. Chem.*, 2000, **275**, 14779–14782.

67. M. Ito, C. X. Yuan, H. J. Okano *et al.*, 'Involvement of the TRAP220 component of the TRAP/SMCC coactivator complex in embryonic development and thyroid hormone action.' *Mol. Cell*, 2000, **5**, 683–693.

68. R. B. Lanz, N. J. McKenna, S. A. Onate *et al.*, 'A steroid receptor coactivator, SRA, functions as an RNA and is present in an SRC-1 complex.' *Cell*, 1999, **97**, 17–27.

69. E. Leygue, H. Dotzlaw, P. H. Watson *et al.*, 'Expression of the steroid receptor RNA activator in human breast tumors.' *Cancer Res.*, 1999, **59**, 4190–4193.

70. V. Cavailles, S. Dauvois, F. L'Horset *et al.*, 'Nuclear factor RIP140 modulates transcriptional activation by the estrogen receptor.' *EMBO J.*, 1995, **14**, 3741–3751.

71. M. B. Kumar, R. W. Tarpey and G. H. Perdew, 'Differential recruitment of coactivator RIP140 by Ah and estrogen receptors. Absence of a role for LXXLL motifs.' *J. Biol. Chem.*, 1999, **274**, 22155–22164.

72. N. Subramaniam, E. Treuter and S. Okret, 'Receptor interacting protein RIP140 inhibits both positive and negative gene regulation by glucocorticoids.' *J. Biol. Chem.*, 1999, **274**, 18121–18127.

73. E. Treuter, T. Albrektsen, L. Johansson *et al.*, 'A regulatory role for RIP140 in nuclear receptor activation.' *Mol. Endocrinol.*, 1998, **12**, 864–881.

74. S. H. Windahl, E. Treuter, J. Ford *et al.*, 'The nuclear-receptor interacting protein (RIP) 140 binds to the human glucocorticoid receptor and modulates hormone-dependent transactivation.' *J. Steroid. Biochem. Mol. Biol.*, 1999, **71**, 93–102.

75. T. N. Collingwood, O. Rajanayagam, M. Adams *et al.*, 'A natural transactivation mutation in the thyroid hormone beta receptor: impaired interaction with putative transcriptional mediators.' *Proc. Natl. Acad. Sci. USA*, 1997, **94**, 248–253.

76. A. Joyeux, V. Cavailles, P. Balaguer *et al.*, 'RIP 140 enhances nuclear receptor-dependent transcription *in vivo* in yeast.' *Mol. Endocrinol.*, 1997, **11**, 193–202.

77. C. H. Lee, C. Chinpaisal and L. N. Wei, 'Cloning and characterization of mouse RIP140, a corepressor for nuclear orphan receptor TR2.' *Mol. Cell. Biol.*, 1998, **18**, 6745–6755.

78. L. N. Wei, X. Hu, D. Chandra *et al.*, 'Receptor-interacting protein 140 directly recruits histone deacetylases for gene silencing.' *J. Biol. Chem.*, 2000, **275**, 40782–40787.

79. R. White, G. Leonardsson, I. Rosewell *et al.*, 'The nuclear receptor co-repressor nrip1 (RIP140) is essential for female fertility.' *Nat. Med*, 2000, **6**, 1368–1374.

80. K. S. Miyata, S. E. McCaw, L. M. Meertens *et al.*, 'Receptor-interacting protein 140 interacts with and inhibits transactivation by, peroxisome proliferator-activated receptor alpha and liver-X-receptor alpha.' *Mol. Cell. Endocrinol.*, 1998, **146**, 69–76.

81. F. M. Chuang, B. L. West, J. D. Baxter *et al.*, 'Activities in Pit-1 determine whether receptor interacting protein 140 activates or inhibits Pit-1/nuclear receptor transcriptional synergy.' *Mol. Endocrinol.*, 1997, **11**, 1332–1341.

82. J. C. Blanco, S. Minucci, J. Lu *et al.*, 'The histone acetylase PCAF is a nuclear receptor coactivator.' *Genes Dev.*, 1998, **12**, 1638–1651.

83. E. Korzus, J. Torchia, D. W. Rose *et al.*, 'Transcription factor-specific requirements for coactivators and their acetyltransferase functions.' *Science*, 1998, **279**, 703–707.

84. A. Imhof, X. J. Yang, V. V. Ogryzko *et al.*, 'Acetylation of general transcription factors by histone acetyltransferases.' *Curr. Biol.*, 1997, **7**, 689–692.

85. V. V. Ogryzko, T. Kotani, X. Zhang *et al.*, 'Histone-like TAFs within the PCAF histone acetylase complex.' *Cell*, 1998, **94**, 35–44.

86. H. Cho, G. Orphanides, X. Sun *et al.*, 'A human RNA polymerase II complex containing factors that modify chromatin structure.' *Mol. Cell. Biol.*, 1998, **18**, 5355–5363.

87. J. C. Chrivia, R. P. Kwok, N. Lamb *et al.*, 'Phosphorylated CREB binds specifically to the nuclear protein CBP.' *Nature*, 1993, **365**, 855–859.

88. R. Ravi, B. Mookerjee, Y. van Hensbergen *et al.*, 'p53-mediated repression of nuclear factor-κ B Rel A via the transcriptional integrator p300.' *Cancer Res.*, 1998, **58**, 4531–4536.

89. D. M. Scolnick, N. H. Chehab, E. S. Stavridi *et al.*, 'CREB-binding protein and p300/CBP-associated factor are transcriptional coactivators of the p53 tumor suppressor protein.' *Cancer Res.*, 1997, **57**, 3693–3696.

90. N. Sang, M. L. Avantaggiati and A. Giordano, 'Roles of p300, pocket proteins, and hTBP in E1A-mediated transcriptional regulation and inhibition of p53 transactivation activity.' *J. Cell. Biochem.*, 1997, **66**, 277–285.

91. M. Paulson, S. Pisharody, L. Pan *et al.*, 'Stat protein transactivation domains recruit p300/CBP through widely divergent sequences.' *J. Biol. Chem.*, 1999, **274**, 25343–25349.

92. S. Gingras, J. Simard, B. Groner *et al.*, 'p300/Cbp is required for transcriptional induction by interleukin-4 and interacts with Stat6.' *Nucleic Acids Res.*, 1999, **27**, 2722–2729.

93. V. Sartorelli, J. Huang, Y. Hamamori *et al.*, 'Molecular mechanisms of myogenic coactivation by p300: direct interaction with the activation domain of MyoD and with the MADS box of MEF2C.' *Mol. Cell Biol.*, 1997, **17**, 1010–1026.

94. P. L. Puri, M. L. Avantaggiati, C. Balsano *et al.*, 'p300 is required for MyoD-dependent cell cycle arrest and muscle-specific gene transcription.' *EMBO J.*, 1997, **16**, 369–383.

95. D. Chakravarti, V. J. LaMorte, M. C. Nelson *et al.*, 'Role of CBP/P300 in nuclear receptor signalling.' *Nature*, 1996, **383**, 99–103.

96. R. Eckner, M. E. Ewen, D. Newsome *et al.*, 'Molecular cloning and functional analysis of the adenovirus E1A-associated 300-kD protein (p300) reveals a protein with properties of a transcriptional adaptor.' *Genes Dev.*, 1994, **8**, 869–884.

97. N. J. McKenna, Z. Nawaz, S. Y. Tsai *et al.*, 'Distinct steady-state nuclear receptor coregulator complexes exist *in vivo.*' *Proc. Natl. Acad. Sci. USA*, 1998, **95**, 11697–11702.

98. M. J. Tetel, P. H. Giangrande, S. A. Leonhardt *et al.*, 'Hormone-dependent interaction between the amino- and carboxyl-terminal domains of progesterone receptor *in vitro* and *in vivo.*' *Mol. Endocrinol.*, 1999, **13**, 910–924.

99. W. L. Kraus and J. T. Kadonaga, 'p300 and estrogen receptor cooperatively activate transcription via differential enhancement of initiation and reinitiation.' *Genes Dev.*, 1998, **12**, 331–342.

100. T. P. Yao, S. P. Oh, M. Fuchs *et al.*, 'Gene dosage-dependent embryonic development and proliferation defects in mice lacking the transcriptional integrator p300.' *Cell*, 1998, **93**, 361–372.

101. Y. Tanaka, I. Naruse, T. Maekawa *et al.*, 'Abnormal skeletal patterning in embryos lacking a single cbp allele: a partial similarity with Rubinstein–Taybi syndrome.' *Proc. Natl. Acad. Sci. USA*, 1997, **94**, 10215–10220.

102. A. L. Kung, V. I. Rebel, R. T. Bronson *et al.*, 'Gene dose-dependent control of hematopoiesis and hematologic tumor suppression by CBP.' *Genes Dev.*, 2000, **14**, 272–277.

103. H. Kawasaki, R. Eckner, T. P. Yao *et al.*, 'Distinct roles of the co-activators p300 and CBP in retinoic-acid-induced F9-cell differentiation.' *Nature*, 1998, **393**, 284–289.

104. F. Petrij, R. H. Giles, H. G. Dauwerse *et al.*, 'Rubinstein–Taybi syndrome caused by mutations in the transcriptional co-activator CBP.' *Nature*, 1995, **376**, 348–351.

105. Y. Oike, A. Hata, T. Mamiya *et al.*, 'Truncated CBP protein leads to classical Rubinstein–Taybi syndrome phenotypes in mice: implications for a dominant-negative mechanism.' *Hum. Mol. Genet.*, 1999, **8**, 387–396.

106. P. Puigserver, Z. Wu, C. W. Park *et al.*, 'A cold-inducible coactivator of nuclear receptors linked to adaptive thermogenesis.' *Cell*, 1998, **92**, 829–839.

107. P. Puigserver, G. Adelmant, Z. Wu *et al.*, 'Activation of PPARγ coactivator-1 through transcription factor docking.' *Science*, 1999, **286**, 1368–1371.

108. R. B. Vega, J. M. Huss and D. P. Kelly, 'The coactivator PGC-1 cooperates with peroxisome proliferator-activated receptor alpha in transcriptional control of nuclear genes encoding mitochondrial fatty acid oxidation enzymes.' *Mol. Cell. Biol.*, 1868, **20**, 1868–1876.

109. M. Monsalve, Z. Wu, G. Adelmant *et al.*, 'Direct coupling of transcription and mRNA processing through the thermogenic coactivator PGC-1.' *Mol. Cell*, 2000, **6**, 307–316.

110. D. Knutti, A. Kaul and A. Kralli, 'A tissue-specific coactivator of steroid receptors, identified in a functional genetic screen.' *Mol. Cell. Biol.*, 2000, **20**, 2411–2422.

111. S. Yeh and C. Chang, 'Cloning and characterization of a specific coactivator, ARA70, for the androgen receptor in human prostate cells.' *Proc. Natl. Acad. Sci. USA*, 1996, **93**, 5517–5521.

112. T. Gao, K. Brantley, E. Bolu *et al.*, 'RFG (ARA70, ELE1) interacts with the human androgen receptor in a ligand-dependent fashion, but functions only weakly as a coactivator in cotransfection assays.' *Mol. Endocrinol.*, 1999, **13**, 1645–1656.

113. C. A. Heinlein, H. J. Ting, S. Yeh *et al.*, 'Identification of ARA70 as a ligand-enhanced coactivator for the peroxisome proliferator-activated receptor gamma.' *J. Biol. Chem.*, 1999, **274**, 16147–16152.

114. S. Yeh, H. Y. Kang, H. Miyamoto *et al.*, 'Differential induction of androgen receptor transactivation by different androgen receptor coactivators in human prostate cancer DU145 cells.' *Endocrine*, 1999, **11**, 195–202.

115. P. W. Hsiao and C. Chang, 'Isolation and characterization of ARA160 as the first androgen receptor N-terminal-associated coactivator in human prostate cells.' *J. Biol. Chem.*, 1999, **274**, 22373–22379.

116. A. M. Moilanen, U. Karvonen, H. Poukka *et al.*, 'A testis-specific androgen receptor coregulator that belongs to a novel family of nuclear proteins.' *J. Biol. Chem.*, 1999, **274**, 3700–3704.

117. J. D. Chen and R. M. Evans, 'A transcriptional corepressor that interacts with nuclear hormone receptors.' *Nature*, 1995, **377**, 454–457.

118. A. J. Horlein, A. M. Naar, T. Heinzel *et al.*, 'Ligand-independent repression by the thyroid hormone receptor mediated by a nuclear receptor co-repressor.' *Nature*, 1995, **377**, 397–404.

119. T. A. Jackson, J. K. Richer, D. L. Bain *et al.*, 'The partial agonist activity of antagonist-occupied steroid receptors is controlled by a novel hinge domain-binding coactivator L7/SPA and the corepressors N-CoR or SMRT.' *Mol. Endocrinol.*, 1997, **11**, 693–705.

120. R. Kurokawa, M. Soderstrom, A. Horlein *et al.*, 'Polarity-specific activities of retinoic acid receptors determined by a co-repressor.' *Nature*, 1995, **377**, 451–454.

121. R. M. Lavinsky, K. Jepsen, T. Heinzel *et al.*, 'Diverse signaling pathways modulate nuclear receptor recruitment of N-CoR and SMRT complexes.' *Proc. Natl. Acad. Sci. USA*, 1998, **95**, 2920–2925.

122. C. L. Smith, Z. Nawaz and B. W. O'Malley, 'Coactivator and corepressor regulation of the agonist/antagonist activity of the mixed antiestrogen, 4-hydroxytamoxifen.' *Mol. Endocrinol.*, 1997, **11**, 657–666.

123. C. D. Laherty, A. N. Billin, R. M. Lavinsky *et al.*, 'SAP30, a component of the mSin3 corepressor complex involved in N-CoR- mediated repression by specific transcription factors.' *Mol. Cell*, 1998, **2**, 33–42.

124. C. W. Wong and M. L. Privalsky, 'Transcriptional repression by the SMRT-mSin3 corepressor: multiple interactions, multiple mechanisms, and a potential role for TFIIB.' *Mol. Cell. Biol.*, 1998, **18**, 5500–5510.

125. Y. Zhang, Z. W. Sun, R. Iratni *et al.*, 'SAP30, a novel protein conserved between human and yeast, is a component of a histone deacetylase complex.' *Mol. Cell*, 1998, **1**, 1021–1031.

126. J. Zhang, M. G. Guenther, R. W. Carthew *et al.*, 'Proteasomal regulation of nuclear receptor corepressor-mediated repression.' *Genes Dev.*, 1998, **12**, 1775–1780.

127. B. Le Douarin, C. Zechel, J. M. Garnier *et al.*, 'The N-terminal part of TIF1, a putative mediator of the ligand-dependent activation function (AF-2) of nuclear receptors, is fused to B-raf in the oncogenic protein T18.' *EMBO J.*, 1995, **14**, 2020–2033.

128. B. Le Douarin, A. L. Nielsen, J. M. Garnier *et al.*, 'A possible involvement of TIF1 α and TIF1 β in the epigenetic control of transcription by nuclear receptors.' *EMBO J.*, 1996, **15**, 6701–6715.

129. B. Le Douarin, J. You, A. L. Nielsen *et al.*, 'TIF1α: a possible link between KRAB zinc finger proteins and nuclear receptors.' *J. Steroid Biochem. Mol. Biol.*, 1998, **65**, 43–50.

130. D. Chen, H. Ma, H. Hong *et al.*, 'Regulation of transcription by a protein methyltransferase.' *Science*, 1999, **284**, 2174–2177.

131. S. S. Koh, D. Chen, Y. H. Lee *et al.*, 'Synergistic enhancement of nuclear receptor function by p160

coactivators and two coactivators with protein methyltransferase activities.' *J. Biol. Chem.*, 2001, **276**, 1089–1098.

132. J. Tang, A. Frankel, R. J. Cook *et al.*, 'PRMT1 is the predominant type I protein arginine methyltransferase in mammalian cells.' *J. Biol. Chem.*, 2000, **275**, 7723–7730.

133. B. D. Darimont, R. L. Wagner, J. W. Apriletti *et al.*, 'Structure and specificity of nuclear receptor-coactivator interactions.' *Genes Dev.*, 1998, **12**, 3343–3356.

134. A. K. Shiau, D. Barstad, P. M. Loria *et al.*, 'The structural basis of estrogen receptor/coactivator recognition and the antagonism of this interaction by tamoxifen.' *Cell*, 1998, **95**, 927–937.

135. D. M. Heery, S. Hoare, S. Hussain *et al.*, 'Core LXXLL motif sequences in CBP, SRC1 and RIP140 define affinity and selectivity for steroid and retinoid receptors.' *J. Biol. Chem.*, 2000, **14**, 14.

136. J. D. Fondell, H. Ge and R. G. Roeder, 'Ligand induction of a transcriptionally active thyroid hormone receptor coactivator complex.' *Proc. Natl. Acad. Sci. USA*, 1996, **93**, 8329–8333.

137. Y. Ren, E. Behre, Z. Ren *et al.*, 'Specific structural motifs determine TRAP220 interactions with nuclear hormone receptors.' *Mol. Cell. Biol.*, 2000, **20**, 5433–5446.

138. J. D. Fondell, M. Guermah, S. Malik *et al.*, 'Thyroid hormone receptor-associated proteins and general positive cofactors mediate thyroid hormone receptor function in the absence of the TATA box-binding protein-associated factors of TFIID.' *Proc. Nat. Acad. Sci. USA*, 1999, **96**, 1959–1964.

139. C. Rachez, Z. Suldan, J. Ward *et al.*, 'A novel protein complex that interacts with the Vitamin D3 receptor in a ligand-dependent manner and enhances VDR transactivation in a cell-free system.' *Genes Dev.*, 1998, **12**, 1787–1800.

140. N. Chiba, Z. Suldan, L. P. Freedman *et al.*, 'Binding of liganded Vitamin D receptor to the Vitamin D receptor interacting protein coactivator complex induces interaction with RNA polymerase II holoenzyme.' *J. Biol. Chem.*, 2000, **275**, 10719–10722.

141. D. Burakov, C. W. Wong, C. Rachez *et al.*, 'Functional Interactions between the Estrogen Receptor and DRIP205, a Subunit of the Heteromeric DRIP Coactivator Complex.' *J. Biol. Chem.*, 2000, **275**, 20928–20934.

142. A. M. Naar, P. A. Beaurang, S. Zhou *et al.*, 'Composite co-activator ARC mediates chromatin-directed transcriptional activation.' *Nature*, 1999, **398**, 828–832.

143. S. Ryu, S. Zhou, A. G. Ladurner *et al.*, 'The transcriptional cofactor complex CRSP is required for activity of the enhancer-binding protein Sp1.' *Nature*, 1999, **397**, 446–450.

144. A. B. Hittelman, D. Burakov, J. A. Iniguez-Lluhi *et al.*, 'Differential regulation of glucocorticoid receptor transcriptional activation via AF-1-associated proteins.' *EMBO J.*, 1999, **18**, 5380–5388.

145. T. G. Boyer, M. E. Martin, E. Lees *et al.*, 'Mammalian Srb/Mediator complex is targeted by adenovirus E1A protein.' *Nature*, 1999, **399**, 276–279.

146. T. A. Nguyen, D. Hoivik, J. E. Lee *et al.*, 'Interactions of nuclear receptor coactivator/corepressor proteins with the aryl hydrocarbon receptor complex.' *Arch. Biochem. Biophys.*, 1999, **367**, 250–257.

147. A. Kobayashi, K. Numayama-Tsuruta, K. Sogawa *et al.*, 'CBP/p300 functions as a possible transcriptional coactivator of Ah receptor nuclear translocator (Arnt).' *J. Biochem. (Tokyo)*, 1997, **122**, 703–710.

148. A. Fujisawa-Sehara, K. Sogawa, M. Yamane *et al.*, 'Characterization of xenobiotic responsive elements upstream from the drug-metabolizing cytochrome P-450c gene: a similarity to glucocorticoid regulatory elements.' *Nucleic Acids Res.*, 1987, **15**, 4179–4191.

149. L. Wu and J. P. Whitlock, Jr., 'Mechanism of dioxin action: receptor-enhancer interactions in intact cells.' *Nucleic Acids Res.*, 1993, **21**, 119–125.

150. H. P. Ko, S. T. Okino, Q. Ma *et al.*, 'Dioxin-induced CYP1A1 transcription *in vivo*: the aromatic hydrocarbon receptor mediates transactivation, enhancer–promoter communication, and changes in chromatin structure.' *Mol. Cell. Biol.*, 1996, **16**, 430–436.

151. J. E. Morgan and J. P. Whitlock, Jr., 'Transcription-dependent and transcription-independent nucleosome disruption induced by dioxin.' *Proc. Nat. Acad. Sci. USA*, 1992, **89**, 11622–11626.

152. H. P. Ko, S. T. Okino, Q. Ma *et al.*, 'Transactivation domains facilitate promoter occupancy for the dioxin-inducible CYP1A1 gene *in vivo*.' *Mol. Cell. Biol.*, 1997, **17**, 3497–3507.

153. R. T. D. Dunn, T. S. Ruh, L. K. Burroughs *et al.*, 'Purification and characterization of an Ah receptor binding factor in chromatin.' *Biochem. Pharmacol.*, 1996, **51**, 437–445.

154. A. Kobayashi, K. Sogawa and Y. Fujii-Kuriyama, 'Cooperative interaction between AhR.Arnt and Sp1 for the drug-inducible expression of CYP1A1 gene.' *J. Biol. Chem.*, 1996, **271**, 12310–12316.

155. N. L. Ge and C. J. Elferink, 'A direct interaction between the aryl hydrocarbon receptor and retinoblastoma protein. Linking dioxin signaling to the cell cycle.' *J. Biol. Chem.*, 1998, **273**, 22708–22713.

156. S. K. Kolluri, C. Weiss, A. Koff *et al.*, 'p27 (Kip1) induction and inhibition of proliferation by the intracellular Ah receptor in developing thymus and hepatoma cells.' *Genes Dev.*, 1999, **13**, 1742–1753.

157. Y. Tian, S. Ke, M. S. Denison *et al.*, 'Ah receptor and NF-κB interactions, a potential mechanism for dioxin toxicity.' *J. Biol. Chem.*, 1999, **274**, 510–515.

158. C. Sakakura, A. Hagiwara, R. Yasuoka *et al.*, 'Amplification and over-expression of the AIB1 nuclear receptor co-activator gene in primary gastric cancers.' *Int. J. Cancer*, 2000, **89**, 217–223.

159. S. Bautista, H. Valles, R. L. Walker *et al.*, 'In breast cancer, amplification of the steroid receptor coactivator gene AIB1 is correlated with estrogen and progesterone receptor positivity.' *Clin. Cancer Res.*, 1998, **4**, 2925–2929.

160. Y. Zhu, C. Qi, S. Jain *et al.*, 'Amplification and overexpression of peroxisome proliferator-activated receptor binding protein (PBP/PPARBP) gene in breast cancer.' *Proc. Nat. Acad. Sci. USA*, 1999, **96**, 10848–10853.

161. S. K. Lee, S. L. Anzick, J. E. Choi *et al.*, 'A nuclear factor, ASC-2, as a cancer-amplified transcriptional coactivator essential for ligand-dependent transactivation by nuclear receptors *in vivo*.' *J. Biol. Chem.*, 1999, **274**, 34283–34293.

162. M. M. Montano, K. Ekena, R. Delage-Mourroux *et al.*, 'An estrogen receptor-selective coregulator that potentiates the effectiveness of antiestrogens and represses the activity of estrogens.' *Proc. Nat. Acad. Sci. USA*, 1999, **96**, 6947–6952.

163. J. D. Graham, D. L. Bain, J. K. Richer *et al.*, 'Nuclear receptor conformation, coregulators, and tamoxifen-resistant breast cancer.' *Steroids*, 2000, **65**, 579–584.

164. U. Mahlknecht and D. Hoelzer, 'Histone acetylation modifiers in the pathogenesis of malignant disease.' *Mol. Med.*, 2000, **6**, 623–644.

165. R. H. Giles, D. J. Peters and M. H. Breuning, 'Conjunction dysfunction: CBP/p300 in human disease.' *Trends Genet.*, 1998, **14**, 178–183.

166. J. Font de Mora and M. Brown, 'AIB1 is a conduit for kinase-mediated growth factor signaling to the estrogen receptor.' *Mol. Cell. Biol.*, 2000, **20**, 5041–5047.

167. S. H. Hong and M. L. Privalsky, 'The SMRT corepressor is regulated by a MEK-1 kinase pathway: inhibition of corepressor function is associated with SMRT phosphorylation and nuclear export.' *Mol. Cell. Biol.*, 2000, **20**, 6612–6625.

J.P. Vanden Heuvel, G.H. Perdew, W.B. Mattes and W.F. Greenlee (Eds.)
Comprehensive Toxicology, Vol. xiv

# 14.2.9
# Convergence of Multiple Nuclear Receptor Signaling Pathways in the Mediation of Xenobiotic-Induced Nongenotoxic Carcinogenesis

DANIEL J. NOONAN, WILLIAM R. HOWARD, XIAOJIE YUAN and MICHELLE L. O'BRIEN
*University of Kentucky, Lexington, KY, USA*

### 14.2.9.1   INTRODUCTION

Peroxisome proliferator-induced rodent hepatocarcinogenesis is a well-established mechanism for nongenotoxic carcinogenesis. The large list of structurally diverse compounds that are able to induce proliferation of peroxisomes in rodent livers includes hypolipidemic drugs, environmental pollutants, analgesics, phthalates and dietary fats.[1,2] Although these compounds appear to not interact directly with DNA, long term exposure to them in rodents results in the development of hepatocellular carcinomas.[2-4] Over the past decade significant progress has been made in the elucidation of how these compounds stimulate rodent hepatocarcinogenesis.[5-8] In this chapter we will review some of this work along with a presentation of evolving data that suggests this process may be substantially more complex than originally hypothesized.

The primary mediator of peroxisome proliferator-induced rodent hepatocarcinogenesis is a member of the steroid/nuclear receptor family of ligand-induced transcription factors. This receptor, the peroxisome proliferator activated receptor alpha (PPARα) has been shown to directly bind several of the classical peroxisome proliferators and in turn stimulate the production of peroxisomal-specific enzymes. Over the past decade the mechanism of PPARα activation and its association with rodent hepatocarcinogenesis has been extensively studied. This chapter will not discuss in any great detail steroid receptor-mediated transcription mechanisms, the PPAR family of receptors or PPAR-specific gene transcription. The readers are referred to Chapter 14.02.3 of this text along with several other recent reviews on these subjects.[9-13] The focus of this chapter will be on accumulated evidence that suggests peroxisome proliferators mediate their toxicity through multiple signaling pathways, the consequences of which can result in serious metabolic disturbances in both rodents and humans. Through the science presented in this chapter it will hopefully become clear that although we have made significant advances in our understanding of how peroxisome prolif-

erators might be altering normal metabolic processes and how PPAR might be involved in this alteration of metabolism, we still have much to learn about the molecular consequences of these alterations.

### 14.2.9.2   NONGENOTOXIC PEROXISOME PROLIFERATOR-INDUCED RODENT HEPATOCARCINOGENESIS

Peroxisomes are ubiquitous cytoplasmic organelles found in animals, plants, fungi and protozoa, and are defined by their role in the metabolism of hydrogen peroxide.[14] Peroxisomes functionally compartmentalize cellular β-oxidation reactions, and the oxidative enzymes found in peroxisomes are involved in a variety of metabolic pathways, including: respiration, lipid metabolism, cholesterol metabolism, and gluconeogenesis. Hess and co-workers[15] were the first to report the proliferation of peroxisomes in rat livers following administration of the hypolipidemic drug clofibrate. Soon after, a variety of other structurally unrelated hypolipidemic agents were identified that possessed peroxisomal proliferation properties, thus suggesting that lipid metabolism was central to peroxisomes and peroxisome proliferation. In rodents, fat metabolism has been directly linked to chemically induced hepatocarcinogenesis. Furthermore, this hepatocarcinogenesis can be closely correlated with the proliferation of liver peroxisomes as induced by the aforementioned hypolipidemic agents (collectively termed peroxisome proliferators).[5-8] Although these peroxisome proliferators appear to act as hepatocarcinogens, they do not show detectable mutagenic or genotoxic activity in suitable test systems.[3] This lack of a direct relationship with mutagenesis compartmentalizes peroxisome proliferator-induced rodent hepatocarcinogenesis into the classification of nongenotoxic carcinogenesis.

## 14.2.9.3 OXIDATIVE STRESS AS A MEDIATOR OF PEROXISOME PROLIFERATOR-INDUCED RODENT HEPATOCARCINOGENESIS

Carcinogenesis can be most simplistically viewed as a 2-step process involving DNA mutation or tumor initiation, and immortalization of the initiation events through tumor promotion (e.g., modulation of cell cycle and/ or apoptosis pathways). The associations between chemically induced peroxisome proliferation and rodent hepatocarcinogenesis brought with it several hypothetical mechanisms for explaining their relationship. These include a substrate overload-perturbation of metabolism hypothesis presented by Lock *et al.*[16] This hypothesis proposes that substrate overload in the form of long chain dicarboxylic fatty acids (LCDCA) results in the activation of peroxisomal enzymes which in turn mediate the growth and proliferation of peroxisomes. This hypothesis is consistent with the role of peroxisome proliferation with lipid metabolism, but provided only weak hypothetical links to tumor initiation and tumor promotion.

Studying nongenotoxic carcinogen induced uncoupling of mitochondrial oxidative phosphorylation, Keller *et al.* demonstrated a direct correlation between tumorigenicity of peroxisome proliferators and their ability to uncouple mitochondrial oxidative phosphorylation.[17] From theirs and other supporting data they proposed that uncoupling mitochondrial oxidative phosphorylation leads to a decline in cellular ATP resulting in an elevation of free fatty acids. Elevated free fatty acids were in turn hypothesized to activate protein kinase C, leading to replicative DNA synthesis and tumor promotion. These data are consistent with linkage of potential tumor promotion pathways and peroxisome proliferator activities, but does not address tumor initiation mechanisms.

The most compelling early hypothesis linking peroxisome proliferators, peroxisomal fat metabolism and rodent hepatocarcinogenesis was that of Reddy and coworkers, who suggested that $H_2O_2$ buildup as a result of excessive peroxisomal-specific β-oxidation of fatty acids results in the production of oxygen and hydroxyl free radicals.[6,8,18,19] This hypothesis, coined the "oxidative stress hypothesis", predicts free radical DNA damage as the principle mechanism for tumor initiation. The free radicals generated as a result of these pathways were also proposed to lead to DNA strand breakage and tumor development. Exactly how this mechanism might lead to tumor immortalization was unclear, but insightfully, these investigators suggested peroxisomal proliferators mediate their effects through a ligand-receptor mechanism, in which a chemical inducer or its metabolite binds to a specific receptor and activates genes linked to peroxisomal-specific oxidation pathways.

## 14.2.9.4 PEROXISOME PROLIFERATOR-ACTIVATED RECEPTOR ALPHA (PPARα) AS A MEDIATOR OF OXIDATIVE STRESS AND PEROXISOME PROLIFERATOR-INDUCED RODENT HEPATOCARCINOGENESIS

The early hypothesis of Reddy and coworkers received experimental ratification in the 1990s with the identification of an intracellular protein that could bind peroxisome proliferators and regulate the transcription of peroxisomal-specific fatty acid metabolism genes. In 1990, Issemann and Green reported the cloning of a mouse steroid hormone receptor family member (peroxisome proliferator-activated receptor: mPPAR) that could be activated by peroxisome proliferators in an *in vitro* transcription assay.[20] The steroid hormone family of receptors consists of a group of ligand-activated DNA transcription factors that bind regulatory sequences upstream of their target gene(s) resulting in the activation or repression of specific gene transcription.[10,21] Subsequently it was shown that PPAR is a small family of genes, with reports of at least α, β and γ isoforms in mouse,[22,23] Xenopus,[24] rat[25] and humans.[26–28] An examination of PPAR regulatable promoters suggests this receptor family is intimately involved in fat metabolism including their breakdown,.[29,30] storage[31] and synthesis.[32] DNA binding by PPAR family members has been shown to be contingent upon both heterodimerization with a member of the retinoid X receptor (RXR) family[33,34] and a restrictively defined DNA sequence motif (most often a 6 base pair imperfect direct repeat).[29,30,35,36] More recent evidence suggests that the specificity of PPAR transcriptional activation may also be contingent upon coregulatory molecules[37–40] and phosphorylation.[41–48]

It soon became apparent that the three isoforms of PPAR performed distinct functions within higher eukaryotes. PPARα was observed to be widely distributed but with greatest expression in highly exergonic tissues (e.g., liver, kidney, heart, and gut).[20,24,49–51] The genes identified as regulated by PPARα provide the major insight into its function. PPARα was observed to bind to and regulate the transcription of genes for peroxisomal-specific β-oxidation of fatty acids,[29,30,35,36] the intracellular transport of lipids,[52] the synthesis of cholesterol[32,53–55] and the synthesis of bile acids.[56,57] The PPARγ isoform on the other hand was found to be more restricted in its tissue distribution with its highest expression in adipose tissue[31,51,58,59] suggesting an important role in the storage of fats. This contention was further supported by the identification of PPARγ-specific regulatory sequences in the promoters of the genes for lipoprotein lipase,[60,61] fatty acid transport protein[55] and fatty acid translocase.[62] Finally, the third member of the PPAR family, PPARβ/δ, is expressed at low levels in a wide range of tissues including heart, adipose, brain, intestine, muscle, spleen, and lung.[51,58,63] Very little is known about its true function in higher eukaryotes, although recent reports have implicated it in brain function,[64] female reproductive processes[65] and colorectal cancer.[66]

With the accumulated evidence for PPARα regulation of peroxisomal-specific fat metabolism, coupled with the oxidative stress hypothesis for peroxisome proliferator induced rodent hepatocarcinogenesis is consistent with the observed effects of peroxisome proliferators. Figure 1 incorporates our current perspectives on PPARα gene regulation into an oxidative stress-based molecular mechanism for hepatocarcinogenesis. This hypothesis proposes that the stimulation of PPARα by peroxisome proliferators can result in the enhanced production of enzymes for the peroxisomal-specific β-oxidation of fatty acids. This event generates $H_2O_2$, which, under normal conditions, is rapidly neutralized into $O_2$ and $H_2O$ by the enzyme catalase. In conditions of chronic exposure to peroxisome proliferators, it is hypothesized that excess and unneutralized $H_2O_2$ would be converted to $O^-$ and $OH^-$ free radicals, both of which are capable of modifying the structure of DNA, proteins and lipids. Oxidative stress-induced mutations to key cell cycle genes, proto-oncogenes, and genes regulating apoptotic events are ultimately proposed to both initiate and immortalize the cellular changes responsible for hepatocarcinogenesis. Perhaps the most compelling data supporting this hypothesis, or at least the PPARα component of the hypothesis, are the results from PPARα null mice.[67–69] Disruption of the gene for PPARα in a transgenic mouse model system specifically eliminates chemical induction of peroxisome proliferation and the susceptibility of these mice to peroxisome proliferator-induced rodent hepatocarcinogen-esis. From these data it can be inferred that PPARα mediated activities are required for the pathology of peroxisome proliferator-induced rodent hepatocarcinogenesis. Logically this would also implicate peroxisomal-specific fat catabolism as an essential component of the carcinogenesis pathway. Beyond these considerations, there is only circumstantial evidence to support an oxidative stress mechanism for tumor initiation and even less evidence for implicating oxidative stress in tumor promotion.

## 14.2.9.5 PEROXISOMES AS MEDIATORS OF ISOPRENOID AND CHOLESTEROL SYNTHESIS

In higher eukaryotes, cholesterol can be derived from both diet and intracellular synthesis. Cellular cholesterol synthesis has been extensively studied over the past 40 years and involves the assembly of 18 two carbon acetyl-CoAs into the 27 carbon cholesterol structure (Figure 2(A)). Although initially regarded as primarily a center for long chain fatty acid β-oxidation, it has become clear that peroxisomes and the enzymes within peroxisomes also play a central role in cholesterol metabolism and steroidigenesis.[70–73] The evolving picture for cholesterol synthesis (Figure 2(B)) appears to require a combination of cytosolic, endoplasmic reticulum (ER) and peroxisomal localized reactions. Enzymes for the conversion of acetyl-CoA into mevalonate have been identified in both cytosol/ER and peroxisomal compartments.[74–76] The observation that HMG-CoA reductase, considered the rate-limiting enzyme in cholesterol synthesis, is localized to both compartments and responds differentially to regulators of cholesterol synthesis[74,77] would suggest that the compartmentalization of these enzymes might also provide a new layer of regulation in cholesterol synthesis. Perhaps one of the more intriguing observations to come out of the new research on subcellular localization of cholesterol synthesis were those of the Krisans *et al.* These investigators, utilizing refined protocols for purification of peroxisomes away from microsomes, identified the majority if not all of the enzyme activity for converting mevalonate into the 15 carbon farnesyl pyrophosphate (FPP), as residing exclusively in the peroxisomes.[71] Interestingly, little if any squalene synthetase activity, responsible for converting two 15 carbon FPPs into the 30 carbon squalene and subsequent cyclization of this structure into lanosterol, can be identified in peroxisomes.

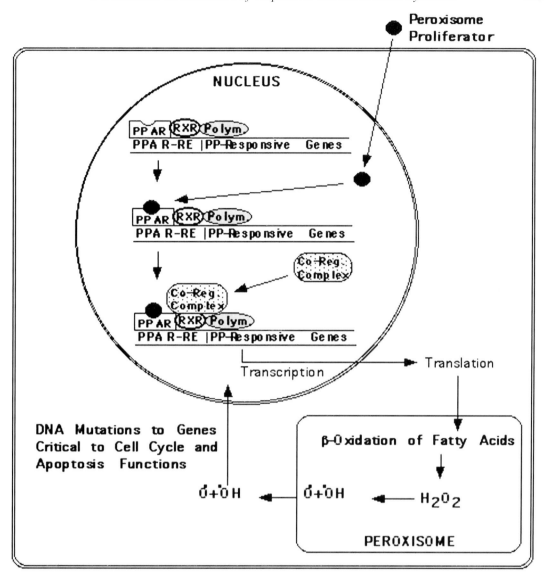

**Figure 1** PPARα as a mediator of oxidative stress and peroxisome proliferator stimulated rodent hepatocarcinogenesis. The current model for peroxisome proliferator stimulated rodent hepatocarcinogenesis has peroxisome proliferators binding the nuclear receptor complex of PPARα and its heterodimer partner RXR. As I have presented it here PPARα and RXR are pre-bound to the promoter region of a peroxisome proliferator-responsive gene. However, the only real evidence for this are *in vitro* studies that suggest PPAR and RXR can form complexes and bind DNA in the absence of exogenously added ligands. Ligand binding of the nuclear receptor complex is thought to facilitate the recruitment of specific co-activators (Co-Reg. Complex) and other proteins required for the activation of RNA polymerase (Polym.). PPARα has been shown to upregulate the transcription of genes involved in the peroxisomal-specific β-oxidation of fatty acids, a process known to generate $H_2O_2$. Chronic stimulation of this system by peroxisome proliferators is hypothesized to saturate available controls for neutralizing the $H_2O_2$, resulting in the formation of mutagenic $O^-$ and $OH^-$ free radicals. Mutations to genes critical to cell cycle control and/or apoptosis by these free radicals is hypothesized to result in the ultimate initiation and immortalization events required for tumor formation.

Finally, enzymes for conversion of lanosterol into cholesterol have been identified in both the cytosolic/ER compartment and the peroxisome. These data suggest that peroxisomes play an essential role in the conversion of mevalonate into FPP, which must subsequently be transported to the cytosolic/ER compartment for conversion to lanosterol. Lanosterol,

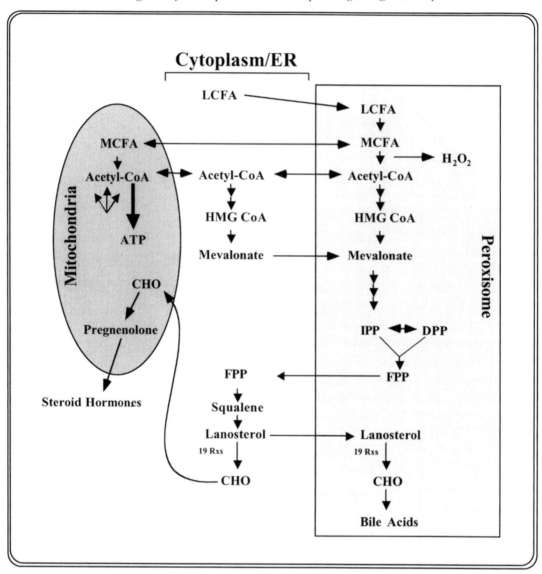

**Figure 2** Peroxisomes play an essential role in the synthesis of cholesterol. Peroxisome deficiency diseases have identified these organelles as essential for conversion of long chain fatty acids (LCFA) into medium chain fatty acids (MCFA). Subsequently, peroxisomes were also shown to contain the enzymes and capacity to completely break down LCFAs into 2 carbon acetyl-CoAs. Classically, cholesterol synthesis has been interpreted to occur primarily in the cytoplasm and endoplasmic reticulum (ER). Recent improvements in the purification and fractionation of cytoplasmic, microsomal and peroxisomal compartments have identified a major and essential role for peroxisomes in the cholesterol biosynthetic pathway. The data at this time suggest synthesis of mevalonate from acetyl-CoA can occur in both the cytosolic/ER and peroxisome compartments. Conversion of mevalonate into the isoprenes, isopentenyl pyrophosphate (IPP) and dimethylallyl pyrophosphate (DPP), as well as their condensation into the 15-carbon farnesyl pyrophosphate (FPP) is thought to occur exclusively in the peroxisomes. Conversely, the condensation of 2 FPPs into squalene and its subsequent cyclization into lanosterol is thought to occur exclusively in the cytosolic/ER compartment. Finally, the subsequent series of reactions converting lanosterol into cholesterol (CHO) have been shown to occur in both compartments. Also noted in this diagram are the essential roles the peroxisomes and mitochondria play in the conversion of cholesterol into bile acids and steroid hormones respectively.

in turn, may be transported back into the peroxisome for conversion to cholesterol or further processed into cholesterol in the cytosolic/ER compartment. The reason for this trafficking and compartmentalization of cholesterol synthesis is yet unclear, but it may serve as a mechanism for regulation of primarily anabolic cholesterol events (e.g., steroid biosynthesis), which utilize mitochondria and other cytoplasmic compartments for side chain cleavage and specific hydroxylations, versus the peroxisomes which appear to be centers for bile acid synthesis and subsequent conjugation with specific amino acids.[78] This latter event is often equated with cholesterol catabolism and excretion. The geographical division of labor while maintaining critical dependencies on both compartments may in turn coordinate signaling events for fat/cholesterol catabolism and fat/cholesterol synthesis, for it can also be conjectured that regulatory pathways influencing cytosolic cholesterol synthesis (e.g., endogenous HMG-CoA reductase modulators) might not necessarily impact peroxisomal cholesterol synthesis in a like manner. The accumulating evidence that peroxisomes are intimately involved in steroidigenesis,[73] ubiquinone synthesis[79,80] and bile acid synthesis[56,78,81] fits well with the concept that peroxisomes compartmentalize the breakdown of fatty acids into acetyl-CoA, the precursor for all of these cholesterol pathway byproducts. Evidence supporting this concept includes: (1) peroxisome proliferators stimulate bile acid synthesis in rodents and humans;[82–86] (2) peroxisome proliferators increase ubiquinone synthesis in rat hepatocytes;[80,87] and (3) in rodents, *in vivo* administration of peroxisome proliferators upregulates both the synthesis of cholesterol byproducts as well as the activity of HMG-CoA reductase.[85] These data, along with corroborative data associating peroxisomal dysfunctions with steroidigenesis (reviewed in Reference 73), strongly implicate peroxisomes in the synthesis of isoprenoids.

## 14.2.9.6 PPARα AS A MEDIATOR OF PEROXISOMAL ISOPRENOID SYNTHESIS

The first suggestions of PPARα involvement in cholesterol synthesis are implied in the identification of PPARα responsive promoters in the rat peroxisomal-specific thiolase gene and the rat mitochondrial HMG CoA synthetase gene.[29,53,54] These two enzymes catalyze the condensation of three molecules of acetyl-CoA into the six carbon 3-hydroxy-3-methyglutaryl coenzyme A (HMG-CoA) molecule, a precursor in the synthesis of cholesterol. Interestingly, the theoretical initial step in cholesterol synthesis is mediated by the same enzyme (thiolase) responsible for the final step in peroxisome-specific fatty acid breakdown. PPARα involvement was further suggested in studies by O'Brien and coworkers[32] wherein statin inhibitors of HMG-CoA reductase, the rate limiting step in cholesterol synthesis, were shown to substantially reduce the transcrip-

tional activity of PPARα (Figure 3(A)). Using an *in vitro* co-transfection assay containing a rPPARα expression construct and an acyl-CoA oxidase response element-driven reporter vector, it was observed that peroxisome proliferator-induced PPARα activity could be repressed by the HMG-CoA reductase inhibitor lovastatin, and that this repression could be overcome by the addition of mevalonate. These investigators went on to further demonstrate that the isoprenoid precursor of cholesterol, farnesol, could, in a stereo-specific manner upregulate PPARα transcriptional activation (Figure 3(B)). These data strongly implicate some cholesterol pathway byproduct in the endogenous activation of PPARα. Interestingly, of the variety of endogenous lipophilic activators of PPARα identified thus far, none have a sterol structure. This would suggest that either a cholesterol-based endogenous activator of PPARα is yet to be identified or that farnesol (most probably in the form of a CoA ester of farnesoic acid) serves a central role in coordinating PPARα activities with the state of lipid metabolism in these cells.

## 14.2.9.7 ISOPRENOIDS AS STIMULATORS OF FARNESOID X-ACTIVATED RECEPTOR (FXR)

The observation that farnesol could serve as an activator of PPARα interestingly overlapped with published data on another member of the steroid/nuclear receptor family, farnesoid X-activated receptor. This receptor, as first reported by Forman *et al.*,[88] was shown not only to share this common activation pathway but also to localize to tissues in which PPARα was abundantly expressed (liver, kidney and gut) and to utilize RXR as a heterodimer partner in transcriptional activation. Although the receptor activation pathways were similar in these respects, FXR was shown to distinguish itself from PPARα by both its activator specificity and the program of genes it regulated (Figure 4). This statement is supported by the observations that: (1) classical peroxisome proliferator activators of PPARα could not activate FXR-mediated transcription; (2) the FXR activating compound, juvenile hormone, could not activate PPARα-mediated transcription; (3) the acyl-CoA-oxidase-derived DNA binding response element motif utilized by PPARα could not function in FXR-mediated transcription events; and (4) the *Drosophila hsp27*-derived DNA binding motif used by FXR[89] could not function in PPARα-mediated transcription

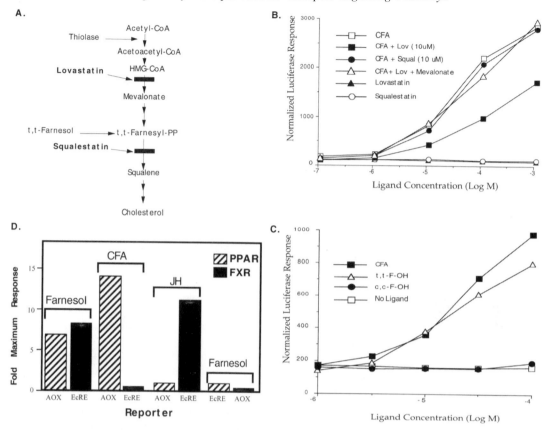

**Figure 3** The cholesterol synthesis pathway produces endogenous activators of PPARα and farnesoid X receptor (FXR). Selective inhibitors of the cholesterol synthesis pathway (**A**) were analyzed for their ability to modulate PPARα-mediated transcription (**B**). rPPARα and β-galactosidase expression plasmids, and an acyl-CoA oxidase promoter region driven luciferase reporter plasmid (AOX-RE) were transfected into CV-1 cells and assayed for the effects Lovastatin (Lov), Squalestatin (Squal) and mevalonate + Lov had on clofibric acid (CFA) stimulation of PPARα-mediated transcription. (**C**) In a similar assay, farnesol was analyzed for its stereo-specific activation of PPARα-mediated transcription. rPPARα and β-galactosidase expression plasmids, and an AOX-RE driven luciferase reporter plasmid were transfected into CV-1 cells and the naturally occurring t,t-farnesol (t,t-F-OH) and a plant derived c,c-farnesol (c,c-F-OH) stereoisomer were assayed for stimulation of PPARα-mediated transcription. Activity is expressed as the average of triplicate luciferase responses normalized to their respective β-galactosidase rate. Finally, in (**D**), rPPARα or FXR mammalian expression plasmids, a β-galactosidase expression plasmid and/or a designated luciferase reporter construct [AOX-RE or an ecdysone receptor response element driven reporter plasmid (EcRE)] were transfected into CV-1 cells and analyzed for t,t-farnesol (F-OH), CFA, and juvenile hormone (JH) stimulation of receptor activities. Activity is expressed as fold stimulation over reporter alone activation for the respective ligands. Adapted with permission from O'Brien *et al.*[33]

events. These data would suggest farnesol serves as a common intermediate for distinct endogenous activators of both PPARα and FXR, but they also suggest that stimulation of PPARα-directed transcription events can impact transcription mediated by FXR.

### 14.2.9.8 CONVERGENCE OF PPARα, FXR AND RETINOID X RECEPTOR (RXR) PATHWAYS

With this as a foundation, O'Brien *et al.* in 1996[32] merged FXR into the PPARα-based mechanism for rodent hepatocarcinogenesis

(Figure 4). In this hypothesis, PPARα activation of oxidative catabolism of fatty acids is proposed to generate excess acetyl-CoA, driving the overexpression of FPP, which subsequently feeds into the activation pathways for both PPARα and FXR. A classical biochemist might necessarily be skeptical about this hypothesis. The cholesterol synthesis pathway is one of the most regulated pathways known, and at first glance this hypothesis would appear counterintuitive. Compounding this is the observation that mevalonate and sterols have been shown to activate a cytoplasmic protease capable of inactivating ER-bound HMG-CoA reductase, the rate-limiting enzyme in farnesol

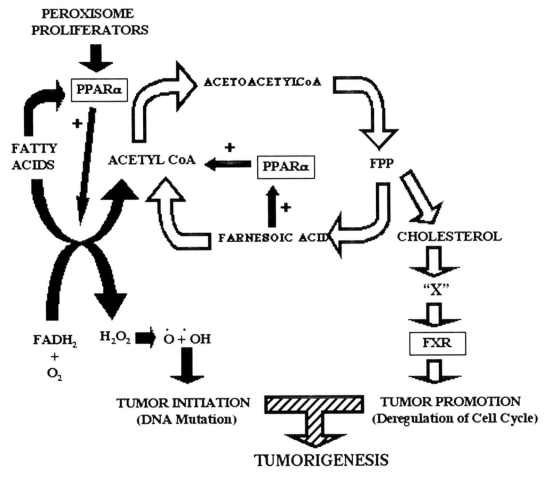

**Figure 4** Marrying FXR into the hypothetical mechanism PPARα-mediation of for peroxisome proliferator-induced rodent hepatocarcinogenesis. The filled in black arrows represent the well-established PPARα component of the "oxidative stress" hypothesis for peroxisome proliferator-induced rodent hepatocarcinogenesis. Chronic peroxisome proliferator stimulation of PPARα results in an increase in β-oxidation of fatty acids, the generation of $H_2O_2$ and the production of acetyl-CoA. Excess and unneutralized $H_2O_2$ results in the generation of $O^-$ and $OH^-$ free radicals capable of mutating DNA and tumor initiation. In an extension of this hypothesis, (white arrows) excess acetyl CoA or acetoacetyl CoA is seen to drive the formation of farnesyl pyrophosphate (FPP). Excess FPP is viewed in this model to have duel consequences on nuclear receptor signaling pathways. One being its conversion to farnesoic acid, which in an autocatalytic fashion can re-activate PPARα, and the other is its conversion to a cholesterol-based activator ("X") of FXR. Upregulation

synthesis.[90] But a closer examination of this biochemical pathway reveals that classical views of cholesterol synthesis, primarily derived from the cytoplasmic and ER localized enzymes and intermediates, may not fully apply here. As described above, it is now clear that HMG-CoA synthesis can also be compartmentalized to the peroxisome and these enzymes and intermediates are, logistically at least, under different constraints and regulatory mechanisms than those used by the pathway found outside of the peroxisome. Yet, it is safe to predict that even the compartmentalization of these processes by peroxisomes would, under normal metabolic situations, be relegated to rigorous molecular controls. Alternatively, the compartmentalization of this pathway together with its precursor fatty acid β-oxidation pathway, might, in a scenario of chronic exposure to PPARα activators, be expected to result in saturation of available controls and the upregulation of FXR activators. The consequence of this alternative upreg ulation of FXR-mediated events is predicted by the hypothesis to result in tumor promotion of the mutagenic environment created by the PPARα-generated oxidative stress. The key to validation of this hypothesis rests in (a) demonstrating PPARα can modulate cholesterol synthesis; (b) identifying a cholesterol-based ligand for FXR; and (c) demonstrating that this FXR ligand can be associated with tumor promotion events.

### 14.2.9.9 PPARα CAN REGULATE THE CHOLESTEROL SYNTHESIS PATHWAY

Although the aforementioned studies by O'Brien *et al.*, demonstrating inhibitors of cholesterol can modulate PPAR-mediated transcription events, are the only direct associations of PPARα with cholesterol synthesis, there are a variety of circumstantial associations of PPARα activators with steroidigenesis. In animal model systems peroxisome proliferators have been reported to cause male and female reproductive dysfunction,[91–95] inhibition of testosterone release from leydig cells,[96] alteration of estradiol secretion and serum levels[93,97–99] and a reduction in the hepatic activity of two major male estrogen-metabolizing enzymes.[99] The important role of PPAR in this event is suggested by studies with PPARα knockout mice wherein it was observed that they express substantially altered levels of sex steroids when placed in a lipid restricted environment.[68] In humans, male patients receiving fibrate drugs report an intriguingly similar impotence[100] to that reported for male workers exposed to estrogens in pharmaceutical plants.[101] Furthermore, this impotence effect reversed after interruption of therapy and reoccurred upon resumption of therapy. Reduced fertility has also been reported among women exposed to peroxisome proliferator class industrial chemicals.[102] It should also be pointed out that impotency in both males and females is substantially higher in the United States than in less industrialized countries where exposure to these commercial peroxisome proliferators is minimal.[101,103] These data imply that peroxisome proliferators, presumably through their activities on PPARα, can seriously alter normal human steroidigenic pathways.

### 14.2.9.10 IDENTIFICATION OF A CHOLESTEROL-BASED LIGAND FOR FXR

The data on farnesol activation of FXR implies that either farnesol or some metabolite of farnesol serves as an endogenous ligand for FXR. This, together with the observation that FXR expression is limited to the classical steroidigenic tissues, leads to the identification of a variety of endogenous activators of FXR. In late 1999, three groups published data demonstrating that bile acids could bind to FXR and stimulate FXR-mediated transcription.[104–106] Furthermore, it was demonstrated that FXR could modulate transcription of the key enzyme in bile acid synthesis, cholesterol 7α hydroxylase (CYP7A),[104,106] as well as the gene for a primary mediator of bile acid transport in the intestine, ileal bile acid binding protein (I-BABP).[104] These data were expanded upon by Howard *et al.*[107] who demonstrated that, in addition to the classical bile acids deoxycholate (DCA) and chenodeoxycholate (CDCA), FXR was transcriptionally active in the presence of the androgen catabolites 5α-androstan-3α-ol-17-one (androsterone) and 5β-androstan-3α-ol-17-one (etiocholanolone), as well as the sterol bronchodilatory drug forskolin (Figure 5A). Furthermore, these investigators presented the intriguing observation that the hydrophilic bile acid ursodeoxycholate (UDCA) could inhibit DCA and CDCA transcriptional activation of FXR in a manner parallel to its ability to antagonize DCA and CDCA induction of apoptosis (Figure 5B). By far, the most efficacious activator of FXR reported thus far was forskolin. Interestingly, although forskolin is classically viewed as an initiator of the adenylate cyclase/protein kinase A (PKA) pathway, PKA inhibition did not inhibit forskolin's activation of FXR, nor was cyclic AMP (cAMP) able to stimulate FXR-mediated transcription. These data suggest forskolin either acts directly as a ligand for FXR, or functions in some other indirect pathway that has yet to be identified.

Data confirming the importance of FXR in bile acid metabolism were generated in studies of FXR knockout mice.[108] Mice lacking FXR were observed to develop normally but could be distinguished from wild type mice by the presence of elevated serum bile acids, cholesterol and triglycerides, as well as increased hepatic cholesterol and triglycerides. Furthermore, it was observed that these mice were no longer able to regulate the expression of both the CYP7A and I-BABP genes. These data cumulatively support the contention that a cholesterol metabolite serves as a primary activator of FXR-mediated transcription events.

If accurate, the oxidative stress model presented above (Figure 3) would predict peroxisome proliferators, as well as PPARα, modulate the production of bile acids. It is well established that peroxisomes are major sites of bile acid synthesis.[81,86,109] The final oxidative

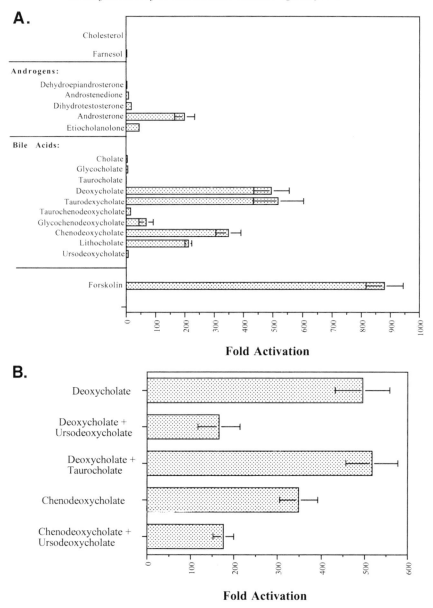

**Fold Activation**

**Figure 5** (**A**) Bile acids, selective androgens and forskolin activate FXR-mediated transcription. FXR and β-galactosidase mammalian expression plasmids along with an EcRE luciferase reporter construct, were transfected into HepG2 cells and analyzed for ligand stimulation of receptor activities. The originally described FXR activator, farnesol, gave approximately threefold increase in activation. Androsterone, etiocholanolone, and the α-hydroxylated diterpenoid forskolin significantly activated FXR-mediated transcription. All compounds were used at 100 μM (their most efficacious concentration) with the exception CDCA and forskolin (10 μM). Activation is expressed as fold increase in luciferase activity for treated versus solvent controls. Normalized luminescence values for triplicate readings of three experiments for each compound or solvent control were subjected to a one-way analysis of variance for verification of significance ($p < 0.05$). Androsterone: 5α-androstan-3α-ol-17-one. Etiocholanolone: 5β-androstan-3α-ol-17-one. Dihydrotestosterone: 5α-androstan-17β-ol-3-one. (**B**) Ursodeoxycholate inhibition of deoxycholate and chenodeoxycholate activation of FXR. Ursodeoxycholate, a very weak activator of FXR, was evaluated for its ability to inhibit deoxycholate and chenodeoxycholate activation of FXR-mediated transcription. A FXR mammalian expression plasmid, a β-galactosidase expression plasmid and an EcRE luciferase reporter construct were transfected into HepG2 cells and analyzed for ligand stimulation of receptor activities. Deoxycholate, ursodeoxycholate and taurocholate were used at 100 μM while chenodeoxycholate was evaluated at 10 μM. Activation is expressed as fold increase in luciferase activity for treated versus solvent controls. Normalized luminescence values for triplicate readings of three experiments for each compound or solvent control were subjected to a one-way analysis of variance for verification of significance ($p < 0.05$). Adapted with permission from Howard *et al.*[109]

steps in the conversion of cholesterol into the primary bile acids cholic acid and chenodeoxycholic acid, as well as the subsequent amidation of these bile acids with taurine and glycine, have been shown to occur in peroxisomes.[78] Furthermore, it has been observed that, in humans and rodents, fibric acid activators of PPARα produce a significant upregulation of cholic acid-based activators of FXR[82–84,86,110] and that fibric acid activation of PPARα modulates the mRNA levels of CYP7A, the key enzyme in the synthesis of bile acids.[56,57,111] Cumulatively, these results support the contention that PPARα is intimately involved in the synthesis of bile acid ligand activators of FXR.

## 14.2.9.11 FXR LIGANDS FUNCTION AS TUMOR PROMOTERS

The original hypothesis of O'Brien *et al.* also suggests that activation of FXR-mediated events plays a tumor promotion role in peroxisome proliferator-induced rodent hepatocarcinogenesis. Although not proven, accumulating evidence appears to strongly support this contention. Bile acids have been suggested to play an important role in the etiology of both rodent and human colon cancers. In rodents, cholic acid has been shown to increase tumor incidence for colon cancer[112,113] and bile acids have been reported to be co-carcinogenic in germ-free rats.[114] Furthermore, in line with the FXR/UDCA studies of Howard *et al.*,[107] UDCA has been shown to decrease tumor incidence in the bile acid facilitated rat model for colon cancer.[112] The underlying molecular mechanism is unclear, but deoxycholic acid has been shown to increase mitotic events of colonic mucosa[115] and preneoplastic lesions in rats fed with cholic acid were found to be less susceptible to apoptotic stimuli.[116,117] In humans, colorectal cancers have been shown to express an increased serum level of deoxycholic acid,[118] reduced bile acid-induced apoptosis of colonic mucosa goblet cells[96] and an impairment of peroxisomal biogenesis.[119] The observations that bile acids: (a) drive AP-1-dependent promoters;[120] (b) decrease transcription of putative tumor inhibiting genes like HLA class I genes;[121] (c) have well established immunosuppressive properties;[122–126] and (d) modulate protein kinase C expression[127–129] would suggest that bile acids function as tumor promoters in human and rat colorectal cancers. When coupled with the observation that FXR and PPARs co-localize to colon,[63,130] these data fit well with the hypothesized tumor promotion model for FXR and strongly suggest that, beyond peroxisome proliferator-induced rodent hepatocarcinogenesis, this model may help to dissect mechanisms of human colorectal and obesity-associated cancers.

## 14.2.9.12 DO PEROXISOME PROLIFERATORS MODULATE THE EXPRESSION OF FXR-REGULATED GENES?

If one accepts the hypothesis that PPARα can upregulate the synthesis of activators of FXR, it would likewise be predicted that xenobiotic activation of PPARα results in the modulation of FXR-regulated genes. Due in part to the limited number of known FXR-regulated genes, the evidence supporting this idea is fairly circumstantial. Peroxisome proliferators have been shown to negatively regulate the expression of CYP7A gene expression in a manner similar to that observed for bile acid activation of FXR.[56,57,104,106,111] Furthermore, this regulation of CYP7A gene expression by PPARα agonists is in the absence of any direct binding of PPARα to the CYP7A promoter region.[111] In reality, this regulatory event may be even more complicated than simply the induction of FXR by bile acids. Recent reports have presented evidence for bile acids stimulating a cascade of nuclear receptor activity initiated by FXR induction of the expression of small heterodimer partner 1 (SHP-1), an atypical member of the nuclear receptor family that lacks a DNA-binding domain. SHP-1 is believed to repress expression of CYP7A by inhibiting the activity of liver receptor homolog 1 (LRH-1), an orphan nuclear receptor that is known to upregulate CYP7A expression.[131] These data would support the idea that peroxisome proliferators mediate their regulation of CYP7A gene repression through their ability to stimulate bile acid activators of FXR. With the recent identification of a variety of other genes containing FXR regulated promoters [e.g., I-BABP,[104] the basolateral sodium taurocholate cotransporter protein (Ntcp),[110] hepatic canicular bile salt export pump (Bsep),[108] and the phospholipid transfer protein (PLTP)[132]] it should be possible, in the near future, to definitively establish whether or not peroxisome proliferator stimulation of PPAR has any downstream effects on FXR mediated gene expression.

### 14.2.9.13 OTHER SIGNALING PATHWAYS CONVERGING WITH PPARα/ FXR/RXR

An interesting facet of at least PPARα and PPARγ transcriptional regulation is their common utilization of a DNA binding response element motif consisting generally of an imperfect direct repeat separated by a single base pair (DR1). Although there appears to be some form of transcription regulatory specificity indigenous to these repeats and their surrounding sequences, PPAR isoforms bind the repeats with similar affinities and it can be demonstrated *in vitro*, that PPARα-specific response elements (such as the one found in the promoter of the acyl-CoA oxidase gene) as well as PPARα-specific ligands (e.g., hypolipidemic drugs) effectively mediate PPARγ-stimulated transcription events.[24,63,133,134] These data would suggest either cell-specific co-regulatory molecules or/and tissue-specific expression of receptors play a significant role in defining PPAR isoform activity. Once again this would be an insignificant consideration for normal metabolic events and also appears to be relatively unimportant in rodent hepatocarcinogenesis, but may be a significant consideration in tissues such as colon, wherein normal expression of alternative PPAR isoforms is relatively high[130,135] and where tumor formation results in a significant elevation in the expression of PPAR isoforms.[66,136,137]

Recently, several other orphan members of the steroid receptor superfamily of genes have been characterized as having overlapping and converging signaling pathways with PPARα and FXR. These include the liver X receptor (LXR),[138] the pregnane X receptor (PXR)[139] and the constitutive androstane receptor (CAR).[140,141] All of these co-localize with FXR and PPARα and are activated by ligands derived from the cholesterol synthesis pathway. Of the three, LXR is perhaps the most intriguing, LXR has been shown to be activated by 6α-hydroxylated bile acids[142] and bile acid metabolism is impaired in mice deleted for the LXR gene.[143] Furthermore, in contrast to FXR,[142] activation of LXR results in the upregulation of transcription of the CYP7A gene.[144] Very little information is available with respect to peroxisome proliferator effects on LXR, but complicating these regulatory pathways even further are the findings of Miyata *et al.*[145] who demonstrated that LXR and PPARα can heterodimerize. The PPARα/ LXR heterodimer was unable to bind classical PPARα or LXR response element motifs, but LXR was able to inhibit the formation of PPARα/ RXR heterodimers. The biological significance of these findings is unclear but they do add a new wrinkle into an already complex story.

Beyond these, there are several other nuclear receptor family members whose tissue distribution and involvement in cholesterol signaling pathways might converge with PPAR and FXR. Although an intriguing concept, at this time there is a lack of significant characterization with respect to overlap with PPARα/ FXR-mediated events to speculate on their integration into these signaling pathways. These receptors, include; the hepatocyte nuclear factor 4 (HNF4), a receptor whose binding site on the CYP7A promoter is intimately linked to PPARα regulation of that promoter;[57] LRH-1, an orphan nuclear receptor directly involved in regulation of CYP7A gene expression (see 02.9.12 above); and chicken ovalbumin upstream promoter transcription factor II (COUP-TFII), an orphan nuclear receptor also implicated in the expression of CYP7A.[146] These data suggest a complex overlapping nuclear receptor network of activity that regulates the ultimate metabolism of synthesized cholesterol.

### 14.2.9.14 DO PEROXISOME PROLIFERATORS FUNCTION AS CARCINOGENS IN HUMANS?

The pleiotrophic effects of peroxisome proliferators appear to be species-specific and the available data suggests humans have a reduced sensitivity to them.[7,9,147–149] Much of the data on humans is based on a limited number of autopsy or epidemiological studies on chronic users of fibric acid drugs, and from *in vitro* experiments with human hepatocytes.[7,150] The results of these studies are inconclusive with respect to long term exposure to peroxisome proliferators but do reflect differences in sensitivity with respect to peroxisome proliferation. Several mechanisms have been proposed to explain the differing sensitivities of humans and rodents to peroxisome proliferators including; (a) relative amounts of PPARα (rat hepatocytes express 10 times more PPARα than found in human hepatocytes);[151–154] (b) differing PPARα regulatory mechanisms (e.g., human hepatocytes may either lack a PPARα positive co-activator protein or contain a unique PPARα-specific co-repressor protein);[23,37–40,155] and/or (c) promoter regions of human and rat PPARα target genes may differ in respect to sequence context and therefore also with respect to their binding and transcriptional activation by PPARα.[156–158]

Although there are published data supporting facets of all of these hypotheses, the bottom line is: (a) fibric acid peroxisome proliferators bind and transcriptionally activate human PPARα; (b) fibric acid peroxisome proliferators modulate human serum cholesterol and lipid levels; and (c) fibric acid peroxisome proliferators are effective lipid lowering agents in humans. Perhaps the major weakness in all of these evaluations analyzing the toxicity of peroxisome proliferators in rodents and humans is the concerted focus on liver function. As mentioned earlier, beyond liver, PPAR and FXR have also been shown to co-localize to kidney and gut. Furthermore, as presented above, human and rodent cancers can be correlated with expression of FXR activators in extrahepatic tissues to which PPARα and FXR co-localize. Therefore, although humans may not experience extensive hepatomegally and hepatocarcinogenesis upon exposure to peroxisome proliferators, it is far too early to exonerate peroxisome proliferators from toxicity and carcinogenicity in humans.

### 14.2.9.15    CONCLUSIONS

In this chapter we have tried to tie some of the current knowledge on xenobiotic activation of PPARα with accumulating evidence supporting its involvement in cholesterol synthesis, bile acid synthesis, and the activation of multiple nuclear receptor-based signaling pathways. Furthermore, a hypothesis was presented that integrated the activation of these signaling pathways with the pathogenesis of rodent and human cancers. Components of the findings, as presented here, challenge classical perspectives concerning PPARα mediation of carcinogenesis pathways. Although this hypothesis is strongly supported by existing experimental data, the complete story, and even the accuracy of the hypothesis as presented here, is a long way from being established. It is hoped that with the future characterization of the various nuclear receptor pathways impacting, and impacted by, cholesterol synthesis, the accuracy of this hypothesis will be resolved.

### 14.2.9.16    REFERENCES

1. D. S. Marsman, R. C. Cattley, J. G. Conway *et al.*, 'Relationship of hepatic peroxisome proliferation and replicative DNA synthesis to the hepatocarcinogenicity of the peroxisome proliferators di(2-ethylhexyl)phthalate and [4-chloro-6-(2,3-xylidino)-2-pyrimidinylthio]acetic acid (Wy-14,643) in rats.' *Cancer Res.*, 1988, **48**, 6739–6744.
2. D. E. Moody, J. K. Reddy, B. G. Lake *et al.*, 'Peroxisome proliferation and nongenotoxic carcinogenesis: commentary on a symposium.' *Fundam. Appl. Toxicol.*, 1991, **16**, 233–248.
3. P. Grasso, M. Sharratt and A. J. Cohen, 'Role of persistent, non-genotoxic tissue damage in rodent cancer and relevance to humans.' *Annu. Rev. Pharmacol. Toxicol.*, 1991, **31**, 253–287.
4. J. Ashby, A. Brady, C. R. Elcombe *et al.*, 'Mechanistically-based human hazard assessment of peroxisome proliferator-induced hepatocarcinogenesis.' *Hum. Exp. Toxicol.*, 1994, **13**(2), S1–117.
5. A. J. Cohen and P. Grasso, 'Review of the hepatic response to hypolipidaemic drugs in rodents and assessment of its toxicological significance to man.' *Food and Cosmetic Toxicology*, 1981, **19**, 585–605.
6. J. K. Reddy and N. D. Lalwani, 'Carcinogenesis by hepatic peroxisome proliferators: evaluation of the risk of hypolipidemic drugs and industrial plasticizers to humans.' *Crit. Rev. Toxicol.*, 1983, **12**, 1–58.
7. C. R. Elcombe, 'Species differences in carcinogenicity and peroxisome proliferation due to trichloroethylene: a biochemical human hazard assessment.' *Archiv. Toxicol. Supplement*, 1985, **8**, 6–17.
8. M. S. Rao and J. K. Reddy, 'Peroxisome proliferation and hepatocarcinogenesis.' *Carcinogenesis*, 1987, **8**, 631–636.
9. J. P. Vanden Heuvel, 'Peroxisome proliferator-activated receptors (PPARS) and carcinogenesis.' *Toxicol. Sci.*, 1999, **47**, 1–8.
10. V. Giguere, 'Orphan nuclear receptors: from gene to function.' *Endocr. Rev.*, 1999, **20**, 689–725.
11. J. C. Fruchart, P. Duriez and B. Staels, 'Molecular mechanism of action of the fibrates.' *J. Soc. Biol.*, 1999, **193**, 67–75.
12. L. Michalik and W. Wahli, 'Peroxisome proliferator-activated receptors: three isotypes for a multitude of functions.' *Curr. Opin. Biotechnol.*, 1999, **10**, 564–570.
13. J. C. Corton, S. P. Anderson and A. Stauber, 'Central role of peroxisome proliferator-activated receptors in the actions of peroxisome proliferators.' *Annu. Rev. Pharmacol. Toxicol.*, 2000, **40**, 491–518.
14. I. Singh, 'Biochemistry of peroxisomes in health and disease.' *Mol. Cell. Biochem.*, 1997, **167**, 1–29.
15. R. Hess, W. Stäubli and W. Riess, 'Nature of the hepatomegalic effect produced by ethyl-chlorophenoxy-isobutyrate in the rat.' *Nature*, 1965, **208**, 856–858.
16. E. A. Lock, A. M. Mitchell and C. R. Elcombe, 'Biochemical mechanisms of induction of hepatic peroxisome proliferation.' *Annu. Rev. Pharmacol. Toxicol.*, 1989, **29**, 145–163.
17. B. J. Keller, D. S. Marsman, J. A. Popp *et al.*, 'Several nongenotoxic carcinogens uncouple mitochondrial oxidative phosphorylation.' *Biochim. Biophys. Acta*, 1992, **1102**, 237–244.
18. J. K. Reddy, M. S. Rao, D. L. Azarnoff *et al.*, 'Mitogenic and carcinogenic effects of a hypolipidemic peroxisome proliferator, [4-chloro-6-(2,3-xylidino)-2-pyrimidinylthio]acetic acid (Wy-14, 643), in rat and mouse liver.' *Cancer Res.*, 1979, **39**, 152–161.
19. J. K. Reddy and D. L. Azarnoff, 'Hypolipidaemic hepatic peroxisome proliferators form a novel class of chemical carcinogens.' *Nature*, 1980, **283**, 397–398.
20. I. Issemann and S. Green, 'Activation of a member of the steroid hormone receptor superfamily by peroxisome proliferators.' *Nature*, 1990, **347**, 645–650.
21. R. M. Evans, 'The steroid and thyroid hormone receptor superfamily.' *Science*, 1988, **240**, 889–895.
22. F. Chen, S. W. Law and B. W. O'Malley, 'Identification of two mPPAR related receptors and evidence for the existence of five subfamily members.' *Biochem. Biophys. Res. Commun.*, 1993, **196**, 671–677.
23. Y. Zhu, C. Qi, C. Calandra *et al.*, 'Cloning and

identification of mouse steroid receptor coactivator-1 (mSRC-1), as a coactivator of peroxisome proliferator-activated receptor γ.' *Gene Expr.*, 1996, **6**, 185–195.

24. C. Dreyer, G. Krey, H. Keller *et al.*, 'Control of the peroxisomal β-oxidation pathway by a novel family of nuclear hormone receptors.' *Cell*, 1992, **68**, 879–887.

25. M. Göttlicher, E. Widmark, Q. Li *et al.*, 'Fatty acids activate a chimera of the clofibric acid-activated receptor and the glucocorticoid receptor.' *Proc. Natl. Acad. Sci. USA*, 1992, **89**, 4653–4657.

26. M. E. Greene, B. Blumberg, O. W. McBride *et al.*, 'Isolation of the human peroxisome proliferator activated receptor γ cDNA: expression in hematopoietic cells and chromosomal mapping.' *Gene Expr.*, 1995, **4**, 281–299.

27. R. Mukherjee, L. Jow, G. E. Croston *et al.*, 'Identification, characterization, and tissue distribution of human peroxisome proliferator-activated receptor (PPAR) isoforms PPARγ2 versus PPARγ1 1 and activation with retinoid X receptor agonists and antagonists.' *J. Biol. Chem.*, 1997, **272**, 8071–8076.

28. T. Sher, H. F. Yi, O. W. McBride *et al.*, 'cDNA cloning, chromosomal mapping, and functional characterization of the human peroxisome proliferator activated receptor.' *Biochemistry*, 1993, **32**, 5598–5604.

29. J. D. Tugwood, I. Issemann, R. G. Anderson *et al.*, 'The mouse peroxisome proliferator activated receptor recognizes a response element in the 5′ flanking sequence of the rat acyl CoA oxidase gene.' *EMBO. J.*, 1992, **11**, 433–439.

30. B. Zhang, S. L. Marcus, F. G. Sajjadi *et al.*, 'Identification of a peroxisome proliferator-responsive element upstream of the gene encoding rat peroxisomal enoyl-CoA hydratase/3-hydroxyacyl-CoA dehydrogenase.' *Proc. Natl. Acad. Sci. USA*, 1992, **89**, 7541–7545.

31. P. Tontonoz, E. Hu, R. A. Graves *et al.*, 'mPPARg2: tissue-specific regulator of an adipocyte enhancer.' *Genes Devel.*, 1994, **8**, 1224–1234.

32. M. L. O'Brien, S. M. Rangwala, K. W. Henry *et al.*, 'Convergence of three steroid receptor pathways in the mediation of nongenotoxic hepatocarcinogenesis.' *Carcinogenesis*, 1996, **17**, 185–190.

33. K. Henry, M. L. O'Brien, W. Clevenger *et al.*, 'Peroxisome proliferator-activated receptor response specificities as defined in yeast and mammalian cell transcription assays.' *Toxicol. Appl. Pharmacol.*, 1995, **132**, 317–324.

34. S. A. Kliewer, K. Umesono, D. J. Mangelsdorf *et al.*, 'Retinoid X receptor interacts with nuclear receptors in retinoic acid, thyroid hormone and Vitamin D$_3$ signalling.' *Nature*, 1992, **355**, 446–449.

35. T. Osumi, J. K. Wen and T. Hashimoto, 'Two *cis*-acting regulatory sequences in the peroxisome proliferator-responsive enhancer region of rat acyl-CoA oxidase gene.' *Biochem. Biophys. Res. Commun.*, 1991, **175**, 866–871.

36. S. A. Kliewer, K. Umesono, D. J. Noonan *et al.*, 'Convergence of 9-*cis* retinoic acid and peroxisome proliferator signalling pathways through heterodimer formation of their receptors.' *Nature*, 1992, **358**, 771–774.

37. Y. Zhu, C. Qi, S. Jain *et al.*, 'Isolation and characterization of PBP, a protein that interacts with peroxisome proliferator-activated receptor.' *J. Biol. Chem.*, 1997, **272**, 25500–25506.

38. Y. Zhu, L. Kan, C. Qi *et al.*, 'Isolation and characterization of peroxisome proliferator-activated receptor (PPAR) interacting protein (PRIP) as a coactivator for PPAR.' *J. Biol. Chem.*, 2000, **275**, 13510–13516.

39. Y. Zhu, C. Qi, Y. Jia *et al.*, 'Deletion of PBP/PPARBP, the gene for nuclear receptor coactivator peroxisome proliferator-activated receptor-binding protein, results in embryonic lethality.' *J. Biol. Chem.*, 2000, **275**, 14779–14782.

40. R. T. Nolte, G. B. Wisely, S. Westin *et al.*, 'Ligand binding and co-activator assembly of the peroxisome proliferator- activated receptor-γ.' *Nature*, 1998, **395**, 137–143.

41. S. Altiok, M. Xu and B. M. Spiegelman, 'PPARγ induces cell cycle withdrawal: inhibition of E2F/DP DNA-binding activity via down-regulation of PP2A.' *Genes Dev.*, 1997, **11**, 1987–1998.

42. H. S. Camp and S. R. Tafuri, 'Regulation of peroxisome proliferator-activated receptor γ activity by mitogen-activated protein kinase.' *J. Biol. Chem.*, 1997, **272**, 10811–10816.

43. M. Adams, M. J. Reginato, D. Shao *et al.*, 'Transcriptional activation by peroxisome proliferator-activated receptor γ is inhibited by phosphorylation at a consensus mitogen-activated protein kinase site.' *J. Biol. Chem.*, 1997, **272**, 5128–5132.

44. E. Hu, J. B. Kim, P. Sarraf *et al.*, 'Inhibition of adipogenesis through MAP kinase-mediated phosphorylation of PPARγ.' *Science*, 1996, **274**, 2100–2103.

45. A. Shalev, C. A. Siegrist-Kaiser, P. M. Yen *et al.*, 'The peroxisome proliferator-activated receptor alpha is a phosphoprotein: regulation by insulin.' *Endocrinology*, 1996, **137**, 4499–4502.

46. B. Zhang, J. Berger, G. Zhou *et al.*, 'Insulin- and mitogen-activated protein kinase-mediated phosphorylation and activation of peroxisome proliferator-activated receptor γ.' *J. Biol. Chem.*, 1996, **271**, 31771–31774.

47. C. E. Juge-Aubry, E. Hammar, C. Siegrist-Kaiser *et al.*, 'Regulation of the transcriptional activity of the peroxisome proliferator-activated receptor alpha by phosphorylation of a ligand-independent trans-activating domain.' *J. Biol. Chem.*, 1999, **274**, 10505–10510.

48. P. Escher and W. Wahli, 'Peroxisome proliferator-activated receptors: insight into multiple cellular functions.' *Mutat. Res.*, 2000, **448**, 121–138.

49. R. Mukherjee, L. Jow, D. J. Noonan *et al.*, 'Human and rat peroxisome proliferator-activated receptors (PPARs) demonstrate similar tissue distribution but different responsiveness to PPAR activators.' *J. Steroid Biochem. Molec. Biol.*, 1994, **51**, 157–166.

50. T. Lemberger, O. Braissant, C. Juge-Aubry *et al.*, 'PPAR tissue distribution and interactions with other hormone-signaling pathways.' *Ann. NY Acad. Sci.*, 1996, **804**, 231–251.

51. O. Braissant, F. Foufelle, C. Scotto *et al.*, 'Differential expression of peroxisome proliferator-activated receptors (PPARs): tissue distribution of PPAR-α, -β, and -γ in the adult rat.' *Endocrinology*, 1996, **137**, 354–366.

52. I. Issemann, R. Prince, J. Tugwood *et al.*, 'A role for fatty acids and liver fatty acid binding protein in peroxisome proliferation?' *Biochem. Soc. Trans.*, 1992, **20**, 824–827.

53. J. A. Ortiz, J. Mallolas, C. Nicot *et al.*, 'Isolation of pig mitochondrial 3-hydroxy-3-methylglutaryl-CoA synthase gene promoter: characterization of a peroxisome proliferator-responsive element.' *Biochem. J.*, 1999, **337**, 329–335.

54. J. C. Rodriguez, G. Gil-Gomez, F. G. Hegardt *et al.*, 'Peroxisome proliferator-activated receptor mediates induction of the mitochondrial 3-hydroxy-3-methyl-

glutaryl-CoA synthase gene by fatty acids.' *J. Biol. Chem.*, 1994, **269**, 18767–18772.

55. G. Martin, K. Schoonjans, A. M. Lefebvre *et al.*, 'Coordinate regulation of the expression of the fatty acid transport protein and acyl-CoA synthetase genes by PPARα and PPARγ activators.' *J. Biol. Chem.*, 1997, **272**, 28210–28217.

56. M. C. Hunt, Y. Z. Yang, G. Eggertsen *et al.*, 'The peroxisome proliferator-activated receptor α (PPARα) regulates bile acid biosynthesis.' *J. Biol. Chem.*, 2000, **275**, 28947–28953.

57. D. D. Patel, B. L. Knight, A. K. Soutar *et al.*, 'The effect of peroxisome-proliferator-activated receptor-alpha on the activity of the cholesterol 7α-hydroxylase gene.' *Biochem. J.*, 2000, **351**, 747–753.

58. C. Dreyer, H. Keller, A. Mahfoudi *et al.*, 'Positive regulation of the peroxisomal beta-oxidation pathway by fatty acids through activation of peroxisome proliferator-activated receptors (PPAR).' *Biol. Cell*, 1993, **77**, 67–76.

59. A. J. Vidal-Puig, R. V. Considine, M. Jimenez-Linan *et al.*, 'Peroxisome proliferator-activated receptor gene expression in human tissues. Effects of obesity, weight loss, and regulation by insulin and glucocorticoids.' *J. Clin. Invest.*, 1997, **99**, 2416–2422.

60. C. E. Robinson, X. Wu, D. C. Morris *et al.*, 'DNA bending is induced by binding of the peroxisome proliferator-activated receptor γ 2 heterodimer to its response element in the murine lipoprotein lipase promoter.' *Biochem. Biophys. Res. Commun.*, 1998, **244**, 671–677.

61. K. Schoonjans, J. Peinado-Onsurbe, A. M. Lefebvre *et al.*, 'PPARα and PPARγ activators direct a distinct tissue-specific transcriptional response via a PPRE in the lipoprotein lipase gene.' *EMBO J.*, 1996, **15**, 5336–5348.

62. K. Motojima, P. Passilly, J. M. Peters *et al.*, 'Expression of putative fatty acid transporter genes are regulated by peroxisome proliferator-activated receptor alpha and gamma activators in a tissue- and inducer-specific manner.' *J. Biol. Chem.*, 1998, **273**, 16710–16714.

63. S. A. Kliewer, B. M. Forman, B. Blumberg *et al.*, 'Differential expression and activation of a family of murine peroxisome proliferator-activated receptors.' *Proc. Natl. Acad. Sci. USA*, 1994, **91**, 7355–7359.

64. P. Kremarik-Bouillaud, H. Schohn and M. Dauca, 'Regional distribution of PPARβ in the cerebellum of the rat (in process citation).' *J. Chem. Neuroanat.*, 2000, **19**, 225–232.

65. H. Lim, R. A. Gupta, W. G. Ma *et al.*, 'Cyclo-oxygenase-2-derived prostacyclin mediates embryo implantation in the mouse via PPARδ.' *Genes Dev.*, 1999, **13**, 1561–1574.

66. R. A. Gupta, J. Tan, W. F. Krause *et al.*, 'Prostacyclin-mediated activation of peroxisome proliferator-activated receptor δ in colorectal cancer.' *Proc. Natl. Acad. Sci. USA*, 2000, **97**, 13275–13280.

67. S. S.-T. Lee, T. Pineau, J. Drago *et al.*, 'Targeted disruption of the a isoform of the peroxisome proliferator-activated receptor gene in mice results in abolishment of the pliotropic effects of peroxisome proliferators.' *Molec. Cell. Biol.*, 1995, **15**, 3012–3022.

68. F. J. Gonzalez, 'Recent update on the PPAR alpha-null mouse.' *Biochimie.*, 1997, **79**, 139–144.

69. F. J. Gonzalez, J. M. Peters and R. C. Cattley, 'Mechanism of action of the nongenotoxic peroxisome proliferators: role of the peroxisome proliferator-activator receptor α.' *J. Natl. Cancer Inst.*, 1998, **90**, 1702–1709.

70. S. K. Krisans, 'The role of peroxisomes in cholesterol metabolism.' *Am. J. Respir. Cell Molec. Biol.*, 1992, **7**, 358–364.

71. S. K. Krisans, 'Cell compartmentalization of cholesterol biosynthesis.' *Ann. NY Acad. Sci.*, 1996, **804**, 142–164.

72. R. J. Wanders and G. J. Romeijn, 'Differential deficiency of mevalonate kinase and phosphomevalonate kinase in patients with distinct defects in peroxisome biogenesis: evidence for a major role of peroxisomes in cholesterol biosynthesis.' *Biochem. Biophys. Res. Commun.*, 1998, **247**, 663–667.

73. M. M. Magalhaes and M. C. Magalhaes, 'Peroxisomes in adrenal steroidogenesis.' *Microsc. Res. Tech.*, 1997, **36**, 493–502.

74. G. A. Keller, M. Pazirandeh and S. Krisans, '3-Hydroxy-3-methylglutaryl coenzyme A reductase localization in rat liver peroxisomes and microsomes of control and cholestyramine-treated animals: quantitative biochemical and immunoelectron microscopical analyses.' *J. Cell Biol.*, 1986, **103**, 875–886.

75. S. L. Thompson and S. K. Krisans, 'Rat liver peroxisomes catalyze the initial step in cholesterol synthesis. The condensation of acetyl-CoA units into acetoacetyl-CoA.' *J. Biol. Chem.*, 1990, **265**, 5731–5735.

76. R. Hovik, B. Brodal, K. Bartlett *et al.*, 'Metabolism of acetyl-CoA by isolated peroxisomal fractions: formation of acetate and acetoacetyl-CoA.' *J. Lipid Res.*, 1991, **32**, 993–999.

77. N. Rusnak and S. K. Krisans, 'Diurnal variation of HMG-CoA reductase activity in rat liver peroxisomes.' *Biochem. Biophys. Res. Commun.*, 1987, **148**, 890–895.

78. K. Solaas, A. Ulvestad, O. Soreide *et al.*, 'Subcellular organization of bile acid amidation in human liver: a key issue in regulating the biosynthesis of bile salts.' *J. Lipid Res.*, 2000, **41**, 1154–1162.

79. E. L. Appelkvist, F. Aberg, Z. Guan *et al.*, 'Regulation of coenzyme Q biosynthesis.' *Mol. Aspects Med.*, 1994, **15**, s37–46.

80. M. Turunen, J. M. Peters, F. J. Gonzalez *et al.*, 'Influence of peroxisome proliferator-activated receptor alpha on ubiquinone biosynthesis.' *J. Mol. Biol.*, 2000, **297**, 607–614.

81. S. K. Krisans, S. L. Thompson, L. A. Pena *et al.*, 'Bile acid synthesis in rat liver peroxisomes: metabolism of 26-hydroxycholesterol to 3 β-hydroxy-5-cholenoic acid.' *J. Lipid Res.*, 1985, **26**, 1324–1332.

82. B. Angelin, K. Einarsson and B. Leijd, 'Biliary lipid composition during treatment with different hypolipidaemic drugs.' *Eur. J. Clin. Invest.*, 1979, **9**, 185–190.

83. K. von Bergmann and O. Leiss, 'Effect of short-term treatment with bezafibrate and fenofibrate on biliary lipid metabolism in patients with hyperlipoproteinaemia.' *Eur. J. Clin. Invest.*, 1984, **14**, 150–154.

84. A. K. Farrants, A. Nilsson, G. Troen *et al.*, 'The effect of retinoids and clofibric acid on the peroxisomal oxidation of palmitic acid and of 3 α,7 α,12 α-trihydroxy-5 β-cholestanoic acid in rat and rabbit hepatocytes.' *Biochem. Biophys. Acta*, 1993, **1168**, 100–107.

85. F. Hashimoto, T. Ishikawa, S. Hamada *et al.*, 'Effect of gemfibrozil on lipid biosynthesis from acetyl-CoA derived from peroxisomal β-oxidation.' *Biochem. Pharmacol.*, 1995, **49**, 1213–1221.

86. C. M. Rodrigues, B. T. Kren, C. J. Steer *et al.*, 'Formation of δ 22-bile acids in rats is not gender specific and occurs in the peroxisome.' *J. Lipid Res.*, 1996, **37**, 540–550.

87. F. Aberg and E. L. Appelkvist, 'Clofibrate and di(2-ethylhexyl)phthalate increase ubiquinone contents without affecting cholesterol levels.' *Acta Biochem. Pol.*, 1994, **41**, 321–329.

88. B. M. Forman, E. Goode, J. Chen *et al.*, 'Identifica-

tion of a nuclear receptor that is activated by farnesol metabolites.' *Cell*, 1995, **81**, 687–693.

89. T. P. Yao, B. M. Forman, Z. Jiang *et al.*, 'Functional ecdysone receptor is the product of EcR and Ultraspiracle genes.' *Nature*, 1993, **366**, 476–479.

90. T. P. McGee, H. H. Cheng, H. Kumagai *et al.*, 'Degradation of 3-hydroxy-3-methylglutaryl-CoA reductase in endoplasmic reticulum membranes is accelerated as a result of increased susceptibility to proteolysis.' *J. Biol. Chem.*, 1996, **271**, 25630–25638.

91. D. Agarwal, S. Eustis, C. Lamb *et al.*, 'Influence of dietary zinc on di(2-ethyl-hexyl) phthalate-induced testicular atrophy and zinc depletion in adult rats.' *Toxicol. Appl. Pharmacol.*, 1986, **84**, 12–24.

92. I. Chu, R. Poom, P. Lecavalier *et al.*, 'Subchronic oral toxicity of di-*n*-octylphthalate (DNOP) and di-(2-ethyl hexyl) phthalate (DEHP) in the rat.' *Toxicologist*, 1995, **15**, 1278.

93. P. Eagon, M. Elm, M. Epley *et al.*, 'Peroxisome proliferators: sex steroid receptors and metabolism in hepatic hyperplasia and neoplasia.' *Toxicologist*, 1996, **15**, 1057.

94. R. Liu, C. Hahn and M. Hutt, 'The direct effect of 11 hepatic peroxisome proliferators (PPs) on rat Leydig cell function *in vitro*.' *Toxicologist*, 1995, **15**, 1291.

95. N. Worrell, W. Cook, C. Thompson *et al.*, 'Effect of mono-(2-ethylhexyl) phthalate on the metabolism of energy yielding substances in rat Sertoli cell-enriched cultures.' *Toxicol. In vitro.*, 1989, **3**, 77–81.

96. H. Garewal, H. Bernstein, C. Bernstein *et al.*, 'Reduced bile acid-induced apoptosis in "normal" colorectal mucosa: a potential biological marker for cancer risk.' *Cancer Res.*, 1996, **56**, 1480–1483.

97. J. C. Corton, C. Bocos, E. S. Moreno *et al.*, 'Rat 17 β-hydroxysteroid dehydrogenase type IV is a novel peroxisome proliferator-inducible gene.' *Mol. Pharmacol.*, 1996, **50**, 1157–1166.

98. J. C. Corton, C. Bocos, E. S. Moreno *et al.*, 'Peroxisome proliferators alter the expression of estrogen metabolizing enzymes.' *Biochimie*, 1997, **79**, 151–162.

99. P. Eagon, N. Chandler, M. Epley *et al.*, 'Di(2-ethylhexyl) phthalate-induced changes in liver estrogen metabolism and hyperplasia.' *In. J. Cancer*, 1994, **58**, 736–743.

100. 'IARC monographs on the evaluation of carcinogenic risks to humans of some pharmaceutical drugs.' *IARC Monogr. Eval. Carcinog. Risks Hum.*, 1996, **66**, 391–344.

101. E. Lamb and S. Bennett, 'Epidemiological studies of male factors in infertility.' *Ann. NY Acad. Sci.*, 1994, **709**, 165–178.

102. M. Sallmen, M.-L. Lindbahm, P. Kyyronen *et al.*, 'Reduced fertility among women exposed to organic solvents.' *Am. J. Industr. Med.*, 1995, **27**, 699–713.

103. G. Bahadur, K. Ling and M. Katz, 'Statistical model reveals demography and time are the main contributing factors to global sperm count changes between 1938 and 1996.' *Human Rep.*, 1996, **11**, 2635–2639.

104. M. Makishima, A. Y. Okamoto, J. J. Repa *et al.*, 'Identification of a nuclear receptor for bile acids.' *Science*, 1999, **284**, 1362–1365.

105. D. J. Parks, S. G. Blanchard, R. K. Bledsoe *et al.*, 'Bile acids: natural ligands for an orphan nuclear receptor.' *Science*, 1999, **284**, 1365–1368.

106. H. Wang, J. Chen, K. Hollister *et al.*, 'Endogenous bile acids are ligands for the nuclear receptor FXR/BAR.' *Mol. Cell*, 1999, **3**, 543–553.

107. W. R. Howard, J. A. Pospisil, E. Njolito *et al.*, 'Catabolites of cholesterol synthesis pathways and forskolin as activators of the farnesoid X-activated nuclear receptor.' *Toxicol. Appl. Pharmacol.*, 2000, **163.**, 195–202.

108. C. J. Sinal, M. Tohkin, M. Miyata *et al.*, 'Targeted disruption of the nuclear receptor FXR/BAR impairs bile acid and lipid homeostasis (in process citation).' *Cell*, 2000, **102**, 731–744.

109. H. Hayashi, K. Fukui and F. Yamasaki, 'Association of the liver peroxisomal fatty acyl-CoA beta-oxidation system with the synthesis of bile acids.' *J. Biochem.(Tokyo)*, 1984, **96**, 1713–1719.

110. B. O. Angelin, I. Bjorkhem and K. Einarsson, 'Effects of clofibrate on some microsomal hydroxylations involved in the formation and metabolism of bile acids in rat liver.' *Biochem. J.*, 1976, **156**, 445–448.

111. M. Marrapodi and J. Y. Chiang, 'Peroxisome proliferator-activated receptor α (PPARα) and agonist inhibit cholesterol 7α-hydroxylase gene (CYP7A1) transcription.' *J. Lipid Res.*, 2000, **41**, 514–520.

112. D. L. Earnest, H. Holubec, R. K. Wali *et al.*, 'Chemoprevention of azoxymethane-induced colonic carcinogenesis by supplemental dietary ursodeoxycholic acid.' *Cancer Res.*, 1994, **54**, 5071–5074.

113. C. K. McSherry, B. I. Cohen, V. D. Bokkenheuser *et al.*, 'Effects of calcium and bile acid feeding on colon tumors in the rat.' *Cancer Res.*, 1989, **49**, 6039–6043.

114. B. S. Reddy, K. Watanabe, J. H. Weisburger *et al.*, 'Promoting effect of bile acids in colon carcinogenesis in germ-free and conventional F344 rats.' *Cancer Res.*, 1977, **37**, 3238–3242.

115. M. J. Wargovich, V. W. Eng, H. L. Newmark *et al.*, 'Calcium ameliorates the toxic effect of deoxycholic acid on colonic epithelium.' *Carcinogenesis*, 1983, **4**, 1205–1207.

116. B. A. Magnuson and R. P. Bird, 'Reduction of aberrant crypt foci induced in rat colon with azoxymethane or methylnitrosourea by feeding cholic acid.' *Cancer Lett.*, 1993, **68**, 15–23.

117. B. A. Magnuson, I. Carr and R. P. Bird, 'Ability of aberrant crypt foci characteristics to predict colonic tumor incidence in rats fed cholic acid.' *Cancer Res.*, 1993, **53**, 4499–4504.

118. E. Bayerdorffer, G. A. Mannes, W. O. Richter *et al.*, 'Increased serum deoxycholic acid levels in men with colorectal adenomas.' *Gastroenterology*, 1993, **104**, 145–151.

119. C. Lauer, A. Volkl, S. Riedl *et al.*, 'Impairment of peroxisomal biogenesis in human colon carcinoma (published erratum appears in Carcinogenesis 1999 Oct; **20**(10), 2037).' *Carcinogenesis*, 1999, **20**, 985–989.

120. F. Hirano, H. Tanada, Y. Makino *et al.*, 'Induction of the transcription factor AP-1 in cultured human colon adenocarcinoma cells following exposure to bile acids.' *Carcinogenesis*, 1996, **17**, 427–433.

121. P. Arvind, E. D. Papavassiliou, G. J. Tsioulias *et al.*, 'Lithocholic acid inhibits the expression of HLA class I genes in colon adenocarcinoma cells. Differential effect on HLA-A, -B and -C loci.' *Mol. Immunol.*, 1994, **31**, 607–614.

122. L. Gianni, F. Di Padova, M. Zuin *et al.*, 'Bile acid-induced inhibition of the lymphoproliferative response to phytohemagglutinin and pokeweed mitogen: an *in vitro* study.' *Gastroenterology*, 1980, **78**, 231–235.

123. R. M. Keane, T. R. Gadacz, A. M. Munster *et al.*, 'Impairment of human lymphocyte function by bile salts.' *Surgery*, 1984, **95**, 439–443.

124. T. D. Feduccia, C. E. Scott-Conner and J. B. Grogan, 'Profound suppression of lymphocyte function in early biliary obstruction.' *Am. J. Med. Sci.*, 1988, **296**, 39–44.

125. D. L. Nelson, D. E. Frazier, Jr., J. E. Ericson *et al.*, 'The effects of perfluorodecanoic acid (PFDA) on humoral, cellular, and innate immunity in Fischer 344

rats.' *Immunopharmacol. Immunotoxicol.*, 1992, **14**, 925–938.

126. Y. Calmus, B. Weill, Y. Ozier *et al.*, 'Immunosuppressive properties of chenodeoxycholic and ursodeoxycholic acids in the mouse.' *Gastroenterology*, 1992, **103**, 617–621.

127. L. M. Brady, D. W. Beno and B. H. Davis, 'Bile acid stimulation of early growth response gene and mitogen-activated protein kinase is protein kinase C-dependent.' *Biochem. J.*, 1996, **316**, 765–769.

128. Y. P. Rao, R. T. Stravitz, Z. R. Vlahcevic *et al.*, 'Activation of protein kinase C α and δ by bile acids: correlation with bile acid structure and diacylglycerol formation.' *J. Lipid Res.*, 1997, **38**, 2446–2454.

129. B. S. Reddy, B. Simi, N. Patel *et al.*, 'Effect of amount and types of dietary fat on intestinal bacterial 7 α-dehydroxylase and phosphatidylinositol-specific phospholipase C and colonic mucosal diacylglycerol kinase and PKC activities during stages of colon tumor promotion.' *Cancer Res.*, 1996, **56**, 2314–2320.

130. J. J. Repa and D. J. Mangelsdorf, 'The Role of Orphan Nuclear Receptors in the Regulation of Cholesterol Homeostasis.' *Annu. Rev. Cell Dev. Biol.*, 2000, **16**, 459–481.

131. B. Goodwin, S. A. Jones, R. R. Price *et al.*, 'A regulatory cascade of the nuclear receptors FXR, SHP-1, and LRH-1 represses bile acid.' *Mol. Cell*, 2000, **6**, 517–526.

132. N. L. Urizar, D. H. Dowhan and D. D. Moore, 'The farnesoid X-activated receptor mediates bile acid activation of phospholipid transfer protein gene expression.' *J. Biol. Chem.*, 2000.

133. C. Juge-Aubry, A. Pernin, T. Favez *et al.*, 'DNA binding properties of peroxisome proliferator-activated receptor subtypes on various natural peroxisome proliferator response elements. Importance of the 5′-flanking region.' *J. Biol. Chem.*, 1997, **272**, 25252–25259.

134. X. Yuan, Ph.D. Dissertation, University of Kentucky, 2000.

135. C. Huin, L. Corriveau, A. Bianchi *et al.*, 'Differential expression of peroxisome proliferator-activated receptors (PPARs) in the developing human fetal digestive tract.' *J. Histochem. Cytochem.*, 2000, **48**, 603–611.

136. R. N. DuBois, R. Gupta, J. Brockman *et al.*, 'The nuclear eicosanoid receptor, PPARγ, is aberrantly expressed in colonic cancers.' *Carcinogenesis*, 1998, **19**, 49–53.

137. M. Lefebvre, B. Paulweber, L. Fajas *et al.*, 'Peroxisome proliferator-activated receptor γ is induced during differentiation of colon epithelium cells.' *J. Endocrinol.*, 1999, **162**, 331–340.

138. D. J. Peet, B. A. Janowski and D. J. Mangelsdorf, 'The LXRs: a new class of oxysterol receptors.' *Curr. Opin. Genet. Dev.*, 1998, **8**, 571–575.

139. S. A. Kliewer, J. T. Moore, L. Wade *et al.*, 'An orphan nuclear receptor activated by pregnanes defines a novel steroid signaling pathway.' *Cell*, 1998, **92**, 73–82.

140. L. B. Moore, D. J. Parks, S. A. Jones *et al.*, 'Orphan nuclear receptors constitutive androstane receptor and pregnane X receptor share xenobiotic and steroid ligands.' *J. Biol. Chem.*, 2000, **275**, 15122–15127.

141. N. Masahiko and P. Honkakoski, 'Induction of drug metabolism by nuclear receptor CAR: molecular mechanisms and implications for drug.' *Eur. J. Pharm. Sci.*, 2000, **11**, 259–264.

142. C. Song, R. A. Hiipakka and S. Liao, 'Selective activation of liver X receptor α by 6α-hydroxy bile acids and analogs.' *Steroids*, 2000, **65**, 423–427.

143. D. J. Peet, S. D. Turley, W. Ma *et al.*, 'Cholesterol and bile acid metabolism are impaired in mice lacking the nuclear oxysterol receptor LXR α.' *Cell*, 1998, **93**, 693–704.

144. G. Wolf, 'The role of oxysterols in cholesterol homeostasis.' *Nutr. Rev.*, 1999, **57**, 196–198.

145. K. S. Miyata, S. E. McCaw, H. V. Patel *et al.*, 'The orphan nuclear hormone receptor LXR α interacts with the peroxisome proliferator-activated receptor and inhibits peroxisome proliferator signaling.' *J. Biol. Chem.*, 1996, **271**, 9189–9192.

146. D. Stroup, M. Crestani and J. Y. Chiang, 'Orphan receptors chicken ovalbumin upstream promoter transcription factor II (COUP-TFII) and retinoid X receptor (RXR) activate and bind the rat cholesterol 7α-hydroxylase gene (CYP7A).' *J. Biol. Chem.*, 1997, **272**, 9833–9839.

147. P. Bentley, I. Calder, C. Elcombe *et al.*, 'Hepatic peroxisome proliferation in rodents and its significance for humans.' *Food Chem. Toxicol.*, 1993, **31**, 857–907.

148. P. R. Holden and J. D. Tugwood, 'Peroxisome proliferator-activated receptor α: role in rodent liver cancer and species differences.' *J. Mol. Endocrinol.*, 1999, **22**, 1–8.

149. A. I. Choudhury, S. Chahal, A. R. Bell *et al.*, 'Species differences in peroxisome proliferation; mechanisms and relevance.' *Mutat. Res.*, 2000, **448**, 201–212.

150. P. Passilly, B. Jannin and N. Latruffe, 'Influence of peroxisome proliferators on phosphoprotein levels in human and rat hepatic-derived cell lines.' *Eur. J. Biochem.*, 1995, **230**, 316–321.

151. J. D. Tugwood, T. C. Aldridge, K. G. Lambe *et al.*, 'Peroxisome proliferator-activated receptors: structures and function.' *Ann. NY Acad. Sci.*, 1996, **804**, 252–265.

152. J. D. Tugwood, P. R. Holden, N. H. James *et al.*, 'A peroxisome proliferator-activated receptor-α (PPARα) cDNA cloned from guinea-pig liver encodes a protein with similar properties to the mouse PPARα: implications for species differences in responses to peroxisome proliferators.' *Arch. Toxicol.*, 1998, **72**, 169–177.

153. D. Auboeuf, J. Rieusset, L. Fajas *et al.*, 'Tissue distribution and quantification of the expression of mRNAs of peroxisome proliferator-activated receptors and liver X receptor-α in humans: no alteration in adipose tissue of obese and NIDDM patients.' *Diabetes*, 1997, **46**, 1319–1327.

154. C. N. A. Palmer, M. H. Hsu, K. J. Griffin *et al.*, 'Peroxisome proliferator activated receptor-α expression in human liver.' *Mol. Pharmacol.*, 1998, **53**, 14–22.

155. S. Westin, R. Kurokawa, R. T. Nolte *et al.*, 'Interactions controlling the assembly of nuclear-receptor heterodimers and co-activators.' *Nature*, 1998, **395**, 199–202.

156. U. Varanasi, R. Chu, Q. Huang *et al.*, 'Identification of a peroxisome proliferator-responsive element upstream of the human peroxisomal fatty acyl coenzyme A oxidase gene (published erratum appears in *J. Biol. Chem.* 1998 Nov 13; **273**(46); 30842).' *J. Biol. Chem.*, 1996, **271**, 2147–2155.

157. N. J. Woodyatt, K. G. Lambe, K. A. Myers *et al.*, 'The peroxisome proliferator (PP) response element upstream of the human acyl CoA oxidase gene is inactive among a sample human population: significance for species differences in response to PPs.' *Carcinogenesis*, 1999, **20**, 369–372.

158. S. C. Hasmall, N. H. James, N. Macdonald *et al.*, 'Species differences in response to diethylhexylphthalate: suppression of apoptosis, induction of DNA synthesis and peroxisome proliferator activated receptor α-mediated gene expression.' *Arch. Toxicol.*, 2000, **74**, 85–91.

J.P. Vanden Heuvel, G.H. Perdew, W.B. Mattes and W.F. Greenlee (Eds.)
Comprehensive Toxicology, Vol. xiv

## 14.3 GENETIC DETERMINANTS OF SUSCEPTIBILITY TO ENVIRONMENTAL AGENTS

# 14.3.1
# Introduction and Overview

## D. J. TOMSO and D. A. BELL

*National Institute for Environmental Health Sciences, Research Triangle Park, NC, USA*

## 14.3.1.1  HUMAN GENETIC VARIATION

Human genetic variation is both complex and self-evident. Characteristics such as appearance, size, and even temperament are influenced by inheritance, and a tremendous diversity of traits is maintained within the human population. It is not surprising that genetic variation has a pronounced impact on other less-obvious characteristics, including human responses to toxic exposures. The rapid development of molecular techniques in the biological sciences, coupled with global efforts to sequence the human genome, has provided a unique opportunity to identify and describe genetic variation as it relates to toxicology. In this chapter, we introduce some of the terminology and concepts required for a study of human genetic variation (see Table 1), while subsequent chapters in this section provide in-depth reviews of current topics. We also include specific examples of the interplay between human genes and toxic exposures, and suggest a framework in which to consider the broader concept of variation as it applies to toxicology.

## 14.3.1.2  VARIATION, GENETIC DISEASE, AND POLYMORPHISM

Genetic differences distinguish individuals within a population. It is possible for unique

**Table 1** Terminology.

| Term | Definition and comments |
| --- | --- |
| Allele | In classical (Mendelian) genetics, an allele refers to a particular version of a gene, typically associated with a phenotype (e.g., blue eyes or wrinkled seed coats). In molecular genetics, the basic concept is the same, except that we can now identify alleles as specific DNA sequences at a given location (and not necessarily within a gene), many of which may lack observable phenotypes. |
| Frequency | The frequency of a polymorphism measures how often it occurs in a population. Frequencies can be measured for any allelic combination of interest (e.g., homozygous variant, heterozygous). |
| Haplotype | A specific combination of alleles that occur on a chromosome and that tend to be inherited together. Depending on context, the DNA segment associated with a haplotype can be small (e.g., an exon) or very large (an entire chromosome). |
| Locus | A locus is a particular region or site within a DNA sequence. |
| Penetrance | Penetrance is a measure of the phenotypic impact of an allele. High-penetrance alleles generally have easily observed phenotypes, while low-penetrance alleles have more subtle effects. Assessing impact can be quite complicated, especially for conditional or exposure-specific phenotypes. |
| Polymorphism | A polymorphism is a common DNA sequence variation within a population. Typically, a polymorphism is represented at 1% or greater frequency, although this determination is complicated by the composition and size of the population examined. Polymorphisms are distinct from rare, disease-causing mutations, which generally occur at much lower frequencies. |
| SNP | A single nucleotide polymorphism, the most common type of polymorphism. SNPs represent alternative base pairs at the same position in a DNA sequence and are distinct from other types of polymorphisms (e.g., insertions, deletions, and rearrangements). |
| Variant | A polymorphism consists of at least two alternative sequences at a particular locus. By convention, the variant sequence is the sequence that occurs less frequently at that site. Multiple variant alleles can exist at a given site. For polymorphisms with high frequency, it may be difficult to establish which alleles are the variants and which is 'normal.' |

genetic changes to occur *de novo* in the germline of a newborn (for example, as a result of mutation in sperm or egg). Such changes are thought to be randomly distributed in the genome and, because they occur only in single individuals, produce no predictable impact on the human population as a whole. Mutations also occur in critical physiological genes and produce diseases, such as cystic fibrosis or hemophilia, which can be inherited from one generation to the next. Disease-causing mutations have tragic health effects but fortunately are relatively rare in human populations, occurring at frequencies of about 1 in 10,000 individuals. In contrast, *polymorphisms* are defined as genetic differences between individuals that occur at a frequency of at least 1% in a population. Although these commonly found variations are generally considered distinct from rare, disease-causing mutations, this separation is sometimes unclear (see Impact and Penetration, below). It is clear, however, that polymorphisms are widespread, take many forms, and have demonstrable effects on human health.

A variety of DNA sequence changes can be considered polymorphisms, including nucleotide substitutions, deletions, insertions, and repeats. The most common type is the *single nucleotide polymorphism*, or *SNP* (commonly pronounced "snip"). As a large proportion of the human genome is noncoding (i.e., it does not generate messenger RNA or encode proteins), a high percentage of polymorphisms (>95%) occur in regions of the genome that have no apparent functional role.[1] These *noncoding polymorphisms* can be quite abundant. In a high density survey of over 4.5 million sequences and 920,752 polymorphisms, the International SNP Map Working Group reported an average spacing of 1.3 kilobases (kb) between SNPs in human chromosomes. Polymorphisms within coding regions were observed at a slightly higher frequency (approximately once every 1.1 kb), although this difference may simply represent higher density of data collection for coding regions. The same group identified a total of 1.4 million SNPs and found, on average, two coding SNPs in every human gene.[1] Other gene-specific analyses of variation, particularly those examining large numbers of individuals, have reported higher overall SNP densities, with some studies reporting SNPs spaced as closely as one every 200 to 500 basepairs (bp).[2–4] In general, studies that compare greater numbers of individuals will detect SNPs that occur at lower frequencies in a population, and consequently will observe SNPs at higher densities. Overall, it is clear that SNPs are very abundant, and that all human genes display some variability.

Analyses in global populations suggest that most polymorphisms are of ancient origin. In modern populations, polymorphisms presumably exist as a consequence of several forces, including mutation, selection, population movement, and random variation. In order for a unique mutation to become a polymorphism (i.e., to increase to a frequency of $\sim 1\%$ or greater in the population), it must be disseminated by a dynamic population process. A mutation that confers a survival advantage can be selected for and preserved within a population. Alternatively, a genetic "founder effect" can occur, in which one or a few individuals carrying a mutation give rise to a large population of descendents (for example, by populating an isolated geographic area). In fact, the current distribution and frequencies of polymorphisms in human groups reflect the historical evolution, migration and interbreeding of human populations. Most high frequency polymorphisms are shared among all people, while some less common polymorphisms (1–10%) are unique to individual ethnic groups.[1,3] Africans, who represent the most ancient population on earth, have accumulated the largest number of unique polymorphisms.[1,3] The abundance of polymorphisms in the human genome allows clear differentiation of individuals as well as population groups. This discrimination forms the basis for DNA-based forensic science. Identifying those polymorphisms that influence human health, especially in the context of complex and subtle responses to toxic exposures, is a formidable challenge for the modern molecular toxicologist.

## 14.3.1.3  GENE/ENVIRONMENT INTERACTIONS

Some highly polymorphic genes or gene families in humans offer an intriguing perspective on polymorphism and its relevance to toxicology. For example, xenobiotic metabolizing enzymes convert many toxic components in smoke and food to nontoxic metabolites.[5]

Presumably, particular configurations and activities of these enzymes have been preserved within the population because of historical exposure to environmental toxins. The genes encoding these enzymes also have a surprisingly high frequency of loss-of-function alleles, ultra high-activity alleles, or highly inducible alleles (frequencies for some alleles approach 50% in the population).[6,7] It is interesting to note that some nontoxic chemicals can be rendered toxic by the same set of enzymes that are, from an evolutionary standpoint, protective. This example illustrates a key concept: it is the interaction between *environment* and *genetics* that determines human response to toxic exposures.

## 14.3.1.4  FUNCTIONAL VERSUS NONFUNCTIONAL POLYMORPHISMS

Polymorphisms begin their environmental interactions at the molecular level. With some exceptions, those that occur in non-coding and non-regulatory regions of the genome have little effect on gene function or phenotype. Furthermore, some polymorphisms in coding regions have no apparent structural effect on proteins. Silent polymorphisms are generally third-position (in the codon) nucleotide changes that do not result in amino acid alterations. Some silent polymorphisms may have subtle effects (for example on codon usage or mRNA stability), but these effects are not well characterized. Likewise, it is conceivable that changes in non-coding regions may have a secondary impact on gene expression, for example by influencing chromatin structure or altering protein-binding sites.

Some polymorphisms that result in altered amino acids may have no apparent functional effect because the variant amino acid is biochemically similar to the common amino acid. Although these situations have not been well-studied, such conservative changes could produce subtle or conditional effects on function. In contrast to noncoding or silent polymorphisms, some changes in the coding regions of genes can directly affect phenotype (i.e., those which nonconservatively alter amino acids or introduce stop codons). An amino acid change can alter a protein's shape, affect its charge, disrupt an active site, or otherwise influence some critical feature of the molecule. Polymorphic alleles with partially or completely deleted coding regions, as well as those with duplicated coding regions, are surprisingly common. These types of alterations can have a significant impact by producing loss-of-function or high-activity phenotypes. For example, one polymorphism in the *CCR5* chemokine receptor gene results in a 32 bp deletion in the coding region. A functional *CCR5* receptor facilitates entry of the HIV virus into human lymphocytes, where it establishes an infection.[8,9] This polymorphism occurs at about 9% frequency in people of European descent, and inheriting two copies of this defective receptor provides resistance to HIV infection.[8,10] Similar coding polymorphisms exist in genes involved in biological responses to toxic exposures (see Genetic Modulation of Exposure and Response, below).

### 14.3.1.5  PENETRANCE, FREQUENCY AND SELECTION

*Penetrance* describes the impact that a particular mutation or polymorphism has on phenotype; those that have a large impact (usually disease mutations) are considered high-penetrance alleles, while those with less impact are considered low-penetrance. The relationship between penetrance and frequency for mutations or polymorphisms is complex. In general it is assumed that genetic variants that have adverse effects, the population frequency will be inverse to penetrance. Conversely, high frequency polymorphisms would be expected to have more modest effects on phenotype. This follows from the general principle of selection, in that variant alleles that reduce survival should remain rare in a population, while those with some benefit should increase in frequency. For example, it is hypothesized that the *CCR5* deletion polymorphism described above may have conferred resistance to other retroviral infections during human evolution. Additionally, no biological disadvantage (e.g., disease) has been attributed to the variant *CCR5* allele to date. Consequently, the variant allele may have been maintained within the population because it confers a benefit without imparting a significant disadvantage. However, the benefit is evident only under specific conditions (exposure to viral pathogens), and a negative biological impact may yet be discovered for the *CCR5* variant allele. Although not strictly within the realm of toxicology, studies of human disease provide an excellent context for an investigation of gene/environment interactions. A classic example is that of the sickle-cell allele, which is common in areas with endemic malaria. Heterozygous individuals are resistant to infection by malarial pathogens, while homozygous variant sickle-cell individuals develop severe anemia. In this case, an allele that is manifestly detrimental in some

conditions is preserved in the population because it is beneficial under other conditions. In these examples, frequency and penetrance are closely related through the process of selection, although in complex ways. In human populations, frequency probably depends on a combination of penetrance and random population effects. Clearly, alleles that result in deadly disease (e.g., are high penetrance, in a negative way) should occur at low frequencies, although the sickle-cell example shows that this is not always the case. Similarly, alleles that are highly beneficial should exist at higher frequencies. Even for disease alleles that have dramatic effects on phenotype, gene-environment interactions can significantly affect frequency. For polymorphisms with subtle or condition-specific effects, the case is even more complicated. There is a general belief that in complex diseases like cancer and hypertension there are hundreds of genes that alter risk in relatively small ways (many low-penetrance alleles, each increasing or reducing risk by a small amount). Each of these genes appears to have many at-risk alleles, with varying levels of effect. Some combinations of these low-impact genetic traits may be especially high-risk, while others may be particularly low-risk. Only when both the environmental factors and the alleles themselves are appropriately considered will the influence of these polymorphisms be fully evident.

It is clear that polymorphisms are very frequent in the genome and very common in the population. Most are totally innocuous, some are associated with disease in ways that seem obvious (detoxification of carcinogens, protection against viruses), and some have subtle effects that are not understood. However, most do not have detectable effects on biological function. Even among polymorphisms in the coding regions of genes, only a fraction will have a measurable effect on the biological function of the gene. In spite of this uncertainty, the sheer number of polymorphisms all but guarantees that some will have a real impact on human health. For the remainder of this chapter, we will provide specific examples of polymorphisms that demonstrably influence some aspect of modern toxicology.

### 14.3.1.6  GENETIC MODULATION OF EXPOSURE AND RESPONSE

In addition to causing acute toxicity to specific organs or cells, toxic agents can damage human health by generating genomic damage. This type of genotoxic stress is particularly significant in complex diseases such as cancer, where long-term effects are thought to modulate the development of disease. As shown in Figure 1, the route from exposure to disease involves many steps, all of which are potentially influenced by the genetics of the affected individual. Uptake and distribution of toxic agents and subsequent metabolism to reactive or non-reactive forms can significantly affect the generation of genotoxic damage (for a review of metabolic enzymes see Reference 7). Once damage has occurred, DNA repair pathways have the capability of removing many types of lesions. As with metabolism genes, genetic variation in DNA repair functions seems to influence the development of complex disease.[11,12] Post-damage functions, such as entry into apoptosis or proliferation, may also play a role in the development of human tumors.

Polymorphisms have been described in a large number of metabolism and DNA repair genes (for example see References 5,12). Evaluating the impact of these polymorphisms has been difficult because the processes involved are complex and multifaceted. Furthermore, as described above, the impact of a particular polymorphism may be evident only under specific conditions, or a given polymorphism may have no effect whatsoever. In spite of this uncertainty, several clear examples have emerged that illustrate the importance of genetic variation as it relates to toxic exposure and response. *N*-acetyltransferases are enzymes that act on aromatic or heterocyclic amine carcinogens and modify their potential toxicity. A second class of toxin-metabolizing enzymes, the glutathione-*S*-transferases, conjugate glutathione to potential carcinogens and facilitate their elimination. Both classes exhibit considerable polymorphism, and some variants have been linked to the development of toxicant-induced cancers (described below, also see References 13–15) Similarly, the human *XRCC1* gene, which encodes a DNA base excision repair enzyme,

is polymorphic.[16,17] Some of the variants of this gene are associated with increased mutation and disease and will be discussed below.

### 14.3.1.6.1 Carcinogen Metabolism Polymorphisms

The metabolism of carcinogens, in particular aromatic and heterocyclic amine compounds, provides perhaps the best example of polymorphism influencing the risk of exposure-induced disease. Both aromatic and heterocyclic amines are present in the environment from several sources, including cigarette smoke and cooked foods. In the cell, many carcinogens are converted first to reactive intermediates by Phase I enzymes (e.g., cytochromes P450, CYPs), then conjugated to water-soluble moieties (e.g., glutathione) by Phase II enzymes to facilitate elimination.[7] The Phase II enzymes include, among others, the *N*-acetyltransferases (NATs) and the glutathione-*S*-transferases (GSTs). Alterations in the activities of these enzymes would be expected to influence the body's ability to activate or detoxify carcinogens. Indeed, specific polymorphisms in both NATs and GSTs have demonstrable impacts, by both molecular and epidemiological criteria, on human health.

The human *NAT1* and *NAT2* genes encode two distinct *N*-acetyltransferases, and both enzymes are capable of the metabolic activation of carcinogens. Polymorphisms in *NAT2* are abundant and are used to classify the population into rapid, intermediate, and slow acetylator phenotypes.[18,19] Variation in these genes has been associated with cancer of the breast, colon, lung, and urinary bladder.[18,20,21] *NAT1* alleles that confer both high and low activity have also been identified,[22] and are likewise associated with altered risk of bladder and colorectal cancer.[23,24] However, of the 19 *NAT1* and 24 *NAT2* alleles thus far identified, only a limited number have been evaluated for

| Organism/Organ Effects | Cellular Effects | DNA Effects | Post-Damage Effects |
|---|---|---|---|
| **Behavior** | **Influx** | **DNA Damage** | **Checkpoints** |
| **Uptake** |   Membrane Structure | **DNA Repair** | **Apoptosis** |
| **Distribution** |   Pumps |   Base Excision | **Proliferation** |
| |   Endocytosis |   Nucleotide Excision | **Immune Evasion** |
| | **Metabolism** |   Break Repair | |
| |   Phase I | **DNA Replication** | |
| |   Phase II | **Translesion Synthesis** | |
| | **Eflux** | | |
| | **Signal Transduction** | | |

**Figure 1** Genetic modulation of exposure-induced disease. This table includes some of the many processes that are important in the development of disease. Some are influenced by known polymorphisms (e.g., metabolism, repair).

their impact on cancer risk. Furthermore, the relationship between *NAT1/NAT2* phenotype and disease is complicated and may involve interaction between the two genes, as well as divergent pathways for different types of cancer (for example see References 25–27).

The glutathione-*S*-transferases comprise a large super family of soluble enzymes, with multiple classes (e.g., alpha (*GSTA*), mu (*GSTM*), pi (*GSTP*), theta (*GSTT*)) and subclasses (e.g., *GSTA1–4, GSTM1–5, GSTT1–2*) represented.[28] Polymorphisms are common in these genes, including nonfunctional alleles that result from the absence or deletion of the entire coding region.[29–31] Two such null alleles, in the *GSTM1* and *GSTT1* genes, are common in human populations.[32–34] For example, an estimated 45% of ethnic whites lack a functional *GSTM1* allele, and 20% of whites lack *GSTT1* activity.[35] The impact of these widespread polymorphisms has not been fully assessed, but it appears that a nonfunctional *GSTM1* allele increases the risk for cancer of the bladder and lung.[6,36] Additionally, molecular studies have suggested that certain combinations of *GST* and *CYP* alleles may result in increased DNA damage in human cells exposed to carcinogens.[37] By implication, individuals with similar genotypes may be at increased risk for developing disease after exposure to cigarette smoke. As with the *NAT* polymorphisms, many *GST* polymorphisms remain uncharacterized, and additional sequence variants undoubtedly remain to be discovered. In both cases, however, the precedent is clear—genetic variation in carcinogen metabolism genes has a significant impact on disease risk.

### 14.3.1.6.2   DNA Repair Polymorphisms

DNA repair systems in human cells operate by removing or circumventing DNA lesions produced by both endogenous and exogenous sources of damage. A diversity of repair pathways, including base excision repair, nucleotide excision repair, recombinational repair, and mismatch repair, have evolved to handle a wide array of lesions. Each of these systems involves multiple, interacting proteins, and significant cross-talk between the various repair pathways also occurs. Mutations in some known or presumed repair genes can cause serious illnesses, including xeroderma pigmentosum, Werner's syndrome, and Cockayne syndrome.[38,39] Alterations in human mismatch repair genes (e.g., *hMSH2*, *hMSH3*, *hMLH1*) are associated with

increased genomic instability and familial dispositions to certain cancers.[40,41] As DNA repair enzymes are directly responsible for removing toxin-induced damage, the potential impact of polymorphisms in these genes is immense. Any variation could significantly influence the efficiency and fidelity of DNA repair pathways.

Although few widespread DNA repair polymorphisms have been identified to date, work in this field is rapidly expanding. A polymorphism in the human *XRCC1* gene, which encodes an enzyme involved in base excision repair, is associated with increased mutation frequencies in DNA from cigarette smokers.[17,42] Polymorphisms in XRCC1 are linked to head and neck, lung, and bladder cancers, again in association with cigarette smoke exposure.[43–45] Polymorphisms in the human XPD repair enzyme have been linked to altered DNA repair capacity *in vitro* and to disease risk.[46,47] Although current examples are limited, the conceptual link between DNA repair and exposure-induced disease is clear. Polymorphism in repair genes, as well as related genes involved in DNA replication and genome maintenance, may yet prove to have a major impact on human health.

### 14.3.1.7   FUTURE RESEARCH

Individuals with certain polymorphisms in xenobiotic metabolism (including carcinogen metabolism), DNA repair, and a large number of other candidate susceptibility genes may be at increased risk for exposure-induced diseases. Existing research has revealed some strong associations between genotype and risk for these diseases, particularly for genes involved in metabolism of foreign compounds. Similarly, a recent comprehensive review of epidemiologic data concluded that DNA repair proficiency and cancer risk were significantly correlated in humans.[11] The authors suggested that a detailed analysis of repair polymorphisms, in conjunction with *in vitro* assays for repair efficiency and large-scale epidemiology studies, would be valuable for the advancement of the field. However, a large number of genes potentially associated with human disease have not been evaluated for polymorphism content. Even relatively well-studied genes such as the *NAT* and *GST* families have many uncharacterized (and probably many other unidentified) polymorphisms.

Identifying polymorphisms is an important first step in understanding the relationship between genetic variation and susceptibility to

**Table 2** Online resources for polymorphism studies and environmental genomics.

| Project | Sponsoring Institution(s) | URL | Comments |
|---------|---------------------------|-----|----------|
| Environmental Genome Project | National Institute of Environmental Health Sciences (NIEHS), National Institutes of Health (NIH) | http://www.niehs.nih.gov/envgenom/ | The ultimate aim of the Environmental Genome Project is to understand the impact and interaction of environmental exposures on human disease. This site contains an overview of the NIEHS Environmental Genome Project, links to many other relevant sites, and funding opportunities. |
| GeneSNPs Public Internet Resource | NIEHS, University of Utah Genome Center | http://www.genome.utah.edu/genesnps/ | This Environmental Genome Project web resource integrates gene, sequence and polymorphism data into individually annotated gene models. The human genes included are related to DNA repair, cell cycle control, cell signaling, cell division, homeostasis and metabolism, and are thought to play a role in susceptibility to environmental exposure. |
| National Center for Biotechnology Information SNP Database | National Center for Biotechnology Information (NCBI), NIH | http://www.ncbi.nlm.nih.gov/SNP/ | This site houses dbSNP, a large and regularly-updated database of human single nucleotide polymorphisms. |
| The SNP Consortium | A consortium of several major biotechnology, pharmaceutical, and computer companies. | http://snp.cshl.org/ | A non-profit foundation organized for the purpose of providing public genomic data. Its mission is to develop up to 300,000 SNPs distributed evenly throughout the human genome and to make the information related to these SNPs available to the public without intellectual property restrictions. |
| HGBASE | Various European public and private organizations | http://hgbase.cgr.ki.se/ | The objective is to provide an accurate, high utility and ultimately fully comprehensive catalog of normal human gene and genome variation, useful as a research tool to help define the genetic component of human phenotypic variation. |

toxic exposures. Global efforts to sequence the human genome have provided a valuable opportunity to collect information on polymorphism. In addition to whole-genome sequencing efforts, several major projects are underway to collect cross-genome data, many with the specific goal of identifying polymorphisms in human populations (see Table 2). To date, these efforts are poorly integrated and only sparsely annotated, with data in diverse formats and from many sources. However, these projects should soon produce databases representing millions of individual polymorphisms, and the tools to evaluate this data are developing at a similar rapid pace.

Once polymorphisms are identified, the enormously complex process of ascertaining their significance begins. In the context of toxicology, the polymorphisms must be evaluated under specific conditions of exposure in order to establish their relationship to human health. Although a formidable challenge, these types of studies also hold the potential for tremendous benefit. For example, knowledge of the impact of a given allele should allow the identification of subgroups within the population who are particularly susceptible to exposure. Preventative, regulatory, or even therapeutic efforts could then be efficiently targeted, and disease averted or ameliorated. On a similar note, it is known that the xenobiotic metabolizing enzymes are important to the metabolism of various therapeutic drugs. Pharmaceutical companies are genotyping individuals in clinical trials in order to understand why drugs are ineffective or toxic in some people. Similar approaches can be envisioned for a wide variety of exposure-related genetic profiling applications.

The large number of genes involved in human responses to toxic challenge, coupled with the high frequency of low-penetrance genotypes, suggest that, as Francis Collins, director of the National Human Genome Research Institute, has said, "Everybody is at risk for something." In fact, it is probable that each of us has some level of inherent genetic risk for many adverse health outcomes. Additionally, each of us likely shares risk-group identity with many people from all ethnic, economic, and geographical groups. Despite this widespread dispersion of polymorphisms, a significant potential for genetic discrimination already exists. Technological advancement and scientific endeavor continue to provide more abundant and reliable information regarding genetic risk, and it is crucial that social and legal development proceed apace. Although we cannot (as yet) change our genotype, we can and should ensure that our environment, lifestyle, and laws incorporate our emerging knowledge of genetic risk in both an ethical and scientifically sound manner.

### 14.3.1.8  REFERENCES

1. R. Sachidanandam, D. Weissman, S. C. Schmidt *et al.*, 'A map of human genome sequence variation containing 1.42 million single nucleotide polymorphisms.' *Nature*, 2001, **409**, 928–933.
2. M. Cargill, D. Altshuler, J. Ireland *et al.*, 'Characterization of single-nucleotide polymorphisms in coding regions of human genes.' *Nat. Genet.*, 1999, **22**, 231–238.
3. M. K. Halushka, J. B. Fan, K. Bentley *et al.*, 'Patterns of single-nucleotide polymorphisms in candidate genes for blood-pressure homeostasis.' *Nat. Genet.*, 1999, **22**, 239–247.
4. D. A. Nickerson, S. L. Taylor, S. M. Fullerton *et al.*, 'Sequence diversity and large-scale typing of SNPs in the human apolipoprotein E gene.' *Genome Res.*, 2000, **10**, 1532–1545.
5. L. W. Wormhoudt, J. N. Commandeur and N. P. Vermeulen, 'Genetic polymorphisms of human *N*-acetyltransferase, cytochrome P450, glutathione-*S*-transferase, and epoxide hydrolase enzymes: relevance to xenobiotic metabolism and toxicity.' *Crit. Rev. Toxicol.*, 1999, **29**, 59–124.
6. D. A. Bell, J. A. Taylor, D. F. Paulson *et al.*, 'Genetic risk and carcinogen exposure: a common inherited defect of the carcinogen-metabolism gene glutathione *S*-transferase M1 (GSTM1) that increases susceptibility to bladder cancer.' *J. Natl. Cancer Inst.*, 1993, **85**, 1159–1164.
7. U. A. Meyer and U. M. Zanger, 'Molecular mechanisms of genetic polymorphisms of drug metabolism.' *Annu. Rev. Pharmacol. Toxicol.*, 1997, **37**, 269–296.
8. S. J. O'Brien and J. P. Moore, 'The effect of genetic variation in chemokines and their receptors on HIV transmission and progression to AIDS.' *Immunol. Rev.*, 2000, **177**, 99–111.
9. N. L. Michael, L. G. Louie, A. L. Rohrbaugh *et al.*, 'The role of CCR5 and CCR2 polymorphisms in HIV-1 transmission and disease progression.' *Nat. Med.*, 1997, **3**, 1160–1162.
10. O. J. Cohen, A. Kinter and A. S. Fauci, 'Host factors in the pathogenesis of HIV disease.' *Immunol. Rev.*, 1997, **159**, 31–48.
11. M. Berwick and P. Vineis, 'Markers of DNA repair and susceptibility to cancer in humans: an epidemiologic review.' *J. Nat. Cancer Inst.*, 2000, **92**, 874–897.
12. H. W. Mohrenweiser and I. M. Jones, 'Variation in DNA repair is a factor in cancer susceptibility: a paradigm for the promises and perils of individual and population risk estimation?' *Mutat. Res.*, 1998, **400**, 15–24.
13. F. F. Kadlubar and A. F. Badawi, 'Genetic susceptibility and carcinogen-DNA adduct formation in human urinary bladder carcinogenesis.' *Toxicol. Lett.*, 1995, **82–83**, 627–632.
14. M. B. Oude Ophuis, E. M. van Lieshout, H. M. Roelofs *et al.*, 'Glutathione *S*-transferase M1 and T1 and cytochrome P4501A1 polymorphisms in relation to the risk for benign and malignant head and neck lesions.' *Cancer*, 1998, **82**, 936–943.

15. M. Sato, T. Sato, T. Izumo *et al.*, 'Genetic polymorphism of drug-metabolizing enzymes and susceptibility to oral cancer.' *Carcinogenesis*, 1999, **20**, 1927–1931.

16. M. R. Shen, I. M. Jones and H. Mohrenweiser, 'Nonconservative amino acid substitution variants exist at polymorphic frequency in DNA repair genes in healthy humans.' *Cancer Res.*, 1998, **58**, 604–608.

17. R. M. Lunn, R. G. Langlois, L. L. Hsieh *et al.*, 'XRCC1 polymorphisms: effects on aflatoxin B1-DNA adducts and glycophorin A variant frequency.' *Cancer Res.*, 1999, **59**, 2557–2561.

18. D. A. Bell, J. A. Taylor, M. A. Butler *et al.*, 'Genotype/ phenotype discordance for human arylamine *N*-acetyltransferase (NAT2) reveals a new slow-acetylator allele common in African-Americans.' *Carcinogenesis*, 1993, **14**, 1689–1692.

19. H. J. Lin, 'Smokers and breast cancer. "Chemical individuality" and cancer predisposition (editorial; comment).' *Jama*, 1996, **276**, 1511–1512.

20. K. F. Ilett, B. M. David, P. Detchon *et al.*, 'Acetylation phenotype in colorectal carcinoma.' *Cancer Res.*, 1987, **47**, 1466–1469.

21. N. P. Lang, M. A. Butler, J. Massengill *et al.*, 'Rapid metabolic phenotypes for acetyltransferase and cytochrome P4501A2 and putative exposure to food-borne heterocyclic amines increase the risk for colorectal cancer or polyps.' *Cancer Epidemiol. Biomarkers Prev.*, 1994, **3**, 675–682.

22. W. W. Weber and K. P. Vatsis, 'Individual variability in *p*-aminobenzoic acid *N*-acetylation by human *N*-acetyltransferase (NAT1) of peripheral blood.' *Pharmacogenetics*, 1993, **3**, 209–212.

23. D. A. Bell, A. F. Badawi, N. P. Lang *et al.*, 'Polymorphism in the *N*-acetyltransferase 1 (NAT1) polyadenylation signal: association of NAT1*10 allele with higher *N*-acetylation activity in bladder and colon tissue.' *Cancer Res.*, 1995, **55**, 5226–5229.

24. D. A. Bell, E. A. Stephens, T. Castranio *et al.*, 'Polyadenylation polymorphism in the acetyltransferase 1 gene (NAT1) increases risk of colorectal cancer.' *Cancer Res.*, 1995, **55**, 3537–3542.

25. P. M. Marcus, P. Vineis and N. Rothman, 'NAT2 slow acetylation and bladder cancer risk: a meta-analysis of 22 case-control studies conducted in the general population.' *Pharmacogenetics*, 2000, **10**, 115–122.

26. S. Mommsen, N. M. Barfod and J. Aagaard, '*N*-Acetyltransferase phenotypes in the urinary bladder carcinogenesis of a low-risk population.' *Carcinogenesis*, 1985, **6**, 199–201.

27. M. C. Yu, P. L. Skipper, K. Taghizadeh *et al.*, 'Acetylator phenotype, aminobiphenyl-hemoglobin adduct levels, and bladder cancer risk in white, black, and Asian men in Los Angeles, California.' *J. Nat. Cancer Inst.*, 1994, **86**, 712–716.

28. D. L. Eaton and T. K. Bammler, 'Concise review of the glutathione *S*-transferases and their significance to toxicology.' *Toxicol. Sci.*, 1999, **49**, 156–164.

29. J. Seidegard, W. R. Vorachek, R.W. Pero *et al.*, 'Hereditary differences in the expression of the human glutathione transferase active on trans-stilbene oxide are due to a gene deletion.' *Proc. Nat. Acad. Sci. USA*, 1988, **85**, 7293–7297.

30. F. Ali-Osman, O. Akande, G. Antoun *et al.*, 'Molecular cloning, characterization, and expression in *Escherichia coli* of full-length cDNAs of three human glutathione *S*-transferase π gene variants. Evidence for differential catalytic activity of the encoded proteins.' *J. Biol. Chem.*, 1997, **272**, 10004–10012.

31. M. A. Watson, R. K. Stewart, G. B. Smith *et al.*, 'Human glutathione *S*-transferase P1 polymorphisms: relationship to lung tissue enzyme activity and population frequency distribution.' *Carcinogenesis*, 1998, **19**, 275–280.

32. R. C. Strange, C. G. Faulder, B. A. Davis *et al.*, 'The human glutathione *S*-transferases: studies on the tissue distribution and genetic variation of the GST1, GST2 and GST3 isozymes.' *Ann. Hum. Genet.*, 1984, **48**, 11–20.

33. S. Pemble, K. R. Schroeder, S. R. Spencer *et al.*, 'Human glutathione *S*-transferase θ (GSTT1): cDNA cloning and the characterization of a genetic polymorphism.' *Biochem. J.*, 1994, **300**, 271–276.

34. S. C. Cotton, L. Sharp, J. Little *et al.*, 'Glutathione *S*-transferase polymorphisms and colorectal cancer: a HuGE review.' *Am. J. Epidemiol.*, 2000, **151**, 7–32.

35. C. L. Chen, Q. Liu and M. V. Relling, 'Simultaneous characterization of glutathione *S*-transferase M1 and T1 polymorphisms by polymerase chain reaction in American whites and blacks.' *Pharmacogenetics*, 1996, **6**, 187–191.

36. J. Brockmoller, I. Cascorbi, R. Kerb *et al.*, 'Combined analysis of inherited polymorphisms in arylamine *N*-acetyltransferase 2, glutathione *S*-transferases M1 and T1, microsomal epoxide hydrolase, and cytochrome P450 enzymes as modulators of bladder cancer risk.' *Cancer Res.*, 1996, **56**, 3915–3925.

37. M. Rojas, I. Cascorbi, K. Alexandrov *et al.*, 'Modulation of benzo[a]pyrene diolepoxide-DNA adduct levels in human white blood cells by CYP1A1, GSTM1 and GSTT1 polymorphism.' *Carcinogenesis*, 2000, **21**, 35–41.

38. A. D. Auerbach and P. C. Verlander, 'Disorders of DNA replication and repair.' *Curr. Opin. Pediatr.*, 1997, **9**, 600–616.

39. J. Nakura, L. Ye, A. Morishima *et al.*, 'Helicases and aging.' *Cell. Mol. Life Sci.*, 2000, **57**, 716–730.

40. C. Schmutte and R. Fishel, 'Genomic instability: first step to carcinogenesis.' *Anticancer Res.*, 1999, **19**, 4665–4696.

41. J. S. Hoffmann and C. Cazaux, 'DNA synthesis, mismatch repair and cancer.' *Int. J. Oncol.*, 1998, **12**, 377–382.

42. S. Z. Abdel-Rahman and R. A. El-Zein, 'The 399Gln polymorphism in the DNA repair gene XRCC1 modulates the genotoxic response induced in human lymphocytes by the tobacco-specific nitrosamine NNK.' *Cancer Lett.*, 2000, **159**, 63–71.

43. E. M. Sturgis, E. J. Castillo, L. Li *et al.*, 'Polymorphisms of DNA repair gene XRCC1 in squamous cell carcinoma of the head and neck.' *Carcinogenesis*, 1999, **20**, 2125–2129.

44. K. K. Divine, F. D. Gilliland, R. E. Crowell *et al.*, 'The XRCC1 399 glutamine allele is a risk factor for adenocarcinoma of the lung.' *Mutat. Res.*, 2001, **461**, 273–278.

45. M. C. Stern, D. M. Umbach, C. H. van Gils *et al.*, 'DNA repair gene XRCC1 polymorphisms, smoking, and bladder cancer risk.' *CEBP*, 2001, **10**.

46. M. Dybdahl, U. Vogel, G. Frentz *et al.*, 'Polymorphisms in the DNA repair gene XPD: correlations with risk and age at onset of basal cell carcinoma.' *Cancer Epidemiol. Biomarkers Prev.*, 1999, **8**, 77–81.

47. R. M. Lunn, K. J. Helzlsouer, R. Parshad *et al.*, 'XPD polymorphisms: effects on DNA repair proficiency.' *Carcinogenesis*, 2000, **21**, 551–555.

J.P. Vanden Heuvel, G.H. Perdew, W.B. Mattes and W.F. Greenlee (Eds.)
Comprehensive Toxicology, Vol. xiv

# 14.3.2
# Molecular Biomarkers for Human Liver Cancer

JOHN D. GROOPMAN,  PETA E. JACKSON,  JIA-SHENG WANG
and THOMAS W. KENSLER
*Johns Hopkins Medical Institutions, Baltimore, MD, USA*

## 14.3.2.1  INTRODUCTION

The use of chemical and biological specific biomarkers for identifying stages in the progression of development of the health effects of environmental agents has the potential for providing important information for critical regulatory, clinical and public health problems.[1,2] Since the development of a paradigm for molecular biomarkers by a committee of the National Research Council over a decade ago, some progress has been made in applying such chemical and biological biomarkers to specific environmental situations that may be hazardous to humans, as exemplified by the studies of aflatoxin and hepatitis B virus as risk factors for human liver cancer. The major goals of these studies are to develop and validate biomarkers that reflect specific exposures and predict disease risk in individuals. Presumably after an exposure each person has a unique response to both dose and time to disease onset. These responses will be affected both by intrinsic (genetic) and by extrinsic (such as dietary) modifiers. It is assumed that biomarkers that reflect the mechanism of action of a specific exposure situation will be strong predictors of an individual's risk of disease. It is also expected that these biomarkers can more clearly classify the status of exposure of individuals, local communities, and larger populations. These studies should also help to elucidate the molecular processes of chemically induced human disease and underlying susceptibility factors. A conceptual model for this work is shown in Figure 1.

### 14.3.2.2 HUMAN LIVER CANCER

Normally the liver functions to maintain homeostasis, by processing nutrients such as dietary amino acids, carbohydrates, lipids, and vitamins. The liver also is involved in the phagocytosis of particulate material in the splanchnic circulation, synthesis of serum proteins; formation of bile and biliary excretion of endogenous products and xenobiotics. This organ is the major site of biotransformation and detoxification of drugs, chemicals, and circulating metabolites in the body.[3,4] Thus, the liver's central role in total health status makes it vulnerable to a wide variety of environmental and occupational toxic insults.

There are two types of human malignant liver disorders associated with occupational and environmental hepatotoxicants; hepatocellular carcinoma (HCC) and hepatic angiosarcoma (HAS). HAS, also described as endothelial cell sarcoma, is a rare malignant tumor and is associated with chronic exposure to the vinyl chloride monomer, arsenic, anabolic steroids, and Thorotrast, a contrast agent that contained colloidal thorium dioxide, an emitter of α-particle ionizing radiation.[5–7] Many of the reported cases of HAS appeared in workers who were exposed chronically to vinyl chloride.[8,9] HAS has also been diagnosed in vineyard workers and others who used arsenicals, including Fowler's solution (1% potassium arsenite), or copper as a pesticide.[10] In addition, there were several case reports that long-term ingestion of arsenic-contaminated well-water could cause HAS.[11]

In contrast to HAS, HCC is one of the leading causes of cancer mortality in the world

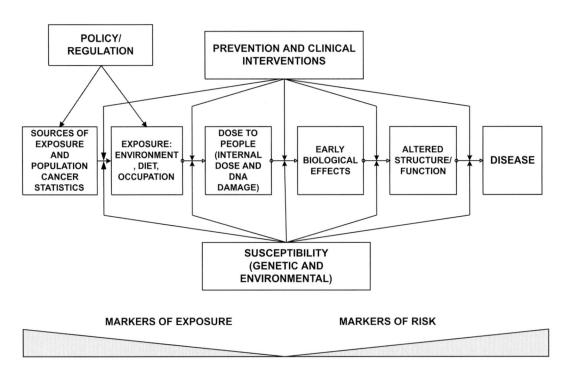

**Figure 1** Model for the molecular epidemiology paradigm.

and in parts of Asia and Africa it is the leading cause of cancer death.[12] Further, its rapidly developing fatal course makes it a difficult disease to cure. In the People's Republic of China, HCC is the third leading cause of cancer mortality and accounts for at least 250,000 deaths per year with an incidence rate in some areas of the country approaching 100 cases per 100,000 per year.[13] In contrast, HCC incidence in the United States is about 1.5 cases per 100,000 per year. Thus, this malignancy varies world-wide by at least 100 fold. In regions of the world with a high rate of HCC, this disease is typically diagnosed much earlier in life than in the U.S. For example, the median onset of liver cancer is 40–45 years of age in some high-risk regions of China.[14]

In the United States and most Western countries, excessive ethanol consumption, hepatitis B virus (HBV) and/or hepatitis C virus (HCV) infections, and possible exposure to chlorinated hydrocarbons have been estimated to account for as many as 75% of HCC cases.[15–17] In addition, a slightly increased risk is considered to be associated with use of oral contraceptives or androgenic-anabolic steroid hormones.[18] A number of epidemiologic studies have also examined the association between occupation and risk of HCC.[19,20] In parts of the US, an excess risk of HCC mortality has been described in certain occupations such as, oil refinery workers, plumbers, pipefitters, butchers and meat cutters, textile workers and longshoremen.[21] Another study reported increase risk of HCC in the chemical or petrochemical industry.[22] A case-control study of HCC in New Jersey found an association with road builders, manufacturers of automobiles and plastics, workers exposed to anesthetics, gas station workers, even after adjustment for level of ethanol consumption.[23] Studies also described increases in HCC in the highway construction industry, especially among workers who were exposed to asphalt,[24] synthetic abrasive manufacture[25] and automobile workers.[26] Many of these studies; however, could not adjust for some of the major viral risk factors, such as the hepatitis viruses.

## 14.3.2.3 VALIDATION STRATEGIES AND METHODOLOGICAL APPROACHES IN MOLECULAR BIOMARKER STUDIES

A number of issues distinguish the problems encountered in molecular biomarker research from those in traditional epidemiology and experimental bioassays. Since a major objective is to assess an individual's exposure status as judged by biomarker measurements, the analytical approaches to measure these biomarkers must be sufficiently sensitive and specific to quantify levels in a limited sample from a single person. Therefore, interpretations of biomarker levels are made on an individual basis and not by using population means and variance. For these markers to be truly valid and valuable, it is the individual and not the group level that must track with specific disease outcome. Thus, the biomarker must span the range from exposure to disease outcome to meet these criteria. Few such biomarkers have heretofore met such stringent criteria.

The problems faced by the field are exemplified by considering the development of chemical-DNA adducts as biomarkers of specific exposures. Since formation of chemical-DNA adducts are thought to be important in the carcinogenesis pathway, it would be expected that the person with higher levels of DNA adduct biomarkers would be at greater risk of disease. Adduct studies, however, may be hindered because samples from target organs are often not available. Hence surrogate markers, such as chemical-protein adducts, and surrogate tissues, such as white blood cells, have been used routinely. The rationale for using surrogate targets is well justified on practical grounds. Interpretation of the data obtained with these markers must be very conservative; however, because the more removed a biomarker becomes from the causal pathway of disease, the less precise it will be to predict disease risk.

Given the difficulty of the studies for comprehensive evaluation of chemical or biological-specific biomarkers, recent efforts have been made to use the levels of a gene or gene product as a surrogate for an exposure when a disease outcome is known. This raises another major issue. It is presumed that if a gene-environment interaction is essential for a disease outcome, if the disease is known and the gene or gene product can be measured, a putative exposure can be inferred with high certainty. These studies are very attractive because of the multitude of genotyping and phenotyping methods now available and the relative ease of designing a case-control study to test the strength of the interaction. In rare instances when the level of exposure is high, the dominance or penetrance of the gene is strong, and the disease outcome is closely linked temporally, this strategy will be effective. Given our present lack of information on many exposures in people, however, it is doubtful that case-control studies that attempt to backtrack from disease and gene to initial exposure will be very accurate. For the molecular biomarkers to be used effectively, very high standards for validation and utilization must be established, such as those used in evaluating serum cholesterol as a risk factor for heart disease. Years of systematic study and multiple validation methods were required before the public could be provided with risk guidelines for personal use. Consequently, we should strive to meet a similar objective for toxicological biomarkers.

### 14.3.2.4 THE MOLECULAR BIOMARKER PARADIGM

In order to enhance the opportunity for biomarkers to guide and assess public health-based strategies for prevention, we have extended the molecular biomarker paradigm by adding molecular interventions along the continuum from exposure to disease (Figure 1). Development of putative biomarkers for environmental agents must be based on specific knowledge of metabolism, product formation, and general mechanisms of action.[27] Validation of any biomarker-effect link requires parallel experimental and human studies. Ideally, an appropriate experimental model is used first to determine the associative or causal role of the marker in the disease or effect pathway and to establish a dose response. The putative marker can then be validated in pilot human studies to establish sensitivity, specificity, accuracy, and reliability for individuals.[28] Results of these studies can be used to assess intra- or interindividual variability, background levels, and relationship of marker to external dose or to disease status, as well as feasibility for use in larger population-based studies. It is important to establish a connection between the biological marker and exposure, effect, or susceptibility. Full interpretation of the information obtained may require prospective epidemiological studies to demonstrate the role of the marker in the overall pathogenesis of the disease or effect.

To implement the strategies described above, we have devised a rational systematic approach for the validation and application of a sub-set of molecular biomarkers, the chemical-specific biomarkers, to human studies. Our model for investigating exposures from, and risk to, environmental carcinogens by parallel investigations in experimental systems and humans is shown in Figure 2. The two endpoints for this process are markers of exposure and markers of risk.

#### 14.3.2.4.1 Exposure Markers

People are exposed to chemical, physical, or biological agents through contaminated air, water, soil, or food.[27] Thus, a person's exposure is the result of proximity to the agent superimposed over many modifying factors. Biomarkers of exposure may be the parent chemical itself, as exemplified by metals such as lead. Frequently, however, it is the metabolic products of the agent that occur in the body that serve as the markers of exposure and provide the "internal dose measure." Carcinogen-DNA and protein adducts are also markers of exposure and in the model shown in Figure 1 are referred to as "biologically effective dose" measures. Finally, altered structure or function, such as those mediated through changes in gene expression, can serve as the exposure marker when the two processes are directly linked.

Ideally, a biomarker of exposure would indicate the presence and magnitude of previous exposure to an environmental agent. In the absence of biomarkers, assessment of exposure typically requires measurement of toxicant levels in the environment and characterization of the individual's presence in, and interaction with, that environment. Toxicants can be measured in air, water, soil, or food by a wide array of analytical methods. In addition to measurement of chemical or physical agents in the ambient environment, exposure can be assessed by use of personal external monitors and questionnaires. Although questionnaires have been extensively used to determine broad dietary exposures to compounds, smoking histories, and genetic backgrounds, except for certain circumstances such as assessing smoking status, this approach is imprecise unless specific chemical agents are already known.

Use of ambient measurements to determine exposure status of individuals is complicated because most environmental contaminations are heterogeneous. It is rare for an agent to be evenly distributed in the environment, so that it is very difficult to extrapolate data from these measurements to an individual's exposure. Therefore, the requirements for the practical development of specific biomarkers to assess exposure must include an ability to integrate multiple routes and fluctuating exposures over time, relate time of exposure to dose, and examine mechanisms in important biological targets. This is important, because safety regulations designed to limit human risk are often set on the basis of ambient exposure determinations. For example, these markers are critically important for verifying that interventions such as dumpsite cleanups have

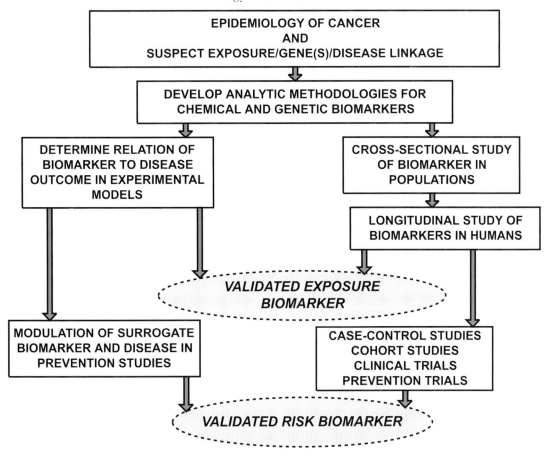

**Figure 2** Validation scheme for molecular biomarker research.

effectively lowered exposure in individuals. Accurate biomarkers of exposure will limit misclassification, which is often the greatest source of error in environmental epidemiology.

### 14.3.2.4.2 Risk Markers

Almost any measurement of an environmental chemical-specific marker in a human sample is going to reflect some index of exposure. The relation between exposure and dose or effect can be represented by a simple linear or a more complex nonlinear response curve. It is apparent from basic toxicological principles that not every measure of exposure will accurately reflect risk of disease outcome. Most biomarkers will only be risk markers if they are involved in the mechanism of disease. Thus, a chemical-DNA adduct biomarker, if linked through experimental studies to specific mutations that are critical for the induction of cancer, might become a validated risk marker in appropriate human investigations. Given the multistage process and long latency of cancer and other chronic human diseases, it is likely

that relatively few chemical-specific biomarkers will prove to be validated risk markers. Rather, certain chemical-specific biomarkers, such as a DNA adduct, in combination with other markers, including metabolic phenotype and/ or DNA repair capacity, will be needed to determine individual risk. Thus, most validated risk markers may turn out to be composites of a group of biomarkers, each of which contributes in some quantifiable way to overall risk. This problem is exemplified by the study of the etiology of human liver cancer where a confluence of two very different fields, chemical carcinogenesis and viral carcinogenesis, have been brought together in order to understand the pathogenesis of the disease.

### 14.3.2.5 VIRAL ETIOLOGY IN HUMAN LIVER CANCER

Major risk factors for HCC in the world are the hepatitis viruses, and in Africa and Asia including China, co-exposure with aflatoxins, which will be described in greater detail in the following sections.[29] The two major hepatitis viruses associated with HCC are Hepatitis B virus (HBV) and Hepatitis C virus (HCV). HBV is a member of the hepadnavirus family, a group of DNA-containing viruses. The genome of HBV is a partially double-stranded circular DNA molecule having 3200 nucleotides.[30] The organization of the HBV genome includes; a nucleocapsid, termed hepatitis B core antigen (HBcAg); envelope glycoprotein, called hepatitis B surface antigen (HBsAg), DNA polymerase (a product of the P gene), and a protein from the X gene, HBeAg, a polypeptide virtually identical to HBcAg. HBV infected hepatocytes are capable of synthesizing and secreting massive quantities of non-infective HBsAg, which appears in serum before the onset of symptoms and peaks during overt disease. HBsAg levels decline to undetectable amounts in 3–6 months following infection. HBeAg, HBV-DNA, and DNA polymerase appear in the serum soon after HBsAg and these are all biomarkers of active viral replication.[31] Persistence of HBeAg is an important clinical indicator and probably marks the progression point towards chronic hepatitis. Anti-HBe is detectable shortly after the disappearance of HBeAg, implying that the acute infection has peaked and this phase of the disease is declining. IgM anti-HBc begins to be measurable in the serum shortly before the onset of symptoms and usually concurrent with the onset of elevated liver serum transaminases. Over several months, the IgM antibody is replaced by IgG anti-HBc, and thus an elevated level of IgM anti-HBc indicates a recent acute infection. Anti-HBs does not rise until the acute disease is over and is usually discerned for a few weeks to several months after the disappearance of HBsAg. During this interval anti-HBc and anti-HBe are the only biomarkers of the disease. Anti-HBs persists for many years, conferring protection and forming the basis for current vaccination strategies.[32,33] Thus, the complex but well-defined pattern of these biomarkers has been important in characterizing the progression of this infection in people.

Hepatitis C virus (HCV) is a small, enveloped single-stranded RNA virus with continued alteration in envelope antigen expression.[34] The genome codes for a single polypeptide of approximately 3010 amino acids in one single open reading frame. This peptide is subsequently processed into a functional protein. HCV RNA is detectable in serum for 1–3 weeks after infection, coincident with elevations in serum transaminase. Circulating HCV RNA persists in many patients despite the presence of neutralizing antibodies, which indicates that the anti-HCV IgG following active infection

does not confer effective protective immunity to subsequent HCV infections. This seriously hampers efforts to develop an effective HCV vaccine. Indeed, HCV has a high rate (more than 50%) of progression to chronic disease and eventual cirrhosis. Currently, HCV is the leading cause of chronic liver disease in many Western countries and Japan.[18]

### 14.3.2.6 AFLATOXIN BIOMARKERS AND THE ETIOLOGY OF HUMAN LIVER CANCER

The molecular epidemiology investigations of aflatoxins probably represents one of the most extensive data sets in the field and this work may serve as a template for future studies of other environmental agents. The aflatoxins are naturally occurring mycotoxins found on foods such as corn, peanuts, various other nuts and cottonseed. They have been demonstrated to be carcinogenic in many animal species including rodents, non-human primates and fish. They were also initially suspected to contribute to human hepatocellular carcinoma (HCC).[35] As a result of nearly 30 years of study, consideration of the accumulated experimental data and epidemiological studies in human populations, led to the classification of aflatoxin B1 (AFB$_1$) as carcinogenic to humans by the International Agency for Research on Cancer in 1993.[36]

#### 14.3.2.6.1 Chemical Carcinogen in Animals

AFB$_1$ has been suspected to contribute to human HCC since the 1960s when its potent activity as a carcinogen in many species of animals, including rodents, nonhuman primates, and fish, was first reported.[35] Generally the liver was the primary target organ; however, under certain circumstances, depending on animal species and strain, dose, route of administration, and dietary factors, significant numbers of tumors have been induced at other sites, such as kidney and colon. Indeed, very few animal species have been found to be resistant to aflatoxin carcinogenesis. Wide cross-species potency including sensitivity of primates provided the justification for suspecting that this agent could contribute to human cancer.

#### 14.3.2.6.2 Suspect Human Carcinogen/ Disease Linkage

On the basis of animal studies, extensive efforts have been made to investigate the association between aflatoxin exposure and risk of HCC in humans. These studies have been hindered by the lack of adequate data on aflatoxin intake, excretion, and metabolism in humans, underlying susceptibility factors such as diet and viral exposure, as well as by the incomplete cancer morbidity and mortality statistics worldwide. These deficiencies provided the impetus to develop biomarker technologies to assess exposure status. Subsequently these biomarkers were applied to traditional and molecular epidemiological investigations to determine their relationships to risk.

The first element in the molecular biomarker paradigm is exposure. The potential for widespread exposure to aflatoxin from the food supply has been well documented and food safety regulations for aflatoxins have been implemented around the world. Commodities most often found to be contaminated are peanuts, various other nuts, cottonseed, corn, and rice.[37] Humans can also be exposed by consumption of products derived from these sources, such as eggs and milk (e.g., aflatoxin $M_1$ (AFM$_1$) from animals that consume contaminated feeds). Requirements for aflatoxin production are relatively nonspecific since molds can produce them on almost any foodstuff. Whereas contamination by molds may be widespread in a given geographical area, the final concentrations in the grain product can vary from less than 1 μg/kg (1 ppb) to more than 12,000 μg/kg (12 ppm).[38] Consequently, measurement of human exposure to aflatoxin by sampling foodstuffs or by dietary questionnaires is extremely imprecise and identification of aflatoxin biomarkers represented a significant advance for accurate assessment of exposure.

During the late 1960s and early 1970s several epidemiological studies conducted in Asia and Africa related estimated dietary intake of aflatoxin with the incidence of HCC. They showed that increased ingestion of aflatoxin corresponded to increased incidence of HCC (reviewed in Groopman[39]). Biomarkers of aflatoxin exposure were not available for these early studies, and the dietary surveys were population-based rather than measures of direct exposure of the persons who developed disease. Biomarkers of other important etiologic agents in HCC such as hepatitis B virus (HBV) were also not available. Thus, although these data provided strong circumstantial association between aflatoxin ingestion and HCC incidence, these findings were not considered sufficient for determining causality.

### 14.3.2.6.3 Identify and Develop Methodologies for Measuring Chemical-Specific Biomarkers

Development of biomarker methods to monitor human exposure to aflatoxins required analytic techniques that were sensitive, specific, and perhaps more important, could be applied to large numbers of samples. Measurement of aflatoxin-DNA and -protein adducts were of major interest because they are direct products of (or surrogate markers for) damage to a critical cellular macromolecular target. The chemical structures of the major aflatoxin macromolecular DNA and protein adducts were known.[40,41] The finding that the major aflatoxin-nucleic acid adduct (AFB-N$^7$-gua) was excreted exclusively in urine of exposed rats[42] spurred interest in using this metabolite. Serum aflatoxin-albumin adduct was also examined as a biomarker of exposure. Because of the longer half-life *in vivo* of aflatoxin-albumin compared to the DNA adduct excreted in urine, the serum albumin adduct apparently reflects exposures over longer time periods. Despite these kinetic differences, it was later found in experimental models that the formation of aflatoxin-DNA adducts in liver, excretion of the urinary aflatoxin nucleic acid adduct, and formation of the serum albumin adduct are highly correlated.[43] Thus, results from the two biomarkers should be comparable. A variety of different analytical methods are available for quantitation of these adducts in biological samples, including chromatography (thin layer and high performance liquid chromatography), immunological assays with specific antibodies or antisera (enzyme-linked immunosorbent assays, radioimmunoassays, and immunohistochemistry visualization in tissues), and instrumentation-based methods (synchronous fluorescence and mass spectroscopy, reviewed in Reference. 37). Each methodology has unique specificity and sensitivity and, depending on the application, the user can choose which is most appropriate. For example, to measure a single aflatoxin metabolite, a chromatographic method can resolve mixtures of aflatoxins into individual compounds, providing that the extraction procedure does not introduce large amounts of interfering chemicals. Antibody-based methods are often more sensitive than chromatography, but immunoassays are less selective because the antibody may cross-react with multiple metabolites.

In work in our laboratories we took advantage of the strengths of antibody and chromatographic separations to develop an immunoaffinity chromatography/high performance liquid chromatography procedure to isolate and measure aflatoxin metabolites in biological samples.[44,45] With this approach, we performed initial validation studies for the dose-dependent excretion of urinary aflatoxin biomarkers in rats after a single exposure to $AFB_1$.[46] A linear relationship was found between $AFB_1$ dose and excretion of the $AFB-N^7$-gua adduct in urine over the initial 24 h period of exposure. In contrast, excretion of other oxidative metabolites such as aflatoxin $P_1$ ($AFP_1$) showed no linear associations with dose. Subsequent studies in rodents that assessed the formation of aflatoxin-macromolecular adducts after chronic administration also support the use of DNA and protein adducts as molecular measures of exposure.[46–48] Recent work from our laboratories have extended the sensitivity of method of analysis of the $AFB-N^7$-gua adduct in rat urine by combining immunoaffinity chromatography clean-up with liquid chromatography/Electrospray ionization tandem mass spectroscopy (LC/ESI-MS/MS).[49] To illustrate the almost 100 fold increase in sensitivity in measuring this DNA adduct in urine the data in Figure 3 illustrates historic and recent dose response studies in the rat to measure the adduct. As the data in open circles show, the detection limit for this adduct is increased considerably by using the LC/ESI-MS/MS technique. Experimental models play a very important role in the development of analytical methods for measuring biomarkers. In studies of xenobiotics, such as aflatoxin, either single or multiple administration doses are used and this greatly diminishes the wide variations in exposure usually encountered in humans and assures that, unless the method is extremely insensitive, all samples will be detected. Unfortunately, extrapolation of the data in these experimental models to humans often neglected to take into account the enormous day-to-day variations that occur in exposure of people to toxins. Further, statistical assumptions of normality used in animal models, where there are few nondetectable values, do not apply to human studies where more than 50% of the values may be nondetectable. Thus, early studies in rodents with aflatoxin biomarkers did not indicate the complexity of future investigations.

### 14.3.2.6.4    Determine Relation of Biomarker to Exposure and Disease in Experimental Animals

In the early 1980s our laboratories started a collaboration to identify effective chemopre-

vention strategies for aflatoxin carcinogenesis. We hypothesized that reduction of aflatoxin-DNA adduct levels by chemopreventive agents would be mechanistically related to and therefore predictive of cancer preventive efficacy. Preliminary studies with a variety of established chemopreventive agents demonstrated that after a single dose of aflatoxin, levels of DNA adducts were reduced.[50] Therefore we next carried out a more comprehensive study using multiple doses of aflatoxin and the chemopreventive agent ethoxyquin to examine effects on DNA adduct formation and removal, and hepatic tumorigenesis in rats.[47] Treatment with ethoxyquin reduced both area and volume of liver occupied by presumptive preneoplastic foci by more than 95%. This same protocol also dramatically reduced binding of $AFB_1$ to hepatic DNA, from 90% initially to 70% at the end of a 2-week dosing period. No differences in residual DNA adduct burdens, however, were discernible several months after dosing. Thus, the apparent efficacy of the intervention depended on the time of analysis.

The experiment was then repeated with several different chemopreventive agents and in all cases aflatoxin-derived DNA and protein adducts were reduced; however, even under optimal conditions, the reduction of the macromolecular adducts always underrepresented the magnitude of tumor burden.[51,52] These macromolecular adducts can track with disease outcome on a population basis, but in the multistage process of cancer the absolute level of adduct provides a necessary but insufficient measure of likelihood of tumor formation. Indeed, it is reasonable to envision a situation where a chemopreventive agent could suppress adduct formation, but through other actions promote tumors, leading to a dichotomous outcome of fewer adducts and more tumors. Finally, because the design of these DNA adduct studies requires serial sacrifice of the animals, it is not possible to track the fate of an individual's adduct burden with tumor outcome. Hence, these investigations could only be used to predict the protective effects of an intervention at the level of the group, but not individual risk of disease.

### 14.3.2.6.5    Molecular Epidemiological Studies of Aflatoxin and Human Liver Cancer

During the 1960s and 1970s several epidemiological studies were conducted in Asia and Africa to determine whether there was an association between high aflatoxin exposure,

## AFB-N7-GUANINE EXCRETION IN RAT URINE

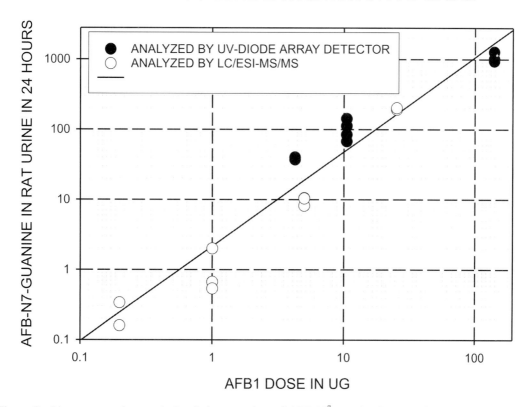

**Figure 3** Linear regression analysis of the excretion of AFB-N[7]-guanine in rat urine. The closed circles represent the data reported in Groopman *et al.*[46] from data collected by immunoaffinity chromatography clean-up and HPLC detection with a diode-array detector. The open circles represents the data reported in Reference 49 that uses an identical isolation scheme, but with LC/ESI-MS/MS detection. The correlation coefficient for data is 0.96 and the limit of detection of the nucleic acid adduct has been decreased by almost 100-fold.

estimated by sampling foodstuffs or by dietary questionnaires, and increased incidence of HCC (reviewed in Reference 39). These investigations showed that increased aflatoxin ingestion from 3 to 222 ng per kg body weight per day corresponded to increased liver cancer incidence, extending from a minimum of 2.0 to a maximum of 35.0 cases per l00,000 population per year. While these early studies could not account for confounding factors such as HBV or HCV infection, this information provided a strong motivation to further investigate the circumstantial relationship between aflatoxin ingestion and liver cancer incidence.

Within the past 20 years, epidemiological studies have been published measuring the correspondence of aflatoxin exposure and liver cancer. In one of the early reports, Bulatao-Jayme and coworkers[53] compared the dietary intakes of confirmed primary liver cancer cases in the Philippines against age-sex-matched controls. By using dietary recall, the frequency and amounts of food items consumed were calculated into units of aflatoxin load per day. It was observed that the mean aflatoxin load

per day of the liver cancer cases was 4.5 times higher than the controls. Alcohol intake as a risk factor was also analysed. It was determined that a synergistic and statistically significant effect on relative risk of HCC with aflatoxin exposure and alcohol intake occurs.

Van Rensburg and collaborators[54] compared the incidence rates of HCC and aflatoxin intake in several regions in Mozambique and Transkei in South Africa. The levels of aflatoxin were determined in randomly collected food samples from each region and again it was found that the mean dietary intake of aflatoxins was significantly correlated with HCC incidence and there was evidence that this was dose-dependent. Another epidemiological study to investigate the relationship between HCC incidence and aflatoxin exposure estimated in dietary samples was conducted in Swaziland.[55] In this study, the prevalence of HBV infection in the population was also estimated from the serum of blood donors and was observed to be high, but varied little across the different geographic regions analyzed. In contrast, there was more than a five-fold variation in aflatoxin exposure between different regions, which was strongly associated with a five-fold range in HCC incidence. Thus, aflatoxin was found to be a more important determinant of the variation in HCC incidence than HBV infection in this population.

In Guangxi province, China, Yeh and colleagues[56,57] examined the interaction between HBV infection and dietary aflatoxin exposure dichotomized for heavy and light contamination. The staple food of people living in this region during the 1960s and 1970s was corn, much of which was determined to have high levels of aflatoxin. In the heavily contaminated areas, aflatoxin content in corn ranged from 53.8 to 303 ppb, while the lightly contaminated regions showed aflatoxin levels in grains of less than 5 ppb. Those individuals who were positive for hepatitis B surface antigen (HBsAg) and had heavy aflatoxin exposure had an incidence of HCC 10-fold higher than did people living in areas with light aflatoxin contamination. People who were HBsAg-negative and who ate diets heavily contaminated with aflatoxin had a rate of HCC comparable to the HBsAg-positive people with light aflatoxin contamination.[57]

Although there is strong evidence for an association between dietary aflatoxin exposure and HCC incidence seen in the above case-control studies, there have been several investigations that do not support this positive association (reviewed in Wild and Hall in Reference 58). A cross-sectional study conducted in China by Campbell and collaborators[59] revealed a non-association between aflatoxin metabolites in human urine and liver cancer disease rates, while HBsAg status was positively correlated. Similar findings were reported for a case-control study in Thailand;[60] there was a positive correlation with HBsAg status, but no increase in risk with recent aflatoxin exposure estimated by dietary intake and measurement of aflatoxin-albumin adducts in serum. Alcohol was also associated with a non-significant elevation in risk.

In general, the most rigorous test of an association between an agent and disease outcome is found in prospective epidemiological studies, in which healthy people are recruited, questionnaires and biological samples taken, and the cohort followed until significant numbers of cases are obtained. A nested study within the cohort can then be designed to match cases and controls. Since the controls were recruited at the same time and with the same health status as the cases, they are better matched than in traditional case-control studies.

There have been two major cohort studies to address the relationship of aflatoxin exposure to HCC incidence reported to date. The first comprises over 18,000 people in Shanghai from whom urine and blood samples were collected.[61,62] After a 7-year follow-up, 50 of the individuals developed liver cancer. These cases were age and residence matched with 267 controls in a nested case-control study to examine the association between markers of aflatoxin exposure and HBV infection and the development of HCC. The data revealed a highly significant increase in the relative risk ($RR = 3.4$) for those liver cancer cases where urinary aflatoxin biomarkers were detected. The relative risk for people who tested positive for HBsAg was 7.3, but individuals with both urinary aflatoxins and positive HBsAg status had a relative risk for developing HCC of about 59. These results strongly supported a causal relationship between two major HCC risk factors, HBV and $AFB_1$ exposure. Finally, when individual aflatoxin metabolites were stratified for liver cancer outcome, the presence of the aflatoxin-nucleic acid adduct ($AFB-N^7$-gua) in urine always resulted in a two- to three-fold elevation in risk of developing HCC. These findings extend the conclusions from rodent chemoprotection studies that monitoring levels of this biomarker may be very useful for assessing risk.

Subsequent cohort studies carried out in Taiwan have also examined the relationship of HBV status, AFB1 exposure and incidence of HCC and have confirmed the results of the Shanghai investigation.[63,64] A nested case-control study with 56 cases of HCC and 220 matched healthy controls from the same cohort of over 15,000 people in Taiwan found that in HBV infected males there was an adjusted odds ratio of 2.8 for detectable compared with non-detectable aflatoxin-albumin adducts and 5.5 for high compared with low levels of aflatoxin metabolites in urine.[63] A second cohort study in Taiwan observed a dose-response relationship between urinary $AFM_1$ levels and HCC in chronic HBV carriers.[64] Similar to the Shanghai data, the HCC risk associated with $AFB_1$ exposure was more striking among the HBV carriers with detectable $AFB-N^7$-gua in urine. In Qidong county, Jiangsu province, People's Republic of China, HCC accounts for 10% of all adult deaths and both HBV and aflatoxin exposures are common.[65] A prospective nested case-control study was initiated through the collection of serum samples from 804 healthy HBsAg positive individuals (728 male, 76 female) aged 30–65 years in 1992. Between the year 1993–1995, 38 of these individuals developed HCC. The serum samples from 34 of these cases were matched by age, gender, residence and time of sampling to 170 controls. AFB$_1$-albumin adduct levels were determined by radioimmunoassay. The relative risk for HCC cases among AFB$_1$-albumin positive individuals was 2.4 (95% C.I:1.2, 4.7).[66] Another nested case-control study[67] was carried out in Taiwan. A cohort of 8068 men was followed-up for three years, and 27 cases of HCC were identified and matched with 120 healthy controls. Serum samples were analyzed for AFB$_1$-albumin adducts by ELISA. The proportion of subjects with a detectable serum AFB$_1$-albumin adducts was higher for HCC cases (74%) than matched controls (66%) with an odds ratio of 1.5. There was also a statistically significant association between detectable levels of AFB$_1$-albumin adduct and HCC risk among men younger than 52 years old, showing a multivariant-adjusted odds ratio of 5.3.

### 14.3.2.6.5.1 Aflatoxin Exposure and HBV Infection in Children

The use of biomarkers in cohort studies has clearly shown the chemical-viral interaction in the induction of HCC. However, aflatoxin exposure in the absence of chronic hepatitis B infection is also etiologically associated with liver cancer. These findings provide the compelling basis to increase efforts both in HBV immunization programs and in the development of concerted programs to lower dietary aflatoxin exposure as means of lowering human cancer risk. These epidemiological studies have also compelled the case to understand if a mechanistic interaction between HBV infection and enhanced aflatoxin metabolism. Early studies in adults infected with HBV and exposed to aflatoxin showed no effect of HBV infection on the levels of aflatoxin biomarkers;[68,69] however, a different picture has emerged in children.

Aflatoxin-albumin adduct levels were measured in serum samples obtained from a group of Gambian children. Aflatoxin-albumin adduct was found in nearly all serum samples collected during a survey performed at the end of the dry season and levels of adduct were generally high (up to 720 pg aflatoxin-lysine equivalent/mg albumin). Higher levels of aflatoxin-albumin adduct were detected in Wollof children than in children of other ethnic groups and marked variation in mean adduct levels between villages was observed. Aflatoxin-albumin adduct levels were up to 3-fold higher in children who were HBsAg positive than in controls. Much lower levels of aflatoxin-

albumin adduct were detected in repeat samples obtained at the end of the rainy season. There was poor correlation between dry and rainy season levels of adduct in individual children. Thus, these studies showed that HBV infection enhanced albumin adduct formation in Gambian children exposed to high levels of aflatoxin.[70]

There have been continuing studies demonstrating that the strongest interaction for HBV infection and aflatoxin exposure biomarkers occurs in children. In a recent cross-sectional analysis of unvaccinated (against HBV) children in rural Gambia, sera from 444 children aged 3–4 yrs, selected to be representative of their communities, were analyzed for aflatoxin-albumin adducts and markers of HBV infection. There was a large inter-individual variation in adduct levels between children (range: 2.2–459 pg AF-lysine eq./mg albumin). Adduct level was strongly correlated with season, with an approximately 2-fold higher mean level in the dry season than the wet. Geometric mean adduct levels in uninfected children, chronic carriers and acutely infected children were 31.6 ($n = 404$), 44.9 ($n = 34$) and 96.9 ($n = 6$) pg/mg respectively. The relationship of adduct level to ethnicity, month of sampling and HBV status was examined in a multiple regression model. Month of obtaining the blood sample ($p = 0.0001$) and HBV status ($p = 0.0023$) each made a highly significant contribution to the model; the high aflatoxin-albumin levels were particularly associated with acute infection. The effect of seasonality on adducts was also observed in a previous study of 347 Gambian adults, however, no correlation between adduct level and HBV status was observed in that population.[71] This difference between children and adults may reflect a more severe effect of HBV infection, particularly acute infection, in childhood on hepatic aflatoxin metabolism. Thus childhood may be a particularly critical period to which intervention strategies should be addressed.[72]

### 14.3.2.6.5.2 Aflatoxin Exposure and Mutations in the p53 Tumor Suppressor Gene

In addition to the evidence from epidemiological studies and the use of biomarkers of biologically effective dose, further support for the involvement of aflatoxin in HCC incidence in certain parts of the world has come from investigations of mutations in the *p53* tumor suppressor gene. The *p53* gene is found mutated in a majority of human cancers and there is a large variation in number and type of mutations between cancers of different tissues.[73,74] Such diversity lends itself to the analysis of the mutational spectrum within the gene with a view to determining information about the etiology of the tumor and potential risk assessment.[75,76] One of the most striking examples of a "molecular fingerprint" in the *p53* gene is a characteristic G→T transversion at the third base of codon 249 observed in liver cancer patients from regions of high aflatoxin exposure.

Initial reports of a specific mutation in the *p53* gene of HCCs in populations exposed to high levels of aflatoxin came from two independent studies in southern Africa and Qidong, China in which a G→T transversion in codon 249 of the *p53* was observed in approximately 50% of HCCs.[77,78] In contrast, no codon 249 mutations were detected in areas with low aflatoxin exposure such as Japan, Europe and the USA.[79]

Since these early findings, numerous investigations to determine the frequency of *p53* mutations, in particular the prevalence of G→T transitions at codon 249, in populations with high and low levels of aflatoxin exposure have been conducted. Li and coworkers[80] analyzed samples from two areas of China and found that 9/20 (45%) HCCs from Qidong, an area of high $AFB_1$ exposure, had a *p53* mutation and all of these were codon 249 mutations. In contrast, in HCCs from Shanghai, an area of intermediate aflatoxin exposure, 3/18 (17%) cases had a *p53* mutation and only 1/3 (33%) was located in codon 249. All but one of the *p53* mutations determined in this study were in found cases positive for HBV infection. A recent study of the HCC cases from the same regions found 7/14 (50%) and 5/10 (50%) samples from Qidong and Shanghai, respectively, to have a *p53* mutation. Of these, 100% (7/7) and 60% (3/5) were codon 249 transversions, respectively.[81] Similar results have been reported for the *p53* mutation in liver tumors from Guangxi Province in China.[82] Senegal is another country that has an extremely high incidence of liver cancer and in which there is a high exposure to dietary aflatoxin. A study conducted on 15 HCC samples from this country detected 10 (67%) cases with a mutation at codon 249 of the *p53* gene, the highest frequency reported to date.[83]

Aguilar *et al.*[84] investigated the mutagenesis of the *p53* gene in non-tumorigenic liver DNA from HCC cases in the United States, Thailand and Qidong, areas of negligible, low and high aflatoxin exposure. They found that the frequency of codon 249 mutations paralleled the level of aflatoxin exposure. Similar observations were made in a comparison of HCCs from Qidong and Beijing, an area of low aflatoxin exposure, in which a codon 249 G→T transversions were seen in 52% (13/25) and 0% (0/9) of HCCs respectively.[85] However, the total frequency of mutations and loss of heterozygosity in the *p53* gene in the 2 regions (both of which have a high incidence of HBV) were similar indicating that alterations in *p53* independent of the 249 mutation may play an important role in HBV-associated HCC.

The implication from these studies in populations of varying exposure to aflatoxin that the G→T mutation at the third base of codon 249 is aflatoxin specific has been supported by studies in bacteria, which have shown that aflatoxin exposure causes almost exclusively G→T transversions.[86] It has also been shown that the aflatoxin-epoxide can bind to codon 249 of *p53* in a plasmid *in vitro*, providing further indirect evidence for a putative role of aflatoxin exposure in *p53* mutagenesis.[87] A recent study has mapped $AFB_1$ adduct formation to codon 249.[88] However, this was not the major site of adduction by AFB1 and adducts were detected in several other codons in exons 7 and 8. It was also observed that the rate of adduct removal from codon 249 was relatively high suggesting that additional mechanisms are involved in mutagenesis at this site.

*In vitro* studies to measure the frequency of *p53* mutations in HepG2 human hepatocarcinoma cells incubated with $AFB_1$ in the presence of rat liver microsomes found that G→T transversions at the third base of codon 249 were preferentially induced.[89,90] G→T and C→A transversions were also observed in adjacent codons in cells exposed to $AFB_1$, although at lower frequencies. These findings also suggest that there is some additional selection of cells with a G→T transversion at the third base of codon 249, perhaps due to an altered function of the mutant 249serine *p53* protein.

While the evidence for a role of aflatoxin in the incidence of codon 249 mutations in *p53* from the above studies is persuasive, the level of exposure to aflatoxin has been classified by factors such as geographic residence rather than measurement of aflatoxin exposure at the individual level. The limitations of this approach have been reviewed by Lasky and Magder.[91] Only a few studies have been conducted to date that measure both aflatoxin adduct levels and *p53* mutational status in HCC cases. The largest and most recent study to assess the relationship between the presence of $AFB_1$ adducts, HBV status and *p53* mutations was reported by Lunn and coworkers[92] who investigated 105 HCC cases in Taiwan. Mutations in the *p53* gene were detected in

28% (29/105) of cases by single strand conformation polymorphism and DNA sequencing revealed that 12 of these were specific codon 249 mutations. Mutant *p53* protein was detected in 35% (37/105) of HCC cases by immunohistochemisty. AFB$_1$-DNA adducts were detected in 66% (69/105) of tumour tissues and were associated with *p53* DNA (odd ratio, OR = 2.9 *p* = 0.082) and protein mutations (OR = 2.9, *p* = 0.054) with only borderline significance. There was no statistically significant association between AFB$_1$-DNA adducts and codon 249 mutations. All of the codon 249 mutations (*n* = 12) occurred in HBsAg-seropositive carriers, resulting in an OR of 10.0 (*p*<0.05), suggesting that HBV may be involved in the selection of these mutations.

### 14.3.2.6.5.3 Detection of p53 Mutations in Human Plasma Samples

While the detection of specific *p53* mutations in liver tumors has provided insight into the etiology of certain liver cancers, the application of these specific mutations to the early detection of cancer offers great promise for prevention.[93] A report by Kirk *et al.*[94] reported for the first time the detection of codon 249 *p53* mutations in the plasma of liver tumor patients from the Gambia; however, the mutational status of the tumors were not known. These authors also reported a small number of cirrhosis patients having this mutation and given the strong relation between cirrhosis and future development of HCC, the possibility of this mutation being an early detection marker needs to be explored.

In a recent paper by Jackson *et al.*,[95] Short Oligonucleotide Mass Analysis (SOMA) was compared with DNA sequencing in 25 HCC samples for specific *p53* mutations. Mutations were detected in 10 samples by SOMA, in agreement with DNA sequencing. Analysis of another 20 plasma and tumor pairs showed 11 tumors containing the specific mutation and this change was detected in six of the paired plasma samples. Four of the plasma samples had detectable levels of the mutation; however, the tumors were negative suggesting possible multiple independent HCCs. Ten plasma samples from healthy individuals were all negative. While more work is required to assess the predictive value of this molecular diagnostic technique, it has potential applications in prevention trials and the early diagnosis of HCC.

In summary, studies of the prevalence of codon 249 mutations in HCC cases from patients in areas of high or low exposure to aflatoxin suggest that a G→T transition at the third base is associated with aflatoxin exposure and *in vitro* data would seem to support this hypothesis. A majority of codon 249 mutations are found in patients with an HBV infection implicating an association. However, in comparisons of codon 249 mutations in regions of high HBV infection but varying levels of AFB$_1$ exposure, the mutation only occurs in areas of high AFB$_1$ exposure. HBV evidently plays an important role in mutagenesis, perhaps by causing preferential selection of cells harboring the mutation. The use of the codon 249 mutation as a marker of exposure to aflatoxin must be done with caution until evidence has been obtained from studies measuring both AFB$_1$ adducts and mutations in the same individual.

### 14.3.2.6.5.4 Aflatoxin Exposure and Mutations in the HPRT Gene

Molecular biomarkers using specific genetic markers are becoming increasingly important tools to identify people who are at highest risk to develop cancer. For many years the investigations of residents of Qidong County, People's Republic of China to examine the combined impact of aflatoxin exposure with other risk factors as contributors to the high liver cancer incidence rates in this region have been very important for the molecular epidemiology field. To further this work, a study was conducted to determine the effects of aflatoxin exposure, as measured by serum aflatoxin albumin adduct levels, on somatic mutation frequency in the human hypoxanthine guanine phosphoribosyl transferase (HPRT) gene. Subjects were assigned as low or high according to a dichotomization around the population mean of aflatoxin albumin adducts. HPRT mutant frequency was determined in individuals by a T cell clonal assay and the samples were categorized as low or high according to mean values. Separate analyses were also conducted for the small set of HBsAg-positive and the larger set of HBsAg-negative individuals, known risk factors for liver cancer. An odds ratio of 19.3 (95% confidence interval: 2.0, 183) was demonstrated for a high HPRT mutation frequency in individuals with high aflatoxin exposure compared to those with low aflatoxin exposure. This association indicates that aflatoxin induced DNA damage in T-lymphocytes, assessed using the validated surrogate albumin-adduct markers, leads to increased mutations reflected as elevated

HPRT gene mutations. This cross-sectional study suggests the potential to use mutation frequency of the HPRT gene as a long-term biomarker of aflatoxin exposure in high-risk populations.[96]

### 14.3.2.7 BIOMARKERS AND LIVER CANCER PREVENTION

Several approaches can be considered for the prevention of liver cancer. A first approach is vaccination against HBV. Unfortunately, many people living in high-risk areas for liver cancer acquire the HBV infection before age three. Thus, an immunization program for total population protection would have to occur over several generations, provided that mutant strains of HBV do not arise, thereby eliminating the utility of current vaccines. Despite these problems, vaccination programs for HBV have been implemented in several areas of Africa and Asia. A second approach for cancer prevention would be the elimination of aflatoxin exposures. Primary prevention of aflatoxin exposures could be accomplished through large expenditures of resources for proper crop storage and handling; however, this approach is not economically feasible in many areas of the world. Secondary prevention measures using chemopreventive agents that block the activation and enhance the detoxification of AFB1 are being investigated in high-risk populations.

Cancer prevention trials that use biomarkers as intermediate endpoints provide the ability to assess the efficacy of promising chemopreventive agents in an efficient manner by reducing sample size requirements as well as the time required to conduct the studies compared to trials that have cancer incidence or mortality as endpoints.[97] These intermediate markers are particularly valuable when investigating chemopreventive agents such as oltipraz that may have an effect at early, preclinical stages of carcinogenesis. The key issue in trials that use biomarkers as the outcome of interest is to have a marker that is directly associated with the evolution or development of neoplasia. These biomarkers are typically organ site-specific genomic, proliferation, and differentiation markers that reflect different intermediary stages of the neoplastic process. As an adjunct to these process-dependent markers it may also be possible to devise markers specific to interventions in selected groups at high risk for carcinogen exposure. These agent-dependent approaches would be based upon knowledge of the etiologic agent(s) in the study population.

It has been shown that several agents can provide some level of protection against aflatoxin-induced liver cancer in experimental systems (Kensler *et al.*[98]). One of the most promising of these agents is oltipraz (4-methyl-5-(2-pyrazinyl)-1,2-dithiole-3-thione) which has been shown to inhibit AFB₁ hepatocarcinogenesis in rats when administered 1 week prior to and throughout carcinogen exposure.[99] This and subsequent studies have also shown that hepatic AFB-DNA adducts, urinary AFB-$N^7$-gua and serum albumin adducts are all reduced in animals given oltipraz during aflatoxin administration, highlighting the utility of these biomarkers in intervention studies.[100]

A double-blind Phase IIa clinical chemoprevention trial with oltipraz was conducted in Qidong, China in 1995.[101] Healthy individuals were randomized into groups receiving 125 mg of oltipraz daily, 500 mg of oltipraz weekly, or placebo. Blood and urine specimens were collected biweekly over the 8-week intervention period and an 8-week follow-up period. Levels of aflatoxin-albumin adducts in serum, and AFM1 and AFB-NAC excreted in the urine were examined as primary biomarker endpoints in the study. There were no consistent changes observed in levels of aflatoxin-albumin adducts in the placebo group or those receiving 125 mg of oltipraz daily. However, there was a significant decline in aflatoxin-albumin levels beginning 1 month into intervention with 500 mg of oltipraz weekly which continued for 1 month after treatment was stopped.[102]

Urinary levels of AFM₁, the primary oxidative metabolite of AFB₁, was reduced by 51% in individuals receiving weekly doses of 500 mg oltipraz as compared to the placebo controls.[102] No significant differences were observed in the levels of AFM₁ in the arm receiving 125 mg of oltipraz daily. In contrast, levels of AFB-NAC, a detoxification product of AFB₁, were increased 2.6-fold in the 125 mg group, but were unchanged in the 500 mg group. An increase in AFB-NAC in the 125 mg group confirms the ability of oltipraz to induce phase 2 enzymes to increase aflatoxin conjugation. The lack of an effect of 500 mg of oltipraz on AFB-NAC probably reflects masking due to diminished substrate formation through the inhibition of cytochrome p450 activities seen in this group (measured as a reduction in AFM1 levels). Overall, these results demonstrate that biomarkers can be used in chemoprevention studies to determine the efficacy of new agents.

### 14.3.2.8 SUMMARY

The long-term goal of the research described herein is the application of molecular biomarkers to the development of preventive interventions for use in human populations at high-risk for cancer. Several of the aflatoxin specific biomarkers have been validated in epidemiological studies and are now being used as intermediate biomarkers in prevention studies. The development of these aflatoxin biomarkers has been based upon the knowledge of the biochemistry and toxicology of aflatoxins gleaned from both experimental and human studies. These biomarkers have subsequently been utilized in experimental models to provide data on the modulation of these markers under different situations of disease risk. In Figure 4, the validation scheme is shown with highlights of the aflatoxin/hepatitis virus/human liver cancer investigations that suggests that this may serve as a template for future investigations. This comprehensive and systematic approach provides encouragement for the development of successful preventive interventions and should serve as a template for the development, validation and application of other chemical-specific biomarkers to reducing the burden of cancer or other chronic diseases.

### 14.3.2.9 ACKNOWLEDGMENTS

Financial support for this work has been provided by grants R01 CA39416, P01 ES06052, NIEHS Center P30 ES03819 and contract N01-CN-25437

### 14.3.2.10 REFERENCES

1. Anonymous, 'Biological markers in environmental health research.' *Environ. Health Perspect.*, 1987, **74**, 3–9.
2. G. N. Wogan, 'Molecular epidemiology in cancer risk assessment and prevention: Recent progress and avenues for future research.' *Environ. Health Perspect.*, 1992, **98**, 167–178.
3. J. J. Gumucio and J. Chianale, in 'The Liver: Biology and Pathobiology', 2nd edn, ed. I. M. Arias, New York, Raven Press, 1988, pp. 931–947.
4. A. M. Rappaport and I. R.Wanless, in 'Diseases of the Liver', 7th edn, Vol I, eds. L. Schiff, E.R. Schiff and J.B. Lippincott, Company, Philadelphia, 1993, pp. 1–41.
5. J. L. Creech and M. N. Johnson, 'Angiosarcoma of the liver in the manufacture of polyvinyl chloride.' *J. Occup. Med.* 1981, **16**, 150–151.
6. H. Falk, L. B. Thomas, H. Popper *et al.*, 'Hepatic angiosarcoma associated with androgenic anabolic steroids.' *Lancet*, 1979, **2**, 1120–1123.
7. H. Falk, G. G. Caldwell, K. G. Ishak *et al.*, 'Arsenic-related hepatic angiosarcoma.' *Am. J. Ind. Med.*, 1981, **2**, 43–50.

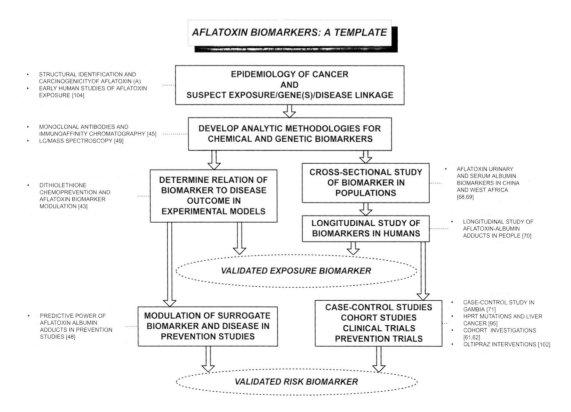

**Figure 4** Aflatoxin as a template for the validation of molecular biomarkers.

8. D. Forman, B. Bennett, J. Stafford *et al.*, 'Exposure to vinyl chloride and angiosarcoma of the liver: A report of the register of cases.' *Br. J. Ind. Med.*, 1985, **42**, 750–753.

9. H. Fald, J. L. Creech, C. W. Heath *et al.*, 'Hepatic disease among workers at a vinyl chloride polymerization plant.' *JAMA*, 1974, **230**, 59–63.

10. H. Popper, L. B. Thomas, N. C. Telles *et al.*, 'Development of hepatic angiosarcoma in man induced by vinyl chloride, thorotrast and arsenic.' *Am. J. Pathol.*, 1978, **92**, 349–376.

11. R. Zaldivar, L. Prunes and G. L. Ghai, 'Arsenic dose in patients with cutaneous carcinomata and hepatic haemangioendothelioma after environmental and occupational exposure.' *Arch. Toxicol.*, 1981, **47**, 154.

12. K. Okuda, M. Kojiro and H. Okuda, in 'Diseases of the Liver', Vol. 2, 7th edn, eds, L. Schiff and E. R. J. B. Lippincott Company, Philadelphia, 1993, pp. 1236–1296.

13. National Cancer Office of the Ministry of Public Health, P.R.C. 'Studies on mortality rates of cancer in China.' Beijing, People's Publishing House, 1980.

14. J. S. Wang, T. Huang, J. Su *et al.*, 'A molecular epidemiological study of hepatocellular carcinoma in Zhuqing village, Fusui County, People's Republic of China.' *Cancer Epidemiology, Biomarkers and Prevention*, 2001, **9**, in press.

15. H. Popper, 'Hepatic cancers in man: quantitative perspectives.' *Environ. Res.*, 1979, **19**, 482–494.

16. H. Popper and L. B. Thomas, 'Environmental liver tumors.' *Kurume Med. J.*, 1979, **26**, 189–204.

17. M. Clolmbo, G. Kuo, Q. L. Choo *et al.*, 'Prevalence of antibodies to hepatitis C virus in Italian patients with hepatocellular carcinoma.' *Lancet*, 1989, **2**, 1006–1008.

18. J. M. Crawford, in 'Robbins Pathologic Basis of Disease', 5th edn, eds, R. S. Cotran, V. Kumar and S. L. Robbins, W. B. Saunders Company, Philadelphia, 1994, pp. 831–896.

19. W. H. Chow, J. K. McLaughlin, W. Zheng *et al.*, 'Occupational risks for primary liver cancer in Shanghai, China.' *Am. J. Ind. Med.*, 1993, **24**, 93–100.

20. Y. M. Hsueh, G. S. Cheng, M. M. Wu *et al.*, 'Multiple risk factors associated with arsenic-induced skin cancer: effects of chronic liver disease and malnutritional status.' *Br. J. Cancer*, 1995, **71**, 109–114.

21. L. Suarez, N. S. Weiss and J. Martin, 'Primary liver cancer death and occupation in Texas.' *Am. J. Ind. Med.*, 1989, **15**, 167–175.

22. R. Hoover and J. F. Fraumeni, 'Cancer mortality in the US counties with chemical industries.' *Environ. Res.*, 1975, **9**, 196–207.

23. A. Stemhagen, J. Slade, R. Altman *et al.*, 'Occupational risk factors and liver cancer. A retrospective case control study of primary liver cancer in New Jersey.' *Am. J. Epidemiol.*, 1983, **117**, 443–454.

24. H. Austin, E. Delzell, S. Grufferman *et al.*, 'Case-control study of hepatocellular carcinoma, occupation and chemical exposures.' *J. Occup. Med.*, 1987, **29**, 665–669.

25. D. H. Wegman and E. Eisen, 'Causes of death among employees of a synthetic abrasive product manufacturing company.' *J. Occup. Med.*, 1981, **11**, 748–753.

26. J. E. Vena, H. A. Sultz, R. C. Fiedler *et al.*, 'Mortality of workers in an automobile engine and parts manufacturing complex.' *Br. J. Ind. Med.*, 1985, **42**, 85–93.

27. J. M. Links, T. W. Kensler and J. D. Groopman, 'Biomarkers and mechanistic approaches in environmental epidemiology.' *Annu. Rev. Public Health*, 1995, **16**, 83–103.

28. B. S. Hulka, 'Epidemiological studies using biological markers: Issues for epidemiologists.' *Cancer Epidemiology Biomarkers and Prevention*, 1991, **1**, 13–20.

29. C. C. Harris and T.-T. Sun, 'Multifactorial etiology of human liver cancer.' *Carcinogenesis*, 1984, **5**, 697–701.

30. J. Y. N. Lau and T. L. Wright, 'Molecular virology and pathogenesis of hepatitis B.' *Lancet*, 1993, **342**, 1335–1340.

31. J. H. Hoofnagle and D. F. Schafer, 'Serologic markers of hepatitis B virus infection.' *Semin. Liver Dis.*, 1986, **6**, 1–10.

32. R. S. Koff, in 'Diseases of the Liver', Vol. 1, 7th edn, eds, L. Schiff and E. R., J.B. Lippincott Company, Philadelphia, 1993, pp. 492–577.

33. J. L. Boyer and A. Reuben, in 'Diseases of the Liver', Vol. 2, 7th edn, eds, L. Schiff and E. R., J.B. Lippincott Company, Philadelphia, 1993, pp. 586–637.

34. O. Weiland and R. Schvarcz, 'Hepatitis C: virology, epidemiology, clinical course, and treatment.' *Scand. J. Gastroenterol.*, 1992, **27**, 337–342.

35. W. F. J. Busby and G. N. Wogan, in 'Chemical Carcinogens', 2nd edn, ed, C. E. Searle, American Chemical Society, Washington DC., 1984, pp. 945–1136.

36. 'Aflatoxins.' *IARC Monogr. Eval. Carcinog. Risks Hum.*, 1993, **56**, pp. 245–395.

37. 'The Toxicology of Aflatoxins: Human Health, Veterinary, and Agricultural Significance.' eds D. L. Eaton and J. D. Groopman, Academic Press, Inc., San Diego 1994.

38. W. O. Ellis, J. P. Smith, B. K. Simpson *et al.*, 'Aflatoxin in food: occurrence, biosynthesis, effects on organisms, detection, and methods of control.' *Crit. Rev. Food Sci. Nutr.*, 1991, **30**, 403–439.

39. J. D. Groopman, P. Scholl and J. S. Wang, 'Epidemiology of human aflatoxin exposures and their relationship to liver cancer.' *Prog. Clin. Biol. Res.*, 1996, **395**, 211–222.

40. J. M. Essigmann, R. G. Croy, A. M. Nadzan *et al.*, 'Structural identification of the major DNA adduct formed by aflatoxin $B_1$ *in vitro*.' *Proc. Nat. Acad. Sci. USA*, 1977, **74**, 1870–1874.

41. G. Sabbioni, P. L. Skipper, G. Büchi *et al.*, 'Isolation and characterization of the major serum albumin adduct formed by aflatoxin $B_1$ *in vivo* in rats.' *Carcinogenesis*, 1987, **8**, 819–824.

42. R. A. Bennett, J. M. Essigmann and G. N. Wogan, 'Excretion of an aflatoxin-guanine adduct in the urine of aflatoxin $B_1$-treated rats.' *Cancer Res.*, 1981, **41**, 650–654.

43. J. D. Groopman, P. DeMatos, P. A. Egner *et al.*, 'Molecular dosimetry of urinary aflatoxin—$N^7$-guanine and serum aflatoxin—albumin adducts predicts chemoprotection by 1,2-dithiole-3-thione in rats.' *Carcinogenesis*, 1992, **13**, 101–106.

44. J. D. Groopman, L. J. Trudel, P. R. Donahue *et al.*, 'High-affinity monoclonal antibodies for aflatoxins and their application to solid-phase immunoassays.' *Proc. Nat. Acad. Sci. USA*, 1984, **81**, 7728–7731.

45. J. D. Groopman, P. R. Donahue, J. Zhu *et al.*, 'Aflatoxin metabolism in humans: Detection of metabolites and nucleic acid adducts in urine by affinity chromatography.' *Proc. Nat. Acad. Sci. USA*, 1985, **82**, 6492–6496.

46. J. D. Groopman, J. A. Hasler, L. J. Trudel *et al.*, 'Molecular dosimetry in rat urine of aflatoxin-$N^7$-guanine and other aflatoxin metabolites by multiple monoclonal antibody affinity chromatography and immunoaffinity/high performance liquid chromatography.' *Cancer Res.*, 1992, **52**, 267–274.

47. T. W. Kensler, P. A. Egner, N. E. Davidson *et al.*, 'Modulation of aflatoxin metabolism, aflatoxin-$N^7$-guanine formation, and hepatic tumorigenesis in rats

fed ethoxyquin: Role of induction of glutathione *S*-transferases.' *Cancer Res.*, 1986, **46**, 3924–3931.

48. P. A. Egner, S. J. Gange, P. M. Dolan *et al.*, 'Levels of aflatoxin-albumin biomarkers in rat plasma are modulated by both long-term and transient interventions with oltipraz.' *Carcinogenesis*, 1995, **16**, 1769–1773.

49. M. Walton, P. Egner, P. F. Scholl *et al.*, 'Modulation of urinary aflatoxin biomarkers by chemopreventive intervention using oltipraz measured with liquid chromatography electrospray-mass spectrometry.' *Chemical Research in Toxicology*, submitted, 2001.

50. T. W. Kensler, P. A. Egner, M. A. Trush *et al.*, 'Modification of aflatoxin $B_1$ binding to DNA *in vivo* in rats fed phenolic antioxidants, ethoxyquin and a dithiothione.' *Carcinogenesis*, 1985, **6**, 759–763.

51. B. D. Roebuck, Y.-L. Liu, A. E. Rogers *et al.*, 'Protection against aflatoxin $B_1$-induced hepatocarcinogenesis in F344 rats by 5-(2-pyrazinyl)-4-methyl-1,2-dithiole-3-thione (oltipraz): Predictive role for short-term molecular dosimetry.' *Cancer Res.*, 1991, **51**, 5501–5506.

52. M. G. Bolton, A. Muñoz, L. P. Jacobson *et al.*, 'Transient intervention with olitpraz protects against aflatoxin-induced hepatic tumorigenesis.' *Cancer Res.*, 1993, **53**, 3499–3504.

53. J. Bulatao-Jayme, E. M. Almero, M. C. Castro *et al.*, 'A case-control dietary study of primary liver cancer risk from aflatoxin exposure.' *Int. J. Epidemiol.*, 1982, **11**(2), 112–119.

54. S. J. van Rensburg, P. Cook-Mozaffari, D. J. Van Schalkwyk *et al.*, 'Hepatocellular carcinoma and dietary aflatoxin in Mozambique and Transkei.' *Brit. J. Cancer*, 1985, **51**(5), 713–726.

55. F. Peers, X. Bosch, J. Kaldor *et al.*, 'Aflatoxin exposure, hepatitis B virus infection and liver cancer in Swaziland.' *Int. J. Cancer*, 1987, **39**(5), 545–553.

56. F. S. Yeh and K.-N. Shen, 'Epidemiology and early diagnosis of primary liver cancer in China.' *Adv. Cancer Res.*, 1986, **47**, 297–329.

57. F. S. Yeh, M. C. Yu, C. C. Mo *et al.*, 'Hepatitis B virus, aflatoxins, and hepatocellular carcinoma in southern Guangxi, China.' *Cancer Res.*, 1989, **49**(9), 2506–2509.

58. A. J. Hall and C. P. Wild, in 'The Toxicology of Aflatoxins: Human Health, Veterinary, and Agricultural Significance.' eds. D. L. Eaton and J. D. Groopman, Academic Press, Inc., San Diego, 1994.

59. T. C. Campbell, J. S. Chen, C. B. Liu *et al.*, 'Nonassociation of aflatoxin with primary liver cancer in a cross-sectional ecological survey in the People's Republic of China (see comments).' *Cancer Res.*, 1990, **50**(21), 6882–6893.

60. P. Srivatanakul, D. M. Parkin, M. Khlat *et al.*, 'Liver cancer in Thailand. II. A case-control study of hepatocellular carcinoma.' *Int. J. Cancer*, 1991, **48**(3), 329–332.

61. R. K. Ross, J. M. Yuan, M. C. Yu *et al.*, 'Urinary aflatoxin biomarkers and risk of hepatocellular carcinoma (see comments).' *Lancet*, 1992, **339**, 943–946.

62. G. S. Qian, R. K. Ross, M. C. Yu *et al.*, 'A follow-up study of urinary markers of aflatoxin exposure and liver cancer risk in Shanghai, People's Republic of China (see comments).' *Cancer. Epidemiol., Biomarkers and Prev.*, 1994, **3**(1), 3–10.

63. L. Y. Wang, M. Hatch, C. J. Chen *et al.*, 'Aflatoxin exposure and risk of hepatocellular carcinoma in Taiwan.' *Int. J. Cancer*, 1996, **67**(5), 620–625.

64. M. W. Yu, J. P. Lien, Y. H. Chiu *et al.*, 'Effect of aflatoxin metabolism and DNA adduct formation on hepatocellular carcinoma among chronic hepatitis B carriers in Taiwan.' *J. Hepatology*, 1997, **27**(2), 320–330.

65. J. S. Wang, G.-S. Qian, A. Zarba *et al.*, 'Temporal patterns of aflatoxin-albumin adducts in hepatitis B surface antigen-positive and antigen-negative residents of Daxin, Qidong county, People's Republic of China.' *Cancer Epidemiol. Biomarkers Prev.*, 1996, **5**, 253–261.

66. S.-Y. Kuang, X. Fang, P.-X. Lu *et al.*, 'Aflatoxin-albumin adducts and risk for hepatocellular carcinoma in residents of Qidong, People's Republic of China.' *Proc. Am. Assoc. Cancer Res.*, 1996, **37**, 1714.

67. M.-W. Yu, C.-J. Chen, L.-W. Wang *et al.*, 'Aflatoxin $B_1$ adduct level and risk of hepatocellular carcinoma.' *Proc. Am. Assoc. Cancer Res.*, 1995, **36**, 1644.

68. J. D. Groopman, A. Hall, H. Whittle *et al.*, 'Molecular dosimetry of aflatoxin-$N^7$-guanine in human urine obtained in the Gambia, West Africa.' *Cancer Epidemiology, Biomarkers Prev.*, 1992, **1**, 221–228.

69. C. P. Wild, G. Hudson, G. Sabbioni *et al.*, 'Correlation of dietary intake of aflatoxins with the level of albumin bound aflatoxin in peripheral blood in the Gambia, West Africa.' *Cancer Epidemiology, Biomarkers Prev.*, 1992, **1**, 229–234.

70. S. J. Allen, C. P. Wild, J. G. Wheeler *et al.*, 'Aflatoxin exposure, malaria and hepatitis B infection in rural Gambian children.' *Trans. R. Soc. Trop. Med. Hyg.*, 1992, **86**, 426–430.

71. C. P. Wild, F. Yin, P. C. Turner *et al.*, 'Environmental and genetic determinants of aflatoxin-albumin adducts in the Gambia.' *Int. J. Cancer*, 2000, **86**, 1–7.

72. P. C. Turner, M. Mendy, H. Whittle *et al.*, 'Hepatitis B infection and aflatoxin biomarker levels in Gambian children.' *Trop. Med. Int. Health*, 2000, **5**(12), 837–841.

73. M. Hollstein, D. Sidransky, B. Vogelstein *et al.*, 'p53 mutations in human cancers.' *Science*, 1991, **253**(5015), 49–53.

74. M. S. Greenblatt, W. P. Bennett, M. Hollstein *et al.*, 'Mutations in the p53 tumor suppressor gene: clues to cancer etiology and molecular pathogenesis.' *Cancer Res.*, 1994, **54**(18), 4855–4878.

75. B. Vogelstein and K. W. Kinzler, 'Carcinogens leave fingerprints (news).' *Nature*, 1992, **355**(6357), 209–210.

76. C. C. Harris, 'p53 tumor suppressor gene: at the crossroads of molecular carcinogenesis, molecular epidemiology, and cancer risk assessment. (Review).' *Environmental Health Perspectives*, 1996, **104**(Suppl 3), 435–439.

77. I. C. Hsu, R. A. Metcalf, T. Sun *et al.*, 'Mutational hotspot in the p53 gene in human hepatocellular carcinomas (see comments).' *Nature*, 1991, **350**(6317), 427–428.

78. B. Bressac, M. Kew, J. Wands *et al.*, 'Selective G to T mutations of p53 gene in hepatocellular carcinoma from southern Africa (see comments).' *Nature*, 1991, **350**(6317), 429–431.

79. C. Challen, J. Lunec, W. Warren *et al.*, 'Analysis of the p53 tumor-suppressor gene in hepatocellular carcinomas from Britain.' *Hepatology*, 1992, **6**(6), 1362–1366.

80. D. Li, Y. Cao, L. He *et al.*, 'Aberrations of p53 gene in human hepatocellular carcinoma from China.' *Carcinogenesis*, 1993, **14**(2), 169–173.

81. A. Rashid, J.-S. Wang, G. S. Qian *et al.*, 'Genetic alterations in hepatocellular carcinomas: association between loss of chromosome 4q and p53 gene mutations.' *Brit. J. Cancer*, 1999, **80**, 59–66.

82. D. Zhuolin, M. Yun, P. Langxing *et al.*, 'A molecular epidemiologic marker of hepatocellular carcinoma

from aflatoxin B1 contaminated area in the southwest of Guangxi.' *Chinese J. Cancer Res.*, 1997, **9**, 163–167.

83. P. Coursaget, N. Depril, M. Chabaud *et al.*, 'High prevalence of mutations at codon 249 of the p53 gene in hepatocellular carcinomas from Senegal.' *Brit. J. Cancer*, 1993, **67**(6), 1395–1397.

84. F. Aguilar, C. C. Harris, T. Sun *et al.*, 'Geographic variation of p53 mutational profile in non-malignant human liver.' *Science*, 1994, **264**(5163), 1317–1319.

85. Y. Fujimoto, L. L. Hampton, P. J. Wirth *et al.*, 'Alterations of tumor suppressor genes and allelic losses in human hepatocellular carcinomas in China (see comments).' *Cancer Res.*, 1994, **54**(1), 281–285.

86. P. L. Foster, E. Eisenstadt and J. H. Miller, 'Base substitution mutations induced by metabolically activated aflatoxin B1.' *Proc. Nat. Acad. Sc. USA*, 1983, **80**(9), 2645–2648.

87. A. Puisieux, S. Lim, J. Groopman *et al.*, 'Selective targeting of p53 gene mutational hotspots in human cancers by etiologically defined carcinogens.' *Cancer Res.*, 1991, **51**(22), 6185–6189.

88. M. F. Denissenko, T. B. Koudriakova, L. Smith *et al.*, 'The p53 codon 249 mutational hotspot in hepatocellular carcinoma is not related to selective formation or persistence of aflatoxin B1 adducts (in process citation).' *Oncogene*, 1998, **17**(23), 3007–3014.

89. F. Aguilar, S. P. Hussain and P. Cerutti, 'Aflatoxin B1 induces the transversion of G→T in codon 249 of the p53 tumor suppressor gene in human hepatocytes.' *Proc. Nat. Acad. Sci. USA*, 1993, **90**(18), 8586–8590.

90. P. Cerutti, P. Hussain, C. Pourzand *et al.*, 'Mutagenesis of the H-ras protooncogene and the p53 tumor suppressor gene.' *Cancer Res.*, 1994, **54**(Suppl. 7), 1934s–1938s.

91. T. Lasky and L. Magder, 'Hepatocellular carcinoma p53 G→T transversions at codon 249: the fingerprint of aflatoxin exposure?. (Review).' *Environ. Health Perspectives*, 1997, **105**(4), 392–397.

92. R. M. Lunn, Y. J. Zhang, L. Y. Wang *et al.*, 'p53 mutations, chronic hepatitis B virus infection, and aflatoxin exposure in hepatocellular carcinoma in Taiwan.' *Cancer Res.*, 1997, **57**(16), 3471–3477.

93. D. Sidransky and M. Hollstein, 'Clinical implications of the *p53* gene.' *Annu. Rev. Med.*, 1996, **47**, 285–301.

94. G. D. Kirk, A.-M. Camus-Randon, M. Mendy *et al.*, 'Ser-249 *p53* mutations in plasma DNA of patients with hepatocellular carcinoma from The Gambia.' *J. Nat. Cancer Inst.*, 2000, **92**, 148–153.

95. P. E. Jackson, G.-S. Qian, M. D. Friesen *et al.*, 'Specific *p53* mutations detected in plasma and tumors of hepatocellular carcinoma patients by electrospray ionization mass spectrometry.' *Cancer Res.*, 2001, **61**, 33–35.

96. S. S. Wang, J. P. O'Neill, G.-S. Qian *et al.*, 'Elevated HPRT mutation frequencies in highly aflatoxin exposed residents of Daxin, Qidong County, People's Republic of China.' *Carcinogenesis*, 1999, **20**, 2181–2184.

97. T. W. Kensler, J. D. Groopman and G. N. Wogan, 'Use of carcinogen-DNA and carcinogen-protein adduct biomarkers for cohort selection and as modifiable end points in chemoprevention trials.' *Principles of Chemoprevention*, 1996, **139**, 237–248.

98. T. W. Kensler, J. D. Groopman and B. D. Roebuck, 'Use of aflatoxin adducts as intermediate endpoints to assess the efficacy of chemopreventive interventions in animals and man.' *Mutat. Res.*, 1998, **402**, 165–172.

99. B. D. Roebuck, Y. L. Liu, A. E. Rogers *et al.*, 'Protection against aflatoxin B1-induced hepatocarcinogenesis in F344 rats by 5-(2-pyrazinyl)-4-methyl-1,2-dithiole-3-thione (oltipraz): predictive role for short-term molecular dosimetry.' *Cancer Res.*, 1991, **51**(20), 5501–5506.

100. P. A. Egner, S. J. Gange, P. M. Dolan *et al.*, 'Levels of aflatoxin-albumin biomarkers in rat plasma are modulated by both long-term and transient interventions with oltipraz.' *Carcinogenesis*, 1995, **16**(8), 1769–1773.

101. L. P. Jacobson, B. C. Zhang, Y. R. Zhu *et al.*, 'Oltipraz chemoprevention trial in Qidong, People's Republic of China: study design and clinical outcomes.' *Cancer Epidemiol. Biomarkers Prev.*, 1997, **6**(4), 257–265.

102. T. W. Kensler, X. He, M. Otieno *et al.*, 'Oltipraz chemoprevention trial in Qidong, People's Republic of China: modulation of serum aflatoxin albumin adduct biomarkers.' *Cancer Epidemiol. Biomarkers Prev.*, 1998, **7**(2), 127–134.

103. J.-S. Wang, X. Shen, X. He *et al.*, 'Protective alterations in Phase 1 & 2 metabolism of aflatoxin B1 by Oltipraz in residents of Qidong, P.R.C.' *J. Nat. Cancer Inst.*, 1999, **91**, 347–353.

104. T. Asao, G. Buchi, M. Abdel-Kader *et al.*, 'Aflatoxins B and G Aflatoxins B and G.' *J. Am. Chem. Soc.*, 1963, **85**, 1706–1707.

J.P. Vanden Heuvel, G.H. Perdew, W.B. Mattes and W.F. Greenlee (Eds.)
Comprehensive Toxicology, Vol. xiv

# 14.3.3
# Inherited Susceptibility and Prostate Cancer Risk

## TIMOTHY R. REBBECK

*University of Pennsylvania School of Medicine, Philadelphia, PA, USA*

## 14.3.3.1  INTRODUCTION

Prostate cancer has the highest incidence of any cancer site in American men. The American Cancer Society estimates that approximately 180,400 men were diagnosed with prostate cancer in 2000, and 31,900 men died of this disease. Only lung cancer accounts for more cancer deaths in men. Prostate cancer remains the leading cause of cancer death in elderly men in the U.S.[1] Unlike lung cancer, the factors involved in prostate cancer etiology are largely unelucidated. Age, race, and family history have been strongly associated with prostate cancer risk.[2] However, few specific exposures or genetic risk factors have been consistently associated with prostate cancer risk.

Molecular and genetic epidemiological studies have recently begun to identify specific genes that may be involved in prostate cancer etiology. Two classes of prostate cancer genes have been inferred to be associated with prostate cancer risk. First, there may be a few genes with allelic variants that confer a high degree of risk to the individual. These genes will be referred to here as "high penetrance" genes. Relatively few individuals in the population are predicted to carry risk-increasing genotypes at these loci. Therefore, the population attributable risk (i.e., the proportion of prostate cancer in the population that may be explained by these genotypes) is low. Because of the large magnitude of effects these genotypes have on cancer risk, one hallmark of high penetrance genes is the creation of a Mendelian (usually autosomal dominant) pattern of cancer.

Second, there may be genes that confer a small to moderate degree of prostate cancer risk to the individual. These "low penetrance" genes may not be associated with Mendelian patterns of prostate cancer, but instead may be associated with sporadic prostate cancer. In addition, it is expected that allelic variability at these genes may be relatively common, and thus may explain a relatively larger proportion of prostate cancer in the population.

In the following sections, an overview of high and low penetrance prostate cancer susceptibility genes is presented, with specific examples of genes belonging to each group. Note that throughout this manuscript, the Mendelian Inheritance in Man number (MIM#) is provided, and additional updated information about the genes described here can be found at the Online MIM (OMIM) at http://www3.ncbi.nlm.nih.gov/Omim.

### 14.3.3.2 HIGH PENETRANCE GENES IN PROSTATE CANCER ETIOLOGY

#### 14.3.3.2.1 Overview

Since the time that prostate cancer was recognized as a familial disease,[3] complex segregation analyses have been undertaken that indicated a rare, autosomal dominant gene segregated in some families to explain patterns of hereditary prostate cancer.[4–6] This putative gene or genes confers a relatively high lifetime prostate cancer penetrance (e.g., 63–

89%, depending on the data set and model applied). However, the frequency of disease causing alleles in the populations studies has been inferred to be 0.3–0.6% in the US and 1.7% in Scandinavia. These figures, and the estimates emanating from these analyses, suggest that only a small proportion (perhaps no more than 10%) of all prostate cancer will occur in men who carry a single gene mutation with an autosomal dominant pattern of inheritance. A number of loci have been identified which explain these hereditary patterns of prostate cancer (summarized in Table 1, and reviewed in Ostrander and Stanford[7]). At least three distinct loci (*HPC1*, *CAPB*, and *PCAP*) have been inferred to lie on chromosome 1, with suggestions that additional loci may lie on this chromosome. However, none of the reports of genetic linkages to specific loci have resulted in cloned prostate cancer susceptibility genes. However, *HPC2/ELAC2* has been identified that may confer prostate cancer risk. These loci are discussed below.

#### 14.3.3.2.2 *CAPB* (MIM# 603688; Chr. 1p36)

One of the loci inferred to lie on chromosome 1p is the *CAPB* locus,[8] named because of its association with both prostate and brain cancers. Gibbs[8] inferred the existence of the *CAPB* locus by genetic linkage analysis of 12 families with aggregations of both prostate and brain cancer, and no evidence for this locus was found in families with prostate cancer only. In contrast, Badzioch[9] and Berry[10] reported no evidence for linkage with brain and prostate cancers in a study nine families that had both prostate and brain cancers. In their larger analysis of 207 families from Europe and North America, only weakly positive and statistically non-significant evidence for linkage to the *CAPB* locus was found.

#### 14.3.3.2.3 *HPC1* (MIM# 601518; Chr. 1q24-q25)

Of the hereditary prostate cancer susceptibility loci identified to date by linkage analysis, the existence of *HPC1* has the greatest support. Using a sample of 91 North American and Scandinavian hereditary prostate cancer families, Smith[11] inferred the existence of a prostate cancer predisposition gene at chromosome 1q24-q25. Those authors estimated that 34% of the families in their sample could be explained by *HPC1*. There were no unusual features of prostate tumors (e.g., age at

**Table 1** Loci associated with a high risk of prostate cancer.

| Locus | Chromosomal location | Independent confirmation? | Reports supporting linkage or cloning | Reports not strongly supporting linkage[*] |
|---|---|---|---|---|
| CAPB | 1p36 | No | Gibbs *et al.* 1999 | Badzioch 2000, Berry 2000 |
| HPC1 | 1q24–25 | Yes | Smith 1996; Cooney 1997; Gronberg 1997; Hsieh 1997<br><br>Neuhausen 1999; Xu 2000 | McIndoe 1997; Eeles 1997; Schleutker 2000; Bergthorsson 2000<br>Suarez 2000; Goode 2000; Gibbs 2000; Berry 2000 |
| HPC2/ ELAC2 | 17p12 | Cloned | Tavtigian 2001 | — |
| HPC20 | 20q13 | No | Berry 2000 | Bock 2001 |
| HPCX | Xq27–28 | Yes | Xu 1999; Schleutker 2000 | Lange 2000; Peters 2001; Bergthorsson 2000 |
| PCAP | 1q42.2–q43 | No | Berthon 1998 | Gibbs 1999; Whittemore 1999; Bergthorsson 2000; Suarez 2000; Berry 2000 |

[*] Includes reports that did not attain a lod score (or its equivalent) of $\geq 3.0$, and/or did not provide statistically significant evidence favoring linkage.

diagnosis, histopathological characteristics of tumors) that distinguished families who were or were not linked to the *HPC1* locus. Subsequent to this report, a series of genetic linkage analyses have been undertaken to confirm evidence for the existence of this locus. A number of studies have provided evidence supporting the existence of *HPC1* as a hereditary prostate cancer susceptibility gene, while other studies have not supported this inference.

McIndoe[12] was unable to confirm the existence of the *HPC1* locus in a sample of American prostate cancer families, and reported no support for this locus as a hereditary prostate cancer susceptibility gene in their sample as a whole or in any subset of families. Eeles[13] also found no support for the existence of *HPC1* in a sample of 60 British pairs of men with prostate cancer and an additional 76 families with hereditary patterns of prostate cancer. Subsequently, a series of studies have failed to confirm the existence of linkage in a wide variety of populations.[14–19]

In contrast, Cooney[20] and to a lesser degree Hsieh[21] reported evidence supporting the existence of *HPC1* in two sets of American families that did not necessarily have strong hereditary patterns of prostate cancer. Families with greater numbers of prostate cancer cases, those with prostate cancer in at least three generations, or those with early mean ages of diagnosis provided even greater support for the existence of this gene. Of note in the study of Cooney[20] were the six African American families, who contributed disproportionately to the evidence for linkage with *HPC1*. Gronberg[22,23] reported evidence favoring linkage to *HPC1* in 40 Swedish families (including

12 from the original report of Smith.[24] The evidence supporting this linkage came predominantly from families with relatively early mean age at diagnosis ( < 65 years). Gronberg[23] estimated that a substantial proportion of Swedish hereditary prostate cancer families (approximately 50%) could be explained by mutations at *HPC1*. Finally, Neuhausen[25] reported evidence supporting the existence of *HPC1* using 41 large hereditary prostate cancer families from Utah. Families with the earliest mean age at diagnosis provided the best support for the existence of *HPC1*.

The strongest evidence for the existence of *HPC1* comes from a joint analysis of 772 hereditary prostate cancer families[26] that included many of the groups that previously individually reported linkage analyses on chromosome 1q24-q25. While the evidence overall for linkage was weakly positive, subsets of families provided much stronger evidence for the existence of *HPC1*. These subsets included families with male-to-male transmission of prostate cancer (i.e., excluding those families that showed potential X-linked patterns of inheritance) and an earlier age at diagnosis.

Until *HPC1* is cloned, its effects on prostate cancer risk cannot be understood well enough to be of clinical utility, nor can the biology of *HPC1*-associated prostate carcinogenesis be elucidated. However, the region surrounding *HPC1* has been reported to undergo both loss of heterozygosity[27] and increases in genomic copy number,[28] suggesting that somatic genetic changes at this locus may be involved in prostate carcinogenesis. While it is highly likely that *HPC1* will eventually be isolated and identified to be a hereditary prostate cancer gene, it also appears that (aside from possibly Scandinavian populations) only a small proportion of prostate cancer will be explained by *HPC1*.

### 14.3.3.2.4 *HPC2/ELAC2* (MIM# 605367; Chr. 17p11)

*HPC2* was originally identified to lie on chromosome 17p11 by a genome-wide linkage analysis using 33 families from Utah (unpublished results). A subsequent analysis in a larger set of families from the same population, however, failed to demonstrate linkage at this locus. However, the information from the original linkage report and the large Utah families available were sufficient to lead to the isolation of the *HPC2* gene. Therefore, the *HPC2* gene was the first putative hereditary prostate cancer gene to be cloned.[29]

*HPC2* is comprised of 24 coding exons and one non-coding exon,[30] and produces a 3 kb mRNA which results in a protein of 826 amino acids. *HPC2* is a metal-dependent hydrolase with homology to the *ELAC* gene family. Portions of the gene are evolutionarily conserved across species as divergent as archaea and humans. The protein encoded by the *HPC2/ELAC2* gene also bears amino acid sequence homology with two other genes, namely *PSO2*, which is involved in DNA interstrand crosslink repair, and *CPSF73*, which encodes the 73 kD subunit of the mRNA 3-prime end cleavage and polyadenylation specificity factor.

A number of germline variants have been identified in *HPC2/ELAC2*. Four of these have been reported to be associated with prostate cancer risk. Two apparently rare mutations have been reported in hereditary prostate cancer families: An *Arg781His* missense mutation and a 1 bp insertion creating a frameshift that results in a *Leu547Stop* nonsense mutation.[29] In addition, two common missense variants have been identified that may be associated with non-hereditary prostate cancer risk. The *Ser217Leu* variant is found in approximately 30% of both U.S. Caucasians and African Americans. The *Ala541Thr* variant occurs more rarely in approximately 3–4% of US Caucasians and 1% of African Americans. Tavtigian[29] and Rebbeck[31] reported that either *Leu217* homozygotes and/or carriers of the *Thr541* allele are associated with prostate cancer. Rebbeck[31] also reported that the *Thr541* variant was associated with elevated prostate specific antigen at the time of diagnosis. These groups have also reported sub-

stantial linkage disequilibrium, such that the *Thr541* variant is always observed on a background of the *Leu217* allele.

### 14.3.3.2.5 *HPC20* (MIM# Unassigned; Chr. 20q13)

Berry[32] reported linkage to hereditary prostate cancer on chromosome 20q13 using 162 American families. Those authors also reported that evidence favoring linkage was strongest in families with less than 5 affected members, a later average age of diagnosis, and no male-to-male transmission. However, Bock[33] could not replicate this result in a sample of 172 American families, although they found suggestive evidence for linkage in the same region as *HPC20* in a subset of 16 African American families. Therefore, the existence of *HPC20* remains in question.

### 14.3.3.2.6 *HPCX* (MIM# 300147; Chr. Xq27-q28)

There is substantial evidence that prostate cancer in some families follows an X-linked pattern of inheritance.[34] In support of this hypothesis, evidence for a prostate cancer susceptibility gene on the X chromosome (denoted *HPCX*) has been found using genetic linkage analysis of 360 North American and Scandinavian families.[35] Schleutker[16] provided no significant support for the existence of *HPCX* in 57 Finnish families overall, but did obtain a logarithm of odds (lod) score of 3.1 in a subset of families with late onset prostate cancer showing male to male transmission. Lange[36] studying 153 families, Peters[37] studying 186 families, and Bergthorsson[14] studying 87 families could not confirm this finding with a high degree of statistical certainty. Both authors concluded that there is suggestive but not conclusive evidence that this locus exists. There remains some evidence that *HPCX* does exist based on patterns of prostate cancer in some families, and the confirmatory study may have not had sufficient statistical power to detect the existence of *HPCX*. Therefore, additional studies of larger numbers of families and study subjects need to be undertaken.

### 14.3.3.2.7 *PCAP* (MIM# 602759, Chr. 1q42.2-q43)

*PCAP* was originally identified in a genetic linkage analysis of French and German hereditary prostate cancer families,[38] and is distinct from *HPC1* despite being located to a nearby chromosomal region. Berthon[38] estimated that approximately 50% of the hereditary pattern of prostate cancer families in their analysis could be explained by *PCAP*. However, Gibbs[8] studied 152 American families with hereditary patterns of prostate cancer, and found no evidence for linkage to chromosome 1q42.2-q43. Even after analyses of family subsets defined by number of prostate cancer cases and mean age of prostate cancer diagnosis in the family, Gibbs[8] found no evidence for linkage in any subset of their families. Similarly, Whittemore,[39] Bergthorsson,[14] Suarez,[40] and Berry[32] also showed no evidence for linkage in European and American prostate cancer families, respectively. Therefore, the existence of *PCAP* remains controversial.

### 14.3.3.2.8 Summary of High Penetrance Genes

While there is substantial suggestive evidence for the existence of high penetrance genes in prostate cancer, only one of these genes (*HPC2*) has been cloned, and this gene appears unlikely to account for a substantial proportion of hereditary prostate cancer. The data available to date suggest that the genetics of hereditary prostate cancer may be substantially more complex than that of breast or colon cancers. It is likely that many high penetrance genes, each of which may explain a small proportion of hereditary prostate cancer, may exist. Because prostate cancer occurs at very high rates in general, the presence of "phenocopies" (i.e., cases of prostate cancer in families that are not caused by the gene of interest) limits the ability of linkage analyses to detect segregating loci. Similarly, some families with apparent patterns hereditary prostate cancer may represent excess aggregation of prostate cancer due to chance alone (possibly exacerbated by the use of prostate cancer screening tools), rather than actual hereditary prostate cancer families caused by the inheritance of mutations in a susceptibility gene. Unlike breast, colon, and other cancers, hereditary patterns of prostate cancer do not have characteristic phenotypic or syndromic features (e.g., early age at diagnosis or characteristic histopathological features) that allow hereditary cancers to be readily distinguished from non-hereditary cancers. These factors are likely to continue to complicate the discovery of hereditary prostate cancer genes.

### 14.3.3.3 LOW PENETRANCE GENES IN PROSTATE CANCER ETIOLOGY

In contrast to the substantial evidence that high penetrance genes are responsible for some hereditary occurrences of prostate cancer (Table 1), molecular epidemiologic studies have yet to provide consistent inferences about the role of low penetrance genes in prostate cancer etiology. However, low penetrance genes have great potential to explain the occurrence of prostate cancer in the general population. First, while prostate cancer often aggregates in families, it does not usually segregate in families in a Mendelian manner. This argues that genes other than the high penetrance class described previously may be acting to predispose some men in the population to develop prostate cancer. In addition, it is likely that non-genetic factors (e.g., exogenous exposures) are associated with prostate cancer risk, and these factors act in concert with genetic susceptibility to predispose some men to develop prostate cancers. The interaction of inherited genotypes and environmental exposures has the potential to explain a relatively large proportion of prostate cancer in the general population.

One rationale for a substantial role of low penetrance genes to explain prostate cancer etiology in the general population is based on the observation that variant alleles of low penetrance genes tend to be common. If these alleles confer prostate cancer risk, and they are carried by a large number of men in the population, the proportion of prostate cancer in the general population that may be explained by this allelic variability could be high. An estimate of the proportion of cancer attributable to a high- and low-penetrance genes can be undertaken using the formula $100\% \times p \times (R-1)/\{p \times (R-1)+1\}$, where $R$ is the risk conferred by the variant, and $p$ is the variant allele frequency in the general population.[41] Assuming a typical high penetrance gene variant confers a large relative risk of, e.g., $R=8$, and has a rare frequency of $p=0.005$, the resulting attributable risk is only approximately 3%. In contrast, an estimate of 20% attributable risk results if we assume a low risk ($R=1.5$) but a common variant genotype frequency ($p=0.5$) for a low penetrance genetic variant. These calculations suggest that a sizable proportion of cancer may be explained by the effect of low penetrance genes. These estimates further support the idea that studies of prostate cancer etiology may need to consider both high penetrance and low penetrance genes to

**Table 2a**    Selected genes reported to confer low- to moderate-risk of prostate cancer: Carcinogen metabolism genes.

| Gene (MIM#)–Chromosome | Positive association (Sample size; Risk estimate[a]) | No association (Sample size) |
|---|---|---|
| CYP2D6 (124030)–22q13.1 | — | Febbo 1998 (571/767); Wadelius 1998 (850) |
| CYP2C19(124020)–10q24.1-q24.3 | — | Wadelius 1998 (850) |
| GSTM1 (138350)–1p13.3 | — | Murata (115/204); Rebbeck 1999 (237/239) |
| | | Autrup 1999 (153/288); Kelada 2000 (276/499) |
| GSTT1 (600436)–22q11.2 | Rebbeck 1999 (237/239; 1.8) | Autrup 1999 (153/288) |
| | Kelada 2000 (276/499; 1.6) | |
| GSTP1 (134660)–11q13 | | Wadelius 1998 (850) |
| | | Shepard 2000 (590/803) |
| | | Autrup 1999 (153/288) |
| NAT1 (108345)–8p23.1-p21.3 | Fukutome (101/97; 2.4) | — |
| NAT2 (243400)–8p23.1-p21.3 | — | Wadelius 1998 (850) |

[a] Case and control sample sizes and a representative Risk or Odds Ratio estimate.

explain genetic susceptibility to develop prostate cancer in the general population.

Table 2 presents examples of low penetrance genes that have been reported to be involved in prostate cancer etiology. These are described in the following sections. While a number of other genes and pathways may be considered in the etiology of prostate cancer, etiological studies of germline variants in these genes have not been undertaken. For example, pathways involved in DNA damage and repair, immune surveillance, the cell cycle, and other genes in hormone and carcinogen metabolism pathways have yet to be formally studied.

### 14.3.3.3.1    Carcinogen Metabolism Genes

Genes involved in carcinogen metabolism may be considered as candidate prostate cancer susceptibility genes because they may determine the rate of somatic genetic mutations in prostate tissue, thus influencing the probability that a cell will undergo neoplastic changes. Genes in this class include the glutathione-*S*-transferases, the *N*-acetyl transferases, or members of the cytochrome P450 multigene family such as *CYP1A1*, *CYP2C9*, *CYP2D6*, and others.

The μ (*GSTM1*), π (*GSTP1*) and θ (*GSTT1*) members of the glutathione-*S*-transferase

**Table 2b**    Selected genes reported to confer low- to moderate-risk of prostate cancer: Steroid hormone metabolism genes.

| Gene (MIM#)–Chromosome | Positive association (Sample size; Risk estimate[a]) | No association (Sample size) |
|---|---|---|
| AR (313700)–Xq11-q12 | | |
|    CAG Repeats | Giovannucci 1997 (587/588;↓ CAG = ↑ Risk) | Edwards 1999 (178/195); Bratt 1999 (190/186) |
| | Ingles 1997 (57/169; 4.6 with < 20 CAG) | |
| | Hsing 2000 (190/304;↓ CAG = ↑ Risk) | |
|    GGN Repeats | Platz 1998 (582/794; (↓ GGN = ↑ Risk) | Edwards 1999 (178/195); Hsing 2000 (190/304) |
| CYP3A4 (124010)–7q22.1 | Paris 1999 (174/116; 1.7) | — |
| CYP17 (202110)–10q24.3 | Lunn 1999 (108/167; 1.7); Wadelius 1999 (178/160; 1.6); Habuchi 2000 (252/333; 2.6); Gsur 2000 (63/126; 2.8) | — |
| SRD5A2 (264600)–2p23 | | |
|    V89L | Nam 2001(158/162; 2.5) | Febbo 1999 (592/799); |
|    A49T | Makridakis 1999(216/261; 7.2) (AA) | Lunn 1999 (108/167) |
| | Makridakis 1999(172/200; 3.6) (HI) | |

[a] Case and control sample sizes and a representative Risk or Odds Ratio estimate.

**Table 2c** Selected genes reported to confer low- to moderate-risk of prostate cancer: Other candidate genes.

| Gene (MIM#)–Chromosome | Positive association (Sample size; Risk estimate[a]) | No association (Sample size) |
|---|---|---|
| BRCA1(113705)–17q21 or BRCA2 (600185)–13q12.3 | — | Lehrer 1998 (60)[c]; Hubert 1999 (87)[c]; Nastuik 1999 (83)[c]; Vazina 2000 (174)[c] |
| BRCA2 (600185)–13q12.3 | Easton 1997[d]; Johannsson 1999[d]; Risch 2000[d]; Gayther 2000[d] | — |
| HPC2 (605367)–17p11.2 | Rebbeck 2000 (359/266; 2.4) Tavtigian 2000 (429/148; 2.0) | Vesprini 2001 (431/513), Xu 2001 (362/182) |
| PSA[c] (176820)–19q13 | Xue 2000 (57/156; 2.9) | — |
| VDR (601769)–12q12-q14 | Taylor 1996 (108/170; 0.3–exon 9) Ingles 1997 (57/169; 4.6–polyA) Habuchi 2000 (222/226; 3.3–intron 8) | Ma 1998 (372/591–exon 8,9) Blazer 2000 (77/183–exon 9, PolyA) |

[a] Case and control sample sizes and a representative Risk or Odds ratio estimate.
[b] Gene Symbol: *APS* or *KLK3* for Kallikrein 3.
[c] *BRCA1/2* mutations were not identified in prostate cancer cases or were identified at a rate comparable to the general population.
[d] Excess relative risk in families with germline BRCA2 mutations.

family have been suggested as prostate cancer susceptibility genes because they are involved in the detoxification or activation of putative prostate carcinogens. Murata[42] reported a statistically non-significant trend towards an effect of the homozygously deleted *GSTM1* and prostate cancer risk. They also reported a potential interaction of *GSTM1* genotype with CYP1A1 genotype conferring an increased prostate cancer risk, and suggested that the effect of these genotypes was greater in early onset cases and in advanced stage disease. Autrup[43] reported no effect of variant genotypes at *GSTM1*, *GSTP1*, or *GSTT1* alone, and only statistically non-significant trends toward an effect with deletion of both *GSTM1* and *GSTT1*. They also reported a non-statistically significant trend toward increased risk in men who smoked and who carried at least one homozygous deletion *GST* genotype. However, these studies had limited statistical power to detect potential main effects of *GST* genotypes or of interactions of these genotypes with smoking. Rebbeck[44] reported that men who carried an active (non-deleted) *GSTT1* was associated with increased prostate cancer risk with an odds ratio of 1.8 (95% CI: 1.2–2.8), and subsequently showed that *GSTT1* interacts with smoking to elicit an effect on prostate cancer risk.[45] Possible explanations for this interaction include metabolism of carcinogenic intermediates by *GSTT1*, which is highly expressed in the prostate and can produce genotoxic effects upon exposure to specific carcinogens, or the response to chronic inflammation associated mediated by *GSTT1*.

It has been widely reported that somatic suppression of *GSTP1* expression by methyla-

tion occurs commonly in prostate tumors.[46] However, germline variants in *GSTP1* have not been consistently associated with prostate cancer risk,[47] Shepard,[48] Autrup.[43] In part, this may be due to the lack of clear functional significance of the *GSTP1* variants studied to date. Alternatively, the lack of an association may be explained by the likely contribution of multiple genotypes interacting with environmental exposures to confer prostate cancer risk if these pathways are in fact involved in prostate carcinogenesis.

The *N*-acetyl-transferase genes *NAT1* and *NAT2* have similarly been proposed as prostate cancer risk factors because of their ability to metabolize putative prostate cancer risk factors such as those found in the diet. Wadelius[47] found no association of *NAT2* and prostate cancer. Fukutome[49] using a small study sample reported an association between *NAT1* genotype and prostate cancer risk. However, the small sample size studied by Fukutome[49] limits the inferences that can be made from this analysis.

Members of the cytochrome P450 family involved in carcinogen metabolism have been proposed as susceptibility genes for prostate cancer. However, few studies relating inherited genetic variation in these pathways with prostate cancer risk have been reported. Two case-control studies have evaluated whether poor metabolizers at *CYP2D6*, a gene involved in the metabolism of a wide variety of carcinogens, were at increased prostate cancer risk.[47,50] Neither study reported an association between *CYP2D6* and prostate cancer. Wadelius[47] also reported no association between prostate cancer and *CYP2C19* genotypes. However, those authors did report a three-fold increase in prostate cancer risk among smokers who were also *CYP2D6* poor metabolizers (OR = 3.1, 95% CI: 1.1–8.9).

The literature therefore provides little consistent evidence that individual carcinogen metabolism genotypes are associated with prostate cancer risk. This is not surprising given the complex, multifactorial nature of prostate cancer etiology. It is noteworthy that the strongest reports of carcinogen metabolism genes and prostate cancer risk have been made among smokers (i.e., interactions of smoking with *CYP2D6* or *GSTT1*). While smoking has not been consistently reported as a prostate cancer risk factor, it is possible that a subset of smokers may be at elevated prostate cancer risk because of an inherited susceptibility to the carcinogens found in cigarette smoke. Additional large-scale studies that can evaluate interactions between carcinogen metabolizing genotypes and specific exposures may provide additional insight into the complex etiology of prostate cancer, and may identify specific pathways by which environmental carcinogens may act in prostate carcinogenesis.

### 14.3.3.3.2  Androgen Metabolism Genes

Testosterone is a major determinant of prostate growth and differentiation. Testosterone is synthesized in the testis leydig cells and, to some degree, catabolized outside of the prostate itself. Testosterone is bioactivated by 5α-reductase to form dihydrotestosterone (DHT) in prostate stromal and basal cells. DHT is the ligand with the highest affinity for the androgen receptor and mediates its activity in prostate epithelial cells to regulate androgen responsive genes.[51] There are numerous lines of evidence that support the role of androgen metabolism in prostate cancer etiology. Circulating levels of androgens have been reported to be higher in populations at increased prostate cancer risk, including African American men,[52] and lower in populations at decreased prostate cancer risk, including Chinese men.[53] Although serum levels of testosterone do not correlate well with prostate cancer risk[54,55] serum levels of DHT and other testosterone metabolites do correlate with prostate cancer risk.[54,55,52] Second, there is abundant clinical evidence that androgens are related to the growth and development of prostate cancers. It is well known that androgen ablation in men with hormone sensitive prostate cancers reduces tumor size, and decreases the associated disease burden. This evidence suggests that the disposition of

testosterone may be important in determining prostate cancer risk.

Testosterone and DHT bioavailabilty are determined by a number of metabolic pathways. These include cytochrome P450 IIIA4 (encoded by *CYP3A4*), 5α-reductase type II (encoded by *SRD5A2*), the 3α- and 3β-hydroxysteroid dehydrogenases (encoded by *HSD3A* and *HSD3B2*), the androgen receptor (encoded by *AR*), and the cytochrome P450's *CYP17* and *CYP19* (aromatase). Each of the major proteins involved in testosterone formation and disposition is regulated by the activity of at least one cloned gene with known polymorphic variation. Of these, only *AR*, *CYP17*, *CYP3A4* and *SRD5A2* have been studied in an epidemiological context (e.g., a case-control association study) to assess their role in prostate cancer etiology. There is also only limited information about exogenous exposures that might interact with these genes in prostate cancer etiology (e.g., endocrine disruptors), and no direct evaluations of cancer risk that includes both these exposures and the genes in these pathways.

#### 14.3.3.3.2.1  Androgen Receptor (AR–MIM# 313700, Chr. Xq12)

The androgen receptor functions as a ligand dependent transcriptional activator in response to androgens. The *AR* gene contains a highly polymorphic *CAG* trinucleotide repeat encoding polyglutamines in its first exon. The length of the *CAG* polymorphism is inversely associated with the degree of transcriptional activation by *AR*.[56,57] Individuals with X-linked spinal and bulbar muscular atrophy (Kennedy's disease) have 40 or more *CAG* repeats and manifest clinical androgen insensitivity.[58] A *GGN* (polyglycine) repeat polymorphism in *AR* has also been identified. These findings suggest that *CAG* or *GGN* repeat length polymorphism may be involved in modifying the development of diseases caused by alterations in androgen signaling.

A number of studies have evaluated the relationship between repeat polymorphisms and prostate cancer risk. Early reports showed positive associations of the *CAG* and *GGN* polymorphisms. Giovannucci[59] studying the Physician's Health Study cohort identified an increase in risk with decreasing *CAG* repeat length. Ingles,[60] also using a cohort-based sample, reported increased prostate cancer risk associated with shorter *CAG* repeat lengths. Hsing[64] reported longer mean *CAG* repeat lengths in Chinese men, and that *CAG* repeats of 23 or greater conferred an increased prostate cancer risk (1.7; 95% CI: 1.1–2.4) in a sample of 190 Chinese prostate cancer cases and 304 controls.

Despite these reports, a number of other studies have been published that cast doubt that *AR* was associated with prostate cancer risk in all populations. Correa-Cerro[61] studied 132 cases and 105 controls in a French/German population, in addition to a set of familial prostate cancer cases, and showed no relationship between the *CAG* repeat polymorphisms and prostate cancer. Bratt[62] studied a population of 190 cases and 186 female controls, and reported no association with the *CAG* repeat length polymorphism, although they did identify associations with endocrine treatment response and age at diagnosis. Edwards[63] also reported no association of *AR* genotypes with prostate cancer risk in an English population. Finally, Lange[65] found no association with *CAG* repeat length and prostate cancer risk, in a sample of men from families with hereditary prostate cancer. Platz[66] and Hsing [64] also studied the *GGN* (polyglycine) repeat in *AR*. Platz[66] also reported that decreased number of *GGN* repeats were associated with an increase in prostate cancer risk in a US population, but Hsing,[64] Correa-Cerro,[61,63] showed no such association in Chinese, French/German, or English populations. These studies provide some evidence that the *CAG* or *GGN* repeat polymorphisms in *AR* are involved in prostate cancer susceptibility, although the inconsistency between reports suggests additional studies are required to better evaluate the conditions under which these polymorphisms may confer risk.

### 14.3.3.3.2.2 CYP17 (MIM# 202110, Chr. 10q24.3)

*CYP17* encodes the enzyme cytochrome P-450c17α, which mediates both 17 α-hydroxylase and 17,20-lyase activity in steroid biosynthesis. A single nucleotide polymorphism in the *CYP17* promoter region creates an Sp1-type (CCACC box) element. This polymorphism (denoted *A2*) has been associated with prostate cancer in a number of studies. Lunn[67] and Wadelius[47] reported associations between the *A2* allele at *CYP17* and prostate cancer (OR = R 1.7, 95% CI: 1.0–3.0 and OR = 1.6, 95% CI: 1.02–2.5, respectively). Gsur[68] also reported an association between men carrying the *A2/A2* genotype and prostate cancer risk (OR = 2.8, 95%CI: 1.02–77.8). This risk was substantially higher in men over age 66 (OR = 8.9, 95%CI = 1.8–49.2). However, the sample studied in Gsur[68] was small (63 cases and 126

controls), and the controls were men who had benign prostatic hyperplasia (BPH). Subsequent to these reports, Habuchi[69] reported that Japanese men who carried the *A1/A1* genotype were at increased prostate cancer risk (OR = 2.6; 95% CI: 1.4–4.8) compared to carriers of the *A2/A2* genotype. They reported no significant association between the *CYP17* genotype and the prostate tumor grade and stage. While this was a Japanese population which might be different in many regards to the Caucasian populations studied by Lunn,[67] Wadelius[47] and Gsur,[68] it is not clear what may have led to the discrepancy between these studies and that of Habuchi[69] in terms of which allele should be considered the risk allele. Because of this discrepancy and the small study samples used to generate some of these results, the inference that *CYP17* is associated with prostate cancer risk must be interpreted with caution.

### 14.3.3.3.2.3 SRD5A2 (MIM# 264600, Chr. 2p23)

The 5α-reductase type 2, encoded by the *SRD5A2* gene, is responsible for the bioactivation of testosterone to dihydrotestosterone (DHT), the primary intracellular mediator of androgenic effects to the prostate, and binds preferentially to the androgen receptor.[70] Germline mutations in *SRD5A2* have been associated with male pseudohermaphrodism. A number of missense variants in *SRD5A2* have also been identified that are not associated with pseudohermaphrodism. These include *C5R*, *P30L*, *A49T*, *V89L*, *T187M*, *R227Q*, *F234L*, and a $(TA)_n$ repeat polymorphism.[71] The majority of these are rare variants, with the most common variants being *A49T*, *V89L*, and $(TA)_n$.[71] The LL genotype of the *V89L* polymorphism is associated with decreased circulating levels of androstenediol glucuronide,[72] and the variant *A49T* enzyme has a markedly increased ability to convert testosterone to DHT than the wild type genotype.[71] If conversion of testosterone to DHT is an important part of prostate carcinogenesis, these data predict that the *V89L* variant may be associated with decreased prostate cancer risk, while the *A49T* variant would be associated with increased prostate cancer risk.

A number of case-control association studies have suggested that these polymorphisms may be associated with prostate cancer risk. Makridakis[73] studied 216 African-American cases and 261 controls, as well as 172 Hispanic cases and 200 controls for the *A49T* variant. They found an elevated risk of clinically significant disease in African-Americans (OR = 7.2; 95%CI: 2.2–27.9) and Hispanics (OR = 3.6, 95%CI: 1.1–12.3). Nam[74] undertook a study of 312 cases and 318 controls (all of whom had undergone prostate biopsy), and reported an association between the *V89L* polymorphism and prostate cancer (OR = 2.5). However, other studies that did not use biopsy controls[67,75] have not shown an association between *V89L* and prostate cancer. Therefore, while there is substantial interest in and some preliminary indication that *SRD5A2* is a prostate cancer susceptibility gene, additional studies will be required to confirm these associations. No relationship between *SRD5A2* and environmental agents in the etiology of prostate cancer has been reported.

### 14.3.3.3.3  Other Pathways

#### *14.3.3.3.3.1  Vitamin D Receptor (MIM# 601769, Chr. 12q12-q14)*

Epidemiological studies have observed that prostate cancer risk varies with latitude, and that UV exposure may in part be associated with prostate cancer risk.[76] The Vitamin D receptor (VDR) mediates Vitamin D-associated cellular differentiation, and a relationship of higher levels of the active hormonal form of serum Vitamin D have been associated with risk of prostate cancer.[77] A number of polymorphisms have been described in the 3′UTR of the VDR that are correlated with mRNA stability.[78] Other polymorphisms in intron 8 and exon 9 have also been reported and appear to be in linkage disequilibrium.[78]

A number of studies have evaluated whether these polymorphisms in VDR are associated with prostate cancer risk. Taylor[79] first reported an altered risk of prostate cancer in 108 radical prostatectomy cases compared with 170 controls associated with a *TaqI* polymorphism in exon 9 (OR = 0.34, 95%CI : 0.16–0.76). Ingles subsequently studied 57 Caucasian prostate cancer cases and 169 controls, and reported that a poly-A microsatellite in the 3′ UTR of VDR was associated with prostate cancer (OR = 4.6, 95%CI: 1.3–15.8) among men who carried at least one long repeat in this poly-A microsatellite. Habuchi[69] reported that the intron 8 polymorphism was significantly associated with risk in a Japanese sample of 222 cases and 226 controls (including a large number of female controls) with an OR = 3.3 (95%CI: 2.1–5.3). However, using a large nested case control sample for 372 prostate cancer cases and 591 controls, Ma[80] reported that the intron 8 and exon 9

polymorphisms were not associated with prostate cancer risk. While there was no association of these polymorphisms overall, men with intron 8 variants and lower than average plasma 25-hydroxy Vitamin D showed a significant association with risk, particularly in a subset of older men. Blazer[81] also reported no association of VDR polymorphisms and prostate cancer risk. However they used a very small sample of 77 cases and 183 controls, and reported an OR effect of 2.5–2.8. This suggests that their study did not have sufficient statistical power to detect potentially interesting effects. While VDR polymorphisms have been studied in multiple settings, it remains unclear whether this gene is associated with prostate cancer risk. In particular, studies that consider not only the genotype but also sun exposure (or its surrogates) may be required to better understand the relationship of this gene with prostate cancer.

#### *14.3.3.3.3.2  BRCA1 (MIM# 133705, Chr. 17q21) and BRCA2 (MIM# 600185, Chr. 13q12.3)*

Deleterious mutations in *BRCA1* and *BRCA2* (*BRCA1/2*) are strongly associated with hereditary breast and ovarian cancer risk. However, germline *BRCA1/2* mutations have not been found to be responsible for hereditary prostate cancer in families.[82,83]

In contrast, an excess risk of prostate cancer has been observed in a number of studies using families with hereditary patterns of breast and ovarian cancer who carry *BRCA2* mutations.[84–87] Excess relative risks of prostate cancer in families who carry these mutations have been estimated to be approximately 2–3-fold higher than expected in the general population. In addition, Risch[88] suggested that prostate cancer occurred in excess among relatives of *BRCA2* mutation carriers whose mutation occurred in the ovarian cancer-cluster region (OCCR) of exon 11. However, it is not clear whether the excess risk of prostate cancer in families may come from heightened cancer awareness and therefore increased detection of otherwise latent prostate cancers because of the existence of a germline *BRCA1/2* mutation in these families.

Other studies have not reported an excess of prostate cancer in individuals or families who carry germline *BRCA1/2* mutations drawn from populations unselected by family history of cancer. In particular, there is no excess of the *BRCA1/2* founder mutations in Jewish men with prostate cancer.[89–92] Therefore, it is not yet clear whether germline *BRCA1/2* mutations are responsible for increased prostate cancer risk outside of the excess rates observed in hereditary breast-ovarian cancer families.

The biological rationale for *BRCA1/2* in the etiology of prostate cancer is not clear. *BRCA1* is a co-activator of the androgen receptor,[93] and therefore may play a role in androgen-dependent carcinogenesis. Studies of *BRCA1/2* function have suggested that these genes may act as prostate tumor suppressor genes,[94] although LOH at *BRCA1/2* has not been consistently observed in prostate tumors.[95,96]

### 14.3.3.3.3  Prostate Specific Antigen (PSA) (KLK3; MIM# 176820, Chr. 19q13)

Elevated prostate specific antigen (PSA) levels are used as biomarkers of benign prostatic hyperplasia (BPH) or prostate cancer presence. Although the designation *PSA* is used here, the gene that encodes *PSA* is correctly denoted *APS* or *KLK3* (after Kallekrein 3). Xue[97] studied *PSA* and *AR* genotype in 57 cases and 156 controls, and reported *PSA* genotype was associated with advanced (but not localized) prostate cancer. In addition, men with both a short *CAG* allele and *PSA* genotype *GG* had a more than 5-fold increase in prostate cancer risk (OR, 5.08; 95% CI: 1.6–16.3), with even stronger associations seen in advanced disease. This is the only published reported to date of *PSA* genotype and prostate cancer risk, and because of the small sample size employed, additional studies will need to be undertaken to confirm these results.

### 14.3.3.3.4  SUMMARY OF LOW PENETRANCE GENES

It is biologically plausible that some candidate low penetrance genes are involved in prostate cancer etiology. Candidate genes have been identified for which functional polymorphisms are available, and many of these genes have been studied in molecular epidemiologic investigations of prostate cancer etiology. A few patterns have emerged from the studies reported thus far. First, few genes have been consistently associated with prostate cancer risk across different studies or populations. In part, this is because the field is relatively new and few large, epidemiologically well-designed studies have been published in this area. In general, disease association studies involving low penetrance genes have often suffered from small or poorly designed samples, which limit the inferences that can be

drawn about the role of these genes in the general population. However, even among well-designed studies, inconsistent results have arisen in part because of differences in the way the data were analyzed. Second, prostate cancer etiology is clearly heterogeneous. As evidenced by the numerous reports of genetic linkage to a series of different genomic loci, the degree of etiologic heterogeneity in prostate cancer etiology may be substantial. Therefore, prostate cancer susceptibility genotypes may be associated with prostate cancer in some groups or populations but not others. Additional research using well-designed epidemiologic studies with large samples that consider interactions among multiple factors (including genotype–genotype and genotype–environment interactions) are required to help elucidate the role of low penetrance genes in prostate cancer etiology. Of particular importance are studies of interactions between these genotypes and specific exposures that may be relevant in prostate carcinogenesis, including those found in the diet.

### 14.3.3.4  IMPLICATIONS FOR PROSTATE CANCER RISK ASSESSMENT

To date, only one potential high penetrance gene, *HPC2*, has been identified that may significantly increase a man's prostate cancer risk. However, it is likely that germline mutations in this gene are rare in the general population, and the specific prostate cancer risk conferred by these mutations is not clear. No other genes have been identified that individually are likely to be appropriately applied in a clinical setting to evaluate prostate cancer risk. In addition, it is unclear at this time how knowledge of genetic information could be used to reduce prostate cancer incidence or mortality. Screening and diagnostic practices have not been modified to incorporate genetic information, and no studies have provided a strong justification for the application of a prostate cancer genetic test for clinical purposes. While the clinical utility of genetic information in prostate cancer is not immediately apparent, it seems possible that clinical risk evaluation may be used as additional information becomes available about prostate cancer genetic risk, and as clinical correlates of this genetic information becomes available. First, knowledge of these factors may be used to guide recommendations for modification of certain exposures (e.g., diet), which may become a part of standard care in the follow-up of men who are found to carry these variants. Second, knowledge of risk modifying factors in these men may lead to tailored cancer screening strategies. This could include recommendations related to the timing or intensity of screening in subsets of men. Finally, knowledge of factors that modify cancer risk may direct research efforts to identify important prostate carcinogenesis pathways. This may in turn lead to the novel use of existing (e.g., hormonal) agents, or the development of new agents that could be applied in chemoprevention or treatment strategies specifically targeted toward men with a specific genetic susceptibility to prostate cancer.

In contrast, low penetrance genes are expected to confer a relatively low degree of risk to the individual. As a result, it is unclear how genetic information at these loci may be of value in clinical risk assessment at this time. If an array of genotypes at multiple loci (possibly in conjunction with exposure information) is identified to have a large effect on a man's prostate cancer risk, then low penetrance genotype information may be clinically valuable. However, substantial additional information is needed about the conditions under which low penetrance genotypes affect prostate cancer risk. To date, relatively few environmental exposures have been consistently associated with prostate cancer risk, and even fewer interactions between inherited genotype and exposures have been reported. However, it is very possible that if low penetrance prostate cancer susceptibility genes are acting in the etiology of prostate cancer, that their effects will be seen in the context of specific environmental exposures. Therefore, future studies of specific carcinogen metabolism genes should explicitly be designed to be able to study these interactions.

## 14.3.3.5 ACKNOWLEDGMENTS

This study was supported by grants from the Public Health Service (CA85074) and the University of Pennsylvania Cancer Center.

## 14.3.3.6 REFERENCES

1. J. L. Stanford, R. A. Stephenson, L. M. Coyle *et al.*, 'Prostate cancer trends 1973–1995, SEER Program, National Cancer Institute. NIH Pub, Bethesda, MD, 1999, **99**, 4543.
2. V. L. Ernster, W. Winkelstein, S. Selvin *et al.*, 'Race, socioeconomic status, and prostate cancer.' *Cancer Treat. Rep.*, 1977, **61**(2), 187–191.
3. G. Morganti, L. Gianferrari, A. Cresseri *et al.*, 'Recherches clinicostatistiques et genetiques sur les neoplasias de la prostate.' *Acta Gen. Med. et Gem.*, 1956, **6**, 304–305.
4. B. S. Carter, T. H. Beaty, G. D. Steinberg *et al.*, 'Mendelian inheritance of familial prostate cancer.' *Proc. Nat. Acad. Sci. USA*, 1992, **89**(8), 3367–3371.
5. H. Gronberg, L. Damber, J. E. Damber *et al.*, 'Segregation analysis of prostate cancer in Sweden: support for dominant inheritance.' *Am. J. Epi.*, 1997, **146**, 552–557.
6. D. J. Schaid, S. K. McDonnell, M. L. Blute *et al.*, 'Evidence for autosomal dominant inheritance of prostate cancer.' *Am. J. Hum. Genet.*, 1998, **62**(6), 1425–1438.
7. E. A. Ostrander and J. L. Stanford, 'Genetics of prostate cancer: too many loci, too few genes.' *Am. J. Hum. Genet.*, 2000, **67**(6), 1367–1375.
8. M. Gibbs, J. L. Stanford, R. A. McIndoe *et al.*, 'Evidence for a rare prostate cancer-susceptibility locus at chromosome 1p36.' *Am. J. Hum. Genet.*, 1999, **64**(3), 776–787.
9. M. Badzioch, R. Eeles, G. Leblanc *et al.*, 'Suggestive evidence for a site specific prostate cancer gene on chromosome 1p36.' The CRC/BPG UK Familial Prostate Cancer Study Coordinators and Collaborators. The EU Biomed Collaborators. *J. Med. Genet.*, 2000, **37**(12), 947–949.
10. R. Berry, J. J. Schroeder, A. J. French *et al.*, 'Evidence for a prostate cancer-susceptibility locus on chromosome 20.' *Am. J. Hum. Genet.*, 2000, **67**, 82–91.
11. R. Smith, D. Freije, J. D. Carpten *et al.*, 'Major susceptibility locus for prostate cancer on chromosome 1 suggested by a genome-wide search.' *Science*, 1996, **274**, 1371–1374.
12. R. A. McIndoe, J. L. Stanford, M. Gibbs *et al.*, 'Linkage analysis of 49 high-risk families does not support a common familial prostate cancer-susceptibility gene at 1q24–25.' *Am. J. Hum. Genet.*, 1997, **61**, 347–353.
13. R. A. Eeles, F. Durocher, S. Edwards *et al.*, 'Linkage analysis of chromosome 1q markers in 136 prostate cancer families.' The Cancer Research Campaign/British Prostate Group U.K. Familial Prostate Cancer Study Collaborators. *Am. J. Hum. Genet.*, 1998, **62**(3), 653–658.
14. J. T. Bergthorsson, G. Johannesdottir, A. Arason *et al.*, 'Analysis of HPC1, HPCX, and PCap in Icelandic hereditary prostate cancer.' *Hum. Genet.*, 2000, **107**(4), 372–375.
15. B. K. Suarez, J. Lin, J. S. Witte *et al.*, 'Replication linkage study for prostate cancer susceptibility genes.' *Prostate*, 2000, **45**(2), 106–114.
16. J. Schleutker, M. Matikainen, J. Smith *et al.*, 'A genetic epidemiological study of hereditary prostate cancer (HPC) in Finland: frequent HPCX linkage in families with late-onset disease.' *Clin. Cancer Res.*, 2000, **6**(12), 4810–4815.
17. E. L. Goode, J. L. Stanford, L. Chakrabarti *et al.*, 'Linkage analysis of 150 high-risk prostate cancer families at 1q24–25.' *Genet. Epi.*, 2000, **18**(3), 251–275.
18. M. Gibbs, J. L. Stanford, G. P. Jarvik *et al.*, 'A genomic scan of families with prostate cancer identifies multiple regions of interest.' *Am. J. Hum. Genet.*, 2000, **67**(1), 100–109.
19. R. Berry, J. J. Schroeder, A. J. French *et al.*, 'Evidence for a prostate cancer-susceptibility locus on chromosome 20.' *Am. J. Hum. Genet.*, 2000, **67**, 82–91.
20. K. A. Cooney, J. D. McCarthy, E. Lange *et al.*, 'Prostate cancer susceptibility locus on chromosome 1q: a confirmatory study.' *J. Nat. Cancer Inst.*, 1997, **89**, 955–999.
21. C. L. Hsieh, I. Oakley-Girvan, R. P. Gallagher *et al.*, 'Re: prostate cancer susceptibility locus on chromo-

some 1q: a confirmatory study.' (Letter) *J. Nat. Cancer Inst.*, 1997, **89**, 1893–1894.

22. H. Gronberg, J. Xu, J. R. Smith *et al.*, 'Early age at diagnosis in families providing evidence of linkage to the hereditary prostate cancer locus (HPC1) on chromosome 1.' *Cancer Res.*, 1997, **57**, 4707–4709.

23. H. Gronberg, J. Smith, M. Emanuelsson *et al.*, 'In Swedish families with hereditary prostate cancer, linkage to the HPC1 locus on chromosome 1q24–25 is restricted to families with early-onset prostate cancer.' *Am. J. Hum. Genet.*, 1999, **65**(1), 134–140.

24. J. R. Smith, D. Freije, J. D. Carpten *et al.*, 'Major susceptibility locus for prostate cancer on chromosome 1 suggested by a genome-wide search.' *Science*, 1996, **274**, 1371–1374.

25. S. L. Neuhausen, J. M. Farnham, E. Kort *et al.*, 'Prostate cancer susceptibility locus HPC1 in Utah high-risk pedigrees.' *Hum. Mol. Genet.*, 1999, **8**(13), 2437–2442.

26. J. Xu, 'Combined analysis of hereditary prostate cancer linkage to 1q24–25: results from 772 hereditary prostate cancer families from the International Consortium for Prostate Cancer Genetics.' *Am. J. Hum. Genet.*, 2000, **66**(3), 945–957.

27. W. D. Dunsmuir *et al.*, 'Allelic imbalance in familial and sporadic prostate cancer at the putative human prostate cancer susceptibility locus, HPC1.' CRC/BPG UK Familial Prostate Cancer Study Collaborators. Cancer Research Campaign/British Prostate Group. *Br. J. Cancer*, 1998, **78**(11), 1430–1433.

28. M. L. Cher, G. S. Bova, D. H. Moore *et al.*, 'Genetic alterations in untreated metastases and androgen-independent prostate cancer detected by comparative genomic hybridization and allelotyping.' *Cancer Res.*, 1996, **56**(13), 3091–3102.

29. S. V. Tavtigian, J. Simard, D. H. F. Teng *et al.*, 'A candidate prostate cancer susceptibility gene at chromosome 17p.' *Nat. Genet.*, 2001, **27**, 172–180.

30. S. V. Tavtigan, J. Simard, J. Rommens *et al.*, 'The complete BRCA2 gene and mutations in chromosome 13q-linked kindreds.' *Nat. Genet.*, 1996, **12**(3), 333–337.

31. T. R. Rebbeck, A. H. Walker, C. Zeigler-Johnson *et al.*, 'Association of HPC2/ELAC2 genotypes and prostate cancer.' *Am. J. Hum. Genet.*, 2000, **67**, 1014–1019.

32. R. Berry, J. J. Schroeder, A. J. French *et al.*, 'Evidence for a prostate cancer-susceptibility locus on chromosome 20.' *Am. J. Hum. Genet.*, 2000, **67**, 82–91.

33. C. H. Bock, J. M. Cunningham, S. K. McDonnell *et al.*, 'Analysis of the prostate cancer-susceptibility locus HPC20 in 172 families affected by prostate cancer.' *Am. J. Hum. Genet.*, 2001, **68**(3), 795–801.

34. K. R. Monroe, M. C. Yu, L. N. Kolonel *et al.*, 'Evidence of an X-linked or recessive genetic component to prostate cancer risk.' *Nat. Med.*, 1995, **1**, 827–829.

35. J. Xu, D. Meyers, D. Freije *et al.*, 'Evidence for a prostate cancer susceptibility locus on the X chromosome.' *Nat. Genet.*, 1998, **20**(2), 175–179.

36. E. M. Lange, H. Chen, K. Brierley *et al.*, 'The polymorphic exon 1 androgen receptor CAG repeat in men with a potential inherited predisposition to prostate cancer.' *Cancer Epidemiol. Biomarkers Prev.*, 2000, **9**(4), 439–442.

37. M. A. Peters, G. P. Jarvik, M. Janer *et al.*, 'Genetic linkage analysis of prostate cancer families to Xq27–28.' *Hum. Hered.*, 2001, **51**(1–2), 107–113.

38. P. Berthon, A. Valeri, A. Cohen-Akenine *et al.*, 'Predisposing gene for early-onset prostate cancer, localized on chromosome 1q42.2–43, *Am. J. Hum. Genet.*, 1998, **62**, 1416-1424.

39. A. S. Whittemore, I. G. Lin, I. Oakley-Girvan *et al.*, 'No evidence of linkage for chromosome 1q42.2–43 in prostate cancer.' *Am. J. Hum. Genet.*, 1999, **65**(1), 254–256.

40. B. K. Suarez, J. Lin, J. K. Burmester *et al.*, 'A genome screen of multiplex sibships with prostate cancer.' *Am. J. Hum. Genet.*, 2000, **66**(3), 933–944.

41. A. M. Lilienfeld and D. E. Lilienfeld, 'Foundations of Epidemiology,' Oxford University Press, New York, NY, 1980, pp. 346–347.

42. M. Murata, T. Shiraishi, K. Fukutome *et al.*, 'Cytochrome P4501A1 and glutathione *S*-transferase M1 genotypes as risk factors for prostate cancer in Japan.' *Jpn. J. Clin. Oncol.*, 1998, **28**(11), 657–660.

43. J. L. Autrup, L. H. Thomassen, J. H. Olsen *et al.*, 'Glutathione *S*-transferases as risk factors in prostate cancer.' *Eur. J. Cancer Prev.*, 1999, **8**(6), 525–532.

44. T. R. Rebbeck, A. H. Walker, J. M. Jaffe *et al.*, 'Glutathione *S*-transferase-μ (GSTM1) and –θ (GSTT1) genotypes in the etiology of prostate cancer.' *Cancer Epidemiol. Biomarkers Prev.*, 1999, **8**(4 Pt 1), 283–287.

45. S. N. Kelada, S. L. R. Kardia, A. H. Walker *et al.*, 'The glutathione *S*-transferase-μ (GSTM1) and θ (GSTT1) genotypes in the etiology of prostate cancer: gene-environment interactions with smoking. *Cancer Epidemiol. Biomarkers Prev.*, 2001, **9**, 1329–1334.

46. W. H. Lee, R. A. Morton, J. I. Epstein *et al.*, 'Cytidine methylation of regulatory sequences near the π-class glutathione *S*-transferase gene accompanies human prostatic carcinogenesis.' *Proc. Nat. Acad. Sci. USA*, 1994, **91**(24), 11733–11737.

47. M. Wadelius, A. O. Andersson, J. E. Johansson *et al.*, 'Prostate cancer associated with *CYP17* genotype.' *Pharmacogenetics*, 1999, **9**(5), 635–639.

48. T. F. Shepard, E. A. Platz, P. W. Kantoff *et al.*, 'No association between the I105V polymorphism of the glutathione *S*-transferase P1 gene (GSTP1) and prostate cancer risk: a prospective study.' *Cancer Epidemiol. Biomarkers Prev.*, 2000, **9**(11), 1267–1268.

49. K. Fukutome, M. Watanabe, T. Shiraishi *et al.*, '*N*-acetyltransferase 1 genetic polymorphism influences the risk of prostate cancer development.' *Cancer Lett.*, 1999, **136**(1), 83–87.

50. P. G. Febbo, P. W. Kantoff, E. Giovannucci *et al*, 'Debrisoquine hydroxylase (CYP2D6) and prostate cancer.' *Cancer Epidemiol. Biomarkers Prev.*, 1998, **7**(12), 1075–1078.

51. S. M. Galbraith and G. M. Duchesne, 'Androgens and prostate cancer: biology, pathology and hormonal therapy.' *Eur. J. Cancer*, 1997, **33**, 545–554.

52. R. K. Ross, L. Bernstein, R. A. Lobo *et al.*, '5-α-reductase activity and risk of prostate cancer among Japanese and US white and black males.' *Lancet*, 1992, **339**, 887–889.

53. D. P. Lookingbill, L. M. Demers, C. Wang *et al.*, 'Clinical and biochemical parameters of androgen action in normal healthy Caucasian versus Chinese subjects.' *J. Clin. Endo. Metab.*, 1991, **72**, 1242–1248.

54. L. J. Vatten, G. Ursin, R. K. Ross *et al.*, 'Androgens in serum and the risk of prostate cancer: a nested case-control study from the Janus serum bank in Norway.' *Cancer Epidemiol. Biomarkers Prev.*, 1997, **6**, 967–969.

55. P. H. Gann, C. H. Hennekens, J. Ma *et al.*, 'Prospective study of sex hormone levels and risk of prostate cancer.' *J. Nat. Cancer Inst.*, 1996, **88**, 1118–1126.

56. N. L. Chamberlain, E. D. Driver and R. L. Miesfeld, 'The length and location of CAG trinucleotide repeats in the androgen receptor *N*-terminal domain affect

transactivation function.' *Nucleic Acids Res.*, 1994, **22**(15), 3181–3186.

57. P. Kazemi-Esfarjani, M. A. Trifiro and L. Pinsky, 'Evidence for a repressive function of the long polyglutamine tract in the human androgen receptor: possible pathogenetic relevance for the (CAG)*n*-expanded neuronopathies.' *Human Mol. Genet.*, 1995, **4**(4), 523–527.

58. A. R. Spada, E. M. Wilson, D. B. Lubahn *et al.*, 'Androgen receptor gene mutations in x-linked spinal and bulbar muscular atrophy.' *Nature*, 1991, **352**(6330), 77–79.

59. E. Giovannucci, M. J. Stampfer, K. Krithivas *et al.*, 'The CAG repeat within the androgen receptor gene and its relationship to prostate cancer.' *Proc. Nat. Acad. Sci. USA*, 1997, **94**(7), 3320–3323.

60. S. A. Ingles, R. K. Ross, M. C. Yu *et al.*, 'Association of prostate cancer risk with genetic polymorphisms in Vitamin D receptor and androgen receptor.' *J. Nat. Cancer Inst.*, 1997, **89**(2), 166–170.

61. L. Correa-Cerro, G. Wohr, J. Haussler *et al.*, '(Cag)nCAA and GGN repeats in the human androgen receptor gene are not associated with prostate cancer in a French–German population.' *Eur. J. Hum. Genet.*, 1999, **7**(3), 357–362.

62. O. Bratt, A. Borg, U. Kristoffersson *et al.*, 'CAG repeat length in the androgen receptor gene is related to age at diagnosis of prostate cancer and response to endocrine therapy, but not to prostate cancer risk.' *Br. J. Cancer*, 1999, **81**(4), 672–676.

63. S. M. Edwards, M. D. Badzioch, R. Minter *et al.*, 'Androgen receptor polymorphisms: association with prostate cancer risk, relapse, and overall survival.' *Int. J. Cancer*, 1999, **84**(5), 458–465.

64. W. Hsing, Y. T. Gao, G. Wu *et al.*, 'Polymorphic CAG and CGN repeat lengths in the androgen receptor gene and prostate cancer risk: a population-based case-control study in China.' *Cancer Res.*, 2000, **60**(18), 5111–5116.

65. E. M. Lange, H. Chen, K. Brierley *et al.*, 'The polymorphic exon 1 androgen receptor CAG repeat in men with a potential inherited predisposition to prostate cancer.' *Cancer Epidemiol. Biomarkers Prev.*, 2000, **9**(4), 439–442.

66. E. A. Platz, E. Giovannucci, D. M. Dahl *et al.*, 'The androgen receptor gene GGN microsatellite and prostate cancer risk.' *Cancer Epidemiol. Biomarkers Prev.*, 1998, **7**(5), 379–384.

67. R. M. Lunn, D. A. Bell, J. L. Mohler *et al.*, 'Prostate cancer risk and polymorphism in 17 hydroxylase (CYP17) and steroid reductase (SRD5A2).' *Carcinogenesis*, 1999, **20**(9), 1727–1731.

68. A. Gsur, G. Bernhofer, S. Hinteregger *et al.*, 'A polymorphism in the CYP17 gene is associated with prostate cancer risk.' *Int. J. Cancer*, 2000, **87**(3), 434–437.

69. T. Habuchi, Z. Liqing, T. Suzuki *et al.*, 'Increased risk of prostate cancer and benign prostatic hyperplasia associated with a CYP17 gene polymorphism with a gene dosage efect.' *Cancer Res.*, 2000, **60**(20), 5710–5713.

70. A. K. Roy and B. Chatterjee, 'Androgen action.' *Crit. Rev. Eukaryot. Gene. Expr.*, 1995, **5**, 157–176.

71. R. K. Ross, G. A. Coetzee, C. L. Pearce *et al.*, 'Androgen metabolism and prostate cancer: establishing a model of genetic susceptibility.' *Eur. Urol.*, 1999, **35**(5–6), 355–361.

72. N. Makridakis, R. K. Ross, M. C. Pike *et al.*, 'A prevalent missense substitution that modulates activity of prostatic steroid 5 α-reductase.' *Can. Res.*, 1997, **57**(6), 1020–1022.

73. N. M. Makridakis, R. K. Ross, M. C. Pike *et al.*, 'Association of missense substitution in SRD5A2 gene with prostate cancer in African-American and Hispanic men in Los Angeles, USA.' *Lancet*, 1999, **354**(9183), 975–978.

74. R. K. Nam, A. Toi, D. Vesprini *et al.*, 'V89L polymorphism of type-2, 5-α reductase enzyme gene predicts prostate cancer presence and progression.' *Urology*, 2001, **57**(1), 199–204.

75. P. G. Febbo, P. W. Kantoff, E. A. Platz *et al.*, 'The V89L polymorphism in the 5α- reductase type 2 gene and risk of prostate cancer.' *Cancer Res.*, 1999, **59**(23), 5878–5881.

76. G. G. Schwartz and B. S. Hulka, 'Is Vitamin D deficiency a risk factor for prostate cancer?' *Anticancer Res.*, 1990, **10**(5A), 1307–1311.

77. E. H. Corder, H.A, Guess, B. S. Hulka *et al.*, 'Vitamin D and prostate cancer: a prediagnostic study with stored sera.' *Cancer Epidemiol. Biomarkers Prev.*, 1993, **2**(5), 467–472.

78. N. A. Morrison, J. C. Qi, A. Tokita *et al.*, 'Prediction of bone density from Vitamin D receptor alleles.' *Nature*, 1994, **20**, **367**(6460), 284–287.

79. J. A. Taylor, A. Hirvonen, M. Watson *et al.*, 'Association of prostate cancer with Vitamin D receptor gene polymorphism.' *Cancer Res.*, 1996, **15**(18), 4108–4110.

80. J. Ma, M. J. Stampfer, P. H. Gann *et al.*, 'Vitamin D receptor polymorphisms, circulating Vitamin D metabolites, and risk of prostate cancer in United States physicians.' *Cancer Epidemiol. Biomarkers Prev.*, 1998, **7**(5), 385–390.

81. D. G. Blazer, D. M. Umbach, R. M. Bostick *et al.*, 'Vitamin D receptor polymorphisms and prostate cancer.' *Mol. Carcinog.*, 2000, **27**(1), 18–23.

82. E. P. Wilkens, D Freije, D. R. Nusskern *et al.*, 'No evidence for a role of BRCA1 or BRCA2 mutations in Ashkenazi Jewish families with hereditary prostate cancer.' *Prostate*, 1999, **1**, **39**(4), 28–284.

83. C. S. Sinclair, R. Berry, D. Schaid *et al.*, 'BRCA1 and BRCA2 have a limited role in familial prostate cancer.' *Cancer Res.*, 2000, **60**(5), 1371–1375.

84. D. F. Easton, L. Steele, P. Fields *et al.*, 'Cancer risks in two large breast cancer families linked to BRCA2 on chromosome 13q12–13.' *Am. J. Hum. Genet.*, 1997, **61**(1), 120–128.

85. O. Johannsson, N. Loman, T. Moller *et al.*, 'Incidence of malignant tumours in relatives of BRCA1 and BRCA2 germ line mutation carriers.' *Eur. J. Cancer*, 1999, **35**(8), 1248–1257.

86. H. A. Risch, J. R. McLaughlin, D. E. Cole *et al.*, 'Prevalence and penetrance of germline BRCA1 and BRCA2 mutations in a population series of 649 women with ovarian cancer.' *Am. J. Hum. Genet.*, 2001, **68**(3), 700–710.

87. S. A. Gayther, K. A. de Foy, P. Harrington *et al.*, 'The frequency of germ-line mutations in the breast cancer predisposition genes BRCA1 and BRCA2 in familial prostate cancer.' *Cancer Res.*, 2000, **60**(16), 4513–4518.

88. N. J. Risch, 'Searching for genetic determinants in the new millennium.' *Nature*, 2000, **405**(6788), 847–856.

89. S. Lehrer, F. Fodor, R. G. Stock *et al.*, 'Absence of 185delAG mutation of the BRCA1 gene and 617delT mutation of the BRCA2 gene in Ashkenazi Jewish men with prostate cancer.' *Br. J. Cancer*, 1998, **78**(6), 771–773.

90. A.Hubert, T. Peretz, O. Manor *et al.*, 'The Jewish Ashkenazi founder mutations in the BRCA1/BRCA2 genes are not found at an increased frequency in Ashkenazi patients with prostate cancer.' *Am. J. Hum. Genet.*, 1999, **65**(3), 921–924.

91. K. L. Nastiuk, M. Mansukhani, M. B. Terry *et al.*, 'Common mutations on BRCA1 and BRCA2 do not

contribute to early prostate cancer in Jewish men.' *Prostate*, 1999, **40**(3), 172–177.

92. A. Vazina, J. Baniel, Y. Yaacobi *et al.*, 'The rate of the founder Jewish mutations in BRCA1 and BRCA2 in prostate cancer patients in Israel.' *Br. J. Cancer*, 2000, **83**(4), 463–466.

93. J. J. Park, R. A. Irvine, G. Buchanan *et al.*, 'Breast cancer susceptibility gene 1 (BRCA1) is a coactivator of the androgen receptor.' *Cancer Res.*, 2000, **60**(21), 5946–5949.

94. S. Fan, J. A. Wang, R. Q. Yuan *et al.*, 'BRCA1 as a potential human prostate tumor suppressor: modulation of proliferation, damage responses and expression of cell regulatory proteins.' *Oncogene*, 1998, **16**(23), 3069–3082.

95. M. Watanabe, T. Shiraishi, T. Muneyuki *et al.*, 'Allelic loss and microsatellite instability in prostate cancers in Japan.' *Oncology*, 1998, **55**(6), 569–574.

96. T. Uchida, C. Wang, T. Sato *et al.*, 'BRCA1 gene mutation and loss of heterozygosity on chromosome 17q21 in primary prostate cancer.' *Int. J. Cancer*, 1999, **84**(1), 19–23.

97. W. Xue, R. A. Irvine, M. C. Yu *et al.*, 'Susceptibility to prostate cancer: interaction between genotypes at the androgen receptor and prostate-specific antigen loci.' *Cancer Res.*, 2000, **60**(4), 839–841.

J.P. Vanden Heuvel, G.H. Perdew, W.B. Mattes and W.F. Greenlee (Eds.)
Comprehensive Toxicology, Vol. xiv

# 14.3.4
# Modeling Genetic Susceptibility to Cancer in the Mouse

## CHERYL LYN WALKER

*The University of Texas M.D. Anderson Cancer Center, Science Park – Research Division, Smithville, TX, USA*

## 14.3.4.1   DEVELOPMENT OF GENETICALLY ENGINEERED MICE: HISTORICAL PERSPECTIVE

### 14.3.4.1.1   Transgenic Mice

Modern techniques for generating genetically engineered mice are founded on studies of developmental genetics in the 1950s (Figure 1). While breeding of various inbred strains of mice with specific characteristics and phenotypes had been widely practiced since the early 1900s, it was not until 1956 with the first successful culture of 1-cell mouse embryos to the blastocyst stage[1] that the possibility of genetically manipulating mouse embryos first became a reality. Development of *in vitro*

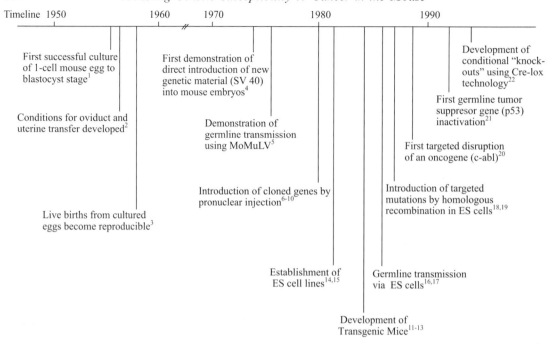

**Figure 1**  Development of techniques for generating genetically engineered mouse models of cancer susceptibility.

culture conditions for mouse embryos occurred concurrently with the development of conditions for routinely transferring mouse eggs into the uterus,[2] making it possible to generate live mice from cultured embryos.[3] These early technical advances were followed in the 1970s and 1980s by genetic manipulations of the germline made possible by the era of molecular biology. The first demonstration that genetic material could be introduced into mouse embryos and incorporated into the genome of embryonic cells were from experiments by Jaenish and Mintz. These pioneers injected purified SV40 DNA into mouse embryos and demonstrated that viral sequences were present in the somatic tissues of the resulting mice.[4] Shortly thereafter germline transmission of exogenous DNA was demonstrated using MoMuLV infected preimplantation embryos.[5] Introduction of the first cloned gene into mice by pronuclear injection was reported by Gordon and Ruddle in 1980 using the herpes simplex TK gene[6] followed by similar reports from Brinster, Jaenish, Mintz and others using TK and beta-globin.[7–10] The era of genetically-defined mouse cancer models dawned soon after with the generation of transgenic mice expressing SV40 T-antigen that developed brain tumors, MTV/myc expressing mice that developed mammary tumors, and mice expressing the c-myc oncogene that developed lymphomas.[11–13]

### 14.3.4.1.2   Embryonic Stem Cells

The next step forward in mouse genetic engineering occurred with the development of pluripotential embryonic stem cell (ES) lines established by the laboratories of Evans and Martin.[14,15] These ES cells were capable of contributing to somatic tissues, as well as the germline, when injected into blastocysts. The establishment of these ES cell lines made it possible to generate transgenic mice in which the exogenous transgene was introduced first into ES cells in culture and then transferred into mouse blastocysts.[16,17] Soon thereafter, targeting of specific genes *in vitro* by homologous recombination in ES cells was developed as a technique for manipulating the mouse genome,[18,19] initially using the HPRT gene as a target. Further advances established the possibility of introducing targeted mutations into oncogenes[20] and this technique was ultimately developed as a means to inactivate tumor suppressor genes with the generation of the first tumor suppressor gene "knockout" mouse deficient for the p53 tumor suppressor gene.[21]

### 14.3.4.1.3   Conditional Knockouts

Recently, conditional knockouts have been developed in which gene inactivation can be selectively targeted to specific tissues. This advance occurred with the development of cre-lox technology,[22] which has made it possible to circumvent many problems, such as embryonic lethality, associated with constitutive germline gene inactivation of tumor suppressor genes. The P1 bacteriophage cre enzyme mediates recombination between two sites containing a specific 34-base pair sequence, termed loci of recombination or "loxP" sites. The result of this recombination is deletion of any intervening sequence lying between these two sites. Genetically engineering loxP sequences into sites flanking a gene, such as a tumor suppressor gene, can be used to generate cre-mediated targeted gene deletion or inactivation *in vivo* or *in vitro*. When transgenic mice expressing cre recombinase are crossed with mice carrying a target gene flanked by loxP sites, the target gene will be deleted in cells of the offspring expressing cre recombinase. This inactivation is conditional on cre expression, and when cre is under the control of a tissue-specific promoter, gene inactivation occurs only in those tissues expressing cre recombinase. Conditional knockouts allow the investigator to study loss of tumor suppressor gene function in specific target tissues, such as the breast, and avoid the problems often encountered with constitutive loss of function, such as embryonic lethality. Germline homozygous deficiency for Brca1, for example, is embryonic lethal[23–26] and Brca1-deficient heterozygotes fail to develop breast cancer. However, conditional inactivation of Brca1 in the mammary gland subsequent to homozygous deletion of a Brca1 tumor suppressor gene flanked by loxP sites ("floxed"), mediated by tissue-specific cre recombinase expression driven by the MMTV or WAP promoter, results in the development of tumors in this organ.[27,28] Similarly, tumor spectrum can also be altered in conditional knockouts. Heterozygous adenomatous polyposis coli (Apc)-deficient mice develop tumors in their small intestine, rather than the colon where tumors develop in humans carrying germline APC defects. Conditional knockout mice driven by an adenovirus cre expression vector delivered colorectally to mice carrying floxed APC alleles develop adenomas in the colon,[29] more closely recapitulating the human disease.

### 14.3.4.1.4 Recoverable Transgenes

The technical developments described above have also made it possible to generate mice (and rats) carrying recoverable transgenes. The bacterial enzyme lacZ has been introduced as a transgene into the germline of mice (Muta-Mouse or Big Blue Mouse). The transgene can be isolated from the tissues of these animals, recovered and expressed in *E. coli*.[30–32] The functionality of this gene can be determined by the ability of bacteria transformed with the recovered transgene to express β-galactosidase. When lacZ mice are exposed to mutagens, the integrity and consequent functionality of the lacZ gene recovered from these mice can be assessed by metabolism of the X-gal substrate by β-galactosidase using an *in vitro* assay. This information can then be extrapolated to quantitate mutation frequency and determine the mutation spectrum at this gene in specific tissues *in vivo*.[33–35] Crossing mice carrying a recoverable transgene with other genetically engineered mice, such as mice deficient in p53 or DNA repair enzymes such as XPA, can modulate mutation frequency and spectrum and can be used to gain valuable insights into the role of specific genes in DNA damage, repair and mutagenesis.[36,37]

### 14.3.4.2 MOUSE MODELS OF GENETIC SUSCEPTIBILITY

#### 14.3.4.2.1 Establishing Models of Genetic Susceptibility to Cancer

Many types of genes have been identified that act as genetic determinants of susceptibility to cancer. Tumor suppressor genes, metabolizing and DNA repair enzymes, and modifier genes are all able to modulate an individual's risk of developing cancer. In some cases, such as genes involved in DNA repair, alterations in gene function may be silent until the organism is challenged in some way, for example by the induction of DNA damage when gene function becomes required. While germline alterations in these types of genes can predispose to cancer, they are not themselves the target for tumorigenesis but rather predispose to alterations in other genes, such as oncogenes or tumor suppressor genes, that initiate the cancer process. This distinction has led to the classification of susceptibility genes as being either "caretakers", which are primarily responsible for the integrity of the genome, "gatekeepers", which are the targets for genetic alterations and initiate tumorigenesis, or "landscapers", that modify the cellular environment in ways that can increase the likelihood of tumor progression.[38] In this context, xenobiotic metabolizing enzyme genes could also be included in the class of "landscapers", as they can modify the dose to tissue of carcinogens, both endogenous and exogenous, and modulate the likelihood of relevant mutations occurring in target genes.

In the case of tumor suppressor genes, germline inactivation of one allele of the tumor suppressor gene is recessive, and tumor development requires the inactivation of the remaining wild-type allele, either spontaneously or induced by carcinogens. However, because the predisposing alteration is carried in the germline, the number of cells at risk is quite high, resulting in a high probability of sustaining a second genetic alteration in the wild-type allele. Consequently, tumors tend to develop in susceptible individuals with a high frequency and multiplicity. For example, a germline defect in the APC gene predisposes to colon cancer with multiple lesions, both adenomas and carcinomas, developing in the colons of susceptible individuals, a phenotype also seen in mice with defective Apc genes.[39] Conversely, additional copies of genes that protect against tumor development, such as $O^6$-methylguanine-DNA-methyltransferase (MGMT), which participates in the repair of $O^6$-methylguanine adducts caused by alkylating agents such as methylnitrosourea (MNU), can confer resistance to tumorigenesis.[40,41] For example, the carcinogen MNU induces mammary tumors that contain GC to AT transitions in the H-ras oncogene,[42,43] and transgenic mice expressing extra copies of MGMT are resistant to NMU-induced mammary carcinogenesis.[44] $O^6$-alkylguanine-DNA-alkyltransferase and MGMT overexpression also protect against liver, lung, colon and skin carcinogenesis and thymic lymphomas induced by alkylating agents,[45–49] while knockout mice in which MGMT has been inactivated are more susceptible to tumor induction.[50] Sensitivity to alkylating agents is consistent with this increased susceptibility, with the $LD_{50}$ of MGMT $-/-$ mice being reduced to 20 mg/kg as compared to MGMT $+/+$ mice which have an $LD_{50}$ of 240 mg/kg.[51]

### 14.3.4.2.2 Animal Cancer Models: Surrogates for Human Disease?

Hereditary forms of human cancer represent the best examples of the genetic basis for this disease. With a few notable exceptions, hereditary human cancers occur as a result of a germline inactivation of a tumor suppressor gene.[52] Numerous mouse models now have been developed in which genes involved in hereditary and sporadic human cancer have been inactivated in the mouse germline (Tables 1 and 2). The extent to which tumor development and disease course recapitulates the human disease varies tremendously between these mouse models.[53] For example, the multiple tumor types that develop in mice lacking the p53 tumor suppressor gene that are either heterozygous ($p53^{+/-}$) or homozygous ($p53^{-/-}$) for the inactive p53 allele closely recapitulates human Li-Fraumeni syndrome in which patients carry a germline p53 mutation.[21] These mice develop primarily lymphomas and soft tissues sarcomas, with tumors arising earlier in $p53^{-/-}$ homozygotes than $p53^{+/-}$ heterozygotes. Carcinomas also occur with a low frequency, principally in heterozygote $p53^{+/-}$ mice, presumably because homozygotes succumb to lymphomas before carcinomas have sufficient time to develop.

Although our current understanding of p53's role as the "guardian of the genome" is consistent with the observed viability of mice homozygous deficient for this gene, the initial report of the viability of $p53^{-/-}$ mice was considered remarkable. Viability of mice homozygous deficient for tumor suppressor genes now appears to be the exception rather than the rule, as most tumor suppressor genes also play key roles in embryonic development. Exceptions to this rule appear to be largely those genes that serve a "caretaker" rather than a "gatekeeper" function. As discussed above, caretaker genes are those that function primarily to protect genomic integrity and prevent mutations in gatekeeper genes that are themselves the targets for tumor development. Caretaker genes such as MSH2 and XPC that function in DNA mismatch and nucleotide excision repair respectively, are required for cells to respond to DNA damage. However, in the absence of such damage these genes do not appear to be necessary for survival and mice with two defective alleles of these genes are viable.

Conversely, many other examples have been reported of tumor suppressor genes that cause human cancers such as breast carcinomas (BRCA), renal cell carcinoma (VHL) and retinoblastoma (RB) when inactivated in the human germline, but which fail to cause these same tumors when inactivated in the mouse. Failure to recapitulate the human disease may be the result of unanticipated redundancies that exist in murine physiology that are not present in humans. For example, humans are susceptible to retinoblastoma when one allele of this gene is inactivated. The failure of mice heterozygous for a defective Rb gene to develop retinoblastoma may reflect the ability of other Rb family members such as p130 and p107 to substitute for Rb in $Rb^{-/+}$ mice in ways that do not occur in humans.[54] This is supported by the finding that heterozygous Rb-deficient mice develop retinoblastoma on a

**Table 1** Tumor suppressor gene knockout mice.

| Gene | Homozygous deficient (−/−) | Heterozygous (−/+) | Human disease | Reference |
|------|---------------------------|--------------------|--------------|-----------|
| p53 | increased and/or accelerated development of multiple tumor types | multiple tumor types | LiFraumeni syndrome, multiple tumor types | 21 |
| APC$^{(min)}$ | embryonic lethal | intestinal tumors (small intestine), mammary carcinomas in females | adenomatous polyposis coli (APC), tumors in large intestine, extra-colonic lesions | 102 |
| APC$^{(716)}$ |  | polyps (small intestine) |  | 115 |
| Apc 1638N | embryonic lethal | tumors in small intestine, extra-colonic lesions |  | 116 |
| NF-1 | embryonic lethal | pheochromo-cytomas (late), myeloid leukemia | juvenile chronic myelogenous leukemia (CML) | 117, 118 |
| NF-2 | embryonic lethal | osteosarcomas, multiple metastatic tumors | benign nervous system tumors | 119 |
| RB family |  |  |  |  |
| RB | embryonic lethal | brain, pituitary tumors (no retinoblastoma) | retinoblastoma (RB) | 120–123 |
| p107 | normal | normal | no involvement in RB | 124 |
| p130 | normal | normal | no involvement in RB | 55 |
| BRCA-1 | germline: embryonic lethal (conditional KO: mammary tumors) | no mammary tumors | breast and ovarian cancer | 11, 25, 26 27, 28 |
| BRCA-2 | embryonic lethal | no mammary tumors | breast cancer | 125, 126 |
| VHL | embryonic lethal | no renal tumors | von Hippel–Lindau disease (hereditary renal cell carcinoma) | 127 |
| p19$^{ARF}$/p16/ INK4a | fibrosarcoma, B-cell lymphoma |  | melanoma | 128 |
| PTEN | embryonic lethal | prostate, skin and colon tumors, T-cell lymphomas | Cowden's disease | 129, 130 |
| WT-1 | embryonic lethal | no nephroblastoma | Wilm's Tumor (nephroblastoma) | 131 |
| E2F-1 | lymphomas, sarcomas, lung tumors |  | NA | 132, 133 |
| p27 | increased susceptibility to carcinogen-induced tumors pituitary tumors | intermediate susceptibility | NA | 134, 135 |

homozygous p107-deficient background.[55,56] Between these two extremes lie mouse tumor models that recapitulate some, but not all aspects of their human disease counterpart. For example, *min* mice defective for the Apc tumor suppressor gene develop tumors of the GI tract similar to APC patients, but polyps rather than carcinomas predominate, and lesions occur predominately in the small intestine, rather than the colon, in these animals.[57] Similarly, mice carrying defective alleles of DNA mismatch repair genes exhibit the same microsatellite instability phenotype seen in their human counterparts, but the frequency of colon tumors, which occur in hereditary nonpolyposis colorectal cancer (HNPCC) patients carrying defective alleles of these genes, varies in these animals.[57,58]

Ideally, the development of mouse models for human cancer should be an iterative process (Figure 2). Characterization of mouse tumor models should include an assessment of the extent to which they model human cancer; tumor histology, latency, progression and metastasis are a few key characteristics that should be evaluated in assessing relevance to the human disease. It is also desirable to determine at the molecular level if additional genetic events that occur in genetically engineered mice during tumor development are similar or different between mice and humans. Similarly, as data becomes available on genetic

**Table 2**　DNA repair deficient mice.

| Gene | Homozygous deficient (−/−) | Heterozygous (+/−) | Human disease | Reference |
|---|---|---|---|---|
| MLH1 (mismatch repair) | lymphomas GI tumors | normal | Hereditary nonpolyposis colon cancer (HNPCC) | 136–138 |
| PMS2/PMS1 (mismatch repair) | lymphomas no GI Tumors | normal | HNPCC | 139 |
| MSH2 (mismatch repair) | lymphomas intestinal tumors secondary to Apc mutations | normal | HNPCC | 140, 141 |
| MSH 6 (mismatch repair) | lymphomas and GI tumors | normal | HNPCC (rare) | 142 |
| XPA (nucleotide excision repair) | liver and UV-induced skin tumors | increased tumor latency | Xeroderma Pigmentosum (XP) group A skin tumors at sun exposed sites | 75, 143, 144 |
| XPC (nucleotide excision repair) | UV-induced skin squamous cell carcinomas, carcinogen-induced liver and lung cancer | normal | XP group C (various types of skin cancer) | 145, 146 |
| CsB (transcription-coupled repair) | susceptibility to UV and DMBA-induced skin cancer | | Cockayne's syndrome (CS) (no skin cancer susceptibility phenotype) | 147 |

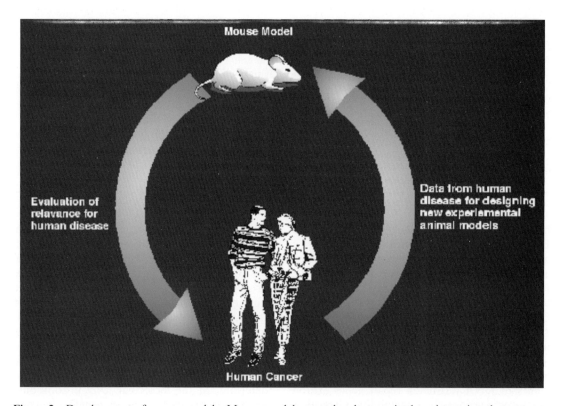

**Figure 2**　Development of mouse models. Mouse models must be characterized to determine the extent to which they recapitulate their cognate human disease. As data emerge regarding the genetic events that occur in specific types of human cancer, this information can be utilized to develop new mouse models that more accurately model the human disease.

alterations that participate in human cancer, this information can be used to design new mouse tumor models. In this way it becomes possible to continuously improve existing mouse tumor models and to develop new mouse models that recapitulate human disease. For example, INK4a/p16/p19ARF null mice fail to develop melanoma, although defects in this gene are clearly linked to the human disease.[59,60] This may be due in part to differences that exist between the melanocyte microenvironment in mice and humans. In humans, melanomas arise in the epidermis, and progression of these lesions into the dermis requires additional genetic alterations, such as N- and H-ras mutations found in nodular human melanomas.[61] Recognizing that in the mouse, melanocytes reside in the dermis rather than epidermis, Chin and colleagues crossed INK4a null mice with transgenic mice expressing H-ras driven by a tyrosinase promoter.[62] Nodular melanomas developed in the offspring of these animals and loss of the wild-type INK4a allele occurred with a high frequency in these tumors, demonstrating the involvement of INK4a in cooperation with H-ras in the genesis of this disease. The Mouse Models of Human Cancer Consortium (MMHCC) is a NIH-supported initiative designed to utilize this approach for developing mouse models for human cancer. The MMHCC is comprised of ~20 groups of investigators involved in the development, characterization and utilization of mouse cancer models. Information on this consortium and resources provided by the MMHCC to the scientific community such as workshops, symposiums and repositories can be obtained at http://mmhcc.nci.nih.gov.

### 14.3.4.3 GENETICALLY ENGINEERED MOUSE MODELS AS ALTERNATIVES FOR THE 2 YEAR BIOASSAY FOR CARCINOGENICITY TESTING

Currently, the National Toxicology Program (NTP) conducts a two-species, both sexes, two year bioassay to assess potential carcinogenicity of chemicals. There is an enormous cost associated with this assay, both in time and resources required to conduct these bioassays and a consequent limitation on the numbers of compounds that can be screened using this approach. Transgenic and knockout mice are now being evaluated as alternative animal models that may prove useful for shortening this evaluation process and/or yielding information regarding mechanisms of action of potential carcinogens.[63,64] Models under eva-luation include TgAC and H-ras-2 transgenic mice and p53$^{(+/-)}$ and XPA$^{(-/-)}$ knockout mice (Figure 3).

#### 14.3.4.3.1 TgAC Mouse

The TgAC mouse carries a mutant H-ras transgene under the control of the fetal zetaglobin promoter.[65] This activated H-ras transgene is expressed in hair follicle cells following dermal carcinogen exposure, leading to the development of skin papillomas in response to chemical carcinogens. Expression of the activated transgene in TgAC mice is dependent on a palindromic repeat that occurred in the zetaglobin promoter during integration of tandem repeats of the transgene,[65–67] resulting in an enhanced suscept-ibility to develop skin papillomas upon carcinogen exposure.[68] Transgene expression correlates with promoter hypomethylation in skin papillomas, consistent with the ability of these animals to detect nongenotoxic carcino-gens.[69] Rearrangement of this promoter sequence can abrogate the responsiveness of these animals to chemical carcinogens, result-ing in a "non-responder" phenotype.[70] Quan-titation of skin papilloma incidence, multiplicity and latency can be used to assess carcinogenic potential using these animals,[71] and the development of readily detectable dermal lesions is an advantage over the conventional bioassay in which most tumors develop in internal organs and are not detect-able until necropsy.

#### 14.3.4.3.2 H-ras-2 Mouse

H-ras-2 mice contain a human H-ras proto-oncogene transgene integrated in multiple copies. Elevated H-ras expression occurs in several tissues as a result of tandem repeats of the transgene and a mutation in the $3'$UTR of the transgene that appears to help stabilize the H-ras transcripts.[72] H-ras-2 transgenics develop spontaneous lung and splenic tumors by 12–18 months of age and additional activating mutations can be detected in the H-ras transgene in carcinogen-induced tumors.[72] Carcinogen exposure decreases tumor latency and increases tumor inci-dence[73,74] and enhanced susceptibility to che-mical carcinogenesis appears to be a result of elevated expression of the carcinogen-activated H-ras transgene that occurs as a consequence of the stabilizing $3'$ mutation.

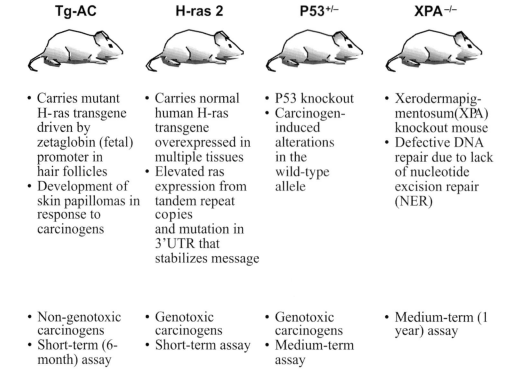

**Figure 3** Genetically engineered mice as alternatives for the NTP 2-year bioassay.

### 14.3.4.3.3 p53 and XPA Knockout Mice

p53 knockout mice have a germline inactivation of one allele of the p53 tumor suppressor gene (p53$^{+/-}$)[21] and XPA knockout mice XPA$^{(-/-)}$ are deficient in the xeroderma pigmentosum (XP) complementation group A DNA repair gene.[75] p53 knockout mice are genetically susceptible to tumor development because they require only a single inactivating mutation in the wild-type p53 gene to inactivate this gene function and induce tumors. Similarly, XPA knockout mice are susceptible to tumor development due to their lack of nucleotide excision repair capability and inability to efficiently repair DNA damage induced by chemical carcinogens, with a resulting increase in mutation frequency and tumor development. The p53 knockout mice are susceptible to tumor development in multiple organs and are particularly sensitive to genotoxic carcinogens.[76] XPA$^{(-/-)}$ mice exhibit an enhanced susceptibility to skin tumors and lymphomas.[75,77]

### 14.3.4.3.4 Model Evaluation

The rapidity with which these different models can be used to assess carcinogenicity and the spectrum of carcinogens detected varies between these animal models. The TgAC, H-ras-2 and p53 mice develop tumors relatively quickly, generally <6 months and may be useful as short-term assays. In contrast, the XPA$^{(-/-)}$ knockout mice develop tumors in ~1 year, suggesting these animals may prove useful as an intermediate term assay. TgAC transgenic mice also appear to be particularly useful for detecting non-genotoxic carcinogens whereas H-ras-2, p53$^{(+/-)}$ and XPA$^{(-/-)}$ mice detect genotoxic carcinogens. These and other inherent differences between the model systems raise questions regarding the ability of these mouse models to fully recapitulate the ability of the 2 year bioassay to predict carcinogenicity.[78] This has led the International Life Sciences Institute's (ILSI) Health and Environmental Sciences Institute (HESI) to organize efforts to evaluate the utility of these genetically engineered mouse models as alternatives for carcinogenesis testing.[79] The Alternatives to Carcinogenicity Testing (ACT) Committee was formed by ILSI/HESI as an international collaborative effort composed of participants from industry, academia and regulatory agencies. Using these 4 animal models, ~20 chemicals for which there is NTP bioassay data, both carcinogens and noncarcinogens, are being evaluated in multiple laboratories under the ACT program. These data, taken together with similar data generated by the

National Institute of Environmental Health Sciences (NIEHS, U.S.), the Central Institute of Experimental Animals (CIEA, Japan) and the National Institute of Public Health and the Environment (RIVM, the Netherlands) are being evaluated to determine the merit of adopting genetically engineered mouse models as alternatives to the conventional rodent bioassay.

### 14.3.4.4 MODELING INTERINDIVIDUAL DIFFERENCES IN GENETIC SUSCEPTIBILITY

#### 14.3.4.4.1 Heterogenous Response of Human Populations

The heterogenous response of human populations exposed to carcinogens and murine strain-specific cancer susceptibility are ultimately determined by the combined effect of mutations or polymorphisms in caretaker and landscaper type susceptibility genes, such as metabolizing and DNA repair enzymes, and tumor suppressor genes (Figure 4). For example, as illustrated in Figure 4 individuals with polymorphisms/mutations in metabolizing genes that result in high Phase I activity and efficient activation of chemical carcinogens or low Phase II detoxication activity of carcinogenic metabolites would be at increased risk for carcinogenesis. Recent estimates suggest that variation in the US population between indi-

viduals due to differences in metabolic activity and DNA repair may be as high as 95–500-fold.[80]

Many polymorphisms in metabolizing genes are associated with increased cancer risk.[81–83] *N*-acetyl transferase 2 (NAT2) for example participates in the metabolism of aromatic amines present in cigarette smoke and heterocyclic amines (HCAs) formed from pyrolysis products from cooked meats.[84] In the colon, metabolism of HCAs to their reactive form by metabolizing enzymes such as NAT2 results in the formation of guanine adducts and subsequent base substitutions, deletions and insertions in DNA leading to the development of colon cancer. Polymorphisms in the NAT2 gene result in rapid, intermediate and slow acetylator phenotypes, with rapid acetylators who consume red meat being at increased risk for the development of colon cancer.[84,85] Conversely, slow acetylators are at increased risk for bladder cancer, as they are less efficient at detoxifying aromatic amines which are bladder carcinogens.[86] Similarly, glutathione-*S*-transferases of the mu and pi class (GSTM1, GSTP1) are responsible for detoxication of polyaromatic hydrocarbons (PAH) and metabolism of lipids and reactive oxygen species.[87] Approximately 20–50% of the population possesses a homozygous deletion, or "GST null variant" genotype, and these GSTP1-deficient individuals are at increased risk for cancer.[88–90] Mutations in DNA repair genes would also modify cancer risk, with defects in

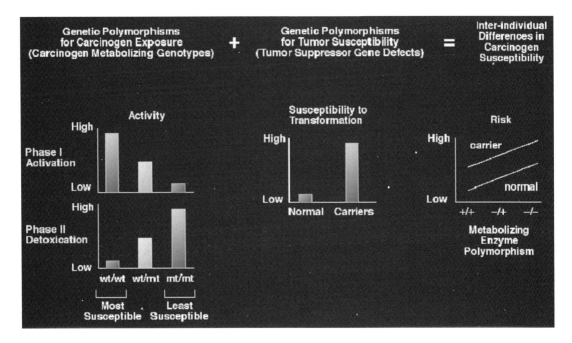

**Figure 4** Genetic basis for heterogeneity in response to carcinogen exposure.

these genes predisposing individuals to mutations as a result of inefficient or erroneous repair of carcinogen-induced DNA damage in other genes involved in tumor development. As discussed above, individuals carrying defects in the mismatch repair gene MSH2 display genomic instability characterized by microsatellite instability and develop hereditary non-polyposis coli cancer (HNPCC)[91–96] and XP-deficient individuals with defective nucleotide excision repair develop several forms of skin cancer, including basal cell carcinoma and melanoma.[97–99]

In addition to genetic defects in caretaker and landscaper susceptibility genes, defects in tumor suppressor genes have a dramatic impact on cancer risk. The number of genes now recognized to act as tumor suppressor genes in humans is quite large. Classically, these genes are defined as those that participate in hereditary as well as spontaneous human cancer syndromes such as retinoblastoma (RB), Li-Fraumeni (p53), and adenomatous polyposis coli (APC). These genes generally follow Knudson's "two-hit" hypothesis in which both alleles of the gene must be inactivated for tumor development to occur.[52] However, there are a few notable exceptions to this rule, such as the Wilms tumor (WT-1) tumor suppressor gene where haploinsufficiency in heterozygous individuals that carry a mutation in one copy of this tumor suppressor gene is sufficient for tumor development in Denys–Drash syndrome[100] or the MET proto-oncogene gene where germline gene activation, rather than loss of function, is the predisposing event for cancer susceptibility and development of hereditary papillary renal cell carcinoma.[101]

Individual cancer risk would be determined by the combination of mutations/polymorphisms in susceptibility genes that influence the probability of mutations occurring at other loci, and germline mutations/polymorphisms in genes that are directly involved in tumorigenesis, such as tumor suppressor genes, or in the case of the H-ras-2 mouse, cellular oncogenes (Figure 4). Individuals with germline tumor suppressor gene mutations/polymorphisms are at increased risk relative to non-carriers (carriers versus normal, respectively in Figure 4) but the magnitude of this increased risk would be modified by their genotype at other loci such as genes for metabolizing enzymes.

### 14.3.4.4.2   Identification of Modifier Genes

In addition to DNA repair genes and metabolizing enzymes, mouse models of genetic susceptibility to cancer are being used to identify other types of modifier genes, which may exist as polymorphic variants within human populations and modulate cancer risk. Modifier genes can effect disease penetrance, tumor latency, multiplicity, and incidence. Because allelic variants of modifier genes may have a high prevalence, i.e., occur at high frequencies in the population, but have low penetrance, i.e., individually contribute only slightly to an increased cancer risk, they may in aggregate have a large effect on cancer susceptibility but be difficult to identify individually in heterogeneous human populations. However, while humans are genetically very heterogeneous, strains of mice that have become inbred (i.e., maintained by brother–sister mating for >20 generations) are genetically homogeneous. Over 200 of these strains of inbred mice are currently available. For many inbred strains of mice, the role of genetic variation in determining carcinogenic outcomes has been recognized for some time, and many examples exist of strain-specific differences in cancer susceptibility. Therefore, an attractive approach to identify modifier genes is to clone these genes from susceptible or resistant strains of mice, and once in hand, use the mouse gene to identify its human homologue.

The success of this approach is based upon the fact that although individual inbred strains are genetically homogeneous, many differences exist between different inbred strains at the DNA level, making it possible to discriminate between alleles of the same gene from different mouse strains. When two strains are crossed, the F1 progeny carry an allele from each parental strain at every gene locus, and thus their entire genome is heterozygous. Using strain-specific polymorphic markers, this heterozygosity can be used to tract alleles from each strain to correlate gain or loss of cancer susceptibility (or resistance) with gain or loss of a specific strain's allele. Using this approach, it becomes possible to localize and even identify modifier genes that modulate cancer susceptibility from different mouse strains. The best example to date of the use of this approach is the identification in the Apc[min] mouse of a modifier gene for colon cancer, Mom-1 (modifier of min). Apc[min] mice carry a mutation in the murine Apc gene involved in the development of colon cancer.[102] Like their human counterparts, mice carrying a defective *Apc* gene (the *min* mutation) develop colon cancer with a high frequency. However, whereas the *min* mutation carried on a C57BL/6 background predisposes to ~30 tumors/mouse, when these mice are crossed with AKR, MA or CAST strains, the F1 progeny have fewer than 10 tumors/mouse.[103] The Mom-1 modifier gene responsible for modulating tumor incidence was mapped to mouse chromosome 4 and later identified as phospholipase Pla2s.[104] This gene is expressed in crypts of the small intestine and hydrolyzes phosphoglycerides to free fatty acids and lysophospholipids and is involved in the production of arachidonic acid.[105]

Similarly, several loci have been identified that can modulate lung cancer risk in mice.[106,107] SWR, A/J, and FBV/N strains of mice are highly susceptible to urethane-induced lung carcinogenesis while DBA, CEH, C57Bl/6(B6) and SM/J are resistant. Urethane exposure in susceptible strains results in 100% tumor incidence and a tumor multiplicity of >10 tumors/mouse while the same exposure in resistant strains produces a <50% tumor incidence and a multiplicity of <1 tumor/animal. When susceptible strains are crossed with resistant strains, susceptibility appears to be a dominant trait and linkage analysis using these crosses has identified several susceptibility and resistance loci.[108–113] Pulmonary adenoma susceptibility (Pas) genes 1, 2 and 3 have been localized to mouse chromosomes 6, 17 and 19 respectively.[108,111] Pas1 appears to be linked to the K-ras protooncogene, which carries a 37 base-pair insertion in the susceptible allele which appears to elevate transcription of this K-ras allele. Resistance genes, Par 1, 2, 3 and 4 have been localized to chromosomes 11, 18, 12 and 6 respectively, with additional susceptibility loci identified on chromosomes 2, 6, 19 and 11 that can modify tumor size and rate of growth.[109,112–114] Although these mouse susceptibility/resistance genes which modulate lung cancer risk have not yet been identified (with the possible exception of K-ras), once these genes are inhand it will be possible to identify their human homologues and determine their contribution to the risk of developing cancer in humans.

## 14.3.4.5 CONCLUSIONS

Cancer has been recognized for over a century as a genetic disease and modeling the genetic events that predispose to cancer in mice has become one of the fundamental focuses of carcinogenesis research. The ability to genetically engineer mice that are predisposed to the development of tumors via expression of exogenous transgenes or inactivation of endogenous tumor suppressor genes has allowed researchers to study mechanisms of tumorigenesis and develop novel strategies for therapeutic interventions and cancer prevention. Mouse models have been developed for many types of human cancer, although the extent to which any given model faithfully recapitulates all aspects of the human disease varies considerably. Many strains of mice are also predisposed naturally to develop specific types of cancer, and strain-specific cancer susceptibility or resistance has also been exploited as an approach to identify mouse genes and their human homologs that can modulate cancer risk.

## 14.3.4.6 REFERENCES

1. W. K. Whitten, 'Culture of tubal mouse ova.' *Nature*, 1956, **177**, 96.
2. A. McLaren and D. Michie, 'Studies on the transfer of fertilized mouse eggs to uterine foster-mothers. I. Factors affecting the implantation and survival of native and transferred eggs.' *J. Exp. Biol.*, 1956, **33**, 394–416.
3. A. McLaren and J. D. Biggers, 'Successsful development and birth of mice clutivated *in vitro* as early embryos.' *Nature*, 1958, **182**, 877–878.
4. R. Jaenisch and B. Mintz, 'Simian virus 40 DNA sequences in DNA of healthy adult mice derived from preimplantation blastocysts injected with viral DNA.' *Proc. Nat. Acad. Sci. USA*, 1974, **71**, 1250–1254.
5. R. Jaenisch, 'Germ line integration and Mendelian transmission of the exogenous Moloney leukemia virus.' *Proc. Nat. Acad. Sci. USA*, 1976, **73**, 1260–1264.
6. J. W. Gordon, G. A. Scangos, D. J. Plotkin *et al.*, 'Genetic transformation of mouse embryos by microinjection of purified DNA.' *Proc. Nat. Acad. Sci. USA*, 1980, **77**, 7380–7384.
7. J. W. Gordon and F. H. Ruddle, 'Integration and stable germ line transmission of genes injected into mouse pronuclei.' *Science*, 1981, **214**, 1244–1246.
8. R. L. Brinster, H. Y. Chen, M. Trumbauer *et al.*, 'Somatic expression of herpes thymidine kinase in mice following injection of a fusion gene into eggs.' *Cell*, 1981, **27**, 223–231.
9. K. Harbers, D. Jahner and R. Jaenisch, 'Microinjection of cloned retroviral genomes into mouse zygotes: integration and expression in the animal.' *Nature*, 1981, **293**, 540–542.
10. E. F. Wagner, T. A. Stewart and B. Mintz, 'The human β-globin gene and a functional viral thymidine kinase gene in developing mice.' *Proc. Nat. Acad. Sci. USA*, 1981, **78**, 5016–5020.
11. R. L. Brinster, H. Y. Chen, A. Messing *et al.*, 'Transgenic mice harboring SV40 T-antigen genes develop characteristic brain tumors.' *Cell*, 1984, **37**, 367–379.
12. T. A. Stewart, P. K. Pattengale and P. Leder, 'Spontaneous mammary adenocarcinomas in transgenic mice that carry and express MTV/myc fusion genes.' *Cell*, 1984, **38**, 627–637.
13. J. M. Adams, A. W. Harris, C. A. Pinkert *et al.*, 'The c-myc oncogene driven by immunoglobulin enhancers induces lymphoid malignancy in transgenic mice.' *Nature*, 1985, **318**, 533–538.

14. M. J. Evans and M. H. Kaufman, 'Establishment in culture of pluripotential cells from mouse embryos.' *Nature*, 1981, **292**, 154–156.

15. G. R. Martin, 'Isolation of a pluripotent cell line from early mouse embryos cultured in medium conditioned by teratocarcinoma stem cells.' *Proc. Nat. Acad. Sci. USA*, 1981, **78**, 7634–7638.

16. E. Robertson, A. Bradley, M. Kuehn et al., 'Germ-line transmission of gene introduced into cultured pluripotential cells by retroviral vector.' *Nature*, 1986, **323**, 445–448.

17. A. Gossler, T. Doetschman, R. Korn et al., 'Transgenesis by means of blastocyst-derived embryonic stem cell lines.' *Proc. Nat. Acad. Sci. USA*, 1986, **83**, 9065–9069.

18. T. Doetschman, R. G. Gregg, N. Maeda et al., 'Targetted correction of a mutant HPRT gene in mouse embryonic stem cells.' *Nature*, 1987, **330**, 576–578.

19. K. R. Thomas and M. R. Capecchi, 'Site-directed mutagenesis by gene targeting in mouse embryo-derived stem cells.' *Cell*, 1987, **51**, 503–512.

20. P. L. Schwartzberg, S. P. Goff and E. J. Robertson, 'Germ-line transmission of a c-abl mutation produced by targeted gene disruption in ES cells.' *Science*, 1989, **246**, 799–803.

21. L. A. Donehower, M. Harvey, B. L. Slagle et al., 'Mice deficient for p53 are developmentally normal but susceptible to spontaneous tumours.' *Nature*, 1992, **356**, 215–221.

22. H. Gu, J. D. Marth, P. C. Orban et al., 'Deletion of a DNA polymerase β gene segment in T cells using cell type-specific targeting.' *Science*, 1994, **265**, 103–106.

23. T. Ludwig, D. L. Chapman, V. E. Papaioannou et al., 'Targeted mutations of breast cancer susceptibility gene homologs in mice: lethal phenotypes of Brca1, Brca2, Brca1/Brca2, Brca1/p53, and Brca2/p53 nullizygous embryos.' *Genes Dev.*, 1997, **11**, 1226–1241.

24. L. C. Gowen, B. L. Johnson, A. M. Latour et al., 'Brca1 deficiency results in early embryonic lethality characterized by neuroepithelial abnormalities.' *Nat. Genet.*, 1996, **12**, 191–194.

25. R. Hakem, J. L. de la Pompa, A. Elia et al., 'Partial rescue of Brca1 (5–6) early embryonic lethality by p53 or p21 null mutation.' *Nat. Genet.*, 1997, **16**, 298–302.

26. C. Y. Liu, A. Flesken-Nikitin, S. Li et al., 'Inactivation of the mouse Brca1 gene leads to failure in the morphogenesis of the egg cylinder in early postimplantation development.' *Genes Dev.*, 1996, **10**, 1835–1843.

27. C. X. Deng and F. Scott, 'Role of the tumor suppressor gene Brca1 in genetic stability and mammary gland tumor formation.' *Oncogene*, 2000, **19**, 1059–1064.

28. X. Xu, K. U. Wagner, D. Larson et al., 'Conditional mutation of Brca1 in mammary epithelial cells results in blunted ductal morphogenesis and tumour formation.' *Nat. Genet.*, 1999, **22**, 37–43.

29. H. Shibata, K. Toyama, H. Shioya et al., 'Rapid colorectal adenoma formation initiated by conditional targeting of the Apc gene.' *Science*, 1997, **278**, 120–123.

30. J. A. Gossen, W. J. de Leeuw, C. H. Tan et al., 'Efficient rescue of integrated shuttle vectors from transgenic mice: a model for studying mutations *in vivo*.' *Proc. Nat. Acad. Sci. USA*, 1989, **86**, 7971–7975.

31. S. W. Kohler, G. S. Provost, A. Fieck et al., 'Spectra of spontaneous and mutagen-induced mutations in the lacI gene in transgenic mice.' *Proc. Nat. Acad. Sci. USA*, 1991, **88**, 7958–7962.

32. M. E. Boerrigter, M. E. Dolle, H. J. Martus et al., 'Plasmid-based transgenic mouse model for studying *in vivo* mutations.' *Nature*, 1995, **377**, 657–659.

33. N. J. Gorelick, 'Overview of mutation assays in transgenic mice for routine testing.' *Environ. Mol. Mutagen.*, 1995, **25**, 218–230.

34. M. E. Dolle, H. J. Martus, M. Novak et al., 'Characterization of color mutants in lacZ plasmid-based transgenic mice, as detected by positive selection.' *Mutagenesis*, 1999, **14**, 287–293.

35. J. Jiao, G. R. Douglas, J. D. Gingerich et al., 'Analysis of tissue-specific lacZ mutations induced by N- nitrosodibenzylamine in transgenic mice.' *Carcinogenesis*, 1997, **18**, 2239–2245.

36. H. Nishino, A. Knoll, V. L. Buettner et al., 'p53 wild-type and p53 nullizygous Big Blue transgenic mice have similar frequencies and patterns of observed mutation in liver, spleen and brain.' *Oncogene*, 1995, **11**, 263–270.

37. H. van Steeg, L. H. Mullenders and J. Vijg, 'Mutagenesis and carcinogenesis in nucleotide excision repair-deficient XPA knock out mice.' *Mutat. Res.*, 2000, **450**, 167–180.

38. K. W. Kinzler and B. Vogelstein, 'Landscaping the cancer terrain.' *Science*, 1998, **280**, 1036–1037.

39. K. H. Goss and J. Groden, 'Biology of the adenomatous polyposis coli tumor suppressor.' *J. Clin. Oncol.*, 2000, **18**, 1967–1979.

40. J. Engelbergs, J. Thomale and M. F. Rajewsky, 'Role of DNA repair in carcinogen-induced ras mutation.' *Mutat. Res.*, 2000, **450**, 139–153.

41. M. Sekiguchi and M. Sanada, 'Alkylation carcinogenesis in mice with altered levels of DNA repair methyltransferase.' *Prog. Exp. Tumor. Res.*, 1999, **35**, 25–36.

42. S. Sukumar, V. Notario, D. Martin-Zanca et al., 'Induction of mammary carcinomas in rats by nitroso-methylurea involves malignant activation of H-ras-1 locus by single point mutations.' *Nature*, 1983, **306**, 658–661.

43. H. Zarbl, S. Sukumar, A. V. Arthur et al., 'Direct mutagenesis of Ha-ras-1 oncogenes by N-nitroso-N-methylurea during initiation of mammary carcinogenesis in rats.' *Nature*, 1985, **315**, 382–385.

44. J. Engelbergs, J. Thomale, A. Galhoff et al., 'Fast repair of O6-ethylguanine, but not O6-methylguanine, in transcribed genes prevents mutation of H-ras in rat mammary tumorigenesis induced by ethylnitrosourea in place of methylnitrosourea.' *Proc. Nat. Acad. Sci. USA*, 1998, **95**, 1635–1640.

45. L. L. Dumenco, E. Allay, K. Norton et al., 'The prevention of thymic lymphomas in transgenic mice by human O6-alkylguanine-DNA alkyltransferase.' *Science*, 1993, **259**, 219–222.

46. N. H. Zaidi, T. P. Pretlow, M. A. O'Riordan et al., 'Transgenic expression of human MGMT protects against azoxymethane-induced aberrant crypt foci and G to A mutations in the K-ras oncogene of mouse colon.' *Carcinogenesis*, 1995, **16**, 451–456.

47. K. Becker, J. Dosch, C. M. Gregel et al., 'Targeted expression of human O(6)-methylguanine-DNA methyltransferase (MGMT) in transgenic mice protects against tumor initiation in two-stage skin carcinogenesis.' *Cancer Res.*, 1996, **56**, 3244–3249.

48. L. Liu, X. Qin and S. L. Gerson, 'Reduced lung tumorigenesis in human methylguanine DNA-methyltransferase transgenic mice achieved by expression of transgene within the target cell.' *Carcinogenesis*, 1999, **20**, 279–284.

49. Y. Nakatsuru, S. Matsukuma, N. Nemoto et al., 'O6-methylguanine-DNA methyltransferase protects against nitrosamine-induced hepatocarcinogenesis.' *Proc. Nat. Acad. Sci. USA*, 1993, **90**, 6468–6472.

50. T. Iwakuma, K. Sakumi, Y. Nakatsuru et al., 'High incidence of nitrosamine-induced tumorigenesis in

mice lacking DNA repair methyltransferase.' *Carcinogenesis*, 1997, **18**, 1631–1635.

51. K. Sakumi, A. Shiraishi, S. Shimizu *et al.*, 'Methylnitrosourea-induced tumorigenesis in MGMT gene knockout mice.' *Cancer Res.*, 1997, **57**, 2415–2418.

52. A. G. Knudson, Jr., 'Mutation and cancer: statistical study of retinoblastoma.' *Proc. Nat. Acad. Sci. USA*, 1971, **68**, 820–823.

53. N. Ghebranious and L. A. Donehower, 'Mouse models in tumor suppression.' *Oncogene*, 1998, **17**, 3385–3400.

54. M. Vooijs and A. Berns, 'Developmental defects and tumor predisposition in Rb mutant mice.' *Oncogene*, 1999, **18**, 5293–5303.

55. M. H. Lee, B. O. Williams, G. Mulligan *et al.*, 'Targeted disruption of p107: functional overlap between p107 and Rb.' *Genes Dev.*, 1996, **10**, 1621–1632.

56. E. Robanus-Maandag, M. Dekker, M. van der Valk *et al.*, 'p107 is a suppressor of retinoblastoma development in pRb-deficient mice.' *Genes Dev.*, 1998, **12**, 1599–1609.

57. J. Heyer, K. Yang, M. Lipkin *et al.*, 'Mouse models for colorectal cancer.' *Oncogene*, 1999, **18**, 5325–5333.

58. T. A. Prolla, A. Abuin and A. Bradley, 'DNA mismatch repair deficient mice in cancer research.' *Semin. Cancer Biol.*, 1996, **7**, 241–247.

59. M. F. Roussel, 'The INK4 family of cell cycle inhibitors in cancer.' *Oncogene*, 1999, **18**, 5311–5317.

60. L. Chin, 'Modeling malignant melanoma in mice: pathogenesis and maintenance.' *Oncogene*, 1999, **18**, 5304–5310.

61. M. Jafari, T. Papp, S. Kirchner *et al.*, 'Analysis of ras mutations in human melanocytic lesions: activation of the ras gene seems to be associated with the nodular type of human malignant melanoma.' *J. Cancer Res. Clin. Oncol.*, 1995, **121**, 23–30.

62. L. Chin, J. Pomerantz, D. Polsky *et al.*, 'Cooperative effects of INK4a and ras in melanoma susceptibility *in vivo*.' *Genes Dev.*, 1997, **11**, 2822–2834.

63. D. Gulezian, D. Jacobson-Ram, C. B. McCullough *et al.*, 'Use of transgenic animals for carcinogenicity testing: considerations and implications for risk assessment.' *Toxicol. Pathol.*, 2000, **28**, 482–499.

64. R. R. Maronpot, 'The use of genetically modified animals in carcinogenicity bioassays.' *Toxicol. Pathol.*, 2000, **28**, 450–453.

65. A. Leder, A. Kuo, R. D. Cardiff *et al.*, 'v-Ha-ras transgene abrogates the initiation step in mouse skin tumorigenesis: effects of phorbol esters and retinoic acid.' *Proc. Nat. Acad. Sci. USA*, 1990, **87**, 9178–9182.

66. K. L. Thompson, B. A. Rosenzweig and F. D. Sistare, 'An evaluation of the hemizygous transgenic Tg.AC mouse for carcinogenicity testing of pharmaceuticals II. A genotypic marker that predicts tumorigenic responsiveness.' *Toxicol. Pathol.*, 1998, **26**, 548–555.

67. R. Honchel, B. A. Rosenzweig, K. L. Thompson *et al.*, 'Loss of palindromic symmetry in Tg.AC mice with a nonresponder phenotype.' *Mol. Carcinog.*, 2001, **30**, 99–110.

68. J. W. Spalding, J. Momma, M. R. Elwell *et al.*, 'Chemically induced skin carcinogenesis in a transgenic mouse line (TG.AC) carrying a v-Ha-ras gene.' *Carcinogenesis*, 1993, **14**, 1335–1341.

69. R. E. Cannon, J. W. Spalding, K. M. Virgil *et al.*, 'Induction of transgene expression in Tg.AC(v-Ha-ras) transgenic mice concomitant with DNA hypomethylation.' *Mol. Carcinog.*, 1998, **21**, 244–250.

70. R. W. Tennant, S. Stasiewicz, J. Mennear *et al.*, 'Genetically altered mouse models for identifying carcinogens.' *IARC Sci. Publ.*, 1999, **146**, 123–150.

71. D. B. Dunson, J. K. Haseman, A. P. van Birgelen *et al.*, 'Statistical analysis of skin tumor data from Tg.AC mouse bioassays.' *Toxicol. Sci.*, 2000, **55**, 293–302.

72. A. Saitoh, M. Kimura, R. Takahashi *et al.*, 'Most tumors in transgenic mice with human c-Ha-ras gene contained somatically activated transgenes.' *Oncogene*, 1990, **5**, 1195–1200.

73. S. Yamamoto, K. Mitsumori, Y. Kodama *et al.*, 'Rapid induction of more malignant tumors by various genotoxic carcinogens in transgenic mice harboring a human prototype c-Ha-ras gene than in control non-transgenic mice.' *Carcinogenesis*, 1996, **17**, 2455–2461.

74. S. Yamamoto, K. Urano, H. Koizumi *et al.*, 'Validation of transgenic mice carrying the human prototype c-Ha-ras gene as a bioassay model for rapid carcinogenicity testing.' *Environ. Health Perspect.*, 1998, **106**(Suppl 1), 57–69.

75. A. de Vries, C. T. van Oostrom, F. M. Hofhuis *et al.*, 'Increased susceptibility to ultraviolet-B and carcinogens of mice lacking the DNA excision repair gene XPA.' *Nature*, 1995, **377**, 169–173.

76. R. W. Tennant, J. Spalding and J. E. French, 'Evaluation of transgenic mouse bioassays for identifying carcinogens and noncarcinogens.' *Mutat. Res.*, 1996, **365**, 119–127.

77. R. J. Berg, A. de Vries, H. van Steeg *et al.*, 'Relative susceptibilities of XPA knockout mice and their heterozygous and wild-type littermates to UVB-induced skin cancer.' *Cancer Res.*, 1997, **57**, 581–584.

78. R. D. Storer, 'Current status and use of short/medium term models for carcinogenicity testing of pharmaceuticals — scientific perspective.' *Toxicol. Lett.*, 2000, **112–113**, 557–566.

79. D. Robinson, 'The International Life Sciences Institute's role in the evaluation of alternative methodologies for the assessment of carcinogenic risk.' *Toxicol. Pathol.*, 1998, **26**, 474–475.

80. F. P. Perera and I. B. Weinstein, 'Molecular epidemiology: recent advances and future directions.' *Carcinogenesis*, 2000, **21**, 517–524.

81. M. Taningher, D. Malacarne, A. Izzotti *et al.*, 'Drug metabolism polymorphisms as modulators of cancer susceptibility.' *Mutat. Res.*, 1999, **436**, 227–261.

82. A. Hirvonen, 'Polymorphisms of xenobiotic-metabolizing enzymes and susceptibility to cancer.' *Environ. Health Perspect.*, 1999, **107**(Suppl 1), 37–47.

83. J. G. Hengstler, M. Arand, M. E. Herrero *et al.*, 'Polymorphisms of *N*-acetyltransferases, glutathione *S*-transferases, microsomal epoxide hydrolase and sulfotransferases: influence on cancer susceptibility.' *Recent Results Cancer Res.*, 1998, **154**, 47–85.

84. D. W. Hein, '*N*-Acetyltransferase genetics and their role in predisposition to aromatic and heterocyclic amine-induced carcinogenesis.' *Toxicol. Lett.*, 2000, **112–113**, 349–356.

85. K. F. Ilett, B. M. David, P. Detchon *et al.*, 'Acetylation phenotype in colorectal carcinoma.' *Cancer Res.*, 1987, **47**, 1466–1469.

86. R. A. Cartwright, R. W. Glashan, H. J. Rogers *et al.*, 'Role of *N*-acetyltransferase phenotypes in bladder carcinogenesis: a pharmacogenetic epidemiological approach to bladder cancer.' *Lancet*, 1982, **2**, 842–845.

87. R. C. Strange, P. W. Jones and A. A. Fryer, 'Glutathione *S*-transferase: genetics and role in toxicology.' *Toxicol. Lett.*, 2000, **112–113**, 357–363.

88. J. Seidegard, R. W. Pero, M. M. Markowitz *et al.*, 'Isoenzyme(s) of glutathione transferase (class μ) as a marker for the susceptibility to lung cancer: a follow up study.' *Carcinogenesis*, 1990, **11**, 33–36.

89. D. A. Bell, J. A. Taylor, D. F. Paulson *et al.*, 'Genetic

risk and carcinogen exposure: a common inherited defect of the carcinogen-metabolism gene glutathione *S*-transferase M1 (GSTM1) that increases susceptibility to bladder cancer.' *J. Nat. Cancer Inst.*, 1993, **85**, 1159–1164.

90. J. E. McWilliams, B. J. Sanderson, E. L. Harris *et al.*, 'Glutathione *S*-transferase M1 (GSTM1) deficiency and lung cancer risk.' *Cancer Epidemiol. Biomarkers Prev.*, 1995, **4**, 589–594.

91. N. Papadopoulos, N. C. Nicolaides, Y. F. Wei *et al.*, 'Mutation of a mutL homolog in hereditary colon cancer.' *Science*, 1994, **263**, 1625–1629.

92. R. Parsons, G. M. Li, M. J. Longley *et al.*, 'Hypermutability and mismatch repair deficiency in RER+ tumor cells.' *Cell*, 1993, **75**, 1227–1236.

93. C. E. Bronner, S. M. Baker, P. T. Morrison *et al.*, 'Mutation in the DNA mismatch repair gene homologue hMLH1 is associated with hereditary nonpolyposis colon cancer.' *Nature*, 1994, **368**, 258–261.

94. R. Fishel, M. K. Lescoe, M. R. Rao *et al.*, 'The human mutator gene homolog MSH2 and its association with hereditary nonpolyposis colon cancer.' *Cell*, 1993, **75**, 1027–1038.

95. F. S. Leach, N. C. Nicolaides, N. Papadopoulos *et al.*, 'Mutations of a mutS homolog in hereditary nonpolyposis colorectal cancer.' *Cell*, 1993, **75**, 1215–1225.

96. R. Kolodner, 'Biochemistry and genetics of eukaryotic mismatch repair.' *Genes Dev.*, 1996, **10**, 1433–1442.

97. J. E. Cleaver, 'Defective repair replication of DNA in xeroderma pigmentosum.' *Nature*, 1968, **218**, 652–656.

98. J. E. Cleaver and K. H. Kraemer, in 'The Metabolic and Molecular Basis of Inherited Disease,' eds. Scriver C. R., Beaudet A. L., Sly W. S. and Valle D., McGraw-Hill, New York, 1995, pp. 4393–4419.

99. J. H. H. Hoeijmakers, in 'DNA Repair Mechnisms: Impact on Human Diseases and Cancer,' ed. Vos J.-H. H., Springer, New York, 1995, pp. 125–151.

100. S. B. Lee and D. A. Haber, 'Wilms tumor and the WT1 gene.' *Exp. Cell Res.*, 2001, **264**, 74–99.

101. L. Schmidt, K. Junker, N. Nakaigawa *et al.*, 'Novel mutations of the MET proto-oncogene in papillary renal carcinomas.' *Oncogene*, 1999, **18**, 2343–2350.

102. A. R. Moser, H. C. Pitot and W. F. Dove, 'A dominant mutation that predisposes to multiple intestinal neoplasia in the mouse.' *Science*, 1990, **247**, 322–324.

103. A. Bilger, A. R. Shoemaker, K. A. Gould *et al.*, 'Manipulation of the mouse germline in the study of Min-induced neoplasia.' *Semin. Cancer Biol.*, 1996, **7**, 249–260.

104. W. F. Dietrich, E. S. Lander, J. S. Smith *et al.*, 'Genetic identification of Mom-1, a major modifier locus affecting Min-induced intestinal neoplasia in the mouse.' *Cell*, 1993, **75**, 631–639.

105. M. MacPhee, K. P. Chepenik, R. A. Liddell *et al.*, 'The secretory phospholipase A2 gene is a candidate for the Mom1 locus, a major modifier of ApcMin-induced intestinal neoplasia.' *Cell*, 1995, **81**, 957–966.

106. A. M. Malkinson, 'The genetic basis of susceptibility to lung tumors in mice.' *Toxicology*, 1989, **54**, 241–271.

107. T. A. Dragani, G. Manenti and M. A. Pierotti, 'Genetics of murine lung tumors.' *Adv. Cancer Res.*, 1995, **67**, 83–112.

108. M. Gariboldi, G. Manenti, F. Canzian *et al.*, 'A major susceptibility locus to murine lung carcinogenesis maps on chromosome 6.' *Nat. Genet.*, 1993, **3**, 132–136.

109. G. Manenti, M. Gariboldi, R. Elango *et al.*, 'Genetic

mapping of a pulmonary adenoma resistance (Par1) in mouse.' *Nat. Genet.*, 1996, **12**, 455–457.

110. G. Manenti, A. Stafford, L. De Gregorio *et al.*, 'Linkage disequilibrium and physical mapping of Pas1 in mice.' *Genome Res.*, 1999, **9**, 639–646.

111. M. F. Festing, A. Yang and A. M. Malkinson, 'At least four genes and sex are associated with susceptibility to urethane-induced pulmonary adenomas in mice.' *Genet. Res.*, 1994, **64**, 99–106.

112. M. Obata, H. Nishimori, K. Ogawa *et al.*, 'Identification of the Par2 (Pulmonary adenoma resistance) locus on mouse chromosome 18, a major genetic determinant for lung carcinogen resistance in BALB/cByJ mice.' *Oncogene*, 1996, **13**, 1599–1604.

113. R. J. Fijneman, S. S. de Vries, R. C. Jansen *et al.*, 'Complex interactions of new quantitative trait loci, Sluc1, Sluc2, Sluc3, and Sluc4, that influence the susceptibility to lung cancer in the mouse.' *Nat. Genet.*, 1996, **14**, 465–467.

114. A. Pataer, M. Nishimura, T. Kamoto *et al.*, 'Genetic resistance to urethan-induced pulmonary adenomas in SMXA recombinant inbred mouse strains.' *Cancer Res.*, 1997, **57**, 2904–2908.

115. M. Oshima, H. Oshima, K. Kitagawa *et al.*, 'Loss of Apc heterozygosity and abnormal tissue building in nascent intestinal polyps in mice carrying a truncated Apc gene.' *Proc. Nat. Acad. Sci. USA*, 1995, **92**, 4482–4486.

116. R. Fodde, W. Edelmann, K. Yang *et al.*, 'A targeted chain-termination mutation in the mouse Apc gene results in multiple intestinal tumors.' *Proc. Nat. Acad. Sci. USA*, 1994, **91**, 8969–8973.

117. C. I. Brannan, A. S. Perkins, K. S. Vogel *et al.*, 'Targeted disruption of the neurofibromatosis type-1 gene leads to developmental abnormalities in heart and various neural crest-derived tissues.' *Genes Dev.*, 1994, **8**, 1019–1029.

118. T. Jacks, T. S. Shih, E. M. Schmitt *et al.*, 'Tumour predisposition in mice heterozygous for a targeted mutation in Nf1.' *Nat. Genet.*, 1994, **7**, 353–361.

119. A. I. McClatchey, I. Saotome, V. Ramesh *et al.*, 'The Nf2 tumor suppressor gene product is essential for extraembryonic development immediately prior to gastrulation.' *Genes Dev.*, 1997, **11**, 1253–1265.

120. T. Jacks, A. Fazeli, E. M. Schmitt *et al.*, 'Effects of an Rb mutation in the mouse.' *Nature*, 1992, **359**, 295–300.

121. A. R. Clarke, C. A. Purdie, D. J. Harrison *et al.*, 'Thymocyte apoptosis induced by p53-dependent and independent pathways.' *Nature*, 1993, **362**, 849–852.

122. E. Y. Lee, C. Y. Chang, N. Hu *et al.*, 'Mice deficient for Rb are nonviable and show defects in neurogenesis and haematopoiesis.' *Nature*, 1992, **359**, 288–294.

123. N. Hu, A. Gutsmann, D. C. Herbert *et al.*, 'Heterozygous Rb-1 delta 20/+mice are predisposed to tumors of the pituitary gland with a nearly complete penetrance.' *Oncogene*, 1994, **9**, 1021–1027.

124. D. Cobrinik, M. H. Lee, G. Hannon *et al.*, 'Shared role of the pRB-related p130 and p107 proteins in limb development.' *Genes Dev.*, 1996, **10**, 1633–1644.

125. S. K. Sharan, M. Morimatsu, U. Albrecht *et al.*, 'Embryonic lethality and radiation hypersensitivity mediated by Rad51 in mice lacking Brca2.' *Nature*, 1997, **386**, 804–810.

126. A. Suzuki, J. L. de la Pompa, R. Hakem *et al.*, 'Brca2 is required for embryonic cellular proliferation in the mouse.' *Genes Dev.*, 1997, **11**, 1242–1252.

127. J. R. Gnarra, J. M. Ward, F. D. Porter *et al.*, 'Defective placental vasculogenesis causes embryonic lethality in VHL-deficient mice.' *Proc. Nat. Acad. Sci. USA*, 1997, **94**, 9102–9107.

128. M. Serrano, H. Lee, L. Chin *et al.*, 'Role of the INK4a locus in tumor suppression and cell mortality.' *Cell*, 1996, **85**, 27–37.

129. A. Di Cristofano, B. Pesce, C. Cordon-Cardo *et al.*, 'Pten is essential for embryonic development and tumour suppression.' *Nat. Genet.*, 1998, **19**, 348–355.

130. V. Stambolic, A. Suzuki, J. L. de la Pompa *et al.*, 'Negative regulation of PKB/Akt-dependent cell survival by the tumor suppressor PTEN.' *Cell*, 1998, **95**, 29–39.

131. J. A. Kreidberg, H. Sariola, J. M. Loring *et al.*, 'WT-1 is required for early kidney development.' *Cell*, 1993, **74**, 679–691.

132. L. Yamasaki, T. Jacks, R. Bronson *et al.*, 'Tumor induction and tissue atrophy in mice lacking E2F-1.' *Cell*, 1996, **85**, 537–48.

133. S. J. Field, F. Y. Tsai, F. Kuo *et al.*, 'E2F-1 functions in mice to promote apoptosis and suppress proliferation.' *Cell*, 1996, **85**, 549–561.

134. M. L. Fero, M. Rivkin, M. Tasch *et al.*, 'A syndrome of multiorgan hyperplasia with features of gigantism, tumorigenesis, and female sterility in p27(Kip1)-deficient mice.' *Cell*, 1996, **85**, 733–744.

135. K. Nakayama, N. Ishida, M. Shirane *et al.*, 'Mice lacking p27(Kip1) display increased body size, multiple organ hyperplasia, retinal dysplasia, and pituitary tumors.' *Cell*, 1996, **85**, 707–720.

136. W. Edelmann, P. E. Cohen, M. Kane *et al.*, 'Meiotic pachytene arrest in MLH1-deficient mice.' *Cell*, 1996, **85**, 1125–1134.

137. W. Edelmann, K. Yang, M. Kuraguchi *et al.*, 'Tumorigenesis in Mlh1 and Mlh1/Apc1638N mutant mice.' *Cancer Res.*, 1999, **59**, 1301–1307.

138. S. M. Baker, A. W. Plug, T. A. Prolla *et al.*, 'Involvement of mouse Mlh1 in DNA mismatch repair and meiotic crossing over.' *Nat. Genet.*, 1996, **13**, 336–342.

139. S. M. Baker, C. E. Bronner, L. Zhang *et al.*, 'Male mice defective in the DNA mismatch repair gene PMS2 exhibit abnormal chromosome synapsis in meiosis.' *Cell*, 1995, **82**, 309–319.

140. N. de Wind, M. Dekker, A. Berns *et al.*, 'Inactivation of the mouse Msh2 gene results in mismatch repair deficiency, methylation tolerance, hyperrecombination, and predisposition to cancer.' *Cell*, 1995, **82**, 321–330.

141. A. H. Reitmair, R. Schmits, A. Ewel *et al.*, 'MSH2 deficient mice are viable and susceptible to lymphoid tumours.' *Nat. Genet.*, 1995, **11**, 64–70.

142. W. Edelmann, K. Yang, A. Umar *et al.*, 'Mutation in the mismatch repair gene Msh6 causes cancer susceptibility.' *Cell*, 1997, **91**, 467–477.

143. A. de Vries, C. T. van Oostrom, P. M. Dortant *et al.*, 'Spontaneous liver tumors and benzo[a]pyrene-induced lymphomas in XPA-deficient mice.' *Mol. Carcinog.*, 1997, **19**, 46–53.

144. H. Nakane, S. Takeuchi, S. Yuba *et al.*, 'High incidence of ultraviolet-B-or chemical-carcinogen-induced skin tumours in mice lacking the xeroderma pigmentosum group A gene.' *Nature*, 1995, **377**, 165–168.

145. A. T. Sands, A. Abuin, A. Sanchez *et al.*, 'High susceptibility to ultraviolet-induced carcinogenesis in mice lacking XPC.' *Nature*, 1995, **377**, 162–165.

146. D. L. Cheo, D. K. Burns, L. B. Meira *et al.*, 'Mutational inactivation of the xeroderma pigmentosum group C gene confers predisposition to 2-acetylaminofluorene-induced liver and lung cancer and to spontaneous testicular cancer in Trp53−/− mice.' *Cancer Res.*, 1999, **59**, 771–775.

147. G. T. van der Horst, H. van Steeg, R. J. Berg *et al.*, 'Defective transcription-coupled repair in Cockayne syndrome B mice is associated with skin cancer predisposition.' *Cell*, 1997, **89**, 425–435.

J.P. Vanden Heuvel, G.H. Perdew, W.B. Mattes and W.F. Greenlee (Eds.)
Comprehensive Toxicology, Vol. xiv

# 14.3.5
# DNA Methylation and Gene Silencing

## JESSICA E. SUTHERLAND and MAX COSTA
*New York University School of Medicine, NY, USA*

## 14.3.5.1   INTRODUCTION

In addition to its involvement in genomic imprinting, methylation of the 5-position of cytosine in the context of the 5'-CG-3' sequence (hereafter referred to as DNA methylation) is involved in embryogenesis, X chromosome inactivation, cellular differentiation, protection of the genome from invading DNA, and the establishment of tissue specific patterns of gene expression. Moreover, as will be discussed, DNA methylation of tumor suppressor genes has been linked to tumor formation.

The CpG dinucleotide is present at only 5–10% of its predicted frequency in the genome of higher eukaryotes and approximately 70–80% of these CpG sites are methylated in humans.[1] In contrast, 5' regulatory regions of genes frequently contain CpG islands. These regions range from 0.5–5 kb in length, are generally unmethylated, and are GC rich, containing numerous CpG sites.[2] CpG islands are prevalent in the promoters of housekeeping genes and to a lesser extent, are also present in 40% of the genes with tissue-specific expression patterns.[1] In normal cells, promoter region CpG islands are generally unmethylated regardless of the transcriptional status of the gene.[2] Exceptions include nontranscribed genes on the inactive X chromosome and imprinted alleles that have methylated promoters. In many tumors, hypermethylation of promoters in tumor suppressor genes has been reported and this event is also associated with transcriptional represssion.[3] This chapter will describe the initiation and/or maintenance of DNA methylation patterns, the association between DNA methylation and transcriptional repression, the association between DNA methylation and carcinogenesis, the techniques used to study DNA methylation, and the ability of certain xenobiotics to alter DNA methylation patterns.

### 14.3.5.2  ESTABLISHMENT OF DNA METHYLATION PATTERNS

DNA methylation is believed to be regulated through three steps including: (1) formation of a new methylation pattern (*de novo* methylation); (2) maintenance of the existing methylated pattern during cell division (maintenance methylation); and (3) erasure of the methylation pattern (demethylation).[4] During embryonic development, methylation patterns are established. Sperm cell DNA is highly methylated whereas oocyte DNA is not. During blastulation, methylation patterns in both paternal and maternal alleles are erased followed by extensive *de novo* methylation of both alleles prior to implantation of the zygote.[5] It is believed that housekeeping genes and imprinted genes escape these processes. Protection from methylation in housekeeping genes may arise from binding of transcription factors to the CpG islands in these genes.[6,7]

*De novo* and maintenance DNA methylation are catalyzed by DNA methyltransferase (DNMT) enzymes.[8] The first of these enzymes to be identified was DNMT1 which has high affinity for hemimethylated DNA substrates and functions during replication to maintain pre-existing methylation patterns.[9,10] DNMT1 has also been reported to perform *de novo* DNA methylation *in vitro*.[2] Three other DNA methyltransferases have been cloned from mammalian cells. DNMT3A and DNMT3B appear to catalyze *de novo* DNA methylation.[11,12] Another enzyme, DNMT2 shares homology with the other DNA methyltransferases but has not been shown to have DNA methyltransferase activity.[13,14]

The importance of DNA methylation during embryonic development was underscored by the finding that targeted mutation of the DNA methyltransferase gene (*Dnmt1*) resulted in embryonic lethality in mice.[15] More recently, it was shown that the DNA methyltransferases DNMT3A and DNMT3B were essential for proper developmental processes in mouse embryonic stem cells.[16]

DNA demethylation may involve passive and active processes and these events are not as well understood as those involved in DNA methylation. Passive demethylation involves the suppression of maintenance DNA methylation (perhaps via interference by chromatin structure), resulting in a 50% decrease in methylation during each round of DNA replication.[17,18] Active demethylation occurs via enzymatic processes and several mechanisms have been postulated. Methylated cytosines can be removed by DNA glycosylases followed by mismatch repair.[19–23] Weiss *et al.* reported excision and repair of the entire CpG site.[24] All of these processes were reported to involve RNA but this remains ambiguous.[25,26] Most recently, Bhattacharya *et al.* identified a mammalian DNA demethylase that cleaves the methyl group from cytosine and releases it as methanol.[27,28] Characterization of the enzymes involved in DNA demethylation remains an active area of investigation.[18]

### 14.3.5.3  DNA METHYLATION AND TRANSCRIPTIONAL REPRESSION

Hypermethylation of CpG sites in promoter regions of genes is associated with transcriptional repression (i.e., gene silencing; See Chapter 14.01.4). There are a few potential mechanisms by which DNA methylation can repress gene expression. First of all, methylation at CpG sites may interfere with binding of transcription factors to their recognition sites in promoters. Binding of several transcription factors (e.g., AP-2, CREB, E2F, c-Myc/Myn, and NF-κ B) has been shown to be inhibited by DNA methylation.[29] However, binding of other transcription factors (i.e., Sp1) is not sensitive to the methylation state of the recognition sequences.[30]

A second mechanism involves the preferential binding of transcriptional repressors to methylated CpG residues. Several proteins (MeCP1, MeCP2, MBD1-4) are capable of specifically binding methylated DNA[31] and have been associated with transcriptional repression.[32–34] A key finding was that MeCP2 recruits histone deacetylases (HDAC1 and HDAC2) via the corepressor protein mSin3A.[35,36] The important implication from these studies is that DNA methylation and chromatin structure are linked. In chromatin, acetylation of lysine residues on the N-terminal tails of histones H4 and presumably H3 is important for limiting interactions between these tails and acidic residues on histones H2A/B in neighboring core particles thus allowing for an open chromatin conformation.[37] Histone deacetylases remove acetyl groups from these lysine residues thereby increasing internucleosomal interactions. This results in a more condensed chromatin conformation that is less accessible to transcriptional machinery.[3,38] MeCP2 is also capable of repressing transcription independently of histone deacetylation. Nan *et al.* demonstrated that MeCP2 repressed transcription of naked DNA in a cell-free system.[34] In transcriptional repressor complexes, MBD2 and MBD3 also associate with histone deacetylases.[39–41]

Further evidence for the link between DNA methylation and histone deacetylation was provided by Eden *et al.* who transfected mouse fibroblasts with *in vitro* methylated and unmethylated thymidine kinase (*tk*) gene constructs.[42] The fraction of chromatin immunoprecipitated with antibodies directed against acetylated histone H4 revealed that, unlike the unmethylated gene, the methylated gene was not enriched for acetylated histones. Treatment of cells with the histone deacetylase inhibitor trichostatin A (TSA) resulted in increased expression of the methylated *tk* gene. TSA treatment also increased the accessibility of the methylated *tk* gene to nucleases suggesting that the inhibition of histone deacetylation restored transcriptionally-competent chromatin even in the presence of DNA methylation.

Recently, a more direct association between DNA methylation and histone deacetylation was demonstrated. The DNA methyltransferase DNMT1 is itself associated with a histone deacetylase (HDAC1) *in vivo*.[43] This discovery allows the possibility that the DNMT1 via its association with HDAC1 directly influences chromatin structure.

Whereas it has been generally accepted that methylation of promoter region CpG sites is associated with transcriptional repression, methylation of downstream CpG sites does not appear to inhibit and in some cases, may up-regulate gene expression.[44] The mechanisms by which methylated downstream CpG sites are elongated during transcription are not well-understood.[45]

## 14.3.5.4 DETECTION OF DNA METHYLATION

Several methods have been developed for detecting cytosine methylation in DNA. These include both gene-specific and non gene-specific approaches. These techniques will be briefly described. Comprehensive details are published elsewhere.[17,46,47]

### 14.3.5.4.1 Gene-Specific Methylation Analysis

In many instances, it is desirable to measure the degree of DNA methylation in a gene of interest (i.e., gene-specific methylation analysis). The use of restriction endonuclease isoschizomers that have different sensitivities to cytosine methylation is a widely-used technique for quantitating DNA methylation in selected regions of specific genes. Briefly, one restriction enzyme (*Hpa* II for example) cuts its recognition site only if the internal cytosine is unmethylated. In contrast, the isoschizomer (e.g., *MspI*) will cleave this restriction site whether or not the internal cytosine is methylated. In the assay, equal amounts of DNA are digested with one or the other of the isoschizomers. The results of each restriction digest are then compared by Southern blotting using the gene of interest as a probe. If cytosine methylation is present at the restriction sites, the bands will be of different sizes resulting in a different digestion pattern for each enzyme. The degree of partial methylation (i.e., some cytosines at each restriction site methylated but others not methylated) can be estimated by comparing the intensities of the bands from each digest to each other. Some drawbacks of this technique are that only cytosines that are part of enzyme recognition sites can be analyzed and hemimethylated DNA cannot usually be detected. In addition, incomplete digestion may potentially lead to erroneous interpretation.

A related technique uses methylation sensitive/insensitive restriction enzymes followed by PCR amplification of the target site. The site will only amplify if the restriction site(s) in the target region are not cleaved. As with the Southern blot approach, incomplete digestion and partial methylation are potential complications.[48]

Other approaches utilize differential base modification followed by gene sequencing. The first approach utilizes the fact that DNA will react with hydrazine at C and T but not at methylated C residues. The location of the C and T residues are then mapped by Maxam–Gilbert sequencing and the methylated cytosines identified by their absence from the sequencing ladder.[49] A second approach is to utilize the fact that methylated cytosines and thymines are sensitive to oxidation by permanganate. Subsequent treatment with piperidine will cleave DNA at these oxidized bases.[46,50] With either of these approaches, further resolution by ligation-mediated PCR of the cleavage products before sequencing has been recommended.[17,46,49] Bisulfite modification of unmethylated cytosines constitutes a third base modification procedure. This method capitalizes on the ability of sodium bisulfite to deaminate all the cytosines in single-stranded DNA to uracil while leaving the methylated cytosines intact.[51] The bisulfite-modified DNA is then amplified by PCR using primers designed to amplify the converted sequence. The uracils are amplified as thymine and only the methylated cytosines remain as cytosine and can be identified by sequencing. Harrison *et al.* have warned against artifacts in this procedure resulting from incomplete bisulfite conversion of cytosines adjacent to methylated CpG sites.[52] Detailed discussions of the potential artifacts arising from this procedure have been published.[17,46]

#### 14.3.5.4.2  Non-Specific Analysis of DNA Methylation

On a genomic basis, the methylated cytosines in DNA can be quantitated by reverse-phase high-performance liquid chromatography (HPLC) following enzymatic digestion of DNA via DNase I and nuclease P1 or snake venom phosphodiesterase and subsequent treatment with alkaline phosphatase.[53–55] This procedure, when coupled with mass spectrometry, results in more specific identification of methylated cytosines.[56]

Thin-layer chromatography (TLC) has also been used to measure genome-wide cytosine methylation. After enzymatic digestion of CCGG sites with the methylation-insensitive restriction enzyme *MspI*, the 5′ phosphate of the internal C is radiolabeled. The DNA is then digested to mononucleotides using nuclease P1, which are subsequently separated on TLC plates.[17]

Genomic DNA methylation levels can also be estimated with methylation acceptance assays. These assays typically employ the bacterial enzyme *SssI* DNA methyltransferase to catalyze *de novo* methylation of cytosines using radiolabeled S-adenosyl-methionine (SAM) as a donor of tritiated methyl groups.[47,57,58] The amount of methyl group incorporation, measured with a scintillation counter, is inversely proportional to the original amount of cytosine methylation in the DNA.

Recently, Oakeley *et al.* described a fluorescence assay for DNA methylation levels.[17,59] Bisulfite-modifed pyrimidine bases are treated with chloracetaldehyde creating ethenocytosine derivatives of methylated cytosines that are fluorescent and can be detected using a fluorimeter.

Monoclonal antibodies directed against methylated cytosine have been used to detect cytosine methylation in genomic DNA that has been denatured and immobilized on a membrane.[60] The technique has also been used to detect DNA methylation in tobacco plant chromosomes by fluorescence microscopy.[17,60]

### 14.3.5.5  DNA METHYLATION AND CANCER

Most human tumors have aberrant DNA methylation patterns compared to normal tissue. Frequently, there is a global loss of DNA methylation in tumors.[61–63] In contrast, hypermethylation of CpG islands and increased DNMT1 levels are often observed in neoplastic cells (Table 1).[64–66] There are at least five pathways by which alterations in cytosine methylation are believed to contribute to carcinogenesis.

First of all, methylated cytosines represent a pre-mutagenic lesion.[67] C to T transition mutations are favored at methylated CpG dinucleotides because of spontaneous deamination of 5-methylcytosine to thymine. A survey of mutations in the human *p53* gene revealed that 28% of the point mutations were due to C to T transitions at CpG nucleotides.[68] Recently, a potential repair enzyme for this lesion (i.e., MBD4) was identified[69] but its role in carcinogenesis remains unclear.[67] Methylated CpG sites also appear to be susceptible to attack by some environmental carcinogens. For example, in the human *p53* gene, guanines flanked by 5′methyl cytosines were preferentially adduced by benzo[a]pyrene diol epoxide (BPDE).[70–72] On the other hand, DNA hypomethylation has also been linked to increased mutagenesis.[73] Murine embryonic stem cells nullizygous for the *Dnmt1* gene and exhibiting a great degree of DNA hypomethylation, had elevated mutation rates (primarily gene deletions) presumably because of a loss of genomic stability (see below).

Altered expression patterns of imprinted genes may also contribute to carcinogenesis.[74,75] Normally, the maternal *IGF2* allele (encoding for insulin-like growth factor II) and the paternal *H19* allele are imprinted. In a murine model, *H19* acts as a putative tumor suppressor gene by regulating the *IGF2* gene in *cis*. Essentially, the maternal *H19* promoter is competing out the *IGF2* gene from a common enhancer so that it cannot be expressed.[76,77] Because the paternal *H19* allele is silenced by DNA methylation, paternal *IGF2* has access to the enhancer and is transcribed. In several

**Table 1** Cancer-related genes silenced by DNA hypermethylation.[48]

---

*Gene*

---

Tumor suppressors:
  p15 INK4B[140]
  p16 INK4A[141,142]
  p73[143]
  ARF/INK4A[144]
  Wilms tumor[145]
  Von Hippel Lindau (VHL)[146]
  Mammary-derived growth inhibitor[147]
  Hypermethylated in cancer (HIC1)[148]
  Retinoblastoma (Rb)[149]

Receptors:
  Estrogen[150]
  Androgen[151]
  Retinoic Acid Beta[152,153]

Invasion/Metastasis suppressors:
  E-cadherin[154]
  Tissue inhibitor metalloproteinase-3 (TIMP-3)[155]
  mts-1[156]
  CD-44[157]

DNA Repair:
  Methylguanine methyltransferase[158]
  hMLH1[159,160]
  BRCA-1[161,162]

Carcinogen Detoxification:
  Glutathione *S*-transferase[163]

Angiogenesis Inhibitors:
  Thrombospondin-1[164]
  TIMP-3[155]

Tumor Antigen:
  MAGE-1[165,166]

---

cancers, biallelic expression of the *IGF2* gene via loss of imprinting has been reported.[77,78] In Wilm's tumor, biallelic hypermethylation of the *H19* gene is frequently observed resulting in a complete loss of *H19* gene expression and reciprocal biallelic expression of *IGF2*.[75,79]

Reduced cytosine methylation is also believed to contribute to the chromosomal instability frequently observed in tumors.[73,80–83] Stability of satellite DNA on chromosomes 1,9, and 16 has been found to be dependent on methylation catalyzed by the DNMT3B enzyme.[67] It has been proposed that DNA methylation functions in suppressing mitotic recombination and contributes to faithful chromosomal segregation during mitosis.[73,84]

In some cancers, specific oncogenes may be hypomethylated. DNA methylation and gene expression were inversely correlated in the antiapoptotic *bcl*-2 gene in B-cell chronic lymphocytic leukemia[85] and in the *k-ras* proto-oncogene in lung and colon carcinomas.[86]

Finally, hypermethylation of CpG islands in the promoter regions of tumor suppressor genes has been frequently linked to carcinogenesis and has received a great deal of research attention.[87–91] Such changes generally lead to transcriptional repression by the mechanisms described in Section 14.03.5.3. The methylation status of many cancer-related genes has been examined using many of the gene-specific detection techniques discussed in Section 14.03.5.4 (see Table 1). In addition, in many instances, treatment of tumor cells *in vitro* with the DNMT1 inhibitors 5-Azacytidine or 5-Azadeoxycytidine reactivated the expression of these genes. In fact, these drugs along with other inhibitors of DNA methylation have been evaluated clinically for treatment of leukemias, myelodysplastic syndromes, and hemoglobinopathies.[92,93]

Another clinical application involves screening for hypermethylated loci in certain genes which may enable early detection of certain cancers.[91] For example, hypermethylation of the *p16* gene promoter has been detected in sputum of patients at high risk for lung cancer and in serum of lung cancer patients and those with hepatocellular carcinoma.[94–96] Importantly, early detection using these biomarkers may result in higher cure rates.

The mechanisms responsible for aberrant DNA methylation patterns observed in neoplastic cells are not well-understood. A singular mechanism will probably not explain the paradoxical phenotype of global hypomethylation and regional hypermethylation present in these cells.

Many reports have focused on the role of increased DNA methyltransferase enzyme activity in tumor cells.[64,65] Recently, Butler *et al.* reported genomic DNA hypomethylation in the *5-MeTase* gene (presumably DNMT1) in tumors and tumor-derived cell lines and hypermethylation of the same gene in normal cells.[97] It has also been reported that overexpression of the *c-fos* gene resulted in morphological transformation of rat fibroblasts and increased levels of DNMT1 expression.[98] On the other hand, DNMT1 expression levels in a variety of tumors have been found to be highly variable.[67] In addition, the current view that DNMT1 functions solely as a maintenance DNA methyltransferase *in vivo* is difficult to reconcile with a presumed role in *de novo* methylation of CpG islands during tumorigenesis. In fact, Eads *et al.* reported that CpG island hypermethylation in four promoters including *p16* was not correlated with expression levels of DNMT1, DNMT3A or DNMT3B.[99] It has been hypothesized that the deregulation of DNA methyltransferase expression may be a by-product and not a cause of tumorigenesis.[67] Further research is needed to further clarify these issues.

Recent experimental evidence suggests that hypermethylation of CpG islands in neoplastic cells may be a result of aberrant cell cycle control. It was shown that DNA methyltransferase 1 (DNMT1) binds to proliferating cell nuclear antigen (PCNA) in newly replicated DNA without effects on its methylating activity. The interaction between DNMT1 and PCNA was interrupted by the cyclin-dependent kinase inhibitor p21. It was proposed that in normal cells, the p21 protein negatively regulates DNMT1 during S-phase, preventing CpG island methylation. The loss of p21 in cancer cells would allow more DNA methylation to occur in early S-phase.[100]

## 14.3.5.6    EFFECTS OF AGING, NUTRITION, AND ENVIRONMENTAL TOXICANTS ON DNA METHYLATION

Increased age is the most important risk factor for most adult cancers.[101,102] Likewise, nutritional factors are believed to contribute to the etiology of many cancers.[103] Exposure to environmental agents is also an important contributor to carcinogenesis. Not surprisingly, all three of these risk factors have been shown to alter DNA methylation patterns.

### 14.3.5.6.1    Aging

It has been known for a long time that overall levels of 5-methylcytosine in cultured normal cells decrease as a function of passage number.[104] However, in several animal models and human tissues, progressive age-related methylation in the coding region of many genes has been reported.[105–108] Coding region DNA methylation may not impact gene expression however as discussed earlier, methylated CpG sites in gene coding regions may be susceptible to increased mutagenesis.

Age-related promoter CpG island methylation was first demonstrated in the estrogen receptor (ER) gene in human colon carcinomas and adenomas.[109] Reduced expression of ER in these cells is associated with hyper-proliferation. Since then, age-related increases in promoter CpG sites in other genes in colon and other tissues have been identified.[101,102,110] It is therefore tempting to speculate that the hypermethylation of promoter CpG islands in tumors arise from normal aging processes. In fact, 19 of 26 CpG islands known to be methylated in primary colon cancer were also methylated in normal colon samples as a function of age.[111] Whether or not age-related promoter methylation alters the physiology of cells or predisposes them to neoplasia has not been established.

### 14.3.5.6.2    Nutrition

Insufficient folate status is positively associated with breast, colorectal, pancreatic, cervical, lung, and brain cancers.[112] Biochemically, folate (as $5'$ methyltetrahydrofolate) mediates the transfer of methyl groups to homocysteine, forming methionine, a precursor of $S$-adenosylmethionine, (SAM), the methyl group donor in DNA methylation reactions. Folate also participates in synthesis of purines and thymidine necessary for nucleotide synthesis. Dietary studies have demonstrated that reduced folate intake by women resulted in hypomethylation of lymphocyte DNA which was reversed by subsequent folate supplementation.[113] Links between folate depletion, DNA hypomethylation and proto-oncogene activation are purely speculative.

Long-term dietary folate depletion in rats initially caused DNA hypomethylation in the coding region of the hepatic *p53* gene. Interestingly, this event was followed by a rebound hypermethylation at the time when liver neoplastic foci were observed.[114] Hypermethylation of tumor suppressor promoter region CpG islands has not been demonstrated following folate depletion. In addition to alteration of DNA methylation, folate depletion also results in disruption of DNA integrity and altered DNA repair that may also constitute pro-carcinogenic changes.[115]

Dietary selenium (Se) intake is protective against some types of cancers[116,117] and epidemiological studies have indicated an increased colon cancer risk in humans in geographical regions where there are low soil Se levels.[118] In the body, Se is converted to monomethylated, dimethylated, and trimethylated forms by enzymes that utilize SAM as the methyl donor. Rats fed Se-deficient diets had reduced global DNA methylation in liver and colon and it has been proposed that this hypomethylation participates in the tumorigenic events associated with low Se status.[119]

### 14.3.5.6.3    Environmental Exposure

Altered DNA methylation patterns may also explain, in part, the carcinogenic properties of some environmental agents. Nickel (Ni) is a potent human and rodent carcinogen.[120] Occupational exposure to Ni is associated with an increased risk of nasal and lung cancers.[121] However, in most assays, Ni is not very mutagenic.[122,123] Instead, it may exert some of its carcinogenic potential via epigenetic mechanisms involving alteration of DNA methylation patterns.

Nickel is clastogenic and generates heterochromatin-specific chromosomal aberrations. In Chinese hamster embryo cells, *in vitro* Ni treatment resulted in decondensation of the heterochromatin in the long (q) arm of the X chromosome.[124] Moreover, Ni exposure also resulted in large deletions in this chromosomal region. Notably, a putative X-linked senescence gene was identified in this region after experiments revealed that restoration of an intact active X chromosome into nickel-immortalized cells containing the deletion restored the senescent phenotype.[125] The senescent phenotype was enhanced after addition of the DNA demethylating agent 5-azacytidine which suggested that silencing of a senescent gene(s) by Ni involved DNA hypermethylation. Others have since confirmed this finding in a different cell system.[126]

The susceptibility of genes to nickel-induced DNA methylation is influenced by their proximity to heterochromatic regions (i.e., they exhibit position effect variegation). The G10 and G12 cell lines were derived from *hprt* V-79 Chinese hamster cell lines and contain the bacterial xanthine guanine phosphoribosyl transferase (*gpt*) transgene which is functionally analogous to the *hprt* gene.[127] Inactivation of *gpt* confers cellular resistance to 6-thioguanine whereas expression of *gpt* enables the cells to grow in a mixture of hypoxanthine, aminopterin, and thymidine. In G12 cells, Ni treatment caused very high levels of *gpt* gene inactivation compared with other known mutagens[122,128,129] whereas Ni did not drastically affect *gpt* gene expression in G10 cells.[122,128] In G12 cells, the *gpt* transgene is inserted near a subtelomeric band of heterochromatin on chromosome 1[130] whereas in G10 cells, it is inserted in a euchromatic region of chromosome 6.[131]

Although Ni exposure resulted in various mutations in the *gpt* gene in G12 cells, the rate of *gpt* gene inactivation in Ni-exposed G12 cells was much higher than would be expected if mutagenic effects were the sole cause of the gene's inactivity.[132] This observation, coupled with the evidence of the heterochromatic position effect variegation, and the knowledge that transgenes are frequently silenced by DNA methylation[133] suggested the involvement of DNA methylation in Ni-induced *gpt*

gene silencing. Digestion with methylation sensitive restriction enzymes revealed that the silenced *gpt* gene was heavily methylated, could be reactivated with 5-azacytidine, and was located in condensed heterochromatin.[131]

Nickel's epigenetic effects upon gene expression may contribute to this metal's carcinogenicity. A model has been proposed whereby nickel induced heterochromatin condensation and DNA hypermethylation may cause silencing of tumor suppressor and/or senescence genes as illustrated in Figure 1.[131]

Given the relationship between DNA methylation and histone deacetylation (as described in Section 14.03.5.3), it was of interest to determine if histone deacetylation is also involved in Ni-induced gene inactivation. DNA methylation has not been detected in the yeast *Saccharomyces cerevisiae*. However, histone acetylation status is an important regulator of gene expression in these cells. In a manner analogous to the position effect observed in the G10/G12 mammalian system, Ni silenced the *URA3* marker gene when it was located in close proximity to the telomeric-silencing element but not when it was further away.[134] Subsequent analyses revealed that Ni decreased the acetylation of lysines in the N-terminal tail of histone H4, an event that would lead to a "closed" chromatin conformation.[135] In this instance, nickel's effects on gene expression and chromatin structure are independent of DNA methylation. It will be interesting to further define the role of histone acetylation status in Ni-exposed mammalian cells that are also capable of undergoing alterations in DNA methylation patterns.

Other xenobiotics may alter DNA methylation patterns in cells. The synthetic estrogen diethylstilbestrol (DES) was found to hypermethylate the *gpt* gene in G12 but not G10 cells albeit at a lower frequency than found for Ni.[136] Many cancer chemotherapeutic drugs have been reported to alter DNA methylation patterns.[137] Given the important role of methylation in the metabolism of arsenic, it was postulated that arsenite exposure might influence DNA methylation levels.[138] However the experimental evidence for this phenomenon is equivocal. Arsenic-induced transformation of a rat liver cell line was accompanied by global DNA hypomethylation and depressed levels of SAM.[138] Dietary arsenite did not significantly alter liver or colon DNA methylation status in rats.[119] Arsenite and arsenate increased cytosine methylation in the p53 promoter region in human lung A549 cells.[139] It is unclear at this time as to whether or not arsenic's carcinogenicity is manifested, at least in part, through alterations in DNA methylation.

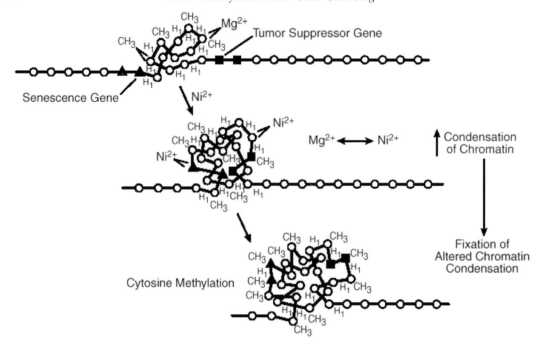

**Figure 1** Model of gene silencing induced by Ni. Nickel-induced increases in heterochromatin condensation and DNA methylation may cause inherited inactivation of critical tumor suppressor and/or senescence genes. Tumor suppressor genes (black squares) and/or senescence genes (black triangles) may become incorporated into heterochromatin that is seeded by nickel-induced DNA condensation (middle panel) and subsequently stabilized by DNA hypermethylation (bottom panel). Reproduced with permission.[131]

The advent of the field of genomics has enabled the comprehensive screening of the effects of a variety of toxicants on gene expression (see Section 5, present volume). This research will probably identify other compounds capable of influencing gene expression via effects upon DNA methylation status. Such endeavors will contribute greatly to our understanding of the effects of altered DNA methylation patterns in environmentally-induced cancers and developmental insults.

## 14.3.5.7 REFERENCES

1. F. Antequera and A. Bird, 'Number of CpG islands and genes in human and mouse.' *Proc. Nat. Acad. Sci. USA*, 1993, **90**, 11995–11999.
2. R. Singal and G. D. Ginder, 'DNA methylation.' *Blood*, 1999, **93**, 4059–4070.
3. S. B. Baylin and J. G. Herman, 'DNA hypermethylation in tumorigenesis: epigenetics joins genetics.' *Trends Genet.*, 2000, **16**, 168–174.
4. S. Tajima and I. Suetake, 'Regulation and function of DNA methylation in vertebrates.' *J. Biochem. (Tokyo)*, 1998, **123**, 993–999.
5. A. Razin and T. Kafri, 'DNA methylation from embryo to adult.' *Prog. Nucleic Acid Res. Mol. Biol.*, 1994, **48**, 53–81.
6. M. Dush, M. Briggs, M. Royce *et al.*, 'Identification of DNA sequences required for mouse APRT gene expression.' *Nucl. Acids Res.*, 1988, **16**, 8509–8525.
7. D. Macleod, R. R. Ali and A. Bird, 'An alternative

promoter in the mouse major histocompatibility complex class II I-Abeta gene: implications for the origin of CpG islands.' *Mol. Cell. Biol.*, 1998, **18**, 4433–4443.
8. T. H. Bestor, 'The DNA methyltransferases of mammals.' *Hum. Mol. Genet.*, 2000, **9**, 2395–2402.
9. M. S. Turker and T. H. Bestor, 'Formation of methylation patterns in the mammalian genome.' *Mutat. Res.*, 1997, **386**, 119–130.
10. T. Bestor, A. Laudano, R. Mattaliano *et al.*, 'Cloning and sequencing of a cDNA encoding DNA methyltransferase of mouse cells. The carboxyl-terminal domain of the mammalian enzymes is related to bacterial restriction methyltransferases.' *J. Mol. Biol.*, 1988, **203**, 971–983.
11. S. Xie, Z. Wang, M. Okano *et al.*, 'Cloning, expression and chromosome locations of the human DNMT3 gene family.' *Gene*, 1999, **236**, 87–95.
12. C. L. Hsieh, '*In vivo* activity of murine de novo methyltransferases, Dnmt3a and Dnmt3b.' *Mol. Cell. Biol.*, 1999, **19**, 8211–8218.
13. M. Okano, S. Xie and E. Li, 'Dnmt2 is not required for de novo and maintenance methylation of viral DNA in embryonic stem cells.' *Nucleic Acids Res.*, 1998, **26**, 2536–2540.
14. C. L. Hsieh, 'Dynamics of DNA methylation pattern.' *Curr. Opin. Genet. Dev.*, 2000, **10**, 224–228.
15. E. Li, T. H. Bestor and R. Jaenisch, 'Targeted mutation of the DNA methyltransferase gene results in embryonic lethality.' *Cell*, 1992, **69**, 915–926.
16. M. Okano, D. W. Bell, D. A. Haber *et al.*, 'DNA methyltransferases Dnmt3a and Dnmt3b are essential for *de novo* methylation and mammalian development.' *Cell*, 1999, **99**, 247–257.
17. E. J. Oakeley, 'DNA methylation analysis: a review of current methodologies.' *Pharmacol. Ther.*, 1999, **84**, 389–400.

18. A. P. Wolffe, P. L. Jones and P. A. Wade, 'DNA demethylation.' *Proc. Nat. Acad. Sci. USA*, 1999, **96**, 5894–5896.

19. J. Jost, 'Nuclear extracts of chicken embryos promote an active demthylation of DNA by excision repair of 5-methyldeoxycytidine.' *Proc. Nat. Acad. Sci. USA*, 1993, **90**, 4684–4688.

20. J. Jost and Y. Jost, 'Transient DNA demethylation in differentiating mouse myoblasts correlates with higher activity of 5-methyldeoxycytidine excision repair.' *J. Biol. Chem.*, 1994, **269**, 10040–10043.

21. J. P. Jost and Y. C. Jost, 'Mechanism of active DNA demethylation during embryonic development and cellular differentiation in vertebrates.' *Gene*, 1995, **157**, 265–266.

22. J. P. Jost, M. Fremont, M. Siegmann *et al.*, 'The RNA moiety of chick embryo 5-methylcytosine-DNA glycosylase targets DNA demethylation.' *Nucleic Acids Res.*, 1997, **25**, 4545–4550.

23. M. Fremont, M. Siegmann, S. Gaulis *et al.*, 'Demethylation of DNA by purified chick embryo 5-methylcytosine-DNA glycosylase requires both protein and RNA.' *Nucleic Acids Res.*, 1997, **25**, 2375–2380.

24. A. Weiss, I. Keshet, A. Razin *et al.*, 'DNA demethylation *in vitro*: involvement of.' *Cell*, 1996, **86**, 709–718.

25. J. F. Swisher, E. Rand, H. Cedar *et al.*, 'Analysis of putative RNase sensitivity and protease insensitivity of demethylation activity in extracts from rat myoblasts.' *Nucleic Acids Res.*, 1998, **26**, 5573–5580.

26. J. P. Jost, M. Siegmann, S. Thiry *et al.*, 'A re-investigation of the ribonuclease sensitivity of a DNA demethylation reaction in chicken embryo and G8 mouse myoblasts.' *FEBS Lett.*, 1999, **449**, 251–254.

27. S. K. Bhattacharya, S. Ramchandani, N. Cervoni *et al.*, 'A mammalian protein with specific demethylase activity for mCpG DNA.' *Nature*, 1999, **397**, 579–583.

28. S. Ramchandani, S. K. Bhattacharya, N. Cervoni *et al.*, 'DNA methylation is a reversible biological signal.' *Proc. Nat. Acad. Sci. USA*, 1999, **96**, 6107–6112.

29. P. H. Tate and A. P. Bird, 'Effects of DNA methylation on DNA-binding proteins and gene expression.' *Curr. Opin. Genet. Dev.*, 1993, **3**, 226–231.

30. M. Harrington, P. Jones, M. Imagawa *et al.*, 'Cytosine methylation does not affect binding of transcription factor Sp1.' *Proc. Nat. Acad. Sci. USA*, 1988, **85**, 2066–2070.

31. B. Hendrich and A. Bird, 'Mammalian methyltransferases and methyl-CpG-binding domains: proteins involved in DNA methylation.' *Curr. Top. Microbiol. Immunol.*, 2000, **249**, 55–74.

32. J. Boyes and A. Bird, 'Repression of genes by DNA methylation depends on CpG density and promoter strength: evidence for involvement of a methyl-CpG binding protein.' *EMBO J.*, 1992, **11**, 327–333.

33. J. Breivik and G. Gaudernack, 'Genomic instability, DNA methylation and natural selection in colorectal carcinogenesis.' *Semin. Cancer Biol.*, 1999, **9**, 245–254.

34. X. Nan, F. J. Campoy and A. Bird, 'MeCP2 is a transcriptional repressor with abundant binding sites in genomic chromatin.' *Cell*, 1997, **88**, 471–481.

35. X. Nan, H. H. Ng, C. A. Johnson *et al.*, 'Transcriptional repression by the methyl-CpG-binding protein MeCP2 involves a histone deacetylase complex.' *Nature*, 1998, **393**, 386–389.

36. P. L. Jones, G. J. Veenstra, P. A. Wade *et al.*, 'Methylated DNA and MeCP2 recruit histone deacetylase to repress transcription.' *Nat. Genet.*, 1998, **19**, 187–191.

37. K. Luger, A. W. Mader, R. K. Richmond *et al.*, 'Crystal structure of the nucleosome core particle at 2.8 A resolution.' *Nature*, 1997, **389**, 251–260.

38. W. L. Cheung, S. D. Briggs and C. D. Allis, 'Acetylation and chromosomal functions.' *Curr. Opin. Cell Biol.*, 2000, **12**, 326–333.

39. H. H. Ng, Y. Zhang, B. Hendrich *et al.*, 'MBD2 is a transcriptional repressor belonging to the MeCP1 histone deacetylase complex.' *Nat. Genet.*, 1999, **23**, 58–61.

40. P. A. Wade, A. Gegonne, P. L. Jones *et al.*, 'Mi-2 complex couples DNA methylation to chromatin remodelling and histone deacetylation.' *Nat. Genet.*, 1999, **23**, 62–66.

41. Y. Zhang, H. H. Ng, H. Erdjument-Bromage *et al.*, 'Analysis of the NuRD subunits reveals a histone deacetylase core complex and a connection with DNA methylation.' *Genes Dev.*, 1999, **13**, 1924–1935.

42. S. Eden, T. Hashimshony, I. Keshet *et al.*, 'DNA methylation models histone acetylation.' *Nature*, 1998, **394**, 842.

43. F. Fuks, W. Burgers, A. Brehm *et al.*, 'DNA methyltransferase Dnmt1 associates with histone deacetylase activity.' *Nat. Genet.*, 2000, **24**, 88–91.

44. M. F. Chan, G. Liang and P. A. Jones, 'Relationship between transcription and DNA methylation.' *Curr. Top. Microbiol. Immunol.*, 2000, **249**, 75–86.

45. P. A. Jones, 'The DNA methylation paradox.' *Trends Genet.*, 1999, **15**, 34–37.

46. T. Rein, M. L. DePamphilis and H. Zorbas, 'Identifying 5-methylcytosine and related modifications in DNA genomes.' *Nucleic Acids Res.*, 1998, **26**, 2255–2264.

47. L. Broday and M. Costa, in 'Current Protocols in Toxicology,' eds. M. Maines, L. Costa, D. Reed, S. Sassa and I. Sipes, John Wiley, New York, NY, 1999, Unit 3.6.

48. R. L. Momparler and V. Bovenzi, 'DNA methylation and cancer.' *J. Cell Physiol.*, 2000, **183**, 145–154.

49. G. P. Pfeifer, S. D. Steigerwald, P. R. Mueller *et al.*, 'Genomic sequencing and methylation analysis by ligation mediated PCR.' *Science*, 1989, **246**, 810–813.

50. E. Fritzsche, H. Hayatsu, G. L. Igloi *et al.*, 'The use of permanganate as a sequencing reagent for identification of 5-methylcytosine residues in DNA.' *Nucleic Acids Res.*, 1987, **15**, 5517–5528.

51. M. Frommer, L. E. McDonald, D. S. Millar *et al.*, 'A genomic sequencing protocol that yields a positive display of 5-methylcytosine residues in individual DNA strands.' *Proc. Nat. Acad. Sci. USA*, 1992, **89**, 1827–1831.

52. J. Harrison, C. Stirzaker and S. J. Clark, 'Cytosines adjacent to methylated CpG sites can be partially resistant to conversion in genomic bisulfite sequencing leading to methylation artifacts.' *Anal. Biochem.*, 1998, **264**, 129–132.

53. K. C. Kuo, R. A. McCune, C. W. Gehrke *et al.*, 'Quantitative reversed-phase high performance liquid chromatographic determination of major and modified deoxyribonucleosides in DNA.' *Nucleic Acids Res.*, 1980, **8**, 4763–4776.

54. J. K. Christman, 'Separation of major and minor deoxyribonucleoside monophosphates by reverse-phase high-performance liquid chromatography: a simple method applicable to quantitation of methylated nucleotides in DNA.' *Anal. Biochem.*, 1982, **119**, 38–48.

55. J. D. Gomes and C. J. Chang, 'Reverse-phase high-performance liquid chromatography of chemically modified DNA.' *Anal. Biochem.*, 1983, **129**, 387–391.

56. A. Razin and H. Cedar, 'Distribution of 5-methylcytosine in chromatin.' *Proc. Nat. Acad. Sci. USA*, 1977, **74**, 2725–2728.

57. J. Wu, J. P. Issa, J. Herman *et al.*, 'Expression of an exogenous eukaryotic DNA methyltransferase gene induces transformation of NIH 3T3 cells.' *Proc. Nat. Acad. Sci. USA*, 1993, **90**, 8891–8895.

58. Y. W. Lee, L. Broday and M. Costa, 'Effects of nickel on DNA methyltransferase activity and genomic DNA methylation levels.' *Mutat. Res.*, 1998, **415**, 213–218.

59. E. J. Oakeley, F. Schmitt and J. P. Jost, 'Quantification of 5-methylcytosine in DNA by the chloroacetaldehyde reaction.' *Biotechniques*, 1999, **27**, 744–746, 748–750, 752.

60. E. J. Oakeley, A. Podesta and J. P. Jost, 'Developmental changes in DNA methylation of the two tobacco pollen nuclei during maturation.' *Proc. Nat. Acad. Sci. USA*, 1997, **94**, 11721–11725.

61. G. A. Romanov and B. F. Vanyushin, 'Methylation of reiterated sequences in mammalian DNAs. Effects of the tissue type, age, malignancy and hormonal induction.' *Biochem. Biophys. Acta*, 1981, **653**, 204–218.

62. M. A. Gama-Sosa, V. A. Slagel, R. W. Trewyn *et al.*, 'The 5-methylcytosine content of DNA from human tumors.' *Nucleic Acids Res.*, 1983, **11**, 6883–6894.

63. A. P. Feinberg, C. W. Gehrke, K. C. Kuo *et al.*, 'Reduced genomic 5-methylcytosine content in human colonic neoplasia.' *Cancer Res.*, 1988, **48**, 1159–1161.

64. T. L. Kautiainen and P. A. Jones, 'DNA methyltransferase levels in tumorigenic and nontumorigenic cells in culture.' *J. Biol. Chem.*, 1986, **261**, 1594–1598.

65. J. P. Issa, P. M. Vertino, J. Wu *et al.*, 'Increased cytosine DNA-methyltransferase activity during colon cancer progression.' *J. Nat. Cancer Inst.*, 1993, **85**, 1235–1240.

66. S. A. Belinsky, K. J. Nikula, S. B. Baylin *et al.*, 'Increased cytosine DNA-methyltransferase activity is target-cell-specific and an early event in lung cancer.' *Proc. Nat. Acad. Sci. USA*, 1996, **93**, 4045–4050.

67. P. M. Warnecke and T. H. Bestor, 'Cytosine methylation and human cancer.' *Curr. Opin. Oncol.*, 2000, **12**, 68–73.

68. P. Hainaut, T. Hernandez, A. Robinson *et al.*, 'IARC Database of p53 gene mutations in human tumors and cell lines: updated compilation, revized formats and new visualisation tools.' *Nucleic Acids Res.*, 1998, **26**, 205–213.

69. B. Hendrich, U. Hardeland, H. H. Ng *et al.*, 'The thymine glycosylase MBD4 can bind to the product of deamination at methylated CpG sites.' *Nature*, 1999, **401**, 301–304.

70. M. F. Denissenko, J. X. Chen, M. S. Tang *et al.*, 'Cytosine methylation determines hot spots of DNA damage in the human P53 gene.' *Proc. Nat. Acad. Sci. USA*, 1997, **94**, 3893–3898.

71. G. P. Pfeifer, M. Tang and M. F. Denissenko, 'Mutation hotspots and DNA methylation.' *Curr. Top. Microbiol. Immunol.*, 2000, **249**, 1–19.

72. G. P. Pfeifer, 'p53 mutational spectra and the role of methylated CpG sequences.' *Mutat. Res.*, 2000, **450**, 155–166.

73. R. Z. Chen, U. Pettersson, C. Beard *et al.*, 'DNA hypomethylation leads to elevated mutation rates.' *Nature*, 1998, **395**, 89–93.

74. B. Tycko, 'Genomic imprinting and cancer.' *Results Probl. Cell Differ.*, 1999, **25**, 133–169.

75. B. Tycko, 'Epigenetic gene silencing in cancer.' *J. Clin. Invest.*, 2000, **105**, 401–407.

76. P. A. Leighton, J. R. Saam, R. S. Ingram *et al.*, 'An enhancer deletion affects both H19 and Igf2 expression.' *Genes Dev.*, 1995, **9**, 2079–2089.

77. S. G. Gray, T. Eriksson and T. J. Ekstrom, 'Methylation, gene expression and the chromatin connection in cancer.' *Int. J. Mol. Med.*, 1999, **4**, 333–350.

78. A. P. Feinberg, 'DNA methylation, genomic imprinting and cancer.' *Curr. Top. Microbiol. Immunol.*, 2000, **249**, 87–99.

79. D. Dao, C. P. Walsh, L. Yuan *et al.*, 'Multipoint analysis of human chromosome 11p15/mouse distal chromosome 7: inclusion of H19/IGF2 in the minimal WT2 region, gene specificity of H19 silencing in Wilms' tumorigenesis and methylation hyper-dependence of H19 imprinting.' *Hum. Mol. Genet.*, 1999, **8**, 1337–1352.

80. N. Kokalj-Vokac, A. Almeida, E. Viegas-Pequignot *et al.*, 'Specific induction of uncoiling and recombination by azacytidine in classical satellite-containing constitutive heterochromatin.' *Cytogenet. Cell Genet.*, 1993, **63**, 11–15.

81. A. Almeida, N. Kokalj-Vokac, D. Lefrancois *et al.*, 'Hypomethylation of classical satellite DNA and chromosome instability in lymphoblastoid cell lines.' *Hum. Genet.*, 1993, **91**, 538–546.

82. C. Lengauer, K. W. Kinzler and B. Vogelstein, 'Genetic instability in colorectal cancers.' *Nature*, 1997, **386**, 623–627.

83. Y. Kondo, Y. Kanai, M. Sakamoto *et al.*, 'Genetic instability and aberrant DNA methylation in chronic hepatitis and cirrhosis-A comprehensive study of loss of heterozygosity and microsatellite instability at 39 loci and DNA hypermethylation on 8 CpG islands in microdissected specimens from patients with hepatocellular carcinoma.' *Hepatology*, 2000, **32**, 970–979.

84. C. Lengauer, K. W. Kinzler and B. Vogelstein, 'DNA methylation and genetic instability in colorectal cancer cells.' *Proc. Nat. Acad. Sci. USA*, 1997, **94**, 2445–2450.

85. M. Hanada, D. Delia, A. Aiello *et al.*, 'bcl-2 gene hypomethylation and high-level expression in B-cell chronic lymphocytic leukemia.' *Blood*, 1993, **82**, 1820–1828.

86. A. P. Feinberg and B. Vogelstein, 'Hypomethylation distinguishes genes of some human cancers from their normal counterparts.' *Nature*, 1983, **301**, 89–92.

87. S. B. Baylin, M. Makos, J. J. Wu *et al.*, 'Abnormal patterns of DNA methylation in human neoplasia: potential consequences for tumor progression.' *Cancer Cells*, 1991, **3**, 383–390.

88. J. G. Herman, A. Merlo, L. Mao *et al.*, 'Inactivation of the CDKN2/p16/MTS1 gene is frequently associated with aberrant DNA methylation in all common human cancers.' *Cancer Res.*, 1995, **55**, 4525–4530.

89. P. A. Jones, 'DNA methylation errors and cancer.' *Cancer Res.*, 1996, **56**, 2463–2467.

90. J. G. Herman, 'Hypermethylation of tumor suppressor genes in cancer.' *Semin. Cancer Biol.*, 1999, **9**, 359–367.

91. J. G. Herman and S. B. Baylin, 'Promoter-region hypermethylation and gene silencing in human cancer.' *Curr. Top. Microbiol. Immunol.*, 2000, **249**, 35–54.

92. C. M. Bender, J. M. Zingg and P. A. Jones, 'DNA methylation as a target for drug design.' *Pharm. Res.*, 1998, **15**, 175–187.

93. M. Lubbert, 'DNA methylation inhibitors in the treatment of leukemias, myelodysplastic syndromes and hemoglobinopathies: clinical results and possible mechanisms of action.' *Curr. Top. Microbiol. Immunol.*, 2000, **249**, 135–164.

94. S. A. Belinsky, K. J. Nikula, W. A. Palmisano *et al.*, 'Aberrant methylation of p16(INK4a) is an early event in lung cancer and a potential biomarker for early diagnosis.' *Proc. Nat. Acad. Sci. USA*, 1998, **95**, 11891–11896.

95. M. Esteller, M. Sanchez-Cespedes, R. Rosell *et al.*,

'Detection of aberrant promoter hypermethylation of tumor suppressor genes in serum DNA from non-small cell lung cancer patients.' *Cancer Res.*, 1999, **59**, 67–70.

96. I. H. Wong, Y. M. Lo, J. Zhang *et al.*, 'Detection of aberrant p16 methylation in the plasma and serum of liver cancer patients.' *Cancer Res.*, 1999, **59**, 71–73.

97. T. L. Butler, P. H. Kay and P. F. Jacobsen, 'Hypomethylation of cytosine 5-methyltransferase in human neoplasms.' *Anticancer Res.*, 2000, **20**, 1435–1438.

98. A. V. Bakin and T. Curran, 'Role of DNA 5-methylcytosine transferase in cell transformation by fos.' *Science*, 1999, **283**, 387–390.

99. C. A. Eads, K. D. Danenberg, K. Kawakami *et al.*, 'CpG island hypermethylation in human colorectal tumors is not associated with DNA methyltransferase overexpression.' *Cancer Res.*, 1999, **59**, 2302–2306.

100. L. S. Chuang, H. I. Ian, T. W. Koh *et al.*, 'Human DNA-(cytosine-5) methyltransferase-PCNA complex as a target for p21WAF1.' *Science*, 1997, **277**, 1996–2000.

101. J. P. Issa, 'Aging, DNA methylation and cancer.' *Crit. Rev. Oncol. Hematol.*, 1999, **32**, 31–43.

102. J. P. Issa, 'CpG-island methylation in aging and cancer.' *Curr. Top. Microbiol. Immunol.*, 2000, **249**, 101–118.

103. I. E. Dreosti, 'Nutrition, cancer and aging.' *Ann. N. Y. Acad. Sci.*, 1998, **854**, 371–377.

104. V. L. Wilson and P. A. Jones, 'DNA methylation decreases in aging but not in immortal cells.' *Science*, 1983, **220**, 1055–1057.

105. T. Ono, R. Tawa, K. Shinya *et al.*, 'Methylation of the c-myc gene changes during aging process of mice.' *Biochem. Biophys. Res. Commun.*, 1986, **139**, 1299–1304.

106. T. Ono, S. Yamamoto, A. Kurishita *et al.*, 'Comparison of age-associated changes of c-myc gene methylation in liver between man and mouse.' *Mutat. Res.*, 1990, **237**, 239–246.

107. H. Ghazi, F. A. Gonzales and P. A. Jones, 'Methylation of CpG-island-containing genes in human sperm, fetal and adult tissues.' *Gene*, 1992, **114**, 203–210.

108. Y. Uehara, T. Ono, A. Kurishita *et al.*, 'Age-dependent and tissue-specific changes of DNA methylation within and around the c-fos gene in mice.' *Oncogene*, 1989, **4**, 1023–1028.

109. J. P. Issa, Y. L. Ottaviano, P. Celano *et al.*, 'Methylation of the oestrogen receptor CpG island links ageing and neoplasia in human colon.' *Nat. Genet.*, 1994, **7**, 536–540.

110. N. Ahuja, Q. Li, A. L. Mohan *et al.*, 'Aging and DNA methylation in colorectal mucosa and cancer.' *Cancer Res.*, 1998, **58**, 5489–5494.

111. M. Toyota and J. P. Issa, 'The role of DNA hypermethylation in human neoplasia.' *Electrophoresis*, 2000, **21**, 329–333.

112. Y. Kim, 'Folate and carcinogenesis: evidence, mechanisms and implications.' *J. Nutr. Biochem.*, 1999, **10**, 66–88.

113. R. A. Jacob, D. M. Gretz, P. C. Taylor *et al.*, 'Moderate folate depletion increases plasma homocysteine and decreases lymphocyte DNA methylation in postmenopausal women.' *J. Nutr.*, 1998, **128**, 1204–1212.

114. I. P. Pogribny, B. J. Miller and S. J. James, 'Alterations in hepatic p53 gene methylation patterns during tumor progression with folate/methyl deficiency in the rat.' *Cancer Lett.*, 1997, **115**, 31–38.

115. S. W. Choi and J. B. Mason, 'Folate and carcinogenesis: an integrated scheme.' *J. Nutr.*, 2000, **130**, 129–132.

116. P. Kneckt, A. Aromaa, J. Maatela *et al.*, 'Serum selenium and subsequent risk of cancer among Finnish men and women.' *J. Nat. Cancer Inst.*, 1990, **82**, 864–868.

117. L. Clark, G. Combs, B. Turnbull *et al.*, 'Effect of selenium supplementation for cancer prevention in patients with carcinoma of the skin: a randomized controlled trial.' *J. Am. Med. Assoc.*, 1996, **276**, 1957–1963.

118. L. Clark, K. Cantor and W. Allaway, 'Selenium in forage crops and cancer mortality in US counties.' *Arch. Environ. Health*, 1991, **46**, 37–42.

119. C. D. Davis, E. O. Uthus and J. W. Finley, 'Dietary selenium and arsenic affect DNA methylation *in vitro* in caco-2 cells and *in vivo* in rat liver and colon.' *J. Nutr.*, 2000, **130**, 2903–2909.

120. Chromium, nickel and welding. IARC, Lyon, FR, 1990, pp. 1–677.

121. R. Doll, J. D. Mathews and L. G. Morgan, 'Cancers of the lung and nasal sinuses in nickel workers: a reassessment of the period of risk.' *Br. J. Ind. Med.*, 1977, **34**, 102–105.

122. B. Kargacin, C. B. Klein and M. Costa, 'Mutagenic responses of nickel oxides and nickel sulfides in Chinese hamster V79 cell lines at the xanthine-guanine phosphoribosyl transferase locus.' *Mutat. Res.*, 1993, **300**, 63–72.

123. G. G. Fletcher, F. E. Rossetto, J. D. Turnbull *et al.*, 'Toxicity, uptake and mutagenicity of particulate and soluble nickel compounds.' *Environ. Health Perspect.*, 1994, **102**(Suppl 3), 69–79.

124. K. Conway and M. Costa, 'Nonrandom chromosomal alterations in nickel-transformed Chinese hamster embryo cells.' *Cancer Res.*, 1989, **49**, 6032–6038.

125. C. B. Klein, K. Conway, X. W. Wang *et al.*, 'Senescence of nickel-transformed cells by an X chromosome: possible epigenetic control.' *Science*, 1991, **251**, 796–799.

126. D. A. Trott, A. P. Cuthbert, R. W. Overell *et al.*, 'Mechanisms involved in the immortalization of mammalian cells by ionizing radiation and chemical carcinogens.' *Carcinogenesis*, 1995, **16**, 193–204.

127. C. B. Klein and T. G. Rossman, 'Transgenic Chinese hamster V79 cell lines which exhibit variable levels of gpt mutagenesis.' *Environ. Mol. Mutagen.*, 1990, **16**, 1–12.

128. C. B. Klein, L. Su, T. G. Rossman *et al.*, 'Transgenic gpt+ V79 cell lines differ in their mutagenic response to clastogens.' *Mutat. Res.*, 1994, **304**, 217–228.

129. Y. W. Lee, C. Pons, D. M. Tummolo *et al.*, 'Mutagenicity of soluble and insoluble nickel compounds at the gpt locus in G12 Chinese hamster cells.' *Environ. Mol. Mutagen.*, 1993, **21**, 365–371.

130. C. Klein and E. Snow, 'Localization of the gpt sequence in transgenic G12 cells by fluorescent in situ hybridization.' *Environ. Mol. Mutagen.*, 1993, **21**, 35.

131. Y. W. Lee, C. B. Klein, B. Kargacin *et al.*, 'Carcinogenic nickel silences gene expression by chromatin condensation and DNA methylation: a new model for epigenetic carcinogens.' *Mol. Cell Biol.*, 1995, **15**, 2547–2557.

132. M. Costa and C. B. Klein, 'Nickel carcinogenesis, mutation, epigenetics, or selection.' *Environ. Health Perspect.*, 1999, **107**, A438–439.

133. M. M. Gebara, C. Drevon, S. A. Harcourt *et al.*, 'Inactivation of a transfected gene in human fibroblasts can occur by deletion, amplification, phenotypic switching, or methylation.' *Mol. Cell Biol.*, 1987, **7**, 1459–1464.

134. L. Broday, J. Cai and M. Costa, 'Nickel enhances telomeric silencing in *Saccharomyces cerevisiae*.' *Mutat. Res.*, 1999, **440**, 121–130.

135. L. Broday, W. Peng, M. H. Kuo *et al.*, 'Nickel

compounds are novel inhibitors of histone H4 acetylation.' *Cancer Res.*, 2000, **60**, 238–241.

136. C. B. Klein and M. Costa, 'DNA methylation, heterochromatin and epigenetic carcinogens.' *Mutat. Res.*, 1997, **386**, 163–180.

137. J. W. Nyce, 'Drug-induced DNA hypermethylation: a potential mediator of acquired drug resistance during cancer chemotherapy.' *Mutat. Res.*, 1997, **386**, 153–161.

138. C. Q. Zhao, M. R. Young, B. A. Diwan *et al.*, 'Association of arsenic-induced malignant transformation with DNA hypomethylation and aberrant gene expression.' *Proc. Nat. Acad. Sci. USA*, 1997, **94**, 10907–10912.

139. M. J. Mass and L. Wang, 'Arsenic alters cytosine methylation patterns of the promoter of the tumor suppressor gene p53 in human lung cells: a model for a mechanism of carcinogenesis.' *Mutat. Res.*, 1997, **386**, 263–277.

140. J. G. Herman, J. Jen, A. Merlo *et al.*, 'Hypermethylation-associated inactivation indicates a tumor suppressor role for p15INK4B.' *Cancer Res.*, 1996, **56**, 722–727.

141. G. A. Otterson, S. N. Khleif, W. Chen *et al.*, 'CDKN2 gene silencing in lung cancer by DNA hypermethylation and kinetics of p16INK4 protein induction by 5-aza 2'deoxycytidine.' *Oncogene*, 1995, **11**, 1211–1216.

142. A. Merlo, J. G. Herman, L. Mao *et al.*, '5' CpG island methylation is associated with transcriptional silencing of the tumour suppressor p16/CDKN2/MTS1 in human cancers.'. *Nat. Med.*, 1995, **1**, 686–692.

143. P. G. Corn, S. J. Kuerbitz, M. M. van Noesel *et al.*, 'Transcriptional silencing of the p73 gene in acute lymphoblastic leukemia and Burkitt's lymphoma is associated with 5' CpG island methylation.' *Cancer Res.*, 1999, **59**, 3352–3356.

144. K. D. Robertson and P. A. Jones, 'The human ARF cell cycle regulatory gene promoter is a CpG island which can be silenced by DNA methylation and down-regulated by wild-type p53.' *Mol. Cell Biol.*, 1998, **18**, 6457–6473.

145. D. Laux, E. Curran, K. Malik *et al.*, 'Hypermethylation of the Wilms tumor suppressor gene CpG island in human breast cancer.' *Proc. Am. Assoc. Cancer Res.*, 1997, **38**, 178.

146. J. Herman, F. Latif, Y. Weng *et al.*, 'Silencing of the VHL tumor-suppressor gene by DNA methylation in renal carcinoma.' *Proc. Nat. Acad. Sci. USA*, 1994, **91**, 9700–9704.

147. H. Huynh, L. Alpert and M. Pollak, 'Silencing of the mammary-derived growth inhibitor (MDGI) gene in breast neoplasms is associated with epigenetic changes.' *Cancer Res.*, 1996, **56**, 4865–4870.

148. N. Ahuja, A. Mohan, Q. Li *et al.*, 'Association between CpG islands methylation and microsatellite instability in colorectal cancer.' *Cancer Res.*, 1997, **57**, 3370–3374.

149. N. Ohtani-Fujita, T. Fujita, A. Aoike *et al.*, 'CpG methylation inactivates the promoter activity of the human retinoblastoma tumor-suppressor gene.' *Oncogene*, 1993, **8**, 1063–1067.

150. A. T. Ferguson, R. G. Lapidus, S. B. Baylin *et al.*, 'Demethylation of the estrogen receptor gene in estrogen receptor-negative breast cancer cells can reactivate estrogen receptor gene expression.' *Cancer Res.*, 1995, **55**, 2279–2283.

151. D. F. Jarrard, H. Kinoshita, Y. Shi *et al.*, 'Methyla-

tion of the androgen receptor promoter CpG island is associated with loss of androgen receptor expression in prostate cancer cells.' *Cancer Res.*, 1998, **58**, 5310–5314.

152. S. Cote and R. L. Momparler, 'Activation of the retinoic acid receptor beta gene by 5-aza-2'-deoxycytidine in human DLD-1 colon carcinoma cells.' *Anticancer Drugs*, 1997, **8**, 56–61.

153. S. Cote, D. Sinnett and R. L. Momparler, 'Demethylation by 5-aza-2'-deoxycytidine of specific 5-methylcytosine sites in the promoter region of the retinoic acid receptor β gene in human colon carcinoma cells.' *Anticancer Drugs*, 1998, **9**, 743–750.

154. J. R. Graff, J. G. Herman, R. G. Lapidus *et al.*, 'E-cadherin expression is silenced by DNA hypermethylation in human breast and prostate carcinomas.' *Cancer Res.*, 1995, **55**, 5195–5199.

155. K. E. Bachman, J. G. Herman, P. G. Corn *et al.*, 'Methylation-associated silencing of the tissue inhibitor of metalloproteinase-3 gene suggest a suppressor role in kidney, brain and other human cancers.' *Cancer Res.*, 1999, **59**, 798–802.

156. E. Tulchinsky, M. Grigorian, T. Tkatch *et al.*, 'Transcriptional regulation of the mts1 gene in human lymphoma cells: the role of DNA-methylation.' *Biochem. Biophys. Acta*, 1995, **1261**, 243–248.

157. N. S. Verkaik, J. Trapman, J. C. Romijn *et al.*, 'Down-regulation of CD44 expression in human prostatic carcinoma cell lines is correlated with DNA hypermethylation.' *Int. J. Cancer*, 1999, **80**, 439–443.

158. X. C. Qian and T. P. Brent, 'Methylation hot spots in the 5' flanking region denote silencing of the O6-methylguanine-DNA methyltransferase gene.' *Cancer Res.*, 1997, **57**, 3672–3677.

159. J. G. Herman, A. Umar, K. Polyak *et al.*, 'Incidence and functional consequences of hMLH1 promoter hypermethylation in colorectal carcinoma.' *Proc. Nat. Acad. Sci. USA*, 1998, **95**, 6870–6875.

160. G. Deng, A. Chen, J. Hong *et al.*, 'Methylation of CpG in a small region of the hMLH1 promoter invariably correlates with the absence of gene expression.' *Cancer Res.*, 1999, **59**, 2029–2033.

161. A. Dobrovic and D. Simpfendorfer, 'Methylation of the BRCA1 gene in sporadic breast cancer.' *Cancer Res.*, 1997, **57**, 3347–3350.

162. D. N. Mancini, D. I. Rodenhiser, P. J. Ainsworth *et al.*, 'CpG methylation within the 5' regulatory region of the BRCA1 gene is tumor specific and includes a putative CREB binding site.' *Oncogene*, 1998, **16**, 1161–1169.

163. M. Esteller, P. G. Corn, J. M. Urena *et al.*, 'Inactivation of glutathione S-transferase P1 gene by promoter hypermethylation in human neoplasia.' *Cancer Res.*, 1998, **58**, 4515–4518.

164. Q. Li, N. Ahuja, P. C. Burger *et al.*, 'Methylation and silencing of the Thrombospondin-1 promoter in human cancer.' *Oncogene*, 1999, **18**, 3284–3289.

165. J. Weber, M. Salgaller, D. Samid *et al.*, 'Expression of the MAGE-1 tumor antigen is up-regulated by the demethylating agent 5-aza-2'-deoxycytidine.' *Cancer Res.*, 1994, **54**, 1766–1771.

166. S. Coral, L. Sigalotti, A. Gasparollo *et al.*, 'Prolonged upregulation of the expression of HLA class I antigens and costimulatory molecules on melanoma cells treated with 5-aza-2'-deoxycytidine (5-AZA-CdR).' *J. Immunother.*, 1999, **22**, 16–24.

J.P. Vanden Heuvel, G.H. Perdew, W.B. Mattes and W.F. Greenlee (Eds.)
Comprehensive Toxicology, Vol. xiv
2002 Elsevier Science BV. All rights reserved.

## 14.4 ALTERATIONS OF CELL SIGNALING BY XENOBIOTICS

# 14.4.1
# Introduction and Overview

## JOHN P. VANDEN HEUVEL
### Penn State University, University Park, PA, USA

## 14.4.1.1  INTRODUCTION

Maintaining growth control is a critical undertaking in the development of any multicellular organism, and this control often goes awry in the disease state. Following xenobiotic treatment, the cell may be subjected to a series of new challenges to the maintenance of homeostasis. The cell must *sense* these homeostatic cues, and signals produced that can affect the outcome of that cell, such as proliferation, apoptosis or differentiation. (Another outcome may be considered an *adaptive response* where enzymes involved in the catabolism of that chemical are induced. The induction of cytochrome P450 enzymes is a prime example of this adaptive response and has been discussed

elsewhere in this volume (14.02.1 Table 6, Reference 1)). The sensing and responding process is defined as cellular *signaling* and the pathways that link the sensors (i.e., receptors) to the effectors (i.e., transcription factors, caspases) of cell fate, the *signal transduction* cascade (see Figure 1).

There have been several key findings in the general area of signaling, the first of which may have been the observation that protein activity can be reversibly altered by phosphorylation (reviewed in Reference 2), in the mid-to-late 1950s. Later discoveries of kinases and phosphatases showed the dynamic nature of this process of protein modification (reviewed in Reference 3). In 1970, Martin Rodbell discovered that signal transmission requires a

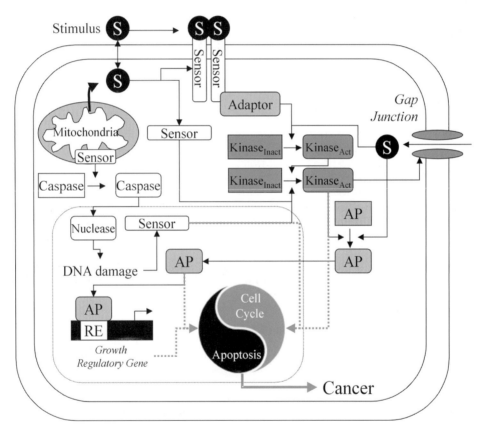

**Figure 1**  Overview of cell signaling events described in Section 4. Xenobiotics can affect signaling in multiple ways within this paradigm, as discussed in the text. A stimulus may include growth factors or reactive oxygen species (ROS, Chapter 4.1). These signals are extracellular or intracellular; gap junctions play essential functions in the communication of signals with adjacent cells (Chapter 4.5). Sensor molecules include cell surface receptors such as the epidermal growth factor receptor, intracellular proteins such as HIF1 (discussed in 4.4) and NFκ B, or enzymes such as protein kinase C. Antioxidants may affect similar signaling intermediates as ROS, but also have unique effects on gene expression (Chapter 4.3). The mitogen and stress activated protein kinase (MAPK, SAPK, Chapter 4.2) cascades are important players in the signal transduction cascade. Ultimately, the signals are propogated into the form of activators proteins (i.e., transcription factors) and caspases. The former are involved in the decision of whether a cell should enter the cell cycle (Chapter 4.6) while the latter act predominantly to initiate apoptosis (Chapter 4.7). Abbreviations used: S, stimulus; Kinase$_{inact/act}$, inactive or active kinase; AP, activator protein; RE, response element. Rectangles depict inactive molecules while rounded rectangles are active. Similar nomenclature and symbols are used throughout this chapter.

small intracellular molecule called guanine triphosphate (GTP) and that the breakdown of GTP to GDP was an integral regulatory step in the control of hormone action.[4] GTP was subsequently found to bind to G-proteins, a family of proteins that serve as intermediaries between incoming signaling molecules, such as hormones and drugs, and the cellular proteins that respond to the signals.[4] Subsequently, several second messengers were discovered as intermediates in the signaling cascade including cyclic adenosine mono-phosphate (cAMP), calcium and diacylglyerol. The pioneers in signal transduction have been receiving much-warranted accolades for their discoveries, most notably the Nobel Prize in Physiology and Medicine. For example, in 1994 Rodbell shared the Nobel Prize with Alfred Gilman for their work on G-proteins and the communication system that regulates cellular activity. In 1998, the pharmacologists Robert Furchgott, Louis Ignarro, and Ferid Murad received this award for their discoveries concerning nitric oxide as a signaling molecule in the cardiovascular system.[5] Finally, the 2000 Nobel Prize recognized the key discoveries of three scientists who have played a significant role in pioneering this frontier. Arvid Carlsson, Paul Greengard, and Eric Kandel made seminal contributions to the current understanding of information processing and signaling within the brain.[6]

In addition to its being fertile ground for the Nobel Prize, the importance of signaling in biology and toxicology is underscored by three key observations. First, many proteins involved in the signal transduction cascade are highly conserved across diverse species. For example, cAMP-dependent protein kinase 1 (Pka-C1) in the fruitfly (*Drosophila melanogaster*) is 82% identical to its human counterpart (PKA).[7] Conservation of protein sequence generally implies that a given protein is playing an important biological role. Correlatively, alteration in this protein's function via a xenobiotic may have dire consequences. Second, dysregulation of several signal transduction pathways has been associated with tumorigenesis (reviewed in Reference 8). Many oncogenes and tumor suppressors genes are signal transduction proteins including ras, myc, fos and jun. Third, there is increasing evidence for a role of the signal transduction cascade in modulation of xenobiotic and hormone receptors. *Cross-talk* implies that signaling may affect the activity of the xenobiotic receptor (and *visa versa*) and the effect of a particular chemical may vary depending on the activity of various signal transduction pathways in a given cell.

Xenobiotics can affect the signal transduction process in a variety of ways. For simplicity, this chapter will be broken down into the major components of signal transduction: the signals, the sensors, and the signaling pathways. How exogenous compounds impinge on each of these components will be discussed. The goal of the present section is to describe the signaling pathways that are often affected by carcinogenic compounds, specifically tumor promoters. In this chapter, the basic framework of the signal transduction cascade will be investigated. In subsequent chapters specific examples of xenobiotic alterations in signal transduction will be portrayed (Chapters 4.2–4.5), as will the ultimate toxic responses of altered cell cycle control (Chapter 4.6) and apoptosis (Chapter 4.7).

### 14.4.1.2 THE SIGNALS

A cell is constantly being bombarded with information from within and without, all of which must be sensed and transduced through the signal transduction cascade. Signaling ligands can be small diffusible molecules, large molecules in the extracellular matrix, non-diffusible surface components of neighboring cells, or intermediates in intracellular intermediary metabolism. Signaling involving small ligands that interact with receptors can be classified as *autocrine* or self-signaling; *paracrine*, signaling between nearby cells, and; *endocrine*, signaling over a long distance, usually via the bloodstream. Signaling involving ligands from a neighboring cell is called *juxtacrine* whereby the signaling and receiving cells must be in contact for the signal to be transmitted. An important aspect of juxtacrine signaling is gap junction intracellular communication (GJIC). Still other signals can come directly from extracellular matrix (ECM) components. Several examples of important signaling molecules are given, with an emphasis on those known to be affected by xenobiotics.

#### 14.4.1.2.1 Growth Factors

*Growth factors* are proteins that bind to receptors on the cell surface, with the primary result of activating a signal transduction cascade, and eventually influencing cellular proliferation, differentiation or apoptosis.* Many growth factors stimulate growth or

---

*Some information came from the "Signal Transduction" website, http://www.way2goal.com/sigpath.html

differentiation of specific cell types, whereas their absence in differentiated cells will result in apoptosis. Most growth factors are polypeptides, and their actions are similar in many respects to that of hormones. Insulin, for example, in addition to its rapid effects on glucose uptake, also acts as a growth factor for several types of cells. Most growth factors were discovered as requirements for cells grown in culture and have cell type-specific effects on proliferation and differentiation. For example, nerve growth factor (NGF) is required for differentiation of neuronal precursor cells in culture. Epidermal growth factor (EGF) is required for proliferation of epithelial cells in culture, and stimulates the growth of many other cell types as well. Additional growth factors include fibroblast growth factor (FGF), platelet-derived growth factor (PDGF), transforming growth factor TGF-β, and insulin-like growth factors (ILGF), required for growth of bone, muscle and other cells. All have specific transmembrane receptors, as will be discussed in 14.4.1.3.

*Cytokines* are a unique family of growth factors. Secreted primarily from leukocytes, cytokines stimulate the humoral and cellular immune responses, as well as the activation of phagocytic cells. Cytokines that are secreted from lymphocytes are deemed *lymphokines*, whereas those secreted by monocytes or macrophages are termed *monokines*. Many of the lymphokines are also known as *interleukins* (ILs), since they are not only secreted by leukocytes but also able to affect the cellular responses of leukocytes. Specifically, interleukins are growth factors targeted to cells of hematopoietic origin, of which at least 18 are known. In addition to ILs, an important immune system-modifying cytokine is tumor necrosis factor (TNF, also called cachectin), produced mainly by activated macrophages.

Since growth factors and cytokines are responsible for a variety of physiological processes, including neoplasia, apoptosis, inflammation, immunity and hematopoiesis, disruption of their expression by xenobiotics could have far-reaching effects. There are several examples of xenobiotics affecting the expression of growth factors and cytokines, and with the advent of high-throughput screening tools, such as microarray analysis, the number of examples will undoubtedly increase. The alteration in growth factor concentrations could have autocrine, endocrine, paracrine or juxtacrine ramifications and can be produced by a xenobiotic at several levels including:[9] (1) *Direct toxicity to growth-factor producing cells.* This is com-monly observed with anti-mitotic agents such as cyclophosphamide, which destroy cytokine producing cells.[9] TCDD (2,3,7,8-tetrachloro-dibenzo-*p*-dioxin) alters the differentiation of T-cells, which partially explains its immuno-toxicity.[10,11] (2) *Alteration of growth factor expression, transcriptionally.* This is a very common means by which chemicals can affect growth factor production and is also where microarrays can be utilized to its fullest extent. Compounds such as cyclosporin, and other immunosuppressive drugs, inhibit the transcriptional regulation of various cytokines in T-cells.[9] Similarly, non-steroidal anti-inflammatory drugs (NSAIDS), including thiazolidinediones and PPARγ agonists, repress TNFα and IL-1 expression.[12,13] In the liver, TCDD affects the expression of the important hepatocyte proliferation inhibitor, TGF-β and induces expression of other factors such as IL-1β and plasminogen activator-inhibitor-2 (PAI-2);[14] these changes may have important consequences in the process of liver tumor promotion.[15] Finally, NF-κB and HIF are important transcriptional modulators of various growth factors including endothelial growth factor (VEGF), platelet-derived growth factor (PDGF) and fibroblast growth factor (FGF), as well as vasodilators produced by enzymes such as inducible nitric oxide synthase (iNOS) and heme oxygenase-1 (HO1).[16] There are numerous xenobiotics that affect NF-κB and HIF activity, as discussed below, that will ultimately affect the production of signaling molecules. (3) *Alteration in growth factor expression, post-transcriptionally.* Alteration in the activity of enzymes involved in growth factor processing is an example of this mechanism of action. Cyclooxygenase (COX) is an inhibitory target of NSAIDs and is an important enzyme in the production of prostaglandins, signaling molecules that have been implicated in carcinogenesis.[17] Cytochrome P450 inducers or inhibitors may also affect hormone, cytokine and growth factor production by altering their steady state levels. For example various AhR ligands affect aromatase (CYP19) activity, the result of which may be altered estrogen and androgen levels.[18] The antiprotozoal drug penta-miidine inhibits the release of IL-1 by altering its processing.[9] Rapamycin does not affect transcription of cytokine genes, but it has been shown to affect cytokine mRNA stability.[9] By whatever means, the alteration in growth factors and hormone concentration is an important mechanism by which xenobiotics cause toxicity, either in the cell of altered production itself or to a cell at a distal site.

### 14.4.1.2.2 Calcium

The divalent cation $Ca^{2+}$ is a major signaling molecule involved in cell cycle regulation, gene expression and apoptosis (reviewed in References 19,20). Although $Ca^{2+}$ is also an important regulator of a variety of metabolic enzymes and proteases, our focus will be on its role in cellular signal transduction and gene expression. The intracellular concentration of calcium is 20,000 fold lower than that found extracellularly and much effort is expended by the cell to maintain this gradient. ATP-dependent pumps transport calcium into the two major sinks, the endoplasmic reticulum (ER) and the extracellular space.[19] Various signals, including xenobiotics, can cause the release of calcium from either or both of these stores (which sink utilized depends on the stimulus), thereby increasing the intracellular concentration from 100 nm to 1–2 μM. Receptor tyrosine kinases and G-coupled receptors increase inositol 3-phosphate ($IP_3$) which stimulates ER release of calcium. Ligand-gated and voltage-dependent ion channels release calcium from the extracellular spaces. Nuclear calcium is also under tight, albeit ill-defined, control that is independent of the $Ca^{2+}$ regulation in the cytosol.[19]

Calcium ionophores such as ionomycin[21] are good examples of xenobiotics that affect the intracellular concentration of $Ca^{2+}$. Toxic agents such as thapsigargin and tributyltin deplete the ER calcium pools, resulting in an opening of the ion channels.[20] As mentioned above, both G-protein and tyrosine kinase receptors utilize calcium as a second messenger; therefore, xenobiotics that affect these receptors will indirectly result in altered intracellular calcium levels and signaling.

### 14.4.1.2.3 Oxygen and Reactive Oxygen Species (ROS)

Although molecular oxygen ($O_2$) is essential to eukaryotic cells, the formation of reactive oxygen species (ROS) is a potential cause of cellular damage. Most ROS are produced in the mitochondria,[16] although other organelles such as the peroxisome may also be involved.[22] In the mitochondria, the electron transport chain adds four electrons to $O_2$, yielding two molecules of water. However, in this process $O_2^{-\bullet}$ (superoxide radical), $H_2O_2$ (hydrogen peroxide) and $OH^\bullet$ (hydroxyl radical) are produced sequentially. These ROS, upon reacting with other molecules, result in the formation of the potent oxidants ONOO- (peroxynitrite) or HOCl (hydrochlorous acid)

or lipid radicals. The cell has developed means of sensing the amount of $O_2$/ROS in the cell (i.e., the redox state) (described below), as well as several systems to directly combat the damaging effects of ROS. The most common mechanisms of protection include superoxide dismutase (SOD), catalase, antioxidant vitamins (VitE, selenium), and the glutathione peroxidase systems.[16]

Xenobiotics that affect the mitochondria are common, and they have the potential to affect the redox state of the cell as well as the production of ROS (reviewed in Reference 23). Uncoupling of oxidative phosphorylation by various substituted naphthoquinones and nitrosamines, as well as adriamycin or paraquat, can produce superoxide free radicals.[23] Furan is metabolized in the mitochondria by a cytochrome P450 enzyme and the subsequent uncoupling metabolite produces ROS.[22] An interesting, albeit controversial, means to produce ROS involves the tumor promoting xenobiotics, the peroxisome proliferators (see Chapter 2.3). These compounds induce the expression of acyl-CoA oxidase (ACO) the first and rate-limiting enzyme in the fatty acid catabolism pathway in the peroxisome. Unlike its mitochondrial counterpart, ACO produces $H_2O_2$ as a byproduct, which may react with other molecules to form more potent ROS. Incidentally, the peroxisome contains a variety of hydrogen peroxide producing enzymes and it also contains the important detoxification enzyme catalase. Therefore, if peroxisome proliferators increase ACO or other $H_2O_2$ generating enzymes to an extent that overwhelms the ability of catalase to degrade, ROS may be liberated into the cytosol. As will be discussed subsequently, ROS and oxidative stress are important by-products of the toxic response and can affect many signaling processes and ultimately cell fate.

### 14.4.1.2.4 Stresses

The concept of *cellular stress* is somewhat diffuse and the initiation of such a signaling event in many cases is not known. For our purposes, stress is defined as any event initiated by a chemical treatment that results in the activation of the stress activated protein kinase (SAPK, also called jun N-terminal kinase, JNK) cascade (discussed in detail in Chapter 4.2). Stress responses are initiated by heat and osmotic shock, inhibition of DNA, RNA, and protein synthesis, and to a lesser extent by growth factors. In addition, the generation of ROS as described above would be considered an *oxidative stress* response. Stress may primarily involve an organelle such as the mitochondria or ER and is indicative of overt damage.[19] Cell shape and contact inhibition may affect actin stress fibers, which in turn may signal through the SAPK cascade. This spectrum of regulators suggests that the SAPK enzymes are transducers of a variety of stress responses. Most of our discussion will focus on oxidative stress and its effects on signaling.

### 14.4.1.3   THE SENSORS

The sensors of cellular homeostasis are, with a few exceptions, *receptors* for the signals described above and the two terms will be used interchangeably. (The concept of a receptor has been described in Chapter 1.3 and soluble receptors have been described extensively in Chapters 2.1–2.8.) The classification of receptors involved in signal transduction, and their basic structure-function relationship is shown in Figure 2 (adapted from Reference 24).

The propagation of the signal transduction cascade starts with the interaction between the signal and the sensor; the result of this bimolecular interaction is a conformational change in the sensor that alters its function. The change in shape of the sensor protein

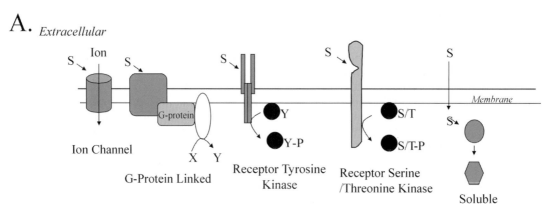

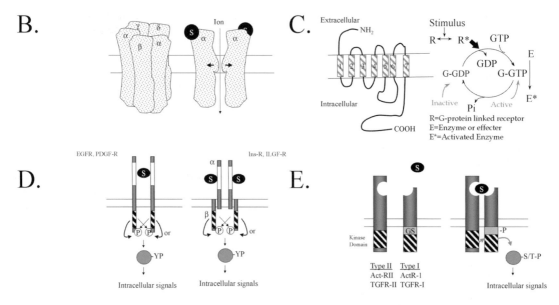

**Figure 2**   Classification and general mechanisms of sensors/receptors. (**A**). Cell surface receptors exist in three basic forms, ion channels, GTP-binding protein linked (GPLR), and kinase linked (which includes protein tyrosine kinase (PTK) and receptors serine/threonine kinases). Receptors/sensor molecules can also exist in the cytoplasm of the cell. (**B**)–(**E**) Basic mechanism of action of ion channel, GPLR, PTK and TGFβ-receptors. See text for details.

results in a variety of functional alterations including: revealing of cryptic motifs, such as those involved in nuclear localization; changing of protein–protein interaction including kinase recognition; or, altering the protein's enzymatic activity, most notably tyrosine phosphorylation. For cell surface receptors, changes such as oligomerization, dimerization and internalization are common following binding of ligand.[25] The *activated* sensor is *coupled* to the signal transduction cascade. In fact is it convenient to consider the activated sensor as a new signal that is able to activate downstream events. For all intents and purposes, signaling is a series of molecular interactions, conformation changes and altered activity.

The biologist studying sensor/receptor molecules owe a debt of gratitude to the pharmacologist and toxicogist. It was only through the use of various toxins and synthetic agonists and antagonists that the characterization of these macromolecules could be performed. For an interesting history of how receptors have been examined at the cellular and molecular level through the use of xenobiotics, the reader is directed to *Principles of Drug Action*.[24]

Although detailed description of the huge number of sensor molecules is beyond the scope of the present chapter, understanding basic receptor structure and function is important for three reasons: First, sensors are the molecules that initiate a biological response from several important chemicals and xenobiotic-induced stresses. Second, disruption of these receptor systems, i.e., through mutation or changes in expression, is an important paradigm in the etiology of cancer. Third, receptors and signal transduction processes are polymorphic (described in Section 14.3). Therefore, by understanding the molecular biology of receptors we can understand the mechanism of action of xenobiotics, examine why cancer cells lose normal growth regulation and last why certain individuals may be more susceptible to disease.

### 14.4.1.3.1   Cell Surface Sensors/Receptors

Cell surface sensors/receptors exist in three basic forms, ion channels, GTP-protein coupled receptors (GPCR), and kinase linked receptors including receptor tyrosine kinases (RTKs) and transforming growth factor β receptors (TGF-β, Tβ R). The basic structure of these signaling molecules is shown in Figure 2 and will be discussed individually.

#### 14.4.1.3.1.1   *Ion Channel Receptors*

Receptors that form channels are generally large, multiprotein complexes. Two varieties exist, potential sensitive (voltage-gated) and chemosensitive (ligand-gated) channels (reviewed in Reference 24). Perhaps the best-characterized receptor system is the nicotinic acetylcholine receptor (AchR), one of the ligand-gated channels. This receptor consists of four distinct subunits (α, β, γ and δ), with one of the subunits present twice (α$_2$β γ δ) arranged around an axis of symmetry (see Figure 2(B)). Each subunit is the product of a separate gene and is 30–40% identical to the other subunits at the amino acid level. The central cavity of this pentamer forms a channel, which at the resting state is impermeable to ions. The α subunits contain the ligand binding domain that are the contact site for agonists such as nicotine and acetylcholine, competitive antagonists such as trimethaphan, and the irreversible antagonist snake α-toxin. Agonist binding results in a conformational change that opens the ion channel to a width of 6.5 Å. The open channel is selective for cations, but they must conform to the width of the channel.

The voltage sensitive channels have a structure that is similar that described for AchR, except the subunits are arranged differently and the stimulus is a change in electrical potential. These channels are extremely selective for a given ion and allow for a rapid and robust flow of ions down a concentration gradient. The best example is perhaps the voltage-sensitive sodium channel, which consists of three subunits (α, β1, β2). The α subunit is by far the most studied since it contains the actual pore of the receptor. The sodium channel α subunit contains up to 32 membrane-spanning segments and is sufficient to form an active channel, without the β subunits. One of the membrane spanning sections contains a series of positively charged amino acids and serves as a sensor of electrical charge. Voltage-dependent gating, or opening of the pore, is initiated by a movement of these positively charged residues toward the cell surface.

As mentioned above, the study of the effects of xenobiotics on ion channel receptors has a long history and is one of the important areas of biological research where toxicology has contributed significantly.[24] The characterization of both ligand- and voltage-gated channels would not have been possible without the availability of selective agonists and antagonist (notably, toxins and toxicants, see Table 1). Also, drugs such as the calcium channel

**Table 1**  Agonists and antagonists for channel forming receptors.

| Receptor | Agonist | Antagonist |
| --- | --- | --- |
| Nicotinic Acetylcholine receptor (muscle) | Acetylcholine, phenyltri-methylammonium | α -toxin, d-tubocurarine |
| Nicotinic Acetylcholine receptor (neuronal) | Acetylcholine, dimethyl-phenylpiperizium | Trimethaphan |
| Muscarinic M1 receptor | Acetylcholine, muscarine | Atropine, pirenzepine |
| Muscarinic M2 receptor | Acetylcholine, muscarine | Atropine, AF-DX116 |
| Muscarinic M3 receptor | Acetylcholine, muscarine | Atropine, hexahydrosila-difenidol |
| Sodium Channel | Acotine, veratridine, α-scorpion toxin | Tetradotoxin, saxitoxin |
| Calcium Channel | Bay K-8644 | Verapamil, diltiazem |

Adapted from Reference 24.

blockers verapamil and diltiazem, are important clinical drugs because of their cardiac properties on ion channels. Pesticides such as dichlorodiphenyltrichloroethane (DDT) have multiple effects on ion channel conductance, including reduction of potassium transport, inactivating sodium channel closure and inhibiting sodium/potassium ATPases.[26]

Another area in which xenobiotics affect the activity of the ion channels is through an indirect process, i.e., by altering the levels of the endogenous signal for these receptors. The most prevalent examples are the organophosphate pesticides and alteration of acetylcholine levels in the synapse of cells (for a review see Reference 26). Certain organophosphorus ester pesticides, including parathione, compete with Ach for binding to, or are poor substrates for, acetylcholinesterase (AChE); AchE is the predominant enzyme in the breakdown of ACh in the synapse and hence the major means by which the signal is attenuated. Other chemicals such as pargyline and isocarboxazid are irreversible inhibitors of this enzyme. Anesthetics may work in a similar fashion by decreasing the conductance of the action potential, i.e., the signal, for voltage-gated channels.

### 14.4.1.3.1.2   G-Protein Linked Receptors

The G-protein linked receptors (GPLR) are a large group of receptors with well over 1000 different members cloned (most with no known ligand). As their name implies, GPLRs utilize guanine triphosphate (GTP)-binding proteins (G-proteins) as essential components of signal transduction. GPLRs have three essential components, the ligand binding subunit (*receptor*), the *G-protein* and the *effector* (enzyme, second messenger). Each of these subunits will be addressed separately.

#### 14.4.1.3.1.2.1   Ligand binding component

Several members of the GPCR family have been characterized in detail and share common structural features (see Figure 2(B)).[24] The amino terminal region contains glycosylation sites and is therefore the extracellular component of the receptor. There are 7 transmembrane regions that give the receptor the characteristic *serpentine* structure. The intracellular loop between the fifth and sixth transmembrane domain appears to be recognition sites for the G-proteins. This loop, as well as the carboxyl terminus, contain phosphorylation sites for cAMP dependent kinase (PKA). Key regions involved in defining ligand-binding specificity are found within the seventh membrane-spanning domain. As discussed below, certain receptor-ligand complexes are coupled to different G proteins ($G_s$ and $G_i$, respectively), via the cytoplasmic (carboxyl) domain of the receptor.

#### 14.4.1.3.1.2.2   GTP-binding protein

There are several types of GTP binding proteins, each of which share common structural features; however, the two predominant forms are $G_s$ and $G_i$ (see Table 2).[24] The general structure of G-proteins is a trimer ($\alpha$, $\beta$ and $\gamma$ subunits) with the $\alpha$ subunit existing in two different forms $\alpha_i$ and $\alpha_s$. The $\alpha$ subunits contain the GTP-binding and hydrolytic (GTPase) domains of the protein. The $\beta$ subunits of the G-proteins appear to be identical in the different forms of G-proteins. The $\gamma$ subunit may function as a membrane anchor for the complex, in particular the $G_i$ complex. In addition to $G_s$ and $G_i$, there are several types of G-proteins, including: $G_o$, found in the central nervous system and may be linked to calcium channels; transducin, the G-protein of photoreceptor cells associated

**Table 2** Basic Properties of $G_s$ and $G_i$ proteins.

| Properties | $G_s$ | $G_i$ |
|---|---|---|
| Enzymatic activity | GTPase activity; Increase adenylate cyclase (AC) activity | GTPase activity; Decrease adenylate cyclase activity |
| Toxin | Cholera toxin | Pertussis toxin |
| Structure | $\alpha_s$, $\beta$ , $\gamma$ | $\alpha_i$, $\beta$, $\gamma$ |
| Molecular mechanism (See Figure 2(B)) | Dissociation of heterotrimer from GPLR upon addition of ligand→GTP for GDP exchange→complex associates with effector→GTPase activated to terminate | Dissociation of heterotrimer from GPLR upon addition of ligand→GTP for GDP exchange→β γ complexes with $\alpha_s$ to form inactive complex; $\alpha_i$ inhibits AC (?)→GTPase activated to terminate |

Adapted from Reference 24.

with cGMP production, and; $G_p$, which is coupled to phospholipase C production. The basic mechanism of action of G proteins is given in Table 2 as well as Figure 2 (C). Once again, toxins have played critical roles in the dissection of the signaling process. The bacterial toxins of *vibrio cholerae* and *bordatella pertissus* covalently modify the G-protein subunit $\alpha_s$ and $\alpha_i$, respectively. Interestingly, the result of both toxins is an increase in cAMP production in the presence of agonist. GDP and GTP analogs such as decoyinine[27] alter the levels of the guanine nucleotides and hence affect G-protein function. The ras family of G-proteins as well as several $\gamma$ subunits of G-proteins are farnesylated and can be inhibited by farnesyltransferase inhibitors such as *N*-acetyl-*S*-farnesyl-L-cysteine (AFC).[28,29]

#### 14.4.1.3.1.2.3 Effectors

The activated $G_s$-proteins (GTP bound) have the ability to start the signaling process, if they are coupled to an effector molecule. The most common effector is adenylate cyclase (AC) which hydrolyses ATP to form cyclic AMP (cAMP) and inorganic phosphate ($P_i$). Other classifications are listed below, with examples of receptors that utilize the effector.

*GPCRs that modulate adenylate cyclase activity*. This class of receptors activate AC, the enzyme leading to the production of cAMP, as the second messenger. Receptors of this class include the β-adrenergic and glucagon receptors. Increases in the production of cyclic adenosine monophosphate (cAMP) leads to an increase in the activity of PKA. In contrast, the α-type adrenergic receptors are coupled to inhibitory G-proteins that repress adenylate cyclase activity upon receptor activation. Forskolin, a diterpenoid isolated from *Coleus forskohlii*, interacts directly with the catalytic subunit of AC, thereby increasing its

activity and intracellular cAMP concentration.[30]

*GPCRs that activate phospolipase C* $(\gamma)$ *PLC-$\gamma$* . These receptors lead to hydrolysis of phosphoinositides (e.g., $PIP_2$) generating the second messengers diacylglycerol (DAG) and inositoltrisphosphate ($IP_3$). DAG serves as the endogenous activator of several protein kinase C (PKC) isoforms (see Chapter 2.1) or may be cleaved by monoacylglycerol or diacylglycerol lipases, liberating fatty acids, most notably arachidonic acid. This particular polyunsaturated fatty acid is a precursor for a variety of signaling molecules such as prostaglandins and leukotrienes. $IP_3$ is capable of liberating calcium from intracellular stores. This class of receptors includes the angiotensin, bradykinin and vasopressin receptors. The aminoglycoside antibiotic, neomycin sulfate, inhibits PLCα activity by binding to inositol phospholipids, thereby removing the substrate from the enzyme.[30]

*Ion channels coupled to G proteins*. G-proteins have been demonstrated to signal through $K^+$ and $Ca^{2+}$ channels. The best characterized example may be the muscarinic receptor activation of $K^+$ channels and its inhibition by pertussis toxin; however, the precise mechanism by which G-proteins affect ion channels is unknown.

*Photoreceptor GPCRs that activate transducin*. This class is coupled to the G-protein transducin that activates a phosphodiesterase that leads to a decrease in the level of cyclic guanine monophosphate (cGMP). The drop in cGMP then results in the closing of a $Na^+/Ca^{2+}$ channel leading to hyperpolarization of the cell.

#### 14.4.1.3.1.3  Receptor Protein Tyrosine Kinase (RTK)

The receptor tyrosine kinases (RTKs) are an important class of membrane receptors that initiate the signal transduction process through an autophosphorylation event on tyrosine residues (reviewed in References 25, 31, [†]). As shown in Figure 2(D), the proteins encoding RTKs contain four major domains: (1) An extracellular ligand-binding domain; (2) An intracellular tyrosine kinase domain; (3) An intracellular regulatory domain; and (4) A transmembrane domain. The domains may be found on a single peptide (epidermal growth factor receptor, EGFR) or in different subunits (α and β, insulin receptor) linked as a disulfide-linked dimer. In this latter case, the α subunit contains the cysteine-rich and ligand-binding domains whereas the β chain contains the kinase domain.

The amino acid sequences of the tyrosine kinase domains of RTKs are highly conserved with those of PKA within the ATP binding and substrate binding regions and are also similar among the RTKs. Autophosphorylation on tyrosine has been demonstrated for receptors from many tissues and may be the result of an intra- (i.e., same chain) or extra- (i.e., with heterodimer) subunit phosphorylation. Incidentally, tyrosine kinases in general are much less prevalent than kinases of serine or threonine with levels of phosphorylation up to 100-fold less.[24]

RTK proteins are classified into families based upon structural features in their extracellular portions, as well as the presence of a *kinase insert* (insertion of non-kinase domain amino acids into the kinase domain). Distinguishing structural features include cysteine-rich domains, immunoglobulin-like domains, leucine-rich domains, Kringle domains, cadherin domains, fibronectin type III repeats, discoidin I-like domains, acidic domains, and EGF-like domains.[25,31] Based upon the presence of these various extracellular domains the RTKs have been sub-divided into at least 15 different families, the most common of which are:

*EGF receptor, NEU/HER2, HER3, erbB.* Contains cysteine-rich sequences in the extracellular domain. The oncogene v-erbB is a truncated form of EGFR that lacks the extracellular ligand binding domains. Lovendustin A is potent and selective inhibitor of EGFR tyrosine kinase that inhibits EGF-mediated smooth muscle cell proliferation.[32]
*Insulin receptor (IR), IGF-1 receptor.* Contains cysteine-rich sequences in the extracellular domain; characterized by disulfide-linked heterotetramers ($\alpha_2\beta_2$) and presence

of fibronectin type III regions in the α and β chains. The β-chains contain the transmembrane domain. Hydroxy-2-naphthalenyl-methylphosphonic acid (HNMPA) is an inhibitor of IR tyrosine and serine autophosphorylation.[33]
*Platelet derived growth factor (PDGF) receptors (α and β), c-kit.* Contain the kinase insert ($\approx 100$ residues) and 5 immunoglobulin-like domains.
*Fibroblast growth factor (FGF) receptors.* Contain 3 immunoglobulin-like domains without kinase insert.
*Neurotrophin receptor family (trkA, trkB, trkC), nerve growth factor receptor.* Contain cysteine-rich and leucine-rich domains in the extracellular portion.
*Vascular endothelial cell growth factor (VEGF) receptor.* Contain 7 immunoglobulin-like domains as well as the kinase insert domain.
*Hepatocyte growth factor (HGF) receptors, MET, RON.* Heterodimeric like the IR class of receptors except that one of the two protein subunits is completely extracellular. Lacks cysteine rich and kinase insert domains.

Upon phosphorylation the RTKs interact with other proteins of the signaling cascade. These other proteins contain a domain of amino acid sequences that are homologous to a domain first identified in the c-Src proto-oncogene, termed SH2 domains (Src homology domain 2).[25,31] Other conserved protein–protein interaction domains identified in many signal transduction proteins include a third domain in c-Src identified as the SH3 domain as well as the PTB (phosphotyrosine binding) domains of *adaptor proteins*. (Adaptor proteins form links between the receptor and various kinase cascades). The interactions of SH2 domain containing proteins with RTKs or receptor-associated tyrosine kinases leads to tyrosine phosphorylation of the SH2 containing proteins and an alteration in their enzymatic activity. Several SH2 containing proteins that have intrinsic enzymatic activity include phospholipase C-γ (PLC-γ), the proto-oncogene c-Ras associated GTPase activating protein (rasGAP), phosphatidylinositol-3-kinase (PI3K), protein phosphatase-1C (PTP1C), as well as members of the Src family of protein tyrosine kinases (PTKs, described below).[25,31]

### 14.4.1.3.1.4 *Others*

The classes of receptors listed above represent a large proportion of growth factor,

---

†Some of this information was derived from "Signal Transduction", *http://www.med.unibs.it/~marchesi/signal.html* and the "Signal Transduction Knowledge Environment", http://www.stke.org/

neurotransmitter and cytokine receptors. In addition, these classes of proteins are frequently mutated in human cancers or are affected by xenobiotics. However, there are several other receptor systems that are important in biology and toxicology that, for the limitation in space, will not be discussed in detail herein. These include receptors for the transforming growth factor β (TGFβR),[34] tumor necrosis factor-α (TNFα),[35] lipoprotein,[36] endothelin[37] and various chemokines of the immune system.[37] Interested readers are directed to the recent reviews referenced.

Perhaps the TGFβ and TNF receptors should be mentioned, due to their importance in the regulation of apoptosis and cell proliferation. The TGF-β superfamily of cytokines signal *via* plasma membrane serine/threonine kinase receptors and cytoplasmic effectors.[34] The mechanism by which TGFβ R initiates the signaling cascade is somewhat reminiscent of the RTKs in that the receptor spans the membrane one time and it contains an intrinsic kinase activity (see Figure 2(E)). However TβR functions as a heterodimer of two subunits, type I and type II. Binding of TGF-β or activin-A to type I receptors requires the presence of the corresponding type II receptor. Type I and type II receptors form heteromeric complexes after ligand binding. The type II receptor kinases phosphorylate and thereby activate the type I receptor cytoplasmic domains. The Smads (their name is a contraction of the *C. elegans* Sma and *Drosophila* Mad proteins) then act as type I receptor–activated signaling effectors, which regulate transcription of select genes in response to ligand.[34]

The discovery that cachectin, a protein known to cause fever and wasting, was identical to TNF provided evidence of the importance of this cytokine, in particular in the immune system.[35] The TNFR ligands are polypeptides that can have both membrane-embedded "pro" as well as cleaved, soluble "mature" forms. The ligand shape is that of an inverted bell that is embraced on three sides at the base by elongated receptor chains forming a symmetric complex of three receptor molecules and three ligands. Although this appears to be very complicated ligand-binding phenomena, this system may enhance regulatory flexibility and complexity. Within the extracellular domains, disulfide bonds form "cysteine-rich domains" (CRDs) that are the hallmark of the TNFR family (i.e., TNFR-1, FasR) and serve as the ligand-binding site. The cytoplasmic domains of TNFRs are modest in length and function as docking sites for signaling molecules. Signaling occurs through two principal classes of cytoplasmic adaptor proteins: TRAFs (TNF receptor–associated factors) and "death domain" (DD) molecules. For the subset of receptors that have DDs, often called "death receptors" (DR), ligand engagement typically causes the association of adaptors such as Fas-associated DD protein (FADD) and TNFR-associated DD protein (TRADD) that ultimately cause caspase activation and cell death. The receptor cytoplasmic tails form a 3:3 internal complex with signaling proteins such as TRAF2 or FADD.[35]

### 14.4.1.3.2 Soluble Sensors/Receptors

Whereas membrane bound receptors require several adapter proteins, kinases and activator proteins to affect gene expression, soluble receptors can effectively bypass many signaling pathways. Soluble receptors interact with their ligands while they reside in the cytoplasm or nucleus. This is not to say that all soluble receptors exert their effects oblivious to the membrane; in fact soluble xenobiotic receptors such as protein kinase C α (PKCα) require association with the membrane to active its kinase function (see Chapter 2.1). Additionally, signaling events have a profound impact on the activity of the soluble receptors and *visa versa*, an event labeled as *cross-talk* between signaling pathways.

Many soluble sensor/receptors have been described in previous chapters. These include nuclear receptors of the steroid/thyroid hormone receptor superfamily, the arylhydrocarbon receptor (AhR), and metal transcription factor (MTF). Many of these proteins are xenobiotic receptors since direct binding of an exogenous chemical has been described. However, they may also serve a role of as a sensor of internal homeostasis. Other cytosolic proteins such as the PAS domain proteins (i.e., HIF-1) as well as activator protein-1 (AP-1) family members (including NF-E2 related factors 1 and 2, Nrf1 and Nrf2) and NFκB are important molecules in the maintaining and responding to the oxygen tension in the cell; however, their role as a true "receptor" is a matter of debate. These sensor molecules (PAS proteins, Nrf1, Nrf2 and NFkB) are a primary focus of the present series of chapters (see Chapter 4.3–4.5). Their mechanism of action can be summarized much in the same way as nuclear receptors or AhR. A signal is received (i.e., ROS) and causes a conformational change in the soluble receptor. This may result in an alteration in protein–protein interactions, activation of kinase activity or revealing a nuclear uptake sequence. These activated receptors are transcription factors that can regulate a specific battery of genes upon dimerizing with other transcription factors. Following the formation of a protein-DNA complex, co-regulators are recruited and gene transcription proceeds.

### 14.4.1.3.3   Oxygen and Reactive Oxygen Sensors

As mentioned above, maintaining oxygen homeostasis is critical to cell survival. In fact, there are challenges of too little (hypoxia) or too much (hyperoxia) oxygen as well as the generation of damaging reactive oxygen species (ROS). It is important that a few terms be clarified[16] before proceeding with the sensors. *Oxygen* will refer to molecule oxygen ($O_2$), whereas *oxidant* will describe the oxidizing potential of both $O_2$ and ROS. *Redox* describes the balance between oxidizing and reducing potentials within the cell and takes into account the presence of both ROS and anti-oxidants.

### *14.4.1.3.3.1   HIF1*

The Per-Arnt-Sim (PAS) family of proteins is an evolutionary ancient group of genes that appear to be involved in the sensing of the internal environment of the cell, in particular oxygen status.[38] In hypoxic cells, regulation of genes involved in protecting against ischemic injury and angiogenesis are controlled by two basic-helix-loop-helix PAS proteins, hypoxia inducible factors-1$\alpha$ (HIF1$\alpha$) and aryl hydrocarbon nuclear translocator (ARNT, or HIF1$\beta$) (see Chapters 4.4 and 2.2 for discussion of structure and function of these molecules). In the presence of normal oxygen tension (*normoxia*), HIF1$\alpha$ rapidly degrades. However, when oxygen tension falls below a critical threshold, HIF1$\alpha$ becomes stabilized and can interact with ARNT to form the HIF1 complex, ultimately regulating gene expression. How HIF1$\alpha$ senses oxygen status is still elusive.[39] There is evidence to support the involvement of a prolyl hydroxylase (PH) enzyme that hydroxylates HIF1$\alpha$ under normoxic conditions, allowing for interaction with von Hipple Lindau protein (pVHL) and subsequent proteasomal degradation.[39] If PH is not the sensor of oxygen concentrations, certainly another heme-containing protein, such as a cytochrome, is involved.[16] Antioxidants, iron chelators and reducing agents stimulate HIF1 under normoxic conditions, showing that redox status affects the stability of the HIF1$\alpha$.[40]

### *14.4.1.3.3.2   NF-$\kappa$B*

NF-$\kappa$B was first identified as a nuclear factor bound to an enhancer element of the immunoglobulin $\kappa$-light chain gene. This protein is a combination of members of the NF-$\kappa$B/Rel family of proteins and also is regulated by an inhibitory subunit (I$\kappa$B) and an I$\kappa$B-kinase (IKK). The basic structure of subunits NF$\kappa$B/Rel and I$\kappa$B will be discussed below followed by how this protein responds to redox status. Much of the following information came from several new review articles on this fast moving field.[16,22,41,42],‡

### *14.4.1.3.3.2.1   NF$\kappa$B/Rel proteins*

Five mammalian members of the NF-$\kappa$B/Rel family have been identified, NF-$\kappa$B1 (p50 and its precursor p105), NF-$\kappa$B2 (p52 and its precursor p100), c-Rel, RelA (p65), and RelB (see Figure 3(A)). These proteins contain a highly conserved Rel homology region (RHR) composed of two immunoglobulin-like domains that extend over 300 residues. The RHR is responsible for dimerization, DNA binding, and interaction with the inhibitory I$\kappa$B proteins (see Figure 3(B)). The p100 and p105 family members contain ankyrin repeats, and when cleaved result in the formation of active p52 and p50 proteins (known as Class 1 members). A region critical for the regulation of NF$\kappa$B activity, the nuclear localization sequence (NLS), is also contained within the RHR and has been the characterized extensively with p65. The class 2 family members (p65, RelB and c-Rel), contain transactivation domains (TAD) in their carboxyl terminus. NF$\kappa$B dimers are most commonly heterodimers of Class 1 and Class 2 components. Heterodimers of p65 and p50 were the first form of NF-$\kappa$B to be identified and are the most abundant in the majority of cell types; even though the term applies to all dimeric forms of the Rel proteins, this particular complex (p50/p65) is the one most often referred to a "NF$\kappa$B". NF-$\kappa$B complexes are regulated through interactions with another member of the family, I$\kappa$B, as discussed below. Studies using knockout animals have examined the role of each of the members of this family, although surprisingly only p65/RelA is essential for survival, due perhaps to compensation between other NF-$\kappa$B members.

---

‡D. M. Rothwarf and M. Karin, 'The NF-$\kappa$B activation pathway: A paradigm in information transfer from the membrane to nucleus' *Science's STKE*: http://stke.sciencemag.org/cgi/content/full/OC_sigtrans/1999/5/re1

A.

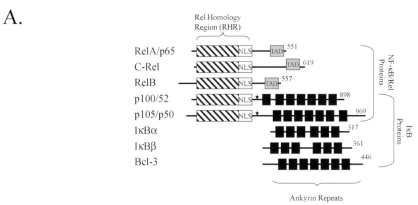

B. C.

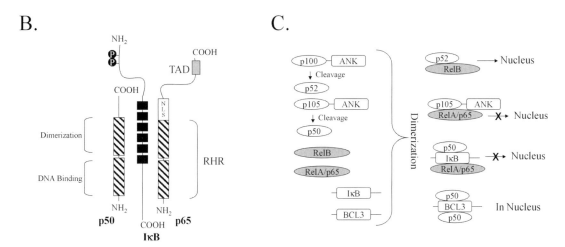

**Figure 3** The Rel family of proteins. (**A**). The basic structure of the Rel family of proteins. Shown are the Rel homology regions (RHR), the nuclear localization sequence (NLS), the transcriptional activation domain (TAD) and the ankyrin repeats. The arrow indicates the site of proteolytic cleavage responsible for the release of the p50 and p52 subunits of NFκB. (**B**). Arrangement of subunits in the p50/RelA-p65/IκB complex, including location of the phosphorylation sites on IB that cause dissociation and degradation. (C) Different combinations of Rel family members and ability to translocate to the nucleus and regulate gene expression.

*14.4.1.3.3.2.2 IκB proteins*

The inhibitory IκB family includes IκBα, IκBα, Bcl-3 as well as the unprocessed p105 and p100. All IκBs contain either six or seven ankyrin repeats, and these stacked helical domains bind to the RHR and mask the NLS of NF-κB. IκBα and IκBβ contain an NH$_2$-terminal regulatory region, in particular serines 32 and 36 and lysines 21 and 22 (IκKα), that is the site for stimulus dependent degradation, a key step in NFκB activation. The binding of IκB to NFκB in the cytosol masks the NLS and blocks nuclear uptake. In addition, IκB can enter the nucleus and bind to NF-κB, thereby enhancing its dissociation from the DNA and causing its export to the cytoplasm by means of a nuclear export sequence (NES) present on IκBα .

*14.4.1.3.3.2.3 Direct oxygen sensing by NFκB*

NFκB is activated by many distinct stimuli, including proinflammatory cytokines such as TNF-α and interleukin 1 (IL-1), T and B cell mitogens, bacteria and bacterial lipopolysaccharide (LPS) as well as physical and chemical stresses. Many of these stimuli may activate various kinase cascades, in particular the IκB kinase (IKK) and SAPK/JNK cascades, and will be discussed in a subsequent section. However, there is some, albeit controversial evidence for direct redox regulation of the NFκB transcription factor. For example, cells lacking functional mitochondrial electron transport show significant repression of NFκB activation from TNFα or IL1.[22] Also, antioxidants such as Vitamin E and overexpression of Cu/Zn superoxide dismutase can block the activation of this transcription factor from most stimuli; however, this is only observed in certain cell types. Exposure of cells to hyperoxia results in nuclear translocation of NFκB. The OxyR and SoxR genes in *Escherichia coli* contain critical cysteine residues that change protein conformation and activity in response to redox status.[16] Whether a similar situation exists for NFκB is not clear but oxidation of Cys-62 in the p50 subunit affects DNA binding.[22]

### 14.4.1.3.3.3    Activator Protein-1 (AP-1) Family

The activation protein-1 (AP-1) transcription factor consists of either Jun homodimers or Fos/Jun heterodimeric complexes.[22] As discussed in Chapter 1.4, fos and jun are basic domain/leucine zipper (bZIP) proteins that bind to tetradecanoyl-phorbol-13-acetate (TPA) response elements (TRE, the palindromic sequence TGA(C/G)TCA). The Jun family includes c-Jun, Jun B and Jun D while the Fos gene family members include c-Fos, Fos B, Fra-1 and Fra-2. In addition, activating transcription factor-1 (ATF) and cAMP-response element binding protein (CREB) can form leucine zipper dimers with Fos and Jun, which exhibit different specificities for AP-1 and CREB-response element (CRE) DNA motifs. NF-E2 related factors 1 and 2 (Nrf1 and Nrf2) are also bZIP proteins that do not heterodimerize with each other, but require bZIP proteins, such as c-Jun or Maf, to bind to the anti-oxidant response element (ARE, discussed in Chapter 4.3).

In many ways, the regulation of AP-1 by redox is a similar story as that of NFκB. That is, it has been known for some time that ROS stimulate the activity of this transcription factor. However, whether this is due to a direct effect of ROS on AP-1 or due to upstream kinase signaling is a matter of debate. Also, both AP-1 and NFκB are integrators of multiple signals in the cell and are a ultimate target of the growth factor signaling pathways. Therefore, it is likely that multiple means of regulation are at play in response to oxidant insult. None-the-less, the case for direct oxidation of AP-1 is more convincing than that of NFκB.[22] Exposure of cells to hydrogen peroxide results in enhanced AP-1 DNA binding; this effect is independent of new protein synthesis, indicating a post-translational modification. Overexpression of thioredoxin in

transient transfection experiments increased AP-1 DNA binding, indicating that cysteine residues are important in function of this protein. Substitution of key cysteine residues in jun (Cys-272) or fos (Cys-154) lead to a loss of redox sensitive regulation. A model has emerged where oxidants modify cysteine residues in fos or jun thereby decreasing DNA binding and transcriptional regulation. Re-activation of the oxidized AP-1 complex can be performed by thioredoxin and redox-factor 1 (REF-1). REF-1 is an endonuclease that may also be involved in DNA damage repair, but is capable of promoting the cycling between the reduced form and the oxidation product of AP-1 as well.[22]

A hypothetical model illustrating the activation of detoxifying enzyme genes by antioxidants and xenobiotics through the Nrf pathway is just beginning to be unraveled. The ROS signal presumably passes through an unknown cytosolic factor. This factor(s) then catalyzes the modification of Nrf2 or its cytosolic partner, inhibitory Nrf2 (INrf2). As a result of the modification of the Nrf2-INrf2 complex, Nrf2 is released and translocates to the nucleus where it heterodimerizes with c-Jun and induces the expression of ARE-regulated genes (Chapter 4.3).

### 14.4.1.4    SIGNAL TRANSDUCTION PATHWAYS

Dissecting the pathways that bridge the signals (growth factors, ROS) to the biological event (cell proliferation, apoptosis) is an extremely arduous task, and describing in detail what is known about the signal transduction cascades is far beyond the scope of this text. The goal of the present section is to emphasize three major points: (1) The signals and sensors can utilize a variety of different signal transduction cascades to regulate cell fate. Which pathway is used depends to a large extent on cell-, development-, sex- and disease state-specific factors; (2) The signal transduction cascades are integrated with one another. Although isolating one pathway for detailed examination is a necessary evil, one must be cognizant of the fact that in the *in vivo* situation, pathways such as the mitogen-activated protein kinase (MAPK), SAPK and PKC are inter-dependent. The "spider-web" analogy applies to these interwoven pathways. Tugging on one strand (i.e., perturbing one cascade) of the spider web (cell) reverberates throughout, and can be sensed by the spider (kinase) sitting on a distant tendril (i.e., another cascade). (3) Although many diverging pathways are activated by a given signal/sensor, often relatively small subsets of activator proteins are ultimately activated. This diverging/converging paradigm allows for a high degree of fine-tuning of the signaling process. For example, phosphorylation of c-jun on different sites by different kinases may result in a protein with distinctive properties when compared to its singly- or un-phosphorylated form.[43] It also explains why mutations or xenobiotic alterations in these convergent activator proteins (such as oncogenes and tumor suppressor genes) are often involved in disease. These three underlying points are illustrated in Figure 4. Representative signal transduction cascades for EGF (Figure (A) and (B)), insulin (C) and TNFα (D) are depicted and include components of each of the pathways to be discussed. Chemical inhibitors and activators of particular pathways that are useful in dissecting the signal transduction process are listed in Table 3.

### 14.4.1.4.1 Growth Factor Signaling

#### 14.4.1.4.1.1 Role of Adapters and Non-receptor Tyrosine Kinases

Following activation of cell surface receptors, further protein–protein interactions are required to transmit this message. In the case of GPCRs, *coupling* is performed by the G-proteins, which are then able to transmit signals to various effectors. The connection between RTKs and signaling events is more complicated and involves proteins of two basic types, kinase-inactive (*adaptor* proteins) or active (non-receptor protein tyrosine kinase, PTKs). RTKs can activate the small guanine nucleotide binding protein Ras (a common early target of cell surface receptors) through the use of nucleotide exchange factors such as Son of Sevenless (SOS).[‡] Ras is a membrane-associated protein that exists in either a GTP-bound active state or GDP-bound inactive state. SOS exhanges GDP for GTP while GTPase activating proteins (GAPs) inactivate Ras. Additional adapter proteins have been

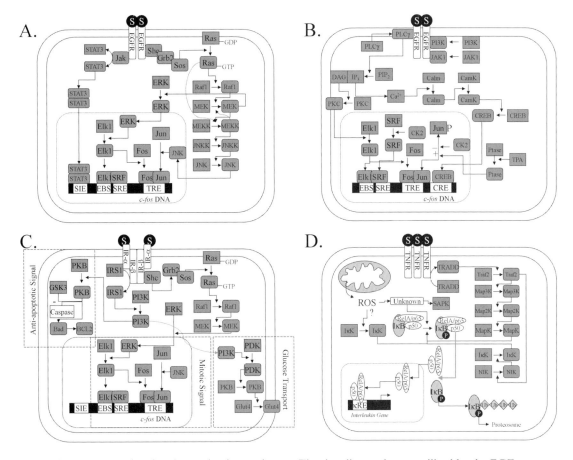

**Figure 4** Representative signal transduction pathways. The signaling pathways utilized by the EGF-receptor (EGFR, panels (A) and (B)), insulin receptor (IR, panel (C)) and the TNFα-receptor (TNFR, panel (D)). See text for details and abbreviations.

**Table 3** Xenobiotics that affect cellular signaling and are useful tools in dissecting signaling pathways.

| Pathway | Chemical | Target (if known) and effect |
|---|---|---|
| Src-1 | Herbimycin A | Inhibits tyrosine kinases by reacting with thiol groups.[44] |
| | | Inhibits src and a variety of other oncogenes. |
| SAPK | Anisomycin | Protein synthesis inhibitor and inducer of SAPK but not ERK[45] |
| MAPK | PD98059 | MEK inhibitor[46] |
| P38 | SB203580 | P38 MAPK inhibitor[47] |
| PI3K | Wortmannin | Covalent PI3K inhibitor[48] |
| | Ly2940002 | PI3K inhibitor[49] |
| Calcium/Calmodulin | KN93 | CamKII inhibitor[50] |
| PKA | H-89 | PKA inhibitor[51] |

identified in this pathway, the most common being growth factor receptor-binding protein 2 (Grb2) and Shc that may couple the receptor to SOS. The Rho family of GTPases (Cdc43 and Rac) have a similar mechanism of action, but may signal through different growth factor dependent pathways than Ras.[52] Cytokine receptors such as IL-1R and TNFR, utilize TNFR-associated factors (TRAF) as adaptor proteins. For example, IL-1R recruits TRAF6, which is then able to recruit a member of the JNK pathway MEKK1.[52]

Another example of receptor-signaling through protein interaction involves the insulin receptor (IR) (see Figure 4(C)).[53] This receptor has intrinsic tyrosine kinase activity but does not directly interact, following ligand-dependent autophosphorylation, with enzymatically active proteins containing SH2 domains (e.g., PI-3K or PLC-γ). Instead, the principal IR substrate is a protein termed IRS-1. IRS-1 contains several motifs that resemble SH2 binding consensus sites for the catalytically active subunit of PI-3K. These domains allow complexes to form between IRS-1 and PI-3K. This model suggests that IRS-1 acts as a docking or adapter protein to couple the IR to SH2 containing signaling proteins.

There are numerous intracellular PTKs that are responsible for phosphorylating a variety of intracellular proteins on tyrosine residues following activation of cellular growth and proliferation signals.[2,25] There are two distinct families of non-receptor PTKs. The archetypal PTK family is related to the Src protein. The v-Src protein is a tyrosine kinase first identified as the transforming protein in Rous sarcoma virus. Subsequently, a cellular homolog was identifed as c-Src. This protein contains SH2 and SH3 domains that are used to associate with RTKs. Activated receptors may recruit

Src, resulting in a conformational change and activation of the kinase.[2]

The second family of PTKs is related to the Janus kinase (Jak) and is utilized by cellular receptors that lack enzymatic activity themselves.[54] This class of receptors includes all of the cytokine receptors, growth hormone, prolactin as well as the CD4 and CD8 cell surface glycoproteins of T cells and the T cell antigen receptor (TCR). Jaks are directly linked to the signal transducers of transcription (STAT) signaling cascade and this system is often called the Jak/STAT pathway. Jaks are constitutively associated with the cytokine receptors. The binding of ligand to the receptor results in activation of Jaks, which in turn phosphorylates the receptor on tyrosine residues. The phosphorylated receptor serves as a docking site for the STAT SH2 domain; The receptor associated STAT is then phosphorylated by Jak.[54] The STATs form homo or heterodimers and regulate transcription by binding to IFN-stimulated response elements (ISRE), γ-IFN activated sequences (GAS).

### 14.4.1.4.1.2 Mitogen Activated Protein Kinase Cascade: MAPK/SAPK

MAP kinases were identified by virtue of their activation in response to growth factor stimulation of cells in culture, hence the name mitogen activated protein kinases. MAP kinases are also called extracellular-signal regulated kinases (ERKs). Although MAP kinase activation was first observed in response to activation of the EGF, PDGF, NGF and insulin receptors, other cellular stimuli include phorbol esters (that function through activation of PKC), thrombin, bombesin and bradykinin (that function through G-proteins) as well as *N*-methyl-D-aspartate (NMDA) receptor activation and electrical stimulation.[2,55] MAP kinases are, however, not the direct

---

‡T. Jun, O. Gjoerup and T. M. Roberts, 'Tangled webs: Evidence of cross-talk between c-raf-1 and akt'. *Science's STKE*: http://stke.sciencemag.org/cgi/content/full/OC_sigtrans;1999/13/pe1

substrates for RTKs nor receptor associated tyrosine kinases but are in fact activated by an additional class of kinases termed MAP kinase kinases (MAPK kinases, MKK) and MAPK kinase kinases (MAPKK kinases, MKKK).[2] One of the MAPKKKs has been identified as the proto-oncogenic serine/threonine kinase, Raf. The three major, distinct groups of MAPKs include: The ERK/MAPK pathway (raf→MEK→MAPK/ERK); the Jun amino-terminal kinases (JNK, or stress activated protein kinase SAPK; MEKK1→SEK1/2→SAPK/Jnk), and the p38 proteins (MLK1→MKK3/6→p38). An over-simplified diagram of the MAPK pathways is shown in Figure 5 and will be discussed in more detail in Chapter 4.2. Ultimate targets of the MAP kinases are several transcriptional regulators e.g., serum response factor (SRF), and the proto-oncogenes Fos, Myc and Jun as well as members of the steroid/thyroid hormone receptor super family of proteins.[55]

### 14.4.1.4.1.3  *Phospholipase/PKC Signaling*

Phospholipases, in particular PLC-γ, are important effector enzymes for membrane bound receptors.[25] PLC-γ contains SH2 domains that function as interacting domains for tyrosine phosphorylated RTKs. This allows PLC-γ to be intimately associated with the signal transduction complexes of the membrane as well as membrane phospholipids that are its substrates. Activation of PLC-γ leads to the hydrolysis of membrane phosphatidylinositol bisphosphate (PIP$_2$) resulting in an increase in intracellular diacylglycerol (DAG) and inositol trisphosphate (IP$_3$) as second messengers. The released IP$_3$ interacts with intracellular membrane receptors leading to an increased release of stored calcium ions (discussed below). The primary signaling molecule affected by the activation of phospholipases is protein kinase C (PKC), which is maximally active in the presence of calcium ion and DAG

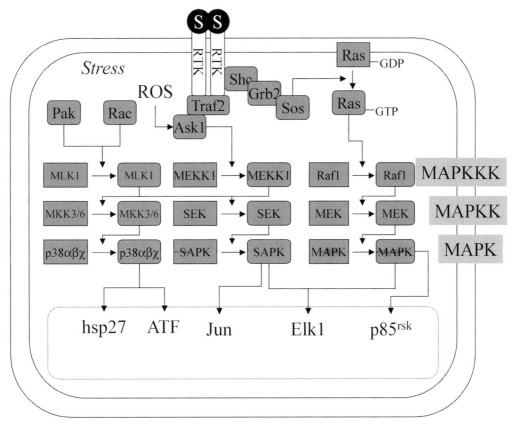

**Figure 5**  The mitogen-activated protein kinase (MAPK) cascade. RTK, and other cell surface receptors and stress (including reactive oxygen species, ROS) can signal through the MAPK cascade. The three major branches of the MAPK pathways are shown, the MAPK, stress-activated protein kinase (SAPK) and the p38 cascades. Each cascade has MAPK-kinase kinase (MAPKKK), MAPK-kinase (MAPKK) and MAPK components. The result of activation of these pathways is the regulation of transcription factors and other proteins that can affect cell fate including heat shock protein 27 (hsp27), activating transcription factor (ATF), the oncogene c-jun, ets family member ELK-1, and p85$^{rsk}$.

(see Chapter 2.1). Although the principal mediators of PKC activity are receptors coupled to activation of PLCγ, phospholipases D and A2 (PLD and PLA2) may also be involved in the sustained activation of PKC through their hydrolysis of membrane phosphatidylcholine (PC).[56] PLD action on PC leads to the release of phosphatidic acid that in turn is converted to DAG by a specific phosphatidic acid phosphomonoesterase. PLA2 hydrolyzes PC to yield free fatty acids and lysoPC both of which have been shown to potentiate the DAG mediated activation of PKC.[56] Of medical significance is the ability of phorbol ester tumor promoters to activate PKC directly. The targets for PKC have been described elsewhere in this volume and include components of the AP-1 transcription factor as well as certain nuclear receptors.

### 14.4.1.4.1.4 *Phosphoinositol Signaling: Phosphatidylinositol-3-Kinase (PI3K)*

PI-3K is activated by various RTKs and receptor-associated PTKs (reviewed in Reference 57). In particular, PI3K associates with and is activated by the PDGF, EGF, insulin, IGF-1, HGF and NGF receptors. This enzyme is a heterodimeric protein containing 85 kDa and 110 kDa subunits. The p85 subunit contains SH2 domains that interact with activated receptors or other receptor-associated PTKs and is itself subsequently tyrosine phosphorylated and activated. The 85 kDa subunit is non-catalytic; however, it does contain a domain homologous to GTPase activating (GAP) proteins. It is the 110 kDa subunit that is enzymatically active as both a lipid and serine/threonine kinase.

PI-3K phosphorylates various phosphatidylinositols at the 3 position of the inositol ring.[57] This activity generates additional substrates for PLC-γ allowing an amplification of the initial signal to include other pathways (i.e., PKC). In addition to PKC, another lipid activated kinase, phosphoinositide-dependent kinase (PDK), is activated by PI3K presumably as a result of its lipid kinase activity. PDK is a serine/threonine kinase that activates and phosphorylates Protein Kinase B (PKB, or Akt) as well as the S6-kinase. PKB in turn has multiple substrates that are important in cell proliferation and apoptosis including BAD, caspases, and glycogen synthase kinase 3 (GSK3). PDK and PKB play an important role in the control of protein synthesis, gluconeogenesis and glycolysis in response to insulin stimulation (see Figure 4(C)).[25,58]

The protein kinase activity of PI3K may serve a role in the regulation of the MAPK pathway. First, PI3K associates with p21-ras and these two proteins appear to enhance each other's activity. Second, the activity of Erk1 was increased in the present of a PI3K mutant that lacked lipid kinase, but retained its protein kinase activity (reviewed in Reference 25). The protein substrates of PI3K have not been conclusively elucidated, as of yet, but there have been several reports of insulin receptor substrate (IRS-1) as being phosphorylated by PI3K.[59]

### 14.4.1.4.1.5 *Calcium Signaling: Calmodulin and CamK*

Calcium is an important second messenger utilized by ion channel receptors, GPLR as well as RTKs to regulate cell cycle, gene expression and apoptosis (reviewed in Reference 19). As stated above, calcium levels can vary dramatically as a result of extracellular and intracellular release. One of the key signaling molecules that responds to intracellular calcium is calmodulin (CaM). CaM is a small protein ($\approx 17$ kD) that contains 4 HLH motifs called "EF hands", each of which is able to bind one molecule of $Ca^{2+}$. The binding occurs cooperatively, the result of which is a conformational change that exposes a hydrophobic pocket the enables CaM to recognize the binding domain of target proteins.

Perhaps the most important action of CaM is to activate a family of serine/threonine kinases, the $Ca^{2+}$/CaM-dependent kinases (CaMK).[19] CamK contain a N-terminal kinase domain, an auto-inhibitory domain, a CaM binding site and a dimerization domain. To activate CaMK, $Ca^{2+}$/CaM associates with its binding site and causes the auto-inhibitory domain to vacate the catalytic site. There are phosphorylation sites for other kinases in the activation loop of CaMKs that suggest other means of regulation in addition to CaM (i.e., CaMKK).

There are three CaMK (I, II, and IV) with different tissue distribution, cellular localization and consensus phosphorylation sites.[19] One substrate they all have the ability to phosphorylate is CREB. CREB is an important transcription factor, that when activated interacts with CRE response elements in target genes. CREB is a basic/leucine zipper protein (bZIP) and is also a substrate for cAMP-dependent kinase (PKA). Genes regulated in this fashion ($Ca^{2+} \rightarrow CaM \rightarrow CaMK \rightarrow CREB$) include c-fos and IL-2. However both CamK and CREB participate in numerous other signaling cascades. For example, CREB heterodimerizes with another bZIP protein, ATF-1, and regulates gene expression from CREs.[19] CCAT-Enhancer binding protein (C/EBP) and serum response factor (SRF) are substrates for CaMK and their activity is modulated by calcium levels. Likewise, nuclear receptors including ROR and COUP-TF are affected by $Ca^{2+}$ and CamK, but it is unclear whether this effect is at the level of the receptor or a co-regulator such as CREB-binding protein (CBP).[19]

### 14.4.1.4.1.6 Cyclic Adenosine Monophosphate (cAMP) Signaling

The GPLR effector enzyme adenylase cyclase produces cAMP as a second messenger. Similar to that of calcium signaling, one of the ultimate targets of cAMP is activation of CREB and regulation of gene expression from CRE (reviewed in Reference 19). However, the kinases used to activate CREB are different, with Protein Kinase A performing this function in the cAMP pathway. Cyclic AMP binds directly to PKA causing a conformational change and release of the active subunit. CREB contains a consensus PKA phosphorylation site (RRP*S*Y, Serine 133), as do other family members (ATF-1 and CREM).[19] The Serine 133 phosphorylated form of the CREB has enhanced transactivation potential, with little effect on DNA binding or heterodimerization with other bZIP members. *In vitro* binding experiments have shown that CBP binds specifically to Ser-133 phosphorylated CREB.

### 14.4.1.4.2 Oxygen Signaling: NFκB as an Example

The signaling of soluble sensors/receptors is fairly straightforward and has been discussed in detail elsewhere in this volume for notable members of this group. In the present section, the focus will be on the oxidant sensor, NFκB. Although there is no full understanding as to how various extracellular and intracellular stimuli trigger NF-κB activation, there are several common features for all of the major activation pathways studied (see Figure 4(D)[16,22,41,42,60,§]). Potent activators, such as TNF-α and IL-1, induce the rapid degradation of the IκBs in a series of well-characterized steps. Inducible phosphorylation of IκB occurs at Ser-32 and Ser-36 (the kinase(s) responsible will be discussed shortly). Phosphorylation leads to the recognition by a component of an Skp1-Cullin-F-box (SCF)-type E3 ubiquitin-protein ligase complex.[6] This complex consequently poly-ubiquitinates IκBα at lysines 21 and 22, which targets IκB for rapid degradation by the 26S proteasome. The dissociation and subsequent degradation of Iκ B results in an unmasking of the NLS of NF-κ B, and translocation of this transcription factor to the nucleus.

Once in the nucleus, the extent of the response from the response element (GGGRNNYYCC where R is a purine, Y pyrimidine)[6] within a target gene may depend on the constituents of the NFκB complex. Only the p65/RelA and c-Rel function as potent transcription activators. Dimers of p50, which lack the TAD, bind to DNA but may mediate repression. RelA recruits histone acetyltransferase activity to the promoter through interactions with p300 and CREB-binding protein. The NFκB-battery of genes include many pro-inflammatory cytokines such as TNFα, IL1β, and IL-8.

Obviously, based on the summary of NFκB signaling described above, a key regulatory step in this process is the phosphorylation of IκB. A cytokine-responsive protein kinase specific for the $NH_2$-terminal regulatory serines of IκB (α and β) has been identified (IκB kinase, IKK).[60] Initially, IKK activity eluted at an apparent molecular size of 700 to 900 kD, suggesting a very large, multicomponent protein complex.[6] By means of protein purification, microsequencing, and molecular cloning, three components of IKK have been identified, IKKα, β and γ. IKKα and IKKβ are both approximately 85-kD proteins, and contain $NH_2$-terminal protein kinase domains and leucine zipper (LZ) and helix-loop-helix (HLH) motifs. These proteins serve as the catalytic subunits of the IKK complex. IKKγ is a 48-kD protein that serves as a regulatory subunit.[60] Other proteins that may associate with this complex include NF-κB-inducing kinase (NIK) and MAPK/ERK kinase kinase-1 (MEKK1).[6]

Finding the adaptor or PTK proteins that link growth factors to IKK has been somewhat elusive. However, a number of different protein kinases activate IKK when overexpressed, including PKCζ and the MAPK kinase kinase (MAPKKK) family members NIK, MEKK1, MEKK2 and TAK1 (reviewed in Reference 6). The ROS sensor in this pathway is also unclear, although it has been suggested that a

---

§D. M. Rothwarf and M. Karin, 'The NF-κB activation pathway: A paradigm in information transfer from the membrane to nucleus' *Science's STKE*: *http://stke.sciencemag.org/cgi/content/full/OC_sigtrans;1999/5/re1*

ROS-induced S-thiolation of critical residues in IKK could activate this enzyme.[22] If this does indeed prove to be the case, IKK could become recognized as the ROS sensor in the cell, as well as a signaling molecule.

#### 14.4.1.4.3 Apoptosis Signaling

The topic of xenobiotic-induced apoptosis will be covered in much detail in Chapter 4.7; however, the signaling cascade is connected to, yet distinct from, that of growth factors and a limited discussion is warranted. According to Strasser *et al.*,[61,62] there are four major components of apoptosis signaling, the *TNFR* and death domain receptors, the *caspases*, *adaptors* that activate the initiator caspase and the *BCL-2* family of proteins (see Figure 6). There are also three distinct types of apoptosis signaling, "death by design", "death by neglect" and "death by stress". Each pathway ultimately results in the activation of an effector caspase (usually caspase 3) but the adaptors and the involvement of BCL-2 family members are different.[52,61,63,64]

In the death by design signaling cascade, TNFR (or Fas/APO-1, CD95), the adaptors Fas-associated death domain protein/mediator of receptor-induced toxicity (FADD/MORT), TNF-R associated death domain protein (TRADD) or receptor interacting protein (RIP) may be utilized. After activation of the death domain (DD)-containing receptor, the adaptors allow caspase aggregation and activation. Certain caspases and adaptors contain domains that allow for aggregation called DEDs. In addition, TRADD and TRAF2 signal through the JNK and NFκB pathway, which was described above, and may influence apoptotic signaling. Caspase-8 and its adaptor FADD are needed for TNFR-induced apoptosis, but they are dispensable for the other two forms of cell death.

Bcl-2 and its homologs are potent modulators of cell death caused by "death by neglect" and "death by stress". Both types of stress require apoptotic protease activating factor 1 (Apaf-1) and caspase-9 to initiate apoptosis. In chemically or stress mediated apoptosis, the release of cytochrome c is observed, which then binds to Apaf-1 and recruits the initiator

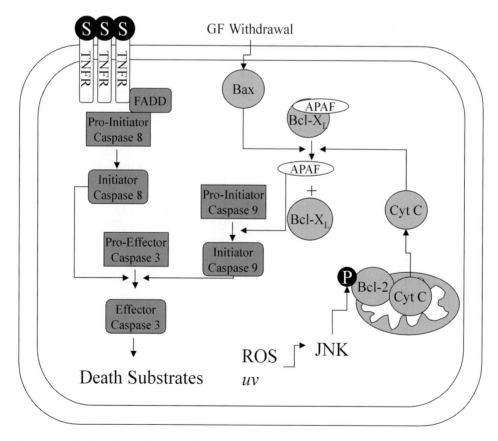

**Figure 6** Apoptosis signaling pathways. Shown are the three major pathways that lead to activation of caspases and death substrates, "Death by Design" (TNFR), "Death by Neglect" (growth factor (GF) withdrawal), and "Death by Stress" (reactive oxygen species, ROS, or ultraviolet light (UV)).

caspase-9. The mechanism of cytochrome c release from the mitochondria is still unclear. In general, it is believed that Bcl-2 and Bcl-X$_L$ prevent, while Bax and Bid promote the release of cytochrome c from the mitochondria.

#### 14.4.1.4.4 Growth Factor Signaling and Gap Junctions

Gap junctions permit the passage of small molecules between cells and have been implicated in many biological processes including cellular growth control (see Chapter 4.5). This important channel is composed of proteins from the connexin gene family. One means by which gap junction intracellular communication (GJIC) is controlled is at the post-translational level. In fact the majority of connexins are phosphoproteins (predominantly phospho-serine) and GJIC can be affected by xenobiotics that alter the activity of kinases/phosphatases.[65] For example, tumor promoters such as phorbol esters inhibit GJIC in a PKC-dependent manner. In addition to PKC, connexins are substrates for other kinase cascades including MAPK and v-Src. Similarly, protein phosphatases such as PP1 and PP2A can regulate connexin activity. The result of phosphate modification of connexins is changes in "gating" but assembly and degradation may also be affected. Based on the importance of GJIC in tumor promotion, connexins may be important non-transcription factor targets of the growth factor and kinase cascades.[65]

### 14.4.1.5 CONCLUSIONS

The signal transduction cascade is a complicated web of intertwined biochemical pathways. Signaling from cell surface receptors or from sensors of the internal cellular environment, are key determinants of the fate of the cell. Important decisions need to be made as a result of the myriad of signaling information being received, including whether to divide, actively die, differentiate, or remain quiescent. Dysregulation of several signal transduction pathways has been associated with tumorigenesis, and hence understanding how a chemical can perturb signaling is an important endeavor. In the present chapter the basic framework of the signaling transduction cascade was put forth. Emphasis was placed on the paradigm of signal→sensor→signal transduction→cell fate, and the major players in each category were discussed. Throughout the remaining chapters in Section 4, specific examples of how xenobiotics can affect cell fate will be described.

### 14.4.1.6 REFERENCES

1. D. J. Waxman, 'P450 gene induction by structurally diverse xenochemicals: central role of nuclear receptors CAR, PXR, and PPAR.' *Arch. Biochem. Biophys.*, 1999, **369**, 11–23.
2. T. Hunter, 'Signaling—2000 and beyond.' *Cell*, 2000, **100**, 113–127.
3. E. G. Krebs and J. A. Beavo, 'Phosphorylation-dephosphorylation of enzymes.' *Annu. Rev. Biochem.*, 1979, **48**, 923–959.
4. M. Rodbell, 'Nobel Lecture. Signal transduction: evolution of an idea.' *Biosci. Rep.*, 1995, **15**, 117–133.
5. T. N. Raju, 'The Nobel chronicles. 1998: Robert Francis Furchgott (b 1911), Louis J. Ignarro (b 1941), and Ferid Murad (b 1936).' *Lancet*, 2000, **356**, 346.
6. P. De Camilli and T. J. Carew, 'Nobel celebrates the neurosciences. Modulatory signaling in the brain.' *Cell*, 2000, **103**, 829–833.
7. D. Kalderon and G. M. Rubin, 'Isolation and characterization of *Drosophila* cAMP-dependent protein kinase genes.' *Genes Dev.*, 1988, **2**, 1539–1556.
8. T. Hunter, 'Oncoprotein networks.' *Cell*, 1997, **88**, 333–346.
9. R. V. House, 'Theory and practice of cytokine assessment in immunotoxicology.' *Methods*, 1999, **19**, 17–27.
10. B. L. Blaylock, S. D. Holladay, C. E. Comment *et al.*, 'Exposure to tetrachlorodibenzo-p-dioxin (TCDD) alters fetal thymocyte maturation.' *Toxicol. Appl. Pharmacol.*, 1992, **112**, 207–213.
11. C. Esser, 'Dioxins and the immune system: mechanisms of interference. A meeting report.' *Int. Arch. Allergy Immunol.*, 1994, **104**, 126–130.
12. P. Tontonoz, L. Nagy, J. G. Alvarez *et al.*, 'PPARγ promotes monocyte/macrophage differentiation and uptake of oxidized LDL.' *Cell*, 1998, **93**, 241–252.
13. L. Nagy, P. Tontonoz, J. G. A. Alvarez *et al.*, 'Oxidized LDL regulates macrophage gene expression through ligand activation of PPARγ .' *Cell*, 1998, **93**, 229–240.
14. T. R. Sutter, K. Guzman, K. M. Dold *et al.*, 'Targets for dioxin: genes for plasminogen activator inhibitor-2 and interleukin-1 β.' *Science*, 1991, **254**, 415–418.
15. J. H. Yang, C. Vogel and J. Abel, 'A malignant transformation of human cells by 2,3,7,8-tetrachlorodibenzo-p-dioxin exhibits altered expressions of growth regulatory factors.' *Carcinogenesis*, 1999, **20**, 13–18.
16. C. T. D'Angio and J. N. Finkelstein, 'Oxygen regulation of gene expression: a study in opposites.' *Mol. Genet. Metab.*, 2000, **71**, 371–380.
17. M. Cuendet and J. M. Pezzuto, 'The role of cyclooxygenase and lipoxygenase in cancer chemoprevention.' *Drug Metabol. Drug Interact.*, 2000, **17**, 109–157.
18. H. J. Drenth, C. A. Bouwman, W. Seinen *et al.*, 'Effects of some persistent halogenated environmental contaminants on aromatase (CYP19) activity in the human choriocarcinoma cell line JEG-3.' *Toxicol. Appl. Pharmacol.*, 1998, **148**, 50–55.
19. S. S. Hook and A. R. Means, 'Ca2+/cam-dependent kinases: from activation to function.' *Annu. Rev. Pharmacol. Toxicol.*, 2001, **41**, 471–505.
20. G. E. Kass and S. Orrenius, 'Calcium signaling and cytotoxicity.' *Environ. Health Perspect.*, 1999, **107**(Suppl 1), 25–35.
21. H. C. Lee, 'Potentiation of calcium- and caffeine-induced calcium release by cyclic ADP-ribose.' *J. Biol. Chem.*, 1993, **268**, 293–299.
22. T. P. Dalton, H. G. Shertzer and A. Puga, 'Regulation of gene expression by reactive oxygen.' *Annu. Rev. Pharmacol. Toxicol.*, 1999, **39**, 67–101.

23. K. B. Wallace and A. A. Starkov, 'Mitochondrial targets of drug toxicity.' *Annu. Rev. Pharmacol. Toxicol.*, 2000, **40**, 353–388.

24. P. Taylor and P. A. Insel, in 'Molecular Basis of Drug Action,' 3rd edn, eds. W. B. Pratt, P. Taylor and A. Goldstein, Churchill Livingstone, New York, 1990, pp. 103–200.

25. J. Schlessinger, 'Cell signaling by receptor tyrosine kinases.' *Cell*, 2000, **103**, 211–225.

26. D. J. Ecobichon, in 'Toxic Effects of Pesticides,' 3rd edn, eds. L. J. Casarett, J. Doull, C. D. Klaassen and M. O. Amdur, Macmillan, New York, 1986, pp. 643–690.

27. R. M. Burch and J. Axelrod, 'Dissociation of bradykinin-induced prostaglandin formation from phosphatidylinositol turnover in Swiss 3T3 fibroblasts: evidence for G protein regulation of phospholipase A2.' *Proc. Nat. Acad. Sci. USA*, 1987, **84**, 6374–6378.

28. C. Volker, P. Lane, C. Kwee et al., 'A single activity carboxyl methylates both farnesyl and geranylgeranyl cysteine residues.' *FEBS Lett.*, 1991, **295**, 189–194.

29. C. Volker, R. A. Miller, W. R. McCleary et al., 'Effects of farnesylcysteine analogs on protein carboxyl methylation and signal transduction.' *J. Biol. Chem.*, 1991, **266**, 21515–21522.

30. A. Laurenza, E. M. Sutkowski and K. B. Seamon, 'Forskolin: a specific stimulator of adenylyl cyclase or a diterpene with multiple sites of action?' *Trends Pharmacol. Sci.*, 1989, **10**, 442–447.

31. S. R. Hubbard and J. H. Till, 'Protein tyrosine kinase structure and function.' *Annu. Rev. Biochem.*, 2000, **69**, 373–398.

32. J. L. Kornyei, X. Li, Z. M. Lei et al., 'Analysis of epidermal growth factor action in human myometrial smooth muscle cells.' *J. Endocrinol.*, 1995, **146**, 261–270.

33. R. Saperstein, P. P. Vicario, H. V. Strout et al., 'Design of a selective insulin receptor tyrosine kinase inhibitor and its effect on glucose uptake and metabolism in intact cells.' *Biochemistry*, 1989, **28**, 5694–5701.

34. J. Massague, 'TGF-β signal transduction.' *Annu. Rev. Biochem.*, 1998, **67**, 753–791.

35. R. M. Locksley, N. Killeen, M. J. Lenardo, 'The TNF and TNF receptor superfamilies: integrating mammalian biology.' *Cell*, 2001, **104**, 487–501.

36. M. M. Hussain, 'Structural, biochemical and signaling properties of the low-density lipoprotein receptor gene family.' *Front. Biosci.*, 2001, **6**, D417–428.

37. D. Giannessi, S. Del Ry and R. L. Vitale, 'The role of endothelins and their receptors in heart failure.' *Pharmacol. Res.*, 2001, **43**, 111–126.

38. Y. Z. Gu, J. B. Hogenesch and C. A. Bradfield, 'The PAS superfamily: sensors of environmental and developmental signals.' *Annu. Rev. Pharmacol. Toxicol.*, 2000, **40**, 519–561.

39. H. Zhu and H. F. Bunn, 'Signal transduction. How do cells sense oxygen?' *Science*, 2001, **292**, 449–451.

40. C. V. Dang and G. L. Semenza, 'Oncogenic alterations of metabolism.' *Trends Biochem. Sci.*, 1999, **24**, 68–72.

41. D. M. Rothwarf, E. Zandi, G. Natoli et al., 'IKK-γ is an essential regulatory subunit of the IκB kinase complex.' *Nature*, 1998, **395**, 297–300.

42. N. D. Perkins, 'The Rel/NF-κ B family: friend and foe.' *Trends Biochem. Sci.*, 2000, **25**, 434–440.

43. P. Cohen, 'The regulation of protein function by multisite phosphorylation—a 25 year update.' *Trends Biochem. Sci.*, 2000, **25**, 596–601.

44. H. Fukazawa, P. M. Li, C. Yamamoto et al., 'Specific inhibition of cytoplasmic protein tyrosine kinases by herbimycin A in vitro.' *Biochem. Pharmacol.*, 1991, **42**, 1661–1671.

45. R. Zinck, M. A. Cahill, M. Kracht et al., 'Protein synthesis inhibitors reveal differential regulation of mitogen-activated protein kinase and stress-activated protein kinase pathways that converge on Elk-1.' *Mol. Cell. Biol.*, 1995, **15**, 4930–4938.

46. D. T. Dudley, L. Pang, S. J. Decker et al., 'A synthetic inhibitor of the mitogen-activated protein kinase cascade.' *Proc. Nat. Acad. Sci. USA*, 1995, **92**, 7686–7689.

47. G. W. Gould, A. Cuenda, F. J. Thomson et al., 'The activation of distinct mitogen-activated protein kinase cascades is required for the stimulation of 2-deoxyglucose uptake by interleukin-1 and insulin-like growth factor-1 in KB cells.' *Biochem. J.*, 1995, **311**, 735–738.

48. M. Ui, T. Okada, K. Hazeki et al., 'Wortmannin as a unique probe for an intracellular signalling protein, phosphoinositide 3-kinase.' *Trends Biochem. Sci.*, 1995, **20**, 303–307.

49. C. J. Vlahos, W. F. Matter, K. Y. Hui et al., 'A specific inhibitor of phosphatidylinositol 3-kinase, 2-(4-morpholinyl)-8-phenyl-4H-1-benzopyran-4-one (LY294002).' *J. Biol. Chem.*, 1994, **269**, 5241–5248.

50. N. Mamiya, J. R. Goldenring, Y. Tsunoda et al., 'Inhibition of acid secretion in gastric parietal cells by the Ca2+/calmodulin-dependent protein kinase II inhibitor KN-93.' *Biochem. Biophys. Res. Commun.*, 1993, **195**, 608–615.

51. T. Chijiwa, A. Mishima, M. Hagiwara et al., 'Inhibition of forskolin-induced neurite outgrowth and protein phosphorylation by a newly synthesized selective inhibitor of cyclic AMP-dependent protein kinase, N-[2-(p-bromocinnamylamino)ethyl]-5- isoquinolinesulfonamide (H-89), of PC12D pheochromocytoma cells.' *J. Biol. Chem.*, 1990, **265**, 5267–5272.

52. R. J. Davis, 'Signal transduction by the JNK group of MAP kinases.' *Cell*, 2000, **103**, 239–252.

53. D. J. Burks and M. F. White, 'IRS proteins and β -cell function.' *Diabetes*, 2001, **50**(Suppl 1), S140–145.

54. K. Imada and W. J. Leonard, 'The Jak-STAT pathway.' *Mol. Immunol.*, 2000, **37**, 1–11.

55. L. Chang and M. Karin, 'Mammalian MAP kinase signalling cascades.' *Nature*, 2001, **410**, 37–40.

56. Y. Nishizuka, 'Protein kinase C and lipid signaling for sustained cellular responses.' *FASEB J.*, 1995, **9**, 484–496.

57. M. A. Krasilnikov, 'Phosphatidylinositol-3 kinase dependent pathways: the role in control of cell growth, survival, and malignant transformation.' *Biochemistry (Mosc)*, 2000, **65**, 59–67.

58. E. Hajduch, G. J. Litherland and H. S. Hundal, 'Protein kinase B (PKB/Akt)—a key regulator of glucose transport?' *FEBS Lett.*, 2001, **492**, 199–203.

59. A. Guilherme, K. Torres and M. P. Czech, 'Cross-talk between insulin receptor and integrin α5 β1 signaling pathways.' *J. Biol. Chem.*, 1998, **273**, 22899–22903.

60. P. A. Baeuerle, 'IκB-NF-κB structures: at the interface of inflammation control.' *Cell*, 1998, **95**, 729–731.

61. A. Strasser, L. O'Connor and V. M. Dixit, 'Apoptosis signaling.' *Annu. Rev. Biochem.*, 2000, **69**, 217–245.

62. L. O'Connor, D. C. Huang, L. A. O'Reilly et al., 'Apoptosis and cell division.' *Curr. Opin. Cell. Biol.*, 2000, **12**, 257–263.

63. B. B. Aggarwal, 'Apoptosis and nuclear factor-κ B: a tale of association and dissociation.' *Biochem. Pharmacol.*, 2000, **60**, 1033–1039.

64. K. Cain and C. Freathy, 'Liver toxicity and apoptosis: role of TGF-β1, cytochrome c and the apoptosome.' *Toxicol. Lett.*, 2001, **120**, 307–315.

65. P. D. Lampe and A. F. Lau, 'Regulation of gap junctions by phosphorylation of connexins.' *Arch. Biochem. Biophys.*, 2000, **384**, 205–215.

J.P. Vanden Heuvel, G.H. Perdew, W.B. Mattes and W.F. Greenlee (Eds.)
Comprehensive Toxicology, Vol. xiv

# 14.4.2
# Mitogen-Activated Protein Kinase (MAPK) Signal Transduction Pathways Regulated by Stresses and Toxicants

JOHN M. KYRIAKIS

*Massachusetts General Hospital and Harvard Medical School, Charlestown, MA, USA*

## 14.4.2.1   INTRODUCTION

The molecular details of mammalian stress-activated signal transduction pathways have only begun to be dissected. This, in spite of the fact that the activation of these pathways exerts a profound impact on the pathology of diseases such as chronic inflammation, athero-sclerosis, and the debilitating effects of diabetes mellitus; as well as the side effects of cancer therapy, not to mention embryonic development, innate and acquired immunity. From the point of view of the toxicologist, stress-activated signaling pathways represent a central mechanism by which the cell and the organism respond to environmental insults. Thus, it is not surprising that understanding these pathways has attracted wide interest; and in the past 10 years dramatic progress has been made. Accordingly, it is now becoming possible to envisage the transition of these findings to the development of novel treatment strategies and to a new appreciation of the molecular basis of toxicity. This chapter will focus on the biochemical components and regulation of mammalian stress-regulated mitogen-activated protein kinase (MAPK) pathways and their activation by environmental stresses and toxicants.

Mitogen-activated protein kinase (MAPK) signal transduction pathways are among the most widespread mechanisms of eukaryotic cell regulation. All eukaryotic cells possess multiple MAPK pathways, each of which is preferentially recruited by distinct sets of stimuli, thereby allowing the cell to respond coordinately to simultaneous divergent inputs. Mammalian MAPK pathways can be activated by a wide variety of different stimuli acting through diverse receptor families, including hormones and growth factors that act through receptor tyrosine kinases (e.g., insulin, EGF, PDGF, FGF) or cytokine receptors (e.g., growth hormone) to vasoactive peptides acting through G-protein coupled, seven-transmembrane receptors (e.g., angiotensin-II, endothelin), TGFβ-related polypeptides, acting through Ser-Thr kinase receptors, as well as inflammatory cytokines of the tumor necrosis factor (TNF) family. MAPKs are also recruited by environmental stresses such as osmotic shock, ionizing radiation, metabolic and genotoxins, oxidants and ischemic injury—stimuli thought not to engage specific receptors. MAPK pathways, in turn, coordinate activation of gene transcription, protein synthesis, cell cycle machinery, cell death and differentiation. Accordingly, these pathways exert a major effect on cell physiology.[1-3]

All MAPK pathways include central three tiered "core signaling modules" (Figure 1) in which MAPKs are activated by concomitant Tyr and Thr phosphorylation within a con-

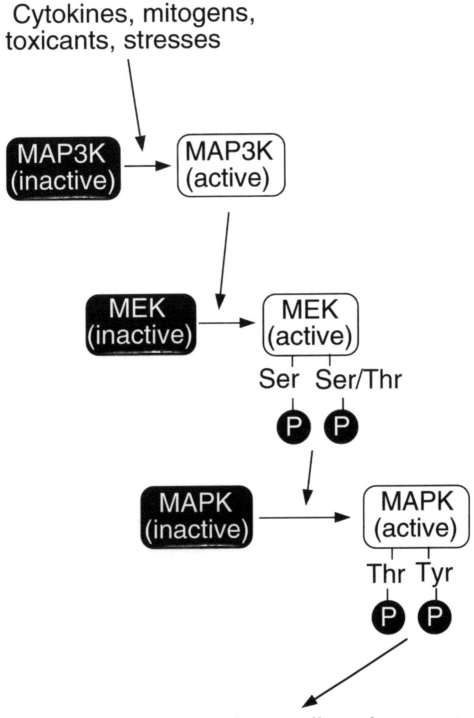

**Figure 1** The MAPK core signaling module. Divergent inputs feed into core MAP3K/ MEK/MAPK core pathways which then recruit appropriate responses.

served Thr-X-Tyr motif in the activation loop of the kinase domain subdomain-VIII. MAPK phosphorylation and activation are catalyzed by a family of dual specificity kinases referred to as MAPK/extracellular signal-regulated kinase (ERK)-kinases (MEKs or MKKs). MEKs, in turn, are regulated by Ser/Thr phosphorylation, also within a conserved motif in kinase domain subdomain-VIII, catalyzed by any of several protein kinase families collectively referred to as MAPK-kinase-kinases (MAP3Ks). MAPK core signaling modules are themselves regulated by a wide variety of upstream activators and inhibitors[1-3].

The notion of multiple parallel MAPK signaling cascades was first appreciated from studies of simple eukaryotes such as the budding yeast *Saccharomyces cerevisiae*. To date, six *S. cerevisiae* MAPK signaling pathways have been identified and have been reviewed elsewhere.[3] The features of yeast MAPK pathways as well as early biochemical studies have revealed common principles shared by all MAPK pathways.

*MAPKs are proline-directed kinases; however, substrate selectivity is often conferred by specific MAPK docking sites present on physiologic substrates*: The proline-directed substrate specificity of the MAP kinases was established using peptide substrates corresponding to the sequences surrounding Thr 669 of the EGF receptor (Glu-Leu-Val-Glu-Pro-Leu-*Thr*-Pro-Ser-Gly-Glu-Ala-Pro-Asn-Gln-Ala-Leu-Leu-Arg) and the site on myelin basic protein (Ala-Pro-Arg-*Thr*-Pro-Gly-Gly-Arg; phosphorylation site underlined) phosphorylated *in vitro* by the 42-kDa insulin- and mitogen-stimulated MAPK extracellular signal-regulated kinase (ERK)-2, a target of the Ras proto oncoprotein. Systematic variation of the sequences surrounding the single Thr phosphoacceptor sites in both peptides established the essential role of the proline immediately carboxyl to the phosphoacceptor site.[4-6] However, this selectivity for proline is not sufficient to account for the high degree of substrate selectivity manifested by different MAPK subgroups. The $K_m$ for native protein substrates is usually several orders of magnitude lower than the $K_m$ for synthetic peptides corresponding in amino acid sequence to the region immediately surrounding the phosphorylation site on these substrates. It has become apparent that most, if not all, of the physiologic substrates of the MAPKs possess specific docking sites, often at considerable distance from the phosphorylation site in the primary sequence, that allow for a strong interaction with select MAPK subfamilies to the exclusion of others.[7-10] In turn, MAPKs themselves possess complementary docking sites that interact with the MAPK binding domains on substrate proteins.[7,10] This confers striking substrate specificity on a family of protein kinases with an otherwise apparently broad *in vitro* substrate profile.

*Signaling components with more than one biological function/signaling component under multiple forms of regulation*: Within MAPK core signaling modules there are instances wherein individual elements can function promiscuously in several pathways. In addition, MAPK pathway components are often subject to regulation by multiple inputs. For example, the *S. cerevisiae* MAP3K Ste11p functions as part of the mating pheromone response pathway and the osmosensing pathway,[3,11] while the mammalian MAP3K MEK-kinase-3 can activate the mitogenic ERK and three stress-activated MAPK pathways.[12-16] Conversely, the yeast osmosensing MEK Pbs2p can be regulated by three MAP3Ks: Ssk2p, Ssk22p and Ste11p while the mammalian stress-regulated MEK stress-activated protein kinase (SAPK)/ ERK-kinase-1 (SEK1) is a putative substrate for at least ten known MAP3Ks.[1,3]

*Drugs and toxins that are considered to affect specific signaling pathways are often nonspecific and can have completely unexpected effects* in vivo *that contrast with their* in vitro *effects*: A recent study of 28 commercially available compounds on 40 different protein kinases revealed a disappointing degree of target specificity for compounds touted as "selective" inhibitors.[17] Moreover, the pyridinyl imidazole SB203580, considered a reliable and specific inhibitor of p38 MAPK, was recently shown to be an inhibitor of the mitogenic MAP3K Raf1 *in vitro*, but a potent *activator* of Raf1 *in vivo*.[18] Thus, when employing such compounds at the bench, or when evaluating toxins *in vivo* and *in vitro*, consideration must be given to the likelihood of multiple intracellular targets. In addition, the *in vitro* effects of a compound may not translate to similar effects *in vivo*.

*A single toxicant may act at several points in the cell to recruit MAPK pathways*: Because many toxicants and drugs can freely migrate to different parts of a cell, they can often initiate signaling pathways contemporaneously at several sites in a cell. For example, the chemotherapeutic genotoxins bleomycin or *cis*-platinum may initiate signaling at sites of DNA damage and through multiple pathways triggered by the generation of reactive oxygen species (ROS).

## 14.4.2.2 THE STRESS-ACTIVATED PROTEIN KINASE (SAPK)/C-JUN-NH$_2$-TERMINAL KINASE (JNK), AND P38 AND ERK5/BIG MAPK-1 (BMK1) PATHWAYS: MAPK SIGNALING PATHWAYS ACTIVATED BY STRESSES, TOXICANTS AND INFLAMMATORY CYTOKINES. CORE PATHWAYS AND THEIR TARGETS

### 14.4.2.2.1 General Considerations

The insulin/mitogen-regulated extracellular signal-regulated kinase (ERK) pathway was the first mammalian MAPK pathway to be identified. This pathway is largely regulated by the monomeric GTPase Ras which recruits MAP3Ks of the Raf family to activate two MEKs: MEK1 and MEK2. These, in turn, activate the ERKs. The biochemistry, biology and regulation of the ERKs have been reviewed exhaustively elsewhere.[2,19,20] As with yeast, multiple parallel mammalian MAPK pathways exist; most of these, in conjunction with the nuclear factor-κB (NF-κB) pathway, are pivotal to stress and inflammatory responses rather than to mitogen responses.

### 14.4.2.2.2 The Stress-activated Protein Kinases/c-Jun NH$_2$-Terminal Kinases (SAPKs), a Family of MAPKs Activated by Environmental Stresses and Inflammatory Cytokines of the Tumor Necrosis Factor Family

Shortly after the identification of the ERKs, it became apparent that mammalian cells possessed several MAPK pathways. The protein synthesis inhibitor cycloheximide can elicit the *in vivo* activation of ribosomal S6 phosphorylation; and, in fact, this strategy was used to activate the p70 ribosomal S6 kinase *in vivo* prior to purification.[20] Cycloheximide also activated a novel Ser/Thr kinase activity that could phosphorylate microtubule-associated protein-2 (MAP2). Purification of this kinase revealed a 54-kDa polypeptide, initially named p54-MAP2 kinase, or p54.[1,22] p54 could be inactivated with Tyr or Ser/Thr phosphatases, indicating that, like the ERKs, the p54 kinase required concomitant Tyr and Ser/Thr phosphorylation for activity;[1,22] moreover, studies with synthetic peptide substrates showed that p54 was proline-directed, but phosphorylated Ser/Thr-Pro motifs with a specificity distinct for that of the p42/p44MAPKs.[21,23]

The specificity of p54 toward native protein substrates clearly differed from that of the ERKs. Most importantly, p54 was able to phosphorylate the c-Jun transcription factor, at a rate at least an order of magnitude greater than that catalyzed by the ERKs, at two sites (Ser63 and Ser73) implicated in regulation of c-Jun and AP-1 *trans* activation function.[24]

To clone the p54 Kyriakis *et al.*, used the amino acid sequence of tryptic peptides derived from purified p54 to design specific PCR primers,[25] whereas Dérijard *et al.*,[26] used a pure PCR strategy employing degenerate primers derived from regions conserved among known MAPKs. Assay of p54 kinase activity immunoprecipitated from extracts of cells subjected to various treatments revealed that, in contrast to the p42/p44 MAPKs, p54 was not strongly activated in most cells by mitogens such as insulin, EGF, PDGF or FGF. Instead, p54 was vigorously activated by a variety of noxious treatments such as heat shock, ionizing radiation, oxidant stress, DNA damaging chemicals (topoisomerase inhibitors and alkylating agents), reperfusion injury, mechanical shear stress and, of course, protein synthesis inhibitors (cycloheximide and anisomycin).[1,25–27]

From a toxicologic perspective, the potent activation of p54 by tunicamycin was conceptually important; tunicamycin inhibits N-linked protein glycosylation and leads to the accumulation of misfolded proteins exclusively within the lumen of the endoplasmic reticulum (ER—the unfolded protein response, a form of ER stress). The ability of an ER-localized perturbation to activate the cytosolic p54 kinase was one of the first clear-cut examples of ER stress-activated signal transduction through a protein kinase cascade.[25]

The p54 is also strongly activated by all the inflammatory cytokines of the tumor necrosis factor (TNF) family (TNF, interleukin-1, CD40 ligand, CD27 ligand, Fas ligand, receptor activator of NF-κB (RANK), RANK ligand, *etc.*) as well as by vasoactive peptides (endothelin and angiotensin-II) (References 25, 28–34). p54 protein has since been renamed; and the nomenclature of this family of kinases is somewhat confusing (Table 1). Two systems are generally accepted: stress-activated protein kinase (SAPK) in reference to the regulation of these kinases by environmental stress and inflammation; and c-Jun-NH$_2$-terminal kinase (JNK) in reference to the phosphorylation by these kinases of the c-Jun amino terminal *trans* activation domain.

The SAPKs are encoded by at least three genes: SAPKα/JNK2, SAPKβ/JNK3 and SAPKγ/JNK1 (References 25, 35, 36) (Table 1).

**Table 1** MAPK nomenclature.

| Name | Alternate names | Substrates |
|------|-----------------|------------|
| ERK1 | p44-MAPK | MAPKAP-K1, MNKs, MSKs, Elk1 |
| ERK2 | p42-MAPK | MAPKAP-K1, MNKs, MSKs, Elk1 |
| SAPK-α | JNK2, SAPK1a | c-Jun, JunD, ATF2, Elk1 |
| SAPK-p54α1 | JNK2β2 | |
| SAPK-p54α2 | JNK2α2 | |
| SAPK-p46α1 | JNK2β1 | |
| SAPK-p46α2 | JNK2α1 | |
| SAPK-β | JNK3, SAPK1b | c-Jun, JunD, ATF2, Elk1 |
| SAPK-p54β1 | JNK3β2 | |
| SAPK-p54β2 | JNK3α2 | |
| SAPK-p46β1 | JNK3β1 | |
| SAPK-p46β2 | JNK3α1 | |
| SAPK-γ | JNK1, SAPK1c | c-Jun, Jun D, ATF2, Elk1 |
| SAPK-p54γ1 | JNK1β2 | |
| SAPK-p54γ2 | JNK1α2 | |
| SAPK-p46γ1 | JNK1β1 | |
| SAPK-p46γ2 | JNK1α1 | |
| p38α | SAPK2a, CSBP1 | MAPKAP-K2/3, MSKs, PRAK, ATF2, Elk1, MEF2A/C |
| p38β | SAPK2b, p38-2 | MAPKAP-K2/3, MSKs, PRAK, ATF2 |
| p38γ | SAPK3, ERK6 | ATF2 |
| p38δ | SAPK4 | ATF2 |

As with all MAPKs, each SAPK isoform contains a characteristic Thr-X-Tyr phosphoacceptor loop in subdomain VIII of the protein kinase catalytic domain. Whereas the ERK sequence is Thr185-Glu-Tyr187, that of the SAPKs is Thr183-Pro-Tyr185. The expression of each SAPK gene is further diversified by differential hnRNA splicing within the catalytic domain at a region spanning subdomains IX and X, resulting in type-1 and -2 SAPKs (α and β JNKs, respectively, Table 1). Additional differential splicing at the extreme carboxyl terminus yields 54- (p54) and 46-kDa (p46) polypeptides (type-2 and -1 JNKs, respectively, Table 1). The significance of the carboxyl terminal isoforms is not clear, but the type-1 and -2 kinases differ modestly in their substrate binding affinities.[8,35,36]

### 14.4.2.2.3 The p38 MAPKs, a Second Stress-activated MAPK Subfamily

The p38 MAPKs are a second mammalian stress-activated MAPK family. p38 (the α isoform) was purified by anti phosphotyrosine immunoaffinity chromatography as a 38-kDa polypeptide that underwent Tyr phosphorylation in response to endotoxin treatment and osmotic shock.[37] cDNA cloning revealed that p38 was the mammalian MAPK homologue most closely related to *HOG1*, the osmosensing MAPK of *S. cerevisiae*. Most notably, the p38s, like Hog1p, contain the phosphoacceptor sequence Thr-Gly-Tyr.[3,37] p38α was also iden-

tified contemporaneously, and independently as a kinase activated by stress and IL-1 that could phosphorylate and activate MAPK-activated protein kinase-activated protein kinase-1 (MAPKAP kinase-2, Section 14.4.2.2.5.1), a novel Ser/Thr kinase implicated in the phosphorylation and activation of the small heat shock protein Hsp27.[38,39]

Finally, p38α was also purified and cloned as the polypeptide receptor for a class of experimental pyridinyl-imidazole antiinflammatory drugs, the cytokine suppressive anti inflammatory drugs (CSAIDs), the most extensively characterized of which are the compounds SB203580 and SB202092.[17,40,41] CSAIDs were originally identified in a screen for compounds that could inhibit the transcriptional induction of TNF and IL-1 during endotoxin shock.[40] By binding and directly inhibiting a subset of the p38s, these compounds block p38-mediated activation of AP-1, a *trans* acting factor required for TNF and IL-1 induction[40] (see below).

With the identification of additional p38 isoforms, four p38 genes are now known (Table 1): the original isoform, here referred to as p38α (also called CSAIDs binding protein (CSBP) and, somewhat confusingly, SAPK2a), p38β (also called SAPK2b, and p38-2), p38γ (also called SAPK3 and ERK6) and p38δ (also called SAPK4).[37–40,42–48]

Only p38α and p38β are inhibited by CSAIDs, p38γ and p38δ are completely unaffected by these drugs *in vitro* or in transfected cells.[46] Subsequent studies have demonstrated that SB203580 and SB202092 are remarkably specific when assayed for inhibition of a variety of protein kinases.[17] The basis for this specificity was revealed in the crystal structure of p38α complexed with the SB203580. In order to accommodate a fluorophenyl moiety present in the SB203580 structure, the amino acid at position 106 must be no larger than Thr.[49,50] Thus, substitution of Thr106 with the corresponding, more bulky residues from p38γ or p38δ (Met, and Pro or Phe, respectively, in both cases) abolishes SB203580 binding. Conversely, if the amino acid of p38γ, p38δ or even SAPKγ which corresponds to p38α Thr106 is replaced with Thr, the resulting mutants display at least partial sensitivity to SB203580.[49,50]

Nevertheless, in spite of the apparent biochemical and cellular selectivity of the CSAIDs, recent studies have shown these compounds to have unexpected nonspecific properties. The proto oncoprotein Raf1 is a MAP3K upstream of the ERKs and downstream of Ras.[2,19] Raf1 has a Thr (Thr321 in rat c-Raf1) at a site corresponding to Thr106 of p38α. Consistent with this, Raf-1 is inhibited by SB203580 and SB202092 *in vitro*, albeit at concentrations higher than that needed to inhibit p38α *in vitro*.[18] However, *in vivo* the Ras/Raf/ERK pathway is not inhibited by SB203580 or SB202092. Surprisingly, the pyridinyl imidazole p38 inhibitors trigger a striking activation of Raf1 *in vivo* (an activation observed on subsequent assay of Raf1 *in vitro*, in the absence of drug, after recovery of the Raf1 from drug-treated cells). This activation is not accompanied by activation of Ras.[18] In an intriguing subsequent finding, the phenylamido derivative compound ZM336372 (*N*-[5-(3-dimethylaminobenzamido)-2-methylphenyl]-4-hydroxybenzamide), a novel *in vitro* Raf inhibitor was shown to have an a effect similar to SB203580 and SB202092. Thus, ZM336372 also inhibits p38s-α and β *in vitro* and *in vivo* in a manner dependent upon Thr106, but is a potent *in vivo* activator of Raf1.[51] The basis for these paradoxical and unexpected findings is unknown, but is indicative of the fact that assertions as to the specificity of a compound *in vitro* require rigorous and comprehensive testing. Moreover, the *in vitro* toxicologic/pharmacologic effects of a drug on MAPK signaling may differ substantially from the *in vivo* effects. Thus, *in vitro* results must be cautiously interpreted when proceeding to *in vivo* investigations.

Like the SAPKs, the p38s are strongly activated *in vivo* by environmental stresses and inflammatory cytokines and are inconsistently activated by insulin and growth factors. In almost all instances, the same stimuli that recruit the SAPKs also recruit the p38s.[1] One exception is ischemia-reperfusion. SAPKs are not activated during ischemia, but rather during reperfusion; whereas the p38s are activated during ischemia and remain active during reperfusion.[1,27,52] The basis for this difference is unknown.

### 14.4.2.2.4 ERK5/Big MAP Kinase-1 (BMK1), a Third Class of Stress-activated MAPK

The novel MEK, MEK5, was cloned by degenerate PCR as part of an effort to identify new MAPK pathways and regulators.[53,54] ERK5, a MEK5 substrate, was cloned as part of a two-hybrid screen that employed MEK5 as bait.[53] ERK5 is a ~90-kDa MAPK of which only one mammalian homologue is known. ERK5 has the sequence Thr-Glu-Tyr in its phosphoacceptor loop. The amino terminal kinase domain of ERK5 is followed by an extensive carboxyl terminal tail of unknown function that contains several consensus proline-rich motifs indicative of binding sites for proteins with SH3 domains.[53]

The stimuli that recruit ERK5 have not been comprehensively characterized; however, ERK5 can be substantially activated by environmental stresses such as oxidant stress (peroxide) and osmotic shock (sorbitol), but not by vasoactive peptides or inflammatory cytokines (TNF).[55] ERK5 is also activated by EGF and may therefore lie downstream of receptor Tyr kinases.[16]

### 14.4.2.2.5 SAPK, p38 and ERK5 Substrates

The substrates of stress-activated MAPKs point to the importance of these MAPKs to the stress response. As with the ERKs, the SAPKs p38s and ERK5 phosphorylate both transcription factors and other protein kinases. Some of the protein kinase substrates (MAPKAP-K2 and -K3 and PRAK) are selectively recruited by stress-activated MAPKs. Others (MNKs and MSKs), however, are activated by both stress- and mitogen-regulated MAPKs and, like the AP-1 transcription factor, integrate both stress and mitogenic signaling pathways.

#### 14.4.2.2.5.1 Protein Kinases

*Mitogen-activated protein kinase-activated protein kinases (MAPKAP-Ks)-2 and -3*: MAPKAP kinase-2 (MAPKAP-K2) and the structurally related MAPKAP-K3 (also called three pathway regulated kinase or 3PK) are a small family of Ser/Thr kinases that consist of an amino terminal regulatory domain and a carboxyl terminal kinase domain.[56–59] They are unrelated to the MAPKAP-K1s/Rsks, targets of the ERKs (reviewed in References 20, 60). Along with p38-regulated and activated kinase (PRAK, see below), MAPKAP-K2 and MAP-KAP-K3 phosphorylate the small heat shock protein Hsp27.[56–59,61,62] Nonphosphorylated Hsp27 normally exists in high molecular weight aggregates that serve as molecular chaperones. Phosphorylation of Hsp27 by MAPKAP kinase-2/3 at Ser15, Ser78, Ser82 and Ser90 coincides with the dissociation of Hsp27 into monomers and dimers, and its redistribution of to the actin cytoskeleton.[61] In peroxide-treated human umbilical vein endothelial cells, redistribution of Hsp27 may participate in eliciting the reorganization of F-actin into stress fibers, thereby affecting cell motility.[61,62] MAPKAP-K2/3-catalyzed phosphorylation of Hsp27 at Ser90 appears necessary for this process and mutation of Ser90 to Ala prevents stimulus-induced changes in Hsp27 oligomerization.[62]

Like p38, MAPKAP-K2 is activated by stresses and inflammatory cytokines.[38,39] MAPKAP-K2 is phosphorylated and activated *in vitro* and *in vivo* by p38α and p38β (but not by p38γ or p38δ)[63] (Figure 2). Activation of MAPKAP kinase 2 is a multistep process. Phosphorylation of Thr25, catalyzed by p38α and p38β, gates subsequent p38-catalyzed phosphorylation of Thr222 and Ser272 in the kinase activation loop. An additional autophosphorylation at Thr334 then results in activation of MAPKAP-K2.[64] Consistent with regulation by p38α and p38β, MAPKAP-K2 activation and Hsp27 phosphorylation are inhibited by CSAIDs.[61,62,64] MAPKAP-K3 can also phosphorylate Hsp27.[58] As with MAPKAP-K2, endogenous MAPKAP-K3 is activated primarily by stresses and inflammatory cytokines and in a manner that can be completely inhibited with CSAIDs, suggesting that p38α and/or p38β are the major MAP-KAP-K3 kinases *in vivo*.[58]

*PRAK*: p38-regulated/activated kinase (PRAK) is a ∼50-kDa Ser/Thr kinase with a similar overall structure to MAPKAP-K2, -K3 and the MAPK-interacting kinases (MNKs, see below). Thus, PRAK consists of an amino terminal regulatory domain and a carboxyl terminal kinase domain. As with MAPKAP-K2 and -K3, PRAK is activated selectively in response to stress and inflammatory cytokines and is not detectably activated by mitogens. PRAK is phosphorylated at Thr182 in the kinase domain activation loop and activated *in vivo* and *in vitro* by p38α and p38β. Consistent with this, PRAK activation can be blocked with CSAIDs.[65] Once activated, PRAK can phosphorylate Hsp27 at the physiologically relevant sites. In-gel kinase assays indicate that, in addition to MAPKAP-Ks-2 and –3, PRAK is a potentially important stress-activated Hsp27 kinase[65] (Figure 2). It should be noted, however, that disruption of *mapkap-k2* suggests that the relative contribution of different kinases to Hsp27 phosphorylation may depend on cell type. Thus, stimulation of Hsp27 phosphorylation by bacterial lipopolysaccharide, a well established response during sepsis, is completely absent in *mapkap-k2−/−* mice.[66] These mice are also resistant to LPS toxicity, showing substantially reduced synthesis of TNF in response to LPS challenge.[66]

*MAPK-interacting kinases (MNKs)*: Most cellular mRNAs contain a 5′ cap structure, the $N^7$-methylguanosine cap, which is essential for translational regulation. The $N^7$-methylguanosine-binding protein eIF-4E, recruits mRNAs onto a scaffold protein eIF-4G, which also binds the RNA helicase eIF-4A. The latter, in collaboration with the RNA binding protein eIF-4B, unwind mRNA secondary structure in the 5′ untranslated segment, thereby facilitating the scanning of the mRNA by the 40S ribosomal complex to the ATG translational initiation site. The complex of eIF-4A, B, G and E is known as eIF-4F (reviewed in Reference 67).

The eIF-4E-eIF-4G interaction, a rate-limiting step in translational initiation, is negatively regulated by the translational repressor protein 4E-binding protein-1 (4E-BP1 also called phosphorylated heat- and acid-stable protein regulated by insulin, PHAS-I). 4E-BP1 is multiply phosphorylated in response to insulin and mitogens resulting in the dissociation of 4E-BP1 from eIF-4E, and the availability of eIF-4E for incorporation into the eIF-4F complex. *In vivo*, the phosphorylation of eIF-4E is strongly inhibited by rapamycin and wortmannin, consistent with the fact that dissociation of 4E-BP1 is regulated by the mammalian target of rapamycin (mTOR) as well as by protein kinases downstream of PI3-kinase (reviewed in References 20, 67). In addition, eIF-4E itself also undergoes a regulatory phosphorylation at Ser209 in response to both insulin/mitogen and environmental stress. This phosphorylation increases the affinity of eIF-4E for the 5′-cap by about three-fold.[67]

MAPK interacting kinases (MNKs)-1 and -2 are two closely related kinases that are probably the physiologically relevant eIF-4E

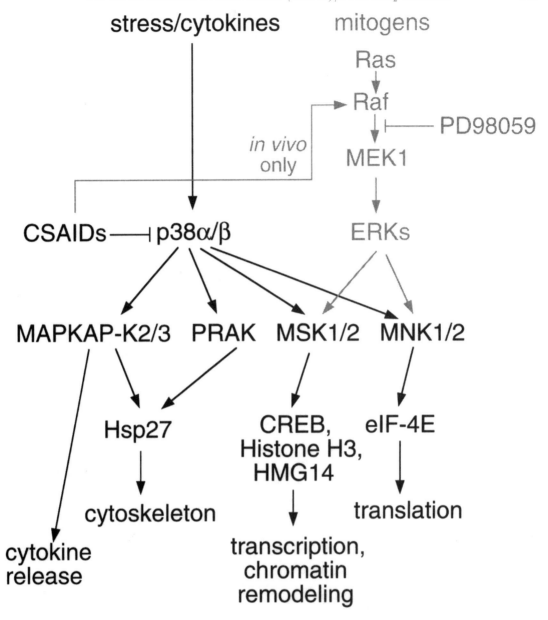

**Figure 2** Regulation of protein kinases by p38s. Note that MSKs and MNKs are also regulated by the ERKs.

Ser209 kinases. As the name implies, MNKs associate *in vivo* with MAPKs and are MAPK substrates *in vitro* and *in vivo*. MNKs are phosphorylated and activated both by ERKs 1/2 (in response to insulin and mitogens) and by the p38s (in response to stress).[68–70] The regulation of the MNKs by both the ERKs and p38s indicates that, as with AP-1 (see below), MNKs are a site of integration of stress and mitogenic signaling pathways (Figure 2).

*MSK1/2*: Mitogen- and stress-activated protein kinases (MSKs)-1 and -2 are a recently identified family of Ser/Thr protein kinases with an overall structure similar to that of the MAPKAP-K1s/Rsks (reviewed in Reference 60)—notably, the MSKs possess two tandem protein kinase domains. MSKs1/2 were identified in an interrogation of databases for DNA sequences homologous to p70 S6 kinase. *In vitro* MSK1 will phosphorylate the synthetic peptide "Crosstide" (Gly-Arg-Pro-Arg-Thr-Ser-Ser-Phe-Ala-Glu-Gly); however, MSK1 purified from unstimulated cells possesses low Crosstide kinase activity. By contrast, the Crosstide kinase activity of MSK1 is activated > 100 fold upon incubation *in vitro* with ERK2 (Figure 2); and MSK1 is activated *in vivo* by mitogens in a manner inhabitable by the MAPK pathway inhibitor PD98059.[71] MSK1 is also a substrate for the p38s. Consistent with this, MSK1 is also activated *in vivo* by environmental stresses (arsenite, UV, peroxide), in a manner inhibited by the CSAID SB203580.[71] MSK1 is also activated by TNF. Here, the activation pattern is more complex. In HeLa cells, TNF activates the p38s (and the SAPKs) as well as the ERKs. However, p38 activation is more rapid, reaching a maximum at 5 min, while ERK activation is comparatively slower, reaching an apparent maximum at 15 min. Accordingly, SB203580 completely blocks early (5 min) TNF activation of MSK1 in HeLa cells. After 15 min, however, the ability of SB203580 to block MSK activation by TNF is only partial and MSK1 can now also be partially inhibited with the ERK pathway inhibitor PD98059[71] (Figure 2).

cAMP response element binding protein (CREB) is a bZIP transcription factor that binds and *trans* activates genes containing the palindromic cAMP response element (CRE-consensus sequence: TGACGTCA). CREB *trans* activating activity can be activated not only by stimuli that elevate cAMP, but by mitogens and stresses. Activation of CREB *trans* activating activity coincides with phosphorylation at Ser133. PKA likely represents the major CREB kinase recruited by cAMP.[72] MSK1 is also a potent CREB kinase *in vitro* and strong evidence indicates that MSK1, and not MAPKAP-K1/Rsk or MAPKAP-K2/3, is the physiologically relevant stress- and mitogen-activated CREB kinase *in vivo*.[71] First, the $K_{cat}$ for MSK1-catalyzed phosphorylation of CREB is nearly 100 fold greater than that for the next most efficient CREB kinase MAPKAP-K1/Rsk. Moreover, the MSK1 polypeptide has a bipartite nuclear localization signal, and is localized exclusively in the nucleus, where CREB resides. Finally, the pattern of pharmacologic inhibition of stress- and mitogen-activated CREB phosphorylation *in vivo* parallels precisely the pattern of pharmacologic inhibition of MSK1 activation. Thus, activation of CREB by stress and mitogens is blocked by the broad specificity kinase inhibitor Ro318220. MSK1 is strongly inhibited *in vitro* by Ro318220 whereas MAPKAP-K2 and MAPKAP-K3 are not. Finally, CREB activation *in vivo* by TNF can be blocked with SB203580 and the ERK inhibitor PD98059 with kinetics that parallel those described above for inhibition of MSK1.[71]

In addition to CREB regulation, MSKs may be involved in the phosphorylation of components involved in chromatin remodeling. Both mitogens and stresses, as they stimulate gene expression, must prompt the loosening of chromatin structure from around target genes.

This provides the transcriptional machinery access to genes that are to be expressed.[73,74] This alteration in chromatin structure involves both the acetylation and phosphorylation of histones. As a result of stimulus-dependent phosphorylation, many transcription factors recruit transcriptional coactivators with intrinsic histone acetyl transferase activity. In addition, protein kinases activated by mitogens and stresses can phosphorylate histones.[73,74] The nucleosomal proteins histone H3 and high mobility group-14 (HMG-14) are prime targets of phosphorylation.[73–75] Agonist-stimulated phosphorylation of histone H3 occurs at Ser10, while HMG-14 is phosphorylated at Ser6. Recent genetic evidence indicates that mitogenic histone H3 phosphorylation is mediated at least in part by the ERK substrate MAPKAP-K1b/Rsk2. Thus, cells derived from patients with Coffin–Lowry syndrome, a loss of function mutation in the *mapkap-k1b/rsk2* gene, display a substantial deficit in mitogen-stimulated histone H3 phosphorylation.[74] However, it is unclear if MAPKAP-K1b/Rsk2 is the only histone H3 kinase *in vivo*. Both mitogens and stresses can stimulate H3 and HMG-14 phosphorylation *in vivo*[75] and MAPKAP-K1s/Rsks are not strongly activated by stresses.[1,20,21,60] Moreover, MAPKAP-K1s/Rsks are poor HMG-14 kinases *in vitro*, and are not inhibited by H89, an inhibitor that blocks HMG-14 kinase activity *in vivo* and *in vitro*. By contrast MSK1 is strongly inhibited by H89 in parallel with *in vivo* inhibition of histone H3 and HMG-14 phosphorylation. In addition, *in vitro*, MSK1 can phosphorylate histone H3 at Ser10, at a rate comparable to or better than that phosphorylated by MAPKAP-K1s/Rsks *in vitro*.[75]

### 14.4.2.2.5.2 *Transcription Factors*

*CHOP/GADD153*: CREB homologous protein (CHOP)/growth arrest and DNA damage-155 (GADD153) is a bZIP transcription factor of the CREB family.[72] CHOP/GADD153 is induced at the mRNA level in response to genotoxic and inflammatory stresses. These treatments also activate the transcriptional regulatory functions of CHOP/GADD153 by triggering phosphorylation of Ser78 and Ser81. CHOP/GADD153 is a transcriptional repressor of certain cAMP-regulated genes and a transcriptional activator of some stress-induced genes. Activation by genotoxins of CHOP/GADD153 mediates, in part, cell cycle arrest at G1/S, a well established consequence of DNA damage. p38α is a likely stress-activated regulator of CHOP/GADD153 func-

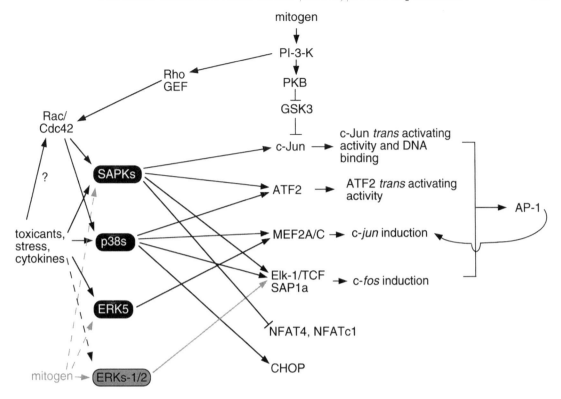

**Figure 3** Regulation of transcription factors by MAPKs. Note that AP-1 regulation involves both the direct phosphorylation of AP-1 components, as well as the transcriptional induction of AP-1 components mediated by distinct transcription factor targets of MAPKs.

tion, given that p38α, and not SAPK or ERK, can phosphorylate CHOP/GADD153 at Ser78 and Ser81 *in vivo* and *in vitro*[76] (Figure 3).

*NFAT4*: Transcription factors of the nuclear factor of activated T cells (NFAT) family are distantly related to Rel/NF-κB (reviewed in Reference 77). In resting cells, NFATs are retained in the cytosol as a consequence of phosphorylation (catalyzed by casein kinase-Iα and, possibly, glycogen synthase kinase (GSK)-3) at 5–6 sites (within NFAT4, this comprises a region spanning amino acids 204–215). This phosphorylation affects NFAT conformation so as to mask the nuclear localization signal. Agonist-induced $Ca^{2+}$ entry recruits the $Ca^{2+}$-dependent phosphatase calcineurin (phosphatase 2B), which dephosphorylates NFATs, exposing their nuclear localization signals and triggering NFAT nuclear translocation. Dephosphorylation of NFATs also enhances DNA binding affinity. NFATs bind and *trans* activate genes with the cognate *cis* acting element (consensus sequence: $_T/^AGGAAAAT$). NFAT sites are often located close to AP-1 sites in many promoters, allowing for the coordinated *trans* activation of numerous proinflammatory genes (IL-2, IL-4, IL-5 and CD40L are examples).[77] Calcineurin is a major target of the immuno-

suppressants FK506 and cyclosporin-A, and, accordingly, inhibition of NFAT activity is an important consequence of FK506 and cyclosporin-A action.[77,78]

Serum factors can substantially suppress the $Ca^{2+}$-mediated nuclear translocation of NFATs. Insofar as casein kinases and GSK3 are not serum-stimulated, the serum effect is likely due to the activation of serum-responsive kinase cascades.[77,79] The SAPKs can phosphorylate the NFAT family member NFAT4 (also called NFATc3) at Ser163 and Ser165. This phosphorylation correlates with an inhibition of stimulus-induced NFAT4 nuclear translocation. Based on this finding, it has been suggested that the SAPK pathway antagonizes NFAT4 action[79] (Figure 3). However, the significance of these results is unclear. Thus, NFAT4 mutants in which the putative SAPK phosphoacceptor sites (Ser163 and Ser165) have been changed to Ala still show serum-stimulated inhibition of nuclear translocation in many instances. Moreover, the SAPK-specific MAP3K MEK-kinase-1 (Section 14.4.2.2.7.2) can block NFAT4 dephosphorylation and activation whether or not the SAPK phosphorylation sites have been mutated to alanine. Dominant inhibitors of the SAPK pathway do not reverse the effect of

MEKK1. Apparently, MEKK1 fosters inhibition of NFAT nuclear translocation by stabilizing the association between NFAT4 and the inhibitory NFAT kinase casein kinase-Iα.[80]

In addition to NFAT4, the SAPKs can phosphorylate NFATc1 (also called NFAT2). In this instance, phosphorylation is stimulated by PMA and ionomycin and occurs at Ser117 and 172. Phosphorylation inhibits or delays the accumulation of NFATc1 in the nucleus by blocking the binding of calcineurin, an essential step in NFATc1 activation.[81] NFATc1 is crucial to the differentiation of $T_H$ cells to the $T_H2$ effector phenotype. Interestingly, disruption of SAPKγ or both SAPKγ and α leads to the preferential accumulation of $T_H2$ cells, consistent with the notion that SAPK-mediated inhibition of $T_H2$ differentiation is mediated through NFATc1.[81-83]

*AP-1*: The SAPKs and p38s are the major Ser/Thr kinases responsible for the recruitment of the activator protein-1 (AP-1) transcription factor in response to stresses.[1,84] ERK5 also plays a role in stress-activated AP-1 regulation.[85] AP-1 is heterodimeric complex, consisting of bZIP transcription factors of the Jun family (c-Jun, JunD) partnered with other Jun transcription factors or with members of the Fos (usually c-Fos) or activating transcription factor (ATF—typically ATF2) family[84] (Figure 3).

The presence of Jun family members enables AP-1 to bind to *cis* acting elements containing the tetradecanoyl phorbol myristate acetate (TPA) response element (TRE—consensus sequence: TGAC/GTCA). ATFs, including ATF2, are members of the CREB subfamily. Thus, AP-1 heterodimers containing ATF transcription factors can bind both to the TRE and to the CRE.[72,84] AP-1 is an important *trans* activator of a number of stress responsive genes including the genes for interleukins-1 and -2, CD40, CD30, TNF and c-Jun itself. In addition, AP-1 participates in the transcriptional induction of proteases and cell adhesion proteins (e.g., E-selectin) important to inflammation.[77,84,86]

AP-1 activation is complex, involving both the direct phosphorylation/dephosphorylation of AP-1 components, as well as the phosphorylation and activation of transcription factors that induce elevated expression of *c-jun* or *c-fos*. Both events can be activated independently by several signaling pathways making AP-1 a site of integration for numerous protein kinase pathways (Figure 3). Thus, c-Jun is phosphorylated in resting cells at a region immediately upstream of the carboxyl terminal the DNA binding domain (Thr231, Thr239, Ser243, Ser249). This phosphorylation inhibits DNA

binding and is catalyzed *in vivo* by glycogen synthase kinase-3 (GSK3).[87] GSK3 is inhibited upon mitogen stimulation by a mechanism dependent upon phosphatidyl inositol-3-kinase (PI-3-K).[88,89] Upon stimulation of cells with AP-1 activators, the c-Jun carboxyl terminal phosphates are removed under conditions in which GSK3 is inactivated (Figure 3).[84,87]

Phosphorylation of c-Jun or ATF2 within their amino terminal *trans* activation domains correlates well with enhanced *trans* activating activity.[24-26,90] c-Jun residues Ser63 and Ser73 in the *trans* activation domain are phosphorylated *in vivo* under conditions wherein the SAPKs are activated and SAPKs can phosphorylate these sites *in vitro* and *in vivo*.[24-26] Immunodepletion of SAPK from cell extracts removes all stress- and TNF-activated c-Jun kinase.[25] Thus, SAPKs are the dominant kinases responsible for stress- and TNF-activated c-Jun phosphorylation (Figure 3). JunD is also phosphorylated at Ser90 and Ser100 by SAPKs, albeit less effectively than is c-Jun. Ser90 and Ser100 of JunD lie within a region of the JunD *trans* activation domain similar to the phosphoacceptor domain of c-Jun.[7]

The SAPKs and p38s can both phosphorylate ATF2 at Thr69 and Ser71 in the *trans* activation domain. Again, these residues are phosphorylated *in vivo* under conditions in which the SAPKs and p38s are activated. Phosphorylation of ATF2 at Thr69 and Ser71 correlates with activation of ATF2 *trans* activating activity[90] (Figure 3). Whether the SAPKs or p38s represent the dominant ATF2 kinases depends on the cell type and stimulus used. During reperfusion of ischemic kidney, for example, the SAPKs are the only detectable ATF2 kinases[91] whereas in KB keratinocytes treated with interleukin-1 (IL-1), the p38s are the major ATF2 kinases.[63]

In response to stress, the SAPKs and p38s also contribute to AP-1 activation by stimulating the transcription of genes encoding AP-1 components[84,85] (Figure 3). The *fos* promoter includes a *cis* acting element, the serum response element (SRE), that mediates the recruitment of a heterodimer of the serum response factor (SRF) and a member of the ternary complex factor (TCF) family (reviewed in Reference 92) (Figure 3). The TCFs include Elk-1 and Sap-1a.[92] The SAPKs, and ERKs, but not p38, can phosphorylate two critical residues in the Elk1 C-terminus (Ser383, Ser389), while the p38s can efficiently phosphorylate the corresponding residues (Ser381 and Ser387) on Sap1a.[9,93-96] Phosphorylation of TCFs enhances their binding to the SRF and thereby triggers *trans* activation at the SRE. By these processes the SAPKs and p38s along with the ERKs can convergently contribute to c-*fos* induction (Figure 3).[9,93–96]

The p38s and ERK5 can also phosphorylate transcription factors of the myocyte enhancer factor-2 MEF2 subgroup of the MCM1-agamous and deficiens-SRF (MADS) box transcription factor family.[85,97,98] MEF2s (MEF2A-D) were originally identified as transcription factors that bound to AT-rich sequences (consensus: CTAAAAATAA) and *trans* activated key genes involved in myoblast differentiation; however, some MEF2s, notably MEF2C, are widely expressed and may regulate additional transcriptional regulatory events,[97,98] (reviewed in Reference 99). MEF2A and C are MAPK substrates, MEF2B and D are not. However, MEF2B and D may act in conjunction with phosphorylated MEF2A/C. Thus, phosphorylation enhances the *trans* activating activity of MEF2A and B or A and D heterodimers.[98] MEF2A is phosphorylated at Thr312 and Thr319 by p38α.[98] MEF2C is a substrate of both the p38s and ERK5; however, the p38s and ERK5 phosphorylate different sets of sites on the MEF2C polypeptide. Thus, p38s phosphorylate Thr293 and Thr300 whereas ERK5 phosphorylates Ser387.[85,97] Of note, all three sites are phosphorylated in response to serum or stress. Thr293/Thr300 phosphorylation is sufficient for p38 activation of MEF2C while Ser387 phosphorylation is sufficient for ERK5 activation of MEF2C.[85,97] A *cis* element for MEF2C is present in the c-*jun* promoter; thus, p38 and ERK5 activation can contribute to the induction of c-*jun* expression[97] that in turn potentiates AP-1 activation. Indeed MEF2A or C, once activated by p38s can *trans* activate c-*jun*[85,97,98] (Figure 3).

### 14.4.2.2.5.3 Regulation of AP-1 by Stress-activated MAPKs is Mediated by Divergent AP-1 Component Transcription Factor Subunits

The different aspects of AP-1 regulation—activation of constituent transcription factor expression and direct phosphorylation/activation of constituent transcription factors—can be independently regulated by several pathways in response to different types of stimuli (Figure 3). Stresses and inflammatory cytokines such as TNF, which preferentially activate the SAPKs and p38s, can recruit AP-1 through the direct phosphorylation of AP-1 components (c-Jun by the SAPKs and ATF2 by both the SAPKs and p38s). In addition, stress pathways can also promote enhanced expression of AP-1 components through recruitment of Elk-1 (mediated by SAPK phosphorylation), which results in elevated c-*fos* expression, and through p38-catalyzed phosphorylation of MEF2A/C (Figure 3). MEF2A/C can bind and *trans* activate the promoter for c-*jun*. Stresses can also modestly recruit PKB/Akt, which can act to inhibit GSK3 thereby blocking its negative regulation of c-Jun DNA binding. Finally, the c-Jun promoter also contains an AP-1 site; thus, c-*jun* expression can be autoregulated by any pathway that activates AP-1 (Figure 3).

### 14.4.2.2.6 MEKs Upstream of the SAPKs and p38s

#### 14.4.2.2.6.1 Activation of the SAPKs by SEK1 and MKK7

The SAPKs are activated upon concomitant phosphorylation at Thr183 and Tyr185. The MEK SAPK/ERK kinase-1 (SEK1, also called MAPK-kinase-4, MKK4; MEK4; JNK kinase-1, JNKK1; and SAPK-kinase-1, SKK1, Table 2) was cloned independently by two groups who employed degenerate PCR to identify novel MAPK signaling components.[100,101]

The homology shared by SEK1 and MEKs-1 and -2 (as well as yeast MEKs) indicated that this kinase lay upstream of MAPKs. It was shown subsequently that SEK1 could phosphorylate and activate all three SAPK isoforms *in vivo* and *in vitro*.[100,101] Dérijard *et al.* also showed that SEK1 could phosphorylate and activate p38 *in vivo* when overexpressed, and *in vitro*.[101] However, the role of SEK1 in p38 activation by SEK1 is unclear and may be cell-

**Table 2**  MEK nomenclature.

| Name | Alternate names | Substrate(s) |
| --- | --- | --- |
| MEK1 | MAPKK1, MKK1 | ERK1, ERK2 |
| MEK2 | MAPKK2, MKK2 | ERK1, ERK2 |
| SEK1 | MKK4, JNK kinase (JNKK)-1, MEK4, SAPK-kinase (SKK)-1 | SAPKs |
| MKK7 | JNKK2, MEK7, SKK4 | SAPKs |
| MKK3 | MEK3, SKK2 | p38s |
| MKK6 | MEK6, SKK3 | p38s |

and stimulus-dependent. Thus, if activated SEK1 concentrations in *in vitro* assays are adjusted to initial rate conditions for SAPK activation, little or no p38 activation is observed.[63,102] Moreover, targeted disruption of *sek1* in mice has no effect on p38 activity in ES cells.[103] However, disruption of *sek1* does impair p38 activation in knockout fibroblasts.[104]

Considerable genetic and biochemical evidence indicated that SEK1 was not the only SAPK-activating MEK. These studies revealed that the spectrum of SAPK activators recruited by different stimuli depended on the stimulus used and on the cell type. Thus, hydroxylapatite fractionation of extracts of osmotically shocked 3Y1 fibroblasts identified a broad peak of SAPK activating activity that was fully resolved from a separate peak of SEK1 immunoreactivity.[105] Likewise, Mono-S chromatography of KB cell extracts showed that IL-1 failed to activate SEK1, but stimulated a broadly eluting peak of SAPK activating activity distinct from SEK1. Similar multiple peaks of SAPK activating activity were observed in extracts of PC-12 cells treated with arsenite or osmotic shock; and in KB cells treated with osmotic shock, UV radiation or anisomycin.[102] In osmotically shocked 3Y1 fibroblasts, SEK1 represented a comparatively minor peak of SAPK activating activity.[105] By contrast, SEK1 was more strongly activated by osmotic shock, UV and arsenite in PC-12 cells.[102] In KB cells, IL-1 failed to activate SEK1; however, SEK1 and other SAPK activators were activated by anisomycin, osmotic shock and UV radiation.[102]

Targeted disruption of *sek1*, while embryonically lethal, results in only a partial ablation of SAPK activation, blocking anisomycin and heat shock activation of SAPK, while leaving osmotic shock and UV activation of SAPK unaffected.[103] Finally, *hemipterous* is a *Drosophila* MEK required for dorsal closure during embryogenesis and deletion/mutagenesis of *hemipterous* is lethal. Although *hemipterous* is significantly homologous to SEK1, SEK1 cannot rescue *Drosophila* mutants wherein *hemipterous* is deficient.[106,107]

MKK7 (also called MEK7, JNKK2 and SKK4, Table 2) was isolated roughly contemporaneously by several laboratories and cDNA sequencing revealed an enzyme with close homology to *hemipterous*.[107–113] Consistent with its structural homology to *hemipterous*, MKK7 can effectively rescue *hemipterous* lethality whereas SEK1 cannot.[107]

MKK7 displays a strong preference for SAPK, even under conditions of high expression. This contrasts with SEK1 that can, under

some conditions, activate p38.[101,103,104,107–113] MKK7 can activate all SAPK isoforms tested equally well.[107–113] MKK7, like SEK1 is subject to alternative hnRNA splicing where the α, β and γ isoforms differ at their amino termini while type 1 and 2 isoforms differ at their carboxyl termini. The β and γ isoforms bind directly to SAPKs via an amino terminal extension not present in the α isoforms. Upon overexpression, the α isoforms exhibit a lower basal activity and higher-fold activation by upstream stimuli.[114]

MKK7 is vigorously activated by TNF and IL-1, conditions which, at best, induce modest SEK1 activation. In contrast, SEK1 and MKK7 are activated with equal potency by osmotic shock, while MKK7 is more weakly activated by anisomycin, a potent SEK1 activator.[103,104,107–113] Thus it is plausible to argue that MKK7 represents at least a substantial portion of the biochemically detected SAPK activating activity present in 3Y1, PC-12 or KB cells subjected to osmotic shock or anisomycin, or in KB cells treated with IL-1 (Figure 4).

While the activation patterns of SEK1 and MKK7 do not to overlap entirely, there is evidence that SEK1 and MKK7 actively cooperate in the activation of SAPK. Thus, whereas SEK/MKK4 is not strongly activated by TNF, TNF cannot activate SAPK in fibroblasts derived from *sek1−/−* early mouse embryos, in spite of the fact that MKK7 is expressed in these cells.[104,115] A possible reason for this anomaly was revealed in biochemical studies of the mechanisms of SEK1 and MKK7 catalysis. As was mentioned above, SEK1 preferentially targets SAPK Tyr185 and only weakly phosphorylates Thr183.[100,116] By contrast, MKK7 preferentially phosphorylates SAPKs at Thr183 rather than Tyr185.[116] Accordingly, when assayed individually at low concentrations, SEK1 and MKK7 are comparatively poor SAPK activators *in vitro*—each stimulating at best a 5–10 fold activation of SAPK. When added together, however, SEK1 and MKK7 synergistically activate SAPK, resulting in >100 fold SAPK activation accompanied by equal phosphorylation of Thr183 and Tyr185[116] *in vitro*. Thus, *in vivo* activation of SAPK by TNF may involve the combined effects of modestly activated SEK1 acting primarily on SAPK Tyr185 and strongly activated MKK7 acting primarily on Thr183. Because both Thr183 and Tyr185 must be phosphorylated for SAPK activation, deletion of *sek1* would, by decreasing sharply SAPK Tyr phosphorylation, substantially compromise SAPK activation by TNF.[116]

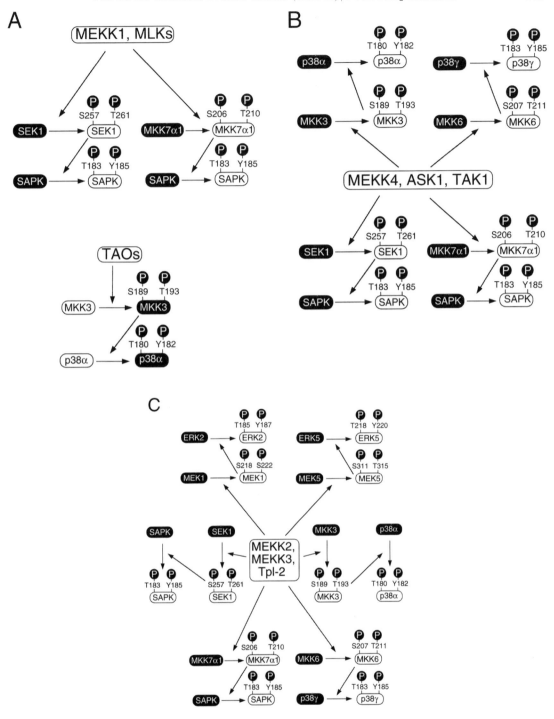

**Figure 4** Representative stress and cytokine-regulated three tiered core signaling pathways. Note the diversity of MAP3K promiscuity. (A) MEKK1, MLK2 and TAOs are relatively selective for the SAPKs; (B) MEKK4, ASK1, and TAK1 regulate the SAPKs and p38s; (C) MEKK2, MEKK3 and Tpl-2 can regulate all MAPK pathways.

### *14.4.2.2.6.2 Activation of the p38s by MKK3 and MKK6*

MKK3 (also called MEK3 and SKK2) and MKK6 (also called MEK6 and SKK3, Table 2) were cloned by degenerate PCR using conserved MEK sequences as templates.[101,117] Independently, MKK6 was cloned by PCR using primers derived from purified protein sequence.[118] Both enzymes are highly selective for p38 and do not activate SAPK or ERK under all conditions tested (Figure 4).[101,117,118]

MKK3 and MKK6 manifest more substantial differences in their substrate selectivity with regard to p38 isoforms than do SEK1 and MKK7 for the various SAPK isoforms. MKK3 preferentially activates p38α and p38β while MKK6 can activate strongly all known p38 isoforms. Similarly, MKK3 appears to be more limited with regard to activation by upstream stimuli. Whereas MKK6 is activated by all known p38 activators, MKK3, like SEK1, is more strongly activated by physical and chemical stresses.[118]

### 14.4.2.2.6.3 Activation of ERK5 by MEK5

MEK5 (Table 2) was identified in a degenerate PCR screen designed to identify novel MEKs.[53] MEK5 is subject to differential hnRNA splicing resulting in four polypeptides. Splicing near the 5' end shifts the reading frame to generate a long (450 amino acids, AA) polypeptide localized in the particulate, cytoskeletal fraction, and a short (359 AA) polypeptide that is cytosolic. A second splicing event substitutes two 10 AA cassettes in a region between subdomains IX and XI, and gives rise to α and β MEK5 isoforms that are analogous to type-1 and -2 SAPKs (Table 1).[25,36,54] Insofar as subdomains in the region spanning subdomains IX and XI are important to substrate binding, α and β MEK5s may differ in their substrate specificities. ERK5 was isolated in a two-hybrid screen for proteins that interact with MEK5.[53] Upon expression with the MAP3Ks Tpl-2 and MEKK3 (Section 14.4.2.2.7.2), MEK5 is activated and can itself activate ERK5 *in vivo* and *in vitro*.[16,119]

### 14.4.2.2.7 Several Divergent Families of MAP3Ks are Upstream of the SAPKs, p38s and ERK5

### 14.4.2.2.7.1 General Considerations

As with all MEKs, the stress-activated MEKs are activated by phosphorylation at two Ser/Thr residues within a conserved region of the P-activation loop of subdomain-VIII of the kinase domain. The SEK1 phosphorylation sites are Ser257 and Thr261,[100,101] those for MKK7 are Ser206 and Thr210 (α isoforms),[107–114] those for MKK3 are Ser189 and Thr193,[101] those for MKK6 are Ser207 and Thr211[117,118] and those for MEK5 are Ser311 and Thr315 (for the long form of MEK5).[53,54]

A large number of MAP3Ks with an intimidating, complex enzymology regulate stress-activated MAPKs. This heterogeneity is consistent with the many different stimuli types that recruit these MAPK pathways. The MAP3Ks upstream of SAPK and p38 fall into three broad protein kinase families: the MEK kinases (MEKKs), the mixed lineage kinases (MLKs) and the thousand and one kinases (TAOs) (Figure 4).

### 14.4.2.2.7.2 The MEKKs

The MEKKs are the most diverse of the eukaryotic MAP3Ks in terms of enzymology, regulation and molecular structure. The common feature of the MEKKs is a conserved catalytic domain homologous to that of *S. cerevisiae* STE11, a MAP3K that regulates both the yeast mating pheromone and osmosensing pathways. Outside of the kinase domains, there is little or no conservation. Some MEKKs can catalyze the activation of multiple different MAPK pathways and interact with a wide array of putative regulatory proteins. Mammalian MEKKs include MEKKs-1–4, apoptosis signal-regulating kinase-1 (ASK1), transforming growth factor-β (TGF-β)-activated kinase-1 (TAK1), and Tpl-2[12,14,120–129] (Table 3). NF-κB-inducing kinase (NIK) is an additional MEKK that is a specific activator of the NF-κB pathway[130] and will not be discussed here (reviewed in References 131, 132).

*MEKK1, a comparatively selective activator of the SAPK pathway*: In an effort to identify MAP3Ks that were upstream of the ERKs, Johnson and colleagues employed a cloning strategy that exploited the existing knowledge of yeast signaling pathways (the mating factor pathway of *S. cerevisiae* in particular) to isolate ~80-kDa fragment of MEKK1 (Table 3). Thus, degenerate PCR primers based on conserved elements of the STE11 sequence and the related *S. pombe* MAP3K *byr2* were used to amplify mammalian homologues of the yeast kinases.[3,120] Full length MEKK1 is a 150-kDa polypeptide that consists of a carboxyl terminal kinase domain (AAs 1221–1493) and an extensive amino terminal domain (AAs 1–1221) that includes two proline-rich segments (AAs 74–149 and 233–291) containing putative binding sites for proteins with SH3 domains, a consensus binding site for 14-3-3 proteins (AAs 239–243) two PH domains (AAs 439–455 and 643–750) and an acid rich motif (AAs 817–1221) which contains two sites for cleavage by cysteine proteases of the proapoptotic caspase family (Asp871, Asp874).[120,124,133,134] Within the kinase domain is a binding site for Ras, the exact location of which has not been clearly

**Table 3** MAP3K nomenclature.

| Name | Alternate names | Substrates/Effectors |
|------|-----------------|----------------------|
| Raf-1 | | MEK1, MEK2 |
| A-Raf | | MEK1, MEK2 |
| B-Raf | | MEK1, MEK2 |
| MEKK1 | | SEK1, MKK7 |
| MEKK2 | | SEK1, MEK1 |
| MEKK3 | | SEK1, MEK1, MKK3, MKK6 |
| MEKK4 | MTK1 | SEK1, MKK3, MKK6 |
| ASK1 | MAPK kinase-kinase-5 (MAPKKK5) | SEK1, MKK3, MKK6 |
| TAK1 | | SEK1, MKK3, MKK6 |
| Tpl-2 (in rats) | Cot (in humans) | MEK1, SEK1 |
| MLK2 | MST | SEK1, MKK7 |
| MLK3 | SPRK, PTK1 | SEK1, MKK7 |
| DLK | MUK, ZPK | SEK1, MKK7 |
| TAO1/2 | PSK (a splicing isoform of TAO2) | MKK3, SEK/MKK7 (PSK only) |

defined.[135] MEKK1 can also bind Rho family GTPases.[136]

MEKK1 is a highly selective activator of the SAPKs (Figure 4(A)). Consistent with this, biochemical studies that indicate that MEKK1 can activate SEK1 *in vitro* and *in vivo*.[137,138] Whereas overexpressed MEKK1 can activate the ERKs and even p38, kinetic studies of MEK1/2 and MKK3/6 activation by MEKK1 suggest that these MEKs are activated with a $K_{cat}$ at least three orders of magnitude lower that that for activation of SEK1.[120,130] In addition to SEK1, MEKK1 can activate MKK7α1, α2, β1 and β2 as well as γ1 *in vivo*; and of the MAP3Ks tested in transfection experiments for MKK7 activation, MEKK1 is the most potent overall *in vivo* MKK7 activator.[114] However, careful *in vitro* biochemical and kinetic analysis of MEKK1 activation of MKK7γ1 *versus* SEK1 indicate that SEK1 is a preferred MEKK1 substrate (other MKK7 isoforms were not tested), with a $K_m$ nearly two orders of magnitude lower than that for activation of MKK7γ1.[139] The basis for this difference in *in vivo* and *in vitro* activity is unclear, and may reflect artifacts of transient expression. Alternatively, *in vivo* yet to be identified scaffold proteins, not employed in *in vitro* assays using purified proteins, may enable significant MKK7 activation by MEKK1 *in vivo*. MEKK1 is significantly activated by TNF,[140] microtubule poisons,[141] oxidant stress[142] and some receptor Tyr kinase agonists.[136]

*MEKKs-2 and -3: more promiscuous MAP3Ks that can activate the SAPK, p38, ERK-1/2 and ERK5 pathways*: Using the same approach as that employed in the cloning of MEKK1, Johnson *et al.* cloned murine MEKKs-2 and -3 (Table 3). Independently, human MEKK3 was cloned by differential display screening.[12,14] MEKKs-2 and -3 are more closely related within their kinase domains to each other (> 90% identity) than they are to MEKK1 (~65% identity). MEKK2 and MEKK3 are smaller polypeptides than MEKK1 (~70 and 71-kDa, respectively); and each has a carboxyl terminal catalytic domain (AAs 362–619 for MEKK2, AAs 368–626 for MEKK3) preceded by amino terminal noncatalytic domains. The amino terminal noncatalytic portions of MEKKs-2 and -3, while significantly similar in primary sequence, contain no motifs suggestive of function or regulation.[12,14]

In contrast to MEKK1 that is strongly selective for the SAPKs, MEKKs-2 and -3 are considerably more promiscuous. In transfection experiments, each can activate both the SAPK and ERK pathways (Figure 4(C)). Consistent with concomitant ERK and SAPK activation, MEKKs-2 and -3 can each activate MEK1 and SEK1 *in vivo* and *in vitro*.[12,14] MEKK3 can also activate MKK7α1 and, to a much lesser extent, MKK7α2, β1 and β2 *in vivo*.[13,114] The reason for the selectivity of MEKK3 for different MKK7 isoforms is unclear. MEKK2 and MEKK3 can also activate the p38s *in vivo*[13,15,143] (and see below); and MEKK3 can activate MKK3 and MKK6 *in vivo* and *in vitro*.[13,15] Chao *et al.* recently demonstrated that MEKK3 could activate MEK5 and the ERK5 *in vivo*.[16]

Stimuli that recruit MEKKs-2 and -3 have not been well characterized. Recent studies indicate that MEKK2 is activated in T cells upon stimulation with antigen presenting cells (APCs).[143] EGF activates MEKK3's *in vitro* kinase activity towards MEK5 and MEKK3's ability to recruit ERK5 *in vivo*.[16] Interestingly, a dominant inhibitory form of MEKK3 (Lys391Trp) blocked EGF activation of

ERK5 but not ERK1.[16] This is likely due to the fact that Raf-1 is the dominant EGF-activated MAP3K for the ERK pathway.[2] Thus, MEKK3 is a putative effector for mitogen/Tyr kinase regulation of all known mammalian MAPK pathways (Figure 4(C)).

*MEKK4 can activate both the SAPKs and the p38s:* MEKK4 (also called MAP three kinase-1, MTK1, Table 3) was isolated by Johnson *et al.* employing again a PCR strategy based on degenerate primers derived from *STE11* and *byr2*.[126] MEKK4 was also cloned independently by Saito *et al.* as a human cDNA that, when expressed in *S. cerevesiae*, could rescue cells missing the genes for all three MAP3Ks of the *HOG1* osmoadaptation pathway (*SSK2*, *SSK22*, and *STE11*).[3,127] MEKK4 is a ~150-kDa polypeptide that consists of a carboxyl terminal kinase domain (AAs 1337–1597) and an extensive amino terminal regulatory region that includes a putative polyproline SH3 binding motif (AAs 27–38),[126,127] a binding site for growth arrest and DNA damage (GADD)-45 family proteins (AAs 147–250, see Section 14.4.2.4.4.3),[144] a putative PH domain (AAs 225–398), and a Cdc42/Rac interaction and binding (CRIB) domain (AAs 1311–1324) that enables GTP-independent binding to Rho subfamily GTPases,[126,127,136] (see Section 14.4.2.3.2.1). The kinase domain of MEKK4 shares ~55% amino acid homology with that of MEKKs-1, -2 and -3. Johnson and colleagues have reported that MEKK4 is selective for the SAPKs and can activate only SEK1 *in vivo* and *in vitro*;[126] however, Saito *et al.* used titered expression of increasing levels of MEKK4 and demonstrated that MEKK4 could activate the SAPKs (via SEK1) and the p38s (via MKK3 and MKK6) with equal potency *in vivo* and *in vitro*.[127] By contrast, Davis and colleagues reported that MEKK4 selectively activated the p38 pathway *in vivo*[114] (Figure 4(B)). The reason for these discrepancies is unclear. MEKK4 may be an effector for stress pathways activated by genotoxins (Section 14.4.2.4.4.3).

*TAK1 can activate both the SAPKs and the p38s:* TGF-β-activated kinase-1 (TAK1) (Table 3) was cloned in a novel screen for mammalian MAP3Ks. This screen employed a mutant strain of *S. cerevisiae* in which the mating factor pathway MAP3K *STE11* was deleted, and a mutant form of the mating factor pathway MEK *STE7*, *STE7*[P368], was expressed. *STE7*[P368] is a gain-of-function mutant that still requires Ste11p for full activity; however, the selectivity of Ste7[P368]p for upstream activators is less stringent than that of wt Ste7p. Thus, constitutively active Raf-1, MEKK1, and, by extension, presum-

ably other MAP3Ks, can substitute for Ste11p in *STE11/STE7*[P368] mutant cells; thus, the *STE11/STE7*[P368] mutant is a powerful reagent for the identification of novel MAP3Ks. Accordingly, TAK1 was cloned as an additional MAP3K that could substitute for Ste11p in *STE11/STE7*[P368] cells.[3,122] TAK1 is a ~60-kDa polypeptide with an amino terminal kinase domain (AAs 30–294) preceded by a short regulatory motif (AAs 1–22) and followed by a carboxyl terminal extension (AAs 295–557) of unknown function.[122] The amino terminal regulatory motif appears to serve an inhibitory role. Thus, full length TAK1 cannot substitute for Ste11p in *STE11/STE7*[P368] cells whereas TAK11–22 is able to substitute.[122] Two-hybrid screening with TAK1 AAs 1–22 as bait identified TAK1 binding proteins (TABs)-1 and -2, two regulatory proteins that bind the TAK1 amino terminal inhibitory motif and are essential for coupling TAK1 to upstream signals (see Section 14.4.2.4.5).[145,146] In over-expression experiments, TAK1 can activate both the SAPKs and p38s (Figure 4(B)). *In vitro* and *in vivo*, TAK1 can phosphorylate and activate SEK1, MKK3 and MKK6. Endogenous TAK1 is activated by TGF-β, IL-1 and TNF.[122,145–148]

*ASK1 can activate both the SAPKs and the p38s:* Apoptosis signal-regulating kinase-1 (ASK1, also called MAPK kinase-kinase-5, MAPKKK5, Table 3) was cloned by Wang *et al.* and, independently, by Ichijo *et al.*, both of whom used a PCR strategy that employed degenerate primers based on conserved elements of Ser/Thr kinase subdomains VI and VIII.[125,128] A ~150-kDa polypeptide, ASK1 consists of a centrally located kinase domain (AAs 677–936), an amino terminal extension (AAs 1–676) that includes segments that bind polypeptides of the TNFR-associated factor (TRAF) family and the redox sensing enzyme thioredoxin.[125,128,149–151] In addition, ASK1 possesses a carboxyl terminal extension (AAs 937–1375) that has also been implicated in TRAF binding.[150,151] ASK1 can activate the SAPKs (via SEK1) and the p38s (via MKKs-3 and -6) *in vivo* and *in vitro*[125,128] (Figure 4). Endogenous ASK1 is activated by oxidant stress, TNF and FasL. Activation of ASK1 by TNF is dependent upon TNF-induced production of reactive oxygen species.[128,149–153] ASK1 is likely to be an important effector coupling these agonists to the SAPKs and p38s.

Under low serum conditions, expression of ASK1 from a Zn-inducible promoter induces apoptosis in several cell lines.[128] The targets recruited by ASK1 (especially if the SAPKs or p38s are involved) that promote apoptosis are unknown. Dominant inhibitory, kinase-dead

ASK1 mutants can block apoptosis stimulated by TNF or oxidant stress.[128,149] Thus, either ASK1 is a direct apoptogenic signaling component recruited by these ligands, or the dominant inhibitory ASK1 is sequestering a common upstream element that regulates both ASK1 and TNF/oxidant-induced apoptosis.

*Tpl-2 can activate the SAPKs, p38s, ERKs1/2 and ERK5: Tumor progression locus-2 (Tpl-2)* is the rat homologue of the human proto oncogene *cot* (Table 3). *Tpl-2* encodes a ∼ 50-kDa protein Ser/Thr kinase with an amino terminal domain, of unknown function, which is truncated in some mRNA splicing isoforms, a central catalytic domain (AAs 139–394) that is significantly homologous to *STE11*, and a carboxyl terminal regulatory domain (AAs 395–467).[121,129] The oncogenic potential of *Tpl-2* is activated in rat thymomas as a result of the additive, spontaneous proviral insertion of the Moloney leukemia virus into the infected cell genome at the *Tpl-2* locus, the consequence of which is positive selection of tumor cells and enhanced tumor progression *in vivo*.[121] Moloney leukemia virus proviral insertions at *Tpl-2* always target the last intron of the gene and elicit the enhanced expression of a carboxyl terminally truncated, constitutively active protein, both through enhanced transcription and mRNA stabilization. It is likely, therefore, that the C-terminal regulatory domain of Tpl-2 exerts a negative effect on Tpl-2 activity.[129]

Transient expression of Tpl-2 activates the ERK pathway in parallel to Raf-1.[154] Subsequently, Ley *et al.*, as well as Tsichlis *et al.* showed that expression of Tpl-2 activated the SAPKs and ERKs with equal potency.[123,129] Ley *et al.* then demonstrated that, consistent with its homology to *STE11*, Tpl-2 was a MAP3K that could directly activate MEK1 and SEK1 *in vivo* and *in vitro*[123] (Figure 4(C)). While overexpression of full length Tpl-2 activates both the SAPKs and ERKs, expression of the oncogenic carboxyl terminally truncated Tpl-2 results in substantially greater SAPK and ERK activation, further supporting the contention that the carboxyl-terminal domain negatively regulates Tpl-2 activity.[129]

A recent study by Chiariello *et al.* has revealed a far broader substrate specificity for Tpl-2. Thus, ectopic expression of Tpl-2 expression activates not only ERK1 and SAPK, but p38γ and ERK5 *in vivo*. Notably, Tpl-2 does not activate p38α or δ *in vivo*. Correspondingly, coexpression with Tpl-2 activates MEK1, MKK6, MEK5 and SEK1. The selective activation of p38γ is curious, insofar as MKK6 can activate all known p38 isoforms *in vitro* and in transfection experiments *in vivo*.[119] Mechanisms of sequestration, such as

scaffold proteins, may isolate Tpl-2 and MKK6 from all p38 isoforms other than γ (Figure 4(C)). Extracellular stimuli that recruit Tpl-2 *in vivo* are unknown.

### 14.4.2.2.7.3  Mixed Lineage Kinases (MLKs) MLK3, MLK2 and DLK: Selective Activators of the SAPKs

The mixed lineage kinases (MLKs) are a small family of protein Ser/Thr kinases that share a general structural configuration wherein an amino terminal kinase domain is followed by one to two leucine zippers, a Cdc42/Rac interaction and binding (CRIB) domain (Section 14.4.2.3.2.1) and a carboxyl terminal proline-rich domain with several consensus SH3 binding motifs. Four MLKs have been identified, MLK1, MLK2 (also called MKN28 cell-derived Ser/Thr kinase, MST, Table 3), MLK3 (also called SH3 domain-containing proline-rich kinase, SPRK or protein Tyr kinase-1, PTK1) and dual leucine zipper kinase (DLK, also called MAPK upstream kinase, MUK or zipper-containing protein kinase, ZPK, Table 3). MLK2 and MLK3, in addition to the common features described above, each contain an SH3 domain amino terminal to their kinase domains.[155–157]

As their names suggest, the kinase domains of the MLKs bear structural similarities to both Ser/Thr and Tyr kinases. However, the MLKs are clearly Ser/Thr specific *in vitro* and *in vivo*. For example, MLK1 contains a Lys residue (Lys129 in the sequence His-Arg-Asp-Leu-Lys) in subdomain VIb that is characteristic of Ser/Thr kinases; however, two Trp residues in MLK1 subdomain IX (Trp192 and Trp199) are highly conserved among Tyr kinases, as is a motif in subdomain XI (Met241-Glu-Asp-Cys-Trp-Asn-Pro-Asp-Pro-His-Pro-Ser-Arg-Pro-Ser-Phe255) which conforms to a conserved region in Tyr kinases (Met - X - X - Cys - Trp - X - X - Asp/Glu - Pro-X-X- Arg-Pro-Ser/Thr-Phe, where X is any amino acid).[155].

MLK3, MLK2 and DLK are potent and, in contrast to the more promiscuous MEKKs, specific activators of the SAPKs *in vivo* in transfection experiments.[114,158–162] ERK, p38 and NF-κB are not activated by MLKs-2, -3 or DLK except under conditions of extreme overexpression (Figure 4(C)). MLK3, MLK2 and DLK can activate SEK1 *in vivo* and *in vitro* to a degree commensurate with that catalyzed by MEKK1.[114,158,159,161,162] In addition, MLK3 and DLK can activate MKK7 *in vivo* in transfection experiments, with the degree of activation dependent upon the MKK7 isoform. Thus, MKK7β1 and β2 are strongly activated *in vivo* by MLK3 and DLK, while α isoforms are more modestly activated.[114] This selectivity is likely due to the structural differences in the amino termini of the α and β isoforms of MKK7 (discussed in Section 14.4.2.2.6.1). By contrast, MLK2 can strongly activate MKK7α1. Indeed, MLK2 is a stronger MKK7α1 kinase than it is a SEK1 kinase.[162].

### 14.4.2.2.7.4    TAO Kinases are Novel MAP3Ks that Regulate the p38s and are Structurally Homologous to both Ste20 and MAP3Ks

A novel family of 1001 amino acid Ser/Thr kinases, the 1001 (TAO) kinases (TAO1 and TAO2), were recently identified in a screen for additional mammalian kinases homologous to *S. cerevisiae* STE20, a proximal kinase thought to be important in the regulation of the Ste11p/Ste7p/Fus3p/Kss1p yeast mating pheromone MAPK core signaling module.[3,163,164] Prostate-derived STE20-like kinase (PSK, Table 3), a possible splicing isoform of TAO2 that includes an extended carboxyl terminal tail, was cloned independently by Moore, *et al.* as a kinase expressed at elevated levels in prostate tumors.[165] TAOs consist of amino terminal kinase domains and very large (700 AA) carboxyl terminal extensions of poorly characterized function. The kinase domains of TAOs are significantly homologous to Ste20p (40% identity) and the germinal center kinases (43% identity with GCK) (Section 14.4.2.3.3), a family of mammalian Ste20-like kinases.[3,163,164,166] However, there is also appreciable identity with the MLK2 kinase domain (33% overall), especially within the substrate binding motif.[161,163]

TAOs-1 and -2 appear to be specific activators of the p38s *in vivo*. The purified kinase domain of TAO1 will directly phosphorylate and activate MKK3 *in vitro* and *in vivo*. Moreover, when expressed in Sf9 cells from a recombinant baculovirus, TAO1 will interact *in vivo* and coimmunoprecipitate with MKK3. Thus, although TAO1 bears significant homology to Ste20s and GCKs, kinases which are thought to regulate MAP3Ks (Section 14.4.2.3.3), TAOs appear to be direct MAP3Ks selective for MKK3[163] (Figure 4(A)). The ability of TAO1 to catalyze directly the activation of MEKs may be due to the kinase domain homology with MLK2 in the substrate binding region.

The high *in vivo* selectivity of the TAOs for MKK3 may be due to the presence of a specific MKK3 binding site. AAs 314–451 of the TAO2 non-catalytic domain selectively binds MKK3 *in vitro* and *in vivo*.[164]

PSK, the TAO2 splicing isoform, appears to activate selectively the SAPK pathway; ERK and p38 are not activated in transfection experiments.[165] This difference in pathway selectivity is curious given that the MKK3 interaction loop found in TAO2[164] is present in PSK.[165] The C-terminal tail unique to PSK may account for the altered specificity of PSK, although this has not been demonstrated.

### 14.4.2.3    UPSTREAM OF MAP3KS

#### 14.4.2.3.1    General Considerations

Sections 14.4.2.2.2–2.7 have described much of what we know of mammalian stress-activated MAP3K/MEK/MAPK core signaling modules. The following five Sections (14.4.2.3.1–3.5) will describe some of the known proximal signaling components thought to couple these core modules to environmental stresses and toxicants.

Monomeric GTPases of the Ras superfamily are potent upstream activators of signal transduction pathways, and Ras is pivotal to the activation of the mitogenic ERK pathway. The activation of Ras by guanine nucleotide exchange factors (GEFs), and the inactivation of Ras by GTPase activating proteins (GAPs) has been extensively reviewed (reviewed in References 167–169).

Ras regulates ERK predominantly through triggering the activation of the MAP3K Raf-1 (reviewed in Reference 2). The regulation of Raf-1 by Ras is a paradigm for MAP3K regulation, and studies of Raf-1 regulation point to the following crucial steps in MAP3K activation: (i) binding to an upstream activating protein accompanied by translocation to the membrane, (ii) phosphorylation and (iii) homo-oligomerization (Figure 5(A)). Thus, Raf activation requires binding to GTP-Ras, an event that results in membrane translocation.[2] At the membrane, Raf is phosphorylated by several protein kinases including Tyr kinases such as Src and Ser/Thr kinases including p21-activated kinase (PAK)-3, an effector for Rac1 and a Rho family GTPase that is itself a Ras target.[170–173] Phosphorylation of Raf-1 results in structural alterations that change the binding to proteins of the 14-3-3 family. These structural changes are necessary for Raf-1 activation.[174] Ras-dependent oligomerization is also necessary for Raf-1 activation.[175] In the same vein, activator protein binding/membrane translocation, phosphorylation and oligomer-

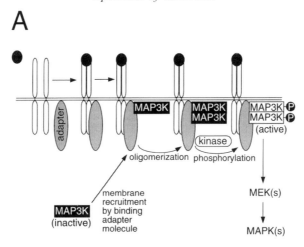

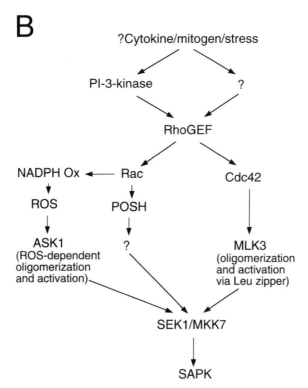

**Figure 5** (A) General principles of MAP3K regulation. The steps illustrated do not necessarily proceed in the order shown; (B) Rho GTPase effectors (POSH, MLK3 and ROS-dependent activation of ASK1) that might regulate the SAPKs. Stimuli that recruit Rac and Cdc42 are not well understood. Exactly which mechanism(s) are employed likely depends on the cell type and the stimulus.

ization may be important in the regulation of stress-activated MAP3Ks; and indeed, regulation by at least some of these mechanisms has been documented for stress-activated MAP3Ks.

### 14.4.2.3.2 Regulation of the SAPKs and p38s by Rho Subfamily GTPases

As with Ras-dependent recruitment of the ERK pathway, at least some stress-activated

MAPK core pathways may be regulated by members of the Rho subfamily of the Ras superfamily. The mammalian Rho subfamily consists of the Rho (RhoA-E), Rac (Rac1 and -2) and Cdc42 (Cdc42Hs, G25K, Tc10 and Chp) groups.[167,176,177] Rho GTPases were originally implicated in the regulation of the actin cytoskeleton.[176] As with other members of the Ras superfamily, Rho subfamily GTPases are active in the GTP-bound state (promoted by GEFs) and inactive in the GDP-bound state (promoted by GAPs). Rac, Cdc42Hs and other Rho subfamily GTPases share a large number of GEFs that belong to the Dbl family. Examples include Ost, a GEF for RhoA and Cdc42Hs; Tiam-1, a GEF for Rac1 and Cdc42Hs; Lbc, a GEF for RhoA; and faciogential dysplasia-1 (FGD1), a GEF selective for Cdc42Hs,[167–169,176] (reviewed in Reference 178).

All Rho family GEFs consist of a conserved Dbl homology (DH) domain that catalyzes GDP dissociation, and a plekstrin homology (PH) domain. Inositol phospholipids, including inositol-3′4′5′-trisphosphate, the product of PI-3-kinase activity, are targeting signals for PH domain-containing polypeptides.[89] Activation of PI-3-kinase requires Ras[179] and, accordingly, Rho family GEFs themselves may be subject to regulation by Ras/PI-3-kinase.[171,178] Indeed mitogen activation of Rac1 requires activation of PI-3-kinase.[171]

GTPase-deficient (Val12), constitutively active mutants of Rac1, Cdc42Hs or Chp are potent activators of the SAPKs and p38s.[177,180–184] RhoA generally does not activate SAPK or p38. Consistent with these findings, expression of the *dbl* proto-oncogene product or FGD1 (Rho GEFs that recruit Rac and Cdc42) also results in strong SAPK activation.[180,185] EGF (through Ras), or transforming alleles of *ras* may signal to the SAPKs via Rac1 inasmuch as Asn17 Rac1 can block the activation of SAPK by these stimuli. In addition, Rac1 is also involved in SAPK activation in response to CD3-CD28 T cell costimulation.[181,186] Cdc42Hs, in contrast, is activated by lysophosphatidic acid, bombesin and, in spreading melanoma cells, engagement of chondroitin sulfate proteoglycan.[176,187]

### 14.4.2.3.2.1 Coupling Rho GTPase Target to SAPK and p38 Core Signaling Pathways

As mentioned above, proteins of the Ras superfamily regulate their targets in part through a direct binding interaction.[168,169] This binding results in membrane translocation and, in the case of Ras-Raf, is a necessary prerequisite for subsequent activation events such as homo-oligomerization and phosphorylation by upstream kinases.[172,173,175] Most, but not all direct targets for Rac1 and Cdc42Hs contain a Cdc42/Rac interaction and binding (CRIB) domain which interacts directly with Rac1 and Cdc42Hs.[188] Several polypeptide species including protein kinases and adapter proteins are candidate Rac1 and Cdc42Hs effectors that couple to the SAPKs and p38s.

*p21-activated kinases (PAKs):* STE20 encodes a *S. cerevisiae* Ser/Thr protein kinase with an amino terminal regulatory domain containing a CRIB motif that binds Cdc42Sc, the yeast Cdc42Hs ortholog.[3] Ste20p is thought to regulate the mating pheromone MAPK core signaling module Ste11p/Ste7p/Fus3p/Kss1p.[3] Genetic epistasis studies provisionally placing Ste20p upstream of the MAP3K Ste11p led to the idea that Ste20-like kinases in mammals might couple Rho GTPases MAPK core signaling modules.[3]

The p21-activated kinases (PAKs, PAK1, also called αPAK; PAK2, also called γPAK or hPAK65; PAK3, also called βPAK; and PAK4, Table 4) are structural and functional orthologs of *S. cerevesiae* Ste20p.[3,189–193] *In vitro* and *in vivo*, the kinase activity of PAKs is activated upon binding GTP-Rac1 or GTP-Cdc42Hs through a process that

**Table 4** GCK/PAK nomenclature.

| Name | Alternate names | Substrates/Effectors |
|---|---|---|
| GCK | | MEKK1, MLK3 |
| GCKR | kinase homologous to Ste20 (KHS) | MEKK1 |
| GLK1 | | ? |
| HPK1 | | MEKK1, MLK3 |
| NIK | NIK (not to be confused with NF-κB-inducing kinase), HPK1/ GCK-related kinase (HGK) | MEKK1 |
| Misshapen | Msn | ? |
| PAK1 | αPAK | ? |
| PAK2 | γPAK, hPAK65 | ? |
| PAK3 | βPAK | Raf-1 |

Tables 1–4 Nomenclature for mammalian MAPK pathway components. Included are commonly accepted names found in the primary literature; ERK pathway nomenclature is also shown, but is not discussed in the text (for review see Marshall, 1995). Not all of the names listed are included in the text.

involves an obligatory autophosphorylation event.[189,190,192] PAK2 can also undergo activation during apoptosis through a mechanism that involves caspase 9-mediated cleavage at Asp212. This proteolytic event removes the amino terminal regulatory domain (including the CRIB motif).[194] The significance of this cleavage-mediated activation is unclear, although PAK2 may mediate some morphologic functions associated with apoptosis such as cell shrinkage.[194]

A major function of the PAKs (PAK1 in particular) is to serve as effectors for Cdc42 and Rac in the regulation of the actin cytoskeleton.[195,196] However, several reports have indicated that constitutively activated forms of PAKs can also activate coexpressed SAPKs and p38s.[183,184,197,198] For example, mutation of Leu107 to Phe, within the CRIB motif of human PAK1, results in a constitutively-active construct that can activate coexpressed SAPK.[198] Moreover, addition to cell-free extracts of *Xenopus* oocytes of a PAK1 mutant wherein the amino terminal regulatory domain has been deleted also results in substantial SAPK activation.[197] However, SAPK activation by coexpressed PAKs has not been universally observed; expression of PAKs does not synergize with Rac or Cdc42 to activate SAPK further.[136] Thus, it is not clear if PAKs are true effectors coupling Rho family GTPases to the SAPKs and p38s.

*MAP3Ks*: Several established stress-sensitive MAP3Ks, including MEKK1,[136] MEKK4,[126,136] MLK2 and MLK3[188,199] bind to and are putative effectors of Rac1 and Cdc42Hs (Figure 5(B)). Consistent with this, the MLKs and MEKK4 possess CRIB motifs. Moreover, MLKs-2 and -3 as well as MEKK1 can bind Rac1 and Cdc42 in a GTP-dependent manner, and there is some evidence that MLK3 is a Cdc42 effector. On the other hand, it is unlikely that any of these MAP3Ks is a direct Rac effector (below). MEKK4 binds Cdc42 only, and in a GTP-independent manner,[136,188,199] making the significance of this interaction unclear. With the exception of MLK3 (below), a functional consequence for GTPase binding to stress-activated MAP3Ks has not been established.

MLK3 and MLK2, two mixed lineage kinase activators of the SAPKs, contain CRIB motifs and can interact directly with Cdc42Hs and Rac1 in a GTP-dependent manner.[188,199] As noted above, there is evidence that MLK3 is a Cdc42Hs effector. As with all MLKs, MLK3 possesses a leucine zipper (AAs 400–487).[156] Leucine zippers frequently mediate protein-protein binding and

homo or heterodimerization. MLK3, when overexpressed, will spontaneously dimerize *in vivo*.[200] The leucine zipper domain of MLK3 is required for homodimerization, insofar as deletion of this domain abrogates *in vivo* MLK3 homodimerization.[200] The ability to homodimerize is critical to MLK3 activity and deletion of this domain inactivates MLK3 even if the kinase domain is intact. While MLK3 will spontaneously homodimerize *in vivo* upon transient expression, coexpression with Cdc42 significantly enhances MLK3 homodimerization suggesting that Cdc42 regulates MLK3 by triggering its homodimerization and activation.[200]

While MLKs 2 and -3 may be Cdc42Hs effectors (below) it is unlikely that they are Rac targets.[199,201] Thus, Phe37Ala-Rac1 is an effector loop mutant of Rac1 that retains its ability to recruit the SAPK pathway *in vivo*.[202] However, neither MLK3 nor MLK2 can interact with Phe37Ala-Rac1 *in vivo* either in COS1 cells or in a yeast two-hybrid assay.[201,202] As is discussed below, Phe37Ala-Rac1 can bind POSH, and POSH is a likely candidate effector coupling Rac1 to the SAPKs[201] (Figure 5(B)).

*POSH*: Plenty of SH3 domains (POSH) is a ~90-kDa polypeptide that consists of four SH3 domains (AAs 139–190, 198–254, 457–511 and 838–892), a domain rich in putative polyproline SH3 binding sites (AAs 368–405), but no CRIB motif. In spite of this, POSH interacts directly with Rac1 (but not Cdc42Hs, Ras or Rho) in yeast two-hybrid assays and a GTP-dependent manner *in vitro*. The Rac binding domain has been localized to AAs 292–362, a domain that appears thus far to be unique to POSH.[201] Transient expression of POSH in COS1 cells results in potent SAPK activation.[201] POSH recruitment by Rac1 effector mutants correlates tightly with the ability of these Rac1 mutants to activate the SAPKs. Thus, in contrast to MLK3 and MLK2, POSH can interact with Phe37Ala-Rac1, a Rac1 effector mutant that cannot elicit lammelopodium formation but does activate coexpressed SAPK.[201,202] However, POSH cannot interact with Tyr40Cys Rac1, a second Rac1 effector mutant that does not activate the SAPKs *in vivo*.[201,202] These results suggest that POSH, but not the MLKs, is a likely effector for Rac1 activation of the SAPKs. The mechanism by which POSH recruits the SAPKs is unclear. It is plausible to speculate that the SH3 domains of POSH may serve a scaffolding function, binding MAP3Ks such as MEKK1 or MLKs, whichcontain SH3 binding sites (Figure 5(B)).

#### 14.4.2.3.4 Regulation of the SAPKs by the Germinal Center Kinases

##### 14.4.2.3.4.1 General Considerations

Germinal center kinase (GCK) is the founding member of a novel family of stress-regulated Ser/Thr protein kinases of which some are selective activators of the SAPKs.[166,203] Those GCKs that recruit the SAPKs may do so in part by binding MAP3Ks and elements upstream of MAP3Ks.[166,204] Thirteen mammalian GCKs have been cloned. In addition, there are *Drosophila*, *C. elegans* and *Dictyostelium* homologues, as well as two *S. cerevisiae* genes with known phenotypes (reviewed in Reference 166). All GCKs possess amino terminal kinase domains that are distantly related to those of the PAKs and Ste20p. The kinase domains are followed by extensive carboxyl terminal regulatory domains (CTDs).[166] While the distant sequence homology between the kinase domains of the GCKs, Ste20p and the PAKs led to the initial placement of the GCKs in the Ste20 family, the different domain arrangement of the GCKs (amino terminal kinase domain, carboxyl terminal regulatory domain, no CRIB motif) indicate that the GCKs should be considered a distinct protein kinase family.[166]

GCKs can be subdivided into at least two, and possibly three groups based on overall sequence conservation and function: Group-I GCKs are structurally similar within both the kinase and CTD regions to GCK itself and are specific upstream activators of the SAPKs. Group-II GCKs are more closely related to the *S. cerevisiae* GCK Sps1p and are activated *in vivo* by extreme stresses. Few downstream effectors of group-II GCKs have been identified.[166] A putative third GCK subgroup consists thus far only of JNK/SAPK inhibitory kinase (JIK).[166,205]

Seven mammalian group-1 GCKs have been identified thus far: GCK itself, GCK-related (GCKR, also called kinase homologous to Ste20, KHS, Table 4), GCK-like kinase (GLK), hematopoietic progenitor kinase-1 (HPK1) and Nck-interacting kinase (NIK, also called HPK1/GCK-like kinase, HGK; The GCK NIK should not be confused with NF-κB inducing kinase,[130] which is also abbreviated NIK), TRAF and Nck interacting kinase (TNIK), and NIK-related kinase (Nrk) (Table 4).[203,206-213]

There is remarkable structural conservation among the CTDs of group-I GCKs, which suggests similar functional and regulatory properties. The CTDs of all group-I GCKs include two or more proline/glutamic acid/ serine/threonine (PEST) motifs and at least two polyproline putative SH3 domain binding sites. Of particular importance, all of the group-I kinases possess a highly conserved ~350 AA region, the GCK homology domain, at the carboxyl-terminal end of the CTD. The GCK homology domain can be subdivided into a somewhat hydrophobic, leucine-rich domain and a 142–152 AA stretch, the carboxyl terminal (CT) motif. The leucine residues in the Leu-rich domains are not organized into leucine zippers, nor are these domains sufficiently hydrophobic for membrane insertion[203,206-213] (reviewed in Reference 166).

GCK, GCKR, GLK and NIK are all activated *in vivo* by TNF. TNIK may also be a TNF effector and the CT extensions of the CTDs of GCK, GCKR and GLK along with that of HPK1 are quite homologous (Figure 8(A)).[166,203,204,206,207,209-211]

Several group-I GCKs (GCK, HPK1 and NIK) can bind MAP3Ks. For example, GCK and NIK can bind MEKK1, HPK1 can bind both MEKK1 and MLK3.[204,206-208] Moreover, GCK, GCKR and the *Drosophila* GCK Misshapen can interact with proteins of the TNF receptor-associated factor (TRAF) family (Section 14.04.2.3.5) (Figure 6).[204,211,214,215] Studies of GCK and GCKR indicate that the CT motif is required for binding TRAFs, and, possibly, for regulating the binding of MAP3Ks.[204,214] The notion that GCKs can both potently activate the SAPKs as well as stably associate with SAPK core signaling module elements and their upstream activators, suggests that the group-I GCKs may coordinately regulate MAPK core signaling modules in conjunction with additional upstream elements.

#### 14.4.2.3.5 Regulation of the SAPKs and p38s by TRAFs, Adapter Proteins that Couple to TNF Family Receptors and to ER Stress Signaling

Upon binding ligand, receptors of the tumor necrosis factor family (the TNFR family) can elicit a wide variety of inflammatory responses and are critical to immune cell development, innate and acquired immunity as well as the pathogenesis of a number of diseases such as arthritis, septic shock and, possibly, type-2 diabetes mellitus. The TNFR superfamily includes TNFR1, TNFR2, the lymphotoxin-β (LT-β) receptors, CD40, CD27, Fas, receptor activator of NF-κB (RANK, also called osteoprotegerin, OPG, or TNF-related activation-induced cytokine, TRANCE, receptor), TNF-related apoptosis inducing ligand recep-

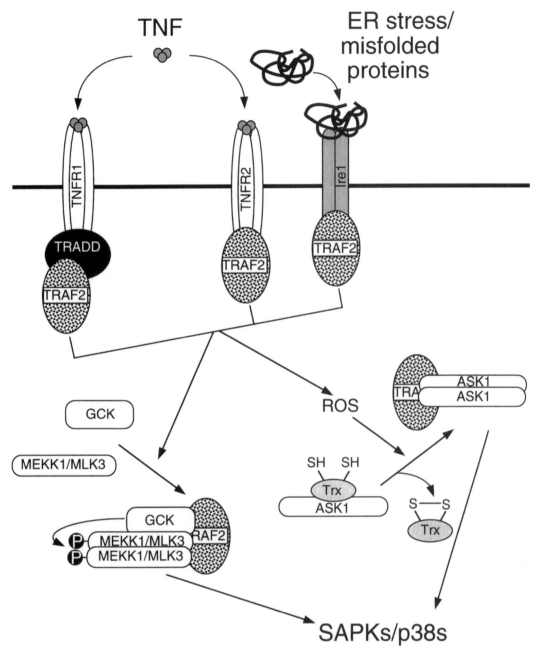

**Figure 6** TRAF2 conveys cytokine and ER stress signals to the SAPKs and p38s. Note that TNF/TRAF2 regulation of ASK1 is mediated in part by ROS. GCK family kinases likely can bind both MEKK1 and MLK3. ASK1, MLK3 and, possibly, MEKK1 are activated by oligomerization, as indicated in the diagram. GCK may also stimulate an obligatory activating autophosphorylation of MEKK1.

tor (TRAILR), CD30, Ox40, 4–1BB, the p75 neurotrophin receptor (p75NTR) and the death receptors (DRs). The receptors for interleukin 1 (IL-1) and the Toll-like receptors (TLRs, which, among other things, include receptors for endotoxin) are structurally completely divergent from the TNFR family; however, their mechanisms of signaling are quite similar to those of TNFR superfamily and, accordingly, these receptors are often grouped with the TNFR superfamily. Several viral genes also encode TNFR family polypeptides; the latent membrane protein-1 (LMP1) of Epstein–Barr virus is an example. With the exception of the IL-1R and the TLRs, most TNFR superfamily receptors share a modest structural similarity within their extracellular domains, in keeping with the structural conservation of many TNF superfamily ligands. By contrast, in spite of the similarities in TNFR superfamily signaling programs and modes of receptor activation, the intracellular extensions of TNFR family receptors are quite divergent.[216–221]

### 14.4.2.3.5.1    The Protein Recruitment Model for TNFR Family Signaling

Receptors of the TNFR family possess no intrinsic enzymatic activity. In the resting state, at least a subset of these receptors (TNF receptors and Fas) exist as preassembled trimers tethered together by an extracellular motif referred to as the pre-ligand assembly domain (PLAD). Upon binding ligand, these receptors assemble into higher order complexes (in the case of receptors held together by PLAD domains, the PLAD–PLAD interactions dissociate and new interactions occur). This further receptor oligomerization is thought to trigger conformational changes in the receptor intracellular extensions that enable the binding of cytosolic signal transducing polypeptides which, in turn, recruit downstream targets.[219,222,223]

*Death domains and signaling*: Several TNFR family receptors (TNFR1, Fas, TRAILR and the DRs, in particular) contain an 82–102 AA extension, the death domain.[201–204] Death domains mediate homotypic and heterotypic protein–protein interactions and are critical for nucleating receptor-effector complexes and implementing several signaling programs including, as the name suggests, apoptosis.[218,219,222] The type-1 TNFR (TNFR1) can recruit all of the known signaling pathways activated by TNF. Signal transduction by TNFR1 is now understood in considerable detail and has become a paradigm for signaling by the rest of the TNFR family. Upon ligand engagement, the TNFR1 death domain binds the death domain of the platform adapter protein TNFR-associated death domain protein (TRADD).[219,222,224] TNFR2, which does not possess a death domain, upon ligand engagement, instead binds directly to TNFR-associated factor (TRAF) proteins through a TRAF domain binding site (below).[225]

TRADD is a critical proximal component in the recruitment by TNFR1 of downstream targets. TRADD consists of a carboxyl terminal death domain and an amino terminal domain, the TRAF interaction domain, which binds TRAF proteins (see below). Overexpression of TRADD can activate many TNF signaling pathways including apoptosis and NF-κB.[225] TRADD overexpression does not activate the SAPKs,[226–228] in spite of the fact that TRADD can recruit TRAF2 and receptor interacting protein (RIP),[229,230] two key upstream activators of the SAPKs and p38s.[204,226–228] The reason for this discrepancy is unknown.

The interaction between the TRADD death domain and that of TNFR1 is also thought to trigger the binding of additional death domain proteins to the TRADD death domain. An example is Fas-associated death domain protein (FADD), an adapter protein that couples TNFR1 to the apoptotic machinery. FADD associates with TRADD through a homotypic interaction between the two proteins' death domains.[229] TRAF proteins are also recruited to the TRADD-TNFR1 complex through an interaction between the TRAF interaction domain of TRADD and the TRAF domains of TRAF2 (below)[218,219,222,229] (Figure 6).

*TRAFs*: The TRAFs are a family of signal transducing adapter proteins that are important for the activation of a number of pathways in response to TNF family ligands and environmental stresses[231] (reviewed in References 218, 219). Six mammalian TRAF polypeptides (TRAFs-1–6) have been identified, all of which consist of two tandem carboxyl terminal TRAF domains (TRAF-N followed by TRAF-C), a central zinc finger motif and, with the exception of TRAF1, an amino terminal RING finger motif.[218,219] The TRAF domains are responsible for binding upstream activators such as TRADD, as well as some TRAF effectors. The function of the Zn finger domains is unclear. The RING fingers are required for the activation of downstream effectors[140,151,214,218,219,226–228,232] (Figure 6).

Details of the structural features required for TRAF binding to upstream proteins are emerging. TRAFs-1 and -2 were identified based on their interactions with the intracellular extension of TNFR2, to which both proteins bind directly.[218,219,225] Since then, the repertoire of polypeptides that bind TRAFs has expanded dramatically and a greater understanding has developed of the structural basis of TRAF proteins binding to their activators/effectors. Crystallographic studies indicate that the TRAF domains of TRAF2 trimerize upon interaction with trimerized upstream activating receptors.[233,234] A subset of TRAF2 interacting proteins contains at least one of two primary sequence motifs to which the TRAF2 TRAF domains bind: Pro/Ser/Ala/Thr-X-Gln/Glu-Glu, the predominant site, and a secondary sequence, Pro-X-Gln-X-X-Asp, where X is any amino acid.[235]

A notable exception to this rule is TRADD, which couples TNFR1 to TRAF2.[224,229]

TRADD does not contain consensus TRAF2 TRAF domain binding sites.[224,235] Thus, other TRAF domain binding motifs likely exist, and these other sites may link TRADD to TRAF2. This leaves open the possibility that TRAF domains could conceivably associate contemporaneously with upstream and downstream polypeptides each with distinct TRAF domain binding motifs. For example, once TRAF2 is bound to TRADD, the regions of the TRAF2 TRAF domains that interact with the consensus TRAF2 binding motifs are free to interact with other proteins. Conversely, TRAF2 associated with a "consensus" TRAF2 binding site on one of its binding partners may have additional sites on its TRAF domains free to associate with other proteins.

### 14.4.2.3.5.2  TRAFs-2, -5, and -6 can Activate the SAPKs and p38s

Transient overexpression of TRAF2 results in potent SAPK and p38 activation.[204,226–228] Deletion of the TRAF2 RING finger domain abrogates completely the ability of TRAF2 to recruit the SAPKs and p38s. Moreover, RING-TRAF2 can act as a dominant inhibitor of TNF activation of the SAPKs and p38s.[204,226–228] Targeted disruption of *traf2*, or transgenic expression of RING-TRAF2, abolishes TNF activation of SAPK; however, IL-1 activation of SAPK is unaffected (p38 was not tested).[236,237] Thus, TRAF2 is a rate limiting signal transducer that couples TNFRs to the SAPKs and, possibly, the p38s. TRAFs-5 and -6, when expressed ectopically, can also strongly activate the SAPKs; however, overexpression of TRAFs-1, 3 and 4 does not result in SAPK activation.[238] Disruption of *traf6* blocks IL-1 and CD40 signaling to SAPK,[239] indicating that TRAF6 is central to these cytokines' signaling to the SAPKs just as TRAF2 is essential to TNF activation of the SAPKs (Figure 6).

### 14.4.2.3.6  GCKs may Collaborate with TRAFs to Activate MAP3Ks Including MEKK1

The group-I GCKs, GCK itself, GCKR, GLK and NIK are activated by TNF *in vivo*;[209,211,240,241] and there is evidence that GCK, GCKR and, possibly, NIK and TNIK are TRAF effectors.[204,208,212,214,215] In particular, GCKR is activated by TNF and upon coexpression with TRAF2 and both GCK and GCKR, binds TRAF2 *in vivo*.[211,214] Further-more, in spite of the observation that TRAF2 and MEKK1 can associate apparently independently of coexpressed GCK family kinases,[140] there is evidence that GCKs, TRAFs and, perhaps additional components TNFR family complexes, collaboratively activate MAP3Ks upstream of the SAPKs and p38s. Thus, (i) antisense GCKR constructs will block TNF activation of SAPK in 293 cells, (ii) some GCKs (GCK, GCKR, TNIK) associate *in vivo* with TRAFs[204,212,214,215] and with MAP3Ks regulated by TRAFs (GCK's interaction with MEKK1[204]), (iii) dominant inhibitory MEKK1 can block GCK, NIK, HPK1 and GCKR activation of SAPK,[207,208,211] and (iv) genetic studies of *Drosophila* implicate GCK family kinases as TRAF effectors.[215]

GCK, GCKR and TNIK can bind TRAF2 *in vivo* in a reaction requiring the TRAF2 TRAF domains. A similar interaction was reported for the interaction between *Drosophila* Misshapen and a *Drosophila* TRAF DTRAF.[204,212,214,215] Coexpresson with TRAF2 activates GCKR; and deletion of the TRAF2 RING domain prevents the mutant TRAF2 from activating coexpressed GCKR.[211] From these results, it is plausible to speculate that the TRAF domains of TRAF2 bind GCKs and GCKR while the RING domains mediate GCK/GCKR activation. Thus, a stable interaction with the TRAF2 RING domain is not an absolute requirement for all TRAF2 effectors, although the RING domain is essential for regulation of TRAF effectors.

Although biochemical and genetic studies suggest that GCKs either relay signals from TRAFs to MAP3Ks or collaborate with TRAFs to activate MAP3Ks,[204,214,215] biochemical evidence also indicates that TRAF2 and MEKK1 interact apparently independently of coexpressed GCKs.[140] How might these conflicting results be reconciled? It is possible, of course, that both mechanisms operate separately and in parallel. Alternatively, the TRAF2-MEKK1 interaction may be comparatively weak, and might be stabilized or potentiated by GCKs once the GCKs themselves are activated by TRAFs. GCKs would not be detected in assays of a TRAF2-MEKK1 interaction.[140] Both the TRAF-MEKK and GCK-MEKK binding results are consistent with the notion that a complex consisting of a group-I GCK and MEKK1 could interact with TRAF2 in such a way as to permit a regulatory interaction between MEKK1 and the TRAF2 RING domain sufficiently stable to detect in TRAF2-MEKK1 coimmunoprecipitation assays.

Another possibility is that both TRAFs and GCKs might contribute contemporaneous, yet independent inputs to MEKK1 activation, requiring the obligate formation of a TRAF-group-I GCK family-MEKK1 complex.

Just what constitutes group-1 GCK and MEKK1 activation is still nebulous. Inasmuch as forced oligomerization of TRAF2 results in activation of the SAPKs, and TRAF2 and MEKK1 can coimmunoprecipitate in a TNF-dependent manner,[140] oligomerization may be key to MEKK1 activation; but this has yet to be demonstrated. The fact that GCK's kinase activity enables maximal recruitment of the SAPKs[204,240] suggests that phosphorylation may also contribute to MEKK1 activation. Support for a role for autophosphorylation in the regulation of MEKK1 comes from the finding that the activity of a truncated MEKK1 construct (either AAs 817–1493 or 1221–1493) requires phosphorylation at Thr1381 and Thr1393 for activity. As with many kinase regulatory phosphoacceptor sites, both of these residues lie within the activation loop just upstream of kinase subdomain VIII. 1–1221 MEKK1 autophosphorylates at Thr 1381, *in vitro* and mutagenesis of either Thr1381 or Thr1393 to Ala abolishes activity.[242,243] What is still unclear is the mechanism by which this autophosphorylation is triggered—specifically, is the autophosphorylation cotranslational? Does the autophosphorylation require additional steps, including priming phosphorylation by upstream kinases such as GCK, or is autophosphorylation triggered by aggregation alone?

## 14.4.2.4   PIECING IT ALL TOGETHER: COUPLING MAP3K / MEK / MAPK CORES TO FIVE STRESS INPUTS: OXIDANT STRESSES, ULTRAVIOLET RADIATION, HEAT STRESS, GENOTOXINS AND ER STRESS

### 14.4.2.4.1   General Considerations

Sections 14.4.2.2.2–2.7 have described much of what we know of mammalian stress-activated MAP3K/MEK/MAPK core signaling modules; Sections 14.4.2.3.1–3.5 have outlined some of the proximal signals that are thought to couple to these cores. The following sections (14.4.2.4.1–4.5) will describe how some of the known proximal signaling components are thought to couple these core modules to environmental stress stimuli. Discussion will be restricted to the SAPKs and p38s insofar as little of MEK5/ERK5 regulation is under-

stood. I will focus on five broad classes of stress: oxidant stresses, ultraviolet radiation, heat stress, genotoxins and endoplasmic reticulum (ER) stress. As described above (Section 14.4.2.3.1), studies of the Ras-ERK pathway suggest that MAP3K regulation involves four basic mechanisms: binding to upstream regulatory proteins, translocation to membrane compartments, oligomerization and phosphorylation. Recent studies indicate that toxicants and environmental stresses recruit MAP3Ks by similar processes.

### 14.4.2.4.2   Oxidant Stress Regulation of SAPK, p38

#### 14.4.2.4.2.1   Overview

Oxidant stress, the effect of oxidizing/reducing agents on cell processes, is a phenomenon associated with a number of pathophysiologic conditions, and is a byproduct of a diverse array of stressful stimuli. Thus, many chemotherapeutic agents, including compounds thought primarily to be DNA damaging agents (*cis*-platinum [CDDP], bleomycin, doxorubicin [dxr], etoposide, β-lapachone [β-lap]), trigger the production of ROS in their target cells. Certain general DNA damaging agents such as methyl methane sulphonate activate the SAPKs via DNA damage-dependent and -independent mechanisms—with the DNA-independent mechanisms likely involving ROS production. Moreover, ionizing radiation (UV) plus older anti-cancer drugs such as nitrogen mustards and cyclophosphamide, as well as the commonly used nonsteroidal chemotherapeutic tamoxifen, can also induce ROS production.[244–248] Isothiocyanates also activate SAPK in a ROS-dependent manner.[249] Conversely, naturally occurring antioxidants, including a subset from green teas (epigallocatechin), block serum-dependent SAPK activation.[250] Interestingly, the ROS-dependent SAPK activation associated with isothiocyanates has been implicated in the antiproliferative/antioncogenic and proapoptotic effects of these compounds[249] while the proliferative effect of serum on vascular smooth muscle is *blocked* by the green tea antioxidant epigallocatechin via inhibition of ROS-dependent activation.[250] Thus, ROS-dependent SAPK activation may have proliferative or antiproliferative effects depending on the cell type. As is noted in Section 14.4.2.5, recent studies implicate the SAPKs in toxicant-dependent apoptosis.

Ultraviolet radiation stimulates the production of ROS, likely by affecting membrane lipid oxidation and by promoting the oxidation of cell surface protein sulfhydryls. This, in turn, may trigger the oligomerization and activation of cell surface receptors; thus any oxidant could also trigger cell surface receptor clustering and signaling (see Section 14.4.2.4.3.2 and Figures 5(B)–7). Proinflammatory cytokines such as TNF and IL-1 also stimulate the production of ROS in reactions catalyzed by dedicated oxidases including the Rac-dependent activation of NADPH oxidase and other mechanisms (Figures 5(B)–7).[176,178,251,252] Notably, TNF and related cytokines also stimulate the production of prostanoids, including leukotrienes. The ω-oxidation of leukotrienes by membrane cyto-chrome p450s has also been associated with the production of biologically active ROS.[252] Finally, the release of cytochrome *c* during cell death induced by treatment with a combination of TNF and cycloheximide, by preventing the transport of electrons from cytochrome complex-III to cytochrome *b* and cytochrome *c* oxidase, has also been associated with the production of ROS which may participate in amplifying proapoptotic signal transduction.[253]

Although it is clear that oxidants can activate the SAPKs and the p38s (and ERK5), very little is known about the mechanisms that govern ROS production, and even less is known of how this ROS production is coupled to regulation of signal transduction. ASK1 is a MAP3K that may serve as a target for ROS signaling to the SAPKs, while MEKK1 may act to protect cells from ROS-induced cell death.

### 14.4.2.4.2.2  *ASK1, Thioredoxin and Regulation by ROS*

Expression of TRAFs-2, 5 or 6 results in vigorous SAPK and p38 activation, and TRAF2 is required for SAPK activation by TNF, while TRAF6 is required for IL-1 activation of SAPK.[204,218,219,226–228,236–239] ASK1 is activated by coexpressed TRAF2, and recombinant ASK1 can interact *in vivo* with recombinant TRAFs-2, -5 and -6. Moreover, endogenous TRAF2 and ASK1 interact *in vivo* in a TNF-dependent manner[150,151,254] (Figure 6). Both the amino and carboxyl terminal non-catalytic domains of ASK1 (AAs 937–1375) interact with TRAF2 (through the TRAF domains).[150,151]

The regulation of ASK1 by TRAF2 is redox dependent (Figure 6). Thus, addition of free radical scavengers prevents the TNF-dependent association of endogenous ASK1 and TRAF2.[151] Yeast two-hybrid screening has revealed that the redox sensing oxidoreductase thioredoxin (Trx) is an endogenous inhibitor of ASK1. This inhibition requires that Trx be in a reduced state. Thus, treatment of cells with oxidant stresses ($H_2O_2$) triggers the dissociation of Trx from ASK1 and the activation of ASK1 *in vivo*;[149] and TNF fosters the dissociation of ASK1 from Trx by a process that can be blocked with free radical scavengers.[149] TNF treatment is known to generate a pulse of reactive oxygen intermediates (ROIs) with kinetics that parallel the activation of ASK1 by TNF.[153,255] Furthermore, this ROS pulse may itself be TRAF2-dependent inasmuch as activation of the SAPKs by coexpressed TRAF2 is partially reversed with free radical scavengers,[227] and expression of TRAF2, but not a mutant TRAF2 construct missing the RING effector domain, triggers the production of ROS.[151]

Recent results suggest that Trx sequesters ASK1 from TRAF2 until TRAF2-dependent ROS production stimulates the dissociation of the ASK1-Trx complex[151] (Figure 6). Trx is abundant in cells and serves as a catalytic antioxidant.[256,257] By contrast, TRAF2 and ASK1 are low abundance polypeptides.[128,225] Expression of Trx at levels in excess of coexpressed TRAF2 and ASK1 completely reverses the ASK1-TRAF2 interaction. The interaction between ASK1 and TRAF6 is also be reversed upon coexpression of excess Trx. The inhibition of the ASK1-TRAF2 interaction can be reversed upon the subsequent administration of oxidant ($H_2O_2$).[151] Taken together, these results suggest that TRAF2 engagement by the TNFR complex triggers the production of ROS, which oxidize Trx causing its dissociation from ASK1, thereby enabling the binding of TRAF2 to the free ASK1 polypeptide. How is it possible, then, to observe an interaction between overexpressed ASK1 and TRAF2, in the absence of excess Trx, without addition of oxidant? It is likely that overexpression of ASK1 and TRAF2 enables the ASK1-TRAF2 interaction against a background of endogenous Trx (i) because the overexpressed ASK1 titers out endogenous Trx, creating a pool of free ASK1 to which TRAF2 can bind and/or (ii), overexpression of TRAF2 generates a ROS pulse *in vivo* resulting in oxidation of endogenous Trx, thereby enabling the ASK1-TRAF2 interaction.[151]

A likely consequence of Trx dissociation from ASK1 is dimerization-dependent ASK1 activation. Upon overexpression, ASK1 spontaneously dimerizes *in vivo*, and TNF promotes the dimerization of endogenous ASK1 by a mechanism that requires ROS and can be reversed with free radical scavengers.[153] Moreover, expressed fusion proteins of ASK1 and DNA gyrase can be forced to dimerize *in vivo* upon administration of the binary DNA gyrase-binding drug coumermycin. This coumermycin-induced dimerization results in substantial activation of coexpressed SAPK.[153] TRAF2 is known to homodimerize *in vivo*;[232–234] and coexpression of ASK1 and TRAF2 enhances the level of recoverable ASK1 homomers in a reaction that requires the TRAF2 RING domain (i.e., is not observed upon expression of TRAF2RING) and is blocked by free radical scavengers[151] (Figure 6).

### 14.4.2.4.2.3  *MEKK1 Disruption Indicates a Role in Protection from Oxidant Stress*

A recent study has demonstrated that MEKK1 suppresses oxidant stress-induced apoptosis in ES cell-derived cardiomyocytes.[142] Thus, *mekk1−/−* ES cells were stimulated to differentiate into embryoid bodies from which cardiomyocytes were isolated and cultured. Cardiomyocytes derived from *mekk1−/−* embryoid bodies were substantially more sensitive to the cytocidal effects of oxidant stress ($H_2O_2$), increasing oxidant stress-induced apoptosis. Coincident with this, oxidant stress- and hypoxia-induced activation of SAPK was abrogated in the *mekk−/−* cells. p38 and ERK activity were unaffected.[142] Thus, MEKK1, through SAPK, apparently acts to protect cardiomyocytes from oxidant stress-induced cytotoxicity. Oxidant stress causes cardiomyocytes to release TNF, triggering a TNF-induced caspase apoptotic cascade. Oxidant stress-induced TNF release was blunted in *mekk1−/−* cells, indicating that MEKK1 negatively regulates the expression of TNF.[142]

### 14.4.2.4.3  Ultraviolet Radiation, Heat Shock and Regulation of the SAPKs and p38s: Unexpected Mechanisms

#### 14.4.2.4.3.1  *Overview*

Among the first activators of the SAPKs to be identified were ultraviolet (UV) radiation and heat shock. In contrast to γ radiation and genotoxins (Section 14.4.2.4.4), UV is generally accepted as a potent SAPK and p38 activator. Both heat shock and UV activation of the SAPKs and p38s proceed by unexpected mechanisms.

#### 14.4.2.4.3.2  *UV Radiation Likely Acts by Triggering the Clustering of Cell Surface Receptors*

Activation of the SAPKs by UV radiation (it must be noted that most studies have employed UV-C, a component of the UV spectrum that does not penetrate the Earth's upper atmosphere) is rapid; and it is generally agreed that UV signaling to protein kinase cascades does not require DNA damage *per se*. Instead UV triggers the clustering and activation of cell surface mitogen and cytokine receptors—a process likely driven by the formation of disulfide bridges between adjacent receptor proteins—thus, any other ROS-producing agent could trigger cell surface receptor clustering.[248,258] Signaling through these receptors then presumably proceeds in the normal fashion. UV also triggers ROS production, possibly through membrane lipid modification or through receptor-mediated recruitment of intracellular oxidases—although the mechanism is far from clear. These ROS may then trigger SAPK and p38 activation through ASK1.

#### 14.4.2.4.3.3  *Heat Shock Regulation of the SAPKs, a Novel Mechanism*

Heat shock activation of the SAPKs appears to bypass the "traditional" three-tiered core module. Although SEK1 is required for heat shock activation of SAPK[103] (Section 14.4.2.2.6.1), SEK1 is not actually activated by heat shock *per se*. Instead heat shock suppresses a Tyr phosphatase that ordinarily inactivates SAPK thereby allowing basal SEK1 Tyr kinase activity (SEK1 and SAPK activity is never entirely absent, even in the resting state) to slowly activate SAPK.[259] A consequence of heat stress is the transcriptional induction of the chaperone protein HSP70. HSP70 expression reactivates this SAPK-targeted Tyr phosphatase resulting in SAPK inactivation.[259]

### 14.4.2.4.4  Genotoxin Regulation of the SAPKs and p38s

#### 14.4.2.4.4.1  *Overview and Background*

Although chemical genotoxins are encountered in the environment, they are also widely employed in conventional chemotherapeutic strategies to battle cancer and other proliferative disorders. High-energy radiation such as γ radiation is also used in the clinic to inflict lethal DNA damage on cancerous cells. While both approaches are, to varying degrees effective, chemical and radiant genotoxins also have numerous side effects stemming from the excessive toxicity of these methods to normal, rapidly growing cells such as GI epithelia, hair follicles, *etc.* Accordingly, considerable effort has been expended to understand how chemical and radiant genotoxins kill cells, as part of an effort to improve treatment approaches and to reduce the effects of environmental toxins.

It is generally accepted that chemical and radiant genotoxins kill cells by triggering apoptosis. In some instances, this death requires the tumor suppressor protein p53; however, inasmuch as many chemotherapies are effective against cancer cells in which p53 function is compromised, p53-independent killing mechanisms must also exist. As noted above (Section 14.4.2.4.2), many chemotherapeutic agents and DNA damaging agents, trigger the production of ROS in their target cells.[244–248] The regulation, through ASK1, of the SAPKs and p38s by ROS is discussed in Section 14.4.2.4.2 and Section 14.4.2.5 will assess the role of the evidence favoring a biological role for the SAPKs in apoptosis. In addition to ROS production, however, several other mechanisms of SAPK regulation by genotoxins have been proposed. These potential mechanisms of SAPK and p38 regulation merit discussion inasmuch as they are controversial at this time, as are observations regarding the spectrum of genotoxins thought to recruit the SAPKs and p38s. In addition, the findings to be discussed point to the complexity of how toxicants recruit MAPK pathways.

### 14.4.2.4.4.2 Genotoxins and SAPK Activation, Controversy over Abl

Activation of the SAPKs by γ radiation and most chemical genotoxins proceeds slowly, if at all, and most investigators believe that recruitment of the SAPKs by these stimuli requires in part DNA damage as a triggering event.[260–263] Thus, for example, the *ataxia telangiectasia mutated* (ATM) gene product, a polypeptide implicated in the cellular response to DNA damage, is necessary for SAPK activation in response to a number of radiant and chemical genotoxins.[264–266] In addition to ATM, the Tyr kinase proto oncoprotein c-Abl has been implicated in signaling from DNA damage to the SAPKs and p38s.

There is considerable disagreement, however, among investigators as to the degree of SAPK activation incurred by different genotoxins and the role played by the Abl Tyr

kinase in genotoxin signaling. Thus, Kharbanda *et al.* observed strong activation of the SAPKs by γ radiation and chemical genotoxins (CDDP, mitomycin-C) which apparently required c-Abl and did not occur in *abl−/−* mouse fibroblasts. These investigators were able to restore genotoxin activation of the SAPKs into *abl−/−* cells upon reintroduction of c-Abl cDNA.[260,262] By contrast, Liu *et al.* observed at best modest SAPK activation by γ radiation and CDDP and substantial activation by mitomycin-C; however, activation of the SAPKs by these stimuli was not compromised in *c-abl−/−* fibroblasts.[261] The reason for this discrepancy is unclear as is the relationship, if any, between ATM and Abl in signaling from DNA damage to the SAPKs. Moreover, these studies do not indicate which MAP3Ks couple the SAPKs to genotoxin signals.

### 14.4.2.4.4.3 Genotoxins and SAPK Activation, Controversy over the Role of MEKK4 and GADD45s

A series of recent studies has potentially identified one MAP3K/MEK/MAPK core pathway recruited by genotoxins. As with the *abl−/−* studies, however, there have been conflicting findings. Results from Saito and co-workers indicate that MEKK4 is recruited by genotoxins that transcriptionally induce proteins of the growth arrest and DNA damage-45 (GADD45) family. GADD genes are a set of transcripts that are rapidly induced by DNA damage and are thought to play a role in coordinating the cellular response to genotoxins.[264,265] GADD153/CHOP, described in Section 14.4.2.2.5.2, is one example. *Gadd45* was one of the first GADD genes to be identified, although its function has remained obscure.

Expression of GADD45 is the culmination of a signaling pathway that requires prior expression of the tumor suppressor protein p53 which *trans* activates the *gadd45* (*gadd45α*) gene. p53 expression is, in turn, driven by the activity of the ATM gene product.[265] This is particularly noteworthy given the observation, cited above, that cell lines derived from AT patients with null or inactivating *ATM* mutations are resistant to SAPK activation by genotoxins but not by TNF.[266]

Yeast two-hybrid screening using MEKK4 as bait identified three *gadd45* homologues (GADD45β and -γ plus GADD45α, the original GADD45) as direct MEKK4 binding proteins.[144] These results were confirmed in experiments in which MEKK4 was overexpressed with recombinant GADD45. In these studies, GADD45γ was shown to be the strongest MEKK4 interactor of the three GADD45 isoforms. A broad segment in the middle of the GADD45 polypeptides (AAs 24–147 of GADD45γ) binds to a small domain (AAs 147–250) at the amino terminus of the MEKK4 polypeptide.[144]

Evidence favoring a role for GADD45s in genotoxin regulation of MEKK4 signaling to the SAPKs and p38s came from the observation that all three *gadd45* genes are transcribed in response to the same genotoxic stimuli shown to upregulate *gadd45α*.[144,264] Moreover, ectopic overexpression of all three GADD45 polypeptides resulted in substantial activation of both p38 and SAPK—a response that was reversed upon coexpression with kinase-dead MEKK4. Kinase-dead MEKK4 could also inhibit activation of p38 by genotoxic stress that transcriptionally induced *gadd45* gene expression. In addition, expression of all three GADD45 polypeptides activates MEKK4 *in vivo*.[144] Most importantly, however, addition of purified, recombinant GADD45 proteins to immunoprecipitated MEKK4 activates MEKK4 *in vitro*, suggesting that the GADD45s are direct MEKK4 activators.[144]

Ectopic overexpression of GADD45 proteins is apoptogenic in HeLa cells, a reaction that may require in part MEKK4 insofar as either kinase-dead MEKK4 or the GADD45 binding domain of MEKK4 (AAs 147–250) significantly blocks GADD45-induced apoptosis, while an in-frame deletion of the GADD45 binding domain of MEKK4 renders kinase-dead MEKK4 incapable of blocking apoptosis.[144]

Taken together, these results implicate GADD45s as activators of the SAPKs, p38s and apoptosis via a MEKK4-dependent mechanism. The requirement for GADD45 transcription also explains the slow kinetics of SAPK and p38 activation by many genotoxins.[260–263] The fact that SAPK activation by genotoxins is ablated in *ATM* null or mutant cells[266] can also be explained by the observation that GADD45 expression (and, hence, subsequent activation of the SAPKs) is also ablated in these cells.[265]

However, although expression of GADD45s is upregulated by stimuli that can activate the SAPKs, missing from the studies implicating MEKK4 in coupling genotoxic stress and GADD45s to the SAPKs are experiments showing that GADD45 proteins are present at times when MEKK4 and the SAPKs are activated. In this regard, recent findings apparently conflict with the observation that DNA damage activates SAPK through GADD45α and, perhaps, other GADD45

homologues. Thus, studies from Karin *et al.*, as well as Holbrook *et al.* indicate that none of the endogenous GADD45 homologues is present at detectable levels when SAPK is fully activated by DNA damaging chemicals (methyl methane sulfonate, MMS).[267,268] Moreover, *gadd45α* knockout cells show adequate SAPK activation by MMS and other genotoxic stresses. Although this experiment does not rule out the possibility that this SAPK activation is mediated by GADD45β or γ, both of which are also induced by MMS, incubation with cycloheximide (employed at concentrations too low to activate SAPK), to block induction of all GADD45 isoforms, also fails to prevent MMS activation of SAPK.[267] In contrast to the results of Saito, Holbrook *et al.* did not observe activation of SAPK by coexpressed GADD45α, β or γ.[268] The results of Holbrook and colleagues may reflect the effects of modest GADD45 expression, whereas greater levels of GADD45 may promote SAPK activation.[144] However, it is unclear if, under physiologic circumstances, such high levels of GADD45 are attained. Even so, Karin and colleagues found that there was little or no significant SAPK activation in fibroblasts (p53 positive—therefore theoretically able to manifest GADD45 induction in response to DNA damage) treated with ionizing radiation sufficient to induce GADD45 expression.[267] It must be kept in mind, however, that while a role for GADD45s in the regulation by genotoxins of the SAPKs (via MEKK4 or otherwise) is controversial, there may be other physiologic circumstances, such as certain developmental stages, during which GADD45s are expressed under conditions in which they can recruit MEKK4, and through MEKK4, the SAPKs and p38s.

### 14.4.2.4.4.4    The Role of SAPK-Mediated FasL Upregulation in Toxicant-induced Apoptosis

Regardless of the route by which genotoxins might activate the SAPKs and p38s, additional findings have provided insight into the mechanism by which the SAPKs promote apoptosis in response to genotoxic stresses, UV irradiation or heat shock. Jurkat cells treated chemotherapeutic genotoxins (etoposide, tenoposide) or UV irradiation rapidly undergo apoptosis coincident with the transcriptional induction and cell surface expression of the death cytokine FasL.[269] The binding of newly expressed FasL to preexisting constitutively expressed Fas triggers an autocrine apoptotic program in response to these genotoxins. The promoter for FasL contains AP-1 and NF-κB *cis* acting elements, and inhibition of AP-1 activation (with dominant inhibitory SAPK to block AP-1 activation or with a dominant negative c-Jun construct, TAM67, in which the *trans* activating δ domain is deleted) prevents FasL induction.[269] These results indicate that genotoxic stresses promote apoptosis at least in part by recruiting the SAPKs which, in turn, activate AP-1 and NF-κB-mediated induction of FasL.[269] As is discussed in Section 14.4.2.5, the SAPKs are probably not relevant to FasL-induced apoptosis *per se*. Still, a toxicant-induced apoptotic pathway could require the SAPKs in part for their ability to induce FasL.

#### 14.4.2.4.5 The ER Stress Response: Commonality with TNFR Signaling, the Role of TRAF2

A recent study has implicated TRAFs in the coupling of endoplasmic stress (ER stress) signaling to the SAPKs.[231] Stress in the ER is induced by environmental perturbations such as oxidants, heat, ionizing radiation, translational inhibition or chemical denaturants such as tunicamycin.[270] Sections 14.4.2.4.3 and 14.4.2.4.4 discuss some possible mechanisms by which oxidants, heat, ionizing radiation can recruit the SAPKs. Many of the stimuli that cause ER stress also lead to the buildup of denatured proteins in the ER lumen. This misfolded protein response adds an additional dimension to stress/toxicant-induced SAPK and p38 signaling. Misfolded proteins are sensed by the Ire1s, a family of mammalian homologues of *S. cerevisiae* IRE1, the product of the inositol auxotrophy gene *IRE1* (also called *ERN1*).[270–274] Ire1s are transmembrane receptor-like proteins with cytosol-facing domains that possess both endoribonuclease and protein Ser/Thr kinase activity, each of which is required for biological function.[270,271,274]

In mammalian cells the ER stress response can be triggered by tunicamycin;[270] and tunicamycin stimulates vigorous SAPK activation.[25] Tunicamycin and other ER stresses lead to the oligomerization and activation of mammalian Ire1p,[273,274] which, in turn activates SAPK.[231] Thus, overexpression of mammalian Ire1α, which is likely sufficient to promote spontaneous Ire1α oligomerization, is capable of activating SAPK; and ER stress activation of SAPK is lost in *ire1α−/−* cells, or in cells expressing kinase-dead Ire1α.[231]

Urano *et al.*, performed a yeast two-hybrid screen using the intracellular extension of Ire1α as bait and, surprisingly, isolated TRAF2. Consistent with the possibility that TRAF2 is a genuine effector for Ire1, ER stress leads to the association of endogenous TRAF2 and endogenous Ire1α. Furthermore, dominant inhibitory TRAF2 can block activation of SAPK by overexpressed Ire1p.[231] Inasmuch as Ire1α activation involves homo-oligomerization, it is likely, that Ire1α oligomerization, in turn, triggers the binding and oligomerization-dependent activation of TRAF2 and TRAF2-dependent signaling.[231,270,272] Thus, TRAFs may serve in a broad range of signaling pathways, transducing not only cytokine signals, but ER stress signals as well (Figure 6).

### 14.4.2.5 BIOLOGICAL ROLES OF STRESS-ACTIVATED MAPKS: A TOXICOLOGIC PERSPECTIVE. INSIGHTS FROM SAPK KNOCKOUTS

Section 14.4.2.4 has provided some information as to how environmental stresses and toxicants can recruit the SAPKs. Until now, however, a genetic link implicating the SAPKs in initiating the biological responses to toxic stress has been missing. A series of SAPK knockout studies has recently revealed a role for the SAPKs in the proapoptotic response to toxicants.

Gene disruption studies have revealed a role for one SAPK isoform in brain apoptosis lending clues as to the biological functions of different SAPK isoforms in response to environmental stresses. Kainic acid is a potent glutamate receptor agonist that rapidly induces siezures similar to those of epilepsy. In addition, kainate causes widespread neuronal apoptosis in the hippocampus, a phenomenon similar to the typical outcome of hippocampal infarcts (stroke). Whereas SAPKα/JNK2 and SAPKγ/JNK1 are ubiquitously expressed, SAPKβ/JNK3 is selectively expressed in brain and heart. Disruption of *sapkβ* in mice leads to a striking resistance to kainate siezures.[275] More importantly, however, deletion of *sapkβ* protected the mutant mice from kainate-induced hippocampal neuronal apoptosis.[275] Hippocampal cell death is a major clinical indicator in stroke and Alzheimer's disease. It will be important to determine if ischemia or Alzheimer's-induced cell death correlates with SAPKβ activation. As was noted above, induction of FasL, through an AP-1-dependent mechanism, has a major role in triggering apoptosis in response to genotoxins.[260] It will also be important to determine if hippocampal cell

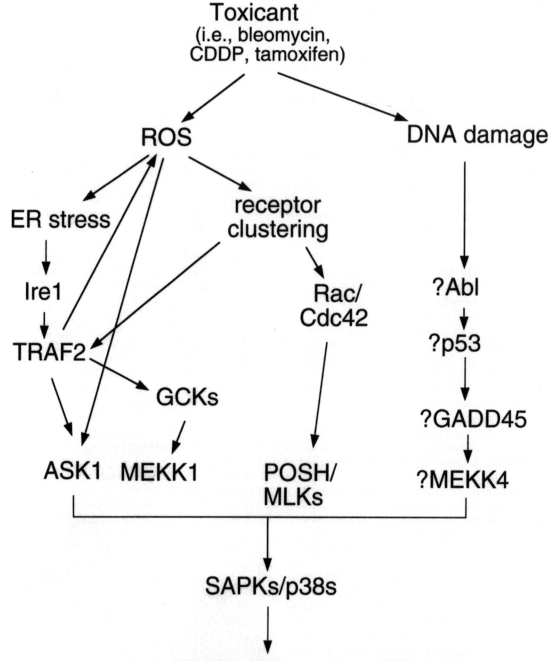

**Figure 7**   Heterogeneity of mechanisms that a single toxicant with multiple intracellular targets uses to recruit the SAPKs and p38s. This diagram is meant to illustrate the many avenues by which a toxic agent can activate MAPKs. This complexity should be appreciated when analyzing toxicant regulation of MAPKs and when considering the pharmaceutical use of toxicants.

death induced by kainate involves upregulation of FasL (Section 14.4.2.4.4.2).

Although stress- or genotoxin-induced expression of FasL (and consequent Fas-driven apoptosis) is apparently SAPK-dependent, it has until recently remained controversial as to whether or not the SAPKs had a more direct role in signaling apoptosis. Recent studies using mouse embryo fibroblasts (MEFs) deficient in SAPKα and γ indicate that the SAPKs are necessary for UV and anisomycin (i.e., environmental stress-induced) apoptosis, but not Fas (i.e., cytokine-driven, developmental) apoptosis.[276]

Two distinct, but overlapping mechanisms for cell death have been identified; both pathways culminate in the activation of the effector caspase caspase 3. The receptor-driven, developmental pathway is activated by receptors such as Fas and TNFR1. In this pathway, receptor engagement recruits death domain adapter proteins (TRADD to TNFR1 and the related Fas-associated death domain protein, FADD, to Fas). TRADD, once at the receptor will itself recruit FADD. Caspases are a family of cysteine proteases that are pivotal to the initiation and execution of apoptosis (reviewed in Reference 277). FADD recruits the initiator caspase, caspase 8. The induced oligomerization of caspase 8 at the receptor complex prompts caspase 8 autoproteolytic activation. Caspase 8, in turn, proteolytically activates procaspase 3. Activation of caspase 3 essentially commits the cell to apoptosis.[277]

The second mechanism of apoptosis is activated by environmental toxicants and stress. This mechanism requires the mitochondrion (see Chapter 4.7). Environmental stresses trigger membrane depolarization and a permeability transition in the mitochondrion leading to the release of cytochrome *c*. In the cytosol, cytochrome *c* serves as an adapter protein to trigger the assembly of an "apoptosome" consisting of the procaspase 9 and the scaffold protein apoptosis-activating factor-1 (Apaf1). Through a poorly understood mechanism that requires ATP and is probably mediated in part by homo-oligomerization, procaspase 9 is then activated and in turn proteolytically activates procaspase 3 thereby committing the cell to apoptosis.[277]

MEFs deficient in both SAPKα and γ undergo apoptosis in response to Fas engagement at a rate indistinguishable from that of wt-MEFs. By contrast, both UV-C- and anisomycin induced apoptosis are markedly inhibited in *sapkα/γ* double knockout cells. Coincident with this, UV-C-stimulated mitochondrial depolarization and release of cytochrome *c* and initiation of the mitochondrial apoptosis pathway are inhibited in these cells.[276] The mechanisms by which UV-activated SAPKs trigger mitochondrial cytochrome *c* release and apoptosis remain to be determined.

## 14.4.2.6 CONCLUDING REMARKS

In considering the regulation of MAPK pathways by toxicants and environmental stresses, two overarching themes emerge. First, from a biochemical standpoint, MAP3K regulation is still poorly understood. Although it is likely that this regulation encompasses the four themes of binding to upstream regulatory proteins, membrane translocation, phosphorylation, and oligomerization; the molecular players involved in the recruitment by stresses and toxicants of specific MAP3Ks is still largely a mystery.

On a more physiologic level, it should be evident from this discussion that, in keeping with the multiple sites of action of toxicants, not to mention the fact that a single toxicant can have multiple sites of action within a cell; a broad heterogeneity of mechanisms may be brought to bear in order to activate MAPK pathways in response to environmental toxins (Figure 7). Thus, for example, a single chemotherapeutic agent such as bleomycin could recruit the SAPKs via mechanisms triggered by DNA damage (Abl-dependent, GADD45/ MEKK4-dependent), ROS (ASK1/Trx) and ER stress (TRAF2, GCKs, MEKK1) (Figure 7). It is critical that this complexity be appreciated when considering the administration of chemotherapeutics, as well as the design of novel treatments, and when dealing with the effects of naturally occurring toxins.

## 14.4.2.7 REFERENCES

1. J. M. Kyriakis and J. Avruch, 'Sounding the alarm: protein kinase cascades activated by stress and inflammation.' *J. Biol. Chem.*, 1999, **271**, 24313–24316.
2. C. J. Marshall, 'Specificity of receptor tyrosine kinase signaling: Transient versus sustained extracellular signal regulated kinase activation.' *Cell*, 1995, **80**, 179–185.
3. I. Herskowitz, 'MAP kinase pathways in yeast: for mating and more.' *Cell*, 1995, **80**, 187–197.
4. J. L. Countaway, I. C. Northwood and R. J. Davis, 'Mechanism of phosphorylation of the epidermal growth factor receptor on threonine 669.' *J. Biol. Chem.*, 1989, **264**, 10828–10835.
5. A. K. Erickson, D. M. Payne, P. A. Martino *et al.*, 'Identification by mass spectrometry of threonine 97 in bovine myelin basic protein as a specific phosphorylation site for mitogen-activated protein kinase.' *J. Biol. Chem.*, 1990, **265**, 19728–19735.
6. I. Clark-Lewis, J. S. Sanghera and S. L. Pelech, 'Definition of a consensus sequence for peptide substrate recognition by p44mpk, the meiosis-activated myelin basic protein kinase.' *J. Biol. Chem.*, 1991, **266**, 15180–15184.
7. T. Kallunki, T. Deng, M. Hibi *et al.*, 'c-Jun can recruit JNK to phosphorylate dimerization partners via specific docking interactions.' *Cell*, 1996, **87**, 929–939.
8. T. Dai, E. Rubie, C. C. Franklin *et al.*, 'Stress-activated protein kinases bind directly to the δ domain of c-Jun in resting cells: implications for repression of c-Jun function.' *Oncogene*, 1995, **10**, 849–855.
9. S.-H. Yang, A. J. Whitmarsh, R. J. Davis *et al.*, 'Differential targeting of MAP kinases to the ETS-

domain transcription factor Elk-1.' *EMBO J.*, 1998, **17**, 1740–1749.

10. T. Tanoue, M. Adachi, T. Moriguchi *et al.*, 'A conserved docking motif in MAP kinases common to substrates, activators and regulators.' *Nat. Cell. Biol.*, 2000, **2**, 110–116.

11. F. Posas and H. Saito, 'Osmotic activation of the HOG MAPK pathway via Ste11p MAPKKK: Scaffold role of Pbs2p MAPKK.' *Science*, 1997, **276**, 1702–1705.

12. J. L. Blank, P. Gerwins, E. M. Elliot *et al.*, 'Molecular cloning of mitogen activated protein/ERK kinase kinases (MEKK) 2 and 3.' *J. Biol. Chem.*, 1996, **271**, 5361–5368.

13. K. Deacon and J. L. Blank, 'MEK kinase 3 directly activates MKK6 and MKK7, specific activators of the p38 and c-Jun NH$_2$-terminal kinases.' *J. Biol. Chem.*, 1999, **278**, 16604–16610.

14. H. Ellinger-Ziegelbauer, K. Brown, K. Kelly *et al.*, 'Direct activation of the stress-activated protein kinase and extracellular signal-regulated protein kinase (ERK) pathways by an inducible mitogen-activated protein kinase/ERK kinase kinase 3 derivative.' *J. Biol. Chem.*, 1997, **272**, 2668–2674.

15. H. Ellinger-Ziegelbauer, K. Kelly and U. Siebenlist, 'Cell cycle arrest and reversion of Ras-induced transformation by a conditionally activated form of mitogen-activated protein kinase kinase kinase 3.' *Mol. Cell. Biol.*, 1999, **19**, 3857–3868.

16. T.-H. Chao, M. Hayashi, R. I. Tapping *et al.*, 'MEKK3 directly regulates MEK5 as part of the big mitogen-activated protein kinase 1 (BMK1) signaling pathway.' *J. Biol. Chem.*, 1999, **274**, 36035–36038.

17. S. P. Davies, H. Reddy, M. Caivano *et al.*, 'Specificity and mechanism of action of some commonly used protein kinase inhibitors.' *Biochem. J.*, 2000, **351**, 95–105.

18. C. A. Hall-Jackson, M. Goedert, P. Hedge *et al.*, 'Effect of SB203580 on the activity of c-Raf *in vitro* and *in vivo*.' *Oncogene*, 1999, **18**, 2047–2054.

19. J. Avruch, X.-f.Zhang and J. M. Kyriakis, 'Raf meets Ras: Completing the framework of a signal transduction pathway.' *Trends Biochem. Sci.*, 1994, **19**, 279–283.

20. J. Avruch, 'Insulin signal transduction through protein kinase cascades.' *Mol. Cell. Biochem.*, 1998, **182**, 31–48.

21. J. M. Kyriakis and J. Avruch, 'pp54 MAP-2 kinase. A novel serine/threonine protein kinase regulated by phosphorylation and stimulated by poly-L-lysine.' *J. Biol. Chem.*, 1990, **265**, 17355–17363.

22. J. M. Kyriakis, D. L. Brautigan, T. S. Ingebritsen *et al.*, 'pp54 microtubule-associated protein-2 kinase requires both tyrosine and serine/threonine phosphorylation for activity.' *J. Biol. Chem.*, 1991, **266**, 10043–10046.

23. N. K. Mukhopadhyay, D. J. Price, J. M. Kyriakis *et al.*, 'An array of insulin-activated, proline-directed (Ser/Thr) protein kinases phosphorylate the p70 S6 kinase.' *J. Biol. Chem.*, 1992, **267**, 3325–3335.

24. B. J. Pulverer, J. M. Kyriakis, J. Avruch *et al.*, 'Phosphorylation of c-jun mediated by MAP kinases.' *Nature*, 1991, **353**, 670–674.

25. J. M. Kyriakis, P. Banerjee, E. Nikolakaki *et al.*, 'The stress-activated protein kinase subfamily of c-Jun kinases.' *Nature*, 1994, **369**, 156–160.

26. B. Dérijard, M. Hibi, I.-H. Wu *et al.*, 'JNK1: A protein kinase stimulated by UV light and Ha-Ras that binds and phosphorylates the c-Jun transactivation domain.' *Cell*, 1994, **76**, 1025–1037.

27. C. M. Pombo, J. V. Bonventre, J. Avruch *et al.*, 'The stress-activated protein kinases are major c-Jun

28. amino-terminal kinases activated by ischemia and reperfusion.' *J. Biol. Chem.*, 1994, **269**, 26546–26551.

28. P. S. Shapiro, J. N. Evans, R. J. Davis *et al.*, 'The seven transmembrane-spanning receptors for endothelin and thrombin cause proliferation of airway smooth muscle cells and activation of the extracellular signal regulated kinase and c-Jun NH$_2$-terminal kinase groups of mitogen-activated protein kinases.' *J. Biol. Chem.*, 1996, **271**, 5750–5754.

29. I. E. Zohn, H. Yu, X. Li *et al.*, 'Angiotensin II stimulates the calcium-dependent activation of c-Jun N-terminal kinase.' *Mol. Cell. Biol.*, 1995, **15**, 6160–6168.

30. I. Berberich, G. Shu, F. Siebelt *et al.*, 'Cross-linking CD40 on B cells preferentially induces stress-activated protein kinases rather than mitogen-activated protein kinases.' *EMBO J.*, 1996, **15**, 92–101.

31. T. A. Bird, J. M. Kyriakis, L. Tyshler *et al.*, 'Interleukin-1 activates p54 mitogen-activated protein (MAP) kinase/stress-activated protein kinase by a pathway that is independent of p21ras, Raf-1 and MAP kinase-kinase.' *J. Biol. Chem.*, 1994, **269**, 31836–31842.

32. J. M. Lenczowski, L. Dominguez, A. M. Eder *et al.*, 'Lack of a role for Jun kinase and AP-1 in Fas-induced apoptosis.' *Mol. Cell. Biol.*, 1997, **17**, 170–181.

33. L. Galibert, M. E. Tometsko, D. M. Anderson *et al.*, 'The involvement of multiple tumor necrosis factor receptor (TNFR)-associated factors in the signaling mechanisms of receptor activator of NF-κB, a member of the TNFR superfamily.' *J. Biol. Chem.*, 1998, **273**, 34120–34127.

34. H. Akiba, H. Nakano, S. Nishinaka *et al.*, 'CD27, a member of the tumor necrosis factor receptor superfamily, activates NF-κB and stress-activated protein kinase/c-Jun N-terminal kinase via TRAF2, TRAF5 and NF-κB-inducing kinase.' *J. Biol. Chem.*, 1998, **273**, 13353–13358.

35. T. Kallunki, B. Su, I. Tsigelny *et al.*, 'JNK2 contains a specificity-determining region responsible for efficient c-Jun binding and phosphorylation.' *Genes Dev.*, 1994, **8**, 2996–3007.

36. S. Gupta, T. Barrett, A. J. Whitmarsh *et al.*, 'Selective interaction of JNK protein kinase isoforms with transcription factors.' *EMBO J.*, 1996, **15**, 2760–2770.

37. J. Han, J.-D. Lee, L. Bibbs *et al.*, 'A MAP kinase targeted by endotoxin and hyperosmolarity in mammalian cells.' *Science*, 1994, **265**, 808–811.

38. J. Rouse, P. Cohen, S. Trigon *et al.*, 'A novel kinase cascade triggered by stress and heat shock that stimulates MAPKAP kinase-2 and phosphorylation of the small heat shock proteins.' *Cell*, 1994, **78**, 1027–1037.

39. N. W. Freshney, L. Rawlinson, F. Guesdon *et al.*, 'Interleukin-1 activates a novel protein kinase cascade that results in the phosphorylation of Hsp27.' *Cell*, 1994, **78**, 1039–1049.

40. J. C. Lee, J. T. Laydon, P. C. McDonnell *et al.*, 'A protein kinase involved in the regulation of inflammatory cytokine biosynthesis.' *Nature*, 1994, **273**, 739–746.

41. P. Cohen, 'The search for physiological substrates of MAP and SAP kinases in mammalian cells.' *Trends Cell. Biol.*, 1997, **7**, 353–361.

42. Y. Jiang, H. Gram, M. Zhao *et al.*, 'Characterization of the structure and function of the fourth member of the p38 group mitogen-activated protein kinases, p38δ.' *J. Biol. Chem.*, 1997, **272**, 30122–30128.

43. C. Lechner, M. A. Zahalka, J. F. Giot *et al.*, 'ERK6, a mitogen-activated protein kinase is involved in C2C12 myoblast differentiation.' *Proc. Nat. Acad. Sci. USA*, 1996, **93**, 4355–4359.

44. S. Mertens, M. Craxton and M. Goedert, 'SAP kinase-3, a new member of the family of mammalian stress-activated protein kinases.' *FEBS Lett.*, 1996, **383**, 273–276.

45. B. Stein, M. X. Yang, D. B. Young *et al.*, 'p38-2, a novel mitogen-activated protein kinase with distinct properties.' *J. Biol. Chem.*, 1997, **272**, 19509–19517.

46. M. Goedert, A. Cuenda, M. Craxton *et al.*, 'Activation of the novel stress-activated protein kinase SAPK4 by cytokines and cellular stresses is mediated by SKK3 (MKK6); comparison of its substrate specificity with that of other SAP kinases.' *EMBO J.*, 1997, **16**, 3563–3571.

47. X. S. Wang, K. Diener, C. L. Manthey *et al.*, 'Molecular cloning and characterization of a novel p38 mitogen-activated protein kinase.' *J. Biol. Chem.*, 1997, **272**, 23688–23674.

48. Y. Jiang, C. Chen, Z. Li *et al.*, 'Characterization of the structure and function of a new mitogen-activated protein kinase (p38β).' *J. Biol. Chem.*, 1998, **271**, 17920–17926.

49. P. A. Eyers, M. Craxton, N. Morrice *et al.*, 'Conversion of SB 203580-insensitive MAP kinase family members to drug-sensitive forms by a single amino-acid substitution.' *Chem. Biol.*, 1998, **5**, 321–328.

50. R. J. Gum, M. M. McLaughlin, S. Kumar *et al.*, 'Acquisition of sensitivity of stress-activated protein kinases to the p38 inhibitor, SB 203580, by alteration of one or more amino acids within the ATP binding pocket.' *J. Biol. Chem.*, 1998, **273**, 15605–15610.

51. C. A. Hall-Jackson, P. A. Eyers, P. Cohen *et al.*, 'Paradoxical activation of Raf by a novel raf inhibitor.' *Chem. Biol.*, 1999, **6**, 559–569.

52. M. A. Bogoyevitch, J. Gillespie-Brown, A. J. Ketterman *et al.*, 'Stimulation of the stress-activated mitogen-activated protein kinase subfamilies in perfused heart. p38/RK mitogen-activated protein kinases and c-Jun N-terminal kinases are activated by ischemia/reperfusion.' *Circ. Res.*, 1996, **79**, 162–173.

53. G. Zhou, Z. Q. Bao and J. E. Dixon, 'Components of a new human protein kinase signal transduction pathway.' *J. Biol. Chem.*, 1995, **270**, 12665–12669.

54. J. M. English, C. A. Vanderbilt, X. Xu *et al.*, 'Isolation of MEK5 and differential expression of alternatively spliced forms.' *J. Biol. Chem.*, 1995, **270**, 28897–28902.

55. J.-I. Abe, M. Kusuhara, R. J. Ulevitch *et al.*, 'Big mitogen-activated protein kinase 1 (BMK1) is a redox-sensitive kinase.' *J. Biol. Chem.*, 1996, **271**, 16586–16590.

56. D. Stokoe, D. G. Campbell, S. Nakielny *et al.*, 'MAPAKP kinase-2; a novel protein kinase activated by mitogen-activated protein kinase.' *EMBO J.*, 1992, **11**, 3985–3994.

57. D. Stokoe, K. Engel, D. G. Campbell *et al.*, 'Identification of MAPKAP kinase 2 as a major enzyme responsible for the phosphorylation of the small mammalian heat shock proteins.' *FEBS Lett.*, 1992, **313**, 307–313.

58. M. M. McLaughlin, S. Kumar, P. C. McDonnell *et al.*, 'Identification of mitogen-activated protein (MAP) kinase-activated protein kinase-3, a novel substrate of CSBP p38 MAP kinase.' *J. Biol. Chem.*, 1996, **271**, 8488–8492.

59. G. Sithanandam, F. Latif, U. Smola *et al.*, '3pK, a new mitogen-activated protein kinase-activated protein kinase located in the small cell lung cancer tumor suppressor gene region.' *Mol. Cell. Biol.*, 1996, **16**, 868–876 (published erratum appears in *Mol. Cell. Biol.*, 1996, **16**, 1880).

60. R. L. Erikson, 'Structure, expression and regulation of protein kinases involved in phoshorylation of ribosomal protein S6.' *J. Biol. Chem.*, 1991, **266**, 6007–6010.

61. J. Huot, F. Houle, F. Marceau *et al.*, 'Oxidative stress-induced actin reorganization mediated by the p38 mitogen-activated protein kinase/heat shock protein 27 pathway in vascular endothelial cells.' *Circ. Res.*, 1997, **80**, 383–392.

62. H. Lambert, S. J. Charette, A. F. Bernier *et al.*, 'HSP27 multimerization mediated by phosphorylation-sensitive intermolecular interactions at the amino terminus.' *J. Biol. Chem.*, 1999, **274**, 9378–9385.

63. A. Cuenda, P. Cohen, V. Buee-Scherrer *et al.*, 'Activation of stress-activated protein kinase-3 (SAPK3) by cytokines and cellular stresses is mediated via SAPKK3 (MKK6); comparison of the specificities of SAPK3 and SAPK2 (RK/p38).' *EMBO J.*, 1997, **16**, 295–305.

64. R. Ben-Levy, I. A. Leighton, Y. N. Doza *et al.*, 'Identification of novel phosphorylation sites required for activation of MAPKAP kinase-2.' *EMBO J.*, 1995, **14**, 5920–5930.

65. L. New, Y. Jiang, M. Zhao *et al.*, 'PRAK, a novel protein kinase regulated by the p38 MAP kinase.' *EMBO J.*, 1998, **17**, 3372–3384.

66. A. Kotlyarov, A. Neininger, C. Schubert *et al.*, 'MAPKAP kinase 2 is essential for LPS-induced TNF-α biosynthesis.' *Nat. Cell. Biol.*, 1999, **1**, 94–97.

67. N. Sonenberg and A.-C. Gingras, 'The mRNA 5′ cap-binding protein eIF-4E and control of cell growth.' *Curr. Opin. Cell. Biol.*, 1998, **10**, 268–275.

68. A. J. Waskiewicz, A. Flynn, C. G. Proud *et al.*, 'Mitogen-activated protein kinases activate the serine/threonine kinases Mnk1 and Mnk2.' *EMBO J.*, 1997, **16**, 1909–1920.

69. R. Fukunaga and T. Hunter, 'MNK1, a new MAP kinase-activated protein kinase, isolated by a novel expression screening method for identifying protein kinase substrates.' *EMBO J.*, 1997, **16**, 1921–1933.

70. A. J. Waskiewicz, J. C. Johnson, B. Penn *et al.*, 'Phosphorylation of the cap-binding protein eukaryotic translation initiation factor 4E by protein kinase Mnk1 *in vivo*.' *Mol. Cell Biol.*, 1999, **19**, 1871–1880.

71. M. Deak, A. D. Clifton, J. Lucocq *et al.*, 'Mitogen-and stress-activated protein kinase-1 (MSK1) is directly activated by MAPK and SAPK2/p38, and may mediate activation of CREB.' *EMBO J.*, 1998, **17**, 4426–4441.

72. J. F. Habener, 'Cyclic AMP-response element binding proteins: a cornucopia of transcription factors.' *Mol. Endocrinol.*, 1990, **4**, 1087–1094.

73. D. N. Chadee, M. J. Hendzel, C. P. Tylipski *et al.*, 'Increased Ser-10 phosphorylation of histone H3 in mitogen-stimulated and oncogene-transformed mouse fibroblasts.' *J. Biol. Chem.*, 1999, **274**, 24914–24920.

74. P. Sassone-Corsi, C. A. Mizzen, P. Cheung *et al.*, 'Requirement of Rsk-2 for epidermal growth factor-activated phosphorylation of histone H3.' *Science*, 1999, **285**, 886–891.

75. A. Thomson, A. Clayton, C. A. Hazzalin *et al.*, 'The nucleosomal response associated with immediate-early gene induction is mediated via alternative MAP kinase cascades: MSK1 is a potential H3/HMG-14 kinase.' *EMBO J.*, 1999, **18**, 4779–4793.

76. X. Wang and D. Ron, 'Stress-induced phosphorylation and activation of the transcription factor CHOP (GADD153) by p38 MAP kinase.' *Science*, 1996, **272**, 1347–1349.

77. A. C. Rao and P. G. Hogan, 'Transcription factors of the NFAT family: regulation and function.' *Annu. Rev. Immunol.*, 1997, **15**, 707–747.

78. S. L. Schreiber and G. R. Crabtree, 'The mechanism of action of cyclosporin A and FK506.' *Immunol. Today*, 1992, **13**, 136–142.

79. C.-W. Chow, M. Rincón, J. Cavanagh *et al.*, 'Nuclear accumulation of NFAT4 opposed by the JNK signal transduction pathway.' *Science*, 1997, **278**, 1638–1641.

80. J. Zhu, F. Shibakashi, R. Price *et al.*, 'Intramolecular masking of nuclear import signal on NFAT4 by casein kinase I and MEKK1.' *Cell*, 1998, **93**, 851–861.

81. C.-W. Chow, C. Dong, R. A. Flavell *et al.*, 'c-Jun NH2-terminal kinase inhibits targeting of the protein phosphatase calcineurin to NFATc1.' *Mol. Cell Biol.*, 2000, **20**, 5227–5234.

82. C. Dong, D. D. Yang, M. Wysk *et al.*, 'Defective T cell differentiation in the absence of Jnk1.' *Science*, 1998, **282**, 2092–2095.

83. C. Dong, D. D. Yang, C. Tournier *et al.*, 'JNK is required for effector T-cell function but not for T cell activation.' *Nature*, 2000, **405**, 91–94.

84. M. Karin, Z.-G. Liu and E. Zandi, 'AP-1 function and regulation.' *Curr. Opin. Cell Biol.*, 1997, **9**, 240–246.

85. Y. Kato, V. V. Kravchenko, R. I. Tapping *et al.*, 'BMK1/ERK5 regulates serum-induced early gene expression through transcription factor MEF2C.' *EMBO J.*, 1997, **16**, 7054–7066.

86. M. A. Read, M. Z. Whitley, S. Gupta *et al.*, 'Tumor necrosis factor-α-induced E-selectin expression is activated by the nuclear factor-κB and c-JUN N-terminal kinase/p38 mitogen-activated protein kinase pathways.' *J. Biol. Chem.*, 1997, **272**, 2753–2761.

87. W. J. Boyle, T. Smeal, L. H. Defize *et al.*, 'Activation of protein kinase C decreases phosphorylation of c-Jun at sites that negatively regulate its DNA binding activity.' *Cell*, 1991, **64**, 573–584.

88. D. A. E. Cross, D. R. Alessi, P. Cohen *et al.*, 'Inhibition of glycogen synthase kinase-3 by insulin mediated by protein kinase B.' *Nature*, 1995, **378**, 785–789.

89. J. Downward, 'Mechanisms and consequences of activation of protein kinase B/Akt.' *Curr. Opin. Cell Biol.*, **10**, 262–267.

90. S. Gupta, D. Campbell, B. Dérijard *et al.*, 'Transcription factor ATF2 regulation by the JNK signal transduction pathway.' *Science*, 1995, **267**, 389–393.

91. H. Morooka, J. V. Bonventre, C. M. Pombo *et al.*, 'Ischemia and reperfusion enhance ATF-2 and c-Jun binding to cAMP response elements and to an AP-1 binding site from the c-Jun promoter.' *J. Biol. Chem.*, 1995, **270**, 30084–30092.

92. R. Treisman, 'Regulation of transcription by MAP kinase cascades.' *Curr. Opin. Cell Biol.*, **8**, 205–215.

93. H. Gille, A. D. Sharrocks and P. E. Shaw, 'Phosphorylation of transcription factor p62TCF by MAP kinase stimulates ternary complex formation at c-fos promoter.' *Nature*, 1992, **358**, 414–417.

94. R. Marais, J. Wynne and R. Treisman, 'The SRF accessory protein Elk-1 contains a growth factor-regulated transcriptional activation domain.' *Cell*, 1993, **73**, 381–393.

95. A. J. Whitmarsh, P. Shore, A. D. Sharrocks *et al.*, 'Integration of MAP kinase signal transduction pathways at the serum response element.' *Science*, 1995, **269**, 403–407.

96. R. Janknecht and T. Hunter, 'Convergence of MAP kinase pathways on the ternary complex factor Sap-1a.' *EMBO J.*, 1997, **16**, 1620–1627.

97. J. Han, Y. Jiang, Z. Li *et al.*, 'MEF2C participates in inflammatory responses via p38-mediated activation.' *Nature*, 1997, **386**, 563–566.

98. M. Zhao, L. New, V. V. Kravchenko *et al.*, 'Regulation of the MEF2 family of transcription factors by p38.' *Mol. Cell Biol.*, 1999, **19**, 21–30.

99. J. N. Francisco and N. Eric, 'MEF2: a transcriptional target for signaling pathways controlling skeletal muscle growth and differentiation.' *Curr. Opin. Cell Biol.*, 1999, **11**, 683–688.

100. I. Sánchez, R. T. Hughes, B. J. Mayer *et al.*, 'Role of SAPK/ERK kinase-1 in the stress-activated pathway regulating transcription factor c-Jun.' *Nature*, 1994, **372**, 794–798.

101. B. Dérijard, J. Raingeaud, T. Barrett *et al.*, 'Independent human MAP kinase signal transduction pathways defined by MEK and MKK isoforms.' *Science*, 1995, **267**, 682–685.

102. R. Meier, J. Rouse, A. Cuenda *et al.*, 'Cellular stresses and cytokines activate multiple mitogen-activated-protein kinase kinase homologues in PC12 and KB cells.' *Eur. J. Biochem.*, 1996, **236**, 796–805.

103. H. Nishina, K. D. Fischer, L. Radvanyi *et al.*, 'Stress-signalling kinase Sek1 protects thymocytes from apoptosis mediated by CD95 and CD3.' *Nature*, 1997, **385**, 350–353.

104. S. Ganiatsas, L. Kwee, Y. Fujiwara *et al.*, 'SEK1 deficiency reveals mitogen-activated protein kinase cascade crossregulation and leads to abnormal hepatogenesis.' *Proc. Nat. Acad. Sci. USA*, 1998, **95**, 6881–6886.

105. T. Moriguchi, H. Kawasaki, S. Matsuda *et al.*, 'Evidence for multiple activators for stress-activated protein kinases/c-Jun amino terminal kinases.' *J. Biol. Chem.*, 1995, **270**, 12969–12972.

106. B. Glise, H. Bourbon and S. Noselli, 'Hemipterous encodes a novel *Drosophila* MAP kinase kinase required for epithelial cell sheet movement.' *Cell*, 1995, **83**, 451–461.

107. P. M. Holland, M. Suzanne, J. S. Campbell *et al.*, 'MKK7 is a stress-activated mitogen-activated protein kinase kinase functionally related to hemipterous.' *J. Biol. Chem.*, 1997, **272**, 24994–24998.

108. C. Tournier, A. J. Whitmarsh, J. Cavanagh *et al.*, 'Mitogen-activated protein kinase kinase 7 is an activator of the c-Jun NH2-terminal kinase.' *Proc. Nat. Acad. Sci. USA*, 1997, **94**, 7337–7342.

109. X. Lu, S. Nemoto and A. Lin, 'Identification of c-Jun NH2-terminal protein kinase (JNK)-activating kinase 2 as an activator of JNK but not p38.' *J. Biol. Chem.*, 1997, **272**, 24751–24754.

110. Z. Wu, J. Wu, E. Jacinto *et al.*, 'Molecular cloning and characterization of human JNKK2, a novel Jun NH2-terminal kinase-specific kinase.' *Mol. Cell Biol.*, 1997, **17**, 7407–7416.

111. Z. Yao, K. Deiner, X. S. Wang *et al.*, 'Activation of stress-activated protein kinases/c-Jun N-terminal kinases (SAPKs/JNKs) by a novel mitogen-activated protein kinase kinase (MKK7).' *J. Biol. Chem.*, 1997, **272**, 32378–32383.

112. T. Moriguchi, F. Toyoshima, N. Masuyama *et al.*, 'A novel SAPK/JNK kinase, MKK7, stimulated by TNFα and cellular stresses.' *EMBO J.*, 1997, **16**, 7045–7053.

113. I. N. Foltz, R. E. Gerl, J. S. Wieler *et al.*, 'Human mitogen-activated protein kinase kinase 7 (MKK7) is a highly conserved c-Jun N-terminal kinase/stress-activated protein kinase (JNK/SAPK) kinase activated by environmental stresses and physiological stimuli.' *J. Biol. Chem.*, **273**, 9344–9351.

114. C. Tournier, A. J. Whitmarsh, J. Cavanagh *et al.*, 'The MKK7 gene encodes a group of c-Jun NH2-terminal kinase kinases.' *Mol. Cell. Biol.*, **19**, 1569–1581.

115. H. Nishina, A. Vaz, P. Billia *et al.*, 'Defective liver formation and liver cell apoptosis in mice lacking the stress signaling kinase SEK1/MKK4.' *Development*, 1999, **126**, 505–516.

116. S. Lawler, Y. Fleming, M. Goedert *et al.*, 'Synergistic activation of SAPK1/JNK1 by two MAP kinase kinases *in vitro*.' *Curr. Biol.*, 1998, **8**, 1387–1390.

117. J. Raingeaud, A. J. Whitmarsh, T. Barett *et al.*, 'MKK3- and MKK6-regulated gene expression is mediated by the p38 mitogen-activated protein kinase signal transduction pathway.' *Mol. Cell. Biol.*, 1996, **16**, 1247–1255.

118. A. Cuenda, G. Alonso, N. Morrice *et al.*, 'Purification and cDNA cloning of SAPKK3, the major activator of RK/p38 in stress- and cytokine-stimulated monocytes and epithelial cells.' *EMBO J.*, 1996, **15**, 4156–4164.

119. M. Chiariello, M. J. Marinissen and J. S. Gutkind, 'Multiple mitogen-activated protein kinase signaling pathways connect the Cot oncoprotein to the c-*jun* promoter and to cellular transformation.' *Mol. Cell. Biol.*, 2000, **20**, 1747–1758.

120. C. A. Lange-Carter, C. M. Pleiman, A. M. Gardner *et al.*, 'A divergence in the MAP kinase regulatory network defined by MEK kinase and Raf.' *Science*, 1993, **260**, 315–319.

121. C. Patriotis, A. Makris, S. E. Bear *et al.*, 'Tumor progression locus 2 (Tpl-2) encodes a protein kinase involved in the progression of rodent T-cell lymphomas and in T-cell activation.' *Proc. Nat. Acad. Sci. USA*, 1993, **90**, 2251–2255.

122. K. Yamaguchi, K. Shirakabi, H. Shibuya *et al.*, 'Identification of a member of the MAPKKK family as a potential mediator of TGF-β signal transduction.' *Science*, 1995, **270**, 2008–2011.

123. A. Salmerón, T. B. Ahmad, G. W. Carlile *et al.*, 'Activation of MEK-1 and SEK-1 by Tpl-2 proto oncoprotein, a novel MAP kinase kinase kinase.' *EMBO J.*, 1996, **15**, 817–826.

124. S. Xu, D. J. Robbins, L. B. Christerson *et al.*, 'Cloning of Rat MEK kinase 1 cDNA reveals an endogenous membrane-associated 195-kDa protein with a large regulatory domain.' *Proc. Nat. Acad. Sci. USA*, 1996, **93**, 5291–5295.

125. X. S. Wang, K. Diener, D. Jannuzzi *et al.*, 'Molecular cloning and characterization of a novel protein kinase with a catalytic domain homologous to mitogen-activated protein kinase kinase kinase.' *J. Biol. Chem.*, 1996, **271**, 31607–31611.

126. P. Gerwins, J. L. Blank and G. L. Johnson, 'Cloning of a novel mitogen-activated protein kinase-kinase-kinase, MEKK4, that selectively regulates the c-Jun amino terminal kinase pathway.' *J. Biol. Chem.*, 1997, **272**, 8288–8295.

127. M. Takekawa, F. Posas and H. Saito, 'A human homolog of the yeast Ssk2/Ssk22 MAP kinase kinase kinases, MTK1, mediates stress-induced activation of the p38 and JNK pathways.' *EMBO J.*, 1997, **16**, 4973–4982.

128. H. Ichijo, E. Nishida, K. Irie *et al.*, 'Induction of apoptosis by ASK1, a mammalian MAPKKK that activates SAPK/JNK and p38 signaling pathways.' *Science*, 1997, **275**, 90–94.

129. J. D. Ceci, C. P. Patriotis, C. Tsatsanis *et al.*, 'Tpl-2 is an oncogenic kinase that is activated by carboxy-terminal truncation.' *Genes Dev.*, 1997, **11**, 688–700.

130. N. L. Malinin, M. P. Boldin, A. V. Kovalenko *et al.*, 'MAP3K-related kinase involved in NF-κB induction by TNF, CD-95 and IL-1.' *Nature*, 1997, **385**, 540–544.

131. D. M. Rothwarf and M. Karin, 'The NF-κB activation pathway: a paradigm in information transfer from membrane to nucleus.' *Science STKE*, 1999, www.stke.org/cgi/content/full/OC_sigtrans;1999/5/re1.

132. E. Zandi and M. Karin, 'Bridging the gap: Composition, regulation and physiological function of the IκB kinase complex.' *Mol. Cell. Biol.*, 1999, **19**, 4547–4551.

133. M. H. Cardone, G. S. Salveson, C. Widmann *et al.*, 'The regulation of anoikis: MEKK-1 activation requires cleavage by caspases.' *Cell*, 1997, **90**, 315–323.

134. G. R. Fanger, C. Widmann, A. C. Porter *et al.*, '14-3-3 proteins interact with specific MEK kinases.' *J. Biol. Chem.*, 1998, **273**, 3476–3483.

135. M. Russell, C. A. Lange-Carter and G. L. Johnson, 'Direct interaction between Ras and the kinase domain of mitogen-activated protein kinase kinase kinase (MEKK1).' *J. Biol. Chem.*, 1995, **270**, 11757–11760.

136. G. R. Fanger, N. L. Johnson and G. L. Johnson, 'MEK kinases are regulated by EGF and selectively interact with Rac/Cdc42.' *EMBO J.*, 1997, **16**, 4961–4972.

137. A. Minden, A. Lin, M. McMahon *et al.*, 'Differential activation of ERK and JNK mitogen-activated protein kinases by Raf-1 and MEKK.' *Science*, 1994, **266**, 1719–1723.

138. M. Yan, T. Dai, J. C. Deak *et al.*, 'Activation of stress-activated protein kinase by MEKK1 phosphorylation of its activator SEK1.' *Nature*, 1994, **372**, 798–800.

139. Y. Xia, Z. Wu, B. Su *et al.*, 'JNKK1 organizes a MAP kinase module through specific and sequential interactions with upstream and downstream components mediated by its amino terminal extension.' *Genes Dev.*, 1998, **12**, 3369–3381.

140. V. Baud, Z.-G. Liu, B. Bennett *et al.*, 'Signaling by proinflammatory cytokines: oligomerization of TRAF2 and TRAF6 is sufficient for JNK and IKK activation and target gene induction via an amino terminal effector domain.' *Genes Dev.*, 1999, **13**, 1297–1308.

141. T. Yujiri, S. Sather, G. R. Fanger *et al.*, 'Role of MEKK1 in cell survival and activation of JNK and ERK pathways defined by targeted gene disruption.' *Science*, 1998, **282**, 1911–1914.

142. T. Minamino, T. Yujiri, P. J. Papst *et al.*, 'MEKK1 suppresses oxidative stress-induced apoptosis of embryonic stem cell-derived cardiac myocytes.' *Proc. Nat. Acad. Sci. USA*, 1999, **96**, 15127–15132.

143. B. C. Schaefer, M. F. Ware, P. Marrack *et al.*, 'Live cell fluorescence imaging of T cell MEKK2: redistribution and activation in response to antigen stimulation of the T cell receptor.' *Immunity*, 1999, **11**, 411–421.

144. M. Takekawa and H. Saito, 'A family of stress-inducible GADD45-like proteins mediate activation of the stress-responsive MTK1/MEKK4 MAPKKK.' *Cell*, 1998, **95**, 521–530.

145. H. Shibuya, K. Yamaguchi, K. Shirakabe *et al.*, 'TAB1: An Activator of the TAK1 MAPKKK in TGF-β Signal Transduction.' *Science*, 1996, **272**, 1179–1182.

146. J. Ninomiya-Tsuji, K. Kishioto, A. Hiyama *et al.*, 'The kinase TAK1 can activate the NIK-IκB as well as the MAP kinase cascade in the IL-1 signalling pathway.' *Nature*, 1999, **398**, 252–256.

147. T. Moriguchi, N. Kuroyanagi, K. Yamaguchi *et al.*, 'A novel kinase cascade mediated by mitogen-activated protein kinase kinase 6 and MKK3.' *J. Biol. Chem.*, 1996, **271**, 13675–13679.

148. K. Shirakabe, K. Yamaguchi, H. Shibuya *et al.*, 'TAK1 mediates the ceramide signaling to stress-activated protein kinase/c-Jun N-terminal kinase.' *J. Biol. Chem.*, 1997, **272**, 8141–8144.

149. M. Saitoh, H. Nishitoh, M. Fujii *et al.*, 'Mammalian thioredoxin is a direct inhibitor of apoptosis signal-regulating kinase (ASK) 1.' *EMBO J.*, 1998, **17**, 2596–2606.

150. H. Nishitoh, M. Saitoh, Y. Mochida *et al.*, 'ASK1 is essential for JNK/SAPK activation by TRAF2.' *Mol. Cell.*, 1998, **2**, 389–395.

151. H. Liu, H. Nishitoh, H. Ichijo *et al.*, 'Activation of apoptosis signal-regulating kinase-1 (ASK1) by TNF receptor-associated factor-2 requires prior dissociation of the ASK1 inhibitor thioredoxin.' *Mol. Cell. Biol.*, 2000, **20**, 2198–2208.

152. H. Y. Chang, H. Nishitoh, X. Yang *et al*, 'Activation of apoptosis signal-regulating kinase 1 (ASK1) by the adapter protein Daxx.' *Science*, 1998, **281**, 1860–1863.

153. Y. Gotoh and J. A. Cooper, 'Reactive Oxygen Species- and Dimerization-induced Activation of Apoptosis Signal-regulating Kinase 1 in Tumor Necrosis Factor-α Signal Transduction.' *J. Biol. Chem.*, 1998, **273**, 17477–17482.

154. C. Patriotis, A. Makris, J. Chernoff *et al.*, 'Tpl-2 acts in concert with Ras and Raf-1 to activate mitogen-activated protein kinase.' *Proc. Nat. Acad. Sci. USA*, 1994, **91**, 9755–9759.

155. D. S. Dorow, L. Devereux, E. Dietzsch *et al.*, 'Identification of a new family of human epithelial protein kinases containing two leucine/isoleucine-zipper domains.' *Eur. J. Biochem.*, 1993, **213**, 701–710.

156. K. A. Gallo, M. R. Mark, D. T. Scadden *et al.*, 'Identification and characterization of SPRK, a novel src-homology 3 domain-containing proline-rich kinase with serine/threonine kinase activity.' *J. Biol. Chem.*, 1994, **269**, 15092–15100.

157. L. B. Holtzman, S. E. Merritt and G. Fan, 'Identification, molecular cloning and characterization of dual leucine zipper bearing kinase.' *J. Biol. Chem.*, 1994, **269**, 30808–30817.

158. A. Rana, K. Gallo, P. Godowski *et al.*, 'The mixed lineage protein kinase SPRK phosphorylates and activates the stress-activated protein kinase activator, SEK1.' *J. Biol. Chem.*, 1996, **271**, 19025–19028.

159. L. A. Tibbles, Y. L. Ing, F. Kiefer *et al.*, 'MLK-3 activates the SAPK/JNK and p38/RK pathways via SEK1 and MKK3/6.' *EMBO J.*, 1996, **15**, 7026–7035.

160. G. Fan, S. E. Merritt, M. Kortenjann *et al.*, 'Dual leucine zipper-bearing kinase (DLK) activates p46SAPK and p38mapk but not ERK2.' *J. Biol. Chem.*, 1996, **271**, 24788–24793.

161. S.-I. Hirai, M. Katoh, M. Terada *et al.*, 'MST/MLK2, a member of the mixed lineage kinase family, directly phosphorylates and activates SEK1, an activator of c-Jun N-terminal kinase/stress-activated protein kinase.' *J. Biol. Chem.*, 1997, **272**, 15167–15173.

162. S.-I. Hirai, K. Noda, T. Moriguchi *et al.*, 'Differential activation of two JNK activators, MKK7 and SEK1, by MKN28-derived nonreceptor serine/threonine kinase/mixed lineage kinase 2.' *J. Biol. Chem.*, 1998, **273**, 7406–7412.

163. M. Hutchinson, K. S. Berman and M. H. Cobb, 'Isolation of TAO1, a protein kinase that activates MEKs in stress-activated protein kinase cascades.' *J. Biol. Chem.*, 1998, **273**, 28625–28632.

164. Z. Chen, M. Hutchinson and M. H. Cobb, 'Isolation of the protein kinase TAO2 and identification of its mitogen-activated protein kinase/extracellular signal-regulated kinase kinase binding domain.' *J. Biol. Chem.*, 1999, **274**, 28803–28807.

165. T. M. Moore, R. Garg, C. Johnson *et al.*, 'PSK, a novel STE20-like kinase derived from prostatic carcinoma that activates the c-Jun N-terminal kinase mitogen-activated protein kinase pathway and regulates actin cytoskeletal organization.' *J. Biol. Chem.*, 2000, **275**, 4311–4322.

166. J. M. Kyriakis, 'Signaling by the germinal center kinase family of protein kinases.' *J. Biol. Chem.*, 1999, **274**, 5259–5262.

167. H. R. Bourne, D. A. Sanders and F. McCormick, 'The GTPase superfamily: Conserved structure and molecular mechanism.' *Nature*, 1991, **349**, 117–127.

168. F. McCormick, 'Activators and effectors of ras p21 proteins.' *Curr. Opin. Genet. Dev.*, 1994, **4**, 71–76.

169. F. McCormick and A. Wittinghofer, 'Interactions between Ras proteins and their effectors.' *Curr. Opin. Biotechnol.*, 1996, **7**, 449–456.

170. J. R. Fabian, I. O. Daar and D. K. Morrison, 'Critical tyrosine residues regulate the enzymatic and biological activity of Raf-1 kinase.' *Mol. Cell. Biol.*, 1993, **11**, 7170–7179.

171. P. T. Hawkins, A. Eguinoa, R. G. Giu *et al.*, 'PDGF stimulates an increase in GTP-Rac via activation of phosphoinositide 3-kinase.' *Curr. Biol.*, 1995, **5**, 393–403.

172. A. J. King, H. Sun, B. Diaz *et al.*, 'The protein kinase Pak3 positively regulates Raf-1 through phosphorylation of serine 338.' *Nature*, 1998, **396**, 180–183.

173. C. S. Mason, C. J. Springer, R. G. Cooper *et al.*, 'Serine and tyrosine phosphorylations cooperate in Raf-1, but not B-Raf activation.' *EMBO J.*, 1999, **18**, 2137–2148.

174. G. Tzivion, Z. Luo and J. Avruch, 'A dimeric 14-3-3 protein is an essential cofactor for Raf kinase activity.' *Nature*, 1998, **394**, 88–92.

175. Z. Luo, G. Tzivion, P. J. Belshaw *et al.*, 'Oligomerization activates c-Raf-1 through a Ras-dependent mechanism.' *Nature*, 1996, **383**, 181–184.

176. A. Hall, 'Rho GTPases and the actin cytoskeleton.' *Science*, 1998, **279**, 509–514.

177. A. Aronheim, Y. C. Broder, A. Cohen *et al.*, 'Chp, a homologue of the GTPase Cdc42Hs, activates the JNK pathway and is implicated in reorganizing the actin cytoskeleton.' *Curr. Biol.*, 1998, **8**, 1125–1128.

178. R. A. Cerione and Y. Zheng, 'The Dbl family of oncogenes.' *Curr. Opin. Cell. Biol.*, 1996, **8**, 216–222.

179. P. Rodriguez-Viciana, P. H. Warne, R. Dhand *et al.*, 'Phosphatidylinositol-3-OH kinase as a direct target of Ras.' *Nature*, 1994, **370**, 527–532.

180. O. A. Coso, M. Chiarello, J.-C. Yu *et al.*, 'The small GTP binding proteins Rac1 and Cdc42 regulated the activity of the JNK/SAPK signaling pathway.' *Cell*, 1995, **81**, 1137–1146.

181. A. Minden, A. Lin, F.-X. Claret *et al.*, 'Selective activation of the JNK signaling cascade and c-Jun transcriptional activity by the small GTPases Rac and Cdc42Hs.' *Cell*, 1995, **81**, 1147–1157.

182. M. F. Olson, A. Ashworth and A. Hall, 'An essential role for Rho, Rac, and Cdc42 GTPases in cell cycle progression through G1.' *Science*, 1995, **269**, 1270–1272.

183. S. Zhang, J. Han, M. A. Sells *et al.*, 'Rho family GTPases regulate p38 mitogen-activated protein kinase through the downstream mediator Pak1.' *J. Biol. Chem.*, 1995, **270**, 23934–23936.

184. S. Bagrodia, B. Dérijard, R. J. Davis *et al.*, 'Cdc42 and PAK-mediated signaling leads to Jun kinase and p38 mitogen-activated protein kinase activation.' *J. Biol. Chem.*, 1995, **270**, 27995–27998.

185. Y. Zheng, D. J. Fischer, M. F. Santos *et al.*, 'The factigenital dysplasia gene product FGD1 functions as a Cdc42Hs-specific guanine-nucleotide exchange factor.' *J. Biol. Chem.*, 1996, **271**, 33169–33172.

186. E. Jacinto, G. Werlen and M. Karin, 'Cooperation between Syk and Rac1 leads to synergistic JNK activation in T lymphocytes.' *Immunity*, 1998, **8**, 31–41.

187. K. M. Eisenmann, M. A. Simpson, P. J. Keely *et al.*, 'Melanoma chondroitin sulphate proteoglycan regulates cell spreading through Cdc42, Ack-1 and p130cas.' *Nat. Cell Biol.*, 1999, **1**, 507–513.

188. P. D. Burbelo, D. Drechsel and A. Hall, 'A conserved binding motif defines numerous candidate target proteins for both Cdc42 and Rac GTPases.' *J. Biol. Chem.*, 1995, **270**, 29071–29074.

189. E. Manser, T. Leung, H. Salihuddin *et al.*, 'A brain serine/threonine protein kinase activated by Cdc42 and Rac1.' *Nature*, 1994, **367**, 40–46.

190. G. A. Martin, G. Bollag, F. McCormick *et al.*, 'A novel serine kinase activated by rac1/CDC42Hs-dependent autophosphorylation is related to PAK65 and STE20.' *EMBO J.*, 1995, **14**, 1970–1978.

191. E. Manser, C. Chong, Z.-S. Zhao *et al.*, 'Molecular cloning of a new member of the p21-Cdc42/Rac-activated kinase (PAK) family.' *J. Biol. Chem.*, 1995, **270**, 25070–25078.

192. M. Teo, E. Manser and L. Lim, 'Identification and molecular cloning of a p21cdc42/rac1-activated serine/threonine kinase that is rapidly activated by thrombin in platelets.' *J. Biol. Chem.*, 1995, **270**, 26690–26697.

193. A. Abo, J. Qu, M. S. Cammarano *et al.*, 'PAK4, a novel effector for Cdc42Hs, is implicated in the reorganization of the actin cytoskeleton and in the formation of filopodia.' *EMBO J.*, 1998, **17**, 6527–6540.

194. N. Lee, H. MacDonald, C. Reinhard *et al.*, 'Activation of hPAK65 by caspase cleavage induces some of the morphological and biochemical changes of apoptosis.' *Proc. Nat. Acad. Sci. USA*, 1997, **94**, 13642–13647.

195. M. A. Sells, U. G. Knaus, S. Bagrodia *et al.*, 'Human p21-activated kinase (Pak1) regulates actin organization in mammalian cells.' *Curr. Biol.*, 1997, **7**, 202–210.

196. M. A. Sells, J. T. Boyd and J. Chernoff, 'p21-activated kinase 1 (Pak1) regulates cell motility in mammalian fibroblasts.' *J. Cell Biol.*, 1999, **145**, 837–849.

197. A. Polverino, J. Frost, P. Yang *et al.*, 'Activation of mitogen-activated protein kinase cascades by p21-activated protein kinases in cell-free extracts of Xenopus oocytes.' *J. Biol. Chem.*, 1995, **270**, 26067–26070.

198. J. L. Brown, L. Stowers, M. Baer *et al.*, 'Human Ste20 homologue hPAK1 links GTPases to the JNK MAP kinase pathway.' *Curr. Biol.*, 1996, **6**, 598–605.

199. K. Nagata, A. Puls, C. Futter *et al.*, 'The MAP kinase kinase kinase MLK2 co-localizes with activated JNK along microtubules and associates with kinesin super-family motor KIF3.' *EMBO J.*, 1998, **17**, 149–158.

200. I. W.-L. Leung and N. Lassam, 'Dimerization via tandem leucine zippers is essential for the activation of the mitogen-activated protein kinase kinase kinase, MLK-3.' *J. Biol. Chem.*, 1998, **273**, 32408–32415.

201. N. Tapon, K. Nagata, N. Lamarche and A Hall, 'A new Rac target POSH is an SH3-containing scaffold protein involved in the JNK and NF-κB signalling pathways.' *EMBO J.*, 1998, **17**, 1395–1404.

202. N. Lamarche, N. Tapon, L. Stowers *et al.*, 'Rac and Cdc42 induce actin polymerization and G1 cell cycle progression independently of p65PAK and the JNK/SAPK MAP kinase cascade.' *Cell*, 1996, **87**, 519–529.

203. P. Katz, G. Whalen and J. H. Kehrl, 'Differential expression of a novel protein kinase in human B lymphocytes: preferential localization in the germinal center.' *J. Biol. Chem.*, 1994, **269**, 16802–16809.

204. T. Yuasa, S. Ohno, J. H. Kehrl *et al.*, 'Tumor necrosis factor signaling to stress-activated protein kinase (SAPK)/ Jun NH$_2$-terminal kinase (JNK) and p38.' *J. Biol. Chem.*, 1998, **273**, 22681–22692.

205. E. Tassi, Z. Biesova, P. P. Di Fiore *et al.*, 'Human JIK, a novel member of the STE20 kinase family that inhibits JNK and is negatively regulated by epidermal growth factor.' *J. Biol. Chem.*, 1999, **274**, 33287–33295.

206. F. Kiefer, L. A. Tibbles, M. Anafi *et al.*, 'HPK1, a hematopoietic protein kinase activating the SAPK/JNK pathway.' *EMBO J.*, 1996, **15**, 7013–7025.

207. M. C.-T. Hu, W. R. Qiu, X. Wang *et al.*, 'Human HPK1, a novel human hematopoietic progenitor kinase that activates the JNK/SAPK kinase cascade.' *Genes Dev.*, 1996, **10**, 2251–2264.

208. Y.-C. Su, J. Han, S. Xu *et al.*, 'NIK is a new Ste20-related kinase that binds NCK and MEKK1 and activates the SAPK/JNK cascade via a conserved regulatory domain.' *EMBO J.*, 1997, **16**, 1279–1290.

209. K. Diener, X. S. Wang, C. Chen *et al.*, 'Activation of the c-Jun N-terminal kinase pathway by a novel protein kinase related to human germinal center kinase.' *Proc. Nat. Acad. Sci. USA*, 1997, **94**, 9687–9692.

210. R. M. Tung and J. Blenis 'A novel human SPS1/STE20 homologue, KHS, activates Jun N-terminal kinase.' *Oncogene*, 1997, **14**, 653–659.

211. C.-S. Shi and J. H. Kehrl, 'Activation of stress-activated protein kinase/c-Jun N-terminal kinase, but not NF-κB, by the tumor necrosis factor (TNF) receptor 1 through a TNF receptor-associated factor 2- and germinal center kinase related-dependent pathway.' *J. Biol. Chem.*, 1997, **272**, 32102–32107.

212. C. A. Fu, M. Shen, B. C. B. Huang *et al.*, 'TNIK, a novel member of the terminal center kinase family that activates the c-Jun N-terminal kinase pathway and regulates the cytoskeleton.' *J. Biol. Chem.*, 1999, **274**, 30729–30737.

213. M. Kanai-Azuma, Y. Kanai, M. Okamoto *et al.*, 'Nrk: a murine X-linked NIK (Nck-interacting kinase)-related kinase gene expressed in skeletal muscle.' *Mech. Dev.*, **89**, 155–160.

214. C.-S. Shi, A. Leonardi, J. Kyriakis *et al.*, 'TNF-mediated activation of the stress-activated protein kinase pathway: TNF receptor-associated factor 2 recruits and activates germinal center kinase related.' *J. Immunol.*, 1999, **163**, 3279–3285.

215. H. Liu, Y.-C. Su, E. Becker *et al.*, 'A Drosophila TNF-receptor-associated factor (TRAF) binds the Ste20 kinase Misshapen and activates Jun kinase.' *Curr. Biol.*, 1999, **9**, 101–104.

216. K. J. Tracey and A. Cerami, 'Tumor necrosis factor, other cytokines and disease.' *Annu. Rev. Cell Biol.*, 1993, **9**, 317–343.

217. C. A. Smith, T. Farrah and R. G. Goodwin, 'The TNF receptor superfamily of cellular and viral proteins: Activation, costimulation and death.' *Cell*, 1994, **76**, 959–962.

218. R. H. Arch, R. W. Gedrich and C. B. Thompson, 'Tumor necrosis factor receptor-associated factors (TRAFs)—a family of adapter proteins that regulates life and death.' *Genes Dev.*, 1998, **12**, 2821–2830.

219. D. Wallach, E. E. Varfolomeev, N. L. Malinin *et al.*, 'Tumor necrosis factor receptor and fas signaling mechanisms.' *Annu. Rev. Immunol.*, 1999, **17**, 331–367.

220. B. Beutler, 'Tlr4: central component of the sole mammalian LPS sensor.' *Curr. Opin. Immunol.*, 2000, **12**, 20–26.

221. B. Beutler, 'Endotoxin, Toll-like receptor 4 and the afferent limb of innate immunity.' *Curr. Opin. Microbiol.*, 2000, **3**, 23–28.

222. P. Vandenabeele, W. Declercq, R. Beyaert *et al.*, 'Two tumour necrosis factor receptors: structure and function.' *Trends. Cell. Biol.*, 1995, **5**, 392–399.

223. F. K. Chan, R. M. Siegel and M. J. Lenardo, 'Signaling by the TNF receptor superfamily and T cell homeostasis.' *Immunity*, 2000, **13**, 419–411.

224. H. Hsu, J. Xiong and D. V. Goeddel, 'The TNF receptor-1-associated protein TRADD signals cell death and NF-κB activation.' *Cell*, 1995, **81**, 495–504.

225. M. Rothe, S. C. Wong, W. J. Henzel *et al.*, 'A novel family of putative signal transducers associated with the cytoplasmic domain of the 75-kDa tumor necrosis factor receptor.' *Cell*, 1994, **78**, 681–692.

226. Z.-G. Liu, H. Hsu, D. V. Goeddel *et al.*, 'Dissection of TNF receptor-1 effector functions: JNK activation is not linked to apoptosis while NF-κB activation prevents cell death.' *Cell*, 1996, **87**, 565–576.

227. G. Natoli, A. Costanzo, A. Ianni *et al.*, 'Activation of SAPK/JNK by TNF receptor-1 through a noncytotoxic TRAF2-dependent pathway.' *Science*, 1997, **275**, 200–203.

228. C. Reinhard, B. Shamoon, V. Shyamala *et al.*, 'Tumor necrosis factor-α-induced activation of c-jun N-terminal kinase is mediated by TRAF2.' *EMBO J.*, 1997, **16**, 1080–1092.

229. H. Hsu, H.-B. Shu, M.-G. Pan *et al.*, 'TRADD-TRAF2 and TRADD-FADD interactions define two distinct TNF receptor 1 signal transduction pathways.' *Cell*, 1996, **84**, 299–308.

230. H. Hsu, J. Huang, H.-B. Shu *et al.*, 'TNF-dependent recruitment of the protein kinase RIP to the TNF receptor-1 signaling complex.' *Immunity*, 1996, **4**, 387–396.

231. F. Urano, X. Wang, A. Bertolli *et al.*, 'Coupling of stress in the ER to activation of JNK protein kinass by transmembrane protein kinase IRE1.' *Science*, 2000, **287**, 664–666.

232. M. Takeuchi, M. Rothe and D. V. Goeddel, 'Anatomy of TRAF2.' *J. Biol. Chem.*, 1996, **271**, 19935–19942.

233. Y. C. Park, V. Burkitt, A. R. Villa *et al.*, 'Structural basis for self-association and receptor recognition of human TRAF2.' *Nature*, 1999, **398**, 533–538.

234. S. M. McWhirter, S. S. Pullen, J. M. Holton *et al.*, 'Crystallographic analysis of CD40 recognition and signaling by human TRAF2.' *Proc. Nat. Acad. Sci. USA*, 1999, **96**, 8408–8413.

235. H. Ye, Y. C. Park, M. Kreishman *et al.*, 'The structural basis for the recognition of diverse receptor sequences by TRAF2.' *Mol. Cell.*, 1999, **4**, 321–330.

236. S. Y. Lee, A. Reichlin, A. Santana *et al.*, 'TRAF2 is essential for JNK but not NF-κB activation and regulates lymphocyte proliferation and survival.' *Immunity*, 1997, **7**, 703–713.

237. W.-C. Yeh, A. Shahinian, D. Speiser *et al.*, 'Early lethality, functional NF-κB activation, and increased sensitivity to TNF-induced cell death in TRAF2-deficient mice.' *Immunity*, 1997, **7**, 715–725.

238. H. Y. Song, C. H. Régnier, C. J. Kirschning *et al.*, 'Tumor necrosis factor (TNF)-mediated kinase cascades: Bifurcation of Nuclear Factor-B and c-jun N-terminal kinase (JNK/SAPK) pathways at TNF receptor-associated factor 2.' *Proc. Nat. Acad. Sci. USA*, 1997, **94**, 9792–9796.

239. M. A. Lomaga, W.-C. Yeh, I. Sarosi *et al.*, 'TRAF6 deficiency results in osteopetrosis and defective interleukin-1, CD40, and LPS signaling.' *Genes Dev.*, 1999, **13**, 1015–1024.

240. C. M. Pombo, J. H. Kehrl, I. Sánchez *et al.*, 'Activation of the SAPK pathway by the human STE20 homologue germinal center kinase.' *Nature*, 1995, **377**, 750–754.

241. Z. Yao, G. Zhou, X. S. Wang *et al.*, 'A novel human STE20-related protein kinase, HGK, that specifically activates the c-Jun N-terminal kinase signaling pathway.' *J. Biol. Chem.*, 1999, **274**, 2118–2125.

242. J. C. Deak and D. L. Templeton, 'Regulation of the activity of MEK kinase 1 (MEKK1) by autopho-sphorylation within the kinase activation domain.' *Biochem. J.*, 1997, **322**, 185–192.

243. T. L. Siow, G. B. Kalmar, J. S. Sanghera *et al.*, 'Identification of two essential phosphorylated threonine residues in the catalytic domain of MEKK1.' *J. Biol. Chem.*, 1997, **272**, 7586–7594.

244. J. D. Hayes, E. M. Ellis, G. E. Neal *et al.*, 'Cellular response to cancer chemotherapreventive agents: contribution of the antioxidant responsive element to the adaptive response to oxidative and chemical stress.' *Biochem. Soc. Symp.*, 1999, **64**, 141–168.

245. A. Yokomizo, M. Ona, H. Nanri *et al.*, 'Cellular levels of thioredoxin associated with drug sensitivity to cisplatin, mitomycin C, doxorubicin, and etoposide.' *Cancer Res.*, 1995, **55**, 4293–4296.

246. U. Gundimeda, Z. H. Chen and R. Gopalakrishna, 'Tamoxifen modulates protein kinase C via oxidative stress in estrogen receptor-negative breast cancer cells.' *J. Biol. Chem.*, 1996, **271**, 13504–13514.

247. C. Ferlin, G. Scambia, M. Marone *et al.*, 'Tamoxifen induces oxidative stress and apoptosis in oestrogen receptor-negative human cancer cell lines.' *Brit. J. Cancer*, 1999, **79**, 257–263.

248. Y. Devary, R. A. Gottlieb, T. Smeal *et al.*, 'The mammalian ultraviolet response is triggered by activation of Src tyrosine kinases.' *Cell*, 1992, **71**, 1081–1091.

249. Y.-R. Chen, W. Wang, A. N. Kong *et al.*, 'Molecular mechanisms of c-Jun N-terminal kinase-mediated apotosis induced by anticarcinogenic isothiocyanates.' *J. Biol. Chem.*, 1998, **273**, 1769–1775.

250. L. H. Lu, S. S. Lee and H. C. Huang, 'Epigallocatechin suppression of proliferation of cascular smooth muscle cells: correlation with c-jun and JNK.' *Brit. J. Pharmacol.*, 1998, **124**, 1227–1237.

251. G. Bonizzi, J. Piette, S. Schoonbroodt *et al.*, 'Reactive oxygen intermediate-dependent NF-κB activation by interleukin 1β requires 5-lipoxygenase or NADPH oxidase activity.' *Mol. Cell. Biol.*, 1999, **19**, 1950–1960.

252. C.-H. Woo, Y.-W. Eom, M.-H. Yoo *et al.*, 'Tumor necrosis factor-α generates reactive oxygen species via a cytosolic phospholipase $A_2$-linked cascade.' *J. Biol. Chem.*, 2000, **275**, 32357–32362.

253. J. Sánchez-Alcázar, E. Schneider, M. A. Martinez *et al.*, 'Tumor necrosis factor-α increases the steady-state reduction of cytochrome *b* of the mitochondrial respiratory chain in metabolically inhibited L929 cells.' *J. Biol. Chem.*, 2000, **275**, 13353–13361.

254. K. P. Hoeflich, W.-C. Yeh, Z. Yao *et al.*, 'Mediation of TNF receptor-associated factor effector functions by apoptosis signal-regulating kinase-1 (ASK1).' *Oncogene*, 1999, **18**, 5814–5824.

255. V. Goossens, J. Grooten, K. De Vos *et al.*, 'Direct evidence for tumor necrosis factor-induced mitochondrial reactive oxygen intermediates and their involvement in cytotoxicity.' *Proc. Nat. Acad. Sci. USA*, 1995, **92**, 8115–8119.

256. A. Holmgren, 'Thioredoxin.' *Annu. Rev. Biochem.*, 1985, **54**, 237–271.

257. A. Holmgren, 'Thioredoxin and glutaredoxin systems.' *J. Biol. Chem.*, 1989, **264**, 13963–13966.

258. C. Rosette and M. Karin, 'Ultraviolet light and osmotic stress: activation of the JNK cascade through miltiple growth factor and cytokine receptors.' *Science*, 1996, **274**, 1194–1197.

259. A. Meriin, J. A. Yaglom, V. L. Gabai *et al.*, 'Protein damaging stresses activate JNK via inhibition of its phosphatase: a novel pathway controlled by HSP72.' *Mol. Cell. Biol.*, 1999, **19**, 2547–2555.

260. S. Kharbanda, R. Ren, P. Pandey *et al.*, 'Activation of the c-Abl tyrosine kinase in the stress response to DNA damaging agents.' *Nature*, 1995, **376**, 785–788.

261. Z.-G. Liu, R. Baskaran, E. T. Lea-Chou *et al.*, 'Three distinct signalling responses by murine fibroblasts to genotoxic stress.' *Nature*, 1996, **384**, 273–276.

262. P. Pandey, J. Raingeaud, M. Kaneki *et al.*, 'Activation of p38 mitogen-activated protein kinase by c-Abl-dependent and -independent mechanisms.' *J. Biol. Chem.*, 1996, **271**, 23755–23779.

263. Y.-R. Chen, C. F. Meyer and T.-H. Tan, 'Persistent activation of c-Jun N-terminal kinase-1 (JNK1) in γ radiation induced apoptosis.' *J. Biol. Chem.*, 1996, **271**, 631–634.

264. A. J. Fornace Jr., I. Alamo Jr. and C. Hollander, 'DNA damage-inducible transcripts in mammalian cells.' *Proc. Nat. Acad. Sci. USA*, 1988, **85**, 8800–8804.

265. M. B. Kastan, Q. Zhan, W. S. el-Diery *et al.*, 'A mammalian cell cycle checkpoint pathway utilizing p53 and GADD45 is defective in ataxia-telangiectasia.' *Cell*, 1992, **71**, 587–597.

266. T. D. Shafman, A. Saleem, J. Kyriakis *et al.*, 'Defective induction of stress-activated protein kinase activity in ataxia-telangiectasia cells exposed to ionizing radiation.' *Cancer Res.*, 1995, **55**, 3242–3245.

267. E. Shaulian and M. Karin, 'Stress-induced JNK activation is independent of gadd45 induction.' *J. Biol. Chem.*, 1999, **274**, 29595–29598.

268. X. Wang, M. Gorospe and N. J. Holbrook, 'gadd45 is not required for activation of c-Jun N-terminal kinase or p38 during acute stress.' *J. Biol. Chem.*, 1999, **274**, 29599–29602.

269. S. Kasibhatla, T. Brunner, L. Genestier *et al.*, 'DNA damaging agents induce expression of Fas ligand and subsequent apoptosis in T lymphosytes via the activation of NF-κB and AP-1.' *Mol. Cell.*, 1998, **1**, 543–551.

270. R. J. Kaufman, 'Stress signaling from the lumen of the endoplasmic reticulum: coordination of gene transcriptional and translational controls.' *Genes Dev.*, 1999, **13**, 1211–1233.

271. K. Mori, W. Ma, M.-J. Gething *et al.*, 'A transmembrane protein with a Cdc2/CDC28-related kinase activity is required for signaling from the ER to the nucleus.' *Cell*, 1993, **74**, 743–756.

272. C. E. Shamu and P. Walter, 'Oligomerization and phosphorylation of the Ire1p kinase during intracellular signaling from the endoplasmic reticulum to the nucleus.' *EMBO J.*, 1996, **15**, 3028–3039.

273. X.-Z. Wang, H. P. Harding, Y. Zhang *et al.*, 'Cloning of mammalian Ire1 reveals diversity in the ER stress response.' *EMBO J.*, 1998, **17**, 5708–5717.

274. W. Tirasophon, A. A. Welihinda and R. J. Kaufman, 'A stress response pathway from the endoplasmic reticulum to the nucleus requires a novel bifunctional protein kinase/endoribonucleuase (Ire1p) in mammalian cells.' *Genes Dev.*, 1998, **12**, 1812–1824.

275. D. D. Yang, C. Y. Kuan, A. J. Whitmarsh *et al.*, 'Absence of excitotoxicity-induced apoptosis in the hippocampus of mice lacking the Jnk3 gene.' *Nature*, 1997, **389**, 865–870.

276. C. Tournier, P. Hess, D. D. Yang *et al.*, 'Requirement of JNK for stress-induced activation of the cytochrome *c*-mediated death pathway.' *Science*, 2000, **288**, 870–874.

277. J. Yuan, 'Transducing signals of life and death.' *Curr. Opin. Cell. Biol.*, 1997, **9**, 247–251.

J.P. Vanden Heuvel, G.H. Perdew, W.B. Mattes and W.F. Greenlee (Eds.)
Comprehensive Toxicology, Vol. xiv

# 14.4.3
# Antioxidant Induction of Genes Encoding Detoxifying Enzymes

AMOS GAIKWAD,  SARAVANAKUMAR DHAKSHINAMOORTHY,
DAVID BLOOM and ANIL K. JAISWAL
*Baylor College of Medicine, Houston, TX, USA*

## 14.4.3.1  INTRODUCTION

Cellular exposure to xenobiotics and drugs leads to electrophilic stress, oxidative stress, cytotoxicity, mutagenicity and carcinogenicity.[1] Antioxidants reduce or neutralize harmful reactive oxygen species (ROS) and protect cells from the adverse effects of electrophilic and oxidative stress.[2–5] Several antioxidants including ascorbic acid, β-carotene, metallothionein, polyamines, melatonin, superoxide dismutase, catalase, 3-(2)-tert-butyl-4-hydroxyanisole (BHA), 3,5-di-tert-butyl-4-hydroxytoluene (BHT), and t-butyl hydroquinone (t-BHQ) are known to play significant roles in the prevention of oxidative stress and its related consequences.[2–12] Exactly how antioxidants play a role in the elimination of ROS is an intriguing question. Many of the antioxidants (e.g., glutathione, thioredoxin, superoxide dis-

mutase and catalase) are scavengers, directly neutralizing the ROS. On the other hand, others work indirectly by inducing mechanisms within cells that increase the endogenous levels of glutathione, prevent the formation of ROS, or facilitate the detoxification of reactive metabolites. In the present chapter, we will focus our discussion on the induction of detoxifying enzymes by antioxidants, and the mechanism of this induction.

## 14.4.3.2  ANTIOXIDANT INDUCTION OF GENES ENCODING DETOXIFYING ENZYMES

The various enzymes that detoxify xenobiotics and drugs have been classified into two-groups based on their mode of action and the effect they exert on cellular growth and function. The phase I enzymes include cytochromes P450, cytochrome P450 reductase, hydroxylases, lipoxygenases, peroxidases and oxidases, which activate chemicals to their mutagenic and carcinogenic metabolic products.[13–14] The phase II enzymes include NAD(P)H:quinone oxidoreductases (NQOs) that compete with cytochromes P450 and P450 reductase and catalyze two-electron reductive metabolism and detoxification of quinones;[15–19] glutathione *S*-transferases (GSTs), which conjugate hydrophobic electrophiles and ROS with glutathione;[20–21] UDP-glucuronosyl transferases (UDP-GT), which catalyze the conjugation of glucuronic acid with xenobiotics and drugs for their excretion;[22] epoxide hydrolase (EH), which inactivates epoxides;[23] γ-glutamylcysteine synthetase (γ-GCS), which plays a role in the glutathione metabolism;[24] ferritin-L gene, which plays an important role in iron storage;[25] and heme oxygenase-1 (HO-1), which catalyzes the first and rate-limiting step in heme catabolism.[26] The phase II enzyme proteins have been shown to protect the cells against toxicity, mutagenicity and carcinogenicity due to exposure to environmental and synthetic chemicals and drugs.[15–26]

Among phase II enzymes, our studies are focused on the NQOs.[15–19,27,28] NQOs (NQO1 and NQO2) are ubiquitously present in all tissues types.[15–19] NQO1−/− mice, lacking a functional NQO1 gene, have no developmental or reproductive defects.[27] However, the NQO1−/− mice are much more sensitive to menadione induced cytotoxicity, when compared to the NQO1+/+ mice. NQO1−/− mice are also more prone to develop benzo(a)pyrene and DMBA induced skin tumors, than wild type mice.[29] NQO2, like NQO1 also plays a role in chemoprevention.[15–16] However, NQO2 uses dihydronicotinamide riboside (NRH) rather than NAD(P)H as an electron donor.[30–31] It has been reported that a point mutation, P187S, results in the loss of NQO1 activity.[32–33] This mutation has been found in colon carcinoma (BE), lung cancer cells (H596) and basal cell carcinoma.[32–34] Studies have also proposed that individuals carrying mutant alleles of NQO1 have an increased risk of developing leukemia in response to benzene and its metabolites.[35–38] The loss of other detoxifying enzymes due to mutations is also reported to be associated with certain types of cancers.[39–42]

The transcription of phase I and II detoxifying enzymes is induced by a variety of chemicals including xenobiotics [e.g., β-naphthoflavone (β-NF)]; antioxidants [e.g., tert-butylhydroquinone (t-BHQ)]; oxidants [e.g., hydrogen peroxide ($H_2O_2$)]; 2,3,7.8-tetrachlorodibenzo-*p*-dioxin (TCDD); heavy metals [e.g., arsenic]; UV light and ionizing radiation.[13–14,16,43–45] Among these, xenobiotics and TCDD are known to induce both phase I and phase II enzyme genes expression.[13–14] In contrast, antioxidants only induce phase II detoxifying enzymes.[16,43–45] A coordinated induction of phase II detoxifying enzyme genes is considered crucial for chemopreventive activity of antioxidants.[15–19] A higher induction of phase I enzymes than phase II enzymes leads to the generation of excess amounts of electrophiles and ROS, which may not be efficiently neutralized/detoxified because of lower amounts of phase II enzymes. This may cause DNA and membrane damage and other adverse effects. On the other hand, a higher or exclusive induction of phase II enzymes, such as in response to antioxidants, will ensure protection to the cells from xenobiotics and drugs.

### 14.4.3.3    ANTIOXIDANT RESPONSE ELEMENT

Several *cis*-elements have been identified in the human NQO1 gene promoter region,[46,47,28] (reviewed in References 15–16) One of these elements, an antioxidant response element (ARE), is located between nucleotides −470 and −447 (Figure 1). This ARE is required for basal expression, as well as the induction of NQO1 in response to β-NF, BHA, tBHQ and $H_2O_2$. Another region, between nucleotides −780 and −365, is required for TCDD induction of NQO1. An activator protein 2 (AP2) element at nucleotide position −157 is essential for cAMP induced expression of NQO1. Yet other elements (between −837 to −560 and 130 to −47) control its basal expression. ARE elements have also been found in the promoter region of the rat and mouse GST Ya subunit genes,[48–52] the rat GST P gene,[53] the human γ-GCS gene,[24] the ferritin-L gene[25] and the human HO-1 gene[54] (Figure 1). Because the AREs of many detoxifying enzymes are highly conserved, it is likely that these genes may be coordinately regulated by a single mechanism involving the ARE. Nucleotide sequence analysis of the NQO1 gene ARE revealed that it contains one perfect and one imperfect AP1 element. These elements are arranged as inverse repeats that are separated by three base pairs and followed by a 'GC' box,[46–47,28,55] (Figure 1). Analysis of the AREs from other genes revealed that they also contain AP1/AP1-like elements arranged as inverse or direct repeats.[55] These repeats are

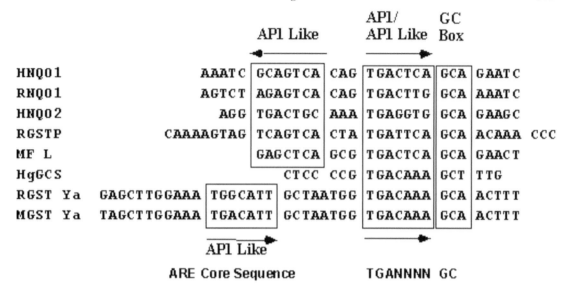

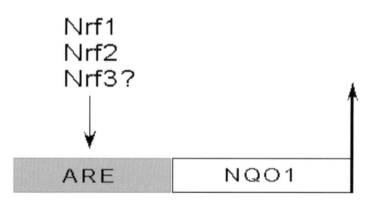

**Figure 1** Alignment of AREs from various genes of detoxifying enzymes. hNQO1, human NAD(P)H:quinone oxidoreductase1 gene ARE; rNQO1, rat NAD(P)H:quinone oxidoreductase1 gene ARE; hNQO2, human NAD(P)H:quinone oxidoreductase2 gene ARE; rGSTP, rat glutathione *S*-transferase P subunit gene ARE; mFL, mouse ferritin L subunit gene ARE; rGST Ya, rat glutathione *S*-transferase Ya gene ARE; mGST Ya, mouse glutathione *S*-transferase Ya gene ARE; hγGCS, human γ-glutamyl cysteine synthetase gene ARE.

separated by either three or eight nucleotides, followed by a GC box.[55] Previous studies revealed that the AP1 element is essential for activation of the collagenase and metallothionein genes in response to TPA.[56] Even though the human NQO1 ARE and other detoxifying enzyme gene AREs contains AP1 and AP1-like elements, they act as unique *cis*-elements. It is the ARE, rather than the AP1 element, that responds to xenobiotics and antioxidants.[28,55] Mutational analysis of the ARE identified GTGAC***GC as the core of the ARE sequence.[57,28,50,55] Additional *cis*-element and nucleotide sequences flanking the core sequence have been shown to contribute to the ARE-mediated expression and induction of detoxifying genes.[25,28,58]

### 14.4.3.4 ARE-BINDING PROTEINS

Several nuclear proteins have been shown to bind to the ARE as either homodimers or heterodimers.[47–55,57–67] Analysis of ARE-nuclear protein complexes have identified a number of nuclear transcription factors including c-Jun, Jun-B, Jun-D, c-Fos, Fra1, Nrf1, Nrf2, YABP, ARE-BP1, MafK, Ah (aromatic hydrocarbon) receptor and the estrogen receptor.[47–55,57–67] Among these transcription factors, c-Jun, Jun-B, Jun-D, c-Fos, Fra1, Nrf1 and Nrf2 and small Maf proteins have been shown to bind to the human NQO1 gene ARE.[47–48,59,64]

NF-E2 related factors 1 and 2 (Nrf1 and Nrf2) positively regulate the ARE-mediated expression and induction of the NQO1 gene in response to antioxidants and xenobiotics.[59] Nrf1 and Nrf2 are b-zip (leucine zipper) proteins that do not heterodimerize with each other. They each require another leucine zipper protein, such as c-Jun or Maf, to bind to the ARE;[59,64,68–69] (Figure 2). Recently, Nrf3, a third member of the Nrf family of transcription factors, was cloned and sequenced,[70] (Figure 2). However, the role of Nrf3 in the regulation of ARE-mediated expression and induction of detoxifying enzyme genes remains unknown. Nrf2−/− mice exhibit significantly reduced NQO1 expression and induction. This demonstrates that Nrf2 plays a physiological (*in vivo*) role in the regulation of NQO1.[65] Recently, we have shown that nuclear transcription factors Nrf1 and Nrf2 heterodimerize with c-Jun, Jun-B and Jun-D proteins. These dimers then bind to the ARE, resulting in expression and coordinated induction of detoxifying enzyme genes in response to antioxidants and xenobiotics.[64] Interestingly, Nrf-Jun association/ heterodimerization and binding to the ARE require unknown cytosolic factor(s).[64] However, the nature, and mode of action of the cytosolic factor(s) remain unknown. Nrf1 and Nrf2 have also been shown to regulate the ARE-mediated expression and induction of other detoxifying enzymes including GST Ya, γ-GCS and HO-1.[54,59,67,71–75]

More recently, a cytosolic factor, Kelch-like ECH-associated protein 1 (Keap1/INrf2), was identified,[76–77] (Figure 3). Under normal conditions, INrf2 retains Nrf2 in the cytoplasm.[76–77] Exposure of cells to antioxidants leads to the release of Nrf2 from INrf2. Nrf2 then translocates to the nucleus resulting in the activation of ARE-mediated gene expression (Figure 4). Analysis of the INrf2 amino acid sequence revealed a BTB/POZ BTB (broad complex, tramtrack, bric-a-brac)/ POZ (poxvirus, zinc finger) domain and a Kelch domain (Figure 3). The role of the BTB/POZ domain remains unknown in INrf2, but in other proteins it has been shown to be responsible for protein–protein interactions. In the *Drosophila* Kelch protein and in PIP, the Kelch domain binds to actin.[78–79] Therefore, it is expected that INrf2 will bind to actin in the cytoskeleton. There is a single report that shows that it is the C-terminal portion of INrf2 that interacts with Nrf2.[76] This portion contains the Kelch domain, but it is unlikely that the Kelch domain alone binds Nrf2. INrf2 appears to be a specific cytosolic inhibitor for Nrf2, because it does not interact with Nrf1 or Nrf3 (Dhakshinamoorthy and Jaiswal, Unpublished). This begs the question, are there cytosolic inhibitors of Nrf1 and Nrf3? If so, they remain unknown at this time.

The small Maf proteins (MafG, MafK, and MafF) are a family of nuclear transcription factors, which can act as either activators or repressors of a number of eukaryotic genes.[80–83] The small Mafs are homologous to viral-Maf (v-Maf) in the DNA binding domain and the leucine zipper domain, but they lack the transactivation domain that is present in v-Maf.[80–82] Small Mafs form homodimers, as well as heterodimers with Nrf2. The homodimers repress, while small Maf-Nrf2

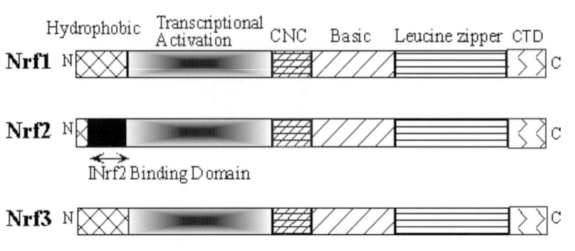

**Figure 2** NF-E2 related factors. The various domains of NF-E2 related factors 1, 2 and 3 (Nrf1, Nrf2 and Nrf3) are shown. The Nrf2 domain that interacts with the cytosolic inhibitor, INrf2, is shown. CNC, cap'n'collar domain; CTD, Carboxy terminal domain.

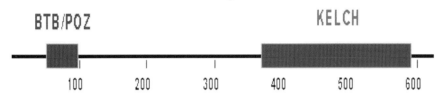

**Figure 3** Rat INrf2 protein. The BTB/POZ and KELCH binding domains are highlighted. The numbers denote amino acids.

heterodimers activate β-globin expression.[83] Analysis of ARE-nuclear protein complexes has revealed the presence of small Mafs as well as Nrf1, Nrf2, Jun, Fos, and Fra.[65, 74] Overexpression of MafG and MafK represses the γ-GCS gene.[74] Recently, Maf homodimers and Maf-Nrf2 heterodimers have been shown to repress the ARE-mediated expression and antioxidant induction of NQO1 and GST Ya genes.[84–85] However, the *in vivo* role of Nrf2-Maf heterodimers in repression of ARE-mediated gene expression is not clear.[84]

In addition to Maf proteins, overexpression of c-Fos or Fra1 also represses ARE-mediated gene expression.[59] Mice lacking c-Fos have increased expression of NQO1 and GST Ya, substantiating a negative role for c-Fos in ARE-mediated gene expression *in vivo*.[86] As with many response elements, it appears that ARE-mediated gene expression is a balance between positive and negative regulatory factors. One theory is that it is always necessary to have a small amount of ROS and other free radicals present to keep the cellular defenses active. Since activation of detoxifying enzymes and other defensive proteins leads to significant reduction in the levels of superoxide and other free radicals, the cell may require negative regulatory factors like the small Mafs and c-Fos to keep the expression of these proteins "in check".

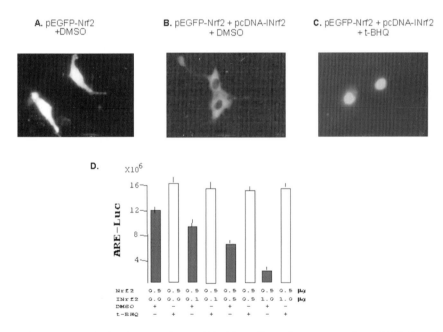

**Figure 4** INrf2 retains Nrf2 in the cytoplasm. (A)–(C). Effect of overexpression of INrf2 and t-BHQ on localization of EGFP-Nrf2. Quail fibroblasts (QT6) cells were transfected with plasmid pEGFP-Nrf2 alone or in combination with pcDNA-INrf2 as shown. The transfected cells were treated with DMSO (A) and (B) or 50μM t-BHQ (C). The cells were processed and visualized for florescence. Files saved in Black and White (D). Effect of overexpression of INrf2 on NQO1 gene ARE-mediated luciferase gene expression and induction in response to tert-butyl hydroquinone (t-BHQ). QT6 cells were co-transfected with reporter plasmid NQO1 gene hARE-tk-CAT and expression plasmids pcDNA-Nrf2 and pcDNA-INrf2 in concentrations as shown. The transfected cells were treated with either DMSO (vehicle control) or t-BHQ (50μM) for 12 h. The cells were analyzed for luciferase activity.

### 14.4.3.5 MECHANISM OF ACTIVATION OF ARE-MEDIATED EXPRESSION AND COORDINATED INDUCTION OF DETOXIFYING GENES

The signal transduction pathway that leads from antioxidants and xenobiotics to INrf2, Nrf1, Nrf2, Jun, Fos, and Mafs, the proteins that regulate ARE-mediated expression of detoxifying genes, remains unknown. Because the metabolism of both antioxidants and xenobiotics results in the creation of superoxides and electrophiles,[87] it is thought that these molecules might act as second messengers, activating ARE-mediated expression of a host of detoxifying enzyme genes including NQO1, GST Ya, $\gamma$-GCS and HO-1. These genes protect the cell against the damaging effects of oxidative stress. For instance, hydroxyl radicals, generated by quinones, induce the expression of the mouse GST gene.[88] However, this effect has not yet been demonstrated using the AREs of other detoxifying enzyme genes. $H_2O_2$, on the other hand, does induce ARE-mediated expression of rat GST Ya, rat NQO1, and human NQO1.[89] Therefore, $H_2O_2$ might play an important part in the induction of detoxifying genes by xenobiotics and antioxidants. Even though electrophiles are thought to be possible messengers in the oxidative stress pathway, their role, if any, has not been demonstrated.

A hypothetical model illustrating the activation of detoxifying enzyme genes by antioxidants and xenobiotics is shown in Figure 5. The ROS signal presumably passes through unknown cytosolic factor(s). This factor(s) then catalyzes the modification of Nrf2, INrf2, or both. This modification is expected because treatment of cells with antioxidants and xenobiotics does not alter the protein levels of Nrf2 or INrf2 (Venugopal et al., unpublished). As a result of the modification of the Nrf2-INrf2 complex, Nrf2 is released. Nrf2 then translocates to the nucleus where it heterodimerizes with c-Jun and induces the expression of NQO1 and other ARE-regulated genes. The levels of c-Jun increase significantly in response to oxidative stress,[90] perhaps to increase expression of ARE-mediated genes at first, and then later, at higher levels or when the stimulus is terminated, to repress their expression. In fact, c-Jun has an ARE in its promoter region, and this has been shown to respond to oxidative stress.[90] Nrf1 is expected to function in a manner similar to Nrf2. Likewise, Jun-B and Jun-D are expected to function similarly to c-Jun. Their participation in ARE-mediated expression of NQO1 has already been demon-

strated.[64] It is possible that others as yet unknown transcription factors heterodimerize with Nrf factors to upregulate ARE-mediated detoxifying enzyme genes. Among these are small Maf (MafK and MafG) proteins that are known to heterodimerize with Nrf2. However, it is not clear if Nrf2-small Maf heterodimers activate or repress ARE-mediated detoxifying enzyme genes expression and induction. Current opinion suggests a negative role for Nrf2-small Maf heterodimers in ARE-mediated gene regulation.[74,84-85] Further work is required to study this postulate.

The identification of the unknown cytosolic factor(s) is crucial for our understanding of how the oxidative stress signal is transmitted from ROS to the transcription factors that induce the expression of detoxifying genes. However, even the nature of this factor(s) remains largely unknown. It could be a kinase, phosphatase, or oxidation-reduction (redox) protein (Figure 5). It could modify Nrf2, INrf2, Maf or even possibly c-Jun. Most likely, the cytosolic factor(s) acts as the cellular sensor of oxidative stress, receiving the signal from superoxides and related ROS, and activating Nrf2. This sequence of events results in the increased expression of ARE-regulated detoxifying genes that protect the cell against the damaging effects of oxidative stress. Recent studies have shown a role for p38 and MEK kinases in the ARE-mediated regulation of detoxifying enzyme genes.[91-92] However, the molecular targets for these kinases remain unknown. Other studies have disputed the involvement of p38 and MEK kinases in ARE-mediated gene expression. Instead, they have shown that PKC phosphorylates Nrf2, thereby inducing ARE-mediated gene expression.[93] There are four PKC sites in Nrf2 that are highly conserved across several species (Figure 6). However, the role of these PKC sites in Nrf2-mediated regulation of detoxifying enzymes remains unknown. Nrf2 also contains a conserved cysteine residue in its DNA binding domain that has been shown to be redox regulated in c-Jun.[94-95] However, the role of this cysteine in Nrf2 regulation of ARE-mediated detoxifying enzymes remains to be determined.

The above studies raise interesting questions. These include: what are the roles of Nrf3 and large Maf proteins (c-Maf and MafB), which contain a transcriptional activation domain, in the expression and induction of NQO1 and other detoxifying enzymes? What is the mechanism of dissociation of Nrf2 from INrf2? Are there cytosolic inhibitors for Nrf1 and Nrf3? What is the process by which Nrf2 and c-Jun heterodimerize and then bind to the

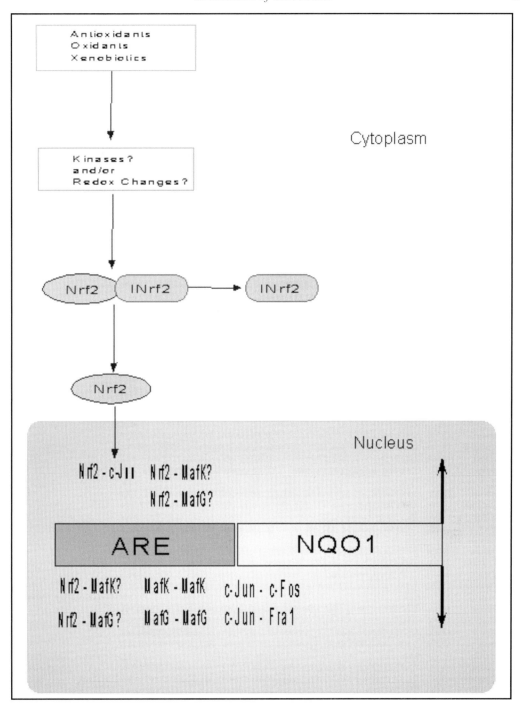

**Figure 5** Model to illustrate the role of NF-E2 related factors, Nrf1 and Nrf2, in ARE-mediated expression and induction of detoxifying enzyme genes in response to antioxidants. Nrf2-c-Jun dimers upregulate ARE-mediated gene expression and induction. MafK-MafK, MafG-MafG, c-Jun-c-Fos and c-Jun-Fra1 dimers negatively regulate ARE-mediated gene expression and induction. Nrf2-MafK and Nrf2-MafG heterodimers bind to the ARE without a clear understanding of their role in upregulation or down regulation of ARE-mediated gene expression. Therefore, these are listed with both positive and negative regulators of ARE.

ARE? Are there transcription factors other than c-Jun that heterodimerize with Nrf2 leading to the upregulation of ARE-mediated gene expression and induction? The answers to these and other questions will help us to understand the mechanism of signal transduction from xenobiotics and antioxidants to the transcription factors that regulate the ARE.

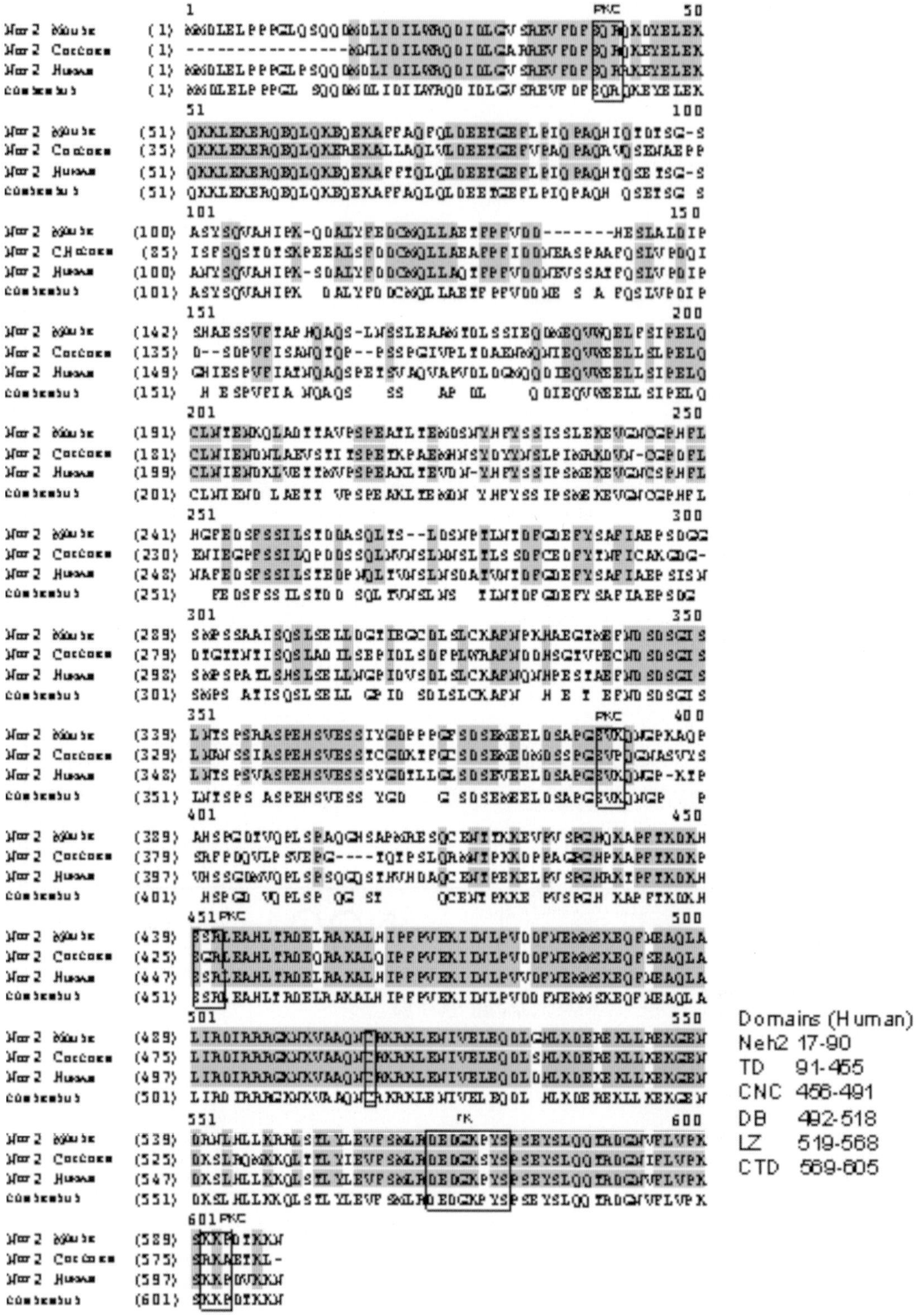

Domains (Human)

Neh2   17-90

TD      91-465

CNC   456-491

DB     492-518

LZ      519-568

CTD   569-605

**Figure 6**  Alignment of amino acid sequence of Nrf2 from human, mouse and chicken. Protein kinase C sites are boxed. The amino acid positions of the various domain are indicated in the right hand margin. Neh2, Nrf2 domain that interacts with INrf2; TD, Transcriptional activation domain; CNC, Cap'n'collar domain; DB, DNA binding domain; LZ, Leucine zipper region; CTD, carboxy terminal domain.

### 14.4.3.6 ACKNOWLEDGMENTS

We are thankful to our colleagues from Baylor College of Medicine, Houston, Texas for valuable suggestions. This work was supported by NIH grants RO1 GM47644, RO1 ES07943 and RO1 CA81057.

### 14.4.3.7 REFERENCES

1. B. Halliwell, 'Antioxidant characterization: methodology and mechanism.' *Biochem. Pharm.*, 1995, **49**, 1341–1348.
2. L. W. Wattenberg, 'Inhibition of carcinogenic and toxic effects of polycyclic aromatic hydrocarbons by phenolic antioxidants and ethoxyquin.' *J. Nat. Cancer Inst.*, 1972, **48**, 1425–1430.
3. B. M. Ulland, J. H. Weisburger, R. S. Yamamoto *et al.*, 'Antioxidants and carcinogenesis: butylated hydroxytoluene, but not diphenyl-*p*-phenylenediamine, inhibits cancer induction by *N*-2-fluorenylacetamide and by *N*-hydroxy-*N*-2-fluorenylacetamide in rats.' *Food Cosmet. Toxicol.*, 1973, **11**, 199–207.
4. R. Kahl, 'Synthetic antioxidants: biochemical actions and interference with radiation, toxic compounds, chemical mutagens and chemical carcinogens.' *Toxicology*, 1984, **33**, 185–228.
5. P. Talalay, 'Mechanisms of induction of enzymes that protect against chemical carcinogenesis.' *Adv. Enzyme Regul.*, 1989, **28**, 149–159.
6. A. Meister, 'Glutathione-Ascorbic acid antioxidant system in animals.' *J. Biol. Chem.*, 1994, **269**, 9397–9400.
7. B. Halliwell and J. M. C. Gutteridge, 'Free Radicals in Biology and Medicine,' 2nd edn, Oxford, Clarendon Press., 1989.
8. B. Halliwell, 'How to characterize a biological antioxidant.' *Free Radic. Res. Commun.*, 1990, **9**, 1–32.
9. H. Sies, 'Oxidative Stress,' New York, Academic Press, 1985.
10. J. M. C. Gutteridge and J. Stocks, 'Caeruloplasmin: Physiological and pathological perspectives.' *Crit. Rev. Clin. Lab. Sci.*, 1981, **14**, 257–329.
11. A. T. Diplock, Vitamin E, in 'Fat-Soluble Vitamins.' ed. Diplock, A. T., London, Heinemann, 1981, pp. 154–224.
12. I. Fridovich, 'Superoxide dismutases. An adaptation to a paramagnetic gas.' *J. Biol. Chem.*, 1989, **264**, 7761–7764.
13. F. P. Guengerich, 'Roles of cytochromes P-450 enzymes in chemical carcinogenesis and cancer chemotherapy.' *Cancer Res.*, 1988, **48**, 2946–2954.
14. R. Synder, 'Cytochrome P450, the oxygen-activating enzyme in xenobiotic metabolism.' *Toxicol. Sci.*, 2000, **58**, 3–4.
15. P. Joseph, T. Xie, Y. Xu and A. K. Jaiswal, 'NAD(P)H:Quinone Oxidoreductase1 (DT diaphorase): expression, regulation, and role in cancer.' *Oncology Research*, 1994, **6**, 525–532.
16. V. Radjendirane, P. Joseph and A. K. Jaiswal, 'Gene expression of DT-diaphorase (NQO1) in cancer cells, in 'Oxidative Stress and Signal Transduction,' eds. Henry J. Forman and Enrique Cadenas. Publisher Chapman and Hall, New York, 1997, pp. 441–475.
17. P. Talalay, J. W. Fahey, W. D. Holtzclaw *et al.*, 'Chemoprotection against cancer by phase 2 enzyme induction.' *Toxicol. Lett.*, 1995, **82–83**, 173–179.
18. R. J. Riley and P. Workman, 'DT-diaphorase and cancer chemotherapy.' *Biochem. Pharmacol.*, 1992, **43**, 1657–1669.
19. L. Ernster, 'DT-diaphorase: A historical review.' *Chem. Scripta.*, 1987, **27A**, 1–13.
20. C. B. Pickett and A. Y. H. Lu, 'Glutathione *S*-transferases: gene structure, regulation and biological function.' *Ann. Rev. Biochem.*, 1989, **58**, 743–764.
21. S. Tsuchida and K. Sato, 'Glutathione transferases and cancer.' *Crit. Rev. Biochem. Mol. Biol.*, 1992, **27**, 337–384.
22. T. Tephly and B. Burchell, 'UDP-glucuronosyl transferase: a family of detoxifying enzymes.' *Trends Pharmacol.*, 1990, **11**, 276–279.
23. F. Oesch, I. Gath, T. Igarashi, H. R. Glatt *et al.*, 'Molecular aspects of monooxygenases and bioactivation of toxic compounds.' eds. E. Arinc, J. B. Schenkman, and E. Hodgson, New York, Plenum, 1991, pp. 447–461.
24. R. T. Mulkahy, M. A. Wartman, H. H. Bailey *et al.*, 'Constitutive and β-naphthoflavone-induced expression of the human γ-glutamylcysteine synthetase heavy subunit gene is regulated by a distal antioxidant response element/TRE sequence.' *J. Biol. Chem.*, 1997, **272**, 7445–7454.
25. W. Wasserman and W. E. Fahl, 'Functional antioxidant responsive elements.' *Proc. Natl. Acad. Sci. USA*, 1997, **94**, 5361–5366.
26. A. M. Choi and J. Alam, 'Heme oxygenase-1: function, regulation, and implication of a novel stress-inducible protein in oxidant-induced lung injury.' *Am. J. Respir. Cell Mol. Biol.*, 1996, **15**, 9–19.
27. V. Radjendirane, P. Joseph, Y. H. Lee *et al.*, 'Disruption of the DT Diaphorase (NQO1) in mice leads to increased menadione toxicity.' *J. Biol. Chem.*, 1998, **273**, 7382–7389.
28. T. Xie, M. Belinsky, Y. Xu *et al.*, 'ARE- and TRE-mediated regulation of gene expression: Response to xenobiotics and antioxidants.' *J. Biol. Chem.*, 1995, **270**, 6894–6900.
29. D. J. L. Long II, R. L. Waikel, X. Wang *et al.*, 'NAD(P)H:quinone oxidoreductase1 deficiency increases susceptibility to benzo(a)pyrene-induced mouse skin carcinogenesis.' *Cancer Res.*, 2000, **60**, 5913–5915.
30. K. B. Wu, R. Knox, X. Z. Sun *et al.*, 'Catalytic properties of NAD(P)H-quinone oxidoreductase-2 (NQO2), a dihydronicotinamide riboside dependent oxidoreductase.' *Arch. Biochem. Biophys.*, 1997, **347**, 221–228.
31. Q. Zhao, X. L. Yang, W. D. Holtzclaw *et al.*, 'Unexpected genetic and structural relationships of a long-forgotten flavoenzyme to NAD(P)H:quinone reductase (DT-diaphorase).' *Proc. Natl. Acad. Sci. USA*, 1997, **94**, 1669–1674.
32. D. Ross, H. Beall, R. D. Traver *et al.*, 'Bioactivation of quinones by DT-diaphorase, molecular, biochemical, and chemical studies.' *Oncology Res.*, 1994, **6**, 493–500.
33. A. M. Rauth, Z. Goldberg and V. Misra, 'DT-diaphorase: Possible roles in cancer chemotherapy and carcinogenesis.' *Oncology Res.*, 1997, **9**, 339–349.
34. A. Clairmont, H. Sies, S. Ramachandran *et al.*, 'Association of NAD(P)H:quinone oxidoreductase (NQO1) null with numbers of basal cell carcinomas: use of a multivariate model to rank the relative importance of this polymorphism and those at other relevant loci.' *Carcinogenesis*, 1999, **20**, 235–1240.
35. N. Rothman, M. T. Smith, R. B. Hayes *et al.*, 'Benzene poisoning, a risk factor for hematological malignancy, is associated with the NQO1 609C→T mutation and rapid fractional excretion of chlorzoxazone.' *Cancer Res.*, 1997, **57**, 2839–2842.
36. J. L. Moran, D. Siegel and D. Ross, 'A potential mechanism underlying the increased susceptibility of

individuals with polymorphism in NAD(P)H:quinone oxidoreductase1 (NQO1) to benzene toxicity.' *Proc. Nat. Acad. Sci. USA*, 1999, **96**, 8150–8155.

37. M. T. Smith, 'Benzene, NQO1, and genetic susceptibility to cancer.' *Proc. Nat. Acad. Sci. USA*, 1999, **96**, 7624–7626.

38. R. A. Larson, Y. Wang, M. Banerjee *et al.*, 'Prevalence of the inactivating 609 C→T polymorphism in the NAD(P)H:quinone oxidoreductase1 (NQO1) gene in patients with primary and therapy-related myeloid leukemia.' *Blood*, 1999, **94**, 803–807.

39. S. C. Cotton, L. Sharp, J. Little *et al.*, 'Glutathione S-transferase polymorphisms and colorectal cancer: a HuGE review.' *Am. J. Epidemiol.*, 2000, **151**, 7–32.

40. H. Autrup, 'Genetic polymorphisms in human xenobiotica metabolizing enzymes as susceptibility factors in toxic response.' *Mutat. Res.*, 2000, **464**, 65–76.

41. J. T. Lear, A. G. Smith, R. C. Strange *et al.*, 'Detoxifying enzyme genotypes and susceptibility to cutaneous malignancy.' *Br. J. Dermatol.*, 2000, **142**, 8–15.

42. C. L. Rock, J. W. Lampe and R. E. Patterson, 'Nutrition, genetics, and risks of cancer.' *Annu. Rev. Public Health*, 2000, **21**, 47–64.

43. T. H. Rushmore and C. B. Pickett, 'Glutathione-S-transferases, structure, regulation and therapeutic implications.' *J. Biol. Chem.*, 1993, **268**, 11475–11478.

44. D. W. Nebert, 'Drug-metabolizing enzymes in ligand-modulated transcription.' *Biochem. Pharmacol.*, 1994, **47**, 25–37.

45. V. Daniel, R. Sharon and A. Bensimon, 'Regulatory elements controlling the basal and drug-inducible expression of glutathione S-transferase Ya subunit gene.' *DNA*, 1989, **8**, 399–408.

46. A. K. Jaiswal, 'Human NAD(P)H:Quinone Oxidoreductase gene structure and induction by dioxin.' *Biochemistry*, 1991, **30**, 10647–10653.

47. Y. Li and A. K. Jaiswal, 'Regulation of human NAD(P)H:quinone oxidoreductase gene: Role of AP1 binding site contained within human antioxidant response element.' *J. Biol. Chem.*, 1992, **267**, 15097–15104.

48. T. H. Rushmore, R. G. King, K. E. Paulson *et al.*, 'Regulation of glutathione S-transferase Ya subunit gene expression: identification of a unique xenobiotic-responsive element controlling inducible expression by planar aromatic compounds.' *Proc. Nat. Acad. Sci. USA*, 1990, **87**, 3826–3830.

49. T. H. Rushmore and C. B. Pickett, 'Transcriptional regulation of the rat glutathione S-transferase Ya subunit gene: characterization of a xenobiotic-responsive element controlling inducible expression by phenolic antioxidants.' *J. Biol. Chem.*, 1990, **265**, 14648–14653.

50. T. H. Rushmore, M. R. Morton and C. B. Pickett, 'The antioxidant responsive element. Activation by oxidative stress and identification of the DNA consensus sequence required for functional activity.' *J. Biol. Chem.*, 1991, **266**, 11632–11639.

51. R. S. Friling, A. Bensimon, Y. Tichauer *et al.*, 'Xenobiotic-inducible expression of murine glutathione S-transferase Ya subunit gene is controlled by an electrophile-responsive element.' *Proc. Nat. Acad. Sci. USA*, 1990, **87**, 6258–6262.

52. R. S. Friling, S. Bergelson and V. Daniel, 'Two adjacent AP-1-like binding sites form the electrophile responsive element of the murine glutathione S-transferase Ya subunit gene.' *Proc. Nat. Acad. Sci. USA*, 1992, **89**, 668–672.

53. A. Okuda, M. Imagawa, Y. Maeda, M. Sakai *et al.*, 'Functional co-operativity between two TPA responsive elements in undifferentiated F9 embryonic stem cells.' *EMBO Journal.*, 1990, **9**, 1131–1135.

54. J. Alam, D. Stewart, C. Touchard *et al.*, 'Nrf2, a Cap'n'Collar transcription factor, regulates induction of the heme oxygenase-1 gene.' *J. Biol. Chem.*, 1999, **274**, 26071–26078.

55. A. K. Jaiswal, 'Antioxidant response element.' *Biochem. Pharmacol.*, 1994, **48**, 439–444.

56. P. Angel and M. Karin, 'The role of Jun, Fos and AP-1 complex in cell proliferation and transformation.' *Biochem. Biophys. Acta.*, 1991, **1072**, 129–157.

57. S. Dhakshinamoorthy, D. J. Long II and A. K. Jaiswal, 'Antioxidant regulation of genes encoding enzymes that detoxify xenobiotics and carcinogens.' *Curr. Topic Cell. Reg.*, 2000, **36**, 201–216.

58. T. Prestera, W. D. Holtzclaw, Y. Zhang *et al.*, 'Chemical and molecular regulation of enzymes that detoxify carcinogens.' *Proc. Natl. Acad. Sci. USA*, 1993, **90**, 2965–2969.

59. R. Venugopal and A. K. Jaiswal, 'Nrf1 and Nrf2 positively and c-Fos and Fra1 negatively regulate the human antioxidant response element-mediated expression of NAD(P)H:quinone oxidoreductase1 gene.' *Proc. Nat. Acad. Sci. USA*, 1996, **93**, 14960–14965.

60. S. Liu and C. B. Pickett, 'The rat liver glutathione S-transferase Ya subunit gene: characterization of the binding properties of a nuclear protein from Hep-G2 cells that has high affinity for the antioxidant response element.' *Biochemistry*, 1996, **35**, 11517–11521.

61. V. Vasliou, A. Puga, C. Chang *et al.*, 'Interaction between the Ah-receptor and proteins binding to the AP-1 like electrophile response element (EpRE) during murine phase II [Ah] battery gene expression.' *Biochem. Pharmcol.*, 1995, **12**, 2057–2068.

62. K. Yoshioka, T..L. Deng, M. Cavigelli *et al.*, 'Antitumor promotion by phenolic antioxidants: Inhibition of AP-1 activity through induction of Fra expression.' *Proc. Nat. Acad. Sci. USA*, 1995, **92**, 4972–4976.

63. M. M. Montano, A. K. Jaiswal and B. S. Katzenellenbogen, 'Transcriptional regulation of the human quinone reductase gene by antiestrogen-liganded estrogen receptor-alpha and estrogen receptor-beta.' *J. Biol. Chem.*, 1998, **273**, 25443–25449.

64. R. Venugopal and A. K. Jaiswal, 'Nrf2 and Nrf1 in association with Jun proteins regulate antioxidant response element-mediated expression and coordinated induction of genes encoding detoxifying enzymes.' *Oncogene*, 1998, **17**, 3145–3156.

65. K. Itoh, T. Chiba, S. Takahashi *et al.*, 'An Nrf2/small maf heterodimer mediates the induction of phase II detoxifying enzyme genes through antioxidant response elements.' *Biochem. Biophys. Res. Commun.*, 1997, **236**, 313–322.

66. M. M. Montano, A. K. Jaiswal and B. S. Katzenellenbogan, 'Transcriptional regulation of the human quinone reductase gene by antiestrogen-liganded estrogen receptor gene by antiestrogen receptor-α and extrogen receptor-β.' *J. Biol. Chem.*, 1998, **39**, 25433–25449.

67. T. Nguyen, H. C. Huang and C. B. Pickett, 'Transcriptional regulation of the antioxidant response element—Activation by Nrf2 and repression by MafK.' *J. Biol. Chem.*, 2000, **275**, 15466–15473.

68. Y. Y. Chan, X. Han and Y. W. Kan, 'Cloning of Nrf1, an NF-E2-related transcription factor, by genetic selection in yeast.' *Proc. Nat. Acad. Sci. USA*, 1993, **90**, 11371–11375.

69. P. Moi, K. Chan, I. Asunis *et al.*, 'Isolation of NF-E2-related factor 2 (Nrf2), a NF-E2 like basic leucine zipper transcriptional activator that binds to the tandem NF-E2/AP1 repeat of β-globin locus control region.' *Proc. Nat. Acad. Sci. USA*, 1994, **91**, 9926–9930.

70. A. Kobayashi, E. Ito, T. Toki *et al.*, 'Molecular cloning and functional characterization of a new cap'n'collar

family transcription factor Nrf3.' *J. Biol. Chem.*, 1999, **274**, 6443–6452.

71. A. K. Jaiswal, 'Nrf1 and Nrf2 regulate ARE-mediated expression and coordinated induction of a battery of genes encoding proteins that protect cells against oxidative stress,' in 'Molecular Medicine' eds. I. Hashimoto *et al.*, 1999, pp. 103–112.

72. J. Jayepaul and A. K. Jaiswal, 'Nrf2 and c-Jun regulation of ARE-mediated expression and induction of γ-glutamylcysteine synthetase heavy subunit gene.' *Biochem. Pharmacol.*, 2000, **59**, 1433–1439.

73. H. Moninova and R. T. Mulcahy, 'Up-regulation of the human g-glutamylcysteine synthetase regulatory subunit gene involves binding of Nrf-2 to an electrophile response element.' *Biochem. Biophys. Res. Commun.*, 1999, **261**, 661–668.

74. A. C. Wild, H. R. Moinova and R. T. Mulcahy, 'Regulation of γ-glutamylcysteine synthetase subunit gene expression by transcription factor Nrf2.' *J. Biol. Chem.*, 1999, **274**, 33627–33636.

75. M. Kwong, Y. W. Kan and J. Y. Chan, 'The CNC basic leucine zipper factor, Nrf1, is essential for cell survival in response to oxidative stress-inducing agents.' *J. Biol. Chem.*, 1999, **274**, 37491–37498.

76. K. Itoh, N. Wakabayashi, Y. Katoh *et al.*, 'Keap1 represses nuclear activation of antioxidant responsive elements by Nrf2 through binding to the amino-terminal Neh2 domain.' *Genes Dev.*, 1999, **13**, 76–86.

77. S. Dhakshinamoorthy and A. K. Jaiswal, 'Functional Characterization of a cytosolic inhibitor of Nrf2 (INrf2): Its role in ARE mediated expression and Antioxidant Induction of the NAD(P)H:quinone Oxidoreductase1 Gene.' Manuscript submitted, 2001.

78. O. Albagli, P. Dhordain, C. Deweindt *et al.*, 'The BTB/POZ domain: a new protein–protein interaction motif common to DNA- and actin-binding proteins.' *Cell Growth and Differentiation*, 1999, **6**, 1193–1198.

79. I. F. Kim, E. Mohammadi, and R. C. C. Huang, 'Isolation and characterization of IPP, a novel human gene coding an actin-binding, KELCH-like protein.' *Gene*, 1999, **228**, 73–83.

80. M. J. Kim, and N. C. Andrews, 'Human MafG is a functional partner for p45 NF-E2 in activating globin gene expression.' *Blood*, 1997, **89**, 3925–3935.

81. K. T. Fujiwara, K. Ashida, H. Nishina *et al.*, 'Two new members of the maf oncogene family, mafK and mafF, encode nuclear b-zip proteins lacking putative transactivator domain.' *Oncogene*, 1993, **8**, 2371–2380.

82. K. Kataoka, M. Noda and M. Nishizawa, 'Maf nuclear oncoprotein recognizes sequences related to an AP-1 site and forms heterodimers with both Fos and Jun.' *Mol. Cell Biol.*, 1994, **14**, 700–712.

83. M. G. Marini, K. Chan, L. Casula *et al.*, 'hMAF, a small human transcription factor that heterodimerizes specifically with Nrf1 and Nrf2.' *J. Biol. Chem.*, 1997, **272**, 16490–16497.

84. S. Dhakshinamoorthy and A. K. Jaiswal, 'Small Maf (MafG and MafK) proteins Negatively Regulate ARE-mediated Expression and Antioxidant Induction of the NAD(P)H: quinone Oxidoreductase1 Gene.' *J. Biol. Chem.*, 2000, **275**, 40134–40141.

85. T. Nguyen, H. C. Huang and C. B. Pickett, 'Transcriptional regulation of the antioxidant response element—activation by Nrf2 and repression by MafK.' *J. Biol. Chem.*, 2000, **275**, 15466–15473.

86. J. Wilkinson IV, V. Radjendirane, G. R. Pfeiffer *et al.*, 'Disruption of c-Fos leads to increased expression of NAD(P)H:quinone oxidoreductase1 and glutathione *S*-transferase.' *Biochem. Biophys. Res. Commun.*, 1998, **253**, 855–858.

87. M. J. De Long, A. B. Santamaria and P. Talalay, 'Role of cytochrome $P_1$–450 in the induction of NAD(P)H:quinone reductase in a murine hepatoma cell line and its mutants.' *Carcinogenesis*, 1987, **8**, 1549–1553.

88. R. Pinkus, L. M. Weiner and V. Daniel, 'Role of oxidants and antioxidants in the induction of AP-1, NF-kB, and glutathione *S*-transferase gene expression.' *J. Biol. Chem.*, 1996, **271**, 13422–13429.

89. L. V. Favreau and C. B. Pickett, 'Transcriptional regulation of the rat NAD(P)H:quinone reductase gene. Identification of regulatory elements controlling basal level expression and inducible expression by planar aromatic compounds and phenolic antioxidants.' *J. Biol. Chem.*, 1991, **266**, 4556–4561.

90. R. Venugopal and A. K. Jaiswal, 'Coordinated induction of c-Jun gene with genes encoding quinone oxidoreductases in response to xenobiotics and antioxidants.' *Biochem. Pharmacol.*, 1999, **58**, 597–603.

91. R. Yu, S. Mandlekar, W. Lei *et al.*, 'p38 mitogen-activated protein kinase negatively regulates the induction of phase II drug-metabolizing enzymes that detoxify carcinogens.' *J. Biol. Chem.*, 2000, **275**, 2322–2327.

92. R. Yu, C. Chen, Y. Y. MO *et al.*, 'Activation of mitogen-activated pathways induced antioxidant response element-mediated gene expression via a Nrf2-dependent mechanism.' *J. Biol. Chem.*, 2000, **275**, 39907–39913.

93. H. C. Huang, T. Nguyen and C. B. Pickett, 'Regulation of the antioxidant response element by protein kinase C-mediated phosphorylation of NF-E2 related factor 2.' *Proc. Nat. Acad. Sci. USA*, 2000, **97**, 12475–12480.

94. S. Xanthoudakis, and T. Curran, 'Identification and characterization of Ref-1, a nuclear protein that facilitates AP-1 DNA binding activity.' *EMBO J.*, 1992, **11**, 653–665.

95. S. Xanthoudakis, G. Miao, F. Wang *et al.*, 'Redox activation of Fos-Jun DNA binding activity is mediated by a DNA repair enzyme.' *EMBO J.*, 1992, **11**, 3323–3335.

## Abbreviations

NQO, NAD(P)H:Quinone Oxidoreductase; NQO1, the dioxin-inducible cytosolic form of NAD(P)H:Quinone Oxidoreductase (also known as DT diaphorase); NQO2, a second cytosolic form of NAD(P)H:Quinone Oxidoreductase; GST Ya, glutathione *S*-transferase Ya subunit; GST P, glutathione *S*-transferase P subunit; ARE, Antioxidant Response Element; β-NF, β-naphthoflavone, BHA, 3-(2)-tert-butyl-4-hydroxyanisole; BHT, 3,5-di-tert-butyl-4-hydroxytoluene; $H_2O_2$, Hydrogen peroxide; ROS, Reactive Oxygen Species.

J.P. Vanden Heuvel, G.H. Perdew, W.B. Mattes and W.F. Greenlee (Eds.)
Comprehensive Toxicology, Vol. xiv

# 14.4.4
# Hypoxia Regulation of Gene Expression through HIF1 Signaling

MARY K. WALKER

*College of Pharmacy, University of New Mexico Health Sciences Center, Albuquerque, NM, USA*

CHRIS A. BRADFIELD and JACQUELINE A. WALISSER

*McArdle Laboratory for Cancer Research, University of Wisconsin-Madison, Madison, WI, USA*

## 14.4.4.1  INTRODUCTION

Oxygen is required by every cell, tissue and organ of the body for oxidative phosphorylation reactions and, as such, complex mechanisms have evolved to ensure oxygen delivery and to maintain oxygen homeostasis. These mechanisms are essential for survival. Conditions leading to the lack of oxygen have severe consequences for an organism, including failure to generate sufficient ATP to maintain essential cellular functions. As a result, sophisticated regulatory mechanisms exist to detect and respond to low oxygen (hypoxic) environments. In keeping with the concept of gene-environment interactions, this section will focus on one mechanism by which vertebrates sense and adapt to a low oxygen environments through the transcriptional activity of the hypoxia inducible factor-1$\alpha$ (HIF1$\alpha$) and how this transcriptional activity can contribute to the physiology and pathology of human diseases.

## 14.4.4.2  RESPONSE TO HYPOXIA

As an environmental cue, hypoxia can stimulate a variety of systemic, local, and cellular responses.[1] In mammalian systems, the systemic response includes the transcriptional upregulation of the gene encoding the peptide hormone erythropoietin (EPO). EPO is a cytokine that stimulates erythropoiesis, thus increasing red blood cell number and the efficiency of $O_2$ transport throughout the body.[2] A second aspect of the systemic response to low oxygen tension is the increase in respiration rate that occurs through dopaminergic input to the carotid body. The upregulation of dopamine is due to the hypoxia-induced transcriptional activation of tyrosine hydroxylase, the rate-limiting enzyme of catecholamine synthesis.[3] At the local level, hypoxic environments can occur during embryogenesis and in a variety of pathophysiological conditions including pulmonary hypertension, myocardial ischemia, pre-eclampsia, and tumor growth. In these processes, hypoxic tissues upregulate the transcription of genes encoding various angiogenic factors, such as vascular endothelial growth factor (VEGF), platelet-derived growth factor (PDGF) and

fibroblast growth factor (FGF),[4–6] as well as vasodilators produced by enzymes such as inducible nitric oxide synthase (iNOS) and heme oxygenase-1 (HO1).[7,8] The upregulation of these factors results in an increased vascular bed density, increased vascular permeability, and increased oxygen availability to the starved tissues. At the cellular level, adaptation to a low oxygen environment occurs as cells switch from oxidative metabolism to glycolysis for energy production. This cellular response is mounted through the transcriptional activation of genes encoding glycolytic enzymes such as aldolase A, phosphoglycerate kinase 1, lactate dehydrogenase A, phosphofructokinase L, and glucose transporters such as GLUT-1.[9–12] At a physiological level, induction of still other genes by hypoxia may mediate tissue-specific responses. For example, cardiac hypoxia transcriptionally induces adrenomedullin (ADM),[13] a vasodilator; endothelin-1 (ET-1),[14] a myocyte hypertrophic factor; and atrial natriuretic peptide (ANP),[15] a diuretic, natriuretic, vasodilator. During cardiac ischemia, these gene products may be important in maintaining cardiac output by decreasing blood pressure and increasing stroke volume. Thus, hypoxia appears to be an important environmental cue for the transcription of a variety of genes involved tissue-specific and global adaptational responses. As will become apparent later in this chapter, persistent activation of these genes by chronic hypoxia can also contribute to the pathogenesis of human disease.

## 14.4.4.3  PAS PROTEINS AND OXYGEN HOMEOSTASIS

The connection between oxygen homeostasis and Per-Arnt/AHR-Sim (PAS) proteins was revealed through studies designed to understand the regulation of hypoxia-induced genes such as *Epo*. Early on, the upregulation of EPO production by hypoxia was shown to be due in large part to increased transcription of the *Epo* gene.[16,17] This regulation was mediated by a hypoxia inducible factor (HIF1), which bound to a hypoxia responsive element (HRE) found in the 3′ untranslated region of the *Epo* gene.[16,18,19] Purification of the HIF1 protein from induced HeLa cells revealed that this transcription factor was composed of two subunits, denoted HIF1α and HIF1β, and that both subunits contained bHLH and PAS domains.[20,21] Amino acid sequence analysis demonstrated that HIF1β was identical to the aryl hydrocarbon receptor nuclear translocator (ARNT) protein, previously shown to be required for aryl hydrocarbon receptor (AHR) signal transduction.[20,21] Protein sequencing and cDNA cloning experiments also demonstrated that the HIF1α subunit was a novel member of the PAS superfamily. The characterization of the HIF1α-ARNT dimer provided the second example of a basic helix-loop-helix (bHLH)-PAS heterodimer that played an important role in sensing and adapting to environmental change. Dimerization of AHR-ARNT (discussed in detail in Chapter 14.2.2) represented the first example of a bHLH-PAS heterodimer that sensed an environmental change, i.e., presence of hydrocarbon pollutants like 2,3,7,8-tetrachlorodibenzo-*p*-dioxin (dioxin), and regulated an adaptational response, i.e., transcriptional induction of phase I and phase II drug metabolizing enzymes by binding to specific DNA sequences termed dioxin-response elements (DRE).

in known target genes, the core consensus sequence for the binding of HIF1α-ARNT dimer has been defined as either 5′-TACGTG-3′ or 5′-RCGTG-3′,[10,23] Methylation interference assays support this core element and indicate contact between the HIF1 dimer and all four guanine residues found in both strands of the *Epo* HRE, i.e., 5′-TACGTGCT-3′.[24] This result is consistent with the consensus data and indicates that contacts between HIF1α and its response element extend beyond the minimal core sequence. The length of the core sequence and the idea that recognition extends beyond the core element is similar to that seen for the AHR-ARNT heterodimer.[25–28] Based upon these similarities, one can predict that the ARNT protein maintains the same half-site specificity within the HRE as it does within the DRE (i.e., the 3′ GTG half-site). This prediction is based upon the observations that the AHR has been shown definitively to bind to the 5′ TNGC half-site and ARNT the 3′ GTG half-site of the core DRE sequence, TNGCGTG.[25,26] It follows that the HIF1α subunit would bind to the 5′ TAC half site of the HRE while ARNT would bind the 3′ GTG. This latter prediction has not been formally tested.

## 14.4.4.4 THE HYPOXIA RESPONSE PATHWAY

The observation that HIF1α heterodimerizes with ARNT, a known partner of the AHR, suggested that hypoxia signaling through HIF1α would share certain features with that of the dioxin-AHR-ARNT signal transduction pathway (Figure 1). Under hypoxic conditions, HIF1α exists in the cytoplasm with its co-chaperone, a dimer of heat shock protein 90 (Hsp90). The hypoxic stimulus causes HIF1α to translocate to the nucleus where it partners with ARNT and this heterodimer subsequently binds to specific DNA sequences called hypoxia response elements (HREs). The bHLH and PAS domains of HIF1α and ARNT confer both DNA binding and dimerization specificity as would be predicted based upon data from the AHR and ARNT studies.[22] Binding of the HIF1 heterodimer to these response elements initiates transcription of a battery of hypoxia-inducible genes involved in mediating a response to a low oxygen environment. Comparison of the core sequences of the HRE with those of the DRE also reveals some similarities and demonstrates that signaling through bHLH-PAS heterodimers follows a structured set of guidelines. Based on functionally-active HREs identified

## 14.4.4.5 FUNCTIONAL DOMAINS OF HIF1α

In addition to the bHLH-PAS domains, the HIF1α protein contains functional domains that confer its ability to respond to hypoxia and its transactivation potential (Figure 2). GAL4 chimeras containing the C-terminal half of HIF1α, in combination with a GAL-UAS driven reporter system, were used to analyze the transactivation function of HIF1α.[11,29,30] The C-terminal HIF1α sequence was demonstrated to function as a potent transactivation domain (TAD) and the ability of this fusion protein to activate transcription was dramatically increased in hypoxic cells. Further deletion analyses lead to the identification of two hypoxia responsive domains (HRDs) within the C-terminal half of HIF1α.[11,29,30] The exact boundaries and identifiers for these domains vary between labs. For simplicity, we will us the definitions of Jiang *et al.*[30] That is, HRD1 lies between residues 531 and 575 and HRD2 lies between residues 786 and 826 of the human HIF1α protein.[30] The HRD sequences were 100% conserved between human and mouse HIF1α, confirming the importance of these regions for protein function. Both HRD1 and HRD2 act as minimal hypoxia activated TADs; however, HRD1 also appears to

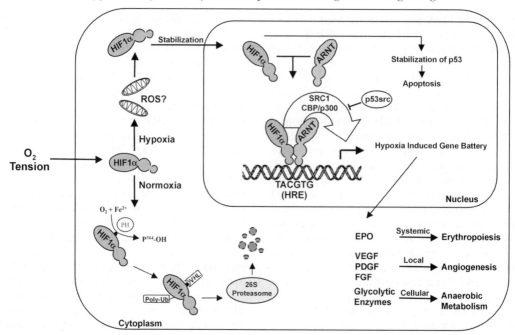

**Figure 1**    Model of hypoxia signaling pathways. HIF1α is differentially regulated by oxygen tension. Under normoxic conditions and in the presence of iron, HIF1α becomes hydroxylated at a specific proline residue ($P^{564}$) located within HRD1. This post-translational modification is mediated by a prolyl-hydroxylase (PH). Once hydroxylated, HIF1α is recognized by the von Hippel–Lindau (VHL) protein complex, polyubiquitinated and recruited to the 26S proteasome for degradation. Under hypoxic conditions, P564 is not hydroxylated and the targeted degradation of HIF1α by the VHL complex is suspended. HIF1α protein stabilization during hypoxia may involve a redox active protein and/or the generation of reactive oxygen species (ROS) from the mitochondria. HIF1α protein is translocated from the cytoplasm to the nucleus where it dimerizes with ARNT. The HIF1α-ARNT heterodimer then binds to hypoxia response elements (HREs) and activates the transcription of downstream target genes. The coactivators, CBP/p300 and SRC1, are involved in this transcriptional activation event. The target genes such as erythropoietin (EPO), vascular endothelial growth factor (VEGF), platelet derived growth factor (PDGF), fibroblast growth factor (FGF) and a series of glycolytic enzymes are involved in the systematic, local and cellular hypoxic responses. HIF1α also is involved in the stabilization of p53 protein and may play a role in the hypoxia-induced apoptosis. See text for details.

respond to low $O_2$ tension by stabilizing the HIF1α protein.[11,29,30] TADs can interact with coactivators, which function in remodeling of chromatin architecture and recruitment of the transcriptional machinery to the gene promoter. The transactivation activity of HIF1α is potentiated by the general coactivator CBP/p300 through direct physical interaction with HIF1α.[31–33] The interaction of p300 with the HRD2 of HIF1α is inhibited by p35src, the product of a gene that is induced by hypoxia via HIF1α mediated transcriptional activa-

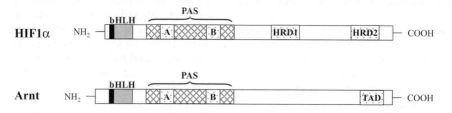

**Figure 2**    Functional domains of HIF1. Towards the amino terminus of both proteins is a region of basic amino acid residues that bind the DNA consensus sequence, 5′ RCGTG 3′, or hypoxia response elements (HRE), in 5′ UTR of genes regulated by hypoxia. Adjacent to the basic residues are amino acids that form a three-dimensional helix-loop-helix motif which serves as the primary dimerization interface between HIF1α and ARNT. The PAS domain consists of two 50 amino acid direct repeats (PAS A and PAS B) that functions as a secondary dimerization interface for HIF1α and ARNT, and functions as a binding site for two molecules of the molecular chaperone, heat shock protein 90 (Hsp90). HIF1α contains two hypoxia responsive domains (HRD1 and HRD2, see text for details) and ARNT contains a transactivation domain (TAD).

tion.[34] This interaction may provide a potential mechanism for feedback downregulation of hypoxic responses. Recent studies suggest that the transcriptional activator, SRC-1, is also recruited to HIF1α.[33]

## 14.4.4.6 REGULATION OF HIF1α BY OXYGEN TENSION

Oxygen concentrations have a direct influence on HIF1α protein stability and, therefore, on the regulation of the hypoxia-stimulated signaling pathway. In contrast to the AHR, HIF1α protein is present in very low amounts or is undetectable in the cell under normal physiological conditions. Experiments using antibodies against HIF1α indicate that HIF1α protein levels rise dramatically in response to hypoxia and agents that mimic hypoxia, like desferrioxamine or $CoCl_2$, whereas ARNT protein levels are relatively unaffected by these treatments.[20,35,36] These observations suggest that the increased transcriptional activity in response to hypoxia is due, at least in part, to increased abundance of HIF1α protein and enhanced binding to HREs in target genes. The absence of detectable HIF1α protein in non-hypoxic cells also suggests that protein stability is under negative regulation as a function of oxygen concentration. It has been observed that under normoxic conditions, the HIF1α protein undergoes rapid degradation via the ubiquitin-proteosome pathway, while under hypoxic conditions the protein is relatively stable.[18,24,37–39] Mapping studies using heterologous protein constructs with various C-terminal deletions of HIF1α have identified the region encompassing the HRD1 (also called the oxygen-dependent degradation domain, ODD) to be responsible for increased protein stability upon hypoxia.[38,39]

### 14.4.4.6.1 Proteasomal Degradation of HIF1α

It is now becoming clear that the proteosomal degradation of HIF1α under normoxic conditions is carried out by a protein complex of the von-Hippel–Lindau (VHL) tumor suppressor protein.[40–44] Initial suggestions that the *VHL* tumor suppressor gene may be involved in regulation of HIF1α came from studies of tumors in patients with VHL disease, an autosomal dominant familial cancer syndrome.[45,46] The majority of tumors seen in VHL disease were highly vascularized due to constitutive expression of VEGF, a hypoxia regulated gene.[47–50] In those tumors lacking the functional *VHL* gene, the increase in

VEGF expression was correlated with the increased expression of HIF1α and HIF2α. Reintroduction of wild-type *VHL* gene into clear cell renal carcinoma cells was sufficient to restore the normal oxygen-dependent regulation of VEGF. The VHL complex of proteins, including Cul2, Elongins B and C and Rbx1 proteins, resembles the well-characterized SCF ubiquitin ligase complexes. VHL directly interacts with the PYI motif contained in the HRD1 domain of HIF1α.[43] This interaction results in multi-ubiquitination within this domain at a critical lysine residue and subsequent targeting to the 26S proteosome.[42,43] It is interesting to note that the functional architecture of the HRD1 encompasses overlapping structures that have the opposing functions of protein degradation versus activation of gene transcription. The finding that tumor suppressor genes function in the degradation pathway of HIF1α may suggest a common function for this class of proteins. Another tumor suppressor gene, p53, has also been shown to interact with HIF1α and limit its hypoxia-induced expression.[51] The p53 protein appears to promote degradation of HIF1α by recruitment to the E3 ubiquitin-protein ligase. Mutation and loss of function of p53 is common finding in a variety of tumors and there is a statistically significant correlation between the presence of mutant p53 and HIF1α overexpression.[52] Therefore, the increased HIF1α activity resulting from a loss of p53 function may contribute to increased VEGF expression and the subsequent increase in angiogenesis and vascularization that is observed in a wide variety of human cancers.

### 14.4.4.6.2 Mechanism of Oxygen Sensing

Early studies of hypoxia-driven gene expression by $pO_2$, iron chelators, and divalent metal ions provided evidence that a heme protein was involved in this process,[2] however, the identification of the oxygen sensing protein has remained somewhat elusive. Recently two laboratories have discovered a heme-dependent cytosolic enzyme, prolyl hydoxylase, which plays a fundamental role in control of HIF-1α ubiquitination.[53,54] In cells essentially lacking ubiquitin degradation, HIF1α accumulated under both normoxic and hypoxic conditions; however, VHL protein could only associate with HIF1α under normoxic conditions, suggesting that a post translational modification occurs to HIF1α under normoxia, allowing for VHL complexation and subsequent ubiquitination.[53] Furthermore, it was shown that association of HIF1α and VHL, when expressed in reticulocyte lysates, occurred in an iron- and oxygen-dependent manner.[54] Using a Gal4-HIF1α fusion protein, it was identified that HIF1α amino acid residues within the HRD1, containing a highly conserved 8-mer sequence (MLAPYIPM), physically bound to VHL.[53,54] Using mass spectrometry analysis, it was shown that Pro[564] was hydroxlated and this proline residue was bound by VHL. The enzyme responsible for this hydroxylation was identified as a prolyl-4-hydroxylase which had a requirement for reduced iron and molecular oxygen.[53,54] Thus, this enzyme may represent a cellular oxygen sensor, because its enzymatic activity is dependent on molecular oxygen as a co-substrate; however, future research must determine whether the affinity of this enzyme for molecular oxygen is sufficiently sensitive to allow graded changes in enzymatic activity as intracellular $pO_2$ changes over a range of physiological levels.

### 14.4.4.6.3   Other Mechanisms of HIF1α Regulation

Other mechanisms have been suggested to account for the oxygen dependent regulation of HIF1α activity. Results from a number of laboratories have indicated that HIF1α activation is dependent on the red-ox state of the cell.[37,55,56] In this regard, it has been shown that Ref-1, a redox sensitive regulator, is necessary for full activity of the HIF1α transactivation domains.[57] It has also been shown that hypoxia-, but not $Co^{++}$- or desferrioxamine-induced HIF1α activation requires the production of reactive oxygen species (ROS) that is dependent upon the mitochondrial electron transport chain.[56,58] By employing *in vitro* methods in combination with various inhibitors of electron transport and ROS generation, this model suggests that hypoxia partially inhibits mitochondrial electron transport, resulting in redox changes in the electron carriers, particularly complex III, that increase the generation of ROS. These oxidants then enter the cytosol and function as second messengers in the signaling pathway leading to HIF1α stabilization. The finding that $Co^{++}$—and desferrioxamine—induce protein stabilization independent of the mitochondria, suggests that induction of HIF1α may employ different pathways and that a common downstream step may link hypoxia-, $Co^{++}$— and desferrioxamine-induced stabilization of HIF1α. Given the knowledge of VHL's involvement in HIF1α degradation, it has been hypothesized that the hypoxic signal conditionally regulates VHL-mediated degra-

dation of HIF1α.[43] It has also been suggested that HIF1α protein stability is conferred through a conformational change that occurs upon binding the ARNT protein.[59] Finally, stabilization of HIF1α expression and increased HIF1 signaling have been observed under *normoxic* conditions by stimuli other than hypoxia, transition metals, or iron-chelators. Insulin and insulin-like growth factor-1 are potent inducers of HIF1α protein expression and downstream gene activation in myoblasts,[60] while thrombin, angiotensin II, and platelet-derived growth factor are potent stimuli for these responses in vascular smooth muscle cells.[61] Taken in sum, these results suggest an extremely complicated regulatory mechanism is at play to stabilize the HIF1α protein and activate downstream gene expression.

To add to this complexity, results from a number of labs have indicated that under certain circumstances, the level of HIF1α mRNA also can be upregulated.[24,62–64] Although this has not been a widely reproduced finding, it should not be discounted and may be an indication that alternative methods of regulation are at play. For example, chronic exposure to hypoxic conditions *in vivo* increased HIF1α mRNA stabilization and transcriptional expression in the chick embryo[64] and rat heart,[65] and in the fetal rat brain.[66] Furthermore, cardiac expression of HIF1α mRNA increased following cardiac ischemia in humans.[67] Taken together these data suggest that the regulation of HIF1α mRNA expression could contribute to the physiological responses *in vivo* during prolonged exposure to hypoxic conditions.

The above discussion highlights the fact that activation of HIF1 signaling is a complex pathway that is regulated by multiple mechanisms. This regulation appears to be cell-type specific and likely depends on the physiological and pathological conditions of a given tissue. Moreover, attempts to model this biology must not only explain how a similar response can be elicited from low $O_2$ tension, ferric ions, cobalt and oxidative stress, but also peptide hormones, such as angiotensin II and insulin. Furthermore, this model also must explain a mechanism by which some responses to hypoxia can be evoked at moderately low levels (8% $O_2$),[68,69] while other responses require more extreme hypoxia (close to 1% $O_2$).[22] Finally, it must also take into account the apparent redundancy of both the α and β class partners (see below, e.g., HIF1α, HIF2α, HIF3α, ARNT and ARNT2), as well as the multiple levels of control that appear to be exerted over this system.

## 14.4.4.7 HOMOLOGUES OF HIF1α- AND β-CLASS PARTNERS

Elucidation of the nucleotide sequence of HIF1α facilitated the discovery of related hypoxia-inducible factors through screens of expressed sequence tag (EST) databases. Two other homologues of HIF1α that also respond to a low oxygen tension are HIF2α (also called endothelial PAS domain protein 1, EPAS1 or MOP2) and HIF3α.[70-76] While amino acid sequence identity indicates that this subset of the bHLH-PAS superfamily is highly related, there does not appear to be redundancy in their functions since their expression profiles are distinct. HIF2α, identified by a number of laboratories, is highly homologous to HIF1α in the bHLH-PAS domains and is able to bind the same DNA elements as the HIF1α-ARNT complex.[71,74-76] In addition, HIF2α contains structural and functional similarity to the HRD1 and HRD2 domains found in HIF1α.[77,78] In contrast to the widespread expression of HIF1α, HIF2α is expressed mainly in endothelial cells and in certain non-endothelial tissues such as the olfactory epithelium and the adrenal gland.[76,79] HIF3α is a newly identified hypoxia inducible factor that shares considerable sequence homology with HIF1α and HIF2α in the basic region, the HLH and PAS domains. It dimerizes with ARNT and this complex activates transcription of reporter genes driven by HRE elements in a heterologous expression system.[73] Less is known about the expression of the 3α homologue, although preliminary evidence from our laboratory suggests high-level expression in the developing trachea and olfactory epithelium (unpublished observation). Interestingly, both sequence analysis and functional analysis suggests that HIF3α contains an HRD1 domain but does not harbor a domain equivalent to HRD2.[73] The ARNT2 and MOP3 homologues of ARNT (described in Chapter 14.2.2) may also play roles as β-class partners of the α-class HIF sensor subunits. The ARNT2 protein shares 81% identity with ARNT in the bHLH-PAS domains (57% overall sequence identity) and thus was predicted to be a second partner of AHR.[70,80] In DNA binding assays, the ARNT2 protein is able to substitute for ARNT, directing HIF1α, HIF2α, and HIF3α to HREs (unpublished results). Coupled with these *in vitro* results, the overlapping developmental profiles of ARNT2 and the α-class HIF proteins, it seems highly likely that these proteins are biologically relevant partners *in vivo*.[79] The third ARNT homologue, MOP3, was originally cloned in EST screens for novel PAS encoding cDNAs.[71,72,81] It has homology with the ARNT protein in both its bHLH and PAS domains (66% and 40% identity, respectively). Although MOP3 and HIF1α are co-expressed in a number of tissues, MOP3 is a fairly weak dimerization partner of the α-class HIFs.[79,82] Thus, it is still unclear if MOP3 plays a significant role in hypoxia signal transduction.

## 14.4.4.8 GENE INACTIVATION IN ANIMAL MODELS OF HYPOXIA SIGNALING

To understand the biological roles of the HIF and ARNT family of PAS proteins in hypoxia signaling, a number of gene inactivation models have been exploited. Murine strains lacking the ARNT protein were the first to be developed.[83-85] These models displayed embryonic lethality between 9.5–10.5 days of gestation. Although some controversy remains as to whether the yolk sac circulation was affected, both ARNT knockout mouse strains display major blocks in developmental angiogenesis, suggesting that this failure is the primary cause of embryonic lethality.[83,84] The phenotype of the ARNT null mice provided important genetic support for the idea that hypoxia is an important signal in normal development and vascularization of the embryo. Use of the Cre-lox technology (see Chapter 5.2) allows the selective disruption of a gene in a tissue specific manner at various developmental stages.[86] Using this technology, recent studies in which the ARNT allele is selectively knocked out in the adult mouse liver suggest that the HIF1α pathway of gene induction is substantially diminished as a result, implying that the response to hypoxia is also compromised.[87] The biological consequences of this impairment in hypoxia signaling in the adult animal are yet to be determined. The role of the HIF subunits in angiogenesis also has support from studies using hepatoma cells that do not express ARNT protein. In these studies, neovascularization is inhibited in transplanted tumors that are devoid of the ARNT protein.[88] An alternate β-class partner for the HIFs is ARNT2. ARNT2 was shown to form functional HIF complexes in tumors derived from ARNT$^{-/-}$ embryonic stem (ES) cell.[89] Studies in ARNT deficient ES cells and hepatocytes have demonstrated that transfected ARNT2 forms functional HIF complexes and can restore hypoxia-induced gene expression.[89] These results suggest that there is some functional redundancy between the ARNT and ARNT2 proteins, and this may be biologically relevant in hypoxia signaling in those tissues where they are co-expressed, such as CNS and kidney.

In keeping with the idea that hypoxia is an important signal for normal development, mice homozygous for disruption at the HIF1α locus (HIF1α$^{-/-}$) display embryonic lethality at day 11 of gestation with neural tube defects, cardiovascular malformation and lack of cephalic vascularization.[90,91] In addition, these mice display a lack of the classic hypoxia responses, such as the upregulation of VEGF, glycolytic enzymes and glucose transporters.[90–92] This phenotype is similar to the ARNT null mice discussed earlier, and is consistent with the role of HIF1 heterodimer in the hypoxia-driven transcriptional activation of a number of genes involved in developmental angiogenesis such as VEGF, FGF and PDGF. Like many null alleles, the phenotype of HIF2α null mice provided a surprising result. The HIF2α$^{-/-}$ embryos die at day 12.5 of gestation due to pronounced bradycardia related to substantially decreased catecholamine levels.[93] This phenotype is consistent with HIF2α's high level of expression in the organ of Zuckerkandl (OZ), the principal source of catecholamine production.[93] Due to the early death of HIF2α homozygous $-/-$ embryo, it is unclear whether HIF2α plays a role in hypoxia responses in later stages of life or if this protein plays a role in angiogenesis as might have been predicted based upon its expression in the vascular endothelium. Conditional knockout of this gene may shed more light on such physiological functions.

## 14.4.4.9　HETEROTYPIC MODELS OF HIF1α SIGNALING

Experiments from a number of laboratories have demonstrated the importance of hetero-dimerization between the a and β-class subunits in the regulation of the systematic, local and cellular responses to hypoxia.[4,9,10,16,19,94] Despite the importance of heterodimerization, recent evidence also suggests that HIFs may signal through heterologous interactions with non-PAS containing proteins. In this regard, HIF1α has also been shown to be involved in the stabilization of p53 protein and may play a role in hypoxia-induced apoptosis[95] (Figure 1). This stabilization appears to be directly related to protein-protein interactions between HIF1α and p53 protein. Such a mechanism of protein stabilization may be a common activity of HIF1α.

## 14.4.4.10　PHYSIOLOGICAL AND PATHOLOGICAL ROLES OF HIF1 IN HUMAN DISEASE

As discussed above, hypoxia induces changes in gene expression that result in tissue-specific and global adaptational responses, but in some cases persistent activation of these genes can become maladaptive and contribute to disease pathogenesis. Recent studies have demonstrated that HIF1 signaling mediates important physiological and pathological responses in a variety of human diseases, including pulmonary hypertension, ischemic heart disease, pre-eclampsia, and cancer.

### 14.4.4.10.1　Hypoxia-induced Pulmonary Hypertension

Pulmonary hypertension results from an increase in vascular resistance within the lung and frequently occurs following long-term exposure to alveolar hypoxia. The hypoxia can result from the presence of chronic obstructive or severe restrictive lung diseases as well as by resident exposure to high altitudes, approximately >4300 m. This chronic hypoxia-induced pulmonary hypertension can lead to right ventricular hypertrophy, right-sided heart failure, and death. Pathologically, hypoxia-induced pulmonary hypertension is associated with pulmonary arterial vasoconstriction, polycythemia, and pulmonary arterial remodeling. The pulmonary arterial remodeling is characterized by a hypertrophy of the medial layer of medium-sized muscular arteries as well as thickening and sometimes duplication of internal and external elastic membranes of those same muscular arteries. In addition, pulmonary arterioles frequently demonstrate medial hypertrophy that can severely narrow the vessel lumen. The combination of pulmonary vasoconstriction, thickened pulmonary vascular walls, and increased circulating RBCs all contribute to increased pulmonary vascular resistance.

The pathophysiological development of hypoxia-induced pulmonary hypertension is mediated, in part, by HIF1α. In animal models, HIF1α mRNA expression was increased in the lung when animals were exposed to either acute or chronic hypoxia;[96,97] in cell culture, HIF1α protein expression was induced by hypoxia in a variety of pulmonary cells, including arterial endothelium, arterial smooth muscle, bronchial epithelium, alveolar macrophages, alveolar epithelium, and microvascular endothelium.[98] The contribution of HIF1α to the pathophysiology of pulmonary hypertension is demonstrated by the pulmonary responses of HIF1α[+/−] heterozygote mice to hypoxia. HIF1α[+/−] heterozygotes develop normally and are indistinguishable from HIF1α[+/+] wildtype animals under normoxic conditions. When HIF1α[+/−] mice were exposed chronically to 10% $O_2$, they exhibited a significant delay in the development of polycythemia, pulmonary hypertension, and right ventricular hypertrophy.[99] Pulmonary histology revealed that HIF1α[+/−] mice exposed to hypoxia had fewer muscularized pulmonary arterioles compared to HIF1α[+/+] and, of those arterioles with a muscle wall, the wall was significantly thinner in lungs from hypoxia-exposed HIF1α[+/−] mice compared to HIF1α wildtype animals. These data demonstrate that reduced levels of HIF1α delay the pulmonary pathophysiological responses to hypoxia.

Studies of the contribution of HIF1α-regulated genes to the pathophysiology of hypoxia-induced pulmonary hypertension have revealed a number of genes that likely play a role, including ET-1[100] and ANP.[101] Currently, a significant number of studies implicate ET-1 as a central player in the pathological progression of hypoxia-induced pulmonary hypertension. ET-1 is a potent vasoconstrictor and has mitogenic properties for both smooth and cardiac muscle cells. These physiological functions of ET-1 are likely important in pulmonary vasoconstriction, pulmonary vascular hypertrophy, and right ventricular hypertrophy observed in hypoxia-induced pulmonary hypertension. The ET-1 proximal promoter contains a HRE binding site that is required for hypoxia-induced ET-1 expression.[102] Hypoxia is a potent inducer of ET-1 in both cultured endothelial cells and cardiac myocytes. *In vivo* exposure of animals to hypoxia induced ET-1 in lungs, pulmonary arteries,[103,104] and heart ventricles.[105] Furthermore, ET-1 mRNA and protein are increased in pulmonary vascular endothelial cells in patients with pulmonary hypertension,[100] while circulating plasma ET-1 protein increases in healthy mountain climbers at high altitudes.[106]

Animal models have further elucidated the role of ET-1 in hypoxia-induced pulmonary hypertension. Transgenic mice expressing the firefly luciferase gene driven by the ET-1 promoter, exhibited luciferase activity in the same tissues as endogenous ET-1 mRNA expression, including lung, heart, aorta, kidney, liver, and spleen.[107] In these transgenic animals, hypoxia increased luciferase activity by six-fold in the lung and two-fold in the heart

with the strongest expression detected in smooth muscle and endothelial cells of pulmonary arteries. Another animal model which has implicated ET-1 in the progression of hypoxia-induced pulmonary hypertension is the Wistar-Kyoto rat. These rats are more sensitive to hypoxia-induced pulmonary hypertension than Fisher 344 rats. Following chronic hypoxia, Wistar-Kyoto rats exhibited a greater increase in pulmonary and circulating levels of ET-1, and exhibited a greater degree of pulmonary vascular remodeling than Fisher 344 rats.[108]

Finally, the most definitive evidence demonstrating a role for ET-1 in hypoxia-induced pulmonary hypertension comes from studies utilizing selective endothelin receptor antagonists. In rats exposed to 10% $O_2$ for 90 min, the highly selective endothelin A receptor antagonist, sitaxsentan, prevented the acute pulmonary vasoconstriction, while the endothelin B receptor antagonist, BQ-788, had no effect.[109] In a chronic hypoxia study where rats were exposed to 10% $O_2$ for two weeks, subsequent treatment with sitaxsentan significantly attenuated the pulmonary hypertension, right ventricular hypertrophy, and pulmonary vascular remodeling.[109] These studies strongly suggest that ET-1 plays a fundamental role in the development and pathological progression of hypoxia-induced pulmonary hypertension. Interestingly, transgenic mice overexpressing ET-1 in the lung exhibited pulmonary fibrosis, but did not develop pulmonary hypertension. The degree of ET-1 overexpression in the lung on these transgenic mice compared to ET-1 levels during hypoxia-induced pulmonary hypertension were not reported. Furthermore, the sensitivity of these mice to hypoxia-induced pulmonary hypertension is not known. It would appear, however, that overexpression of pulmonary ET-1 alone is not sufficient for the development of pulmonary hypertension, and other factors, which may include other hypoxia-inducible genes, such as nitric oxide synthase and ANP, are necessary for the pathological development of hypoxia-inducible pulmonary hypertension.

### 14.4.4.10.2  Ischemic Heart Disease

Ischemic heart disease results when myocardial oxygen demand exceeds oxygen delivery from the coronary blood supply. Ischemic heart disease frequently is called coronary heart disease, because in human disease myocardial ischemia most commonly results from atherosclerotic plaques, which narrow the coronary arteries and reduce oxygen delivery to the myocardium. Thus, ischemic heart disease results in localized regions of myocardial hypoxia. Transient myocardial hypoxia can cause angina pectoris, while severe myocardial hypoxia can result in myocardial infarction and sudden death.

The regulation of the physiological responses to myocardial ischemia likely involves HIF1α. In animal models, expression of HIF1α mRNA is increased in cardiomyocytes exposed to hypoxic conditions in culture[63,65,110] as well as *in vivo*.[64,65] In humans, heart biopsy samples demonstrated that myocardial expression of HIF1α mRNA increased following myocardial ischemia, as a result of coronary artery disease and myocardial infarction.[67] These same samples also demonstrated that HIF1α protein expression was restricted to the areas of ischemia or infarction, but was not detectable in non-ischemic or non-infarcted myocardium. The HIF1α$^{-/+}$ mice would provide an excellent experimental model for determining whether HIF1α contributes to physiological and pathological responses to myocardial ischemia.

The induction of the HIF1-regulated gene, VEGF, plays a beneficial role in the physiological responses to ischemic heart disease. VEGF is an endothelial cell mitogen that stimulates both the *de novo* formation of new blood vessels (vasculogenesis) and the branching of existing blood vessels (angiogenesis). VEGF induction is viewed as a beneficial adaptational tissue response to myocardial ischemia, because it stimulates the formation of collateral coronary blood vessels which reduces localized myocardial hypoxia and helps to prevent myocardial infarction. The VEGF promoter contains an HRE binding site which is required for hypoxia-induced VEGF expression,[4] while the 3′ UTR of the VEGF mRNA confers hypoxia-induced mRNA stabilization.[5] VEGF mRNA and protein are highly inducible in primary cardiac myocytes and cardiac explant cultures under hypoxic conditions.[111,112] Similarly, VEGF is inducible *in vivo* in the heart when hypoxia is induced locally by coronary artery occlusion or globally by exposure to reduced $O_2$ percentage or injection of transition metals such as cobalt and manganese.[113–117] Hypoxia-induced VEGF subsequently stimulates the formation coronary blood vessels[112] and increases collateral blood flow.[113] In humans with coronary artery disease there is a direct correlation between the amount of blood flow through collateral coronary vessels and the local concentration of VEGF in those vessels.[118] Furthermore, circulating VEGF increases by 2.5 fold within 10 days of an acute myocardial infarction.[119]

The degree of collateral coronary vessel growth stimulated by VEGF varies considerably among individuals with coronary artery disease and the explanation for this difference is not clear. It has been demonstrated that individuals with coronary artery disease differed in the ability of hypoxia to induce VEGF and this difference was directly correlated with the number of collateral coronary vessels detected by angiography.[120] Patients with no collateral vessels had a significantly reduced responsiveness to hypoxia-induced VEGF expression in circulating monocytes, a primary source of circulating VEGF, when compared to patients with collateral vessels. Furthermore, this correlation was independent of age, gender, hypertension, hypercholesterolemia, cigarette smoking, diabetes, family history, history of myocardial infarction, prior coronary artery bypass, use of β-blockers, and the number of diseased coronary vessels.[120] Decreased responsiveness to hypoxia-induced VEGF expression has also been described in an animal model and was determined to be age-related. Following hypoxic exposure of vascular smooth muscle cells in culture, HIF1α protein, HRE binding, and induction of VEGF mRNA were significantly reduced in smooth muscle of older animals (4–5 years) when compared to younger animals (6–8 months).[121]

Given that the humans exhibit considerable heterogeneity in their responsiveness to hypoxia-induced VEGF expression and that considerable benefit can be derived from VEGF-induced neovascularization, there is great interest in the therapeutic use of VEGF for individuals with coronary artery disease.[122] Therapeutic administration of recombinant VEGF protein or VEGF DNA has demonstrated significant physiological benefits in animal models of coronary artery disease as well as in humans. In animal models, intracoronary administration of recombinant VEGF protein increased collateral blood flow in coronary arteries,[115] while intramyocardial injection of plasmid VEGF DNA or an adenoviral vector containing VEGF DNA[123,124] resulted in coronary neovascularization and increased collateral blood flow. In clinical trials intracoronary administration of recombinant VEGF protein[125] or myocardial injection of plasmid VEGF DNA[126,127] to patients with end-stage coronary artery disease have demonstrated new collateral vessels, improved myocardial perfusion, and a reduction the frequency of angina.[122] Thus, in ischemic heart disease increased HIF1 signaling and induction of VEGF, in particular, can slow the disease progression.

### 14.4.4.10.3 Pre-Eclampsia

Pre-eclampsia, often called pregnancy-induced hypertension, is a pregnancy-specific disease that is characterized by hypertension, edema, and proteinuria. It occurs at a frequency of 7–10% of pregnancies and represents the primary cause of maternal morbidity and mortality in developed countries.[128] In about 4% of the women with pre-eclampsia, the disease progresses to a convulsive phase known as eclampsia. Following parturition, maternal blood pressure returns to normal and the disease state resolves. The etiology of pre-eclampsia is complex and not fully understood; however, it has been shown that the pathogenesis involves uterine ischemia. This ischemia likely results from (1) inadequate invasion of fetal trophoblasts into the uterine decidua, resulting in shallower implantation; and (2) failure of uterine spiral arterioles to remodel to larger, less resistant vessels, leading to placental hypoxia.

The regulation of placental development is dependent on hypoxia gradients between the embryo and maternal myometrium.[129] During the first 10 weeks of human gestation, there is very limited maternal blood flow to the placenta and the embryo develops within a hypoxic environment. This hypoxic environment stimulates embryonic trophoblasts to proliferate.[129–131] From 10–12 weeks of human gestation, maternal blood flow to the placenta increases, exposing the embryonic trophoblasts to higher oxygen tension,[132] which stimulates them to differentiate and invade into the myometrium. Thus, in the developing placenta, hypoxia stimulates trophoblast proliferation and subsequent normoxia promotes trophoblast differentiation and invasion.[129,130] In pre-eclamptic patients, there is a reduction in the degree of trophoblast invasion, suggesting that oxygen tension fails to increase, trophoblasts do not detect or are unable to response to an oxygen increase.[133]

This hypoxic regulation of placental development and pathology of pre-eclampsia are mediated, in part, by HIF1 signaling.[131] Factors regulating hypoxia responses, including ARNT, HIF1α, HIF2α, and VHL, are detected in the developing placenta.[79,131,134–138] The expression of HIF2α and VHL mRNA are detected in proliferating trophoblasts, but rapidly downregulate as the trophoblasts are exposed to increased oxygen tension of maternal blood flow.[138] Other investigators have shown that HIF1α mRNA and protein are expressed at relatively high levels in trophoblasts at 5–8 weeks of gestation, but their expression rapidly decreases between 9 and 15 weeks,[133] when placental $pO_2$ levels increase and trophoblasts begin to invade. The regulation of trophoblasts invasion by HIF-1α has been demonstrated in trophoblast explant culture.[131] Trophoblast explants proliferate when exposed to 3% $O_2$, but do not express markers of an invasive phenotype, such as α1 integrin or MMMP2, nor do explant cells invade into the underlying matrix. In contrast, treatment of these hypoxic explant cultures with antisense HIF-1α oligonucleotides inhibits trophoblast proliferation, increases markers of invasion, and increases the number of cell invading the underlying matrix,[131] demonstrating that in human trophoblasts HIF-1α regulates trophoblast differentiation and invasion.

The identity of specific HIF1-regulated genes that contribute to the pathophysiology of pre-eclampsia remains somewhat elusive, but studies have shown that one candidate proteins may be transforming growth factor-β3 (TGFβ3). TGFβ3 is regulated directly by the HIF1 complex; however, the temporal and spatial expression of TGFβ3 in the placenta parallels the expression of HIF-1α. TGFβ3 expression is high in proliferating embryonic trophoblasts at 6–8 weeks gestation and this expression rapidly decreases in differentiated trophoblasts at 10–12 weeks gestation when placental oxygen tension increases.[133] Furthermore, in trophoblast explant cultures, inhibition of HIF-1α signaling reduces TGFβ3 expression. In these same cultures, treatment with antisense TGFβ3 inhibits trophoblast proliferation and stimulates trophoblast invasion.[139] Of clinical importance to pre-eclampsia, TGFβ3 is overexpressed in the placentas of pre-eclamptic patients and *in vitro* treatment of these placentas with antisense TGFβ3 restores the invasive capacity of the trophoblasts.[139]

Further evidence for the involvement of bHLH-PAS proteins and TGFβ3 in placental development and pre-eclampsia come from gene inactivation models. ARNT, a dimer partner for HIF-1α, is required for normal placental development as has been demonstrated in mice harboring null alleles for this gene.[84,83,140] Placentas from $Arnt^{-/-}$ embryos exhibit reduced trophoblast proliferation rates, poor trophoblast invasion of the uterine decidua, and a lack of fetal blood vessel development. In addition, placental TGFβ3 is downregulated in the absence of ARNT and cultures of $Arnt^{-/-}$ trophoblast cells are not responsive to the induction of TGFβ3 by hypoxia.[140] These studies provide additional evidence that HIF1 signaling and TGFβ3 are important for trophoblast proliferation and placental development. In addition, while placentas from $Arnt^{-/-}$ embryos do not exhibit an alteration of VEGF expression, they do exhibit reduced expression of VEGF receptor 2, which may further contribute to the failure of trophoblast invasion and formation of fetal blood vessels.[85] Thus, the inability of placentas from *Arnt*[−/−] embryos to respond to hypoxia represents a new *in vivo* model for studying pre-eclampsia. Interestingly, neither the *Hif1α* or *Hif2α* homozygous null animals exhibit placental defects,[90,91,93] suggesting that the presence of one protein may compensate for the absence of the other. Thus, one might expect that an *HIF1α*[−/−] × *Hif2α*[−/−] cross would exhibit abnormal placental development. Finally, in addition to the bHLH-PAS proteins, homozygous null mouse embryos for *VHL* also exhibit abnormal placental development and die between gestation days 10.5 and 12.5.[135] Placentas from *VHL*[−/−] embryos exhibit poor trophoblast invasion of the uterine decidua and a lack of fetal blood vessel development, a phenotype that is very similar to the genetic deletion of *Arnt*. Thus, loss of VHL, a protein which prevents HIF-1α degradation under hypoxic conditions, represents another *in vivo* model for pre-eclampsia in which the embryonic trophoblasts appear unable to respond to their hypoxic environment.

#### 14.4.4.10.4    Cancer

Tumors cannot grow beyond the limit of diffusion of $O_2$, glucose, or other nutrients without neovascularization and this occurs by a process of angiogenesis, the branching and in-growth of new blood vessels from preexisting vessels in adjacent normal tissue. In many cancers, the extent of vascularization positively correlates with tumor cell invasion and metastasis, and inversely with patient survival. Thus, as a tumor grows the microenvironment becomes hypoxic and the tumor recruits blood vessels by releasing angiogenic factors that induce proliferation and migration of surrounding endothelial cells.

The induction of HIF1α and or disruption of factors that negatively regulate HIF1α expression, such as p53 and VHL, represent pathological mechanisms by which tumors can undergo neovascularization and continue to grow. There are numerous studies in the literature that have shown that the tumor expression of HIF-1α correlates positively with tumor invasiveness and negatively with patient prognosis. Expression of HIF1α protein above levels detected in normal tissue has been detected in preneoplastic lesions of the mammary gland, colon, and prostate.[52]

Furthermore, overexpression of HIF1α is associated with aggressive tumors in mammary gland,[141] may be associated with invasion in gliomas,[142] and is an unfavorable prognostic marker in cervical cancer[143] and oropharyngeal cancer.[144]

In some cancers, the overexpression of HIF1α or increase in HIF-dependent transcription can be related to the loss of the tumor suppressors, p53 or VHL, as discussed earlier in this chapter. Elevated expression of p53 promotes degradation of HIF1α[51] and inhibits HIF1-inducible transcription.[145] In contrast, loss of p53 protein in human colon cancer cells increases HIF1α expression, enhances HIF1-dependent transcription of VEGF following hypoxia, and promotes neovascularization and growth of tumors when injected into nude mice.[51] Similarly, aggressive prostate cancer, which has lost p53 function, is associated with an elevated inducibility of HIF1-dependent genes as compared to less aggressive prostate cancers.[146] Similarly, loss of VHL function in tumors is associated with increased expression of HIF1α, HIF2α, and VEGF.[49,50] Thus, loss of tumor suppressors and increased ability to induce HIF1-dependent angiogenesis may result in the selection for more highly metastatic tumors. The dependence of tumor growth on induction of angiogenesis through an HIF1-dependent mechanism represents a promising strategy for cancer gene therapy.[147,148]

### 14.4.4.11    EXPERIMENTAL METHODS FOR HYPOXIA IMAGING

As has been discussed above, localized oxygen gradients contribute to normal embryonic angiogenesis and placental development, and represent a fundamental component of pulmonary hypertension, ischemic heart disease, pre-eclampsia, and cancer. Thus, methods have been developed that qualitatively and quantitatively can determine the degree of tissue hypoxia in both a spatial and temporal manner. These methods can be used not only to study the extent of tissue hypoxia in embryonic and disease processes, but also can be useful tools for when investigators experimentally induce hypoxia to determine both molecular and cellular responses to changes in oxygen tension *in vivo*. These tools have been applied most frequently in the former case and less so in the latter. In the latter case investigators commonly assume that whole animal exposure to reduced oxygen percentage or reduce partial pressure of oxygen in a hypobaric chamber will result in hypoxia in the target organ of interest. However, it is important to remember that hypoxia induces numerous molecular and cellular changes with some responses occurring within seconds and some responses lasting upwards of months. These changes will alter ventilation rate, heart rate, vascular dilation, and overall tissue perfusion, and so the degree of hypoxia will vary within a single tissue, among tissues, and with time. Thus, hypoxia imaging methods can be used to confirm the occurrence of hypoxia, determine the spatial localization of hypoxia with a single tissue, and allow the comparison of the degree of hypoxia among tissues.

#### 14.4.4.11.1 Nitroimidazoles

A group of compounds that undergo differential intracellular metabolism depending on the oxygen tension within a tissue is the nitroimidazoles. Nitroimidazoles are sufficiently lipophilic to passively enter cells. They then undergo a single electron nitroreduction forming a radical anion in both aerobic and anaerobic cells.[149] In the presence of oxygen, the anion is rapidly reoxidized to the parent compound and can diffuse out of the cell. In tissues with reduced oxygen tension, the anion is not reoxidized to the parent compound, but undergoes three additional electron reductions to produce a bioreactive hydroxylamine, which covalently binds to cellular macromolecules.[150,151] The rate of anion oxidation is dependent on the intracellular oxygen concentration and, thus, the retention of the hydroxylamine can distinguish between normoxic and hypoxic tissues. Additionally, nitroimidazoles only bind in viable cells and thus can distinguish between viable, hypoxic cells and necrotic, hypoxic cells. The rates of [3]H-misonidazole binding increased linearly in murine liver, spleen, heart, brain, kidney, and tumors as tissue $pO_2$ decreased from 20% to approximately 0.1%.[150] While under normoxic conditions, minimal binding occurred in all tissues, except liver, *in vivo*, suggesting that the liver $pO_2$ is significantly lower than other tissues perhaps as a consequence of arterial and portal blood mixing.

In cancer biology it has been recognized that tumor hypoxia can be associated with resistance to therapy, increased metastases, and poor prognosis. Nitroimidazoles were first used experimentally to sensitize hypoxic tumors to radiation or chemical therapy, but more recently have been used to identify the degree of tumor hypoxia and allow for the design of specific treatment strategies that increase tumor oxygenation or directly kill hypoxic

tumor cells.[152] In addition, nitroimidazoles are being used to study the relationships between tumor hypoxia and cell proliferation, apoptosis, and metastases. Various nitroimidazole derivatives and experimental approaches have been used from immunohistochemistry, scintillation counting, and flow cytometry to whole body, noninvasive imaging. Human squamous cell carcinomas have been studied using a biotin-conjugated nitroimidazole derivative, pimonidazole, and the degree of hypoxia determined using a monoclonal antibody and immunohistochemical techniques.[153] Using flow cytometry, quantification of the binding of pimonidazole in squamous cell carcinomas, human glioma cells, and cervical carcinoma xenografts was highly correlated with the hypoxic fraction of cells measured by a comet assay, a measure of radiation induced DNA single-strand breaks that directly correlates with radio-biologically hypoxic cells.[154] Noninvasive imaging of hypoxic hepatomas also has been conducted using a radioactive 18F-labeled nitroimidazole derivative and positron emission tomography,[155] allowing for the assessment of tumor hypoxia without invasive surgery.

Nitroimidazoles have been used to study hypoxia during embryonic development as well as in disease states other than cancer. Pimonidazole and another nitroimidazole derivative, EF5, have been used to visualize areas of hypoxia in developing mouse[156] and rat embryos,[157] respectively, using immunohistochemistry. Early in organogenesis hypoxic microenvironments were detected in the neural tube, somites, and heart, and expression of HIF1α and VEGF were co-localized within these hypoxic regions in the mouse embryo,[156] suggesting that hypoxia may act as a signal for embryonic blood vessel development. The presence of hypoxia also has been demonstrated by increased nitroimidazole binding in the arterial wall and heart of animal models of coronary artery disease,[158,159] while EF5 binding and expression of HIF1α, HIF2α, and VEGF were co-localized in the brain following cerebral artery occlusion.[160] Thus, nitroimidazoles can be used qualitatively or quantitatively to measure the spatial localization and degree of hypoxia in experimental animal models as well as in human disease.

#### 14.4.4.11.2 Electron Paramagnetic Resonance Oximetry

A relatively new method to measure tissue oxygen tension directly uses implantation of oxygen sensitive probes and use of electron paramagnetic resonance (EPR). Lithium phthalocyanin (1μm diameter), an inert paramagnetic crystal, is implanted in the tissue of interest and the EPR spectra is obtained. The width of the L-band EPR spectra directly is correlated with pO$_2$, has a very fast response time to changes in tissue pO$_2$, and can measure tissue pO$_2$ as low as 0.01%.[161] Using this method, it was demonstrated that different anesthetics and routes of administration alter cerebral pO$_2$ in a reproducible manner.[162] Intramuscular injection of either pentobarbital or ketamine/xylazine decreased cerebral pO$_2$ to a greater degree than inhalation of isoflurane or halothane. This method also has been used to study brain ischemia in animal models of stroke[163,164] and in chronic acclimation to hypoxia.[164] Future research is beginning to use injectable, water soluble oxygen sensing probes and EPR to image tissue pO$_2$ *in vivo* in a three-dimensional manner.

## 14.4.4.12 ACKNOWLEDGMENTS

MKW thanks Darlene Gabaldon for her technical assistance. This work was supported by NIH grants R01-ES10433 and R01-ES09804 to MKW; R01-ES05703 and P30-CA07175 to CAB; and a Post-doctoral Fellowship from the Natural Sciences and Engineering Research Council of Canada to JW.

## 14.4.4.13 REFERENCES

1. K. Guillemin and M. A. Krasnow, 'The hypoxic response: huffing and HIFing.' *Cell*, 1997, **89**, 9–12.
2. M. A. Goldberg, S. P. Dunning and H. F. Bunn, 'Regulation of the erythropoietin gene: evidence that the oxygen sensor is a heme protein.' *Science*, 1988, **242**, 1412–1415.
3. M. F. Czyzyk-Krzeska, D. A. Bayliss, E. E. Lawson and D. E. Millhorn, 'Regulation of tyrosine hydroxylase gene expression in the rat carotid body by hypoxia.' *J. Neurochem.*, 1992, **58**, 1538–1546.
4. J. A. Forsythe, B. H. Jiang, N. V. Iyer *et al.*, 'Activation of vascular endothelial growth factor gene transcription by hypoxia-inducible factor 1.' *Mol. Cell Biol.*, 1996, **16**, 4604–4613.
5. A. P. Levy, N. S. Levy, S. Wegner *et al.*, 'Transcriptional regulation of the rat vascular endothelial growth factor gene by hypoxia.' *J. Biol. Chem.*, 1995, **270**, 13333–13340.
6. K. Kuwabara, S. Ogawa, M. Matsumoto *et al.*, 'Hypoxia-mediated induction of acidic/basic fibroblast growth factor and platelet-derived growth factor in mononuclear phagocytes stimulates growth of hypoxic endothelial cells.' *Proc. Natl. Acad. Sci. USA*, 1995, **92**, 4606–4610.
7. S. L. Archer, K. A. Freude and P. J. Shultz, 'Effect of graded hypoxia on the induction and function of inducible nitric oxide synthase in rat mesangial cells.' *Circ. Res.*, 1995, **77**, 21–28.

8. P. J. Lee, B. H. Jiang, B. Y. Chin *et al.*, 'Hypoxia-inducible factor-1 mediates transcriptional activation of the heme oxygenase-1 gene in response to hypoxia.' *J. Biol. Chem.*, 1997, **272**, 5375–5381.
9. G. L. Semenza, P. H. Roth, H. M. Fang *et al.*, 'Transcriptional regulation of genes encoding glycolytic enzymes by hypoxia-inducible factor 1.' *J. Biol. Chem.*, 1994, **269**, 23757–23763.
10. G. L. Semenza, B. H. Jiang, S. W. Leung *et al.*, 'Hypoxia response elements in the aldolase A, enolase 1, and lactate dehydrogenase A gene promoters contain essential binding sites for hypoxia-inducible factor 1.' *J. Biol. Chem.*, 1996, **271**, 32529–32537.
11. H. Li, H. P. Ko and J. P. Whitlock, 'Induction of phosphoglycerate kinase 1 gene expression by hypoxia. Roles of Arnt and HIF1α.' *J. Biol. Chem.*, 1996, **27**, 21262–21267.
12. J. M. Gleadle and P. J. Ratcliffe, 'Induction of hypoxia-inducible factor-1, erythropoietin, vascular endothelial growth factor, and glucose transporter-1 by hypoxia: evidence against a regulatory role for Src kinase.' *Blood*, 1997, **89**, 503–509.
13. S. Cormier-Regard, D. B. Egeland, V. J. Tannoch *et al.*, 'Differential display: Identifying genes involved in cardiomyocyte proliferation.' *Mol. Cell. Biochem.*, 1997, **172**, 111–120.
14. H. Kagamu, T. Suzuki, M. Arakawa *et al.*, 'Low oxygen enhances endothelin-1 (ET-1) production and responsiveness to ET-1 in cultured cardiac myocytes.' *Biochem. Biophys. Res. Commun.*, 1994, **202**, 1612–1618.
15. Y. F. Chen, J. Durand and W. C. Claycomb, 'Hypoxia stimulates atrial natriuretic peptide gene expression in cultured atrial cardiocytes.' *Hypertension*, 1997, **29**, 75–82.
16. G. L. Semenza, M. K. Nejfelt, S. M. Chi *et al.*, 'Hypoxia-inducible nuclear factors bind to an enhancer element located 3' to the human erythropoietin gene.' *Proc. Nat. Acad. Sci. USA*, 1991, **88**, 5680–5684.
17. M. A. Goldberg, C. C. Gaut and H. F. Bunn, 'Erythropoietin mRNA levels are governed by both the rate of gene transcription and posttranscriptional events.' *Blood*, 1991, **77**, 271–277.
18. G. L. Semenza and G. L. Wang, 'A nuclear factor induced by hypoxia via de novo protein synthesis binds to the human erythropoietin gene enhancer at a site required for transcriptional activation.' *Mol. Cell. Biol.*, 1992, **12**, 5447–5454.
19. V. Ho, K. Acquaviva, E. Duh *et al.*, 'Use of a marked erythropoietin gene for investigation of its *cis*-acting elements.' *J. Biol. Chem.*, 1995, **27**, 10084–10090.
20. G. L. Wang, B.-H. Jiang, E. A. Rue *et al.*, 'Hypoxia-inducible factor 1 is a basic-helix-loop-helix-PAS heterodimer regulated by cellular O$_2$ tension.' *Proc. Nat. Acad. Sci. USA*, 1995, **92**, 5510–5514.
21. G. L. Wang and G. L. Semenza, 'Purification and characterization of hypoxia-inducible factor 1.' *J. Biol. Chem.*, 1995, **270**, 1230–1237.
22. B.-H. Jiang, G. L. Semenza, C. Bauer *et al.*, 'Hypoxia-inducible factor 1 levels vary exponentially over a physiologically relevant range of O$_2$ tension.' *Am. J. Physiol.*, 1996, **271**, C1172–C1180.
23. R. H. Wenger and M. Gassmann, 'Oxygen(es) and the hypoxia-inducible factor-1.' *Biol. Chem.*, 1997, **37**, 609–616.
24. G. L. Wang and G. L. Semenza, 'Characterization of hypoxia-inducible factor 1 and regulation of DNA binding activity by hypoxia.' *J. Biol. Chem.*, 1993, **268**, 21513–21518.
25. S. G. Bacsi, S. Reisz-Porszasz and O. Hankinson, 'Orientation of the heterodimeric aryl hydrocarbon (dioxin) receptor complex on its asymmetric DNA

recognition sequence.' *Mol. Pharmacol.*, 1995, **47**, 432–438.

26. H. I. Swanson, W. K. Chan and C. A. Bradfield, 'DNA binding specificities and pairing rules of the Ah receptor, ARNT, and SIM proteins.' *J. Biol. Chem.*, 1995, **270**, 26292–26302.

27. H. I. Swanson and J.-H. Yang, 'Mapping the protein/DNA contact sites of the Ah receptor and Ah receptor nuclear translocator.' *J. Biol. Chem.*, 1996, **27**, 31657–31665.

28. B. N. Fukunaga and O. Hankinson, 'Identification of a novel domain in the aryl hydrocarbon receptor required for DNA binding.' *J. Biol. Chem.*, 1996, **271**, 3743–3749.

29. C. W. Pugh, J. F. O'Rourke, M. Nagao *et al.*, 'Activation of hypoxia-inducible factor-1; definition of regulatory domains within the alpha subunit.' *J. Biol. Chem.*, 1997, **272**, 11205–11214.

30. B. H. Jiang, J. Z. Zheng, S. W. Leung *et al.*, 'Transactivation and inhibitory domains of hypoxia-inducible factor α. Modulation of transcriptional activity by oxygen tension.' *J. Biol. Chem.*, 1997, **272**, 19253–19260.

31. Z. Arany, L. E. Huang, R. Eckner *et al.*, 'An essential role for p300/CBP in the cellular response to hypoxia.' *Proc. Natl. Acad. Sci. USA*, 1996, **93**, 12969–12973.

32. P. J. Kallio, K. Okamoto, S. O'Brien *et al.*, 'Signal transduction in hypoxic cells: inducible nuclear translocation and recruitment of the CBP/p300 coactivator by the hypoxia-inducible factor-1α.' *EMBO J.*, 1998, **17**, 6573–6586.

33. P. Carrero, K. Okamoto, P. Coumailleau *et al.*, 'Redox-regulated recruitment of the transcriptional coactivators CREB-binding protein and SRC-1 to hypoxia-inducible factor 1α.' *Mol. Cell Biol.*, 2000, **20**, 402–415.

34. S. Bhattacharya, C. L. Michels, M. K. Leung *et al.*, 'Functional role of p35srj, a novel p300/CBP binding protein, during transactivation by HIF-1.' *Genes Dev.*, 1999, **13**, 64–75.

35. W. K. Chan, R. Chu, S. Jain *et al.*, 'Baculovirus expression of the Ah receptor and Ah receptor nuclear translocater. Evidence for additional dioxin responsive element-binding species and factors required for signaling.' *J. Biol. Chem.*, 1994, **26**, 26464–26471.

36. K. Gradin, J. McGuire, R. H. Wenger *et al.*, 'Functional interference between hypoxia and dioxin signal transduction pathways: competition for recruitment of the Arnt transcription factor.' *Mol. Cell Biol.*, 1996, **16**, 5221–5231.

37. S. Salceda and J. Caro, 'Hypoxia-inducible factor 1α (HIF-1α) protein is rapidly degraded by the ubiquitin-proteasome system under normoxic conditions.' *J. Biol. Chem.*, 1997, **272**, 22642–22647.

38. P. J. Kallio, W. J. Wilson, S. O'Brien *et al.*, 'Regulation of the hypoxia-inducible transcription factor 1α by the ubiquitin-proteasome pathway.' *J. Biol. Chem.*, 1999, **274**, 6519–6525.

39. L. E. Huang, J. Gu, M. Schau *et al.*, 'Regulation of hypoxia-inducible factor 1α is mediated by an O₂-dependent degradation domain via the ubiquitin-proteasome pathway.' *Proc. Nat. Acad. Sci. USA*, 1998, **95**, 7987–7992.

40. P. H. Maxwell, M. S. Wiesener, G. W. Chang *et al.*, 'The tumour suppressor protein VHL targets hypoxia-inducible factors for oxygen-dependent proteolysis.' *Nature*, 1999, **399**, 271–275.

41. W. Krek, 'VHL takes HIF's breath away.' *Nature Cell Biology*, 2000, **2**, E121–E123.

42. M. E. Cockman, N. Masson, D. R. Mole *et al.*, 'Hypoxia inducible factor-α binding and ubiquity-lation by the von Hippel–Lindau tumor suppressor protein.' *J. Biol. Chem.*, 2000, **275**, 25733–25741.

43. K. Tanimoto, Y. Makino, T. Pereira *et al.*, 'Mechanism of regulation of the hypoxia-inducible factor-1α by the von Hippel–Lindau tumor suppressor protein.' *EMBO J.*, 2000., **19**, 4298–4309.

44. T. Kamura, S. Sato, K. Iwai *et al.*, 'Activation of HIF1α ubiquitination by a reconstituted von Hippel–Lindau (VHL) tumor suppressor complex.' *Proc. Natl. Acad. Sci. USA*, 2000, **97**, 10430–10435.

45. F. Latif, K. Tory, J. R. Gnarra *et al.*, 'Identification of the von Hippel–Lindau disease tumor suppressor gene.' *Science*, 1993, **260**, 1317–1320.

46. J. R. Gnarra, D. R. Duan, Y. Weng *et al.*, 'Molecular cloning of the von Hippel–Lindau tumor suppressor gene and its role in renal carcinoma.' *Biochimica et Biophysica Acta*, 1996, **1242**, 201–210.

47. W. G. Kaelin, Jr. and E. R. Maher, 'The VHL tumour-suppressor gene paradigm.' *Trends Genetics*, 1998, **14**, 423–426.

48. J. R. Gnarra, S. Zhou, M. J. Merrill *et al.*, 'Post-transcriptional regulation of vascular endothelial growth factor mRNA by the product of the VHL tumor suppressor gene.' *Proc. Nat. Acad. Sci. USA*, 1996, **93**, 10589–10594.

49. O. Iliopoulos, A. P. Levy, C. Jiang *et al.*, 'Negative regulation of hypoxia-inducible genes by the von Hippel–Lindau protein.' *Proc. Nat. Acad. Sci. USA*, 1996, **93**, 10595–10599.

50. G. Siemeister, K. Weindel, K. Mohrs *et al.*, 'Reversion of deregulated expression of vascular endothelial growth factor in human renal carcinoma cells by von Hippel–Lindau tumor suppressor protein.' *Cancer Res.*, 1996, **56**, 2299–2301.

51. R. Ravi, B. Mookerjee, Z. M. Bhujwalla *et al.*, 'Regulation of tumor angiogenesis by p53-induced degradation of hypoxia-inducible factor 1α.' *Genes Dev.*, 2000, **14**, 34–44.

52. H. Zhong, A. M. De Marzo, E. Laughner *et al.*, 'Overexpression of hypoxia-inducible factor 1α in common human cancers and their metastases.' *Cancer Res.*, 1999, **59**, 5830–5835.

53. M. Ivan, K. Kondo, H. Yang *et al.*, 'HIFα Targeted for VHL-Mediated Destruction by Proline Hydroxylation: Implications for O₂ Sensing.' *Science*, 2001, **292**, 464–468.

54. P. Jaakkola, D. R. Mole, Y. M. Tian *et al.*, 'Targeting of HIF-α to the von Hippel–Lindau Ubiquitylation Complex by O₂-Regulated Prolyl Hydroxylation.' *Science*, 2001, **292**, 468–472.

55. L. E. Huang, Z. Arany, D. M. Livingston *et al.*, 'Activation of hypoxia-inducible transcription factor depends primarily upon redox-sensitive stabilization of its α subunit.' *J. Biol. Chem.*, 1996, **271**, 32253–32259.

56. N. S. Chandel, E. Maltepe, E. Goldwasser *et al.*, 'Mitochondrial reactive oxygen species trigger hypoxia-induced transcription.' *Proc. Nat. Acad. Sci. USA*, 1998, **95**, 11715–11720.

57. D. Lando, I. Pongratz, L. Poellinger *et al.*, 'A redox mechanism controls differential DNA binding activities of hypoxia-inducible factor (HIF) 1α and the HIF-like factor.' *J. Biol. Chem.*, 2000, **275**, 4618–4627.

58. N. S. Chandel, D. S. McClintock, C. E. Feliciano *et al.*, 'Reactive oxygen species generated at mitochondrial complex III stabilize hypoxia-inducible factor-1α during hypoxia: a mechanism of O₂ sensing.' *J. Biol. Chem.*, 2000, **275**, 25130–25138.

59. P. J. Kallio, I. Pongratz, K. Gradin *et al.*, 'Activation of hypoxia-inducible factor 1α: posttranscriptional regulation and conformational change by recruitment

of the Arnt transcription factor.' *Proc. Nat. Acad. Sci. USA*, 1997, **94**, 5667–5672.

60. E. Zelzer, Y. Levy, C. Kahana *et al.*, 'Insulin induces transcription of target genes through the hypoxia-inducible factor HIF-1α/ARNT.' *EMBO J.*, 1998, **17**, 5085–5094.

61. D. E. Richard, E. Berra and J. Pouyssegur, 'Non-hypoxic pathway mediates the induction of hypoxia-inducible factor 1α in vascular smooth muscle cells.' *J. Biol. Chem.*, 2000, **275**, 26765–26771.

62. G. L. Wang and G. L. Semenza, 'Desferrioxamine induces erythropoietin gene expression and hypoxia-inducible factor 1 DNA-binding activity: implications for models of hypoxia signal transduction.' *Blood*, 1993, **82**, 3610–3615.

63. A. Ladoux and C. Frelin, 'Cardiac expressions of HIF-1α and HLF/EPAS, two basic loop helix/PAS domain transcription factors involved in adaptative responses to hypoxic stresses.' *Biochem. Biophys. Res. Commun.*, 1997, **240**, 552–556.

64. T. F. Catron, M. A. Mendiola, S. M. Smith *et al.*, 'Hypoxia regulates avian cardiac Arnt and HIF-1α mRNA expression.' *Biochem. Biophys. Res. Commun.*, 2001, **28**, 602–607.

65. F. Jung, L. A. Palmer, N. Zhou and R. A. Johns, 'Hypoxic regulation of inducible nitric oxide synthase via hypoxia inducible factor-1 in cardiac myocytes.' *Circ. Res.*, 2000, **86**, 319–325.

66. C. Royer, J. Lachuer, G. Crouzoulon *et al.*, 'Effects of gestational hypoxia on mRNA levels of Glut3 and Glut4 transporters, hypoxia inducible factor-1 and thyroid hormone receptors in developing rat brain.' *Brain Research*, 2000, **856**, 119–128.

67. S. H. Lee, P. L. Wolf, R. Escudero *et al.*, 'Early expression of angiogenesis factors in acute myocardial ischemia and infarction.' *N. Engl. J. Med.*, 2000, **342**, 626–633.

68. A. Minchenko, T. Bauer, S. Salceda *et al.*, 'Hypoxic stimulation of vascular endothelial growth factor expression *in vitro* and *in vivo*.' *Laboratory Investigation*, 1994, **71**, 374–379.

69. K. U. Eckardt, S. T. Koury, C. C. Tan *et al.*, 'Distribution of erythropoietin producing cells in rat kidneys during hypoxic hypoxia.' *Kidney International*, 1993, **43**, 815–823.

70. K. Hirose, M. Morita, M. Ema *et al.*, 'cDNA cloning and tissue-specific expression of a novel basic helix-loop-helix/PAS factor (Arnt2) with close sequence similarity to the aryl hydrocarbon receptor nuclear translocator (Arnt).' *Mol. Cell. Biol.*, 1996, **16**, 1706–1713.

71. J. B. Hogenesch, W. K. Chan, V. H. Jackiw *et al.*, 'Characterization of a subset of the basic-helix-loop-helix-PAS superfamily that interacts with components of the dioxin signaling pathway.' *J. Biol. Chem.*, 1997, **272**, 8581–8593.

72. M. Ikeda and M. Nomura, 'cDNA cloning and tissue-specific expression of a novel basic helix-loop-helix/PAS protein (BMAL1) and identification of alternatively spliced variants with alternative translation initiation site usage.' *Biochem. Biophys. Res. Commun.*, 1997, **23**, 258–264.

73. Y. Gu, S. M. Moran, J. B. Hogenesch *et al.*, 'Molecular characterization and chromosomal localization of a third α-class hypoxia inducible factor subunit, HIF3α.' *Gene. Expr.*, 1998, **7**, 205–213.

74. M. Ema, S. Taya, N. Yokotani *et al.*, 'A novel bHLH-PAS factor with close sequence similarity to hypoxia-inducible factor 1α regulates the VEGF expression and is potentially involved in lung and vascular development.' *Proc. Nat. Acad. Sci. USA*, 1997, **94**, 4273–4278.

75. I. Flamme, T. Frohlich, M. von Reutern *et al.*, 'HRF,

76. H. Tian, S. L. McKnight and D. W. Russell, 'Endothelial PAS domain protein 1 (EPAS1), a transcription factor selectively expressed in endothelial cells.' *Genes Dev.*, 1997, **11**, 72–82.

77. J. F. O'Rourke, Y. M. Tian, P. J. Ratcliffe *et al.*, 'Oxygen-regulated and transactivating domains in endothelial PAS protein 1: comparison with hypoxia-inducible factor-1α.' *J. Biol. Chem.*, 1999, **274**, 2060–2071.

78. M. S. Wiesener, H. Turley, W. E. Allen *et al.*, 'Induction of endothelial PAS domain protein-1 by hypoxia: characterization and comparison with hypoxia-inducible factor-1α.' *Blood*, 1998, **92**, 2260–2268.

79. S. Jain, E. Maltepe, M. M. Lu *et al.*, 'Expression of ARNT, ARNT2, HIF1α, HIF2α and Ah receptor mRNAs in the developing mouse.' *Mech. Dev.*, 1998, **73**, 117–123.

80. G. Drutel, M. Kathmann, A. Heron *et al.*, 'Cloning and selective expression in brain and kidney of ARNT2 homologous to the Ah receptor nuclear translocator (ARNT).' *Biochem. Biophys. Res. Commun.*, 1996, **22**, 333–339.

81. S. Takahata, K. Sogawa, A. Kobayashi *et al.*, 'Transcriptionally active heterodimer formation of an Arnt-like PAS protein, Arnt3, with HIF-1a, HLF, and clock.' *Biochem. Biophys. Res. Commun.*, 1998, **248**, 789–794.

82. J. B. Hogenesch, Y. Z. Gu, S. Jain *et al.*, 'The basic-helix-loop-helix-PAS orphan MOP3 forms transcriptionally active complexes with circadian and hypoxia factors.' *Proc. Nat. Acad. Sci. USA*, 1998, **95**, 5474–5479.

83. E. Maltepe, J. V. Schmidt, D. Baunoch *et al.*, 'Abnormal angiogenesis and responses to glucose and oxygen deprivation in mice lacking the protein ARNT.' *Nature*, 1997, **386**, 403–407.

84. K. R. Kozak, B. Abbott and O. Hankinson, 'ARNT-deficient mice and placental differentiation.' *Dev. Biol.*, 1997., **191**, 297–305.

85. B. D. Abbott and A. R. Buckalew, 'Placental defects in ARNT-knockout conceptus correlate with localized decreases in VEGF-R2, Ang-1, and Tie-2.' *Dev. Dyn.*, 2000, **21**, 526–538.

86. A. Nagy, 'Cre recombinase: the universal reagent for genome tailoring.' *Genesis: J. Genet. Dev.*, 2000, **26**, 99–109.

87. S. Tomita, C. J. Sinal, S. H. Yim *et al.*, 'Conditional disruption of the aryl hydrocarbon receptor nuclear translocator (Arnt) gene leads to loss of target gene induction by the aryl hydrocarbon receptor and hypoxia-inducible factor 1α.' *Mol. Endocrin.*, 2000, **14**, 1674–1681.

88. P. H. Maxwell, G. U. Dachs, J. M. Gleadle *et al.*, 'Hypoxia-inducible factor-1 modulates gene expression in solid tumors and influences both angiogenesis and tumor growth.' *Proc. Nat. Acad. Sci. USA*, 1997, **94**, 8104–8109.

89. E. Maltepe, B. Keith, A. M. Arsham *et al.*, 'The role of ARNT2 in tumor angiogenesis and the neural response to hypoxia.' *Biochem. Biophys. Res. Commun.*, 2000, **27**, 231–238.

90. N. V. Iyer, L. E. Kotch, F. Agani *et al.*, 'Cellular and developmental control of $O_2$ homeostasis by hypoxia-inducible factor 1α.' *Genes Dev.*, 1998, **12**, 149–162.

91. H. E. Ryan, J. Lo and R. S. Johnson, 'HIF-1 alpha is required for solid tumor formation and embryonic vascularization.' *EMBO J.*, 1998, **17**, 3005–3015.

92. P. Carmeliet, Y. Dor, J. M. Herbert *et al.*, 'Role of

a putative basic helix-loop-helix-PAS-domain transcription factor is closely related to hypoxia-inducible factor-1 α and developmentally expressed in blood vessels.' *Mech. Dev.*, 1997, **63**, 51–60.

HIF-1α in hypoxia-mediated apoptosis, cell proliferation and tumour angiogenesis.' *Nature*, 1998, **394**, 485–490.

93. H. Tian, R. E. Hammer, A. M. Matsumoto *et al.*, 'The hypoxia-responsive transcription factor EPAS1 is essential for catecholamine homeostasis and protection against heart failure during embryonic development.' *Genes Dev.*, 1998, **12**, 3320–3324.

94. G. L. Semenza, 'Hypoxia-inducible factor 1: master regulator of O$_2$ homeostasis.' *Curr. Opin. Genet. Dev.*, 1998, **8**, 588–594.

95. W. G. An, M. Kanekal, M. C. Simon *et al.*, 'Stabilization of wild-type p53 by hypoxia-inducible factor 1α.' *Nature*, 1998, **392**, 405–408.

96. L. A. Palmer, G. L. Semenza, M. H. Stoler *et al.*, 'Hypoxia induces type II NOS gene expression in pulmonary artery endothelial cells via HIF-1.' *Am. J. Physiol. Lung Cell Mol. Physiol.*, 1998, **274**, L212–L219.

97. C. M. Wiener, G. Booth and G. L. Semenza, '*In vivo* expression of mRNAs encoding hypoxia-inducible factor 1.' *Biochem. Biophys. Res. Commun.*, 1996, **22**, 485–488.

98. A. Y. Yu, M. G. Frid, L. A. Shimoda *et al.*, 'Temporal, spatial, and oxygen-regulated expression of hypoxia-inducible factor-1 in the lung.' *Am. J. Physiol.*, 1998, **275**, L818–L826.

99. A. Y. Yu, L. A. Shimoda, N. V. Iyer *et al.*, 'Impaired physiological responses to chronic hypoxia in mice partially deficient for hypoxia-inducible factor 1α.' *J. Clin. Invest.*, 1999, **103**, 691–696.

100. M. Fukuchi and A. Giaid, 'Endothelial expression of endothelial nitric oxide synthase and endothelin-1 in human coronary artery disease. Specific reference to underlying lesion.' *Laboratory Investigation*, 1999, **79**, 659–670.

101. Y. Hirata, E. Suzuki, H. Hayakawa *et al.*, 'Role of endogenous ANP in sodium excretion in rats with experimental pulmonary hypertension.' *Am. J. Physiol.*, 1992, **26**, H1684–H1689.

102. J. Hu, D. J. Discher, N. H. Bishopric *et al.*, 'Hypoxia regulates expression of the endothelin-1 gene through a proximal hypoxia-inducible factor-1 binding site on the antisense strand.' *Biochem. Biophys. Res. Commun.*, 1998, **245**, 894–899.

103. T. S. Elton, S. Oparil, G. R. Taylor *et al.*, 'Normobaric hypoxia stimulates endothelin-1 gene expression in the rat.' *Am. J. Physiol.*, 1992, **263**, R1260–R1264.

104. H. Li, S. J. Chen *et al.*, 'Enhanced endothelin-1 and endothelin receptor gene expression in chronic hypoxia.' *J. Appl. Physiol.*, 1994, **77**, 1451–1459.

105. J. P. Loennechen, V. Beisvag, I. Arbo *et al.*, 'Chronic carbon monoxide exposure *in vivo* induces myocardial endothelin-1 expression and hypertrophy in rat.' *Pharmacol. Toxicol.*, 1999, **85**, 192–197.

106. S. Goerre, M. Wenk, P. Bartsch *et al.*, 'Endothelin-1 in pulmonary hypertension associated with high-altitude exposure.' *Circulation*, 1995, **91**, 359–364.

107. C. R. Aversa, S. Oparil, J. Caro *et al.*, 'Hypoxia stimulates human preproendothelin-1 promoter activity in transgenic mice.' *Am. J. Physiol.*, 1997, **273**, L848–L855.

108. J. I. Aguirre, N. W. Morrell, L. Long *et al.*, 'Vascular remodeling and ET-1 expression in rat strains with different responses to chronic hypoxia.' *Am. J. Physiol. Lung Cell Mol. Physiol.*, 2000, **278**, L981–L987.

109. R. G. Tilton, C. L. Munsch, S. J. Sherwood *et al.*, 'Attenuation of pulmonary vascular hypertension and cardiac hypertrophy with sitaxsentan sodium, an orally active ET(A) receptor antagonist.' *Pulm. Pharmacol. Therap.*, 2000, **13**, 87–97.

110. S. V. Nguyen and W. C. Claycomb, 'Hypoxia regulates the expression of the adrenomedullin and HIF-1 genes in cultured HL-1 cardiomyocytes.' *Biochem. Biophys. Res. Commun.*, 1999, **265**, 382–386.

111. A. Ladoux and C. Frelin, 'Hypoxia is a strong inducer of vascular endothelial growth factor mRNA expression in the heart.' *Biochem. Biophys. Res. Commun.*, 1993, **195**, 1005–1010.

112. X. Yue and R. J. Tomanek, 'Stimulation of coronary vasculogenesis/angiogenesis by hypoxia in cultured embryonic hearts.' *Dev. Dyn.*, 1999, **216**, 28–36.

113. T. Matsunaga, D. C. Warltier, D. W. Weihrauch *et al.*, 'Ischemia-induced coronary collateral growth is dependent on vascular endothelial growth factor and nitric oxide.' *Circulation (Online)*, 2000, **102**, 3098–3103.

114. A. P. Levy, N. S. Levy, J. Loscalzo *et al.*, 'Regulation of vascular endothelial growth factor in cardiac myocytes.' *Circ. Res.*, 1995, **76**, 758–766.

115. S. Banai, D. Shweiki, A. Pinson *et al.*, 'Upregulation of vascular endothelial growth factor expression induced by myocardial ischaemia: implications for coronary angiogenesis.' *Cardiovasc. Res.*, 1994, **28**, 1176–1179.

116. E. Hashimoto, T. Ogita, T. Nakaoka *et al.*, 'Rapid induction of vascular endothelial growth factor expression by transient ischemia in rat heart.' *Am. J. Physiol.*, 1994, **267**, H1948–H1954.

117. L. Jingjing, B. Srinivasan, X. Bian *et al.*, 'Vascular endothelial growth factor is increased following coronary artery occlusion in the dog heart.' *Mol. Cell. Biochem.*, 2000, **214**, 23–30.

118. M. Fleisch, M. Billinger, F. R. Eberli *et al.*, 'Physiologically assessed coronary collateral flow and intracoronary growth factor concentrations in patients with 1- to 3-vessel coronary artery disease.' *Circulation*, 1999, **10**, 1945–1950.

119. A. Kranz, C. Rau, M. Kochs *et al.*, 'Elevation of vascular endothelial growth factor-A serum levels following acute myocardial infarction. Evidence for its origin and functional significance.' *J. Mol. Cell Cardiol.*, 2000, **32**, 65–72.

120. A. Schultz, L. Lavie, I. Hochberg *et al.*, 'Interindividual heterogeneity in the hypoxic regulation of VEGF: significance for the development of the coronary artery collateral circulation.' *Circulation*, 1999, **100**, 547–552.

121. A. Rivard, L. Berthou-Soulie, N. Principe *et al.*, 'Age-dependent defect in vascular endothelial growth factor expression is associated with reduced hypoxia-inducible factor 1 activity.' *J. Biol. Chem.*, 2000, **275**, 29643–29647.

122. S. B. Freedman and J. M. Isner, 'Therapeutic angiogenesis for ischemic cardiovascular disease.' *J. Mol. Cell Cardiol.*, 2001, **33**, 379–393.

123. C. A. Mack, S. R. Patel, E. A. Schwarz *et al.*, 'Biologic bypass with the use of adenovirus-mediated gene transfer of the complementary deoxyribonucleic acid for vascular endothelial growth factor 121 improves myocardial perfusion and function in the ischemic porcine heart.' *J. Thorac. Cardiovasc. Surg.*, 1998, **115**, 168–176.

124. H. Su, R. Lu and Y. W. Kan, 'Adeno-associated viral vector-mediated vascular endothelial growth factor gene transfer induces neovascular formation in ischemic heart.' *Proc. Nat. Acad. Sci. USA*, 2000, **97**, 13801–13806.

125. R. C. Hendel, T. D. Henry, K. Rocha-Singh *et al.*, 'Effect of intracoronary recombinant human vascular endothelial growth factor on myocardial perfusion: evidence for a dose-dependent effect.' *Circulation*, 2000, **101**, 118–121.

126. J. F. Symes, D. W. Losordo, P. R. Vale *et al.*, 'Gene

therapy with vascular endothelial growth factor for inoperable coronary artery disease.' *Ann. Thorac. Surg.*, 1999, **68**, 830–836.

127. D. W. Losordo, P. R. Vale, J. F. Symes *et al.*, 'Gene therapy for myocardial angiogenesis: initial clinical results with direct myocardial injection of phVEGF165 as sole therapy for myocardial ischemia.' *Circulation*, 1998, **98**, 2800–2804.

128. J. M. Roberts and D. W. Cooper, 'Pathogenesis and genetics of pre-eclampsia.' *Lancet*, 2001, **357**, 53–56.

129. O. Genbacev, Y. Zhou, J. W. Ludlow *et al.*, 'Regulation of human placental development by oxygen tension.' *Science*, 1997, **277**, 1669–1672.

130. O. Genbacev, R. Joslin, C. H. Damsky *et al.*, 'Hypoxia alters early gestation human cytotrophoblast differentiation/invasion *in vitro* and models the placental defects that occur in preeclampsia.' *J. Clin. Invest.*, 1996, **97**, 540–550.

131. I. Caniggia, H. Mostachfi, J. Winter *et al.*, 'Hypoxia-inducible factor-1 mediates the biological effects of oxygen on human trophoblast differentiation through TGFβ(3).' *J. Clin. Invest.*, 2000, **105**, 577–587.

132. F. Rodesch, P. Simon, C. Donner *et al.*, 'Oxygen measurements in endometrial and trophoblastic tissues during early pregnancy.' *Obstet. Gynecol.*, 1992, **80**, 283–285.

133. I. Caniggia, J. Winter, S. J. Lye *et al.*, 'Oxygen and placental development during the first trimester: implications for the pathophysiology of pre-eclampsia.' *Placenta*, 2000, **21**, Suppl-30.

134. L. A. Carver, J. B. Hogenesch and C. A. Bradfield, 'Tissue specific expression of the rat Ah-receptor and ARNT mRNAs.' *Nucl. Acids Res.*, 1994, **22**, 3038–3044.

135. J. R. Gnarra, J. M. Ward, F. D. Porter *et al.*, 'Defective placental vasculogenesis causes embryonic lethality in VHL-deficient mice.' *Proc. Nat. Acad. Sci. USA*, 1997, **94**, 9102–9107.

136. G. Luo, Y. Z. Gu, S. Jain *et al.*, 'Molecular characterization of the murine Hif-1 α locus.' *Gene Expr.*, 1997, **6**, 287–299.

137. G. Tscheudschilsuren, S. Hombach-Klonisch, A. Kuchenhoff *et al.*, 'Expression of the arylhydrocarbon receptor and the arylhydrocarbon receptor nuclear translocator during early gestation in the rabbit uterus.' *Toxicol. Appl. Pharmacol.*, 1999, **160**, 231–237.

138. O. Genbacev, A. Krtolica, W. G. Kaelin *et al.*, 'Human cytotrophoblast expression of the von Hippel–Lindau protein is downregulated during uterine invasion in situ and upregulated by hypoxia *in vitro*.' *Dev. Biol.*, 2001, **23**, 526–536.

139. I. Caniggia, S. Grisaru-Gravnosky, M. Kuliszewsky *et al.*, 'Inhibition of TGF-β 3 restores the invasive capability of extravillous trophoblasts in preeclamptic pregnancies.' *J. Clin. Invest.*, 1999, **103**, 1641–1650.

140. D. M. Adelman, M. Gertsenstein, A. Nagy *et al.*, 'Placental cell fates are regulated *in vivo* by HIF-mediated hypoxia responses.' *Genes Dev.*, 2000, **14**, 3191–3203.

141. R. Bos, H. Zhong, C. F. Hanrahan *et al.*, 'Levels of hypoxia-inducible factor-1 α during breast carcinogenesis.' *J. Nat. Cancer Inst.*, 2001, **93**, 309–314.

142. D. Zagzag, H. Zhong, J. M. Scalzitti *et al.*, 'Expression of hypoxia-inducible factor 1α in brain tumors: association with angiogenesis, invasion, and progression.' *Cancer*, 2000, **88**, 2606–2618.

143. P. Birner, M. Schindl, A. Obermair *et al.*, 'Overexpression of hypoxia-inducible factor 1α is a marker for an unfavorable prognosis in early-stage invasive cervical cancer.' *Cancer Res.*, 2000, **60**, 4693–4696.

144. D. M. Aebersold, P. Burri, K. T. Beer *et al.*, 'Expression of hypoxia-inducible factor-1α: a novel predictive and prognostic parameter in the radiotherapy of oropharyngeal cancer.' *Cancer Res.*, 2001, **61**, 2911–2916.

145. M. V. Blagosklonny, W. G. An, L. Y. Romanova *et al.*, 'p53 inhibits hypoxia-inducible factor-stimulated transcription.' *J. Biol. Chem.*, 1998, **273**, 11995–11998.

146. K. Salnikow, M. Costa, W. D. Figg *et al.*, 'Hyperinducibility of hypoxia-responsive genes without p53/p21-dependent checkpoint in aggressive prostate cancer.' *Cancer Res.*, 2000, **60**, 5630–5634.

147. G. U. Dachs and G. M. Tozer, 'Hypoxia modulated gene expression: angiogenesis, metastasis and therapeutic exploitation.' *Eur. J. Cancer*, 2000, **36**, 1649–1660.

148. X. Sun, J. R. Kanwar, E. Leung *et al.*, 'Gene transfer of antisense hypoxia inducible factor-1α enhances the therapeutic efficacy of cancer immunotherapy.' *Gene. Therapy*, 2001, **8**, 638–645.

149. A. J. Franko, 'Misonidazole and other hypoxia markers: metabolism and applications.' *Int. J. Rad. Oncol. Biol. Phys.*, 1986, **12**, 1195–1202.

150. D. J. Van Os-Corby, C. J. Koch *et al.*, 'Is misonidazole binding to mouse tissues a measure of cellular pO$_2$?' *Biochem. Pharmacol.*, 1987, **36**, 3487–3494.

151. A. Nunn, K. Linder and H. W. Strauss, 'Nitroimidazoles and imaging hypoxia.' *Eur. J. Nuclear Med.*, 1995, **22**, 265–280.

152. J. M. Brown and A. J. Giaccia, 'The unique physiology of solid tumors: opportunities (and problems) for cancer therapy.' *Cancer Res.*, 1998, **58**, 1408–1416.

153. J. A. Raleigh, S. C. Chou, E. L. Bono *et al.*, 'Semiquantitative immunohistochemical analysis for hypoxia in human tumors.' *Int. J. Rad. Oncol. Biol. Phys.*, 2001, **49**, 569–574.

154. P. L. Olive, R. E. Durand, J. A. Raleigh *et al.*, 'Comparison between the comet assay and pimonidazole binding for measuring tumour hypoxia.' *Brit. J. Cancer*, 2000, **83**, 1525–1531.

155. S. M. Evans, A. V. Kachur, C. Y. Shiue *et al.*, 'Noninvasive detection of tumor hypoxia using the 2-nitroimidazole [18F]EF1.' *J. Nuclear Med.*, 2000, **41**, 327–336.

156. Y. M. Lee, C. H. Jeong, S. Y. Koo *et al.*, 'Determination of hypoxic region by hypoxia marker in developing mouse embryos *in vivo*: a possible signal for vessel development.' *Dev. Dyn.*, 2001, **220**, 175–186.

157. E. Y. Chen, M. Fujinaga and A. J.Giaccia, 'Hypoxic microenvironment within an embryo induces apoptosis and is essential for proper morphological development.' *Teratology*, 1999, **60**, 215–225.

158. W. L. Rumsey, B. Patel, B. Kuczynski *et al.*, 'Potential of nitroimidazoles as markers of hypoxia in heart.' *Adv. Exp. Med. Biol.*, 1994, **34**, 263–270.

159. T. Bjornheden, M. Evaldsson and O. Wiklund, 'A method for the assessment of hypoxia in the arterial wall, with potential application *in vivo*.' *Arterioscler. Thromb. Vasc. Biol.*, 1996, **16**, 178–185.

160. H. J. Marti, M. Bernaudin, A. Bellail *et al.*, 'Hypoxia-induced vascular endothelial growth factor expression precedes neovascularization after cerebral ischemia.' *Am. J. Pathol.*, 2000, **156**, 965–976.

161. K. J. Liu, P. Gast, M. Moussavi *et al.*, 'Lithium phthalocyanine: a probe for electron paramagnetic resonance oximetry in viable biological systems.' *Proc. Nat. Acad. Sci. USA*, 1993, **90**, 5438–5442.

162. K. J. Liu, P. J. Hoopes, E. L. Rolett *et al.*, 'Effect of anesthesia on cerebral tissue oxygen and cardiopulmonary parameters in rats.' *Adv. Exp. Med. Biol.*, 1997, **41**, 33–39.

163. K. J. Liu, G. Bacic, P. J. Hoopes *et al.*, 'Assessment of cerebral pO$_2$ by EPR oximetry in rodents: effects of anesthesia, ischemia, and breathing gas.' *Brain Res.*, 1995, **685**, 91–98.

164. E. L. Rolett, A. Azzawi, K. J. Liu *et al.*, 'Critical oxygen tension in rat brain: a combined (31)P-NMR and EPR oximetry study.' *Am. J. Physiol. Regul. Integr. Comp. Physiol.*, 2000, **279**, R9–R16.

J.P. Vanden Heuvel, G.H. Perdew, W.B. Mattes and W.F. Greenlee (Eds.)
Comprehensive Toxicology, Vol. xiv

# 14.4.5
# Epigenetic-Toxicant Induced Modulated Gap Junctional Intercellular Communication

## JAMES E. TROSKO and CHIA-CHENG CHANG
*Michigan State University, East Lansing, MI, USA*

## 14.4.5.1 INTRODUCTION

"Nothing in biology makes sense except in the light of evolution". The use of this quotation by T. Dobhanzsky[1] serves to frame the philosophical basis for this examination of how chemicals classified as "epigenetic toxicants" work to cause a wide variety of diseases. Specifically, the working hypothesis for the mechanistic basis of "epigenetic" toxicants is that gap junctional intercellular communication is absolutely required for the homeostatic regulation of the five phenotypic options of a metozoan cell, namely the control of (a) cell proliferation; (b) cell differentiation; (c) apoptosis; (d) adaptive responses of differentiated cells; and (e) senescence. Clearly, in single cell organisms, adaptive survival is dependent on the ability of a single cell, due to a random mutation in a population, to resist a lethal environmental change for the population. In effect, single cell organisms survive by symmetrical cell proliferation and, in the absence of nutrient depletion and extreme temperature, uncontrolled cell proliferation.

However, during that transition from a single cell organism to the first metazoan, cell differentiation, apoptosis, adaptive responses of the differentiated cells and senescence appeared at the same time as new genes, including those of the connexin genes (genes coding for proteins that compose the gap junctions).[2] It should also be noted that two other unique cellular characteristics appeared, namely, the ability of certain cells to act as totipotent or pluripotent stem cells and the ability of these stem cells to divide either symmetrically or asymmetrically.[3]

As soon as two or more cells cooperated by joining together, the ability to differentiate gave them new selective advantages for survival. However, that union also put restrictions on uncontrolled cell division. Contact inhibition[4] is that phenotype associated with the control of cell proliferation. With the appearance of these unique cellular functions occurring at the time the gene for the connexin (among many others) appeared, the philosophical question can be raised, "Was the appearance of these new cellular functions, 'causal' or coincidental, with the appearance of the connexin gene in the metazoans?"

As will be elaborated below, most normal cells (exceptions being stem cells, blood cells, highly terminally differentiated cells), capable of proliferation, are suppressed from uncontrolled cell proliferation by contact inhibition. Contact inhibition in these cells is associated with functional gap junctional intercellular communication (GJIC).[5,6] Cells with functional GJIC are capable of differentiation,[7] apoptosis[8] and adaptive responses in differentiated cells.[9] On the other hand, cancer cells, which lack GJIC,[10] do not contact inhibit,[11] do not terminally differentiate,[12] have abnormal apoptotic functions[13] and do not have terminally differentiated adaptive responses,[14] nor do they senesce (i.e., they are "immortal"). It is as though cancer cells have reverted back to the single cell state in evolution. This brief overview should give meaning to Dobzhansky's statement, in that the evolutionary appearance of a ubiquitous membrane structure, the gap junction, which allows contiguous cells to communicate directly from cytoplasm to cytoplasm in order to control vital cellular functions, provides a major function that determines the phenotypes of a multicellular organism. In other words, the gap junction, its genes and its function must be evaluated "in light of evolution".

## 14.4.5.2  HOMEOSTASIS OF THE BIOLOGICAL HIERARCHY

Toxicology as an interdisciplinary science had its primary origin in trying to determine whether organisms, exposed to various agents, might be affected by death, birth defects, cancer, reproductive dysfunction, neurological disorders, genetic diseases, cardiovascular problems, and caustic effects to the eyes or skin, *etc*. Only after these practical issues were determined, did the desire to examine the underlying mechanisms by which physical agents, such as radiations, solid particles, and chemical toxins and toxicants, appear in this field of study. Both short and long-term bioassays on various unicellular and multicellular organs, as well as epidemiological studies, contributed to our understanding of the "toxicities" of these exposures and determination of the LD50 (lethal dose 50%) or NOEL's (no observable effect level).

Much misunderstanding of the potential mechanisms by which these agents can be toxic comes from the reductionalistic approaches used to study the presumptive mechanisms of toxicity. Again, a philosophical approach to understanding how toxic agents can induce a variety of diseases must be considered before the detailed scientific observations can be interpreted. The multicellular organism, such as the human being, consists of a highly orchestrated, cybernetic hierarchy of interacting levels.[15–17] From the bottom of the hierarchy, the atomic/molecular level, consisting of atoms and molecules such as free $Ca^{++}$ or DNA molecules, to the top consisting of the brain/consciousness/ behavior, is linked by the cellular components (mitochondria; chromosomes; endoplasmic reticulum; plasma membranes, *etc.*), different cell types (totipotent; pluripotent stem cells, progenitor cells; terminally-differentiated cells), different tissues, different organs, different organ systems, all interacting with each other and the environment. Unique genes inherited in each individual ("nature") determine the ability of the specific individual hierarchy to respond to environmental influences ("nurture").

The ability of a single fertilized egg (containing the entire gene repertoire) to proliferate to generate the 100 trillion cells needed to made an individual, while at the same time orchestrate the differentiation of the various cell-, tissue- and organ-types, requires the delicate orchestration of all cell functions: symmetric cell proliferation; asymmetrical cell division for generation of stem cells and differentiated cells; programmed cell death for removal of cells/tissues at one stage of development in order for the generation of new cell/tissue types; adaptive responses of the various terminally differentiated cells, such as contractions of muscle cells or induction of insulin from pancreatic beta cells; and ultimately the senescence of some cells. Homeostasis is that regulatory, cybernetic mechanisms existing at all levels of the biological hierarchy[17] (enzyme feedback inhibition systems; contact inhibition to control cell proliferation; growth factor and hormone regulation of differentiation; death signals to trigger apoptosis and genetic and external environmental factors controlling senescence).

Homeostasis at the cellular level is mediated by several highly integrated signaling mechanisms (Figure 1). Extracellular communication, via secreted hormones, growth factors, cytokines, and neurotransmitters of distal cells in certain tissues, can trigger intra-cellular communication in targeted cells, which bind to the extra-cellular messages, causing a variety of signal transduction to modulate transcriptional, translational or posttranslational expression of the genome. In addition, these extra-cellular-triggered intra-cellular changes can modify the gap junctional inter-cellular communication between contiguous cells. Cells in multicellular organisms also receive intracellular signals from the cell adhesion molecules of the contiguous cells as well as from the extra-cellular matrix to which they are bound. The net effect of each of the specific extra- and inter-cellular communication signals will determine if the cell will remain quiescent, proliferate, differentiate, apoptose, adaptively respond if it is terminally differentiate, or go through senescence.

### 14.4.5.3 GAP JUNCTIONAL INTERCELLULAR COMMUNICATION IN THE HOMEOSTATIC CONTROL OF CELLULAR FUNCTIONS

Clearly, the human genome contains many genes not found in single cell organisms. Therefore, it is fair to ask, "Why should the connexin genes be considered the critical genes needed to maintain the multicellular phenotypes not found in single cell organisms?" The connexin genes do not, by themselves, suppress growth in contact inhibited cells; they do not induce specific types of differentiation; they clearly are not the genes involved in whether a given cell apoptoses; and they do not trigger, by themselves, whether a group of muscle cells contract in synchrony or of a group of neuroendocrine secret hormones in synchrony. The answer seems to be, by themselves, they are a necessary but insufficient contributor to these important multicellular functions. However, without the functional gap junctional communication, these other factors can only

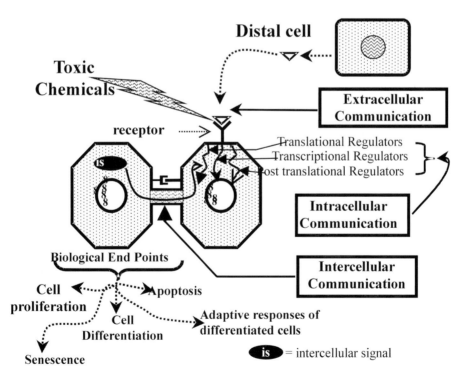

**Figure 1** An integrative scheme of our research proposal. Endogenous and exogenous extra-cellular signals, which can trigger various intra-cellular signal transduction mechanisms, can either increase or decrease gap junctional inter-cellular communication between cells in a multicellular organism. Cell proliferation, differentiation, apoptosis and adaptive responses of differentiated cells can occur as a consequence of the modulation of GJIC.

contribute to control of cell proliferation, differentiation, adaptive responses, apoptosis or senesence if other highly specialized substitutes have appeared. For example, free existing cells such as the hemopoietic stem cell is under growth control by negative growth regulators probably secreted from the differentiated cells of that lineage that bind to receptors that suppress cell cycle genes even though these cells do not express their connexin genes.

The membrane bound structural proteins comprising the intercellular channels, the gap junctions, are encoded by an evolutionary-conserved family of genes referred to as connexins.[2] Currently, about 20 connexin genes have been cloned from a number of vertebrate species.[18] These connexins oligerimize into single-membrane channels (hemi-channels) called connexons, which self-align with the corresponding hemi-channel in the extracellular space between contiguous cells to complete the intercellular channel. Each connexon consists of six connexins to form a central pore (see Figure 2).[19] These connexons can be constructed of a single type of connexin (homomeric) or multiple connexins if the cell expresses more than one type (heteromeric). Many of these single channels aggregate to form the gap junction, the size and number of which are dependent on the dynamic state of the tissue. Because GJIC can occur between certain heterologous cell types,[20] there can be coupling between homotypic, heterotypic or heteromeric connexons.[18,20]

The functions of gap junctions, in general, provide coordinated or synchronized electrotonic responses of groups of cells in excitable tissues (i.e., contraction of cardiac muscles) or metabolic cooperation in non-excitable tissues (sharing of signals needed for growth control, differentiation or adaptive responses to hormones, *etc.*).[20] Moreover, pattern formation during embryonic development depends on

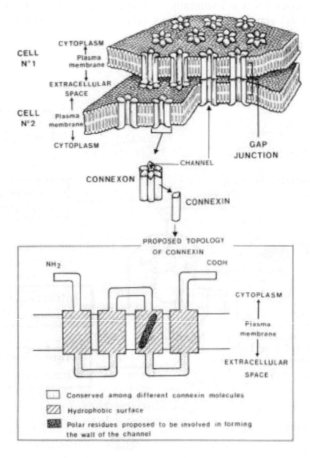

**Figure 2**   Schematic of gap junction and the role of connexins. Connexins oligerimize into single-membrane channels (hemi-channels) called connexons, which self-align with the corresponding hemi-channel in the extracellular space between contiguous cells to complete the intercellular channel. Connexins are transmembrane proteins with regions of conservation among the 20 members of this family. Used with permission from: H. Yamasaki, "Gap junctional intercellular communication and carcinogenesis", *Carcinogenesis*, 1990, **11**, 1051–1058.

sequestering groups of cells for one type of differentiation from other groups requiring another type of differentiation. These channels provide direct exchange of ions and small regulatory molecules (i.e., c-AMP) and molecular substrates (nucleosides, amino acids, sugars, *etc.*).[21] There is also a cell type and tissue specificity of which connexin gene is expressed and functional.[5] Some cells in a tissue, such as a pluripotent stem cell of the human kidney epithelial, human breast epithelial pluripotent stem cell,[22,23] do not express any connexins and do not have functional GJIC. On the other hand, other cells express only one type, and others express at least three types (rat liver epithelial express Cx26, Cx43 and Cx45) while the rat hepatocytes express Cx26 and Cx32. Although the rat liver epithelial cells and hepatocytes coexist side by side, there seems to be no connexon coupling between the two cell types.[24] While the transfer of ions and small molecular weight molecules (below 1200 Da) is passive diffusion, there does seem to be connexin-dependent differences in the transfer of certain molecules based either on shape or molecular weight of the molecules.[25]

## 14.4.5.4 CRISIS IN PARADIGMS: GENOTOXICITY AS A PARALYZING CONCEPT

The field of toxicology was basically concerned, in the beginning, with the potential of toxins and toxicants killing organisms, making organisms sick, causing birth defects or somatic and genetic diseases. As the field evolved, the underlying mechanisms were the focus of toxicological research. With the introduction of the paradigm of "carcinogen as mutagen",[26] and the subsequent experimental evidence that genetic diseases were associated with inherited mutated genes, tumor cells were associated with mutated proto-oncogenes or tumor suppressor genes,[27] as well as many chemicals associated with carcinogenesis were mutagens in various short term "genotoxicity" assays,[28] the primary emphasis in evaluating potential toxicities of chemicals was placed on testing for "genotoxicity". While there is absolutely no doubt that mutation of genes and chromosomes can and do contribute to various human diseases, what has been generally ignored in the field of Toxicology is the fact that chemicals can be toxic via mechanisms independent of their "genotoxicity".[29–31] In general, an agent can interact

with cells to damage genomic DNA that, then, can be repaired in an error free manner, repaired in an error prone manner or not repaired at all. In the latter cases, this can either lead to the death of the cell or to a point- or gene-mutation or chromosome mutation, such as a translocation. Error prone-DNA replication can also lead to either gene mutations without any template DNA damage[32] or amplified chromosome replication.[33]

However, some agents can induced cell killing without DNA damage (e.g., inhibition of critical enzyme function; destruction of membrane integrity; inducing apoptosis genes without genotoxic chemicals (serum withdrawal; dexamethasone-treatment, *etc.*). The most ignored mechanism of toxicity has been the modulation of gene expression or "epigenetic toxicity".[31,34–35] By epigenetic toxicity, the modulation of the expression of genes at the transcriptional level (turning on or off of genes at inappropriate times during development; during the cell cycle; during differentiation; during apoptosis; or during the adaptive responses of differentiated cells) can lead to different "toxic consequences".[36,37] Epigenetic toxicity could also occur at the translational level via altering the stability of the genetic message (mRNA) or by causing alteration in the splicing of the mRNA. Moreover, epigenetic toxicity can also occur by posttranslational processing of proteins (e.g., phosphorylation, nitrosylation, ubiquitization of proteins).

## 14.4.5.5 DISRUPTION OF GJIC AS AN ADAPTIVE OR MALADAPTIVE (TOXICOLOGICAL) MECHANISM

Clearly, if, after fertilization of the egg, the social community of cells always remained coupled after the expression of the connexin genes early in the developmental process,[38] growth would have stopped and further development would have ceased. Turning on or off of specific connexin genes allowed specific differentiation to occur. Modulation of the connexin genes at the posttranslational levels by normal endogenous hormones, growth factors, neurotransmitters and cytokines allowed some cells to proliferate, others to die by apoptosis and others to perform specific adaptive responses in the terminally-differentiated cells.[31] Wound repair depends on surviving cells in the damaged tissue to uncouple, allowing contact-inhibited cells in G0 to enter the cell cycle and to proliferate in order to heal the damaged or removed tissue.[39]

Many toxic chemicals, such as TCDD, DDT, peroxisome proliferators, *etc.*, which are not genotoxic, can modulate signal transduction-induced expression of genes and inhibition of gap junctional intercellular communication,[17,31] to bring about either birth defects, tumor promotion, reproductive, immune modulation and/or neurotoxicities without damaging nuclear DNA or killing cells. While the detailed biochemical (signal transduction signals) or molecular (modified histone acetylation/deacetylation;[40] methylation/demethylation)[41] mechanisms have not yet been delineated for these compounds, evidence showing that they modulate cell proliferation, differentiation or apoptosis implicates the involvement of gap junctions' intercellular communication. Indeed, one of the first series of observations linking epigenetic toxicities to this class of non-mutagenic toxicants has been their ability to modulate GJIC.[42]

Even more dramatic is the idea that many so-classified "genotoxic" carcinogens might not be mutagenic because of misinterpretation of the short-term tests for genotoxicity.[29,30] The fact that many of these "carcinogens" can have epigenetic potential, especially structure/function relationships more closely linked to their ability to be associated with the induction of rodent tumors,[43,44] might suggest that much of the confusing data generated by assuming chemical "carcinogens" must be mutagenic has been misinterpreted because of the "carcinogen as mutagen" paradigm. Only recently, with more data showing that toxic chemicals might be associated with the altered expression of genes (i.e., imprinting errors[45]) or global and specific gene methylation changes in tumor cells, as well as more evidence linking the modulation of GJIC with teratogenesis (thalidamide),[46] tumor promotion (anthracenes),[44,45] reproductive toxicants (DDT as an endocrine disruptor, neurotoxicants (DDT, Dieldrin, toxaphene).[47,48]

One of the major implications of chemicals working via epigenetic mechanisms, including the modulation of gap junctional intercellular communication, comes from the field of carcinogenesis. Early in this field, the concept of the multistage process of carcinogenesis was conceived when it was observed that after exposure of an animal to a no effect level of a chemical[49] or physical agent,[50] chronic and regular posttreatment with an agent such as phorbol esters or TPA having no genotoxic potential, brought about the appearance of skin papillomas. This implied that the multistage process of carcinogenesis also consisted of multi-mechanisms because the initial exposure of the animal's skin induced an "irreversible" event. The subsequent posttreatment with a non-genotoxic chemical had to be applied (a) at a high enough level to exceed a threshold; (b) in a regular, not irregular, fashion and for a long period of time.[51] Stoppage of the TPA treatment could lead to the regression of the papillomas.[52] Exposure of the papillomas to known genotoxic agents could bring about the conversion of the benign papillomas to carcinomas.[53]

These empirical observations led to the operational concepts of "initiation", "promotion", and "progression".[54,55] While these operational concepts do not imply any underlying molecular mechanism, their general characteristics suggest that the "initiation" phase might be due to a DNA damage/mutagenic mechanism because of its irreversibility. On the other hand, because of a threshold concentration of the "promoting" agent,[52,56-62] its non-genotoxic nature (e.g., TCCD, phenobarbital; polybrominated biphenyls; peroxisome proliferators; DDT; saccharin, *etc.*),[34] and its potential interruptibility or reversibility,[63] one could speculate that these chemicals worked via an epigenetic mechanism. The fact that papillomas could be irreversibly converted to carcinomas might suggest a genotoxic event. However, it cannot be ruled out at this point that a stable epigenetic event could be responsible for both the initiation and progression phases.

Based on the foregoing observations, one hypothesis concerning the potential mechanisms of the multistep, multi-mechanism process of the initiation-promotion-concept of carcinogenesis involves the modulation of gap junctional intercellular communication as an epigenetic mechanism. This hypothesis integrates all of the predominant theories of carcinogenesis, namely, the "stem cell theory",[64-67] "clonal origin of tumorigenesis",[68,69] "cancer as a disease of differentiation",[70,71] "Oncogeny as partially blocked ontogeny",[72] the "initiation/promotion/progression" theory of carcinogenesis;[73] the "oncogene/ tumor suppressor gene" theory;[75] and the mutation/epigenetic theory of carcinogenesis.[73] While each individual theory has an element of strength, no one theory gives an adequate picture of what Potter has stated:[72] "Cancer cells should not be seen as exclusively a problem in cell proliferation, but rather as a problem combining the processes of proliferation and differentiation, hence the phrase introduced in 1968: 'oncogeny is blocked ontogeny'. Cancer tissues resemble foetal tissues in many ways but they differ from foetal tissue in being unable to 'recapitulate the total programme leading to an orchestrated collection of organism-serving cells' that are programmed 'to make the organ as adaptive as possible to the range of environmental variations in which it evolved'". Consequently, what is to follow is the proposal that the reversible and stable down-regulation of GJIC during the exposure of a pluri-potent stem cell to the ultimate conversion to a neoplastic, invasive and metastatic cell will help to integrate these different theories of carcinogenesis into one unified theory.

It must be remembered that cells within a tumor appear to be clonally derived, have little growth control or have lost "contact inhibition",[11] do not terminally differentiate under normal circumstances, and appear to be "immortalized". The seminal observation that only a few "contact-insensitive" cells in a population of primary cells can be neoplastically-transformed after exposure to a carcinogen[74] allows us to start the unified hypothesis. If one assumes that in every tissue there exist a few stem cells, one must start with some characteristics of stem cells. By definition, a stem cell must have both the capacity to divide symmetrically (to increase the stem cell numbers) and to divide asymmetrically in order to start the differentiation lineage with one daughter cell (Figure 3) and to maintain "stem cell renewal" with the other daughter. These stem cells would be "primitive", in the sense that they would not express many genes

associated with the differentiated state, including drug metabolizing enzymes.[3]

Another characteristic of stem cells would be their state of "mortality". It has been generally assumed that the carcinogenic process had to first convert a mortal cell to an immortal cell during the early phase of carcinogenesis.[75] In fact, the recent studies on telomerase have been interpreted to support this hypothesis. "Immortalizing" viruses, used to isolate a few immortalized "cells from populations of primary cells, as well as the isolation of spontaneously-immortalized" cells from primary cultures that have experienced "a crisis" period when most of the cells have gone through "senescence", have been viewed as converting a few normal "mortal" cells to the "immortalized" state. The alternative view of these observations is that the stem cell is, by definition, "immortal". It only becomes "mortal" when it is induced to differentiate.[3] Evidence in support of this idea comes from the fact that normal human breast epithelial stem cells express telomerase activity,[76] whereas when they differentiate the telomerase activity is reduced. Based on these observations, the *initiation* process of carcinogenesis is one that "blocks the terminal differentiation of a stem cell" or "oncogeny as partially blocked ontogeny".[72] The major implication of this idea is that the initiation process involves the blockage of a stem cell's ability to divide asymmetrically but to allow it to divide symmetrically to produce two cells that can

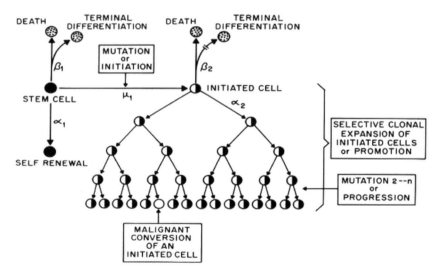

**Figure 3**  The initiation/promotion/progression model of carcinogenesis. $\beta_1$ = rate of terminal differentiation and death of stem cell; $\beta_2$ = rate of death, but not of terminal differentiation of the initiated cell $(-\|\rightarrow)$; $\alpha_1$ = rate of cell division of stem cells; $\alpha_2$ = rate of cell division of initiated cells; $\mu_1$ = rate of the molecular event leading to initiation (i.e., possibly mutation); $\mu_2$ = rate at which second event occurs within an initiated cell. (From Trosko *et al.*, In: *Modern Cell Biology*; Vol. 7, 'Gap Junctions' eds. E. L. Hertzberg and R. G. Johnson, pp. 435–448, 1998; with permission from Alan R. Liss, Inc., New York).

partially, but not terminally, differentiate. It would be able to divide symmetrical to produce more cells like itself and remain "immortal" (Figure 3).

In addition, human stem cells appear not to have expressed gap junctions or functional GJIC. The toti-potent stem cell or fertilized egg, as well as the very early blastula cells, seem to lack expressed connexin genes or do not have functional GJIC.[7] Only after early evidence of development of pattern formation/differentiation, do the connexin genes get expressed and functional GJIC take place.[7] Given that there seems to be a similarity between stem cells (primitive phenotypes; contact-insensitive; no functional GJIC) and malignant tumor cells (they are partially, but not terminally, differentiated; contact-insensitive; no function heterologous or homologous GJIC),[23] it might seem reasonable to question why normal stem cells are not "malignant". The answer seems to be that it is relatively easy for the non-expressed, non gap junctional communicating-stem cells to be induced to express their connexins and to have functional GJIC,[23] whereas it is difficult to restore functional GJIC in tumor cells (either by transfection of connexin genes, by exposure to oncogene-inhibiting drugs or chemicals; by activation of tumor suppressor genes).[73,77] These observations would allow the integration of the "stem cell theory" with the "ontogeny as partially blocked theory", and the initiation phase of the carcinogenic process, as well as cancer as a "disease of differentiation".

Clearly, with the experimental evidence that several genetic/epigenetic events must occur before a normal cell can become an invasive, metastatic cell.[78] Given the relatively low rate of mutations in a given cell and the number of critical genes (oncogenes; tumor suppressor genes; genes for anti-invasiveness and anti-metastasis) needed to convert the normal stem cell to the malignant cell, the most efficient manner to bring about these changes is to clonally multiply the first initiated cell to a large mass by stimulating proliferation in the non-terminally differentiating initiated cell and by preventing the apoptosis of these initiated cells.[78] Both of these processes are triggered by epigenetic chemicals, but not by genotoxic compounds (except only indirectly when genotoxic-induced cell death by necrosis causes compensatory hyperplasia of surviving initiated cells). Genotoxic agents would, for example, not be expected to block apoptosis or directly stimulate DNA synthesis or cell division. They would do just the opposite. Moreover, non-genotoxic chemicals, or chemicals that inhibit GJIC and act as tumor promoters,

block apoptosis and or can stimulate cell proliferation.[5]

The tumor promotion phase, as one that increases the number of initiated cells by cell proliferation and blockage of apoptosis, is done by many non-genotoxic chemicals at non-cytotoxic levels by reversibly blocking GJIC and altering signal transduction and gene expression at the posttranslational, translational or transcriptional levels.[5] These are all epigenetic mechanisms. At this point, the cell is viewed as having reached the *progression* phase of carcinogenesis. During this clonal expansion phase of initiated cells, additional mutations/stable epigenetic events occur which ultimately allows one of those clonally-expanded initiated cell to acquire a stable phenotype that prevents the cell from having functional GJIC. It no longer needs endogenous (hormones, growth factors) or exogenous tumor promoting chemicals to continuously inhibit GJIC. Instead, the activation of certain oncogenes (i.e., Ha-Ras)[79] and the loss of tumor suppressor genes can stably down regulate GJIC. The HeLa cell does not have function GJIC because the connexin genes have been stably down regulated, not mutated or deleted. This is seen when the Cx43 gene can be induced by micro-cell chromosome transfer[81] or exposure to demethylation conditions.[77] In addition, treatment of ras-transformed cells with non-genotoxic chemicals, such as lovastatin or hexamethylene bisacetamide or caffeic acid phenethylene ester, can cause the restoration of normal connexin 43 phosphorylation (a posttranslational epigenetic event).[81–86] Evidence in support of the role gap junctions in carcinogenesis also comes from the opposite set of observations, namely the treatment of non-communicating cancer cells, or the simultaneous treatment of promoter exposed initiated cells, with chemicals that up-regulate GJIC, can act as chemotherapeutic or chemopreventive agents.[5]

### 14.4.5.6   CONCLUSIONS: ENDOGENOUS AND EXOGENOUS CHEMICALS AS EPIGENETIC TOXICANTS

Although it has been long known that endogenous chemicals, such as hormones, growth factors and neurotransmitters, acted as epigenetic agents, in that they could induce signal transduction mechanisms, alter gene expression and bring about cellular phenotypic changes (e.g., cause cells to proliferate, differentiate, apoptose), it had not been formally introduced into the field of toxicology until recently as a significant mechanism of toxicity.

DNA damage and mutagenesis, as well as necrosis, had been considered the primary mechanisms by which chemicals could have toxic potentials. Even today, some view epigenetic toxicology as restricted only to altered gene expression at the transcriptional level.[84] However, with the clear demonstration that many chemicals that can induce teratogenesis (thalidomide), promote tumors (phenobarbital, PBB, TCCD), disrupt reproductive capacities (gossypol), or be neurotoxicants (DDT) are not genotoxic or not even cytotoxic at doses where they are "epigenetic toxicants".[35] In fact, these chemicals can block gap junctional intercellular communication in a dose dependent fashion, showing real threshold-like effects and demonstrate reversibility. The fact that these chemicals do not block gap junctional communication directly but via some indirect induced membrane-triggered effects (e.g., membrane fluidity changes, induced signal transduction mechanisms, *etc.*) which can modify the expression of the connexin genes, connexin message stability, connexin protein translocation, assembly and/or function of the connexins, suggests that, since GJIC can be modulated within seconds or few minutes, the concept of "epigenetic toxicity" must be viewed broader that just transcriptional modulation. A holistic view is that these cells, by early modulation of connexins, can lead to altered gene expression at the transcriptional level, to trigger cell-cycle related genes for cell proliferation, induced cell differentiation, apoptosis, or senescence genes, or simply to allow differentiated cells to perform adaptive functions.

Since GJIC has been implicated in the homeostatic control of cell proliferation, differentiation, apoptosis and adaptive functions of differentiated cells, modulation of GJIC at the transcriptional, translational or posttranslational levels at the inappropriate times during embryogenesis, fetal development, maturation and aging of the multicellular metazoan can have toxicological effects on the individual. Too much or too little cell proliferation, differentiation, apoptosis or adaptive responses of differentiated cells due to disrupted GJIC by non-genotoxic or non-cytotoxic, but epigenetic chemicals, can explain my know chemical toxicities. In effect, even though birth defects, cancer, cataracts, immune modulation, reproductive- and neuro-toxicities are very different clinical disease endpoints, it now seems clear they can all share a common underlying, shared mechanisms, namely, modulated gap junctional intercellular communication. The now voluminous scientific literature, involving human genetic connexin mutations, knockout connexin mice, and experimental animals exposed to chemicals capable of modulating GJIC, should redirect our attention to the importance of using modulation of GJIC as a very important biomarker of toxicity. This has been dramatically-illustrated by a study, using a method that models the properties of a large population of molecules to represent the "universe of chemicals", that showed "inhibition of GJIC is strongly linked to the carcinogenic process in rodents, to cellular but not system toxicity, to biological phenomena that may involve inflammatory processes and to developmental effects".[85]

## 14.4.5.7 REFERENCES

1. T. Dobzhansky, 'Nothing in biology makes sense except in the light of evolution.' *Amer. Biol.*, 1973, **35**, 125–129.
2. R. Bruzzone, T. W. White and D. L. Paul, 'Connection with connexins: The molecular basis of direct intercellular signaling.' *Eur. J. Biochem.*, 1996, **238**, 1–27.
3. J. E. Trosko, C. C. Chang, M. R. Wilson *et al.*, 'Gap junctions and the regulation of cellular functions of stem cells during development and differentiation.' *Methods*, 2000, **20**, 245–264.
4. E. M. Levine, Y. Becker, C. W. M. Boone *et al.*, 'Contact inhibition, macromolecular synthesis and polyribosomes in cultured human diploid fibroblasts.' *Proc. Nat. Acad. Sci. USA*, 1965, **53**, 350–355.
5. J. E Trosko and R. J. Ruch, 'Cell–cell communication in carcinogenesis.' *Front. Biosci.*, 1998, **3**, 208–236.
6. H. Yamasaki and C. C. G. Naus, 'Role of connexin genes in growth control.' *Carcinogenesis*, 1996, **17**, 1199–1213.
7. C. Lo, 'The role of gap junction membrane channels in development.' *J. Bioenerg. Biomembr.*, 1996, **28**, 379–385.
8. M. R. Wilson, T. W. Close and J. E. Trosko, 'Cell population dynamics (Apoptosis, mitosis and cell–cell communication) during disruption of homeostasis.' *Exp. Cell. Res.*, 2000, **254**, 257–268.
9. P. Meda, R. Bruzzone, M. Chanson *et al.*, 'Gap junctional coupling modulates secretion of exocrine pancreas.' *Proc. Nat. Acad. Sci. USA*, 1987, **84**, 4901–4904.
10. H. Yamasaki, M. Hollstein, M. Mesnil *et al.*, 'Selective lack of intercellular communication between transformed and nontransformed cells a common property of chemical and oncogene transformation of BALB/c3T3 cells.' *Cancer Res.*, 1987, **47**, 5658–5664.
11. C. Borek and L. Sachs, 'The difference in contact inhibition of cell replication between normal cells and cells transformed by different carcinogens.' *Proc. Nat. Acad. Sci. USA*, 1966, **56**, 1705–1711.
12. A. Block, 'Induced cell differentiation in cancer therapy.' *Cancer Treat. Rep.*, 1984, **68**, 199–206.
13. C. B. Thompson, 'Apoptosis in the pathogenesis and treatment of disease.' *Science*, 1995, **267**, 1456–1462.
14. V. R. Potter, 'Cancer as a problem in intercellular communication, Regulation by growth-inhibiting factors.' *Prog. Nucleic. Acid Res. Molec. Biol.*, 1983, **29**, 161–173.
15. H. Brody, 'A systems view of man: Implications for medicine, science and ethics.' *Perspect. Biol. Med.*, 1973, **17**, 71–92.

16. V. R. Potter, 'Probabilistic aspects of the human cybernetic machine.' *Perspect. Biol. Med.*, 1974, **17**, 164–183.

17. J. E. Trosko, 'Hierarchical and cybernetic nature of biological systems and their relevance to homeostatic adaptation to low-level exposures to oxidative stress-inducing agents.' *Environ. Health Perspect.*, 1998, **106**, 331–337.

18. J. X. Jiang and D. A. Goodenough, 'Heteromeric connexons in lens gap junction channels.' *Proc. Nat. Acad. Sci. USA*, 1996, **93**, 1287–1291.

19. G. Perkins, D. A. Goodenough and G. Sosinsky, 'Three-dimensional structure of the gap junction connexon.' *Biophys. J.*, 1997, **72**, 533–544.

20. J. C. Saez, V. M. Berthoud, A. P. Moreno *et al.*, 'Gap junctions: Multiplicity of controls in differentiated and undifferentiated cells and possible functional implications.' *Ad. Second Messeng. Phosphoprotein Res.*, 1993, **27**, 163–198.

21. G. S. Goldberg, P. D. Lampe and B. J. Nicholson, 'Selected transfer of endogenous metabolites through gap junctions composed of different connexins.' *Nature Cell Biol.*, 1999, **1**, 457–459.

22. C. C. Chang, J. E. Trosko, M. H. El-fouly *et al.*, 'Contact insensitivity of a subpopulation of normal human fetal kidney epithelial cells and of human carcinoma cell lines.' *Cancer Res.*, 1987, **47**, 1634–1645.

23. C. Y. Koa, K. Nomata, C. S. Oakley *et al.*, 'Two types of normal human breast epithelial cells derived from reduction mammoplasty: Phenotypic characterization and response to SV40 transfection.' *Carcinogen.*, 1995, **16**, 531–538.

24. M. Mesnil, J.-M. Fraslin, C. Piccoli *et al.*, 'Cell-contact but not junctional communication (dye coupling) with biliary epithelial cells is required for hepatocytes to maintain differentiated functions.' *Exp. Cell Res.*, 1987, **173**, 524–533.

25. T. W. White and D. L. Paul, 'Genetic diseases and gene knockouts reveal diverse connexin functions.' *Ann. Rev. Physiol.*, 1999, **61**, 283–310.

26. B. N. Ames, W. E. Durston, E. Yamasaki *et al.*, 'Carcinogens are mutagens: A simple test system combining liver homogenates for activation and bacteria for detection.' *Proc. Nat. Acad. Sci. USA*, 1993, **70**, 2281–2285.

27. H. Varmus and R. H Weinberg (eds.), *Genes and the Biology of Cancer*, Scientific 1993, American Library, New York.

28. R. W. Tennant, B. H. Margolin, M. D. Shelby *et al.*, 'Prediction of chemical carcinogenicity in rodents from *in vitro* genetic toxicity assays.' *Science*, 1987, **236**, 933–941.

29. J. E. Trosko, 'A failed paradigm: Carcinogenesis is more than mutagenesis.' *Mutagenesis*, 1988, **3**, 363–366.

30. J. E. Trosko, 'Challenge to the simple paradigm that "carcinogens" are "mutagens" and to the *in vitro* assays used to test the paradigm.' *Mutation Res.*, 1997, **373**, 245–249.

31. J. E. Trosko, C. C. Chang, B. L. Upham *et al.*, 'Epigenetic toxicology as toxicant-induced changes in intracellular signalling leading to altered gap junctional intercellular communication.' *Toxicol. Lett.*, 1998, **102–103**, 71–78.

32. C. C. Chang, J. A. Boezi, S. T. Warren *et al.*, 'Isolation and characterization of a UV-sensitive hyper-mutable aphidicolin-resistant Chinese hamster cell line.' *Somatic. Cell Genet.*, 1981, **7**, 235–253.

33. Y. Huang, C. C. Chang and J. E. Trosko, 'Aphidicolin-induced endoreduplication in Chinese hamster cells.' *Cancer Res.*, 1983, **43**, 1361–1364.

34. J. E. Trosko and C. C. Chang, 'Nongenotoxic mechanisms in carcinogenesis: Role of inhibited inter-cellular communication.' Banbury Report 31: Carcinogen Risk Assessment: New Directions in the Qualitative and Quantitative Aspects, Cold Spring Harbor Laboratory Press, Cold Spring Harbor, NY, 1988.

35. J. E. Trosko, C. C. Chang, B. V. Madhukar *et al.*, 'Modulators of gap junction function: The scientific basis of epigenetic toxicology.' *In vitro Toxicol.*, 1990, **3**, 9–26.

36. J. E. Trosko, C. C. Chang and B. V. Madhukar, '*In vitro* analysis of modulators of intercellular communication: Implications for biologically-based risk assessment models for chemical exposure.' *Toxicol. In Vitro*, 1990, **4**, 635–643.

37. J. E. Trosko, B. V. Madhukar and C. C. Chang, 'Endogenous and exogenous modulation of gap junctional intercellular communication: Toxicological and pharmacological implications.' *Life Sci.*, 1993, **53**, 1–19.

38. B. S. B. Yancey, S. Biswal and J.-P. Revel, 'Spatial and temporal patterns of distribution of gap junction protein connexin43 during mouse gartrulation and organogenesis.' *Develop.*, 1992, **114**, 203–212.

39. M. Saitoh, M. Oyamada, Y. Oyamada *et al.*, 'Changes in the expression of gap junction proteins (connexins) in hamster tongue epithelium during wound healing and carcinogenesis.' *Carcinogen.*, 1997, **18**, 1319–1328.

40. S. Y. Archer and R. A. Hodin, 'Histone acetylation and cancer.' *Curr. Opin. Genet. Devel.*, 1999, **9**, 171–174.

41. K. D. Robertson and P. A. Jones, 'DNA methylation: Past, present and future.' *Carcinogenesis*, 2000, **21**, 461–467.

42. L. P. Yotti, J. E. Trosko and C. C. Chang, 'Elimination of metabolic cooperation in Chinese hamster cells by a tumor promoter.' *Science*, 1979, **206**, 1089–1091.

43. A. M. Rummel, J. E. Trosko, M. R. Wilson *et al.*, 'Polycyclic aromatic hydrocarbon on gap junction intercellular communication and stimulated MAPK activity.' *Toxicol. Sci.*, 1999, **49**, 232–240.

44. B. L. Upham, L. M. Weis and J. E. Trosko, 'Modulated gap junctional intercellular communication as a biomarker of PAH's epigenetic toxicity in structure/function relationship.' *Environ. Health Perspect.*, 1998, **106**, 975–981.

45. M. Nakao and H. Sasaki, 'Genomic imprinting: Significance in development and diseases and the molecular mechanisms.' *J. Biochem.*, 1996, **120**, 467–473.

46. S. Nicolai, H. Sies and W. Stahl, 'Stimulation of gap junctional intercellular communication by thalidamide and thalidomide analogs in human skin fibroblasts.' *Biochem. Pharmacol.*, 1997, **53**, 1553–1557.

47. G. Tsushimoto, J. E. Trosko, C. C. Chang *et al.*, 'Cytotoxic, mutagenic and tumor promoting properties of DDT, Lindane and Chlodane on Chinese hamster cells *in vitro*.' *Arch. Environ. Cont. Toxicol.* 1983, **12**, 721–730.

48. J. E. Trosko, C. Jone and C. C. Chang, 'Inhibition of gap junctional-mediated intercellular communication *in vitro* by Aldrin, Dieldrin and Toxaphene: A possible cellular mechanism for their tumor-promoting and neurotoxic effects.' *Mol. Toxicol.*, 1987, **1**, 83–93.

49. I. Berenblum, 'A speculative review: The probable nature of promoting action and its significance in the understanding of the mechanism of carcinogenesis.' *Cancer Res.*, 1954, **14**, 471–477.

50. R. J. M. Fry, J. B. Storer and R. L. Ullrich, 'Radiation Toxicology: Carcinogenesis,' The Scientific Basis of Toxicity Assessment, Elsevier/North Holland Biomedical Press, New York, 1980.

51. R. K. Boutwell, 'The function and mechanism of

promoters of carcinogenesis.' *CRC Crit. Rev. Toxicol.*, 1974, **2**, 419–443.

52. A. K. Verma and R. K. Boutwell, 'Effects of dose and duration of treatment with a tumor-promoting agent, TPA, on mouse skin carcinogenesis.' *Carcinogenesis*, 1980, **1**, 271–276.

53. H. Hennings, R. Shores, M. L. Wenk *et al.*, 'Malignant conversion of mouse skin tumours is increased by tumor initiators and unaffected by tumor promoters.' *Nature*, 1983, **304**, 607–609.

54. I. B. Weinstein, S. Cattoni-Celli, P. Kirschmeier *et al.*, 'Multistage carcinogenesis involves multiple genes and multiple mechanisms.' *J. Cell. Physiol.*, 1984, **3**, 127–137.

55. J. E. Trosko, C. C. Chang, B. V. Madhukar *et al.*, 'Intercellular communication: A paradigm for the interpretation of the initiation/promotion/progression model of carcinogenesis,' Chemical Induction of Cancer, Birkhauser Publisher, Boston, 1995.

56. K. L. Kolaja, D. E. Stevenson, E. F. Walborg *et al.*, 'Dose dependence of phenobarbital promotion of preneoplastic hepatic lesions in F344 rats and B6C3F1 mice: Effects on DNA synthesis and apoptosis.' *Carcinogenesis*, 1996, **17**, 947–954.

57. H. C. Pitot, T. L. Goldsworthy, S. Moran *et al.*, 'A method to quantitate the relative initiating and promoting potencies of hepatocarcinogenic agents in their dose-response relationship to altered hepatic foci.' *Carcinogenesis*, 1987, **8**, 1491–1499.

58. E. Deml and D. Oesterle, 'Dose response of promotion by polychlorinated biphenyls and chloroform in rat liver foci bioassay.' *Arch. Toxicol.*, 1987, **60**, 209–211.

59. R. M. David, M. R. Moore, M. A. Cifone *et al.*, 'Chronic peroxisome proliferation and hepatomegaly associated with the hepatocellular tumorigenesis of Di(2-ethylhexyl)phthalate and the effects of recovery.' *Toxicol. Sci.*, 1999, **50**, 195–205.

60. G. M. Williams, M. J. Iatropoulos, C. X. Wang *et al.*, 'Nonlinearities in 2-acetylaminofluorene exposure responses for genotoxic and epigenetic effects leading to initiation of carcinogenesis in rat liver.' *Toxicol. Sci.*, 1998, **45**, 152–161.

61. E. W. Carney, J. W. Crissman, A. B. Liberacki *et al.*, 'Assessment of adult and neonatal reproductive parameters in Sprague-Dawley rats exposed to prylene glycol monomethyl ether vapors for two generations.' *Toxicol. Sci.*, 1999, **50**, 249–258.

62. J. G. Teeguarden, Y. P. Dragan, J. Singh *et al.*, 'Quantitative analysis of dose- and time-dependent promotion of four phenotypes of altered hepatic foci by 2,3,7,8-tetrachlorodibenzo-*p*-dioxin in female Sprague–Dawley rats.' *Toxicol. Sci.*, 1999, **51**, 211–224.

63. H. C. Pitot and Y. P. Dragan, 'Facts and theories concerning the mechanisms of carcinogenesis.' *FASEB*, 1991, **5**, 2280–2286.

64. J. E. Till, 'Stem cells in differentiation and neoplasia.' *J. Cell Physiol. Suppl.*, 1982, **1**, 3–11.

65. S. Kondo, 'Carcinogenesis in relation to the stem cell-mutation hypothesis.' *Differentiation*, 1983, **24**, 1–8.

66. M. F. Greaves, 'Differentiation-linked leukemogenesis in lymphocytes.' *Science*, 1986, **234**, 697–704.

67. R. N. Buick and M. N. Pollak, 'Perspectives on clonogenic tumor cells, stem cells and oncogenes.' *Cancer Res.*, 1984, **44**, 4909–4918.

68. P. C. Nowell, 'The clonal evolution of tumor cell populations.' *Science*, 1976, **194**, 23–28.

69. P. J. Fialkow, 'Clonal origin of human tumors.' *Annu. Rev. Med.*, 1979, **30**, 135–143.

70. C. L. Markert, 'Neoplasia: A disease of cell differentiation.' *Cancer Res.*, 1968, **28**, 1908–1914.

71. G. B. Pierce, 'Neoplasms, differentiation and mutations.' *Am. J. Pathol.*, 1974, **77**, 103–118.

72. V. R. Potter, 'Phenotypic diversity in experimental hepatomas: The concept of partially blocked ontogeny.' *Br. J. Cancer*, 1978, **38**, 1–23.

73. J. E. Trosko, C. C. Chang, B. V. Madhukar *et al.*, 'Oncogenes, tumor suppressor genes and intercellular communication in the "oncogeny as partially blocked ontogeny hypothesis".' New Frontiers in Cancer Causation. Taylor and Francis Publishers, Wash. DC, 1993.

74. S. Nakano, H. Ueo and S. A. Bruce, 'A contact-insensitive subpopulation in Syrian hamster cell cultures with a greater susceptibility to chemically induced neoplastic transformation.' *Proc. Nat. Acad. Sci. USA*, 1985, **82**, 5005–5009.

75. H. Land, L. F. Parada and R. A. Weinberg, 'Tumorigenic conversion of primary embryo fibroblasts requires at least two cooperating oncogenes.' *Nature*, 1983, **304**, 596–602.

76. W. Sun, K.-S. Kang, I. Morita *et al.*, 'High susceptibility of a human breast epithelial cell type with stem cell characteristics to telomerase activation and immortalization.' *Cancer Res.*, 1999, **59**, 6118–6123.

77. T. J. King, L. Fukushima, T. A. Donlon *et al.*, 'Correlation between growth control, neoplastic potential and endogenous connexin 43 expression in HeLa cell lines: implications for tumor progression.' *Carcinogenesis*, 2000, **21**, 311–315.

78. J. E. Trosko and C. C. Chang, 'Role of stem cells and gap junctional intercellular communication in human carcinogenesis.' *Radiat. Res.*, **155**, 175–180.

79. A. W. de Feijter, J. S. Ray, C. M. Weghorst *et al.*, 'Infection of rat liver epithelial cells with V-Ha-ras: Correlation between oncogene expression, gap junctional communication and tumorigenicity.' *Molec. Carcinogen.*, 1990, **3**, 54–67.

80. H. L. de Feijter-Rupp, T. Hayashi, G. H. Kalimi *et al.*, 'Restored gap junctional communication in non-tumorigenic HeLa-normal human fibroblast hybrids.' *Carcinogenesis*, 1998, **19**, 747–754.

81. R. J. Ruch, B. V. Madhukar, J. E. Trosko *et al.*, 'Reversal of ras-induced inhibition of gap junctional intercellular communication, transformation and tumorigenesis by lovastatin.' *Molec. Carcinogen.*, 1993, **7**, 50–59.

82. T. Ogawa, T. Hayashi, S. Kyoizumi *et al.*, 'Upregulation of gap junctional intercellular communication by hexamethylene bisacetamide in cultured human peritoneal mesothelial cells.' *Lab. Invest.*, 1999, **79**, 1511–1520.

83. H.-K. Na, M. R. Wilson, K.-S. Kang *et al.*, 'Restoration of gap junctional intercellular communication by caffeic acid phenethyl ester (CAPE) in a ras-transformed rat liver epithelial cell line.' *Cancer Lett.*, 2000, **157**, 31–38.

84. A. P. Wolffe and M. A. Matzke, 'Epigenetics: Regulation through repression.' *Science*, 1999, **286**, 481–486.

85. H. S. Rosenkranz, N. Pollack and A. R. Cunningham, 'Exploring the relationship between the inhibition of gap junctional intercellular communication and other biological phenomena.' *Carcinogenesis*, 2000, **21**, 1007–1011.

J.P. Vanden Heuvel, G.H. Perdew, W.B. Mattes and W.F. Greenlee (Eds.)
Comprehensive Toxicology, Vol. xiv

# 14.4.6
# Environmental Influences on Cell Cycle Regulation

CORNELIS J. ELFERINK

*University of Texas Medical Branch, Galveston, TX, USA*

## 14.4.6.1  INTRODUCTION

All forms of earthly life are ultimately composed of cells, all of which arise from the division of preexisting cells. Controlled division of cells is crucial to every biological process from development to reproduction. The uncontrolled division of cells, of course, can lead to cancer. Cytological studies in the early twentieth century detailed the processes of mitosis and cell division. However, explaining in molecular terms the complex set of events underlying cell division, from DNA replication to chromosome condensation to the formation of the spindle body and cytokinesis, came much later starting with the identification and characterization of DNA in 1953 by James Watson and Francis Crick.[1] The regulatory scheme that ordered these events remained a mystery until quite recently. Only in the last couple of decades with the advent of sophisticated genetic and biochemical techniques are we beginning to unravel the innermost workings of the cell, including the cell cycle.

The two most significant sets of events in a cell's life span are replication of nuclear DNA and the dual proliferative processes of mitosis (or meiosis) and cytokinesis. The former takes place during the S (synthesis) phase, and the latter is restricted to the M (mitosis) phase. Separating these are the gap phases. The G1 phase follows M phase and precedes S phase; the G2 phase follows S phase and precedes M phase. Interphase is the period comprising G1, S and G2 between mitoses. In the typical dividing mammalian cell, G1 phase lasts approximately 12 hr, S phase 6–8 hr, G2 phase 3–6 hr, and M phase about 30 min, although the exact length of each phase varies with the cell type and growth conditions. Cells may also exit the cell cycle entering a quiescent state designated G0 and reside there for an indefinite period (Figure 1).

The arbitrary division of the cell cycle into these phases begs the question as to how a cell is informed of what phase it currently resides in, and what determines whether it should make the transition into the next chronological phase. Mechanisms that ensure DNA stability are central to this question. The emerging evidence suggests that proteins known as cyclins, cyclin-dependent kinases (CDKs), and the CDK inhibitors (CKIs) may be the crucial indicators and determinants of the cell's position in the cycle. Cellular surveillance pathways that monitor successful completion of cell cycle events and the integrity of the cell are capable of delaying cell cycle progression in response to DNA damage or other events

and are refered to as *checkpoints*. Checkpoint function often involves a delay in activation or inactivation of a particular cyclin-CDK complex.

This chapter examines cell division from the standpoint of cell cycle control with an emphasis on the regulatory components involved, and explores how a very diverse range of environmental signals can influence these cell cycle regulatory components. These signals take the form of normal growth promoting factors (mitogens), as well as viral pathogens, genotoxic and nongenotoxic chemicals, and radiation.

## 14.4.6.2  MOLECULAR BIOLOGY OF THE CELL CYCLE

### 14.4.6.2.1  The Cell Cycle

During a typical mammalian cell cycle the start of G1 phase reveals low cyclin levels, and though CDKs are present, no cyclin-CDK complexes can form. The D-cyclin level rises as G1 progresses and they form complexes preferentially with CDK4 and CDK6 (Figure 2). Dephosphorylation by CDC25A and phosphorylation by cyclin H-CDK7 of specific residues activates the kinases, which begin phosphorylating the Retinoblastoma tumor suppressor protein (pRb). This initiates a process that leads to pRb dissociating from E2F. Freed from pRb's restraint, E2F induces the production of a number of proteins that prepare the cell for DNA synthesis. When overexpressed, E2F can drive quiescent cells into S phase,[2–4] highlighting its central role in regulating the G1-to-S phase transition. Active repression by the pRb–E2F complex (for review see Reference 5), as opposed to a block in E2F-mediated transcriptional activation, is required to enforce G1 arrest by certain antiproliferative signals. In cells treated with TGF-β, for example, introduction of an E2F mutant containing only the DNA-binding domain, which is unable to bind pRb, blocks promoter accessibility and repression by the pRb–E2F complex and triggers S phase entry, even in cells that retain hypophosphorylated pRb.[6] However, such cells cannot complete the cell cycle unless they are also supplied with additional cyclin E-CDK2, implying that at physiologic levels, E2F requires CDK2 activity to stimulate cell proliferation. Among the E2F-regulated genes are cyclins E and A, which are both required to catalyze the G1/S transition in normal cells.[7–13] The ability of E2F to induce cyclin E, which in turn regulates CDK2 to enforce pRb phosphorylation, creates a

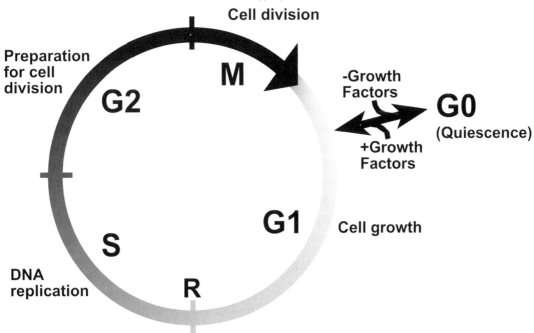

**Figure 1** Diagram depicting the cell cycle of a typical mammalian cell. DNA replication and cell division, occurring in S phase and M phase, respectively, constitute the most significant sets of events in a cell's lifespan. Separating these are the gap phases, G1 and G2. G1 represents a period of cell growth, and G2 is distinguished as an interval in which the cell prepares for cell division. The G0 phase is a quiescent, non-growing state, and may be viewed as being outside the cell cycle. Entry into and exit from G0 occurs at a point during G1 in response to the presence or absence of growth factors. The restriction (R) point denotes the G1-to-S phase transition, and is defined as the point in the cell cycle beyond which the cell is committed to completing cell division and no longer depends on growth factors.

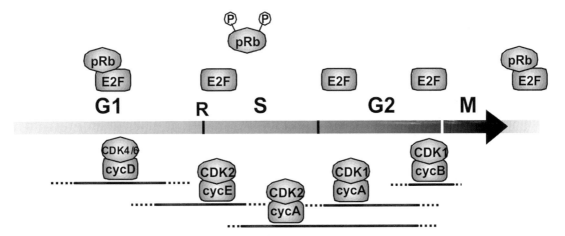

**Figure 2** Schematic illustration of the sequential periodic activation of distinct cyclin-CDK complexes as a function of cell cycle progression. Timed activation of the cyclin-CDK complexes results from the periodic expression of the cyclins, and the actions of CKIs, kinases and phosphatases (not depicted in this figure). The primary CDK target is pRb. In early G1 phase, pRb is in the hypophosphorylated state and functions to silence E2F-mediated gene expression. During mid-to-late G1 phase, cyclin D- and cyclin E-CDK complexes phosphorylate pRb at numerous serine and threonine residues (the two shown are meant to represent pRb hyperphosphorylation and do not reflect the actual number). Hyperphosphorylation results in pRb release from E2F and the induction of genes required for S phase progression and DNA synthesis. During S and G2 phases, the hyperphosphorylation of pRb is maintained by the cyclin A-CDK complexes. Only in M phase, with the degradation of cyclin A, is pRb restored to the hypophosphorylated form and able to repress E2F activity once more.

positive feedback loop that helps contribute to the irreversibility of the G1/S transition.

During S phase the cyclin A level peaks and cyclin A-CDK2 complexes appear (Figure 2). These kinase complexes maintain pRb hyperphosphorylation—as cyclins D and E are by now degraded[14,15]—to sustain DNA replication.[16,17] As S phase gives way to G2 phase, cyclin A-CDK1 complexes also assemble while cyclin A-CDK2 remains active, each phosphorylating substrates in preparation for mitosis.[8,18] Induction of mitosis however, depends on cyclin B, and cyclin B-CDK1 complex formation,[19,20] since cyclin A is degraded by late G2/M.[21,22] Cyclin B-CDK1 activity increases only when cyclin H-CDK7 phosphorylates the CDK1 T-loop threonine and CDC25C phosphatase activity exceeds Wee1 kinase activity at the CDK1 threonine residues 14 and 15. A positive feedback loop between CDC25C and cyclin B-CDK1 ensures activation of the entire pool of cyclin B-CDK1, thus triggering M phase by phosphorylating a variety of targets important in the mitotic process.

Most cells contain a spindle checkpoint that arrests cells in mitosis until all chromosomes are attached properly to the spindle. The critical transition from metaphase to anaphase and the separation of sister chromatids is monitored by the gene products Mad (mitotic arrest defective) proteins, Mad1–3p, the Bub (budding uninhibited by benomyl) proteins, Bub1–3p, and Mps1.[23] Progress through this checkpoint witnesses the destruction of cyclins A and B via ubiquitinated-proteolysis by APC (anaphase promoting complex). As the levels of these cyclins are depleted, CDK1 activity is lost and the impetus for cell division is removed. By this time the original cell has divided into two progeny cells, and the cell cycle begins anew so long as growth factors are present. At any time during this process the cell can be brought to a halt if CKIs conspire to block the ability of the active cyclin-CDK complex to phosphorylate its substrates.

### 14.4.6.2.2 Cyclins and Cyclin-Dependent Kinases

When quiescent (G0) cells enter the cell cycle in response to mitogenic stimulation, genes encoding D-type cyclins (D1, D2, and D3) are induced. The D-type cyclins are very labile, with a half-life of less than 30 min, and expression levels decline rapidly upon withdrawal of mitogens. The D-type cyclins are the regulatory subunits that assemble with their catalytic partners, CDK4 and CDK6, as cells progress through G1 phase.[24,25] Assembled cyclin D-CDK complexes enter the nucleus where they are phosphorylated by a CDK-activating kinase (CAK, composed of cyclin H and CDK7) before they can phosphorylate protein substrates. Indeed, the major function of the D cyclin pathway is to provide a link between mitogenic cues and the potentially autonomous cell cycle machinery which, in vertebrate cells, is composed primarily of CDK2 and cdc2/CDK1 and their associated regulators. Constitutive activation of the D cyclin pathway can reduce or overcome certain mitogen requirements for cell proliferation and thereby contribute to oncogenic transformation.[25,26] In contrast, insertion of antisense oligonucleotides to the cyclin D1 gene or antibodies to the cyclin D1 protein into fibroblasts during mid-G1 phase blocks progress into S phase, although this effects is not observed in cells close to the G1/S boundary.[27,28] This evidence attests to the crucial role for the D-type cyclins in the mid-late G1 phase. This interval also harbors the restriction (R) point, arguably the most important checkpoint in the cell cycle, since progress beyond the R point commits the cell to a round of DNA replication and the subsequent mitosis. The most recognized function of cyclin D-dependent kinases is phosphorylation of pRb (Figure 3). The generally accepted view is that cyclin D-dependent kinases initiate pRb phosphorylation in mid-G1 phase after which cyclin E-CDK2 becomes active in late G1 phase, peaking at the G1/S phase transition. Cyclin E-CDK2 completes the process of hyperphosphorylating pRb.[29–35] Because cyclin E-CDK2 can induce S phase without inactivating pRb, this implies that the kinase must have an additional role, distinct from pRb phosphorylation and E2F activation.[36,37]

### 14.4.6.2.3 The CDK Inhibitors

Mitogen-dependent progression through G1 and initiation of DNA synthesis in S phase during the mammalian cell cycle are cooperatively regulated by the CDKs whose activities are in turn constrained by CDK inhibitors (CKIs). CKIs that govern these events have been assigned to one of two families based on their structures and CDK targets. The first class includes the INK4 proteins (inhibitors of CDK4), so named for their ability to specifically inhibit the catalytic subunits of CDK4 and CDK6. Four such proteins (p16 INK4a,[38] p15 INK4b,[39] p18 INK4c,[40,41] and p19 INK4d)[41,42] bind only to CDK4 and CDK6

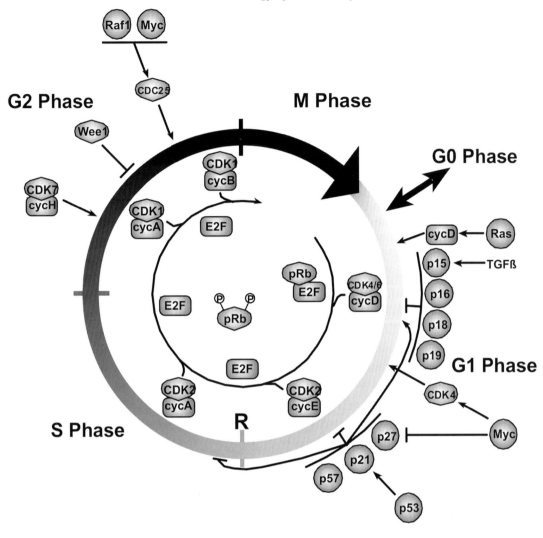

**Figure 3** Diagram depicting a generic cell cycle and its regulation by the various control proteins. Arrows denote an activation or upregulation of the target protein, and bars indicate an inhibition or downregulation of the target protein. For the purpose of clarity, many details have been omitted; for example, while the regulatory actions of cyclin H-CDK7, Wee1 and CDC25 are shown to impact G2 phase exclusively, they are believed to act on the G1 cyclin-CDK complexes as well. The proteins p15, p16, p18, p19, and p21, p27, p57, comprise two groups of CKIs with distinct properties (discussed elsewhere in the chapter).

but not to other CDKs or to D-type cyclins. The INK4 proteins can be contrasted with more broadly acting inhibitors of the Cip/Kip family whose actions affect the activities of cyclin D-, E-, and A-dependent kinases. The latter class includes p21Cip1,[43–48] p27Kip1,[49–51] and p57Kip2.[52,53] They all contain characteristic motifs within their amino-terminal region that enable them to bind both to cyclin and CDK subunits.[54–59] Based largely on *in vitro* experiments and *in vivo* over-expression studies, the Cip/Kip CKIs were initially thought to interfere with the activities of cyclin D-, E-, and A-dependent kinases. More recent work has revealed that although the Cip/Kip proteins are

potent inhibitors of cyclin E- and A-dependent CDK2, they act as positive regulators of cyclin D-dependent kinases[60,61] (see Figure 3 and Section 2.5).

#### 14.4.6.2.4 Regulation of the G1/S Transition

The cyclin D- and E-dependent kinases contribute sequentially to pRb phosphorylation, negating its ability to repress the E2F family of transcription factors that activate numerous genes required for entry into S phase. Amongst these are the genes for cyclin E and A, as well as a bank of gene products that regulate nucleotide metabolism and DNA synthesis.[5] In addition, a noncatalytic function of cyclin D-CDK4 complexes is sequestration of CKIs, including p27Kip1 and p21Cip1.[62] Most p27Kip1 in proliferating cells is complexed to cyclin D-dependent kinases.[50,51] In quiescent cells, the levels of p27Kip1 are relatively high, whereas p21Cip1 levels are low but usually increase in response to mitogenic signals during G1 phase progression. Titration of unbound p27Kip1 and p21Cip1 molecules into higher order complexes with cyclin D-dependent kinases relieves cyclin E-CDK2 from Cip/Kip constraint, thereby facilitating cyclin E-CDK2 activation later during G1 and S phase. The levels of untethered Cip/Kip proteins may set an inhibitory threshold for activation of cyclin E– and A-CDK2 synthesized later in G1 phase. Once the process of Cip/Kip sequestration lowers the effective CKI level below the threshold, cyclin E-CDK2 can facilitate its own activation by phosphorylating p27Kip1 on a specific threonine residue (Thr-187) to trigger its degradation.[63–65] It is worth noting that the two functions of cyclin D-CDK complexes—pRb phosphorylation and CKI sequestration—are compatible, because the Cip/Kip proteins do not inhibit the pRb kinase activity of cyclin D-CDK complexes. Intriguingly, all of the cyclin D-CDK pRb kinase activity in proliferating cells is found in complexes containing Cip/Kip proteins.[60,61,66,67] Although p27Kip1 can inhibit recombinant cyclin D-CDK complexes *in vitro*, it is more effective in antagonizing the activity of cyclin E-CDK2.[50,51] Direct attempts to measure the specific activity of cyclin D-CDK4–p27/p21 complexes *in vitro* showed no significant inhibition at 1:1:1 stoichiometry.[60,67,68] This does not preclude the possibility that inactive complexes containing higher ratios of p21 or p27 to cyclin D-CDK might form under certain physiologic circumstances, either blocking their activation by CAK[69,70] or inhibiting them directly. Nonetheless, this contrasts with the properties of complexes containing CDK2, which are inhibited *in vitro* by equimolar concentrations of p21.[71]

### 14.4.6.2.5  The Cip/Kip CKIs are Positive Regulators of Cyclin D-CDK Function

Not only are cyclin D-CDK complexes resistant to Cip/Kip inhibition, but also their activation is actually facilitated by their interactions with these CKIs. This was first demonstrated by LaBaer and coworkers,[60] who, in recognizing that Cip/Kip proteins bind both to cyclin and CDK subunits, directly tested the possibility that the CKIs actively promoted cyclin D-CDK assembly. Their studies showed that both p21 and p27 promoted interactions between D-type cyclins and their CDK partners *in vitro*, primarily by stabilizing the complexes. *In vivo*, p21Cip1 stimulated the assembly of enzymatically active CDK4 by entering into higher order complexes with cyclin D. All three Cip/Kip family members directed the accumulation of cyclin D-CDK complexes in the cell nucleus. Cheng *et al.*[61] subsequently observed that assembly of cyclin D1/D2-CDK4 complexes was impaired in primary mouse embryo fibroblast (MEF) strains taken from animals lacking the p21 gene, the p27 gene, or both. The level of cyclin D1 and of assembled complexes was reduced >10-fold in MEFs from p21/p27 double-null mice, and immunoprecipitable cyclin D-CDK4-dependent pRb kinase activity was no longer detected. Forced expression of cyclin D1 did not restore complex formation in the double-null MEFs, however, reintroduction of p21 or p27 into the cells restored cyclin D-CDK4 complex formation. The Cip/Kip proteins also promote activation of cyclin D-CDK complexes by directing the heteromeric complex to the cell nucleus, and by increasing the stability of the D cyclins.[60,61]

### 14.4.6.2.6  CDK Inhibition by the INK4 CKIs

The idea that Cip/Kip proteins act as positive regulators of cyclin D-CDK complexes has important implications for how we interpret the effects of other molecules that govern G1 progression, the INK4 family of CKIs being a notable example. The signals that lead to the synthesis of INK4 proteins remain poorly understood, but it is clear that p15 INK4b is induced by TGF-β and contributes to its ability to induce G1-phase arrest.[39,72,73] p16 INK4a accumulates progressively as cells age, possibly being induced by a senescence timer,[74–78] whereas p18 INK4c and p19 INK4d are focally expressed during fetal development and may have roles in terminal differentiation.[78–81] Overexpressed INK4 proteins arrest the cell cycle in G1 phase in a manner that depends on the integrity of pRb.[40,82–84] The simplest interpretation is that pRb phosphorylation is directly antagonized by INK4-mediated inhibition of cyclin D-dependent kinases, thereby maintaining pRb in a growth-suppressive, hypophosphorylated state. One view is that INK4 proteins compete with Cip/Kip proteins (together with D cyclin) for CDK4,[85] consistent with the idea that CDK4 partitions between INK4 and Cip/Kip-bound states.[85,86] Increased amounts of INK4 proteins would alter the distribution of CDK4 in favor of INK4-CDK4 complexes and thereby reverse the normal sequestration of Cip/Kip proteins that occurs during G1 phase. The release of the Cip/Kip proteins can lead to cyclin E-CDK2 inhibition and to G1-phase arrest.[73] Conditional induction of p16 and displacement of cyclin D1 from CDK4 and CDK6 also shunts p21 into complexes with CDK2.[86,87]

That INK4-mediated growth arrest depends on mobilization of Cip/Kip proteins from complexes containing cyclin D-CDK4 to cyclin E-CDK2 at first seems to be at odds with previous observations that the ability of INK4 proteins to induce arrest is lost in cells lacking pRb function. However, loss of pRb also leads to increases in both the amount and total activity of cyclin E-CDK2,[88,89] which may be sufficient to overcome p27 both by titration and by increased p27 destruction.[64,65] The dependency of pRb-positive cells on D cyclins might similarly reflect the ability of cyclin D-CDK complexes to sequester Cip/Kip proteins. Therefore, in normal cells in which pRb function is intact, both pRb phosphorylation and Cip/Kip sequestration by cyclin D-dependent kinases contribute to cyclin E-CDK2 activation and S phase entry, whereas in cells lacking pRb, both of the G1 cyclin-dependent kinases (i.e., CDK4/6 and CDK2) are deregulated.

The introduction of catalytically inactive versions of CDK2 that assemble with cyclins E and A, results in G1 arrest. However, overexpression of analogous forms of CDK4 that assemble into inactive complexes with D cyclins do not halt the cell cycle.[90] This apparent paradox might potentially be explained by the realization that catalytically inactive CDK4 is as effective as the wild-type protein in sequestering Cip/Kip members. Regardless of whether the ectopic overexpression of either wild-type or catalytically inactive CDK4 is enforced, cyclin E-CDK2 becomes active and may be sufficient to catalyze pRb phosphorylation on its own.[91] The clear implication is that under circumstances where cyclin D-CDK4 complexes are overexpressed, their ability to sequester Cip/Kip proteins is predominant. When the inhibitory actions of Cip/Kip family members on cyclin E- and A-CDK2 are completely neutralized, cyclin D-dependent kinases may no longer be required for cell cycle progression. MEFs lacking p27 and p21 did not exhibit aberrant cell cycles despite the fact that overall cyclin D-dependent kinase activity was reduced below the assay

limit of detectability.[61] In turn, cyclin E-CDK2 activity was significantly elevated, and pRb was phosphorylated during the cell cycle, even on sites that are preferentially recognized by cyclin D-dependent kinases. As expected, such cells were also highly resistant to arrest by p16 but not p27. Recent observations that replacement of cyclin D1 with cyclin E can completely rescue the developmental defects observed in D1-deficient animals are compatible with the idea that cyclin E is a downstream target of cyclin D1, even in cell lineages that normally rely strongly on D1 function.[92]

### 14.4.6.2.7  A Role for the Cip/KipCKIs in Cancer

Although disruption of the pRb pathway—loss of pRb or p16 function, or overexpression of cyclin D or CDK4—appears to be a requisite event in human cancer, complete loss of Cip/Kip function has not been observed. Similarly, cyclin E amplification is rare, and gain-of-function mutations involving cyclin E have not been found.[25] This may be a consequence of the Cip/Kip proteins acting both positive and negative to regulators of the cyclin D- and E-dependent kinases, respectively. Low expression of p27 protein occurs frequently in many types of human tumors. This reduction correlates strongly with tumor aggression and connotes a poor prognosis,[93] as first demonstrated for breast and colon carcinoma and more recently for other tumor types.[94–96] Low levels of p27 are likely to be causally related to tumorigenesis, as evidenced by studies performed in p27 heterozygous mice that are haploinsufficient for tumor suppression.[97] These animals develop tumors at greatly increased frequency after exposure to chemical carcinogens or X-rays, even though the arising tumor cells retain the functional p27 allele.

### 14.4.6.3  MYC AND RAS: MITOGEN-RESPONSIVE SIGNALING

### 14.4.6.3.1  Myc

A classical paradigm for oncogene cooperation is the transformation of primary rodent fibroblasts by myc and ras oncogenes. Studies suggest that cooperation between ras and myc may be largely due to their complementary and synergistic action on CDK function, and offers an example of how extracellular signals (mitogenic growth factors) impinge upon the cell cycle.[98] The myc family of proto-oncogenes includes three evolutionarily conserved genes c-,

N- and L-myc, which encode related proteins. c-Myc expression is strictly dependent on mitogenic signals, is suppressed by growth-inhibitory signals and inducers of differentiation, and is important for proliferation and apoptosis in response to the appropriate stimuli.[99] Constitutive expression of Myc reduces growth-factor requirement, prevents growth arrest by a variety of growth-inhibitory signals, and can block cellular differentiation.[100,101] Myc up-regulates and prevents inhibition of cyclin E/CDK2 activity through multiple pathways (Figure 3). First, by inactivating the CKI p27Kip1 involving the induction of a p27-sequestering protein(s), probably CDK4, which may also target p21Cip1 and p57Kip2. Second, by inducing both cyclin E and CDC25A transcription. Third, by facilitating p27Kip1 degradation following CDK2-mediated phosphorylation of threonine-187 in the CKI.[63,102–105]

### 14.4.6.3.2 Ras

Mitogen stimulated activation of the Ras protein, or its effector Raf, induces cell cycle progression. Ras transduces mitogenic stimuli in response to tyrosine-kinase receptors, and its function is required in G1 for passage through the R-point.[106,107] Ras activity is required for the phosphorylation of pRb in response to mitogenic signaling, and functional inactivation of Ras induces G1 arrest in pRb-positive, but not in pRb-negative cells.[108] The mitogenic signal mediated by Ras is propogated through Raf, mitogen-activated protein kinase kinase 1 (MEK1), and the mitogen-activated protein kinases (p42/p44MAPK), to positively regulate cyclin D1 expression and/or the degradation of p27.[107,109,110] Assembly of newly synthesized cyclin D1 with CDK4 depends on the same kinase cascade,[107,111] but the substrate(s) of MAPK phosphorylation that helps mediate this process is as yet unknown. Ectopically expressed cyclin subunits do not efficiently associate with CDKs in the absence of mitogenic signals,[29] demonstrating that the ERKs act post-translationally to regulate cyclin D-CDK assembly. Turnover of D-type cyclins depends on a separate Ras signaling pathway involving phosphatidylinositol 3-kinase (PI3K) and Akt (protein kinase B), which negatively regulates the phosphorylation of cyclin D1 on a single threonine reside (Thr-286) by glycogen synthase kinase-3β (GSK-3β).[112] Inhibition of this signaling pathway leads to cyclin D1 phosphorylation, enhances its nuclear export, and accelerates its ubiquitin-dependent proteasomal degradation in the cytoplasm, shortening the half-life of cyclin D1 to as little as 10 min. Hence, D-type cyclins act as growth factor sensors, with cyclin D transcription, assembly, nuclear transport, and turnover each being mitogen-dependent steps.

### 14.4.6.3.3 Synergism Between Myc and Ras Signaling Pathways

The mechanisms underlying the effects of Ras on p27 remain to be dissected. It is unclear, for instance, whether Ras acts in concert with CDK2-induced degradation of p27[64] or through a different pathway. In any case, the action of Ras on p27 may add to that of Myc,[63,103] both contributing to the derepression of cyclin E/CDK2. The activities of Myc and Ras/Raf may also complement at the level of CDC25A, which might be activated directly by the Raf kinase.[102] The expected result is a synergistic enhancement of cyclin E/CDK2 activity by Myc and Ras, as was shown using adenovirus-mediated expression of the two proteins.[98] Given that the CDK4 gene is a transcriptional target for c-Myc,[107] and Ras facilitates expression of cyclin D1, cooperativity between Myc and Ras also exists at the level of cyclin D-CDK4 by promoting pRb phosphorylation and sequestration of the Cip/Kip CKIs.[60]

### 14.4.6.4 RETINOBLASTOMA PROTEIN

#### 14.4.6.4.1 pRb is a Major Target for CDK Activity

The retinoblastoma tumor suppressor protein (pRb) is a 105 kDa nuclear phosphoprotein encoded by the *RB1* gene. Although the pRb protein level is quite stable during phases of the cell cycle, it is subject to regulation by phosphorylation in a cell cycle-dependent manner. In G0 and early G1 pRb is in a hypophosphorylated form. As the cell traverses the restriction (R) point in late G1 phase and enters S phase, the protein acquires phosphate groups on multiple serine and threonine residues. Phosphatase activity concomitant with reemergence into G1 after mitosis removes the phosphates from hyperphosphorylated pRb. Multiple cyclin-CDK complexes are responsible for pRb phosphorylation as the cell cycle advances. CDK1, CDK2, CDK4 and CDK6 in conjunction with their respective cyclins all phosporylate pRb, the exception being cyclin B-CDK1 which governs mitotic events. Clearly pRb can be regarded as a crucial cell cycle protein for it to be a substrate for so many cyclin-CDKs over a protracted period of the cell cycle. By learning the role pRb plays in cell cycle regulation, we begin to see the purpose behind various kinase complexes keeping pRb hyperphosphorylated.

### 14.4.6.4.2 pRb as a Regulator of Cell Growth

The introduction of the wild-type *RB1* gene into certain *RB1*-deficient tumor cells induces growth arrest. These observations suggest that pRb is a general suppressor of cell proliferation. It is noteworthy that certain viral oncoproteins such as E1A, E7 and SV40 physically associate with pRb; moreover they form complexes almost exclusively with the hypophosphorylated form of pRb using a common motif LXCXE (X designates a variable amino acid) in the viral proteins. The ability of these viral oncoproteins to transform and immortalize cells depends on the LXCXE motif. Studies examining the binding properties of LXCXE motif proteins to pRb identified two proteins closely related to pRb, p107 and p130. Subsequent studies revealed that LXCXE motif protein binding to pRb (and the related proteins) requires two nonadjacent regions in pRb (amino acids 394–571 and 649–772) that form the "pocket" domain. An intact pocket domain is essential for pRb to inhibit proliferation.

pRb has been observed to bind to a large number of molecules including both stimulatory and repressive transcription factors, kinases and phosphatases. Recent estimates are around 50 different proteins,[113] although this number is likely to grow. Most of these proteins bind to the pocket domain. Among the cellular partners for pRb, its regulation of E2F offers the clearest picture of how pRb controls the cell cycle. E2F was initially identified as a mammalian DNA-binding protein required to activate the adenovirus E2 promoter.[114] E2F is a family of transcription factors (E2F1–6) that bind DNA as heterodimers with the DP transcription factors (DP1–3). It is widely anticipated that the different E2F heterodimers will be found to regulate different subsets of E2F target genes. E2F-1, E2F-2 and E2F-3 bind almost exclusively to pRb, E2F-5 associates with p130 and E2F-4 binds primarily with p107 and p130, but will also bind pRb.[5] In general p130/E2F complexes are found primarily in quiescent (G0) or differentiated cells and p107/E2F being most prevalent in S phase cells. pRb/E2F complexes are detect in quiescent and differentiated cells, but are most evident in cycling cells as they transition through G1 and into S phase. E2F transcriptional activity is required to express genes important for entry into S phase and DNA replication. A primary role for pRb is to restrict E2F transcriptional activity in G1 phase cells. Early models of E2F action proposed that pRb binding to the activation domain located in the C-terminus of E2F prevented E2F-mediated gene expression by masking the activation domain. More recent observations have shown that pRb/E2F complexes can also actively repress transcription.[115,116] Evidence is emerging that pRb, in response to certain environmental cues, may effect cell cycle control by means distinct from the direct inhibition of E2F (see Section 14.4.6.6).

### 14.4.6.5 P53 AND ATM: SURVEYORS OF GENOMIC STABILITY

#### 14.4.6.5.1 p53

The ability of cells to maintain genomic integrity is vital for cell survival and proliferation. Lack of fidelity in DNA replication and maintenance can result in deleterious mutations leading to cell death or, in organisms, cancer. Exposure to several common sources of genotoxic stress; including oxidative stress, ionizing radiation, UV radiation and the genotoxic compound benzo[a]pyrene (BP), elicit cell cycle checkpoint responses. Like pRb, the p53 protein exhibits the properties of both a tumor suppressor and a growth suppressor. Wild-type p53 can prevent the transformation of healthy cells by oncogenes and inhibit growth in cells derived from tumors.[117,118] Although viable, mice that are nullizygous for wild-type p53 are extremely susceptible to the development of malignant tumors,[119] and the p53 gene is mutated in a large number of human cancers. This attests to p53 activity as a critical component in the cell's DNA-damage response pathway causing cells to arrest in both G1 and G2 allowing the cell an opportunity for DNA repair before cell cycle progression resumes. Exposure to DNA damaging agents triggers a pronounced increase in the level of p53, and its role in response to DNA damage is underscored by the observation that p53-null cells fail to arrest following irradiation.[120] p53 triggers G1 cell cycle arrest indirectly by acting as a transcription factor inducing the p21Cip1 gene.[45] In turn, p21Cip1 functions as an inhibitor of the G1 cyclin-CDK complexes, particularly the cyclin E-CDK2 complex, thus preventing pRb hyperphosphorylation and entry into S phase (Figure 3).

Recent studies examining the DNA damage-induced G2 checkpoint have identified crosstalk between p53 and the Chk1/Chk2 proteins[121,122] (Figure 4). Both Chk1 and Chk2 are kinases functionally conserved throughout evolution. Upon DNA damage, Chk1 and Chk2 were shown to phosphorylate several serine residues on p53 including serine-20 which is involved in p53 stabilization. Flatt *et al.*[123] demonstrated that p53 prevents exit from G2 after genotoxic stress involving a mechanism dependent upon p21Cip1 and pRb that facilitated an initial inhibition of cyclin B-CDK1 activity, followed by down-regulation of both cyclin B and CDK1 expression. The reliance on pRb suggests that cyclin B and CDK1 expression may be regulated by E2F. Interestingly, p53-mediated induction of p21Cip1 down-regulated Chk1 gene expression by a mechanism that also requires pRb. Chk1 (and Chk2) suppresses cyclin B-CDK1 activity indirectly, by phosphorylating serine-216 on CDC25C, which in turn prevents CDC25C from dephosphorylating threonine-14 and -15 on CDK1 necessary to activate the kinase. The p53-dependent down-regulation of Chk1 is perceived as a means to resume passage through G2 once DNA repair is complete.

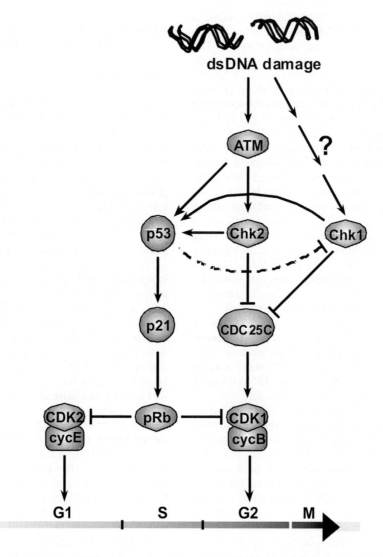

**Figure 4** Pathways leading from DNA damage to cell cycle checkpoint control in G1 and G2 phases. Arrows denote an activation or upregulation of the target protein, and bars indicate an inhibition or downregulation of the target protein. Depicted is the p53-mediated p21Cip1- and pRb-dependent inhibition of G1 and G2 in response to double-stranded—and single-stranded—DNA damage. Also shown is the ATM-mediated response to double-stranded DNA damage involving both p53 and Chk2, and the crosstalk between p53 and Chk1 and 2. The dashed line denotes the pRb-dependent downregulation of Chk1.

The evidence suggests that a common mechanism operates in p53-mediated G1 and G2 cell cycle arrest, which is geared towards suppressing CDK activity and reversing pRb hyperphosphorylation.

In addition to cell cycle arrest, p53 can also induce apoptosis in numerous cell lines. This gave rise to the notion that p53-dependent cell cycle arrest occurs following reparable DNA damage, but if damage is too severe, p53 induces an apoptotic response thus ridding the organism of potentially harmful cells. This has earned p53 the mantle of "guardian of the genome".[124] p53-dependent apoptosis appears to be a response independent of the cell cycle regulators and falls outside the scope of this chapter.

### 14.4.6.5.2   ATM

Several human heritable cancer-prone syndromes known to alter DNA stability have been found to have defects in checkpoint surveillance pathways. *Ataxia telangiectasia* (AT) is an autosomal disease caused by recessive mutations in both copies of the Ataxia telangiectasia mutated (ATM) gene. The disease is associated with a high risk for cancer (particularly leukemias and lymphomas), immune system abnormalities, radiation sensitivity and genetic instability.[125] Studies in AT cells suggest that the ATM protein plays a role in G1 and G2 phase DNA-damage response pathways in conjunction with the p53 protein.[126]

ATM-dependent activity in cells occurs in response to double-strand breaks that result from ionizing radiation (but not UV radiation), endogenous oxidative DNA damage and environmental mutagens, as well as breaks created during meiosis and immunologically important gene rearrangements. Thus one might view ATM, like p53, as a guardian of the genome. Recent studies have provided evidence that ATM operates via the phosphorylation of signaling molecules that link the detection of DNA damage to the maintenance of genomic stability by modulation of the cell cycle progression (Figure 4). ATM acts very early in this signal transduction pathway because the ATM kinase is required for the immediate response to damage.[127,128] Loss of functional ATM protein decreases the fidelity with which DNA lesions are monitored and results in the accumulation of mutations. These mutations might be repaired if a second level of monitoring, such as the p53 pathway, is activated to induce cell cycle arrest.

ATM is involved in the regulation of multiple cell cycle checkpoints (G1/S, S, and G2/M) after DNA damage. Kastan *et al.*[129] demonstrated the importance of p53 in activating the G1/S checkpoint after ionizing irradiation. Subsequently, evidence was obtained demonstrating a direct interaction between ATM and p53.[127] ATM is activated in response to DNA damage, and in turn, activates p53 at least in part by phosphorylating p53 on serine-15.[130,131] Phosphorylation at this site may contribute to the increased half-life of p53 by facilitating its dissociation from MDM2, a protein that promotes p53 proteolysis. As a consequence, p53 is able to induce p21Cip1 gene expression to trigger G1 cell cycle arrest by inhibiting G1 CDK-dependent pRb hyperphosphorylation.

In addition to the G1/S checkpoint control, ionizing radiation also induces S phase and G2/M phase checkpoints in an ATM-dependent mechanism. ATM phosphorylates the Chk2, and possibly Chk1 proteins in response to radiation-induced DNA damage.[132] Both Chk1 and 2 phosphorylate CDC25C on serine-216, leading to inhibition of the CDC25C's ability to dephosphorylate and thereby activate cyclin B-CDK1 complexes. The decrease in cyclin B-CDK1 activity prevents the G2 transition into mitosis. Studies in both fission and budding yeast with the ATM homologs (Mec1/Rad3) and Chk2 homologs (Rad53/Cds1) suggest that Chk2 may also function to inhibit transition through S phase in response to DNA damage.

### 14.4.6.6   ENVIRONMENTAL STRESSORS INFLUENCING THE CELL CYCLE

#### 14.4.6.6.1   Viral Pathogens

Viruses have been key instruments in cancer research for the last 20 years. While originally viewed as unusual agents that cause cancer in animals but otherwise of little relevance to humans, studies on viral carcinogenesis have unlocked many of the mysteries of cell growth control. Tumor viruses played two roles: first, as tools for the discovery and dissection of cell signaling and growth control pathways and; second, as newly identified causative agents of human neoplasia. The tumor viruses are broadly classified as RNA or DNA tumor viruses. The retroviral RNA tumor viruses share the unusual characteristic that soon after infection, the viral RNA genome is reverse transcribed into a double-stranded DNA copy which is then integrated into the host cell genome. Oncogenesis by transforming retro-viruses probably involves proviral integration upstream of a proto-oncogene yielding a chimeric virus-cell transcript. Recombination during a subsequent round of replication may then result in incorporation of the cellular gene into the viral genome.[133] Retroviruses can, but generally do not contain a viral oncogene derived from a previously infected host cell, instead they activate cellular proto-oncogenes upon proviral integration by a mechanism termed "proviral insertional mutagenesis". Proviral insertion introduces a strong promoter and enhancer sequences into the gene locus, resulting in modified gene expression.[134] These retroviruses lack oncogenes, do not transform cells in culture and induce tumors with long latent periods *in vivo*. Since these viruses are replication-competent, with time they can infect numerous cells in the host target organ increasing the likelihood of provirus insertion near a cellular oncogene such as c-Myc, H-Ras, K-Ras and D-type cyclins.[135] Altered expression of such a growth regulatory gene may bestow upon the cell a selective growth advantage ensuring its survival by increasing the time available to acquire additional genetic changes.

DNA tumor viruses target tumor suppressor proteins such as pRb and p53. The oncogenes of the small DNA tumor viruses (simian virus 40, papilloma virus, adenovirus) were found to be of viral, not cellular, origin and to be essential for both viral replication and cellular transformation.[136–138] This is in striking contrast to the cellular-derived oncogenes carried by retroviruses. Studies of the DNA viruses led to the discovery of cellular tumor suppressor genes, which are distinct from the oncogenes but critically important in cell cycle control and cancer development. Their meager genetic content renders the DNA tumor viruses dependent on the host cell machinery to replicate the viral DNA. However, virally encoded onco-proteins trigger resting cells (G0) to enter S phase, thus providing the enzymes and cellular environment needed for viral DNA replication. Simian virus 40 (SV40) large T-antigen is a multifunctional protein capable of usurping cell cycle control largely by binding to the cellular tumor suppressor proteins p53, and members of the pRb family (pRb, p107, p130) (Figure 5). Sequestration of pRb and p53 prevents these proteins from exercising their normal role in cell cycle control thus stimulating viral DNA replication, forming the basis for cellular transformation into a cancer cell. The adenovirus E1A and E1B proteins and papillomavirus E7 and E6 proteins likewise target pRb and p53, respectively.[139–143]

The E1A, E7, and large T antigen proteins all contain an LXCXE motif necessary for forming a complex with pRb (or pRb-related proteins p107 and p130) thus inhibiting pRb activity. Oncoprotein binding abrogates pRb function presumably by displacing its normal cellular partners such as E2F from the pocket domain, achieving the same effects as hyper-phosphoylation of pRb by the cyclin-CDK complexes. The tumor viruses thus bypass the regulatory pathways upstream of pRb and deceive the host cell into inappropriately proceeding through pRb-regulated check-points, most notably the G1/S restriction point. Virally induced entry into S phase in the absence of appropriate mitogenic cues however, induces a p53-mediated cell cycle arrest or apoptosis, conceivably by the mechanism involving E2F-1 and p14ARF.[144–147] p53-specific viral oncoproteins overcome this obstacle by directly binding p53 and inhibiting its function reminiscent of the inhibition of pRb, although this does not rely on a single "pocket" in p53 as is the case with pRb. Accordingly, each oncoprotein counters p53's effects in a different fashion. E1B binds to the transcriptional activation domain, thereby acting as a transcriptional repressor of p53.[148] Large T antigen binds to the DNA-binding domain, preventing p53 from binding DNA and from fulfilling either its stimulatory or repressive activities as a transcription factor.[149] Large T antigen also appears to stabilize p53.[150] E6 is similar to large T antigen in that it also binds the DNA-binding domain, but differs in that E6 targets p53 for ubiquitin-mediated degradation with a concomitant decrease in the half-life of p53.[142,151] Not only does E6 abrogate p53-mediated G1 arrest, it also prevents the host cell from mounting a p53-directed response to DNA damage. Consequently, the infected cell runs an increased risk of acquiring DNA mutations hastening the process towards carcinogenic transformation. This enhanced genomic instability may explain why papilloma virus is a more potent tumor agent than either adenovirus or SV40.

Another strategy for bypassing both INK4 and Cip/ Kip-mediated inhibition is exemplified by the Kaposi's sarcoma virus (KSHV or human herpes virus-8), which encodes a variant D-type cyclin (K cyclin) (Figure 5). This protein forms complexes with CDK6[152] that are resistant to inhibition by either p16 or p27.[153] The cyclin K-CDK6 complex phosphorylates p27 on Thr-187, triggering its degradation,[154] and in cells expressing cyclin K, a catalytically inactive form of CDK6 but not of CDK2 blocks these effects. Importantly, cyclin K-CDK6 is not stably bound by either

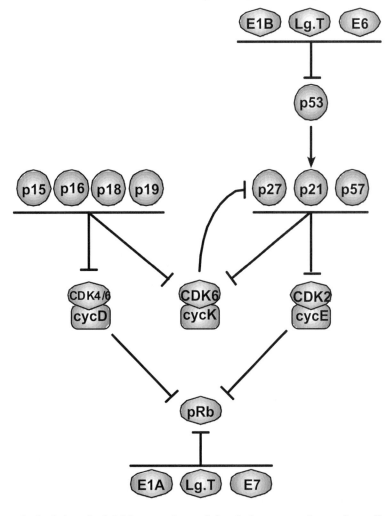

**Figure 5** Schematic depicting the inhibitory actions of the viral oncoproteins on key cell cycle regulators. Arrows denote an activation or upregulation of the target protein, and bars indicate an inhibition or downregulation of the target protein. The "X" through the bars is intended to show that the cyclin K-CDK6 complex is resistant to inhibition by the INK4 and Cip/Kip CKIs. Overall, the viral oncoproteins facilitate the cell's advance into S phase and avoid apoptosis.

p21 or p27, indicating that the holoenzyme is not antagonized by Cip/Kip inhibitors, does not require a Cip/Kip protein for assembly, and does not titrate the Cip/Kip proteins to activate cyclin E-CDK2. Instead, by inactivating p27 through direct phosphorylation, overexpression of cyclin K-CDK6 can bypass a p27-induced cell cycle block. Although cyclin K-CDK6 complexes are themselves resistant to Cip/Kip inhibition, stable phosphorylation-resistant p27 mutants containing an alanine for Thr-187 substitution prevent S-phase entry in KSHV-infected cells. Therefore, cyclin K-CDK6 fulfills only part of the role of cyclin E-CDK2, with the latter still being required for S-phase entry. In short, cyclin K-CDK6 helps promote a cellular environment in which unopposed cyclin E-CDK2 and A-CDK2 can enforce S phase entry and progression.

### 14.4.6.6.2 Radiation-induced DNA Damage

#### 14.4.6.6.2.1 *Ultraviolet Radiation*

Exposure to UV light induces a number of cellular changes, including generation of DNA lesions. As is the case for ionizing radiation (IR) and reactive oxygen species (ROS), UV radiation induces the stress proteins (p53 and p21Cip1), and the initiation of cell cycle checkpoint arrest in cycling cells.[155] UV radiation is divided into three clases based on wavelength; UV-A (400–320 nm), UV-B (320–290 nm), and UV-C (290–100 nm). UV-A and UV-B are the more biologically relevant, since UV-C is largely absorbed by ozone in the upper atmosphere. The primary UV-induced DNA lesion is the cross-linking of adjacent pyrimidines refered to as a pyrimidine dimer.

Formation of the dimer is a UV-reversible process, but the equilibrium strongly favors dimer formation.[156] Thymine–thymine dimers are the most common pyrimidine dimers formed upon UV exposure, with cytosine–cytosine and cytosine–thymine dimers also occurring. Less common is the formation of thymine glycols and protein-DNA crosslinking. UV radiation also generates DNA damage indirectly due to production of reactive oxygen species (ROS), including superoxide ($O_2^{.-}$), the hydroxyl radical ($.OH$), and hydrogen peroxide ($H_2O_2$), all of which rapidly react with each other and surrounding biomolecules.

### 14.4.6.6.2.2  Ionizing Radiation

IR damages all components of the cell and is known to produce more than 100 distinct DNA adducts.[157] IR damages DNA through direct and indirect mechanisms. Direct damage to DNA occurs as a result of the interaction of radiation energy with DNA. This can result in the generation of a variety of lesions, including the generation of abasicdeoxyribose sites in DNA that are produced as a consequence of destabilization of the *N*-glycosidic bond, and generation of both signle stranded (ss) DNA and double stranded (ds) DNA breaks. Indirect DNA damage comes from the interaction of DNA with reactive species formed by IR.[158] Water is the predominant cellular constituent and more than 80% of the energy in IR deposited in the cell results in the ejection of electrons from water. Subsequent reactions can result in the formation of ROS.

### 14.4.6.6.2.3  Reactive Oxygen Species

Even though oxygen is an absolute requirement for aerobic life, it can damage biological molecules, including DNA. Normal cellular metabolism, as well as the metabolism of a variety of xenobiotics, produces an array of ROS that are highly reactive and can readily damage DNA. Under conditions of oxidative stress, cycling cells will exhibit cell cycle checkpoint responses.[159] Ames and Shigenaga[160] have estimated that roughly 20,000 lesions occur per day per human genome because of oxidative damage of DNA. These include ssDNA and dsDNA breaks, DNA-protein cross-links, and a wide variety of base and sugar modifications. Amongst the ROS, the hydroxyl radical ($.OH$) is extremely reactive and is the product of Fenton chemistry following reduction of $H_2O_2$, which is produced by a wide variety of intracellular events, particularly in normal oxidative electron transport in the mitochondria. In addition, $.OH$ can be generated in the human body after exposure to a myriad exogenous substances including cigarette tars, dietary components high in fat and low in plant fiber, ethanol, asbestos fibers, and IR. Superoxides are somewhat less reactive than $.OH$, but can still damage DNA appreciably, estimated at 10,000 lesions per genome per day.

### 14.4.6.6.3  Genotoxic Chemicals

The environment harbors an enormous number of both natural and man-made substances that have genotoxic properties. Most of these substances have the capacity to covalently modify DNA. The polycyclic aromatic hydrocarbons (PAHs) comprise a family of compounds that modify DNA with bulky lesions and, because of their prevalence in the environment, pose a significant human health hazard. Benzo[a]pyrene (BP) represents a well characterized member of the PAHs. BP is generated during combustion of many organic substances such as coal, cigarettes and gasoline, and is a classical complete carcinogen.[161] BP requires biotransformation by the *CYP1A1* gene product, cytochrome P450IA1, to the active carcinogenic form, benzo[a]pyrene-7,8-diol-9,10-epoxide (BPDE). BPDE can covalently attach to DNA to form a variety of adducts. DNA repair to remove these adducts triggers cell cycle arrest in cycling cells, although the precise mechanism by which this occurs remains poorly understood. BPDE-DNA adduct formation correlates with p53 protein expression, consistent with the protective role p53 plays following DNA damage.[162] BP carcinogenicity is lost however, in mice nullizygous for the aryl hydrocarbon receptor (AhR), confirming that AhR-mediated induction of the *Cyp1a1* gene is required to effect biotransformation of BP to the genotoxic BPDE.[163] It is important to note that there are many other classes of environmental agents known to modify DNA, including agents such as aflatoxin and the aromatic heterocyclic amines.

### 14.4.6.6.4  Nongenotoxic Chemicals

The impact of nongenetoxic chemicals on cell cycle checkpoint function are not yet clearly understood. A number of chemicals found in the environment, compounds such as the dioxins and polychlorinated biphenyls are potent tumor promoters without being genotoxic. Paradoxically, dioxins, and in particular the congener 2,3,7,8-tetrachlorodibenzo-*p*-

dioxin (TCDD), is capable of inducing a G1 arrest response.[164–166] The TCDD-induced arrest is dependent on aryl hydrocarbon receptor (AhR) activity.

### 14.4.6.6.4.1 *The Aryl Hydrocarbon Receptor*

The AhR is a ligand-activated transcription factor the binds TCDD.[167] The unliganded AhR is a cytosolic protein which translocates into the nucleus following ligand binding, whereupon it partners with the Arnt protein and binds to AhR DNA recognition sites upstream of target genes, including several genes encoding for the drug metabolizing enzymes *CYP1A1, 1A2, 1B1* and glutathione S-transferase Ya.[168,169] Interest in the AhR dates back to the mid-70s due to its involvement TCDD toxicity.[168] The AhR also binds BP and several other structurally diverse natural compounds.[170–172] Whether any of these constitute physiologically relevant endogenous ligands requires further study however, but the structural diversity amongst the ligands attests to the promiscuous nature of AhR ligand binding.

### 14.4.6.6.4.2 *AhR-mediated G1 Arrest*

Studies on growth rates of AhR-positive and AhR-negative mouse and rat hepatoma cell lines revealed that the AhR modulates G1 cell cycle progression.[164,173] The pronounced TCDD induced G1 arrest in the rat 5L hepatoma cell line involves an AhR-mediated induction of the cyclin-dependent kinase inhibitor, p27Kip1.[174] Likewise, TCDD is known to inhibit DNA synthesis in primary hepatocytes and hepatocyte proliferation *in vivo* following partial hepatectomy.[175,176] Elferink and coworkers recently showed that the AhR and hypophosphorylated pRb directly interact through a LXCXE motif in the AhR, and that the interaction contributes to AhR transcriptional activation and TCDD-induced G1 arrest consistent with pRb functioning as a transcriptional coactivator.[166,177]

A model for how TCDD induces a G1 arrest (Figure 6) proposes that the AhR induces p27Kip1 protein expression thus preventing pRb phosphorylation and keeping E2F repressed. Since hypophosphorylated pRb can function as an AhR coactivator,[166] a positive feedback loop is established sustaining AhR transcriptional activity-conditional upon the presence of an AhR agonist. In contrast, transition through the G1/S checkpoint relies on CDK4/6 and CDK2-mediated pRb hyperphosphorylation to derepress E2F regulated gene expression. In turn, E2F facilitates its own transcriptional activity by controlling expression of the cyclin E gene,[10–13,178] thereby increasing cyclin E-CDK2 activity and hastening pRb hyperphosphorylation. Hence, E2F

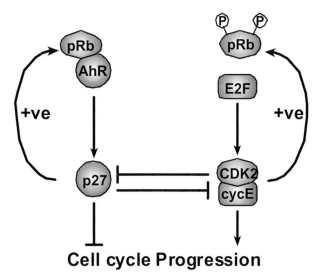

**Cell cycle Progression**

**Figure 6** The diagram groups the components of the growth stimulatory and inhibitory positive feedback mechanisms discussed in the text. AhR-mediated expression of p27Kip1 inhibits cell cycle progression by repressing CDK2 activity thus preventing pRb phosphorylation. E2F regulated expression of cyclin E facilitates CDK2 activity, inhibition of p27Kip1 and pRb hyperphosphorylation, to promote entry into S phase and cell cycle progression. The phosphorylation status of pRb influences the activity of both pathways. Not shown are the numerous other signaling molecules such as Ras, the cyclin D-CDK4/6 complex and pRb phosphatases that also influence the activity of these molecular components.

activity also establishes a positive feedback mechanism driving entry into S phase. Collectively, the opposing actions of p27Kip1 and cyclin E-CDK2 function as a "binary switching mechanism" where G1/S phase transition appears to require not only the emancipation of E2F transcriptional activity, but also AhR inactivation to terminate synthesis of p27Kip1. As a regulatory component common to both pathways, pRb hyperphosphorylation meets both endpoints simultaneously. Puga *et al.*[165] proposed an alternative mechanism for the AhR-dependent G1 arrest response, which invokes the formation of a ternary complex between E2F, pRb and the AhR that represses E2F regulated gene expression. Repression of E2F activity by pRb has been shown to involve recruitment of histone deacetylase 1 by binding with the LXCXE motif in the deacetylase,[179] promoting histone deacetylation and formation of a transcriptionally less active chromatin conformation. However, the existence of the E2F/pRb/AhR ternary complex has not been demonstrated, nor has the AhR been shown to possess histone deacetylase activity. Discerning between a p27Kip1-mediated mechanism and one involving a ternary complex will require further study.

#### 14.4.6.6.4.3   Ras Activity Suppresses AhR Function

Hyperphosphorylation of pRb may also explain why oncogenic Ras suppresses AhR activity in MCF-10A human breast cancer cells.[180] Activation of Ras and the downstream MAPK cascade in response to mitogens induces cyclin D1 expression[107] triggering the activation of cyclin D-CDK4/6. CDK4/6 directly phosphorylates pRb but also sequesters p27Kip1 away from the cyclin E-CDK2 complex, thus increasing CDK2 activity and further pRb phosphorylation.[181] The implication is that the inhibitory effect of Ras on AhR activity may be due to pRb hyperphosphorylation, thereby precluding pRb from coactivating AhR transcriptional activity.

#### 14.4.6.6.4.4   Is the AhR a Tumor Suppressor or an Oncoprotein?

It is evident that AhR action can impinge directly on cell cycle regulation in response to TCDD and probably numerous other nongenotoxic AhR ligands, including a putative endogenous AhR ligand. However, BP is also an AhR agonist, and in fact, requires AhR-mediated expression of *CYP1A1* to render BP

carcinogenic by biotransformation to BPDE.[163] Hence, the AhR paradoxically appears to contribute to processes that lead to cell cycle arrest, and promote cell proliferation. Still more puzzling is the observation that TCDD can induce cell cycle arrest, while simultaneously be classified as a potent tumor promoter based on data from animal studies.[168] Given that TCDD is a persistent environmental contaminant, it is possible that its action as a tumor promoter reflects sustained expression of *CYP1A1* increasing the likelihood of a deleterious DNA lesion due to a *CYP1A1* generated genotoxic metabolite. Alternatively, the AhR may contribute more directly to a proliferative response in light of the recent report that the AhR, in conjunction with RelA, can induce c-Myc gene expression in human breast cells.[182]

### 14.4.6.7   SUMMARY

The information obtained from studies of the molecular events governing cell cycle control has revealed much about the interwoven mechanisms that exist to compensate for dismantled controls. The primary role of the D-type cyclins is to act as growth factor sensors, converting environmental signals into the sparks that ignite the cell cycle. In normal cells, pRb phosphorylation, and Cip/Kip sequestration and degradation activate cyclin E-CDK2, thereby propelling cells into S phase and subsequent completion of the cell cycle in the absence of further mitogenic cues. The phosphorylation-triggered proteolysis of cyclin E that occurs as cells transit S phase, coupled with degradation of cyclin A by G2/M, resets the system to the ground state. This reestablishes a requirement for the D-type cyclins D and mitogen dependence in the ensuing G1 phase to maintain cell cycling, otherwise the cell exits the cell cycle by becoming quiescent (G0).

Cell cycling however, places an enormous emphasis on maintaining genomic stability and checkpoints provide the cell with opportunities to stall progress in the event of DNA damage until repairs are complete. Conversely, cells may abort the process entirely preferring to exit from the cell cycle and enter quiescence, or undergo apoptosis, depending on the condition of the cell and the environment it occupies. Our understanding of the cell cycle machinery has also provided us with a clearer picture of how environmental insults in the form of viral pathogens, genotoxic and nongenotoxic chemicals, and radiation can corrupt cell cycle control to bring about cancer. These same insights may also help in the design of more efficacious therapeutic strategies for the treatment of cancer and other diseases, as well as provide improved risk assessments for specific subpopulations of individuals genetically predisposed to certain diseases.

## 14.4.6.8 REFERENCES

1. J. D. Watson and F. H. C. Crick, 'Molecular structure of nucleic acid. A structure for deoxyribose nucleic acid.' *Nature*, 1953, **171**, 737–738.
2. D. G. Johnson, J. K. Schwarz, W. D. Cress *et al.*, 'Expression of transcription factor E2F1 induces quiescent cells to enter S phase.' *Nature*, 1993, **365**, 349–352.
3. X. Q. Qin, D. M. Livingston, W. G. Kaelin, Jr. *et al.*, 'Deregulated transcription factor E2F-1 expression leads to S-phase entry and p53-mediated apoptosis.' *Proc. Nat. Acad. Sci. USA*, 1994, **91**, 10918–19022.
4. B. Shan and W. H. Lee, 'Deregulated expression of E2F-1 induces S-phase entry and leads to apoptosis.' *Mol. Cell. Biol.*, 1994, **14**, 8166–8173.
5. N. Dyson, 'The regulation of E2F by pRB-family proteins.' *Genes Dev.*, 1998, **12**, 2245–2262.
6. H. S. Zhang, A. A. Postigo and D. C. Dean, 'Active transcriptional repression by the Rb-E2F complex mediates G1 arrest triggered by 16 INK4a, TGFβ, and contact inhibition.' *Cell*, 1999, **97**, 53–61.
7. F. Girard, U. Strausfeld, A. Fernandez *et al.*, 'Cyclin A is required for the onset of DNA replication in mammalian fibroblasts.' *Cell*, 1991, **67**, 1169–1179.
8. M. Pagano, R. Pepperkok, F. Verde *et al.*, 'Cyclin A is required at two points in the human cell cycle.' *EMBO J.*, 1992, **11**, 961–971.
9. F. Zindy, E. Lamas, X. Chenivesse *et al.*, 'Cyclin A is required in S phase in normal epithelial cells.' *Biochem. Biophys. Res. Commun.*, 1992, **182**, 1144–1154.
10. J. A. Knoblich, K. Sauer, L. Jones *et al.*, 'Cyclin E controls S phase progression and its down-regulation during Drosophila embryogenesis and is required for the arrest of cell proliferation.' *Cell*, 1994, **77**, 107–120.
11. P. K. Jackson, S. Chevalier, M. Philippe *et al.*, 'Early events in DNA replication require cyclin E and are blocked by p21CIP1.' *J. Cell Biol.*, 1995, **130**, 755–769.
12. M. Ohtsubo, A. M. Theodoras, J. Schumacher *et al.*, 'Human cyclin E, a nuclear protein essential for the G1 to S phase transition.' *Mol. Cell. Biol.*, 1995, **15**, 2612–2624.
13. U. P. Strausfeld, M. Howell, P. Descombes *et al.*, 'Both cyclin A and cyclin E have S-phase promoting (SPF) activity in Xenopus egg extracts.' *J. Cell Sci.*, 1996, **109**, 1555–1563.
14. B. E. Clurman, R. J. Sheaff, K. Thress *et al.*, 'Turnover of cyclin E by the ubiquitin-proteasome pathway is regulated by CDK2 binding and cyclin phosphorylation.' *Genes Dev.*, 1996, **10**, 1979–1990.
15. K.-A. Won and S. I. Reed, 'Activation of cyclin E/CDK2 is coupled to site-specific autophosphorylation and ubiquitin-dependent degradation of cyclin E.' *EMBO J.*, 1996, **15**, 4182–4193.
16. J. W. Ludlow, J. Shon, J. M. Pipas *et al.*, 'The retinoblastoma susceptibility gene product undergoes cell cycle-dependent dephosphorylation and binding to and release from SV40 large T.' *Cell*, 1990, **60**, 387–396.
17. J. W. Ludlow, 'Interactions between SV40 large-tumor antigen and the growth suppressor proteins pRB and p53.' *FASEB J.*, 1993, **7**, 866–871.
18. D. S. Peeper, L. L. Parker, M. E. Ewen *et al.*, 'A- and B-type cyclins differentially modulate substrate specificity of cyclin-cdk complexes.' *EMBO J.*, 1993, **12**, 1947–1954.
19. W. G. Dunphy, L. Brizuela, D. Beach *et al.*, 'The Xenopus cdc2 protein is a component of MPF, a cytoplasmic regulator of mitosis.' *Cell*, 1988, **54**, 423–431.
20. J. C. Labbe, J. P. Capony, D. Caputet *et al.*, 'MPF from starfish oocytes at first meiotic metaphase is a heterodimer containing one molecule of cdc2 and one molecule of cyclin B.' *EMBO J.*, 1989, **10**, 3053–3058.
21. R. W. King, R. J. Deshaies, J.-M. Peters *et al.*, 'How proteolysis drives the cell cycle.' *Science*, 1996, **274**, 1652–1659.
22. J.-M. Peters, 'SCF and APC: The Yin and Yang of cell cycle regulated proteolysis.' *Curr. Opin. Cell Biol.*, 1998, **10**, 759–768.
23. K. G. Hardwick, 'The spindle checkpoint.' *Trends Genet.*, 1998, **14**, 1–4.
24. C. J. Sherr, 'Mammalian G1 cyclins.' *Cell*, 1993, **73**, 1059–1065.
25. C. J. Sherr, 'Cancer cell cycles.' *Science*, 1996, **274**, 1672–1677.
26. R. A. Weinberg, 'The retinoblastoma protein and cell cycle control.' *Cell*, 1995, **81**, 323–330.
27. D. E. Quelle, R. A. Ashmun, S. E. Shurtleff *et al.*, 'Overexpression of mouse D-type cyclins accelerates G1 phase in rodent fibroblasts.' *Genes Dev.*, 1993, **7**, 1559–1571.
28. V. Baldin, J. Lukas, M. J. Marcote *et al.*, 'Cyclin D1 is a nuclear protein required for cell cycle progression in G1.' *Genes Dev.*, 1993, **7**, 812–821.
29. H. Matsushime, D. E. Quelle, S. A. Shurtleff *et al.*, 'D-type cyclin-dependent kinase activity in mammalian cells.' *Mol. Cell. Biol.*, 1994, **14**, 2066–2076.
30. M. Meyerson and E. Harlow, 'Identification of a G1 kinase activity for cdk6, a novel cyclin D partner.' *Mol. Cell. Biol.*, 1994, **14**, 2077–2086.
31. S. Mittnacht, J. A. Lees, D. Desai *et al.*, 'Distinct subpopulations of the retinoblastoma protein show a distinct pattern of phosphorylation.' *EMBO J.*, 1994, **13**, 118–127.
32. M. Kitagawa, H. Higashi, H. K. Jung *et al.*, 'The consensus motif for phosphorylation by cyclin D1-Cdk4 is different from that for phosphorylation by cyclin A/E-Cdk2.' *EMBO J.*, 1996, **15**, 7060–7069.
33. S. A. Ezhevsky, H. Nagahara, A. M. Vocero-Akbani *et al.*, 'Hypophosphorylation of the retinoblastoma protein (pRb) by cyclin D:Cdk4/6 complexes results in active pRb.' *Proc. Nat. Acad. Sci. USA*, 1997, **94**, 10699–10704.
34. A. S. Lundberg and R. A. Weinberg, 'Functional inactivation of the retinoblastoma protein requires sequential modification by at least two distinct cyclin-CDK complexes.' *Mol. Cell. Biol.*, 1998, **18**, 753–761.
35. J. W. Harbour, R. X. Luo, A. Dei Santi *et al.*, 'Cdk phosphorylation triggers sequential intramolecular interactions that progressively block Rb functions as cells move through G1.' *Cell*, 1999, **98**, 859–869.
36. D. Resnitzky and S. I. Reed, 'Different roles for cyclins D1 and E in regulation of the G1-to-S transition.' *Mol. Cell. Biol.*, 1995, **15**, 3463–3469.
37. J. Lukas, T. Herzinger, K. Hansen *et al.*, 'Cyclin E-induced S phase without activation of the Rb/E2F pathway.' *Genes Dev.*, 1997, **11**, 1479–1492.
38. M. Serrano, G. J. Hannon D. Beach, 'A new regulatory motif in cell cycle control causing specific inhibition of cyclin D/CDK4.' *Nature*, 1993, **366**, 704–707.
39. G. J. Hannon and D. Beach, 'p15 INK4b is a

potential effector of TGFβ-induced cell cycle arrest.'
*Nature*, 1994, **371**, 257–261.

40. K.-L. Guan, C. W. Jenkins, Y. Li *et al.*, 'Growth
suppression by p18, a p16 INK4/MTS1- and p15
INK4b/MTS2–related CDK6 inhibitor, correlates
with wild-type pRb function.' *Genes Dev.*, 1994, **8**,
2939–2952.

41. H. Hirai, M. F. Roussel, J. Kato *et al.*, 'Novel INK4
proteins, p19 and p18, are specific inhibitors of the
cyclin D-dependent kinases CDK4 and CDK6.' *Mol.
Cell. Biol.*, 1995, **15**, 2672–2681.

42. F. K. M. Chan, J. Zhang, L. Chen *et al.*, 'Identification of human/mouse p19, a novel CDK4/CDK6
inhibitor with homology to p16 ink4' *Mol. Cell. Biol.*,
1995, **15**, 2682–2688.

43. Y. Gu, C. W. Turek and D. O. Morgan, 'Inhibition of
CDK2 activity *in vivo* by an associated 20K regulatory subunit.' *Nature*, 1993, **366**, 707–710.

44. J. W. Harper, G. R. Adami, N. Wei *et al.*, 'The p21
cdk-interacting protein Cip1 is a potent inhibitor of
G1 cyclin-dependent kinases.' *Cell*, 1993, **75**, 805–816.

46. Y. Xiong, G. J. Hannon, H. Zhang *et al.*, 'p21 is a
universal inhibitor of cyclin kinases.' *Nature*, 1993,
**366**, 701–704.

47. V. Dulic, W. K. Kaufmann, S. J. Wilson *et al.*, 'p53-
Dependent inhibition of cyclin-dependent kinase
activities in human fibroblasts during radiation-
induced G1 arrest.' *Cell*, 1994, **76**, 1013–1023.

48. A. Noda, Y. Ning, S. F. Venable *et al.*, 'Cloning of
senescent cell-derived inhibitors of DNA synthesis
using an expression screen.' *Exp. Cell. Res.*, 1994, **211**,
90–98.

49. K. Polyak, J.-Y. Kato, M. J. Solomon *et al.*, 'p27
Kip1, a cyclin-cdk inhibitor, links transforming
growth factor-b and contact inhibition to cell cycle
arrest.' *Genes Dev.*, 1994a, **8**, 9–22.

50. K. Polyak, M.-H. Lee, H. Erdjument-Bromage *et al.*,
'Cloning of p27 Kip1, a cyclin-dependent kinase
inhibitor and a potential mediator of extracellular
antimitogenic signals.' *Cell*, 1994b, **78**, 59–66.

51. H. Toyoshima and T. Hunter, 'p27, a novel inhibitor
of G1 cyclin/cdk protein kinase activity, is related to
p21.' *Cell*, 1994, **78**, 67–74.

52. M. H. Lee, I. Reynisdottir and J. Massague, 'Cloning
of p57 Kip2, a cyclin-dependent kinase inhibitor with
unique domain structure and tissue distribution.'
*Genes Dev.*, 1995, **9**, 639–649.

53. S. Matsuoka, M. Edwards, C. Bai *et al.*, 'p57 KIP2 , a
structurally distinct member of the p21 CIP1 cdk
inhibitor family, is a candidate tumor suppressor
gene.' *Genes Dev.*, 1995, **9**, 650–662.

54. J. Chen, P. K. Jackson, M. W. Kirschner *et al.*,
'Separate domains of p21 involved in the inhibition of
cdk kinase and PCNA.' *Nature*, 1995, **374**, 386–388.

55. J. Chen, P. Saha, S. Kornbluth *et al.*, 'Cyclin-binding
motifs are essential for the function of p21 Cip1.'
*Mol. Cell. Biol.*, 1996, **16**, 4673–4682.

56. M. Nakanishi, R. S. Robetorge, G. R. Adami *et al.*,
'Identification of the active region of the DNA
synthesis inhibitory gene p21 Sdi1/CIP1/WAF1.'
*EMBO J.*, 1995, **14**, 555–563.

57. E. Warbrick, D. P. Lane, D. M. Glover *et al.*, 'A
small peptide inhibitor of DNA replication defines the
site of interaction between the cyclin-dependent
kinase inhibitor p21 WAF1 and proliferating cell
nuclear antigen.' *Curr. Biol.*, 1995, **5**, 275–282.

58. J. Lin, C. Reichner, X. Wu *et al.*, 'Analysis of wild-
type and mutant p21 WAF-1 gene activities.' *Mol.
Biol. Cell*, 1996, **16**, 1786–1793.

59. A. A. Russo, P. D. Jeffrey, A. K. Patten *et al.*,
'Crystal structure of the p27Kip1 cyclin-dependent
kinase inhibitor bound to the cyclin A-cdk2 complex.'
*Nature*, 1996, **382**, 325–331.

60. J. LaBaer, M. D. Garrett, L. F. Stevenson *et al.*, 'New
functional activities for the p21 family of CDK
inhibitors.' *Genes Dev.*, 1997, **11**, 847–862.

61. M. Cheng, P. Olivier, J. A. Diehl *et al.*, 'The p21 Cip1
and p27 Kip1 CDK "inhibitors" are essential
activators of cyclin D-dependent kinases in murine
fibroblasts.' *EMBO J.*, 1999, **18**, 1571–1583.

62. C. J. Sherr and J. M. Roberts, 'Inhibitors of
mammalian G1 cyclin-dependent kinases.' *Genes
Dev.*, 1995, **9**, 1149–1163.

63. D. Müller, C. Bouchard, B. Rudolph *et al.*, 'Cdk2-
dependent phosphorylation of p27 facilitates its myc-
induced release from cyclin E/cdk2 complexes.'
*Oncogene*, 1997, **15**, 2561–2576.

64. R. Sheaff, M. Groudine, M. Gordon *et al.*, 'Cyclin E-
CDK2 is a regulator of p27Kip1.' *Genes Dev.*, 1997,
**11**, 1464–1478.

65. J. Vlach, S. Hennecke and B. Amati, 'Phosphoryla-
tion-dependent degradation of the cyclin-dependent
kinase inhibitor p27 Kip1.' *EMBO J.*, 1997, **16**, 5334–
5344.

66. T. J. Soos, H. Kiyokawa, J. S. Yan *et al.*, 'Formation
of p27-CDK complexes during the human mitotic cell
cycle.' *Cell Growth Differ.*, 1996, **7**, 135–146.

67. S. W. Blain, E. Montalvo and J. Massague, 'Differential interaction of the cyclin-dependent kinase
(Cdk) inhibitor p27 Kip1 with cyclin A-Cdk2 and
cyclin D2-Cdk4.' *J. Biol. Chem.*, 1997, **272**, 25863–
25872.

68. H. Zhang, G. J. Hannon and D. Beach, 'p21-
containing cyclin kinases exist in both active and
inactive states.' *Genes Dev.*, 1994, **8**, 1750–1758.

69. A. Koff, M. Ohtsuki, K. Polyak *et al.*, 'Negative
regulation of G1 in mammalian cells: inhibition of
cyclin E-dependent kinase by TGF-β.' *Science*, 1993,
**260**, 536–539.

70. J. Kato, M. Matsuoka, K. Polyak *et al.*, 'Cyclic
AMP-induced G1 phase arrest mediated by an
inhibitor (p27 Kip1) of cyclin-dependent kinase-4
activation.' *Cell*, 1994, **79**, 487–496.

71. L. Hengst and S. Reed, 'Complete inhibition of Cdk/
cyclin by one molecule of p21Cip1.' *Genes Dev.*, 1998,
**12**, 3882–3888.

72. I. Reynisdottir, K. Polyak, A. Iavarone *et al.*, 'Kip/
Cip and Ink4 cdk inhibitors cooperate to induce cell
cycle arrest in response to TGF-β.' *Genes Dev.*, 1995,
**9**, 1831–1845.

73. I. Reynisdottir and J. Massague, 'The subcellular
locations of p15 INK4b and p27 Kip1 coordinate
their inhibitory interactions with cdk4 and cdk2.'
*Genes Dev.*, 1997, **11**, 492–503.

74. D. A. Alcorta, Y. Xiong, D. Phelps *et al.*, 'Involvement of the cyclin-dependent kinase inhibitor p16
(INK4a) in replicative senescence of normal human
fibroblasts.' *Proc. Nat. Acad. Sci. USA*, 1996, **93**,
13742–13747.

75. E. Hara, R. Smith, D. Parry *et al.*, 'Regulation of p16
CDKN2 expression and its implications for cell
immortalization and senescence.' *Mol. Cell. Biol.*,
1996, **16**, 859–867.

76. I. Palmero, B. McConnell, D. Parry *et al.*, 'Accumulation of p16INK4a in mouse fibroblasts as a function
of replicative senescence and not of retinoblastoma
gene status.' *Oncogene*, 1997, **15**, 495–503.

77. M. Serrano, A.W. Lin, M.E. McCurrach *et al.*,
'Oncogenic ras provokes premature cell senescence
associated with accumulation of p53 and p16 INK4a.'
*Cell*, 1997, **88**, 593–602.

78. F. Zindy, D. E. Quelle, M. F. Roussel *et al.*,
'Expression of the p16 INK4a tumor suppressor
versus other INK4 family members during mouse
development and aging.' *Oncogene*, 1997a, **15**, 203–
211.

79. L. Morse, D. Chen, D. Franklin *et al.*, 'Induction of cell cycle arrest and B cell terminal differentiation by CDK inhibitor p18(INK4c) and IL-6.' *Immunity*, 1997, **6**, 47–56.

80. F. Zindy, H. Soares, K.-H. Herzog *et al.*, 'Expression of INK4 inhibitors in cyclin D-dependent kinases during mouse brain development.' *Cell Growth Differ.*, 1997b, **8**, 1139–1150.

81. D. E. Phelps, K. M. Hsiao, Y. Li *et al.*, 'Coupled transcriptional and translational control of cyclin-dependent kinase inhibitor p18INK4c expression during myogenesis.' *Mol. Cell. Biol.*, 1998, **18**, 2334–2343.

82. J. Koh, G. H. Enders, B. D. Dynlacht *et al.*, 'Tumour-derived p16 alleles encoding proteins defective in cell cycle inhibition.' *Nature*, 1995, **375**, 506–510.

83. J. Lukas, D. Parry, L. Aagaard *et al.*, 'Retinoblastoma protein-dependent cell cycle inhibition by the tumor suppressor p16.' *Nature*, 1995, **375**, 503–506.

84. R. H. Medema, R. E. Herrera, F. Lam *et al.*, 'Growth suppression by p16 Ink4 requires functional retinoblastoma protein.' *Proc. Nat. Acad. Sci. USA*, 1995, **92**, 6289–6293.

85. D. A. Parry, D. Mahony, K. Wills *et al.*, 'Cyclin D-CDK subunit arrangement is dependent on the availability of competing INK4 and p21 class inhibitors.' *Mol. Cell. Biol.*, 1999, **19**, 1775–1783.

86. B. B. McConnell, F. J. Gregory, F. J. Stott *et al.*, 'Induced expression of p16 INK4a inhibits both CDK4- and CDK2-associated kinase activity by reassortment of cyclin-CDK-inhibitor complexes.' *Mol. Cell. Biol.*, 1999, **19**, 1981–1989.

87. J. Mitra, C. Y. Dai, K. Somasundaram *et al.*, 'Induction of p21 WAF1/Cip1 and inhibition of Cdk2 mediated by the tumor suppressor p16 INK4a.' *Mol. Cell. Biol.*, 1999, **19**, 3916–3928.

88. R. E. Herrera, V. P. Sah, B. O. Williams *et al.*, 'Altered cell cycle kinetics, gene expression, and G1 restriction point regulation in Rb-deficient fibroblasts.' *Mol. Cell. Biol.*, 1996, **16**, 2402–2407.

89. R. K. J. Hurford, D. Cobrinik, M.-H. Lee *et al.*, 'pRB and P107/p130 are required for the regulated expression of different sets of E2F responsive genes.' *Genes Dev.*, 1997, **9**, 2364–2372.

90. S. Van den Heuvel and E. Harlow, 'Distinct roles for cyclin-dependent kinases in cell cycle control.' *Science*, 1993, **262**, 2050–2054.

91. H. Jiang, H. S. Chou and L. Zhu, 'Requirement of cyclin E-cdk2 inhibition in p16 INK4a-mediated growth suppression.' *Mol. Cell. Biol.*, 1998, **18**, 5284–5290.

92. Y. Geng, W. Whoriskey, M. Y. Park *et al.*, 'Rescue of cyclin D1 deficiency by knockin cyclin E.' *Cell*, 1999, **97**, 767–77.

93. J. Tsihlias, L. Kapusta and J. Slingerland, 'The prognostic significance of altered cyclin-dependent kinase inhibitors in human cancer.' *Annu. Rev. Med.*, 1999, **50**, 401–423.

94. C. Catzavelos, N. Bhattacharya, Y. C. Ung *et al.*, 'Decreased levels of the cell-cycle inhibitor p27Kip1 protein: Prognostic implications in primary breast cancer.' *Nat. Med.*, 1997, **3**, 227–230.

95. M. Loda, B. Cukor, S. W. Tam *et al.*, 'Increased proteasome-dependent degradation of the cyclin-dependent kinase inhibitor p27 in aggressive colorectal carcinomas.' *Nat. Med.*, 1997, **3**, 231–234.

96. P. L. Porter, K. E. Malone, P. J. Heagerty *et al.*, 'Expression of cell-cycle regulators p27Kip1 and cyclin E, alone and in combination, correlate with survival in young breast cancer patients.' *Nat. Med.*, 1997, **3**, 222–225.

97. M. L. Fero, E. Randel, K. E. Gurley *et al.*, 'The murine gene p27 Kip1 is haplo-insufficient for tumour suppression.' *Nature*, 1998, **396**, 177–180.

98. G. Leone, J. DeGregori, R. Sears *et al.*, 'Myc and Ras collaborate in inducing accumulation of active cyclin E/Cdk2 and E2F.' *Nature*, 1997, **387**, 422–426.

99. M. Henriksson and B. Lüscher, 'Proteins of the Myc network: essential regulators of cell growth and differentiation.' *Adv. Cancer Res..*, 1996, **68**, 109–182.

100. M. G. Alexandrow, M. Kawabata, M. Aakre *et al.*, 'Overexpression of the c-Myc oncoprotein blocks the growth-inhibitory response but is required for the mitogenic effects of transforming growth factor β1.' *Proc. Nat. Acad. Sci. USA*, 1995, **92**, 3239–3243

101. H. Hermeking, J. O. Funk, M. Reichert *et al.*, 'Abrogation of p53-induced cell cycle arrest by c-Myc: evidence for an inhibitor of p21WAF1/CIP1/SDI1.' *Oncogene*, 1995, **11**, 1409–1415.

102. K. Galaktionov, X. Chen and D. Beach, 'Cdc25 cell-cycle phosphatase as a target of c-myc.' *Nature*, 1996, **382**, 511–517.

103. I. Perez-Roger, D. L. C. Solomon, A. Sewing *et al.*, 'Myc activation of cyclin E/CDK2 kinase involves induction of cyclin E gene transcription and inhibition of p27Kip1 binding to newly formed complexes.' *Oncogene*, 1997, **14**, 2373–238.

104. H. Hermeking, C. Rago, M. Schuhmacher *et al.*, 'Identification of CDK4 as a target of c-MYC.' *Proc. Nat. Acad. Sci. USA*, 2000, **97**, 2229–2234.

105. E. Santoni-Rugiu, J. Falck, N. Mailand *et al.*, 'Involvement of Myc activity in a G1/S-promoting mechanism parallel to the pRb/E2F pathway.' *Mol. Cell. Biol.*, 2000, **20**, 3497–3509.

106. S. Dobrowolski , M. Harter, D. W. Stacey, 'Cellular ras activity is required for passage through multiple points of the G0/G1 phase in BALB/c 3T3 cells.' *Mol. Cell. Biol.*, 1994, **14**, 5441–5449.

107. D. S. Peeper, T. M. Upton, M. H. Ladha *et al.*, 'Ras signalling linked to the cell cycle machinery by the retinoblastoma protein.' *Nature*, 1997, **386**, 177–181.

108. S. Mittnacht , H. Paterson, M. F. Olson *et al.*, 'Ras signalling is required for inactivation of the tumour suppressor pRb cell-cycle control protein.' *Curr. Biol.*, 1997, **7**, 219–221.

109. H. Aktas, H. Cai and G. M. Cooper, 'Ras links growth factor signaling to the cell cycle machinery via regulation of cyclin D1 and the cdk inhibitor p27 Kip1.' *Mol. Cell. Biol.*, 1997, **17**, 3850–3857.

110. A. Sewing, B. Wiseman, A. C. Lloyd *et al.*, 'High-intensity Raf signal causes cell cycle arrest mediated by p21Cip1.' *Mol. Cell. Biol.*, 1997, **17**, 5588–5597.

111. M. Cheng, V. Sexl, C. J. Sherr *et al.*, 'Assembly of cyclin D-dependent kinase and titration of p27 Kip1 regulated by mitogen-activated protein kinase kinase (MEK1).' *Proc. Nat. Acad. Sci. USA*, 1998, **95**, 1091–1096.

112. J. A. Diehl, M. Cheng, M. F. Roussel *et al.*, 'Glycogen synthase kinase-3β regulates cyclin D1 proteolysis and subcellular localization.' *Genes Dev.*, 1998, **12**, 192–229.

113. G. Mulligan and T. Jacks, 'The retinoblastoma gene family: cousins with overlapping interests.' *Trends Genet.*, 1998, **14**, 223–229.

114. I. Kovesdi, R. Reichel and J. R. Nevins, 'E1A transcription induction: enhanced binding of a factor to upstream promoter sequences.' *Science*, 1986, **231**, 719–722.

115. P. A. Hamel, R. M. Gill, R. A. Phillips *et al.*, 'Transcriptional repression of the E2-containing promoters EIIaE, c-myc, and RB1 by the product of the RB1 gene.' *Mol. Cell. Biol.*, 1992, **12**, 3431–3438.

116. S. J. Weintraub, C. A. Prater and D. C. Dean, 'Retinoblastoma protein switches the E2F site from

positive to negative element.' *Nature*, 1992, **358**, 259–261.

117. C. A. Finlay, P. W. Hinds and A. J. Levine, 'The p53 proto-oncogene can act as a suppressor of transformation.' *Cell*, 1989, **57**, 1083–1093.

118. S. J. Baker, S. Markowitz, E. R. Fearon *et al.*, 'Suppression of human colorectal carcinoma cell growth by wild-type p53.' *Science*, 1990, **249**, 912–915.

119. L. A. Donehower, M. Harvey, B. L. Slagle *et al.*, 'Mice deficient for p53 are developmentally normal but susceptible to spontaneous tumours.' *Nature*, 1992, **356**, 215–221.

120. M. B. Kastan, Q. Zhan, W. S. El-Deiry *et al.*, 'A mammalian cell cycle checkpoint pathway utilizing p53 and GADD45 is defective in ataxia-telangiectasia.' *Cell*, 1992, **71**, 587–597.

121. S. Y. Shieh, J. Ahn, K. Tamai *et al.*, 'The human homologs of checkpoint kinases Chk1 and Cds1 (Chk2) phosphorylate p53 at multiple DNA damage-inducible sites.' *Genes Dev.*, 2000, **14**, 289–300.

122. V. Gottifredi, O. Karni-Schmidt, S. Shieh *et al.*, 'p53 down-regulates CHK1 through p21 and the retinoblastoma protein.' *Mol. Cell. Biol.*, 2001, **21**, 1066–1076.

123. P. M. Flatt, L. J. Tang, C. D. Scatena *et al.*, 'p53 regulation of G(2) checkpoint is retinoblastoma protein dependent.' *Mol. Cell. Biol.*, 2000, **20**, 4210–4223.

124. D. P. Lane, 'p53, guardian of the genome.' *Nature*, 1992, **358**, 15–16.

125. K. Savitsky, A. Bar-Shira, S. Gilad *et al.*, 'A single ataxia telangiectasia gene with a product similar to PI-3 kinase.' *Science*, 1995, **268**, 1749–1753.

126. X. Lu and D. P. Lane, 'Differential induction of transcriptionally active p53 following UV or ionizing radiation: defects in chromosome instability syndromes?' *Cell*, 1993, **75**, 765–778.

127. K. K. Khanna, K. E. Keating, S. Kozlov *et al.*, 'ATM associates with and phosphorylates p53: mapping the region of interaction.' *Nat. Genet.*, 1998, **20**, 398–400.

128. R. S. Tibbetts, K. M. Brumbaugh, J. M. Williams *et al.*, 'A role for ATR in the DNA damage-induced phosphorylation of p53.' *Genes Dev.*, 1999, **13**, 152–157.

129. M. B. Kastan, O. Onyekwere, D. Sidransky *et al.*, 'Participation of p53 protein in the cellular response to DNA damage.' *Cancer Res.*, 1991, **51**, 6304–6311.

130. S. Banin, L. Moyal, S. Shieh *et al.*, 'Enhanced phosphorylation of p53 by ATM in response to DNA damage.' *Science*, 1998, **281**, 1674–1677.

131. C. E. Canman, D. S. Lim, K. A. Cimprich *et al.*, 'Activation of the ATM kinase by ionizing radiation and phosphorylation of p53.' *Science*, 1998, **281**, 1677–1679.

132. S. Matsuoka, M. Huang and S. J. Elledge, 'Linkage of ATM to cell cycle regulation by the Chk2 protein kinase.' *Science*, 1998, **282**, 1893–1897.

133. A. Telesnitsky and S. P. Goff, 'Retroviruses', eds. J. M. Coffin, S. H. Hughes and H. E. Varmus, Cold Spring Harbor Laboratory, NY, 1997, pp. 121–160.

134. H. J. Kung, C. Boerkoel and T. H. Carter, 'Retroviral mutagenesis of cellular oncogenes: a review with insights into the mechanisms of insertional activation.' *Curr. Top. Microbiol. Immunol.*, 1991, **171**, 1–25.

135. N. Rosenberg and P. Jolicoeur, 'Retroviruses', eds. J. M. Coffin, S. H. Hughes and H. E. Varmus, Cold Spring Harbor Laboratory, NY, 1997, pp. 475–585.

136. C. N. Cole, 'Fields Virology', 3rd edn, eds. B. N. Fields, D. M. Knipe, P. M. Howley *et al.*, Philadelphia, PA, 1996, **2**, 1997–2025.

137. P. M. Howley. 'Fields Virology', 3rd edn, eds. B. N.

138. T. Shenk. 'Fields Virology', 3rd edn, eds. B. N. Fields, D. M. Knipe, P. M. Howley *et al.*, Philadelphia, PA, 1996, **2**, 2111–2148.

139. N. Dyson, P. M. Howley, K. Munger *et al.*, 'The human papilloma virus-16 E7 oncoprotein is able to bind to the retinoblastoma gene product.' *Science*, 1989, **243**, 934–937.

140. P. Whyte , N. M. Williamson and E. Harlow, 'Cellular targets for transformation by the adenovirus E1A proteins.' *Cell*, 1989, **56**, 67–75.

141. P. R. Yew and A. J. Berk, 'Inhibition of p53 transactivation required for transformation by adenovirus early 1B protein.' *Nature*, 1992, **357**, 82–85.

142. M. S. Lechner, D. H. Mack, A. B. Finicle *et al.*, 'Human papillomavirus E6 proteins bind p53 *in vivo* and abrogate p53-mediated repression of transcription.' *EMBO J.*, 1992, **11**, 3045–3052.

143. J. W. Ludlow, C. L. Glendening, D. M. Livingston *et al.*, 'Specific enzymatic dephosphorylation of the retinoblastoma protein.' *Mol. Cell. Biol.*, 1993, **13**, 367–372.

144. S. Bates, A. C. Phillips, P. A. Clark *et al.*, 'p14ARF links the tumour suppressors RB and p53.' *Nature*, 1998, **395**, 124–125.

145. F. J. Stott, S. Bates, M. C. James *et al.*, 'The alternative product from the human CDKN2A locus, p14(ARF), participates in a regulatory feedback loop with p53 and MDM2.' *EMBO J.*, 1998, **17**, 5001–5014.

146. J. Pomerantz, N. Schreiber-Agus, N. J. Liegeois *et al.*, 'The Ink4a tumor suppressor gene product, p19Arf, interacts with MDM2 and neutralizes MDM2's inhibition of p53.' *Cell*, 1998, **92**, 713–723.

147. Y. Zhang, Y. Xiong and W.G. Yarbrough, 'ARF promotes MDM2 degradation and stabilizes p53: ARF-INK4a locus deletion impairs both the Rb and p53 tumor supression pathways.' *Cell*, 1998, **92**, 725–734.

148. P. R. Yew, X. Liu and A. J. Berk, 'Adenovirus E1B oncoprotein tethers a transcriptional repression domain to p53.' *Genes Dev.*, 1994, **8**, 190–202.

149. J. Bargonetti, I. Reynisdottir, P. N. Friedman *et al.*, 'Site-specific binding of wild-type p53 to cellular DNA is inhibited by SV40 T antigen and mutant p53.' *Genes Dev.*, 1992, **6**, 1886–1898.

150. W. Deppert, M. Haug and T. Steinmayer, 'Articles-Modulation of p53 protein expression during cellular transformation with simian virus 40.' *Mol. Cell. Biol.*, 1987, **7**, 4453–4463.

151. M. S. Lechner and L. A. Laimins, 'Inhibition of p53 DNA binding by human papillomavirus E6 proteins.' *J. Virol.*, 1994, **68**, 4262–4273.

152. Y. Chang, P. S. Moore, S. J. Talbot *et al.*, 'Cyclin encoded by KS herpes virus.' *Nature*, 1996, **382**, 410.

153. C. Swanton, D. J. Mann, B. Fleckenstein *et al.*, 'Herpes viral cyclin/Cdk6 complexes evade inhibition by CDK inhibitor proteins.' *Nature*, 1997, **390**, 184–187.

154. M. Ellis, Y. P. Chew, L. Fallis *et al.*, 'Degradation of p27 Kip cdk inhibitor triggered by Kaposi's sarcoma virus cyclin-cdk6 complex.' *EMBO J.*, 1999, **18**, 644–653.

155. R. M. Tyrrell, 'Activation of mammalian gene expression by the UV component of sunlight—from models to reality.' *Bioessays*, 1996, **18**, 139–148.

156. R. B. Setlow, 'The photochemistry, photobiology, and repair of polynucleotides.' *Prog. Nucleic Acid Res. Mol. Biol.*, 1968, **8**, 257–295.

157. F. Hutchinson, 'Chemical changes induced in DNA by ionizing radiation.' *Prog. Nucleic Acid Res. Mol. Biol.*, 1985, **32**, 115–154.

158. J. F. Ward, 'DNA damage produced by ionizing radiation in mammalian cells: identities, mechanisms of formation, and reparability.' *Prog. Nucleic Acid Res. Mol. Biol.*, 1988, **35**, 95–125.

159. Q. M. Chen, J. C. Bartholomew, J. Campisi *et al.*, 'Molecular analysis of H₂O₂-induced senescent-like growth arrest in normal human fibroblasts: p53 and Rb control G1 arrest but not cell replication.' *Biochem. J.*, 1998, **332**, 43–50.

160. B. N. Ames and M. K. Shigenaga, 'Oxidants are a major contributor to aging.' *Ann. N.Y. Acad. Sci.*, 1992, **663**, 85–96.

161. IARC, 'in IARC monographs on the evaluation of the carcinogenic riak of chemicals to man,' International Agency for Research on Cancer, Lyon, France 1973, **3**, 91–136.

162. M. Ramet, K. Castren, K. Jarvinen *et al.*, 'p53 protein expression is correlated with benzo[a]pyrene-DNA adducts in carcinoma cell lines.' *Carcinogenesis*, 1995, **16**, 2117–2124.

163. Y. Shimizu, Y. Nakatsuru, M. Ichinose *et al.*, 'Benzo[a]pyrene carcinogenicity is lost in mice lacking the aryl hydrocarbon receptor.' *Proc. Nat. Acad. Sci. USA*, 2000, **97**, 779–782.

164. C. Weiss, S. K. Kolluri, F. Kiefer *et al.*, 'Complementation of Ah receptor deficiency in hepatoma cells: negative feedback regulation and cell cycle control by the Ah receptor.' *Exp. Cell Res.*, 1996, **226**, 154–163.

165. A. Puga, S. J. Barnes, T. P. Dalton *et al.*, 'Aromatic hydrocarbon receptor interaction with the retinoblastoma protein potentiates repression of E2F-dependent transcription and cell cycle arrest.' *J. Biol. Chem.*, 2000, **275**, 2943–2950.

166. C. J. Elferink, N.-L. Ge, A. Levine. 'Maximal Ah receptor activity depends on an interaction with the retinoblastoma protein.' *Mol. Pharmacol.*, 2001, **59**, 664–673.

167. J. V. Schmidt and C. A. Bradfield, 'Ah receptor signaling pathways.' *Annu. Rev. Cell Dev. Biol.*, 1996, **12**, 55–89.

168. A. Poland and J. C. Knudson, '2,3,7,8-Tetrachlorodibenzo-*p*-dioxin and related aromatic hydrocarbons: examination of the mechanism of toxicity.' *Annu. Rev. Pharmacol. Toxicol.*, 1982, **22**, 517–554.

169. J. P. Whitlock Jr., 'Mechanistic aspects of dioxin action.' *Chem. Res. Toxicol.*, 1993, **6**, 754–763.

170. L. F. Bjeldanes, J.-Y. Kim, K. R. Grose *et al.*, 'Aromatic hydrocarbon responsiveness-receptor agonists generated from indole-3-carbinol *in vitro* and *in vivo*: comparisons with 2,3,7,8-tetrachlorodibenzo-*p*-dioxin.' *Proc. Nat. Acad. Sci. USA*, 1991, **88**, 9543–9547.

171. C. J. Sinal and J. R. Bend, 'Aryl hydrocarbon receptor-dependent induction of cyp1a1 by bilirubin in mouse hepatoma hepa 1c1c7 cells.' *Mol. Pharmacol.*, 1997, **52**, 590–599.

172. C. M. Schaldach, J. Riby, L. F. Bjeldanes, 'Lipoxin A4: a new class of ligand for the Ah receptor.' *Biochemistry*, 1999, **38**, 7594–7600.

173. Q. Ma and J. P. Whitlock Jr., 'The aromatic hydrocarbon receptor modulates the Hepa 1c1c7 cell cycle and differentiated state independently of dioxin.' *Mol. Cell. Biol.*, 1996, **16**, 2144–2150.

174. S. K. Kolluri, C. Weiss, A. Koff *et al.*, 'p27Kip1 induction and inhibition of proliferation by the intracellular Ah receptor in developing thymus and hepatoma cells.' *Genes Dev.*, 1999, **13**, 1742–1753.

175. J. W. Bauman, T. L. Goldsworthy, C. S. Dunn *et al.*, 'Inhibitory effects of 2,3,7,8-tetrachlorodibenzo-*p*-dioxin on rat hepatocyte proliferation induced by 2/3 partial hepatectomy.' *Cell Proliferation*, 1995, **28**, 437–451.

176. D. R. Huskha and W. F. Greenlee, '2,3,7,8-Tetrachlorodibenzo-*p*-dioxin inhibits DNA synthesis in rat primary hepatocytes.' *Mutation Research*, 1995, **333**, 89–99.

177. N.-L. Ge and C. J. Elferink, 'A direct interaction between the aryl hydrocarbon receptor and retinoblastoma protein: linking dioxin signaling to the cell cycle.' *J. Biol. Chem.*, 1998, **273**, 22708–22713.

178. K. Ohtani, J. DeGregori, J. R. Nevins, 'Regulation of the cyclin E gene by transcription factor E2F1.' *Proc. Nat. Acad. Sci. USA*, 1995, **92**, 12146–12150.

179. A. Brehm, E. A. Miska, D. J. McCance *et al.*, 'Retinoblastoma protein recruits histone deacetylase to repress transcription.' *Nature*, 1998, **391**, 597–601.

180. J. J. Reiners Jr., C. L. Jones, N. Hong *et al.*, 'Downregulation of aryl hydrocarbon receptor function and cytochrome P450 1A1 induction by expression of Ha-ras oncogenes.' *Mol. Carcinogenesis*, 1997, **19**, 91–100.

181. C. J. Sherr and J. M. Roberts, 'CDK inhibitors: positive and negative regulators of G1-phase progression.' *Genes Dev.*, 1999, **13**, 1501–1512.

182. D. W. Kim, L. Gazourian, S. A. Quadri *et al.*, 'The RelA NF-κB subunit and the aryl hydrocarbon receptor (AhR) cooperate to transactivate the c-myc promoter in mammary cells.' *Oncogene*, 2000, **19**, 5498–5506.

J.P. Vanden Heuvel, G.H. Perdew, W.B. Mattes and W.F. Greenlee (Eds.)
Comprehensive Toxicology, Vol. xiv

# 14.4.7
# Biochemical Signals that Initiate Apoptosis

## JOHN D. ROBERTSON and STEN ORRENIUS
*Karolinska Institutet, Stockholm, Sweden*

## 14.4.7.1  INTRODUCTION

Since the late 1980s, the scientific community has witnessed an explosion in cell death research. Much of the interest in this area of research is linked to the increased understanding that, at the molecular and cellular level, cell death is often as active a process as cell division, insofar as it requires the activity of different genes.[1] The term most commonly applied to indicate active cell death—apoptosis—was first introduced by Kerr and colleagues in 1972 as a means to discern a morphologically distinct form of cell death that occurred physiologically.[2] It was con-trasted with necrosis, which they associated with acute cell injury. The first biochemical evidence of apoptosis was reported by Skalka *et al.*,[3] who reported the appearance of oligonucleosomal-length DNA fragments in irradiated lymphocytes, an effect that was later attributed to increased endonuclease activity.[4] Following relatively little movement in the field during most of the 1980s, advances over the last 10–13 years have been moving at breakneck speed and a role for apoptotic cell death in embryonic development, the removal of damaged cells in the adult animal, and the onset of different diseases, including neurodegenerative disease and cancer, has been demonstrated.

## 14.4.7.2   THE APOPTOTIC PROGRAM

Exposure to various chemicals can result in cell damage and death. Whether a cell survives or dies in the presence of a chemical insult is often determined by proliferative status, repair enzyme capacity, and the ability to induce proteins that either promote or inhibit the cell death process. Tight control of these parameters must be maintained in order to prevent dysregulation of cell death, which can lead to a number of pathological conditions. Much of our understanding about cell death and its involvement in disease relates to its role in regulating tissue turnover or homeostasis. Tissue homeostasis occurs when a balance is achieved between cell renewal and cell death so that no net change in cell number is present. Normal homeostatic cell deletion is controlled, at least in part, by apoptosis.[2,4,5] Apoptosis, a biochemically and morphologically distinct form of cell death, is an active process associated with cell shrinkage, nuclear and cytoplasmic condensation, release of cytochrome *c* and other proteins from mitochondria, caspase activation, plasma membrane blebbing, phosphatidylserine externalization on the plasma membrane, DNA fragmentation, and the formation of apoptotic bodies that can be taken up and degraded by neighboring cells[6–8] (Figure 1). Apoptosis-associated nuclear condensation is usually accompanied by the activation of nucleases that first degrade chromosomal DNA into large 50–300 kilobase subunits and then into smaller units of ~180 base pairs.[4] Because plasma membrane integrity is maintained during apoptosis, which prevents the leakage of cytosolic contents into the extracellular domain, this form of cell death is normally not associated with an inflammatory response. In contrast to apoptosis, necrosis is a passive form of cell death associated with inflammation, often resulting from an overwhelming cellular insult that causes cell and organelle swelling, breakdown of the plasma membrane, release of lysosomal enzymes, and spillage of the cell contents into the extracellular milieu[9] (Figure 1). While apoptosis and necrosis can occur simultaneously, studies suggest that intracellular energy levels are important determinants of the mode of cell death.[10,11] Leist *et al.*[11] demonstrated that different aspects of the cell death program could be halted by manipulating intracellular levels of ATP. Specifically, apoptosis induced by established pro-apoptotic stimuli, e.g., staurosporine or anti-CD95 agonistic antibody, could be switched to necrosis when ATP levels were reduced, an effect believed to be linked to reduced caspase

activity and/or formation of the apoptosome complex (discussed below).[12]

In recent years, this laboratory and others have focused heavily on characterizing molecular and cellular mechanisms involved in triggering, and subsequently activating, apoptosis induced by various stimuli. It is becoming increasingly evident that different chemical agents, e.g., etoposide, staurosporine, glucocorticoids, inhibitors of the 26S proteasome, and various environmental contaminants, can exert their toxicity via apoptotic cell signaling.[13–16] The first part of this chapter will focus on the latest molecular trends in the area of chemical-induced apoptosis with particular attention given to mitochondrial participation, caspase activity, and the Bcl-2 family of proteins, while the second part will provide specific examples of environmental toxicants that are reported to exert their toxicity, at least in part, by apoptosis. Finally, the chapter will conclude with a general discussion of common themes and potential future directions in the area of chemical-induced apoptosis.

### 14.4.7.2.1   Mitochondrial Participation

Despite the seminal observation that Bcl-2 (discussed below), a potent anti-apoptotic protein, was present in the mitochondrial membrane,[17] these organelles were believed originally not to be involved in apoptotic cell signaling since cells lacking mitochondrial DNA (mtDNA), and therefore a functional respiratory chain, could undergo apoptosis.[18] Moreover, given that these cells were protected by Bcl-2 overexpression, the authors concluded that neither the induction of apoptosis nor the cytoprotective effect of Bcl-2 was dependent upon mitochondrial participation. In contrast, a subsequent study using *Xenopus* egg extract supported a role for mitochondrial participation since a membrane fraction containing mitochondria was required for the induction of nuclear apoptosis.[19] Other researchers have since confirmed the participation of mitochondria in effecting apoptosis.[20] Using a cell-free system consisting of HeLa cell extract, Liu *et al.*[21] made the important observation that cytochrome *c* (Figure 2), liberated from mitochondria during extract preparation, was able to stimulate pro-caspase-3 (discussed below) activation in the presence of dATP. Both *in vitro* and *in vivo* studies have since demonstrated the release of cytochrome *c* from mitochondria during apoptosis,[22,23] although the precise mechanism controlling this event is unknown. Once in the cytosol, cytochrome *c* interacts with apoptotic protease-activating

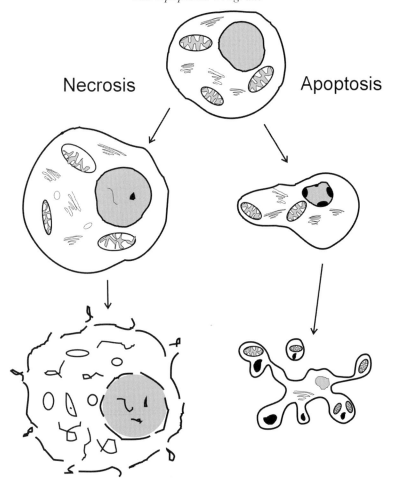

**Figure 1** Schematic representation of apoptosis and necrosis. Necrosis is characterized by cell swelling and lysis, whereas apoptosis is associated with cell shrinkage and ultimate formation of apoptotic bodies that can be engulfed and degraded by macrophages.

factor-1 (Apaf-1) and pro-caspase-9 forming the apoptosome complex[24] (Figure 3). The result is the cleavage and activation of pro-caspase-9 and other pro-caspases that are responsible for the executive stages of apoptotic cell death.

The mitochondrion consists of two membranes: the inner and outer membranes. The inner membrane, which is rich in cristae, contains the molecular components of the electron transport chain (Figure 2). Oxidation of mitochondrial respiratory chain substrates results in the formation of a proton electrochemical gradient across the inner mitochondrial membrane. This gradient is exploited by the $F_0F_1$-ATP-synthase to produce ATP from ADP and $P_i$. Cytochrome $c$ is a component of the electron transport chain and supports oxidative phosphorylation by shuttling electrons between complexes III (cytochrome bc1) and IV (cytochrome oxidase). Apo-cytochrome $c$ is encoded by a nuclear gene and synthesized

in the cytosol. It is imported into the mitochondrial intermembrane space in an unfolded configuration where it receives a heme group.[25] Covalent attachment of this heme group stimulates a conformational change and holo-cytochrome $c$ subsequently assumes its functional role as a component of the electron transport chain. It is important to note that only holo-cytochrome $c$, and not apo-cytochrome $c$, is able to stimulate pro-caspase-9 activation.[22] Moreover, observations made by this laboratory indicate that both reduced and oxidized cytochrome $c$ are able to activate pro-caspase-9 with similar efficiencies, an effect that may be related to the speed with which oxidized cytochrome $c$ is reduced *in vitro*.[26]

As already mentioned, the precise molecular mechanisms controlling cytochrome $c$ release from mitochondria in the presence of a pro-apoptotic stimulus are not clear, although different models have been proposed. It was previously suggested that the induction of

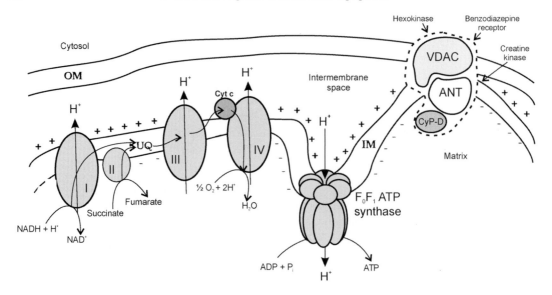

**Figure 2** Schematic representation of the mitochondrial respiratory chain and the permeability transition pore.

mitochondrial permeability transition (MPT), resulting in the loss of membrane potential ($\Delta\psi$) and opening of the permeability transition pore (PTP), stimulated the release of cytochrome $c$, and that this effect could be blocked by the MPT inhibitor cyclosporin A.[27] The PTP consists of both inner and outer mitochondrial membrane proteins, such as adenine nucleotide translocator (ANT) and voltage-dependent anion channel (VDAC), respectively, and is formed at contact sites between these two membranes[25] (Figures 2 and 3). $Ca^{2+}$-dependent opening of this pore—MPT—results in the loss of $\Delta\psi$, uncoupling of oxidative phosphorylation, mitochondrial swelling, and rupture of the outer membrane, leading to the release of cytochrome $c$. However, ample evidence from more recent studies suggests that while MPT is likely to be *a* mechanism responsible for cytochrome $c$ release, it is no longer regarded as *the* mechanism.[28] In particular, Martinou et al.[29] showed that mitochondria of neuronal growth factor NGF-deprived sympathetic neurons undergoing apoptosis released cytochrome $c$ and reduced in size, yet remained intact and resumed normal function when re-incubated with NGF. A more recent study from our laboratory demonstrated that etoposide stimulated cytochrome $c$ release from isolated mitochondria, despite the presence of 1 mM EGTA (a known inhibitor of MPT) in the reaction buffer.[30] Alternatively, it may be that the release of cytochrome $c$ involves a permeabilization of the outer membrane regulated by the pro-apoptotic Bcl-2 family proteins Bax or Bid, proteins known to induce cytochrome $c$

release.[31–34] Specifically, von Ahsen et al.[33] demonstrated the ability of recombinant Bax and cleaved or truncated Bid (tBid) to stimulate *complete* cytochrome $c$ release that was unaffected by cyclosporin A and did not result in mitochondrial depolarization or alterations in ultrastructure as assessed by electron microscopy. These data pointed to the possibility that cytochrome $c$ release involves a specific pore in the outer membrane that may be formed *de novo* by Bax or Bid. A variation of this model has been proposed by Tsujimoto's group who demonstrated the ability of recombinant Bax and Bak (also a pro-apoptotic Bcl-2 family protein) proteins to hasten the opening of VDAC in liposomes and induce changes in $\Delta\psi$, whereas Bcl-$X_L$ (discussed below) was able to bind and close VDAC directly.[35] A subsequent and complementary study reported the ability of certain Bcl-2 family members, i.e., Bid and Bik, to stimulate apoptosis in a fashion different from that of other pro-apoptotic Bcl-2 family proteins, such as Bax and Bak.[36] Specifically, Bid and Bik stimulated cytochrome $c$ release without inducing changes in $\Delta\psi$ or interacting directly with VDAC.

Normally, cytochrome $c$ is preferentially anchored to the inner membrane (Figure 2) by an interaction with the mitochondrial lipid cardiolipin. Recently, evidence was provided that cardiolipin mediates the targeting of tBid to mitochondria through a previously unknown three-helix domain in Bid. These findings implicate cardiolipin and/or certain pro-apoptotic members of the Bcl-2 family of proteins in the pathway for cytochrome $c$ release.[37] In fact, the involvement of cardiolipin in the release of cytochrome *c* was also documented since elevated levels of mitochondrial phospholipid hydroperoxide glutathione peroxidase in cells undergoing apoptosis completely suppressed the release of cytochrome *c*, presumably by preventing the oxidation of cardiolipin to cardiolipin hydroperoxide.[38] Taken together, there is reason to believe that cytochrome *c* release may occur by a two-step process, wherein this protein is first liberated from cardiolipin and then released via specific or non-specific pores/channels in the outer mitochondrial membrane.

A different molecule that is reportedly released from mitochondria in the presence of pro-apoptotic stimuli is apoptosis-inducing factor (AIF).[39,40] AIF, a protein that shares considerable homology with oxidoreductases from vertebrates and non-vertebrates, releases

from mitochondria upon apoptosis stimulation and translocates to the nucleus where it is believed to trigger chromatin condensation and DNA fragmentation. The presence of AIF in the cytosol is also sufficient to induce phosphatidylserine externalization on the plasma membrane and loss of $\Delta\psi$. Yet despite the resemblance of these effects to established apoptotic criteria, AIF-induced cell death is not blocked by caspase inhibitors.

Thus, abundant evidence supports the involvement of mitochondria in regulating apoptosis. Specifically, it is clear that mitochondria are able to release pro-apoptotic proteins into the cytosol. Among these proteins, the best characterized is cytochrome *c*, which participates in the activation of the caspase cascade. What is not as clear are the mechanisms responsible for triggering cytochrome *c* release.

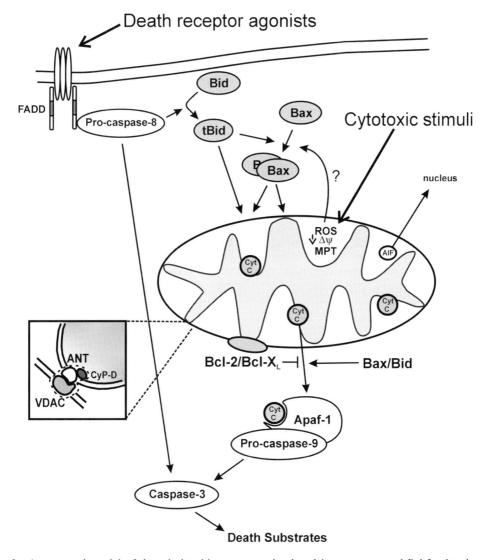

**Figure 3** A proposed model of the relationships among mitochondria, caspases, and Bcl-2-related proteins involved in regulating apoptotic cell death.

Although an attractive model to explain this phenomenon involves the formation of membrane pores/channels, conflicting evidence exists as to the size and nature of these putative conduits.

### 14.4.7.2.2  Caspases

A flood of interest has surrounded a growing family of cysteine-aspartate proteases (caspases) that are intimately, and perhaps inextricably, involved in apoptotic cell signaling and share sequence similarity with the *Caenorhabditis elegans* cell death protease (Ced-3).[41–43] The precise way in which caspases execute cell death remains unknown largely because of the variety of intracellular protein substrates for these enzymes that have been identified and because of their ability to translocate across subcellular compartments upon activation.[44,45] Nonetheless, a role for caspase activity in drug-induced apoptosis is supported by a recent study by Lemaire *et al.*[46] showing a shift from apoptotic to necrotic cell death in the presence of peptide caspase inhibitors.

Since the discovery of interleukin-1β-converting enzyme (ICE or caspase-1), which has no apparent role in cell death, thirteen additional caspases have been identified (Table 1).[41,42,47] All caspases are present constitutively and synthesized in inactive, precursor forms (30–50 kDa) that must be proteolytically cleaved in order to be activated.[42] Each pro-caspase consists of a pro-domain, a large ($\sim 20$ kDa) and small ($\sim 10$ kDa) subunit (Figure 4).[48] Cleavage and subsequent heterodimerization of the larger and smaller subunits result in caspase activation, and all caspases contain the conserved active site pentapeptide sequence QACXG (Figure 4). An active caspase can subsequently activate additional pro-caspases like itself and/or different pro-caspases, since aspartic acid cleavage sites exist between the prodomain of a caspase and its subunits.[49] Caspases are commonly divided into two groups based upon the primary structure of their $NH_2$-terminal prodomain: initiator (caspase-2, -8, -9 and -10) and effector (caspase-3, -6 and -7) enzymes. Initiator pro-caspases are activated with the help of adaptor molecules, e.g., Apaf-1 and Fas-associated death domain (FADD/MORT1) that bring these proteases into close proximity, permitting auto-processing. Once activated, initiator caspases are responsible for cleaving and activating effector pro-caspases. Effector caspases, in turn, cleave various proteins leading to morphological and biochemical features characteristic of apoptosis.[50]

While a role for effector caspases in the executive stages of apoptosis is clear, the specific protein substrates that must be cleaved in order to carry out this process remain unknown. Of the more than 100 caspase substrates that have been identified,[51] merely a handful has been studied extensively for their importance in the apoptotic process. Among these substrates are those that are involved in regulating genomic integrity (PAK2, DNA-PK, PARP, and ICAD/DFF-45) and those that are involved in maintenance of the cytoskeleton and nuclear scaffold (lamin, NuMa, gelsolin, and acinus). DNA-dependent protein kinase (DNA-PK) and poly(ADP-ribose) polymerase (PARP) were shown to be caspase-3 substrates that are involved in the repair of damaged DNA.[52] Another caspase-3

**Table 1**  A list of the 14 known mammalian caspases categorized based upon the primary structure of their pro-domain. All caspases, except*, are reported to participate in the apoptotic process. Although¶ contain a long pro-domain and thus are classified as Class I caspases, it remains unclear at what level within the caspase cascade that these enzymes operate. Consequently,¶ are not currently considered initiator caspases (see text).

| *Class I*<br>*(Long Pro-domain)* | *Class II*<br>*(Short Pro-domain)* |
|---|---|
| Caspase-1 (ICE)* | Caspase-3 (CPP32, apopain, Yama) |
| Caspase-2 (Nedd2,ICH-1) | Caspase-6 (Mch2) |
| Caspase-4 (ICE$_{rel}$-II, TX, ICH-2)* | Caspase-7 (Mch3, ICE-LAP3, CMH-1) |
| Caspase-5 (ICE$_{rel}$-III, TY)* | Caspase-14 (MICE)* |
| Caspase-8 (MACH, FLICE,Mch5) | |
| Caspase-9 (ICE-LAP6, Mch6) | |
| Caspase-10 (Mch4) | |
| Caspase-11 (ICH-3)¶ | |
| Caspase-12 (ICH-4)¶ | |
| Caspase-13 (ERICE)* | |

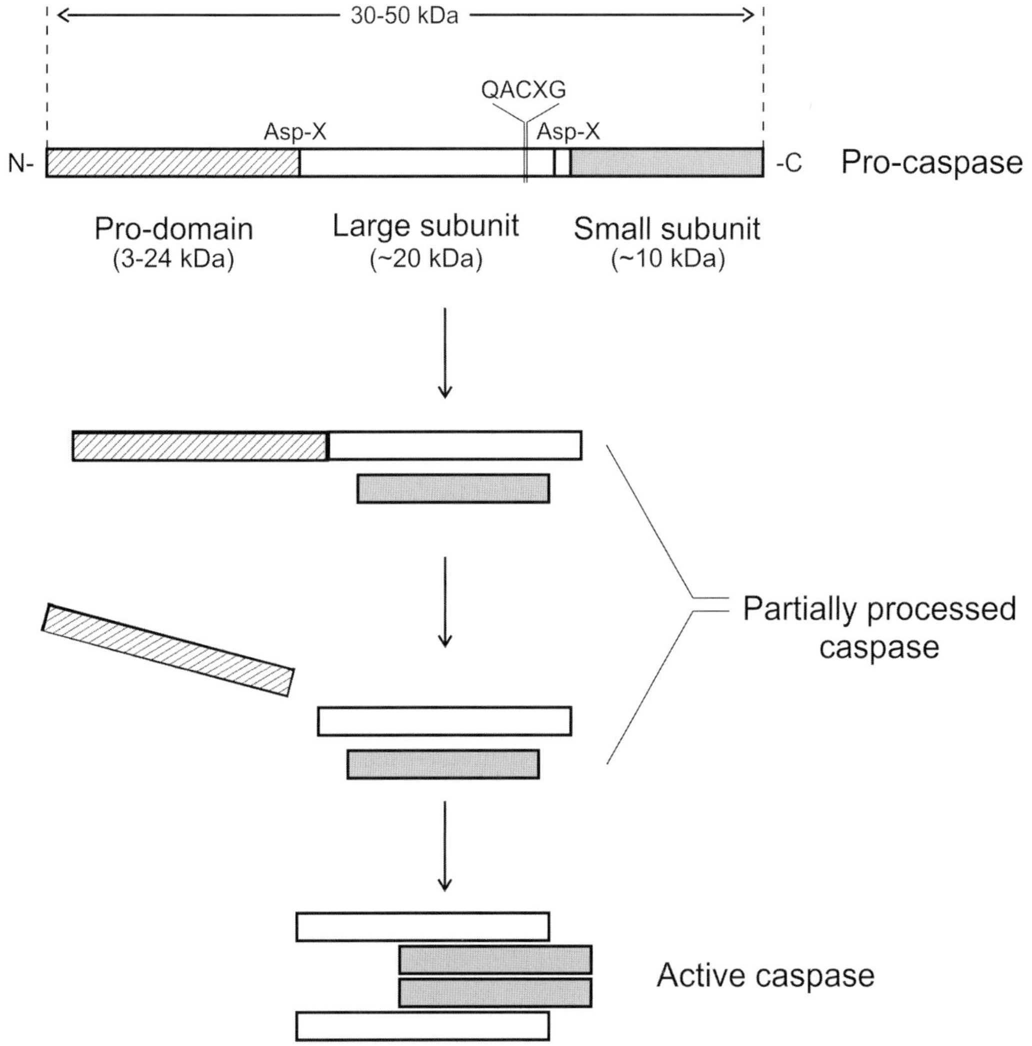

**Figure 4** Caspase pro-enzyme organization and processing to generate a fully active enzyme.

(and -7) substrate that has received considerable attention recently because of its ability to prevent nuclear changes associated with apoptosis is inhibitor of caspase-activated deoxyribonuclease/DNA fragmentation factor-45 (ICAD/DFF-45). This molecule is the inhibitory counterpart of the endonuclease CAD/DFF-40.[53,54] Normally, CAD or DFF-40 is present in an inactive form bound to its inhibitory molecule. It is believed that caspase-3 stimulates the dissociation and degradation of ICAD/DFF-45, which, in turn, liberates CAD/DFF-40 resulting in the induction of nuclease activity, nuclear condensation, and DNA fragmentation.[53–56] DFF-45 knockout mice are refractory to DNA fragmentation and nuclear condensation in response to various apoptotic stimuli, suggesting that the inhibitory molecule to DFF-40 is required for normal function of this enzyme.[57] Conversely, in cells deficient of caspase-3 activity, e.g.,

MCF-7 cells, apoptosis is not associated with CAD/DFF-40 activation and DNA fragmentation. Two protein substrates that can be cleaved by caspase-6 and have received attention for their role in maintaining nuclear structural integrity are lamin and nuclear mitotic apparatus protein (NuMA). In particular, Rao et al.[58] demonstrated that introducing mutant lamin into cells delayed DNA fragmentation for 12 h and completely eliminated nuclear condensation. However, that lamin and NuMa are specific caspase-6 substrates has been challenged, since in some models caspase-3 (and other non-caspase proteases) can also cleave these proteins—at distinct sites—without altering DNA fragmentation or chromatin condensation patterns.[59–61] Gelsolin, a protein involved in actin metabolism, was identified as a substrate for caspase-3 in cells stimulated with CD95(Fas/Apo-1) (see below) and cleaved gelsolin is able to sever actin filaments—an effect that was involved in the detachment of adherent cells in culture and in nuclear condensation.[62] A different study by Tsujimoto's group reported caspase-3-mediated cleavage of gelsolin *in vivo*.[63] This same group was also responsible for identifying and characterizing the nuclear factor acinus that must be cleaved by caspase-3 and is required for chromatin condensation in response to various stimuli.[64]

Different initiator pro-caspases signal via distinct pathways. For instance, pro-caspase-9 is involved in apoptosis induced by cytotoxic stimuli, whereas pro-caspase-8 is recruited during death receptor activation (Figure 3). Moreover, and as mentioned previously, initiator pro-caspases require the assistance of adaptor molecules to become activated. Pro-caspase-9 requires the cytosolic adaptor molecule Apaf-1. In the presence of different chemical stimuli, mitochondria release specific proteins into the cytosol (discussed above). Among these molecules are cytochrome *c*, AIF, and adenylate kinase-2.[21,65,66] Once in the cytosol, cytochrome *c* has been shown to activate Apaf-1, which then oligomerizes, binds, and activates pro-caspase-9 in the presence of dATP.[12,67] Active caspase-9 can signal downstream resulting in the activation of effector pro-caspase-3 and -7. Cleavage of pro-caspase-9 is inhibitable by the caspase inhibitor z-VAD-fmk, whereas the release of cytochrome *c* is unaffected by caspase inhibitors in many models of chemical-induced apoptosis.[68] A possible exception involves cytochrome *c* release induced by DNA-damaging agents, such as the chemotherapeutic drug etoposide, which appear to activate nuclear caspases that, in turn, either directly or indirectly target mitochondria to stimulate cytochrome *c* release.[30,69]

Although pro-caspase-8 is primarily associated with receptor-mediated apoptosis, it bears mentioning because of its ability to activate all known caspases *in vitro* and because of a report indicating that Bid, a caspase-8 substrate, was cleaved during chemical-induced apoptosis.[68,70] Pro-caspase-8 activation requires the cytoplasmic adaptor molecule FADD/MORT1, which is essential for CD95(Fas/APO-1)-mediated forms of apoptosis.[71] Traditionally, when Fas ligand binds to its receptor, FADD/MORT1 is recruited to the receptor through a death effector domain (DED). This allows binding of pro-caspase-8, auto-processing, and activation.[72–74] Once activated, caspase-8 can signal downstream activating the effector pro-caspase-3,[75] seemingly bypassing pro-caspase-9 and mitochondria. However, recent studies demonstrate the ability of mitochondria to significantly amplify Fas-mediated apoptosis by releasing cytochrome *c*.[76,77] Considerable evidence indicates that Bid (a pro-apoptotic Bcl-2 family protein) is cleaved by active caspase-8, yielding tBid (Figure 3). tBid, the active form of Bid, was shown to target mitochondria and stimulate the release of cytochrome *c*.[78,79] While the study by Bossy-Wetzel and Green[77] confirmed the ability of caspase-8 (and caspase-3) to signal through mitochondria by cleaving Bid, they also demonstrated the ability of caspase-6 and caspase-7, which were unable to cleave Bid, to stimulate cytochrome *c* release. However, this ability of caspase-6 and caspase-7 to stimulate cytochrome *c* release was indirect, since the presence of cytosol was essential. This finding suggests that a key caspase substrate, in addition to Bid, may exist in the cytosol and that when cleaved can also act on mitochondria stimulating the release of cytochrome *c*. Thus, recent evidence supports a role for mitochondria in apoptotic cell signaling induced by chemical stimuli. However, the relationship between caspase-8 and chemical-induced apoptosis is uncertain and requires further study.

### 14.4.7.2.3 Bcl-2 Protein Family

Original insight into the molecular nature of apoptotic cell death came from studies with the nematode *C. elegans*. Specifically, three different apoptotic loci were identified. These included two pro-apoptotic, Ced-3 and Ced-4, and one anti-apoptotic, Ced-9, proteins.[1,80] The mammalian equivalents of Ced-3 and Ced-9 are caspases (discussed above) and Bcl-2-related proteins, respectively. The mammalian equivalent to Ced-4 is Apaf-1.[67] It is clear from work with *C. elegans* that Ced-4 and Ced-3 can bind one another, and Ced-4 binding to Ced-3 is believed to be required for Ced-3 activation.[81] Moreover, it is known that Ced-4 and Ced-3 can interact with Ced-9. This led to the notion that Ced-9 may exert its anti-apoptotic effect in *C. elegans* by antagonizing the pro-apoptotic effect(s) of the Ced-4/Ced-3 complex, maintaining Ced-3 in its inactive or precursor form.[81]

*B-cell lymphoma-2* (*bcl-2*) was the first mammalian homologue of the *ced* gene family discovered, and its corresponding protein was shown to provide cell survival following cytokine withdrawal in a cytokine-dependent hematopoietic cell line.[82] Additional proteins, both pro- and anti-apoptotic, with sequence similarity to Bcl-2 have since been identified and taken together these proteins make up the Bcl-2 family of proteins.[83–85] Bcl-2 and Bcl-$X_L$, which share 47% amino acid homology, are two anti-apoptotic members of the Bcl-2 protein family found in mitochondrial, endoplasmic reticular, or nuclear membranes.[17,86] Bax, a pro-apoptotic member of the Bcl-2 family, is present as a monomer in the cytosol or loosely attached to membranes prior to an apoptotic stimulus and requires oligomerization for its activation.[87] In contrast, Bid—also a pro-apoptotic Bcl-2 family protein—requires caspase-8-mediated cleavage for its activation. It is important to note that these are not the only members of the Bcl-2 family of proteins but are four of the best characterized and most relevant in terms of chemical-induced apoptosis (Table 2).

All members of the Bcl-2 family contain up to four conserved domains denoted Bcl-2 homology domains (BH1 to BH4).[88,89] Bcl-2 and Bcl-$X_L$ possess all four domains. Bax, which resembles Bcl-2, possesses BH1, BH2, and BH3 domains. Evidence suggests that the β-helical BH3 domain is the critical 'death domain' among the pro-apoptotic proteins, which are related to the *C. elegans* death-driving protein Egl-1.[90,91] Moreover, a recent report (discussed above) indicates that BH3-only proteins promote apoptosis *via* a mechanism that is distinct from other pro-apoptotic Bcl-2 family proteins.[36]

A precise role for the anti-apoptotic Bcl-2 family proteins remains elusive. However, different cytoprotective functions have been ascribed to Bcl-2, including regulation of intracellular $Ca^{2+}$ homeostasis,[92,93] modulation of antioxidant pathways, and redistribution of glutathione to the nucleus—though

**Table 2** A list of anti- and pro-apoptotic members of the Bcl-2 family of proteins. All proteins are of mammalian origin except § and * which are viral and *C. elegans* proteins, respectively.

| Anti-apoptotic proteins | Pro-apoptotic proteins |
| --- | --- |
| Bcl-2 | Bax |
| Bcl-$X_L$ | Bcl-$X_S$ |
| Bcl-w | Bcl-$G_{L/S}$ |
| Boo | Bak |
| Mcl-1 | Bok |
| A1 | Bik |
| NR-13 | Blk |
| BHRF-1§ | Hrk |
| LMW5-HL§ | NIP3 |
| KS-Bcl-2§ | Nix |
| E1B-19K§ | BimL |
| ORF16§ | Bad |
| Ced-9* | Bid |
| | Egl-1* |

recently the ability of this protein to prevent mitochondrial changes that are characteristic of apoptosis has attracted the most interest. In particular, the release of cytochrome *c* is inhibited by elevated levels of Bcl-2 and/or Bcl-$X_L$, perhaps by binding and closing VDAC, a component of the PTP. On the other hand, there is evidence suggesting that Bcl-$X_L$, but not Bcl-2, can interact with and sequester Apaf-1,[94] which is required for apoptosome formation and the activation of caspase-9 (discussed above). However, these results have been challenged more recently because of an inability to show a physiologically significant level of interaction between Apaf-1 and Bcl-$X_L$.[95,96] Reports describing the 3-dimensional structure of Bcl-$X_L$ also reveal information about its function.[97] Specifically, this protein possesses structural similarity with the pore-forming domains of different bacterial toxins.[98,99] Thus, it may be that Bcl-$X_L$'s capacity to form pores in lipid membranes is instrumental for its anti-apoptotic function insofar as it maintains ion homeostasis across mitochondrial membranes.

There is accumulating evidence that pro-apoptotic Bax and tBid proteins translocate to mitochondria in the presence of a death stimulus, where they exert their pro-apoptotic effect (Figure 3). Bax is reported to integrate into the mitochondrial membrane as an oligomer in the presence of a pro-apoptotic stimulus.[99] Recent evidence suggests that a conformational change in, or removal of, 20 amino acids at the $NH_2$-terminal of this protein is a critical step in targeting Bax for the mitochondrial membrane.[100] Once integrated into the mitochondrial membrane, Bax is believed to elicit a pro-apoptotic response by stimulating the release of cytochrome *c*,[31,32] an effect capable of being blocked by Bcl-2 or Bcl-$X_L$ but not by caspase inhibition in most systems. Bid is a "BH3 domain only" protein that shares some structural similarity with the *C. elegans* death promoting protein Egl-1. Like Bax, Bid exists in an inactive state prior to an apoptotic stimulus and undergoes post-translational modification during its activation. Initial studies with Bid showed that an intact BH3 domain was essential for its ability to bind Bcl-2 and/or Bax and to promote cell death. As mentioned above, this model has been further developed recently by the understanding that activation of Bid involves cleavage by caspase-8 and insertion of this product, tBid, into the outer mitochondrial membrane.[101] Recently, other proteases, like granzyme B and lysosomal proteases, have been reported to cleave Bid, resulting in an amplication of their pro-apoptotic effect by mitochondrial recruitment.[102,103]

Although there is little argument about the pro-apoptotic potential of Bax and Bid and the anti-apoptotic potential of Bcl-2 and Bcl-X$_L$, the mechanisms responsible for eliciting such effects are unclear and require further study.

### 14.4.7.3 ENVIRONMENTAL TOXICANTS: SPECIFIC EXAMPLES

In recent years, this laboratory and others have focused on the ability of a variety of environmental toxicants to influence apoptosis.[104] While toxicologists have traditionally associated cell death with necrosis, it is becoming increasingly evident that different noxious stimuli can exert their toxicity via apoptotic cell death. In many cases, the precise influence an environmental contaminant has on apoptosis is dose-dependent, as well as cell type- and cell cycle-dependent.

#### 14.4.7.3.1 Aryl Hydrocarbon Receptor Agonists

2,3,7,8-Tetrachlorodibenzo-*p*-dioxin (TCDD), polychlorinated biphenyls (PCBs) and polyaromatic hydrocarbons (PAHs) are well-known environmental toxicants that bind to the ligand-activated transcription factor called aryl hydrocarbon receptor (AHR) (see Chapter 14.2.2). AHR is normally found in the cytosol bound to two molecules of Hsp90 that are sloughed upon ligand binding. This allows AHR to translocate to the nucleus where it forms a dimer with the aryl hydrocarbon receptor nuclear translocator (ARNT).[105] AHR/ARNT heterodimers interact with DNA elements of target genes encoding proteins involved in phase I and II biotransformation, cell proliferation and differentiation, and apoptosis.[106–111]

TCDD is recognized as a highly potent environmental toxicant. Adverse effects associated with TCDD exposure include wasting syndrome, immune, reproductive and developmental toxicities, skin lesions, teratogenesis, and carcinogenesis.[112] One of the most pronounced developmental toxicities associated with TCDD is thymic atrophy. This effect could involve a targeting of pro-thymocyte development, immature thymocytes, or supportive thymic epithelial cells that produce growth factors. Over a decade ago, McConkey *et al.*[113] reported that immature thymocytes were induced to undergo apoptosis in the presence of TCDD. This conclusion was, in part, because of similarities in internucleoso-mal DNA fragmentation patterns between cells treated with TCDD or the glucocorticoid methylprednisolone, an agent known to induce thymocyte apoptosis and, in part, because of the disappearance of methylprednisolone-sensitive thymocytes in young rats treated with TCDD.[114] Subsequently, the notion of TCDD-induced thymocyte apoptosis was challenged by Comment *et al.*[115] when they were unable to show a loss of cell viability in TCDD-treated thymocyte cultures. Moreover, the role of apoptosis in TCDD-induced thymic atrophy was further disputed because the concentration of this chemical needed to stimulate apoptosis *in vitro* was twice that needed to induce thymic atrophy *in vivo*.[116] The ability of lower doses of TCDD to induce apoptosis *in vivo* led to the notion that the pro-apoptotic effect of this chemical may be indirect, perhaps targeting thymic epithelial cells that, as already mentioned, are required for support.[116,117] To further complicate matters, more recent studies demonstrated the inability of the anti-apoptotic protein Bcl-2 to protect against thymic atrophy induced by TCDD in transgenic mice.[118,119] In the same study, the presence of Bcl-2 was able to rescue thymocytes from dexamethasone-induced apoptosis suggesting, in contrast to previous reports, that the mechanism for TCDD-induced thymic atrophy differs from that of glucocorticoids and does not involve apoptosis.[120,121]

While it is traditionally believed that TCDD exerts its toxicity via AHR,[122] there is recent evidence suggesting that AHR may not be important for TCDD toxicity in certain cell types.[123,124] Using the human leukemic lymphoblastic T cell line L-MAT, Hossain *et al.*[124] demonstrated the ability of TCDD to induce morphological and biochemical changes characteristic of apoptosis. These changes were inhibitable by the caspase inhibitor z-*Asp*. Furthermore, these changes did not require AHR, since functional AHR is not expressed in L-MAT cells. Moreover, transfecting these cells with human AHR had no effect on TCDD-induced apoptosis. Lastly, a decrease in Bcl-2 protein was observed prior to apoptosis leading to the speculation that removal of this protein is an important step in the pro-apoptotic response to TCDD in this system.

There is additional evidence indicating that TCDD may cause thymocyte apoptosis by up-regulating the production of soluble FasL, since thymocytes from Fas-deficient (*lpr*) or FasL-defective (*gld*) mice were resistant to TCDD-induced toxicity.[125] It should be noted that the notion of Fas-FasL involvement in toxicant-induced apoptosis is not without precedent. In particular, Richburg *et al.*[126]

demonstrated a decreased sensitivity of *gld* mice to germ cell apoptosis induced by the Sertoli cell-specific toxicant mono-(2-ethylhexyl)phthalate. Thus, it may become increasingly apparent that different types of toxicants—especially those that do not generate an oxidative stress and/or directly damage mitochondria—exert their effect, at least in part, by exploiting the Fas signaling system.

In conclusion, the cellular mechanisms mediating TCDD-induced thymic atrophy remain obscure. While there is some evidence supporting TCDD-induced apoptosis, using both *in vitro* and *in vivo* models, additional studies are needed to further characterize TCDD-mediated thymic atrophy given the multitude of potential target sites and therefore alternative mechanisms. Although TCDD was selected here as a model toxicant, other toxic chemicals, including PCBs and PAHs, that are capable of interacting with AHR are reported to induce apoptosis.[127–130] The ability of these chemicals to interact with AHR suggests that they may also share similar cell death mechanisms. Therefore, future studies designed to delineate the respective abilities of dioxins, PCBs, and PAHs to induce cell death, perhaps by apoptosis, should help characterize the importance of AHR in this process.

### 14.4.7.3.2 Metals

#### 14.4.7.3.2.1 Cadmium

Cadmium (Cd) is widely used in the manufacture of batteries, alloys, and pigments used in inks, dyes, paints, enamel and plastic. Although present in food, dietary intake of Cd is normally quite low and gastrointestinal absorption is minimal. The primary risk of human exposure is by inhalation of cadmium particles or fumes during industrial operations. Moreover, cigarette smoke represents a significant source of exposure. Widespread cadmium exposure occurred in Japan, known as Itai-Itai, when consumption of cadmium-contaminated rice led to renal dysfunction and osteomalacia.

The half-life of cadmium in the body is between 7 and 30 years, and a primary toxic effect of chronic Cd exposure involves kidney damage. Cd-induced nephrotoxicity is believed to involve the formation of a cadmium-metallothionein complex (CdMT) in the liver. Normally, the CdMT complex resides in the liver and is only released if this organ is damaged or the pool saturated. If transported to the kidney, CdMT is endocytosed, degraded by lysosomes,[131,132] and free $Cd^{2+}$ is liberated

resulting in damage to the proximal tubule.[133] Although Cd is not a Fenton metal, the mechanism of $Cd^{2+}$ toxicity seems to involve its ability to generate hydroxyl radicals in the presence of metallothioneins[134] and/or to damage mitochondria.[135] This suggests that $Cd^{2+}$ is able to diplace other redox active Fenton metals from metallothioneins and/or depress the level of endogenous thiol antioxidants. Alternatively, cadmium is reported to inactivate thioltransferase (glutaredoxin)-mediated deglutathionylation of protein-SSG in cells, which might account for this metal's ability to alter the intracellular redox status.[136]

The ability of $Cd^{2+}$ to induce biochemical and morphological characteristics of apoptosis has been demonstrated using *in vivo* and *in vitro* models,[137–140] and recently several studies characterizing the molecular framework for this process have appeared in the literature. Ishido *et al.*[141] demonstrated the ability of zinc to inhibit $Cd^{2+}$-induced apoptosis in porcine kidney LLC-PK$_1$ cells. These data, in part, support a previous report indicating that $Cd^{2+}$ was able to stimulate calcium-dependent endonuclease activity in isolated nuclei, an effect blocked by the presence of zinc.[142] However, micromolar concentrations of zinc were able to inhibit $Cd^{2+}$-induced apoptosis, whereas millimolar concentrations of zinc are required to inhibit $Cd^{2+}$-stimulated endonuclease activity, suggesting an alternate mechanism for the anti-apoptotic effect of zinc. The same study reported that zinc stimulates DNA synthesis and the production of Bcl-2, which they propose as important aspects of zinc's anti-apoptotic capacity. Lastly, these authors report that caspase activity is not associated with $Cd^{2+}$-induced apoptosis since caspase inhibitors were unable to rescue cells treated with $Cd^{2+}$. In contrast, a more recent report indicates that cadmium induces apoptosis by a caspase-dependent pathway in human U937 cells.[143] In this study, the authors show that different caspase inhibitors suppressed DNA fragmentation in the order of z-VAD-fmk > caspase-8 inhibitor > caspase-3 inhibitor. Consistent with these data, they also show that Bid, an established caspase-8 substrate, is cleaved to its active form in cadmium-treated cells.

Therefore, apoptosis seems to play a role in $Cd^{2+}$-mediated toxicity, although cell signaling mechanisms governing this effect are unclear. Because of its ability to disrupt $\Delta\psi$ and stimulate apoptosis, it would be interesting to see whether the release of cytochrome *c* occurs following $Cd^{2+}$ exposure. However, since caspases do not appear to be activated by $Cd^{2+}$, it would seem at this time that cytochrome $c$ is not involved. Nevertheless, additional studies are needed to further characterize the nature of mitochondrial involvement in $Cd^{2+}$-induced apoptosis.

### 14.4.7.3.2.2  Methylmercury

Methylmercury (MeHg) is highly toxic and was the culprit in a poisoning that occurred in the 1960s in Japan, known as Minimata disease. It was also involved in a poisoning in Iraq where grain that had been treated with MeHg-containing fungicide was used in bread production. The Japanese incident occurred when industrial emissions that contained inorganic mercury sedimented at the bottom of Minimata Bay. Microorganisms biotransformed the inorganic mercury into methyl and dimethyl mercury and because of the lipophilic nature of these forms of mercury, they entered the food chain when fish consumed organisms that had accumulated MeHg. Local residents who consumed the contaminated fish developed a neurological syndrome, and histological data revealed that areas of the brain controlling vision, including parts of the cerebral cortex, were targeted.[144]

Kunimoto[144] was one of the first to demonstrate the ability of MeHg to induce apoptosis in cerebellar neurons in primary culture. At doses up to 0.3 μM, MeHg was able to induce apoptosis as measured by DNA fragmentation. Concentrations of MeHg that exceeded 0.3 μM induced necrosis in this system. In a different study by Sarafian et al.,[145] overexpression of Bcl-2 in the neural cell line GT1–7 offered protection against MeHg-induced apoptosis. Protection by Bcl-2 correlated with reduced ROS production, which may be important in eliciting a pro-apoptotic response to this chemical.[145]

More recent studies have examined mechanisms involved in MeHg-induced apoptosis in the immune system.[146] Treated T cells exhibit reductions in the level of glutathione (GSH), change in $\Delta\psi$, induction of MPT, ROS generation, and cardiolipin oxidation.[147] Moreover, 8 h after 2.5 μM MeHgCl treatment, a 9-fold increase in mitochondrial cytochrome $c$ release is observed. Since ROS are capable of triggering apoptosis in different systems,[148] it seems likely that the interaction of MeHg with mitochondrial membranes results in uncoupling of mitochondria, production of ROS, depletion of GSH, and MPT induction that potentiates the release of cytochrome $c$. In fact, Shenker et al.[149] showed that MeHgCl caused a significant release of cytochrome $c$ and the activation of the caspase

cascade, whereas $HgCl_2$, while also stimulating caspase-3 activity, did not trigger cytochrome $c$ release—an effect that might be related to an increase in Bcl-2 induced by this particular mercurial species.

Therefore, MeHg-induced toxicity seems to occur by apoptosis in neuronal and immune systems. Although the involvement of caspases in MeHg-induced apoptosis has not been tested, the probability is high given the release of cytochrome $c$, an established apoptogen and component of the apoptosome.[150] The ability of Bcl-2 to protect against MeHg-mediated apoptosis supports a role for mitochondria in this form of toxicity since this protein is present in the outer mitochondrial membrane and is able to inhibit cytochrome $c$ release. Additional studies focusing on events downstream of cytochrome $c$ release are needed in order to further characterize pro-apoptotic cell signaling in response to MeHg.

### 14.4.7.3.3  Organotin Compounds

Recent studies in our laboratory have focused on mechanisms involved in tributyltin (TBT)- and triphenyltin (TPT)-induced apoptosis.[151–153] These chemicals have been used as agricultural fungicides and in anti-fouling paints used for watercraft. Several different types of toxicity have been associated with exposure to these compounds, including neural and renal damage, thymic atrophy, and impaired immune response.

Although a specific intracellular target for TBT-induced toxicity is unknown, this chemical triggers a rise in the level of intracelluar $Ca^{2+}$, cytoskeletal modifications, and disruption of normal mitochondrial function, including cessation of ATP production.[154–157] Stridh et al.[151] demonstrated the ability of low doses of TBT ($< 5$ μM) and TPT ($< 1$ μM) to activate pro-caspase-3-like enzymes in human T lymphocyte Jurkat and HUT-78 cell lines. Moreover, inhibition of caspase activity successfully blocked the apoptotic effect of these compounds. Higher doses of these chemicals induced necrosis. The key determinant controlling the mode of cell death induced by TBT was the intracellular level of ATP.[153] Since TBT targets mitochondria and disrupts ATP production, cells become more reliant on glycolytic ATP. If the level of intracellular ATP is maintained, mitochondrial swelling is observed followed by the release of cytochrome $c$ and pro-caspase activation. If, however, ATP is depleted, mitochondrial swelling is absent and pro-caspase activation is completely blocked. Furthermore and interestingly, higher doses of these compounds were able to inhibit caspase activity induced by anti-Fas treatment, which may be related to the ability of TBT to bind to essential thiol groups and thereby block caspase activity.

In summary, recent studies in our laboratory have demonstrated the ability of low doses of organotin compounds to induce apoptosis *in vitro* by disrupting mitochondrial integrity and stimulating caspase activity. However, the ability of these compounds to induce apoptosis *in vivo* is not known and therefore additional studies are needed to address this question.

#### 14.4.7.3.4 Dithiocarbamates

This broad class of molecules has been used as agricultural fungicides, herbicides, and insecticides. They are also used medically in chemotherapy, alcohol deterrent drugs, and antidotes against metal poisoning. Dithiocarbamates (DCs) are also used in cellular and molecular biology studies, where they are reported to have both anti- and pro-oxidant effects.[158,159]

Recently, studies have focused on the ability of this class of molecules to influence apoptotic cell death. For instance, our laboratory and others have demonstrated the ability of pyrrolidine dithiocarbamate (PDTC) to inhibit apoptosis in a variety of systems.[160,161] While others have proposed that this effect may be related to the antioxidant properties of PDTC, our laboratory showed that the ability of PDTC to inhibit etoposide-induced apoptosis in thymocytes correlated with an increase in glutathione disulfide (GSSG), an effect indicative of a change in the cellular redox level to a more oxidized state.[162,163] Moreover, this effect was accompanied by a decrease in pro-caspase-3 activation, which was found to involve the formation of a mixed disulfide between DC disulfide and cysteine residue(s) on pro-caspase-3.[162] Additional *in vitro* experiments revealed that GSSG, by itself, was sufficient to inhibit pro-caspase-3 processing, suggesting that formation of active caspase-3 is dependent on the maintenance of thiol redox status in a more reduced state. However, longer (> 6 h) incubations with PDTC stimulated apoptosis and were also accompanied by an increase in GSSG levels. The ability of PDTC to induce apoptosis was dependent upon the transport of Cu ions into the cell, since the presence of a Cu chelator, bathocuproinedisulfonic acid, abrogated the pro-apoptotic effect of PDTC.[163] These data were further supported by the presence of 8-fold higher levels of intracellular Cu after PDTC treatment alone.

A subsequent study in our laboratory was performed to characterize the mechanism involved in PDTC-induced apoptosis.[164] Results showed that DC toxicity involves a Cu-catalyzed oxidation to DC disulfides, which are subsequently reduced by GSH, yielding GSSG. As previously mentioned, DC disulfides are able to inhibit pro-caspase-3 processing in the short term but become pro-apoptotic at longer time-points. An explanation for this could be that DC disulfides are metabolized over time by various enzymatic pathways to isothiocyanates, sulfoxides, or sulfones and thus lose their anti-apoptotic capacity.[162] Alternatively, a more recent study demonstrated the ability of PDTC to directly stimulate the release of cytochrome *c* from isolated mitochondria. Moreover, the possible involvement of receptor-mediated pathways was ruled out by time-course analyses of caspase-3 *versus* caspase-8 activation.[165]

Therefore, the ability of dithiocarbamates to influence apoptosis seems clear. However, the precise nature of their involvement remains somewhat obscure. Future studies should focus on the paradoxical issue of dithiocarbamates possessing both anti- and pro-apoptotic effects. Does metabolism, in fact, account for the lost anti-apoptotic capacity of DC disulfides over time or is some other pro-apoptotic pathway operative? Lastly, *in vivo* data currently do not exist and are needed to address the physiological significance of these *in vitro* reports.

#### 14.4.7.3.5 Benzene

Benzene is a widespread environmental pollutant present in cigarette smoke, gasoline emissions, and incomplete combustion products. It is a potent hematotoxicant capable of inducing aplastic anemia and acute myeloid leukemia. Benzene is metabolized in the liver by cytochrome P-450 2E1, yielding catechol, hydroquinone, phenol, and benzenetriol as byproducts. In turn, these metabolites are able to migrate to, and accumulate in, the bone marrow where they are converted by peroxidases to highly active quinones, the ultimate toxic metabolites of benzene.[166,167] NAD(P)H:-quinone oxidoreductase (NQO1) can detoxify quinones by performing two electron reductions, which prevents covalent binding to, and modification of, other cellular constituents[168] and/or redox cycling.[169]

A primary mechanism by which benzene metabolites exert their toxicity involves their ability to redox cycle and produce ROS, including superoxide anion, hydrogen peroxide and, ultimately, the highly reactive hydroxyl radical.[170] Evidence supports the ability of catechol and hydroquinone to exert their toxicities by stimulating apoptosis, whereas phenol does not.[171] Specifically, Moran et al.[171] treated human promyelocytic leukemia cells (HL-60) and CD34+ human bone marrow progenitor cells with catechol and hydroquinone and observed the respective abilities of these compounds to induce apoptosis, whereas Bratton et al.,[172] using the same cell line, reported the ability of 2,3,5-tris(glutathion-S-yl)hydroquinone to deplete intracellular GSH and ultimately activate caspase-3-like proteases. A different study,[173] using HL-60 cells transfected with NQO1, demonstrated the ability of this enzyme to decrease the level of benzene metabolite-DNA adducts and offer modest protection from apoptosis. A more recent study demonstrated the ability of Bcl-2 overexpression to block apoptosis induced by benzene metabolites (1,4-benzoquinone and 1,4-hydroquinone) in HL-60 cells.[174] However, Bcl-2 was not able to prevent ROS production stimulated by these metabolites of benzene, indicating that Bcl-2's cytoprotective effect lies downstream of this event. Strikingly, 8-hydroxydeoxyguanosine, a marker of oxidative DNA damage, was more efficiently removed in neo-transfected cells than in Bcl-2-overexpressing cells suggesting that Bcl-2, in addition to blocking apoptosis, is able to suppress DNA repair enzymes and thus potentially allow the accumulation of gene mutations.

Taken together, these reports support a role for apoptosis in benzene metabolite-mediated toxicity. The nature of the pro-apoptotic effect of these compounds seems to involve their ability to redox cycle and produce ROS. While Bcl-2 was unable to block ROS production as compared to neo-transfected cells, this protein effectively prevented apoptosis, although the specific site of interaction of Bcl-2 in this system is unknown. Finally, whether benzene metabolites are also able to induce apoptosis *in vivo* is unknown and should be studied.

### 14.4.7.4 CONCLUDING REMARKS

A central theme among different forms of apoptosis induced by chemical agents is the involvement of mitochondria (Figure 3). Specifically, it seems that the release of cytochrome *c*, if not a universal event, occurs in a majority of chemical-induced apoptotic models. Once released, cytochrome *c* is able to activate Apaf-1, which oligomerizes in the presence of dATP and recruits pro-caspase-9. Active caspase-9, in turn, cleaves and activates effector pro-caspases, which are responsible for carrying out

biochemical and morphological changes characteristic of apoptosis. Although the release of cytochrome *c* can be promoted by Bax and/or tBid integration into the mitochondrial membrane and prevented by the presence of Bcl-2 and Bcl-X$_L$, fundamental questions remain as to the mechanisms controlling these events. What signal is responsible for triggering Bax translocation to the mitochondria? How is cytochrome *c* released and how do Bcl-2 and Bcl-X$_L$ inhibit this event? Which caspase substrate(s) must be cleaved in order to execute apoptosis? What is the function of nuclear Bcl-2?

Much of our understanding about the molecular mechanisms of apoptosis originates from genetic studies performed on the nematode *C. elegans*, where it was determined that specific gene products are required in order to execute, and protect against, this form of cell death.[175] Since those original discoveries, numerous studies have revealed that similar apoptotic pathways are present in other organisms, including mammals, and may be involved in the onset of various diseases. Many of these studies have used, and continue to use, cytotoxic drugs, e.g., etoposide and dexamethasone, ultraviolet radiation, growth factor withdrawal, overexpression of endogenous pro- or anti-apoptotic proteins, or death receptor agonists in order to characterize and manipulate critical steps in the apoptotic program.

As mentioned previously, an interest in apoptosis is emerging within the field of toxicology as more researchers begin to examine the ability of various environmental contaminants to induce this form of cell death. Evidence of this is reflected by an increasing number of studies reporting the ability of established pollutants to exert their toxicity, in part, by promoting apoptosis. Despite the limited scope of this chapter with respect to the number of chemicals discussed versus the number of xenobiotics present in the environment, there is evidence to suggest a general role for apoptosis in environmental contaminant-induced toxicity.

With the exception of TCDD-induced toxicity, the other examples discussed here are reported to exert their apoptotic effect, at least in part, by generating ROS resulting in mitochondrial damage (Figure 3). Given the number of reports indicating that mitochondria are involved in apoptotic cell signaling induced by various stimuli, it could be that a common mechanism among environmental toxicants capable of damaging mitochondria is their ability to stimulate the release of cytochrome *c* into the cytosol, subsequently triggering the caspase cascade. While recent studies with MeHg and organotin compounds support a role for cytochrome *c* in apoptosis induced by these compounds, overall there is a paucity of information available addressing the participation of mitochondria in biochemical and molecular mechanisms of environmental contaminant-induced apoptosis.

In many instances, the answer to the question of whether a particular environmental toxicant exerts its effect by stimulating apoptosis seems to depend on the experimental system employed. Consequently, *in vitro* and *in vivo* data are often in disagreement. This is evident in the case of TCDD where it is still unclear what role apoptosis plays in the ability of this chemical to cause thymic atrophy. As mentioned previously, the dose of TCDD that is needed to stimulate apoptosis *in vitro* is two times higher than that needed to induce thymic atrophy *in vivo*.[116] However, dose-related issues are not unique to TCDD as evidenced by the ability of a relatively high dose of TBT to actually protect against pro-caspase activation and apoptosis in cytosols isolated from Fas-treated cells, whereas a lower dose of this toxicant is a potent inducer of MPT and cytochrome *c* release from mitochondria. Moreover, depending on the duration of exposure to PDTC, this compound can exert either a pro- or anti-apoptotic effect in treated thymocytes. The same is true for $H_2O_2$ in Jurkat cells and menadione in HepG2 cells.[176,177] Therefore, in some cases the experimental outcome seems to depend considerably on the model and/or dose used. This, combined with an overall shortage of mechanistic studies characterizing environmental toxicant-induced apoptosis, is an indication that it is still too early to draw any substantive conclusions as to the precise role apoptosis plays in cell death stimulated by these agents.

While uncertainties regarding the ability of different xenobiotics to induce apoptosis may involve limitations inherent in the methods used to evaluate apoptosis, it seems more likely that such confusion arises from the absence of a uniformly accepted definition of this form of cell death, as well as an incomplete characterization of the apoptotic pathway. Recently, it was proposed that apoptosis be defined based on the activation of caspases coupled with the presence of distinct morphological criteria.[42] Although there are reports of caspase-independent apoptosis, e.g., $Cd^{2+}$,[141] the prevailing trend within the literature indicates that caspase activation is an essential step in this form of cell death induced by cytotoxic stimuli. Furthermore, many reports of caspase-independent apoptosis were performed using the synthetic, broad-spectrum inhibitor z-VAD-fmk. Even though z-VAD-fmk is capable of inhibiting all known caspases,[178] this does not preclude the possibility that other, yet-to-be-identified caspases are operant in chemical-induced apoptosis—a conceivable scenario given the rate at which this family of proteases has grown in recent years.

It is important to note that, although this chapter focuses on the ability of environmental chemicals to exert their toxicity by inducing apoptosis, a number of environmental toxicants, e.g., nitrosamines and aflatoxin $B_1$, are mutagenic carcinogens that are able to stimulate tumorigenesis by forming DNA adducts. Recent studies have focused on the mutation of the prototypic tumor suppressor gene p53, which can result in the circumvention of a "death signal" and allow the retention of cells that should otherwise have been removed.[179,180] p53, normally expressed in response to DNA damage and/or hypoxia, arrests cells in the G1 phase of the cell cycle, allowing DNA repair or apoptosis if damage is too severe. Current evidence indicates that p53 mutations are present in most human cancers and can result from infection by Hepatitis B virus or exposure to aflatoxin $B_1$. Other studies have examined the ability of established tumor promoters, e.g., PMA, DES, and DDT, to inhibit apoptosis induced by DNA-damaging stimuli. In particular, Wright *et al.*[181] demonstrated the ability of 10 different tumor promoters to block apoptosis, as determined by DNA fragmentation, in a variety of carcinoma cell lines. In this respect, the presence of mutant p53 and/or exposure to environmental chemicals that can suppress apoptosis may be important in the development of neoplastic cells and/or tumorigenesis.

In conclusion, despite the feverish pace at which basic apoptotic research is progressing, relatively little is known about the specific involvement of apoptosis in environmental pollutant-induced toxicity. In that regard, this field is still in its infancy and ample opportunity exists for characterizing the nature of this form of apoptosis. Furthermore, there is considerable evidence indicating that certain cell types are more prone to undergo apoptosis, which should assist researchers in determining potential target sites for investigation. A benefit of the explosion in apoptosis research involves the availability of sensitive, reliable, and specific techniques that allow for thorough investigation of apoptotic mechanisms. Taken together, the tools and knowledge are available to pursue a more detailed understanding of the role that apoptosis plays in cell death induced by cytotoxic chemicals.

## 14.4.7.5 REFERENCES

1. H. M. Ellis and H. R. Horvitz, 'Genetic control of programmed cell death in the nematode *C. elegans.*' *Cell*, 1986, **44**, 817–829.
2. J. F. R. Kerr, A. H. Wyllie and A. R. Currie, 'Apoptosis: a basic biological phenomenon with wide-ranging implications in tissue kinetics.' *Br. J. Cancer*, 1972, **26**, 239–257.
3. M. Skalka, J. Matyasova and M. Cejkova, 'DNA in chromatin of irradiated lymphoid tissues degrades *in vivo* into regular fragments.' *FEBS Lett.*, 1976, **72**, 271–275.
4. A. H. Wyllie, 'Glucocorticoid-induced thymocyte apoptosis is associated with endogenous endonuclease activation.' *Nature*, 1980, **284**, 555–556.
5. J. F. R. Kerr and B. V. Harmon, 'Apoptosis: the molecular basis of cell death,' Cold Spring Harbor Laboratory Press, New York, 1991.
6. M. R. Wilson, 'Apoptosis: unmasking the executioner.' *Cell Death Differ.*, 1998, **5**, 646–652.
7. R. T. Allen, W. J. Hunter and D. K. Agrawal, 'Morphological and biochemical characterization and analysis of apoptosis.' *J. Pharmacol. Toxicol. Methods*, 1997, **37**, 215–228.
8. B. Fadeel, S. Orrenius and B. Zhivotovsky, 'Apoptosis in human disease: a new skin for the old ceremony?' *Biochem. Biophys. Res. Commun.*, 1999, **266**, 699–717.
9. B. F. Trump and I. K. Berezesky, 'Calcium-mediated cell injury and cell death.' *FASEB J.*, 1995, **9**, 219–228.
10. P. Nicotera, M. Leist and E. Ferrando-May, 'Intracellular ATP, a switch in the decision between apoptosis and necrosis.' *Toxicol. Lett.*, 1998, **102–103**, 139–142.
11. M. Leist, B. Single, A. F. Castoldi *et al.*, 'Intracellular ATP concentration: a switch deciding between apoptosis and necrosis.' *J. Exp. Med.*, 1997, **185**, 1481–1486.
12. P. Li, D. Nijhawan, I. Budihardjo *et al.*, 'Cytochrome *c* and dATP-dependent formation of Apaf-1/caspase-9 complex initiates apoptotic protease cascade.' *Cell*, 1997, **91**, 479–489.
13. S. H. Kaufmann, 'Cell death induced by topoisomerase-targeted drugs: more questions than answers.' *Biochem. Biophys. Acta*, 1998, **1400**, 195–211.
14. S. T. Harkin, G. M. Cohen and A. Gescher, 'Modulation of apoptosis in rat thymocytes by analogs of staurosporine: lack of direct association with inhibition of protein kinase C.' *Mol. Pharmacol.*, 1998, **54**, 663–670.
15. J. Beyette, G. G. Mason, R. Z. Murray *et al.*, 'Proteasome activities decrease during dexamethasone-induced apoptosis of thymocytes.' *Biochem. J.*, 1998, **332**, 315–320.
16. J. D. Robertson, K. Datta, S. S. Biswal *et al.*, 'Heat-shock protein 70 antisense oligomers enhance proteasome inhibitor-induced apoptosis.' *Biochem. J.*, 1999, **344**, 477–485.
17. D. Hockenbery, G. Nuñez, C. Milliman *et al.*, 'Bcl-2 is an inner mitochondrial membrane protein that blocks programmed cell death.' *Nature*, 1990, **348**, 334–336.
18. M. D. Jacobson, J. F. Burne, M. P. King *et al.*, 'Bcl-2 blocks apoptosis in cells lacking mitochondrial DNA.' *Nature*, 1993, **361**, 365–369.
19. D. D. Newmeyer, D. M. Farschon and J. C. Reed, 'Cell-free apoptosis in Xenopus egg extracts: inhibition by Bcl-2 and requirement for an organelle fraction enriched in mitochondria.' *Cell*, 1994, **79**, 353–364.
20. D. R. Green and J. C. Reed, 'Mitochondria and apoptosis.' *Science*, 1998, **281**, 1309–1312.
21. X. Liu, C. N. Kim, J. Yang *et al.*, 'Induction of apoptotic program in cell-free extracts: requirement for dATP and cytochrome *c.*' *Cell*, 1996, **86**, 147–157.
22. J. Yang, X. Liu, K. Bhalla *et al.*, 'Prevention of apoptosis by Bcl-2: release of cytochrome *c* from mitochondria blocked.' *Science*, 1997, **275**, 1129–1132.
23. R. M. Kluck, E. Bossy-Wetzel, D. R. Green *et al.*, 'The release of cytochrome *c* from mitochondria: a primary site for Bcl-2 regulation of apoptosis.' *Science*, 1997, **275**, 1132–1136.
24. P. Li, D. Nijhawan, I. Budihardjo *et al.*, 'Cytochrome *c* and dATP dependent formation of Apaf-1/caspase-9 complex initiate an apoptotic protease cascade.' *Cell*, 1997, **91**, 479–489.
25. M. Crompton, 'The mitochondrial permeability transition pore and its role in cell death.' *Biochem. J.*, 1999, **341**, 233–249.
26. M. B. Hampton, B. Zhivotovsky, A. F. Slater *et al.*, 'Importance of the redox state of cytochrome *c* during caspase activation in cytosolic extracts.' *Biochem. J.*, 1998, **329**, 95–99.
27. J. C. Yang and G. A. Cortopassi, 'Induction of the mitochondrial permeability transition causes release of the apoptotgenic factor cytochrome *c.*' *Free Radical Biol. Med.*, 1998, **24**, 624–631.
28. V. Gogvadze, J. D. Robertson, B. Zhivotovsky *et al.*, 'Cytochrome *c* release occurs via $Ca^{2+}$-dependent and $Ca^{2+}$-independent mechanisms that are regulated by Bax.' *J. Biol. Chem.*, 2001, **276**, 19066–19071.
29. I. Martinou, S. Desagher, R. Eskes *et al.*, 'The release of cytochrome *c* from mitochondria during apoptosis of NGF-deprived sympathetic neurons is a reversible event.' *J. Cell. Biol.*, 1999, **144**, 883–889.
30. J. D. Robertson, V. Gogvadze, B. Zhivotovsky *et al.*, 'Distinct pathways for stimulation of cytochrome *c* release by etoposide.' *J. Biol. Chem.*, 2000, **275**, 32438–32443.
31. R. Eskes, B. Antonsson, A. Osen-Sand *et al.*, 'Bax-induced cytochrome *c* release from mitochondria is independent of the permeability transition pore but highly dependent on $Mg^{2+}$ ions.' *J. Cell. Biol.*, 1998, **143**, 217–224.
32. D. M. Finucane, E. Bossy-Wetzel, N. J. Waterhouse *et al.*, 'Bax-induced caspase activation and apoptosis via cytochrome *c* release from mitochondria is inhibitable by Bcl-xL.' *J. Biol. Chem.*, 1999, **274**, 2225–2233.
33. O. von Ahsen, C. Renken, G. Perkins *et al.*, 'Preservation of mitochondrial structure and function after Bid- or Bax-mediated cytochrome *c* release.' *J. Cell. Biol.*, 2000, **150**, 1027–1036.
34. T.-H. Kim, Y. Zhao, M. J. Barber *et al.*, 'Bid-induced cytochrome *c* release is mediated by a pathway independent of mitochondrial permeability transition pore and Bax.' *J. Biol. Chem.*, 2000, **275**, 39474–39481.
35. S. Shimizu, M. Narita and Y. Tsujimoto, 'Bcl-2 family proteins regulate the release of apoptogenic cytochrome *c* by the mitochondrial channel VDAC.' *Nature*, 1999, **399**, 483–487.
36. S. Shimizu and Y. Tsujimoto, 'Proapoptotic BH3-only Bcl-2 family members induce cytochrome *c* release, but not mitochondrial membrane potential loss, and do not directly modulate voltage-dependent anion channel activity.' *Proc. Nat. Acad. Sci. USA*, 2000, **97**, 577–582.
37. M. Lutter, M. Fang, X. Luo *et al.*, 'Cardiolipin provides specificity for targeting of tBid to mitochondria.' *Nat. Cell Biol.*, 2000, **2**, 754–761.

38. K. Nomura, H. Imai, T. Koumura *et al.*, 'Mitochondrial phospholipid hydroperoxide glutathione peroxidase inhibits the release of cytochrome *c* from mitochondria by suppressing the peroxidation of cardiolipin in hypoglycaemia-induced apoptosis.' *Biochem. J.*, 2000, **351**, 183–193.

39. S. A. Susin, N. Zamzami, M. Castedo *et al.*, 'Bcl-2 inhibits the mitochondrial release of an apoptogenic protease.' *J. Exp. Med.*, 1996, **184**, 1331–1341.

40. S. A. Susin, N. Zamzami, M. Castedo *et al.*, 'The central executioner of apoptosis: multiple connections between protease activation and mitochondria in Fas/APO-1/CD95- and ceramide-induced apoptosis.' *J. Exp. Med.*, 1997, **186**, 25–37.

41. E. S. Alnemri, D. J. Livingston, D. W. Nicholson *et al.*, 'Human ICE/CED-3 protease nomenclature.' *Cell*, 1996, **87**, 171.

42. N. A. Thornberry and Y. Lazebnik, 'Caspases: enemies within.' *Science*, 1998, **28**, 1312–1316.

43. A. Samali, B. Zhivotovsky, D. Jones *et al.*, 'Apoptosis: cell death defined by caspase activation.' *Cell Death Differ.*, 1999, **6**, 495–496.

44. D. W. Nicholson and N. A. Thornberry, 'Caspases: killer proteases.' *Trends. Biochem. Sci.*, 1997, **22**, 299–306.

45. B. Zhivotovsky, A. Samali, A. Gahm *et al.*, 'Caspases: their intracellular localization and translocation during apoptosis.' *Cell Death Differ.*, 1999, **6**, 644–651.

46. C. Lemaire, 'Inhibition of caspase activity induces a switch from apoptosis to necrosis.' *FEBS Lett.*, 1998, **425**, 266–270.

47. S. Wang, M. Miura, Y. K. Jung *et al.*, 'Murine caspase-11, an ICE-interacting protease, is essential for the activation of ICE.' *Cell*, 1998, **92**, 501–509.

48. M. Whyte and G. Evan, 'Apoptosis. The last cut is the deepest.' *Nature*, 1995, **376**, 17–18.

49. E. A. Slee, M. T. Harte, R. M. Kluck *et al.*, 'Ordering the cytochrome *c*-initiated caspase cascade: hierarchical activation of caspases-2, -3, -6, -7, -8, and -10 in a caspase-9-dependent manner.' *J. Cell Biol.*, 1999, **144**, 281–292.

50. V. Cryns and J. Yuan, 'Proteases to die for.' *Genes Dev.*, 1998, **12**, 1551–1570.

51. D. W. Nicholson, 'Caspase structure, proteolytic substrates, and function during apoptotic cell death.' *Cell Death Differ.*, 1999, **6**, 1028–1042.

52. L. Casciola-Rosen, D. W. Nicholson, T. Chong *et al.*, 'Apopain/CPP32 cleaves proteins that are essential for cellular repair: A fundamental principle of apoptotic death.' *J. Exp. Med.*, 1996, **183**, 1957–1964.

53. M. Enari, H. Sakahira, H. Yokoyama *et al.*, 'A caspase-activated DNase that degrades DNA during apoptosis, and its inhibitor ICAD.' *Nature*, 1998, **391**, 43–50.

54. X. Liu, H. Zou, C. Slaughter *et al.*, 'DFF, a heterodimeric protein that functions downstream of caspase-3 to trigger DNA fragmentation during apoptosis.' *Cell*, 1997, **89**, 175–184.

55. S. Mitamura, H. Ikawa, N. Mizuno *et al.*, 'Cytosolic nuclease activated by caspase-3 and inhibited by DFF-45.' *Biochem. Biophys. Res. Commun.*, 1998, **243**, 480–484.

56. D. Tang and V. J. Kidd, 'Cleavage of DFF-45/ICAD by multiple caspases is essential for its function during apoptosis.' *J. Biol. Chem.*, 1998, **273**, 28549–28552.

57. J. Zhang, X. Liu, D. C. Schere *et al.*, 'Resistance to DNA fragmentation and chromatin condensation in mice lacking the DNA fragmentation factor 45.' *Proc. Nat. Acad. Sci. USA*, 1998, **95**, 12480–12485.

58. L. Rao, D. Perez and E. White, 'Lamin proteolysis facilitates nuclear events during apoptosis.' *J. Cell Biol.*, 1996, **135**, 1441–1455.

59. H. Hirata, A. Takahashi, S. Kobayashi *et al.*, 'Caspases are activated in a branched protease cascade and control distinct downstream processes in Fas-induced apoptosis.' *J. Exp. Med.*, 1998, **187**, 587–600.

60. D. J. McConkey, 'Calcium-dependent, interleukin 1β-converting enzyme inhibitor-insensitive degradation of lamin B1 and DNA fragmentation in isolated thymocyte nuclei.' *J. Biol. Chem.*, 1996, **271**, 22398–22406.

61. B. Zhivotovsky, A. Gahm and S. Orrenius, 'Two different proteases are involved in the proteolysis of lamin during apoptosis.' *Biochem. Biophys. Res. Commun.*, 1997, **233**, 96–101.

62. S. Kothakota, T. Azuma, C. Reinhard *et al.*, 'Caspase-3-generated fragment of gelsolin: effector of morphological change in apoptosis.' *Science*, 1997, **278**, 294–298.

63. S. Kamada, H. Kusano, H. Fujita *et al.*, 'A cloning method for caspase substrates that uses the yeast two-hybrid system: cloning of the antiapoptotic gene gelsolin.' *Proc. Nat. Acad. Sci. USA*, 1998, **95**, 8532–8537.

64. S. Sahara, M. Aoto, Y. Eguchi *et al.*, 'Acinus is a caspase-3-activated protein required for apoptotic chromatin condensation.' *Nature*, 1999, **401**, 168–172.

65. S. A. Susin, H. K. Lorenzo, N. Zamzami *et al.*, 'Molecular characterization of mitochondrial apoptosis-inducing factor.' *Nature*, 1999, **397**, 441–446.

66. C. Köhler, A. Gahm, T. Noma *et al.*, 'Release of adenylate kinase 2 from the mitochondrial intermembrane space during apoptosis.' *FEBS Lett.*, 1999, **447**, 10–12.

67. H. Zou, W. J. Henzel, X. Liu *et al.*, 'Apaf-1, a human protein homologous to *C. elegans* CED-4, participates in cytochrome *c*-dependent activation of caspase-3.' *Cell*, 1997, **90**, 405–413.

68. X. M. Sun, M. MacFarlane, J. Zhuang *et al.*, 'Distinct caspase cascades are initiated in receptor-mediated and chemical-induced apoptosis.' *J. Biol. Chem.*, 1999, **274**, 5053–5060.

69. A. D. Tepper, E. de Vries, W. J. van Blitterswijk *et al.*, 'Ordering of ceramide formation, caspase activation, and mitochondrial changes during CD95- and DNA damage-induced apoptosis.' *J. Clin. Invest.*, 1999, **103**, 971–978.

70. S. M. Srinivasula, M. Ahmad, T. Fernandes-Alnemri *et al.*, 'Molecular ordering of the Fas-apoptotic pathway: the Fas/APO-1 protease Mch5 is a CrmA-inhibitable protease that activates multiple Ced-3/ICE-like cysteine proteases.' *Proc. Nat. Acad. Sci. USA*, 1996, **93**, 14486–14491.

71. A. Strasser and K. Newton, 'FADD/MORT1, a signal transducer that can promote cell death or cell growth.' *Int. J. Biochem. Cell. Biol.*, 1999, **31**, 533–537.

72. M. Muzio, A. M. Chinnaiyan, F. C. Kischkel *et al.*, 'FLICE, a novel FADD-homologous ICE/CED-3-like protease, is recruited to the CD95 (Fas/APO-1) death-inducing signaling complex.' *Cell*, 1996, **85**, 817–827.

73. M. P. Boldin, T. M. Goncharov, Y. V. Goltsev *et al.*, 'Involvement of MACH, a novel MORT1/FADD-interacting protease, in Fas/APO-1- and TNF receptor-induced cell death.' *Cell*, 1996, **85**, 803–815.

74. J. P. Medema, C. Scaffidi, F. C. Kischkel *et al.*, 'FLICE is activated by association with the CD95 death-inducing signaling complex (DISC).' *EMBO J.*, 1997, **16**, 2794–2804.

75. H. R. Stennicke, J. M. Jurgensmeier, H. Shin *et al.*, 'Pro-caspase-3 is a major physiologic target of caspase-8.' *J. Biol. Chem.*, 1998, **273**, 27084–27090.

76. T. Kuwana, J. J. Smith, M. Muzio *et al.*, 'Apoptosis induction by caspase-8 is amplified through the mitochondrial release of cytochrome *c*.' *J. Biol. Chem.*, 1998, **273**, 16589–16594.

77. E. Bossy-Wetzel and D. R. Green, 'Caspases induce cytochrome *c* release from mitochondria by activating cytosolic factors.' *J. Biol. Chem.*, 1999, **274**, 17484–17490.

78. H. Li, H. Zhu, C. J. Xu *et al.*, 'Cleavage of BID by caspase 8 mediates the mitochondrial damage in the Fas pathway of apoptosis.' *Cell*, 1998, **94**, 491–501.

79. X. Luo, I. Budihardjo, H. Zou *et al.*, 'Bid, a Bcl2 interacting protein, mediates cytochrome *c* release from mitochondria in response to activation of cell surface death receptors.' *Cell*, 1998, **94**, 481–490.

80. R. E. Ellis, D. M. Jacobson and H. R. Horvitz, 'Genes required for the engulfment of cell corpses during programmed cell death in *Caenorhabditis elegans*.' *Genetics*, 1991, **129**, 79–94.

81. A. M. Chinnaiyan, K. O'Rourke, B. R. Lane *et al.*, 'Interaction of CED-4 with CED-3 and CED-9: a molecular framework for cell death.' *Science*, 1997, **275**, 1122–1126.

82. D. L. Vaux, S. Cory and J. M. Adams, 'Bcl-2 gene promotes haemopoietic cell survival and cooperates with c-myc to immortalize pre-B cells.' *Nature*, 1988, **335**, 440–442.

83. B. Fadeel, B. Zhivotovsky and S. Orrenius, 'All along the watchtower: on the regulation of apoptosis regulators.' *FASEB J.*, 1999, **13**, 1647–1657.

84. A. Strasser, D. C. Huang and D. L. Vaux, 'The role of the bcl-2/ced-9 gene family in cancer and general implications of defects in cell death control for tumourigenesis and resistance to chemotherapy.' *Biochem. Biophys. Acta*, 1997, **1333**, F151-F178.

85. D. T. Chao and S. J. Korsmeyer, 'BCL-2 family: regulators of cell death.' *Annu. Rev. Immunol.*, 1998, **16**, 395–419.

86. S. Krajewski, S. Tanaka, S. Takayama *et al.*, 'Investigation of the subcellular distribution of the bcl-2 oncoprotein: residence in the nuclear envelope, endoplasmic reticulum, and outer mitochondrial membranes.' *Cancer Res.*, 1993, **53**, 4701–4714.

87. Y.-T. Hsu and R. J. Youle, 'Bax in murine thymus is a soluble monomeric protein that displays differential detergent-induced conformations.' *J. Biol. Chem.*, 1998, **273**, 10777–10783.

88. J. M. Adams and S. Cory, 'The Bcl-2 protein family: arbiters of cell survival.' *Science*, 1998, **281**, 1322–1326.

89. A. Gross, J. M. McDonnell and S. J. Korsmeyer, 'BCL-2 family members and the mitochondria in apoptosis.' *Genes Dev.*, 1999, **13**, 1899–1911.

90. A. Kelekar and C. B. Thompson, 'Bcl-2-family proteins: the role of the BH3 domain in apoptosis.' *Trends Cell Biol.*, 1998, **8**, 324–330.

91. B. Conradt and H. R. Horvitz, 'The *C. elegans* protein EGL-1 is required for programmed cell death and interacts with the Bcl-2-like protein CED-9.' *Cell*, 1998, **93**, 519–529.

92. J. Hacki, L. Egger, L. Monney *et al.*, 'Apoptotic crosstalk between the endoplasmic reticulum and mitochondria controlled by Bcl-2.' *Oncogene*, 2000, **19**, 2286–2295.

93. L. Zhu, S. Ling, X. D. Yu *et al.*, 'Modulation of mitochondrial $Ca^{2+}$ homeostasis by Bcl-2.' *J. Biol. Chem.*, 1999, **274**, 33267–33273.

94. Y. Hu, M. A. Benedict, D. Wu *et al.*, 'Bcl-XL interacts with Apaf-1 and inhibits Apaf-1-dependent caspase-9 activation.' *Proc. Nat. Acad. Sci. USA*, 1998, **95**, 4386–4391.

95. K. Moriishi, D. C. Huang, S. Cory *et al.*, 'Bcl-2 family members do not inhibit apoptosis by binding the caspase activator Apaf-1.' *Proc. Nat. Acad. Sci. USA*, 1999, **96**, 9683–9688.

96. G. Hausmann, L. A. O'Reilly, R. van Driel *et al.*, 'Pro-apoptotic apoptosis protease-activating factor 1 (Apaf-1) has a cytoplasmic localization distinct from Bcl-2 or Bcl-x(L).' *J. Cell. Biol.*, 2000, **149**, 623–634.

97. S. W. Muchmore, M. Sattler, H. Liang *et al.*, 'X-ray and NMR structure of human Bcl-xL, an inhibitor of programmed cell death.' *Nature*, 1996, **381**, 335–341.

98. S. L. Schendel, Z. Xie, M. O. Montal *et al.*, 'Channel formation by antiapoptotic protein Bcl-2.' *Proc. Nat. Acad. Sci. USA*, 1997, **94**, 5113–5118.

99. A. Gross, J. Jockel, M. C. Wei *et al.*, 'Enforced dimerization of BAX results in its translocation, mitochondrial dysfunction and apoptosis.' *EMBO J.*, 1998, **17**, 3878–3885.

100. I. S. Goping, A. Gross, J. N. Lavoie *et al.*, 'Regulated targeting of BAX to mitochondria.' *J. Cell Biol.*, 1998, **143**, 207–215.

101. A. Gross, X. M. Yin, K. Wang *et al.*, 'Caspase cleaved BID targets mitochondria and is required for cytochrome *c* release, while BCL-XL prevents this release but not tumor necrosis factor-R1/Fas death.' *J. Biol. Chem.*, 1999, **274**, 1156–1163.

102. J. A. Heibein, I. S. Goping, M. Barry *et al.*, 'Granzyme B-mediated cytochrome *c* release is regulated by the Bcl-2 family members Bid and Bax.' *J. Exp. Med.*, 2000, **192**, 1391–1402.

103. V. V. Stoka, B. Turk, S. L. Schendel *et al.*, 'Lysosomal protease pathways to apoptosis: cleavage of bid, not Pro-caspases, is the most likely route.' *J. Biol. Chem.*, 2001, **276**, 3149–3157.

104. D. J. van den Dobbelsteen, S. Orrenius and A. F. G. Slater, 'Environmental toxicants and apoptosis.' *Comments Toxicology*, 1997, **5**, 511–528.

105. F. J. Gonzalez and P. Fernandez-Salguero, 'The aryl hydrocarbon receptor: studies using AHR-null mice.' *Drug Metab. Dispos.*, 1998, **26**, 1194–1198.

106. J. C. Rowlands and J. A. Gustafsson, 'Aryl hydrocarbon receptor-mediated signal transduction.' *Crit. Rev. Toxicol.*, 1997, **27**, 109–134.

107. P. M. Fernandez-Salguero, J. M. Ward, J. P. Sundberg *et al.*, 'Lesions of aryl-hydrocarbon receptor-deficient mice.' *Vet. Pathol.*, 1997, **34**, 605–614.

108. J. V. Schmidt, G. H. Su, J. K. Reddy *et al.*, 'Characterization of a murine Ahr null allele: involvement of the Ah receptor in hepatic growth and development.' *Proc. Nat. Acad. Sci. USA*, 1996, **93**, 6731–6736.

109. N. L. Ge and C. J. Elferink, 'A direct interaction between the aryl hydrocarbon receptor and retinoblastoma protein. Linking dioxin signaling to the cell cycle.' *J. Biol. Chem.*, 1998, **273**, 22708–22713.

110. D. W. Nebert, 'Drug-metabolizing enzymes in ligand-modulated transcription.' *Biochem. Pharmacol.*, 1994, **47**, 25–37.

111. A. Hoffer, C. Y. Chang and A. Puga, 'Dioxin induces transcription of fos and jun genes by Ah receptor-dependent and -independent pathways.' *Toxicol. Appl. Pharmacol.*, 1996, **141**, 238–247.

112. M. J. DeVito and L. S. Birnbaum, 'Dioxins: model chemicals for assessing receptor mediated toxicity.' *Toxicology*, 1995, **102**, 115–123.

113. D. J. McConkey, P. Hartzell, S. K. Duddy *et al.*, '2,3,7,8-Tetrachlorodibenzo-*p*-dioxin kills immature thymocytes by $Ca^{2+}$-mediated endonuclease activation.' *Science*, 1988, **242**, 256–259.

114. D. J. McConkey and S. Orrenius, '2,3,7,8-Tetrachlorodibenzo-*p*-dioxin (TCDD) kills glucocorticoid-sensitive thymocytes *in vivo*.' *Biochem. Biophys. Res. Commun.*, 1989, **160**, 1003–1008.

115. C. E. Comment, B. L. Blaylock, D. R. Germolec *et al.*, 'Thymocyte injury after *in vitro* chemical expo-

sure: potential mechanisms for thymic atrophy.' *J. Pharmacol. Exp. Ther.*, 1992, **262**, 1267–1273.

116. A. E. Silverstone, D. E. Frazier Jr., N. C. Fiore *et al.*, 'Dexamethasone, β-estradiol, and 2,3,7,8-tetrachlorodibenzo-*p*-dioxin elicit thymic atrophy through different cellular targets.' *Toxicol. Appl. Pharmacol.*, 1994, **126**, 248–259.

117. A. B. Kamath, H. Xu, P. S. Nagarkatti *et al.*, 'Evidence for the induction of apoptosis in thymocytes by 2,3,7,8-tetrachlorodibenzo-*p*-dioxin *in vivo*.' *Toxicol. Appl. Pharmacol.*, 1997, **142**, 367–377.

118. J. E. Staples, N. C. Fiore, D. E. Frazier Jr. *et al.*, 'Overexpression of the anti-apoptotic oncogene, bcl-2, in the thymus does not prevent thymic atrophy induced by estradiol or 2,3,7,8-tetrachlorodibenzo-*p*-dioxin.' *Toxicol. Appl. Pharmacol.*, 1998, **151**, 200–210.

119. Z. W. Lai, N. C. Fiore, P. J. Hahn *et al.*, 'Differential effects of diethylstilbestrol and 2,3,7,8-tetrachlorodibenzo-*p*-dioxin on thymocyte differentiation, proliferation, and apoptosis in bcl-2 transgenic mouse fetal thymus organ culture.' *Toxicol. Appl. Pharmacol.*, 2000, **168**, 15–24.

120. J. A. Cidlowski, K. L. King, R. B. Evans-Storms *et al.*, 'The biochemistry and molecular biology of glucocorticoid-induced apoptosis in the immune system.' *Recent Prog. Hor. Res.*, 1996, **51**, 457–491.

121. S. C. Chow, R. Snowden, S. Orrenius *et al.*, 'Susceptibility of different subsets of immature thymocytes to apoptosis.' *FEBS Lett.*, 1997, **408**, 141–146.

122. J. E. Staples, F. G. Murante, N. C. Fiore *et al.*, 'Thymic alterations induced by 2,3,7,8-tetrachlorodibenzo-*p*-dioxin are strictly dependent on aryl hydrocarbon receptor activation in hemopoietic cells.' *J. Immunol.*, 1998, **160**, 3844–3854.

123. J. J. Reiners Jr. and R. E. Clift, 'Aryl hydrocarbon receptor regulation of ceramide-induced apoptosis in murine hepatoma 1c1c7 cells.' *J. Biol. Chem.*, 1999, **274**, 2502–2510.

124. A. Hossain, S. Tsuchiya, M. Minegishi *et al.*, 'The Ah receptor is not involved in 2,3,7,8-tetrachlorodibenzo-*p*-dioxin-mediated apoptosis in human leukemic T cell lines.' *J. Biol. Chem.*, 1998, **273**, 19853–19858.

125. A. B. Kamath, I. Camacho, P. S. Nagarkatti *et al.*, 'Role of Fas-Fas ligand interactions in 2,3,7,8-tetrachlorodibenzo-*p*-dioxin (TCDD)-induce immunotoxicity: increased resistance of thymocytes from Fas-deficient (lpr) and Fas ligand-defective (gld) mice to TCDD-induced toxicity.' *Toxicol. Appl. Pharmacol.*, 1999, **160**, 141–155.

126. J. H. Richburg, A. Nanez, L. R. Williams *et al.*, 'Sensitivity of testicular germ cells to toxicant-induced apoptosis in gld mice that express a nonfunctional form of Fas ligand.' *Endocrinology*, 2000, **141**, 787–793.

127. C. T. Lutz, G. Browne and C. R. Petzold, 'Methylcholanthrene causes increased thymocyte apoptosis.' *Toxicology*, 1998, **128**, 151–167.

128. R. I. Near, R. A. Matulka, K. K. Mann *et al*, 'Regulation of preB cell apoptosis by aryl hydrocarbon receptor/transcription factor-expressing stromal/adherent cells.' *Proc. Soc. Exp. Biol. Med.*, 1999, **222**, 242–252.

129. V. M. Salas and S. W. Burchiel, 'Apoptosis in Daudi human B cells in response to benzo[a]pyrene and benzo[a]pyrene-7,8-dihydrodiol.' *Toxicol. Appl. Pharmacol.*, 1998, **151**, 367–376.

130. P. K. Narayanan, W. O. Carter, P. E. Ganey *et al.*, 'Impairment of human neutrophil oxidative burst by polychlorinated biphenyls: inhibition of superoxide dismutase activity.' *J. Leukoc. Biol.*, 1998, **63**, 216–224.

131. K. Cain and D. E. Holt, 'Studies of cadmium-thionein induced nephropathy: time course of cadmium-thionein uptake and degradation.' *Chem. Biol. Interact.*, 1983, **43**, 223–237.

132. C. D. Klaassen, S. Choudhuri, J. M. McKim Jr. *et al.*, '*In vitro* and *in vivo* studies on the degradation of metallothionein.' *Environ. Health. Perspect.*, 1994, **102**(Suppl. 3), 141–146.

133. M. Piscator, 'Long-term observations on tubular and glomerular function in cadmium-exposed persons.' *Environ. Health Perspect.*, 1984, **54**, 175–179.

134. P. O'Brien and H. J. Salacinski, 'Evidence that the reactions of cadmium in the presence of metallothionein can produce hydroxyl radicals.' *Arch. Toxicol.*, 1998, **72**, 690–700.

135. D. Nigam, G. S Shukla and A. K. Agarwal, 'Glutathione depletion and oxidative damage in mitochondria following exposure to cadmium in rat liver and kidney.' *Toxicol. Lett.*, 1999, **106**, 151–157.

136. C. A. Chrestensen, D. W. Starke and J. J. Mieyal, 'Acute cadmium exposure inactivates thioltransferase (Glutaredoxin), inhibits intracellular reduction of protein-glutathionyl-mixed disulfides, and initiates apoptosis.' *J. Biol. Chem.*, 2000, **275**, 26556–26565.

137. F. Thévenod and J. M. Friedmann, 'Cadmium-mediated oxidative stress in kidney proximal tubule cells induces degradation of Na$^{(+)}$/K$^{(+)}$-ATPase through proteasomal and endo-/lysosomal proteolytic pathways.' *FASEB J.*, 1999, **13**, 1751–1761.

138. T. Hamada, A. Tanimoto, S. Iwai *et al.*, 'Cytopathological changes induced by cadmium-exposure in canine proximal tubular cells: a cytochemical and ultrastructural study.' *Nephron.*, 1994, **68**, 104–111.

139. A. Tanimoto, T. Hamada and O. Koide, 'Cell death and regeneration of renal proximal tubular cells in rats with subchronic cadmium intoxication.' *Toxicol. Pathol.*, 1993, **21**, 341–352.

140. M. Ishido, C. Tohyama and T. Suzuki, 'c-myc is not involved in cadmium-elicited apoptotic pathway in porcine kidney LLC-PK1 cells.' *Life Sci.*, 1999, **63**, 1195–1204.

141. M. Ishido, T. Suzuki, T. Adachi *et al.*, 'Zinc stimulates DNA synthesis during its antiapoptotic action independently with increments of an antiapoptotic protein, Bcl-2, in porcine kidney LLC-PK(1) cells.' *J. Pharmacol. Exp. Ther.*, 1999, **290**, 923–928.

142. R. D. Lohmann and D. Beyersmann, 'Effects of zinc and cadmium on apoptotic DNA fragmentation in isolated bovine liver nuclei.' *Environ. Health Perspect.*, 1994, **102**(Suppl 3), 269–271.

143. M. Li, T. Kondo, Q. L. Zhao *et al.*, 'Apoptosis induced by cadmium in human lymphoma U937 cells through Ca$^{2+}$-calpain and caspase-mitochondria-dependent pathways.' *J. Biol. Chem.*, 2000, **275**, 39702–39709.

144. M. Kunimoto, 'Methylmercury induces apoptosis of rat cerebellar neurons in primary culture.' *Biochem. Biophys. Res. Commun.*, 1994, **204**, 310–317.

145. T. A. Sarafian, L. Vartavarian, D. J. Kane *et al.*, 'bcl-2 expression decreases methyl mercury-induced free-radical generation and cell killing in a neural cell line.' *Toxicol. Lett.*, 1994, **74**, 49–55.

146. B. J. Shenker, T. L. Guo and I. M. Shapiro, 'Low-level methylmercury exposure causes human T-cells to undergo apoptosis: evidence of mitochondrial dysfunction.' *Environ. Res.*, 1998, **77**, 149–159.

147. B. J. Shenker, T. L. Guo, I. O *et al.*, 'Induction of apoptosis in human T-cells by methyl mercury: Temporal relationship between mitochondrial dysfunction and los of reductive reserve.' *Toxicol. Appl. Pharmacol.*, 1999, **157**, 23–35.

148. K. T. Lin, J. Y. Xue, F. F. Sun *et al.*, 'Reactive oxygen species participate in peroxynitrite-induced

apoptosis in HL-60 cells.' *Biochem. Biophys. Res. Commun.*, 1997, **230**, 115–119.

149. B. J. Shenker, T. L. Guo and I. M. Shapiro, 'Mercury-induced apoptosis in human lymphoid cells: evidence that the apoptotic pathway is mercurial species dependent.' *Environ. Res.*, 2000, **84**, 89–99.

150. P. Li, D. Nijhawan, I. Budihardjo *et al.*, 'Cytochrome *c* and dATP-dependent formation of Apaf-1/caspase-9 complex initiates an apoptotic protease cascade.' *Cell*, 1997, **91**, 479–489.

151. H. Stridh, S. Orrenius and M. B. Hampton, 'Caspase involvement in the induction of apoptosis by the environmental toxicants tributyltin and triphenyltin.' *Toxicol. Appl. Pharmacol.*, 1999, **156**, 141–146.

152. H. Stridh, M. Kimland, D. P. Jones *et al.*, 'Cytochrome *c* release and caspase activation in hydrogen peroxide- and tributyltin-induced apoptosis.' *FEBS Lett.*, 1998, **429**, 351–355.

153. H. Stridh, E. Fava, B. Single *et al.*, 'Tributyltin-induced apoptosis requires glycolytic adenosine trisphosphate production.' *Chem. Res. Toxicol.*, 1999, **12**, 874–882.

154. S. Mizuhashi, Y. Ikegaya and N. Matsuki, 'Cytotoxicity of tributyltin in rat hippocampal slice cultures.' *Neurosci. Res.*, 2000, **38**, 35–42.

155. S. C. Chow, G. E. Kass, M. J. McCabe Jr. *et al.*, 'Tributyltin increases cytosolic free $Ca^{2+}$ concentration in thymocytes by mobilizing intracellular $Ca^{2+}$, activating a $Ca^{2+}$ entry pathway, and inhibiting $Ca^{2+}$ efflux.' *Arch. Biochem. Biophys.*, 1992, **298**, 143–149.

156. E. Corsini, B. Viviani, M. Marinovich *et al.*, 'Role of mitochondria and calcium ions in tributyltin-induced gene regulatory pathways.' *Toxicol. Appl. Pharmacol.*, 1997, **145**, 74–81.

157. S. C. Chow and S. Orrenius, 'Rapid cytoskeleton modification in thymocytes induced by the immunotoxicant tributyltin.' *Toxicol. Appl. Pharmacol.*, 1994, **127**, 19–26.

158. Y. Izumi, K. Ogino, T. Murata *et al.*, 'Effects of diethyldithiocarbamate on activating mechanisms of neutrophils.' *Pharmacol. Toxicol.*, 1994, **74**, 280–286.

159. G. M. Bartoli, A. Muller, E. Cadenas *et al.*, 'Antioxidant effect of diethyldithiocarbamate on microsomal lipid peroxidation assessed by low-level chemiluminescence and alkane production.' *FEBS Lett.*, 1983, **164**, 371–374.

160. S. Verhaegen, A. J. McGowan, A. R. Brophy *et al.*, 'Inhibition of apoptosis by antioxidants in the human HL-60 leukemia cell line.' *Biochem. Pharmacol.*, 1995, **50**, 1021–1029.

161. H. Albrecht, J. Tschopp and C. V. Jongeneel, 'Bcl-2 protects from oxidative damage and apoptotic cell death without interfering with activation of NF-κB by TNF.' *FEBS Lett.*, 1994, **351**, 45–48.

162. C. S. Nobel, D. H. Burgess, B. Zhivotovsky *et al.*, 'Mechanism of dithiocarbamate inhibition of apoptosis: thiol oxidation by dithiocarbamate disulfides directly inhibits processing of the caspase-3 proenzyme.' *Chem. Res. Toxicol.*, 1997, **10**, 636–643.

163. C. S. Nobel, M. Kimland, B. Lind *et al.*, 'Dithiocarbamates induce apoptosis in thymocytes by raising the intracellular level of redox-active copper.' *J. Biol. Chem.*, 1995, **270**, 26202–26208.

164. M. J. Burkitt, H. S. Bishop, L. Milne *et al.*, 'Dithiocarbamate toxicity toward thymocytes involves their copper-catalyzed conversion to thiuram disulfides, which oxidize glutathione in a redox cycle without the release of reactive oxygen species.' *Arch. Biochem. Biophys.*, 1998, **353**, 73–84.

165. F. Della Ragione, V. Cucciolla, A. Borriello *et al.*, 'Pyrrolidine dithiocarbamate induces apoptosis by a cytochrome *c*-dependent mechanism.' *Biochem. Biophys. Res. Commun.*, 2000, **268**, 942–946.

166. M. T. Smith, 'Overview of benzene-induced aplastic anaemia.' *Eur. J. Haematol. Suppl.*, 1996, **60**, 107–110.

167. D. J. Thomas, A. Sadler, V. V. Subrahmanyam *et al.*, 'Bone marrow stromal cell bioactivation and detoxification of the benzene metabolite hydroquinone: comparison of macrophages and fibroblastoid cells.' *Mol. Pharmacol.*, 1990, **37**, 255–262.

168. D. Ross, D. Siegel, N. W. Gibson *et al.*, 'Activation and deactivation of quinones catalyzed by DT-diaphorase. Evidence for bioreductive activation of diaziquone (AZQ) in human tumor cells and detoxification of benzene metabolites in bone marrow stroma.' *Free Radic. Res. Commun.*, 1990, **8**, 373–381.

169. C. Lind, P. Hochstein and L. Ernster, 'DT-diaphorase as a quinone reductase: a cellular control device against semiquinone and superoxide radical formation.' *Arch. Biochem. Biophys.*, 1982, **216**, 178–185.

170. A. Yardley-Jones, D. Anderson and D. V. Parke, 'The toxicity of benzene and its metabolism and molecular pathology in human risk assessment.' *Br. J. Ind. Med.*, 1991, **48**, 437–444.

171. J. L. Moran, D. Siegel, X. M. Sun *et al.*, 'Induction of apoptosis by benzene metabolites in HL60 and CD34+ human bone marrow progenitor cells.' *Mol. Pharmacol.*, 1996, **50**, 610–615.

172. S. B Bratton, S. S. Lau and T. J. Monks, 'The putative benzene metabolite 2,3,5-tris(glutathion-*S*-yl)hydroquinone depletes glutathione, stimulates sphingomyelin turnover, and induces apoptosis in HL-60 cells.' *Chem. Res. Toxicol.*, 2000, **13**, 550–556.

173. J. Wiemels, J. K. Wiencke, A. Varykoni *et al.*, 'Modulation of the toxicity and macromolecular binding of benzene metabolites by NAD(P)H:Quinone oxidoreductase in transfected HL-60 cells.' *Chem. Res. Toxicol.*, 1999, **12**, 467–475.

174. M. L. Kuo, S. G. Shiah, C. J. Wang *et al.*, 'Suppression of apoptosis by Bcl-2 to enhance benzene metabolites-induced oxidative DNA damage and mutagenesis: a possible mechanism of carcinogenesis.' *Mol. Pharmacol.*, 1999, **55**, 894–901.

175. M. M. Metzstein, G. M. Stanfield and H. R. Horvitz, 'Genetics of programmed cell death in *C. elegans*: past, present and future.' *Trends Genet.*, 1998, **14**, 410–416.

176. M. B. Hampton and S. Orrenius, 'Dual regulation of caspase activity by hydrogen peroxide: implications for apoptosis.' *FEBS Lett.*, 1997, **414**, 552–556.

177. A. Samali, H. Nordgren, B. Zhivotovsky *et al.*, 'A comparative study of apoptosis and necrosis in HepG2 cells: oxidant-induced caspase inactivation leads to necrosis.' *Biochem. Biophys. Res. Commun.*, 1999, **255**, 6–11.

178. M. Garcio-Calvo, E. P. Peterson, B. Leiting *et al.*, 'Inhibition of human caspases by peptide-based and macromolecular inhibitors.' *J. Biol. Chem.*, 1998, **273**, 32608–32613.

179. S. Prost, C. O. C. Bellamy, D. S. Cunningham *et al.*, 'Altered DNA repair and dysregulation of p53 in IRF-1 null hepatocytes.' *FASEB J.*, 1998, **12**, 181–188.

180. W. R. Kobertz, D. Wang, G. N. Wogan *et al.*, 'An intercalation inhibitor altering the target selectivity of DNA damaging agents: synthesis of site-specific aflatoxin B1 adducts in a p53 mutational hotspot.' *Proc. Nat. Acad. Sci. USA*, 1997, **94**, 9579–9584.

181. S. C. Wright, J. Zhong and J. W. Larrick, 'Inhibition of apoptosis as a mechanism of tumor promotion.' *FASEB J.*, 1994, **8**, 654–660.

J.P. Vanden Heuvel, G.H. Perdew, W.B. Mattes and W.F. Greenlee (Eds.)
Comprehensive Toxicology, Vol. xiv

## 14.5 EMERGING TECHNOLOGIES AND PREDICTIVE ASSAYS

# 14.5.1
# Introduction and Overview

## WILLIAM B. MATTES
*Pharmacia, Kalamazoo, MI, USA*

## 14.5.1.1  INTRODUCTION

Paracelsus would be pleased. Dogma in all of the biomedical sciences, especially toxicology, has been a fragile thing in the last half century. While Paracelsus had to confront a standard medical practice based on teachings established over 14 centuries earlier, modern day iconoclasts are lucky to find scientific tenets that have survived for more that 50 years. In fact, the principles of experimentation, chemical toxicity, dose, and chemotherapy that Paracelsus himself established are some of the few concepts that have survived over the centuries.

To a large part the rapid changes in biomedical scientific thinking can be directly linked to the rapid changes in the technology of scientific exploration. The widespread application of radioisotopic tracers, X-ray crystallography, ultracentrifugation, electrophoresis, and protein sequencing transformed biochemistry in the latter part of the 20th century. Likewise, electron and fluorescence microscopy, monoclonal antibodies, and tissue culture changed cell biology, while restriction endonucleases, molecular cloning, DNA sequencing and polymerase chain reaction

(PCR) created the field of molecular biology as we know it today. Toxicology has drawn on all these innovations, although as a field it still leans heavily on its roots in pathology studies in whole animal models.

Toxicology of the 21st century to come will have similarities with the toxicology of the 19th and 20th centuries, but the differences will be levels of detail and sophistication driven by the new technologies. Just as rat studies have been distinguished from mouse studies, studies with one genotyped strain will be distinguished from those of another genotyped strain. The results will be similarly interpreted in terms of how a particular agent or chemical will affect one human or animal population versus another. The parameters examined in future studies will be correlated with database resources such that their predictive value will be greatly amplified compared to current practice. Focused *in vitro* studies, based upon the understanding of the target organ-tissue-cell and the biochemical context of that target, will be more commonplace. If there *is* a universal theme, it will be that toxicity will be approached as an interaction between the molecules of the toxicant, the molecules of the target cell, the cells of the target tissue and the tissue in the organism, with an eye toward homeostatic mechanisms at all levels. Overall, the game, as Paracelsus defined it, will be the same, but the players will be different.

Predicting the technologies that will be central to toxicology 50 years from now would be rather pretentious. However, it is the goal of this chapter, and this volume, to highlight and describe technologies that, while in a developing stage, have demonstrated results and power that identify them as some of the new players in toxicology. Some of these, such as metabonomics and laser capture microdissection have literally arisen in the last few years. The others, such as transgenic animal models, bioinformatics, gene expression profiling, proteomics, data mining, and systems modeling represent approaches that have been developing over some time. While a few of these technologies have already seen application in toxicology, all of them will have important roles in the toxicology of this century.

## 14.5.1.2 BIOINFORMATICS

### 14.5.1.2.1 Background

While it continues to be viewed as one of the more arcane "new" technologies, ironically, bioinformatics is one of the oldest.[1] It dates from the mid 1960s, when the combination of automated protein sequencing and newly developed digital computers prompted Margerett Dayhoff and others to explore computational analyses of protein sequence and evolution.[2,3] While "bioinformatics" is often used as an umbrella term referring to applications from database mining to microarray data analysis, it is truly rooted in the analysis of protein and DNA sequences.

By themselves polypeptide and polynucleotide sequences are no more informative than random collections of letters. To have value for the cell or the researcher they must be "decoded": genomic DNA must be transcribed to a complementary RNA, exons and introns must be identified, a "mature" messenger RNA formed, a protein coding region identified, a start codon and reading frame established, and the amino acid sequence of the protein determined by the genetic code. The polypeptide produced will then fold based upon the characteristics of the amino acid sequence, and/or be targeted for a certain subcellular location based upon the presence of a signalling peptide. Finally, the functional characteristics of this protein, and its interactions with other cell components, will depend upon the peptide motifs it contains and their arrangement in three dimensional space. A prime goal of bioinformatics is to trace theses cellular steps *in silico*, using a combination of algorithms and data based upon established cellular and molecular biology.

### 14.5.1.2.2 Algorithms

Some appreciation of the algorithmic approaches in bioinformatics is useful for understanding the basis and reliability of the many different applications. At the simplest level is the comparison of the sequence with a "code" database. Examples of this approach is the identification of restriction endonuclease recognition sites in a DNA sequence and the prediction of the amino acid sequence of a polypeptide coded for by the DNA (Figure 1). Needless to say these comparisons are relatively straightforward, and given the specificity of restriction endonucleases and the robustness of the genetic code, the results hardly warrant experimental confirmation. The pairwise alignment of two reasonably similar DNA sequences is also reasonably straightforward (Figure 2(a)). The pairwise alignment of two protein sequences presents a somewhat different problem.[4] Knowledge of both the characteristics of amino acids and protein evolution tells us that some amino acids are quite similar in function (e.g., alanine and valine), and often the substitution of one for the other at a particular site in a given protein can be tolerated without dramatic changes in function. Thus these comparisons make use of substitution matrices that determine scores for different substitutions: for any given sequence position the highest score is given if the same amino acid is in both proteins, a lower score is given for a conserved amino acid difference (e.g., alanine versus valine), while a penalty score is given for a truly disparate amino acid comparison (e.g., aspartate versus tryptophan) (Figure 2(b)). Gaps in such alignment also elicit penalty values, and ultimately comparing the sequences for two proteins that are not very similar may produce more than one reasonable alignment. In this last case a conclusion of similarity may depend upon other considerations.

### 14.5.1.2.3 Databases

The other key element in bioinformatics is the availability of data for computational comparison for analysis, i.e., relevant databases. The late 1980s saw the development of not only public sequence databases, but also

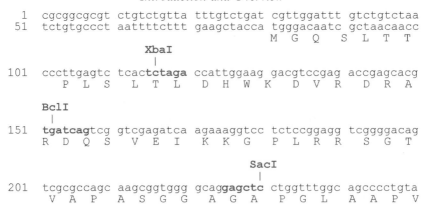

**Figure 1**  Simple comparison of a sequence with the genetic code.

the Internet and the World Wide Web. These two technologies came together to create the current model of myriad public databases readily accessible from any Internet-connected computer. While several databases will be mentioned here, the reader should note that every year the first issue of *Nucleic Acids Research* is devoted to a review of a large number and variety of publicly accessible databases.[5]

First and foremost to be considered are the databases maintained at the three sites of the International Nucleotide Sequence Database Collaboration. These are GenBank (maintained by the National Center for Biotechnology Information (NCBI) in Bethesda, Maryland), the DNA Database of Japan (DDBJ, Mishima, Japan) and the European Bioinformatics Institute (EBI) databases at Hinxton, England, part of the European Molecular Biology Laboratory (EMBL). These sites (see Table 1) are the central repository for publicly available sequence information, and while they contain essentially identical information, they maintain separate formats and

different tools for accessing the information. It should be noted that this data is contributed from a variety of sources, and should be contrasted with edited ("curated") databases, where the entries are checked for accuracy and annotated with information from several sources. Such databases include SwissProt, LocusLink, and GeneCards (Table 1).

Other sites host databases that focus beyond sequence information. Thus the Mouse Genome Informatics (MGI) site, the Saccharomyces Genome Database (SGD), the Sanger Centre, and the Stanford Genomic Resources focus on genomic and chromosomal information (Table 1). The Kyoto Encyclopedia of Genes and Genomes (KEGG) and What Is There (WIT) offer metabolic pathway information, while three dimensional protein structure files are maintained at the Protein Data Bank (PDB). PROSITE, BLOCKS, PRINTS, and PFAM represent examples of databases of protein motifs and paterns. Lastly two publicly accessible sites, UniGene and TIGR, maintain database covering ESTs and it is this resource that demands special description.

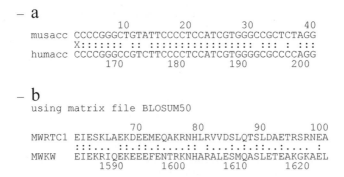

**Figure 2**  (A) Pairwise alignment of two DNA sequences; (B) Pairwise alignment of two peptide sequences, using a substitution matrix.

**Table 1** Internet Sites for Bioinformatics.

| | |
|---|---|
| *Link Gateways* | |
| Research Tool Sites for the Molecular Biology | http://www.yk.rim.or.jp/~aisoai/molbio.html |
| CMS-SDSC Molecular Biology Resource | http://restools.sdsc.edu/ |
| | |
| *Main Databases* | |
| EBI Home Page | http://www.ebi.ac.uk/ebi_home.html |
| EMBL – European Molecular Biology Laboratory | http://www.embl-heidelberg.de/ |
| National Center for Biotechnology Information (NCBI): | http://www.ncbi.nlm.nih.gov/ |
| DNA Data Bank of Japan | http://www.ddbj.nig.ac.jp/ |
| | |
| *Curated Databases* | |
| LocusLink | http://www.ncbi.nlm.nih.gov/LocusLink/ |
| GeneCards human genes, proteins and diseases (Weizmann) | http://bioinformatics.weizmann.ac.il/cards/ |
| SWISS-PROT | http://www.expasy.ch/sprot/ |
| | |
| *Genomic Databases* | |
| Mouse Genome Informatics | http://www.informatics.jax.org/ |
| Whitehead Institute | http://www-genome.wi.mit.edu/ |
| Saccharomyces Genome Database | http://genome-www.stanford.edu/Saccharomyces/ |
| The Sanger Centre | http://www.sanger.ac.uk/ |
| Stanford Genomic Resources | http://genome-www.stanford.edu/ |
| | |
| *Metabolic Pathways* | |
| Kyoto Encyclopedia of Genes and Genomes | http://www.genome.ad.jp/kegg/ |
| WIT Home Page | http://wit.mcs.anl.gov/WIT2/ |
| | |
| *Protein Structures, Motifs, and Patterns* | |
| PDB Home Page | http://www.rcsb.org/pdb/ |
| Blocks WWW Server | http://www.blocks.fhcrc.org/ |
| Pfam | http://www.sanger.ac.uk/Software/Pfam/ |
| PRINTS DATABASE | http://bmbsgi11.leeds.ac.uk/bmb5dp/prints.html |
| | |
| *EST Databases* | |
| The Institute for Genomic Research | http://www.tigr.org/index.html |
| UniGene | http://www.ncbi.nlm.nih.gov/UniGene/ |
| The I.M.A.G.E. Consortium | http://image.llnl.gov/ |
| | |
| *Multiple Alignment Tools* | |
| CLUSTALW at EBI | http://www2.ebi.ac.uk/clustalw/ |
| MACAW (download site) | ftp:/ /ncbi.nlm.nih.gov/pub/macaw/ |
| | |
| *Gene Regulation Predictions* | |
| Transcription Regulatory Regions Database Main Page | http://wwwicg.bionet.nsc.ru/trrd/ |
| Molecular Informatics Resource for the Analysis of GeneExpression | http://www.ifti.org/ |
| | |
| *Protein Structure Searching* | |
| Molecular Modeling Database | http://www.ncbi.nlm.nih.gov/Structure/MMDB/mmdb.shtml |
| VAST | http://www.ncbi.nlm.nih.gov/Structure/VAST/vastsearch.html |

## 14.5.1.2.4 ESTs

Expressed Sequence Tags (ESTs) were conceived as means to gain useful genomic information from automated DNA sequencing without waiting for the completion of whole genome projects.[6] In this approach hundreds to thousands of clones from collections (libraries) of complementary DNA (cDNA) are isolated, and then submitted to automated, partial, single-pass sequences from either end. A cDNA library represents a snapshot of expressed gene transcripts from the tissue or cell at the time the RNA was isolated and cloned, and complete sequencing of a large number of clones would yield a picture not only of the coding regions of the genome, but also of the genes expressed under the particular cellular conditions. The EST strategy was developed as a shortcut, where the partial sequences would serve as a resource for database mining. Thus ESTs are "identified" by their homology to known genes: the EST is first translated in all reading frames, and the six sequences are compared against a database of known protein sequences from all species. ESTs with no homology to known genes may represent novel expressed genes and represent an approach to gene discovery However, given the single-pass nature of the sequence information collected, accuracy is a serious concern, and can interfere with the unambiguous identification of any given EST. ESTs are maintained as part of the overall nucleotide databases, but can be uniquely searched at dbEST (Table 1). Additionally, ESTs, along with well-characterized sequences, can be computationally organized into gene-oriented clusters. Two such cluster databases are Uni-Gene,[7] maintained by NCBI, and the TIGR Gene Index,[8,9] maintained by the Institute for Genome Research (TIGR). These resources allow the researcher to determine if any given EST has already been identified as a gene fragment. The UniGene database also serves as a link to the physical cDNA clone through the IMAGE (Integrated Molecular Analysis of Genomes and their Expression) Consortium;[7] once obtained from a licensed distributor the clone's sequence can be confirmed and extended, essentially converting the EST into a full-fledged cDNA sequence.

### 14.5.1.2.5   Applications in Toxicology

#### 14.5.1.2.5.1   Finding New Proteins Based on Similarity to a Known Protein

One of the most common uses of the growing database of sequence information, particularly EST data is searching for new proteins that have sequence similarity to a known protein. The most common approach is to compare a known protein against the EST database using a tool such as BLAST (Basic Local Alignment Search Tool, Table 1).[10] BLAST will then return a listing of similar sequences and the alignment between them and the test sequence. "Similarity" can be a somewhat subjective determination if the sequences are of very different lengths and do not share long stretches of highly similar sequences. Given this caveat, though, this exercise has proved fantastically productive. Thus with the first report of EST sequencing, 337 new genes were identified in brain mRNA.[6] Most recently the dbEST was used to identify potential human counterparts of yeast genes coding for critical steps of oxidative phosphorylation.[11] It is pertinent to consider the different definitions applied to similar proteins, and the implications for their function: *homologues* are presumed to share a common protein ancestor and

may have similar function, *orthologues* are the same protein separated by a speciation event, probably with the same function, *paralogues* are similar proteins within a species and these may have similar function. The members of the basic helix-loop-helix PAS superfamily discovered by Bradfield and coworkers[12] with database searching are examples of paralogues.

#### 14.5.1.2.5.2   Identification, for any Given "Tox Target" Protein, of Similar Targets in Other Species

A common approach in toxicology is that of predicting toxic responses in one species (usually human) based upon results in a model species. Given this model, a logical application would be the search across species for orthologues of a protein identified as a toxicologic "target". Remarkably, this application has not been reported. However, with the increasing amount of rat, mouse, and even monkey sequences this application should begin to see greater use.

#### 14.5.1.2.5.3   Prediction of Function for a Novel Protein

A common experimental situation is that a gene may be identified as having a certain biological role, yet the exact biochemical function of the protein it codes for is unknown. An recent example of this may be found in the discovery that the B subunit of bacterial cytolethal distending toxin (CDT), which causes cell cycle arrest and cell death, bears sequence resemblance to type I deoxyribonucleases.[13] These studies invariably start with a simple comparison of the unknown protein sequence with databases of known proteins. A more detailed and thorough functional prediction can be obtained with an alignment of the novel protein with several related proteins of known functions. Examples of such multiple alignment programs include Multalin,[14] MACAW[15] and Clustal.[16] Thus the Clustal alignment of CDT subunit B with deoxyribonucleases identified residues in the former protein that were conserved across all DNases and were known to be essential for catalytic activity.

In a similar fashion, valuable inference about a protein's cellular role can be made based upon that protein's subcellular localization. Ever since the proposal that such localization was targeted by a signal peptide on the N-terminus of a nascent protein,[17] considerable effort has been made to analyze signal peptide sequences and predict them based on primary amino acid sequence.[18]

Thus, when two forms of mouse deoxyguanosine kinase were cloned, they could be predicted to be mitochondrial and cytoplasmic based upon the presence or absence of a mitochondrial import signal sequence.[19] Several good Web sites exist for analyzing proteins for signal peptides (Table 1), and of particular note is the PSORT WWW Server, which will predict the probable localization into a number of cellular organelles.

Far less common for a novel protein is the availability of a three-dimensional crystal structure. If one was so fortunate, the Vector Analysis Search Tool (VAST)[20] would allow one to search the database of known protein structures for similar structural motifs in other proteins. VAST is one of the tools offered through NCBI's Molecular Modeling Database (MMDB) (Table 1).

#### 14.5.1.2.5.4 *Prediction of Regulation for a Novel Gene*

Yet another opportunity for discovery comes when sequence information 5′ upstream of the mRNA start site is obtained for a gene. Given the paradigm of gene regulation by transcription factor binding to specific sequence motifs,[21] insight into the control and tissue-specific expression of a given gene can be gained by comparing these upstream sequences against a database of transcription factor binding sites. Thus, Spandidos and co-workers found phorbol ester-inducible AP-1 binding sites in the promoter regions for the human H-ras oncogene.[22] Such databases have evolved from Ghosh's early TFD (Transcription Factor Database)[23] to TRRD (Transcription Regulatory Region Database)[24] and EPD (Eucaryotic Promoter Database)[25] (Table 1), and the tools for using these database have evolved from SignalScan[26] to a wide variety of tools[27] including those that compare sequences against nucleotide distribution matrices.[28]

#### 14.5.1.2.5.5 *Establishing Evolutionary Relationships Among Several Similar Proteins*

This application is rooted in the very earliest developments of bioinformatics, at a time when sequences for cytochrome *c* from several species became available.[2,3] Indeed the process of creating an alignment of multiple protein sequences was usually linked to creating a hierarchical tree describing the relatedness of different sequences, and such trees could be used to infer phylogenetic relationships.[29]

While seemingly peripheral to toxicology, this is the one approach in bioinformatics that has had the most dramatic impact on the field. With the visionary suggestion that cytochrome P450s be named and categorized based upon their evolutionary relationships, Nebert and his colleagues gave lasting order to a field of research.[30] The remarkable aspect of this system, and its further updates[31–33] is that it did away with confusing, subjective nomenclature and replaced it with one based entirely upon computer-determined alignments of protein sequences. It is this kind of impact that the field of bioinformatics promises, that of automating difficult analyses and offering an objective understanding of biological molecules.

### 14.5.1.3 TRANSGENIC AND KNOCKOUT ANIMAL MODELS

#### 14.5.1.3.1 Background

Genetically modified (GM) animals have become such a familiar part of the scientific landscape that they are described in most college biology texts. Nonetheless, the sophistication of the models currently possible, and the ease at which they can be generated points to a truly growing impact in the field of toxicology. The power of genetic manipulation of cellular pathways, formerly readily available only to those studying single cell organisms, is now applicable to whole animal models, and in a way that will complement the other advances in understanding gene expression.

#### 14.5.1.3.2 Basic Methodology

Key to appreciating the growing power of the GM animal models is an appreciation of the change in approach over the last few years. Classically, "transgenic" animals are developed by microinjection of the transgene DNA into fertilized eggs. The fertilized eggs are then implanted into pseudo-pregnant mice and offspring are screened for the presence of the transgene. The approach has proved to be extremely powerful, but it is limited to the introduction of genetic information. Furthermore, the transgene is integrated into the genome randomly, and usually in multiple copies, so that its expression and effects are dependent upon many factors, including the number of integrated copies, its chromosomal context, and the efficacy of its promoter and regulatory elements in that context. On the other hand, this approach has been successful in species other than mice.

By contrast, with the development of targeted gene disruptions, insertions, and replacements, modifications are specifically integrated at a desired genomic site by homologous recombination. A vector with targeting sequences is introduced into murine embryonic stem (ES) cells. After selection and screening for desired recombinant clones, the ES cells are injected into blastocysts, which are implanted into a foster mother. The offspring will be chimeric animals that can ultimately be crossed to give homozygous and heterozygous animals. The approach avoids the introduction of modifications outside of the gene of interest, but at this time is limited to mice.

Key to the technology of targeted gene manipulations was the determination of conditions for isolating and maintaining ES cells,[34] and the development of a means to select for homologous recombinants and eliminate heterologous recombinants.[4,35,36] The former innovation has developed to the point that ES cells are available from several commercial sources (e.g., Genome Systems, Inc., Jackson Laboratoris, and the American Type Culture Collection). The latter development is built upon a careful vector design strategy. The vector is designed with two selectable markers, commonly neomycin resistance (*neo*) and the herpes simplex virus thymidine kinase (*tk*) gene. The neo gene is flanked by the sequences bearing homology to the gene of interest, whereas the *tk* gene lies outside of these sequences. Heterologous recombination will result in the insertion of the entire vector into a non-target site (Figure 3). The resulting cell will be neomycin resistant, but sensitive to the thymidine analog ganciclovir, which is recognized only by the HSV thymidine kinase. On the other hand, homologous recombination will result in the insertion of only the homologous sequences and the *neo* gene they flank (Figure 4). These cells will be resistant to both neomycin and ganciclovir. Of course, cells containing no vector will be sensitive to both drugs.

### 14.5.1.3.3  New Tools

The basic approach outlined above can readily generate mice strains with disruptions at specific genomic sites. However, several innovations have allowed for more sophisticated manipulations. The earliest of these was the application of the bacteriophage P1 Cre-Lox system to targeted gene disruption.[37] In P1-infected bacteria, the system operates to precisely excise the phage DNA from bacterial DNA: the Cre recombinase carries out DNA recombination between the two 34 bp loxP sites that flank the integrated P1 genome. In the target gene modification system, the vector is constructed such that sequences to be excised are flanked by *loxP* sequences (i.e., they are "floxed"). Excision of these sequences will then take place if the Cre recombinase is co-expressed in those cells. A functionally similar system makes use of the yeast flp/FRT recombinase.[38] Both systems have the advantage of allowing complete removal of transgenic sequences, such as the Neo gene, from targeted site. Thus a otherwise completely normal gene carrying a precise point mutation can be generated, giving the capability to analyze the effects of specific polymorphisms.

Two component control of sequence excision further opens up the possibility of regulating this recombination at the development, temporal, and anatomical level. For example, liver-specific knockouts can be generated by crossing a mouse bearing the floxed gene with one carrying the Cre recombinase under control of a liver-specific promoter (Figure 5). An inducible knockout can be developed by introducing a third component, e.g., the reverse tetracycline controlled transactivator (rtTA).[39] A tissue specific promoter drives the transcrip-

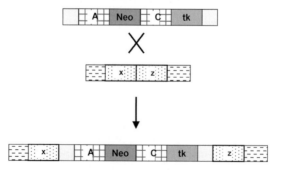

**Figure 3**  Heterologous recombination of a dual-selectable vector with a non-target sequence. The recombinant has neomycin resistance (*neo*) but sensitivity to ganciclovir (HSV-*tk*).

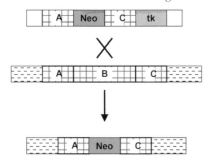

**Figure 4** Homologous recombination of a dual-selectable vector with target sequence. The recombinant has both neomycin resistance (*neo*) and resistance to ganciclovir (HSV-*tk*).

tion of a construct that fuses part of the *E. coli* tetracycline resistance gene to viral transcriptional activator. If Cre recombinase is under control of a minimally active promoter containing tetO sequences, it will be expressed only when the rtTA, in the presence of tetracycline (or its analogue docycline), binds to the tetO sequence and the viral activator enhances Cre transcription. Cre recombinase would then function on the third component, the floxed gene of interest. Of course, any gene of interest could be put under control of the tetO-viral promoter system to render it inducible by doxycycline. Thus positive and negative control of gene expression is afforded in an *in vivo* model with precisely controlled genetic background.

### 14.5.1.3.4 Applications in Toxicology

#### 14.5.1.3.4.1 *Mutagenicity Models*

One of the earliest applications of transgenic animals in toxicology was the development of models for mutagenicity, where bacteriophage lambda sequences were introduced as "neutral" genes that could be recovered and tested for mutation load.[40,41] These models have been extensively reviewed,[42] and the differences between mutation rates and spectra at these loci and endogenous loci offer insight into the way the cell handles transgenic sequences.[43–45] Recently, gene targeting has been used to develop mouse strains heterozygous for the endogenous thymidine kinase locus[46] or the adenosine phosphoribosyl transferase (APRT) locus,[47] allowing for selection of point mutations in these gene in cells recovered from animals treated with mutagens. Fish containing lambda sequences have also been developed,[48] allowing the investigation of mutation mechanisms in non-rodent species. The transgenic fish models also offer the possibility of monitoring environmental contamination using *in vivo* mutagenesis as an endpoint.

#### 14.5.1.3.4.2 *Carcinogenicity Models*

Animals with genetic modifications also been explored for some time as models for short-term carcinogenicity testing.[49] These include the Tg.AC transgenic mouse which carries a v-Ha-ras oncogene, with point mutations and under the control of a zeta globin promoter,[50] the rasH2 mouse, which carries a mutated human c-H-ras oncogene (see Reference 51) and the p53 +/− heterozygous knockout mouse model.[52] While the merits of these models are still being evaluated, each appears to have unique mechanistic characteristics. Tg.AC mice develop papillomas and squamous cell carcinomas (SCC) following dermal application of tumor promoters, in a time frame (20 weeks) where control animals do not show abnormalities. The Tg.AC transgene is not expressed in normal skin, but is expressed in

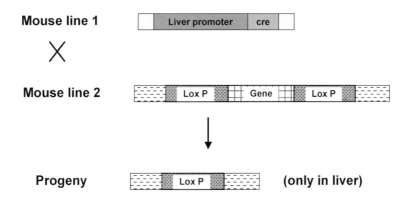

**Figure 5** Cre-loxP system for conditional, targeted gene disruption. The *cre* recombinase is under the control of a promoter expressed only in liver, hence the sequence flanked by *loxP* sites is excised only in liver.

the papillomas and SCC, leading to the suggested use of this model as a short-term test for non-genotoxic carcinogens and tumor promoters. On the other hand, Tg.AC mice do not develop generally tumors at traditional target sites with orally administered compounds, complicating their use as a generalized model. The rasH2 model has a good response rate,[53] with a number of genotoxic and non-genotoxic carcinogens yielding tumors in six months; control animals or those treated with non-carcinogens did not show significant tumor formation. On the other hand, this model displays some important false positives and false negatives.[54] The p53 +/− mice were developed with an eye toward the prevalence of p53 mutations in human cancer; indeed they are analogous to humans with Li-Fraumeni syndrome who are heterozygous for a normal p53 gene and are pre-disposed to cancer. As with the other models, these mice develop tumors when exposed to known carcinogens over a relatively short time frame (six months). In contrast to the other models, though, p53 +/− mice appear to respond exclusively to chemicals with known genotoxic activity.

### 14.5.1.3.4.3   Models Defining Roles of Metabolic Pathways

Xenobiotic and endogenous small molecules are more often than not subject to a variety of enzymatic conversions *in vivo*. Understanding the role of different pathways in these conversions can allow the prediction of metabolic interactions between compounds, as well as the effects of genetic variations among the human population in the activity of these pathways. The conventional approach to determining the role of different enzymes or pathways for metabolism of a particular compound is to examine the effects of specific enzyme inhibitors on that compound's conversion in isolated *in vitro* systems. Alternatively, a number of purified enzymes can be screened for activity towards the compound. The application of targeted gene modification allows the construction of a model where the role of a specific enzyme *in vivo* may be dissected. Thus, both CYP2E1 and CYP1A2 have been hypothesized to play a key role in the metabolism of acetaminophen to toxic metabolites.[55] Mice homozygous for a targeted knockout in the Cyp2E1 gene are indeed more resistant than wild type mice to acetaminophen, requiring higher doses to achieve the same extent of liver necrosis.[56] CYP1A2 knockout mice are also, more resistant to acetaminophen than wild type mice, though more sensitive than the

CYP2E1 knockout mice. However, the double-null Cyp2E1 negative/Cyp1A2 negative strain, was extremely resistant to this drug, indicating that while CYP2E1 plays a primary role in acetaminophen metabolism, CYP1A2 is also relevant *in vivo*. By contrast, CYP1A2 null mice were found to be equally sensitive to 4-aminobiphenyl (4-ABP) carcinogens is as wild type mice,[57] indicating that this enzyme is not essential to the production of carcinogenic metabolites of 4-ABP *in vivo*. Finally, CYP1B1 null mice were constructed and found to be completely resistant to tumor induction by dimethylbenzanthracene,[58] demonstrating that this case, a single enzyme is crucial for carcinogenesis. As more models are constructed with targeted deletions in metabolic enzymes a more complete picture of *in vivo* metabolic pathways will develop.

### 14.5.1.3.4.4   Models Defining Roles of Specific Receptors

Another application for genetically modified animal models is the dissection of cellular signaling pathways, particularly in regards to ligand-binding receptors.[59] An example for this is the PPARα knockout mouse model.[60,61,62] PPARα is a nuclear hormone receptor in steroid receptor superfamily that after binding its ligand functions as a transcription factor regulating hepatic lipid metabolism (e.g., fatty acid β oxidation). While leukotriene B4 appears to be a natural ligand, PPARα also binds a class of compounds referred to as peroxisomal proliferators (PP) (clofibrate, phthalates) and in this context appears to induce a series of responses that lead to liver tumors in rats. The role for PPARα in this process is confirmed by the observation that in the homozygous animals lacking this receptor, of di(2-ethylhexyl)phthalate (DEHP) treatment does not induce peroxisome formation, hepatomegaly or liver lesions.[63,64]

### 14.5.1.4   TRANSCRIPT PROFILING

#### 14.5.1.4.1   Background

Ever since the development of the Northern blot[65] it has been routine to examine the levels of various mRNA transcripts in different tissues or under different biological conditions. Central to this endeavor is the thinking that the levels of certain transcripts are regulated by mechanisms that respond to the cellular environment. In the toxicology arena, this has led to an investigation of the transcript levels for both single and multiple genes under conditions of toxicity.[66–69] These experiments give insight into the response of specific pathways under stress.

The concept of gene expression profiling, or transcript profiling, extends the paradigm of measuring transcript levels to that of measuring the levels of hundreds of transcripts, with an eye toward determining not only the response of certain genes, but also overall patterns of gene expression and relationships between the levels of different transcripts. Thus, while one goal may be to infer a toxicological mechanism from the response of a particular transcript, and perhaps develop a diagnostic assay based on that transcript, another goal gaining in popularity is that of inferring mechanism from the profile of collective response for hundreds of genes.[70,71] This latter goal is based on the hypothesis that the number of "toxicologically relevant" gene expression patterns is manageable[71] and these patterns can be used to classify unknown toxicants. Indeed this paradigm was proposed originally for arrays of reporter gene assays.[72]

### 14.5.1.4.2 "Open" Systems

#### 14.5.1.4.2.1 *Differential Display*

Key to an understanding of transcript profiling is the concept of open and closed systems. An "open" system is a means of examining differentially expressed transcripts that does *not* require prior sequence information. Differential display is just such a system: using a PCR primer set with sequence ambiguities, a subset of the entire transcript population is amplified and electrophoretically separated into discrete bands.[73] Comparison of this display with a comparably prepared sample representing a different experimental condition allows identification of differentially expressed transcripts. In practice, though, the identification step requires isolation of the band, cloning and sequencing of the fragment, and comparison of the sequence with sequence databases. Furthermore, screening of the entire transcript population requires the use of several different PCR primer sets. Nonetheless, the technique is a powerful one for identifying both known and novel genes regulated by altered cellular environments,[74–77] and can be applied to organisms that do not have extensive sequence information in the databanks.[78,79] The limitations of this technique can, to a large part, be overcome by automation and utilization of amplified restriction fragment polymorphism (AFLP) technol-ogy.[78,79] A step is incorporated in the differential display process where the cDNA is first digested with a restriction enzyme, ligated to adapters, and then amplified with PCR primer sets designed to target certain sequences. The presence and length of the resulting gel bands are predictable from the sequence of the transcript, the target sequence of the restriction enzyme, as well as the target sequence of the PCR primer set. Shimkets and coworkers[12] took this logic one step further, and coupled automated separation and sizing of fluorescent differential display/AFLP products with a gene database query such that known transcripts could be immediately identified from the size of the of product and the restriction enzyme used. Fragments from novel genes could be identified by conventional cloning and sequencing. It should be noted that open approaches such as differential display will be particularly useful in monitoring transcript splice variants,[80] and their role in the regulation of gene expression.

#### 14.5.1.4.2.2 *SAGE*

Another open system approach is dubbed Serial Analysis of Gene Expression (SAGE),[81] and makes use of short (9–12 base) sequence "tags", that are isolated, concatenated, cloned, and sequenced. By comparison with sequence databases the abundance of both known and unknown transcripts may be determined. The drawback with this approach is its sequence-intense nature, and the fact that 9–12 bases may not clearly identify a transcript,[82] and do not offer much information for novel transcripts.

### 14.5.1.4.3 "Closed" Systems

#### 14.5.1.4.3.1 *Nylon-membrane Arrays*

A "closed" system for transcript profiling is one that monitors a pre-determined set of sequences. Of course, the Northern blot is the fore-runner of this concept, with later developments being "reverse dot-blots" that monitored the levels of several transcripts simultaneously.[67,83] In the latter approach, cDNA fragments corresponding to different transcripts are bound to a nylon membrane in a grid pattern. RNA is converted to labeled cDNA and hybridized to the membrane, and the intensity of label at any given spot or grid location gives the abundance of that transcript in the original RNA pool. Such membrane-based arrays of cDNAs are commercially available and remain a flexible and cost-effective way to monitor a number of selected gene simultaneously.[84]

### 14.5.1.4.3.2 High-density cDNA and Oligonucleotide Arrays

The field of gene expression profiling, as it is generally thought of today, was certainly born with the development of the glass-slide cDNA microarray[85,86] and its seminal application to examining gene expression "on a genomic scale".[87] The basic concept was similar to that of the "reverse dot-blot", but incorporated several technological innovations. First, known cDNA fragments, prepared from PCR amplification of cloned yeast genes, were arrayed onto coated slides using a robotic device that could print 6400 spots in an 18 mm$^2$ area. Probe material was prepared from RNA using a reverse transcriptase reaction and fluorescently-labeled nucleotides; RNA from an "experimental" condition was labeled with one dye (Cy3) while the RNA from a control sample was labeled with a second dye (Cy5). The two probe samples were then competitively hybridized to the cDNA microarray. Finally, the fluorescence at each spot on the slide was determined with a confocal laser scanning microscope; the ratio of the intensities for the two dyes at each spot gives the ratio of abundance for that transcript in the experimental sample compared to the control sample. These basic features continue to be standard practice in cDNA microarray technology.

Paralleling the development of cDNA microarrays was the development of oligonucleotide microarrays.[88] This approach is quite different from that of the cDNA microarrays; rather than affixing a pre-existing DNA to a surface, the oligonucleotide is built *in situ* using photochemical synthesis. With photolithographic masks oligonucleotides of different sequence may be synthesized in parallel in separate 25 μm$^2$ areas, yielding a chip with thousands of probes (Figure 6). As designed by Affymetrix these GeneChip® arrays make use of several (16–20) 20-mer probe pairs for each RNA being monitored, a pair consisting of an oligonucleotide complementary to the RNA, and another with one mismatched base in the sequence. The mismatch probe for each pair serves as an internal control for hybridization specificity. As with the cDNA microarray approach, RNA samples are converted to a cDNA, but using an oligo-dT primer that also incorporates a bacteriophage T7 RNA polymerase promoter sequence. This allows subsequent production of "complementary RNA" in the presence of fluorescent-labeled nucleo-

tides. Modifications of the labeling procedure allows its use with bacterial RNA lacking polyadenylation sequences.[89] As with the cDNA microarrays, the labeled material is hybridized to the array and the array analyzed with a confocal laser scanning microscope. Unlike the approach with the cDNA microarrays, only one labeled sample is hybridized at a time. The analysis of spot intensities over all the probe pairs for an RNA allows quantification of the abundance of the transcript in the total sample.[88]

Yet another approach combines features of the two other microarray configurations. Arrays of oligonucleotides either pre-synthesized[90] or synthesized *in situ*[91] can be created on nylon or glass surfaces using essentially an ink-jet printing process. Using a competitive hybridization approach, even rare transcripts could be monitored with a single 60-mer oligonucleotide. A particular strength of the ink-jet approach is the ability to create different array configurations rather simply, allowing rapid optimization of probe oligonucleotide sequence as well as creation of many custom arrays.

### 14.5.1.4.4  Data Analysis: Paradigms and Methods

The Brown lab ushered the concept of examining hundreds of transcript changes for patterns of coordinately regulated genes.[87] The coordinate regulation for several known genes could be mechanistically explained by their role in metabolic pathways, as well as their control by common transcription factors. However, for genes with unknown functions, coordinate regulation with known genes suggests common cellular roles. This paradigm of identifying common patterns in transcript changes has driven much of the efforts in data analysis.

The first and foremost challenge in transcript profiling data analysis is the handling of large data sets (at least by biologists' standards).[92–94] Ideally, these data sets have not only the thousands of differential expression values, but also include functional and/or pathway information for the genes examined[94] to allow later functional grouping. Currently, this step is complicated by the limited (but growing) amount of functional annotation for database entries[95–98] and the need for standard gene and gene function nomenclature (i.e., gene ontology).[99,100,101] This issue will be discussed further later, but it must be emphasize that the systems in place for classifying species, and cataloging enyzmes, are in an early stage of development for genomic data.

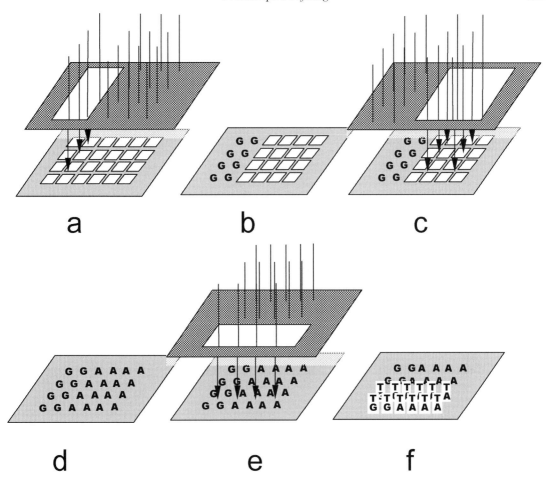

**Figure 6** High density oligonucleotide microarray synthesis via the Affymetrix photolithographic process. A masked light beam (a) directs a single base addition onto a given area of the chip (b). Changing masks (c), (d) and (e) results in single base addition to new areas of the chip and onto growing oligonucleotide chains. The final result is an array with oligonucleotides of different sequence at different positions of the chip, as dictated by the combinations of masks.

The next challenge is that of determining patterns of gene expression relevant to the experimental hypotheses. Even if the question is only what genes are truly regulated under a single experimental condition, data analysis is in order. Thus, microarray experiments, despite their expense, must be replicated,[102] with statistical methods applied to determine random variability. A variation on this approach is described by Hughes *et al.*[103] who developed a database of over 300 experiments, and used this "compendium" and a gene-specific error model to identify significant transcript changes under any given single condition. In terms of identifying trends in groups of transcripts, a variety of clustering methods has been described, including the original hierarchical clustering,[104] interactive clustering,[105] k-means clustering,[106] percolation clustering,[107] self-organizing maps,[108,109]

"gene shaving",[110] and analysis with support vector machines.[111] Many of these methods have been reviewed,[92,93,112] several have been incorporated into commercially-available software packages, and all remain valuable approaches to data analysis.

### 14.5.1.4.5 Applications in Toxicology: Paradigms and Methods

Pennie has noted that gene expression profiling can be applied to toxicology (i.e., toxicogenomics) in two ways: mechanistic research or predictive toxicology.[113] By no means are these exclusive endeavors; mechanistic information in toxicology forms the basis for the development of predictive models. Furthermore, defining the "mechanism of toxicity" in an *in vitro* model may allow prediction of *in vivo* toxicity. However, one might view another division in the application of gene expression profiling to toxicology: discovery of transcript regulation as opposed to classification of toxicity by patterns of transcript regulation. The first makes fewer assumptions about the relationship between gene regulation and toxicity, indeed it views toxicogenomics as a hypothesis-generating tool.[114] The latter keys off of the "fundamental assumption of toxicogenomics" that "there is no toxicologically relevant outcomes *in vitro* or *in vivo*, with the possible exception of rapid necrosis, that do not require differential gene expression.".[71] Both approaches have limitations and examples of success.

### 14.5.1.4.5.1 Discovery of Transcript Regulation by Toxicogenomics

As noted above transcript profiling may be used to identify regulated transcripts with the intent of both understanding the etiology of toxicity as well as developing predictive assays. Pennie and coworkers have developed arrays of selected genes for just this purpose[113] and have applied these arrays to identifying transcripts uniquely regulated in chloroform treatment of hepatoma cells,[115] using dimethyl formamide as a reference, non-hepatotoxic compound. They identified 47 transcripts meeting their criteria for altered regulation, and confirmed that IL-8 protein, as well as mRNA was upregulated under these conditions. Similarly, Rodi *et al.*[116] identified 300 transcripts regulated in rat liver by chronic phenobarbital treatment. Jelinsky and Samson[117] identified 401 transcripts regulated in yeast after treatment with the alkylating agent methyl methanesulfonate; in addition to transcripts associated with DNA repair many of the regulated transcripts are involved in protein turnover and synthesis, suggesting a cellular program to replace alkylated proteins. Transcript profiling was also used to correlate histopathological changes in cultured cells with specific transcript changes[84] after oxidative stress treatment of mesothelial cells. The authors of this study used this data to propose a model for KbrO3-induced carcinogenicity, in which p53 is activated, growth control is dysregulated, and oxidant-induced DNA damage is fixed as mutations. Finally, cDNA microarrays were used to examine the transcriptional response of cells to ionizing radiation:[118] 48 mRNAs were identified as IR-responsive, including 30 not previously thought to respond to DNA damage. Extending these studies to other cell lines containing

various levels of p53 activity allowed dissection of pathways that were p53-dependent. The common theme in all of these studies is that the experimental results were combined with other biological information concerning the responsive transcripts to create a holistic picture of the cellular response to the toxicant.

### 14.5.1.4.5.2 Classification of Toxicity by Patterns of Transcript Regulation

As noted above, one approach to the use of transcript profiling is the classification of toxicants by their transcript profile, not unlike the use of patterns of 2D-protein electrophoresis to identify toxicants.[119] Key to this use is the assumption that "a common set of changes unique to a class of toxicants, termed a toxicant signature" can be determined.[70] This application could serve both in a mechanistic mode by identifying the etiology of toxicity for a novel compound, as well as a predictive mode if the transcript profile were conducted with an *in vitro* model. A criticism of this paradigm, though, is that several modes of toxicity (e.g., acetaminophen hepatotoxicity) do not operate through gene expression pathways.[114] Nonetheless, this approach of pattern matching "guilt by association" has been extremely successful for identifying novel gene and pharmacological drug function in yeast. Marton and coworkers[120] identified response pathways modulated by FK506 independent of of the "target" calcineurin. Similarly, drugs and or genetic deletions affecting the ergosterol synthesis pathway in yeast all elicited gene expression profiles with certain commonalities,[121] while experiments conducted on 300 yeast deletion strains or drug treatments allowed the identification of a novel target for the drug dyclonine.[103] On the other hand, a series of experiments exploring "toxicity" by exposing yeast to "stresses" including temperature shock, hydrogen peroxide, menadione, diamide, dithiothreitol, and osmotic shock produced transcript profiles with a number of common responses, subsequently dubbed the Environmental Stress Response.[122] Finally, a recent report has demonstrated that clustering of transcript profiles alone from treated HepG2 hepatoma cells could be used to discriminate two classes of drugs.[123] An important observation of this study (reminiscent of the gene-error model described above)[103] was that certain transcripts in the initial profiles needed to be excluded for effective clustering. The authors thus argue that toxicant signatures can only be effectively determined with an experimentally chosen transcript set, as opposed to using an array monitoring thousands of genes.

### 14.5.1.4.5.3 *Looking Ahead*

It still remains to be seen how transcript profiling can be most effectively used in toxicology. As noted, the concept of using global patterns of expression as fingerprints diagnostic for different mechanism of toxicity is not new: rather this was "advanced" for proteomics studies several year prior to the development of cDNA microarrays.[119] To some extent the concept hinges on the expectation that "the number of toxicologically relevant gene expression patterns is small".[71] For both gene expression profiling and proteomics there is a paucity of published evidence that a signature pattern achieved a classification more relevant than that achieved by standard histopathology for a toxicity induced by a compound with a *novel* structure. Of course this lack of evidence doesn't discriminate between the failure of the hypothesis or the lack of a sufficiently large and diverse database of signature patterns.

On the other hand, discovery of regulated transcripts has utility in and of itself, as noted above. To the extent that these transcripts are understood in the context of regulation by other stimuli or signals they can be informative in terms of the mechanism of toxicity. Transcript changes also have the potential for biomarker analysis, especially if they can be monitored in an accessible tissue such as peripheral blood.[124,125] Transcript profiles could also be useful in the context of comparing regulatory networks from one species to another, but this begs two other questions: (1) how might regulatory networks be rigorously defined and stored in a database, and (2) what sequence information exists for species of toxicological interest? The former question is somewhat addressed in databases such as the KEGG,[97] but this is an evolving task for signal transduction and transcriptional regulation pathways. Furthermore, such pathway databases do not yet address cell type or species specificity in pathways. Detailed databases for this information will allow full utilization of transcript profiling data in one species and extrapolation to others. The latter question is extremely pertinent in that while the human genome has been sequenced, and the mouse genome is not far behind, sequence information for the rat is still limited and that for the dog is virtually non-existent. Until robust sequence information exists for these species

as well, the full promise of transcript profiling in toxicology will not be met.

### 14.5.1.5 PROTEOMICS

#### 14.5.1.5.1 Background

While "proteomics" is a new term, coined in 1994,[126] the concept of profiling all the expressed proteins in a particular tissue, in a given organism, and under a given environmental condition (e.g., xenobiotic treatment), is not. In fact the idea of a "Human Protein Index" dates back almost two decades.[127] At the heart of this approach is the principle that collectively the amounts of all the individual proteins in a cell or tissue, as well as their modification status, serves as a true state function for that cell or tissue. In essence, the protein complement, i.e., the proteome, defines what that cell or tissue really is. By contrast, the genome represents a blueprint for what the cell or tissue *can* be. The "transcriptome", the collective amounts of all the mRNAs in the cell are revealed by transcript profiling gives a valuable clue into the state of the tissue or cell at any given time. Ultimately, though, it is the collection of enzymes, transcription factors, phosphorylated and de-phosphorylated signal transduction proteins, cytoskeletal components, glycoproteins, etc. that determine if the tissue is proliferating, healthy, diseased or responding to a xenobiotic compound.

Proteomics as it is practiced today is a mixture of both the old and the new, and is still evolving. It makes use of the two dimensional gel technology developed over 25 years ago, coupled with recent advances in mass spectrometry (MS). Ironically, the other key element in modern proteomics is the DNA sequence databases derived from genomic sequencing efforts—these provide the source for virtual peptide map databases that can be queried with results from MS analysis of gel spots. And even with their limitations, together these technologies have made proteomics an extremely productive endeavor.

#### 14.5.1.5.2 Basic Methodology

##### 14.5.1.5.2.1 *Two-dimensional Gel Electrophoresis*

At the core of current proteomics is two-dimensional polyacrylamide gel electrophoresis (2DGE), similar to that as originally described.[128–130] In the first dimension proteins are separated by isoelectric focusing (IEF) on the basis of their isoelectric point (pI). A protein sample suspended in a denaturant solution (e.g., urea and the detergent NP-40) is mixed with carrier ampholytes and applied to a low percentage poly-acrylamide gel. When current is applied the ampholytes set up a stable pH gradient; the proteins will migrate to the pH that matches their pI and at that point (lacking a net charge) their migration will stop. When isoelectric focusing is complete, that gel is place along the top of a standard SDS slab gel.[131] The electrophoresis at this point is such that the proteins migrate out of the IEF gel and into the SDS gel and then are separated according to molecular weight (Mw). Generally this separation, by pI in the first dimension and Mw in the second dimension can resolve 1000–2000 spots.[132]

Key to any analytical technology is reproducibility and comparability between sample analysis. Since one gel runs only one sample (in traditional 2DGE technology), it is critical to assure identical conditions from one gel to the next. An important advance in this regard was the ISO-DALT system for running multiple gels simultaneously.[133,134] Immobilized pH gradient strips (IPG), containing fixed ampholytes addressed many of the reproducibility issues seen with IEF gels. The IPG strips are physically robust and provide a highly stable gradient due to the polymer backing.[135] These systems are bundled in commercial packages for both large and mini-format 2DGE by both Bio-Rad and Amersham Pharmacia Biotech.

### 14.5.1.5.2.2   Visualization

Visualizing the separated proteins has traditionally been done by staining with Coomassie Brilliant Blue R-250 (CBB). While relatively insensitive (50 ng protein/mm$^2$) its staining intensity is linear over a reasonably wide protein concentration range, making densitometric quantification of the protein spots on the stained gel feasible. Silver stains offer a far more sensitive alternative (down to 0.01 ng protein/mm$^2$), and are linear in their staining intensity over a wide protein range. However, they are more variable and problematic in practice, and are not necessarily compatible with downstream peptide digestion or mass spectrometry analysis, as described below. Recently, fluorescent protein stains have been introduced that are at least as sensitive as the silver, are compatible with downstream analysis and offer a linear response to protein concentration over 3 orders of magnitude.[136] The disadvantage here is the requirement for a fluorescent imager and the elimination of

simple human visual inspection as an option. Of course if the protein sample is radiolabeled then the gel may be visualized with autoradiography and/or phosphoimaging. This approach is attractive in that autoradiography allows for human inspection, and phosphoimaging is quantitative over several orders of magnitude; on the other hand radiolabeled samples are not always feasible (especially with animal studies) and introduce the complications of radioisotope handling and waste.

### 14.5.1.5.2.3   Evaluation of Differential Expression

The next step in the traditional proteomic process is the simplest to describe, and in practice can be the most laborious to execute: gels from different samples need to be evaluated for differentially expressed protein spots. For staining techniques such as CBB, silver, and autoradiography this can certainly be done by eye, but practically this will allow the identification of only a few spots that are clearly differentially expressed. If one is interested in a global pattern of differential expression, the gel image must be processed by image analysis software that will align the patterns in different gels, and quantitate spot intensity normalized to a value such as overall staining. Commercial software packages are available from Bio-Rad, Phoretix, and Genome Systems. Image analysis can allow the identification of protein spots by comparison of co-ordinates with spots in the image database that have been previously identified by MS or other analysis. Finally pattern databases themselves have long been viewed as a means for assessing biological status and identifying responses to stimulii.[119] Databases of gel images, with some identification are available at many Web sites (e.g., the NCI Image Meta-Database at http://www-lecb.ncifcrf.gov/2dwgDB/ and SWISS-2DPAGE at http://www.expasy.ch/ch2d/ch2d-top.html) and continue to increase in scope and sophistication.[137]

### 14.5.1.5.2.4   Mass Spec ID

Depending upon its size and quality an image database may be useful in identifying protein spots in any given gel based upon their position. This, of course, presupposes that the database already includes a large number of positively identified spots. Thus at some point in a proteomic project novel protein spots must be identified, and in current terms this invariably involves mass spectrometric analysis. The role of mass spectrometry in proteomics has been reviewed[138] (see also Witzmann, this volume), but common practice involves: (1) excision of the spot from the gel; (2) digestion of the protein with a sequence-specific protease such as trypsin; (3) analysis of the exact peptide masses by mass spectrometry; and (4) comparison of all the peptide masses with those predicted for digestion of all known protein sequences. This final step is a critical one made possible by genomic as well as protein sequencing, databases of predicted peptide fragments and software to carry out the matching.[139] The approach lends itself to automation and several companies offer complete setup for both the hardware and software. Of course, the identification can be ambiguous in some cases, and in others can fail because the protein is not represented in protein or genomic databases, or represents a mutant protein. In such cases, de novo sequencing of the protein using MS approaches is possible.[140]

### 14.5.1.5.3 New Tools

The current approach to proteomics, as described above, has great strength, but also clear liabilities. The procedures are truly laborious, although with the introduction of commercially available packages and the opportunities for automation,[141] part of this issue can be addressed. Even so, the scale of operation required for reproducible gel and gel image analysis, database development and maintenance, and robotic MS analysis requires a commitment that is often beyond even relatively large organizations.[142] Other issues remain, such as the fact that in its present form the procedure has a limited range of sensitivity.[143]

On the other hand, many new tools in the proteomics workshop address some of the issues with the current technology. Thus the issue of matching spots between gels for two differentially treated samples can be addressed by difference gel electrophoresis (DIGE), where two different samples are tagged with two different fluorescent dyes, mixed, and then separated on the same gel system.[144,145] The fluorescent tags allow exact comparison between samples for any given spot, allow sensitive detection, have a wide dynamic range, and are compatible with downstream analysis. DIGE has been applied to acetaminophen-induced hepatotoxicity and determined protein changes not previously described.[145] The drawback with DIGE at this time is the high cost and proprietary hardware and software required.

The sensitivity of traditional proteomic methods is of particular concern when considering changes in transcription factors and signaling proteins. These proteins are invariably at low abundance in the cell, yet are more often than not the key players in a cellular response to external stimuli. A recent approach to examine signal transduction pathways couples 2DGE methodology with immunodetection of phosphorylated proteins.[146] This has been applied to signal tranduction pathways following growth factor stimulation[147,148] and adapted to a mini-gel format to study insulin responses in adipocytes.[149] Alternatively, cytosolic extracts can separated into a phosphoprotein fraction prior to 2DGE, as described by Yagame and coworkers.[150] This technology promises to offer toxicologists a proteomics look with a different focus.

Finally, several efforts have been taken to eliminate the 2DGE step from proteomics.[151] One approach makes use of isotope-coded affinity tags (ICAT) to label peptides from different samples to allow their comparison with subsequent MS analysis.[152] Another approach binds proteins to different "chips" that vary in their adsorptive properties, and then analyzes the bound proteins using surface-enhanced laser desorption/ionization (SELDI).[153,154] While this technique does not permit immediate identification of differentially expressed peptides, it is commercially available in a format that lends itself to general use. In analogy to their use for nucleic acid analysis, microarrays of proteins or antibodies have been explored and discussed.[151,155]

### 14.5.1.5.4 Applications in Toxicology

Given that proteomics, at least in the form of 2DGE, has been in use for over two decades, it has been applied to hundreds of studies examining the biological effect of xenobiotic chemicals. Toxico-proteomics has been reviewed[119,156,157] (see also this volume) and one needs only to note that it has been recently used to identify molecular pathways regulated by lovastatin,[158] explore the molecular basis for a drug-induced hepatomegaly,[159] determine proteins modified in the liver and kidney by jet fuel exposure,[160] and develop both a mechanism and biomarker for cyclosporin A nephrotoxicity.[161] Proteomics serves as an essential complement to transcript profiling, neither replaces the other. At the current time both technologies have strengths and limitations. Some of the current limitations with proteomics are essential technical, and one can expect that with as the new tools noted above are further developed, and others are added, proteomic analysis in one form or another will become routine in toxicological studies.

### 14.5.1.6 LASER CAPTURE MICRODISSECTION

#### 14.5.1.6.1 Background

In the early days of molecular biology, and even recently, the molecular toxicologist faced a conundrum in considering applying molecular tools to toxicological problems. Given the sensitivity of tools such as Northern blots and Coomassie Blue-stained protein gels, large amounts of tissue were need for a signal. Pathologists would be quick (and right) to note that the relevant lesion might be occurring in only a small subset of cell types within that tissue, and that molecular changes in this small subset may be masked by the signal of all the surrounding tissue. As techniques such as PCR were developed, that allowed focused analyses in single cells,[162] examinations of cell populations isolated by flow sorting or microdissection. Thus a study examining microdissected samples demonstrated a 6.7-fold higher expression of the c-myc gene in liver tissue of immature mice when compared to tissue from adult mice),[163] under conditions where the samples were free of confounding microlesiosns. With the introduction of real-time quantitative RT-PCR the molecular analyses of such studies were somewhat simplified.[164] However, these techniques required careful manipulations with a micro-needle and tended to be laborious.

Laser capture microdissection (LCM) (or Laser Capture Microscopy) offers the prospect of bridging the worlds of the pathologist and the molecular toxicologist, and allowing molecular analyses of discrete cell populations. LCM was developed at the National Cancer Institute for microanalysis of tumor tissue, and was quickly commercialized as an instrument marked by Arcturus Engineering (Mountain View, CA, http://www.arctur.com). The principle and practice is truly simple[165,166] (Figure 7). A slide with a histological section is placed on the LCM microscope stage, and the area of interest is selected. The instrument places a transparent ethylene vinyl acetate film in apposition to the specimen. At this point the operator can move the slide with a joystick to bring the cell or cells into the center of the optical field. An infrared laser beam is then focused onto the desired area, and when it is pulsed it transiently melts the film above the target area. The melted polymer enter the voids in the tissue around the cells, then solidifies and fixes the target area to the film, essentially embedding the sample in a fashion analogous to histological embedding. When the film is lifted off of the section, the embedded cells are removed. The remaining tissue is undamaged and can be further dissected. The film itself is fixed onto the bottom of a rigid plastic cap that then be used to cap a standard 0.5 ml microcentrifuge tube, which could contain reagents for subsequent processing of the captured material. The size of the focused laser beam can be 30, 15 or 7.5 μm, allowing capture of even single cells, while up to about 6000 cells can be captured by means of multiple (up to 3000) "shots". Standard protocols for LCM fixation, sectioning, and molecular analyses may be found at the NIH LCM web page (*http://dir.nichd.nih.gov/lcm/lcm.htm*), and recently protocols for molecular anlysis have been described for laser-assisted microdissection of formalin-fixed, paraffin-embedded tissue.[167] Despite its appearance as a new field, LCM has a considerable maturity in its methodology and practice.

LCM is not without some drawbacks. First and foremost is the cost ($100,000–$150,000) which restricts its use to well-funded laboratories or core facilities. The sample lacks

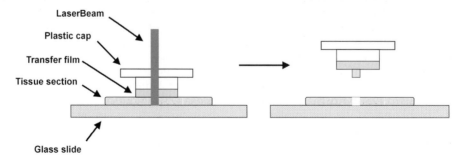

**Figure 7** Simplified cartoon of laser capture microdissection. The area of tissue illuminated by the laser beam is embedded in the transfer film and adheres to the plastic cap.

coverslipping and mounting medium, so its visualization is less than perfect. And while single cell capture is quite possible with the equipment, in practice many molecular analyses (e.g., proteomics and microarrays) do indeed require several thousand cells, and practically this will take time to acquire with the instrument.

### 14.5.1.6.2 Applications

#### 14.5.1.6.2.1 RT-PCR

Given its power to analyze RNA from single cells, real-time quantitative RT-PCR analysis is a natural adjunct to LCM. Real-time RT-PCR was first described for laser-assisted cell picking, and used to demonstrate dose-dependent upregulation of TNF-α in lung alveolar macrophages from rats challenged with endotoxin.[164] Kidney is a particularly diverse tissue and not surprisingly a natural target for LCM/RT-PCR; indeed Kohda and coworkers could isolate pure populations of proximal convoluted tubules, proximal straight tubules, and thick ascending limbs from renal sections, and determine specific mRNA changes in these cells in response to ischemia-reperfusion.[168] Similarly Nagasawa and coworkers determined with real-time RT-PCR that in experimental glomerulonephritis induced with intravenous anti-Thy1.1 monoclonal antibody TGF-beta1 mRNA increased over time in LCM-isolated glomeruli.[169] And in a study of ethyl carbamte-induced lung tumors, individual carcinomas isolated with LCM were found to have 12-fold less TGF-beta RII mRNA when compared with adjacent normal bronchioles.[170]

#### 14.5.1.6.2.2 Genetics

PCR can also be applied to examine genetic changes in various tissue micro-regions isolated by LCM. The caveolin-1 gene promoter was found to be hypermethylated in prostate epithelial cells from tumor tissue, but not normal tissue, using LCM to obtain cells from paraffin-embedded sections.[171] In a study examining fiber atrophy in age-related sarcopenia, LCM was used to isolate individual muscle fibers from animals of different ages. Analysis of mitochondrial DNA (mt DNA) showed that mt DNA deletions co-localized with fiber atrophy, fiber splitting, and oxidative damage.[172]

#### 14.5.1.6.2.3 Proteomics

If enough cells are collected, standard proteomic techniques may be used to analyze LCM samples. Emmert-Buck and coworkers[173] analyzed 50,000 cells from normal and tumor esophageal tissue, using 2-dimensional gel electrophoresis to identify 12 spot differences. These could be analyzed with MS to determine that cytokeratin 1 was overexpressed in the tumor cells, while annexin I was underexpressed. A comparison of the microdissected tumor epithelial cells versus normal epithelial cells versus stroma versus whole-tissue sections demonstrated that the 2D gel pattern of the whole tissue section truly represented a mix of the three pure populations. A similar proteomic study of prostate tumor tissue again demonstrated the utility of microdissection.[174] A natural extension of this approach is coupling LCM with the highly sensitive SELDI approach to proteomics (described above). In a study of cancer progression, Paweletz and coworkers[175] were able to achive reproducible SELDI scans with 1500 cells, and demonstrate patterns indicative of tumor stage in both esophageal and protate tumor cells. Clearly pure cell population offer the idea sample for proteomics.[176]

#### 14.5.1.6.2.4 Microarrays

Pure population of cells represent the ideal sample for transcript profiling experiments as well as proteomics, and it is not surprising that one of the early applications of LCM was the analysis of mRNA expression patterns.[177] Neighboring neurons of different subtypes (1000 cells per set) were isolated with LCM. Total RNA was extracted, and to provide enough material for microarray analysis, amplified about $10^6$ fold using a T7 RNA polymerase amplification strategy. Microarray analysis using cDNAs for 477 genes demonstrated distinct expression patterns for large- and small-sized neurons in the dorsal root ganglia. A similar strategy has been described LCM-derived samples analyzed using high-density oligonucleotide (i.e., Affymetrix) arrays.[178]

#### 14.5.1.6.2.5 Immunoassay

Another analysis tool applied to LCM-procured cell population is that of immunoassay. This has been used to quantitate the number of prostate-specific antigen (PSA) molecules/cell in human prostate tissue from fixed and stained frozen sections.[179] The automated (1.5 h) sandwich chemiluminescent immunoassay technique should be applicable to a wide range of studies and represents one more tool to consider when using LCM.

### 14.5.1.7 METABONOMICS IN TOXICOLOGY

Metabonomics is another of the "-omic" technology which, like genomics and proteomics, attempts to look at changes in hundreds and thousands of components simultaneously to achieve a global perspective of the organism's status. The components examined in this technology are the small molecular weight molecules produced by metabolism. The analytical tools used are high resolution nuclear magnetic resonance (NMR) coupled with pattern recognition algorithms. The samples analyzed are typically urine, blood or another biofluid. The assumptions made in these experiments are similar to those made for transcript profiling and proteomics: toxicants by definition disrupt cellular status and the overall balance of cellular components. In the case of metabonomics, the assumption translates to one where toxicant treatment disrupts the balance of endogenous biochemicals in cellular metabolic pathways, which is reflected in a perturbation in circulating blood metabolites, which will then produce an altered biomolecular trace in urine. As with transcript profiling and proteomics, the technology can focus on a specific component or global patterns, or "fingerprints". Not surprisingly, metabonomics offers similar challenges, but also some unique capabilities, and several good reviews are available[180,181] (see also Robertson, this volume).

#### 14.5.1.7.1 Basic Methodology

##### 14.5.1.7.1.1 Sample Collection

While the focus in metabonomics is often on the NMR technology itself, it is relevant to consider the experimental sample and its collection. While biofluids such as plasma, serum, cerebrospinal fluid (CSF), semen, and bile have all been analyzed with NMR,[182] urine is by far the preferred and common sample. It is easy to analyze and can be automatically collected over a time course by non-invasive (non-stressful) means. Urine's relative lack of proteins and enzymes minimize analyte lability. It can be obtained *prior* to dosing studies and hence can serve to identify inter-animal varia-

bility. It has been argued that because of this characteristic, fewer animals need to be studied at any given dose (see Robertson, this volume). As far as handling urine, samples need only be collected cold to minimize bacterial contamination. Procedures for obtaining spectra from whole tissue samples using magic angle spinning-NMR has been described,[183] yet for all the reasons cited urine remains the sample of choice.

##### 14.5.1.7.1.2 NMR

An essential component of the metabonomics technology is, of course, high-field NMR equipment. Proton ($^1$H) NMR spectroscopy is capable of detecting individual atoms in soluble proton-containing molecules with molecular weights of approximately 20 kD or less, so it is well suited to aqueous samples. However, it was the development of superconducting magnets in the 1970s, that allowed the production of high-field ($>400$ MHz) spectrometers, which in turn were able to report NMR spectrum on whole urine. A 600 MHz spectrometer are capable of detecting thousands of resonances in the urine from molecules at concentrations in the mid micromolar range or higher with a data acquisition time of a few minutes. Such an NMR spectrum is the raw data for metanomics experiments, and can be used to identify individual components by their characteristic chemical shift. Alternatively, the spectrum can be used as a whole in a pattern recognition strategy to define a "metabolic state".

##### 14.5.1.7.1.3 Pattern Recognition

Pattern recognition (PR) techniques have been applied to proton NMR spectrum collected from urine as a means for addressing the complexity of the information.[184–186] An NMR spectrum may be considered as an *n*-dimensional object with the dimensions being the concentrations of individual metabolites. NMR data can be reduced prior to PR by focusing on a few metabolites[184,185] or dividing the spectra into discrete "bins".[181,186,187] One of the better approaches to PR was found to be principal components analysis (PCA), where principle components (PC) are new variables created from linear combinations of the starting variables. The first PC accounts for the maximum amount of variance and each successive PC accounts for less of the remaining variance in the data. PCA may be viewed as a dimension reduction technique in that points in multidimensional space are projected onto a space of fewer dimensions. In fact, this is carried out by means of algebraic transformations with appropriate weighting coefficients, and reduction to two or three PCs allows visual discrimination between data sets. If this clustering is carried out without regard to toxicological mechanism, it is truly "unsupervised" in mathematical terms. Supervised approaches can be taken to develop predictive models, obviously a goal in metabonomics.[187]

### 14.5.1.7.2 Applications in Toxicology

High-field NMR analysis is resource-intensive not just in the nature of the instrumentation but also in the commitment of skilled personnel involved. Only a few laboratories (notably the Nicholson group at the University of London) have invested in the technology and published results. Nonetheless, these results suggest that this technology will play an important role in toxicological investigation, and will be more commonplace in large organizations and core facilities.

#### 14.5.1.7.2.1 Marker Discovery

Obviously high-resolution NMR can identify a component in complex mixtures if the resonances of that component are sufficiently resolved from those of other components. Identifying a single component for later focus as a marker for toxicity was one of the earliest uses of the technology, and, while not strictly "metabonomics", plays an important role. Thus increased urinary citrate and acetate served as a marker for *p*-aminophenol nephrotoxicity.[188] Nephrotoxins damaging the proximal tubules could be distinguished by aminoaciduria and glycosuria while those damaging renal papilla resulting in increased urinary trimethylamine *N*-oxide.[189] On the other hand uranyl nitrate nephrotoxicity was marked by high concentrations of 3-D-hydroxybutyrate (HB) in the urine.[190] Cadmium-induced testicular damage could be detected as increased creatinine excretion,[191,192] and a number of hepatotoxins (but not allyl alcohol) caused in increased excretion of taurine.[193]

#### 14.5.1.7.2.2 Trajectories

Perhaps the most unique and intriguing capability of metabonomics is to examine individual animal responses to treatment over a full time course. By applying PCA to such time points "trajectories" of the principle components (i.e., the reduced data) can be observed as marking the time dependent response to, and recovery from, acute toxic treatment. In an early study, time course NMR data provided the best discrimination between model liver and kidney toxins of different classes.[184] Similarly, using time course data, an acute dose of 2-bromoethanamine (BEA), a medullary toxin, produced a different trajectory of principle components than did $HgCl_2$, a proximal tubular toxin.[194,195] The trajectory for onset of BEA toxicity was different from that for its recovery phase, in contrast to the outward trajectory (onset) for $HgCl_2$ treatment, which mapped a space similar to its inward trajectory (recovery phase). The discriminatory power of such trajectories has been confirmed for other hepatotoxins and nephrotoxins.[181,196] These trajectories also allow inter-animal variation to be detected, and allow determination of time points at which the impact of such variability on the analysis is minimized.[181,194]

#### 14.5.1.7.2.3 Mechanism Discrimination and Prediction

As noted above, trajectories of principal components can discriminate between toxicants of different classes. This observation begs the question whether such analysis can be predictive of toxicant mechanism, akin to the classification of toxicants by mRNA or protein expression patterns. This classification of mechanism was suggested by some of the early metabonomics experiments,[184,185] and explored using a soft independent modelling of class analogy (SIMCA) model.[187] The dataset in this study consisted of the first three principal components derived from the NMR data for 183 samples from animals treated with one of 11 different liver and kidney toxins. The model was able to correctly predict the class of toxicant for 58 out of 61 test samples. Clearly this approach holds promise,[180] and it remains to be seen if a large database of such patterns will allow the classification for novel compounds.

#### 14.5.1.7.2.4 Screening

Because of its non-invasive nature, NMR analysis of urine samples is well suited to a compound screening mode.[197] The proposal is that an expert system could, provide from data collected in this context three levels of information: (1) classification of the individual test animal as "normal" or deviating from the control population; (2) classification of the test article as "toxic" based on the PCA for the treated animals indicating a deviation from the control population, and/or matching a pattern for known classes of toxicity; and (3) biomarker identification. The first level of information is particularly intriguing, since the only major assumption made is that a well maintained population of test animals should have a reasonably similar urine metabolite profile. The other levels of information, of course, depend upon a well-established database and validated criteria. Nonetheless, the ability to gain information with a limited number of animals is attractive, particularly if mice are used, where compound requirements would be minimized (Robertson, this volume).

### 14.5.1.8   DATA MINING

#### 14.5.1.8.1   Background

"Data Mining" is a term that is often used in regards to very different applications. Certainly one common denominator to these applications is the effort to make sense of large data sets.[198] While large data sets are not new in biology or medicine, the availability of this data through inexpensive storage and retrieval, and the ability to further generate it through automated instrumentation is relatively recent. Furthermore, the combination of massive genomic databases, through genome sequencing, protein structure identification, gene expression profiling and pharmacogenetic screening makes the consideration of data mining imperative. Adding to this diverse dataset is the rapidly developing availability of electronic published literature,[199] effectively making all publicly disclosed biomedical information electronically accessible. Identifying non-intuitive patterns and relationships in these data will be the challenge of the next century. This will be possible only with computer-driven techniques such as data mining.

Data mining represents one step in the process of "knowledge discovery in databases" (KDD), where the latter may be defined as "the nontrivial process of identifying valid, novel, potentially useful, and ultimately understandable patterns in data".[200] Data mining (DM) is the step that applies "data analysis and discovery algorithms" to identify patterns and models. While the development of appropriate databases and data mining approaches have just recently been appreciated in gene

expression profiling,[93] these techniques are widely appreciated, developed and applied in the business sector.[201]

#### 14.5.1.8.2   Basic Methodology

Knowledge discovery in databases is an overall process that has been described[202] as having several discrete steps:

(1) Definition of the goals of the KDD process.
(2) Creation or selection of a data set.
(3) Data cleaning and preprocessing.
(4) Data reduction and projection.
(5) Selection of data mining methods.
(6) Exploratory analysis and model/hypothesis selection.
(7) Data mining.
(8) Interpretation/evaluation.
(9) Utilization.

The overall process is, in fact, interactive and requires the input of both a skilled computer scientist and an expert in the field generating the data (i.e., a "data owner").[198] Clearly defining the goals of the project requires both kinds of expertise. Thus, if the task is to use examples of relationships to learn or develop a model that predicts relationships, the task is *predictive* (or *supervised*). On the other hand, a common goal in gene expression experiments is to detect co-regulated genes; in this case it is not known what to predict, and the task is *descriptive* and computer approaches are described as *unsupervised.*

Likewise, steps 2 and 3, the creation or selection of a data set, requires consideration. On one hand, a large and diverse data set is generally desirable.[203] In fact, all too often models are constructed in toxicology based on tests with a limited group of compounds, usually ones that have well-demonstrated activity in the tests. Ideally the data set should include a complete spectrum of values for any attribute,[204] such as chemical structure, biological activity, or gene function. On the other hand, biological data can be generated in different laboratories by different methodologies and protocols. By the old maxim "garbage-in, garbage-out" such data could be inherently inconsistent and poison the model building process. Data cleaning, checking for inconsistencies and correcting errors is also a time consuming task in the process. Obviously, toxicological data is open to error, but chemical structures also can be incorrectly entered in databases, and data for mixtures of compounds may not be suitably included with that for pure compounds. Genomic data offers similar concerns: as one example, sequencing errors are frequent, and as another example, description of biological "function" for a gene may be derived from a flawed or artificial experimental system. Furthermore, this step requires eyes trained not only for experimental error, but also for formatting issues such as the presence of text characters (e.g., tildes, commas) in numerical data. Thus, careful attention to the quality, quantity, and characteristics of the data set is an essential pre-requisite to successful data mining.

Preprocessing may also entail conversion of data types into the forms appropriate for the project. As a example, histopathological observations may need to be reduced to present/not-present codes. An essential step for data mining of diverse chemical structures may be the reduction of those structures to fragments, allowing a predictive data mining exercise to learn the model from substructures, and then make predictions from new examples that contain certain of those substructures. This approach was indeed first used in the CASE/MultiCASE programs.[205,206] Similarly one could envision describing the transcriptional promoter for a gene as a group of elements (e.g., transcription factor binding sites) with various physical arrangements and properties. An environmental example might include describing a water body in terms of inflow and outflow streams with various configurations and rates.

Step 4 of the KDD process requires selection of the data features important for the model development. This could mean selecting chemical properties, biological endpoints, or other parameters. "Projection" is the process by which the data is formatted into a machine-readable form. In this process we can refer to the transformed data as either "propositional" or "relational".[200] Propositional data is generally data that can be simple compiled in a tabular form. Relational data is that which would have been represented with several interconnected tables, or similarly involved data structure. An example of the latter would be compound descriptions, with variable amounts of atom numbers, types, partial charges, bond connectivities, etc. A biological example would be a "gene" description, with variable numbers of exons, transcript variants, polymorphisms, tissue expression, and so forth. This step is again often overlooked in its importance and time consumption, and is most problematic when taking data from unstructured sources (e.g., text). Of course, if the data is drawn from well-structured databases, then automatic transformation into the format required for data mining is possible.

Selecting the Data Mining method (step 5) is intrinsically linked with the goal of the endeavor. Thus if goal entails unsupervised learning, then a method such as clustering may be selected. On the other hand, if the goal is to develop a predictive model, based on a database of examples, then two other options are available. In one case, a model may be developed to predict numerical values, while in the model is designed to predict simply a feature or attribute. The former task in called a regression, while the latter called a classification, with respective examples being Quantitative Structure Activity Relationship (QSAR) models and simple. Structure Activity Relationship (SAR) models. It should be noted that semi-quantiative models can be constructed by classification if the numerical values are "binned", e.g., high, medium, low, and that for many biological endpoints such binning may be more faithful to the actual accuracy of the data.

Steps 6 and 7 involve selecting a data mining model, search strategy, and system such as PROGOL, C4.5, CASE or MultiCASE. For details on these steps the reader is referred to a recent review.[200]

Some mention should be made of evaluating the model, step 8, which in predictive models is often referred to as "validating" the model. Taking the example of SAR models, this step often takes the approach of partitioning the data set into multiple subsets: all but one are combined into a learning data set, which is then used to predict the activity of the remaining subset. Obviously, the smaller the entire data set, the more likely that any given subset will contain unique examples (e.g., chemical structures), that will be poorly predicted by the training set.

#### 14.5.1.8.3 Applications in Toxicology

##### 14.5.1.8.3.1 *Structure-Activity Relationship Models*

Certainly the most mature application of data mining to toxicology is that of predictive structure-activity relationship models.[207] One of the earliest models was developed by Enslein in 1982 to estimate carcinogenicity,[208] and evolved into the TOPKAT program. The model is developed with substructure descriptors and is based on linear discriminant analysis, setting up equations to separate the positive and negative structures. This approach has been developed, as well, for predicting *Salmonella* mutagenicity,[209] developmental toxicity,[210,211] rat oral acute toxicity,[212] and guinea pig dermal sensitization.[213] Current models can be based on sub-models for different chemical classes,[211,213] and include an algorithm to indicate if the "test" chemical structure lies outside the chemical space covered by the model. A slightly different approach was taken by Kopman in the development of the CASE program (Computer-automated structural evaluation):[205] again example molecules are split into structural fragments, but these are then evaluated statistically for a contribution to activity (biophores) or a subtraction from activity (biophobes). CASE develops a QSAR model and thus can predict potency. With the development of the MultiCASE program[206,214] multiple biophores and modulating fragments are included in the prediction. MultiCASE models have been developed to investigate and predict Salmonella mutagenicity,[215] activity in the mouse lymphoma *tk* mutation assay,[216] and potency as a mouse carcinogen;[217] other endpoints as well have been incorporated into a Tox II program to identify overall toxicity for a novel structure.[218] These models have been reviewed[219,220] and compared to other models;[215,221] in the Salmonella mutagenicity model both systems were about 75% concordant with the actual experimental results for the test structure. The DEREK system[222] offers another SAR model approach; this is a rule-based system that again focuses on substructures, and models have been developed for Salmonella mutagenicity,[223] metabolic fate,[223] mouse carcinogenicity,[221] and skin sensitization.[224] Finally, a machine learning approach using inductive logic programming was used to develop models for mutagenicity[225] and rat carcinogenicity.[226] With new algorithmic approaches and with an expanding quality data set one can expect SAR/QSAR data mining models to continue to evolve.[200,221]

### 14.5.1.8.3.2  *Other Applications*

While SAR models, especially of carcinogenicity and mutagenicity have dominated data mining efforts in toxicology, the development of other large data sets will definitely set the stage for other explorations. As noted above, a variety of data mining models have been applied to microarray data analysis;[93] similar models have also been used to compare the gene expression patterns for the 60 cell lines in the NCI drug discovery screen with their response to chemotherapeutic agents.[227,228] The vast amount and variety of genomic and

biologic data available in public databases[95] begs a number of data mining efforts. Andrade and Bork[229] describe how information retrieval and information extraction techniques can be used to identify relevant journals in literature databases and extract appropriate keywords from the abstracts in this subset for annotating genetic loci. This approach could certainly be extended to topics in the toxicology literature. An example is the study of trends in occupational exposures Symanski and colleagues conducted[230] where they compiled about 700 sets of data from the journal literature and other sources. Other large data sets that could be subject for data mining are those from multiple environmental, epidemiological, and clinical studies. While the various testing programs such as the NTP offer public data sets of toxicity data for chemicals of environmental concern, new drug registrations also generate a vast amount of pre-clinical and clinical toxicity data. The more that the classically-trained computer scientist takes on a prominent role in molecular biology and toxicology departments, the more that new knowledge will be gleaned from these diverse data sets.

### 14.5.1.9  REFERENCES

1. J. B. Hagen, 'The origins of bioinformatics.' *Nat. Rev. Genet.*, 2000, **1**, 231–236.
2. M. O. Dayhoff, 'Computer aids to protein sequence determination.' *J. Theor. Biol.*, 1965, **8**, 97–112.
3. M. O. Dayhoff, 'Computer analysis of protein evolution.' *Sci. Am.*, 1969, **22**, 86–95.
4. S. Scherer and R. W. Davis, 'Replacement of chromosome segments with altered DNA sequences constructed *in vitro*.' *Proc. Nat. Acad. Sci. USA*, 1979, **76**, 4951–4955.
5. A. D. Baxevanis, 'The Molecular Biology Database Collection: an updated compilation of biological database resources.' *Nucleic Acids Res.*, 2001, **29**, 1–10.
6. M. D. Adams, J. M. Kelley, J. D. Gocayne *et al.*, 'Complementary DNA sequencing: expressed sequence tags and human genome project.' *Science*, 1991, **252**, 1651–1656.
7. G. Miller, R. Fuchs and E. Lai, 'IMAGE cDNA clones, UniGene clustering, and ACeDB: an integrated resource for expressed sequence information.' *Genome. Res.*, 1997, **7**, 1027–1032.
8. F. Liang, I. Holt, G. Pertea *et al.*, 'An optimized protocol for analysis of EST sequences.' *Nucleic Acids Res.*, 2000, **28**, 3657–3665.
9. J. Quackenbush, F. Liang, I. Holt *et al.*, 'The TIGR gene indices: reconstruction and representation of expressed gene sequences.' *Nucleic Acids Res.*, 2000, **28**, 141–145.
10. S. F. Altschul, W. Gish, W. Miller *et al.*, 'Basic local alignment search tool.' *J. Mol. Biol.*, 1990, **215**, 403–410.
11. A. Rotig, I. Valnot, C. Mugnier *et al.*, 'Screening human EST database for identification of candidate

genes in respiratory chain deficiency.' *Mol. Genet. Metab.*, 2000, **69**, 223–232.

12. J. B. Hogenesch, W. K. Chan, V. H. Jackiw *et al.*, 'Characterization of a subset of the basic-helix-loop-helix-PAS superfamily that interacts with components of the dioxin signaling pathway.' *J. Biol. Chem.*, 1997, **272**, 8581–8593.

13. M. Lara-Tejero and J. E. Galan, 'A bacterial toxin that controls cell cycle progression as a deoxyribonuclease I-like protein.' *Science*, 2000, **290**, 354–357.

14. F. Corpet, 'Multiple sequence alignment with hierarchical clustering.' *Nucleic Acids Res.*, 1988, **16**, 10881–10890.

15. G. D. Schuler, S. F. Altschul and D. Lipman, 'A workbench for multiple sequence alignment construction and analysis.' *Proteins Struct. Funct. Genet.*, 1991, **9**, 180–190.

16. D. G. Higgins, J. D. Thompson and T. J. Gibson, 'Using CLUSTAL for multiple sequence alignments.' *Methods Enzymol.*, 1996, **266**, 383–402.

17. G. Blobel, 'Unidirectional and bidirectional protein traffic across membranes.' *Cold Spring Harb. Symp. Quant. Biol.*, 1995, **60**, 1–10.

18. R. M. Stroud and P. Walter, 'Signal sequence recognition and protein targeting.' *Curr. Opin. Struct. Biol.*, 1999, **9**, 754–759.

19. T. G. Petrakis, E. Ktistaki, L. Wang *et al.*, 'Cloning and characterization of mouse deoxyguanosine kinase. Evidence for a cytoplasmic isoform.' *J. Biol. Chem.*, 1999, **274**, 24726–24730.

20. J. F. Gibrat, T. Madej and S. H. Bryant, 'Surprising similarities in structure comparison.' *Curr. Opin. Struct. Biol.*, 1996, **6**, 377–385.

21. D. S. Latchman, 'Eukaryotic Transcription Factors,' Academic Press, New York, 1991.

22. D. A. Spandidos, R. A. Nichols, N. M. Wilkie *et al.*, 'Phorbol ester-responsive H-ras1 gene promoter contains multiple TPA-inducible/AP-1-binding consensus sequence elements.' *FEBS Lett.*, 1988, **240**, 191–195.

23. D. Ghosh, 'A relational database of transcription factors.' *Nucleic Acids Res.*, 1990, **18**, 1749–1756.

24. T. Heinemeyer, E. Wingender, I. Reuter *et al.*, 'Databases on transcriptional regulation: TRANS-FAC, TRRD and COMPEL.' *Nucleic Acids Res.*, 1998, **26**, 362–367.

25. R. C. Perier, V. Praz, T. Junier *et al.*, 'The eukaryotic promoter database (EPD).' *Nucleic Acids Res.*, 2000, **28**, 302–303.

26. D. S. Prestridge and G. Stormo, 'SIGNAL SCAN 3.0: new database and program features.' *Comput. Appl. Biosci.*, 1993, **9**, 113–115.

27. K. Frech, K. Quandt and T. Werner, 'Software for the analysis of DNA sequence elements of transcription.' *Comput. Appl. Biosci.*, 1997, **13**, 89–97.

28. K. Quandt, K. Frech, H. Karas *et al.*, 'MatInd and MatInspector: new fast and versatile tools for detection of consensus matches in nucleotide sequence data.' *Nucleic Acids Res.*, 1995, **23**, 4878–4884.

29. W. M. Fitch and E. Margoliash, 'Construction of phylogenetic trees.' *Science*, 1967, **155**, 279–284.

30. D. W. Nebert, M. Adesnik, M. J. Coon *et al.*, 'The P450 gene superfamily: recommended nomenclature.' *DNA*, 1987, **6**, 1–11.

31. D. W. Nebert, D. R. Nelson, M. Adesnik *et al.*, 'The P450 superfamily: updated listing of all genes and recommended nomenclature for the chromosomal loci.' *DNA*, 1989, **8**, 1–13.

32. D. R. Nelson, T. Kamataki, D. J. Waxman *et al.*, 'The P450 Superfamily: Update on new sequences, gene mapping, accession numbers, early trivial names of enzymes, and nomenclature.' *DNA Cell. Biol.*, 1993, **12**, 1–51.

33. D. R. Nelson, L. Koymans, T. Kamataki *et al.*, 'P450 superfamily: update on new sequences, gene mapping, accession numbers and nomenclature.' *Pharmacogenetics*, 1996, **6**, 1–42.

34. E. Robertson, A. Bradley, M. Kuehn *et al.*, 'Germ-line transmission of genes introduced into cultured pluripotential cells by retroviral vector.' *Nature*, 1986, **323**, 445–448.

35. S. L. Mansour, K. R. Thomas and M. R. Capecchi, 'Disruption of the proto-oncogene int-2 in mouse embryo-derived stem cells: a general strategy for targeting mutations to non-selectable genes.' *Nature*, 1988, **336**, 348–352.

36. M. R. Capecchi, 'Altering the genome by homologous recombination.' *Science*, 1989, **244**, 1288–1292.

37. B. Sauer, 'Manipulation of transgenes by site-specific recombination: use of Cre recombinase.' *Methods Enzymol.*, 1993, **225**, 890–900.

38. S. M. Dymecki, 'Flp recombinase promotes site-specific DNA recombination in embryonic stem cells and transgenic mice.' *Proc. Nat. Acad. Sci. USA*, 1996, **93**, 6191–6196.

39. A. Kistner, M. Gossen, F. Zimmermann *et al.*, 'Doxycycline-mediated quantitative and tissue-specific control of gene expression in transgenic mice.' *Proc. Nat. Acad. Sci. USA*, 1996, **93**, 10933–10938.

40. S. W. Kohler, G. S. Provost, A. Fieck *et al.*, 'Spectra of spontaneous and mutagen-induced mutations in the lacI gene in transgenic mice.' *Proc. Nat. Acad. Sci. USA*, 1991, **88**, 7958–7962.

41. B. C. Myhr, 'Validation studies with Muta Mouse: a transgenic mouse model for detecting mutations *in vivo*.' *Environ. Mol. Mutagen.*, 1991, **18**, 308–315.

42. J. A. Heddle, S. Dean, T. Nohmi *et al.*, 'In vivo transgenic mutation assays.' *Environ. Mol. Mutagen.*, 2000, **35**, 253–259.

43. L. Cosentino and J. A. Heddle, 'Differential mutation of transgenic and endogenous loci *in vivo*.' *Mutat. Res.*, 2000, **454**, 1–10.

44. V. E. Walker, J. L. Andrews, P. B. Upton *et al.*, 'Detection of cyclophosphamide-induced mutations at the Hprt but not the lacI locus in splenic lymphocytes of exposed mice.' *Environ. Mol. Mutagen.*, 1999, **34**, 167–181.

45. J. J. Monroe, K. L. Kort, J. E. Miller *et al.*, 'A comparative study of *in vivo* mutation assays: analysis of hprt, lacI, cII/cI and as mutational targets for *N*-nitroso-*N*-methylurea and benzo[a]pyrene in Big Blue mice.' *Mutat. Res.*, 1998, **421**, 121–136.

46. V. N. Dobrovolsky, D. A. Casciano and R. H. Heflich, 'Development of a novel mouse tk+/− embryonic stem cell line for use in mutagenicity studies.' *Environ. Mol. Mutagen.*, 1996, **28**, 483–489.

47. H. Vrieling, S. Wijnhoven, P. van Sloun *et al.*, 'Heterozygous Aprt mouse model: detection and study of a broad range of autosomal somatic mutations *in vivo*.' *Environ. Mol. Mutagen.*, 1999, **34**, 84–89.

48. R. N. Winn, M. B. Norris, K. J. Brayer *et al.*, 'Detection of mutations in transgenic fish carrying a bacteriophage lambda cII transgene target.' *Proc. Nat. Acad. Sci. USA*, 2000, **97**, 12655–12660.

49. D. Gulezian, D. Jacobson-Ram, C. B. McCullough *et al.*, 'Use of transgenic animals for carcinogenicity testing: considerations and implications for risk assessment.' *Toxicol. Pathol.*, 2000, **28**, 482–499.

50. J. W. Spalding, J. E. French, R. R. Tice *et al.*, 'Development of a transgenic mouse model for carcinogenesis bioassay: evaluation of chemically induced skin tumors in Tg.AC mice.' *Toxicol. Sci.*, 1999, **49**, 241–254.

51. S. Yamamoto, Y. Hayashi, K. Mitsumori *et al.*, 'Rapid carcinogenicity testing system with transgenic

mice harboring human prototype c-HRAS gene.' *Lab. Anim. Sci.*, 1997, **47**, 121–126.

52. J. F. Mahler, N. D. Flagler, D. E. Malarkey *et al.*, 'Spontaneous and chemically induced proliferative lesions in Tg.AC transgenic and p53-heterozygous mice.' *Toxicol. Pathol.*, 1998, **26**, 501–511.

53. R. R. Maronpot, K. Mitsumori, P. Mann *et al.*, 'Interlaboratory comparison of the CB6F1-Tg rasH2 rapid carcinogenicity testing model.' *Toxicology*, 2000, **146**, 149–159.

54. T. Koujitani, K. Yasuhara, T. Usui *et al.*, 'Lack of susceptibility of transgenic mice carrying the human c-Ha-ras proto-oncogene (rasH2 mice) to phenolphthalein in a 6-month carcinogenicity study.' *Cancer. Lett.*, 2000, **152**, 211–216.

55. J. L. Raucy, J. M. Lasker, C. S. Lieber *et al.*, 'Acetaminophen activation by human liver cytochromes P450IIE1 and P450IA2.' *Arch. Biochem. Biophys.*, 1989, **271**, 270–283.

56. H. Zaher, J. T. Buters, J. M. Ward *et al.*, 'Protection against acetaminophen toxicity in CYP1A2 and CYP2E1 double-null mice.' *Toxicol. Appl. Pharmacol.*, 1998, **152**, 193–199.

57. S. Kimura, M. Kawabe, J. M. Ward *et al.*, 'CYP1A2 is not the primary enzyme responsible for 4-aminobiphenyl-induced hepatocarcinogenesis in mice.' *Carcinogenesis*, 1999, **20**, 1825–1830.

58. J. T. Buters, S. Sakai, T. Richter *et al.*, 'Cytochrome P450 CYP1B1 determines susceptibility to 7, 12-dimethylbenz[a]anthracene-induced lymphomas.' *Proc. Nat. Acad. Sci. USA*, 1999, **96**, 1977–1982.

59. F. J. Gonzalez, P. Fernandez-Salguero, S. S. Lee *et al.*, 'Xenobiotic receptor knockout mice.' *Toxicol. Lett.*, 1995, **82–83**, 117–121.

60. S. S. Lee, T. Pineau, J. Drago *et al.*, 'Targeted disruption of the α isoform of the peroxisome proliferator-activated receptor gene in mice results in abolishment of the pleiotropic effects of peroxisome proliferators.' *Mol. Cell. Biol.*, 1995, **15**, 3012–3022.

61. J. C. Corton, P. J. Lapinskas and F. J. Gonzalez, 'Central role of PPAR α in the mechanism of action of hepatocarcinogenic peroxisome proliferators.' *Mutat. Res.*, 2000, **448**, 139–151.

62. F. J. Gonzalez, J. M. Peters and R. C. Cattley, 'Mechanism of action of the nongenotoxic peroxisome proliferators: role of the peroxisome proliferator-activator receptor α.' *J. Nat. Cancer. Inst.*, 1998, **90**, 1702–1709.

63. J. M. Ward, J. M. Peters, C. M. Perella *et al.*, 'Receptor and nonreceptor-mediated organ-specific toxicity of di(2-ethylhexyl)phthalate (DEHP) in peroxisome proliferator-activated receptor α-null mice.' *Toxicol. Pathol.*, 1998, **26**, 240–246.

64. J. M. Peters, R. C. Cattley and F. J. Gonzalez, 'Role of PPAR α in the mechanism of action of the nongenotoxic carcinogen and peroxisome proliferator Wy-14,643.' *Carcinogenesis*, 1997, **18**, 2029–2033.

65. J. C. Alwine, D. J. Kemp and G. R. Stark, 'Method for detection of specific RNAs in agarose gels by transfer to diazobenzyloxymethyl-paper and hybridization with DNA probes.' *Proc. Nat. Acad. Sci. USA*, 1977, **74**, 5350–5354.

66. B. F. Trump and I. K. Berezesky, 'The role of altered [Ca$^{2+}$]i regulation in apoptosis, oncosis, and necrosis.' *Biochim. Biophys. Acta.*, 1996, **1313**, 173–178.

67. C. L. Parfett, R. Pilon, J. W. Barlow *et al.*, 'Altered Abundance of Multiple, Oxidant-Inducible mRNA Species in C3H/10T1/2 Cells Following Exposure to Asbestos or Reactive Oxygen Species.' *In vitro Toxicol.*, 1996, **9**, 403–417.

68. N. Tygstrup, S. A. Jensen, B. Krog *et al.*, 'Expression of liver-specific functions in rat hepatocytes following

sublethal and lethal acetaminophen poisoning.' *J. Hepatol.*, 1996, **25**, 183–190.

69. N. Tygstrup, S. A. Jensen, B. Krog *et al.*, 'Expression of liver functions following sub-lethal and non-lethal doses of allyl alcohol and acetaminophen in the rat.' *J. Hepatol.*, 1997, **27**, 156–162.

70. E. F. Nuwaysir, M. Bittner, J. Trent *et al.*, 'Microarrays and toxicology: the advent of toxicogenomics.' *Mol. Carcinog.*, 1999, **24**, 153–159.

71. S. Farr and R. T. Dunn II, 'Concise review: gene expression applied to toxicology.' *Toxicol. Sci.*, 1999, **50**, 1–9.

72. M. D. Todd, M. J. Lee, J. L. Williams *et al.*, 'The CAT-Tox (L) assay: a sensitive and specific measure of stress-induced transcription in transformed human liver cells.' *Fundam. Appl. Toxicol.*, 1995, **28**, 118–128.

73. P. Liang and A. B. Pardee, 'Differential display of eukaryotic messenger RNA by means of the polymerase chain reaction.' *Science*, 1992, **257**, 967–971.

74. L. Robinson, A. Panayiotakis, T. S. Papas *et al.*, 'ETS target genes: Identification of Egr1 as a target by RNA differential display and whole genome PCR techniques.' *Proc. Nat. Acad. Sci. USA*, 1997, **94**, 7170–7175.

75. A. Bhattacharjee, V. R. Lappi, M. S. Rutherford *et al.*, 'Molecular dissection of dimethylnitrosamine (DMN)-induced hepatotoxicity by mRNA differential display.' *Toxicol. Appl. Pharmacol.*, 1998, **150**, 186–195.

76. A. E. Kegelmeyer, C. S. Sprankle, G. J. Horesovsky *et al.*, 'Differential display identified changes in mRNA levels in regenerating livers from chloroform-treated mice.' *Mol. Carcinog.*, 1997, **20**, 288–297.

77. D. R. Crawford and K. J. Davies, 'Modulation of a cardiogenic shock inducible RNA by chemical stress: adapt73/PigHep3.' *Surgery*, 1997, **121**, 581–587.

78. Y. Habu, S. Fukada-Tanaka, Y. Hisatomi *et al.*, 'Amplified restriction fragment length polymorphism-based mRNA fingerprinting using a single restriction enzyme that recognizes a 4-bp sequence.' *Biochem. Biophys. Res. Commun.*, 1997, **234**, 516–521.

79. A. Fischer, H. Saedler and G. Theissen, 'Restriction fragment length polymorphism-coupled domain-directed differential display: a highly efficient technique for expression analysis of multigene families.' *Proc. Nat. Acad. Sci. USA*, 1995, **92**, 5331–5335.

80. R. Sorek and M. Amitai, 'Piecing together the significance of splicing.' *Nat. Biotechnol.*, 2001, **19**, 196.

81. V. E. Velculescu, L. Zhang, B. Vogelstein *et al.*, 'Serial analysis of gene expression.' *Science*, 1995, **270**, 484–487.

82. J. Stollberg, J. Urschitz, Z. Urban *et al.*, 'A quantitative evaluation of SAGE.' *Genome. Res.*, 2000, **10**, 1241–1248.

83. J. M. Almendral, D. Sommer, H. Macdonald-Bravo *et al.*, 'Complexity of the early genetic response to growth factors in mouse fibroblasts.' *Mol. Cell. Biol.*, 1988, **8**, 2140–2148.

84. L. M. Crosby, K. S. Hyder, A. B. DeAngelo *et al.*, 'Morphologic Analysis Correlates with Gene Expression Changes in Cultured F344 Rat Mesothelial Cells.' *Toxicol. Appl. Pharmacol.*, 2000, **169**, 205–221.

85. M. Schena, D. Shalon, R. W. Davis *et al.*, 'Quantitative monitoring of gene expression patterns with a complementary DNA microarray.' *Science*, 1995, **270**, 467–470.

86. D. Shalon, S. J. Smith and P. O. Brown, 'A DNA microarray system for analyzing complex DNA samples using two-color fluorescent probe hybridization.' *Genome. Res.*, 1996, **6**, 639–645.

87. J. L. DeRisi, V. R. Iyer and P. O. Brown, 'Exploring the metabolic and genetic control of gene expression on a genomic scale.' *Science*, 1997, **278**, 680–686.

88. D. J. Lockhart, H. Dong, M. C. Byrne *et al.*, 'Expression monitoring by hybridization to high-density oligonucleotide arrays.' *Nat. Biotechnol.*, 1996, **14**, 1675–1680.

89. A. de Saizieu, U. Certa, J. Warrington *et al.*, 'Bacterial transcript imaging by hybridization of total RNA to oligonucleotide arrays.' *Nat. Biotechnol.*, 1998, **16**, 45–48.

90. D. I. Stimpson, P. W. Cooley, S. M. Knepper *et al.*, 'Parallel production of oligonucleotide arrays using membranes and reagent jet printing.' *BioTechniques*, 1998, **25**, 886–890.

91. T. R. Hughes, M. Mao, A. R. Jones *et al.*, 'Expression profiling using microarrays fabricated by an ink-jet oligonucleotide synthesizer.' *Nat. Biotechnol.*, 2001, **19**, 342–347.

92. M. Q. Zhang, 'Large-scale gene expression data analysis: a new challenge to computational biologists.' *Genome. Res.*, 1999, **9**, 681–688.

93. D. E. Bassett, M. B. Eisen and M. S. Boguski, 'Gene expression informatics—it's all in your mine.' *Nat. Genet.*, 1999, **21**, 51–55.

94. O. Ermolaeva, M. Rastogi, K. D. Pruitt *et al.*, 'Data management and analysis for gene expression arrays.' *Nat. Genet.*, 1998, **20**, 19–23.

95. D. L. Wheeler, C. Chappey, A. E. Lash *et al.*, 'Database resources of the National Center for Biotechnology Information.' *Nucleic Acids Res.*, 2000, **28**, 10–14.

96. K. D. Pruitt and D. R. Maglott, 'RefSeq and LocusLink: NCBI gene-centered resources.' *Nucleic Acids Res.*, 2001, **29**, 137–140.

97. M. Kanehisa and S. Goto, 'KEGG: kyoto encyclopedia of genes and genomes.' *Nucleic Acids Res.*, 2000, **28**, 27–30.

98. M. Rebhan, V. Chalifa-Caspi, J. Prilusky *et al.*, 'GeneCards: a novel functional genomics compendium with automated data mining and query reformulation support.' *Bioinformatics*, 1998, **14**, 656–664.

99. C. V. Jongeneel, 'The need for a human gene index.' *Bioinformatics*, 2000, **16**, 1059–1061.

100. J. White, H. Wain, E. Bruford *et al.*, 'Promoting a standard nomenclature for genes and proteins.' *Nature*, 1999, **402**, 347.

101. M. Ashburner, C. A. Ball, J. A. Blake *et al.*, 'Gene ontology: tool for the unification of biology. The Gene Ontology Consortium.' *Nat. Genet.*, 2000, **25**, 25–29.

102. M. L. Lee, F. C. Kuo, G. A. Whitmore *et al.*, 'Importance of replication in microarray gene expression studies: statistical methods and evidence from repetitive cDNA hybridizations.' *Proc. Nat. Acad. Sci. USA*, 2000, **97**, 9834–9839.

103. T. R. Hughes, M. J. Marton, A. R. Jones *et al.*, 'Functional discovery via a compendium of expression profiles.' *Cell*, 2000, **102**, 109–126.

104. M. B. Eisen, P. T. Spellman, P. O. Brown *et al.*, 'Cluster analysis and display of genome-wide expression patterns.' *Proc. Nat. Acad. Sci. USA*, 1998, **95**, 14863–14868.

105. L. J. Heyer, S. Kruglyak and S. Yooseph, 'Exploring expression data: identification and analysis of coexpressed genes.' *Genome. Res.*, 1999, **9**, 1106–1115.

106. S. Tavazoie, J. D. Hughes, M. J. Campbell *et al.*, 'Systematic determination of genetic network architecture.' *Nat. Genet.*, 1999, **22**, 281–285.

107. R. Sasik, T. Hwa, N. Iranfar *et al.*, 'Percolation clustering: a novel approach to the clustering of gene expression patterns in Dictyostelium development.' *Pac. Symp. Biocomput.*, 2001, 335–347.

108. P. Toronen, M. Kolehmainen, G. Wong *et al.*, 'Analysis of gene expression data using self-organizing maps.' *FEBS Lett.*, 1999, **451**, 142–146.

109. P. Tamayo, D. Slonim, J. Mesirov *et al.*, 'Interpreting patterns of gene expression with self-organizing maps: methods and application to hematopoietic differentiation.' *Proc. Nat. Acad. Sci. USA*, 1999, **96**, 2907–2912.

110. T. Hastie, R. Tibshirani, M. B. Eisen *et al.*, ' ''Gene shaving'' as a method for identifying distinct sets of genes with similar expression patterns.' *Genome Biol.*, 2000, **1**, 1–21.

111. M. P. Brown, W. N. Grundy, D. Lin *et al.*, 'Knowledge-based analysis of microarray gene expression data by using support vector machines.' *Proc. Nat. Acad. Sci. USA*, 2000, **97**, 262–267.

112. G. Sherlock, 'Analysis of large-scale gene expression data.' *Curr. Opin. Immunol.*, 2000, **12**, 201–205.

113. W. D. Pennie, J. D. Tugwood, G. J. Oliver *et al.*, 'The principles and practice of toxigenomics: applications and opportunities.' *Toxicol. Sci.*, 2000, **54**, 277–283.

114. M. R. Fielden and T. R. Zacharewski, 'Challenges and limitations of gene expression profiling in mechanistic and predictive toxicology.' *Toxicol. Sci.*, 2001, **60**, 6–10.

115. P. R. Holden, N. H. James, A. N. Brooks *et al.*, 'Identification of a possible association between carbon tetrachloride-induced hepatotoxicity and interleukin-8 expression.' *J. Biochem. Mol. Toxicol.*, 2000, **14**, 283–290.

116. C. P. Rodi, R. T. Bunch, S. W. Curtiss *et al.*, 'Revolution through genomics in investigative and discovery toxicology.' *Toxicol. Pathol.*, 1999, **27**, 107–110.

117. S. A. Jelinsky and L. D. Samson, 'Global response of Saccharomyces cerevisiae to an alkylating agent.' *Proc. Nat. Acad. Sci. USA*, 1999, **96**, 1486–1491.

118. S. A. Amundson, M. Bittner, Y. Chen *et al.*, 'Fluorescent cDNA microarray hybridization reveals complexity and heterogeneity of cellular genotoxic stress responses.' *Oncogene*, 1999, **18**, 3666–3672.

119. N. L. Anderson, J. Taylor, J. P. Hofmann *et al.*, 'Simultaneous measurement of hundreds of liver proteins: application in assessment of liver function.' *Toxicol. Pathol.*, 1996, **24**, 72–76.

120. M. J. Marton, J. L. DeRisi, H. A. Bennett *et al.*, 'Drug target validation and identification of secondary drug target effects using DNA microarrays.' *Nat. Med.*, 1998, **4**, 1293–1301.

121. G. F. Bammert and J. M. Fostel, 'Genome-wide expression patterns in *Saccharomyces cerevisiae*: comparison of drug treatments and genetic alterations affecting biosynthesis of ergosterol.' *Antimicrob. Agents. Chemother.*, 2000, **44**, 1255–1265.

122. A. P. Gasch, P. T. Spellman, C. M. Kao *et al.*, 'Genomic expression programs in the response of yeast cells to environmental changes.' *Mol. Biol. Cell*, 2000, **11**, 4241–4257.

123. M. E. Burczynski, M. McMillian, J. Ciervo *et al.*, 'Toxicogenomics-based discrimination of toxic mechanism in HepG2 human hepatoma cells.' *Toxicol. Sci.*, 2000, **58**, 399–415.

124. J. P. Vanden Heuvel, G. C. Clark, C. L. Thompson *et al.*, 'CYP1A1 mRNA levels as a human exposure biomarker: use of quantitative polymerase chain reaction to measure CYP1A1 expression in human peripheral blood lymphocytes.' *Carcinogenesis*, 1993, **14**, 2003–2006.

125. T. Nakamoto, I. Hase, S. Imaoka *et al.*, 'Quantitative RT-PCR for CYP3A4 mRNA in human peripheral lymphocytes: induction of CYP3A4 in lymphocytes and in liver by rifampicin.' *Pharmacogenetics*, 2000, **10**, 571–575.

126. M. R. Wilkins, J. C. Sanchez, A. A. Gooley *et al.*, 'Progress with proteome projects: why all proteins expressed by a genome should be identified and how to do it.' *Biotechnol. Genet. Eng. Rev.*, 1996, **13**, 19–50.

127. N. G. Anderson and L. Anderson, 'The Human Protein Index.' *Clin. Chem.*, 1982, **28**, 739–748.

128. J. Klose, 'Protein mapping by combined isoelectric focusing and electrophoresis of mouse tissues. A novel approach to testing for induced point mutations in mammals.' *Humangenet.*, 1975, **26**, 231–243.

129. G. A. Scheele, 'Two-dimensional gel analysis of soluble proteins. Charaterization of guinea pig exocrine pancreatic proteins.' *J. Biol. Chem.*, 1975, **250**, 5375–5385.

130. P. H. O'Farrell, 'High resolution two-dimensional electrophoresis of proteins.' *J. Biol. Chem.*, 1975, **250**, 4007–4021.

131. U. K. Laemmli, 'Cleavage of structural proteins during the assembly of the head of bacteriophage T4.' *Nature*, 1970, **227**, 680–685.

132. P. Z. O'Farrell, H. M. Goodman and P. H. O'Farrell, 'High resolution two-dimensional electrophoresis of basic as well as acidic proteins.' *Cell*, 1977, **12**, 1133–1141.

133. N. L. Anderson and N. G. Anderson, 'Analytical techniques for cell fractions. XXII. Two-dimensional analysis of serum and tissue proteins: multiple gradient-slab gel electrophoresis.' *Anal. Biochem.*, 1978, **85**, 341–354.

134. N. G. Anderson and N. L. Anderson, 'Analytical techniques for cell fractions. XXI. Two-dimensional analysis of serum and tissue proteins: multiple isoelectric focusing.' *Anal. Biochem.*, 1978, **85**, 331–340.

135. B. Bjellqvist, K. Ek, P. G. Righetti *et al.*, 'Isoelectric focusing in immobilized pH gradients: principle, methodology and some applications.' *J. Biochem. Biophys. Methods*, 1982, **6**, 317–339.

136. K. Berggren, E. Chernokalskaya, T. H. Steinberg *et al.*, 'Background-free, high sensitivity staining of proteins in one- and two-dimensional sodium dodecyl sulfate-polyacrylamide gels using a luminescent ruthenium complex.' *Electrophoresis*, 2000, **21**, 2509–2521.

137. J.-C. Jean-Charles Sanchez, V. Converset, C. Hoogland *et al.*, 'The mouse SWISS-2D PAGE database: a tool for proteomics study of diabetes and obesity.' *Proteomics*, 2001, **1**, 136–163.

138. M. Wilm, 'Mass spectrometric analysis of proteins.' *Adv. Protein. Chem.*, 2000, **54**, 1–30.

139. A. Shevchenko, O. N. Jensen, A. V. Podtelejnikov *et al.*, 'Linking genome and proteome by mass spectrometry: large-scale identification of yeast proteins from two dimensional gels.' *Proc. Nat. Acad. Sci. USA*, 1996, **93**, 14440–14445.

140. A. Shevchenko, I. Chernushevich, M. Wilm *et al.*, 'De Novo peptide sequencing by nanoelectrospray tandem mass spectrometry using triple quadrupole and quadrupole/time-of-flight instruments.' *Methods. Mol. Biol.*, 2000, **146**, 1–16.

141. P. A. Binz, M. Muller, D. Walther *et al.*, 'A molecular scanner to automate proteomic research and to display proteome images.' *Anal. Chem.*, 1999, **71**, 4981–4988.

142. N. G. Anderson, A. Matheson and N. L. Anderson, 'Back to the future: The human protein index HPI and the agenda for post-proteomic biology.' *Proteomics*, 2001, **1**, 3–12.

143. S. P. Gygi, G. L. Corthals, Y. Zhang *et al.*, 'Evaluation of two-dimensional gel electrophoresis-based proteome analysis technology.' *Proc. Nat. Acad. Sci. USA*, 2000, **97**, 9390–9395.

144. M. Unlu, M. E. Morgan and J. S. Minden, 'Difference gel electrophoresis: a single gel method for detecting changes in protein extracts.' *Electrophoresis*, 1997, **18**, 2071–2077.

145. R. Tonge, J. Shaw, B. Middleton *et al.*, 'Validation and development of fluorescence two-dimensional differential gel electrophoresis proteomics technology.' *Proteomics*, 2001, **1**, 377–396.

146. S. Birkelund, L. Bini, V. Pallini *et al.*, 'Characterization of Chlamydia trachomatis l2-induced tyrosine-phosphorylated HeLa cell proteins by two-dimensional gel electrophoresis.' *Electrophoresis*, 1997, **18**, 563–567.

147. J. Godovac-Zimmermann, V. Soskic, S. Poznanovic *et al.*, 'Functional proteomics of signal transduction by membrane receptors.' *Electrophoresis*, 1999, **20**, 952–961.

148. V. Soskic, M. Gorlach, S. Poznanovic *et al.*, 'Functional proteomics analysis of signal transduction pathways of the platelet-derived growth factor β receptor.' *Biochemistry*, 1999, **38**, 1757–1764.

149. A. Ducret, C. Desponts, S. Desmarais *et al.*, 'A general method for the rapid characterization of tyrosine-phosphorylated proteins by mini two-dimensional gel electrophoresis.' *Electrophoresis*, 2000, **21**, 2196–2208.

150. H. Yagame, T. Horigome, T. Ichimura *et al.*, 'Differential effects of methylmercury on the phosphorylation of protein species in the brain of acutely intoxicated rats.' *Toxicology*, 1994, **92**, 101–113.

151. R. E. Jenkins and S. R. Pennington, 'Arrays for protein expression profiling: Towards a viable alternative to two-dimensional gel electrophoresis?' *Proteomics*, 2001, **1**, 13–29.

152. S. P. Gygi, B. Rist, S. A. Gerber *et al.*, 'Quantitative analysis of complex protein mixtures using isotope-coded affinity tags.' *Nat. Biotechnol.*, 1999, **17**, 994–999.

153. H. Kuwata, T. T. Yip, M. Tomita *et al.*, 'Direct evidence of the generation in human stomach of an antimicrobial peptide domain (lactoferricin) from ingested lactoferrin.' *Biochem. Biophys. Acta*, 1998, **1429**, 129–141.

154. M. Merchant and S. R. Weinberger, 'Recent advancements in surface-enhanced laser desorption/ionization-time of flight-mass spectrometry.' *Electrophoresis*, 2000, **21**, 1164–1177.

155. A. Lueking, M. Horn, H. Eickhoff *et al.*, 'Protein microarrays for gene expression and antibody screening.' *Anal. Biochem.*, 1999, **270**, 103–111.

156. S. Steiner and N. L. Anderson, 'Expression profiling in toxicology—potentials and limitations.' *Toxicol. Lett.*, 2000, **112–113**, 467–471.

157. S. Steiner and F. A. Witzmann, 'Proteomics: applications and opportunities in preclinical drug development.' *Electrophoresis*, 2000, **21**, 2099–2104.

158. S. Steiner, C. L. Gatlin, J. J. Lennon *et al.*, 'Proteomics to display lovastatin-induced protein and pathway regulation in rat liver.' *Electrophoresis*, 2000, **21**, 2129–2137.

159. S. J. Newsholme, B. F. Maleeff, S. Steiner *et al.*, 'Two-dimensional electrophoresis of liver proteins: characterization of a drug-induced hepatomegaly in rats.' *Electrophoresis*, 2000, **21**, 2122–2128.

160. F. A. Witzmann, R. L. Carpenter, G. D. Ritchie *et al.*, 'Toxicity of chemical mixtures: proteomic analysis of persisting liver and kidney protein alterations induced by repeated exposure of rats to JP-8 jet fuel vapor.' *Electrophoresis*, 2000, **21**, 2138–2147.

161. L. Aicher, D. Wahl, A. Arce *et al.*, 'New insights into cyclosporine: A nephrotoxicity by proteome analysis.' *Electrophoresis*, 1998, **19**, 1998–2003.

162. D. A. Rappolee, A. Wang, D. Mark *et al.*, 'Novel method for studying mRNA phenotypes in single or small numbers of cells.' *J. Cell. Biochem.*, 1989, **39**, 1–11.

163. P. M. Fernandez, L. J. Pluta, R. Fransson-Steen *et al.*, 'Reverse transcription-polymerase chain reaction-based methodology to quantify differential gene expression directly from microdissected regions of frozen tissue sections.' *Mol. Carcinogen.*, 1997, **20**, 317–326.

164. L. Fink, W. Seeger, L. Ermert *et al.*, 'Real-time quantitative RT-PCR after laser-assisted cell picking.' *Nat. Med.*, 1998, **4**, 1329–1333.

165. N. L. Simone, R. F. Bonner, J. W. Gillespie *et al.*, 'Laser-capture microdissection: opening the microscopic frontier to molecular analysis.' *Trends Genet.*, 1998, **14**, 272–276.

166. S. Curran, J. A. McKay, H. L. McLeod *et al.*, 'Laser capture microscopy.' *Mol. Pathol.*, 2000, **53**, 64–68.

167. K. Specht, T. Richter, U. Muller *et al.*, 'Quantitative gene expression analysis in microdissected archival formalin-fixed and paraffin-embedded tumor tissue.' *Am. J. Pathol.*, 2001, **158**, 419–429.

168. Y. Kohda, H. Murakami, O. W. Moe *et al.*, 'Analysis of segmental renal gene expression by laser capture microdissection.' *Kidney Int.*, 2000, **57**, 321–331.

169. Y. Nagasawa, M. Takenaka, Y. Matsuoka *et al.*, 'Quantitation of mRNA expression in glomeruli using laser-manipulated microdissection and laser pressure catapulting.' *Kidney Int.*, 2000, **57**, 717–723.

170. Y. Kang, J. M. Mariano, J. Angdisen *et al.*, 'Enhanced tumorigenesis and reduced transforming growth factor-β type II receptor in lung tumors from mice with reduced gene dosage of transforming growth factor-β1.' *Mol. Carcinog.*, 2000, **29**, 112–126.

171. J. Cui, L. R. Rohr, G. Swanson *et al.*, 'Hypermethylation of the caveolin-1 gene promoter in prostate cancer.' *Prostate*, 2001, **46**, 249–256.

172. J. Wanagat, Z. Cao, P. Pathare *et al.*, 'Mitochondrial DNA deletion mutations colocalize with segmental electron transport system abnormalities, muscle fiber atrophy, fiber splitting, and oxidative damage in sarcopenia.' *FASEB J.*, 2001, **15**, 322–332.

173. M. R. Emmert-Buck, J. W. Gillespie, C. P. Paweletz *et al.*, 'An approach to proteomic analysis of human tumors.' *Mol. Carcinog.*, 2000, **27**, 158–165.

174. D. K. Ornstein, J. W. Gillespie, C. P. Paweletz *et al.*, 'Proteomic analysis of laser capture microdissected human prostate cancer and *in vitro* prostate cell lines.' *Electrophoresis*, 2000, **21**, 2235–2242.

175. C. P. Paweletz, J. W. Gillespie, D. K. Ornstein *et al.*, 'Rapid protein display profiling of cancer progression directly from human tissue using a protein biochip.' *Drug Dev. Res.*, 2000, **49**, 34–42.

176. N. L. Simone, C. P. Paweletz, L. Charboneau *et al.*, 'Laser capture microdissection: Beyond functional genomics to proteomics.' *Mol. Diagn.*, 2000, **5**, 301–307.

177. L. Luo, R. C. Salunga, H. Guo *et al.*, 'Gene expression profiles of laser-captured adjacent neuronal subtypes.' *Nat. Med.*, 1999, **5**, 117–122.

178. H. Ohyama, X. Zhang, Y. Kohno *et al.*, 'Laser capture microdissection-generated target sample for high-density oligonucleotide array hybridization.' *BioTechniques*, 2000, **29**, 530–536.

179. N. L. Simone, A. T. Remaley, L. Charboneau *et al.*, 'Sensitive immunoassay of tissue cell proteins procured by laser capture microdissection.' *Am. J. Pathol.*, 2000, **156**, 445–452.

180. J. K. Nicholson, J. C. Lindon and E. Holmes, ''Metabonomics': understanding the metabolic responses of living systems to pathophysiological stimuli via multivariate statistical analysis of biological NMR spectroscopic data.' *Xenobiotica*, 1999, **29**, 1181–119

181. D. G. Robertson, M. D. Reily, R. E. Sigler *et al.*, 'Metabonomics: evaluation of nuclear magnetic resonance (NMR) and pattern recognition technology for rapid *in vivo* screening of liver and kidney toxicants.' *Toxicol. Sci.*, 2000, **57**, 326–337.

182. J. C. Lindon, J. K. Nicholson and J. R. Everett, 'NMR spectroscopy of biofluids.' *Ann. R. NMR S.*, 1999, **38**, 1–88.

183. N. J. Waters, S. Garrod, R. D. Farrant *et al.*, 'High-resolution magic angle spinning (1)H NMR spectroscopy of intact liver and kidney: optimization of sample preparation procedures and biochemical stability of tissue during spectral acquisition.' *Anal. Biochem.*, 2000, **282**, 16–23.

184. K. P. Gartland, C. R. Beddell, J. C. Lindon *et al.*, 'Application of pattern recognition methods to the analysis and classification of toxicological data derived from proton nuclear magnetic resonance spectroscopy of urine.' *Mol. Pharmacol.*, 1991, **39**, 629–642.

185. M. L. Anthony, B. C. Sweatman, C. R. Beddell *et al.*, 'Pattern recognition classification of the site of nephrotoxicity based on metabolic data derived from proton nuclear magnetic resonance spectra of urine.' *Mol. Pharmacol.*, 1994, **46**, 199–211.

186. W. el-Deredy, 'Pattern recognition approaches in biomedical and clinical magnetic resonance spectroscopy: a review.' *NMR. Biomed.*, 1997, **10**, 99–124.

187. E. Holmes, A. W. Nicholls, J. C. Lindon *et al.*, 'Development of a model for classification of toxin-induced lesions using 1H NMR spectroscopy of urine combined with pattern recognition.' *NMR Biomed.*, 1998, **11**, 235–244.

188. K. P. Gartland, F. W. Bonner, J. A. Timbrell *et al.*, 'Biochemical characterisation of para-aminophenol-induced nephrotoxic lesions in the F344 rat.' *Arch. Toxicol.*, 1989, **63**, 97–106.

189. K. P. Gartland, F. W. Bonner and J. K. Nicholson, 'Investigations into the biochemical effects of region-specific nephrotoxins.' *Mol. Pharmacol.*, 1989, **35**, 242–250.

190. M. L. Anthony, K. P. Gartland, C. R. Beddell *et al.*, 'Studies of the biochemical toxicology of uranyl nitrate in the rat.' *Arch. Toxicol.*, 1994, **68**, 43–53.

191. J. K. Nicholson, D. P. Higham, J. A. Timbrell *et al.*, 'Quantitative high resolution 1H NMR urinalysis studies on the biochemical effects of cadmium in the rat.' *Mol. Pharmacol.*, 1989, **36**, 398–404.

192. J. Gray, J. K. Nicholson, D. M. Creasy *et al.*, 'Studies on the relationship between acute testicular damage and urinary and plasma creatine concentration.' *Arch. Toxicol.*, 1990, **64**, 443–450.

193. S. M. Sanins, J. K. Nicholson, C. Elcombe *et al.*, 'Hepatotoxin-induced hypertaurinuria: a proton NMR study.' *Arch. Toxicol.*, 1990, **64**, 407–411.

194. E. Holmes, F. W. Bonner, B. C. Sweatman *et al.*, 'Nuclear magnetic resonance spectroscopy and pattern recognition analysis of the biochemical processes associated with the progression of and recovery from nephrotoxic lesions in the rat induced by mercury(II) chloride and 2-bromoethanamine.' *Mol. Pharmacol.*, 1992, **42**, 922–930.

195. E. Holmes, J. K. Nicholson, F. W. Bonner *et al.*, 'Mapping the biochemical trajectory of nephrotoxicity by pattern recognition of NMR urinanalysis.' *NMR Biomed.*, 1992, **5**, 368–372.

196. B. M. Beckwith-Hall, J. K. Nicholson, A. W. Nicholls *et al.*, 'Nuclear magnetic resonance spectroscopic and principal components analysis investigations into

biochemical effects of three model hepatotoxins.' *Chem. Res. Toxicol.*, 1998, **11**, 260–272.

197. E. Holmes and J. P. Shockcor, 'Accelerated Toxicity Screening Using NMR and Pattern Recognition-Based Methods.' *Curr. Opin. Drug Discovery Dev.*, 2000, **3**, 72–78.

198. P. Smyth, 'Data mining: data analysis on a grand scale?' *Stat. Methods. Med. Res.*, 2000, **9**, 309–327.

199. R. J. Roberts, H. E. Varmus, M. Ashburner *et al.*, 'Building a "GenBank" of the published literature.' *Science*, 2001, **291**, 2318–2319.

200. C. Helma, E. Gottmann and S. Kramer, 'Knowledge discovery and data mining in toxicology (In Process Citation).' *Stat. Methods. Med. Res.*, 2000, **9**, 329–358.

201. I. H. Witten and E. Frank,'Data Mining,' Morgan Kaufmann Publishers, San Fransisco, CA, 2000.

202. U. Fayyad, G. Piatetsky-Shapiro and P. Smmyth, 'The KDD process for extracting useful knowledge from volumes of data.' *Comm. ACM.*, 1996, **39**, 27–34.

203. M. Liu, N. Sussman, G. Klopman *et al.*, 'Estimation of the optimal data base size for structure-activity analyses: the Salmonella mutagenicity data base.' *Mutat. Res.*, 1996, **358**, 63–72.

204. M. Liu, N. Sussman, G. Klopman *et al.*, 'Structure-activity and mechanistic relationships: the effect of chemical overlap on structural overlap in data bases of varying size and composition.' *Mutat. Res.*, 1996, **372**, 79–85.

205. G. Klopman, 'Artificial intelligence approach to structure-activity studies: Computer automated structure evaluation of biological activity of organic molecules.' *J. Am. Chem. Soc.*, 1984, **106**, 7315–7321.

206. G. Klopman, 'MultiCASE: A hierarchical computer automated structure evaluation program.' *Quantit. Struct. Act. Relat.*, 1992, **11**, 176–184.

207. C. Hansch, D. Hoekman, A. Leo *et al.*, 'The expanding role of quantitative structure-activity relationships (QSAR) in toxicology.' *Toxicol. Lett.*, 1995, **79**, 45–53.

208. K. Enslein and P. N. Craig, 'Carcinogenesis: a predictive structure-activity model.' *J. Toxicol. Environ. Health*, 1982, **10**, 521–530.

209. K. Enslein, T. R. Lander, M. E. Tomb *et al.*, 'Mutagenicity (Ames): a structure-activity model.' *Teratog. Carcinog. Mutagen.*, 1983, **3**, 503–513.

210. K. Enslein, T. R. Lander and J. R. Strange, 'Teratogenesis: a statistical structure-activity model.' *Teratog. Carcinog. Mutagen.*, 1983, **3**, 289–309.

211. V. K. Gombar, K. Enslein and B. W. Blake, 'Assessment of developmental toxicity potential of chemicals by quantitative structure-toxicity relationship models.' *Chemosphere*, 1995, **31**, 2499–2510.

212. K. Enslein, T. R. Lander, M. E. Tomb *et al.*, 'A predictive model for estimating rat oral LD50 values.' *Toxicol. Ind. Health*, 1989, **5**, 261–387.

213. K. Enslein, V. K. Gombar, B. W. Blake *et al.*, 'A quantitative structure-toxicity relationships model for the dermal sensitization guinea pig maximization assay.' *Food Chem. Toxicol.*, 1997, **35**, 1091–1098.

214. G. Klopman, 'The MultiCASE program II. Baseline activity identification algorithm (BAIA).' *J. Chem. Inf. Comput. Sci.*, 1998, **38**, 78–81.

215. E. Zeiger, J. Ashby, G. Bakale *et al.*, 'Prediction of Salmonella mutagenicity.' *Mutagenesis*, 1996, **11**, 471–484.

216. B. Henry, S. G. Grant, G. Klopman *et al.*, 'Induction of forward mutations at the thymidine kinase locus of mouse lymphoma cells: evidence for electrophilic and non-electrophilic mechanisms.' *Mutat. Res.*, 1998, **397**, 313–335.

217. A. R. Cunningham, H. S. Rosenkranz, Y. P. Zhang *et al.*, 'Identification of "genotoxic" and "non-genotoxic" alerts for cancer in mice: the carcinogenic potency database.' *Mutat. Res.*, 1998, **398**, 1–17.

218. G. Klopman and H. S. Rosenkranz, 'Toxicity estimation by chemical substructure analysis: the TOX II program.' *Toxicol. Lett.*, 1995, **79**, 145–155.

219. H. S. Rosenkranz, A. R. Cunningham, Y. P. Zhang *et al.*, 'Development, characterization and application of predictive-toxicology models.' *SAR. QSAR. Environ. Res.*, 1999, **10**, 277–298.

220. H. S. Rosenkranz, A. R. Cunningham, Y. P. Zhang *et al.*, 'Applications of the case/multicase SAR method to environmental and public health situations.' *SAR. QSAR. Environ. Res.*, 1999, **10**, 263–276.

221. A. M. Richard, 'Structure-based methods for predicting mutagenicity and carcinogenicity: are we there yet?' *Mutat. Res.*, 1998, **400**, 493–507.

222. D. M. Sanderson and C. G. Earnshaw, 'Computer prediction of possible toxic action from chemical structure; the DEREK system.' *Hum. Exp. Toxicol.*, 1991, **10**, 261–273.

223. N. Greene, P. N. Judson, J. J. Langowski *et al.*, 'Knowledge-based expert systems for toxicity and metabolism prediction: DEREK, StAR and METEOR.' *SAR. QSAR. Environ. Res.*, 1999, **10**, 299–314.

224. M. D. Barratt and J. J. Langowski, 'Validation and subsequent development of the DEREK skin sensitization rulebase by analysis of the BgVV list of contact allergens.' *J. Chem. Inf. Comput. Sci.*, 1999, **39**, 294–298.

225. R. D. King, S. H. Muggleton, A. Srinivasan *et al.*, 'Structure-activity relationships derived by machine learning: the use of atoms and their bond connectivities to predict mutagenicity by inductive logic programming.' *Proc. Nat. Acad. Sci. USA*, 1996, **93**, 438–442.

226. R. D. King and A. Srinivasan, 'Prediction of rodent carcinogenicity bioassays from molecular structure using inductive logic programming.' *Environ. Health Perspect.*, 1996, **104**(Suppl. 5), 1031–1040.

227. U. Scherf, D. T. Ross, M. Waltham *et al.*, 'A gene expression database for the molecular pharmacology of cancer.' *Nat. Genet.*, 2000, **24**, 236–244.

228. D. T. Ross, U. Scherf, M. B. Eisen *et al.*, 'Systematic variation in gene expression patterns in human cancer cell lines.' *Nat. Genet.*, 2000, **24**, 227–235.

229. M. A. Andrade and P. Bork, 'Automated extraction of information in molecular biology.' *FEBS Lett.*, 2000, **476**, 12–17.

230. E. Symanski, L. L. Kupper and S. M. Rappaport, 'Comprehensive evaluation of long-term trends in occupational exposure: Part 1. Description of the database.' *Occup. Environ. Med.*, 1998, **55**, 300–309.

J.P. Vanden Heuvel, G.H. Perdew, W.B. Mattes and W.F. Greenlee (Eds.)
Comprehensive Toxicology, Vol. xiv

# 14.5.2
# Bioinformatics

## WILLIAM B. MATTES
*Pharmacia, Kalamazoo, MI, USA*

## 14.5.2.1  BACKGROUND

Just what *is* bioinformatics? This is a question posed by both biologists and computer scientists alike and it highlights the breadth of the field. As a buzzword, the word "bioinformatics" is certainly applied rather freely. In some biology circles it might seem that any computer application outside of word processing and spreadsheets is "bioinformatics", while computer scientists might use the word to describe statistical analysis of biological data database construction and maintenance, database mining, automated capture of biological data, and analysis of genetic and protein sequence data. The latter viewpoint would rightly define the field as the intersection of computer science, information technology, and biology. Several books are now available on the subject[1–3] and two definitions therein offer perspective: "bioinformatics is the science of using information to understand biology",[3] and represents "a new way to do hard, meaningful work".[1]

While it continues to be viewed as one of the more arcane "new" technologies, ironically, bioinformatics is one of the oldest.[4] It dates from the mid 1960s (Table 1), when the combination of automated protein sequencing and newly developed digital computers prompted Margerett Dayhoff and others to explore computational analyses of protein sequence and evolution.[5,6] Its application to nucleic acid data required the introduction of tools such as restriction endonucleases, cloning, and DNA sequencing, which allowed the isolation and sequencing of specific DNA fragments (Table 1). With the introduction of automated DNA sequencing, and the development of public databases for storing the rapidly accumulating amounts of DNA sequence information, the stage was set for the development of the field we know it as today. And while "bioinformatics" is often used as an umbrella term referring to applications from database mining to microarray data analysis, it is truly rooted in the analysis of protein and DNA sequences.

By themselves polypeptide and polynucleotide sequences are no more informative than random collections of letters (Figure 1). To have value for the cell or the researcher they must be "decoded"—genomic DNA must be transcribed to a complementary RNA, exons and introns must be identified, a "mature" messenger RNA formed, a protein coding region identified, a start codon and reading frame established, and the amino acid sequence of the protein determined by the genetic code (Figure 2). The polypeptide produced will then fold based upon the characteristics of the amino acid sequence and/or be targeted for a certain subcellular location based upon the presence of a signaling peptide. Finally, the functional characteristics of this protein, and its interactions with other cell components, will depend upon the peptide motifs it contains and their arrangement in 3-dimensional space. A prime goal of bioinformatics is to trace theses cellular steps and events *in silico*, using a combination of algorithms and data based upon established cellular and molecular biology.

**Table 1**  Historical Background of Bioinformatics.

| | |
|---|---|
| 1962 | Computer analysis of protein sequences |
| 1965 | Margaret Dayhoff's *Atlas of Protein Sequence and Structure* |
| 1968 | Completion of genetic code |
| 1970 | Type II restriction enzymes |
| 1973 | Molecular cloning |
| 1977 | DNA sequencing |
| 1981 | Concept of sequence motifs proposed |
| 1982 | GenBank |
| 1985 | FASTP/FASTN rapid database searching introduced |
| 1987 | Automated DNA sequencing |
| 1990 | BLAST: rapid database searching introduced |
| 1991 | EST: expressed sequence tag sequencing |
| 1996 | Yeast genome completely sequenced |
| 1999 | *Drosphilla* genome completely sequenced |
| 2000 | Draft of the human genome |

# Raw sequence

```
cgcggcgcgtctgtctgttatttgtctgatcgttggatttgtctgtctaa
tctgtgccctaattttctttgaagctaccatgggacaatcgctaacaacc
cccttgagtctcactctagaccattggaaggacgtccgagaccgagcacg
tgatcagtcggtcgagatcaagaaaggtcctctccggaggtcggggacag
tcgcgccagcaagcggtggggcaggagctcctggtttggcagcccctgta
```

**Figure 1** A "raw" DNA sequence.

## 14.5.2.2 ALGORITHMS

Some appreciation of the algorithmic approaches in bioinformatics is useful for understanding the basis and reliability of the many different applications. At the simplest level is the comparison of the sequence with a "code" database. Examples of this approach is the identification of restriction endonuclease recognition sites in a DNA sequence and the prediction of the amino acid sequence, of a polypeptide coded for by the DNA (Figure 2). Needless to say these comparisons are relatively straightforward, and given the specificity of restriction endonucleases and the robustness of the genetic code, the results hardly warrant experimental confirmation. The pairwise alignment of two reasonably similar DNA sequences is also reasonably straightforward (Figure 3(A)). The pairwise alignment of 2 protein sequences presents a somewhat different problem.[7] Knowledge of both the characteristics of amino acids and protein evolution tells us that some amino acids are quite similar in function (e.g., alanine and valine), and often the substitution of one for the other at a particular site in a given protein can be tolerated without dramatic changes in function. Thus these comparisons make use of

substitution matrices that determine scores for different substitutions: for any given sequence position the highest score is given if the same amino acid is in both proteins, a lower score is given for a conserved amino acid difference (e.g., alanine versus valine), while a penalty score is given for a truly disparate amino acid comparison (e.g., aspartate versus tryptophan) (Figure 3(B)). Gaps in such alignment also elicit penalty values and ultimately comparing the sequences for 2 proteins that are not very similar may produce more than one reasonable alignment. In this last case a conclusion of similarity may depend upon other considerations.

Certainly the level of sophistication in algorithms increases with the complexity of the task involved. Calculations of a polypeptide's physical properties, such as hydrophobicity, isolectric point, and secondary structure, are examples of calculations that must take into account several factors; a similar problem is the identification of "signal sequences",[8,9] where the length, composition, and position of the subsequence must be considered. In the case of determining sites in genomic sequences with the potential for regulating transcription, the patterns recognized by transcription factors are usually heterogeneous; in this case the

# Notated sequence

**Figure 2** A DNA sequence annotated by comparison with the genetic code and with a table of restriction enzyme recognition sites.

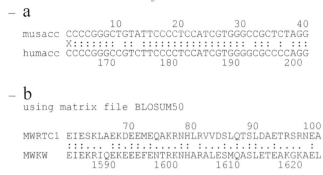

Figure 3   (A) Pairwise alignment of two DNA sequences; (B) Pairwise alignment of two peptide sequences, using a substitution matrix.

genomic sequences are best compared to a matrix that scores the fit for any given sequence position[10] (see Table 2). Finally, with genomic sequences for a number of organisms, including human, the challenge is to predict genes complete with transcription initiation, intron splice, and termination sites. A number of approaches to this have been described, including GENSCAN,[11] Genie,[12] GeneWise,[13] and Otto,[14] with varying degrees of success.[14,15]

### 14.5.2.3   DATABASES

The other key element in bioinformatics is the availability of data for computational comparison for analysis, i.e., relevant databases. The late 1980s saw the development of not only public sequence databases, but also the Internet and the World Wide Web. These 2 technologies came together to create the current model of myriad public databases readily accessible from any Internet-connected computer. Often the challenge is being aware a certain resource exists, and, to this end, gateway Internet sites (Table 3) can be very useful. While several databases will be mentioned here, the reader should note that every year the first issue of *Nucleic Acids Research* is devoted to a review of a large number and variety of publicly accessible databases.[16]

First and foremost to be considered are the databases maintained at the 3 sites of the International Nucleotide Sequence Database Collaboration (INSDBC). These are GenBank (maintained by the National Center for Biotechnology Information [NCBI] in Bethesda, Maryland), the DNA Database of Japan (DDBJ, Mishima, Japan) and the European Bioinformatics Institute (EBI) databases at Hinxton, England, part of the European Molecular Biology Laboratory (EMBL). Theses sites (see Table 3) are the central repository for publicly available nucleotide and protein sequence information, and while they contain essentially identical sequence information, they maintain separate formats and different tools for accessing the information. Each of these sites also maintains other genomic information integrated with the sequence information. Thus NCBI not only offers nucleotide and protein sequence information, but also links taxonomy and related sequences for any given entry.[17]

It should be noted that the data in the "main" databases are contributed from a variety of sources, and should be contrasted with edited ("curated") databases, where the entries are checked for accuracy and annotated with information from several sources. Such databases include SwissProt, LocusLink,[18] and GeneCards[19] (Table 3). In fact, if comprehensive, gene-centered information is desired, as opposed to simple sequence information, these are the databases to consult. They offer canonical gene names, information on function motifs for the protein, and links to bibliographic information. LocusLink and GeneCards offer a link to the OMIM database

Table 2   Trasnfac/MatInspector Matrix Table for AP-1 Transcription Factor Binding Site.

| Position in sequence | 1 | 2 | 3 | 4 | 5 | 6 | 7 | 8 | 9 |
|---|---|---|---|---|---|---|---|---|---|
| A weighting | 4.79 | 0 | 0 | 11.28 | 1.28 | 0 | 1.46 | 12.15 | 0.89 |
| C weighting | 2.3 | 0 | 0 | 0 | 4.71 | 0 | 10.7 | 0 | 1.22 |
| G weighting | 3.06 | 0 | 9.83 | 0 | 6.16 | 0 | 0 | 0 | 5.91 |
| T weighting | 1.3 | 12.15 | 2.32 | 0.87 | 0 | 2.15 | 0 | 10 | 2.17 |
| *IUPAC description* | *N* | *T* | *G* | *A* | *S* | *T* | *C* | *A* | *G* |

**Table 3** Internet Sites for Databases.

*Link Gateways*

| | |
|---|---|
| Research Tool Sites for the Molecular Biology | http://www.yk.rim.or.jp/~aisoai/molbio.html |
| CMS-SDSC Molecular Biology Resource | http://restools.sdsc.edu/ |

*Main Databases*

| | |
|---|---|
| EBI Home Page | http://www.ebi.ac.uk/ebi_home.html |
| EMBL–European Molecular Biology Laboratory | http://www.embl-heidelberg.de/ |
| National Center for Biotechnology Information (NCBI): | http://www.ncbi.nlm.nih.gov/ |
| DNA Data Bank of Japan | http://www.ddbj.nig.ac.jp/ |

*Curated Databases*

| | |
|---|---|
| LocusLink | http://www.ncbi.nlm.nih.gov/LocusLink/ |
| GeneCards human genes, proteins and diseases (Weizmann) | http://bioinformatics.weizmann.ac.il/cards/ |
| SWISS-PROT | http://www.expasy.ch/sprot/ |
| Online Mendelian Inheritance in Man | http://www.ncbi.nlm.nih.gov/Omim/ |
| Online Mendelian Inheritance in Animals | http://www.angis.org.au/Databases/BIRX/omia/ |
| RefSeq | http://www.ncbi.nlm.nih.gov/RefSeq/ |

*Genomic Databases*

| | |
|---|---|
| Mouse Genome Informatics | http://www.informatics.jax.org/ |
| Whitehead Institute | http://www-genome.wi.mit.edu/ |
| Saccharomyces Genome Database | http://genome-www.stanford.edu/Saccharomyces/ |
| The Sanger Centre | http://www.sanger.ac.uk/ |
| Stanford Genomic Resources | http://genome-www.stanford.edu/ |

*Metabolic Pathways*

| | |
|---|---|
| Kyoto Encyclopedia of Genes and Genomes | http://www.genome.ad.jp/kegg/ |
| WIT Home Page | http://wit.mcs.anl.gov/WIT2/ |

*Protein Structures, Motifs, and Patterns*

| | |
|---|---|
| PDB Home Page | http://www.rcsb.org/pdb/ |
| Blocks WWW Server | http://www.blocks.fhcrc.org/ |
| Pfam | http://www.sanger.ac.uk/Software/Pfam/ |
| PROSITE | http://www.expasy.ch/prosite/ |
| PRINTS DATABASE | http://bmbsgi11.leeds.ac.uk/bmb5dp/prints.html |

*Gene Expression Databases*

| | |
|---|---|
| BodyMap Home Page | http://bodymap.ims.u-tokyo.ac.jp/ |
| Human Transcript Database | http://www.hgsc.bcm.tmc.edu/HTDB/ |
| HumanGene Expression Index | http://www.hugeindex.org/ |
| Mouse Gene Expression Menu | http://www.informatics.jax.org/menus/ expression_menu.shtml |

*Homology Databases*

| | |
|---|---|
| HomoloGene | http://www.ncbi.nlm.nih.gov/HomoloGene/ |

*EST Databases*

| | |
|---|---|
| The Institute for Genomic Research | http://www.tigr.org/index.html |
| UniGene | http://www.ncbi.nlm.nih.gov/UniGene/ |
| The I.M.A.G.E. Consortium | http://image.llnl.gov/ |

(Online Mendelian Inheritance in Man, Table 3) that curates bibliographic information on the involvement of a particular gene in human disease. An analogous site maintains information on Mendelian Inheritance in Animals (Table 3).

Other sites host databases that focus beyond sequence information. Thus the Mouse Genome Informatics (MGI) site, the Saccharomyces Genome Database (SGD), the Sanger Centre, and the Stanford Genomic Resources host chromosomal information, gene mapping data, polymorphism data and strain genotypes

(Table 3). The Kyoto Encyclopedia of Genes and Genomes (KEGG) and What Is There (WIT) offer metabolic pathway information, while 3-dimensional protein structure files are maintained at the Protein Data Bank (PDB). PROSITE, BLOCKS, PRINTS, and PFAM represent examples of databases of protein motifs and patterns. Sites such as BodyMap, the Human Transcript database, the HumanGene Expression Index, and Mouse Genome Informatics offer expression data in different tissues. Lastly 2 publicly accessible sites, UniGene and TIGR, maintain database covering

Expressed Sequence Tags (ESTs) and it is this resource that demands special description.

## 14.5.2.4   EXPRESSED SEQUENCE TAGS (ESTS)

ESTs were conceived as means to gain useful genomic information from automated DNA sequencing without waiting for the completion of whole genome projects.[20] In this approach hundreds to thousands of clones from collections (libraries) of complementary DNA (cDNA) are isolated, and then submitted to automated, partial, single-pass sequences from either end. A cDNA library represents a snapshot of expressed gene transcripts from the tissue or cell at the time the RNA was isolated and cloned, and complete sequencing of a large number of clones would yield a picture not only of the coding regions of the genome, but also of the genes expressed under the particular cellular conditions. The EST strategy was developed as a shortcut, where the partial sequences would serve as a resource for database mining. Thus ESTs are "identified" by their homology to known genes: the EST is first translated in all reading frames, and the six sequences are compared against a database of known protein sequences from all species. ESTs with no homology to known genes may represent novel expressed genes and represent an approach to gene discovery However, given the single-pass nature of the sequence information collected, accuracy is a serious concern, and can interfere with the unambiguous identification of any given EST. ESTs are maintained as part of the overall nucleotide databases, but can be uniquely searched at dbEST (Table 3). Additionally, ESTs, along with well characterized sequences, can be computationally organized into gene-oriented clusters. Two such cluster databases are Uni-Gene,[21] maintained by NCBI, and the TIGR Gene Index,[22,23] maintained by The Institute for Genome Research (TIGR). These resources allow the researcher to determine if any given EST has already been identified as a gene fragment. It should be noted that as new sequence information is accumulated clusters may be rearranged and individual EST sequences may be grouped into different clusters. The UniGene database also serves as a link to the physical cDNA clone through the IMAGE (Integrated Molecular Analysis of Genomes and their Expression) Consortium;[21] once obtained from a licensed distributor the clone's sequence can be confirmed and extended, essentially converting the EST into a full fledged cDNA sequence.

## 14.5.2.5   SEQUENCES

### 14.5.2.5.1   Sequence Identifiers

"The ras gene" does not correlate with a unique DNA sequence. In fact, the phrase "K-ras" is associated with no fewer that 123 sequence entries! This example highlights the unique issues with naming and identifying nucleotide and protein sequences. At one time "Locus" roughly corresponding to a genetic locus name, was actually used as an unique identifier and as a mnemonic for the function and origin of the sequence. Thus HUMFOS and MUSJUNC referred to the sequence for the human *fos* gene and the mouse *c-Jun* gene, respectively. However, as new or different functions for a given "gene" were discovered the locus name became volatile, and another identifier was clearly required. The International Nucleotide Sequence Database Collaborators introduced the "accession number" to identify a sequence with an intentional absence of biological connotation. The accession number has remained a gold standard for sequence identification, but even it has deficiencies. It uniquely identifies a sequence *record*, but if the sequence in that record is updated through correction or addition, the actual sequence identified by an accession number can change. Recently the International Nucleotide Sequence Database Collaborators have introduced an accession number that includes a version number, and this should be the preferred method of citing sequences.

NCBI also has introduced a unique identifier for each submitted *sequence*, the genInfo identifier (gi). This "GI number" is indeed unique for any *sequence* and if a sequence is revised or replaced, the new sequence receives a new GI number, and a protein sequence will have a different GI number than its corresponding mRNA.

NCBI has also embarked on a endeavor to create a database, RefSeq,[18] of "curated" entries for mRNAs and proteins with complete sequences. The RefSeq database (Table 3) is a non-redundant set of reference sequences, including constructed genomic contigs, mRNAs, and proteins. RefSeq records are identified with unique accession number prefixes: NM_ for mRNA, NP_ for proteins. Additional feature annotation may be added, and, in some instances the sequence has been modified relative to the original GenBank sequence from which it was derived. RefSeq accession numbers that begin with the prefix XM_(mRNA) and XP_(protein) are model reference sequences produced by NCBI's Genome Annotation project. Some RefSeq entries are "predicted", others "provisional" (includes some initial quality checking), and others "reviewed". As might be expected, "reviewed" RefSeq records can reflect the addition of a variety of cross-referenced information. In cases where splice variant information is well known, it is reflected in RefSeq entries. RefSeq represents an effort to move from the archival (and somewhat chaotic) nature of GenBank to a repository of non-redundant information.

### 14.5.2.5.2 Sequence File Formats

Critical to any discussion of bioinformatics and sequence analysis is a consideration of the different formats commonly used to store sequences and that serve as input to various software tools. The simplest, but least commonly used format is that of the raw sequence (e.g., Figure 1). The most common format used as input for analysis tools is the "Fasta" format (Figure 4A), developed in conjunction with the FASTA sequence analysis package.[24] This format consists of a single description line

A

```
>MMU10101 Mus musculus bcl-x long (bcl-x long) mRNA, complete cds.
atgtctcagagcaaccgggagctggtggtcgactttctctcctacaagctttcccagaaa
ggatacagctggagtcagtttagtgatgtcgaagagaataggactgaggccccagaagaa
actgaagcagagagggagacccccagtgccatcaatggcaacccatcctggcacctggcg
gatagcccggccgtgaatggagccactggccacagcagcagtttggatgcgcgggaggtg
attccatggcagcagtgaagcaagcgctgagagaggcaggcgatgagtttgaactgcgg
taccggagagcgttcagtgatctaacatcccagcttcacataaccccagggaccgcgtat
cagagctttgagcaggtagtgaatgaactctttcgggatggagtaaactggggtcgcatc
gtggccttttctcctttggcggggcactgtgcgtggaaagcgtagacaaggagatgcag
gtattggtgagtcggattgcaagttggatggccacctatctgaatgaccacctagagcct
tggatccaggagaacggccgctgggacactttgtggatctctacgggaacaatgcagca
gccgagagccggaaaggccaggagcgcttcaaccgctggttcctgacgggcatgactgtg
gctggtgtggttctgctgggctcactcttcagtcggaagtga
```

B

```
LOCUS        MMU10101     702 bp    mRNA           ROD       30-NOV-1995
DEFINITION   Mus musculus bcl-x long (bcl-x long) mRNA, complete cds.
ACCESSION    U10101
NID          g506647
VERSION      U10101.1  GI:506647
SOURCE       mouse.
  ORGANISM   Mus musculus
             Eukaryota; Metazoa; Chordata; Craniata; Vertebrata; Mammalia;
             Eutheria; Rodentia; Sciurognathi; Muridae; Murinae; Mus.
REFERENCE    1  (bases 1 to 702)
  AUTHORS    Fang,W., Rivard,J.J., Mueller,D.L. and Behrens,T.W.
  TITLE      Cloning and molecular characterization of mouse bcl-x in B and T
             lymphocytes
  JOURNAL    J. Immunol. 153 (10), 4388-4398 (1994)
  MEDLINE    95052604
FEATURES             Location/Qualifiers
     source          1..702
                     /organism="Mus musculus"
                     /db_xref="taxon:10090"
                     /cell_line="22D6"
                     /clone_lib="pre-B lymphocyte cDNA library"
     gene            1..702
                     /gene="bcl-x long"
BASE COUNT      166 a    166 c    226 g    144 t
ORIGIN
        1 atgtctcaga gcaaccggga gctggtggtc gactttctct cctacaagct ttcccagaaa
       61 ggatacagct ggagtcagtt tagtgatgtc gaagagaata ggactgaggc cccagaagaa
      121 actgaagcag agagggagac ccccagtgcc atcaatggca acccatcctg gcacctggcg
      ....
      661 gctggtgtgg ttctgctggg ctcactcttc agtcggaagt ga
//
```

**Figure 4** (A) FASTA sequence format; (B) GenBank sequence format.

starting with a > character, followed by subsequent lines of sequence without line number. Another common format is that developed at the NCBI, the GenBank format (Figure 4B). As can be seen, this format provides a great deal of annotation. Other formats include the EMBL format, which also carries copious annotation, and the GCG format, used for the GCG Wisconsin Package software. Not all software will recognize multiple formats, so access to a conversion utility (see the BCM Sequence Utilities, Table 5) is useful. It should also be emphasized that almost universally, sequence analysis software requires sequence files to be "*text files*"; manipulating and saving a sequence file using a word processor will almost certainly render it useless for subsequent analysis. This can be avoided if the file is "Saved As, Text File", or if the file is manipulated and edited using a plain text editor (such as the Windows Notepad).

### 14.5.2.6 TOOLS

In every human endeavor there has been a dynamic between tool making and tool using; usually necessity is the mother of invention (tool making), but if an appropriate tool exists most tool users choose not to re-invent it. Fortunately for today's molecular toxicologist a large number of robust and powerful tools for bioinformatics already exist. These software tools range from crude, home-brewed programs with idiosyncratic interfaces, to Web-based software with the familiar browser interface, to complete integrated packages that run a wide variety of programs on a stand-alone computer. A large number of public domain programs that carry out a wide variety of tasks can be obtained from two common archives (Table 5), including some that provide integration of a number of tasks. With a bit of care a local workstation or small network can be outfitted with a suite of public domain programs[2,3] to provide a full menu of applications. On the other hand, for a large number of users where standardized training in the tools is a requirement, commercially available integrated packages (Table 3) offer the advantage of a uniform interface with vendor technical support. Furthermore, these integrated packages allow the analysis of proprietary sequences on local computer systems, and generally have the capability to carry out all but the most sophisticated tasks. The range and capabilities of Web-based tools (Table 5) often exceed that of integrated packages, and while they can involve insecure submission over the Internet, they (generally) do not

**Table 4** Integrated Commercial Bioinformatics Tools.

| Package | Operating system | Contact |
|---------|-----------------|---------|
| GCG Wisconsin Package | UNIX | http://www.gcg.com/ |
| Vector NTI | Mac, Windows | http://www.informaxinc.com/ |
| LaserGene | Mac, Windows | http://www.dnastar.com/ |
| MacVector | Mac | http://www.gcg.com/products/macvector.html |
| Omiga | Windows | http://www.gcg.com/products/omiga.html |
| DoubleTwist[a] | Internet-based | http://www.doubletwist.com/ |
| Bionavigator[b] | Internet-based | http://www.bionavigator.com/ |

[a] DoubleTwist offers some analyses free-of-charge. Access is through a browser supporting a secure connection.
[b] Bionavigator offers the first 100 units of analysis free-of-charge; subsequent charges are for units of analysis time. Access is through a browser supporting a secure connection.

require purchase or installation. As with any tool, more sophisticated tasks require more expertise with the tool, and an understanding of its strengths and limitations. However, the reward for the effort in gaining familiarity and experience with the many bioinformatic tools available is the ability to extend one's research far beyond the bench and lab.

### 14.5.2.6.1 Sequence Alignment

#### 14.5.2.6.1.1 Sequence to Sequence (Pairwise) Comparison

Basic to almost all of the analyses for both nucleic acid and protein sequence is the concept and practice of sequence alignment. As noted above under algorithms, an alignment between two nucleic acid sequences can be as simple as adjusting the two sequences until some sub-sequences match and then noting the identities at other positions. However, if two sequences share common subsequences, but are of different lengths, with divergent sequences separating conserved

sequences, the problem of a pairwise alignment becomes more difficult. Most programs start with "local" alignments—alignments of portions of the sequences. Two common programs are LALIGN and LFASTA,[25] available at the FASTA Web site (Table 5), and as programs for standalone computers. Both of these will accommodate gaps to achieve the best alignment (Figure 5). BLAST (Basic Local Alignment Search Tool)[26] is a program that carries out local alignments, but most commonly is used to take a query sequence and search a database of sequences to find those that have the best alignments with the query sequence. A version of BLAST that compares two sequences is available at the BLAST web site (Table 5). Another tool for examining both local and global sequence similarities is the "dot plot", available in most integrated packages (Table 4) as well as via the Internet (Table 5). In this approach the two sequences are aligned on the two perpendicular axes of a graph. For any given position on one sequence an identity with a position on the other sequence is scored as a dot in the 2-dimensional plot, with the coordinates being given by the

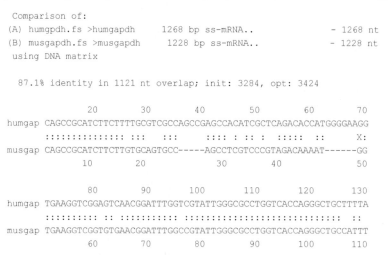

**Figure 5** A portion of the local alignment between the mRNA sequence for human GAPDH (glyceraldehyde 3-phosphate dehydrogenase) and mouse GAPDH. The LFASTA program was used with default parameters.

**Table 5**   Internet Sites for Bioinformatics Tools.

| | |
|---|---|
| *Nucleotide Sequence Analysis* | |
| Webcutter 2 restriction enzyme site mapper | http://www.firstmarket.com/cutter/cut2.html |
| Virtual Genome Center DNA sequence translation | http://alces.med.umn.edu/webtrans.html |
| Rockefeller University Toolkit | http://cs.rockefeller.edu/index.php3?page = toolkit/ |
| Molecular Biology Computing Services— Washington Univ. | http://stl.wustl.edu/service/ |
| BCM Sequence Utilities (conversion, translation, repeats, digests) | http://searchlauncher.bcm.tmc.edu/seq-util/ seq-util.html |
| Primer3 PCR primer design | http://www-genome.wi.mit.edu/cgi-bin/primer/ primer3_www.cgi |
| | |
| *Public Domain Software* | |
| Catalogue of Molecular Biology Programs | http://www.ebi.ac.uk/biocat/ |
| IUBio Archive for Biology | http://iubio.bio.indiana.edu/ |
| *Protein Sequence Analysis* | |
| ExPASy Proteomics tools | http://www.expasy.ch/tools/ |
| SAPS | http://ulrec3.unil.ch/software/SAPS_form.html |
| TGrease | http://alpha10.bioch.virginia.edu/fasta/ |
| *Sequence Alignment* | |
| BLAST | http://www.ncbi.nlm.nih.gov/BLAST/ |
| FASTA Programs at the Univ. of Virginia | http://alpha10.bioch.virginia.edu/fasta/ |
| Nucleotide Dot Plot Page | http://www.molgen.uc.edu/analyze/nucdot.htm |
| *Multiple Alignment Tools* | |
| CLUSTALW at EBI | http://www2.ebi.ac.uk/clustalw/ |
| MULTALIN | http://www.toulouse.inra.fr/multalin.html |
| MACAW (download site) | ftp:/ /ncbi.nlm.nih.gov/pub/macaw/ |
| PHYLIP | http://evolution.geneitcs.washington.edu/phylip.html |
| *Database Searching* | |
| BLAST | http://www.ncbi.nlm.nih.gov/BLAST/ |
| Washington University BLAST | http://blast.wustl.edu/ |
| BCM Search Launcher | http://searchlauncher.bcm.tmc.edu/ |
| EMBL Database Searching and Analysis Tools | http://www.ebi.ac.uk/Tools/ |
| *Functional Prediction* | |
| Blocks WWW Server | http://www.blocks.fhcrc.org/ |
| Pfam | http://www.sanger.ac.uk/Software/Pfam/ |
| PRINTS DATABASE | http://bmbsgi11.leeds.ac.uk/bmb5dp/prints.html |
| ProfileScan Server | http://www.isrec.isb-sib.ch/software/ PFSCAN_form.html |
| Center for Biological Sequence Analysis Prediction Server | http://www.cbs.dtu.dk/services/ |
| PSORT WWW Server | http://psort.nibb.ac.jp/ |
| Mitochondria Project (MITOP) | http://mips.gsf.de/proj/medgen/mitop/ |
| Scansite | http://cansite.bidmc.harvard.edu/cantley85.phtml |
| PhosphoBase | http://www.cbs.dtu.dk/databases/ PhosphoBase/ |
| | |
| *Gene Regulation Predictions* | |
| Transcription Regulatory Regions Database Main Page (TRDD) | http://wwwicg.bionet.nsc.ru/trrd/ |
| Molecular Informatics Resource for the Analysis of GeneExpression (MIRAGE) | http://www.ifti.org/ |
| *Structure Predictions* | |
| MFold RNA and DNA folding | http://bioinfo.math.rpi.edu/~mfold/rna/ |
| CASP Structure Prediction Project | http://predictioncenter.llnl.gov/ |
| *Protein Structure Searching* | |
| Molecular Modeling Database | http://www.ncbi.nlm.nih.gov/Structure/MMDB/ mmdb.shtml |
| VAST | http://www.ncbi.nlm.nih.gov/Structure/VAST/ vastsearch.html |
| | |
| *PERL Programming Tools* | |
| PERL.COM | http://www.perl.com |
| Bioinformatics.org | http://bioinformatics.org/ |
| bio.perl.org—Main page | http://bio.perl.org/ |

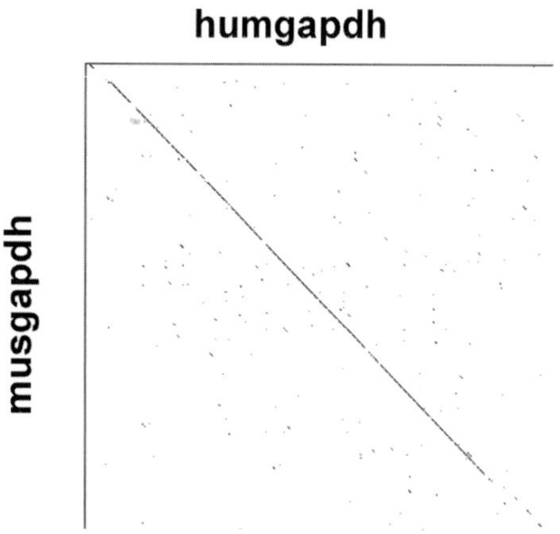

## humgapdh

**musgapdh**

**Figure 6** The "dot-plot" alignment between the mRNA sequence for human GAPDH (glyceralde-hyde 3-phosphate dehydrogenase) and mouse GAPDH. The WINDOT program was used with default parameters. Note that gaps that are also present in the local alignment in Figure 4.

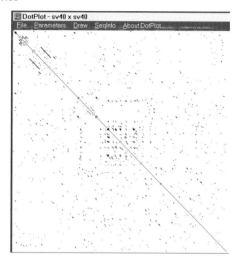

**Figure 7** A dot-plot alignment between a portion of SV40 viral genome and itself. Areas of repeated sequence are seen as small diagonal lines offset from the main diagonal.

positions on the two sequences. Thus two identical sequences will be marked by a solid diagonal line, while similar, but not identical, sequences will be marked by gaps (Figure 6). Comparing a sequence with itself can reveal areas of repeats as diagonals offset from the main diagonal (Figure 7). Obviously, compar-ing an mRNA sequence against genomic sequence will reveal exon-intron boundaries. Finally, as noted for BLAST, pairwise align-ments form the basis for searching sequence databases for entries similar in sequence to a query sequence.

### 14.5.2.6.1.2  Multiple Sequence Analysis

Key to an understanding of relationships between multiple proteins, mRNAs or DNAs is the tool of multiple sequence analysis. Align-ments of multiple peptide sequences can reveal conserved motifs, which may be vital to function. The alignment of a novel protein with a group of proteins of similar sequence and known function can indicate functional motifs (conserved peptide sub-sequences)[27] main-tained in the new protein and give clues to its overall function. Many multiple alignment tools are available through World Wide Web access, (Table 5) although some (such as MACAW, described later) run only on specific operating systems, and a number of these systems have recently been comparatively reviewed.[28]

Most multiple alignment tools start with alignments of pairs of sequences. The degrees of similarity for these pairwise alignments (i.e., the scores as noted in *Algorithms* above) are then used to guide a "progressive alignment" where

the two most similar sequences are aligned and then more dissimilar sequences are then pro-gressively added to the alignment. Examples of such multiple alignment programs include Multalin[29] and Clustal.[30] The latter program is commonly used and is often included in multi-functional packages. It should be emphasized, though, that the process of multiple alignment is such that different programs may not produce the same alignment with their default settings. The ultimate judge of the quality of a multiple alignment must be the researcher, with an eye toward what makes sense, and a "good" alignment may involve considerable user invol-vement by way of adjusting parameters for the multiple alignment software and or the align-ment strategy. Figure 8 displays an example of a Clustal multiple alignment, but required two steps: (1) an alignment of the SRC, MATK, and c-ABL sequences; and (2) an alignment of the JAK2 sequence with the first alignment. That the alignment is "good" is indicated by the proper alignment of the tyrosine kinase catalytic domain.

Another approach is taken by the MACAW (Multiple Alignment Construction and Analy-sis Workbench) program, which simply searches for "blocks" of aligned sequence segments.[31] Block boundaries may be evalu-ated and adjusted by the user, and the assessment of whether or not a block is linked into a multiple alignment is left up to the user, aided by a graphical interface. Thus the process forces user input into the alignment. An example of the MACAW alignment can be seen in Figure 9. This alignment covers the same region as that in Figure 8. Because

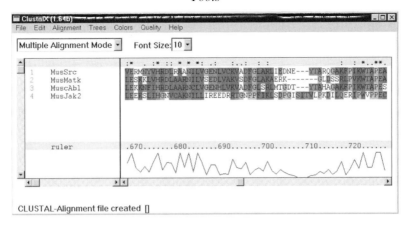

**Figure 8** A Clustal alignment of murine Src (accession # NM_009271), Matk (NM_010768), c-abl (L10656), and Jak2 (NM_008413) amino acid sequences. The region shown is part of the "TyrKc; Region: Tyrosine kinase, catalytic domain" peptide motif.

blocks, by definition, do not contain gaps, the MACAW alignment does not readily highlight the KFPIKWTAPE motif common between the SRC, MATK, and c-ABL sequences (as found by Clustal). On the other hand, other blocks of homology are found with a default MACAW alignment and are not found with a default Clustal alignment, so again the researcher is well to look at any alignment with a careful eye, and consider alignments with several parameter settings and perhaps using different programs.

### 14.5.2.6.1.3 *Phylogenetic Analysis*

One application of multiple sequence alignments, especially with protein alignments, is the inference of evolutionary relationships, i.e., phylogeny. Often these relationships are displayed with a tree diagram, with the branches

and branch points being determined by the relatedness of various sequences (Figure 10). However, key assumptions are made in any phylogenetic analysis, and the reader is encouraged to consider a recent review of the subject.[32] Software tools to conduct phylogenetic analysis exist as part of comprehensive packages, as well as dedicated programs, the most well known being the PHYLIP package. As noted below, phylogenetic analysis plays a key role in the naming conventions for xenobiotic-metabolizing enzymes.

### 14.5.2.6.2 **Database Searching**

### 14.5.2.6.2.1 *Homology to a Single Sequence*

A critical application of sequence alignment is that of database searching, where a nucleotide or protein sequence is sequentially aligned

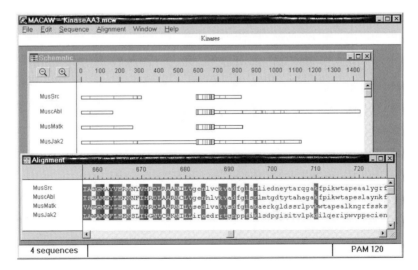

**Figure 9** A MACAW alignment of murine Src, Matk, c-abl, and Jak2 amino acid sequences.

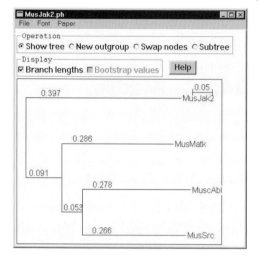

**Figure 10**  A PHYLIP format tree depicting the sequence relatedness determined in the Clustal alignment shown in Figure 8.

with *every* member of a database of sequences. One of the early successful approaches was that taken by the FASTA program.[24] The FASTA algorithm used four steps to determine a score for a pairwise alignment, the first being a rapid scan for short regions of consecutive identities ("words") with an identification of the 10 best diagonal regions. Subsequent steps rescore the diagonals and develop a score for the optimized alignment. In this way the program compares a sequence against a database of sequences and the highest scoring alignments are returned. In the simplest use a protein sequence is compared against a database of protein sequences to find other proteins with similar sequence. A more sophisticated (and computationally expensive) application involves taking a novel mRNA sequence, translating it in all six possible reading frames, and comparing each of these against a protein database to determine a possible protein product. One more variation of the program compares a protein sequence to a database of

DNA sequences, translating each target sequence into each of six possible reading frames "on-the-fly". The FASTA suite of programs (Table 6) continues to be a powerful tool[33] and is available both through a World Wide Web interface (Table 5) and as a stand-alone, local program for many operating system.

The BLAST suite of programs was introduced in 1990[26] as a more rapid approach to database searching. As with FASTA, a query sequence is broken up into "words", and the database is searched for matches to these "words". If a match is found, i.e., between a "word" in the query and a "word" in the target sequence, an attempt is made to extend the alignment in either direction until the score of such an alignment drops below a threshold. Alignments that are generated are called High-scoring Segment Pairs (HSPs). The maximally scoring HSPs for any given target are stored and their scores compared with those for the alignments with other targets. Ultimately the BLAST program returns a list of highest scoring (most similar) target sequences along with a value (the E value) for the number of times that such an alignment might have occurred by chance. Thus a high score and a low E-value (e.g., $1 \times 10^{-33}$) would indicate that the match between the query and the target sequence was significant. Originally, BLAST alignments did not allow gaps, but the recent version BLAST 2.0[34] does. The BLAST suite of programs (Table 6) has become the *de facto* standard for database searching, and as with the FASTA suite, is available both through a World Wide Web interface (Table 5) and as a stand-alone, local program for many operating system. It is also available through email server (blast @ncbi.nlm.nih.gov) where queries can be submitted as an email message, and the results returned as email. (Information on the email server may be obtained by sending an email to blast@ncbi.nlm.nih.gov with just the word

**Table 6**   FASTA and BLAST suites of Database Search Tools.

| Program | Query sequence | Target database |
|---------|----------------|-----------------|
| FASTA   | DNA or Protein | Corresponding type |
| FASTX   | DNA (TRN)      | Protein |
| TFASTX  | Protein        | DNA (TRN) |
| blastn  | DNA            | DNA |
| blastp  | Protein        | Protein |
| blastx  | DNA (TRN)      | Protein |
| tblastn | Protein        | DNA (TRN) |
| tblastx | DNA (TRN)      | DNA (TRN) |

TRN means the database is dynamically translated in all reading frames for each alignment.

HELP in the message). An example of a BLAST search can be found in the following Applications section.

A common database search is one carried out with a new mRNA or protein sequence of unknown function (such as might be identified with differential display),[35] with the question being what protein with *known* function is it identical or similar to. The strategy would then be to use blastp to compare the protein (or blastx to compare the mRNA) against a database of known proteins. Another approach is to compare the sequence for a gene product of *known* function with an EST database, the question being whether similar, but previously undescribed, gene products may be found. An example of such a search will be given in the Applications section.

#### 14.5.2.6.2.2 *Homology to a Motif*

As noted above, multiple alignments of related proteins can reveal motifs, or locally conserved regions of peptide sequence that often are crucial for the shared functions of the aligned proteins. Identifying such motifs in a novel protein can give clues to its function. Thus the search strategy is to compare the novel protein sequence against profile databases, where a sequence profile is a quantitative or qualitative method of describing a motif. BLOCKS, PRINTS, Pfam and Prosite are the most common databases used, and can be accessed at their web sites (Table 3). Several other Web servers (such as the CD-Search engine at the NCBI Blast Web site) access these databases.

#### 14.5.2.6.3 **Nucleotide Sequence Analysis**

When the research project involves collecting and manipulating DNA sequences by cloning, sequencing, or PCR, a variety of tools become critical for success. For cloning, sub-cloning, and Southern blot analysis, a tool to predict restriction enzyme cleavage sites in vectors and known sequences is needed (see Figure 2), and ideally one with the ability to generate maps of such sites based upon the DNA fragment sizes obtained after such an enzyme digest. These operations are relatively simple, but the software tool must accommodate an ever-increasing list of restriction enzymes. If the sequence being determined is a genomic (as opposed to mRNA, or cDNA) sequence, then one requires a software tool to determine open reading frames (ORFs—regions of sequence which when translated into amino acids lack a stop

codon) (again, see Figure 2). Such a tool is also required when mRNA sequences are being determined, as they serve as a check for sequence accuracy; if the mRNA sequence does not have an open reading frame, i.e., contains a stop odon, then the sequence most likely contains an error.

Another useful computational analysis is the design of optimal oligonucleotide primers for polymerase chain reaction (PCR).[36] Such analysis is offered by several commercial packages, such as Oligo, and Primer Premier, but can also be conducted through World Wide Web access to the Primer program at the Whitehead Institute (Table 5). Key to these analyses is selection of oligonucleotides with high specificity to the target sequence and minimal proclivity for hairpin structures and primer-primer dimers.

#### 14.5.2.6.4 **Protein Sequence Analysis**

As noted above, polypeptide sequences were the subject of some of the first computational analyses. Physical properties such as molecular weight and isoelectric point represent rather simple calculations available in the integrated packages as well as through various Web sites (ExPASy Proteomics tool site, SAPS, and the FASTA web site, Table 5). Prediction of protease cleavage sites is also useful, especially in analysis of proteomics data. Calculations of local hydrophobicity (e.g., the TGrease program)[37] make use of the relative hydrophobicity of individual amino acids. The program calculates a moving average across the protein sequence, and the resulting plot offers insight into exposed regions of the protein. Prediction of potential secondary structure poses a somewhat more challenging problem. Various approaches have been taken, one example being that of the PredictProtein program,[38] where the protein is first aligned with similar sequences in the Swiss–Prot database, then the multiple alignment is analyzed via a neural network for probable structures.

#### 14.5.2.6.5 **Function Prediction**

#### 14.5.2.6.5.1 *Proteins*

Generally an endeavor more interesting than determining physical properties for a protein is that of developing hypotheses as to its function. In this regard, bioinformatics can provide an enormous insight, based upon the primary sequence of the protein, and the modular nature of proteins. Here the key observation is that a protein is the sum of its parts, small peptide patterns that carry out different functions, such as protein tyrosine kinase catalytic domain noted in Figure 8. These conserved short sequence patterns are determined from multiple alignments of related proteins, and are termed *motifs*, and may be used to define protein families. Motif databases, such as Blocks, PROSITE, Pfam, and PRINTS (Table 3), may be used as resources to compare a novel protein sequence against. Thus the function of a previously unknown protein may be conjectured on the basis of the functional motifs it contains.

Similarly, "signal sequences"[8,9] are known to direct nascent polypeptides to their ultimate subcellular localization. Thus a powerful clue to protein function can be the identification of such motifs predicted from a translated mRNA sequence. Programs for predicting such sequences can be found at the Mitochondria Project and the Center for Biological Sequence Analysis Prediction Server (Table 5). Of particular note is the PSORT WWW Server (Table 5) which carries out sequential analysis for a number of signal sequences and delivers an "overall" prediction for subcellular localization.

### *14.5.2.6.5.2  Nucleic Acids*

Functional information is coded not only in polypeptide sequences, but also in genomic nucleotide patterns. Specifically, the transcriptional activity of any given gene is usually determined by a committee of proteins that binds to sequences along it and modulates the activity of RNA polymerase. These regulatory proteins (transcription factors) often operate in specific tissue and or developmental contexts (such as the erythroid-specific GATA-1 transcription factoe),[39,40] or confer specific responsiveness (such as members of the estrogen receptor family). Thus hypotheses may be developed as to the expression pattern (i.e., tissue distribution or hormone-responsiveness) of a given gene from the transcription factor binding sites found in its genomic sequence. As noted in the *Algorithms* section such sites are usually heterogeneous and their determination best involves comparison to a matrix that scores the fit for any given sequence position[10] (see Table 2). Two Internet sites that offer such analysis are the Transcription Regulatory Regions Database (TRDD) and the Molecular Informatics Resource for the Analysis of GeneExpression (MIRAGE) (Table 5).

### 14.5.2.6.6  Structure Prediction

### *14.5.2.6.6.1  Nucleic Acids*

While the double helical structure of DNA is all too familiar, the secondary structure of single-stranded RNA and DNA is not. However, the same forces that stabilize the DNA double helix also drive the folding of single stranded polynucleotides into double-stranded regions. Such folding has implications not only for RNA stability, but also for the efficiency in which a given RNA region is reverse-transcribed[41] prior to RT-PCR. Using energy rules for formation of loops, hairpins and double-stranded regions, Zuker developed the MFold program to determine optimal and suboptimal secondary structures for a given RNA or DNA.[42-44] The user should bear in mind though, that, because of uncertainties in the folding model and the folding energies, the "correct" folding may not be the "optimal" folding determined by the program. He/she may therefore want to view many optimal and suboptimal structures within a few percent of the minimum energy. MFold is available both through Web servers (Table 5), as well as a stand-alone program running on UNIX machines, and as part of the GCG Wisconsin package.

### *14.5.2.6.6.2  Proteins*

Despite the fact that much can be hypothesized in terms of functionality from a protein's amino acid sequence, it is that sequence arranged in a 3-dimensional structure that ultimately determines a protein's behavior and function. Unfortunately, while the SWISS–PROT database contains 98,739 curated protein sequence entries and 471,750 entries for translated mRNA sequences, the PDB repository contains only 15435 structures (all data as of June 2001). Thus there is a huge gap between sequence and structure, prompting a search for the "Holy Grail" of structure prediction from amino acid sequence. As noted above, several tools exist for identifying motifs and potential secondary structure, but these fall short of a full 3D prediction. Rather, structure prediction efforts have taken three approaches: homology modeling, threading, and *ab-initio* prediction. The first relies on using a structural template from known structures of similar sequence. Threading uses the polypeptide sequence of the novel protein, computes models based on existing structures, and then evaluates the fit of each amino acid for the novel protein in each structure. Finally, *ab-initio* methods build structures from the "ground up" with molecular modeling principles. Not surprisingly, the search for the "Holy Grail" has led to a competition—the Community Wide Experiment on the Critical Assessment of Techniques for Protein Structure Prediction (CASP, Table 5), which is held every 2 years. The latest, CASP 4, highlights the progress that has been made in the field, as well as the limitations of some of the current approaches.

### 14.5.2.6.7 Pathway Prediction

Proteins do not carry out their functions in isolation, but as part of larger metabolic or signaling pathways. Indeed, inferring a proteins place in such a pathway often gives clues to its functions. Thus the identification of conserved "gene clusters" through the comparison of bacterial genomes allows the prediction of pathways.[45] In a similar fashion, phylogenetic analysis of COGs (clusters of orthologous genes) can place predicted gene within metabolic pathways.[46] Extending this approach to grouping proteins by evolution, correlated mRNA expression patterns and patterns of domain fusion allowed a general function to be assigned to previously uncharacterized yeast proteins.[47] With the finalization of the human and mouse genomes this sort of analysis may be extended to higher eurcaryotes. Paralleling such genome wide comparisons are the efforts to predict signaling pathways: Yaffe and coworkers used an approach that predicted potential phosphorylation targets for any given protein tyrosine kinase, establishing a network of kinases and substrates.[48] Expansion of these efforts offer possibilities for predictive databases that may be used to guide molecular and cellular experiments.

### 14.5.2.6.8 Predictions from Genomic Data

With complete sequences for *C. elegans* and *Drosophila*, and the draft sequence of the human genome,[14,49] in hand, the task is to determine what information can be divined. First off, the exon-intron structure of unknown genes must be predicted. This task was refined during the annotation of the *Drosophila* genome, notably using programs such as Genie,[12] GenWise,[13] and GENSCAN,[11] and extended to annotation of the human genome.[50] Such analysis can be augmented with microarray data[51] to allow identification of splice variants[52] and tissue-specific expression. These predictions allow comparisons between species in terms of transcriptional machinery,[50] DNA repair,[53] and apoptotic mechanisms.[54] Genomic sequence data also allows comparisons of transcriptional control elements; such elements may be identified through comparisons of individual genes,[55] or inferred through clustering of sequence elements using microarry expression data.[56] Again, these efforts offer possibilities for databases that may be used to guide molecular and cellular experiments.

### 14.5.2.6.9 Custom Tools and Programming

No matter how many software tools are available through integrated packages, or the Internet, sooner or later a bioinformatics research project will involve tasks that are best handled with custom programs. Such tasks are often as simple as changing file formats, or manipulating large numbers of text files. To this end it behooves the serious bioinformatics researcher to develop proficiency in one or more programming languages. Of particular note is Perl (Practical Extraction and Report Language), which is a relatively new language, but one that has become a preferred standard in the bioinformatics community.[57] Lincoln Stein even maintains that "PERL saved the human genome project".[58] Even without this accolade Perl offers a programming language that readily interfaces with the Internet, is maintained on almost every operating system (e.g., Windows, Macintosh and UNIX ), is supported for biological applications by a large user base (bio.perl.org, Table 5), and is free. Furthermore, because "programs" written in Perl are actually text "scripts" they can be easily critiqued, and modified by other users. Several good tutorials are also available at the Perl.com web site.

### 14.5.2.7 APPLICATIONS IN TOXICOLOGY

#### 14.5.2.7.1 Finding New Proteins Based on Similarity to Known Proteins

One of the most common uses of the growing database of sequence information, particularly EST data, is searching for new proteins that have sequence similarity to a known protein. The usual approach is to compare a known protein against the EST database using a tool such as BLAST (Table 1).[26] BLAST will return a listing of similar sequences and the alignment between them and the test sequence. "Similarity" can be a somewhat subjective determination if the sequences are of very different lengths and do not share long stretches of highly similar sequences. Given this caveat, though, this exercise has proved fantastically productive. Thus with the first report of EST sequencing, 337 new genes were identified in brain mRNA.[20] Most recently the dbEST was used to identify potential human counterparts of yeast genes coding for critical steps of oxidative phosphorylation.[59] It is pertinent to consider the different definitions applied to similar proteins, and the implications for their function: (1) *homologues* are presumed to share a common protein ancestor and may have similar function; (2) *orthologues* are the same protein separated by a speciation event, probably with the same function; (3) *paralogues* are similar proteins within a species and these may have similar function. The members of the basic helix-loop-helix PAS superfamily discovered by Bradfield and coworkers[60] with database searching are examples of paralogues.

#### 14.5.2.7.2 Identification, for any Given "Tox Target" Protein, of Similar Targets in Other Species

A common approach in toxicology is that of predicting toxic responses in one species (usually human) based upon results in a model species. Given this model, a logical application would be the search across species for orthologues of a protein identified as a toxicologic "target". Remarkably, this application has not been commonly reported. However, with the increasing amount of rat, mouse, and even monkey sequences, this application should begin to see greater use.

An example of this approach can be given here. Cap43 was a transcript identified as inducible by nickel subsulfide in human lung A549 cells.[61] To study its regulation in a rat cell line, a corresponding rat mRNA sequence would be desirable. The first step in this process might be to examine the LocusLink entry for Cap43. Searching for Cap43 at the LocusLink site (Table 3) reveals that this transcript is identical to Drg1, a transcript upregulated in human colon carinoma cells,[62] and RTP, a transcript upregulated by homocysteine in human umbilical vein endothelial cells[63] (Figure 11). This example should demonstrate the unreliability entailed in depending upon annotation for inferences as to gene function. The official gene symbol and name is NDRG1: N-myc downstream regulated gene (LocusID: 10397) and the Web page contains a link to the protein reference

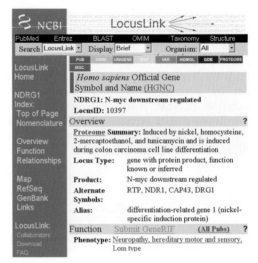

**Figure 11**    The LocusLink result for a search for the gene "Cap43".

sequence (RefSeq) with accession number NP_006087. Accessing the BLAST Web page and choosing TblastN under "Translating BLAST: Protein query—Translated db" brings up the TBLASTN entry page. One enters "NP_06087" (without quotes) in the "Search" box, and selects "est_others" in the Choose Database drop-down box. Under Options for advanced blasting one selects *Rattus norvegicus* in the drop-down box next to "select from", and then submits the search. The results for this search can be seen in Figure 12(A)–(C). First, a table of the best scoring "hits" can be seen in Figure 12(A); the E Values of 2e-43 or less indicate these alignments clearly did not occur by chance, and a glance at the best alignment (Figure 12(C)) confirms that this rat sequence is most certainly the homolog of human NDRG1 (or at least a portion of it). The graph of Blast hits (Figure 12(B)) gives an indication of which are the longest sequences and would direct an effort to join the individual ESTs into a composite, full-length, virtual rat NDRG1 sequence. Such a sequence could be used for PCR primer design, or alternatively the clones for individual ESTs could be obtained from the IMAGE consortium and used as probes.

#### 14.5.2.7.3 Prediction of Function for a Novel Protein

A common experimental situation is that a gene may be identified as having a certain biological role, yet the exact biochemical function of the protein it codes for is unknown. A recent example of this may be found in the discovery that the B subunit of bacterial

A

```
TBLASTN 2.2.1 [Apr-13-2001]

Reference:
Altschul, Stephen F., Thomas L. Madden, Alejandro A. Schäffer,
Jinghui Zhang, Zheng Zhang, Webb Miller, and David J. Lipman (1997),
"Gapped BLAST and PSI-BLAST: a new generation of protein database search
programs", Nucleic Acids Res. 25:3389-3402.

Query= gi|5174657|ref|NP_006087.1|
        (394 letters)

Database: GenBank non-mouse and non-human EST entries
          2,589,715 sequences; 1,196,695,995 total letters

                                                         Score      E
Sequences producing significant alignments:             (bits)   Value
gi|11214483|gb|BF283413.1|BF283413  EST448004 Rat Gene Index...    409   e-114
gi|8083026|gb|AW917279.1|AW917279   EST348583 Rat gene index,...   377   e-105
gi|1503348|dbj|C06572.1|C06572      C06572 Rat pancreatic islet ... 249   2e-66
gi|4289657|gb|AA998409.1|AA998409   UI-R-C0-ia-e-02-0-UI.s1 U...    229   2e-60
gi|13893341|gb|BG671242.1|BG671242  DRNBNE03 Rat DRG Library...    172   2e-43
gi|11673239|gb|BF563509.1|BF563509  UI-R-C4-ale-h-01-0-UI.r1...    141   5e-34
gi|8512551|gb|BE120446.1|BE120446   UI-R-CA0-bam-e-08-0-UI.s1...   119   2e-27
gi|8506330|gb|BE114225.1|BE114225   UI-R-CA0-axo-e-12-0-UI.s1...   119   2e-27
gi|11671236|gb|BF561506.1|BF561506  UI-R-C0-ia-e-02-0-UI.r1 ...     94   1e-19
gi|7172621|gb|AW530207.1|AW530207   UI-R-C4-ale-h-01-0-UI.s1 ...    88   9e-18
gi|7172619|gb|AW530205.1|AW530205   UI-R-C4-ale-g-11-0-UI.s1 ...    82   4e-16
```

B

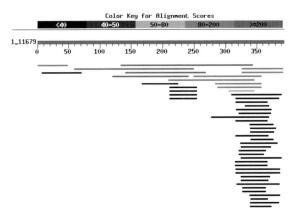

C

**Alignments**

```
>gi|11214483|gb|BF283413.1|BF283413 EST448004 Rat Gene Index, normalized rat, Rattus norvegicus
cDNA
 Rattus norvegicus cDNA clone RGIEB70 3' sequence.
 Length = 656

 Score =  409 bits (1052), Expect = e-114
 Identities = 200/211 (94%), Positives = 207/211 (97%)
 Frame = -3

Query: 134 IIGMGTGAGAYTLTRFALNNPEMVEGLVLINVNPCAEGWMDWAASKISGWTQALPDMVVS 193
           +IGMGTGAGAY LTRFALNN EMVEGLVL+NVNPCAEGWMDWAASKISGWTQALPDMVVS
Sbjct: 654 VIGMGTGAGAYILTRFALNNXEMVEGLVLMNVNPCAEGWMDWAASKISGWTQALPDMVVS 475

Query: 194 HLFGKEEMQSNVEVVHTYRQHIVNDMNPGNLHLFINAYNSRRDLEIERPMPGTHTVTLQC 253
           HLFGKEE+ SNVEVVHTYRQHI+NDMNP NLHLFI+AYNSRRDLEIERPMPGTHTVTLQC
Sbjct: 474 HLFGKEEIHSNVEVVHTYRQHILNDMNPSNLHLFISAYNSRRDLEIERPMPGTHTVTLQC 295

Query: 254 PALLVVGDSSPAVDAVVECNSKLDPTKTTLLKMADCGGLPQISQPAKLAEAFKYFVQGMG 313
           PALLVVGD+SPAVDAVVECNSKLDPTKTTLLKMADCGGLPQISQPAKLAEAFKYFVQGMG
Sbjct: 294 PALLVVGDNSPAVDAVVECNSKLDPTKTTLLKMADCGGLPQISQPAKLAEAFKYFVQGMG 115

Query: 314 YMPSASMTRLMRSRTASGSSVTSLDGTRSRS 344
           YMPSASMTRLMRSRTASGSSVTSL+GTRSRS
Sbjct: 114 YMPSASMTRLMRSRTASGSSVTSLEGTRSRS 22
```

**Figure 12** Results from a BLAST search for ESTs similar to the human NDRG1 transcript. The program TBLASTN was used to compare the protein sequence (RefSeq NP_006087) against the EST database, translating each sequence on the fly before the comparison. The search was limited to *Rattus norvegicus* sequences. (A) table of significant alignments; (B) graphical depiction of the distribution of Blast "hits" along the query sequence; (C) one of the reported alignments. The actual table of alignments is much larger, as is the number of reported alignments. In an email report (i.e., text-based) only (A) and (C) would be returned.

cytolethal distending toxin (CDT), which causes cell cycle arrest and cell death, bears sequence resemblance to type I deoxyribonucleases.[64] These studies invariably start with a simple comparison of the unknown protein sequence with databases of known proteins. A more detailed and thorough functional prediction can be obtained with an alignment of the novel protein with several related proteins of known functions. Thus the Clustal alignment of CDT subunit B with deoxyribonucleases identified residues in the former protein that were conserved across all DNases and were known to be essential for catalytic activity.

In a similar fashion, valuable inference about a protein's cellular role can be made based upon that protein's subcellular localization. Ever since the proposal that such localization was targeted by a signal peptide on the N-terminus of a nascent protein,[8] considerable effort has been made to analyze signal peptide sequences and predict them based on primary amino acid sequence.[9] Thus, when two forms of mouse deoxyguanosine kinase were cloned, they could be predicted to be mitochondrial and cytoplasmic based upon the presence or absence of a mitochondrial import signal sequence.[65] Several good Web sites exist for analyzing proteins for signal peptides (Table 1), and of particular note is the PSORT WWW Server, which will predict the probable localization into a number of cellular organelles.

Far less common for a novel protein is the availability of a 3-dimensional crystal structure. If one was so fortunate, the Vector Analysis Search Tool (VAST)[66] would allow one to search the database of known protein structures for similar structural motifs in other proteins. VAST is one of the tools offered through NCBI's Molecular Modeling Database (MMDB) (Table 1).

### 14.5.2.7.4 Prediction of Regulation for a Novel Gene

Yet another opportunity for discovery comes when sequence information 5′ upstream of the mRNA start site is obtained for a gene. Given the paradigm of gene regulation by transcription factor binding to specific sequence motifs,[67] insight into the control and tissue-specific expression of a given gene can be gained by comparing these upstream sequences against a database of transcription factor binding sites. Thus, Spandidos and coworkers found phorbol ester-inducible AP-1 binding sites in the promoter regions for the human H-ras oncogene.[68] Such databases have

evolved from Ghosh's early TFD (Transcription Factor Database)[69] to TRRD (Transcription Regulatory Region Database)[70] and EPD (Eucaryotic Promoter Database)[71] (Table 5), and the tools for using these database have evolved from SignalScan[72] to a wide variety of tools[73] including those that compare sequences against nucleotide distribution matrices.[10]

### 14.5.2.7.5 Establishing Evolutionary Relationships Among Several Similar Proteins

This application is rooted in the very earliest developments of bioinformatics, at a time when sequences for cytochrome *c* from several species became available.[5,6] Indeed the process of creating an alignment of multiple protein sequences was usually linked to creating a hierarchical tree describing the relatedness of different sequences, and such trees could be used to infer phylogenetic relationships.[74] While seemingly peripheral to toxicology, this is the one approach in bioinformatics that has had the most dramatic impact on the field. With the visionary suggestion that cytochrome P450s be named and categorized based upon their evolutionary relationships, Nebert and his colleagues gave lasting order to a field of research.[75] The remarkable aspect of this system, and its further updates[76–78] is that it did away with confusing, subjective nomenclature and replaced it with one based entirely upon computer-determined alignments of protein sequences. It is this kind of impact that the field of bioinformatics promises, that of automating difficult analyses and offering an objective understanding of biological molecules.

### 14.5.2.8 FUTURE DIRECTIONS AND CHALLENGES

#### 14.5.2.8.1 Gene Index

Despite the opportunities and promise with genomic and bioinformatic resources, several challenges remain. First and foremost of these is the expansion of current gene annotation and indexing efforts.[79] Of course a key element in this process is the establishment of standard and meaningful nomenclature for genes and proteins.[80,81] As noted in the Applications section, emphasis on resources such as LocusLink and GeneCards will be useful in ensuring compliance with standard nomenclature. A critical extension to toxicology and biological sciences in general will be a nomenclature that crosses species, or is simple to translate across species.

### 14.5.2.8.2 Gene Ontology

Another challenge for the bioinformatics and biology community will be an unambiguous description of function for any given gene product. The Gene Ontology Consortium[82] represents just such an effort, although it parallels several other public and private efforts. The challenge will be to arrive at consensus ontological terms that again can be applied across species, a database system that allows easy access to such descriptions, and the flexibility to update a gene's ontology as the biology develops.

### 14.5.2.8.3 Cross-species Annotation and Comparison

Finally, the challenge remains of cross-indexing homologs across species. Much can be learned from such indices[83,84] in terms of the similarities of given proteins and gene regulatory regions. Homologene at the NCBI offers both an automated (i.e., computational) and curated approach to such a database, however, any computational approach to homology is complicated by the existence of *paralogues* (similar proteins within a species and these may have similar function). Nonetheless, Homologene is a valuable resource and remains the only public effort at cross-species indexing.

The future, though, promises that these challenges will be met. With the extension of sequencing efforts to other species (e.g., dog, monkey, rabbit), the value of bioinformatic approaches will only increase. Furthermore, one can expect the development of databases that compare not just gene, but pathways, across species, and allow true cross-species comparison of toxicological responses. Bioinformatics will not remain a arcane tool in toxicology, but will take its place with the standard *in vivo* study as essential in safety assessment and prediction.

### 14.5.2.9 REFERENCES

1. *Bioinformatics: A practical guide to the analysis of genes and proteins*, eds., A. D. Baxevanis and B. F. F. Ouellette, John Wiley and Sons, Inc., New York, 1998.
2. *Bioinformatics: Methods and Protocols*, eds., A. Misener and S. A. Krawetz, Humana Press, Totowa, NJ, 2000.
3. C. Gibas and P. Jambeck, *Developing Bioinformatics Computer Skills*, O'Reily, Cambridge, MA, 2001.
4. J. B. Hagen, 'The origins of bioinformatics.' *Nat. Rev. Genet.*, 2000, **1**, 231–236.
5. M. O. Dayhoff, 'Computer aids to protein sequence determination.' *J. Theor. Biol.*, 1965, **8**, 97–112.
6. M. O. Dayhoff, 'Computer analysis of protein evolution.' *Sci. Am.*, 1969, **22**, 86–95.
7. S. Scherer and R. W. Davis, 'Replacement of chromosome segments with altered DNA sequences constructed *in vitro*.' *Proc. Nat. Acad. Sci. USA*, 1979, **76**, 4951–4955.
8. G. Blobel, 'Unidirectional and bidirectional protein traffic across membranes.' *Cold Spring Harb. Symp. Quant. Biol.*, 1995, **60**, 1–10.
9. R. M. Stroud and P. Walter, 'Signal sequence recognition and protein targeting.' *Curr. Opin. Struct. Biol.*, 1999, **9**, 754–759.
10. K. Quandt, K. Frech, H. Karas *et al.*, 'MatInd and MatInspector: new fast and versatile tools for detection of consensus matches in nucleotide sequence data.' *Nucleic Acids Res.*, 1995, **23**, 4878–4884.
11. C. Burge and S. Karlin, 'Prediction of complete gene structures in human genomic DNA.' *J. Mol. Biol.*, 1997, **268**, 78–94.
12. M. G. Reese, D. Kulp, H. Tammana *et al.*, 'Genie—gene finding in *Drosophila melanogaster*.' *Genome. Res.*, 2000, **10**, 529–538.
13. E. Birney and R. Durbin, 'Using GeneWise in the Drosophila annotation experiment.' *Genome. Res.*, 2000, **10**, 547–548.
14. J. C. Venter, M. D. Adams, E. W. Myers *et al.*, 'The sequence of the human genome.' *Science*, 2001, **291**, 1304–1351.
15. M. G. Reese, G. Hartzell, N. L. Harris *et al.*, 'Genome annotation assessment in *Drosophila melanogaster*.' *Genome. Res.*, 2000, **10**, 483–501.
16. A. D. Baxevanis, 'The Molecular Biology Database Collection: an updated compilation of biological database resources.' *Nucleic Acids Res.*, 2001, **29**, 1–10.
17. D. L. Wheeler, C. Chappey, A. E. Lash *et al.*, 'Database resources of the National Center for Biotechnology Information.' *Nucleic Acids Res.*, 2000, **28**, 10–14.
18. K. D. Pruitt and D. R. Maglott, 'RefSeq and LocusLink: NCBI gene-centered resources.' *Nucleic Acids Res.*, 2001, **29**, 137–140.
19. M. Rebhan, V. Chalifa-Caspi, J. Prilusky *et al.*, 'GeneCards: a novel functional genomics compendium with automated data mining and query reformulation support.' *Bioinformatics*, 1998, **14**, 656–664.
20. M. D. Adams, J. M. Kelley, J. D. Gocayne *et al.*, 'Complementary DNA sequencing: expressed sequence tags and human genome project.' *Science*, 1991, **252**, 1651–1656.
21. G. Miller, R. Fuchs and E. Lai, 'IMAGE cDNA clones, UniGene clustering, and ACeDB: an integrated resource for expressed sequence information.' *Genome. Res.*, 1997, **7**, 1027–1032.
22. F. Liang, I. Holt, G. Pertea *et al.*, 'An optimized protocol for analysis of EST sequences.' *Nucleic Acids Res.*, 2000, **28**, 3657–3665.
23. J. Quackenbush, F. Liang, I. Holt *et al.*, 'The TIGR gene indices: reconstruction and representation of expressed gene sequences.' *Nucleic Acids Res.*, 2000, **28**, 141–145.
24. W. R. Pearson and D. J. Lipman, 'Improved tools for biological sequence comparision.' *Proc. Nat. Acad. Sci. USA*, 1988, **85**, 2444–2448.
25. K. M. Chao, W. R. Pearson and W. Miller, 'Aligning two sequences within a specified diagonal band.' *Comput. Appl. Biosci.*, 1992, **8**, 481–487.

26. S. F. Altschul, W. Gish, W. Miller *et al.*, 'Basic local alignment search tool.' *J. Mol. Biol.*, 1990, **215**, 403–410.

27. R. F. Doolittle, 'Similar amino acid sequences: chance or common ancestry?' *Science*, 1981, **214**, 149–159.

28. P. Briffeuil, G. Baudoux, C. Lambert *et al.*, 'Comparative analysis of seven multiple protein sequence alignment servers: clues to enhance reliability of predictions.' *Bioinformatics*, 1998, **14**, 357–366.

29. F. Corpet, 'Multiple sequence alignment with hierarchical clustering.' *Nucleic Acids Res.*, 1988, **16**, 10881–10890.

30. D. G. Higgins, J. D. Thompson and T. J. Gibson, 'Using CLUSTAL for multiple sequence alignments.' *Methods Enzymol.*, 1996, **266**, 383–402.

31. G. D. Schuler, S. F. Altschul and D. Lipman, 'A workbench for multiple sequence alignment construction and analysis.' *Proteins Struct. Funct. Genet.*, 1991, **9**, 180–190.

32. M. A. Hershkovitz and D. D. Leipe, in 'Bioinformatics: A Practical Guide to the Analysis of Genes and Proteins, 1st edn.', eds., A. D. Baxevanis and B. E. F. Ouellete, Wiley-Liss, Inc, New York, 1988.

33. W. R. Pearson, T. Wood, Z. Zhang *et al.*, 'Comparison of DNA sequences with protein sequences.' *Genomics*, 1997, **46**, 24–36.

34. S. F. Altschul, T. L. Madden, A. A. Schaffer *et al.*, 'Gapped BLAST and PSI-BLAST: a new generation of protein database search programs.' *Nucleic Acids Res.*, 1997, **25**, 3389–3402.

35. P. Liang and A. B. Pardee, 'Differential display of eukaryotic messenger RNA by means of the polymerase chain reaction.' *Science*, 1992, **257**, 967–971.

36. W. B. Mattes, in 'PCR Protocols in Molecular Toxicology, 1st edn.', ed., J. P. Vanden Heuvel, CRC Press, Boca Raton, 1988.

37. J. Kyte and R. F. Doolittle, 'A simple method for displaying the hydropathic character of a protein.' *J. Mol. Biol.*, 1982, **157**, 105–132.

38. B. Rost, C. Sander and R. Schneider, 'PHD–an automatic mail server for protein secondary structure prediction.' *Comput. Appl. Biosci.*, 1994, **10**, 53–60.

39. T. Evans, M. Reitman and G. Felsenfeld, 'An erythrocyte-specific DNA-binding factor recognizes a regulatory sequence common to all chicken globin genes.' *Proc. Nat. Acad. Sci. USA*, 1988, **85**, 5976–5980.

40. L. I. Zon, S. F. Tsai, S. Burgess *et al.*, 'The major human erythroid DNA-binding protein (GF-1): primary sequence and localization of the gene to the X chromosome.' *Proc. Nat. Acad. Sci. USA*, 1990, **87**, 668–672.

41. L. Pallansch, H. Beswick, J. Talian *et al.*, 'Use of an RNA folding algorithm to choose regions for amplification by the polymerase chain reaction.' *Anal. Biochem.*, 1990, **185**, 57–62.

42. M. Zuker, 'On finding all suboptimal foldings of an RNA molecule.' *Science*, 1989, **244**, 48–52.

43. J. A. Jaeger, D. H. Turner and M. Zuker, 'Improved predictions of secondary structures for RNA.' *Proc. Nat. Acad. Sci. USA*, 1989, **86**, 7706–7710.

44. J. A. Jaeger, D. H. Turner and M. Zuker, 'Predicting optimal and suboptimal secondary structure for RNA.' *Methods Enzymol.*, 1990, **183**, 281–306.

45. R. Overbeek, M. Fonstein, M. D'Souza *et al.*, 'The use of gene clusters to infer functional coupling.' *Proc. Nat. Acad. Sci. USA*, 1999, **96**, 2896–2901.

46. M. Y. Galperin and E. V. Koonin, 'Who's your neighbor? New computational approaches for functional genomics.' *Nat. Biotechnol.*, 2000, **18**, 609–613.

47. E. M. Marcotte, M. Pellegrini, M. J. Thompson *et al.*, 'A combined algorithm for genome-wide prediction of protein function.' *Nature*, 1999, **402**, 83–86.

48. M. B. Yaffe, G. G. Leparc, J. Lai *et al.*, 'A motif-based profile scanning approach for genome-wide prediction of signaling pathways.' *Nat. Biotechnol.*, 2001, **19**, 348–353.

49. E. S. Lander, L. M. Linton, B. Birren *et al.*, 'Initial sequencing and analysis of the human genome.' *Nature*, 2001, **409**, 860–921.

50. R. Tupler, G. Perini and M. R. Green, 'Expressing the human genome.' *Nature*, 2001, **409**, 832–833.

51. D. D. Shoemaker, E. E. Schadt, C. D. Armour *et al.*, 'Experimental annotation of the human genome using microarray technology.' *Nature*, 2001, **409**, 922–927.

52. R. Sorek and M. Amitai, 'Piecing together the significance of splicing.' *Nat. Biotechnol.*, 2001, **19**, 196.

53. R. D. Wood, M. Mitchell, J. Sgouros *et al.*, 'Human DNA repair genes.' *Science*, 2001, **291**, 1284–1289.

54. L. Aravind, V. M. Dixit and E. V. Koonin, 'Apoptotic molecular machinery: vastly increased complexity in vertebrates revealed by genome comparisons.' *Science*, 2001, **291**, 1279–1284.

55. G. G. Loots, R. M. Locksley, C. M. Blankespoor *et al.*, 'Identification of a coordinate regulator of interleukins 4, 13, and 5 by cross-species sequence comparisons.' *Science*, 2000, **288**, 136–140.

56. A. Brazma, I. Jonassen, J. Vilo *et al.*, 'Predicting gene regulatory elements in silico on a genomic scale.' *Genome. Res.*, 1998, **8**, 1202–1215.

57. C. Gibas and P. Jambeck, 'Developing Bioinformatics Computer Skills,' O'Reily, Cambridge, MA, 2001.

58. L. Stein, 'How PERL saved the human genome project.' *Perl J.*, 1996, **1**, 8–13.

59. A. Rotig, I. Valnot, C. Mugnier *et al.*, 'Screening human EST database for identification of candidate genes in respiratory chain deficiency.' *Mol. Genet. Metab.*, 2000, **69**, 223–232.

60. J. B. Hogenesch, W. K. Chan, V. H. Jackiw *et al.*, 'Characterization of a subset of the basic-helix-loop-helix-PAS superfamily that interacts with components of the dioxin signaling pathway.' *J. Biol. Chem.*, 1997, **272**, 8581–8593.

61. D. Zhou, K. Salnikow and M. Costa, 'Cap43, a novel gene specifically induced by Ni2+ compounds.' *Cancer Res.*, 1998, **58**, 2182–2189.

62. B. N. Van, W. N. Dinjens, M. P. Diesveld *et al.*, 'A novel gene which is up-regulated during colon epithelial cell differentiation and down-regulated in colorectal neoplasms.' *Lab. Invest.*, 1997, **77**, 85–92.

63. K. Kokame, H. Kato and T. Miyata, 'Homocysteine-respondent genes in vascular endothelial cells identified by differential display analysis. GRP78/BiP and novel genes.' *J. Biol. Chem.*, 1996, **271**, 29659–29665.

64. M. Lara-Tejero and J. E. Galan, 'A bacterial toxin that controls cell cycle progression as a deoxyribonuclease I-like protein.' *Science*, 2000, **290**, 354–357.

65. T. G. Petrakis, E. Ktistaki, L. Wang *et al.*, 'Cloning and characterization of mouse deoxyguanosine kinase. Evidence for a cytoplasmic isoform.' *J. Biol. Chem.*, 1999, **274**, 24726–24730.

66. J. F. Gibrat, T. Madej and S. H. Bryant, 'Surprising similarities in structure comparison.' *Curr. Opin. Struct. Biol.*, 1996, **6**, 377–385.

67. D. S. Latchman, 'Eukaryotic Transcription Factors,' Academic Press, New York, 1991.

68. D. A. Spandidos, R. A. Nichols, N. M. Wilkie *et al.*, 'Phorbol ester-responsive H-ras1 gene promoter contains multiple TPA-inducible/AP-1-binding consensus sequence elements.' *FEBS Lett.*, 1988, **240**, 191–195.

69. D. Ghosh, 'A relational database of transcription factors.' *Nucleic Acids Res.*, 1990, **18**, 1749–1756.

70. T. Heinemeyer, E. Wingender, I. Reuter *et al.*, 'Databases on transcriptional regulation: TRANSFAC, TRRD and COMPEL.' *Nucleic Acids Res.*, 1998, **26**, 362–367.

71. R. C. Perier, V. Praz, T. Junier *et al.*, 'The eukaryotic promoter database (EPD).' *Nucleic Acids Res.*, 2000, **28**, 302–303.

72. D. S. Prestridge and G. Stormo, 'SIGNAL SCAN 3.0: new database and program features.' *Comput. Appl. Biosci.*, 1993, **9**, 113–115.

73. K. Frech, K. Quandt and T. Werner, 'Software for the analysis of DNA sequence elements of transcription.' *Comput. Appl. Biosci.*, 1997, **13**, 89–97.

74. W. M. Fitch and E. Margoliash, 'Construction of phylogenetic trees.' *Science*, 1967, **155**, 279–284.

75. D. W. Nebert, M. Adesnik, M. J. Coon *et al.*, 'The P450 gene superfamily: recommended nomenclature.' *DNA*, 1987, **6**, 1–11.

76. D. W. Nebert, D. R. Nelson, M. Adesnik *et al.*, 'The P450 superfamily: updated listing of all genes and recommended nomenclature for the chromosomal loci.' *DNA*, 1989, **8**, 1–13.

77. D. R. Nelson, T. Kamataki, D. J. Waxman *et al.*, 'The P450 Superfamily: Update on new sequences, gene mapping, accession numbers, early trivial names of enzymes, and nomenclature.' *DNA Cell. Biol.*, 1993, **12**, 1–51.

78. D. R. Nelson, L. Koymans, T. Kamataki *et al.*, 'P450 superfamily: update on new sequences, gene mapping, accession numbers and nomenclature.' *Pharmacogenetics*, 1996, **6**, 1–42.

79. C. V. Jongeneel, 'The need for a human gene index.' *Bioinformatics*, 2000, **16**, 1059–1061.

80. J. White, H. Wain, E. Bruford *et al.*, 'Promoting a standard nomenclature for genes and proteins.' *Nature*, 1999, **402**, 347.

81. H. M. Wain, E. Bruford, A. Duncanson *et al.*, 'Nomenclature: Genes, Weights and Measures, Animals, Elements, and Planets.' *Radiat. Res.*, 2000, **154**, 1–2.

82. M. Ashburner, C. A. Ball, J. A. Blake *et al.*, 'Gene ontology: tool for the unification of biology. The Gene Ontology Consortium.' *Nat. Genet.*, 2000, **25**, 25–29.

83. W. Makalowski, J. Zhang and M. S. Boguski, 'Comparative analysis of 1196 orthologous mouse and human full-length mRNA and protein sequences.' *Genome. Res.*, 1996, **6**, 846–857.

84. W. Makalowski and M. S. Boguski, 'Evolutionary parameters of the transcribed mammalian genome: an analysis of 2820 orthologous rodent and human sequences.' *Proc. Nat. Acad. Sci. USA*, 1998, **95**, 9407–9412.

J.P. Vanden Heuvel, G.H. Perdew, W.B. Mattes and W.F. Greenlee (Eds.)
Comprehensive Toxicology, Vol. xiv

# 14.5.3
# Interpretation of Toxicogenomic Data using Genetically-Altered Mice

J. CHRISTOPHER CORTON,  LESLIE RECIO
*CIIT Centers for Health Research, Research Triangle Park, NC, USA*

STEVEN P. ANDERSON
*GlaxoSmithKline Research and Development, Research Triangle Park, NC, USA*

## 14.5.3.1   INTRODUCTION

There is intense interest in the toxicological community to use functional genomics approaches to better understand chemical-induced toxicity. Fueled by massive sequencing efforts that have resulted in completion of the genomes of humans and a growing number of model organisms, evolving techniques are allowing for the simultaneous analysis of messenger RNA (mRNA) levels of hundreds or thousands genes. These techniques include DNA arrays, created by printing or synthesizing DNA fragments complimentary to specific genes onto silicon, glass, or nylon.[1] These arrays are probed with the labeled cDNA from chemically treated tissue to determine genome-wide mRNA levels (sometimes called transcriptomics). The global analyses of levels of proteins or endogenous and exogenous metabolites (referred to as proteomics[2,3] and metabonomics,[4] respectively) is lagging somewhat behind transcriptomics, but these approaches will be increasingly used by the toxicologist. The data generated from these functional genomics approaches and interpreted with the help of bioinformatics tools will be used to integrate the changes observed at the tissue, cellular and molecular levels of organization to create comprehensive biologically-based models of chemical action.

Toxicologists working in the area of functional genomics have created an exciting new scientific discipline many have called toxicogenomics.[5] Toxicogenomics is used to understand the toxicological significance of chemical exposure by studying perturbations in cellular components on a genome-wide scale. This application of functional genomics approaches is now allowing us an unprecedented view of the global changes that occur after chemical exposure. In the past toxicologists viewed the biological world from the perspective of one gene or signal transduction pathway thought to be relevant to toxicity. We are now beginning a new phase of scientific discovery in which the toxicogenomicist has the opportunity to view the effect of chemical exposure on all genes within a genome. As chemicals can induce both adverse as well as adaptive effects that allow the organism to cope with the chemical stressor, the challenge for the toxicogenomicist now is to correctly interpret the large number of genomic changes anticipated after chemical exposure.

This review will specifically focus on how genetically-altered mice can be used to help the toxicogenomicist interpret the biological significance of global changes in cellular components that occur after chemical exposure. We will describe the types of genetically-altered mice available to the toxicogenomicist and will end with a brief discussion of the prospects for a coordinated and systematic effort to increase the speed, accuracy and detail of molecular mechanisms of chemical action.

## 14.5.3.2 STRATEGIES FOR INTERPRETATION OF GLOBAL CHANGES IN COMPONENTS OF THE GENOME

The universality of the genetic code, the high degree of conservation of key biochemical reactions, the evolutionary conservation of key developmental processes all emphasize the close relationship of humans to species that are morphologically quite distinct and evolutionarily distantly related.[6] In theory, primates should provide the best animal models of human disease because they are so closely related to us. However, because they are comparatively long lived and have less fecundity than rodents, breeding experiments are more difficult to organize; thus, primates are not well suited to experimentation. Instead, mice have been the most widely used animal models of human disease because of their short lifespan (2–3 years), short generation time (90 days), and high fecundity. Our genetic knowledge of the laboratory mouse is secondary only to our knowledge of our own genetics, and the phenotypes of many mouse mutants have been characterized[7] (http://www.informatics.jax.-org/). A general comparison of the mouse and human genomes reveals similar gene numbers and genome size in mice and man.[6] Gene family organization is very similar for individual families, although recent duplications have led to differences in gene numbers within these families. At the DNA level, coding sequences are generally 70–90% homologous; polypeptides are even more similar (75–95%). Likewise, gene expression is very similar between the two species, although for orthologous genes, there are sometimes differences in the choice of alternative promoters and patterns of splicing, or in imprinting. Almost all genes in humans have functional homologs in the mouse. In many cases, these homologs are located in syntenic regions of the corresponding mouse chromosome.

The power of mouse genetics for functional studies lies in the multitude of sophisticated tools that are available to the geneticist. From replacing one nucleotide in a specific gene to replacing whole chromosomes, current technologies make it possible to engineer almost any genetic change in mice. Specific genes are routinely inactivated by homologous recombination in embryonic stem cells. At the opposite end of the spectrum, transgenic mice are generated by inserting functional mouse or human genes into the mouse genome. These technologies for generating novel phenotypes, complement, rather than supersede, the more classical genetic approaches such as generating mutants by exposing mice to chemical mutagens. With the end of the mouse genome sequencing project in sight, these approaches will bring the mouse to the forefront of experimental genetics and functional genomics and provide the proper tools for understanding mechanisms of toxicity.

While genetically engineered mice are already being exploited by toxicologists, there is a growing hope that genetically engineered mice can be useful in identifying consistent patterns of gene expression associated with specific types of toxic insult. Perhaps in the not-so-distant future, we will become sufficiently proficient in integrating this type of data as to allow us to "read the transcriptome" in a manner analogous to a pathologist's interpreting lesions on a microscopic slide. Our ability to do so will largely depend upon our ability to separate adaptive (those not directly related to the mechanism of toxicity) from maladaptive (toxic) responses. One example would be determining that changes in the protein products of the genes mediating the hypolipidemic effects of peroxisome proliferator drugs are not mechanistically responsible for the hepatocellular mitogenesis driving the hepatic carcinogenicity of these chemicals. At present, our ability to separate "cause" from "effect" is woefully inadequate, but linking gene expression fingerprints to toxic mechanisms promises to greatly facilitate this process.

Generating genetically altered mice with either increased resistance or sensitivity to chemical exposure can help to interpret changes observed by genome-wide expression scanning. Figure 1 illustrates the relationship between dose and time of chemical exposure and phenotypic changes associated with exposing wild type and genetically altered mice to a hypothetical chemical. There are several imaginable scenarios compared to the wild type condition, ranging from complete insensitivity to hypersensitivity. Adaptive responses may parallel, or may occur independently of toxic responses. Genes that are linked to toxicity in the wild-type mice can be confirmed in chemically-exposed genetically-altered mice by further associating those expression changes with parallel increases or decreases in toxicity.

One of the most promising uses for genetically-altered mice is to test hypotheses of the role of specific pathways or genes in mediating chemical toxicity from patterns of chemically-induced gene expression changes (Figure 2). Bioinformatics tools can be used to group genes into clusters based on their coordinated regulation over time or dose.[8,9] The clustered genes most closely linked to toxicity can be used to define the mechanism of chemical action in two ways.

Clusters of genes can be used to make predictions of phenotypic changes that may be linked to toxicity. Work in a number of models has shown that coordinately regulated genes are often playing similar functional roles in the cell, e.g., alteration in components of a metabolic pathway. This is especially true with yeast in which transcription is tightly regulated (e.g., reference 10). The known function of genes that cluster with toxicity could be used to predict phenotypic changes occurring during chemical exposure linked to or a manifestation of toxicity. In this context clustering approaches have been used to predict the function of uncharacterized genes that fall into the same functional cluster.[11]

Cluster analysis can also be used to identify early events that precede altered gene expression. The nucleotide sequence of the promoters of the clustered genes can be compared to identify common sequences that may determine coordinate regulation.[12,13] A shared sequence can be compared to previously identified response elements recognized by one or more transcription factors. If the sequence is not characterized, *in vitro* approaches such as electrophoretic mobility shift assays or DNA affinity purification can be used to identify transcription factors that recognize the sequence. Once a transcription factor controlling response genes has been identified, events leading to activation or repression of that transcription factor can be characterized. This approach has been effective in identifying response elements that regulate batteries of yeast genes.[14] This approach will be increasingly used to identify response elements and associated signal transduction pathways that mediate toxicity when the mouse genome sequence is completed.

Hypotheses generated from clustering approaches can be tested by further iterations of expression scanning in appropriate genetically-altered mice. For example, repeating the chemical exposure in mice that lack a key transcription factor involved in chemical responses may reveal either increased or decreased toxicity, depending on the role of that transcription factor in either damage or repair, respectively. As the number of genetically-altered mouse strains available to the researcher increases, this type of analysis in which appropriate mutant mice are used in successive rounds of refining chemical mechanisms will be increasingly used.

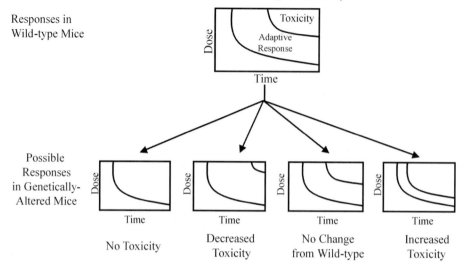

**Figure 1** Altered responsiveness to chemical toxicity in genetically-altered mice. The relationship between dose and time of chemical exposure and two phenotypic responses is illustrated as a two-dimensional landscape for a hypothetical chemical. Gene expression changes that occur over a range of doses or times of chemical exposure could be used to make correlations between phenotypic changes and altered expression of batteries of genes. Confirmation that those genes are clearly associated with chemically-induced toxicity can be made in genetically-altered mice that exhibit differences in responsiveness to chemical exposure. If genes that were identified in the wild-type experiments associated with toxicity are truly linked to toxicity, expression of those genes should track with the altered toxicity exhibited in the genetically-altered mice. This type of approach of following up observations in genetically-altered mice allows definition of genes associated with toxicity as well as those genes more closely associated with adaptive responses.

### 14.5.3.3  TYPES OF GENETICALLY-ALTERED MICE AVAILABLE TO THE TOXICOGENOMICIST

#### 14.5.3.3.1  Transgenic Mice

Transgenic mice contain a foreign gene, the transgene, typically introduced into the mouse germ line by microinjection of DNA into fertilized mouse eggs. The transgene usually consists of a reporter gene or a cDNA under control of a general or tissue-specific promoter.[15] The DNA is integrated randomly into chromosomal DNA, usually at a single site, although rarely two sites of integration are found in a single animal. The level of transgene expression depends on the site of insertion, so the tissue distribution of expression will vary among mice originating from different insertion events. Mice generated using this

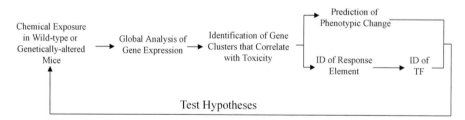

**Figure 2** Coupling genetically-altered mice with global analysis of gene expression to increase mechanistic understanding of chemical action. Using bioinformatics approaches global analysis of gene expression in chemically exposed wild-type mice will lead to the identification of clusters of genes that correlate with toxicity. The known or putative function of those gene products can lead to a prediction of phenotypic changes associated with chemical toxicity. Comparison of the sequence of the promoter regions of the clustered genes can be used to identify common response elements found within the promoter elements that determine coordinate expression. Candidate response elements can be tested for functional significance by assessing interaction with transcription factors that coordinate the expression of those clustered genes. Genetically-altered mice can be used to test hypotheses of the functional significance of the transcription factors in regulating the clustered genes or in gene products that determine phenotypic responses linked to toxicity.

approach most frequently harbor gain of function mutations leading to either constitutive or tissue-specific overexpression. In other cases, transgenic mice have been created which express dominant negative forms of proteins that abolish function of the wild-type protein. Other transgene constructs express ribozymes or cDNA in the antisense direction that abolish expression of the endogenous wild-type protein. Finally, of particular interest to most toxicologists, it is also possible to generate transgenic rats.[16]

Inconveniently, constitutive expression of some transgenes in neonatal mice kills the developing embryo prior to parturition. One way to circumvent this is to induce the gene post-natally, in a reversible fashion if need be, allowing the mouse to develop to term without expressing the fatal transcript. An advantage to this protocol is that each individual mouse serves as its own "control" animal, because the animal can be studied both before and after induction of the gene of interest. Mice that express a gene in a tissue-specific and inducible manner have been created using the binary tetracycline regulatory system.[17] The reversed tetracycline-controlled transactivator (rtTA) is a fusion protein composed of a mutant version of the Tn10 tetracycline repressor from *Escherichia coli* fused to a C-terminal portion of protein 16 from the Herpes Simplex virus that

functions as a strong transcriptional activator (Figure 3). When the rtTA gene under control of a tissue-specific promoter is expressed in the presence of the antibiotic doxycycline (but not in its absence), the rtTA protein binds to the tetracycline operator (tetO) controlling expression of the inducible transgene and activates transcription of the open reading frame from a minimal human cytomegalovirus promoter. Thus when doxycycline is added to the drinking water of the mice, the open reading frame will be expressed only in those cells in which rtTA is expressed by means of the tissue-specific promoter.

Conversely, it is also possible to shut down the otherwise constitutive expression of a transgene using the tetracycline-controlled transactivator (tTA) system.[17] In this system, the transgene and the tetracycline repressor are under control of a tissue-specific promoter on two separate constructs expressed in the same cell-types. The transgene promoter contains a tetO engineered in a location that when bound by the tetracycline receptor will block expression of the transgene. In the absence of doxycycline the tetracycline repressor is expressed but does not bind to the tetO, allowing expression of the transgene. In the presence of doxycycline the tetracycline repressor binds to the tetO, shutting down transgene expression.

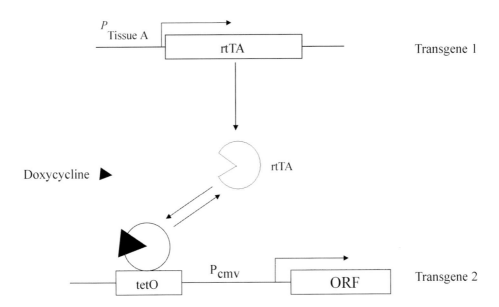

**Figure 3** Inducible expression of a transgene using the reverse tetracycline controlled transactivator (rtTA) system. The rtTA system consists of a fusion protein composed of a mutant version of the Tn10 tetracycline repressor fused to the transcriptional activation domain of protein 16 from Herpes Simplex Virus. The rtTA gene is placed under control of a tissue-specific promoter (transgene 1). When the rtTA gene is expressed in the presence of the antibiotic doxycycline, the rtTA protein binds to the tetracycline operator (tetO) controlling expression of the open reading frame (ORF; transgene 2) leading to expression from a minimal human cytomegalovirus ($P_{cmv}$).

#### 14.5.3.3.2  Targeted Mouse Mutants

Microinjecting DNA into fertilized oocytes is technically difficult and not suitable for production on an industrial scale. As an alternative, foreign DNA can be transferred into embryonic stem cells (ES cells). ES cells are derived from 3.5 to 4.5 day postcoitum embryos and arise from the inner cell mass of the blastocyst. A major advantage of using ES cells is that they can be maintained as undifferentiated pluripotent cells in culture indefinitely. A number of strategies are available for creating genomic alterations in these cells. When genetically-altered ES cells are injected into a host blastocyst, they have the capacity to contribute to all tissues, including the germline, of the developing mouse, where they can transmit the foreign gene to their progeny. The total number of gene knockouts that have been made to date is approaching 5000 with the number of laboratories applying this technology steadily rising. This large number of mouse mutants has provided valuable information about the role of the corresponding genes in development and adult homeostasis.

Certainly, the most powerful genetic use of ES cells is in gene targeting.[18,19] In this approach, targeting vectors are constructed that are specifically integrated at a desired genomic location by homologous recombination in embryonic stem cells (Figure 4). These cells are subsequently injected into blastocysts where they contribute to the developing embryo. Chimeric mice are back-crossed to wild-type mice to obtain heterozygotes that are mated to each other to obtain homozygote

knockouts. The gene target is usually disrupted by introduction of a neomycin resistance marker within the coding region of the gene resulting in deletion of part of the gene. Mouse mutations may be studied in the heterozygous state (to assess potential gene dosage effects) or the homozygous state (for analysis of the null phenotype). The phenotypic changes observed in these knockout mice are expected to provide information on the function of the respective genes in wild-type animals.

Targeted gene inactivation in mice has significantly accelerated our ability to determine the role of the gene products in normal physiology. At the same time, these types of manipulations have also demonstrated the extreme complexity of genetic regulation and gene function. Targeted inactivation of many genes results in no recognizable alteration in phenotype. On the simplest level, there are three possible explanations for this: (1) genes that belong to a family or are involved in parallel pathways impacting the same cellular function are often functionally redundant; (2) many genes have multiple functions; and (3) genes do not act in isolation; susceptibility to a given toxic insult may depend on the presence of several susceptibility alleles, and the variation in alleles at one locus might affect the trait only if there is a certain allele at another locus. Thus, inactivating a gene does not always reveal hints as to its function. In addition, several other complications may arise, including an early embryonically lethal phenotype that can prevent the study of the role of the gene product in the adult. The phenotype can also affect multiple cellular lineages making it difficult to determine functions in different cell

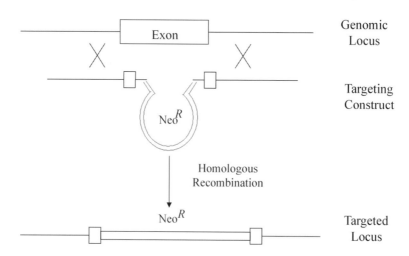

**Figure 4**  Functional inactivation of genes by homologous recombination. Targeting constructs are made in which some or all of the gene's exons are deleted and replaced with the neomycin resistance cassette (NeoR). The construct is transfected into embryonic stem cells and cells resistant to neomycin are selected. Cells which carry a disrupted targeted locus are identified and used to make chimeric mice.

types or tissues. Fortunately, we now have at our disposal a battery of well-characterized tools to circumvent these problems by allowing spatial and temporal restrictions to the genome alterations.

### 14.5.3.3.3 Conditional Knockouts

The most popular concept of a conditional gene knockout is the strategy that combines homologous and site-specific recombination.[20–22] Enzymes known as site-specific recombinases recognize specific DNA recombination sites and either delete or invert the intervening sequences depending on the orientation of the sites. The Cre recombinase of the P1 bacteriophage and the FLP recombinase of yeast have been the choice for experiments in mice because they require only specific short (34 bp) consensus sequences to catalyze recombination.

The general strategy for conditional knockouts is to clone two recognition sites for a site-specific recombinase flanking an essential exon in a way that does not alter the normal function of the gene (Figure 5). Mice carrying this altered allele in homozygous form (floxed mice; *flanked by lox* sites) can then be established without complications of an altered phenotype. To create the conditional knockout, mice possessing these recombinase sites are mated with mice that express the recombinase in a tissue-specific manner. Recombination will occur only in those tissues that express

the recombinase. A drawback here is that the gene deletion could still occur in the developing embryo and if essential for development, will prevent establishing lines of tissue-specific null mice. There are now a growing number of commercially available mouse strains which express the Cre recombinase in a tissue-specific manner. These commercially available mice can be used in conjunction with a mouse line that possesses a floxed gene of interest to make a number of tissue-specific knockouts.

The ability of the Cre recombinase to excise DNA can also be used to turn on a foreign gene.[22] As in the last example, one mouse strain is a conventional transgenic mouse line with Cre targeted to a tissue or cell type. The second mouse strain in this case possesses a transgene whose 5' regulatory elements controlling expression are separated from the coding region with a floxed stop sequence. Recombination, resulting in excision of the stop signal, occurs only in those cells expressing the Cre recombinase, which allows the transgene to be transcribed. In other cell types not expressing Cre the transgene remains inactive.

In addition to creating knockouts in essentially any cell type or tissue, a knockout can also be designed to be inducible at essentially any desired time point throughout development or in the adult. The inducible knockout strategy utilizes a combination of the Cre technology and the rtTA system discussed above. In order to make the knockout inducible, a Cre recombinase transgene is expressed

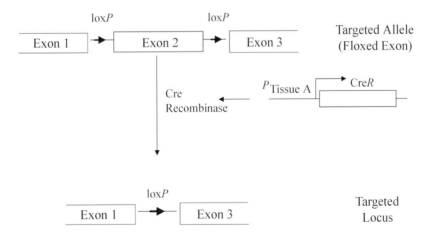

**Figure 5** Tissue-specific functional inactivation of genes using the Cre-lox system. Targeted homologous recombination described in Figure 4 is used to create a targeted allele containing sites recognized by the Cre-recombinase called loxP sites. The loxP sites are usually engineered into introns where they leave intact the normal function of the gene and flank the exon to be deleted said to be floxed (*flanked by lox*P sites). Mice containing these targeted genes are mated with mice that express the Cre recombinase expressed from a tissue-specific promoter (e.g., tissue A). Only in those cells expressing the Cre recombinase will recombination between the loxP sites occur resulting in deletion of the exon.

under the control of the rtTA system (Figure 6). A tissue-specific promoter drives rtTA expression and upon addition of doxycycline, rtTA will bind to a tetO sequence within the Cre recombinase promoter resulting in expression. In those cells in which rtTA and Cre are expressed, the gene will undergo recombination and inactivation in the presence of doxycyclin.

### 14.5.3.3.4   Systematic Inactivation of Gene Function in the Mouse

Possessing a knockout for every gene, together with a unique sequence identifier will be ideal for many biologists. On a large scale, gene trap mutagenesis carried out by a number of organizations hopes to achieve this aim.[23] The gene to be trapped inactivates endogenous genes and consists of a drug resistance marker gene (e.g., puromycin *N*-acetyl transferase, PURO) under control of a phosphoglycerase kinase (PGK) promoter. Instead of carrying a polyA addition signal, the transferase gene contains a consensus exon donor site at the end of its coding region. After transfection of ES cells, the DNA construct integrates into a gene at a site that contains at least one downstream exon and polyA addition site, producing a functional mRNA encoding the transferase. Growth of ES cells in the presence of puromycin selects for vector insertion into genes. The random amplification of cDNA

ends (RACE) strategy allows for isolation of sequences from the gene into which the DNA was inserted. Lexicon (Woodlands, TX) has a growing library of these sequence tag sites referred to as Omnibank sequence tags (OST) which are maintained in a proprietary database. This database of tagged and inactivated genes will enable customers to purchase the embryonic stem cell clone carrying the desired gene disruption obviating the need to genetically engineer the disrupted gene. The ES cell clone can then immediately be injected into blastocytes, significantly reducing the time to obtain mutant null mice. The German Human Genome Project is funding a similar ES cell insertional mutagenesis program at GSF (Neuherberg, Germany). Web sites at each of these locations have searchable databases to find genes that have been disrupted.

### 14.5.3.3.5   Phenotype-Driven Mutagenesis Strategies in the Mouse

The "reverse genetics" (i.e., moving from genotype to phenotype) approaches discussed above will allow the systematic production of mouse mutants for any gene that has been cloned. Complementary to this are more traditional "forward genetics" approaches, where the investigator first generates a "phenotype" (mutant mouse) and then proceeds to identify the genotypic alteration that underlies

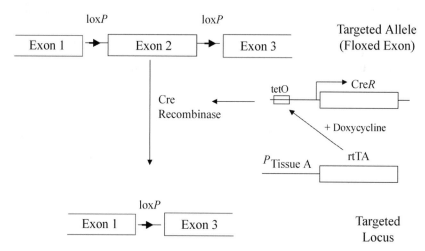

**Figure 6**   The inducible Cre-lox system. Transgenic mice are created which express the reverse tetracycline controlled transactivator (rtTA) protein from a tissue-specific promoter and a Cre recombinase gene under control of the tetracycline operator. These mice are mated to mice that contain an exon flanked by loxP sites in a targeted gene. When mice are exposed to doxycycline, the rtTA protein binds to the tetO resulting in expression of the Cre recombinase. The recombinase then removes sequences between the loxP sites of the targeted locus. Recombination only occurs in those cells which express the rtTA and Cre recombinase proteins and are assessable to doxycycline.

that phenotype. The study of mouse mutants that develop diseases similar to those in humans but have mutations in different genes is essential for our understanding of the molecular mechanisms involved in the pathogenesis of multigenic diseases. The full power of genetic analysis of gene function requires the availability of multiple alleles of the same gene, including alleles of different strength and gain of functional alleles. Many of the clinically relevant human diseases are the result of a partial, but not complete loss of gene function.

A number of large-scale systematic screening efforts to identify mouse mutants by chemical mutagenesis is now being carried out in a number of laboratories.[24] The alkylating agent ethyl nitrosourea (ENU) is currently the most powerful mutagen used to produce germline mutations in mice. The mutations recovered after ENU mutagenesis are mainly point mutations and/or small intragenic lesions resulting in hypomorphic (partial loss of function), gain of function, as well as complete loss of function mutations. Protocols are available that allow very efficient germ cell mutagenesis in mice. The frequency of mutant recovery is about 1/1000 for a specific locus that can be scored phenotypically. ENU mutagenizes premieotic granule stem cells allowing for the production of a large number of F1 founder animals from a single treated male, minimizing the number of animals required and the handing of ENU. A drawback of this approach is that mutations induced by ENU will not be tagged molecularly. Although this is a disadvantage with respect to the cloning of the responsible genes, the availability of point mutations will be important for a more detailed functional analysis of many genes. It is anticipated that the advances currently made in the field of genomics, particularly the production of high-resolution genetic, physical, and transcript maps will reduce the difficulties inherent in cloning genes mutated after ENU treatment.

One approach for an ENU phenotype-based mutagenesis experiment is to screen the entire genome for a new dominant or recessive mutation.[25] Mutagenized males are crossed with wild-type partners and the offspring from this cross are evaluated for new dominant phenotypes. To detect new recessive phenotypes, it is necessary to make the offspring homozygous for the mutations by either intercrossing the F1 offspring or backcrossing F1 females with their mutagenized male parent. In addition, each recovered mutation can be mapped using standard mitotic procedures. These mating steps, and the screening of

prodigy for mutant phenotypes, are required for each step, involve a considerable effort, and a substantial commitment to the maintenance of large breeding colonies of mice. Nonetheless, a genomic scan for new ENU-induced mutations is considered to be the best approach for screening for a single phenotype throughout the genome. These ENU mutagenesis protocols will likely lead to a comprehensive database of phenotypic abnormalities resulting from single or multiple point mutations. This information has the potential of being extremely valuable to the toxicologist in allowing a systematic search for phenotypes in the ENU mutagenized mice that mimic those caused by chemical toxicity. Subsequent mapping experiments could allow identification of key genes whose alteration in function results in the chemical-induced phenotype.

Unless a toxic insult is so egregious that it leads to acute death, different strains of mice almost always respond differently to a given toxic insult. Researchers can take advantage of this genetic variability by surveying susceptibility in various strains of mice and, through selective breeding, identify discretely measurable (i.e., quantitative or digital traits), rather than qualitative (analog, or continuously varying) phenotypes. After identifying susceptible and resistant strains the next step is to map the genetic loci that segregate with the phenotype. In mice, there are several recombinant inbred (RI) strains available that have been developed by crossing two standard inbred strains, then brother × sister mating for 20+ generations to produce a set of several new strains in which the parental genes have recombined.[26] After testing susceptibility to a given toxin (i.e., phenotyping) for a subset containing many RI strains it may be possible to identify genetic markers associated with this trait, thereby mapping some of the "quantitative trait loci" (QTLs) for a given trait. A significant disadvantage of this approach is the large number of animals that must be crossed and phenotyped in order to narrow a susceptibility locus down to a reasonably small region of a chromosome. Furthermore, suitable RI strains are not always available. Nonetheless, these techniques have proven themselves extremely powerful, even if they are cumbersome and expensive. For example, between 10 and 12 loci that appear to be associated with lung tumor susceptibility in the mouse have been mapped.[27] It is likely that the time to identify candidate disease genes will be significantly shortened by using a combination of expression profiling and linkage of the trait to a rapidly growing database of single nucleotide polymorphisms (SNP).[28]

### 14.5.3.4   FUTURE PROSPECTS FOR THE TOXICOGENOMICIST

In the not too distant future, it may be possible for toxicologists to use two types of information derived from traditional toxicity tests to identify candidate target genes responsible for chemical-induced phenotypes (Figure 7). First, gene expression profiles derived from appropriate tissues of chemically-exposed wild-type or genetically-altered mice could theoretically be compared to gene expression databases from mice exposed to well-characterized chemicals either or from tissues from untreated genetically-altered mice. This approach could result in the identification of not only what chemical class the unknown chemical falls into, but the specific pathway that the chemical alters. Second, a set of phenotypic changes induced by the chemical could be queries against phenotypes from characterized mouse mutants. These two types of information will be described in greater detail below.

One of the most intriguing prospects for the use of genomic information in toxicology is the creation of databases useful for categorizing chemicals according to mechanism of action. Comparison of gene expression profiles of individual chemicals from many mechanism of action classes (e.g., peroxisome proliferators, estrogenic chemicals, cytotoxic chemicals)

will allow the identification of common sets of genes whose expression is consistently linked to particular disease outcomes. When the database becomes sufficiently comprehensive, transcript profiles of new chemicals with unknown properties could be compared to profiles in databases allowing new chemicals to be provisionally placed into one or more mechanism-of-action classes. Additional more directed studies could be undertaken to confirm or refute the predicted mechanism-of-action of the new chemical. This type of analysis could potentially simplify the tests needed to characterize toxicity and could substantially reduce time and resources needed to determine the potential toxicity of each new chemical. A substantial matrix of data on many chemicals with known exposure-disease outcomes will need to be obtained to maximize the likelihood of detecting true positives and minimize false negatives. This will require the evaluation of gene expression profiles of structurally-related chemicals not causing disease, as well as those known to cause disease with varying potency. This approach to predicting toxicity has been recently discussed.[5,29]

Beyond their use to predict chemical toxicity, databases of genomic information might someday be used to identify specific targets of chemicals allowing for precise determination of mechanism-of-action. Indications that this is a

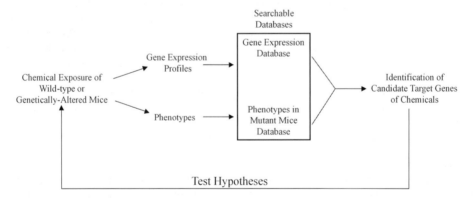

**Figure 7**   Accelerating the identification of targets of toxicants using genetically-altered mice. In the near future, two types of searchable databases will aid the toxicologist in understanding the mechanism of chemical-induced effects. One is a gene expression database that will be maximally effective when it contains two types of information: gene expression changes that occur after exposure to diverse chemicals from different mechanism of action classes (e.g., peroxisome proliferators, estrogenic chemicals, cytotoxic chemicals, *etc.*) and gene expression differences between wild-type and knockout mice. Gene expression in knockout mice will reveal the genetic networks that are perturbed in the absence of the functional protein. The second type of database useful for understanding chemically induced toxicity is one in which various altered phenotypes in mutant mice have been annotated in a database that is directly searchable. This type of database is currently being constructed at a number of different institutions. Starting with conventional toxicity experiments, gene expression profiles and altered phenotypes in target tissues can be compared to databases of gene expression and phenotypes leading to identification of candidate target genes that may mediate the effects of a chemical. The target genes hypothesized to be involved in chemical action can then be tested in appropriate genetically-altered mice.

feasible goal came from recent work carried out by Friend and colleagues who constructed a compendium of expression profiles from approximately 300 mutants and chemically-treated strains of yeast.[11] An important finding from these studies is that the expression profiles obtained from chemical or drug-treated cells are similar to profiles from strains in which the genes encoding targets of the chemicals are mutated. This indicates that the cell responds to the two disparate perturbations (i.e., inactivation of the gene or the protein product of the gene) by regulating the same set of genes. This type of strategy could theoretically be used to create a database of gene expression profiles in toxicologically important tissues from genetically altered mice. Once databases are sufficiently robust it may be possible for toxicologists to compare an expression profile of a chemical under scrutiny to hundreds or thousands of these profiles in web-based databases to identify likely gene or pathway targets that could be confirmed experimentally. Although many years off, this type of analysis may revolutionize toxicology by shifting research from a chemical category-based orientation to one in which mechanism of action can be precisely defined.

There is also hope that the spectrum of adverse or adaptive phenotypic responses observed in chemically-treated mice could be screened against databases of mouse genetic and phenotypic data. An important mouse database is the Mouse Genome Database housed at the Jackson Laboratory which includes a mouse locus catalog describing existing mouse mutants and extensive map information describing their locations. To simplify the search for mutant mice, the International Mouse Strain Resource has recently been developed which is mirrored at two web sites, the Jackson Laboratory and the MRC, Harwell. A great variety of phenotype data are likely to accumulate on large numbers of mutant mouse strains creating the need for phenotype databases that can be linked with mapping, mutagenesis, and gene expression databases.

To provide a genetic resource to the toxicology community, mutant mice must be readily available. Although many mutant stocks are available from commercial vendors or from the Jackson Laboratory, additional distribution centers are required. To meet this demand, the National Institute of Environmental Health Sciences (NIEHS) has funded Mutant Mouse Resource Centers at a number of locations in the United States to serve as stock archives and regional distribution centers. The European Mutant Mouse Archive (EMMA) consists of a number of mutant mouse repositories.

## 14.5.3.5 SUMMARY

Transgenic mice are proving to be powerful tools in mechanistically-based toxicology research. Through the inactivation of key genes by specific targeting, the role of a specific gene in regulating cellular responses to a wide variety of stressors can be assessed in the intact animal given toxicologically relevant exposures. Transgenic animals have been used to define genes that have a central role in the metabolic activation/detoxication of xenobiotics, regulate cellular responses to chemical carcinogens, and mediate the spectrum of biological responses induced by ligand-receptor interactions.[30] The incorporation of recoverable target genes in the genome of mice has been used to examine tissue specific mechanisms of mutagenicity that can be involved in tumor development.[31] Transgenic mice with altered oncogenes or tumor suppressor genes are being tested for being appropriate short-term cancer models.[32]

Transgenic mice are also being used to identify genes and specific genetic mutations that confer genetic susceptibility of humans to the toxic effects of xenobiotics. The enzyme systems involved in the metabolic bioactivation and detoxication of xenobiotics are significant determinants of interindividual variability and risk in response to the toxic effects of xenobiotic exposures. Polymorphisms in DNA repair systems are also being linked to cancer susceptibility. Animal models with specific single nucleotide polymorphisms linked to human genetic susceptibility can be constructed in mice and directly tested for their sensitivity to various agents thereby defining the significance of specific polymorphisms as determinants of human risk. These models can mimic aspects of human responsiveness and susceptibility providing the toxicologist with "humanized" mouse models to directly test hypotheses regarding human genetic susceptibilities.[30]

In this chapter we have outlined approaches for the construction of a new generation of mouse models that exhibit cell type or tissue type specific gene expression. Models that can be induced to express altered gene products in a time and tissue specific manner will be valuable in focusing attention on relevant target tissues. Genes that can be induced to express normal or mutant proteins during specific periods of embryonic development will aid in the identification in critical target proteins as well as windows of sensitivity linked to adverse outcomes. Integrating genomic studies with transgenic mice to identify genetic pathways altered by toxic agents will likely significantly increase the depth of and the pace of our understanding of chemical interactions with biological systems.

## 14.5.3.6 ACKNOWLEDGMENTS

We thank Becky Staben for creating the figures.

## 14.5.3.7 REFERENCES

1. J. Khan, M. L. Bittner, Y. Chen *et al.*, 'DNA microarray technology: the anticipated impact on the study of human disease.' *Biochem. Biophys. Acta*, 1999, **1423**, M17–M28.
2. G. Chambers, L. Lawrie, P. Cash *et al.*, 'Proteomics: a new approach to the study of disease.' *J. Pathol.*, 2000, **192**, 280–288.
3. M. J. Cunningham, 'Genomics and proteomics: the new millennium of drug discovery and development.' *J. Pharmacol. Toxicol. Methods*, 2000, **44**, 291–300.
4. J. K. Nicholson, J. C. Lindon and E. Holmes, '"Metabonomics": understanding the metabolic responses of living systems to pathophysiological stimuli via multivariate statistical analysis of biological NMR spectroscopic data.' *Xenobiotica*, 1999, **29**, 1181–119.
5. E. F. Nuwaysir, M. Bittner, J. Trent *et al.*, 'Microarrays and toxicology: the advent of toxicogenomics.' *Mol. Carcinog.*, 1999, **24**, 153–159.
6. F. Fougerousse, P. Bullen, M. Herasse *et al.*, 'Human-mouse differences in the embryonic expression patterns of developmental control genes and disease genes.' *Hum. Mol. Genet.*, 2000, **9**, 165–173.
7. P. Denny and M. J. Justice, 'Mouse as the measure of man?' *Trends Genet.*, 2000, **16**, 283–287.
8. D. R. Gilbert, M. Schroeder and H. J. van, 'Interactive visualization and exploration of relationships between biological objects.' *Trends. Biotechnol.*, 2000, **18**, 487–494.
9. G. Sherlock, 'Analysis of large-scale gene expression data.' *Curr. Opin. Immunol.*, 2000, **12**, 201–205.
10. K. M. Kuhn, J. L. DeRisi, P. O. Brown *et al.*, 'Global and specific translational regulation in the genomic response of *Saccharomyces cerevisiae* to a rapid transfer from a fermentable to a nonfermentable carbon source.' *Mol. Cell. Biol.*, 2001, **21**, 916–927.
11. T. R. Hughes, M. J. Marton, A. R. Jones *et al.*, 'Functional discovery via a compendium of expression profiles.' *Cell*, 2000, **102**, 109–126.
12. M. Q. Zhang, 'Promoter analysis of co-regulated genes in the yeast genome.' *Comput. Chem.*, 1999, **23**, 233–250.
13. T. G. Wolfsberg, A. E. Gabrielian, M. J. Campbell *et al.*, 'Candidate regulatory sequence elements for cell cycle-dependent transcription in Saccharomyces cerevisiae.' *Genome. Res.*, 1999, **9**, 775–792.
14. M. T. Geraghty, D. Bassett, J. C. Morrell *et al.*, 'Detecting patterns of protein distribution and gene expression in silico.' *Proc. Nat. Acad. Sci. USA*, 1999, **96**, 2937–2942.
15. M. H. Hofker and M. Breuer, 'Generation of transgenic mice.' *Methods. Mol. Biol.*, 1998, **110**, 63–78.
16. J. Heideman, 'Transgenic rats: a discussion.' *Biotechnology*, 1991, **16**, 325–332.
17. A. Kistner, M. Gossen, F. Zimmermann *et al.*, 'Doxycycline-mediated quantitative and tissue-specific control of gene expression in transgenic mice.' *Proc. Nat. Acad. Sci. USA*, 1996, **93**, 10933–10938.
18. M. R. Capecchi, 'Altering the genome by homologous recombination.' *Science*, 1989, **244**, 1288–1292.
19. R. P. Woychik, M. L. Klebig, M. J. Justice *et al.*, 'Functional genomics in the post-genome era.' *Mutat. Res.*, 1998, **400**, 3–14.
20. B. Sauer, 'Manipulation of transgenes by site-specific recombination: use of Cre recombinase.' *Methods Enzymol.*, 1993, **225**, 890–900.
21. S. M. Dymecki, 'Flp recombinase promotes site-specific DNA recombination in embryonic stem cells and transgenic mice.' *Proc. Nat. Acad. Sci. USA*, 1996, **93**, 6191–6196.
22. C. G. Lobe and A. Nagy, 'Conditional genome alteration in mice.' *Bioessays*, 1998, **20**, 200–208.
23. F. Cecconi and B. I. Meyer, 'Gene trap: a way to identify novel genes and unravel their biological function.' *FEBS Lett.*, 2000, **480**, 63–71.
24. M. Hrabe de Angelis and R. Balling, 'Large scale ENU screens in the mouse: genetics meets genomics.' *Mutat. Res.*, 1998, **400**, 25–32.
25. M. J. Justice, 'Capitalizing on large-scale mouse mutagenesis screens.' *Nat. Rev. Genet.*, 2000, **1**, 109–115.
26. B. A. Taylor, in 'Recombinant inbred strains,' eds. M. F. Lyon, S. Rastan and S. D. M. Brown, Genetic Variants and Strains of the Laboratory Mouse, Vol. 2. Oxford University Press, Oxford, BA, 1996, pp. 1597–1659
27. M. F. Festing, L. Lin, T. R. Devereux *et al.*, 'At least four loci and gender are associated with susceptibility to the chemical induction of lung adenomas in A/J × BALB/c mice.' *Genomics*, 1998, **53**, 129–136.
28. A. Grupe, S. Germer, J. Usuka *et al.*, 'In silico mapping of complex disease-related traits in mice.' *Science*, 2001, **292**, 1915–1918.
29. J. C. Corton and A. J. Stauber, 'Toward construction of a transcript profile database predictive of chemical toxicity.' *Toxicol. Sci.*, 2000, **58**, 217–219.
30. F. J. Gonzalez, 'The use of gene knockout mice to unravel the mechanisms of toxicity and chemical carcinogenesis.' *Toxicol. Lett.*, 2001, **120**, 199–208.
31. S. W. Dean, T. M. Brooks, B. Burlinson *et al.*, 'Transgenic mouse mutation assay systems can play an important role in regulatory mutagenicity testing *in vivo* for the detection of site-of-contact mutagens.' *Mutagenesis*, 1999, **14**, 141–151.
32. D. Gulezian, D. Jacobson-Ram, C. B. McCullough *et al.*, 'Use of transgenic animals for carcinogenicity testing: considerations and implications for risk assessment.' *Toxicol. Pathol.*, 2000, **28**, 482–499.

J.P. Vanden Heuvel, G.H. Perdew, W.B. Mattes and W.F. Greenlee (Eds.)
Comprehensive Toxicology, Vol. xiv

# 14.5.4
# Gene Expression Analysis in the Microarray Age

WILLIAM D. PENNIE

*Syngenta Central Toxicology Laboratory, Alderley Park, Macclesfield, Cheshire, SK10 4TJ, UK*

## 14.5.4.1 INTRODUCTION

Experimental toxicologists have at their disposal a wide range of techniques and technologies that can be harnessed to explore the mechanisms by which chemicals may exert undesirable effects on living systems. Recent technological innovations, such as microarray and proteomic technologies, represent nothing short of a revolution in terms of their ability to characterize an unprecedented number of endpoints simultaneously. Such powerful tools bring their own challenges in terms of data management and interpretation. When used in concert with established investigative techniques, as part of a holistic approach to toxicology, these technologies are already contributing significantly to our understanding of basic toxicology mechanisms, and may in the future contribute to the faster development of safer drugs, agrochemicals and chemical products.

### 14.5.4.1.1 The "Genomics Revolution"

The last decade has seen major developments in large-scale genome sequencing and in the development of technical platforms to support this endeavor. There are a considerable, and growing, number of sequencing programs which aim to characterize the genomes of many organisms at varying levels of resolution. At the time of writing in excess of 300 sequencing programs are active, with 60 complete genomes, such as that of yeast,[1] resolved at the nucleotide level;[2] see *http://igweb.integratedgenomics.com/GOLD/*.

The potential benefits of achieving genome closure have been widely discussed in the popular and scientific press with predictions that genetic characterization of individual risk of disease (and likely responses to therapeutic agents) will become standard medical procedures within a relatively short time frame.[3] As well as characterizing individual susceptibility to disease or drugs, it is possible that genomics information could help predict an individual's likely reaction to chemical exposure. With this aim, the Environmental Genome Project (EGP) aims to resequence DNA samples from a broad demographic range in the USA, assembling a database of polymorphic variability that could be useful in characterizing the genetic component of susceptibility to specific environmental agents.[4]

The genome sequencing programs have adopted a "clone by clone" approach and in the process substantial libraries of sequenced genomic DNA have been generated.[5] The availability of the sequence information for many thousands of genes has facilitated the development of a number of technologies, which profile changes in genes (and gene products) at a scale that will greatly facilitate unraveling how many biological processes are regulated. While these sequenced clones are of great value (for example in the measurement of the transcriptional activity of individual genes), detailed annotation of the location and structure of all genes will require a combination of computational analysis of the primary sequence, coupled with an experimental approach to validate expression from the gene.[6] New technologies such as ink-jet based microarray technology have been applied to improve the validation of gene transcripts on a genomic scale.[7] Extending gene identification to functional characterization will be greatly facilitated by the availability of full length coding (cDNA) clones, as illustrated by the three year pilot study being run at the German Cancer Research Center, which aims to clone full length coding sequences from many hundreds of genes and make them publicly available.[8] A similar initiative is underway at the National Institutes of Health as part of the Cancer Genome Anatomy project.

### 14.5.4.1.2 Gene Regulation and Function; The "Omics" Technologies

The unraveling of the genome will increase our understanding of the numbers and types of genes, the proteins they encode, and the interactions between the products of multiple gene families. In mechanistic toxicology research it is clearly our long-term aim to understand how different classes of biomole-cules interact in response to toxicant exposure, and the conditions by which these interactions may or may not give rise to a phenotypic change (toxicity). These different groups of biomolecules can be broadly characterized as;

- Genome; the chromosomal DNA organization, including nucleotide sequence.
- Transcriptome; the messenger RNA from actively transcribed genes.
- Proteome; the entire protein complement of a biological sample
- Metabonome; the constituent metabolites in a biological sample.

Potentially, xenobiotics can have a primary mode of action that affects any of these classes of biomolecule although it is unlikely that any resultant phenotypic change, with the possible exception of necrosis, can take place without measurable alterations of all compartments downstream of the genome. Some of the processes by which changes in one compartment can influence others are outlined simplistically in Figure 1. Such changes, and their consequences, can be characterized by a range of established (or evolving) technologies; *Genomics* research involves the characterization (at the level of DNA sequencing or chromosome mapping) of the DNA sequence of an organism and provides the primary information used for a diverse range of technologies and applications. The technologies to accomplish such large-scale enterprises are evolving rapidly, as demonstrated by the recent announcement that Celera Genomics (www.celera.com) has characterized 95% of the murine genome (approx 9.3 billion base pairs) in approximately seven months of effort.

The investigation of variable, or polymorphic, regions of genes in an attempt to characterize their potential role in idiosyncratic responsiveness is a distinct genomics discipline, and is often referred to as *pharmacogenomics* or *pharmacogenetics*. The associated technologies allow the mapping of single nucleotide polymorphisms (SNPs), regions of variability in gene structure, for specific chromosomal regions. SNP maps are being generated in both the academic (http://snp.cshl.org) and commercial (www.celera.com) domains with approx. 3 million individual SNPs now characterized. Clearly, genetic polymorphism is a critical consideration in exploring inter-individual differences in response to toxic insults and in identifying susceptible sub-populations. For this reason, attempts to map common polymorphisms that may be associated with sensitivity (or resistance) to chemical insults are of particular interest to toxicologists, and the characterization of polymorphic variability[9,10] is likely to drive the evolution of *toxicogenetics*.

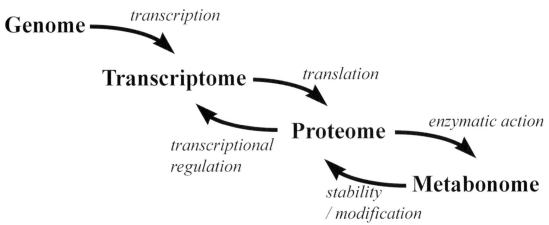

**Figure 1** Simplified overview of classes of biomolecules that can be profiled with the technologies under discussion. Changes in the components of one category of molecule will often impact other categories, e.g., gene expression change may affect the level of a particular enzyme which may then alter the levels of a metabolite.

The term *functional genomics* refers to a host of technologies that enable the biological role of genes to be investigated. Examples of functional approaches include the well-established technologies of transgenic or knockout animal models. More recently the use of heterologous expression systems have allowed a higher throughput approach to assessing gene function. In such systems genes of interest are expressed in a host organism (such as yeast) to monitor the phenotypic effects of gain of gene function. Comparable approaches for deleting genes in simple organisms also exist, and while these are primarily commercial enterprises using proprietary technology, initiatives in the academic domain continue to develop. For example, the *C. elegans* Gene Knockout Project (http://elegans.bcgsc.bc.ca/) is a worldwide consortium whose ultimate goal is to produce null alleles of all known genes in the *C. elegans* genome as a model system. Such approaches are taken one step further with *comparative* genomics, an approach that leverages functional and structural information from one biological system across all other biological systems. This is likely to be a significant growth area in molecular toxicology research given its potential to help in extrapolating mechanistic data from laboratory species to potential risk to man.

*Proteomics* technologies[11] can characterize differential protein expression and/or modifications that may lead to changes in the *activity* of gene products. The traditional approach to proteomics profiling involves two stages. In the first stage the individual proteins in a biological sample are separated and resolved by two-dimensional gel electrophoresis. By comparing the resolved gels from control and treated samples, differentially expressed or modified proteins associated with the treatment can be noted. In the second stage, the identity of individual protein spots of interest is established by either "mapping" their resolved locations based on historic data (or reference samples), or by excising the protein spot from the gel and identifying the protein by peptide cleavage and sequencing, or analysis of mass spectra. There exist many publicly accessible databases on protein structure and function that can assist proteomics analyses;[12] (see for example www.proteome.com).

*Metabonomics* allows assessment of changes in the chemical environment, resulting perhaps from alterations in enzyme production or activity in various biochemical pathways. In this approach the NMR spectra of biofluids or tissues are profiled and used in statistical analysis, typically Principal Component Analysis (PCA), to allow relationships between different compounds or xenobiotics to be estimated. Such associations are made on the basis of reducing an entire NMR spectra result to a single point, plotted in multidimensional space, where the distances between individual points on such a graph are proportional to the statistical similarity between the parent data sets; i.e., spectra of similarly acting compounds will "cluster" together. Metabonomics approaches have been shown to distinguish toxicities (e.g., hepatotoxicity from nephrotoxicity) by NMR spectra analysis of the urine of treated rats with 98% accuracy.[13] In these studies this approach is also able to characterize subtle differences in metabolism that enable the different strains of animal used to be distinguished on the basis of metabonomic analysis. It is pertinent to note that metabonomics is able to distinguish changes in metabolism at dose or time points before the onset of clinical or pathological indicators of toxicity.[14]

Many laboratories, including our own, have utilized new technologies that measure gene expression in research and investigative toxicology. Such approaches have become commonly known as "toxicogenomics" a term synonymous with the large-scale measurement of gene expression change that results as a consequence of xenobiotic exposure. Gene expression analysis has been profoundly impacted by the development of toxicogenomics and it could be argued that a new paradigm for the analysis of gene expression changes has evolved as a result of the development of the microarray technologies which facilitate this approach.

## 14.5.4.2 TRANSCRIPT PROFILING

Until recently, researching gene regulation events during toxicological processes tended to focus on the study of small numbers of genes in some detail. For example, the regulation of the alcohol-inducible cytochrome P4502E1 has been characterized at high resolution, including transcriptional and translation mechanisms by which enzymatic activity is regulated.[15] Such studies help identify specific factors involved in modulating individual gene expression and thus characterize the interplay between discrete signaling and regulatory pathways, allowing the role of transcriptional regulation in the development of toxicity to be investigated.[16] In other approaches the levels and distribution of the mRNA molecules from specific genes can be determined, and used to localize the activity of particular genes[17] or to characterize particular responses, as in the use of cytokine fingerprinting to investigate immune response.[18] The advent of microarray technology (the major technique used in transcript profiling) alters this working model by allowing the researcher to measure changes in the transcriptional regulation of *thousands* of genes simultaneously but without necessarily understanding the mechanism by which expression of any of the genes takes place, or indeed what the consequences of altered expression may be.

DNA microarrays permit the quantitative comparison of the expression levels of thousands of individual genes in different biological samples, thus facilitating comparisons, for instance, between normal tissue and diseased tissue, or between control cell lines and toxicant-treated cell lines. The construction of microarrays involves the immobilization of DNA sequences (either cDNA sequences or oligonucleotides), corresponding to the coding sequence of genes of interest, on a solid support such as a glass slide or nylon membrane.[19] The mRNA prepared from cells or tissues can be labeled and hybridized (usually in the form of a reverse transcribed cDNA copy) to a microarray and visualized using phosphorimager scanning or other appropriate methodologies. Subsequent analysis using appropriate software allows determination of the extent of hybridization of the labeled probes to the corresponding arrayed cDNA or oligonucleotide spots, and a comparison of control with test samples permits quantitative assessment of changes in gene expression associated with treatment.[20]

Many microarrays have been developed by commercial vendors, pharmaceutical or agrochemical companies and academic institutions[21] including "custom" cDNA microarrays designed specifically to measure genes of potential toxicological relevance. The platforms vary from those comprising several hundred to those harboring tens of thousands of probe sequences. Researchers employing microarray technology are thus faced with a number of technical platforms with variations in the chemistry used to immobilize the cDNA (or oligonucleotide) sequences and in the extraction, labeling and detection of the sample RNA.[22] There are a number of public domain repositories for cDNA sequences, which can be used, should an individual researcher want (or need) to construct their own array design. It should be noted that the error rate for some of these clone sets is perceived to be high; error rates of 40% have been reported.[23] Such issues can be overcome by securing sequence verified clones, or by individually sequencing public domain material.

Microarrays can be designed to be either broad coverage "discovery" style arrays, which include an unbiased selection of gene sequences (which may include probes for genes of as yet uncharacterized function), or "hypothesis-driven" where the arrays are designed to focus on genes of relevance to a particular discipline such as toxicology.[24] In this regard, an often-voiced criticism of microarray usage in mammalian systems is that of incomplete coverage, i.e., only a defined subset of genes are profiled by a given technical platform. Platforms with complete coverage of simpler genomes, such as yeast or *C. elegans*, have been developed, and mammalian platforms approaching full genome coverage are likely to become available within the next few years. *Open* approaches to measuring differential gene expression have existed for many years (e.g., differential display polymerase chain reaction)[25] but they are generally low throughput, and require considerable follow-up work to confirm the identity of differentially expressed genes. The SAGE technique (serial analysis of gene expression) has widespread usage in many laboratories[26] and the technology has matured to allow transcript profiling of very small amounts of biological material.[27] In addition SAGE expression databases have been established to allow the analysis of gene expression events associated with endpoints such as carcinogenesis.[28,29] Sophisticated high throughput sequencing technologies, such as MPSS (massively parallel signature sequencing)[30] can also be applied to resolve differences in transcriptional status between different samples. Broad technical platforms may, however, not be desirable (or practical) for certain applications, and individual laboratories have thus developed a number of hypothesis-based array designs. Strategies for the development of these arrays may involve categorizing genes into tissue (or developmental stage) specific classes and/or designing platforms that comprehensively cover specific gene families. It has been suggested that statistical approaches[31] can be employed to help identify genes most likely to be involved in particular physiological processes.[32] An example of a microarray designed to profile a particular tissue or developmental stage is the mouse testis array constructed by the US Environmental Protection MicroArray Consortium (EPAMAC), where a non-redundant list of 950 testis-expressed genes was assembled, and representative sequence-verified public domain clones were used for the array construction.[32] Small scale microarray design can also be mechanism-based, allowing the researcher to focus on discrete classes of genes (e.g., immediate early genes, metabolizing enzymes *etc*.), and such microarrays may be very modest in scale.[33]

### 14.5.4.2.1 Toxblot: A Custom Toxicology Microarray

Our laboratories have a program of in house custom toxicology array design and manufacture; *ToxBlot*. A custom approach allows the selection of the genes to be represented on the array, permitting arrays to be focused on areas of particular interest to mechanistic or investigative toxicology research programs (Table 1). For the construction of the *ToxBlot* arrays, some 13,000 sequences of human or murine origin were identified as being of potential

relevance to various forms of toxicity or normal cell regulatory or signaling pathways. In house cDNA clone sets, of both public domain and proprietary origin, were used as the source material for array construction. Following selection, PCR amplification, purification and (where appropriate) sequence verification, the representative cDNA sequences are immobilized on nylon membranes (Figure 2). To aid in the interpretation of differential gene expression results we have assembled a database on gene function, distribution and known allelic variations for each represented gene. In this regard public domain databases such as the GeneCards[34] system developed at the Weizmann Institute or the Kyoto Encyclopedia of Genes and Genomes (KEGG) databases (http://www.genome.ad.jp/kegg/) can provide valuable supplementary information.

### 14.5.4.2.2 Applications of Transcript Profiling

The obvious toxicology application of these technologies is the investigation of the regulation events that underpin the development of an adverse biological response. An example of such an endpoint, non-genotoxic carcinogenesis, is usually assayed by using a long-term cancer bioassay in rodents,[35] which is expensive in terms of both time and animal usage. Toxicogenomics applications may assist the identification of surrogate markers for the development of this phenotype in cultured cells. For example, the exposure of rodent hepatocytes to the non-genotoxic carcinogen, phenobarbital, has been studied by using microarray and gel-based expression technologies, with the result that in excess of 300 genes were found whose expression is modulated by this reagent.[36] Similarly, oligonucleotide-based arrays have been used to characterize the gene expression changes associated with exposure to damaging dosages of particular compounds such as acetaminophen.[37] In such studies it is important to distinguish mechanism-related responses specific for the compound being studied more general stress responses of xenobiotic exposure. The ability of transcript profiling to distinguish between two distinct classes of compounds (DNA damaging agents and non-steroidal anti-inflammatory drugs) has recently been demonstrated by researchers at the RW Johnson Pharmaceutical Research Institute. Their approach involved the analysis of a prototypic single compound from each compound class, to look for consistent diagnostic expression changes while avoiding background noise in the system (resulting from

**Table 1**  Sample gene categories used in the construction of the *ToxBlot* in-house toxicology microarrays. Approximately 200 broad classes of genes were identified with potential relevance to toxicology. Each class may contain many individual genes. Incorporation of genes of interest into a custom cDNA technical platform may be limited by the availability of proprietary or public domain clones.

| *Metabolism* | *Oxidoreductases* | *Signal transduction and regulation* |
|---|---|---|
| DNA metabolism | Acting on a peroxide | Extracellular messengers |
| Protein metabolism | as acceptor | receptors |
| Lipid metabolism | (peroxidases) | Intracellular signalling |
| Monosaccharide | Carbon-sulfur lyases | Transcription factors |
| Metabolismpolysaccharide and | | |
| Glycoconjugate metabolism | | |
| Amino acid biosynthesis | *Protein modification and maintenance* | *Immunity and defence* |
| Amino acid degradation | Proteases | Antigen processing and |
| Urea metabolism | Protease inhibitors | presentation |
| Energy metabolism | Chaperones | Inflammatory response |
| Cofactor metabolism | | Defence and resistance |
| Sulphate metabolism | *Localized and structural proteins* | |
| Antioxidant metabolism | Secreted and extracellular | |
| Drug and Xenobiotic | Cytoskeleton | *Environmental responses* |
| metabolism | Organelle | Metal tolerances |
| Secondary metabolism | | |
| | *Growth and Development* | *Others* |
| *Adhesion and molecular* | Cell division | Phosphotransferase system |
| *recognition* | Reproduction | VDP-glucuronyl transferases |
| Adhesion | Differentiation and proliferation | Chemokine receptors |
| Antigen recognition | Apoptosis | a-ketoglutarate dH complex |
| | Aging and senescence | Glutathione cycle enzymes |
| | | Proteoglycans |
| *Electron transfer* | *Membrane transport* | Orphan receptors |
| Cytochromes | Channels and porins | Opioid receptors |
| Flavoproteins | Transporters and pumps | Development |
| | | Epithelial cell transporters |
| *Kinesis* | | Ecological interactions |
| Motility | | |

**Figure 2**  Scanned phosphorimage view of a processed cDNA microarray. The extent of hybridization at each individual "spot" can be compared across different experiments and used as a quantitative estimate for changes in gene expression (see main text for more detail). This microarray is based on nylon membrane technology and the transcriptional regulation of approximately 13,000 individual genes can be assessed simultaneously with this platform.

biological variability or inherent noise in the technical platform). Computer-based prediction tools were then employed to expand the list of consistent gene expression events to those genes that could accurately distinguish DNA damaging agents from the anti-inflammatory drugs in a 100 compound "learning set".[38] Such experiments can be viewed as generating specific "candidate" gene lists, which can be used to formulate hypotheses as to the mechanism by which the xenobiotic induces its effect. Many toxic endpoints may be amenable to this approach, and combinatorial approaches such as using cell lines developed from transgenic or knockout animals,[39] could provide powerful insights into the role of specific genes. Clearly, proving the causative involvement of any of these candidates requires detailed follow-up work, most probably employing more traditional "functional genomics" and biochemical approaches (e.g., gene knockout systems, transcriptional "reporter" assays, site-directed mutagenesis, small molecule inhibitors *etc.*).

An alternative to mechanistic research is the application of transcript profiling to enhance our ability to *predict* potential toxicity issues. Treatment of test systems with known reference toxicants (with similar toxic endpoint, mechanism, chemical structure, target organ, *etc.*) could permit the identification of diagnostic gene expression patterns. Such pattern recognition could facilitate the discovery and subsequent validation of biomarkers useful for application in higher throughput technologies to characterize or detect specific toxicity endpoints.

The development of the reference data sets to allow the "pattern recognition" approach to toxicology is likely to require the application of complex computer algorithms and statistical approaches. For example, statistical clustering techniques have been applied to microarray data to analyze the temporal patterns of gene expression associated with serum-responsiveness and wound repair,[40] and to distinguish cancerous tissue from normal tissues and cell lines.[41] A number of resources exist in both the academic and commercial sectors to assist in managing reference compound gene expression data sets to facilitate "pattern recognition".[42] Scientists at The National Human Genome Research Institute have developed one example, the software platform ArrayDB. The system facilitates the storage, retrieval and analysis of microarray data along with information linking some 15,000 genes to public domain sequence and pathway databases.[43] In addition there are collaborative efforts involving pharmaceutical, agrochemical and chemi-

cal industries, such as the ILSI HESI (International Life Sciences Institutes—Health and Environmental Sciences Institute) Subcommittee on Application of Genomics and Proteomics to Mechanism Based Risk Assessment to generate publicly accessible datasets of this type using reference toxicants (see www.ilsi.org). The construction of such baseline datasets will undoubtedly assist in determining the normal variability in levels of individual genes or gene products, which will in turn assist in distinguishing "background noise" from adaptive or mechanistically important change.

### 14.5.4.3 GENE REGULATION STUDIES: RESOURCES, TOOLS, DATABASES

The variety of microarray platforms raises challenges in comparing datasets generated on different platforms. Notwithstanding technical differences that exist between platforms (for example in microarray design, probe chemistry, data analysis *etc.*) these technical platforms may not output the actual data in a consistent format. To address this issue, standard data format standards for exchanging and annotating microarray data are being proposed. An example of such a format is Gene Expression Markup Language (GEML[TM]), developed by a number of organizations including Rosetta Informatics. GEML is designed to separate data reporting and collection from the methodology used to report or collect the data, to be independent of any particular software system or database for gene expression analysis and to serve as a general-purpose data format for exchanging expression data among different programs and databases. In principle GEML format supports multiple data collection methodologies, including cDNA and oligonucleotide based microarrays together with other methodologies such as SAGE and TaqMan. The intention of enterprises such as the GEML initiative is that the exchange format employed accommodates current and future industry data formats, is suitable for easy sharing of data and transmission and that the format is public domain, facilitating the support and development of public access and proprietary data analysis tools. The development of such a "common language" is perhaps the single biggest bottleneck to be overcome before public domain data repositories for microarray experimental data will be feasible.

A number of institutions, in both the public and private sector, are developing such data repositories for gene expression data. An example of such an endeavor is the Stanford

Microarray Database (SMD), which stores both raw and normalized data from more than 1000 microarray experiments. This platform also allows access to analysis and visualization tools and electronic links to published microarray experiments (http://genome-www4.stanford.edu/MicroArray/SMD/). Increasingly such databases are designed to host transcriptional data from multiple technical platforms (cDNA microarray, oligonucleotide-based) and non-microarray data such as Northern/filter hybridization and SAGE. The Gene Expression Omnibus database launched by the NCBI is one such example (http://www.ncbi.nlm.nih.gov/geo). Increasingly, the code for publicly funded database or analysis software is released as "open source" and this approach can only be constructive in facilitating the rapid development of these useful tools. As such databases evolve it is likely to become increasingly important that functional data other than transcriptional changes (e.g., biochemical endpoints, proteomics, growth rates *etc*.) are captured within the datasets. In this regard scientists at Harvard University are developing on their Express DB database, a relational model for yeast expression genomics, to build an integrated database for functional genomics; the Biomolecule Interaction, Growth and Expression Database (BIGED).[44]

The tissue-specific distribution of gene expression is also being characterized by large-scale cDNA library sequencing approaches where individual tissues (at various stages of development) are used to construct cDNA libraries. The sequencing of such libraries will reveal, within various technical constraints, which genes were transcriptionally active in the source material, and the number of times the cDNA for an individual gene is encountered can be used as a crude estimate of message abundance. In this way electronic "Northern blots", comparing genes expressed in particular compartments[45] can be performed by querying databases of this type on the distribution and frequency of particular cDNAs (representing individual genes), and may also be used to identify transcripts with tissue specific functions. The "BodyMap" database (http://bodymap.ims.u-tokyo.ac.jp) is an endeavor by a number of Japanese laboratories to characterize both human and murine transcription "maps" derived by such analysis[46] and other laboratories have reported similar data mining approaches.[47] A number of publicly accessible databases of great value to gene regulation research are available (Table 2). In addition, many laboratories and user groups host public access software analysis tools that facilitate expression analysis. Prominent examples include;

Stanford Biomedical Informatics    http://classify.stanford.edu

Classification of Expression Array (Cleaver) internet site provides a suite of analysis tools for microarray data. All analysis is done on the host server and tools are provided for processes such as data clustering and visualization of analyzed data.

Lawrence Berkley National Lab    http://rana.lbl.gov

Software (including open source) is made freely available for registered non-profit researchers. Software tools include those for microarray image analysis (e.g., spot finding and analysis), and data analysis, clustering and visualization tools.[48]

Univ of California, Irvine    http://www.genomics.uci.edu/software.html

A number of tools are made available for clustering analysis, the management of array data (including the merging of multiple data sets), conversion of microarray data between different technical platforms *etc*.

NHGRI, NIH    http://genome.nhgri.nih.gov/arraydb/

The National Human Genome Research Institute (NHGRI) has developed a software suite, ArrayDB, which allows analysis of gene expression changes at the level of single or multiple experiments. In addition the design of each microarray platform can be included and the software links to external databases on gene function, metabolic pathways *etc*. allowing the researcher to quickly gain an overview of the general biology of the genes which respond in an individual experiment.

### 14.5.4.4 CHALLENGES AND OPPORTUNITIES

The careful application of transcript profiling technologies could also assist research associated with safety assessment by:

● An enhanced ability to extrapolate accurately between experimental animals and humans in the context of risk assessment.

● A more detailed appreciation of molecular mechanisms of toxicity.

**Table 2** Selected publicly-accessible database resources for genomics, proteomics and gene expression research.

| Database | Location | Contents |
|---|---|---|
| ALFRED | http://alfred.med.yale.edu/alfred/index.asp | DNA polymorphisms database |
| AllGenes | http://www.allgenes.org/ | Human mouse gene index integrating gene, transcript and protein annotation |
| BodyMap | http://bodymap.ims.u/ | Expression data; both human and mouse |
| EMBL nucleotide sequence repository | http://www.ebi.ac.uk/embl.html | Nucleotide sequence database |
| ENZYME | http://www.expasy.ch/enzyme/ | Enzyme nomenclature conventions |
| GenAtlas | http://www.citi2.fr/GENATLAS/ | Atlas of human genes and markers |
| Genbank | http://www.ncbi.nlm.nih.gov/ | Sequence database collaboration |
| GOLD | http://igweb.integratedgenomics.com/GOLD/ | Catalogue of genome sequencing projects |
| HGMD | http://www.uwcm.ac.uk/uwcm/mg/hgmd0.html | Published mutations associated with human disease |
| HUNT | http://www.hri.co.jp/HUNT | Annotated human full length cDNA sequences |
| KEGG | http://www.genome.ad.jp/kegg | Biochemical pathway mining |
| MGD | http://www.informatics.jax.org/ | Mouse genetics/ genomics resources |
| Mouse Atlas and Gene Expression Database | http://genex.hgu.mrc.ac.uk/ | Spatially-mapped murine gene expression data |
| Proteome Analysis Database | http://www.ebi.ac.uk/proteome/ | Proteomics resource |
| Stanford Microarray Database | http://genome-www.stanford.edu/microarray | Microarray experiment data repository |
| SWISSPROT | http://www.expasy.ch/sprot | Annotated protein sequences |
| TRRD | http://wwwmgs.bionet.nsc.ru/mgs/dbases/trrd4 | Database of eukaryotic gene regulatory sequences |
| Tumour Gene Families Database | http://www.tumor-gene.org/tgdf.html | Database of information on genes implicated in carcinogenesis |
| UNIGENE | http://www.ncbi.nlm.nih.gov/UniGene/ | Non-redundant gene clusters |
| WormBase | http://www.wormbase.org/ | *C. elegans* resource |

- Facilitating more-rapid screens for compound toxicity.
- The provision of new research leads.

In the context of mechanism-based research it is probably best to regard results obtained using transcript profiling technologies as the springboard to more detailed and focused investigations (using other experimental approaches) that would confirm or refute the significance of the observed changes. Concern has been voiced already that a potential problem is the misinterpretation, or over-interpretation, of such high-volume data analyses particularly in the context of determining product safety. It must be recognized that the interaction of xenobiotics with biological systems will in many instances result in some changes in gene expression, even under circumstances where such interactions are benign with respect to adverse effects. It is therefore important that emerging data from these technologies is followed through with traditional approaches and analyzed fully to establish if the measured changes are background noise, adaptive, beneficial or potentially harmful. Such expression changes clearly do not compromise previously characterized NOAELs of exposure *under circumstances where there are no physiological or pathological indicators of adverse effect*. These are points on which it is critically important to foster the development of consensus and common understanding across the constituencies of industry, academia and regulators.

The attainment of common ground through collaboration involving the generation, sharing and publication of suitable, high-quality, pilot data should be the prime goal for scientists and institutions engaged in researching the new technology and its appropriate application towards toxicology. It is in the interests of all stakeholders to seek opportunities for investing resources in those initiatives that promote the development of a sound gene expression database. Such a database would enable the technologies to be evaluated from an informed perspective, assisting those engaged in assessing chemical safety, including regulatory agencies, to place these data into context with other, more conventional data. In this regard, it is encouraging to see the emergence of academia/industry consortia, such as the ILSI/HESI sponsored genomics subcommittee, with commitment to begin building and populating the databases needed for pattern recognition. Finally, it is important for both scientists and regulators to appreciate what the technologies can and cannot do, and to work together to define their appropriate use in regulatory processes and discussions.

## 14.5.4.5 REFERENCES

1. J. L. DeRisi, V. R. Iyer and P. O. Brown, 'Exploring the metabolic and genetic control of gene expression on a genomic scale.' *Science*, 1997, **278**(5338), 680–686.
2. A. Bernal, U. Ear and N. Kyrpides, 'Genomes OnLine Database (GOLD): a monitor of genome projects world-wide.' *Nuc. Acids Res.*, 2001, **29**(1), 126–127.
3. F. S. Collins and V. A. McKusick, 'Implications of the Human Genome Project for medical science.' *JAMA*, 2001, **285**(5), 540–544.
4. K. Olden and J. Guthrie, 'Genomics: implications for toxicology.' *Mutat. Res.*, 2001, **473**(1), 3–10.
5. P. Little, 'The end of all human DNA maps?' *Nat. Genet.*, 2001, **27**(3), 229–230.
6. D. D. Shoemaker, E. E. Schadt, C. D. Armour *et al.*, 'Experimental annotation of the human genome using microarray technology.' *Nature*, 2001, **409**(6822), 922–927.
7. C. B. Burge, 'Chipping away at the transcriptome.' *Nat. Genet.*, 2001, **27**(3), 232–234.
8. S. Wiemann, B. Weil, R. Wellenreuther *et al.*, 'Toward a Catalog of Human Genes and Proteins: Sequencing and Analysis of 500Novel Complete Protein Coding Human cDNAs.' *Genome Res.*, 2001, **11**(3), 422–435.
9. F. P. Guengerich, 'The environmental genome project: functional analysis of polymorphisms.' *Environ. Health Perspect.*, 1998, **106**, 365–368.
10. D. J. Wang, J.-B. Fan, C.-J. Berno *et al.*, 'Large-scale identification, mapping and genotyping of single-nucleotide polymorphisms in the human genome.' *Science*, 1998, **280**, 1077–1082.
11. N. L. Anderson, J. Taylor, J. P. Hoffmann *et al.*, 'Simultaneous measurement of hundreds of liver proteins: application in assessment of liver function.' *Toxicol. Pathol.*, 1996, **24**(1), 72–76.
12. M. C. Costanzo, J. D. Hogan, M. E. Cusick *et al.*, 'The yeast proteome database (YPD) and Caenorhabditis elegans proteome database (WormPD): comprehensive resources for the organization and comparison of model organism protein information.' *Nucleic Acids Res.*, 2000, **28**(1), 73–76.
13. E. Holmes, A. W. Nicholls, J. C. Lindon *et al.*, 'Chemometric models for toxicity classification based on NMR spectra of biofluids.' *Chem. Res. Toxicol.*, 2000, **13**(6), 471–478.
14. D. G. Robertson, M. D. Reily, R. E. Sigler *et al.*, 'Metabonomics: evaluation of nuclear magnetic resonance (NMR) and pattern recognition technology for rapid *in vivo* screening of liver and kidney toxicants.' *Toxicol. Sci.*, 2000, **57**(2), 326–337.
15. R. F. Novak and K. J. Woodcroft, 'The alcohol-inducible form of cytochrome P450 (CYP 2E1): role in toxicology and regulation of expression.' *Arch. Pharm. Res.*, 2000, **23**(4), 267–282.
16. J. L. Stevens, H. Liu, M. Halleck *et al.*, 'Linking gene expression to mechanisms of toxicity.' *Toxicol. Lett.*, 2000, **112–113**, 479–486.
17. C. N. Kind, 'The application of *in situ* hybridisation and immuno-cytochemistry to problem resolution in drug development.' *Toxicol. Lett.*, 2000, **112–113**, 487–492.
18. R. J. Vandebriel, H. Van Loveren and C. Meredith, 'Altered cytokine (receptor) mRNA expression as a tool in immunotoxicology.' *Toxicology*, 1998, **130**(1), 43–67.
19. D. D. Bowtell, 'Options available—from start to finish—for obtaining expression data by microarray.' *Nat. Genet.*, 1999, **21**(Suppl. 1), 25–32.
20. P. O. Brown and D. Botstein, 'Exploring the new world of the genome with DNA microarrays.' *Nat. Genet.* 1999, **21**(Suppl. 1), 33–37.
21. E. F. Nuwaysir, M. Bittner, J. Trent *et al.*, 'Microarrays and toxicology: The advent of Toxicogenomics.' *Mol. Carcinogenesis*, 1999, **24**, 153–159.
22. S. E. Wildsmith and F. J. Elcock, 'Microarrays under the microscope.' *Mol. Pathol.*, 2001, **54**(1), 8–16.
23. R. G. Halgren, M. R. Fielden, C. J. Fong *et al.*, 'Assessment of clone identity and sequence fidelity for 1189 IMAGE cDNA clones.' *Nucl. Acids Res.*, 2001, **29**(2), 582–588.
24. W. D. Pennie, J. D. Tugwood, G. J. Oliver *et al.*, 'The principles and practice of toxigenomics: applications and opportunities.' *Toxicol. Sci.*, 2000, **54**(2), 277–283.
25. P. Liang and A. B. Pardee, 'Differential display. A general protocol.' *Mol. Biotechnol.*, 1998, **10**(3), 261–267.
26. V. E. Velculescu, L. Zhang, B. Vogelstein *et al.*, 'Serial analysis of gene expression.' *Science*, 1995, **270**(5235), 484–487.
27. V. E. Velculescu, B. Vogelstein and K. W. Kinzler, 'Analysing uncharted transcriptomes with SAGE.' *Trends Genet.*, 2000, **16**(10), 423–425.
28. A. Lal, A. E. Lash, S. F. Altschul *et al.*, 'A public database for gene expression in human cancers.' *Cancer Res.*, 1999, **59**(21), 5403–5407.
29. A. H. van Kampen, B. D. van Schaik, E. Pauws *et al.*, 'USAGE: a web-based approach towards the analysis of SAGE data.' *Bioinformatics*, 2000, **16**(10), 899–890.
30. S. Brenner, M. Johnson, J. Bridgham *et al.*, 'Gene expression analysis by massively parallel signature

sequencing (MPSS) on microbead arrays.' *Nat. Biotechnol.*, 2000, **18**(6), 630–634.

31. M. J. Cunningham, S. Liang, S. Fuhrman *et al.*, 'Gene expression microarray data analysis for toxicology profiling.' *Ann. NY Acad. Sci.*, 2000, **919**, 52–67.

32. J. C. Rockett, J. Christopher Luft, J. Brian Garges *et al.*, 'Development of a 950-gene DNA array for examining gene expression patterns in mouse testis.' *Genome Biol.*, 2001, **2**(4).

33. M. Bartosiewicz, S. Penn and A. Buckpitt, 'Applications of gene arrays in environmental toxicology: fingerprints of gene regulation associated with cadmium chloride, benzo(a)pyrene, and trichloroethylene.' *Environ. Health Perspect.*, 2001, **109**(1), 71–74.

34. M. Rebhan, V. Chalifa-Caspi, J. Prilusky *et al.*, 'GeneCards: integrating information about genes, proteins and diseases.' *Trends Genet.*, 1997, **13**(4), 163.

35. R. S. Chabra, J. E. Huff, B. S. Schwetz *et al.*, 'An overview of prechronic and chronic toxicity/carcinogenicity experimental study designs and criteria used by the National Toxicology Program.' *Environ. Health Perspect.*, 1990, **86**, 313–321.

36. C. P. Rodi, R. T. Bunch, S. W. Curtiss *et al.*, 'Revolution through genomics in investigative and discovery toxicology.' *Toxicol. Pathol.*, 1999, **27**(1), 107–110.

37. T. P. Reilly, M. Bourdi, J. N. Brady *et al.*, 'Expression profiling of acetaminophen liver toxicity in mice using microarray technology.' *Biochem. Biophys. Res. Commun.*, 2001, **282**(1), 321–328.

38. M. E. Burczynski, M. McMillian, J. Ciervo *et al.*, 'Toxicogenomics-based discrimination of toxic mechanism in HepG2 human hepatoma cells.' *Toxicol. Sci.*, 2000, **58**(2), 399–415.

39. B. Ryffel, 'Impact of knockout mice in toxicology.' *Crit. Rev. Toxicol.*, 1997, **27**(2), 135–154.

40. V. R. Iyer, M. B. Eisen, D. T. Ross *et al.*, 'The transcriptional program in the response of human fibroblasts to serum.' *Science*, 1999, **283**(5398), 83–87.

41. U. Alon, N. Barkai, D. A. Notterman *et al.*, 'Broad patterns of gene expression revealed by clustering analysis of tumor and normal colon tissues probed by oligonucleotide arrays.' *Proc. Nat. Acad. Sci. USA*, 1999, **96**(12), 6745–6750.

42. D. E. Bassett Jr., M. B. Eisen and M. S. Boguski, 'Gene expression informatics—it's all in your mine.' *Nat. Genet.*, 1999, **21**(Suppl 1), 51–55.

43. O. Ermolaeva, M. Rastogi, K. D. Pruitt *et al.*, 'Data management and analysis for gene expression arrays.' *Nat. Genet.*, 1998, **20**(1), 19–23.

44. J. Aach, W. Rindone and G. M. Church, 'Systematic management and analysis of yeast gene expression data.' *Genome Res.*, 2000, **10**(4), 431–445.

45. K. Okubo, K. Itoh, A. Fukushima *et al.*, 'Monitoring cell physiology by expression profiles and discovering celltype-specific genes by compiled expression profiles.' *Genomics*, 1995, **30**(2), 178–186.

46. K. Okubo and K. Matsubara, 'Complementary DNA sequence (EST) collections and the expression information of the human genome.' *FEBS Lett.*, 1997, **403**(3), 225–229.

47. R. Pires Martins, R. E. Leach and S. A. Krawetz, 'Whole-body gene expression by data mining.' *Genomics*, 2001, **72**(1), 34–42.

48. M. B. Eisen, P. T. Spellman, P. O. Brown *et al.*, 'Cluster analysis and display of genome-wide expression patterns.' *Proc. Nat. Acad. Sci. USA*, 1998, **95**(25), 14863–14868.

J.P. Vanden Heuvel, G.H. Perdew, W.B. Mattes and W.F. Greenlee (Eds.)
Comprehensive Toxicology, Vol. xiv

# 14.5.5
# Proteomic Applications in Toxicology

## FRANK A. WITZMANN
*Indiana University School of Medicine, Indianapolis, IN, USA*

### 14.5.5.1 INTRODUCTION

With the sequencing of the human genome complete and its final annotation on the near horizon, we have reached the end of the genome era. The genomic information gleaned from these efforts is static and by itself does not reveal the complexity of cellular dynamics such as metabolic, structural, and signaling activities. Though the genome is identical in each individual somatic cell of an organism, gene expression is specific to each cell type and can be altered quantitatively and qualitatively in cellular dysfunction, injury or disease, following administration of pharmaceuticals or xenobiotics, or in response to other epigenetic phenomena. As a result, a new approach called "functional genomics" has gained prominence. This involves global and high throughput analysis of the mRNA and protein products expressed by the genome to correlate their expression profiles with specific effects of disease and altered cell function. However, functional genomics, at the level of mRNA gene transcripts (the *transcriptome*),[1] has been shown to correlate poorly with corresponding profiles at the protein level.[2-6] This strongly suggests that mRNA degradation, alternative splicing, co- and posttranslational modification, and posttranscriptional regulation of gene expression are all fundamental cellular processes that render extrapolations from mRNA to protein and cellular function potentially unreliable. As a result, growing evidence suggests that protein profiling[7] of the expressed genes in tissues and individual cells, e.g., protein signatures,[8] is likely to lead to a better understanding of cellular regulation and provide insights into the molecular effects of xenobiotics and disease. This approach called *proteomics* can be defined as the global analysis of the expressed genome. The term proteomics was initially coined at the 1994 Siena Two-Dimensional Electrophoresis Meeting by Marc Wilkins and published a year later.[9] Proteomics' roots are embedded in the union of two-dimensional electrophoretic separation of complex protein mixtures, mass spectrometric analysis of peptides, and bioinformatics. As these three approaches independently reached a critical level of development, their union was inevitable and their subsequent exploitation has been explosive.

This chapter is intended to give the reader an introduction into the what, why and how of proteomics, provide relevant examples of instances where this approach has been applied successfully in examining chemical toxicity, and offer a useful but not exhaustive review of relevant literature concerning this timely topic.

### 14.5.5.2 PROTEOMIC METHODOLOGIES

Despite its limitations,[10] the principle core technology of protein separation still lies in the two-dimensional electrophoretic (2DE) separation of the complex mixture of thousands of proteins represented in a living cell. Rather than an emerging technology, 2DE is an approach with a 26 year history. The analysis of protein expression profiles using a 2DE approach combining separation by isoelectric focusing and denaturing sodium-dodecyl sulfate electrophoresis, was first published in 1975 independently by Klose,[11] O'Farrell,[12] and Scheele.[13] Global protein analysis using 2DE was proposed early the next decade[14-15] and technical improvements making it a large-scale, quantitative and qualitative analytical tool have continued to the present.[16] In conjunction with 2DE, enzymatic cleavage of separated proteins, mass-spectrometric analysis of the resulting peptides, and online peptide-mass database comparison for protein identification comprise the typical "first-generation proteomics" platform.

In response to 2DE's shortcomings, efforts to streamline by automation[17] or completely replace the traditional approach with more straight-forward, rapid, and sensitive technologies with greater dynamic range in protein detection and identification are ongoing. One of these "second generation proteomics" technologies (see Section 14.5.5.2.5) includes isotope-coded affinity tag (ICAT) peptide labeling[18] that overcomes the need for running gels, is very sensitive, and has greater dynamic range. The development of non-redundant protein arrays from cDNA expression libraries[19] or surface-enhanced laser desorption/ionization (SELDI)[20] that enable less global, high-throughput screening are also examples of proteomic evolution. Finally, though with limitations of its own, the direct analysis of proteins in complex mixtures via multi-dimensional liquid chromatography and tandem mass spectrometry[21] portends to be useful in building proteomic databases rapidly.

A concise description of the various components of both first- and second-generation proteomic technology available to the toxicologist will be addressed in this section. For detailed accounts of all aspects of this approach, the reader should consult any of several useful books and reviews that have been published on this subject, several in cookbook form.[22-27]

#### 14.5.5.2.1 Two-dimensional Electrophoresis

2DE protein separation typically resolves 1000–2000 proteins in a single tissue sample[28] and up to 10,000 on a large format gel[29] when moderately sensitive stains are used. Proteins are separated based on their charge, a function of their constituent acidic and basic amino acids, by isoelectric focusing (IEF) in the first-dimension and then by molecular mass via sodium dodecyl sulfate–polyacrylamide gel electrophoresis (SDS-PAGE or DALT) in the second dimension. When combined in this manner, these two separation techniques produce a unique two-dimensional protein pattern representative of molecular phenotype. The 2DE pattern can then be subjected to image analysis where each protein spot is quantified based on its optical density and alterations in abundance, charge, and molecular mass can be determined relative to a standard pattern. Changes in abundance suggest up- or down-regulation of gene expression or altered protein turnover rates. Modifications in charge and/or molecular mass indicate post-translational protein modifications such as phosphorylation, glycosylation, prenylation, *etc.*, or genetic polymorphisms. Of particular relevance to the toxicologist are the detectable charge modifications associated with conjugation, chemical-adduct formation, or amino acid substitutions resulting from point mutations in the genome.

Completion of the 2DE procedure (2D separation and detection) results in a gel pattern such as the one illustrated in Figure 1.

##### 14.5.5.2.1.1 Sample Preparation

In the analysis of differential protein expression resulting from toxic exposure, optimal sample solubilization[30] is critical and unnecessary sample handling and artifactual alteration of the isolated proteins must be avoided. Generally, one should use a one-step solubilization technique that involves physical disruption of tissues/cells/organelles via ground-glass or Teflon homogenization, or ultrasonication in a medium that simultaneously solubilizes, denatures and reduces the sample proteins, breaks noncovalent and disulfide linkages, removes interfering substances such as nucleic acids, and inhibits the actions of intracellular proteases.[31] Prepared in this way, proteins can be separated as the monomers encoded by the genome while the covalent posttranslational modifications that typify certain proteins and cellular conditions remain unaltered.

A typical solubilization medium contains 7–9 M urea, 4% nonionic or zwitterionic detergent such as 3-[(3-cholamidopropyl)dimethylammonio]-1-propanesulfate (CHAPS) or Igepal CA-630 ([octylphenoxy]polyethoxyethanol), and 50–100 mM dithiotreitol (DTT) or tributylphosphine as a reducing agent. When immobilized pH gradient (IPG) strips are used for 1st dimension separation (see Section 14.5.5.2.1.2.1.2), protein hydrophobicity becomes a problem. In this case, combining urea and thiourea in a two- to four-fold ratio during solubilization and IPG strip reswelling (Section 14.5.5.2.1.2.1.2) prevents protein adsorption to the IPG strip and loss of resolvable proteins.[32]

##### 14.5.5.2.1.2 First-Dimension Isoelectric Focusing

After solubilization (and prefractionation if desired), protein samples are separated according to charge by either one of two common techniques, isoelectric focusing in tube gels with low acrylamide concentrations or immobilized pH-gradient electrophoresis on gel strips.

###### 14.5.5.2.1.2.1 Isoelectric focusing (IEF)

In IEF, charge separation is accomplished by adding carrier ampholytes, a mixture of aliphatic polyaminopolycarboxylic acids with various pH ranging anywhere from 2 to 11, to a low concentration polyacrylamide gel tube. This creates a relatively linear environment of increasing $H^+$ concentration in which a protein, placed at the basic end of the IEF gel, migrates until it reaches a pH where its surface charge equals 0 and its electrophoretic migration ceases. This is the isoelectric point (pI), a unique characteristic of each protein that is solely dependent on amino acid composition and modifications therein. This powerful separation technique enables sharp resolution of complex protein mixtures but suffers from batch-to-batch variability, gradient instability that leads to significant inter- and even intra-laboratory variation, poor resolution at basic pH due to cathodic drift, electroendosmosis, and physical fragility. However, at narrow ranges of pH (4–7) and moderate protein loading (100–200 μg), IEF remains an exceptional first dimension separative tool.

###### 14.5.5.2.1.2.2 Immobilized pH gradient electrophoresis (IPG)

To overcome the limitations of IEF, dried gel strips copolymerized with Immobilines[TM]

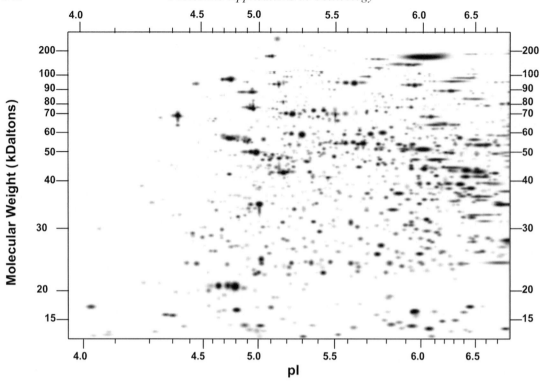

**Figure 1**  Representative proteome profile of mouse liver on a 2DE gel pattern. First dimension separation was accomplished by IEF using pH 4–8 ampholytes and second dimension separation on a 1.5 mm thick slab with an 11–17% polyacrylamide gradient. Proteins were visualized using colloidal Coomassie blue staining. Molecular weight and pI calibrations are estimates based on the calculated MW and pI of protein spots identified in the pattern (via EXPASY Compute pI/MW Tool, http://www.expasy.ch/tools/pi_tool.html).

were developed. These IPG strips consist of immobilized pH gradients[33] with several distinct advantages over conventional IEF. In contrast to IEF separations for 2DE, IPG pI resolution up to 0.01 pH unit has been reported[33] in highly stable gradients where cathodic drift is minimal. Because each IPG strip is cast with a GelBond backing, it has improved physical stability that simplifies handling and avoids gel breakage. Gel sample loading has also been optimized by taking advantage of the commercially available IPG strip shipped in dehydrated form. By rehydrating (reswelling) the IPG strip with sample buffer[34] problematic sample precipitation at the application point is avoided, control of quantitative sample loading is improved, and if necessary >10 mg protein can be loaded to improve detection of low abundance proteins. Recent improvements in pH range and alkaline stability have produced IPG strips with functional immobilized gradients of pH 3–12.[35]

### 14.5.5.2.1.3  *Second Dimension SDS-PAGE*

After first-dimension IEF or IPG, the gel or strip is placed on a polyacrylamide gel slab

(DALT) for separation by molecular mass, producing a 2D gel pattern (so-called IEF-DALT or IPG-DALT). The procedures involved in SDS slab gel electrophoresis vary according to apparatus available with most 2D systems limited to 2–12 slab gels per run. Some large-scale systems support 20 gels per run[36–37] and these are particularly useful in toxicology studies where the effect of several dose groups on protein expression are being compared. A highly parallel slab gel system significantly reduces gel-to-gel variability, improves gel imaging and quantitative/qualitative protein analysis, and is an essential component of automated, large scale, high-throughput proteomics systems.[38]

Sodium dodecyl sulfate (SDS) is an anionic detergent that denatures proteins and forms anionic complexes with constant net negative charge/mass ratio. Proteins treated with both SDS and a reducing agent like dithiotreitol (DTT) can be separated exclusively by molecular weight. Most 2DE systems use a tris-glycine buffer system similar to that described by Laemmli.[39] Based on its sieving effect,[40] acrylamide significantly affects protein mobility particularly when one alters acrylamide concentrations. Incorporation of acrylamide gradients optimizes resolution of heteroge-neous mixtures of low, high and intermediate molecular weight proteins typically found in target tissues such as liver, kidney, *etc.*

## 14.5.5.2.2 Protein Detection and Quantification Methods

Vital to an effective proteomics platform is the ability to quantitatively detect the resolved protein pattern in a manner that is both accurate and reproducible. Numerous methods can be used to stain/visualize/detect proteins separated by 2DE and all have their limita-tions. Those amenable to automation will be the methods of greatest utility to proteomics in the future.

### 14.5.5.2.2.1 *Protein Staining*

Historically, one of the first dyes with widespread use for electrophoretic protein staining was Coomassie Brilliant Blue (CBB) R-250 because its staining intensity is relatively linear over a wide range of protein concentra-tion[41] and thus excellent for densitometric protein quantification. CBB-stained proteins are particularly useful for proteomic studies because they are compatible with mass spectro-metry of their tryptic digests (Section 14.5.5.2.3.2). However, CBB R-250 is rather insensitive ($50 \, ng/mm^2$) so that low abundance proteins are often undetectable. By using a colloidal CBB G-250 staining protocol,[42] protein detection can be greatly improved as it nearly reaches the sensitivity of silver stain (e.g., 0.5–1 ng).

To overcome CBB R-250's lack of sensitiv-ity, protein electrophoretic silver stains were perfected.[43] Silver staining can be very sensitive ($0.01 \, ng/mm^2$) and linear over a 40–50 fold concentration range (up to $2 \, ng/mm^2$) com-pared to a linear range of 10–200 ng for CBB.[44] However, mass spectrometry of tryptic digests from silver stained proteins is problematic[44] and requires the additional step of silver removal (destaining) with potential protein loss. Furthermore, slight variations in the reagents can cause major differences in staining intensity and background thus creating quan-tification problems in large-scale experiments with very large numbers of gels.

Recent modifications in silver stain techni-que have rendered it compatible with MALDI-MS and ESI-MS/MS[45] and formulations are now commercially available. A few other staining techniques that are sensitive, rapid, and compatible with peptide mass spectro-metry are reversible metal chelate stains,[46] an imidazole-zinc negative stain[47] that requires no in-gel protein fixation, and fluorescent SYPRO Ruby protein gel stain.[48] The development of spot-cutting robotic systems compatible with both visible and fluorescent detection and the need for high-throughput in proteomics makes SYPRO staining even more advantageous.

### 14.5.5.2.2.2 *Radiolabeling*

Separated proteins can also be detected and quantified by radiolabeling (*in vivo* or *in vitro*) and autoradiography, mainly using sensitive and flexible phosphor imaging.[49] In addition to providing a very sensitive means for detecting resolved proteins, this approach has the advantage of enabling double-label experi-ments[50] where various cellular/biochemical events can be identified simultaneously using different radiolabels. As with all radioisotopic techniques, disadvantages of this approach involve storage/disposal difficulties, the cost of phosphor imaging equipment, and the specificity of phosphor screens.

### 14.5.5.2.2.3 *2D Protein Image Analysis*

Successful application of 2DE in toxicology relies on accurate differential protein analysis. This necessitates precise comparison of indivi-dual protein levels in control versus exposed/treated groups to properly describe the dynamics of injury and repair processes and to distinguish the magnitude of various chemi-cal effects. Therefore, optimized protein separation, detection, and identification meth-ods are central to this application and to proteomics in general. There is currently no practical approach to routinely and precisely measure the absolute amounts of each protein present in a typical 2D gel. However, one can accurately quantify protein abundances rela-tive to a reference pattern. Most proteomic investigations use CBB or silver staining, or both, for densitometric protein quantification. As a result, relative protein abundance can be determined and quantitative differences in protein expression statistically compared using any one of a number of commercially available 2D gel analysis systems.

Most 2D gel image analysis packages com-monly consist of (a) scanning a stained 2DE pattern via laser densitometry or photodiode array; (b) digitizing the image; (c) processing the image (e.g., background and streak sub-traction and mathematical morphology opera-tions to detect and optimize each spot); and (d) creating a reference pattern to which all other patterns are then matched to create a set of essentially superimposable 2D protein patterns.[51] Lists of protein spot abundances can then be subjected to statistical analysis to search for quantitative, qualitative, and positional (charge) alterations in the proteome patterns (see Figure 2).

### 14.5.5.2.3   Protein Identification

After 2DE separation and image analysis, any or all of the proteins in the gel pattern can be selected for identification. In large proteomic projects, the ultimate goal is to identify all expressed proteins.[9] In less ambitious studies, only those proteins altered in response to some experimental manipulation/disease state or those detected in greatest abundance are assigned for identification. Regardless of project scope, protein identification is becoming less problematic as a result of technical developments in both traditional approaches that are slow and labor intensive as well as contemporary techniques that are fast, less

expensive and appropriate for high-throughput applications.[52]

#### 14.5.5.2.3.1   Immunoblotting

The most traditional method of protein identification involves Western blotting and immunostaining.[53] Proteins on 2DE gels are transferred electrophoretically onto a suitable membrane (e.g., nylon, nitrocellulose or polyvinylidene fluoride [PVDF]) and the membrane is probed with a primary antibody that binds to the desired membrane-bound protein. If the desired protein is present, it is detected by application of a secondary antibody incorporating either a radiolabel or an enzyme that yields a colored product upon addition of an appropriate substrate. While this approach can be extremely precise, it is labor intensive and requires significant optimization for each protein. Furthermore, non-specific binding of either primary or secondary antibodies can be misleading. Because immunoblotting searches for known proteins with available antibodies, it is unsuitable for global protein identification. This technique's greatest utility may relate to

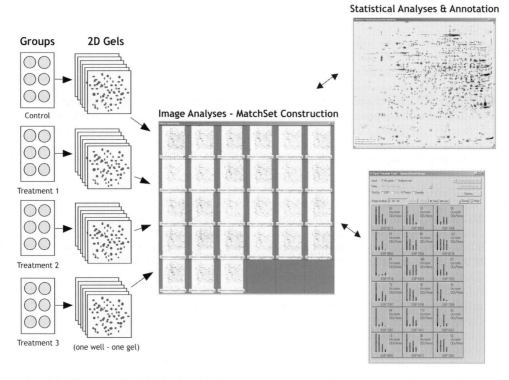

**Figure 2**   This illustrates the initial gel analysis component of a typical *in vitro* toxicoproteomics approach used in the author's laboratory. Groups are represented by 6-well plates in which cells, e.g., hepatocytes, are cultured and solubilized, after which their proteins resolved on separate 2D gels, stained and scanned for image analysis. The effect of various *in vitro* exposures can be compared to a control group by assembling scanned gel images into match sets using a 2D gel analysis system such as PDQuest[TM] (Bio-Rad, Hercules CA) and statistical comparisons of protein spot abundance data.

the selective use of antibodies whose epitopes recognize posttranslationally modified proteins such as phosphoproteins[54] or other covalent adducts[55] to characterize epigenetic phenomena.

### 14.5.5.2.3.2 *Mass Spectrometric Peptide Mass Fingerprinting*

With vast electronic protein sequence databases containing an ever increasing fraction of the proteins coded by organismal genomes, and mass spectrometric techniques that now enable the routine analysis of proteins and peptides with extremely high sensitivity and excellent precision, high-throughput protein identification via peptide mass fingerprinting is quite practical. This approach has been reviewed thoroughly[56] and will be described briefly in this section. Essentially, proteins separated on a 2D gel are cut from the gels, proteolytically fragmented into their constituent peptides, resulting peptide masses are detected by mass spectrometry, and the resulting masses are compared to theoretically derived databases of peptide masses generated by calculating all possible mass fragments (using a variety of fragmentation techniques) for all known protein sequences.[57–61]

#### 14.5.5.2.3.2.1 *Preparation of peptide fragments*

Though several different approaches to generating measurable peptides are possible, the enzymatic in-gel digestion of proteins using a suitable protease such as trypsin is most common. Trypsin is an endopeptidase that cleaves peptide bonds at the C-terminal side of lysine (K) or arginine (R) residues. This protease, along with Lys-C, has been the preferred enzyme for proteomic applications mainly because it generates peptides in the 800–2500 kDa range, optimal for contemporary mass spectrometric techniques. Nonenzymatic cyanogen bromide cleavage as well as other proteases such as ArgC, AspN, GluC (bicarbonate), GluC (phosphate), and chymotrypsin, also work. Various peptide mass databases are designed to reflect each unique cleavage point.

A detailed protocol of various peptide fragment preparation techniques (in-gel and on-membrane) in recipe form has been published recently.[62] Briefly, a gel plug no larger than the stained protein spot is excised either manually or robotically. The gel piece is then washed to remove detergents and other con-

founding materials followed by reduction and alkylation of cysteine residues. The gel is dried and rehydrated by the addition of buffered protease, the protein is digested (often overnight), and the peptides are extracted from the gel piece(s). After extraction, the peptides can be concentrated, desalted and eluted using microcolumn chromatography such as the ZipTip C18 from Millipore. This in-gel approach is particularly advantageous to large-scale protein characterization because each step can be automated.

Electroblotted proteins can also be analyzed in this way. However, robotic systems are less useful in this regard because membrane pieces tend to float on the aqueous solutions making protein digestion more problematic. Immobilon[TM]-CD, a cationic, hydrophilic, charged-modified PVDF membrane; Immobilon[TM]-PSQ, a hydrophobic PVDF membrane with a pore size of 0.22 μm; or Immobilon[TM]-NC membrane, a combination of nitrocellulose and cellulose acetate, can all be used.

Following the generalized steps presented above, the peptide mixture is ready for mass analysis.

#### 14.5.5.2.3.2.2 *Matrix-assisted laser desorption/ionization (MALDI) mass spectrometry*

Another important technical development that has advanced proteomics is the rapid improvement in the performance of mass spectrometers.[63] At the heart of mass spectrometry (MS) is the production of gas-phase ions from solid or liquid states and the accurate determination of their mass. Proteomics-related advances are the direct result of improvements in both ionization technique (matrix-assisted laser desorption and nanoelectrospray) and mass analyzers (linear and reflectron time-of-flight [TOF], various quadrupole configurations, and ion-traps). The principles of ionization and mass analysis for peptide and protein analysis, and application of MALDI and electrospray ionization (ESI) to protein identification and sequencing have been reviewed previously.[64] This section will summarize the relevant aspects of these techniques.

In conventional MALDI-MS analysis, concentrated peptide sample mixtures are combined with a crystalline matrix (e.g., alpha-cyano-4-hydroxycinnamic acid) whose role is to segregate the peptides, absorb the laser light, and transfer the energy to the analyte molecules. The analytes (peptide masses) are ionized by simple protonation by the photo-excited matrix and this leads to the formation of the typical $[M + X]+$ type ion species in what is called a "soft ionization" mechanism. Ionization renders the intact peptides into the gaseous phase. Various matrices are available and the concentration of analyte relative to matrix is typically kept low for efficient, uniform ionization and sensitive detection with high mass accuracy ($\pm$ 20 ppm). Ionized peptides are accelerated in a magnetic field down a time-of-flight tube, detected by a linear detector, and the flight time from ionization to detection is calculated. Because the flight time is dependent solely on momentum, the root of mass/charge ($m/z$), with accurate calibration absolute analyte masses can be calculated easily. The resulting spectrum of masses can then be saved for online database searching (Figure 3).

Despite improved instrument mass accuracy, ambiguous identifications via MALDI-MS do occur occasionally and require additional information such as peptide sequence data. Some MALDI instruments incorporate a reflectron to effectively increase the TOF free-flight path, thereby increasing resolution and improving mass accuracy. This technology also allows researchers to study molecular structure of ions (peptide sequence) via post-source decay (PSD), in which ionized fragments decompose further in the flight tube and the secondary products provide additional information about the structure of the original ion.[65] PSD-derived sequence information is used to resolve ambiguous identifications at picomole quantities and can be used to determine the qualitative nature of post-translational modifications on individual peptides.

MALDI-MS application in proteomic analysis is particularly functional due to its speed, accuracy, chemical tolerance, solid phase ionization, and its successful automation (multiple targeting). Perhaps one of the most important considerations is cost. Several just-for-proteomics MALDI-MS instruments are currently available from a number of vendors and all are relatively inexpensive relative to high-end tandem and multi-quadrupole mass spectrometers. Another unique and potentially automated feature of MALDI MS lies in its capacity to ionize electroblotted proteins/pep-

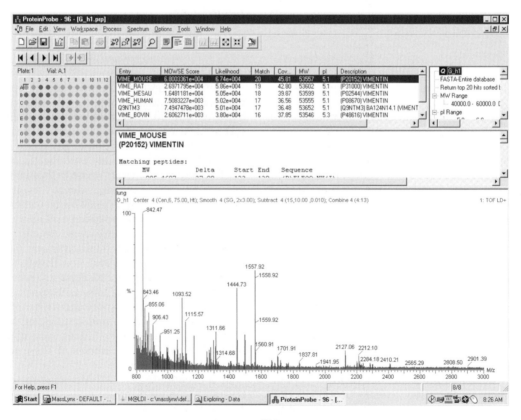

**Figure 3** Screen capture obtained from ProteinProbe™ software display, depicts the mass spectrum of a protein digested by trypsin and detected by MALDI-MS. The search engine automatically searches the monoisotopic peptide masses against databases of known protein sequences. Online searching of the optimized spectrum enables rapid protein identification and subsequent databasing of results.

tides directly on membranes using a scanning IR-MALDI-MS. Future development of this approach may very well include on-blot digestion and direct peptide mass analysis via scanning IR-MALDI-MS.[17]

### 14.5.5.2.3.2.3 Electrospray ionization mass spectrometry (ESI-MS)

The alternative to solid phase peptide MALDI-MS is ESI-MS of liquid phase analytes. ESI separation[66] is based on very low flow rates of peptide eluents from high pressure liquid chromatography (HPLC) columns for LC/MS or capillary electrophoresis (CE) columns for CE/MS into a nanoelectrospray ion source that evaporates the solvent and creates multiply charged ions in a droplet cloud. Various voltages are applied to the quadrupole electrodes to direct the ions toward the mass analyzer and along the way trap and eject them according to their mass-to-charge ratios. The linear quadrupole mass analyzer thus acts as a mass filter by varying potentials on the quadrupole rods to pass only a selected mass-to-charge ratio. Ions are ejected in order of increasing mass-to-charge ratio and detected by the ion detector resulting in a precise mass spectrum.

One of the advantages of ESI relates to the multiply charged species it generates. Because the average charge state increases in an approximately linear fashion with molecular weight, the true molecular mass of an ion can be calculated since more than one charge state is observed. Furthermore, multiple quadrupoles can be added in tandem (MS/MS) to make multiple ion interpretation more robust and facilitate peptide sequence analysis. For example, in a triple quadrupole instrument, peptide ion masses are measured by the first quadrupole, a peptide of interest is sent to the second quadrupole where it is fragmented by collision-induced dissociation (CID), and the fragment masses measured at sub-picomole levels in the third quadrupole known as the "sequence tag" method.[67] For peptide mass mapping, the mass information from several peptides derived from the same protein is required for protein identification by database searching. In contrast, the CID spectrum of a single peptide can contain enough information for unambiguous identification of a protein. It is a direct result of this technical development that it is now possible to identify the components of even relatively complex protein mixtures without prior purification of individual proteins by using two-dimensional chromatography methods coupled to ESI-MS/MS.[21]

In contrast to filtering the ions, a third type of mass spectrometer, the quadrupole ion trap mass spectrometer, uses three electrodes to trap ions in a small volume. The mass analyzer consists of a ring electrode separating two hemispherical electrodes. A mass spectrum is obtained by changing the electrode voltages to eject the ions from the trap. The high resolving power and efficiency of the ion trap allows for high-resolution precursor ion selection and subsequent highly sensitive MS/MS analysis for peptide sequence analysis,[68] plus it enables the investigator to "hold" the ions without losing them as one would in a multiple quadrupole instrument.

Various combinations of the ionization, filtering, trapping, and detection systems are possible and lead to substantially greater precision of mass measurement. Whatever the means the investigator chooses (or is constrained) to employ, the resultant peptide masses and peptide sequences derived from the spectra are then suitable for submission to a database for searching, protein identification, and characterization of post-translational modifications.

### 14.5.5.2.3.3 Protein Sequence and Structure Databases

The annotated protein sequence and peptide mass information in expanding databases is the last crucial element in the proteomics approach. In high-throughput proteome projects where large numbers of protein spots must be analyzed and identified within a short period, the present identification method of choice seems to be peptide mass finger-printing via MALDI-TOF or ESI-MS, and/or MALDI-PSD or ESI-MS/MS sequence analysis. Peptide mass spectra and partial sequence data obtained from highly resolved protein spots are of little immediate use without relating them to known gene and protein sequences. To identify a 2DE protein spot unambiguously, one must search all available, up-to-date peptide mass, MS/MS sequence tag and/or MS/MS fragment-ion databases. These sources are all based on a constantly expanding volume of protein sequence data contained in various protein databases such as SWISS-PROT, PIR, PRF, PDB (115 databases are presented in the 2000 Database issue of Nucleic Acids Research). Currently, several such databases are accessible via the WWW and are listed in Table 1.

#### 14.5.5.2.4  Detecting Posttranslational Modifications

This section will deal briefly with posttranslational modifications (phosphorylation, glycosylation, prenylation, acetylation, adduct formation, *etc.*) and their relevance to proteomics[69] in general and toxicoproteomics specifically.

Translation of mRNA to a polypeptide chain is often preceded by RNA splicing or followed by proteolytic cleavage so that the final protein product has little semblance to that predicted by the genome. In contrast, a protein may be synthesized exactly as coded but specific amino acid residues in the chain may be modified chemically to transform its physicochemical characteristics in a manner not necessarily predicted by the original gene sequence.[70] A good example of this can be found in the alteration of pI (and frequently electrophoretic mobility in terms of mass) by chemical modification of a protein's constituent amino acids,[71] alterations that can be detected, quantified by 2DE,[72] and identified by mass spectrometry.[73–75] However, MS analysis of posttranslational modifications is not a trivial task. First, due to low stoichiometric modification of proteins, extremely high sensitivity of detection for the modified peptides is essential. Second, identification of the modification requires the isolation and analysis of *the specific peptide* that contains the modified residue(s).

Natural cellular processes such as signaling systems, enzymatic regulation (activation/deactivation), intracellular translocation, and structural associations and dissociations are characterized by *in vivo* posttranslational modification of proteins. Protein phosphorylation is essential in cell regulation,[76] phosphoproteins are therefore numerous, and their detection and proteomic analysis can be conducted effectively with[73] or without[56,77] 2DE. Furthermore, it has recently been postulated that more than half of all proteins are glycoproteins.[78] Posttranslational modifications are also the direct consequence of chemical intoxication (pharmaceutical or otherwise), frequently when normal conjugation systems are overwhelmed by chemical exposure and reactive intermediates form protein adducts. Identification of the specific adduct poses significant challenges, but with CID spectrum analysis (Section 14.5.5.2.3.2.3), it is possible.

Accordingly, one aspect of the power of proteomic analysis is its ability to enable the detection, quantification, and identification of xenobiotically induced protein modifications (e.g., in databases such as FindMod[79] *http://www.expasy.ch/tools/findmod*). As will be illustrated later in this chapter, this is a feature of great importance for toxicological research.

**Table 1**  Web-based protein identification resources.

| Database | URL |
|---|---|
| ATLAS | http://speedy.mips.biochem.mpg.de/mips/programs/atlas.html |
| BLAST | *http://www.ncbi.nlm.nih.gov/BLAST/* |
| CombSearch | *http://cuiwww.unige.ch/∼hammerl4/combsearch/* |
| EXPASY | |
|    AACompIdent Tool | *http://www.expasy.ch/tools/aacomp/* |
|    MultiIdentTool | *http://www.expasy.ch/tools/multiident/* |
|    PeptIdent Tool | *http://www.expasy.ch/tools/peptident.html* |
|    TagIdent | *http://www.expasy.ch/tools/tagident.html* |
| FAST3 | *http://www2.ebi.ac.uk/fasta3/* |
| Mascot | http://www.matrixscience.com |
| MassSearch | http://cbrg.inf.ethz.ch |
| MOWSE | *http://srs.hgmp.mrc.ac.uk/cgi-bin/mowse* |
| OWL | *http://www.bis.med.jhmi.edu/Dan/proteins/owl.html* |
| PepFrag, ProFound | *http://prowl.rockefeller.edu/* |
| PeptideSearch | *http://www.mann.embl-heidelberg.de/Services/PeptideSearch/* |
| ProteinProspector | |
|    MS-Fit | http://prospector.ucsf.edu/ucsfhtml3.4/msfit.htm |
|    MS-Tag | http://prospector.ucsf.edu/ucsfhtml3.4/mstagfd.htm |
|    MS-Seq | http://prospector.ucsf.edu/ucsfhtml3.4/msseq.htm |
|    MS-Pattern | *http://prospector.ucsf.edu/ucsfhtml3.4/mspattern.htm* |
| SEQUEST | http://thompson.mbt.washington.edu/sequest |

#### 14.5.5.2.5 Relevant Emerging Technologies

Several new methods are now available for the analysis of protein expression. Each addresses the fundamental limitations of 2DE-based proteomics, e.g., sensitivity and throughput, characteristics essential to the pharmaceutical industry for screening putative drugs, though none of these methods is likely to completely replace 2DE-based proteomic application in toxicology just yet.

##### 14.5.5.2.5.1 *Protein Microarray Technology*

Development of protein chip (array) technology ranges from (1) spotted-protein arrays derived from cDNA libraries where for instance, >25,000 proteins can be screened for antibody-antigen interaction;[80] to (2) printed protein arrays for protein function studies on genome-wide level;[81] to (3) ligand arrays being developed by Ciphergen Biosystems (Palo Alto, CA, USA) known as the SELDI (surface enhanced laser desorption/ionization) ProteinChip[TM] Array (http://www.ciphergen.com/F1_tech2.html). While none of these approaches represents a truly dedicated methodology to enable the global determination of differential protein expression by a toxicological target tissue as a snapshot in time, they do address improvements in the important issues of speed/throughput, cost, and sensitivity. For instance, in SELDI technology, spots up to 1 mm in diameter on each chip can contain either a chemical (ionic, hydrophobic, hydrophilic, *etc.*) or biochemical (antibody, receptor, DNA, *etc.*) surface designed to capture proteins of interest. Application of a complex protein mixture onto the chip results in selected protein binding. This is followed by a wash step to remove unbound proteins, after which the chip is analyzed in a SELDI-TOF-MS instrument to determine the molecular weights of the bound proteins. It is claimed that this system enables the separation, quantification, identification, and characterization of previously characterized proteins at femtomole levels. However, because it is based on analysis of already characterized proteins, the ProteinChip[TM] is much like a very highly parallel ELISA (Enzyme Linked ImmunoSorbent Assay).

##### 14.5.5.2.5.2 *Isotope Coded Affinity Tags*

To address the limitations of 2DE regarding sensitivity of protein detection, Gygi *et al.*[18] have developed isotope-coded affinity tags (ICATs) to quantify relative differences in protein expression in different biological samples. This approach is based on stable isotope dilution techniques using chemical reagents called isotope-coded affinity tags (ICATs) combined with tandem mass spectrometry. The ICAT reagent consists of a biotin affinity tag (to isolate the peptides), a linker to incorporate stable isotopes, and a group that reacts with cysteines. In essence, the technique involves reacting cellular protein mixtures from different experimental conditions with light and heavy ICATs. The mixtures are combined, proteolytically digested to yield peptides, labeled peptides isolated via biotin-tag, peptides separated and quantified by HPLC (relative quantification based on ratio of peptide pairs), and proteins identified by MS/MS. Even very low abundance proteins can thus be accurately quantified *and* identified in the same sample mixture. Because this technique accurately compares relative abundance of proteins at very low levels of expression, its utility in toxicology seems extremely promising.

##### 14.5.5.2.5.3 *Fluorescence 2D Difference Gel Electrophoresis (DIGE)*

A technical approach based on the work of Unlu,[82–83] now refined and commercially available, enables the 2DE separation of up to three different protein samples on a single 2D gel. This is accomplished by labeling each of three different protein samples (tissues, cells, organelles, or body fluids) with each of three different fluorescent dyes (e.g., Cy2 Cy3 and Cy5), mixing, co-separating via 2DE, detecting each fluor with a special scanner, and analyzing the overlayed images. Because each protein migrates to the same *x, y* coordinate position in each sample gel, the merged images rapidly and easily display any protein abundance differences between the three samples, simplifying and improving comparative accuracy.

This approach addresses the problems of gel-to-gel variability inherent in most 2DE analysis systems. DIGE also has the advantage of combining improved throughput/reduced experimentation time, MS-compatibility, sub-nanogram sensitivity, linear response to variation in protein concentration over five orders of magnitude, and a broad dynamic range. Though this approach has obvious toxicological applications, two of its shortcomings include the need for expensive fluorescence imaging reagents, instrumentation and software and the necessity for dividing sample proteins for fluor labeling. In many toxicological proteomics applications (e.g., *in vitro* exposures, tissue micropunches, or laser dissected samples) sample size is already minimal and protein concentration at low detection limits. Dividing the sample in two or three undermines the sensitivity of the approach. Nonetheless, DIGE has already proven its utility in recent toxicologic applications[84] and likely will continue to be an important approach in toxicoproteomics.

### 14.5.5.2.5.4 Proteomics on a Chip Technology

As a result of rapid developments in the area of microfabrication of analytical devices, various aspects of the proteomics process are now being carried out using microfluidic devices. This huge downscaling of the front-end elements of the proteomic approach such as protein extraction, fractionation and digestion can be coupled seamlessly to MS analysis and protein identification. Extremely high-throughput, completely automated, high sensitivity, and relatively inexpensive proteomics will be the result. Recently reviewed,[85] several of these promising developments are still in the proof-of principle stage but their eventual implementation bodes well for the advance of proteomics in all applications.

### 14.5.5.3 TOXICOPROTEOMICS

It is generally agreed that the biological effects of most chemical exposures, including efficacious and adverse effects of drug administration, reveal themselves in target cells as protein modifications resulting from either altered gene expression or epigenetic mechanisms, or both. Therefore, *in vivo* and *in vitro* toxicologic exposure systems and proteomics (i.e., toxico-proteomics) represent a more powerfully relevant analytical combination of conventional techniques than any other available to the modern toxicologist. This notion is supported by the increasing frequency with which scientific papers addressing or including this approach are appearing in the proteomic and toxicological literature.

The growing trend in protein analytical applications in toxicology is likely a direct result of the utility toxicologists now find in the informative power of proteomics. Prior to the proteomics and functional genomics revolution, while protein separation techniques were adequate, protein identification was slow and tedious. Laboratories that lacked image analysis capabilities labored over black and white photos of stained gels or autoradiographs and

toxicologists found little informational gain for all the technical effort required.

Application of global protein analysis in toxicology thus reflects the technical advances made throughout the last three decades such that today, proteomics is used as a sensitive means to screen for chemical toxicity, to investigate toxic mechanisms, and to develop biomarkers of toxic exposure and effect. As a consequence of the very recent developments in proteomic core technologies, (e.g., mass spectrometry and protein bioinformatics), toxicoproteomics is now positioned to revolutionize both toxicity testing and drug development.

This section will present a brief historical perspective of 2DE applications in toxicology and give a few representative citations of the more than 300 published manuscripts that detail efforts in which protein analysis based on 2DE, conventional proteomics, or the emerging proteomic technologies has been used to study the biological effects of xenobiotic chemicals and to understand the protein molecular correlates of toxicologic endpoints.

The first applications of comprehensive protein analysis involved the visual analysis of stained gels or autoradiographs as in the analysis of protein alterations associated with thioacetamide-induced liver injury.[86] It was soon clear that visual inspection of complex 2D gel patterns was inadequate, and so Lemkin *et al.*[87] published the first computerized image analysis of asbestos exposed phagocyte 2D protein patterns. 2DE technology as applied to studying multiple dose groups was greatly facilitated by the first real attempt to increase throughput; the development of large-scale 2DE technology[36-37] in which 20 individual samples could be separated simultaneously under identical running conditions. This approach eliminated gel–gel variability as a confounding factor and is therefore still used successfully for similar applications in many laboratories today. Improved gel apparatus and correspondingly better intra- and inter-run gel uniformity aided the application of computerized 2D protein pattern imaging analysis systems[88-92] to multi-gel experiments that in turn enabled the development of toxicologically-relevant protein databases[93-95] to chronicle the effects of chemical exposures. A typical toxicoproteomics approach incorporating these principles is illustrated in Figure 4.

#### 14.5.5.3.1 Examples of *In Vivo* Studies

Proteomic approaches have been used to study a broad range of toxicants in a variety of test systems in diverse models and target

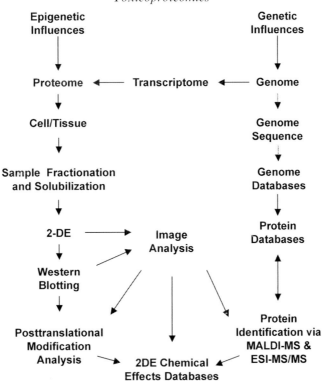

**Figure 4** Typical elements of a proteomics approach that enable one to generate proteome profiles and identify proteins for inclusion in various databases.

tissues. The nature of the protein analysis approach has been diverse as well. For instance, in the examination of peroxisome proliferators (PP) such as Wy-14,643 ([4-chloro-6-(2,3-xylidino)-2-pyrimidinylthio]acetic acid) and the phthalate-ester plasticizer di-(2-ethylhexyl)-phthalate (DEHP) in rodents, early proteomic studies reported alterations in protein expression without extensive image analysis and with either no or limited protein identification.[96–97] Ensuing studies combined sophisticated image analysis and Western blot techniques to provide statistical analyses of protein alterations, identify a limited set of altered proteins, and describe patterns of protein alteration indicative specific chemical exposure.[98–102] The combined results of these efforts provided evidence supporting the proposed receptor-based mechanism regulating PP, showed species-specific changes in proteins that were independent of the PP response, and demonstrated expression alterations in complex sets of proteins that could be implicated in carcinogenesis. More recently, the use of contemporary proteomics has resulted in a better understanding of the range of chemicals involved in this response, specific enzyme systems up- and down-regulated by exposure, and their effect on mechanisms of cell death.[103–106]

A toxicoproteomic approach (*in vivo*) has been used to study other hepatotoxic agents,[73,107–109] the nephrotoxicity of metals,[110–111] drugs,[112] and jet fuel,[113–114] pulmonary toxicity via analysis of lung[115–116] and bronchoalveolar lavage fluid (BALF),[117] immunotoxicity,[118] reproductive toxicity,[119–120] cardiac toxicity,[121–122] and neurotoxicity.[123–124] A rat serum 2D protein database also has been made available for potential use in toxicology studies.[125]

As a final example, proteomics has been used to investigate the effects of exposure to various types of radiation, such as ultraviolet,[126–127] ionizing,[128–130] and electromagnetic radiation.[131–132]

### 14.5.5.3.2  Examples of *In Vitro* Studies

Advances in *in vitro* technology have made this approach a valuable tool in solving toxicological problems. Despite their advantages and disadvantages, *in vitro* test systems form a natural combination with proteomics (Figure 2) and thus numerous examples of this productive articulation can be found in the literature.

Primary hepatocyte cell culture, perhaps the most applicable of test systems, has been used

to investigate the effects of peroxisome proliferators,[104,106] isoproterenol[133] and cell-sig-
naling modulators.[134–135] Other tissues lend
themselves to this approach as well. For
example, the effect of cadmium was studied
in primary proximal tubule cell culture,[136]
copper toxicity was studied in a minnow
epithelial cell line,[137] and epidermal exposure
to a tumor promoter was simulated in explant
culture[138] and in cultured keratinocytes.[139]
Immune cells[140–141] and pancreatic cells[142] are
good candidates for this application as well.

As an alternative to primary cell culture,
tissue slice technology has been greatly
improved[143] and though only a few studies
have combined proteomics with this novel
approach,[144–146] its potential application in
toxico- and pharmacoproteomics seems promising.

### 14.5.5.3.3  Chemically-induced Charge Modifications

As introduced in Section 14.5.5.2.4, 2DE is
particularly well-suited to enable the detection
and quantification of posttranslational protein
modifications that occur naturally and when
detoxification systems are overwhelmed by
chemical exposure and intermediates react
covalently with neighboring proteins.[147] As
illustrated in Figure 5, addition of negatively
charged protein adducts results in a leftward
shift in pI. In this relationship, the greater the
extent of chemical modification of an individual protein, the greater its detectable charge
heterogeneity. Calculation of the charge modification index (CMI)[72] enables one to accurately quantify the extent of adduct formation
and conduct statistical analysis on the empirical expression of these phenomena.

Numerous studies have exploited this capability by observing changes in charge heterogeneity as indicative of acetaminophen
toxicity,[73,148–150] halothane hepatitis,[151] and
cocaine toxicity,[152] and a few others have
used the CMI calculation to accurately quantify and compare putative adduct formation.[72,99,153–157]

### 14.5.5.3.4  Limitations

Because detection sensitivity is an important
issue in global protein analysis and proteomics
has been criticized for its lack thereof, one
approach that addresses this issue is tissue
fractionation.[158] A recent review[159] accurately
points out the advantages of analyzing subcellular organelles, macromolecular structures

and multiprotein complexes by proteomics and
developing subcellular proteomic databases.
Development of mammalian organellar proteome databases has begun[160–161] and should
prove to be invaluable to toxicoproteomics.
Several toxicoproteomic studies have used a
pre-fractionation approach to focus more
precisely on the compartment chemical effect
and characterize protein alterations with
greater specificity and sensitivity.[110,162]

Another limitation of 2DE is the instability
of the alkaline end of the IEF gel. Due to
cathodic drift during focusing runs, proteins
with pI > 7 are often poorly resolved. Unfortunately for toxicoproteomics, this includes
many of the cytochrome P450 monooxygenases
and glutathione S-transferases, some of the
most important proteins in toxicology. One
solution has been to resolve these proteins by
non-equilibrium pH-gradient electrophoresis
(NEPHGE).[156,163] The most effective solution
to this problem has been the development of
first-dimension IPG strips (Section
14.5.5.2.1.2.1.2). Strips with broad and very
alkaline pH gradients of 4–12 have been
reported[164] and when commercially available
with consistent performance, the benefits to
toxicoproteomics will be considerable.

### 14.5.5.4  CONCLUSION AND FUTURE DIRECTIONS

Proteome analysis has gained a prominent
place in modern biotechnology, not simply as a
complementary approach to genomics but
rather as a strategic and comprehensive discipline that can reveal the molecular anatomy
and physiology of a cell and cellular responses
to injury, disease, and repair. Though numerous toxicological studies have used proteomic
technologies to varying degrees, its potential
impact has yet to be realized. To broaden its
demonstrated utility in toxicology and other
life sciences, proteomics faces several technical
challenges.

Small scale, focused applications of proteomics to specific research questions (by "small"
laboratories) can be accomplished via access to
suitably equipped biotechnology facilities.
Large scale and broadly applied proteomics
requires reasonable cost, reasonably high-
throughput, precision, reproducibility, and
above all comprehensiveness to identify all
resolvable proteins that link the genome and
proteome. To achieve this objective innovative
automation and continued improvement and
uniformity in separation technology are critical. The implementation of large format IPG
first-dimension separation media with broad

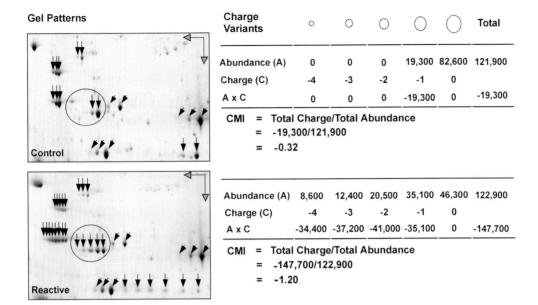

**Figure 5** Quantification of charge modification using the calculable charge modification index (CMI).[72] The upper gel pattern illustrates a portion of a 2DE gel from normal rat liver. Proteins indicated by arrows are those with post-translational modifications and resolved as charge heterogeneities (additional spots with more acidic pI). The lower pattern illustrates the same gel region but from a rat exposed to a chemical that generated a reactive intermediate. The additional arrows indicate new charge variants on some proteins generated by binding of the reactive intermediate exclusively on endoplasmic reticular proteins. Using methods described in this chapter, the additional charge variants can be cut from the gel and the nature of post-translational modification studied using MS/MS. Adapted from author's published manuscript.[155]

pH ranges and large-format slab-gels will improve resolution of proteomic separations. Increased sensitivity of protein detection is necessary so that ultimately a few copies of a protein expressed in a cell can be quantified, and this must be matched by a corresponding increase in the sensitivity of mass spectrometric techniques. Improved fluorescent protein stains and prefractionation of cell/tissue samples by novel centrifugation techniques that ultra-subfractionate cellular compartments will lower the limits of protein detection and enhance the characterization of low-abundance and compartment specific proteins.

Finally, it will be essential for toxicologists to generate both mRNA and protein expression data (e.g., toxicogenomics *and* toxicoproteomics) from the same test systems to decipher the dynamic relationship between the transcriptome and proteome.[165] No definitive examples of such a comparison exist in the toxicologic literature and a comprehensive understanding of the intricacies of cell function during and after chemical exposure requires that we carefully observe changes in both message and gene product. The technological developments presented here together with continued refinement will enable proteomics to reach its full potential and become an essential tool in routine toxicology.

## 14.5.5.5 REFERENCES

1. V. E. Velculescu, L. Zhang, W. Zhou *et al.*, 'Characterization of the yeast transcriptome.' *Cell*, 1997, **88**, 243–251.
2. L. Anderson and J. Seilhamer, 'A comparison of selected mRNA and protein abundances in human liver.' *Electrophoresis*, 1997, **18**, 533–537.
3. I. Humphery-Smith, S. J. Cordwell and W. P. Blackstock, 'Proteome research: complementarity and limitations with respect to the RNA and DNA worlds.' *Electrophoresis*, 1997, **18**, 1217–1242.
4. P. A Haynes, S. P. Gygi, D. Figeys *et al.*, 'Proteome analysis: biological assay or data archive?' *Electrophoresis*, 1998, **19**, 1862–1871.
5. S. P. Gygi, Y. Rochon, B. R. Franza *et al.*, 'Correlation between protein and mRNA abundance in yeast.' *Mol. Cell Biol.*, 1999, **19**, 1720–1730.
6. A. O. Gramolini, G. Karpati and B. J. Jasmin, 'Discordant expression of utrophin and its transcript in human and mouse skeletal muscles.' *J. Neuropathol. Exp. Neurol.*, 1999, **58**, 235–244.
7. S. Steiner and N. L. Anderson, 'Expression profiling in toxicology—potentials and limitations.' *Toxicol. Lett.*, 2000, **112–113**, 467–471.
8. R. A. VanBogelen, E. E. Schiller, J. D. Thomas *et al.*, 'Diagnosis of cellular states of microbial organisms using proteomics.' *Electrophoresis*, 1999, **20**, 2149–2159.
9. M. R. Wilkins, J. C. Sanchez, A. A. Gooley *et al.*, 'Progress with proteome projects: why all proteins expressed by a genome should be identified and how to do it.' *Biotechnol. Genet. Eng. Rev.*, 1996, **13**, 19–50.
10. P. A. Haynes and J. R. Yates III, 'Proteome profiling—pitfalls and progress.' *Yeast*, 2000, **17**, 81–87.

11. J. Klose, 'Protein mapping by combined isoelectric focusing and electrophoresis of mouse tissues. A novel approach to testing for induced point mutations in mammals.' *Humangenetik*, 1975, **26**, 231–243.

12. P. H. O'Farrell, 'High resolution two-dimensional electrophoresis of proteins.' *J. Biol. Chem.*, 1975, **250**, 4007–4021.

13. G. A. Scheele, 'Two-dimensional gel analysis of soluble proteins. Characterization of guinea pig exocrine pancreatic proteins.' *J. Biol. Chem.*, 1975, **250**, 5375–5385.

14. B. F. Clark, 'Towards a total human protein map.' *Nature*, 1981, **292**, 491–492.

15. N. G. Anderson and L. Anderson, 'The Human Protein Index.' *Clin. Chem.*, 1982, **28**, 739–748.

16. N. G. Anderson and N. L. Anderson, 'Twenty years of two-dimensional electrophoresis: Past, present and future.' *Electrophoresis*, 1996, **17**, 443–453.

17. P. A. Binz, M. Muller, D. Walther *et al.*, 'A molecular scanner to automate proteomic research and to display proteome images.' *Anal. Chem.*, 1999, **71**, 4981–4988.

18. S. P. Gygi, B. Rist, S. A. Gerber *et al.*, 'Quantitative analysis of complex protein mixtures using isotope-coded affinity tags.' *Nat. Biotechnol.*, 1999, **17**, 994–999.

19. D. J. Cahill, E. Nordhoff, J. O'Brien *et al.*, in 'Proteomics—From Protein Sequence to Function,' eds., S. R. Pennington, M. J. Dunn, BIOS Scientific Publishers, Oxford, UK, 2001, p. 1.

20. H. Kuwata, T. T. Yip, C. L. Yip *et al.*, 'Bactericidal domain of lactoferrin: detection, quantitation, and characterization of lactoferricin in serum by SELDI affinity mass spectrometry.' *Biochem. Biophys. Res. Commun.*, 1998, **245**, 764–773.

21. A. J. Link, J. Eng, D. M. Schieltz *et al.*, 'Direct analysis of protein complexes using mass spectrometry.' *Nat. Biotechnol.*, 1999, **17**, 676–682.

22. M. R. Wilkins, K. L. Williams, R. D. Appel *et al.*, 'Proteome research: New Frontiers in Functional Genomics,' Springer-Verlag, Berlin, 1997.

23. A. J. Link, '2-D Proteome Analysis Protocols,' Humana Press, Totowa NJ, 1999.

24. R. Kellner, F. Lottspeich and H. E. Meyer, 'Micro-characterization of Proteins,' 2nd edn., Wiley-VCH, Berlin, 1999.

25. T. Rabilloud, 'Proteome Research: Two-dimensional Gel Electrophoresis and Identification Methods,' Springer-Verlag, Berlin, 2000.

26. R. M. Kamp, D. Kyriakidis and T. Choli-Papado-poulou, 'Proteome and Protein Analysis,' Springer-Verlag, Berlin, 2000.

27. S. R. Pennington and M. J. Dunn, 'Proteomics—From Protein Sequence to Function,' BIOS, Oxford, UK, 2001.

28. P. Z. O'Farrell, H. M. Goodman and P. H. O'Farrell, 'High resolution two-dimensional electrophoresis of basic as well as acidic proteins.' *Cell*, 1977, **12**, 1133–1141.

29. J. Klose and U. Kobalz, 'Two-dimensional electrophoresis of proteins: an updated protocol and implications for a functional analysis of the genome.' *Electrophoresis*, 1995, **16**, 1034–1059.

30. B. Herbert, 'Advances in protein solubilisation for two-dimensional electrophoresis.' *Electrophoresis*, 1999, **20**, 660–663.

31. T. Rabilloud, 'Solubilization of proteins in 2-D electrophoresis. An outline.' *Methods Mol. Biol.*, 1999, **112**, 9–19.

32. T. Rabilloud, 'The use of thiourea to increase the solubility of membrane proteins in two-dimensional electrophoresis.' *Electrophoresis*, 1998, **19**, 758–760.

33. B. Bjellqvist, K. Ek, P. G, Righetti *et al.*, 'Isoelectric focusing in immobilized pH gradients: principle, methodology and some applications.' *J. Biochem. Biophys. Methods*, 1982, **6**, 317–339.

34. J. C. Sanchez, V. Rouge, M. Pisteur *et al.*, 'Improved and simplified in-gel sample application using reswelling of dry immobilized pH gradients.' *Electrophoresis*, 1997, **18**, 324–327.

35. A. Görg, C. Obermaier, G. Boguth *et al.*, 'Recent developments in two-dimensional electrophoresis with immobilized pH gradients: Wide pH gradients up to pH 12, longer separation distances and simplified procedures.' *Electrophoresis*, 1999, **20**, 712–717.

36. N. G. Anderson and N. L. Anderson, 'Analytical techniques for cell fractions. XXI. Two-dimensional analysis of serum and tissue proteins: multiple isoelectric focusing.' *Anal. Biochem.*, 1978, **85**, 331–340.

37. N. L. Anderson and N. G. Anderson, 'Analytical techniques for cell fractions. XXII. Two-dimensional analysis of serum and tissue proteins: multiple gradient-slab gel electrophoresis.' *Anal. Biochem.*, 1978, **85**, 341–354.

38. B. J. Walsh, M. P. Molloy, K. L. Williams, 'The Australian Proteome Analysis Facility (APAF): assembling large scale proteomics through integration and automation.' *Electrophoresis*, 1998, **19**, 1883–1890.

39. U. K. Laemmli, 'Cleavage of structural proteins during the assembly of the head of bacteriophage T4.' *Nature*, 1970, **227**, 680–685.

40. H. Langen, D. Roder, J. F. Juranville *et al.*, 'Effect of protein application mode and acrylamide concentration on the resolution of protein spots separated by two-dimensional gel electrophoresis.' *Electrophoresis*, 1997, **18**, 2085–2090.

41. J. A. Lott, V. A. Stephan and K. A. Pritchard Jr., 'Evaluation of the Coomassie Brilliant Blue G-250 method for urinary protein.' *Clin. Chem.*, 1983, **29**, 1946–1950.

42. V. Neuhoff, N. Arold, D. Taube *et al.*, 'Improved staining of proteins in polyacrylamide gels including isoelectric focusing gels with clear background at nanogram sensitivity using Coomassie Brilliant Blue G-250 and R-250.' *Electrophoresis*, 1988, **9**, 255–262.

43. C. R. Merril, D. Goldman, S. A. Sedman *et al.*, 'Ultrasensitive stain for proteins in polyacrylamide gels shows regional variation in cerebrospinal fluid proteins.' *Science*, 1981, **211**, 1437–1438.

44. C. Scheler, S. Lamer, Z. Pan *et al.*, 'Peptide mass fingerprint sequence coverage from differently stained proteins on two-dimensional electrophoresis patterns by matrix assisted laser desorption/ionization-mass spectrometry (MALDI-MS).' *Electrophoresis*, 1998, **19**, 918–927.

45. J. X. Yan, R. Wait and T. Berkelman, 'A modified silver staining protocol for visualization of proteins compatible with matrix-assisted laser desorption/ionization and electrospray ionization-mass spectrometry.' *Electrophoresis*, 2000, **21**, 3666–3672.

46. W. F. Patton, M. J. Lim and D. Shepro, 'Protein detection using reversible metal chelate stains.' *Methods Mol. Biol.*, 1999, **112**, 331–339.

47. L. Castellanos-Serra, W. Proenza, V. Huerta *et al.*, 'Proteome analysis of polyacrylamide gel-separated proteins visualized by reversible negative staining using imidazole-zinc salts.' *Electrophoresis*, 1999, **20**, 732–737.

48. K. Berggren, E. Chernokalskaya, T. H. Steinberg *et al.*, 'Background-free, high sensitivity staining of proteins in one- and two-dimensional sodium dodecyl sulfate-polyacrylamide gels using a luminescent ruthenium complex.' *Electrophoresis*, 2000, **21**, 2509–2521.

49. A. J. Link, 'Autoradiography of 2-D gels.' *Methods Mol. Biol.*, 1999, **112**, 285–290.

50. K. H. Lee and M. G. Harrington, 'Double-label analysis.' *Methods Mol. Biol.*, 1999, **112**, 291–295.

51. R. D. Appel and D. F. Hochstrasser, 'Computer analysis of 2-D images.' *Methods Mol. Biol.*, 1999, **112**, 363–381.

52. R. Aebersold and S. D. Patterson, in 'Proteins: Analysis and Design,' ed., R. H. Angeletti, Academic Press, San Diego, 1998.

53. D. Egger and K. Bienz, 'Protein (western) blotting.' *Mol. Biotechnol.*, 1994, **1**, 289–305.

54. A. Gatti, X. Wang and P. J. Robinson, 'Protein kinase C-α is multiply phosphorylated in response to phorbol ester stimulation of PC12 cells.' *Biochem. Biophys. Acta*, 1996, **1313**, 111–118.

55. T. G. Myers, E. C. Dietz, N. L. Anderson *et al.*, 'A comparative study of mouse liver proteins arylated by reactive metabolites of acetaminophen and its non-hepatotoxic regioisomer, 3′-hydroxyacetanilide.' *Chem. Res. Toxicol.*, 1995, **8**, 403–413.

56. R. Aebersold and D. R. Goodlett, 'Mass spectrometry in proteomics.' *Chemical Reviews*, 2001, ASAP Article 10.1021/cr990076h.

57. W. J. Henzel, T. M. Billeci, J. T. Stults *et al.*, 'Identifying proteins from two-dimensional gels by molecular mass searching of peptide fragments in protein sequence databases.' *Proc. Nat. Acad. Sci. USA*, 1993, **90**, 5011–5015.

58. P. James, M. Quadroni, E. Carafoli *et al.*, 'Protein identification by mass profile fingerprinting.' *Biochem. Biophys. Res. Commun.*, 1993, **195**, 58–64.

59. M. Mann, P. Hojrup and P. Roepstorff, 'Use of mass spectrometric molecular weight information to identify proteins in sequence databases.' *Biol. Mass Spectrom.*, 1993, **22**, 338–345.

60. D. J. C. Pappin, P. Hojrup and A. J. Bleasby, 'Rapid identification of proteins by peptide-mass fingerprinting.' *Curr. Biol.*, 1993, **3**, 327–332.

61. J. R. Yates III, S. Speicher, P. R. Griffin *et al.*, 'Peptide mass maps: a highly informative approach to protein identification.' *Anal. Biochem.*, 1993, **214**, 397–408.

62. P. L. Courchesne and S. D. Patterson, 'Identification of proteins by matrix-assisted laser desorption/ionization mass spectrometry using peptide and fragment ion masses.' *Methods Mol. Biol.*, 1999, **112**, 487–511.

63. A. L. Burlingame, R. K. Boyd and S J Gaskell, 'Mass spectrometry.' *Anal. Chem.*, 1998, **70**, 647R–716R.

64. J. R. Yates III, 'Mass spectrometry and the age of the proteome.' *J. Mass Spectrom.*, 1998, **33**, 1–19.

65. P. R. Griffin, M. J. MacCoss, J. K. Eng *et al.*, 'Direct database searching with MALDI-PSD spectra of peptides.' *Rapid Commun. Mass Spectrom.*, 1995, **9**, 1546–1551.

66. P. A. Haynes, N. Fripp and R. Aebersold, 'Identification of gel-separated proteins by liquid chromatography-electrospray tandem mass spectrometry: comparison of methods and their limitations.' *Electrophoresis*, 1998, **19**, 939–945.

67. A. Shevchenko, O. N. Jensen, A. V. Podtelejnikov *et al.*, 'Linking genome and proteome by mass spectrometry: large-scale identification of yeast proteins from two dimensional gels.' *Proc. Nat. Acad. Sci. USA*, 1996, **93**, 14440–14445.

68. J. A. Loo and H. Muenster, 'Magnetic sector-ion trap mass spectrometry with electrospray ionization for high sensitivity peptide sequencing.' *Rapid Commun. Mass Spectrom.*, 1999, **13**, 54–60.

69. A. A. Gooley and N. H. Packer, in 'Proteome Research: New Frontiers in Functional Genomics,' eds. M. R. Wilkins, K. L. Williams, R. D. Appel *et al.*, Springer-Verlag, Berlin, 1997, p. 65.

70. A. J. Link, K. Robison and G. M. Church, 'Comparing the predicted and observed properties of proteins encoded in the genome of *Escherichia coli* K-12.' *Electrophoresis*, 1997, **18**, 1259–1313.

71. E. Gianazza, 'Isoelectric focusing as a tool for the investigation of post-translational processing and chemical modifications of proteins.' *J. Chromatog. A*, 1995, **705**, 67–87.

72. N. L. Anderson, D. C. Copple, R. A. Bendele *et al.*, 'Covalent protein modifications and gene expression changes in rodent liver following administration of methapyrilene: a study using two-dimensional electrophoresis.' *Fundam. Appl. Toxicol.*, 1992, **18**, 570–580.

73. Y. Qiu, L. Z. Benet and A. L. Burlingame, 'Identification of the hepatic protein targets of reactive metabolites of acetaminophen *in vivo* in mice using two-dimensional gel electrophoresis and mass spectrometry.' *J. Biol. Chem.*, 1998, **273**, 17940–17953.

74. R. Hoffmann, S. Metzger, B. Spengler *et al.*, 'Sequencing of peptides phosphorylated on serines and threonines by post-source decay in matrix-assisted laser desorption/ionization time-of-flight mass spectrometry.' *J. Mass Spectrom.*, 1999, **34**, 1195–1204.

75. J. Abraham, J. Kelly, P. Thibault *et al.*, 'Post-translational modification of p53 protein in response to ionizing radiation analyzed by mass spectrometry.' *J. Mol. Biol.*, 2000, **295**, 853–864.

76. E. H. Fischer, 'Cellular regulation by protein phosphorylation: a historical overview.' *Biofactors*, 1997, **6**, 367–374.

77. A. L. Zarlinga, S. B. Ficarrod, F. M. Whited *et al.*, 'Phosphorylated peptides are naturally processed and presented by major histocompatibility complex class I molecules *in vivo*.' *J. Exp. Med.*, 2000, **192**, 1755–1762.

78. R. Apweiler, H. Hermjakob and N. Sharon, 'On the frequency of protein glycosylation, as deduced from analysis of the SWISS-PROT database.' *Biochem. Biophys. Acta*, 1999, **1473**, 4–8.

79. M. R. Wilkins, E. Gasteiger, A. A. Gooley *et al.*, 'High-throughput mass spectrometric discovery of protein post-translational modifications.' *J. Mol. Biol.*, 1999, **289**, 645–657.

80. L. J. Holt, K. Bussow, G. Walter *et al.*, 'By-passing selection: direct screening for antibody-antigen interactions using protein arrays.' *Nucleic Acids Res.*, 2000, **28**, e72–e72.

81. G. MacBeath and S. L. Schreiber, 'Printing proteins as microarrays for high-throughput function determination.' *Science*, 2000, **289**, 1760–1763.

82. M. Unlu, M. E. Morgan and J. S. Minden, 'Difference gel electrophoresis: a single gel method for detecting changes in protein extracts.' *Electrophoresis*, 1997, **18**, 2071–2077.

83. M. Unlu, 'Difference gel electrophoresis.' *Biochem. Soc. Trans.*, 1999, **27**, 547–549.

84. R. Tonge, J. Shaw and B. Middleton, 'Validation and development of fluorescence 2D-differential gel electrophoresis proteomics technology.' *Proteomics*, 2001, in press.

85. D. Figeys and D. Pinto, 'Proteomics on a chip: promising developments.' *Electrophoresis*, 2001, **22**, 208–216.

86. A. M. Gressner, 'Two-dimensional electrophoretic analysis of ribosomal proteins from chronically injured liver.' *J. Clin. Chem. Clin. Biochem.*, 1979, **17**, 541–545.

87. P. Lemkin, L. Lipkin, C. Merril *et al.*, 'Protein abnormalities in macrophages bearing asbestos.' *Environ. Health Perspect.*, 1980, **34**, 75–89.

88. L. E. Lipkin and P. F. Lemkin, 'Data-base techniques for multiple two-dimensional polyacrylamide gel electrophoresis analyses.' *Clin. Chem.*, 1980, **26**, 1403–1412.

89. N. L. Anderson, J. Taylor, A. E. Scandora *et al.*, 'The TYCHO system for computer analysis of two-dimensional gel electrophoresis patterns.' *Clin. Chem.*, 1981, **27**, 1807–1820.

90. P. F. Lemkin, L. E. Lipkin and E. P. Lester, 'Some extensions to the GELLAB two-dimensional electrophoretic gel analysis system.' *Clin. Chem.*, 1982, **28**, 840–849.

91. T. Pun, D. F. Hochstrasser, R. D. Appel *et al.*, 'Computerized classification of two-dimensional gel electrophoretograms by correspondence analysis and ascendant hierarchical clustering.' *Appl. Theor. Electrophor.*, 1988, **1**, 3–9.

92. J. L. Garrels, 'The QUEST system for quantitative analysis of two-dimensional gels.' *J. Biol. Chem.*, 1989, **264**, 5269–5282.

93. N. L. Anderson, J. Taylor, J. P. Hofmann *et al.*, 'Simultaneous measurement of hundreds of liver proteins: application in assessment of liver function.' *Toxicol. Pathol.*, 1996, **24**, 72–76.

94. C. S. Giometti, J. Taylor and S. L. Tollaksen, 'Mouse liver protein database: a catalog of proteins detected by two-dimensional gel electrophoresis.' *Electrophoresis*, 1992, **13**, 970–991.

95. N. L. Anderson, R. Esquer-Blasco, J. P. Hofmann *et al.*, 'A two-dimensional gel database of rat liver proteins useful in gene regulation and drug effects studies.' *Electrophoresis*, 1991, **12**, 907–930.

96. T. Watanabe, N. D. Lalwani and J. K. Reddy, 'Specific changes in the protein composition of rat liver in response to the peroxisome proliferators ciprofibrate, Wy-14,643 and di-(2-ethylhexyl)phthalate.' *Biochem. J.*, 1985, **227**, 767–775.

97. P. J. Wirth, M. S. Rao, R. P. Evarts *et al.*, 'Coordinate polypeptide expression during hepatocarcinogenesis in male F-344 rats: comparison of the Solt-Farber and Reddy models.' *Cancer Res.*, 1987, **47**, 2839–2851.

98. C. S. Giometti, J. Taylor, M. A. Gemmell *et al.*, 'A comparative study of the effects of clofibrate, ciprofibrate, WY-14,643, and di-(2-ethylhexyl)-phthalate on liver protein expression in mice.' *Appl. Theor. Electrophor.*, 1991, **4–5**, 101–107.

99. F. A. Witzmann, B. M. Jarnot, D. N. Parker *et al.*, 'Modification of hepatic immunoglobulin heavy chain binding protein (BiP/Grp78) following exposure to structurally diverse peroxisome proliferators.' *Fundam. Appl. Toxicol.*, 1994, **23**, 1–8.

100. N. L. Anderson, R. Esquer-Blasco, F. Richardson *et al.*, 'The effects of peroxisome proliferators on protein abundances in mouse liver.' *Toxicol. Appl. Pharmacol.*, 1996, **137**, 75–89.

101. F. Witzmann, M. Coughtrie, C. Fultz *et al.*, 'Effect of structurally diverse peroxisome proliferators on rat hepatic sulfotransferase.' *Chem. Biol. Interact.*, 1996, **99**, 73–84.

102. C. S. Giometti, S. L. Tollaksen, X. Liang *et al.*, 'A comparison of liver protein changes in mice and hamsters treated with the peroxisome proliferator Wy-14,643.' *Electrophoresis*, 1998, **19**, 2498–2505.

103. U. Edvardsson, M. Bergstrom, M. Alexandersson *et al.*, 'Rosiglitazone (BRL49653), a PPARγ-selective agonist, causes peroxisome proliferator-like liver effects in obese mice.' *J. Lipid Res.*, 1999, **40**, 1177–1184.

104. S. Chevalier, N. Macdonald, R. Tonge *et al.*, 'Proteomic analysis of differential protein expression in primary hepatocytes induced by EGF, tumour necrosis factor alpha or the peroxisome proliferator nafenopin.' *Eur. J. Biochem.*, 2000, **267**, 4624–434.

105. C. S. Giometti, X. Liang, S. L. Tollaksen *et al.*, 'Mouse liver selenium-binding protein decreased in abundance by peroxisome proliferators.' *Electrophoresis*, 2000, **21**, 2162–2169.

106. N. Macdonald, K. Barrow, R. Tonge *et al.*, 'PPARα-dependent alteration of GRP94 expression in mouse hepatocytes.' *Biochem. Biophys. Res. Commun.*, 2000, **277**, 699–704.

107. A. Arce, L. Aicher, D. Wahl *et al.*, 'Changes in the liver protein pattern of female Wistar rats treated with the hypoglycemic agent SDZ PGU 693.' *Life Sci.*, 1998, **63**, 2243–2250.

108. M. Fountoulakis, P Berndt, U. A. Boelsterli *et al.*, 'Two-dimensional database of mouse liver proteins: changes in hepatic protein levels following treatment with acetaminophen or its nontoxic regioisomer 3-acetamidophenol.' *Electrophoresis*, 2000, **21**, 2148–2161.

109. S. J. Newsholme, B. F. Maleeff, S. Steiner *et al.*, 'Two-dimensional electrophoresis of liver proteins: characterization of a drug-induced hepatomegaly in rats.' *Electrophoresis*, 2000, **21**, 2122–2128.

110. F. A. Witzmann, C. D. Fultz, R. A. Grant *et al.*, 'Differential expression of cytosolic proteins in the rat kidney cortex and medulla: preliminary proteomics.' *Electrophoresis*, 1998, **19**, 2491–2497.

111. F. A. Witzmann, C. D. Fultz, R. A. Grant *et al.*, 'Regional protein alterations in rat kidneys induced by lead exposure.' *Electrophoresis*, 1999, **20**, 943–951.

112. L. Aicher, D. Wahl, A. Arce *et al.*, 'New insights into cyclosporine A nephrotoxicity by proteome analysis.' *Electrophoresis*, 1998, **19**, 1998–2003.

113. F. A. Witzmann, M. D. Bauer and A. M. Fieno, 'Proteomic analysis of the renal effects of simulated occupational jet fuel exposure.' *Electrophoresis*, 2000, **21**, 976–984.

114. F. A. Witzmann, R. L. Carpenter, G. D. Ritchie *et al.*, 'Toxicity of chemical mixtures: proteomic analysis of persisting liver and kidney protein alterations induced by repeated exposure of rats to JP-8 jet fuel vapor.' *Electrophoresis*, 2000, **21**, 2138–2147.

115. F. A. Witzmann, M. D. Bauer and A. M. Fieno, 'Proteomic analysis of simulated occupational jet fuel exposure in the lung.' *Electrophoresis*, 1999, **20**, 3659–3669.

116. M. W. Lame, A. D. Jones, D. W. Wilson *et al.*, 'Protein targets of monocrotaline pyrrole in pulmonary artery endothelial cells.' *J. Biol. Chem.*, 2000, **275**, 29091–29099.

117. R. Wattiez, C. Hermans, C. Cruyt *et al.*, 'Human bronchoalveolar lavage fluid protein two-dimensional database: study of interstitial lung diseases.' *Electrophoresis*, 2000, **21**, 2703–2712.

118. J. L. Pipkin, W. G. Hinson, S. J. James *et al.*, 'The relationship of p53 and stress proteins in response to bleomycin and retinoic acid in the p53 heterozygous mouse.' *Biochim. Biophys. Acta*, 1999, **1450**, 164–176.

119. J. B. LaBorde, J. L. Pipkin Jr, W. G. Hinson *et al.*, 'Retinoic acid-induced stress protein synthesis in the mouse.' *Life Sci.*, 1995, **56**, 1767–1778.

120. R. Gandley, L. Anderson and E. K. Silbergeld, 'Lead: male-mediated effects on reproduction and development in the rat.' *Environ. Res.*, 1999, **80**, 355–363.

121. M. Toraason, W. Moorman, P. I. Mathias *et al.*, 'Two-dimensional electrophoretic analysis of myocardial proteins from lead-exposed rabbits.' *Electrophoresis*, 1997, **18**, 2978–2982.

122. D. Arnott, K. L. O'Connell, K. L. King *et al.*, 'An integrated approach to proteome analysis: identification of proteins associated with cardiac hypertrophy.' *Anal. Biochem.*, 1998, **258**, 1–18.

123. C. Charriaut-Marlangue, F. Dessi and Y. Ben-Ari, 'Use of two-dimensional gel electrophoresis to characterize protein synthesis during neuronal death in cerebellar culture.' *Electrophoresis*, 1996, **17**, 1781–1786.

124. H. Yagame, T. Horigome, T. Ichimura *et al.*, 'Differential effects of methylmercury on the phosphorylation of protein species in the brain of acutely intoxicated rats.' *Toxicology*, 1994, **92**, 101–113.

125. I. Eberini, I. Miller, M. Gemeiner *et al.*, 'A web site for the rat serum protein study group.' *Electrophoresis*, 1999, **20**, 3599–3602.

126. E. S. Robinson, T. P. Dooley and K. L. Williams, 'UV-induced melanoma cell lines and their potential for proteome analysis: a review.' *J. Exp. Zool.*, 1998, **282**, 48–53.

127. O. Weinreb, M. A. van Boekel, A. Dovrat *et al.*, 'Effect of UV-A light on the chaperone-like properties of young and old lens alpha-crystallin.' *Invest. Ophthalmol. Vis. Sci.*, 2000, **41**, 191–198.

128. C. S. Giometti, S. L. Tollaksen, M A. Gemmell *et al.*, 'The analysis of recessive lethal mutations in mice by using two-dimensional gel electrophoresis of liver proteins.' *Mutat. Res.*, 1990, **242**, 47–55.

129. S. C. Prasad, V. A. Soldatenkov, M. R. Kuettel *et al.*, 'Protein changes associated with ionizing radiation-induced apoptosis in human prostate epithelial tumor cells.' *Electrophoresis*, 1999, **20**, 1065–1074.

130. S. Chen, L. Cai, X. Li *et al.*, 'Low-dose whole body irradiation induces alteration of protein in mouse splenocytes.' *Toxicol. Lett.*, 1999, **105**, 141–152.

131. J. L. Pipkin, W. G. Hinson, J. F. Young *et al.*, 'Induction of stress proteins by electromagnetic fields in cultured HL-60 cells.' *Bioelectromagnetics*, 1999, **20**, 347–357.

132. S. Nakasono and H. Saiki, 'Effect of ELF magnetic fields on protein synthesis in *Escherichia coli* K12.' *Radiat. Res.*, 2000, **154**, 208–216.

133. J. L. Pipkin, W. G. Hinson, J. L. Hudson *et al.*, 'The effect of isoproterenol on nuclear protein synthesis in electrostatically sorted rat hepatocytes.' *Cytometry*, 1980, **1**, 212–221.

134. G. Mellgren, T. Bruland, A. P. Doskeland *et al.*, 'Synergistic antiproliferative actions of cyclic adenosine 3′,5′-monophosphate, interleukin-1β, and activators of $Ca^{2+}$/calmodulin-dependent protein kinase in primary hepatocytes.' *Endocrinology*, 1997, **138**, 4373–4383.

135. K. E. Fladmark, B. T. Gjertsen, S. O. Doskeland *et al.*, 'Fas/APO-1(CD95)-induced apoptosis of primary hepatocytes is inhibited by cAMP.' *Biochem. Biophys. Res. Commun.*, 1997, **232**, 20–25.

136. J. Liu, K. S. Squibb, M. Akkerman *et al.*, 'Cytotoxicity, zinc protection, and stress protein induction in rat proximal tubule cells exposed to cadmium chloride in primary cell culture.' *Ren. Fail.*, 1996, **18**, 867–882.

137. B. M. Sanders, J. Nguyen, L. S. Martin *et al.*, 'Induction and subcellular localization of two major stress proteins in response to copper in the fathead minnow Pimephales promelas.' *Comp. Biochem. Physiol. C Pharmacol. Toxicol. Endocrinol.*, 1995, **112**, 335–343.

138. C. J. Molloy and J. D. Laskin, 'Specific alterations in keratin biosynthesis in mouse epidermis *in vivo* and in explant culture following a single exposure to the tumor promoter 12-*O*-tetradecanoylphorbol-13-acetate.' *Cancer Res.*, 1987, **47**, 4674–4680.

139. C. J. Molloy and J. D. Laskin, 'Altered expression of a mouse epidermal cytoskeletal protein is a sensitive marker for proliferation induced by tumor promoters.' *Carcinogenesis*, 1992, **13**, 963–968.

140. L. Mascarell, J. R. Frey, F. Michel *et al.*, 'Increased protein synthesis after T cell activation in presence of cyclosporin A.' *Transplantation*, 2000, **70**, 340–348.

141. R. Joubert-Caron, J. P. Le Caer, F. Montandon *et al.*, 'Protein analysis by mass spectrometry and sequence database searching: a proteomic approach to identify human lymphoblastoid cell line proteins.' *Electrophoresis*, 2000, **21**, 2566–2575.

142. N. E. John, H. U. Andersen, S. J. Fey *et al.*, 'Cytokine- or chemically-derived nitric oxide alters the expression of proteins detected by two-dimensional gel electrophoresis in neonatal rat islets of Langerhans.' *Diabetes*, 2000, **49**, 1819–1829.

143. A. R. Parrish, A. J. Gandolfi and K. Brendel, 'Precision-cut tissue slices: applications in pharmacology and toxicology.' *Life Sci.*, 1995, **57**, 1887–1901.

144. R. K. Yip, P. T. Kelly *et al.*, '*In situ* protein phosphorylation in hippocampal tissue slices.' *J. Neurosci.*, 1989, **9**, 3618–3630.

145. F. A. Witzmann, C. D. Fultz, J. F. Wyman, 'Two-dimensional electrophoresis of precision-cut testis slices: toxicologic application.' *Electrophoresis*, 1997, **18**, 642–646.

146. A. T. Drahushuk, B. P. McGarrigle, K. E. Larsen *et al.*, 'Detection of CYP1A1 protein in human liver and induction by TCDD in precision-cut liver slices incubated in dynamic organ culture.' *Carcinogenesis*, 1998, **19**, 1361–1368.

147. N. R. Pumford and N. C. Halmes, 'Protein targets of xenobiotic reactive intermediates.' *Annu. Rev. Pharmacol. Toxicol.*, 1997, **37**, 91–117.

148. J. B. Bartolone, R. B. Birge, K. Sparks *et al.*, 'Immunochemical analysis of acetaminophen covalent binding to proteins. Partial characterization of the major acetaminophen-binding liver proteins.' *Biochem. Pharmacol.*, 1988, **37**, 4763–4774.

149. T. G. Myers, E. C. Dietz, N. L. Anderson *et al.*, 'A comparative study of mouse liver proteins arylated by reactive metabolites of acetaminophen and its non-hepatotoxic regioisomer, 3′-hydroxyacetanilide.' *Chem. Res. Toxicol.*, 1995, **8**, 403–413.

150. S. J. Bulera, R. B. Birge, S. D. Cohen *et al.*, 'Identification of the mouse liver 44-kDa acetaminophen-binding protein as a subunit of glutamine synthetase.' *Toxicol. Appl. Pharmacol.*, 1995, **134**, 313–320.

151. N. Frey, U. Christen, P. Jeno *et al.*, 'The lipoic acid containing components of the 2-oxoacid dehydrogenase complexes mimic trifluoroacetylated proteins and are autoantigens associated with halothane hepatitis.' *J. Chem. Res. Toxicol.*, 1995, **8**, 736–746.

152. F. M. Ndikum-Moffor, J. W. Munson, N. K. Bokinkere *et al.*, 'Immunochemical detection of hepatic cocaine-protein adducts in mice.' *Chem. Res. Toxicol.*, 1998, **11**, 185–192.

153. N. L. Anderson, D. C. Copple, R. A. Bendele *et al.*, 'Covalent protein modifications and gene expression changes in rodent liver following administration of methapyrilene: a study using two-dimensional electrophoresis.' *Fundam. Appl. Toxicol.*, 1992, **18**, 570–580.

154. F. C. Richardson, S. C. Strom, D. M. Copple *et al.*, 'Comparisons of protein changes in human and rodent hepatocytes induced by the rat-specific carcinogen, methapyrilene.' *Electrophoresis*, 1993, **14**, 157–161.

155. F. A. Witzmann, C. D. Fultz, R. S. Mangipudy *et al.*, 'Two-dimensional electrophoretic analysis of compartment-specific hepatic protein charge modification induced by thioacetamide exposure in rats.' *Fundam. Appl. Toxicol.*, 1996, **31**, 124–132.

156. F. A. Witzmann, D. A. Daggett, C. D. Fultz *et al.*, 'Glutathione *S*-transferases: two-dimensional electrophoretic protein markers of lead exposure.' *Electrophoresis*, 1998, **19**, 1332–1335.

157. F. A. Witzmann, R. L. Carpenter, G. D. Ritchie *et al.*, 'Toxicity of chemical mixtures: proteomic analysis of persisting liver and kidney protein alterations induced by repeated exposure of rats to JP-8 jet fuel vapor.' *Electrophoresis*, 2000, **21**, 2138–2147.

158. G. L. Corthals, V. C. Wasinger, D. F. Hochstrasser *et al.*, 'The dynamic range of protein expression: a challenge for proteomic research.' *Electrophoresis*, 2000, **21**, 1104–1115.

159. E. Jung, M. Heller, J.-C. Sanchez *et al.*, 'Proteomics meets cell biology: The establishment of subcellular proteomes.' *Electrophoresis*, 2000, **21**, 3369–3377.

160. E. Jung, C. Hoogland, D. Chiappe *et al.*, 'The establishment of a human liver nuclei two-dimensional electrophoresis reference map.' *Electrophoresis*, 2000, **21**, 3483–3487.

161. R. S. Taylor, C. C. Wu, L. G. Hays *et al.*, 'Proteomics of rat liver Golgi complex: minor proteins are identified through sequential fractionation.' *Electrophoresis*, 2000, **21**, 3441–3459.

162. M. A. Kaderbhai, T. K. Bradshaw, R. B. Freedman *et al.*, 'Alterations in the enzyme activity and polypeptide composition of rat hepatic endoplasmic reticulum during acute exposure to 2-acetylaminofluorene.' *Chem. Biol. Interact.*, 1982, **39**, 279–299.

163. F. A. Witzmann and D. N. Parker, 'Hepatic protein pattern alterations following perfluorodecanoic acid exposure in rats.' *Toxicol. Lett.*, 1991, **57**, 29–36.

164. A. Görg, G. Boguth, C. Obermaier *et al.*, 'Two-dimensional electrophoresis of proteins in an immobilized pH 4-12 gradient.' *Electrophoresis*, 1998, **19**, 1516–1519.

165. S. W. Burchiel, C. M. Knall, J. W. Davis II *et al.*, 'Analysis of Genetic and Epigenetic Mechanisms of Toxicity: Potential Roles of Toxicogenomics and Proteomics in Toxicology.' *Toxicol. Sci.*, 2001, **59**, 193–195.

J.P. Vanden Heuvel, G.H. Perdew, W.B. Mattes and W.F. Greenlee (Eds.)
Comprehensive Toxicology, Vol. xiv

# 14.5.6
# Mechanistically-Based Assays for the Identification and Characterization of Endocrine Disruptors

TIM R. ZACHAREWSKI, KIRSTEN C. FERTUCK,
MARK R. FIELDEN and JASON B. MATTHEWS
*Michigan State University, East Lansing, MI, USA*

## 14.5.6.1   INTRODUCTION

Experimental and epidemiological evidence suggests that exposure to hormonally active synthetic and natural chemicals, commonly referred to as endocrine disruptors, may cause adverse health effects in humans and wild-life.[1,2] These documented or suspected effects include hormone-dependent cancers,[3,4] reproductive tract abnormalities,[5] compromised reproductive fitness,[6] impaired cognitive ability[7,8] and developmental abnormalities.[9–14] Despite skepticism and contradictory re-examination of the data,[15–19] mechanistically feasible hypotheses have suggested a link between exposure to endocrine disruptors and adverse health effects in humans and wildlife.[3,12,20,21] The greatest concern is that exposure to endocrine disruptors during critical periods of development may contribute to the manifestation of adverse effects in adults. In response to growing concern, the Safe Drinking Water Act Amendment (05 1996—Bill number S.1316) and the Food Quality Protection Act (1996—Bill number P.L. 104–170) were introduced, which require the United States Environmental Protection Agency (US EPA) to test all chemicals for estrogenic, androgenic and thyroid-like endocrine disrupting activity.

Endocrine disruptors have been defined as any exogenous substances that cause adverse health effects in an intact organism or its progeny, secondary to changes in endocrine function.[22] Suspected endocrine disruptors encompass a wide range of compounds including natural products (e.g., phytoestrogens, mycotoxins), environmental pollutants (e.g., polyaromatic hydrocarbons, dioxins), pharmaceuticals, pesticides and industrial chemicals (Figure 1). These chemicals do not necessarily share structural similarity to endogenous steroids,[23] and therefore can only be identified using functional assays.[24] In general, the activity of estrogenic, androgenic and thyroid-like chemicals is mediated by their respective nuclear receptors[25] (Figure 2), although other mechanisms of action such as interactions with binding globulins,[26–30] inhibition of steroidogenic enzymes[31–33] and binding to membrane receptors[34,35] or other nuclear receptors[36–38] cannot be excluded.

Nuclear receptors comprise a large superfamily of transcription factors that regulate target genes in response to small hydrophobic ligands such as steroids, retinoids, and thyroid hormones. Members of this superfamily play important roles in development and differentiation, and have been implicated in a number of diseases. They are modular proteins consisting of functionally independent and conserved domains. The zinc finger-type DNA-binding domain (DBD) facilitates association with specific DNA response elements in the regulatory region of target genes. A ligand binding domain is essential for interaction with ligands, and for the transmission of the ligand signal to the transcriptional machinery. Transcriptional activity of nuclear receptors is mediated by at least two separate activation domains, the ligand-independent activation function 1 (AF-1) near the N-terminus and the ligand-dependent activation function 2 (AF-2) region in the ligand binding domain near the C-terminus. Although the mechanism by which the AF-2 region transmits ligand signals to the basal transcriptional machinery is poorly understood, several proteins that interact with the AF-2 region in a ligand-dependent manner have been identified.[39] These proteins, collectively termed cofactors, function as coactivators or corepressors to induce or inhibit gene

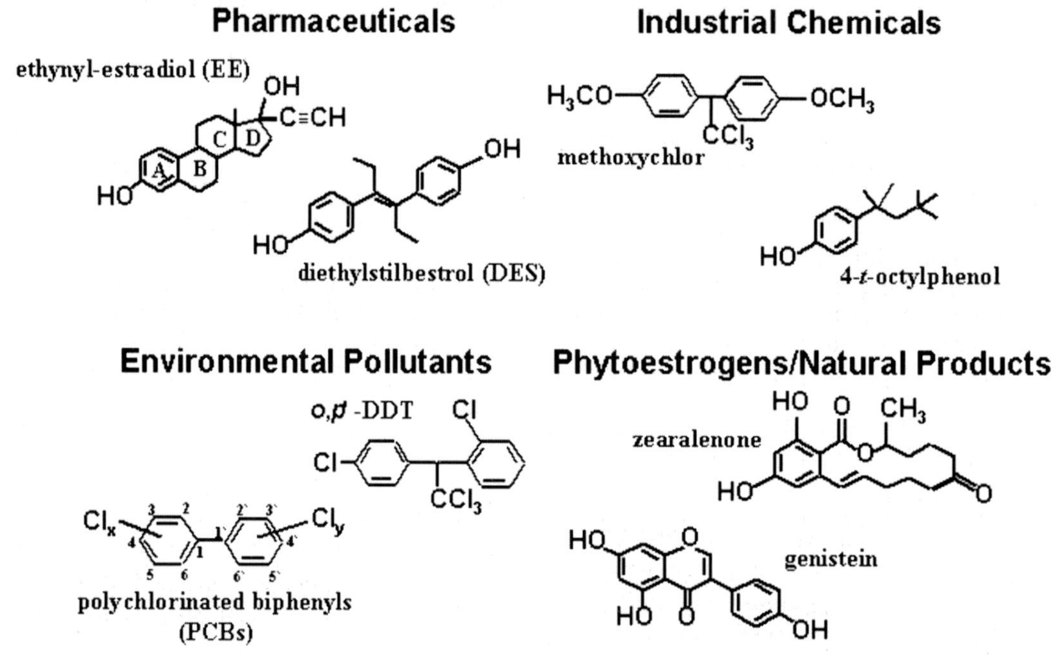

**Figure 1** Structural diversity of selected estrogenic endocrine disruptors.

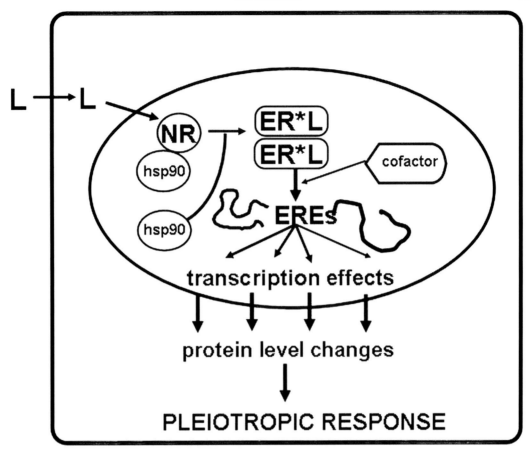

**Figure 2** Proposed mechanism of action of estrogen receptors (ER). The estrogenic substance or ligand (L) binds to the ER causing a conformational change in protein structure and the dissociation of heat shock protein 90 (hsp90). Ligand occupied ERs then undergo homodimerization and the resulting ER homodimer complexes possess high affinity for specific DNA sequences referred to as estrogen response elements (EREs), located in the regulatory region of estrogen-responsive genes. Once bound to the ERE, the ER homodimer complex recruits cofactors that induce changes in gene expression thereby altering the level of proteins that are critical for homeostasis and development.

expression, respectively. Following ligand binding and subsequent receptor dimerization, conformational changes occur that expose critical residues within the AF-2 region that are essential for transcriptional cofactor recruitment. Functional and structural studies have shown that coactivators interact with the AF-2 region via short leucine motifs (i.e., LXXLL, also known as the NR-box) to transduce the ligand signal to the transcriptional machinery. Several methodologies that are amenable to high throughput screening for endocrine disruptors exploit the mechanism of action of nuclear receptors.[24,40]

The US EPA has proposed a tiered program to screen and test chemicals for estrogenic, androgenic and thyroid-like activities.[1] Tier 1 examines several different endpoints using a number of *in vitro* and *in vivo* assays and model species. Following data collection, a weight of evidence approach is then used to determine if further *in vivo* testing (Tier 2) is warranted in order to assess the potential of endocrine disruption in humans. Although data from Tier 1 *in vivo* testing will be given priority, discrepancies between *in vitro* and *in vivo* assay results will likely be encountered, thus further complicating endocrine disruption assessments. In addition, concerns have been expressed regarding the validation of structure-activity relationships, the need for reference chemicals, problems encountered with reproducing endocrine disruption data, and the lack of objective criteria for assessing the predictive value of an assay.[41] Moreover, it is unclear whether Tier 1 testing will be sufficient to detect endocrine disruption elicited by selective receptor modulators[42,43] or that result from divergent mechanisms that converge or cross-talk. Consequently, in order to assess the ability of natural and synthetic chemical to elicit an endocrine disrupting effect following chronic or subchronic exposure, comprehensive strategies must be developed that investigate potential activities at the molecular, cellular and tissue level within the context of the whole organism and its genome, proteome and metabonome. This chapter presents some of the approaches that attempt to address these issues.

### 14.5.6.2  *IN VITRO* ASSAYS FOR THE IDENTIFICATION OF ENDOCRINE-DISRUPTING COMPOUNDS

#### 14.5.6.2.1  Competitive Binding Assays

*In vitro* estrogen receptor (ER) competitive binding assays have been well established and extensively used to investigate ER-ligand interactions. All competitive binding assays involve the displacement of a receptor-bound probe molecule by a test compound. To perform a competitive binding assay, a cytosolic extract from an ER-rich tissue or recombinant source of ER, such as bacteria, yeast, and SF9 insect cells, is prepared. The ER is incubated with an appropriate concentration of labeled probe (ligand) and increasing concentrations of unlabeled competitor. The probe is traditionally radiolabeled, however fluorescently labeled high affinity ligands have also been used.[44] Separation of receptor-bound from free ligand is achieved using a variety of methods including dextran-coated charcoal (DCC),[45] hydroxylapatite (HAP)[46] or by filteration.[47] Blair and coworkers have used rat uterine cytosolic ER to analyze the competitive binding profiles of 188 natural and xenochemicals.[48] Our laboratory has also examined the ability of several natural and synthetic chemicals to compete for binding to recombinantly expressed ERs from a variety of vertebrate species including human, mouse, chicken, anole, *Xenopus*, and rainbow trout.[49] Several differences in the ability of certain chemicals to interact with ERs among different species were observed (Figure 3).

Scintillating microplates have also been developed and used to investigate interactions of a variety of estrogenic compounds with recombinant human ERα ligand binding domain expressed in yeast.[50] In high throughput screening applications, the ease of the scintillation proximity assay (SPA) (Amersham Pharmacia Biotech) and FlashPlates[TM] scintillating microplates (NEN Life Sciences) has largely replaced the use of filtration methods by eliminating the separation and addition of scintillation cocktail steps[51] (Figure 4). In SPA, the target (receptor) is immobilized to a solid support that contains a scintillant. When the probe (i.e., radiolabeled ligand) binds to the target, the radioisotope is immobilized sufficiently close to the solid support to activate the scintillant, thus releasing a signal relative to the amount of ligand bound or displaced. In contrast, free radiolabeled ligands in solution are too far away from the solid support to activate the scintillant and are not detected. A similar strategy is used for the FlashPlates[TM]. Scintillating microplates are available precoated with antibody, glutathione, streptavidin or nickel chelate enabling a variety of tagged proteins to be captured on the plates. In addition, sterile scintillating microplates are available, which allow for a variety of cell-based applications, including receptor binding.

Shortcomings of the competitive binding assay include the inability to distinguish between receptor agonists and antagonists and the possibility that high concentrations of competing ligand may lead to an increase in non-specific binding.[40] However, the assay is amenable to high throughput formats and can be extended to investigate direct interactions between a number of ligands and nuclear receptors. Binding assays are extremely versatile and are part of the standard analysis of drug discovery programs and can be fully automated and performed by laboratory robots in high throughput drug screens.

#### 14.5.6.2.2  Coactivator-Estrogen Receptor Interactions

Other methods that investigate ER-ligand interactions take advantage of the ligand induced conformational change that occurs upon ligand binding and the subsequent interaction with coactivator proteins. Functional and structural studies have demonstrated that these coactivators interact with leucine motifs in the AF-2 region in an agonist-dependent manner to transmit ligand signals to the basal transcriptional machinery.[52-54] Several different coactivators have been described and isolated.[55,56] Many estrogenic endocrine disruptors (EEDs) exhibit a reduced ability to induce ER-mediated gene expression in reporter gene assays than would be expected based on their affinity for the ER.[57,58] The poor efficacy of these partial agonists may be due to the EED-induced conformational changes, which result in compromised interactions with coactivators and reduced transcriptional activity. Alternatively, binding of EEDs may cause a suboptimal agonist conformation of the receptor that does not facilitate stable interactions with coactivators and thus does not efficiently transmit the ligand signal to the basal transcriptional machinery. The crystal structure for the genistein-ERα complex shows that the critical helix 12 is positioned intermediate between agonist and antagonist conformations.[59] Therefore, analyzing the interactions between coactivators and ERs offer another method to further investigate potential EEDs and to further investigate the mechanism of action of existing EEDs.

Methods that have been used to investigate ligand induced ER-coactivator interactions include yeast two-hybrid assays, GST pull down assays, and surface plasmon resonance. Nishikawa and coworkers used all three methods to investigate the EED-induced interactions between the ER and several different coactivators.[60] In yeast two-hybrid assays, a rank order in the interaction of the ER-17β-estradiol (E2) complex with a variety of coactivators was observed, with TIF2 exhibiting the strongest interactions. In addition, certain EEDs induced differential abilities to interact with different coactivators, although the differences were small.[60] Disadvantages of the yeast two-hybrid assay include inability to identify antagonists, potential false positives, the insensitivity of some yeast assays to many estrogenic chemicals and the ability of yeast to effectively pump out chemicals thus reducing the effective concentration within the cell. Nevertheless, recent advances in high throughput yeast two-hybrid screening makes this an exciting assay with enormous possibility.[61,62]

GST pull down assays are a classic *in vitro* method to investigate protein–protein interactions. These assays are relatively simple and are routinely performed to investigate receptor-coactivator interactions.[60,63] Disadvantages of these assays include the use of radioactivity, and the qualitative nature of the results. An alternative to this classic approach is surface plasmon resonance, commercially referred to as Biacore (Biacore, New Jersey, USA).[64] This emerging technology allows the investigation of protein–protein interactions in real time, which allows for the determination of kinetic affinity constants providing a more informative examination of the interaction. The technology is based on fixing one of the two proteins to a sensor chip via nickel nitrilotriacetic acid metal affinity technology using a hexahistidine purified fusion protein or using antibody epitopes, such as GST or protein A, as well as other methods.[60,64,65] The plasmon biosensor detects the binding to the immobilized partner in real time. The time-dependent measurements provide kinetic information that is difficult to obtain by other methods.[64] Although the approach is currently not set up for high throughput screening and requires expensive specialized instrumentation, it offers the tremendous advantages of providing quantitative information regarding specific endocrine disruptor:receptor:coactivator interactions.

### 14.5.6.2.3 Future Considerations

Computational methods such as quantitative structure activity relationships (QSARs) and 3D-QSARs offer the potential for screening large databases of chemicals for novel EEDs.[66] QSARs and 3D-QSARs attempt to correlate spatially-localized features across a number of molecules with a biological activity, while 3D-QSARs also consider the requirement for ligand binding and interactions which may occur in the receptor binding site. 3D-QSARs are commonly used in the pharmaceutical industry, and have been successfully incorporated in toxicology. QSAR and 3D-QSAR approaches have been developed to identify and assess ligands or substrates for ERα,[67,68] ERβ,[68] androgen receptor,[69] Ah receptor,[70] and cytochrome P-450 isozymes.[71]

Similar to QSARs, structure-based drug design is used routinely in the pharmaceutical industry. It offers a powerful alternative to screening large databases by utilizing computer aided ligand docking software and existing 3D protein structure data to predict protein–ligand interactions. Docking of substrates into the active site of a protein of known or modeled structure can be used to explain protein–substrate interactions, and possibly predict protein-substrate interactions. Specific inhibitors of HIV-1 proteases,[72] thymidylate synthase,[73] and dihydrofolate reductase[74] have been successfully identified and developed using structure-based drug design approaches. Ligand docking studies can also be extended to screen large databases of small chemicals to identify novel putative ligands. This strategy has been used to screen databases of over 100,000 compounds to identify HIV protease ligands using the tool SLIDE, which considers both ligand and protein side-chain flexibility in its predictions.[75] In addition, SLIDE has been used to identify potential ligands for the human progesterone receptor (PR), human uracil-DNA glycosylase, and dihydrofolate reductase.[76] SLIDE could also be extended to examine other receptor-endocrine disruptor interactions, since the crystal structures of both ER isoforms, ERα[52,77] and ERβ,[59] as

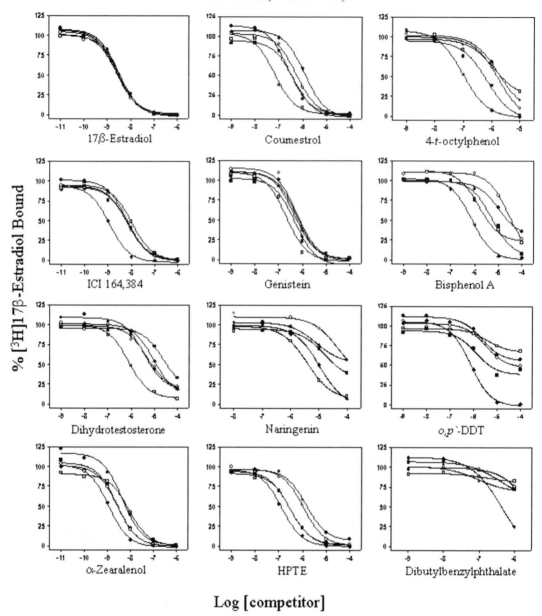

**Figure 3**    Representative competitive binding curves of selected test chemicals to ● GST-hERα (human) ○ GST-mERα (mouse), ■ GST-cERα (chicken), □ GST-aER (anole), and ◇ GST-rtER (rainbow trout) fusion proteins. An aliquot of partially purified GST-ERdef fusion proteins was incubated with 2.5 nM [$^3$H]17β-estradiol and increasing concentrations of unlabeled test chemical and incubated for 24 h at 4°C as described previously.[58] The results are from a representative experiment that was repeated at least two times. Standard deviations for points on graph ranged between 5% and 15% of the mean. 17β-estradiol exhibited a similar ability to compete for binding to the ERs from the different species examined. However, several differences in the affinity of the different ERs for several EEDs were observed. This figure is from Reference.[49]

well as the androgen receptor(AR)[78] and thyroid receptor (TR)[79] are available.

Nanotechnology is a relatively new technology that offers potential applications in several scientific fields, including the examination of biomolecular interactions.[80] The term "nanotechnology" refers to the use of materials with nanoscale dimensions, with sizes ranging from

1 to 100 nanometers. Nanomechanical cantilevers have been used to detect DNA hybridization of the exact complementary to ssDNA sequences of varying lengths.[81,82] Moreover, this technology has been used to detect protein:protein interactions between protein A and rabbit IgG,[81] and biotin and neutravidin.[82]. These studies demonstrate the applic-

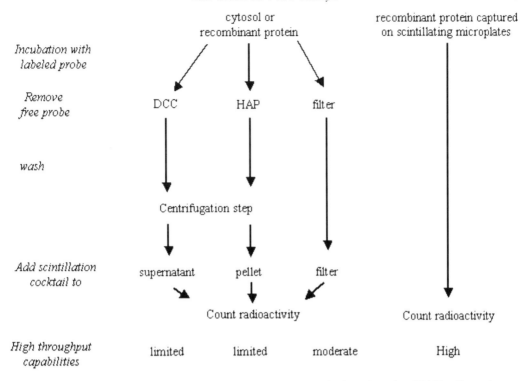

**Figure 4** Comparison of the dextran-coated charcoal (DCC), hydroxylapatite (HAP), filter plate, and scintillating microplate receptor binding assay methods. Both the traditional methods of DCC and HAP are relatively labor intensive and have limited high-throughput capabilities. The filter plate method reduces the amount of hands on time but still requires the removal of free probe and the addition of scintillation cocktail. However, the filter plate assays can be done in 96-well format and thus have moderately high-throughput capabilities. The scintillating microplate format is very well suited for high-throughout screening, since only the incubation of probe with the receptor, which has been captured on the microplates, is required. The removal of free probe, the wash steps, and the addition of scintillation cocktail are not necessary using this methodology.

ability of nanomechanical transduction to detect biomolecular interactions, and offers an exciting alternative screening strategy to detect potential EEDs with high-throughput capabilities.

## 14.5.6.3 CELL-BASED *IN VITRO* ASSAYS FOR THE IDENTIFICATION OF ENDOCRINE-DISRUPTING COMPOUNDS

Cell based *in vitro* assays have several key advantages in the identification of the endocrine disrupting potential of a chemical. These include simplicity, rapidity, relatively low cost, and potential for high throughput. Below, the major types of commonly used and emerging assays are discussed. This section focuses on assays that detect the modulation of gene expression in response to test compound exposure; however several other types of assays are noted (Table 1).

### 14.5.6.3.1 Reporter Gene-based Assays

#### 14.5.6.3.1.1 Recombinant Yeast Screens

Many assays have been developed that use transformed yeast (*Saccharomyces cerevisiae*) containing a hormone receptor of interest and a hormone response element linked to an easily measured reporter gene. Assays that use recombinant strains of yeast are attractive for many reasons, the first of which is the ease of culture and of generating stable transformants containing receptor, reporter, and other genetic elements of interest. Another advantage is that yeast lack endogenous steroid and thyroid hormones and receptors that may confound the interpretation of results, yet possess accessory proteins compatible with the transactivation of genes originating from a variety of species. As well, yeast cells have limited metabolic capability, so that in cases where both parent chemical and common biotransformation products are available, the active compound can be identified.[40,83,84] A further advantage is that yeast cells are amenable to high-throughput screening, which will be discussed further in a later section.

Recombinant yeast-based assays typically involve stably transforming all or part of the receptor of interest into yeast, along with a reporter gene under the control of one or more hormone response elements.[83,85] To date the majority of assays in yeast, like other types of assays examining interactions of compounds with endocrine signaling pathways, have focused on interactions with the ER. $EC_{50}$ values reported for E2 tend to be higher than those in mammalian systems, but detection limits tend to be lower because of high responsiveness and low background.[40] The yeast human ERα screen has been used to examine estrogenic activity of various environmental compounds as single chemicals,[83,85,86] and as mixtures,[87–89] while antiestrogenic effects have been studied by co-treating ER-expressing yeast with both E2 and the compound of interest.[90] Interactions of various compounds with rainbow trout ER have also been studied in yeast.[91] As well, one group of investigators used three separate yeast strains transformed with either the human ER, AR, or PR and their respective response element[84] to examine the activity of various drugs and environmental compounds. Interestingly, no compounds were identified that were able to interact with the PR, while another study identified several PR antagonists.[92] A yeast screen assay has also been developed for identifying AR ligands and aromatase inhibitors concurrently.[93] This yeast strain was cotransformed with human aromatase and mouse AR expression vectors, along with an androgen-responsive reporter. Treatment of the triple transformant with a suspected androgen induced AR-dependent activation of the reporter gene, while activity of a suspected aromatase inhibitor could be detected by co-treating the cells with the test compound and an aromatizable androgen; antiaromatase activity inhibited enzymatic formation of estrogen, and was detected as AR-dependent activation of the reporter gene.

Disadvantages associated with using yeast-based assays include low or selective permeability of the yeast cell wall, differences in chemical transport, metabolism, and bioavailability differences of test compounds compared to more complex organisms, and the high sensitivity to toxic effects of certain chemicals.[40,94,95] As well, several investigators have reported that compounds that typically produce antagonistic ER-mediated effects are often detected as agonists in yeast-based reporter gene assays.[96–98] It has been suggested that the absence of observed antagonism is due to the lack of appropriate endogenous repressor proteins in yeast.

### 14.5.6.3.1.2 *Transient Transfection Assays*

Transient transfection of hormone receptors and corresponding hormone-regulated reporter genes into mammalian cells, which addresses some of the concerns regarding the use of yeast assays to infer responses in more complex organisms.[40,99] Immortal cell lines from several species, including human, are available, and primary cultures can also be transfected effectively. Introducing receptor expression vectors into cells of the same species presumably has enhanced predictive value, since the appropriate transcription factors and coactivators/ corepressors are present.

Transfection assays typically fall into two categories. In the first, a reporter gene-containing plasmid under the transcriptional control of a hormone-responsive element is introduced by one of many available methods, along with a corresponding hormone receptor expression vector if the endogenous receptor content is insufficient to allow effective transactivation. The second type involves the introduction of chimeric (multiple species) receptor/reporter vectors. One commonly used chimeric system employs the Gal4-HEG0 construct, consisting of the yeast Gal4 DNA binding domain upstream of the D, E, and F domains (ligand binding region) of the human ERα, along with a reporter vector construct, composed of five repeats of the 17mer Gal4 response element followed by the luciferase reporter gene (17m5-G-Luc). This construct has been used to identify estrogenic responses of several environmentally relevant compounds and mixtures.[100–104] The chimeric system has the advantage of being regulated through a yeast DNA binding protein (Gal4) with no known mammalian homologue, so that all luciferase activity is attributed to the chimeric receptor. A potential disadvantage is that the N-terminal regions of the receptor have been deleted, and important regulatory regions may be missing. For example, the AF-1 domain of the ER, located in the N-terminal A/B region, is known to be important in transcriptional activation by E2 and other ER ligands on some promoters.[105,106]

Assays evaluating the estrogenic potential of chemicals commonly use MCF-7 human mammary carcinoma cells[85,107] or HeLa human cervical carcinoma cells.[108,109] ER cDNA clones from relevant wildlife species can also be introduced. Meanwhile several groups have transiently transfected human AR and androgen-regulated reporter genes into various cell lines including monkey kidney CV1 cells,[110,111] human prostate cancer LNCaP cells,[112] and mouse vas deferens epithelial cells.[113] In addition, one study of antiandrogenic effects used Chinese hamster ovary (CHO) cells transfected with human AR and luciferase reporter vectors.[114] Antiandrogenic environmental pollutants were able to cause a decrease in the response induced by the synthetic AR agonist R1881. As well, experiments involving the transient transfection of either the A or the B form of the PR and a progestin-responsive luciferase reporter have been performed in HeLa and CV1 cells, as well as in the HepG2 human hepatocarcinoma cell line.[115] Together, these studies show that many combinations of cell and receptor type and species of origin can be employed for the rapid and effective detection of compounds interfering with steroid receptor pathways.

### 14.5.6.3.1.3  Stably Transfected Cell Assays

Recently there has been a sharp increase in the development and use of stably transfected cell lines for examining endocrine-interacting effects. While these recombinant receptor-reporter cell lines have similar advantages and drawbacks compare to the transient transfection assays mentioned above, a key improvement, of course, is that stable transfectants obviate the need for repeatedly introducing the constructs of interest. Clearly, having the cells themselves replicate the necessary DNA segments not only represents a savings in time and materials, but also eliminates the variability inherent in the transient transfection procedure. In addition, reporter gene expression induced from response elements incorporated into the genome of the cells may more accurately reflect endogenous responses compared to those induced on episomal plasmid DNA.

The first estrogen-responsive cell lines to have been stably transfected with recombinant receptor and/or reporter genes were reported by Pons et al.[116] and Kramer et al.[117] (pVit-tk-Luc in MCF-7), and by Balaguer et al.[118] (Gal4-HEG0 and 17m5-G-Luc in HeLa). While a low level of reporter gene induction was characteristic of these cell lines (2- to 10-fold), responsiveness has greatly improved in more recently developed cell lines. For example, Legler et al.[119] recently developed the T47D.Luc cell line, in which T47D human breast adenocarcinoma cells, known to contain endogenous ERα and ERβ, were stably transfected with the pERE-tata-Luc reporter gene vector, and a 100-fold maximum induction of luciferase activity by E2 was reported. The improved responsiveness was attributed to the three sequential estrogen response elements (EREs), the minimal promoter (TATA box alone, without flanking regions that can cause higher background activation), and improvements in the luciferase gene, as well as vigorous serum-stripping to remove estrogenic compounds present in the serum. The use of T47D cells, which express high levels of ERs and transcription factors, also appeared to have been beneficial. Interestingly, the potencies observed for E2 and other estrogens were generally lower than those reported in transient transfection assays, but comparable to those observed in the MCF-7 cell proliferation assay (see table in Reference 119). Notably, this assay does not discriminate between responses mediated through ERα and ERβ. These investigators also attempted the stable introduction of pEREtata-Luc into ECC-1 human endometrial carcinoma cells and the chimeric Gal4-HEG0/17m5-G-Luc into Hepa.1c1c7 mouse hepatoma cells, but to date have observed only constitutive, non-estrogen-inducible luciferase expression.

Balaguer et al.[120] recently reported the generation of several other stably transfected cell lines, which were used to evaluate the estrogenic activity of sewage treatment plant effluent. MELN cells were created from the stable transfection of MCF-7 cells, which are ERα-positive, with the pERE-β-globin-Luc construct. HeLa cells by contrast do not express endogenous ERs, and these cells, after introduction of a luciferase reporter, were stably transfected with either ERα or ERβ under the control of an SV40 promoter to obtain the HELNα and HELNβ cell lines, respectively. The HGELN line was similarly created by stably introducing the chimeric Gal4-HEG0/17m5-G-Luc into HeLa cells. Comparison of responses of these four cell lines to E2 revealed that the MELN and HELNα appeared to be most sensitive, with the HELNβ being slightly less sensitive (attributed to the slightly lower affinity of E2 for ERβ), and the HGELN the least sensitive based on calculated $EC_{50}$ values. A stably transfected cell line containing an estrogen-responsive luciferase reporter was similarly developed in $GH_4C_1$ rat pituitary cells.[99]

Rogers and Denison[121] recently developed the first stably transfected human ovarian cell line from BG-1 cells, which were found to express endogenous ERα and EBβ mRNA but showed no evidence of either protein. Known as BG1Luc4E, this cell line was developed by introducing the pGudLuc7ere construct, which contains four EREs linked to the MMTV promoter and luciferase gene, into BG-1 ovarian carcinoma cells. When characterized, the BG1Luc4Es cell line was found to be comparable in sensitivity and responsiveness to the T47D.Luc cell line described above. These investigators have also developed analogous vectors responsive to androgen and retinoids (pGudLuc7are and pGudLuc7rare), which have been confirmed to be responsive when transiently transfected into human prostate LNCaP cells and MCF-7 cells respectively, and the production of stable transfectants is likely in progress.

Finally, a cell line known as MCF7-ERE was recently developed by Miller *et al.*[122] This cell line was made by incorporating the pERE-PGK-GFP plasmid (two sequential EREs coupled to a human phosphoglycerate kinase promoter linked to an enhanced green fluorescent protein gene) into MCF-7 cells. This is the first report of a stably transfected cell line expressing E2-inducible GFP, and while a maximal fold induction value was not reported, it appeared to have been approximately 7.5-fold compared to solvent-treated controls. Sensitivity was found to be comparable to that of the T47D.Luc cell line.

Overall, stably transfected cell lines show promise because of the ease of handling and reduction in reagent costs compared with transient transfection assays. Several sensitive and responsive cell lines have recently become available, allowing responses with different cell and receptor types to be investigated. Use of several complementary cell lines, including those from environmentally relevant species and specific cell types not yet developed, will increase the predictive power of these assays.

### 14.5.6.3.1.4   Advances in Reporter Gene Use

Some of the advantages of using reporter genes is that assay results are quantitative, can be obtained rapidly, and that the same detection procedure can be used for several different test systems. For example, the cell lines developed by Rogers and Denison described above, when complete, should allow parallel determination of a compound's ability to interact with the ER, AR, and RAR, all using the same protocol for measuring luciferase activity. By contrast, using endogenous gene expression as a marker, discussed in a later section, requires the labor-intensive separate monitoring of one or several transcripts in each cell type.

The use of reporter genes in toxicology has been reviewed in detail[123] (Figure 5). Commonly used reporter genes include firefly luciferase, β-galactosidase, and chloramphenicol acetyltransferase, but several investigators have recently reported progress in developing reporter genes with various enhancements. For example, several researchers have undertaken the task of engineering green fluorescent protein of jellyfish (*Aequorea victoria*) to enhance desirable qualities, and various colors and intensities are now available.[124–126] In addition, the sequential use of firefly (*Photinus pyralis*) and Renilla (*Renilla reniformis*) luciferase to measure receptor-induced expression as well as expression of a constitutive normalization vector can now be achieved using a cost-effective, nonproprietary assay buffer.[127] As well, Olesen *et al.*[128] provide an extensive report and protocols for recent improvements to luciferase, β-galactosidase, β-glucuronidase, and placental alkaline phosphatase chemiluminescent systems. It is anticipated that ongoing research will continue to provide improvements in signal stability and detection, and improve adaptability to high-throughput assay formats.

### 14.5.6.3.1.5   Advanced in High-Throughput Screening

Because of the remarkably high number of compounds for which testing for endocrine-interacting effects is desired, several researchers have recently reported advances in increasing the throughput of *in vitro* assays described in this section. In fact, a major advantage of many of the assays described above is their potential compatibility with a high-throughput format. Information obtained from these rapid screens can then aid in prioritization to determine which chemicals warrant fast tracking into Tier 1 screening.

Many investigators now use 96-well microtiter plates as a higher-density alternative to conventional single dish or 6-, 12-, or 24-well culture plate methods. This has met with some success, and though reduced fold induction was been reported,[85] the 96-well format has been used routinely in several assays of diverse types including transiently transfected MCF-7 cells[129] and stably transfected $GH_4C_1$ cells,[99] mentioned above, as well as in a carp hepatocyte assay.[130] Even more interesting is that several groups[128,131,132] have recently been investigating the feasibility of using 384- and even 1536-well plate formats. Major concerns currently include shearing of cells during pipeting, evaporation causing significant volume loss, and inability to detect signals of low intensity, but continued progress is anticipated.

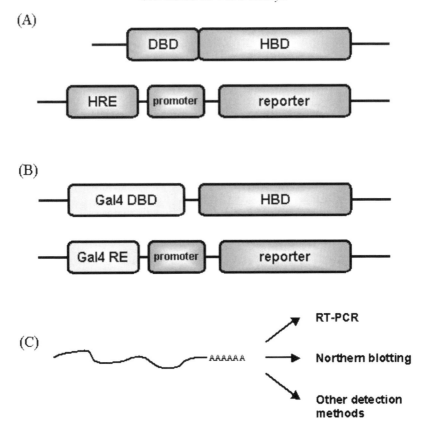

**Figure 5** Several common methods of determining *in vitro* receptor-mediated gene expression. (A) A simple system involving a hormone receptor (HR; both the DNA binding domain (DBD) and hormone binding domain (HBD) are necessary) as well as a reporter linked to one or more hormone response elements (HREs); (B) A more complex chimeric system designed to minimize non-HR-mediated activation, in which the HBD of the receptor of interest is linked to the yeast Gal4 DNA binding domain, and incorporated either transiently or stably into the cell of interest along with a Gal4 response element-driven reporter gene; (C) Measurement of expression of endogenous hormone-regulated genes can be accomplished by several methods.

### 14.5.6.3.2 Monitoring of Endogenous Gene Expression

The E-SCREEN assay, in which the induction or inhibition of MCF-7 cell proliferation by compounds of interest is monitored, has identified some compounds that mimic or interfere with the proliferative effect characteristic of endogenous estrogen.[133] An advantage of this system is that, unlike in transfected cells, only the endogenous cellular components exist. This may enhance the predictive value of the assay. One major disadvantage, however, is that a positive result does not necessarily give information about the mechanism of the effect, since proliferation can be induced not only by estrogen but by other compounds such as epidermal growth factor.[134] A further disadvantage, important in the screening of compounds for diverse effects, is that presently such a test is limited to the detection of estrogens, as analogous test systems do not currently exist for monitoring interference with other pathways. An alternative that has been utilized recently is the monitoring of expression of endogenous genes in response to compounds of interest. Again this, unlike transfected cell assays, monitors genes that are associated with full endogenous promoters and that are present in their natural copy number and in their native context. As well, there is no mixing of cellular components from different species. Another advantage is that expression of genes from more than one pathway can be monitored simultaneously, for example some that act through the ER and some that are associated with estrogen-like effects not mediated through either ER isoform.

Several studies have examined endogenous gene expression in MCF-7 cells, which is of particular interest since, as noted above, these cells have also been used in complementary transient and stable transfection studies and in cell proliferation assays. Earlier studies monitored the expression of one or a small number

**Table 1**  Summary of cell-based *in vitro* assays for endocrine disruptors and their advantages and limitations.

| Assay type | Gene(s) under study | Method of measuring response | Assay advantages | Assay limitations |
|---|---|---|---|---|
| Transformed yeast | Exogenous | Reporter gene activity | • Ease of generating and maintaining transformants<br>• Ease of measuring reporter gene response<br>• No endogenous steroid/thyroid pathways<br>• Limited metabolic capability<br>• Amenable to high-throughput | • Low similarity to cells of mammals or other organisms of interest<br>• Agonist response to several known antagonists<br>• Barrier properties of cell wall<br>• Highly sensitive to toxicity by certain chemicals |
| Transiently trans-fected cells | Exogenous | Reporter gene activity | • Can use primary cultures; many immortal cell lines also available<br>• Ease of measuring reporter gene response<br>• Use of chimeric system ensures specificity of response | • Time and reagent costs associated with transient transfection<br>• Chimeric receptor may lack important functional domains<br>• Not easily adapted to high-throughput format |
| Stably transfected cells | Exogenous | Reporter gene activity | • Same as for transient transfection, above<br>• In addition, no need to perform transient transfection | • Must use immortal cell line<br>• Currently relatively few cell lines available<br>• Some cell lines exhibit poor inducibility<br>• Not easily adapted to high-throughput format |
| Measurement of endogenous gene expression | Endogenous | RT-PCR, Northern blot, slot blot, *etc.* | • Any cell type can be used<br>• Intact, endogenous regulatory elements<br>• Expression of many genes can be detected simultaneously | • Not easily adapted to high-throughput format<br>• Labor intensive |

of E2-inducible gene products such as the PR or pS2 by northern blotting or RT-PCR,[135–138] and several investigators have recently refined these test systems. For example, Kledal et al.[139] have recently discussed the important features of marker genes (high sensitivity, strong dose-dependence, and several-fold difference in expression levels between exposed and unexposed samples), and comment on the use of competitive PCR to detect changes in marker gene expression. Similarly Jorgensen et al.,[140] used four marker genes (pS2, TGFβ3, monoamine oxidase A, and α-1-antichymotrypsin) selected from a prior compilation of approximately 100 estrogen-regulated genes. Notably, they also made vigorous attempts to detect ERβ mRNA in MCF-7 cells, yet were unable to find any evidence for its expression, and therefore concluded that all of their observed results were mediated through ERα. The investigators also added that a key criterion for choosing an estrogen-regulated gene as a marker is the reversal of the effect upon addition of an ER antagonist. Changes in the expression of the four genes that were selected were monitored by competitive PCR and quantified by phosphorimaging of bands resolved by polyacrylamide gel electrophoresis. The authors also noted that similar assays to detect effects mediated through ERβ and the AR could be readily developed from prostate-derived cell lines.

Zajchowski et al.[141] recently reported the use of several diverse cell lines, in addition to MCF-7 cells, for monitoring endogenous gene expression. This study was aimed at developing a suite of assays that could be used to identify selective ER modulators (SERMs) with diverse mechanisms as evidenced by their gene expression profiles. This combination of assays allowed discrimination between E2, tamoxifen, raloxifene, and the pure ER antagonist ICI 164,384 based on northern and slot blots for known estrogen-responsive genes (cathepsin D, growth hormone, prolactin, progesterone receptor, pS2, TGFα, insulin-like growth factor binding protein 1, corticosteroid binding globulin, amphiregulin, and thyrotropin-releasing hormone receptor). Many cell lines were used, including breast, ovarian, pituitary, and liver lines, and the differentiation of various types of SERMs was possible based on characteristic gene expression profiles.

In addition, cells from several wildlife species have been used to detect endogenous expression of estrogen-regulated genes. For example, estrogen-inducible vitellogenin mRNA has been successfully detected in trout hepatocytes by slot blot[91] and dot blot,[142] and in *Xenopus*

(frog) hepatocytes by semi-quantitative RT-PCR.[143] Many other assays have been developed which use similar conditions but measure vitellogenin protein rather than mRNA levels. One example is the CARP-HEP assay, which detects vitellogenin in male carp hepatocytes by ELISA in a 96-well format.[130]

## 14.5.6.4 THE USE OF MICROARRAYS TO SCREEN FOR NOVEL ENDOCRINE DISRUPTORS

Microarrays have quickly emerged as the premier tool for enabling genome-wide analysis of gene expression. With microarrays, the level of mRNA expression for hundreds to tens of thousands of genes can be measured simultaneously in a single experiment (Figure 6). By contrast, northern blot and RT-PCR methods allow for the quantitation of only a few genes at most in any one experiment. With knowledge of gene expression in cells, tissues, organs or whole organisms under a variety of physiological and pathological states, new understanding of the molecular basis of physiology, disease and toxicity can be acquired. As a result, these new technologies are influencing drug discovery and preclinical safety in the biotechnology and pharmaceutical industry. Toxicologists are also promoting genomic expression technologies as a superior alternative to traditional rodent bioassays to identify and assess the safety of chemicals, including potential EEDs and drug candidates.[144,145] Ultimately, toxicogenomics (the interdisciplinary field of genomics, bioinformatics and toxicology) is expected to accelerate drug development and aid risk assessment for drugs, agrochemicals, and industrial chemicals.

It has been proposed that each chemical that acts through a particular mechanism of action will induce a unique and characteristic gene expression profile under a given set of conditions.[145] Microarray experiments applied to tumor samples, for example, have demonstrated the potential of gene expression profiling to accurately classify disease phenotypes based on gene expression alone.[146–148] Therefore, it is expected that gene expression profiling can be used to classify chemicals by the similarity of their expression profiles compared to expression profiles obtained from known chemicals with defined mechanisms of action. Although the cell-based assays previously discussed allow for the rapid identification of chemicals that can act through a known target or pathway, such as the ER,

**A**                                              **B**

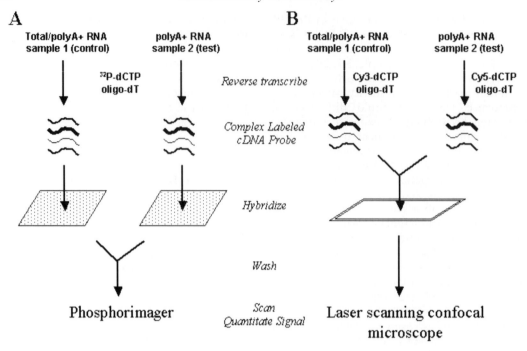

**Figure 6** Overview of microarray methodologies for highly parallel quantitation of gene expression. In this technique, an entire total or polyA+ RNA population from a tissue source of interest is reverse transcribed with an oligo(dT) primer in the presence of (A) radiolabeled nucleotides to generate a complex labeled cDNA probe, such that the abundance of individual mRNA transcripts is reflected in the cDNA product. The labeled probes are then hybridized to an excess amount of double-stranded denatured target cDNA arrayed on a solid support, such as a nylon membrane. The arrayed cDNA contains previously cloned and partially sequenced genes and expressed sequences from a cDNA library. The microarrays are washed to remove unbound probe before being scanned on a phosphorimager to detect and quantitate radioisotopic signal intensity. The intensity of the hybridization signal is proportional to the abundance of cDNA derived from the mRNA population. Two microarrays (e.g., control and test) can then be compared to determine the presence and relative abundance of hundreds to thousands of mRNA transcripts in a single hybridization experiment. (B) An alternative method, made popular by Patrick Brown and colleagues, is based on competitive hybridization of fluor-labeled probes to cDNA arrayed at high density on glass slides.[167,168] In this method, control and test mRNA samples are reverse transcribed in the presence of an oligo(dT) primer and one of two modified nucleotides that fluoresce at a characteristic wavelength (e.g., Cy3-dCTP or Cy5-dCTP). The control and test fluor-labeled cDNA probes are mixed and compete for hybridization to their complementary sequence arrayed on the glass slide. After unbound probe is washed off, the microarray is scanned with a laser scanning confocal microscope. The relative fluorescent signal at each wavelength is proportional to the mRNA abundance in each sample. The relative abundance of mRNA is represented by a ratio of the fluorescent intensity at each wavelength, which indicates the fold change in gene expression in the test sample relative to the control.

large-scale comparison of gene expression profiles has the potential to determine mechanisms of action of uncharacterized chemicals without prior knowledge of potential targets or toxicities. Due to the various mechanisms by which chemicals can disrupt the endocrine system, screening for a mechanism of action, based on expression profiles, rather than screening against a single activity provides a more comprehensive assessment of the potential for EEDs to disrupt the endocrine system. Here we will discuss the possible application of gene expression profiling using microarrays for predicting mechanisms of action and identifying novel EEDs.

#### 14.5.6.4.1  Pairwise-Conditional Expression Analysis Identifies Novel Biomarkers

Biomarkers are mechanistically-based experimental endpoints that are designed to efficiently detect and characterize chemical exposures. Biomarkers have been frequently used to detect and characterize xenobiotics for hormone-like activity. For example, an early biomarker that was developed for detecting estrogen exposure *in vivo* is the egg yolk protein, vitellogenin.[149] This protein is normally expressed in female oviparous vertebrates in response to endogenous circulating estrogen. Its detection in male oviparous vertebrates, however, is diagnostic of exogenous estrogen exposure since males do not normally express vitellogenin due to their low levels of endogenous estrogen. Induction of both vitellogenin have been used to detect chemicals with estrogenic activity *in vivo* and in cultured hepatocytes.[91,130,142,143,150] Many biochemical, hormonal and physiological endpoints are also used to screen for potential EEDs in rodents, including agonists and antagonists of estrogen, androgen, progesterone, dopamine and thyroid hormones.[151]

Despite the wide use of proposed Tier 1 screening assays for EEDs, they remain costly, time consuming, and only identify a limited spectrum of biological activities. By contrast, molecular biomarkers such as mRNA provide a more rapid and sensitive marker that can detect a wider variety of biological activities. Microarrays provide a powerful tool to identify a larger number of molecular biomarkers of exposure to EEDs and other xenobiotics. The first example of microarrays being used to study hormonal gene regulation *in vivo* was by Feng *et al.*,[152] who used fluor-based microarrays to study hepatic gene expression in hypothyroid mice treated with or without thyroid hormone ($T_3$). Hypothyroid mice were used to provide a model system with low levels of endogenous $T_3$, thus increasing the sensitivity to detect subtle changes in gene expression in response to $T_3$ treatment. They were able to verify changes in the expression of genes previously known to be affected by $T_3$, as well as identify many novel genes that were both positively (14 genes) and negatively (41 genes) regulated by $T_3$. This approach could also be utilized for other hormone agonists and antagonists. With a knowledge of biomarkers for exposure to hormone agonists and antagonists, higher throughput assays can now be utilized to screen chemicals for a number of potential endocrine modulating activities *in vivo*. To increase their predictive value, these biomarkers must be validated in other tissues or other *in vivo* and *in vitro* model systems. In addition, the establishment of potential biomarkers should be correlated to adverse physiological effects, since changes in gene expression do not always imply toxicity.

Microarray-based screening is also valuable when studying xenobiotics with unknown mechanisms of action, since identifying novel biomarkers associated with toxicity can often give valuable insight into mechanisms of action and may explain pathogenesis. For example, Holden *et al.*[153] used radiolabel-based microarrays to discover genes associated with hepatotoxicity. In their model system, cultured human hepatoma cells (HepG2) were exposed to carbon tetrachloride ($CCl_4$) for 8 h. Gene expression was compared to cells treated with dimethyl formamide, a chemical not implicated in hepatotoxicity, rather than vehicle solvent. While 47 genes were changed in expression, interleukin-8 (IL-8) was further studied by northern blot and ELISA. The increase in both IL-8 mRNA and protein expression was found to correspond to the time-dependent decrease in cell viability, thus implicating IL-8 in hepatotoxicity. However, it is unclear whether IL-8-induction causes cell damage, or is induced in response to cell damage. Therefore, it would be of interest to determine if the response of IL-8 is specific to $CCl_4$ or if its expression is associated with other hepatotoxicants. Microarrays have also been used for identifying human genes responsive to other physical and chemical stresses, such as irradiation-induced apoptosis,[154,155] DNA-damaging agents and anti-inflammatory drugs,[156] metals, and combustion by-products.[157–159]

However, transcriptional responses to chemical stressors in human cells are complicated by cellular heterogeneity and genotype, which present further challenges to interpreting gene expression changes in human populations. Another potential drawback of gene-based biomarkers is their lack of specificity, since genes are often regulated by more than one signaling pathway, and failure of a biomarker to respond to exposure does not necessarily indicate absence of effect. Having multiple markers of exposure that are correlated with adverse effects, in addition to replication and secondary verification of microarray results using alternative experimental means, are vital for identifying robust predictive biomarkers of endocrine disrupting chemicals and xenobiotic exposure.

### 14.5.6.4.2 Multi-conditional Expression Analysis Predicts Mechanism of Action

Measuring changes in gene expression over time or across increasing doses provides information on the kinetics and coordination of gene expression during the dynamic processes of cellular homeostasis. Analyzing gene expression data across multiple samples can also reveal underlying similarities among different conditions, thus producing correlates of gene behavior that can be used to predict and diagnose cellular responses to exogenous chemicals. In order to extract ordered subsets of information from disordered sets of multiconditional gene expression data, a number of multivariate methods have been applied. The most common method for clustering multi-conditional gene expression data is hierarchical clustering, which was made popular by the work of Eisen _et al._[160] Principal component analysis[161] and partitioning methods, such as k-means clustering[162] and self-organizing maps (SOMs)[163] have also been applied to multi-conditional gene expression data sets. In general, these methods attempt to find order among disordered data sets by grouping similar objects together. Grouping genes based on similarity of expression across multiple conditions is desirable for a number of reasons. For example, two genes of similar expression characteristics may be coordinately regulated and therefore involved in a similar function and/or under the same regulatory control. In addition to grouping genes, samples can be clustered based on the expression of all genes on the microarray. Gene expression profiles induced by chemicals that share a similar mechanism of action can be correlated under certain conditions (Figure 7).[156,164,165] Therefore, it should be possible to predict a potential mechanism of action for a chemical of undefined toxicity based on correlated expression to chemicals of known mechanism. This approach can be used to identify potential EEDs based on large-scale comparison of expression profiles induced by chemicals of known mechanism, such as estradiol, testosterone, $T_3$, and other hormones that modulate the endocrine system.

The overall process of hierarchical cluster analysis begins with defining the objects to be clustered by a measure of similarity. The Pearson correlation coefficient is a common metric that quantifies how well two variables vary together. The correlation coefficient $r$ is always between $-1$ and 1, where 1 indicates an identical profile, 0 indicates the two profiles are independent, and $-1$ indicates the profiles are inversely related. The advantage of this statistic is that it captures similarity in shape without emphasis on magnitude as it is invariant to scale. Correlation measures are often transformed into Euclidean distances prior to clustering. This transformation accounts for similarity with all other genes, rather by than a single pairwise comparison. It is also less sensitive to random fluctuations in expression measurement such that two genes that exhibit poor pairwise correlation may still be similar by virtue of their correlation with all other genes. A matrix of pairwise distances is clustered and visualized using a dendrogram, which is similar to a phylogenetic tree. To begin clustering, each object (i.e., gene or sample) is represented by a single cluster. The two most similar objects are then merged into a new pseudo-object. A new distance measure is calculated for the pseudo-object and merged with the next most similar object. This process is repeated in an iterative fashion until only one pseudo-object remains, which represents the root of the tree. By the process of clustering, objects are organized into branches, such that branches of the tree that are adjacent are more closely related, and the length of the tree branch reflects the degree of similarity between objects. Finally, the branches of the tree are ordered and colored to indicate graphically the nature of the relationships within and between the branches (Figure 7).

Both the genes and the experimental conditions can be hierarchically clustered in a two-dimensional dendrogram. In this manner, treatment conditions that induce similar expression profiles across all genes on the microarray are closer together on the tree in the first dimension, and presumably mechanistically related, while genes that are similar in expression profile across all conditions are closer together on the tree in the other dimension, and perhaps functionally related. This approach can be used to identify treatment conditions and to illustrate relationships between multiple conditions and gene expression so that subsets of genes, rather than a single biomarker, can be used as predictors of cellular response to EEDs and other xenobiotics.

Hierarchical clustering has been suggested to be inappropriate for gene expression data since phylogenetic trees, or dendrograms, are best applied to situations of true hierarchical descent, such as evolution. However, gene expression does not follow a hierarchy but rather is characterized by distinct mechanisms. When hierarchical clustering is applied, the data is forced into associations at some level, regardless of the relationship between the data. As two branches of the tree are joined into one, they become less similar and eventually meaningless. As an alternative to imposing a hierarchical descent on the data, partitioning methods, such as SOMs and k-means clustering, can divide data set into similar, but distinct groups. Hierarchical clustering can then be applied to each partitioned set of genes.

One of the first applications demonstrating the utility of gene expression profiling to characterize mechanisms of toxicity was by Marton _et al._[164] Using cDNA microarrays containing essentially every open reading frame (ORF) in the yeast genome (6000+), Marton _et al._ were able to show that the gene expression profile induced by the immunosuppressant FK506 was highly correlated with the expression profile induced by the mechanistically

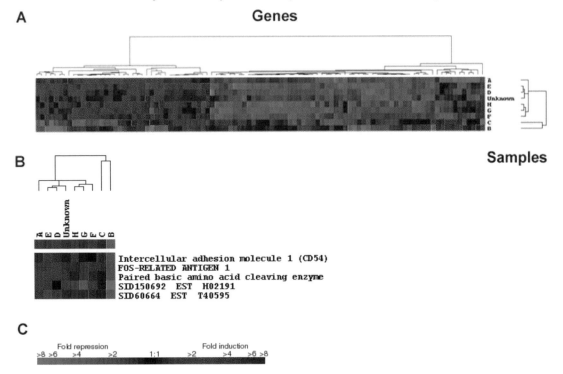

**Figure 7** Hierarchical cluster analysis of genes and samples. (A) This hypothetical data set demonstrates how hierarchical clustering of both genes, across the horizontal axis, and samples (i.e., array experiments), along the vertical axis, illustrates the relationship between gene expression and different experimental conditions. Each column represents a gene and each row is a single microarray experiment measuring gene expression for 100 genes. In this example, each of the nine rows represent an experiment comparing an untreated sample to a sample treated with one of nine different chemicals, including eight chemicals of known mechanism (labeled A to H) and one chemical of unknown mechanism. Genes are hierarchically clustered to indicate the similarity between genes in their expression profile across the nine samples. The length of each branch indicates the relative similarity between branches of the tree. Likewise, the samples are also hierarchically clustered to indicate similarity in expression profile induced by each of the nine chemicals. From the sample dendrogram in A, the expression profile induced by the unknown sample appears most similar to the expression profile induced by chemicals D and E. Note that the similarity between the unknown and chemicals D and E are the same, since both D and E can be reordered about their common node. Therefore, the unknown chemical may act with a similar mechanism of action as D and E. The pink colored branch of the tree is highlighted in B. (B) A close up of a cluster within the dendrogram indicates the genes within that cluster and their pattern of expression across the nine samples. (C) The color hue and intensity represents the fold change in gene expression, as indicated in the color bar. Green indicates repression of gene expression in the test sample relative to the control sample. Red indicates induction of gene expression in the test sample relative to the control sample. Black indicates no change (i.e., ratio of 1) and grey indicates no data point available for that gene on its respective array.

similar immunosuppressant cyclosporin A. These expression signatures were not, however, correlated with the expression profile induced by other unrelated drugs. These experiments demonstrate the principle that chemicals can induce characteristic and unique gene expression profiles. Therefore, expression profiles may be used to identify and classify unknown chemicals with respect to mechanism of action. The use of yeast as a model system allowed Marton *et al.* to also verify the drug target (i.e., calcineurin) by correlating the expression profile of the immunosuppressants to the expression profile induce by genetic disruption

of the target gene. It was also demonstrated that off-target effects exist for FK506 since a distinct profile of gene expression was produced in a calcineurin mutant strain treated with the immunosuppressant. Although the toxicity of FK506 is primarily mechanism-based, these effects point to potential causes of unwanted side effects.

The studies by Marton *et al.* support the use of expression profiling to distinguish mechanisms of action of unknown EEDs, even in simple model systems. However, a large reference database of expression profiles induced by a number of diverse hormones and other EEDs of known mechanism is required. The utility of this approach has been demonstrated in yeast. Using a "compendium" of yeast expression profiles from over 300 diverse mutants and chemical treatments, Hughes *et al.*[165] were able to identify a previously unknown drug target for the commonly used topical anesthetic dyclonine. This was accomplished by matching gene expression profiles caused by uncharacterized perturbations (e.g., gene disruption or chemical inhibition) with a large set of reference profiles corresponding to disturbances of known cellular pathways. Hierarchical clustering and correlation measures were used to match similar expression profiles. When the dyclonine-induced expression profile was compared to the compendium, it was found to be correlated ($r = 0.82$) with the profile resulting from genetic disruption of the ergosterol pathway, specifically *erg2*. The human gene with the greatest sequence similarity to the *erg2* protein was found to be the sigma receptor, which is known to bind a number of neuroactive drugs and other inhibitory compounds that target both *erg2p* and the sigma receptor. Despite the use of yeast as a model system, a potential mechanism of action for dyclonine can be inferred for mammalian systems.

Gene expression profiling could also be extended to identify regulatory motifs, such as hormone response elements, that govern the response to hormones and other EEDs. For example, Gasch *et al.*[166] used yeast to identify regulatory motifs that play a role in the transcriptional response to a variety of physical and chemical perturbations, including heat shock, osmotic shock, nitrogen and amino acid depletion, oxidative stress, and others. When the gene expression profiles induced by a panel of environmental stresses were compared by hierarchical clustering, a set of ~900 genes were found to show a similar response to almost all of the environmental changes. Their regulation, however, was dependent on many signaling systems that acted in a condition-specific and gene-specific manner, rather than being controlled by a single stress-sensing pathway. To determine what factors governed the response to stress, the promoters of subclusters of stress-responsive genes were analyzed for common regulatory elements. It was found that many stress-responding genes contained Msn2 and/or Msn4p binding sites. By examining the expression profile of yeast mutants null for *msn2* or *msn4*, or over expressing MSN2 or MSN4, it was determined that these transcription factors play a role in regulating expression of a subset of responsive genes following environmental stress. Other

responsive genes were not affected by MSN2/4 and are thought to be under control of other independent signaling pathways. These elaborate set of studies demonstrate the complexity at which cells can detect and respond to unique forms of physical and chemical stressors, and underscores the utility of transcriptional responses to diagnose cellular perturbations and decipher the mechanisms of action of chemicals.

### 14.5.6.5  REFERENCES

1. U.S. EPA, 'Endocrine Disruptor Screening Program,' http://www.epa.gov/scipoly/oscpendo/, 2000.
2. National Research Council, 'Hormonally-active Agents in the Environment,' http://www.nap.edu/books/0309064198/html/, 2000, pp. 1–430.
3. D. L. Davis, H. L. Bradlow, M. Wolff *et al.*, 'Medical hypothesis: xenoestrogens as preventable causes of breast cancer.' *Environ. Health Perspect.*, 1993, **101**, 372–377.
4. M. S. Wolff, G. W. Collman, J. C. Barrett *et al.*, 'Breast cancer and environmental risk factors: epidemiological and experimental findings.' *Annu. Rev. Pharmacol. Toxicol.*, 1996, **36**, 573–596.
5. A. Giwercman and N. E. Skakkebaeck, 'The human testis—an organ at risk.' *Int. J. Androl.*, 1992, **15**, 373–375.
6. E. Carlsen, A. Giwercman, N. Keiding *et al.*, 'Evidence for decreasing quality of semen during past 50 years.' *Brit. Med. J.*, 1992, **305**, 609–613.
7. R. F. Seegal, 'Epidemiological and laboratory evidence of PCB-induced neurotoxicity.' *Crit. Rev. Toxicol.*, 1996, **26**, 709–737.
8. H. A. Tilson and P. R. Kodavanti, 'Neurochemical effects of polychlorinated biphenyls: an overview and identification of research needs.' *Neurotoxicology*, 1997, **18**, 727–743.
9. T. Colborn, F. S. vom Saal and A. M. Soto, 'Developmental effects of endocrine-disrupting chemicals in wildlife and humans.' *Environ. Health Perspect.*, 1993, **101**, 378–384.
10. T. Colborn and C. Clement, 'Chemically-induced Alterations in Sexual and Functional Development: The Wildlife/Human Connection,' Princeton Scientific Publishing Co., Princeton, 1992.
11. U.S. EPA. 'Special report on environmental endocrine disruption: an effects assessment and analysis.' U.S. Environmental Protection Agency, Washington, D.C., 1997.
12. L. Guillette Jr., S. Arnold and J. McLachlan, 'Ecoestrogens and embryos—Is there a scientific basis for concern.' *Animal Reprod. Sci.*, 1996, **42**, 13–24.
13. R. J. Kavlock, G. P. Daston, C. DeRosa *et al.*, 'Research needs for the risk assessment of health and environmental effects of endocrine disruptors: a report of the U.S. EPA-sponsored workshop.' *Environ. Health Perspect.*, 1996, **104**(Suppl. 4), 715–740.
14. D. M. Sheehan, 'Herbal medicines, phytoestrogens and toxicity: risk-benefit considerations.' *Proc. Soc. Exp. Biol. Med.*, 1998, **217**, 379–385.
15. R. L. Cooper and R. J. Kavlock, 'Endocrine disruptors and reproductive development: a weight-of-evidence overview.' *J. Endocrinol.*, 1997, **152**, 159–166.
16. H. Fisch, E. T. Goluboff, J. H. Olson *et al.*, 'Semen analyses in 1283 men from the United States over a

25-year period: no decline in quality.' *Fertil. Steril.*, 1996, **65**, 1009–1014.

17. N. Krieger, M. S. Wolff, R. A. Hiatt *et al.*, 'Breast cancer and serum organochlorines: a prospective study among white, black and asian women.' *J. Nat. Cancer Inst.*, 1994, **86**, 589–599.

18. S. H. Safe, 'Environmental and dietary estrogens and human health: is there a problem?' *Environ. Health Perspect.*, 1995, **103**, 346–351.

19. S. H. Safe, 'Endocrine disruptors and human health: is there a problem? An update.' *Environ. Health Perspect.*, 2000, **108**, 487–493.

20. R. M. Sharpe, 'Declining sperm counts in men—is there an endocrine cause?' *J. Endocrinol.*, 1993, **136**, 357–360.

21. J. Toppari, J. C. Larsen, P. Christiansen *et al.*, 'Male reproductive health and environmental xenoestrogens.' *Environ. Health Perspect.*, 1996, **104**(Suppl. 4), 741–803.

22. Environmental and Climate Research Programme of DG XII of the European Commission, 'Proceedings of the European Workshop on the Impact of Endocrine Disruptors on Human Health and Wildlife,' Weybridge, UK, 1996.

23. J. A. Katzenellenbogen, 'The structural pervasiveness of estrogenic action.' *Environ. Health Perspect.*, 1995, **103**(Suppl. 7), 99–101.

24. J. A. McLachlan, 'Functional toxicology: a new approach to detect biologically active xenobiotics.' *Environ. Health Perspect.*, 1993, **101**, 386–387.

25. R. Evans, 'The steroid and thyroid hormone receptor superfamily.' *Science*, 1988, **240**, 889–895.

26. H. Adlercreutz, K. Hockerstedt, C. Bannwart *et al.*, 'Effect of dietary components, including lignans and phytoestrogens, on enterohepatic circulation and liver metabolism of estrogens and on sex hormone binding globulin (SHBG).' *J. Steroid. Biochem.*, 1987, **27**, 1135–1144.

27. A. Brouwer, W. S. Blaner, A. Kukler *et al.*, 'Study of the mechanism of interference of 3,3′,4,4′-tetrachlorobiphenyl with the plasma retinol-binding proteins in rodents.' *Chem.–Biol. Interact.*, 1988, **68**, 203–217.

28. A. Brouwer and K. J. Van den Berg, 'Binding of a metabolite of 3,4,3′,4′-tetrachlorobiphenyl to transthyretin reduces serum vitamin A transport by inhibiting the formation of the protein complex carrying both retinol and thryoxin.' *Toxicol. Appl. Pharmacol.*, 1986, **85**, 301–312.

29. H. Hodgert Jury, T. R. Zacharewski and G. L. Hammond, 'Interactions between human plasma sex hormone-binding globulin and xenobiotic ligands.' *J. Steroid. Biochem. Mol. Biol.*, 2000, **75**, 167–176.

30. M. C. Lans, E. Klasson-Wehler, M. Willemsen *et al.*, 'Structure-dependent, competitive interaction of hydroxy-polychlorobiphenyls, -dibenzo-*p*-dioxins and -dibenzofurans with human transthyretin.' *Chem.–Biol. Interact.*, 1993, **88**, 7–21.

31. S. Chen, Y. C. Kao and C. A. Laughton, 'Binding characteristics of aromatase inhibitors and phytoestrogens to human aromatase.' *J. Steroid. Biochem. Mol. Biol.*, 1997, **61**, 107–115.

32. Y. C. Kao, C. Zhou, M. Sherman *et al.*, 'Molecular basis of the inhibition of human aromatase (estrogen synthetase) by flavone and isoflavone phytoestrogens: A site-directed mutagenesis study.' *Environ. Health Perspect.*, 1998, **106**, 85–92.

33. M. H. Kester, S. Bulduk, D. Tibboel *et al.*, 'Potent inhibition of estrogen sulfotransferase by hydroxylated PCB metabolites: a novel pathway explaining the estrogenic activity of PCBs.' *Endocrinology*, 2000, **141**, 1897–1900.

34. A. K. Loomis and P. Thomas, 'Effects of estrogens and xenoestrogens on androgen production by Atlantic croaker testes *in vitro*: evidence for a nongenomic action mediated by an estrogen membrane receptor.' *Biol. Reprod.*, 2000, **62**, 995–1004.

35. A. Revelli, M. Massobrio and J. Tesarik, 'Nongenomic actions of steroid hormones in reproductive tissues.' *Endocr. Rev.*, 1998, **19**, 3–17.

36. B. Blumberg, W. Sabbagh Jr., H. Juguilon *et al.*, 'SXR, a novel steroid and xenobiotic-sensing nuclear receptor.' *Genes Dev.*, 1998, **12**, 3195–3205.

37. H. Masuyama, Y. Hiramatsu, M. Kunitomi *et al.*, 'Endocrine disrupting chemicals, phthalic acid and nonylphenol, activate Pregnane X receptor-mediated transcription.' *Mol. Endocrinol.*, 2000, **14**, 421–428.

38. L. B. Moore, D. J. Parks, S. A. Jones *et al.*, 'Orphan nuclear receptors constitutive androstane receptor and pregnane X receptor share xenobiotic and steroid ligands.' *J. Biol. Chem.*, 2000, **275**, 15122–15127.

39. S. Westin, M. G. Rosenfeld and C. K. Glass, 'Nuclear receptor coactivators.' *Adv. Pharmacol.*, 2000, **47**, 89–112.

40. T. Zacharewski, '*In vitro* assays used to assess estrogenic substances.' *Environ. Sci. Tech.*, 1997, **31**, 613–623.

41. J. Ashby, 'Validation of *in vitro* and *in vivo* methods for assessing endocrine disrupting chemicals.' *Toxicol. Pathol.*, 2000, **28**, 432–437.

42. H. U. Bryant and W. H. Dere, 'Selective estrogen receptor modulators: an alternative to hormone replacement therapy.' *Proc. Soc. Exp. Biol. Med.*, 1998, **217**, 45–52.

43. A. Negro-Vilar, 'Selective androgen receptor modulators (SARMs): a novel approach to androgen therapy for the new millennium.' *J. Clin. Endocrinol. Metab.*, 1999, **84**, 3459–3462.

44. R. Bolger, T. E. Wiese, K. Ervin *et al.*, 'Rapid screening of environmental chemicals for estrogen receptor binding capacity.' *Environ. Health Perspect.*, 1998, **106**, 551–557.

45. S. Stoessel and G. Leclercq, 'Competitive binding assay for estrogen receptor in monolayer culture: measure of receptor activation potency.' *J. Steroid. Biochem.*, 1986, **25**, 677–682.

46. S. C. Laws, S. A. Carey, W. R. Kelce *et al.*, 'Vinclozolin does not alter progesterone receptor (PR) function *in vivo* despite inhibition of PR binding by its metabolites *in vitro*.' *Toxicology*, 1996, **112**, 173–182.

47. K. G. Coleman, B. S. Wautlet, D. Morrissey *et al.*, 'Identification of CDK4 sequences involved in cyclin D1 and p16 binding.' *J. Biol. Chem.*, 1997, **272**, 18869–18874.

48. R. M. Blair, H. Fang, W. S. Branham *et al.*, 'The estrogen receptor relative binding affinities of 188 natural and xenochemicals: structural diversity of ligands.' *Toxicol. Sci.*, 2000, **54**, 138–153.

49. J. Matthews, T. Celius, R. Halgren *et al.*, 'Differential estrogen receptor binding of estrogenic substances: a species comparison.' *J. Steroid. Biochem. Mol. Biol.*, 2000, **74**, 223–234.

50. J. Haggblad, B. Carlsson, P. Kivela *et al.*, 'Scintillating microtitration plates as platform for determination of [3H]-estradiol binding constants for hER-HBD.' *Biotechniques*, 1995, **18**, 146–151.

51. S. A. Sundberg, 'High-throughput and ultra-high-throughput screening: solution- and cell-based approaches.' *Curr. Opin. Biotechnol.*, 2000, **11**, 47–53.

52. A. K. Shiau, D. Barstad, P. M. Loria *et al.*, 'The structural basis of estrogen receptor/coactivator recognition and the antagonism of this interaction by tamoxifen.' *Cell*, 1998, **95**, 927–937.

53. B. D. Darimont, R. L. Wagner, J. W. Apriletti *et al.*, 'Structure and specificity of nuclear receptor-coactivator interactions.' *Genes Dev.*, 1998, **12**, 3343–3356.

54. D. M. Heery, E. Kalkhoven, S. Hoare *et al.*, 'A signature motif in transcriptional co-activators mediates binding to nuclear receptors.' *Nature*, 1997, **387**, 733–736.

55. C. K. Glass, D. W. Rose and M. G. Rosenfeld, 'Nuclear receptor coactivators.' *Curr. Opin. Cell. Biol.*, 1997, **9**, 222–232.

56. C. K. Glass and M. G. Rosenfeld, 'The coregulator exchange in transcriptional functions of nuclear receptors.' *Genes Dev.*, 2000, **14**, 121–141.

57. G. G. Kuiper, J. G. Lemmen, B. Carlsson *et al.*, 'Interaction of estrogenic chemicals and phytoestrogens with estrogen receptor β.' *Endocrinology*, 1998, **139**, 4252–4263.

58. J. Matthews and T. Zacharewski, 'Differential binding affinities of PCBs, HO-PCBs, and aroclors with recombinant human, rainbow trout (*Oncorhynchus mykiss*), and green anole (*Anolis carolinensis*) estrogen receptors, using a semi-high throughput competitive binding assay.' *Toxicol. Sci.*, 2000, **53**, 326–339.

59. A. C. Pike, A. M. Brzozowski, R. E. Hubbard *et al.*, 'Structure of the ligand-binding domain of oestrogen receptor β in the presence of a partial agonist and a full antagonist.' *EMBO J.*, 1999, **18**, 4608–4618.

60. J. Nishikawa, K. Saito, J. Goto *et al.*, 'New screening methods for chemicals with hormonal activities using interaction of nuclear hormone receptor with coactivator.' *Toxicol. Appl. Pharmacol.*, 1999, **154**, 76–83.

61. P. Uetz, L. Giot, G. Cagney *et al.*, 'A comprehensive analysis of protein-protein interactions in *Saccharomyces cerevisiae*.' *Nature*, 2000, **403**, 623–627.

62. G. Cagney, P. Uetz and S. Fields, 'High-throughput screening for protein-protein interactions using two-hybrid assay.' *Methods Enzymol.*, 2000, **328**, 3–14.

63. L. Johansson, A. Bavner, J. S. Thomsen *et al.*, 'The orphan nuclear receptor SHP utilizes conserved LXXLL-related motifs for interactions with ligand-activated estrogen receptors.' *Mol. Cell. Biol.*, 2000, **20**, 1124–1133.

64. M. Malmqvist, 'BIACORE: an affinity biosensor system for characterization of biomolecular interactions.' *Biochem. Soc. Trans.*, 1999, **27**, 335–340.

65. E. Treuter, L. Johansson, J. S. Thomsen *et al.*, 'Competition between thyroid hormone receptor-associated protein (TRAP) 220 and transcriptional intermediary factor (TIF) 2 for binding to nuclear receptors. Implications for the recruitment of TRAP and p160 coactivator complexes.' *J. Biol. Chem.*, 1999, **274**, 6667–6677.

66. J. D. McKinney, A. Richard, C. Waller *et al.*, 'The practice of structure activity relationships (SAR) in toxicology.' *Toxicol. Sci.*, 2000, **56**, 8–17.

67. C. L. Waller, D. L. Minor and J. D. McKinney, 'Using three-dimensional quantitative structure-activity relationships to examine estrogen receptor binding affinities of polychlorinated hydroxybiphenyls.' *Environ. Health Perspect.*, 1995, **103**, 702–707.

68. W. Tong, R. Perkins, L. Xing *et al.*, 'QSAR models for binding of estrogenic compounds to estrogen receptor α and β subtypes.' *Endocrinology*, 1997, **138**, 4022–4025.

69. C. L. Waller, B. W. Juma, L. E. J. Gray *et al.*, 'Three-dimensional quantitative structure—activity relationships for androgen receptor ligands.' *Toxicol. Appl. Pharmacol.*, 1996, **137**, 219–227.

70. C. L. Waller and J. D. McKinney, 'Three-dimensional quantitative structure-activity relationships of dioxins and dioxin-like compounds: model validation and Ah receptor characterization.' *Chem. Res. Toxicol.*, 1995, **8**, 847–858.

71. C. L. Waller, M. V. Evans and J. D. McKinney, 'Modeling the cytochrome P450-mediated metabolism of chlorinated volatile organic compounds.' *Drug Metab. Dispos.*, 1996, **24**, 203–210.

72. A. Wlodawer and J. W. Erickson, 'Structure-based inhibitors of HIV-1 protease.' *Annu. Rev. Biochem.*, 1993, **62**, 543–585.

73. B. K. Shoichet, R. M. Stroud, D. V. Santi *et al.*, 'Structure-based discovery of inhibitors of thymidylate synthase.' *Science*, 1993, **259**, 1445–1450.

74. D. A. Gschwend, W. Sirawaraporn, D. V. Santi, I. D. Kuntz, 'Specificity in structure-based drug design: identification of a novel, selective inhibitor of *Pneumocystis carinii* dihydrofolate reductase.' *Proteins*, 1997, **29**, 59–67.

75. V. Schnecke and L. A. Kuhn, 'Database screening for HIV protease ligands: The influence of binding-site conformation and representation on ligand selectivity.' In Seventh International Conference on Intelligent Systems for Molecular Biology. AAAI Press, 1999.

76. V. Schnecke and L. A. Kuhn, 'Virtual screening with solvation and ligand-induced complementarity.' *Drug Discov. Dev.*, 2000, **20**, 171–190.

77. A. M. Brzozowski, A. C. Pike, Z. Dauter *et al.*, 'Molecular basis of agonism and antagonism in the oestrogen receptor.' *Nature*, 1997, **389**, 753–758.

78. P. M. Matias, P. Donner, R. Coelho *et al.*, 'Structural evidence for ligand specificity in the binding domain of the human androgen receptor. Implications for pathogenic gene mutations.' *J. Biol. Chem.*, 2000, **275**, 26164–26171.

79. R. L. Wagner, J. W. Apriletti, M. E. McGrath *et al.*, 'A structural role for hormone in the thyroid hormone receptor.' *Nature*, 1995, **378**, 690–697.

80. R. F. Service, 'Issues in nanotechnology.' *Science*, 2000, **290**, 1524–1531.

81. J. Fritz, M. K. Baller, H. P. Lang *et al.*, 'Translating biomolecular recognition into nanomechanics.' *Science*, 2000, **288**, 316–318.

82. G. Wu, H. Ji, K. Hansen *et al.*, 'Origin of nanomechanical cantilever motion generated from biomolecular interactions.' *Proc. Nat. Acad. Sci. USA*, 2001, **98**, 1560–1564.

83. S. F. Arnold, M. K. Robinson, A. C. Notides *et al.*, 'A yeast estrogen screen for examining the relative exposure of cells to natural and xenoestrogens.' *Environ. Health Perspect.*, 1996, **104**, 544–548.

84. K. W. Gaido, S. L. Leonard, S. Lovell *et al.*, 'Evaluation of chemicals with endocrine modulating activity in a yeast-based steroid hormone receptor gene transcription assay.' *Toxicol. Appl. Pharmacol.*, 1997, **143**, 205–212.

85. D. M. Klotz, B. S. Beckman, S. M. Hill *et al.*, 'Identification of environmental chemicals with estrogenic activity using a combination of *in vitro* assays.' *Environ. Health Perspect.*, 1996, **104**, 1084–1089.

86. B. M. Collins, J. A. McLachlan and S. F. Arnold, 'The estrogenic and antiestrogenic activities of phytochemicals with the human estrogen receptor expressed in yeast.' *Steroids*, 1997, **62**, 365–372.

87. K. Graumann, A. Breithofer and A. Jungbauer, 'Monitoring of estrogen mimics by a recombinant yeast assay: synergy between natural and synthetic compounds?' *Sci. Total Environ.*, 1999, **225**, 69–79.

88. V. A. Baker, P. A. Hepburn, S. J. Kennedy *et al.*, 'Safety evaluation of phytosterol esters. Part 1. Assessment of oestrogenicity using a combination of *in vivo* and *in vitro* assays.' *Food Chem. Toxicol.*, 1999, **37**, 13–22.

89. J. Payne, N. Rajapakse, M. Wilkins *et al.*, 'Prediction and assessment of the effects of mixtures of four xenoestrogens.' *Environ. Health Perspect.*, 2000, **108**, 983–987.

90. D. Q. Tran, C. F. Ide, J. A. McLachlan *et al.*, 'The anti-estrogenic activity of selected polynuclear aromatic hydrocarbons in yeast expressing human estrogen receptor.' *Biochem. Biophys. Res. Commun.*, 1996, **229**, 102–108.

91. F. Petit, P. Le Goff, J. P. Cravedi *et al.*, 'Two complementary bioassays for screening the estrogenic potency of xenobiotics: recombinant yeast for trout estrogen receptor and trout hepatocyte cultures.' *J. Mol. Endocrinol.*, 1997, **19**, 321–335.

92. D. Q. Tran, D. M. Klotz, B. L. Ladlie *et al.*, 'Inhibition of progesterone receptor activity in yeast by synthetic chemicals.' *Biochem. Biophys. Res. Commun.*, 1996, **229**, 518–523.

93. P. Mak, F. D. Cruz and S. Chen, 'A yeast screen system for aromatase inhibitors and ligands for androgen receptor: yeast cells transformed with aromatase and androgen receptor.' *Environ. Health Perspect.*, 1999, **107**, 855–860.

94. H. R. Andersen, A.-M. Andersson, S. F. Arnold *et al.*, 'Comparison of short-term estrogenicity tests for identification of hormone-disrupting chemicals.' *Environ. Health Perspect.*, 1999, **107**(Suppl. 1), 89–108.

95. L. E. Gray Jr., W. R. Kelce, T. Wiese *et al.*, 'Endocrine Screening Methods Workshop report: detection of estrogenic and androgenic hormonal and antihormonal activity for chemicals that act via receptor or steroidogenic enzyme mechanisms.' *Reprod. Toxicol.*, 1997, **11**, 719–750.

96. C. R. Lyttle, P. Damian-Matsumura, H. Juul *et al.*, 'Human estrogen receptor regulation in a yeast model system and studies on receptor agonists and antagonists.' *J. Steroid Biochem. Mol. Biol.*, 1992, **42**, 677–685.

97. N. G. Coldham, M. Dave, S. Sivapathasundaram *et al.*, 'Evaluation of a recombinant yeast cell estrogen screening assay.' *Environ. Health. Perspect.*, 1997, **105**, 734–742.

98. J. W. Liu, E. Jeannin and D. Picard, 'The anti-estrogen hydroxytamoxifen is a potent antagonist in a novel yeast system.' *J. Biol. Chem.*, 1999, **380**, 1341–1345.

99. J. S. Edmunds, E. R. Fairey and J. S. Ramsdell, 'A rapid and sensitive high throughput reporter gene assay for estrogenic effects of environmental contaminants.' *Neurotoxicology*, 1997, **18**, 525–532.

100. T. Zacharewski, K. Berhane, B. E. Gillesby, B. K. Burnison, 'Detection of estrogen- and dioxin-like activity in pulp and paper mill black liquor and effluent using *in vitro* recombinant receptor/reporter gene assays.' *Environ. Sci. Technol.*, 1995, **29**, 2140–2146.

101. M. R. Fielden, I. Chen, B. Chittim *et al.*, 'Examination of the estrogenicity of 2,4,6,2′,6′-pentachlorobiphenyl (PCB 104), its hydroxylated metabolite 2,4,6,2′,6′-pentachloro-4-biphenylol (HO-PCB 104), and a further chlorinated derivative, 2,4,6,2′,4′,6′-hexachlorobiphenyl (PCB 155).' *Environ. Health Perspect.*, 1997, **105**, 1238–1248.

102. J. H. Clemons, L. M. Allan, C. H. Marvin *et al.*, 'Evidence of estrogen- and TCDD-like activities in crude and fractionated extracts of PM10 air particulate material using *in vitro* gene expression assays.' *Environ. Sci. Technol.*, 1998, **32**, 1853–1860.

103. T. R. Zacharewski, J. Clemons, M. D. Meek *et al.*, 'Examination of the alleged *in vitro* and *in vivo* estrogenic activities of eight commercial phthalate esters.' *Toxicol. Sci.*, 1998, **46**, 282–293.

104. M. Fielden, Z. Wu, C. Sinal *et al.*, 'Estrogen receptor and aryl hydrocarbon receptor-mediated activities of a coal-tar creosote.' *Environ. Toxicol. Chem.*, 2000, **19**, 1262–1271.

105. S. Ali, D. Metzger, J. M. Bornert *et al.*, 'Modulation of transcriptional activation by ligand-dependent phosphorylation of the human oestrogen receptor A/B region.' *EMBO J.*, 1993, **12**, 1153–1160.

106. M. K. El-Tanani and C. D. Green, 'Two separate mechanisms for ligand-independent activation of the estrogen receptor.' *Mol. Endocrinol.*, 1997, **11**, 928–937.

107. G. Sathya, W. Li, C. M. Klinge *et al.*, 'Effects of multiple estrogen responsive elements, their spacing, and location on estrogen response of reporter genes.' *Mol. Endocrinol.*, 1997, **11**, 1994–2003.

108. P. M. Druege, L. Klein-Hitpass, S. Green *et al.*, 'Introduction of estrogen-responsiveness into mammalian cell lines.' *Nucleic Acids Res.*, 1986, **14**, 9329–9337.

109. M. D. Shelby, R. R. Newbold, D. B. Tully *et al.*, 'Assessing environmental chemicals for estrogenicity using a combination of *in vitro* and *in vivo* assays.' *Environ. Health Perspect.*, 1996, **104**, 1296–1300.

110. Z. X. Zhou, M. Sar, J. A. Simental *et al.*, 'A ligand-dependent bipartite nuclear targeting signal in the human androgen receptor. Requirement for the DNA-binding domain and modulation by NH2-terminal and carboxyl-terminal sequences.' *J. Biol. Chem.*, 1994, **269**, 13115–13123.

111. V. E. Quarmby, J. A. Kemppainen, M. Sar *et al.*, 'Expression of recombinant androgen receptor in cultured mammalian cells.' *Mol. Endocrinol.*, 1990, **4**, 1399–1407.

112. N. Warriar, N. Page, M. Koutsilieris *et al.*, 'Antiandrogens inhibit human androgen receptor-dependent gene transcription activation in the human prostate cancer cells LNCaP.' *Prostate*, 1994, **24**, 176–186.

113. A. Dassouli, C. Darne, S. Fabre *et al.*, '*Vas deferens* epithelial cells in subculture: a model to study androgen regulation of gene expression.' *J. Mol. Endocrinol.*, 1995, **15**, 129–141.

114. A. M. Vinggaard, E. C. Joergensen and J. C. Larsen, 'Rapid and sensitive reporter gene assays for detection of antiandrogenic and estrogenic effects of environmental chemicals.' *Toxicol. Appl. Pharmacol.*, 1999, **155**, 150–160.

115. Z. Nawaz, G. M. Stancel and S. M. Hyder, 'The pure antiestrogen ICI 182,780 inhibits progestin-induced transcription.' *Cancer Res.*, 1999, **59**, 372–376.

116. M. Pons, D. Gagne, J. C. Nicolas *et al.*, 'A new cellular model of response to estrogens: a bioluminescent test to characterize (anti) estrogen molecules.' *Biotechniques*, 1990, **9**, 450–459.

117. V. J. Kramer, W. G. Helferich, A. Bergman *et al.*, 'Hydroxylated polychlorinated biphenyl metabolites are anti-estrogenic in a stably transfected human breast adenocarcinoma (MCF7) cell line.' *Toxicol. Appl. Pharmacol.*, 1997, **144**, 363–376.

118. P. Balaguer, A. Joyeux, M. S. Denison *et al.*, 'Assessing the estrogenic and dioxin-like activities of chemicals and complex mixtures using *in vitro* recombinant receptor-reporter gene assays.' *Can. J. Physiol. Pharmacol.*, 1996, **74**, 216–222.

119. J. Legler, C. E. van den Brink, A. Brouwer *et al.*, 'Development of a stably transfected estrogen receptor-mediated luciferase reporter gene assay in the human T47D breast cancer cell line.' *Toxicol. Sci.*, 1999, **48**, 55–66.

120. P. Balaguer, F. Francois, F. Comunale *et al.*, 'Reporter cell lines to study the estrogenic effects of xenoestrogens.' *Sci. Total Environ.*, 1999, **233**, 47–56.

121. J. M. Rogers and M. S. Denison, 'Recombinant cell bioassays for endocrine disruptors: development of a stably transfected human ovarian cell line for the detection of estrogenic and anti-estrogenic chemicals.' *In vitro Mol. Toxicol.*, 2000, **13**, 67–82.

122. S. Miller, D. Kennedy, J. Thomson *et al.*, 'A rapid and sensitive reporter gene that uses green fluorescent protein expression to detect chemicals with estrogenic activity.' *Toxicol. Sci.*, 2000, **55**, 69–77.

123. J. Clemons and T. Zacharewski, in 'Molecular Biology of the Toxic Response,' eds., A. Puga and K. Wallace, Taylor and Francis, Philadelphia, PA, 1999, pp. 187–204.

124. W. Chiu, Y. Niwa, W. Zeng *et al.*, 'Engineered GFP as a vital reporter in plants.' *Curr. Biol.*, 1996, **6**, 325–330.

125. A. Crameri, E. A. Whitehorn, E. Tate *et al.*, 'Improved green fluorescent protein by molecular evolution using DNA shuffling.' *Nat. Biotechnol.*, 1996, **14**, 315–319.

126. J. Haseloff, 'GFP variants for multispectral imaging of living cells.' *Methods Cell. Biol.*, 1999, **58**, 139–151.

127. B. W. Dyer, F. A. Ferrer, D. K. Klinedinst *et al.*, 'A noncommercial dual luciferase enzyme assay system for reporter gene analysis.' *Anal. Biochem.*, 2000, **282**, 158–161.

128. C. E. Olesen, Y. X. Yan, B. Liu *et al.*, 'Novel methods for chemiluminescent detection of reporter enzymes.' *Methods Enzymol.*, 2000, **326**, 175–202.

129. G. D. Charles, M. J. Bartels, T. R. Zacharewski *et al.*, 'Activity of benzo[a]pyrene and its hydroxylated metabolites in an ERα reporter gene assay.' *Toxicol. Sci.*, 1999, **14**, 207–216.

130. J. M. Smeets, I. van Holsteijn, J. P. Giesy *et al.*, 'Estrogenic potencies of several environmental pollutants, as determined by vitellogenin induction in a carp hepatocyte assay.' *Toxicol. Sci.*, 1999, **50**, 206–213.

131. A. M. Maffia III, I. Kariv and K. R. Oldenburg, 'Miniaturization of a mammalian cell-based assay: luciferase reporter gene readout in a 3 microliter 1536-Well Plate.' *J. Biomol. Screen.*, 1999, **4**, 137–142.

132. M. Berg, K. Undisz, R. Thiericke *et al.*, 'Miniaturization of a functional transcription assay in yeast (human progesterone receptor) in the 384- and 1536-well plate format.' *J. Biomol. Screen.*, 2000, **5**, 71–76.

133. A. M. Soto, C. Sonnenschein, K. L. Chung *et al.*, 'The E-SCREEN assay as a tool to identify estrogens: an update on estrogenic environmental pollutants.' *Environ. Health Perspect.*, 1995, **103**(Suppl. 7), 113–122.

134. B. D. Gehm, J. M. McAndrews, V. C. Jordan *et al.*, 'EGF activates highly selective estrogen-responsive reporter plasmids by an ER-independent pathway.' *Mol. Cell. Endocrinol.*, 2000, **159**, 53–62.

135. F. E. May, M. D. Johnson, L. R. Wiseman *et al.*, 'Regulation of progesterone receptor mRNA by oestradiol and antioestrogens in breast cancer cell lines.' *J. Steroid. Biochem.*, 1989, **33**, 1035–1041.

136. M. J. Pilat, M. S. Hafner, L. G. Kral *et al.*, 'Differential induction of pS2 and cathepsin D mRNAs by structurally altered estrogens.' *Biochemistry*, 1993, **32**, 7009–7015.

137. M. Hirota, Y. Furukawa and K. Hayashi, 'Expression of pS2 gene in human breast cancer cell line MCF-7 is controlled by retinoic acid.' *Biochem. Int.*, 1992, **26**, 1073–1078.

138. L. Ren, M. A. Marquardt and J. J. Lech, 'Estrogenic effects of nonylphenol on pS2, ER and MUC1 gene expression in human breast cancer cells-MCF-7.' *Chem.–Biol. Interact.*, 1997, **104**, 55–64.

139. T. J. Kledal, M. Jorgensen, F. Mengarda *et al.*, 'New methods for detection of potential endocrine disruptors.' *Andrologia*, 2000, **32**, 271–278.

140. M. Jorgensen, B. Vendelbo, N. E. Skakkebaek *et al.*, 'Assaying estrogenicity by quantitating the expression levels of endogenous estrogen-regulated genes.' *Environ. Health Perspect.*, 2000, **108**, 403–412.

141. D. A. Zajchowski, K. Kauser, D. Zhu *et al.*, 'Identification of selective estrogen receptor modulators by their gene expression fingerprints.' *J. Biol. Chem.*, 2000, **275**, 15885–15894.

142. M. Islinger, S. Pawlowski, H. Hollert *et al.*, 'Measurement of vitellogenin-mRNA expression in primary cultures of rainbow trout hepatocytes in a non-radioactive dot blot/RNAse protection-assay.' *Sci. Total Environ.*, 1999, **233**, 109–122.

143. W. Kloas, I. Lutz and R. Einspanier, 'Amphibians as a model to study endocrine disruptors: II. Estrogenic activity of environmental chemicals *in vitro* and *in vivo*.' *Sci. Total Environ.*, 1999, **225**, 59–68.

144. C. A. Afshari, E. F. Nuwaysir and J. C. Barrett, 'Application of complementary DNA microarray technology to carcinogen identification, toxicology, and drug safety evaluation.' *Cancer Res.*, 1999, **59**, 4759–4760.

145. E. F. Nuwaysir, M. Bittner, J. Trent *et al.*, 'Microarrays and toxicology: the advent of toxicogenomics.' *Mol. Carcinog.*, 1999, **24**, 153–159.

146. C. M. Perou, S. S. Jeffrey, M. van de Rijn *et al.*, 'Distinctive gene expression patterns in human mammary epithelial cells and breast cancers.' *Proc. Nat. Acad. Sci. USA*, 1999, **96**, 9212–9217.

147. A. A. Alizadeh, M. B. Eisen, R. E. Davis *et al.*, 'Distinct types of diffuse large B-cell lymphoma identified by gene expression profiling.' *Nature*, 2000, **403**, 503–511.

148. U. Alon, N. Barkai, D. A. Notterman *et al.*, 'Broad patterns of gene expression revealed by clustering analysis of tumor and normal colon tissues probed by oligonucleotide arrays.' *Proc. Nat. Acad. Sci. USA*, 1999, **96**, 6745–6750.

149. C. Purdom, P. Hardiman, V. Bye *et al.*, 'Estrogenic effects of effluents from sewage treatment works.' *Chem. Ecol.*, 1994, **8**, 275–285.

150. J. P. Sumpter and S. Jobling, 'Vitellogenesis as a biomarker for estrogenic contamination of the aquatic environment.' *Environ. Health Perspect.*, 1995, **103**(Suppl. 7), 173–178.

151. J. C. O'Connor, S. R. Frame, L. B. Biegel *et al.*, 'Sensitivity of a Tier I screening battery compared to an *in utero* exposure for detecting the estrogen receptor agonist 17 β-estradiol.' *Toxicol. Sci.*, 1998, **44**, 169–184.

152. X. Feng, Y. Jiang, P. Meltzer *et al.*, 'Thyroid hormone regulation of hepatic genes *in vivo* detected by complementary DNA microarray.' *Mol. Endocrinol.*, 2000, **14**, 947–955.

153. P. R. Holden, N. H. James, A. N. Brooks *et al.*, 'Identification of a possible association between carbon tetrachloride-induced hepatotoxicity and interleukin-8 expression.' *J. Biochem. Mol. Toxicol.*, 2000, **14**, 283–290.

154. S. A. Amundson, M. Bittner, Y. Chen *et al.*, 'Fluorescent cDNA microarray hybridization reveals complexity and heterogeneity of cellular genotoxic stress responses.' *Oncogene*, 1999, **18**, 3666–3672.

155. A. J. Fornace Jr., S. A. Amundson, M. Bittner *et al.*, 'The complexity of radiation stress responses: analysis by informatics and functional genomics approaches.' *Gene. Expr.*, 1999, **7**, 387–400.

156. M. E. Burczynski, M. McMillian, J. Ciervo *et al.*, 'Toxicogenomics-based discrimination of toxic mechanism in HepG2 human hepatoma cells.' *Toxicol. Sci.*, 2000, **58**, 399–415.

157. S. S. Nadadur, M. C. Schladweiler and U. P. Kodavanti, 'A pulmonary rat gene array for screening

altered expression profiles in air pollutant-induced lung injury.' *Inhal. Toxicol.*, 2000, **12**, 1239–1254.

158. M. A. Hossain, C. M. Bouton, J. Pevsner *et al.*, 'Induction of vascular endothelial growth factor in human astrocytes by lead. Involvement of a protein kinase C/activator protein-1 complex-dependent and hypoxia-inducible factor 1-independent signaling pathway.' *J. Biol. Chem.*, 2000, **275**, 27874–27882.

159. H. Sato, M. Sagai, K. T. Suzuki *et al.*, 'Identification, by cDNA microarray, of A-raf and proliferating cell nuclear antigen as genes induced in rat lung by exposure to diesel exhaust.' *Res. Commun. Mol. Pathol. Pharmacol.*, 1999, **105**, 77–86.

160. M. B. Eisen, P. T. Spellman, P. O. Brown *et al.*, 'Cluster analysis and display of genome-wide expression patterns.' *Proc. Nat. Acad. Sci. USA*, 1998, **95**, 14863–14868.

161. S. Raychaudhuri, J. Stuart and R. Altman, 'Principal component analysis to summarize microarray experiments: application to sporulation time series.' *Pacific Symposium on Biocomputing*, 2000, pp. 452–463.

162. S. Tavazoie, J. D. Hughes, M. J. Campbell *et al.*, 'Systematic determination of genetic network architecture.' *Nat. Genet.*, 1999, **22**, 281–285.

163. P. Tamayo, D. Slonim, J. Mesirov *et al.*, 'Interpreting patterns of gene expression with self-organizing maps: methods and application to hematopoietic differentiation.' *Proc. Nat. Acad. Sci. USA*, 1999, **96**, 2907–2912.

164. M. J. Marton, J. L. DeRisi, H. A. Bennett *et al.*, 'Drug target validation and identification of secondary drug target effects using DNA microarrays.' *Nat. Med.*, 1998, **4**, 1293–1301.

165. T. R. Hughes, M. J. Marton, A. R. Jones *et al.*, 'Functional discovery via a compendium of expression profiles.' *Cell*, 2000, **102**, 109–126.

166. A. P. Gasch, P. T. Spellman, C. M. Kao *et al.*, 'Genomic expression programs in the response of yeast cells to environmental changes.' *Mol. Biol. Cell.*, 2000, **11**, 4241–4257.

167. M. Schena, D. Shalon, R. W. Davis *et al.*, 'Quantitative monitoring of gene expression patterns with a complementary DNA microarray.' *Science*, 1995, **270**, 467–470.

168. J. DeRisi, L. Penland, P. O. Brown *et al.*, 'Use of a cDNA microarray to analyse gene expression patterns in human cancer.' *Nat. Genet.*, 1996, **14**, 457–460.

J.P. Vanden Heuvel, G.H. Perdew, W.B. Mattes and W.F. Greenlee (Eds.)
Comprehensive Toxicology, Vol. xiv

# 14.5.7
# Metabonomic Technology as a Tool for Rapid Throughput In Vivo Toxicity Screening

DONALD G. ROBERTSON, MICHAEL D. REILY
*Pfizer Global Research and Development, Ann Arbor MI, USA*

JOHN C. LINDON, ELAINE HOLMES and
JEREMY K. NICHOLSON
*University of London, London, UK*

## 14.5.7.1    HOW IT ALL BEGAN: INTRODUCTION

An emerging approach enabling the realization of rapid-throughput *in vivo* toxicology is the technology of metabonomics.[1–4] Metabonomics combines the techniques of high resolution nuclear magnetic resonance (NMR) coupled with pattern recognition technology to rapidly evaluate the metabolic "status" of an animal from a peripheral sample such as urine. This ability to evaluate animal physiology non-invasively allows for serial evaluation of the metabolic consequences of toxicity such that onset and regression of toxicity can be obtained from a single animal and the findings correlated with traditional measures of toxicity (clinical and histopathology) if desired. Toxicants, by definition, disrupt the normal composition and flux of endogenous biochemicals in, or through, key intermediary cellular metabolic pathways. These disruptions alter, either directly or indirectly, the blood that percolates through the target tissues. This altered blood can then produce altered urine either directly or indirectly producing characteristic biomolecular traces in urine after filtration by the kidney. The diagnostic utility of any one trace biomolecule is limited due to the number of variables affecting it along its route to the urine and by the commonality of biochemical processes disrupted by toxicants. However, if a significant number of trace molecules are monitored, the overall pattern, or "fingerprint" produced may be more consistent and predictive than any one marker. Obtaining such comprehensive biochemical information is made possible using high-field NMR equipment, which is an essential component of the technology. Proton ($^1$H) NMR spectroscopy is capable of detecting individual atoms in soluble proton-containing molecules with a molecular weight of approximately 20 kD or less. Standard flow probes used with 600 MHz spectrometer are capable of detecting thousands of resonances in the urine from molecules at concentrations $> 50\,\mu$M or higher with a data acquisition time of a few minutes. Incorporation of cryoprobe technology will lower this limit significantly.

NMR spectra serve as the raw input for pattern recognition analyses, which simplifies complex multivariate data into 2- or 3-dimensions that can be readily understood and evaluated. Acquiring this wealth of information non-invasively enables the potential to develop relatively high-throughput *in vivo* toxicity screening technologies. The principles of the technique have been described in detail[5–8] and several excellent reviews are available.[1–3] Additionally, a comprehensive review of NMR analysis of biofluids has recently been published.[9] An extensive literature exists on the use of metabonomics to evaluate nephrotoxicants[10–14] and hepatic toxicants[7] or both.[4]

The potential for the technology as a screening tool is now obvious. The advent of combinatorial chemistry and high-throughput efficacy screening have pushed the envelope of how rapidly a new drug can be brought to clinical testing. The result of this increase in activity at the front end of the drug discovery and development process has placed an increased burden further down the pipeline. A primary bottleneck is at the *in vitro/in vivo* interface, where initial high-throughput results are tested in *in vivo* models, assessed for pharmacokinetic attributes, and safety determined. This is one place where metabonomic technology can play the vital role of evaluating *in vivo* toxicity of a compound in a model with minimal compound requirements and in a relatively rapid throughput fashion. Metabonomic technology has application for use in generating biomarkers of both efficacy and toxicity as well, which may be of even greater significance than its use as a screening tool. These "downstream" applications of metabonomic technology are significantly more resource-intensive. *In vivo* screening applications can be thought of as a first-pass analysis without the requirement of analytical identification of every peak in the spectrum. The results of these early analyses will undoubtedly highlight data that can be further mined for gems such as biomarkers and mechanistic insights.

This review is intended to serve as a primer for toxicologists on the practical application of this technology as a screening tool and is divided into 2 complementary sections. The first section provides basic information on the application and use of nuclear magnetic resonance spectroscopy with the assumption that most toxicologists have only limited knowledge about this analytical technique. A full review of the science of magnetic resonance is beyond the scope of this work so the focus of the discussion will be on the practical application of NMR to toxicological samples. The second section deals with the application of metabonomic technology from a toxicologist's perspective. It is hoped that this review will provide those interested in metabonomics with the basic information they need to assess the applicability of this exciting new technology.

### 14.5.7.2 MAGNETIC MUSINGS: ANALYTICAL NMR SPECTROSCOPY AND METABONOMICS

#### 14.5.7.2.1 Principles of NMR Spectroscopy

##### 14.5.7.2.1.1 Introduction

NMR spectroscopy has been used to investigate small molecules of chemical and biological interest since its use was first commercialized in the 1950s. However, rigorous application of NMR to directly investigate large biomolecules or complicated biological mixtures was hindered until the development of routinely available NMR spectrometers with proton observe frequencies of 500 MHz or greater. Indeed the first report of an NMR spectrum of whole urine was not published until the early 1980s.[15] NMR spectrometers are designated according to the frequency at which protons resonate which determines the magnetic field strength, since the field and frequency are linearly related. Thus, a 500 MHz spectrometer utilizes an 11.7 Tesla (T) magnet, whereas a 900 MHz spectrometer would use a 21.0 T magnet (1 T = $10^4$ Gauss). Development of superconducting magnets in the 1970s spawned the affordable production of these "high-field" spectrometers in the mid 1980s. As a direct consequence of these developments, entirely new areas of study have emerged, including biomolecular NMR, flow NMR, and biofluid NMR which can be considered fields in their own right. The latter two developments are pertinent to this discussion. The term "biofluid NMR" can be defined as the analytical application of high resolution NMR to unprocessed, or largely unprocessed, biological fluids to determine composition.

Different biofluids have significantly different properties that directly impact their behavior in a high resolution NMR experiment and the results that can be obtained. NMR studies of plasma, serum, cerebrospinal fluid (CSF), semen, bile and urine have all been reported.[9] Plasma, serum, seminal fluid and CSF are all rich in protein components. This presents some major challenges for high resolution NMR, in that these very large molecules have undesirable NMR properties. As a rough approximation, the width of an NMR peak (line width) increases 1 Hz for each kilodalton of molecular weight. Thus, in a spectrometer with 0.1 Hz resolving power, anything above 40 or so kilodaltons ceases to be a high resolution NMR target and very large molecules essentially cause broad band envelopes. Often, these broad lines can be edited out by choosing an appropriate NMR experiment. However, an additional complication brought on by the presence of proteins in these fluids is that they often bind to some of the smaller molecules, rendering them partially or completely unobservable. As a biofluid for metabonomics-based screening, urine has some distinct advantages from a both a physiological and an NMR perspective. First, urine is the only extensively studied biofluid that is primarily composed of water and small molecules, minimizing the need to account for the dynamical interactions between analytes and macromolecules. Second, the relative lack of proteins in urine is also beneficial since precautions against protein binding or enzymatic decomposition of potential analytes during the data acquisition is minimized. Finally, urine is the least viscous all the biofluids widely studied by NMR,[9] reducing viscosity-related spectral broadening and making fluid transfer and washout in flow systems less problematic.

### 14.5.7.2.1.2 High Resolution NMR versus MRI and MRS

Toxicologists are generally more familiar with the more "clinical" incarnations of NMR, namely magnetic resonance imaging (MRI) and *in vivo* magnetic resonance spectroscopy (MRS), than with the high resolution methods used in biofluids NMR. Therefore, a brief comparison of the techniques is in order. Basically, both operate on identical physical principles with MRI providing an image based on differentiating the properties of nuclei (usually protons) in different tissues and MRS providing a spectrum of the molecular components of a tissue. Like biofluid NMR, these technologies have become commonplace because of revolutionary changes in hardware (e.g., magnet and sample container design). As is often the case with applications that migrate from the chemical to the clinical setting, MRI/MRS and high resolution NMR have distinct practitioners, literature and jargon. Because of the obvious negative implications, the term "nuclear" has disappeared from the MRI and MRS as these techniques found their way into the clinic, although the magnetically active nuclei of interest in most biological applications are all stable (non-radioactive). Another major terminology difference is that MRI/MRS equipment are normally referred to as scanners, whereas high resolution NMR instruments are referred to as spectrometers. Outside of the divergent vernacular of the two techniques, the major differences have to do with the sample itself. In MRI and MRS, the sample is an intact animal or organ and in high resolution NMR, the sample is typically dissolved in a homogeneous solution or at least a mobile fluid. Since the sample must be surrounded by a homogeneous magnetic field, design considerations are significantly different for the two techniques. Magnet design for MRS/MRI equipment has focused on magnets with large horizontal bores (20–1000 cm) for easy access, whereas high resolution NMR magnet design has maintained the traditional narrow (5–8 cm), vertical bores. Because of the larger volumes and ethical/safety considerations, the highest attainable magnetic field strength is significantly smaller for MRI/MRS scanners (up to 7T) than for high resolution NMR spectrometers (up to 21T for currently available commercial spectrometers). This, combined with the often unfavorable solution characteristics of intact systems, provides a powerful advantage for high resolution NMR over MRS in metabonomics analysis. High resolution studies of small tissue samples in very high magnetic fields have been reported using the technique of magic angle spinning (MAS).[16] This technique, also developed originally for chemical analysis overcomes some of the unfavorable aspects of intact tissues for NMR analysis and holds great promise, but currently is rather low throughput.

### 14.5.7.2.1.3 The Experiment

Many atomic nuclei have a property known as "spin" which when combined with the positive charge of the nucleus gives rise to a magnetic moment. In a magnetic field this magnet moment can take certain discrete values or "spin states" and transitions between spin states can be induced by radio frequency irradiation. This is the property known as nuclear magnetic resonance. The $^1H$ isotope of hydrogen (proton) has a very large magnetic moment and this, coupled with its near 100% natural abundance and its diverse chemical distribution, makes it by far the most useful for biofluids NMR and as such this discussion will focus primarily on proton ($^1H$) NMR. As a point of reference, another useful nucleus, $^{13}C$, has a natural sensitivity only 0.25 that of a proton and since it only comprises about 1% of all the naturally occurring carbon isotopes ($^{12}C$ is the most abundant and is not NMR active), $^{13}C$ NMR is much less sensitive. Thus the overall sensitivity of carbon NMR is only about 0.25% that of proton NMR. Other nuclei found naturally in biological systems and which give rise to NMR spectra are $^{13}P$ and $^{15}N$ at 20% and 0.03% the sensitivity of a proton, respectively.

Application of an external static magnetic field to a sample containing NMR-active nuclei results in a series of quantum-mechanically allowed spin states corresponding to the allowed discrete orientation of the nuclear spin angular momentum with the magnetic field. Most NMR-active nuclei of interest in biological applications can adopt only two spin states ($+1/2$ or $-1/2$) that represent alignment with and against the applied field. Each of these states has a slightly different energy associated with it (imagine holding a bar magnet in a larger magnetic field and the noting the different amounts of energy required to hold it in one direction as opposed to the other) and thus the lower energy level will have a slight excess of the population of the nuclei in it at equilibrium. This disproportionation between spin state results in a net macroscopic magnetization, $M_0$, in the sample that is defined by the Boltzmann distribution:

$$M_0 \sim N^+/N^- = \exp[-(E^+ - E^-)/kT] \quad (1.0)$$

where $N^+$, $N^-$ and $E^+$ and $E^-$ are the population and energy values of the $+1/2$ and $-1/2$ spin state, respectively, k is the Boltzmann constant and the T is the absolute temperature.[17]

In NMR spectroscopy, $E^+$, $E^-$ or $\Delta E$ is small, falling within the radio frequency (RF) range of the electromagnetic spectrum ($10^7$–$10^9$ Hz) and its exact value is different for each nucleus and is proportional to the applied magnetic field. It is this latter fact that, more than any other single thing, has driven the development of higher strength magnets for NMR. Furthermore, and very significantly, $\Delta E$ depends on the exact chemical environment that a nucleus is in. Thus it is possible to differentiate all of the protons in a single molecule, for example as arising from an aromatic ring or an aliphatic group. In a simple NMR experiment, a sample containing analyte molecule(s) in solution is enclosed in a glass tube and placed in a strong, highly homogenous magnetic field. A single high powered ($\sim 10^2$ watts) pulse of energy at the appropriate RF wavelength, having a power distribution which covers the whole frequency range for proton transitions, is applied to the sample, inducing transitions between spin states. The subsequent return to equilibrium (relaxation) of the nuclei produces a damped oscillating voltage that encodes the location and intensity of all of like nuclei within the sample and constitutes the measured signal in an NMR experiment. These time domain signals are amplified, recorded and converted to the frequency domain (usually via fast Fourier transformation) to yield an NMR spectrum, the basic analytical commodity provided by NMR.

### 14.5.7.2.1.4    Information Content of NMR Spectra

The NMR spectrum contains peaks (or resonances) corresponding to every atom (of the nucleus selected for observation) contained in the sample at frequencies, or chemical shifts, that depend on the exact chemical environment of the atom. For example, the proton NMR spectrum of a mixture of 3-nitrotyrosine and citrate in $D_2O$ will contain signals corresponding to the 10 types of carbon-bound protons in that molecule, each with a chemical shift characteristic of its environment (Figure 1). Furthermore, the broadness of a line and its temperature or pH dependence can reveal information about conformational equilibrium, chemical exchange, electrostatic interactions (like hydrogen bonds) and ionization states. It should be noted that the two protons in the $CH_2$ group in citrate have different

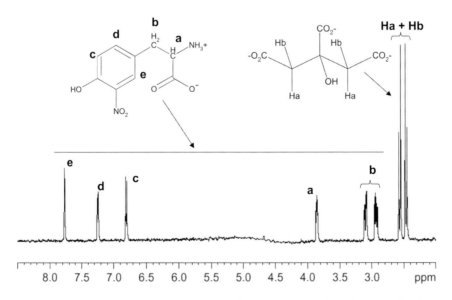

**Figure 1**  The 600 MHz NMR spectrum of 1 mM 3-nitrotyrosine and 1 mM citrate dissolved in 50 mM sodium phosphate buffer, pH 7.4. The chemical shift dependence on the molecular environment is illustrated by the peak assignments. Each molecule produces a characteristic pattern or fingerprint unique to that molecule. In aqueous solvents, the amino and carboxylate protons are generally exchange broadened and not observed.

chemical shifts because they are in different chemical environments. This is also true for the $CH_2$ group of the 3-nitro. Unlike other forms of spectroscopy, NMR is unique in that the intensity of each resonance is proportional to the molar concentration of the corresponding atom, independent of the molecule that atom is part of. In other words, extinction coefficients are identical for a given nucleus, making quantitation relatively trivial. The fact that NMR is an almost completely non-selective technique is a strong attribute as well as a potential pitfall. As an illustration of this latter point, a typical urine $^1H$ NMR spectrum is dominated by the resonances from hundreds of small biomolecules with molecular weights less than 500 and present in concentrations $> 50$ $\mu M$ (Figure 2). Obviously, in order to be a useful analytical tool, methods must be employed to extract necessary information from data that are this complicated.

### 14.5.7.2.1.5   Sensitivity

The detection limits of any spectroscopic method with a finite signal-noise ratio is dependent on two factors: those that influence the signal levels and those that influence noise levels. For a given instrument design, the former is dependent primarily on the Boltzmann distribution and the number of observable nuclear spins in the receiver coil (e.g., the sample size). For protons, room temperature Boltzmann factors are only 1.000080 at 500 MHz (0.008% more protons in the low energy state compared with the high energy state). This small difference limits the number of transitions that can be induced and consequently the signal voltage that is detectable. Thus, NMR spectroscopy is inherently insen-

sitive because the very small energy difference between ground and excited states (Equation 1.0). As it turns out, $M_0$ and thus the signal voltage is directly proportional to the applied magnetic field and the number of observed nuclei within the sample and inversely proportional to the sample temperature. While it is beyond the scope of this chapter to discuss in detail all of the factors influencing NMR sensitivity (for an excellent discussion on sensitivity optimization see Freeman),[18] there are several factors that are crucial in its practical optimization. Important amongst these, only the use of higher applied magnetic fields ($B_0$) provides the anticipated gain in signal-noise ratio. Increasing the sample size by increasing the receiver coil volume will also provide substantial gains but these can be mediated by several factors. First, it is crucial that the receiver coil be as close to the sample as possible to maximize the filling factor, approximated as the (sample volume)/(coil volume). For many biological applications, the actual amount of sample is limited, and so any gains from a larger coil volume are lost if that sample has to be diluted to match it. Also, the ability to achieve a magnetic field homogeneous enough for high resolution NMR ($\sim 1$ part per billion) across the entire sample becomes a challenge for large volumes. If cost and commercial availability were not an issue, ideally one would match the probe volume to the anticipated sample volume within the constraints outlined above. For example, a large coil ($> 500$ $\mu l$) probe might be chosen for dog or primate urine analysis, whereas the best probe for small rodent urine or bile or CSF samples would have a smaller coil volume (60–120 $\mu l$). However, both cost and availability *are* real world constraints; a conventional or flow probe costs about $30,000

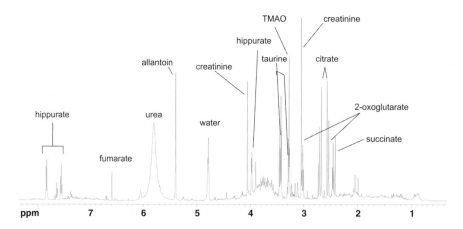

**Figure 2**   600 MHz NMR spectrum of a urine sample from a control rat highlighting common urinary analytes.

and there are a limited number of off-the-shelf configurations. With this in mind, the smaller coil volume probe would be the best choice if a single probe had to be used to analyze all of the samples in the example above. The narrow range of temperatures useful for biofluids analysis essentially eliminates this as a useful signal enhancement variable, at least for liquids.

The factors that contribute to noise in the spectrometer arise predominantly from imperfections and thermal agitation of electrons in the receiver circuitry. These factors are significant and are constantly being optimized by the instrument manufacturers. A typical 500 MHz NMR spectrometer as delivered in 1990 produced a proton S/N on a standard sample of about 450:1 with a conventional 5 mm diameter $^1$H-observe probe. In 1998, a brand new 500 MHz spectrometer produced a S/N of 800:1 on the same system, almost double that of the older system with no change in field strength or sample size. With equipment widely used today, given the time constraints of a reasonable turnaround, good quantitative results can be obtained for components in urine with concentrations as low as 50 micromolar. The development of hypercooled detection hardware (cryoprobe technology), where the detector coil and preamplifier are cooled to about 20°K to reduce thermal noise, promises to extend this limit of detection into the low micromolar range.[19]

### 14.5.7.2.1.6 *Resolution Enhancement and Spectral Assignment*

An underlying tenet in NMR spectroscopy is that one must know what gives rise to a particular peak in order to make use of that peak. Although this may seem obvious, we have seen that peak assignment may not always be a trivial problem with biofluids due to the vast number of components and many spectral peaks present. Fortunately, this tenet is also not always strictly true. Nonetheless, there will always be a need to assign individual NMR peaks buried in a sea of others. Because of the importance of the spectral assignment requirement, many methods have been developed to facilitate the labor of making spectral assignments in NMR. These approaches have spectral simplification as a fundamental objective and fall into two categories, spin manipulation and physical on-line separation.

Spin manipulation is accomplished by multipulse NMR experiments wherein the single excitation pulse is replaced by a series of pulses designed to measure a particular attribute of the nucleus or interactions with other nuclei and includes two-dimensional (2D) NMR and one-dimensional (1D) editing experiments. Development of new pulse sequences has been fertile ground for NMR spectroscopists since the earliest days of NMR and literally hundreds of pulse sequences have been reported over the years. Indeed, there are dozens of pulse sequences that can be immensely helpful for the deconvolution and assignment of raw biofluids (Table 1), including correlation experiments to reduce spectral complexity based on the coupling of nuclear spins[9,20] or editing of components based on nuclear relaxation[21] or molecular diffusion.[22] Although highly reproducible, the 2D approaches suffer from low throughput—a typical 2D $^1$H-$^1$H correlation spectrum of urine may take 30 min to several hours to acquire, dependent on which experiment is used.

An entirely different approach has been to couple liquid chromatography (LC) to a flow NMR spectrometer[9,23–24] providing a real time means of sample fractionation prior to analysis. In LC-NMR, a biofluid is injected onto a chromatography system, such as an HPLC, and NMR spectra are recorded as the eluent flows through the spectrometer. The analytes need not have UV chromophores to be detected (because of the non-selective nature of the NMR spectrometer) nor does separation need to be complete, since the inherent frequency resolution of the NMR spectrum makes it possible to analyze many components in a single spectrum. The biofluid complexity problem is thus reduced chromatographically. Disadvantages of LC-NMR techniques include being slow and difficult to reproduce. In fact most applications of LC-NMR and the related LC-NMR-MS technique are for the identification of drug metabolites in biofluids.

In the case of urine spectra, many of the major and commonly observed components can be identified by manual or automated inspection, based on reported chemical shifts. Nearly a hundred components representing 263 resolved peaks commonly found in urine, plasma, serum, CSF and semen have been tabulated.[9] The methods described above are important tools for identifying what remains as is necessary. Fortunately for applications such as screening, where changes in molecular concentrations from one sample to another are a primary piece of information, assignments are not absolutely necessary. Here, pattern recognition (PR) can be used to identify a characteristic change without the absolute necessity of knowing what is changing. As opposed to an approach based on the assignment of each individual peak, pattern

**Table 1** Basic pulse sequences useful for biofluids applications.

| Pulse sequence | Application | Reference |
|---|---|---|
| CPMG | Edit signals from large molecules out of a spectrum, leaving only signals from small molecule analytes | 45 |
| NOESY1D[a] WET[b] DPFGSE[c] | Provides1D spectra with elimination of large solvent resonances | a—7 b—46 c—47 |
| COSY | Provides 2D NMR spectra, usually plotted as intensity contours, with normal 1D spectrum on the diagonal, as off-diagonal peaks at positions indicating chemical shifts of spin-coupled nuclei (nuclei close in chemical bond terms) | |
| TOCSY | TOCSY is like COSY but identifies chains of spin-coupled nuclei. Identification of biofluid components based on association of peaks with others in the same molecule | 48 |
| DE-TOCSY | Diffusion edited version of TOCSY. Edits out signals from small molecules in biofluids leaving only macromolecules. | 49 |
| DOSY | Diffusion edited 2D NMR method for spectroscopically separating individual components based on their diffusion constants | 50 |
| HMQC HSQC | Obtaining $^{13}$C chemical shifts for analyte CH, CH$_2$ and CH$_3$ groups | 51 |
| HMBC | Like HMQC/HSQC but allows long range $^1$H connections of quaternary carbons | 52 |

recognition of simple 1D biofluid NMR data has the ability to dramatically reduce the dimensionality of the data using multivariate analytical techniques. Indeed, successes in this area have provided the impetus to look at NMR spectroscopy as a potential tool for rapid-throughput *in vivo* toxicity screening.

### 14.5.7.2.2 Pattern Recognition Analysis of NMR Data

#### *14.5.7.2.2.1 Introduction*

Pattern recognition provides a means to identify and interpret meaningful regularities in noisy or complex data sets. As has been seen, such tools will be necessary for successful analysis of biofluid NMR data, especially in a screening environment. Even when relatively small numbers of spectra are to be investigated, as when tracking down biomarkers, pattern recognition approaches can establish whether there is indeed a change in the data that deserves further investigation. The pattern recognition process, specifically as it pertains to analysis of NMR spectral data, and the various pattern recognition techniques that are useful for biomedical NMR applications have been recently been reviewed.[25] By far the most widely used method, principal components analysis (PCA), has been described extensively in the literature.[7–8,12–13,26–29]

#### *14.5.7.2.2.2 Principal Component Analysis*

Principal component modeling can be performed at two levels; either an unsupervised approach to cluster similar samples together and aid in visualization, or in a supervised learning approach to model preconceived groups of samples.[30] A convincing correlation with accepted methodologies is a necessary first step in integration of any new technology. Unsupervised PCA has proven to be a valuable tool for correlating changes in the urine NMR spectra to acute toxicity as measured by both clinical chemistry and histopathology.[4] Ultimately, supervised approaches such as SIMCA[14] or probabilistic neural networks[31] will have to be employed to support a genuine screening paradigm. In SIMCA, a principal component (PC) model is generated which defines each training set class. SIMCA is merely multi-class PCA and as such a working knowledge of PCA is valuable. Such supervised approaches will require a large training set such as will be produced by the COMET consortium (see below).

PCA provides simplified graphical representations of complex data sets containing hundreds of variables by using the first two or three principal components as the axes.[32] PCA generates linear combinations of all variables in a sample data set, in this case a series of NMR spectra that have been reduced in size along the frequency axis. This reduction is accomplished by integrating discrete frequency

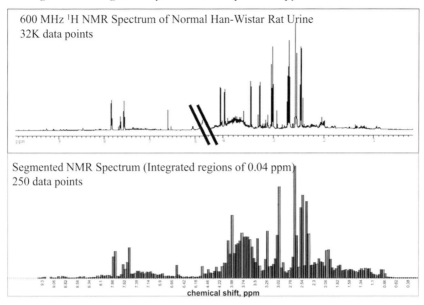

**Figure 3**  Generating the pattern recognition input data from an NMR spectrum. The original data contains 32 K data points and is reduced in this case to ca. 250 data points by integrating over 0.04 ppm regions, or buckets. This both makes the file size tractable and reduces the effect of small variations in chemical shift due to temperature and pH differences from sample to sample.

regions to generate a segmented form of the NMR spectrum as depicted in Figure 3. The next step requires calculation of a new set of independent indices, or principal components (PCs) which are uncorrelated (orthogonal) to each other and reflect the variance in the original data. Furthermore, these new PCs are computed in order of decreasing variance. Thus, the first PC accounts for the maximum amount of variance and each successive PC accounts for less of the remaining variance in the data. This concept can be easily visualized if one limits the number of original variables to 3 as is illustrated in Figure 4.

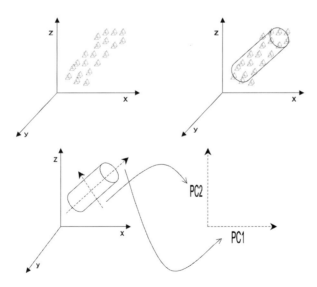

**Figure 4**  A graphical representation of the process involved in computing principal components. In this example, x, y and z are 3 integrated regions in an NMR spectrum. If we plot the values of the these integrals from several NMR spectra, we have a cluster of points in the 3 dimensional chart (upper left) which can captured in the cylinder shown (upper right). The PCs can then be computed from the raw data as depicted in the lower part of the figure. Note that PC1 is the vector that defines the longest axis, or highest variability, in the data. Subsequent PCs are calculated so that they are orthogonal to each other and describe decreasing amounts of variance in the data.

As a simple example, a data set comprised of ¹H NMR spectra of urine samples from one control group and two toxin-treated (para-aminophenol, PAP and α-naphthylisothiocyante, ANIT) test groups is illustrated in Figure 5(A). If PCA is performed on only the control and ANIT samples the resulting PC map is as in Figure 5(B). If the PAP samples are included in the PC analysis and the new data set reprocessed, the PC map is as in Figure 5(C). In the graphical representations, toxicity- or physiologic-induced variations in spectral patterns are indicated by separation from the control region. The magnitude of biochemical changes and correlated severity of toxicity is directly proportional to the distance from the control centroid within any one PCA, but not necessarily across PCAs for different compounds. Additionally, separation of clusters within the PCA is indicative of different NMR spectral patterns, which may involve differences in one or more analytes.

### 14.5.7.2.2.3    *Effects of Xenobiotics*

As shown in the examples above, when exposed to a toxin, some biochemical processes in an organism will be disrupted and others will be initiated in response to the insult and the resulting by-products of these processes will show up as changes in the urine or other biofluid. These changes can be in the form of increases or decreases in ordinary biofluid components or the appearance of endogenous products not normally observed. Additionally, the toxic agent itself, its metabolites or the vehicle it was delivered with can make significant contributions to the NMR spectrum.

Obviously, these xenobiotic contributions, while important diagnostics of detoxification processes, are not direct indicators of toxicity per se and must be excluded from the pattern recognition analysis. In practice, this is easily accomplished by deleting regions of the spectrum that contain peaks from the interfering material. This is nicely illustrated by comparing control animals before and after treatment with a vehicle containing the common constituent polyethylene glycol (PEG) as in Figure 6. These data show that the separation of the two clusters is due solely to the presence of the PEG NMR resonance and its elimination results in a homogenous group. Not all vehicles are so imposing on the NMR spectra. For example in Figure 5(B) the control data represent samples

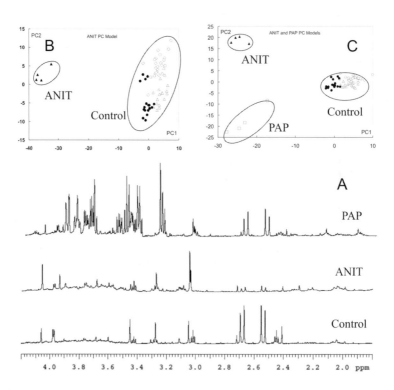

**Figure 5**    PC plots and examples of the corresponding NMR data for three groups of control rats and rats treated with ANIT and PAP (N = 4). (A) A ¹H NMR spectral region showing changes induced by PAP (top trace) and ANIT (middle trace) relative to control (bottom trace); (B) The results when only ANIT and control data are used as input; (C) PC plot when PAP spectral data are included in the analysis. The circles represent normal, ANIT and PAP-induced toxicity models.

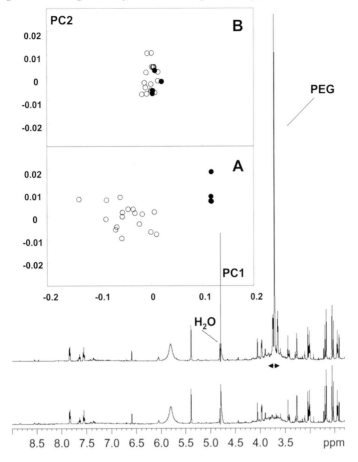

**Figure 6** The effect of PEG on the NMR spectrum and PCA clustering of control rat urine. The NMR spectra were recorded before (bottom) and after (top) dosing a rat with PEG vehicle. Note the dominant resonance from PEG at 3.7 ppm. The insets show the PC plot of 21 samples predose (closed circles) and post dose (open circles) using as input for the calculations: (A) the usual NMR spectral range; and (B) excluding the small PEG associated region indicated by the double arrow in the NMR spectrum.

collected from three different vehicle treatments, corn-oil, saline or naïve (no treatment). There is a small but distinct separation between groups of animals that received nothing, and those that received IP injections of either saline or corn oil (Figure 7). Yet, the small differences in the urine spectra from these rats are indiscernible by visual inspection of the NMR spectra and they are dwarfed by the toxic effects mentioned earlier (compare Figures 5(B), 5(C) and 7).

### 14.5.7.2.3 NMR Logistics (Experimental Considerations)

#### 14.5.7.2.3.1 Equipment

Setting up an NMR facility for biofluid NMR analysis requires careful consideration of many factors, including cost and space, and the exact equipment needed will depend on what types of analyses are desired.

The NMR spectrometer itself is the single most critical (and expensive) component required for establishing an in-house metabonomic capability. For reasons of sensitivity and spectral dispersion, it is important to utilize the highest field strength feasible, and many laboratories have chosen to standardize at 600 MHz observation frequencies for [1]H NMR. Lower observation frequencies will allow classification, but it becomes more difficult and time-consuming to identify the biochemical effects. The magnet should be the self-shielded variety, minimizing the space requirements and greatly reducing the distance from which necessary accessories (e.g., liquid handling robotics and chromatography systems) can be located. The footprint of a shielded system is dramatically smaller than a conventional magnet. The spectrometer should be equipped with a minimum of two R.F.

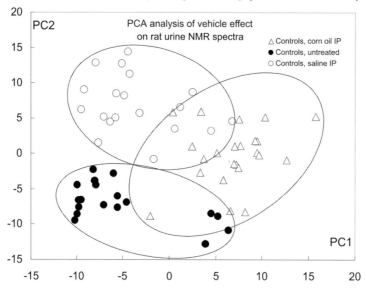

**Figure 7**   PC plots of three groups of control rats. These are the same control samples as used in the analysis in Figure 5(B) and 5(C), with PCs computed only on the control samples. It interesting to note that PCA is sensitive enough to pick up changes due to IP injection of corn oil and even saline.

excitation channels (one broadband), both of which should have the capability of producing shaped pulses; this in conjunction with an appropriate probe will provide all of the multipulse and multidimensional capabilities required. The spectrometer should also have a $Z$ axis gradient amplifier capable of producing a magnetic field gradient across the sample of 30 gauss/cm with rapid recovery time ($< 10\,\mu$s), enabling newer solvent suppression techniques and diffusion editing experiments to be employed. Finally, the system should have a sample temperature control unit capable of maintaining a uniform temperature across the sample of $+/- 0.1°$C or better.

Another important consideration is to provide a stable environment for the system. The room should be isolated from unnecessary traffic, vibrations and rf noise and be outfitted to maintain a consistent temperature $+/- 1°$C. It should also be free of large amount of ferric metals, particularly those that have any chance of moving during the course of an experiment. Periodic cryogen filling (weekly for nitrogen and bimonthly to quarterly for liquid helium) is facilitated by plumbed-in liquid nitrogen and gaseous He with nearby access doors large enough to allow ready access to portable liquid He dewars. The Pfizer Ann Arbor laboratory is built on a raised aluminum floor with aluminum-lined pits to contain the NMR magnets. This places the top of the magnet within easy reach for cryogen fills and when loading the systems using conventional probes. Alternatively, and probably more commonly, one must have a high ceiling and aluminum or

wooden ladders to access the top of the magnet.

The sample probe should be a Z-gradient, proton observe flow cell design with a proton sensitivity of no less than 60:1 (measured on a single scan of the standard sample, the anomeric proton of 2 mM sucrose) and a second coil capable of broad band excitation (for pulsing $^{13}$C, $^{15}$N, $^{31}$P, *etc.*). For urine samples from rats or larger species, it is suggested that a flow probe with an active volume of 120 µl (250 µl total volume) which produces a sucrose sensitivity of 100:1 be used. For mice, a smaller probe with an active volume of 60 µl should be considered due to the smaller sample volumes obtained. In the case of mice, this smaller sample volume is somewhat offset by the highly concentrated urine typical for that species.

The system should be coupled to a liquids handing device capable of holding deep well 96-well polyethylene plates and maintaining them at a temperature of 4–10°C while they are in queue for analysis. Sample handling equipment should be capable of delivering and withdrawing aqueous samples to and from the probe at a minimum rate of 1–2 ml/min without cavitation. Care should be taken to leave a few millimeters clearance between the bottom of the sample tube or well and the tip of the probe to avoid aspirating any settled particulate—a needle with a side orifice is advisable. An iso-osmotic wash buffer should be utilized to clean the probe between samples. For follow-up analysis on individual urine samples, it is desirable to have a binary gradient-capable HPLC system and a mass spectrometer on line, with integrated software to control all of these components.

### 14.5.7.2.3.2 Sample Requirements and Processing Considerations

Because large numbers of samples will be analyzed in a screening paradigm, sample preparation requirements should be kept to a minimum, although several factors can potentially affect the outcome of the NMR results. The flow probe design requires that samples be free of solids to avoid clogging and therefore samples should clarified via gravity precipitation or centrifugation prior to placing into 96 well plates. Osmolarity can affect the efficiency of energy transfer from the probe to the sample and the pH can influence the chemical shifts of molecules with ionizable groups such as amines and carboxylic acids. Both of these effects can be partially normalized by diluting urine using a strong buffer (0.2 M sodium phosphate, pH 7.4) in a 1:2 (buffer:urine) ratio. As an added precaution against bacterial contamination, it is advisable to make a prior addition of sodium azide to the tube in which urine is be collected. Finally, an internal chemical shift standard, TSP (sodium 2,2′,3,3′deutero-3-trimethylsilylpropionate gives a single resonance which defines 0 ppm in an NMR spectrum), and a lock solvent ($D_2O$) should be added. This is efficiently done by adding 1.0 mM solution of TSP dissolved in $D_2O$ to the buffered sample to a final $D_2O$ concentration of 5–10%. It should be noted that normal precipitation of various amounts of calcium phosphates after dilution with buffer is generally observed, but does not cause a problem if the care is taken in setting up the needle depth on the liquids handling equipment. A minimum total final sample volume of approximately 500 μl or 340 μl is necessary for 120 μl and 60 μl flow probes, respectively. Samples should be stored at −20°C after dilution and maintained at 10°C while they are in queue on the liquids handling device.

### 14.5.7.2.3.3 Data Acquisition and Analysis

NMR spectra should be recorded under conditions as close as possible to those in the database that is used to construct the toxicity models. This requires selecting standard conditions for sample preparation and data collection.[33] The sample should be aspirated into the delivery needle at a flow rate that is as fast as possible without cavitation (typically 1–2 ml/min). Each sample is loaded into the NMR flow cell and allowed to come to probe temperature, which should be calibrated and maintained at 27°C. It is recommended that a minimum of 32 pulses be acquired for each spectrum, the obvious trade-off is data collection time versus signal-to-noise. Once the data are collected and safely archived, they are processed using fast Fourier transformation (FFT). After FFT, the data should be phased and their baselines flattened if necessary.

Finally, the data are reduced in size by summing the intensity of all data points over 0.04 ppm regions. The size of these regions is arbitrary, but should be small enough to capture individual peaks and large enough to reduce the affect of small variations in chemical shift. Regions devoid of endogenous peaks, regions containing xenobiotic metabolites and the region 6.0–4.5 ppm containing urea and water resonances should normally be excluded from all of the reduced data. The resulting integrated spectral regions are the input for subsequent statistical analysis as described above.

## 14.5.7.3 URINE, URINE, EVERYWHERE...: TOXICOLOGIC CONSIDERATIONS

### 14.5.7.3.1 Sample Requirements

Almost any biological fluid can be used for metabonomic investigations. Plasma, whole blood, milk, saliva, seminal fluid, aqueous humor, *etc.*, can all be used successfully. Some samples suit toxicological applications more readily than others. Lindon *et al.*[3,9] reviewed the characteristics of many biofluids with regard to their suitability to NMR analyses. While certain fluids lend themselves to intriguing possibilities for mechanistic investigations, any fluid used for rapid throughput screening application, by definition, needs to be easily obtained and available in sufficient quantity. With this in mind, the samples of choice for screening applications are blood (whole blood/plasma/serum) and urine.

### 14.5.7.3.2 Urine as the Sample of Choice

#### 14.5.7.3.2.1 Urine versus Blood

While blood or blood fractions are the standard for comparison with standard clinical laboratory and pharmacokinetic analyses, urine has significant advantages for screening applications. Firstly, while simple, blood withdrawal from an animal is still an invasive procedure requiring animal handling and venipuncture, both of which can induce some level of stress and trauma to the animal. While this trauma can be controlled for and minimized, it can not be eliminated and necessarily introduces some level of variability due to sample collection techniques. More importantly, there are significant limitations with how often blood can be withdrawn. While we all remember graduate school days when collecting samples at 3 AM was viewed as a rite of passage, the necessity of collecting samples at such times *routinely* will eventually induce resentment in even the most dedicated technical groups. Additionally, even if it were possible from a logistics standpoint, there are limitations to the number of times blood can successfully be withdrawn from a rat due to both practical and ethical considerations.[34,35] The limitations are even more profound in mice.

### 14.5.7.3.2.2   Temporal Sampling

The use of urine can tremendously simplify multiple time point sample concerns. Metabolism cages enable continuous collection of sample, such that an entire temporal profile of exposure can be easily obtained after administration of a test compound. This can be done while the animal is left unattended obviating the need for off-hour sample collections. Not only does this simplify technical logistics, it eliminates handling-associated stress.

### 14.5.7.3.2.3   Sample Variability

The biggest advantages of continuous sampling are that nothing is missed and sample variability is reduced or at least is put in context. While it is obvious as to why nothing would be missed with continual monitoring, why there would be a reduction in variability is less obvious. Figure 8(a) presents individual animal data from animals treated with a single IP dose of 0.5 ml of $CCl_4$ with urine collected pretest and continually for 96 h after dosing. Serum was collected for transaminase analyses from the same animals 24, 48 and 96 h after dosing. The serum ALT versus time profiles are presented in Figure 8(b). Both the serum profile data and subsequent pathology evaluation clearly showed hepatic centrilobular lesions in all animals. It is readily apparent that if a single serum sample had been taken

for transaminase measurements only at termination (a typical approach for screening studies) 3 of 4 animals would have normal ALT levels—a clear misrepresentation of the actual results. If serum samples had been obtained only at the 24-h time point, 75% of the animals would have appeared affected. This is better, but still not an accurate representation of the toxicity results. Serum ALT levels are seldom used in isolation for toxicity assessment and, in this case, multiple time-point sampling did identify that all animals were affected by drug treatment. However, standard clinical chemistry data can generate only limited "snapshots" of toxicity.

In contrast, when the experiment was continually monitored using NMR-based metabonomic technology as depicted in Figure 8(a) it became evident that 2 of the 4 animals (a and d) had quite disparate responses, as indicated by the solid trajectory lines in Figure 8(a). Two animals (b and c) responded similarly, with maximal effect 24 h after dose, and subsequent regression towards control (dotted trajectory lines in Figure 8(a)). Animal a, however, responded relatively slowly to the compound with normal ALT levels 24 and 48 h after dose, but moderate to marked elevation 96 h post dose. Animal d had an immediate and marked response 24 h after dose, but surprisingly was within the normal range by 96 h after dose. It is clear from this simple example that evaluating toxicity in terms of animals exhibiting similar responses, regardless of the time after dose, rather than temporally (e.g., hours after dose) may significantly aid interpretation of resultant pathology and perhaps aid understanding etiology. From a screening perspective, this is critical, because it means that fewer animals need be studied at a given dose. Individual animal response is readily apparent and need not be compensated for by increasing the "N" size at discrete sampling times. This along with the fact that the entire time-course of toxicity from onset, through peak effect to regression can be studied non-invasively in a single animal, dramatically reduces costs and compound requirements and aids in the overall goal of animal use reduction.

### 14.5.7.3.2.4   Sample Trajectory

The ability to monitor changes of urinary profiles over time within an individual animal is demonstrated in Figure 9 which presents individual animal PCA trajectories for animals treated with a single intraperitoneal dose of 0.5 ml/kg $CCl_4$ and followed for 4 or 10 days. It is readily apparent that animals responded simi-

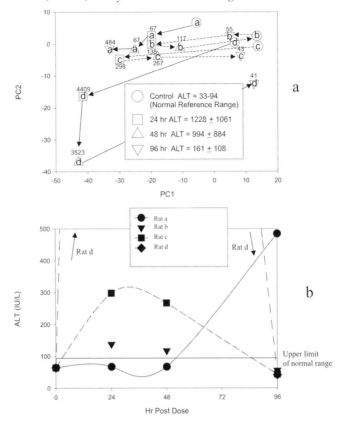

**Figure 8** (a) PC plot of urine samples collected from rats treated with a single dose of 0.5 ml/kg CCl₄. Different symbols represent different time periods with the letter inside each symbol representing an individual animal. Dotted trajectory lines are from two similarly responding animals (b and c) and solid trajectory lines are from two animals with disparate responses (a and d). Numbers above symbols represent concurrently determined serum ALT activities (IU/L). Mean ALT levels are indicated in the legend box. (b) individual animal serum ALT levels over time. Rats a–d are the same in both figures.

larly after treatment, although not necessarily at the same time. The severity of toxicity (as determined by concurrent serum ALT determination) was proportional to the distance from the pretest data point for an individual animal, although absolute distance from pretest was not necessarily proportional to ALT across animals. However, PCA evaluation of urine NMR spectra is not simply an alternative means to measure ALT, it represents the totality of hepatic dysfunction (within the analytical limits of the instrumentation) and the resultant disruption of urinary analytes. Therefore, it should not be expected that ALT would be proportional to PCA distance across animals, but rather PCA distance is proportional to hepatic disruption and/or dysfunction. While ALT is usually a reliable indicator of hepatic necrosis, it is a leakage enzyme and it is not a biomarker of hepatic function. The PCA data give a more comprehensive evaluation. A more analogous situation would be to compare the PCA data to a normalized average of all possible serum liver function tests.

However, even that would fall short, since many changes in urine may be due to secondary or tertiary events that would not necessarily be expressed in serum clinical chemistry analyses.

### 14.5.7.3.2.5 Biochemical Implications of Trajectory Analysis

Figure 9 demonstrates that the trajectories of change in the PCA data were strikingly similar across animals identically treated with 0. 5 ml/kg CCl₄. This makes sense since each point represents a collapsed snapshot of the biochemical makeup of the urine. The similarity in trajectories suggests that there is similarity in the biochemical make up of the urine produced which further suggest a common physiologic (or toxicologic) response to the toxicant. The composition of urine analytes has been used to differentiate between parenchymal and biliary damage induced by hepatic toxicants[7] and trajectory analyses can be used to determine

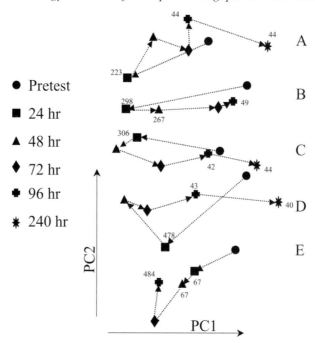

**Figure 9** Individual animal PC trajectories from rats treated with 0.5 ml CCl₄. The trajectories have been spread out in the vertical dimension for the sake of clarity. Numbers above symbols represent concurrently determined individual animal serum ALT activity (IU/L). Note: The PC2 scale is arbitrary and has been offset to emphasize the individual trajectories.

the time course and resolution of various pathophysiologic processes.[12]

#### 14.5.7.3.2.6  *Is Urine an "Appropriate" Sample?*

Despite these advantages, legitimate questions can be raised as to the relationship of the biochemical profile of urine to the toxicity being investigated. Is urine, as a waste product, a valid sample for evaluating the onset and regression of toxicity within in animal? The short answer is yes. Toxicology has a long tradition of urine collection via pans or poorly designed metabolism cages that produce markedly contaminated samples of questionable value. Based on this premise, it can be easily understood why there may be reluctance on the toxicologist's part to consider urine as a useful sample for screening. Modern well-designed metabolism cages reduce or eliminate fecal, water and feed contamination eliminating this source of sample variability. In well-controlled studies, properly collected samples have surprisingly reproducible NMR spectra on a day to day basis as shown by the spectra in Figure 10.

The physiology of any living organism is a dynamic process involving thousands of inter-related biochemical pathways that exist in a state of homeostatic balance at any one period of time. Changes in any one pathway (e.g., due to a toxicologic insult) disrupt the homeostatic balance causing numerous changes, some of which are obviously related to the toxic insult. Other changes, secondary or tertiary to the insult are not so obvious. Obvious links can be directly measured (e.g., fasting induced keto-nuria), without the need for metabonomic technology. However, the complexity of biochemical processes dictates that most changes reflected in the urine, at least at this point in time, are at best poorly understood, and in most cases not understood at all. In our experience, we see many consistent changes in urine analytes associated with a particular toxin. For example α-naphthalisothiocyanate (ANIT) induces decreased urinary citrate, succinate and 2-oxoglutarate within 24 h after administration to rats.[4,7] While it is tempting to speculate as to the meaningfulness of decreases in these Krebs intermediates, at this point in time the most we can say is that the significance of these findings is unknown. Additionally, these particular changes, while not the only changes, are also produced by the nephrotoxin 2-bromoethylamine[5] and certain vascular toxins.[36] Clearly then, changes in any one of these analytes can not be viewed as a specific marker for any particular toxicity. Rather it is the pattern of numerous changes

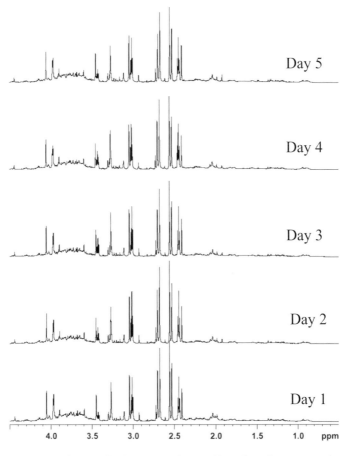

**Figure 10** Urine NMR spectra from a single untreated rat collected on five consecutive days demonstrating very little change in sample constituents with time.

that becomes unique such that the urine spectral patterns of BEA and ANIT can be easily separated.[4]

### 14.5.7.3.3 Species Choice

Any species that produces a bodily fluid (i.e., anything) can be used in toxicologic applications of metabonomic technologies. Practically, certain species are easier to use than others.

#### 14.5.7.3.3.1 Rodents

##### 14.5.7.3.3.1.1 Sample size

Rats are the species of choice for metabonomic studies since they produce an ample amount of urine within a 24-h period and metabolism cages are commercially available allowing fairly clean collection of urine (see below). A typical Sprague–Dawley rat weighing from 200 to 250 g eliminates approximately 13.5 ml of urine over 24 h. Similarly weighted Wistar rats produce slightly more urine, but have higher interanimal variability (Table 2). Rates of urine production vary somewhat with the light cycle, but generally enough urine is produced over every 8-h period to easily allow metabonomic evaluations in these temporal increments. This is important for mechanistic studies where early and rapid changes in animal physiology are important for understanding the toxicity.[6,10] Mice are an attractive choice for screening assays, because their small size allows for reduced compound requirements, an extremely important consideration for any potential screening procedure. Table 2 demonstrates that mice (20–25 g) while producing, on average, less than 1 ml of urine/24 h, still almost always produce sufficient urine for metabonomic evaluation ( > 300 μl).

**Table 2**    Urine production in rats and mice.

| Species | Strain | n | Weight range (g) | Collection period (h) | Volume (ml) Mean ± SD | Volume range (ml) |
|---------|--------|---|------------------|----------------------|------------------------|--------------------|
| Rat | Sprague–Dawley | 30 | 200–250 | 8 | 4.0 ± 1.2 | 2–6 |
| Rat | Sprague–Dawley | 30 | 200–250 | 24 | 13.5 ± 2.5 | 9–18 |
| Rat | Wistar | 36 | 200–250 | 24 | 15.5 ± 5.7 | 5–29 |
| Mouse | B6C3F1 | 30 | 20–25 | 24 | 0.81 ± 0.45 | 0.36–2.24 |

*14.5.7.3.3.1.2    Strain and Age considerations*

Figure 11 shows that, urine from different species can be clearly separated by metabonomic analyses. This is not terribly surprising, since it would be anticipated that different species would have both qualitative and quantitative differences in urine analyte profiles. The question then becomes how discriminating can this new technology be? Holmes *et al.*[2] demonstrated that different strains of rats produced clearly definable PCA patterns. The technique has also been used to phenotype two closely related mouse strains.[37] Unmistakably, metabonomics is a powerful and precise tool for elucidating subtle differences in urine analyte profiles. These differences must be compensated for if cross strain evaluations are to be considered. This is no different than the extrapolation necessary when trying to compare toxicity responses across strains using conventional methods such as clinical chemistry or histopathologic assessment. However,

predictive metabonomics technology will rely heavily on a database of spectral data that will probably be generated in a single strain. To date, most metabonomic data has been generated in either the Sprague–Dawley or Wistar rat. The COMET consortium (see below) is using the Sprague–Dawley (Crl:CD(SD)IGS BR) rat as its model for developing its database. It remains to be conclusively demonstrated that the urine biochemical profile produced in one strain would represent the same mechanism/severity information indicated by a database generated in another strain. In all likelihood, this will probably be true, but there may be exceptions.

The aging rat goes through a variety of physiological changes that are reflected by physical and biochemical changes within the animal. Not surprisingly the urinary profile of a lean 200 g rat that is undergoing a period of rapid growth is quite different from a plump 600 g animal that is entering its golden days.[9] This became quite apparent in an anecdotal

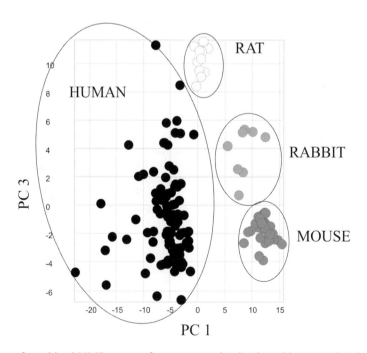

**Figure 11**    PC plot of combined NMR spectra from untreated animals and humans showing clear separation of all species.

incident at the Pfizer Ann Arbor laboratory. In early studies, the only metabolism cages available for metabonomic experiments were for rats > 300 g in size. Correspondingly, rats for these studies were 12–13 weeks of age. Subsequently, cages capable of handling smaller rats were obtained, and the lab reverted to testing 7–8 week old rats, which was the norm for the laboratory. Initial evaluation of the PCA data revealed that the spectral pattern of urine from 8-week-old untreated rats could be almost completely resolved from 13-week-old rats (Figure 12). This observation underscores the need for comparing data only among similarly aged rats. Unfortunately the study mentioned above was not designed to explore age relationships, and currently studies are underway to define the nature and extent of age related changes in urine spectral patterns.

### 14.5.7.3.3.2 Non Rodents

Relatively little has been published on the use of metabonomic technology in species other than the rat.[3] This will surely change as the technology expands in use. Preliminary data from rabbits (Figure 11) and dogs (data not shown) suggests that these species will be quite amenable to metabonomic analyses. The principal limitations to use of species other than rodents are practical rather than scientific. Urine collection for large animals using simple catch pans produces samples that are frequently heavily contaminated with feed, feces and water. Such samples are useless for metabonomic evaluation. Short, timed collec-

tions in which animals are removed from feed and water make for cleaner samples (as long as fecal contamination can be avoided), but this may be impractical in some situations. Metabolism cages for large animals are not commonly available, but appropriately designed custom metabolism cages for large animals can be obtained (e.g., Allentown Caging Equipment Co., Allentown NJ). Cystocentesis can be used in any situation where collection in metabolism cages is not practical. However, except for terminal samples, the technique does require appropriate veterinary skill and the frequency of collection is limited. Likewise catheterization takes appropriate veterinary skill (particularly for females) and can be subject to bacterial contamination.

### 14.5.7.3.3.3 Human Urine

Clearly the most-important samples are those obtained from the end-users of pharmaceutical products. Ironically the volume of metabonomic data from human samples is second only to rats with regard to the published literature. NMR spectra of urine have long been used to successfully diagnose inborn errors of metabolism and disease progression. Lindon *et al.*[3] lists numerous historical applications of NMR technology to the diagnosis of human disease. An interesting application of metabonomic technology to human disease are experiments conducted by Foxall *et al.*[20] and Le Moyec *et al.*,[38] as published by Lindon *et al.*[3] In these experiments pattern recognition evaluation of NMR

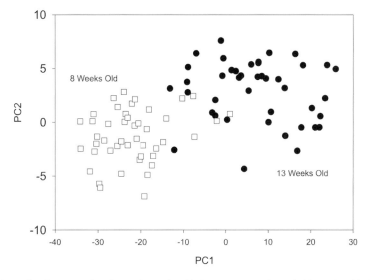

**Figure 12** PC plot of urine samples from 8 week old (open squares) and 13-week-old (solid circles) rats showing distinct clustering of samples from animals differing in age by only four weeks.

spectra from urine collected from patients after kidney transplant was investigated. The technique could distinguish patients undergoing graft rejection from patients experiencing cyclosporin (an anti-rejection drug) toxicity. Clearly the application of NMR technology to human biofluid analysis has a storied history. With regard to metabonomic analyses, the ability to readily transfer the technology from preclinical animal models to human clinical trials is one of the most desirable features of the approach. The ability to monitor humans for biomarkers of toxicity and efficacy from the same peripheral sample is extremely appealing.

### 14.5.7.3.4 Sample Collection

The discussion below describes "real-world" application of metabonomics to toxicology screening in a pharmaceutical company. These observations should be viewed as approaches that have been successfully employed in industry, but should not be seen at the only approach or even the best approach. The sequence of sample collection and handling processes as performed by the Pfizer laboratory is diagrammed in Figure 13. Further details are explained below.

#### *14.5.7.3.4.1 Collection Options*

##### *14.5.7.3.4.1.1 Metabolism cage considerations*

There are several vendors that supply appropriate metabolism cages for metabonomic analyses. Nalgene® cages supplied by several vendors including Harvard Apparatus (Holliston, MA) and VWR/Scientific Products (Plainfield, NJ) are ideal for this work. One advantage of these cages is that adapter kits can be purchased to convert the cages to mouse metabolism cages. A primary concern for collection of samples is that they need to be collected cold (0–4°C). This can be accomplished simply by placing collection vessels on ice. However, ice needs to be changed periodically (i.e., nights and weekends) making this option unappealing for routine use. A step up from collection on ice involves construction of jacketed collection containers that can be hooked in series to refrigerated water circulators. A picture of such a set-up is presented in Figure 14. The use of such a rig, allows constant, unattended, cooled, constant temperature sample collection. However, there are limitations to the number of jacketed vessels that can be handled in such a fashion. Experience suggests that a maximum of 4 can be hooked in series using a standard laboratory refrigerated water circulator. Therefore, 3

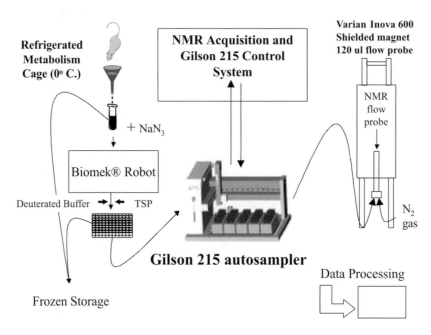

**Figure 13** Diagrammatic representation of sample flow within the Pfizer Ann Arbor laboratories. Samples are collected in metabolism cages while maintained at 0° into tubes containing sodium azide. Samples can be then frozen or further diluted with buffer, internal standard and D₂O and distributed into 96 well plates using a BIOMEK® robot. The plates can then be frozen or proceed to the Gilson NMR fluidics workstation for analysis using the Varian Inova 600 system. Collected raw spectra are processed and PCA conducted.

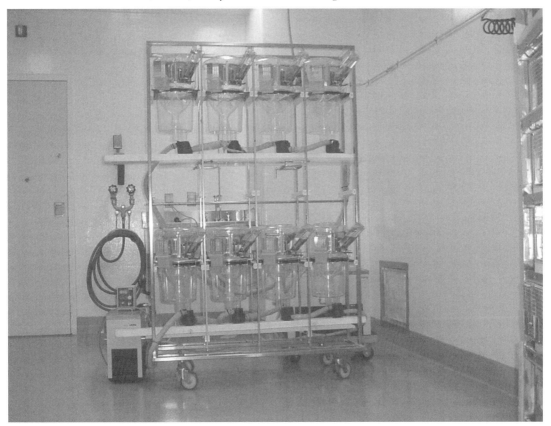

**Figure 14** Photograph of a custom built cooling apparatus used for collecting cooled urine samples. Refrigerated water circulators can be seen to the left and right rear of the rack of metabolism cages. Each circulator cools a bank of four glass jackets which contain the urine collection tubes located under the individual metabolism cages.

circulators would be required for collection of a standard rack of 12 cages. An interesting side note is that the plumbing of such "Rube Goldberg" contrivances can adversely affect room temperature. In the Pfizer Ann Arbor laboratories' original construction of the apparatus pictured in Figure 14, the apparatus and associated tubing dropped room temperature by 8–10 degrees when in operation. Subsequent insulation of all tubing alleviated the problem. Perhaps the best, but most expensive, approach to the cold collection requirement is the use of refrigerated metabolism racks. These racks are generally available from the same vendors that supply the caging. An example of such a rack is pictured in Figure 15. These racks allow timed collection of up to 12 fractions at any desired interval from 10 sec to 30 h. This unit is more compact and easier to set-up than the jury-rigged device mentioned above, while maintaining the ability for unattended collection of cooled samples. Additionally, the unit allows collection of timed fractions with no operator

intervention, which would be useful for mechanistic studies as indicated above. A significant caveat for their use is their cost and the time required to obtain the cages once ordered (currently at least several months).

### 14.5.7.3.4.1.2 Other urine collection procedures

While metabolism cage collection is preferred for the reasons stated above, any urine collection procedure that is capable of collecting clean (non-bacterially contaminated) urine in sufficient volume (e.g., $> 100\,\mu l$) can be utilized. Addition of sodium azide or other "NMR friendly" (i.e., without $^1$H NMR peaks) bacterial-static compound should be considered since samples will frequently thaw and come to room temperature at some time during processing or analysis. A final concentration of approximately 0.1% $NaN_3$ is adequate for this purpose.

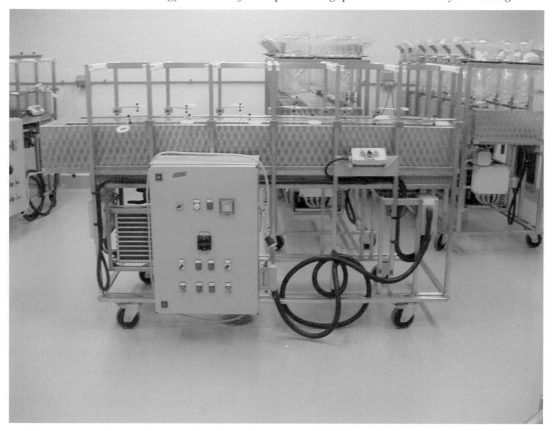

**Figure 15**    Photograph of a refrigerated metabolism cage rack capable of simultaneously collecting samples from 12 cages. Racks with cages installed can be seen in the background.

#### 14.5.7.3.4.2    *Sample Handling and Processing*

One of the advantages of using NMR as an analytical tool, is that little sample preparation is required prior to analysis. At some point in the processing (either prior to or after plate preparation, samples need to be centrifuged to remove particulates. This is critical for rodent samples, which contain a significant sediment concentration. It is also important that pH be controlled. After sample collection, urine can be diluted in buffer to minimize pH variations and $D_2O$ and TSP added as described earlier. While these additions are simple, when dealing with the hundreds of samples typically generated in a metabonomics experiment, this repetitive sample handling can be time-consuming and a source of error if the dilutions are conducted manually. The simple nature of these dilutions makes them ideal for robotic processing. In practice, Pfizer laboratories use a Biomek® 2000 robot with a refrigerated side-loader to dilute the samples and fraction them out into multiple deep-well 96-well format plates. This enables preparation of extra plates for back-up purposes. The 96-well format is ideal format for storing samples as plates

requires little freezer space. Additionally, the plates can then be directly used by the NMR fluidics system.

### 14.5.7.4    PROVE IT!: METABONOMICS IN PRACTICE

The examples given below document some of the uses for which metabonomic technology has been successfully applied.

#### 14.5.7.4.1    The Hepatotoxicity of Allyl Alcohol

Allyl alcohol is a documented hepatotoxin typically producing elevated transaminases and periportal necrosis within 24 h of dosing.[39,40] Metabonomic technology was used to examine the temporal effects of allyl alcohol when administered as a single oral dose to groups of male Wistar rats $(N=4)$ at doses of 0 (control), 12 (low dose) and 120 (high dose) mg/kg. Twenty-four hour urine samples were collected for the 24-h period prior to dosing (pretest) and daily for the next four days. Serum samples were collected for clinical chemistry analyses 24, 48 and 96 h after dose (24 and 96 h post-dose for control). The animals were sacrificed 96 h after dose and the livers examined microscopically. The clinical chemistry data are summarized in Table 3. Statistically significant effects on mean clinical chemistry parameters were restricted to the high dose, with marked increases in transaminases and lesser increases in ALP and total bilirubin 24 h after dose, gradually returning towards control levels by 96 h after dose. Total protein was also moderately decreased at the 24 h time point. These effects were consistent with the histopathology described below. No hepatic histopathology was evident in any control or low dose-treated animal. In the high dose group, hepatic pathology was evident in all four animals characterized by minimal to marked epithelial hyperplasia (4/4), mild to moderate fibrosis (3/4), minimal to mild single cell necrosis (3/4) and moderate disseminated coagulative necrosis (1/4).

The metabonomic data for this experiment is summarized as a PC plot in Figure 16. These data clearly demonstrate the onset and regression of hepatic toxicity with allyl alcohol. Severity, as determined by distance from the pretest cluster, was maximal 24–48 h after dose, regressing towards control by 96 h which is consistent with the clinical chemistry parameters. However, the utility of the data does not end there. Although there appeared to be a very slight increase in mean transaminase and ALP levels at the low dose, these changes were not statistically significant and did not correlate to any histopathology. In typical safety evaluations, findings such as these would probably be discounted. However, on an individual animal basis a single low dose animal that had slightly elevated ALP (260–280 IU/l) at all post-dose time points was readily identified (left side of Figure 16).

Additionally, the only high dose animal having serum chemistry data consistent with a moderate biliary effect (ALP = 730 IU/l, bilirubin = 1.1 mg/dl) had a trajectory distinct from the other three animals (trajectory lines in Figure 16). The data were suggestive of a metabolic profile consistent with necrosis and elevated transaminases for samples plotting to the southeast corner of the plot and hepatobiliary toxicity including elevated ALP and bilirubin for samples plotting to the southwest corner of the plot.

### 14.5.7.4.2 Vascular Toxicity

Vasculitis is of growing concern in the development of novel therapeutics. As a class, the vasculitides encompass a variety of clinical and pathologic conditions and are, in general, empirically defined and poorly understood because of the relative rarity of the disorders, the lack of specificity of associated signs and symptoms, the inaccessibility of certain target tissues for pathologic examination, and the absence of efficient non-invasive diagnostic tests.[41] The lack of biomarkers means that currently histopathology is required to definitively assess vasculitis. Therefore, even if vasculitis is suspected in the class, sufficient drug needs to be synthesized to conduct fairly extensive whole animal studies. Further, no decision can be reached until after histopathologic confirmation, a frequently time-consuming process. Phosphodiesterase Type IV (PDE4) inhibitors are being developed for a variety of immunologically mediated indications including asthma and chronic obstructive pulmonary disease. PDE inhibitors are well-documented inducers of vasculitis in rats.[42,43] CI-1018, a PDE4 inhibitor, was found to be associated with mesenteric vasculitis in rats.[44]

**Table 3** Serum chemistry parameters after allyl alcohol administration (Mean ± SE).

| Dose (mg/kg) | Timepoint (h) | ALT (IU/l) | AST (IU/l) | ALP (IU/l) | Total bilirubin[b] (mg/dl) | Total protein[b] (g/dl) |
|---|---|---|---|---|---|---|
| 0 | 24 | 40 ± 1.4 | 79 ± 2.8 | 230 ± 12.2 | ND | ND |
| | 48 | ND[a] | ND | ND | ND | ND |
| | 96 | 39 ± 1.7 | 73 ± 0.9 | 221 ± 13.2 | ND | ND |
| 12 | 24 | 39 ± 0.9 | 94 ± 7.1 | 250 ± 11.6 | 0.10 ± 0.0 | 5.55 ± 0.1 |
| | 48[b] | 39 ± 1.0 | 90 ± 1.3 | 245 ± 10.3 | 0.10 ± 0.0 | 5.45 ± 0.1 |
| | 96 | 39 ± 1.7 | 104 ± 29.8 | 240 ± 14.8 | 0.18 ± 0.1 | 5.93 ± 0.1 |
| 120 | 24 | 1741 ± 617* | 2289 ± 600* | 407 ± 96* | 0.38 ± 0.2 | 4.75 ± 0.1 |
| | 48[b] | 299 ± 108.9 | 515 ± 166 | 464 ± 108 | 0.30 ± 0.2 | 5.18 ± 0.1 |
| | 96 | 47 ± 4.5 | 90 ± 13.8 | 190 ± 14.2 | 0.15 ± 0.0 | 5.68 ± 0.1 |

* $= p < 0.05$.
[a] ND = Not Done.
[b] Statistical comparisons not performed (no concurrent control).

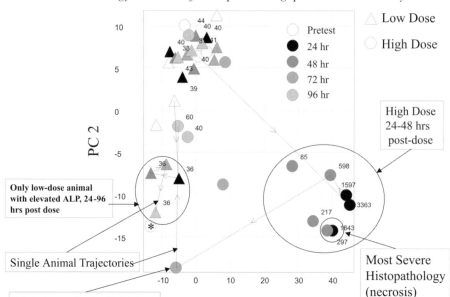

**Figure 16** PC plot of samples obtained from rats treated with a single oral 12 mg/kg (low—triangles) or 120 mg/kg (high—circles) dose of allyl alcohol. Numbers above samples are concurrently determined individual animal serum ALT values (IU/L). Markedly elevated transaminases and a southeast trajectory of all high dose animals characterized initial response with subsequent regression toward control. A single low dose animal and a high dose animal with secondary evidence of biliary involvement (elevated ALP and bilirubin) tracked towards the southeast corner of the plot before regressing towards control (trajectory lines). The asterisk represents an outlier control sample.

Based on these findings an attempt was made to evaluate the utility of metabonomics to assess CI-1018 induced vasculitis in rats.[36] Rats were administered CI-1018 at doses up to 3000 mg/kg for a period of up to 4 days. Urine (24 h samples) was collected from all animals pretest and daily for metabonomic analysis. Eight of 32 CI-1018 treated animals (across doses) were found to have vascular injury of varying severity at doses ≥750 mg/kg. The metabonomic data is summarized as a principal components plot in Figure 17. Principal component analysis produced a clear pattern separation (as indicated by the dotted circle in Figure 17) between 5 of 8 animals with lesions and 27 of 28 animals without lesions. The three animals with lesions that were not identified in the PC plot, had the most minimal lesions, with one animal dosed for only two days. During the course of the experiment, a single high dose animal was found dead on the last day. It had displayed clinical signs of hypoactivity, reduced feces and red stained muzzle the day prior to its death. One other animal, while not displaying clinical signs of toxicity lost a significant amount of weight (16 g) over the four days of dosing. The observed clinical signs and weight loss are correlates of moderate to marked

vasculitis typically observed with PDE4 inhibitors. The spectra of the 72 and 96 h samples from these two animals were dominated by ketone bodies such as β-hydroxybutyrate, acetylacetone, acetate, and acetone. Correspondingly, these samples clustered distinctly separate from both the control cluster and the other animals with lesions but without clinical signs or weight loss (solid circle in Figure 17). Although the dead animal was not examined the other animal was found to have marked mesenteric vasculitis and mild hepatic vasculitis. These findings are consistent with an inappetence-induced ketonuria secondary to the vasculitis. These data demonstrate the utility of metabonomic technology for evaluating a novel toxicity for which existing biomarkers do not exist. The data also demonstrate the potential of metabonomic data to aid clinical and morphologic evaluations. If the initial promise of these data is confirmed, metabonomic technology represents a novel approach for dealing with the thorny problem of vasculitis in drug development. Perhaps even more exciting, is the potential for generating novel biomarkers of vasculitis in the preclinical setting that can subsequently be transferred to the clinical setting.

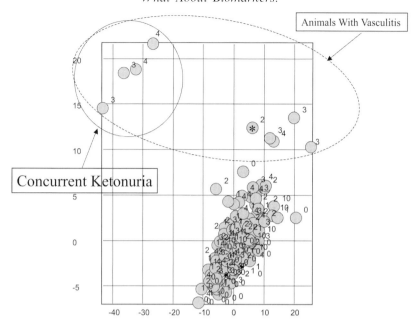

**Figure 17** PC plot of samples from animals treated with 250–3000 mg/kg CI-1018 for up to four days. Black circles (only partially visible behind large cluster) represent control samples and gray circles represent samples from treated animals (collapsed across doses). Numbers above symbols represent day of sample collection with 0 being a pretest sample. The large dotted circle encompasses samples from 5 of 8 animals that had vascular lesions (some animals had more than one sample above the line) from animals without lesions. The asterisk denotes a sample from a high dose animal that had no apparent vasculitis. The smaller solid circle encompasses samples from animals that had marked vasculitis and concurrent ketonuria.

### 14.5.7.5 WHAT'S IT ALL MEAN?: METABONOMICS AS A SCREENING TECHNOLOGY

Although the data from the two examples cited above are too limited to draw definitive conclusions, the utility of the metabonomics approach should be readily apparent. Not only can onset, severity and reversal of toxicity be monitored from a peripheral sample, the data is meaningful even at the individual animal level. The technique appears to be as sensitive as traditional clinical chemistry indices and allows for ready identification of idiosyncratic responses to drug treatment. Additionally, target tissues and even sub-localization within tissues can potentially be determined. These data, in conjunction with the temporal sequence of events, can provide mechanistic insights into the etiology of observed lesions. Taken together these data demonstrate the enormous potential this technology has in a screening environment, where blood sampling and histopathology would be too time and resource intensive to conduct on a routine basis. Theoretically, most of the technical aspects of the technique conducted after urine sample collection can be automated to a significant extent. The ability to maximize the significance of individual animal data suggests that group sizes for such screens could be relatively small, thereby decreasing drug requirements. If mice are used in the screening approach, the potential exists for conducting *in vivo* screens with as little as 50–500 mg of test material. Although even this quantity of drug may seem substantial early in drug development, the utilization of this technology represents a significant leap in our ability to decrease toxicity-induced late-stage attrition of lead compounds.

### 14.5.7.6 WHAT ABOUT BIOMARKERS?

Although not the topic of this discussion, the potential for biomarker generation is a significant attraction of metabonomic technology. Arguably it may represent the most important contribution this technology will make to the area of pharmaceutical development. By definition, if a toxicity, efficacy or physiological change can reproducibly be associated with a specific PC pattern, a biomarker must exist. However, we may need to rethink the concept of biomarkers, since a PC cluster may be caused by a pattern of analyte changes within the sample and may not be due to changes in a single molecule. The pattern then becomes the biomarker. Additionally, the analytical changes may be secondary or tertiary to the initiating event, be it toxicity or efficacy. Obviously such a biomarker would be less desirable than a mechanistically linked biomarker. Another major drawback is that while validation for a single analyte as a biomarker is difficult enough, validating a pattern change as a biomarker expands the difficulty exponentially. However, in many cases any biomarker (e.g., for vasculitis) would represent a significant advancement from current practices.

### 14.5.7.7   IT SEEMS LIKE A LOT OF WORK: THE COMET CONSORTIUM

Even a basic understanding of metabonomics technology rapidly leads to the realization that in order for pattern recognition evaluation of NMR spectra to work as a screening tool for identifying the severity and localization of a lesion, a pattern has to be established. Unfortunately, one test compound does not a pattern make. In order to reliably assess the toxicity of novel compounds, a database of numerous compounds inducing numerous types of physiologic and toxicologic morphofunctional changes needs to be generated. The generation of such a database represents a fairly Herculean task, well beyond the capabilities of any one academic or industrial institution. With this in mind, the *C*onsortium *O*n *M*etabonomics *T*echnology (COMET) has been formed under the auspices of Imperial College in London devoted to the task of generating just such a database. The current make-up of the consortium is indicated in Table 4. The consortium is seen as a joint industry/academia research initiative into this promising technology. The initial goals of the consortium are to generate a database for hepatic and renal toxicity with subsequent targets to be determined later. As of this writing, the legal agreements have been finalized by all parties and the first sets of data have been generated by members. So far, the project is on schedule and initial results are very promising. As a research consortium, publication of relevant findings can be anticipated in coming years.

### 14.5.7.8   NOW WHAT?: CONCLUSION

The potential for metabonomic technology is enormous and an in depth discussion of all possible applications is beyond the scope of a single review. Currently the realization of this potential is underway and the literature on novel applications will undoubtedly expand exponentially in the next few years. Biological fluids represent a vast reservoir of information that can be sampled and assessed using metabonomics technology. Clearly one of our biggest hurdles in the near future will be designing approaches to collect, collate and comprehend this wealth of information. However, the task for this technology will be no greater than the task set before toxciogenomic and proteomic initiatives. Metabonomics will not replace these sister technologies, but should serve as an extension of them, aiding in placing data gleaned from these approaches in proper context. One can easily envision a combined genomic/proteomic/metabonomic tactical approach to addressing etiology and pathology from the gene through the protein to the phenotype. This "grand unification" of technologies will be an exciting development to look for in the near future.

### 14.5.7.9   REFERENCES

1. J. K. Nicholson, J. C. Lindon and E. Holmes, '*Metabonomics*: Understanding the metabolic responses of living systems to pathophysiological stimuli via multivariate statistical analysis of biological NMR spectroscopic data.' *Xenobiotica*, 1999, **29**, 1181–1189.
2. E. Holmes and J. P. Shockcor, 'Accelerated toxicity screening using NMR and pattern recognition-based methods.' *Curr. Opin. Drug Discovery Dev.*, 2000, **3**, 72–78.

**Table 4**   COMET Membership (as of March 2001).

Imperial College of Science, Technology and Medicine (London)
Dupont Pharmaceuticals Inc.
Lilly Research Laboratories, Eli Lilly and Co.
Novo Nordisk A/S
Pharmacia Corporation
Pfizer Global Research and Development
F. Hoffmann-La Roche AG

3. J. Lindon, J. K. Nicholson, E. Holmes *et al.*, 'Metabonomics: Metabolic processes studied by NMR spectroscopy of biofluids.' *Concepts Magn. Reson.*, 2000, **12**, 289–320

4. D. G. Robertson, M. D. Reily, R. E. Sigler *et al.*, 'Metabonomics: Evaluation of nuclear magnetic resonance (NMR) and pattern recognition technology for rapid *in vivo* screening of liver and kidney toxicants.' *Toxicol. Sci.*, 2000, **57**, 326–337.

5. E. Holmes, F. W. Bonner, B. C. Sweatman *et al.*, 'Nuclear magnetic resonance spectroscopy and pattern recognition analysis of the biochemical processes associated with the progression of and recovery from nephrotoxic lesions in the rat induced by mercury(II) chloride and 2-bromoethanamine.' *Mol. Pharmacol.*, 1992a, **42**, 922–930.

6. M. L. Anthony, K. P. R. Gartland, C. R. Beddell *et al.*, 'Studies of the biochemical toxicology of uranyl nitrate in the rat.' *Arch. Toxicol.*, 1994, **68**, 43–53.

7. B. M. Beckwith-Hall, J. K. Nicholson, A. W. Nicholls *et al.*, 'Nuclear magnetic resonance spectroscopic and principle component analysis investigations into biochemical effects of three model hepatotoxins.' *Chem. Res. Toxicol.*, 1998, **11**, 260–272.

8. E. Holmes, A. W. Nicholls, J. C. Lindon *et al.*, 'Development of a model for classification of toxin-induced lesions using $^1$H NMR spectroscopy of urine combined with pattern recognition.' *NMR Biomed.*, 1998, **11**, 235–244.

9. J. C. Lindon, J. K. Nicholson and J. R. Everett, 'NMR spectroscopy of biofluids.' *Annu. Reports NMR Spectros.*, 1999, **38**, 1–88.

10. K. P. R. Gartland, F. W. Bonner and J. K. Nicholson, 'Investigations into the biochemical effects of region-specific nephrotoxins.' *Mol. Pharmacol.*, 1989, **35**, 242–250.

11. K. P. R. Gartland, F. W. Bonner, J. A. Timbrell *et al.*, 'Biochemical characterization of para-aminophenol-induced nephrotoxic lesions in the F344 rat.' *Arch. Toxicol.*, 1989b, **63**, 97–106.

12. E. Holmes, J. K. Nicholson, F. W. Bonner *et al.*, 'Mapping the biochemical trajectory of nephrotoxicity by pattern recognition of NMR urinalysis.' 1992, *NMR Biomed.*, **5**, 368–372.

13. M. L. Anthony, B. C. Sweatman, C. R. Beddell *et al.*, 'Pattern recognition classification of the site of nephrotoxicity based on metabolic data derived from proton nuclear magnetic resonance spectra of urine.' *Mol. Pharmacol.*, 1994, **46**, 199–211.

14. E. Holmes, J. K. Nicholson, A. W. Nicholls *et al.*, 'The identification of novel biomarkers of renal toxicity using automatic data reduction techniques and PCA of proton NMR spectra of urine.' *Chemom. Intell. Lab. Syst.*, 1998, **44**(1,2), 245–255.

15. K. Matsushita, K. Yoshikawa and A. Ohsaka, '1H NMR of human urine: An application of NMR spectroscopy in clinical pathology.' 1982, *JEOL News (Analytical Instrumentation)*, 1982, **18A**(2), 2–4.

16. N. J. Waters, S. Garrod, R. D. Farrant *et al.*, 'High-resolution magic angle spinning 1H NMR spectroscopy of intact liver and kidney: Optimization of sample preparation procedures and biochemical stability of tissue during spectral acquisition.' *Anal. Biochem.*, 2000, **282**(1), 16–23.

17. A. Abragam, 'Principles of Nuclear Magnetism,' Oxford University Press, 1961.

18. R. Freeman, 'A Handbook of Nuclear Magnetic Resonance,' Longman Scientific and Technical/John Wiley and Sons, 1988.

19. D. J. Russell, C. E. Hadden, G. E. Martin *et al.*, 'A comparison of inverse-detected heteronuclear NMR performance: conventional versus cryogenic microprobe performance.' *J. Nat. Prod.*, 2000, **63**(8), 1047–1049.

20. P. J. D. Foxall, J. A. Parkinson, I. H. Sadler *et al.*, 'Analysis of biological fluids using 600 MHz proton NMR spectroscopy: Application of homonuclear two-dimensional J-resolved spectroscopy to urine and blood plasma for spectral simplification and assignment.' *J. Pharm. Biomed. Anal.*, 1993, **11**(1), 21–31.

21. E. Holmes, P. J. D. Foxall and J. K. Nicholson, 'Proton NMR analysis of plasma from renal failure patients: evaluation of sample preparation and spectral-editing methods.' *J. Pharm. Biomed. Anal.*, 1990, **8**(8–12), 955–958.

22. J. C. Lindon, M. Liu and J. K. Nicholson, 'Diffusion coefficient measurement by high resolution NMR spectroscopy: Biochemical and pharmaceutical applications.' *Rev. Anal. Chem.*, 1999, **8**(1–2), 23–66.

23. J. P. Shockcor, S. H. Unger, I. D. Wilson *et al.*, 'Combined HPLC, NMR spectroscopy and ion-trap mass spectrometry with application to the detection and characterization of xenobiotic and endogenous metabolites in human urine.' *Anal. Chem.*, 1996, **68**(24), 4431–4435.

24. U. G. Sidelmann, U. Braumann, M. Hofmann *et al.*, 'Directly coupled 800 MHz HPLC-NMR spectroscopy of urine and its application in the identification of the major phase II metabolites of tolfenamic acid.' *Anal. Chem.*, 1997, **69**(4), 607–612.

25. W. El-Deredy, 'Pattern recognition approaches in biomedical and clinical NMR spectroscopy: A review.' *NMR in Biomed.*, 1997, **10**, 99–124.

26. A. W. Nicholls, J. K. Nicholson, J. N. Haselden *et al.*, 'A metabonomic approach to the investigation of drug-induced phospholipidosis: An NMR spectroscopy and pattern recognition study.' *Biomarkers*, 2000, **5**, 410–423.

27. J. L. Griffin, L. A. Walker, J. Troke *et al.*, The initial pathogenesis of cadmium-induced renal toxicity.' *FEBS Lett.*, 2000, **478**(1,2), 147–150.

28. M. A Warne, E. M. Lenz, D. Osborn *et al.*, 'An NMR-based metabonomic investigation of the toxic effects of 3-trifluoromethyl-aniline on the earthworm eisenia veneta.' *Biomarkers*, 2000, **5**(1), 56–72.

29. E. Holmes, A. W. Nicholls, J. C. Lindon *et al.*, 'Chemometric models for toxicity classification based on NMR spectra of biofluids.' *Chem. Res. Toxicol.*, 2000, **13**(6), 471–478.

30. S. Kruse, O. M. Kvalheim, G. Gadeholt *et al.*, 'Multivariate analysis of proton NMR spectra of serum from rabbits. Monitoring progressive growth of implanted VX-2 carcinoma.' *Chemom. Intell. Lab. Syst.*, 1991, **11**(2), 191–6.

31. E. Holmes, J. K. Nicholson and G. Tranter, 'Metabonomic characterization of genetic variations in toxicological and metabolic responses using probabilistic neural networks.' *Chem. Res. Toxicol.*, 2001, **14**(2), 182–191.

32. K. R. Beebe, R. J. Pell and M. B. Seasholtz, 'Chemometrics: A Practical Guide,' 1998, John Wiley and Sons (Wiley-Interscience).

33. B. C. M. Potts, A. J. Deese, G. J. Stevens *et al.*, 'NMR of biofluids and pattern recognition: Assessing the impact of NMR parameters on the principal component analysis of urine from rat and mouse.' *J. Pharmaceutical Biomed. Anal.*, 2001, in press.

34. M. W. McGuill and A. N. Rowan, 'Biological effects of blood loss: Implications for sampling volumes and techniques.' *ILAR News*, 1989, **4**, 5–20.

35. K. Diehl, R. Hull, D. Morton *et al.*, 'A good practice

guide to the administration of substances and removal of blood including routes and volumes.' *J. Appl. Toxicol.*, 2001, **21**, 15–23.

36. D. G. Robertson, M. D. Reily, M. Albassam *et al.*, 'Metabonomic assessment of vasculitis in rats.' *Cardiovasc. Toxicol.*, 2001 in Press.

37. C. L. Gavaghan, E. Holmes, E. Lenz *et al.*, 'An NMR-based metabonomic approach to investigate the biochemical consequences of genetic strain differences: Application to the C57BL10J and Alpk:ApfCD mouse.' *FEBS Let.*, 2000, **484**, 169–174.

38. L. Le Moyec, A. Pruna, M. Eugene *et al.*, 'Proton nuclear magnetic resonance spectroscopy of urine and plasma in renal transplantation.' *Nephron.*, 1993, **65**, 433–439.

39. R. A. Branch, R. Cotham, R. Johnson *et al.*, 'Periportal localization of lorazepam glucuronidation in the isolated perfused rat liver.' *J. Lab. Clin. Med.*, 1983, **102**, 805–812.

40. J. Wojcicki, A. Hinek and L. Samochoweic, 'The protective effect of pollen extracts against allyl alcohol damage.' *Arch. Imunol. Ther. Exp. Warsz.*, 1985, **33**, 841–849.

41. C. Fieschi, M. Rasura, A. Anzini *et al.*, 'Central nervous system vasculitis.' *J. Neurol. Sci.*, 1998, **153**, 159–171.

42. J. L. Larson, M. V. Pino, L. E. Geiger *et al.*, 'The toxicity of repeated exposures to rolipram, a type IV phosphodiesterase inhibitor, in rats.' *Pharmacol. Toxicol.*, 1996, **78**, 44–49.

43. A. Nyska, R. A. Herberts, P. C. Chan *et al.*, 'Theophylline-induced mesenteric periarteritis in F344/N rats.' *Arch. Toxicol.*, 1998, **72**, 731–737.

44. L. A. Dethloff, D. G. Pegg and A. L. Metz, 'Preclinical toxicity studies of the phosphodiesterase IV inhibitor CI-1018 in rats.' *Toxicol. Sci.*, 1999, **48**(1S), 1502a.

45. S. Meiboom and D. Gill, 'Modified spin-echo method for measuring nuclear relaxation times.' *Rev. Sci. Instrum.*, 1958, **29**, 688–691.

46. S. H. Smallcombe, S. L Patt and P. A. Keifer, 'WET solvent suppression and its applications to LC NMR and high-resolution NMR spectroscopy.' *J. Magn. Reson., Ser. A*, 1995, **117**(2), 295–303.

47. T. L. Hwang and A. J. J. Shaka, 'Water supression that works. Excitation sculpting using arbitrary waveforms and pulsed field gradients.' *Magn. Reson. Ser.*, 1995, **112**, 275–279.

48. A. Bax and D. Davis, 'MLEV-17-based two-dimensional homonuclear magnetization transfer spectroscopy.' *J. Magn. Reson.*, 1985, **65**, 355–360.

49. M. Liu, J. K. Nicholson, J. A. Parkinson *et al.*, 'Measurement of biomolecular diffusion coefficients in blood plasma using two-dimensional $^1$H-$^1$H diffusion-edited total-correlation NMR spectroscopy.' *Anal. Chem.*, 1997, **69**(8), 1504–1509.

50. C. S. Johnson Jr., 'Diffusion ordered nuclear magnetic resonance spectroscopy: Principles and applications.' *Prog. Nucl. Magn. Reson. Spectrosc.*, 1999, **34**(3,4), 203–256.

51. A. Bax, R. H. Griffey and B. L. Hawkins, 'Correlation of proton and nitrogen-15 chemical shifts by multiple quantum NMR.' *J. Magn. Reson.*, 1983, **55**, 301–315.

52. A. Bax and M. F. Summers, 'Proton and carbon-13 assignments from sensitivity-enhanced detection of heteronuclear multiple-bond connectivity by 2D multiple quantum NMR.' *J. Am. Chem. Soc.*, 1986, **108**(8), 2093–2094.

J.P. Vanden Heuvel, G.H. Perdew, W.B. Mattes and W.F. Greenlee (Eds.)
Comprehensive Toxicology, Vol. xiv

# 14.5.8
# Data Mining in Toxicology

CHRISTOPH HELMA

*University of Freiburg, Freiburg, Germany*

## 14.5.8.1  INTRODUCTION

The present situation in biological and medical sciences is characterized by the availability of massive amounts of experimental data. This has two main reasons, which are both related to the developments in computer hardware and software. First, it is possible to generate massive amounts of data with the help of modern automated laboratory techniques. Second, it becomes cheaper and easier to store and retrieve this data.* In most cases these databases are used to retrieve experimental data in a convenient way and to circumvent the consultation of the original literature. But these databases provide also the empirical

---

A collection of links to toxicity databases on the Internet can be found at http://helma.informatik.uni-freiburg.de/db links/.

evidence, which may be the basis for the discovery of new scientific theories.

The derivation of theories from experimental data is traditionally a process depending on the experience, skill and intuition of the individual researcher. But even the best expert works under limitations: Humans cannot handle large amounts of data, have troubles detecting complex relationships, cannot keep working for a long time and tend to make errors in an unpredictable way, *etc*. Therefore many efforts have been made to use computational methods in this process. In Artificial Intelligence research Data Mining techniques were developed for the identification of useful and understandable patterns in databases.

This review will focus on the application of Knowledge Discovery in Databases (KDD), Data Mining (DM) and Machine Learning (ML) in toxicology. We will discuss primarily methods which are capable of providing new insights and theories. Methods which work well for predictive purposes, but fail to provide models which are interpretable in terms of toxicological knowledge (e.g., many connectionist and multivariate approaches), will not be covered here. As the author works primarily on the application of symbolic Machine Learning to Structure Activity Relationships (SARs) and Predictive Toxicology, there might be a bias towards these areas, although I have tried to report other applications as objectively as possible. The following chapter consists of two main sections: The first part will be an introduction to Data Mining, to introduce the terminology, describe the basic techniques and summarize present implementations of Data Mining algorithms. The second section will present a survey about current applications of Data Mining in Toxicology.

### 14.5.8.2   DATA MINING TECHNIQUES

This section shall provide a nontechnical introduction to Data Mining. The book "Data Mining"[1] by I. H. Witten and E. Frank provides an excellent introduction into this area and it is well readable also for non-computer scientists. A recent review[2] covers Data Mining applications in toxicology. Another recommended reading is "Advances in Knowledge Discovery and Data Mining" by U. M. Fayyad *et al.*[3]

#### 14.5.8.2.1   Concepts and Definitions

First we will have to clarify the meaning of Data Mining (DM) and its relation to other terms frequently used in this area, namely Knowledge Discovery in Databases (KDD) and Machine Learning (ML). Common definitions[1,4–7] in the Data Mining Community are:

- *Knowledge Discovery* is the non-trivial process of identifying valid, novel, potentially useful, and ultimately understandable patterns in data.
- *Data Mining* is a step in the Knowledge Discovery process that consists of the application of statistics, machine learning and database techniques to the data.
- *Machine Learning* is the study of computer algorithms that improve automatically through experience.

This means, that Knowledge Discovery is the process of supporting humans in their enterprise to make sense of massive amounts of data, Data Mining is the application of techniques to achieve this goal and Machine Learning is one of the Data Mining techniques suitable for this task. Other Data Mining techniques originate from diverse fields, such as statistics, visualization, and database research. The focus in this article will be primarily on Data Mining techniques based on Machine Learning. In practice many of these terms are not used in their strict sense. In this article we will also use sometimes the popular term Data Mining, when we mean Knowledge Discovery in Databases or Machine Learning.

#### 14.5.8.2.2   The Knowledge Discovery Process

We will illustrate the Knowledge Discovery process with an hypothetical example, the extraction of Structure Activity Relationships (SARs) from experimental data. The whole project will consist of the following steps:

(1) Definition of the goal of the project and the purpose of the SAR models (e.g., predictions for untested compounds, scientific insight into toxicological mechanisms).

(2) Creation or selection of the dataset (e.g., by performing experiments, downloading data).

(3) Checking the dataset for mistakes and inconsistencies and perform corrections.

(4) Selection of the features that are relevant to the project and transformation of the data into a format, which is readable by Data Mining programs.

(5) Selection of the Data Mining technique.

(6) Exploratory application and optimization of the Data Mining tools to see if they provide useful results.

(7) Application of the selected and optimized Data Mining technique to the dataset.

(8) Interpretation of the derived model and evaluation of its performance.

(9) Application of the derived model, e.g., to predict the activity of untested compounds.

These tasks closely resemble the nine steps of the Knowledge Discovery process as described by Fayyad[6] (Table 1).

The typical Knowledge Discovery setting involves several iterations over these steps. Human intervention is an essential component of the KDD process. Although most research has been focused on the Data Mining step of the process (and the present article will not make an exception), the other steps are at least equally important. In practical applications, the data cleaning and preprocessing step is by far the most laborious and time consuming task in the Knowledge Discovery process (and therefore often neglected).

In the following paragraphs, we will present a few general criteria that are useful for describing and choosing Data Mining systems on a general level. The additional benefit of this description will be the explanation of some basic concepts in Data Mining. First we will discuss the structure of the data, which can be used by Data Mining programs. Then we will differentiate between predictive and descriptive methods and classification versus regression problems. After discussing the model types frequently used as output from Data Mining programs, we will present the basic search strategy of Data Mining algorithms.

### 14.5.8.2.3 Data Representation

Before feeding the data into a Data Mining program we have to transform it into a computer understandable form. From a computer scientists point of view there are two basic data representations relevant to Data Mining, both will be illustrated with examples: Table 2 shows a table with physico-chemical properties of chemical compounds. For every compound there are a fixed number of parameters or features available, therefore it is

possible to represent the data in a single table. In this table, each row represents an example and each column an attribute. We call this type of representation *propositional*.

Let us assume we want to represent chemical structures by identifying atoms and the connections (bonds) between them. It is obvious, that this type of data does not fit into a single table, because each compound may have a different number of atoms and bonds. Instead we may write down the atoms and the relations (bonds) between them as in Figure 1. This is called a *relational* representation. Other biologically relevant structures (e.g., genes, proteins) may be represented in a similar manner.

Many Data Mining techniques can use only propositional data. This is a pity because toxicology deals with complex structures and relations. There are, however, techniques that allow the transformation of relational to propositional data, a process which is called *propositionalization*.[8] In the case of chemical structures we may, for example, count the occurrences of certain structural features and write the results into a table. There are, of course, more sophisticated procedures for propositionalizations, some examples can be found in Section 14.5.8.3.1.3.

### 15.5.8.2.4 Data Mining Tasks

When preparing a dataset for Data Mining the learning task is in most cases very obvious. In the SAR case, for example, we want to learn to predict toxicity from chemical structures. If we know the learning task and tell it the program we have an example for *predictive* Data Mining or *supervised* Machine Learning, because the learning task is defined by the user. The algorithm has to use the given examples to learn a model that predicts unseen examples.

Sometimes, however, we do not know what to predict. This is often the case when analyzing the huge amount of data generated by gene expression experiments. Here, the primary aim is to understand the data and to

**Table 1** The Knowledge Discovery process according to Fayyad.[6]

1. Definition of the goals of the KDD process
2. Creation or selection of a data set
3. Data cleaning and preprocessing
4. Data reduction and projection
5. Selecting Data Mining methods
6. Exploratory analysis and model/hypothesis selection
7. Data Mining
8. Interpretation/evaluation
9. Utilization

**Table 2** Example of a propositional representation of chemical compounds using molecular properties.

| CAS | logP | HOMO | LUMO | ... | MUTAGEN |
|---|---|---|---|---|---|
| 100–01–6 | 1.47 | −9.42622 | −1.01020 | ... | 1 |
| 100–40–3 | 3.73 | −9.62028 | 1.08193 | ... | 0 |
| 100–41–4 | 3.03 | −9.51833 | 0.37790 | ... | 0 |
| ... | ... | ... | ... | ... | ... |
| 99–59–2 | 1.55 | −9.01864 | −0.98169 | ... | 1 |
| 999–81–5 | −1.44 | −9.11503 | −4.57044 | ... | 0 |

detect regularities (e.g., coregulated genes), without a reference to a certain endpoint. In this case we have to use *descriptive* Data Mining or *unsupervised* learning. The various clustering techniques, that are presently used to analyze gene expression data are one example for an unsupervised Data Mining technique. Several other Data Mining techniques (e.g., association rules) can be used for unsupervised learning.[1]

#### 14.5.8.2.5 Prediction Tasks

In toxicology we can differentiate between two basic types of effects: those with a threshold (the majority of toxicological effects) and those without (e.g., carcinogenicity, mutagenicity). For the last category it is possible to differentiate between active and inactive compounds, whereas for other endpoints a number (e.g., $LD_{50}$) is used to indicate the toxic

```
% Trichloromethane [67-66-3]

atom('67-66-3','67-66-3_1').
element('67-66-3_1',c).

atom('67-66-3','67-66-3_2').
element('67-66-3_2',cl).

atom('67-66-3','67-66-3_3').
element('67-66-3_3',cl).

atom('67-66-3','67-66-3_4').
element('67-66-3_4',cl).

atom('67-66-3','67-66-3_5').
element('67-66-3_5',h).

bond('67-66-3','67-66-3_1_2').
connected('67-66-3_1','67-66-3_2','67-66-3_1_2').
bond_type('67-66-3_1_2',single).

bond('67-66-3','67-66-3_1_3').
connected('67-66-3_1','67-66-3_3','67-66-3_1_3').
bond_type('67-66-3_1_3',single).

bond('67-66-3','67-66-3_1_4').
connected('67-66-3_1','67-66-3_4','67-66-3_1_4').
bond_type('67-66-3_1_4',single).

bond('67-66-3','67-66-3_1_5').
connected('67-66-3_1','67-66-3_5','67-66-3_1_5').
bond_type('67-66-3_1_5',single).
```

**Figure 1** Example of a relational representation of Trichloromethane. This molecular representation is written in Prolog, a programming language frequently used in Data Mining. Traditional computational chemical formats use connection tables, which are just another form to represent the same data.

potency. If it is sufficient to distinguish between different classes of compounds (e.g., active and inactive) we have, in computer science terms, a *classification* problem. If we want to predict numerical values we need to use *regression* techniques. Most Data Mining techniques focus on classification. Again it is possible to transform a regression problem into a classification problem. We use categories of activities (e.g., low, medium, high) instead of the activities themselves. Although this looks very much like a workaround, discretization may even improve performance, especially when dealing with noisy biological data.

### 14.5.8.2.6 Model Types

What kind of output can we obtain from a Data Mining tool? I will present here two of the most popular (and in my opinion well understandable) types of models: *rules* and *trees*. Figure 2 depicts an example set of rules derived from PART, a rule learner implemented in the WEKA workbench.[1] It is quite obvious how to read it: If a compound fulfills the criteria of the first rule it is classified as positive, if it fulfills the criteria of the second rule it is classified as negative. Both rules test for the presence or absence of molecular substructures (see Section 3.1.3, '~' indicates aromatic bonds). The numbers in brackets summarize the correctly/incorrectly-classified compounds in the training set.

Figure 3 shows an example of a *decision tree* generated by C4.5,[9] one of most popular Data Mining algorithms. It may require a little more explanation, but is equally well understandable. If a compound contains a N=O fragment (usually a nitro group) it is classified as active. Using this method, 121 compounds were

```
c=c-n-o no AND
n=o yes AND
c-c=c-c=c no AND
o-c~c~c~c-n-o no AND
o-c-c-c-o no: 1 (73.0/4.0)

c-c-c-c yes AND
c-c-c-c-c-o-c-c-c no AND
o=c-n-c=o no AND
c-c~c~c~c-n no AND
c-c~c~c~c-c no: 0 (76.0/4.0)
```

**Figure 2**  Two exemplary rules extracted from a PART decision list.

classified correctly, 13 incorrectly. If it does not contain such a fragment, it is checked if the HOMO is higher or lower than −8.6 eV, followed by the ring strain energy and the presence of O−C=O fragments. Both models are simplified; in practice they will be more complex. Of course there are methods available, which can generate more sophisticated models involving for example regression to predict numerical values (Table 3).

### 14.5.8.2.7 Data Mining Algorithms

Finally we may ask how the algorithms generate the models. The details are quite complicated and vary from algorithm to algorithm. A detailed explanation requires an understanding of relational databases, logic and statistics; please refer to the specialized textbooks[1,3] for a detailed treatment. I will explain the basic procedure with a simplified example using the decision tree example of Figure 3. Let's assume, we have a data set with active and inactive compounds together with a set of descriptors, in our case molecular substructures and physico-chemical properties. The first task is to identify criteria that provide the best discrimination between active and inactive compounds. There are a lot of possibilities and trying them all would be hard and boring for humans. Fortunately computers are very good at such a work. So, the program finds, that checking the presence or absence of a N=O substructure provides the best discrimination. This test will be the first node of the tree.

This criteria works quite well for positive classifications, because 121 compounds are classified correctly and only 13 incorrectly. But for the negative examples, the situation is not satisfying. There are still too many positive examples within the negatives. Therefore the algorithm will start again looking for another test. In our case it comes up asking, if the HOMO is higher or lower than −8.6, and this will be the next node in the tree. This procedure will be repeated until (almost) all examples are correctly classified.

In practice the search strategies and capabilities vary considerably between the individual algorithms and we will present a few of them in the next section. Table 3 summarizes the Data Mining algorithms, which have been applied in toxicology so far. An overview about software packages containing implementations of these algorithms can be found at: http://www.kdnuggets.com/.

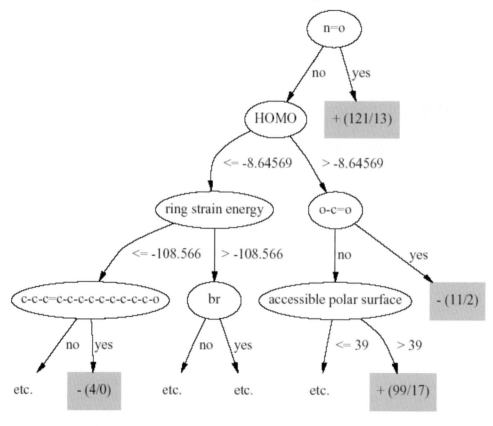

**Figure 3** Example of a C4.5 decision tree.

### 14.5.8.3 APPLICATIONS OF DATA MINING IN TOXICOLOGY

Presently the main application area of Data Mining techniques in toxicology is the detection of *Structure Activity Relationships* (SARs), but the basic principles of Knowledge Discovery and Data Mining are applicable also to other areas (e.g., toxicogenomics). SARs are based on the assumption, that biological effects are determined by the chemical structure of a compound (Figure 4). The development of SAR models has a long and successful history in medicinal chemistry, but the application of SAR techniques to toxicological problems has started much more recently. Although the objectives, the prediction of biological activities from chemical structures, seem to be

**Table 3** Data Mining systems that have been applied in toxicology.

| System | Prediction Task | Data Representation | Model Type |
|---|---|---|---|
| Golem[75] | classification | relational | rules |
| Progol[77] | classification | relational | rules |
| ICL[78] | classification | relational | rules |
| RL[70,79] | classification | propositional | rules |
| C4.5[9] | classification | propositional | trees |
| Tilde[80] | classification | relational | trees |
| M5[81] | regression | propositional | trees |
| CART[36] | classification, regression | propositional | trees |
| S-CART[37,38] | classification, regression | relational | trees |
| MONKEI[82] | classification | propositional | rules |
| CASE[25] | classification, regression | relational, propositional | fragments lin. regression |
| MULTICASE[26] | classification, regression | relational, propositional | fragments lin. regression |
| Warmr[83] | descriptive | relational | DATALOG queries |
| bottom-up propositionalization[27] | descriptive | relational | fragments |
| RDM-Fragments[8] | descriptive | relational | fragments |

Chemical Structure            SAR-Model            Toxic Effect

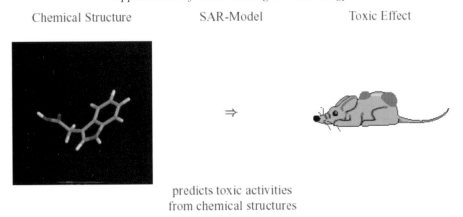

predicts toxic activities
from chemical structures

**Figure 4**  The concept of Structure-Activity Relationships (SARs).

closely related, there are substantial differences between medical chemistry and toxicology.

SAR techniques in pharmaceutical research rely on the hypothesis, that all molecules contain the basic functionality required for activity (i.e., they are *congeneric*), but vary in a secondary feature responsible for variations of their potency. The therapeutic agent (e.g., receptor, active site of an enzyme) is usually well known and some knowledge about the biological mechanisms is also available. For the majority of toxic effects there is, on the other hand, no unique cellular target. As a consequence, chemicals with very different structures (*non-congenerics*) may cause the same toxic effect. Toxicity assays are in many cases very time consuming and expensive. Therefore it is not possible to perform toxicity tests especially for SAR studies, and SAR models have to use already available databases.

Traditional SAR approaches[10] are limited to *congeneric* series of compounds and therefore not applicable to many toxicological problems. Recent applications of SARs in toxicology are either knowledge-based or rely on some kind of data-mining technique.[11–13]

This section will start with the discussion of methodological issues which apply to all Data Mining systems. The application part will deal to a large extend with CASE/MultiCASE, a program developed especially for predictive toxicology and the application of symbolic Machine Learning for the detection of SARs. The last part will cover various other applications of Data Mining techniques in toxicology.

### 14.5.8.3.1  Methodological Considerations

Several methodological issues are relevant to all applications of Data Mining in toxicology. The topics presented in this section refer in most cases to steps 2, 3 and 8 in the KDD process as presented in Section 14.5.8.2.2.

#### 14.5.8.3.1.1  Data Selection

It was already mentioned, that the toxicological data miner has to rely in most cases on experimental results, which were generated for other purposes than building SAR models. A researcher in this area has therefore little flexibility in choosing datasets; usually he/she has to use what is already available. Nevertheless there are several issues concerning size and composition of the dataset, which should be considered when applying Data Mining to toxicological databases.

*Size of the Learning Set.* Most publicly available toxicological databases are, in Data Mining terms, rather small and contain typically a few hundred compounds. It is therefore hard to estimate the optimal size of a learning set, because too few examples are available for systematic experiments. The only public dataset which is large enough for such investigations are results from the *Salmonella* mutagenicity assay (Ames test). Using multiple sampling and model building,[14] found that the optimal predictivity per unit cost was obtained from databases with approximately 350 compounds. Larger databases improved the predictivity of the derived models, but not very dramatically.

But the predictivity depends not only on the size of the learning data, but also on its composition. If the goal is to make predictions for the "chemical universe", the learning set should be as diverse as possible. In CASE/MultiCASE (Section 14.5.8.3.2) the program issues a warning, when it has to predict the activity of a compound containing a substructure which was not present in the learning set.

This makes it possible to estimate the general suitability of a SAR model, to make predictions for a large number of compounds and record the warning messages. Several researchers used this approach by making predictions for 5000 representative compounds.[14–17] The main result was that the information content (related to the number of predictions without warning) of a model is critically related to its predictivity,[2,14] which in turn, is determined mainly by the size of the learning set,[15] especially if the size of the data base is smaller than 350 compounds. The estimation of predictivity will be discussed in Section 14.5.8.3.1.5.

Another important finding in this area are the results from Reference 18. They combined the dataset from the National Toxicology Program (NTP), containing predominantly industrial chemicals, with data from the Food and Drug Administration (FDA), containing mainly pharmaceuticals, which led to a significant improvement of the resulting model. When combining datasets from different sources, it is crucial, that both results were determined by similar protocols and evaluated by identical criteria. Adding results from the general literature to the NTP carcinogenicity data, does not lead to improvements, because the experimental protocols differ too much.[19]

*Ratio of Active to Inactive Compounds.* As toxicological data sets were not generated to meet the needs of an SAR models, they are frequently skewed towards active chemicals. This may affect the performance of SAR models, because many Data Mining systems work best with equal frequencies of positive and negative examples. Therefore databases with predominantly active compounds can be supplemented with physiological chemicals, which are assumed to lack toxic activity (e.g., sugars, amino acids, lipids) to obtain a 1:1 distribution.[20–23]

### 14.5.8.3.1.2   Quality of Datasets

Many SAR modelers take the validity of chemical structures and experimental results for granted and seldom perform a quality check of their data. Checking the integrity of the learning set is the most laborious task in Knowledge Discovery, because automated routines are seldom available and a lot of human interventions and expert knowledge is required.

*Chemical Structures.* In many cases toxicological databases will contain chemical structures, otherwise it is necessary to download them from the Internet or enter them manually. The first task is to remove mixtures and compounds with undefined structures (e.g., polymers). Then the validity of the remaining structures have to be checked. Formal errors may be detected, by applying various computational chemistry programs to the structures and record error messages. But the final check still has to be performed by a chemist. Our experience shows, that in real datasets a high proportion of structures have to be edited, e.g., 142 from 444 structures for the preparation of the dataset for the Predictive Toxicology Challenge (http://www.informatik.uni-freiburg.de/~ml/ptc/, see Section 14.5.8.3.3.1). The generation of "correct" structures is, however, impossible in all cases, because conformations depend on environmental factors (e.g., pH). It is therefore more important to find a *consistent* representation for all compounds, which give sensible results, when chemical properties are calculated. Several strategies to achieve this goal are discussed in Reference 24.

*Toxicological Endpoints.* Quality assurance for toxicological data involves similar tasks as for chemical structures. Here it is necessary to look at the protocols behind the actual results, to detect, e.g., unusual administration routes, strains or experimental protocols. Again, this task cannot be automated at present and requires a toxicological expert.

Another point, which has to be considered in this context, is the validity and reproducibility of the experimental data itself. Unfortunately the reproducibility of the majority of toxicity assays is unknown and hard to assess, because too few replicate experiments are available. Our group for example has tried to assess the reproducibility of rodent carcinogenicity assays, by comparing results from the (standardized) National Toxicology Program with (nonstandardized) results from the general literature.[19] The main result was, that both datasets had less than 60% concordance for carcinogen/non-carcinogen classifications. It is presently impossible to decide, if this value represents biological variability, or if it is caused by different protocols (although taking in account details of the protocols did not improve the concordance significantly). Such investigations are not only necessary to evaluate the validity of toxicity data, but also to provide a benchmark for the derived SAR model, because Data Mining cannot provide accurate models from inaccurate data.

Poor reproducibility of certain endpoints may also be an explanation for a common observation within the SAR community: That it is harder to predict complex toxicological endpoints (e.g., carcinogenicity) than those involving comparably simple mechanisms (e.g., mutagenicity), even if both datasets are reduced to the same size.[20,23] My hypothesis is, that complex mechanisms lead to higher biological variability and thus to noisier data. This has the consequence, that SAR models derived from such datasets are inevitably inaccurate, not because they cannot account for the complexity of mechanisms, but because they are based on inaccurate data.

### 14.5.8.3.1.3 Preprocessing

Both, chemical and toxicological data, require preprocessing before the application of Data Mining. On the chemical side it will be often necessary to calculate properties at the molecular (e.g., reactivity indices, lipophilicity) or atomic (e.g., electronegativity) level. The choice of descriptors to be calculated is still a trial and error process, especially for complex endpoints without *a priori* knowledge of detailed biochemical mechanisms. It will also depend on the capability of the Data Mining system to be used, e.g., if it can use relational data or not. I do not want to go into details of computational chemistry, but want to stress, that one should become familiar with the demands of the applied programs (e.g., regarding the presence of disconnected structures, representation of nitro groups) and prepare the dataset accordingly.

A very useful concept, especially for *non-congeneric* datasets, is to use structural fragments as descriptors for a molecule. The idea is to decompose every compound in the dataset into all possible linear substructures (neglecting hydrogen) up to a certain length or frequency. Figure 5 shows the fragments of Trichloroethene as an example. These fragments may be used as descriptors for chemical structures and Data Mining algorithms have to find the fragments, that are present in active compounds, but not in inactive ones (and vice versa). This approach was first used in the CASE/MultiCASE programs.[25–26] Recently it was reformulated in the context of Inductive Logic Programming and Relational Data Mining.[27–30] In this framework fragment generation is already a Data Mining tool, because only frequent fragments and those discriminating between active and inactive compounds are generated. It is also tool for the propositionalization of relational data, because the presence and absence of each fragment can be represented in a single table. Kramer and De Raedt[30] were able to gain significant improvements in various toxicity domains using fragment generation in conjunction with symbolic Machine Learning.

Data preprocessing is also necessary for toxicological data. The original data in carcinogenicity experiments, for example, are histopathological observations. Using statistical criteria they are condensed to positive and negative responses in each organ, which are further summarized to sex and species specific responses. These may be further abstracted to obtain a classification for "rodent carcinogenicity". The Data Miner has to decide at which level of abstraction he/she wants to build the model. In the carcinogenicity case it is sensible to generate separate models for each sex/species group[18] to account for the different susceptibilities and to combine these predictions in a later step, if necessary. Unfortunately there are presently not enough data available to develop models for organ specific effects (maybe with exception of the liver).[19]

### 14.5.8.3.1.4 Data Representation

*Chemical Structures and Properties.* In computational chemistry, chemical structures are represented either in line notations (e.g., SMILES)[31] or connection tables (e.g., MDL Molfiles).[32] Both representations can be easily transformed into a format suitable for Machine Learning systems (e.g., Prolog or Relational Databases, Figure 1). Additional properties may be added at the molecular level (e.g., reactivity indices, lipophilicity) or for each atom or bond (e.g., charges, bond orders). It is even possible to represent compounds as a connection of functional groups[33] or to represent the 3 dimensional structure completely in Prolog or Relational Databases.[34] A representation in Relational Database formalisms allows experimentation with various sets of descriptors. It is possible to use different sets of descriptors depending on the scope of the study (e.g., functional groups for computer aided

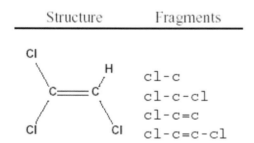

| Structure | Fragments |
| --- | --- |
| | cl-c |
| | cl-c-cl |
| | cl-c=c |
| | cl-c=c-cl |

**Figure 5**  Structural fragments for Trichloroethene.

drug design, pharmacophores for mechanistic studies, connectivity indices for the mass screening of combinatorial chemistry databases). Although the possibilities of a representation of chemical structures as logic programs sound promising, they have been rarely applied in toxicology studies. The most common representation of compounds uses a fairly primitive atom-bond representation, similar to Table 1.

Only recently several attempts have been made to formulate concepts used in other computational chemistry applications as logic programs. Dehaspe *et al.*[35] tried to find frequent substructures in chemical compounds using the WARMR algorithm. Pfahringer *et al.*[33] intended to develop an abstract representation of chemical structures based on functional groups, while Finn *et al.*[34] developed a pharmacophore description based on 3D geometries, which may be also useful in toxicity applications. Another very promising approach is fragment generation, which was covered in the previous section. The representation of chemical structures and properties as logic program provides powerful capabilities, and we hope that this area will attract more interest in the near future.

*Toxic Properties and Background Knowledge.* In a SAR study, toxic properties are the variables to be predicted. The degree of intended accuracy may vary considerably. Sometimes it is sufficient to classify compounds, e.g., to distinguish between mutagenic and non-mutagenic compounds. In most cases quantitative values (e.g., lethal concentrations $LC_{50}$) are of interest and sometimes the prediction of specific effects (e.g., target organ specific toxicity) is required. Most Machine Learning programs can deal with classifications, only a few (M5, CART, S-CART[9,36–39]) are able to learn quantitative values. One workaround for this shortcoming is to classify the examples according to their activity (e.g., strong/medium/weak). This is in many cases legitimate because quantitative experimental toxic activities are often very unreliable and a prediction of numerical values would provide a false impression of accuracy. Another solution was proposed by King and Srinivasan:[40] They used *Inductive Logic Programming* (ILP) to discover indicator variables (attributes), which were subsequently used in classical Quantitative Structure Activity Relationship analysis with linear regression techniques.

The ability of symbolic Machine Learning systems to use background knowledge is very advantageous in the representation of toxic endpoints. It is for example possible to use databases with very specific effects (e.g., results individual mutagenicity assays), provide abstractions (e.g., for bacterial mutagenicity, mammalian mutagenicity) and learn models at different levels of abstractions. It is also possible to provide current theoretical concepts (e.g., structural alerts) as background knowledge and see, if the learning algorithm uses them, or not. Despite the promising prospects, this approach has been rarely used in the toxicity domain. The only example, we are aware of, is the work by King and Srinivasan,[41] who used structural alerts and rules for mutagenicity to predict carcinogenicity.

### 14.5.8.3.1.5 Validation of Data Mining Results

Many SAR studies still validate their results, by applying their model to the learning set and report the percentage of correctly classified instances. This is not appropriate, because the examples have been already seen by the model and a high concordance may indicate an overfitting to the data. It is therefore necessary to use a separate test set, to estimate the real performance. In most cases, toxicological databases contain too few examples, to separate them into a learning and a test set. It is therefore necessary to use *cross-validation*. This means, that the complete dataset is partitioned into *n* subsets (folds) of approximately the same size. One fold is removed as test set from the learning set and the remaining examples are used to learn a model. Using this model, predictions are made for the test set. As the test set was not contained in the learning set, these predictions give a better estimation of the model's capabilities. The whole procedure is repeated *n* times, until predictions for all examples are available. These predictions are combined to estimate the overall performance of the SAR model.[42]

In practical applications 10-fold cross-validation is most frequently used. Smaller datasets require a Leave-One-Out procedure, where one example is removed from the learning set. The model is trained with the remaining compounds to obtain a prediction for the removed compound. This procedure is repeated until all compounds have been predicted.

### 14.5.8.3.1.6 Combination of Models

It is often desirable to combine multiple models to obtain a better overall predictivity. First steps into this direction were started approximately 15 years ago with the *Carcino-*

*genicity Prediction and Battery Selection Approach* (CPBS)[43–44] based on Bayes' Theorem. Similar approaches were used for

- combining qualitative and quantitative CASE/MultiCASE predictions
- combining species and sex specific models
- combining SAR models based on different experimental databases
- combining SAR models with results from short term assays.

The application of these techniques resulted in significant improvements of the predictive performance. It is for example advantageous to learn separate models for the carcinogenicity in male/female rats/mice and to combine the results for a prediction of rodent carcinogenicity.[18]

### 14.5.8.3.2   CASE/MultiCASE

With more than 150 published references, the CASE/Multi-CASE system is the most extensively used predictive toxicology system. It was designed to deal especially with *non-congeneric* databases and situations without *a priori* knowledge of biochemical mechanisms. The objective is to find structural entities discriminating active molecules from inactives and to enable the (re)discovery of structural features responsible for toxic action. CASE was developed by G. Klopman[25] and later expanded to MultiCASE.[26,45] A separate program (TOX II)[46] uses the knowledge from CASE/MultiCASE to make predictions for molecules without experimental data. Recent reviews cover the methodology[23] and the application[47] of CASE/MultiCASE.

#### 14.5.8.3.2.1   CASE

The development of a CASE[25] model starts with a training set, containing chemical structures and biological activities. Each molecule is split into all possible linear subunits with 2–10 non-hydrogen atoms (the size of fragment varies in CASE related publications). All fragments belonging to an active molecule are labeled as active, while those belonging to an inactive molecule are labeled as inactive. Most of the fragments are unrelated to the observed activity and the program has to discard them. CASE takes a statistically significant deviation from random distribution as indication that the fragment is relevant for biological activity. Structures contributing to activity are prevalent in active molecules and are called *biophores*. Those preventing activity can be found predominantly in inactive chemicals and are called *biophobes*.

Once the system has been trained with a particular data base it is possible to predict the activity of unknown compounds. The entry of an unknown molecule will lead to the generation of all inherent substructures and these will be compared to the previously identified biophores and biophobes. Based on the presence of these descriptors the probability of activity or inactivity will be predicted. In addition to predicting classifications (qualitative prediction) CASE has the ability to predict the toxic potency based on a *Quantitative Structure Activity Relationship* (QSAR) model (quantitative prediction). This is derived by a classical multivariate statistical analysis based. Biophores, biophobes and calculated values of lipophilicity (logP, logP$^2$) are incorporated within a regression equation in a forward stepwise manner until no significant improvement is observed between actual and calculated values. The coefficient of each of the fragments selected by the regression analysis is a measure of the contribution to activity/inactivity.

#### 14.5.8.3.2.2   MultiCASE

In CASE all relevant fragments are given a certain weight, but no provision is made that correlated fragments do not reinforce each other in the prediction of activity. Furthermore CASE does not distinguish between fragments causing activity and those modulating activity. MultiCASE[26] tries to solve these problems by performing a multiple CASE type analysis.

First MultiCASE identifies only one biophore, the substructure with the highest probability of being responsible for activity. Molecules containing this biophore are removed from the data set and the remaining compounds are submitted to a new analysis. This procedure is repeated until the entire set is eliminated or until the remaining data cannot be explained by statistically significant descriptors. For a new structure the likelihood to exhibit activity is determined by the occurrence of biophores and their probabilities for contributing to activity. A compound without biophores is classified as inactive by default.

For predictions of toxic potencies the molecules of the training set are separated into subclasses according to the presence of each of the biophores. For each of the subclasses a QSAR analysis is performed to identify *modulators* increasing or decreasing the activity of the associated biophore. Modulators can be molecular fragments, but also two-dimensional (2D) distance descriptors, calculated electronic indices (molecular orbital energies, electron densities) and calculated transport parameters (lipophilicity, water solubility). For quantitative predictions of an unknown compound MultiCASE will search for the presence of a biophore. If it finds one, it will look for the presence of modulators associated with that biophore to predict the potency of the compound.

The application of CASE/MultiCASE results in four individual predictions corresponding to CASE/MultiCASE probabilities for activities and CASE/MultiCASE predictions of toxic potencies. As they reflect different aspects of the structural basis for activity, they may be combined in an overall prediction using Bayes theorem. The CASE/MultiCASE models may provide conclusive as well as inconclusive predictions. If the system encounters a fragment which was not contained in the training set, the prediction is labeled with a warning, because the unknown fragment may be an unrecognized biophore, biophobe or modulator.

### 14.5.8.3.2.3 *Applications, Problems and Limitations*

The obvious application of CASE/MultiCASE is the development of SAR models for different endpoints and the prediction of the biological activity of untested compounds. Application areas can be found in research, design of new compounds, priority setting for experimental testing and regulation.[47] As CASE/MultiCASE provides rationales for its predictions, it is also possible to study toxicological mechanisms and to discover new knowledge in toxicological databases. A survey of CASE/MultiCASE applications can be found in References 2 and 47. CASE/MultiCASE is delivered as monolithic commercial product without source code and possibilities for extensions and adaptations. Its applicability is therefore restricted to problems that can be formulated in the original CASE context. It cannot be applied to other toxicology data mining problems, e.g., in environmental or epidemiological databases.

Although the huge number of CASE/MultiCASE related publications is impressive, all of them originate from the same group. The only effort to validate the methodology independently was conducted by an Italian group who successfully reimplemented the CASE system[48–51] and applied it for mechanistic studies in the carcinogenicity domain. For SAR applications the major limitation of CASE/MultiCASE is in our opinion its inflexibility in respect to structural descriptors. CASE/MultiCASE is conceptually limited to two dimensional structures of organic molecules and a few predefined global properties and it cannot use relative positions of biophores and modulators. It is impossible to experiment with other types of of descriptors (e.g., local properties of atoms, 3-dimensional structures, other global parameters). Such a flexibility would be desirable for the optimization for a particular application domain (e.g., high throughput screening, computer aided drug design or mechanistic studies) or a special group of compounds (e.g., anorganics).

Another limitation of CASE and Multi-CASE is the linearity of their regression models, which may not be appropriate for the complex relationships encountered in toxicology.

### 14.5.8.3.3 **Symbolic Machine Learning**

The goal of symbolic Machine Learning systems is to derive understandable and interpretable models from examples and background knowledge and to apply them for predictive and explanatory purposes. Machine Learning is therefore very well suited for learning SARs from toxicity databases. The examples, background knowledge and final descriptions may be formulated as logic programs which makes them relatively easy to interpret for domain experts. The application of symbolic Machine Learning to toxicity databases was pioneered by Bahler and Bristol[52] who used C4.5[9] to predict carcinogenicity from biological and chemical information. Since that time more than 20 applications of Machine Learning algorithms to toxicological problems have been published.

### 14.5.8.3.3.1 *Applications*

The application of symbolic Machine Learning to toxicological problems has started only a few years ago. Therefore the range of endpoints is much narrower than for the CASE/MultiCASE system, which is in use for more than 15 years. A summary of symbolic Machine Learning applications can be found in Reference 2.

Several comparative investigations[53–56] demonstrated that the performance of the individual algorithms is generally comparable.

There are also no major differences to non-explanatory techniques, like artificial neural networks or multivariate techniques. This demonstrates, that it is possible to obtain interpretable models without loosing predictive accuracy. The past Predictive Toxicology Challenges[57–59] have also demonstrated, that Data Mining algorithms can outperform human experts when it comes to predicting toxic effects of compounds with unknown activity.

The explanatory power of symbolic Machine Learning and the potential to detect new knowledge has been claimed by many authors. Researchers such as Bahler and Bristol[52] have rediscovered many expert heuristics (e.g., utility of microbial assays for certain compounds, importance of subchronic toxicity in some organs) in carcinogenicity risk assessment. Also, as outlined in Reference 60, others found structural features associated with the mutagenicity of heterocyclic aromatic amines, which were explainable as hydrophobic, electronic and steric effects. In another experiment, the authors in Reference 41 identified structural alerts for carcinogenicity, many of them were similar to those identified by Ashby[61] with his expert knowledge. Dzerosky *et al.*[56] were also able to rediscover expert knowledge about biodegradation. Sometimes the results are unexpected and counterintuitive. The authors of Reference 62 found that negative results in certain genotoxicity assays (Sister Chromatid Exchanges and Chromosomal Aberrations) are an indication of carcinogenicity, which conflicts with the hypothesis that genetic damage leads to cancer. Such findings require further investigations and may lead to the formulation of new hypotheses.

Despite these success stories the situation is far from ideal. In many cases domain experts were unable to interpret models[56,59] derived from Machine Learning. In our opinion this lack of interpretability does not originate from ML algorithms, but from an inadequate representation of chemical and biological data and background knowledge and from a poor visualization of the resulting models (Section 14.5.8.3.3.2).

### 14.5.8.3.3.2 *Problems and Limitations*

The main problems with the application of symbolic Machine Learning in toxicological Data Mining do not arise from the framework itself, which is very powerful and flexible, but from the concrete application to toxicological problems. The majority of applications were performed by computer scientists, who do not have the domain knowledge to formulate an adequate representation of chemical and toxicological data. Therefore ML applications are focused towards a few "standard" datasets, which are used in a very uncritical way. As chemists and toxicologists lack the ability to handle and improve ML programs appropriately, a close interaction between the disciplines is needed. Computational chemists are required to choose and calculate chemical properties and assist in their representation for learning. Toxicologists should provide background knowledge about molecular mechanisms, assess the quality of biological data, define their requirements for an interpretable SAR model and formulate the learning problem in collaboration with the other experts. Computer scientists should adopt their methods to the peculiarities of the application domain. In our opinion the major demand for research lies in the following areas:

- Quality control of toxicity data
- Representation of chemical structures and properties
- Representation and visualization of SAR models
- Adaptation of Machine Learning techniques to deal with the uncertainty of toxicological data
- Optimization of Machine Learning algorithms for the utilization of structural information

### 14.5.8.3.3.3 *Other Applications in Toxicology*

To the best of our knowledge there are very few applications of Knowledge Discovery in other toxicological areas than the detection of Structure Activity Relationships.

*Activity-Activity Relationships (AAR).* In analogy to Structure-Activity Relationships Activity-Activity Relationships (AARs) may be defined as the derivation of a biological activity from other biological activities. In toxicology this makes sense for predicting important effects, which are hard/expensive to measure (e.g., carcinogenicity) from a series of cheap short term tests (e.g., genotoxicity assays). This type of analysis is also useful in answering scientific questions, e.g., which biological mechanisms are involved in a certain toxic endpoint.

Historically this type of analysis originated in the 1970s, when alternatives to the rodent carcinogenicity assays (e.g., the *Salmonella* mutagenicity assay)[63] were developed. Later it became clear, that the predictivity of this assay is much lower (slightly more than 60%)[64] than originally claimed (approximately 75%). This resulted in the development of new test systems, but they were also unable to obtain a better performance.[65] One solution in this situation was to combine different assays to a battery of short-term tests. In most cases the selection of assays depends on vague mechanistic considerations without empirical evidence. One noteworthy exception is the *Carcinogenicity Prediction and Battery Selection Method* (*CPBS*).[43,44] This is a method for analyzing large, often sparsely filled databases containing short term test results, which have often a marginal representation of non-carcinogens. The analytical tools are based on Bayes' decision analysis, cluster analysis, dynamic programming and multi-objective decision making. The CPBS was applied to the Gene-Tox Database from the US. EPA and the carcinogenicity results from the National Toxicology Program (NTP),[66–68] and many CPBS techniques were applied in investigations using CASE/MultiCASE (Section 14.5.8.3.2).

Other AAR approaches were based on symbolic Machine Learning: Bahler and Bristol[52] applied the C4.5 decision tree learning program[9] to predict rodent carcinogenicity from short term assays, organ specific toxicity and structural alerts. Lee *et al.*[69] applied the RL (Rule Learner) program[70] to predict rodent carcinogenicity from organ specific toxicity.

*Epidemiological and Clinical Studies*. In epidemiological investigations[71] applied C4.5 to study the influence of exposure to air polluted by coal mining and other environmental and social factors on acute respiratory diseases of children in Slovakia. The researchers of Reference 72 used C4.5 to investigate causal relationships between exposure to diesel exhaust and human lung cancer. Similarly, the authors of Reference 73 used RL to investigate geographical and seasonal variations of Datura intoxications.

In clinical studies Reference 74 applied rule induction to discover drug side effects in clinical databases. Spiehler[75] used various AI programs (Expert 4, BEAGLE and Knowledge Maker) to develop rules for the interpretation of morphine-involved cases and to identify the parameters, which are most predictive in estimating the outcome of acute morphine intoxication.

These few examples demonstrate, that the applicability of Knowledge Discovery in Toxicology goes beyond the detection of Structure Activity Relationships. Knowledge Discovery is applicable to all areas where large amounts of empirical data are available and expert knowledge alone is not sufficient to find meaningful relationships. We see potential applications mainly in toxicological epidemiology, the analysis of clinical trials, decision support in clinical toxicology and the analysis of ecotoxicological field data. Another possible application would be to derive toxicologically relevant knowledge from the constantly growing genome and proteome databases.

### 14.5.8.4 CONCLUSION

Up to now Knowledge Discovery in Databases and Data Mining have been applied in Toxicology primarily to identify Structure-Activity Relationships. The most important techniques are CASE/MultiCASE and symbolic Machine Learning. While CASE/Multi-CASE has been used extensively since the 1980s, symbolic Machine Learning algorithms have been applied to toxicological problems approximately 10 years later. Machine Learning techniques are very flexible, because they are not tailored for a specific problem. Therefore they are also applicable for the analysis of other types of data in toxicology (e.g., genetic, epidemiological and environmental data). Although their flexibility is a major advantage over CASE/MultiCASE, improper use of chemical and biological data will prohibit the discovery of useful knowledge. The successful application of symbolic Machine Learning in Toxicology requires the close interaction of Chemists, Toxicologists and Computer Scientists. Further research is required to adapt the algorithms to the specific learning problems in this area, to develop improved representations of chemical and biological data and to enhance the interpretability of the SAR models for toxicologists.

### 14.5.8.5 REFERENCES

1. I. H. Witten and E. Frank, 'Data Mining,' Morgan Kaufmann Publishers, San Francisco, California, 2000.
2. C. Helma, E. Gottmann and S. Kramer, 'Knowledge discovery and data mining in toxicology.' *Stat. Methods Med. Res.*, 2000, **9**, 329–358.
3. U. M. Fayyad, G. Piatesky-Shaprio, P. Smyth *et al.*, 'Advances in Knowledge Discovery and Data Mining,' AAAI/MIT Press, 1996.
4. U. Fayyad, G. Piatetsky-Shapiro and P. Smmyth, 'From Data Mining to Knowledge Discovery: An overview,' in 'Advances in Knowledge Discovery and Data Mining,' eds., U. M. Fayyad, G. Piatesky-Shaprio, P. Smyth *et al.*, AAAI Press, Menlo Park, Calif., 1996, pp. 1–30.
5. U. Fayyad and R. Uthurusamy, 'Data mining and knowledge discovery in databases.' *Commun. ACM*, 1996, **39**, 24–26.
6. U. Fayyad, G. Piatetsky-Shapiro and P. Smmyth, 'The KDD process for extracting useful knowledge from volumes of data.' *Commun. ACM*, 1996, **39**, 27–34.

7. T. M. Mitchell, 'Machine Learning,' The McGraw-Hill Companies, Inc., 1997.

8. S. Kramer, N. Lavrac and P. Flach, 'Propositionalization Approaches to Relational Data Mining,' Springer Verlag, Berlin Heidelberg New York, 2001.

9. J. R. Quinlan, 'C4.5: Programs for Machine Learning,' Morgan Kaufmann, San Mateo, CA, 1993.

10. C. Hansch, D. Hoekman, A. Leo *et al.*, 'The expanding role of quantitative structure-activity relationships (QSAR) in toxicology.' *Toxicol. Lett.*, 1995, **79**, 45–53.

11. A. M. Richard, 'Structure-based methods for predicting mutagenicity and carcinogenicity: Are we there yet?' *Mutation Res.*, 1998, **400**, 493–507.

12. R. Benigni, 'Predicting chemical carcinogenesis in rodents: The state of art in light of a comparative exercise.' *Mutation Res.*, 1995, **334**, 103–113.

13. A. M. Richard, 'Application of SAR methods to non-congeneric data bases associated with carcinogenicity and mutagenicity: Issues and approaches.' *Mutation Res.*, 1994, **305**, 73–97.

14. M. Liu, N. Sussman, G. Klopman and H. S. Rosenkranz, 'Estimation of the optimal data base size for structure-activity analyses: The *Salmonella* mutagenicity data base.' *Mutation Res.*, 1996, **358**, 63–72.

15. N. Takihi, Y. P. Zhang, G. Klopman *et al.*, 'Development of a method to assess the informational content of structure-activity data bases.' *Qual. Assur.*, 1993, **2**, 255–264.

16. N. Takihi, H. S. Rosenkranz and G. Klopman, 'Identification of chemicals for testing in the rodent cancer bioassay.' *Qual. Assur.*, 1993, **2**, 232–243.

17. N. Takihi, Y. P. Zhang, G. Klopman *et al.*, 'An approach for evaluating and increasing the informational content of mutagenicity and clastogenicity data bases.' *Mutagenesis*, 1993, **8**, 257–264.

18. E. J. Matthews and J. F. Contrera, 'A new highly specific method for predicting the carcinogenic potential of pharmaceuticals in rodents using enhanced MCASE QSAR-ES software.' *Regul. Toxicol. Pharmacol.*, 1998, **28**, 242–264.

19. E. Gottmann, S. Kramer, B. Pfahringer *et al.*, 'Data quality in predictive toxicology: Reproducibility of rodent carcinogenicity experiments.' *Environ. Health Perspect.*, 2001, **109**, 509–514.

20. B. Henry, S. G. Grant, G. Klopman *et al.*, 'Induction of forward mutations at the thymidine kinase locus of mouse lymphoma cells: Evidence for electrophilic and non-electrophilic mechanisms.' *Mutation Res.*, 1998, **397**, 313–335.

21. W. L. Yang, G. Klopman and H. S. Rosenkranz, 'Structural basis of the *in vivo* induction of micronuclei.' *Mutation Res.*, 1992, **272**, 111–124.

22. A. Labbauf, G. Klopman and H. S. Rosenkranz, 'Dichotomous relationship between DNA reactivity and the induction of sister chromatid exchanges *in vivo* and *in vitro*.' *Mutation Res.*, 1997, **377**, 37–52.

23. H. S. Rosenkranz, A. R. Cunningham, Y. P. Zhang *et al.*, 'Development, characterization and application of predictive-toxicology models.' *SAR QSAR Environ. Res.*, 1999, **10**, 277–298.

24. C. Helma, S. Kramer, B. Pfahringer *et al.*, 'Data quality in predictive toxicology: Identification of chemical structures and calculation of chemical properties.' *Environ. Health Perspect.*, 2000, **108**, 1029–1033.

25. G. Klopman, 'Artificial intelligence approach to structure-activity studies: Computer automated structure evaluation of biological activity of organic molecules.' *J. Am. Chem. Soc.*, 1984, **106**, 7315–7321.

26. G. Klopman, 'MultiCASE: A hierarchical computer automated structure evaluation program.' *Quantit. Struct. Act. Relat.*, 1992, **11**, 176–184.

27. S. Kramer and E. Frank, 'Bottom-up propositionalization,' in 'Work-In-Progress Rep. 10th International Conference on Inductive Logic Programming (ILP-2000)', 2000, pp. 156–162.

28. S. Kramer, E. Frank and C. Helma, 'Fragment generation and support vector machines for inducing SARs.' *SAR QSAR Environ. Res.*, in press.

29. L. De Raedt and S. Kramer, 'The level-wise version space algorithm and its application to molecular fragment finding.' in '17th International Joint Conference on Artificial Intelligence' (IJCAI-01), volume submitted, 2001.

30. S. Kramer and L. De Raedt, 'Feature construction with version spaces for biochemical applications', in '18th International Conference on Machine Learning' (ICML-01), in press.

31. D. Weininger, 'SMILES, a chemical language and information system 1. Introduction and encoding rules.' *J. Chem. Inf. Comput. Sci.*, 1988, **28**, 31–36.

32. A. Dalby, J. G. Nourse, W. D. Hounshell *et al.*, 'Description of several chemical structure file formats used by computer programs developed at Molecular Design Limited.' *J. Chem. Inf. Comput. Sci.*, 1992, **32**, 244–255.

33. B. Pfahringer, E. Gottmann, C. Helma *et al.*, 'Efficiency/representational issues in toxicological knowledge discovery,' in 'Predictive Toxicology of Chemicals: Experiences and Impact of AI Tools,' AAAI Press, Menlo Park, California, 1999, pp. 86–89.

34. P. Finn, S. Muggleton, D. Page *et al.*, 'Pharmacophore discovery using the inductive logic programming system Progol'. *Machine Learning*, 1998, **30**, 241–271.

35. L. Dehaspe, H. Toivonen and R. D. King, 'Finding frequent substructures in chemical compounds,' in '14th International Conference on Knowledge Discovery and Data Mining', AAAI Press, Menlo Park, CA, 1998.

36. L. Breiman, J. H. Friedman, R. A. Olshen *et al.*, 'Classification and Regression Trees.' The Wadsworth Statistics/Probability Series. Wadsworth International Group, Belmont, CA, 1984.

37. S. Kramer, 'Structural regression trees,' in 'Proc. 13th National Conference on Artificial Intelligence' (AAAI-96), 1996.

38. S. Kramer, 'Relational Learning versus Propositionalization: Investigations in Inductive Logic Programming and Propositional Machine Learning.' Ph.D. thesis, Vienna University of Technology, 1999.

39. S. Kramer and G. Widmer, 'Inducing Classification and Regression Trees in First-Order Logic,' Springer Verlag, Berlin Heidelberg, New York, 2001.

40. R. D. King and A. Srinivasan, 'The discovery of indicator variables for QSAR using inductive logic programming.' *J. Comput. Aided Mol. Des.*, 1997, **11**, 571–580.

41. R. D. King and A. Srinivasan, 'Prediction of rodent carcinogenicity bioassays from molecular structure using inductive logic programming.' *Environ. Health Perspect.*, 1996, **104**(Suppl. 5), 1031–1040.

42. R. Kohavi, 'A Study of Cross-validation and Bootstrap for Accuracy Estimation and Model Selection.' in ed., C. S. Mellish, 'Proc. 14th International Joint Conference on Artificial Intelligence,' (IJCAI-95), Morgan Kaufmann, San Mateo, CA, 1995, pp. 1137–1143.

43. V. Chankong, Y. Y. Haimes, H. S. Rosenkranz *et al.*, 'The carcinogenicity prediction and battery selection (CPBS) method: A Bayesian approach.' *Mutation Res.*, 1985, **153**, 135–166.

44. J. Pet-Edwards, Y. Y. Haimes, V. Chankong *et al.*, 'Risk Assessment and Decision Making Using Test Results: The Carcinogenicity Prediction and Battery

Selection Approach,' Plenum Press, New York and London, 1989.

45. G. Klopman, 'The MultiCASE program II. Baseline activity identification algorithm (BAIA).' *J. Chem. Inf. Comput. Sci.*, 1998, **38**, 78–81.

46. G. Klopman and H. S. Rosenkranz, 'Toxicity estimation by chemical substructure analysis: the TOX II program.' *Toxicol. Lett.*, 1995, **79**, 145–155.

47. H. S. Rosenkranz, A. R. Cunningham, Y. P. Zhang *et al.*, 'Applications of the case/multicase sar method to environmental and public health situations.' *SAR QSAR Environ. Res.*, 1999, **10**, 263–276.

48. D. Malacarne, R. Pesenti, M. Paolucci *et al.*, 'Relationship between molecular connectivity and carcinogenic activity: A confirmation with a new software program based on graph theory.' *Environ. Health Perspect.*, 1993, **101**, 332–342.

49. D. Malacarne, M. Taningher, R. Pesenti *et al.*, 'Molecular fragments associated with non-genotoxic carcinogens, as detected using a software program based on graph theory: Their usefulness to predict carcinogenicity.' *Chem. Biol. Interact.*, 1995, **97**, 75–10.

50. A. Perrotta, D. Malacarne, M. Taningher *et al.*, 'A computerized connectivity approach for analyzing the structural basis of mutagenicity in *Salmonella* and its relationship with rodent carcinogenicity.' *Environ. Mol. Mutagen.*, 1996, **28**, 31–50.

51. M. Taningher, D. Malacarne, A. Perrotta *et al.*, 'Computer-aided analysis of mutagenicity and cell transformation data for assessing their relationship with carcinogenicity.' *Environ. Mol. Mutagen.*, 1999, **33**, 226–239.

52. D. Bahler and D. W. Bristol, 'The induction of rules for predicting chemical carcinogenesis in rodents.' *Intel. Sys. Mol. Biol.*, 1993, **1**, 29–37.

53. R. King, M. Sternberg and A. Srinivasan, 'Relating chemical activity to structure: An examination of ILP successes.' *New Generation Computing*, 1995, **13**, 411–433.

54. A. Srinivasan, S. Muggleton, R. D. King *et al.*, 'Theories for mutagenicity: A study of first-order and feature based induction.' *Artificial Intelligence*, 1996, **85**, 277–299.

55. S. Kramer, B. Pfahringer and C. Helma, 'Mining for causes of cancer: Machine learning experiments at various levels of detail,' in 'Proc. 3rd International Conference on Knowledge Discovery and Data Mining (KDD-97)', AAAI Press, Menlo Park, CA, 1997.

56. S. Dzeroski, H. Blockeel, B. Kompare *et al.*, 'Experiments in predicting biodegradability', in 'Proc. 9th International Conference on Inductive Logic Programming,' Springer, Berlin Heidelberg New York, 1999, pp. 80–91.

57. D. W. Bristol, J. T. Wachsman and A. Greenwell, 'The NIEHS predictive toxicology evaluation project.' *Environ. Health Perspect.*, 1996, **104**(Suppl. 5), 1001–1010.

58. A. Srinivasan, S. H. Muggleton, R. D. King *et al.*, 'The predictive toxicology evaluation challenge,' in 'Proc. 15th International Joint Conference on Artificial Intelligence (IJCAI-97),' Morgan Kaufmann, San Mateo, CA, 1997.

59. A. Srinivasan, R. D. King and D. W. Bristol, 'An assessment of submissions made to the predictive toxicology evaluation challenge,' in 'Proc. 16th International Joint Conference on Artificial Intelligence (IJCAI-99)', Morgan Kaufmann, San Francisco, CA, 1999, pp. 270–275.

60. R. D. King, S. H. Muggleton, A. Srinivasan *et al.*, 'Structure activity relationships derived by machine learning: the use of atoms and their bond connectivities to predict mutagenicity by inductive logic programming.' *Proc. Nat. Acad. Sci. USA*, 1996, **93**, 438–442.

61. J. Ashby and D. Paton, 'The influence of chemical structure on the extent and sites of carcinogenesis for 522 rodent carcinogens and 55 different human carcinogen exposures.' *Mutation Res.*, 1993, **286**, 3–74.

62. Y. Lee, B. G. Buchanan, D. M. Mattison *et al.*, 'Learning rules to predict rodent carcinogenicity of non-genotoxic chemicals.' *Mutation Res.*, 1995, **328**, 127–149.

63. B. N. Ames, W. E. Durston, E. Yamasaki *et al.*, 'Carcinogens are mutagens: A simple test system combining liver homogenates for activation and bacteria for detection.' *Proc. Nat. Acad. Sci. USA*, 1973, **70**, 2281–2285.

64. H. S. Rosenkranz, N. Takihi and G. Klopman, 'Structure activity-based predictive toxicology: An efficient and economical method for generating non-congeneric data bases.' *Mutagenesis*, 1991, **6**, 391–394.

65. G. Klopman and H. S. Rosenkranz, 'Quantification of the predictivity of some short-term assays for carcinogenicity in rodents.' *Mutation Res.*, 1991, **253**, 237–240.

66. J. Pet-Edwards, V. Chankong, H. S. Rosenkranz *et al.*, 'Application of the carcinogenicity prediction and battery selection method to the Gene-Tox data-base.' *Mutation Res.*, 1985, **153**, 187–200.

67. F. K. Ennever and H. S. Rosenkranz, 'Evaluating batteries of short-term genotoxicity tests.' *Mutagenesis*, 1986, **1**, 293–298.

68. B. S. Kim and B. H. Margolin, 'Predicting carcinogenicity by using batteries of dependent short-term tests.' *Environ. Health Perspect.*, 1994, **102**(Suppl. 1), 127–130.

69. Y. Lee, B. G. Buchanan, G. Klopman *et al.*, 'The potential of organ specific toxicity for predicting rodent carcinogenicity.' *Mutation Res.*, 1996, **358**, 37–62.

70. S. H. Clearwater and F. J. Provost, 'RL4: A tool for knowlege-based induction', in 'Proc. Tools for Artificial Intelligence 90,' IEEE Computer Society Press, 1990, pp. 24–30.

71. B. Kontic and S. Dzeroski, 'Perspectives of machine learning in epidemiological studies,' in 'Proc. International Symposium on Enviromental Epidemiology in Central and Eastern Europe,' International Institute for Rural and Environmental Health, Bratislava, Slovakia, 1997, pp. 27–30.

72. L. A. Cox, 'Does diesel exhaust cause human lung cancer?' *Risk Analysis*, 1997, **17**, pp. 807–829.

73. J. Aronis, F. Provost and B. Buchanan, 'Exploiting background knowledge in automated discovery,' in 'Proc. 2nd International Conference on Knowledge Discovery and Data Mining,' (KDD-96) eds., E. Simoudis, J. W. Han and U. Fayyad, AAAI Press, Menlo Park, Calif., 1996, pp. 355–359.

74. S. Tsumoto, 'Discovery of knowledge about drug side effects in clinical databases based on rough set model,' in 'Predictive Toxicology of Chemicals: Experiences and Impact of AI Tools,' AAAI Press, Menlo Park, California, 1999, pp. 100–104.

75. V. R. Spiehler, 'Computer-assisted interpretation in forensic toxicology: Morphine-involved death.' *J. Forensic Sci.*, 1987, **32**, 906–916.

# Appendix I

ADDITIONAL WEB RESOURCES

**Receptor Binding (Chapter 14.1.3)**
Analyzing Radioligand Binding Data
http://www.graphpad.com/www/radiolig/radiolig.htm

Pharmacology Guide to Receptors
http://science.glaxowellcome.com/science/pharm_guide/index.htm

The Merck Manual
http://www.merckhomeedition.com/home.html

Kinetics & Ligand-Binding Resources on the Web
http://www.med.umich.edu/biochem/enzresources/

**Gene Expression (Chapter 14.1.4)**
Signal Transduction Knowledge Environment
http://stke.sciencemag.org/

Classification of Transcription factors
http://transfac.gbf.de/TRANSFAC/cl/cl.html

Protein Kinase Resource
http://www.sdsc.edu/Kinases/pk_home.html

Nuclear Receptor Resource
http://nrr.georgetown.edu/nrr/nrr.html

TFSEARCH: Searching Transcription Factor Binding Sites
http://www.etl.go.jp/etl/cbrc/research/db/TFSEARCH.html

CMS Molecular Biology Resource
http://restools.sdsc.edu/

**Xenobiotic Receptors (Chapter 14.2.1)**
Classification of Transcription factors
http://transfac.gbf.de/TRANSFAC/cl/cl.html

Protein Kinase Resource
http://www.sdsc.edu/Kinases/pk_home.html

Nuclear Receptor Resource
http://nrr.georgetown.edu/nrr/nrr.html

PPAR Resource Page
http://ppar.cas.psu.edu

TFSEARCH: Searching Transcription Factor Binding Sites
http://www.etl.go.jp/etl/cbrc/research/db/TFSEARCH.html

## DNA Methylation (Chapter 14.3.5)

A vast amount of information about DNA methylation, chromatin structure, histones, and nucleosomes exists on the internet. A few recommended sites are:

The DNA Methylation Society: http://dnamethsoc.server101.com/

The DNA Methylation Database: http://www.methdb.de/

The Chromatin Structure & Function Page: http://www.cstone.net/~jrb7q/chrom.html

The Histone Sequence Database: http://genome.nhgri.nih.gov/histones

The Nuclesome Core Particle — Richmond Group: http://www.mol.biol.ethz.ch/richmond/

## Gap Junctions (Chapter 14.4.5)

Google Website
http://www.google.com/search?q = gap+junction

Gap Junction and Stem Cell Home Page
www.phd.msu.edu/trosko

Image Library-Gap Junctions
http://dilbert.scripps.edu/gallery/gap_jnctn.html

Gap Junction Page
http://plaza.snu.ac.kr/~kangpub/Gjic.htm

## Cell Cycle (Chapter 14.4.6)

The Cell Cycle Mitosis Tutorial
http://www.biology.arizona.edu/cell_bio/tutorials/cell_cycle/main.html

The Eukaryotic Cell Cycle and Cancer
http://www.ndsu.nodak.edu/instruct/mcclean/plsc431/cellcycle/cellcycl1.htm

# Subject Index

629